Graduate Texts in Mathematics 239

T0255201

Graduate Texts in Mathematics

(continued after index)

Henri Cohen

Number Theory

Volume I:
Tools and Diophantine
Equations

 Springer

Henri Cohen
Université Bordeaux I
Institut de Mathématiques de Bordeaux
351, cours de la Libération
33405, Talence cedex
France
Henri.Cohen@math.u-bordeaux1.fr

Mathematics Subject Classification (2000): 11-xx 11-01 11Dxx 11Rxx 11Sxx

ISBN-13:   978-1-4419-2390-5   eISBN-13:   978-0-387-49923-9

Printed on acid-free paper.

9 8 7 6 5 4 3 2 1

springer.com

Preface

This book deals with several aspects of what is now called "explicit number theory," not including the essential algorithmic aspects, which are for the most part covered by two other books of the author [Coh0] and [Coh1]. The central (although not unique) theme is the solution of Diophantine equations, i.e., equations or systems of polynomial equations that must be solved in integers, rational numbers, or more generally in algebraic numbers. This theme is in particular the central motivation for the modern theory of arithmetic algebraic geometry. We will consider it through three of its most basic aspects.

The first is the *local* aspect: the invention of p-adic numbers and their generalizations by K. Hensel was a major breakthrough, enabling in particular the simultaneous treatment of congruences modulo prime powers. But more importantly, one can do *analysis* in p-adic fields, and this goes much further than the simple definition of p-adic numbers. The local study of equations is usually not very difficult. We start by looking at solutions in *finite fields*, where important theorems such as the Weil bounds and Deligne's theorem on the Weil conjectures come into play. We then *lift* these solutions to local solutions using *Hensel lifting*.

The second aspect is the *global* aspect: the use of number fields, and in particular of class groups and unit groups. Although local considerations can give a considerable amount of information on Diophantine problems, the "local-to-global" principles are unfortunately rather rare, and we will see many examples of failure. Concerning the global aspect, we will first require as a prerequisite of the reader that he or she be familiar with the standard basic theory of number fields, up to and including the finiteness of the class group and Dirichlet's structure theorem for the unit group. This can be found in many textbooks such as [Sam] and [Marc]. Second, and this is less standard, we will always assume that we have at our disposal a computer algebra system (CAS) that is able to compute rings of integers, class and unit groups, generators of principal ideals, and related objects. Such CAS are now very common, for instance Kash, magma, and Pari/GP, to cite the most useful in algebraic number theory.

The third aspect is the theory of zeta and L-functions. This can be considered a *unifying theme*[1] for the whole subject, and it embodies in a beautiful way the local and global aspects of Diophantine problems. Indeed, these functions are defined through the local aspects of the problems, but their analytic behavior is intimately linked to the global aspects. A first example is given by the Dedekind zeta function of a number field, which is defined only through the splitting behavior of the primes, but whose leading term at $s = 0$ contains at the same time explicit information on the unit rank, the class number, the regulator, and the number of roots of unity of the number field. A second very important example, which is one of the most beautiful and important conjectures in the whole of number theory (and perhaps of the whole of mathematics), the Birch and Swinnerton-Dyer conjecture, says that the behavior at $s = 1$ of the L-function of an elliptic curve defined over $\mathbb{Q}$ contains at the same time explicit information on the rank of the group of rational points on the curve, on the regulator, and on the order of the torsion group of the group of rational points, in complete analogy with the case of the Dedekind zeta function. In addition to the purely *analytical* problems, the theory of L-functions contains beautiful results (and conjectures) on *special values*, of which Euler's formula $\sum_{n \geqslant 1} 1/n^2 = \pi^2/6$ is a special case.

This book can be considered as having four main parts. The first part gives the tools necessary for Diophantine problems: equations over finite fields, number fields, and finally local fields such as $\mathfrak{p}$-adic fields (Chapters 1, 2, 3, 4, and part of Chapter 5). The emphasis will be mainly on the theory of $\mathfrak{p}$-adic fields (Chapter 4), since the reader probably has less familiarity with these. Note that we will consider function fields only in Chapter 7, as a tool for proving Hasse's theorem on elliptic curves. An important tool that we will introduce at the end of Chapter 3 is the theory of the Stickelberger ideal over cyclotomic fields, together with the important applications to the Eisenstein reciprocity law, and the Davenport–Hasse relations. Through Eisenstein reciprocity this theory will enable us to prove Wieferich's criterion for the first case of Fermat's last theorem (FLT), and it will also be an essential tool in the proof of Catalan's conjecture given in Chapter 16.

The second part is a study of certain basic Diophantine equations or systems of equations (Chapters 5, 6, 7, and 8). It should be stressed that even though a number of general techniques are available, each Diophantine equation poses a new problem, and it is difficult to know in advance whether it will be easy to solve. Even without mentioning *families* of Diophantine equations such as FLT, the congruent number problem, or Catalan's equation, all of which will be stated below, proving for instance that a specific equation such as $x^3 + y^5 = z^7$ with x, y coprime integers has no solution with $xyz \neq 0$ seems presently out of reach, although it has been proved (based on a deep theorem of Faltings) that there are only finitely many solutions; see [Dar-Gra]

---

[1] Expression due to Don Zagier.

and Chapter 14. Note also that it has been shown by Yu. Matiyasevich (after a considerable amount of work by other authors) in answer to Hilbert's tenth problem that there cannot exist a general algorithm for solving Diophantine equations.

The third part (Chapters 9, 10, and 11) deals with the detailed study of analytic objects linked to algebraic number theory: Bernoulli polynomials and numbers, the gamma function, and zeta and L-functions of Dirichlet characters, which are the simplest types of L-functions. In Chapter 11 we also study p-adic analogues of the gamma, zeta, and L-functions, which have come to play an important role in number theory, and in particular the Gross–Koblitz formula for Morita's p-adic gamma function. In particular, we will see that this formula leads to remarkably simple proofs of Stickelberger's congruence and the Hasse–Davenport product relation. More general L-functions such as Hecke L-functions for Grössencharacters, Artin L-functions for Galois representations, or L-functions attached to modular forms, elliptic curves, or higher-dimensional objects are mentioned in several places, but a systematic exposition of their properties would be beyond the scope of this book.

Much more sophisticated techniques have been brought to bear on the subject of Diophantine equations, and it is impossible to be exhaustive. Because the author is not an expert in most of these techniques, they are not studied in the first three parts of the book. However, considering their importance, I have asked a number of much more knowledgeable people to write a few chapters on these techniques, and I have written two myself, and this forms the fourth and last part of the book (Chapters 12 to 16). These chapters have a different flavor from the rest of the book: they are in general not self-contained, are of a higher mathematical sophistication than the rest, and usually have no exercises. Chapter 12, written by Yann Bugeaud, Guillaume Hanrot, and Maurice Mignotte, deals with the applications of Baker's explicit results on linear forms in logarithms of algebraic numbers, which permit the solution of a large class of Diophantine equations such as Thue equations and norm form equations, and includes some recent spectacular successes. Paradoxically, the similar problems on elliptic curves are considerably less technical, and are studied in detail in Section 8.7. Chapter 13, written by Sylvain Duquesne, deals with the search for rational points on curves of genus greater than or equal to 2, restricting for simplicity to the case of hyperelliptic curves of genus 2 (the case of genus 0—in other words, of quadratic forms—is treated in Chapters 5 and 6, and the case of genus 1, essentially of elliptic curves, is treated in Chapters 7 and 8). Chapter 14, written by the author, deals with the so-called super-Fermat equation $x^p + y^q = z^r$, on which several methods have been used, including ordinary algebraic number theory, classical invariant theory, rational points on higher genus curves, and Ribet–Wiles type methods. The only proofs that are included are those coming from algebraic number theory. Chapter 15, written by Samir Siksek, deals with the use of Galois representations, and in particular of Ribet's level-lowering theorem and Wiles's and Taylor–Wiles's theorem proving the modularity conjecture. The main application is to equations of "abc" type, in other words, equations of the form $a + b + c = 0$ with a, b, and c highly composite, the "easiest" application of this method being the proof of FLT. The author of this chapter has tried to hide all the sophisticated mathematics and to present the method as a black box that can be used without completely understanding the underlying theory. Finally, Chapter 16, also written by the author, gives the complete proof of Catalan's conjecture by P. Mihăilescu. It is entirely based on notes of Yu. Bilu, R. Schoof, and especially of J. Boéchat and M. Mischler, and the only reason that it is not self-contained is that it will be necessary to assume the validity of an important theorem of F. Thaine on the annihilator of the plus part of the class group of cyclotomic fields.

Warnings

Since mathematical conventions and notation are not the same from one mathematical culture to the next, I have decided to use systematically unambiguous terminology, and when the notations clash, the French notation. Here are the most important:

– We will systematically say that a is strictly greater than b, or greater than or equal to b (or b is strictly less than a, or less than or equal to a), although the English terminology a is greater than b means in fact one of the two (I don't remember which one, and that is one of the main reasons I refuse to use it) and the French terminology means the other. Similarly, positive and negative are ambiguous (does it include the number 0?) Even though the expression "x is nonnegative" is slightly ambiguous, it is useful, and I *will* allow myself to use it, with the meaning $x \geqslant 0$.
– Although we will almost never deal with noncommutative fields (which is a contradiction in terms since in principle the word field implies commutativity), we will usually not use the word field alone. Either we will write explicitly commutative (or noncommutative) field, or we will deal with specific classes of fields, such as finite fields, $\mathfrak{p}$-adic fields, local fields, number fields, etc., for which commutativity is clear. Note that the "proper" way in English-language texts to talk about noncommutative fields is to call them either skew fields or division algebras. In any case this will not be an issue since the only appearances of skew fields will be in Chapter 2, where we will prove that finite division algebras are commutative, and in Chapter 7 about endomorphism rings of elliptic curves over finite fields.
– The GCD (respectively the LCM) of two integers can be denoted by (a, b) (respectively by $[a, b]$), but to avoid ambiguities, I will systematically use the explicit notation $\gcd(a, b)$ (respectively $\operatorname{lcm}(a, b)$), and similarly when more than two integers are involved.

- An open interval with endpoints a and b is denoted by (a, b) in the English literature, and by $]a, b[$ in the French literature. I will use the French notation, and similarly for half-open intervals $(a, b]$ and $[a, b)$, which I will denote by $]a, b]$ and $[a, b[$. Although it is impossible to change such a well-entrenched notation, I urge my English-speaking readers to realize the dreadful ambiguity of the notation (a, b), which can mean either the ordered pair (a, b), the GCD of a and b, the inner product of a and b, or the open interval.
- The trigonometric functions $\sec(x)$ and $\csc(x)$ do not exist in France, so I will not use them. The functions $\tan(x)$, $\cot(x)$, $\cosh(x)$, $\sinh(x)$, and $\tanh(x)$ are denoted respectively by $\operatorname{tg}(x)$, $\operatorname{cotg}(x)$, $\operatorname{ch}(x)$, $\operatorname{sh}(x)$, and $\operatorname{th}(x)$ in France, but for once to bow to the majority I will use the English names.
- $\Re(s)$ and $\Im(s)$ denote the real and imaginary parts of the complex number s, the typography coming from the standard TEX macros.

Notation

In addition to the standard notation of number theory we will use the following notation.

- We will often use the practical self-explanatory notation $\mathbb{Z}_{>0}$, $\mathbb{Z}_{\geqslant 0}$, $\mathbb{Z}_{<0}$, $\mathbb{Z}_{\leqslant 0}$, and generalizations thereof, which avoid using excessive verbiage. On the other hand, I prefer not to use the notation $\mathbb{N}$ (for $\mathbb{Z}_{\geqslant 0}$, or is it $\mathbb{Z}_{>0}$?).

- If a and b are nonzero integers, we write $\gcd(a, b^\infty)$ for the limit of the ultimately constant sequence $\gcd(a, b^n)$ as $n \to \infty$. We have of course $\gcd(a, b^\infty) = \prod_{p | \gcd(a,b)} p^{v_p(a)}$, and $a / \gcd(a, b^\infty)$ is the largest divisor of a coprime to b.
- If n is a nonzero integer and $d \mid n$, we write $d \| n$ if $\gcd(d, n/d) = 1$. Note that this is *not* the same thing as the condition $d^2 \nmid n$, except if d is prime.
- If $x \in \mathbb{R}$, we denote by $\lfloor x \rfloor$ the largest integer less than or equal to x (the *floor* of x), by $\lceil x \rceil$ the smallest integer greater than or equal to x (the *ceiling* of x, which is equal to $\lfloor x \rfloor + 1$ if and only if $x \notin \mathbb{Z}$), and by $\lfloor x \rceil$ the nearest integer to x (or one of the two if $x \in 1/2 + \mathbb{Z}$), so that $\lfloor x \rceil = \lfloor x + 1/2 \rfloor$. We also set $\{x\} = x - \lfloor x \rfloor$, the *fractional part* of x. Note that for instance $\lfloor -1.4 \rfloor = -2$, and not -1 as almost all computer languages would lead us to believe.
- For any α belonging to a field K of characteristic zero and any $k \in \mathbb{Z}_{\geqslant 0}$ we set
$$\binom{\alpha}{k} = \frac{\alpha(\alpha - 1) \cdots (\alpha - k + 1)}{k!}.$$
In particular, if $\alpha \in \mathbb{Z}_{\geqslant 0}$ we have $\binom{\alpha}{k} = 0$ if $k > \alpha$, and in this case we will set $\binom{\alpha}{k} = 0$ also when $k < 0$. On the other hand, $\binom{\alpha}{k}$ is *undetermined* for $k < 0$ if $\alpha \notin \mathbb{Z}_{\geqslant 0}$.

- Capital italic letters such as K and L will usually denote number fields.
- Capital calligraphic letters such as $\mathcal{K}$ and $\mathcal{L}$ will denote general $\mathfrak{p}$-adic fields (for specific ones, we write for instance $K_{\mathfrak{p}}$).
- Letters such as $\mathbb{E}$ and $\mathbb{F}$ will always denote finite fields.
- The letter $\mathbb{Z}$ indexed by a capital italic or calligraphic letter such as $\mathbb{Z}_K$, $\mathbb{Z}_L$, $\mathbb{Z}_{\mathcal{K}}$, etc., will always denote the ring of integers of the corresponding field.
- Capital italic letters such as A, B, C, G, H, S, T, U, V, W, or lowercase italic letters such as f, g, h, will usually denote polynomials or formal power series with coefficients in some base ring or field. The coefficient of degree m of these polynomials or power series will be denoted by the corresponding letter indexed by m, such as A_m, B_m, etc. Thus we will always write (for instance) $A(X) = A_d X^d + A_{d-1} X^{d-1} + \cdots + A_0$, so that the ith elementary symmetric function of the roots is equal to $(-1)^i A_{d-i}/A_d$.

Acknowledgments

A large part of the material on local fields has been taken with little change from the remarkable book by Cassels [Cas1], and also from unpublished notes of Jaulent written in 1994. For p-adic analysis, I have also liberally borrowed from work of Robert, in particular his superb GTM volume [Rob1]. For part of the material on elliptic curves I have borrowed from another excellent book by Cassels [Cas2], as well as the treatises of Cremona and Silverman [Cre2], [Sil1], [Sil2], and the introductory book by Silverman–Tate [Sil-Tat]. I have also borrowed from the classical books by Borevich–Shafarevich [Bor-Sha], Serre [Ser1], Ireland–Rosen [Ire-Ros], and Washington [Was]. I would like to thank my former students K. Belabas, C. Delaunay, S. Duquesne, and D. Simon, who have helped me to write specific sections, and my colleagues J.-F. Jaulent and J. Martinet for answering many questions in algebraic number theory. I would also like to thank M. Bennett, J. Cremona, A. Kraus, and F. Rodriguez-Villegas for valuable comments on parts of this book. I would especially like to thank D. Bernardi for his thorough rereading of the first ten chapters of the manuscript, which enabled me to remove a large number of errors, mathematical or otherwise. Finally, I would like to thank my copyeditor, who was very helpful and who did an absolutely remarkable job.

It is unavoidable that there still remain errors, typographical or otherwise, and the author would like to hear about them. Please send e-mail to

Henri.Cohen@math.u-bordeaux1.fr

Lists of known errors for the author's books including the present one can be obtained on the author's home page at the URL

http://www.math.u-bordeaux1.fr/~cohen/

Table of Contents

Volume II

Part III. Analytic Tools

1. Introduction to Diophantine Equations

1.1 Introduction

The study of *Diophantine equations* is the study of solutions of polynomial equations or systems of equations in integers, rational numbers, or sometimes more general number rings. It is one of the oldest branches of number theory, in fact of mathematics itself, since its origins can be found in texts of the ancient Babylonians, Chinese, Egyptians, and Greeks. One of the fascinations of the subject is that the problems are usually easy to state, but more often than not very difficult to solve, and when they can be solved, they sometimes involve extremely sophisticated mathematical tools.

Perhaps even more importantly, mathematicians must often invent or extensively develop entirely new tools to solve the number-theoretical problems, and these become in turn important branches of mathematics per se, which often have applications in completely different problems from those from which they originate.

1.1.1 Examples of Diophantine Problems

Let me give five examples. The first and most famous is "Fermat's last theorem" (FLT), stating that for $n \geqslant 3$, the curve $x^n + y^n = 1$ has no rational points other than the ones with x or y equal to 0 (this is of course equivalent to the usual statement).[1]

In the nineteenth century, thanks in particular to the work of E. Kummer and P.-G. Lejeune-Dirichlet, the theorem was proved for quite a large number

---

[1] Incidentally, this is the place to destroy the legend concerning this statement, which has produced an enormous number of "Fermatists" claiming to have found an "elementary" proof that Fermat may have found himself: Fermat made this statement in the margin of his copy of the book by Diophantus on number theory (at the place where Diophantus discusses Pythagorean triples, see below), and claimed to have found a marvelous proof and so on. However, he wrote this statement when he was young, never claimed it publicly, and certainly never imagined that it would be made public, so he forgot about it. It *may* be possible that there does exist an elementary proof (although this is unlikely), but we can be positively sure that Fermat did not have it, for otherwise he would at least have challenged his English colleagues, as was the custom at that time.

of values of n, including all $n \leqslant 100$. Together with the theory of quadratic forms initiated by A.-M. Legendre and especially by C. F. Gauss, one can without exaggeration say that this single problem gave rise to algebraic number theory (rings, ideals, prime ideals, principal ideals, class numbers, units, Dirichlet series, L-functions, etc.). As is well known, although these methods were pushed to the extreme in the twentieth century, they did not succeed in solving the problem completely. The next progress on FLT came from algebraic geometry thanks to the work of G. Faltings, who proved the so-called Mordell conjecture, which in particular implies that for a *fixed* $n \geqslant 3$ the number of solutions to the Fermat equation is finite. However, it was only thanks to the work of several mathematicians starting with Y. Hellegouarch and G. Frey, and culminating with the work of K. Ribet, then finally of A. Wiles (helped for a crucial part by R. Taylor), that the problem was finally completely solved using completely different tools from those of Kummer (and even Faltings): elliptic curves, Galois representations, and modular forms. Although these subjects were not initiated by FLT, their development was certainly accelerated by the impetus given by FLT. In particular, thanks to the work of Wiles, the complete proof of the Taniyama–Shimura–Weil conjecture was obtained a few years later by C. Breuil, B. Conrad, F. Diamond, and R. Taylor. This latter result can be considered in itself a more important (and certainly a more useful) theorem than FLT.

A second rather similar problem whose history is slightly different is *Catalan's conjecture*. This states that when n and m are greater than or equal to 2, the only solutions in nonzero integers x and y of the equation $x^m - y^n = 1$ come from the equality $3^2 - 2^3 = 1$. This problem can be naturally attacked by the standard methods of algebraic number theory originating in the work of Kummer. However, it came as a surprise that an elementary argument due to Cassels (see Theorem 6.11.5) shows that the "first case" is impossible, in other words that if $x^p - y^q = 1$ with p and q primes then $p \mid y$ and $q \mid x$. The next important result, due to R. Tijdeman using Baker's theory of linear forms in logarithms of algebraic numbers, was that the total number of quadruplets (m, n, x, y) satisfying the required conditions is finite. Note that the proof of this finiteness result is completely different from Faltings's proof of the corresponding one for FLT, and in fact in the latter his result did not imply the finiteness of the number of triples (x, y, n) with $n \geqslant 3$ and $xy \neq 0$ such that $x^n + y^n = 1$.

Until the end of the 1990s the situation was quite similar to that of FLT before Wiles: under suitable conditions on the nondivisibility of the class number of cyclotomic fields, the Catalan equation was known to have no nontrivial solutions. It thus came as a total surprise that in 1999 P. Mihăilescu proved that if Catalan's equation $x^p - y^q = 1$ with p and q odd primes has a solution then p and q must satisfy the so-called double Wieferich condition $p^{q-1} \equiv 1 \pmod{q^2}$ and $q^{p-1} \equiv 1 \pmod{p^2}$. These conditions were known before him, but he completely removed the conditions on class numbers. The last step was again taken by Mihăilescu in 2001, who finished the proof of Catalan's conjecture. His proof was improved and simplified by several people, including in particular Yu. Bilu and H. W. Lenstra. The remarkable thing about the final proof is that it uses *only* algebraic number theory techniques on cyclotomic fields. However, it uses a large part of the theory, including the relatively recent theorem of F. Thaine, that has had some very important applications elsewhere. It does not use any computer calculations, while the initial proof did.

A third example is the *congruent number problem*, stated by Diophantus in the fourth century A.D. The problem is to find all integers n (called congruent numbers) that are equal to the area of a *Pythagorean triangle*, i.e., a right-angled triangle with all three sides rational. Very simple algebraic transformations show that n is congruent if and only if the Diophantine equation $y^2 = x^3 - n^2 x$ has rational solutions other than those with $y = 0$. The problem was in an "experimental" state until the 1970s; more precisely, one knew the congruent or noncongruent nature of numbers n up to a few hundred (and of course of many other larger numbers). Remarkable progress was made on this problem by J. Tunnell in 1980 using the theory of modular forms, and especially of modular forms of half-integral weight. In effect, he completely solved the problem, by giving an easily checked criterion for n to be a congruent number, assuming a weak form of the Birch–Swinnerton-Dyer conjecture, see Theorem 6.12.4. This conjecture (for which a prize of 1 million U.S. dollars has been offered by the Clay foundation) is probably one of the most important, and also one of the most beautiful, conjectures in all of mathematics in the twenty-first century.

A fourth important example is the Weil conjectures. These have to do with the number of solutions of Diophantine equations *in finite fields*. Indeed, one of the main themes of this book is that to study a Diophantine equation it is essential to start by studying it in finite fields. Let us give a simple example. Let $N(p)$ be the number of solutions modulo p of the equation $y^2 = x^5 - x$. Then $|N(p) - p|$ can never be very large compared to p, more precisely $|N(p) - p| < 4\sqrt{p}$, and the constant 4 is best possible. This result is already quite nontrivial, and the general study of the number of points on *curves* culminated with work of A. Weil in 1949 proving that this phenomenon occurs for all (nonsingular) curves and many other results besides. It was then natural to ask the question for surfaces, and more generally varieties of any dimension. This problem (in a very precise form, which in particular implied excellent bounds on the number of solutions) became known as the Weil conjectures. A general strategy for solving these conjectures was put forth by Weil himself, but the achievement of this goal was made possible only by an amazing amount of work by numerous people. It included the creation of modern algebraic geometry by A. Grothendieck and his students (the famous EGA and SGA treatises). The Weil conjectures were finally solved by P. Deligne in the early 1970s, exactly following Weil's strategy, but using all the tools developed since.

As a last example we mention Waring's problem. One of its forms (by far not the only one) is the following: given an integer $k \geqslant 2$, find the smallest integer $g(k)$ such that any nonnegative integer can be represented as a sum of $g(k)$ nonnegative kth powers. It has been known since J.-L. Lagrange that any integer is a sum of 4 squares, and that integers congruent to 7 modulo 8 are not the sum of 3 squares, so that $g(2) = 4$. It was proved by D. Hilbert that $g(k)$ is finite (this can be proved with not too much difficulty from Lagrange's result, but is still not completely trivial: try it as an exercise). However, the major advances on this problem were made by G. H. Hardy and J. Littlewood, who invented the *circle method* in order to treat the problem. One of the important aspects of the circle method is the so-called *singular series*, which regroups all the arithmetic information obtained by studying the problem modulo p for each prime p. The other major advances were made by I.-M. Vinogradov using the theory of *trigonometric sums*. Both the circle method and trigonometric sums have found universal application in the branch of number theory called "additive number theory," and also in other branches of number theory. To finish this example, we note that Waring's problem as given above (as already mentioned, there are other versions) is completely solved. Perhaps surprisingly, when one compares it with FLT for example, the hardest cases are not for large k but for small k: the most difficult is $k = 4$, solved only in the 1980s by R. Balasubramanian, J.-M. Deshouillers, and F. Dress, see [BDD]. For the record, we have $g(2) = 4$, $g(3) = 9$, and $g(4) = 19$.

Additive number theory forms a large part of what is usually called "analytic number theory" because many sophisticated analytic techniques come into play. Analytic number theory will *not* be studied in this book, with the exception of a few basic results such as the prime number theorem and Dirichlet's theorem on primes in arithmetic progression. The expression "analytic methods" used in the third part of this book (Chapters 9 to 11) refer to the study of Bernoulli polynomials, gamma and L-functions, integral transforms, summation formulas, and the like. We refer for instance to [Ell] and [Iwa-Kow] among many others for excellent expositions of analytic number theory.

1.1.2 Local Methods

As is explicit or implicit in all of the examples given above (and in fact in all Diophantine problems), it is essential to start by studying a Diophantine equation *locally*, in other words prime by prime (we will see later precisely what this means). Let p be a prime number, and let $\mathbb{F}_p \simeq \mathbb{Z}/p\mathbb{Z}$ be the prime finite field with p elements. We can begin by studying our problem in $\mathbb{F}_p$ (i.e., modulo p), and this can already be considered as the start of a local study.

This is sometimes sufficient, but usually not. In that case, keeping the same prime p, we will see that there are two *totally different* ways to refine the study of the equation.

The first is to consider it modulo p^2, p^3, and so on, i.e., to work in $\mathbb{Z}/p^2\mathbb{Z}$, $\mathbb{Z}/p^3\mathbb{Z}$,... An important discovery, made by K. Hensel in the beginning of the twentieth century, is that it is possible to regroup all these rings with zero divisors into a single object, called the p-adic integers, and denoted by $\mathbb{Z}_p$, which is an integral domain. Not only do we have the benefit of being able to work conveniently with all the congruences modulo p, p^2, p^3,... simultaneously, but we have the added benefit of having *topological properties,* which add a considerable number of tools that we may use, in particular *analytic methods* (note that this type of limiting construction is very frequent in mathematics, with the same type of benefits). When we say that we study our Diophantine problem locally at p, this means that we study it in $\mathbb{Z}_p$, or in the field of fractions $\mathbb{Q}_p$ of $\mathbb{Z}_p$. We will devote the entirety of Chapter 4 to the study of p-adic numbers and their generalizations. The reason for the word "local" will become clear when we study p-adic numbers.

A second way to refine the study of our equation, which is explicit for example in Weil's estimates and conjectures, is to study our equations in the finite fields $\mathbb{F}_{p^2}$, $\mathbb{F}_{p^3}$, etc. (Note that usually this does not bring any information for equations over $\mathbb{Q}$, since in that case only local methods are useful.) At this point, recall that the main theorem on finite fields (which we will recall, with proof, in Chapter 2) is that for any prime power $q = p^n$ there exists up to isomorphism exactly one finite field $\mathbb{F}_{p^n}$ of that cardinality, and all finite fields have this form. They are of course *not* isomorphic to $\mathbb{Z}/p^2\mathbb{Z}$, $\mathbb{Z}/p^3\mathbb{Z}$,... since the latter are not even fields. We will come back to the structure of finite fields in the text. Once again, we can use a limiting process of a slightly different kind so as to put all these finite fields of characteristic p together: this leads to $\overline{\mathbb{F}_p}$, the algebraic closure of $\mathbb{F}_p$. In this case we of course do not say that we study it locally, but simply over $\overline{\mathbb{F}_p}$.

Let us give simple but typical examples of all this. Consider first the Diophantine equation $x^2 + y^2 = 3$ to be solved in rational numbers or equivalently, the Diophantine equation $x^2 + y^2 = 3z^2$ to be solved in rational integers. We may assume that x and y are coprime (exercise). Looking at the equation modulo 3, i.e., in the field $\mathbb{F}_3$, we see that it has no solution (x^2 and y^2 are congruent to 0 or 1 modulo 3; hence $x^2 + y^2$ is congruent to 0 modulo 3 if and only if x and y are both divisible by 3, excluded by assumption). Thus, our initial Diophantine equation does not have any solution.

We are here in the case of a *quadratic* Diophantine equation. It is crucial to note that this type of equation can *always* be solved by local methods. In other words, either we can find a solution to the equation (often helped by the local conditions), or it is possible to prove that the equation does not have any solutions using positivity conditions together with congruences as above (or equivalently, real and p-adic solubility). This is the so-called *Hasse principle*, a nontrivial theorem (see Theorem 5.3.3) that is valid for a *single homogeneous quadratic* Diophantine equation, but is in general not true for higher-degree equations or for systems of equations.

Consider now the Diophantine equation $x^3 + y^3 = 1$ to be solved in nonzero rational numbers, or equivalently, the Diophantine equation $x^3 + y^3 = z^3$ to be solved in nonzero rational integers. Once again we may assume that x, y, and z are pairwise coprime. It is natural to consider once more the problem modulo 3. Here, however, the equation has nonzero solutions (for example $1^3 + 1^3 \equiv 2^3 \pmod 3$). We must go up one level, and consider the equation modulo $9 = 3^2$ to obtain a partial result: since it is easily checked that an integer cube is congruent to -1, 0, or 1 modulo 9, if we exclude the possibility that x, y, or z is divisible by 3 we see immediately that the equation does not have any solution modulo 9, hence no solution at all. Thus we have proved that if $x^3 + y^3 = z^3$, then one of x, y, and z is divisible by 3. This is called solving the first case of FLT for the exponent 3. To show that the equation has no solutions at all, even with x, y, or z divisible by 3, is more difficult and *cannot* be shown by congruence conditions alone (see Sections 6.4.5 and 6.9). Indeed Proposition 6.9.11 tells us that the equation $x^3 + y^3 = z^3$ has a solution with $xyz \neq 0$ in every p-adic field, hence modulo p^k for any prime number p and any exponent k (and it of course has real solutions). Thus, the Hasse principle clearly fails here since the equation does not have any solution in rational integers with $xyz \neq 0$. When this happens, it is necessary to use additional *global* arguments, whose main tools are those of algebraic number theory developed by Kummer et al. in the nineteenth century, and in particular class and unit groups, which are objects of a strictly *global* nature.

1.1.3 Dimensions

An important notion that has come to be really understood only in the twentieth century is that of *dimension*. It is not our purpose here to define it precisely,[2] but to give a feeling of its meaning. We stick to the algebraic and/or arithmetic case, since the topological or analytic case is simpler.

Consider first the classical (algebraic) situation, say over the complex numbers $\mathbb{C}$. A *point* is clearly of dimension 0, and more generally a finite set of points defined by a system of algebraic equations has dimension 0. Similarly, a curve (for example defined by a single equation in two variables $f(x, y) = 0$ in affine coordinates) has dimension 1 (note however that a *complex* curve has dimension 1 over $\mathbb{C}$ but has dimension 2 over $\mathbb{R}$), and so on with surfaces which have by definition dimension 2, or arbitrary varieties of higher dimension.

Consider now the *arithmetic* situation, say over the integers $\mathbb{Z}$. If $f(x, y)$ is a polynomial in two variables with integer coefficients, we can of course

---

[2] in the language of schemes, it is the maximal length of an ascending chain of irreducible subschemes.

consider the curve $f(x, y) = 0$ as defining a complex curve of dimension 1. But when we consider the Diophantine equation $f(x, y) = 0$, then as we have seen, it is essential to consider it also modulo p and more generally in the p-adic fields $\mathbb{Q}_p$ for every prime p (including the prime "at infinity," which gives the field $\mathbb{R}$). Thus, as a Diophantine equation, $f(x, y) = 0$ should not be seen as a curve (i.e., of dimension 1), but in fact as a surface, called an *arithmetic surface*. In other words, the ring $\mathbb{Z}$ must be considered to be of dimension 1 (its points being the prime numbers p together with 0 corresponding to the prime at infinity), and any system of equations considered as a system of Diophantine equations over $\mathbb{Z}$ should be considered to have one additional dimension compared to its ordinary complex dimension. See Exercise 4 for an illustration.

One of the goals of the modern theory of *arithmetic geometry* is to extend to arithmetic surfaces and more generally to arithmetic varieties of any dimension results known for ordinary surfaces and varieties.

Using these notions, we can quite naturally put a hierarchy on the objects that naturally occur in algebraic number theory.

- **Finite fields**. These are the simplest objects, not only because they are finite (finite rings and groups are extremely difficult to study; see Exercise 3) but because they have a very simple structure, which we will recall in detail in the text. They occur as *residue fields* (we will see the meaning of this later, but $\mathbb{Z}/p\mathbb{Z}$ is a typical example).
- **Local fields**. Local fields of characteristic 0 are the p-adic fields $\mathbb{Q}_p$, the real numbers $\mathbb{R}$, and their finite extensions, which are the $\mathfrak{p}$-adic fields $K_{\mathfrak{p}}$ and the field $\mathbb{C}$. There are also local fields of nonzero characteristic, which we will not consider in this book.
- **Global fields, and rings of dimension** 1. Global fields are the field of rational numbers $\mathbb{Q}$, its finite extensions (i.e., number fields), and in nonzero characteristic the fields $\mathbb{F}_q(X)$ and their finite extensions, which are the function fields of curves. The corresponding rings of integers of these global fields ($\mathbb{Z}$, $\mathbb{Z}_K$, $\mathbb{F}_q[X]$, etc.) are of dimension 1.
- Any object of higher dimension will be called a curve, surface, etc. Be careful with the terminology: when we speak of a *curve*, it usually means a variety of dimension 1 over the base field, but if we consider it over $\mathbb{Z}$, it then becomes an arithmetic *surface*, hence of dimension $2 = 1 + 1$. Another possible confusion is that a complex curve is a real manifold of dimension 2, i.e., a surface, here because $2 = 2 \cdot 1$.

The reason and necessity of using this language cannot be clearly understood without a course in modern algebraic geometry, but nevertheless it is a good thing to have in mind, since it explains the utmost importance of the objects that we are going to study.

1.2 Exercises for Chapter 1

1. The following problem seems similar to the congruent number problem, but is much simpler. Show that for any integer n there exists a (not necessarily right-angled) triangle with rational sides a, b, and c and area n (recall that the area is given by $n^2 = s(s-a)(s-b)(s-c)$, where $s = (a+b+c)/2$). Try to give a complete parametrization of such triangles (I do not know the answer to this latter question).

2. Let us say that a triangle is *almost equilateral* if it satisfies the following two conditions: its vertices have integer coordinates, and the lengths of its sides are three consecutive integers $a-1$, a, and $a+1$. Show that such a triangle exists if and only if a has the form $a = (2+\sqrt{3})^k + (2-\sqrt{3})^k$ for $k \geqslant 1$ (you will first need to know the solution to the Pell–Fermat equation, see Proposition 6.3.16).

3. This exercise is mainly to emphasize that finite fields are really simple objects. Let $G(n)$ (respectively $R(n)$, $F(n)$) be the number of groups (respectively rings with nonzero unit, fields) of order n up to isomorphism. Compute $G(n)$ for $1 \leqslant n \leqslant 11$ (you will need a little group theory for this), $F(n)$ for $1 \leqslant n \leqslant 100$ (using the theory recalled in the next chapter), and $R(n)$ for as many consecutive values of n starting at $n = 1$ as you can. In the same ranges compute the number $G_a(n)$ of abelian groups, and $R_c(n)$ the number of commutative rings.

4. The goal of this exercise is to illustrate the fact that $\mathbb{Z}[X]$ has dimension 2.

   (a) Let I be a nonzero ideal of $\mathbb{Z}[X]$. Prove that there exists a primitive polynomial $G(X) \in \mathbb{Z}[X]$ such that $\mathbb{Q}I = G(X)\mathbb{Q}[X]$, and that $I \subset G(X)\mathbb{Z}[X]$. We set $J = I/G(X)$, which is again a nonzero ideal of $\mathbb{Z}[X]$, and it contains I.

   (b) Prove that there exists $n \in \mathbb{Z}_{\geqslant 1}$ such that $J \cap \mathbb{Z} = n\mathbb{Z}$ (you must prove that $n \neq 0$).

   (c) From now on, assume that I is a prime ideal. Prove that either $J = \mathbb{Z}[X]$, or that J is a prime ideal.

   (d) If $J = \mathbb{Z}[X]$ prove that $G(X)$ is irreducible in $\mathbb{Q}[X]$, and conversely that if $G(X)$ is irreducible then $I = G(X)\mathbb{Z}[X]$ is a prime ideal.

   (e) Finally, we assume from now on that $J \neq \mathbb{Z}[X]$, or equivalently, that $G(X) \notin I$, hence that J is a prime ideal. Prove that $J \subset I$, hence that $I = J$, and deduce that $G(X) = 1$.

   (f) Prove that the integer n defined above is a prime number p, and in particular that $n \neq 1$.

   (g) Prove that there exists a polynomial $H(X) \in \mathbb{Z}[X]$ such that $I = p\mathbb{Z}[X] + H(X)\mathbb{Z}[X]$, and that the reduction $\overline{H}(X)$ in $\mathbb{F}_p[X]$ is either 0 or is irreducible in $\mathbb{F}_p[X]$. In particular, you must show that $\overline{H}(X)$ cannot be a nonzero constant. Conversely, if $\overline{H}(X)$ is irreducible in $\mathbb{F}_p[X]$ prove that the above ideal I is a prime ideal.

   (h) Conclude that the (nonzero) prime ideals of $\mathbb{Z}[X]$ have three types: first the ideals $I = G(X)\mathbb{Z}[X]$ with $G(X) \in \mathbb{Z}[X]$ irreducible and primitive, second the ideals $I = p\mathbb{Z}[X]$ for p prime, and third the prime ideals $p\mathbb{Z}[X] + H(X)\mathbb{Z}[X]$ with $\overline{H}(X)$ irreducible in $\mathbb{F}_p[X]$.

   The ascending chains of prime ideals

   $$\{0\} \subset p\mathbb{Z}[X] \subset p\mathbb{Z}[X] + H(X)\mathbb{Z}[X] \text{ and } \{0\} \subset H(X)\mathbb{Z}[X] \subset p\mathbb{Z}[X] + H(X)\mathbb{Z}[X]$$

   mean that the dimension of $\mathbb{Z}[X]$ is equal to 2.

Part I

Tools

2. Abelian Groups, Lattices, and Finite Fields

This chapter introduces a number of necessary tools for the rest of the book, at different levels. The theory of finitely generated abelian groups, including the elementary divisor theorem and the structure theorem, as well as the theory of finite fields, should be known at the undergraduate level, but experience shows that this is not always the case, so for completeness we will give all the important proofs. Note that the theory of finitely generated abelian groups extends almost completely verbatim to finitely generated modules over a principal ideal domain; we will have the occasion to use this more general setting over the ring of p-adic integers $\mathbb{Z}_p$.

At a deeper level in this chapter we will also describe important results on the number of solutions of systems of polynomial equations over finite fields, culminating with the Weil conjectures proved by Deligne. Finally, we also include a section on lattices, seen mainly from the point of view of the LLL algorithm, which will be the main tool that we will use in applications to Diophantine equations.

2.1 Finitely Generated Abelian Groups

A set G is an abelian group if and only if it is a $\mathbb{Z}$-module. We will use indifferently both terms, but we will usually use abelian group when we want to emphasize group-theoretic properties, while we will use $\mathbb{Z}$-module when considering bases or generating families.

2.1.1 Basic Results

Lemma 2.1.1. *Let G be a finitely generated torsion-free abelian group generated by $x_1, \ldots, x_n$, and assume that G cannot be generated by fewer than n elements. Then there is no nontrivial relation $\sum_{1 \leqslant i \leqslant n} a_i x_i = 0$ with $a_i \in \mathbb{Z}$.*

Proof. Assume the contrary, and among all sets of n generators and all such relations on them, choose one for which $\sum_{1 \leqslant i \leqslant n} |a_i|$ is the smallest. We distinguish two cases:

– If at least two of the a_i are nonzero, then permuting subscripts and changing signs if necessary we may assume that $a_1 \geqslant a_2 > 0$. Clearly $x_1, x_1 + x_2$,
$x_3, \ldots, x_n$ still generate G, and the relation between these generators is
$(a_1 - a_2)x_1 + a_2(x_1 + x_2) + \cdots + a_n x_n = 0$, and the corresponding sum
of absolute values of coefficients is thus strictly smaller than the preceding
one, a contradiction.
– If only one of the a_i, say a_1, is nonzero, the relation is $a_1 x_1 = 0$, and
since G is torsion-free $x_1 = 0$; hence G is generated by the $n - 1$ elements
$x_2, \ldots, x_n$, a contradiction.   □

Corollary 2.1.2. *Any finitely generated torsion-free $\mathbb{Z}$-module is free.*

Proof. Indeed, choose a generating system (x_i) having the smallest number
of elements. By the lemma, it is a $\mathbb{Z}$-basis of G.   □

Theorem 2.1.3 (Elementary divisor theorem I). *Let G be a finitely generated torsion-free (hence free) abelian group, and let H be a subgroup of G.
There exists a basis $x_1, \ldots, x_n$ of G and strictly positive integers $m_1, \ldots, m_r$
for some $r \leqslant n$ such that $m_i \mid m_{i+1}$ for $1 \leqslant i \leqslant r - 1$ and such that the $m_i x_i$
for $1 \leqslant i \leqslant r$ form a basis for H. In addition, if H has finite index in G then
$r = n$.*

Proof. We can assume H nontrivial, otherwise we can choose $r = 0$.
For the moment let $y_1, \ldots, y_n$ be any basis of G. For any nonzero $h = \sum_{1 \leqslant i \leqslant n} a_i y_i \in H$, set $d(h) = \gcd(a_1, \ldots, a_n)$. I claim that this does not
depend on the chosen basis, but only on h: indeed, any other $\mathbb{Z}$-basis is given
in terms of the initial one by an $n \times n$ integral matrix P whose inverse is also
integral, in other words that has determinant ± 1 (the group of such matrices
is denoted by $\mathrm{GL}_n(\mathbb{Z})$). If A is the column vector of the (a_i), the new coefficients are given by the vector $P^{-1}A$. Clearly the GCD of the coefficients of
A divides that of $P^{-1}A$, and since $A = PP^{-1}A$, the converse is also true;
hence the GCD is the same, proving our claim.

Choose for h a nonzero element of H for which $d(h)$ is as small as possible, and choose a basis $y_1, \ldots, y_n$ of G for which the corresponding sum
$\sum_{1 \leqslant i \leqslant n} |a_i|$ is as small as possible. If two of the a_i were nonzero, as in the
proof of Lemma 2.1.1, we could decrease $\sum_{1 \leqslant i \leqslant n} |a_i|$ by modifying the basis,
a contradiction. Thus only one a_i is nonzero, and after permuting subscripts
we may assume that $a_1 = m_1 > 0$ is the only nonzero coefficient.

Now let $z = \sum_{1 \leqslant i \leqslant n} b_i y_i$ be any element of H. Then we obtain successively

– $m_1 \mid b_1$ since otherwise $0 < b_1 - cm_1 < m_1$ for the Euclidean quotient c of
b_1 by m_1 would give $d(z - ch) \leqslant b_1 - cm_1 < m_1 = d(h)$, a contradiction.
– $\sum_{2 \leqslant i \leqslant n} b_i y_i = z - (b_1/m_1)h \in H$ clearly.
– For each i, $m_1 \mid b_i$, for otherwise $t = m_1 y_1 + \sum_{2 \leqslant i \leqslant n} b_i y_i \in H$ would have
$d(t) < m_1 = d(h)$.

Finally, let G_1 be the group generated by $y_2, \ldots, y_n$, let $H_1 = G_1 \cap H$, and choose $x_1 = y_1$. We have proved that

$$G = \mathbb{Z}x_1 \oplus G_1 \quad \text{and} \quad H = \mathbb{Z}m_1 x_1 \oplus H_1 \,,$$

and that the coefficients of the elements of H_1 on $y_2, \ldots, y_n$, hence on any $\mathbb{Z}$-basis of G_1, are divisible by m_1.

We now repeat the process on G_1 and H_1 instead of G and H, and by the last remark, we obtain $m_1 \mid m_2$, and we continue until H is exhausted, proving the main part of the theorem. If $r < n$, then the mx_n for $m \in \mathbb{Z}$ belong to distinct cosets of H, so H does not have finite index in G, proving the last point.   □

Corollary 2.1.4. *With the notation of the theorem, if we denote by $\overline{x}$ the class of an element of G in G/H, we have*

$$G/H = \bigoplus_{1 \leqslant i \leqslant r} (\mathbb{Z}/m_i \mathbb{Z})\overline{x_i} \oplus \bigoplus_{r < i \leqslant n} \mathbb{Z}\overline{x_i} \,.$$

Proof. Clear. Note that this is an *equality*, not only an isomorphism.   □

Corollary 2.1.5. *Any subgroup of a finitely generated free abelian group is a finitely generated free abelian group of lower dimension.*

Proof. Also clear, H being free on the $m_i x_i$ for $1 \leqslant i \leqslant r$ and $r \leqslant n$.   □

In a different direction we have the following result, which can also be proved directly (see Exercise 1).

Corollary 2.1.6. *Let $V \in \mathbb{Z}^n$ be a column vector of n globally coprime integers. There exists an integral matrix $A \in \mathrm{GL}_n(\mathbb{Z})$ (in other words with determinant ± 1) having V as first column.*

Proof. In the proposition, we let $G = \mathbb{Z}^n$ and $H = \mathbb{Z}V$. There exists a basis $A_1, \ldots, A_n$ of G and $d \in \mathbb{Z}_{\geqslant 1}$ such that dA_1 is a basis of H. In particular, $V \in dk A_1$ for some $k \in \mathbb{Z}$, and since the coefficients of V are globally coprime it follows that $d = 1$ and $k = \pm 1$; hence $\pm V = A_1$ is the first column of the matrix of the A_i, which is in $\mathrm{GL}_n(\mathbb{Z})$.   □

We now come to the general structure theorem for finitely generated abelian groups.

Theorem 2.1.7 (Elementary divisor theorem II). *Let G be a finitely generated abelian group. There exist elements $x_1, \ldots, x_n$ of G and positive integers $m_1, \ldots, m_r$ for some $r \leqslant n$ such that $m_i > 1$ for $1 \leqslant i \leqslant r$, $m_i \mid m_{i+1}$ for $1 \leqslant i \leqslant r - 1$, $m_i x_i = 0$ for $1 \leqslant i \leqslant r$, and such that every element of G can be written uniquely in the form $\sum_{1 \leqslant i \leqslant n} a_i x_i$ with $0 \leqslant a_i < m_i$ when $1 \leqslant i \leqslant r$. Furthermore, n, r, and the m_i are unique.*

Proof. To prove existence, let $y_1, \ldots, y_N$ be any generators of G, and let $G^* = \bigoplus_{1 \leqslant i \leqslant N} \mathbb{Z} Y_i \simeq \mathbb{Z}^N$ be the free abelian group on N generators Y_i. There is a natural surjection from G^* to G sending Y_i to y_i for all i, and if H^* is its kernel, we have a natural isomorphism $G \simeq G^*/H^*$, so under this isomorphism we can identify G with G^*/H^*. Since G^* is free, we can apply the above theorem and corollary to G^* and H^*, obtaining generators X_i and integers M_i for $1 \leqslant i \leqslant R$. If we denote by x_i the image of X_i in G, we thus have

$$G = \bigoplus_{1 \leqslant i \leqslant R} (\mathbb{Z}/M_i\mathbb{Z})x_i \oplus \bigoplus_{r < i \leqslant N} \mathbb{Z} x_i \ .$$

We can evidently suppress from this equality all the components with $M_i = 1$, and if we call m_i the remaining M_i (in the same order), all the conditions of the theorem are satisfied, proving existence.

To prove uniqueness, assume that we have a second such representation, where we add $'$ to all the letters. We first prove that $n = n'$. Indeed, assume for instance that $n > n'$, and let p be a prime dividing m_1 if $r > 0$, and let p be any prime otherwise. Using the first representation, we have a surjection from G to $(\mathbb{Z}/p\mathbb{Z})^n$ sending $\sum a_i x_i$ to the vector of a_i modulo p, which makes sense since $p \mid m_i$ for all i. Since the (x_i') generate G, it follows that $(\mathbb{Z}/p\mathbb{Z})^n$ must be generated by the images of the x_i', which is absurd since there exist at most $n' < n$ such images.

Now for any $m > 0$, consider the group $mG = \{mx, \ x \in G\}$. We can obtain a representation as above by replacing the x_i by the mx_i, m_i by $m_i/\gcd(m_i, m)$, and deleting the mx_i for which $m_i \mid m$. It follows that for a fixed i, m_i is uniquely defined by the property that m_i is the smallest $m > 1$ for which a canonical representation of mG as above uses at most $n - i$ generators. $\square$

As in Corollary 2.1.4, we can restate the existence part of the theorem by writing

$$G = \bigoplus_{1 \leqslant i \leqslant r} (\mathbb{Z}/m_i\mathbb{Z})x_i \oplus \bigoplus_{r < i \leqslant n} \mathbb{Z} x_i \ .$$

Corollary 2.1.8. *Any subgroup of a finitely generated abelian group is finitely generated.*

Proof. Once again, we use a finitely generated free abelian group G^* and a surjective map from G^* to G. If H is a subgroup of G, denote by H^* the inverse image of H by this map. By Corollary 2.1.5, H^* is finitely generated, and the images of a finite set of generators of H^* generate H. $\square$

We now easily deduce the structure theorem for finite abelian groups:

Theorem 2.1.9. *Let G be a finite abelian group. There exist unique integers $m_i > 1$ for $1 \leqslant i \leqslant k$ such that $m_i \mid m_{i+1}$ for $1 \leqslant i < k$, and nonunique elements $g_i \in G$ such that*

$$G = \bigoplus_{1 \leqslant i \leqslant k} (\mathbb{Z}/m_i\mathbb{Z})g_i \,,$$

so that in particular $G \simeq \bigoplus_{1 \leqslant i \leqslant k}(\mathbb{Z}/m_i\mathbb{Z})$.

Proof. Indeed, if G is finite it is finitely generated. We have seen above as a consequence of Theorem 2.1.7 that any such group can be written

$$G = \bigoplus_{1 \leqslant i \leqslant r} (\mathbb{Z}/m_i\mathbb{Z})g_i \oplus \bigoplus_{r < i \leqslant k} \mathbb{Z}g_i \,,$$

for some $g_i \in G$ and $m_i > 1$ such that $m_i \mid m_{i+1}$ for $1 \leqslant i < r$. If $r < k$, then G contains copies of $\mathbb{Z}$; hence it is infinite. Thus, if G is finite we must have $r = k$, proving the theorem. $\qquad\square$

Finally, note that there is a matrix version of the elementary divisor theorem, called the *Smith normal form*. Recall that a matrix is *unimodular* if it is an element of $\mathrm{GL}_k(\mathbb{Z})$, i.e., an integral square matrix with determinant equal to ± 1 (not only $+1$).

Theorem 2.1.10 (Smith normal form). *Let A be a square integral matrix with nonzero determinant. There exist two unimodular matrices U and V and a diagonal integral matrix D with strictly positive diagonal entries such that $D = UAV$, and if $D = (d_{i,j})$ then $d_{i,i} \mid d_{i+1,i+1}$ for all $i \leqslant k - 1$.*

Proof. We apply Theorem 2.1.3 to $G = \mathbb{Z}^k$ and H the group of $\mathbb{Z}$-linear combinations of columns of A considered as elements of $\mathbb{Z}^k$. We leave to the reader to check that we thus obtain the present theorem (Exercise 2). $\qquad\square$

Note that D is unique but U and V are not (for instance if A is the identity matrix I, then $D = I$ and we can take any matrix U and $V = U^{-1}$).

To finish this section, recall the following.

Definition 2.1.11. *Let G be a group and $g \in G$. The set E of elements $e \in \mathbb{Z}$ such that $g^e = 1$ is of the form $k\mathbb{Z}$ for a unique $k \geqslant 0$. If $k = 0$ we say that g has infinite order, otherwise we call k the order of g in G. It is thus characterized by the following: $g^k = 1$, and $g^n = 1$ if and only if $k \mid n$.*

Proposition 2.1.12. *Let G be an abelian group and let $g \in G$ be an element of finite order $k = k_1 k_2$ with k_1 and k_2 coprime. There exist g_1 and g_2 in G of respective orders k_1 and k_2 such that $g = g_1 g_2$.*

Proof. Since k_1 and k_2 are coprime there exist integers u_1 and u_2 such that $u_1 k_1 + u_2 k_2 = 1$. We set $g_1 = g^{u_2 k_2}$ and $g_2 = g^{u_1 k_1}$. It is clear that $g_1 g_2 = g$, and furthermore by definition $g_1^n = 1$ if and only if $g^{u_2 k_2 n} = 1$ if and only if $k \mid u_2 k_2 n$ if and only if $k_1 \mid u_2 n$, and since $u_1 k_1 + u_2 k_2 = 1$, k_1 and u_2 are coprime, hence $k_1 \mid n$, so that g_1 has order k_1, and similarly g_2 has order k_2. $\qquad\square$

2.1.2 Description of Subgroups

Given a finite abelian group G, it is often useful to enumerate the subgroups of G. This can easily be done using the *Hermite normal form* of a matrix. Recall the following definition.

Definition 2.1.13. *An $n \times n$ matrix M is said to be in (upper triangular) Hermite normal form (HNF for short) if M is upper triangular with nonnegative integral entries, the diagonal entries $m_{i,i}$ of M are strictly positive, and the nondiagonal entries $m_{i,j}$ with $j > i$ are such that $0 \leqslant m_{i,j} < m_{i,i}$.*

Note that this definition can easily be generalized to nonsquare matrices (see Definition 6.2.1), but here we do not need this generality.

It is easy to show that if A is an integral matrix with nonzero determinant, there exists a unimodular matrix U such that $H = AU$ is in HNF, and H (and therefore U) is unique. More generally, if A is an $n \times k$ matrix with $n \leqslant k$ of maximal rank n, there exists a unimodular matrix U and a matrix H in HNF such that $AU = (H|0)$, concatenation of H with $k - n$ zero columns, and H is unique (but not U if $k > n$); see Exercise 3 and Proposition 6.2.2.

The HNF is useful in many contexts, essentially of algorithmic nature. Its relevance here is the following result.

Theorem 2.1.14. *Let G be a finite abelian group, and using the notation of Theorem 2.1.9, write*

$$G = \bigoplus_{1 \leqslant i \leqslant k} (\mathbb{Z}/m_i\mathbb{Z})g_i \ .$$

Denote by D be the $k \times k$ diagonal matrix whose diagonal entries are the integers m_i and by E the row vector whose entries are the generators g_i. The subgroups G' of G are in one-to-one correspondence with left divisors M of D in HNF, i.e., integral matrices M in HNF such that $M^{-1}D$ has integral entries. The correspondence is as follows:

(1) *If $M = (m_{i,j})_{1 \leqslant i,j \leqslant k}$ is such an HNF matrix, the subgroup G' is generated by $E' = EM$ with relations given by the columns of the matrix $M^{-1}D$.*

(2) *Conversely, if G' is a subgroup of G generated by a row of elements E', there exists an integer matrix P such that $E' = EP$, and the corresponding HNF matrix M is the HNF of the matrix $(P|D)$ obtained by concatenation of the matrices P and D.*

(3) *In this correspondence, we have $|G'| = |G| / \det(M)$, or equivalently, $|G/G'| = [G : G'] = \det(M)$.*

Proof. By definition, the following sequence is exact:

$$1 \longrightarrow \bigoplus_{i=1}^{k} m_i\mathbb{Z} \longrightarrow \mathbb{Z}^k \overset{\phi}{\longrightarrow} G \longrightarrow 1 \ ,$$

where

$$\phi(x_1, \ldots, x_k) = \prod_{1 \leqslant i \leqslant k} x_i g_i \ .$$

Let $(\varepsilon_i)_{1 \leqslant i \leqslant k}$ be the canonical basis of $\mathbb{Z}^k$, and let Λ be the subgroup of $\mathbb{Z}^k$ defined by $\Lambda = \bigoplus_i m_i \varepsilon_i$ (this is a *lattice*, see Section 2.3.1). We thus have a canonical isomorphism $G \simeq \mathbb{Z}^k/\Lambda$, obtained by sending the ith generator g_i of G to the class of ε_i.

Subgroups of $\mathbb{Z}^k/\Lambda$ have the form Λ'/Λ, where Λ' is a lattice such that $\Lambda \subset \Lambda' \subset \mathbb{Z}^k$. By existence and uniqueness of the HNF of a matrix of maximal rank, such a lattice Λ' can be uniquely defined by a matrix M in HNF such that the columns of this matrix express a $\mathbb{Z}$-basis of Λ' on the ε_i. The condition $\Lambda' \subset \mathbb{Z}^k$ means that M has integer entries, and the condition $\Lambda \subset \Lambda'$ means that $M^{-1}D$ also has integer entries, since it is the matrix that expresses the given basis of Λ in terms of that of Λ'. In terms of generators, this correspondence translates into the equality $E' = EM$. Furthermore, if 0_G denotes the unit element of G then $E'X = 0_G$ if and only if $EMX = 0_G$; hence $MX = DY$, or $X = M^{-1}DY$, and so if G' is the subgroup of G corresponding to Λ'/Λ, it is given in terms of generators and relations by $(EM, M^{-1}D)$, proving (1).

For (2), we note that the entries of ED are equal to 0_G; hence if $E'' = E(P|D)$, we have simply added some $1_{G'}$'s to the generators of G'. Thus, the group can be defined by the generators E'' and the matrix of relations of maximal rank $(P|D)$, hence also by (E'', M), where M is the HNF of this matrix.

For (3), we know that $M^{-1}D$ expresses a basis of Λ in terms of a basis of Λ'; hence

$$|G'| = |\Lambda'/\Lambda| = \det(M^{-1}D) = |G| / \det(M) \ .$$

$\square$

Example. The matrix M corresponding to the subgroup $\{0_G\}$ of G is $M = D$, and the matrix corresponding to the subgroup G of G is $M = I_k$, the $k \times k$ identity matrix.

Remark. Thanks to the above theorem the algorithmic *enumeration* of subgroups of a finite abelian group is reduced to the enumeration of the integral left divisors of a diagonal matrix. This is considerably more technical, and since the present book is not primarily algorithmic in nature we refer to [Coh1] Section 4.1.10 for complete details on the subject.

2.1.3 Characters of Finite Abelian Groups

First, an important notation. Here and in the rest of this book we use the symbol ζ_n for a *primitive* nth root of unity, either viewed as an (abstract) algebraic number (see Chapter 3), or as an element of $\mathbb{C}$ (for instance $\exp(2i\pi/n)$), or sometimes of other fields such as $\mathfrak{p}$-adic fields. If $d \mid n$, it is understood that we choose ζ_d such that $\zeta_d = \zeta_n^{n/d}$.

Definition 2.1.15. *Let G be a finite abelian group. A character of G is a group homomorphism from G to the multiplicative group $\mathbb{C}^*$ of nonzero complex numbers. The group of characters of G is called the* dual group *of G and denoted by $\widehat{G}$. The character sending all elements of G to 1, which is the unit element of $\widehat{G}$, is called the trivial character.*

Let $\chi \in \widehat{G}$. If $|G| = n$, then for every $g \in G$ we have $\chi(g)^n = \chi(g^n) = \chi(1) = 1$. It follows that any character takes values in the unit circle of complex numbers of modulus 1, more precisely in the group of nth roots of unity, which we denote by μ_n.

Proposition 2.1.16. *Let G be a finite abelian group. The dual group $\widehat{G}$ is noncanonically isomorphic to G (hence has the same cardinality).*

Proof. By the structure theorem for finite abelian groups (Theorem 2.1.9) we know that

$$G = \bigoplus_{1 \leqslant i \leqslant k} (\mathbb{Z}/m_i\mathbb{Z})g_i \simeq \bigoplus_{1 \leqslant i \leqslant k} (\mathbb{Z}/m_i\mathbb{Z})$$

for certain integers m_i and $g_i \in G$. On the other hand, we clearly have $\widehat{G_1 \oplus G_2} \simeq \widehat{G_1} \oplus \widehat{G_2}$. It follows that to prove the first part of the proposition it is sufficient to prove it for finite cyclic groups. But such a group is isomorphic to $\mathbb{Z}/m\mathbb{Z}$ for some m, and characters of $\mathbb{Z}/m\mathbb{Z}$ are simply determined by the image of the class of 1, which can be any mth root of unity. Thus we have (canonically) $\widehat{\mathbb{Z}/m\mathbb{Z}} \simeq \mu_m$, and (noncanonically) $\mu_m \simeq \mathbb{Z}/m\mathbb{Z}$, proving the result. $\qquad \square$

Remark. It follows from the proof that characters of a finite abelian group G can be described very concretely. We write $G = \bigoplus_{1 \leqslant i \leqslant k}(\mathbb{Z}/m_i\mathbb{Z})g_i$ as in Theorem 2.1.9, and for each i we choose some $a_i \in \mathbb{Z}/m_i\mathbb{Z}$. We then define

$$\chi_{a_1,\ldots,a_k}\left(\sum_{1 \leqslant i \leqslant k} x_i g_i\right) = \prod_{1 \leqslant i \leqslant k} \zeta_{m_i}^{a_i x_i} \,,$$

where the ζ_{m_i} are fixed primitive m_ith roots of unity in $\mathbb{C}$. Even more explicitly, since all the m_i divide $m_k = m$, we fix $\zeta = \zeta_m$ and choose $\zeta_{m_i} = \zeta^{m/m_i}$, so that

$$\chi_{a_1,\ldots,a_k}\left(\sum_{1 \leqslant i \leqslant k} x_i g_i\right) = \zeta^S \quad \text{with} \quad S = \sum_{1 \leqslant i \leqslant k} a_i x_i(m/m_i) \,,$$

so the value of the character χ can be represented by the integer S.

Corollary 2.1.17. *Let G be a finite abelian group and H a subgroup of G. Any character of H can be extended to exactly $[G : H]$ characters of G. In particular, the natural restriction map from $\widehat{G}$ to $\widehat{H}$ is surjective.*

Proof. Let f be the above restriction map. The kernel of f is the group of characters χ of G that are trivial on H, in other words the characters of G/H. It follows by the proposition that the cardinality of the image of f is equal to

$$\frac{|\widehat{G}|}{|\widehat{G/H}|} = \frac{|G|}{|G/H|} = |H| = |\widehat{H}| \; ;$$

hence f is surjective, as claimed, and the number of preimages of a character by the restriction map is equal to $|\operatorname{Ker}(f)| = [G : H]$. $\qquad\square$

Remarks. (1) Since we have proved surjectivity using a counting argument, the above reasoning cannot be applied to infinite abelian groups. See Proposition 4.4.43 for a generalization using Zorn's lemma.

(2) The importance of the above corollary is not so much the exact number of extensions of a character, but the simple fact that such extensions exist. For instance:

Corollary 2.1.18. *If g is not the unit element of G there exists $\chi \in \widehat{G}$ such that $\chi(g) \neq 1$.*

Proof. If $n > 1$ is the order of g we set $\chi(g^k) = \zeta_n^k$, which defines a character such that $\chi(g) \neq 1$ on the subgroup H of G generated by g, and we extend χ to G using the above corollary. $\qquad\square$

Corollary 2.1.19. *The natural map $a \mapsto (\chi \mapsto \chi(a))$ gives a canonical isomorphism from G to the dual of its dual.*

Proof. By the preceding corollary this map is injective, and since both groups have the same cardinality it is an isomorphism. $\qquad\square$

One of the most important properties of characters is their orthogonality properties as follows.

Proposition 2.1.20. *Let G be a finite abelian group and let K be a commutative field.*

(1) *If χ_1 and χ_2 are distinct group homomorphisms from G to K^* then*

$$\sum_{g \in G} \chi_1(g)\chi_2^{-1}(g) = 0 \;,$$

or equivalently, if χ is not the constant homomorphism equal to 1 then

$$\sum_{g \in G} \chi(g) = 0 \;.$$

(2) *In the special case $K = \mathbb{C}$ then if g_1 and g_2 are* distinct *elements of G we have*

$$\sum_{\chi \in \hat{G}} \chi(g_1 g_2^{-1}) = 0 \,,$$

or equivalently, if g is not the unit element of G then

$$\sum_{\chi \in \hat{G}} \chi(g) = 0 \,.$$

Proof. The two statements of (1) are clearly equivalent, as are those of (2). Set $S = \sum_{g \in G} \chi(g)$. Let $h \in G$ such that $\chi(h) \neq 1$. Then

$$\chi(h)S = \sum_{g \in G} \chi(h)\chi(g) = \sum_{g \in G} \chi(hg) = \sum_{g' \in G} \chi(g') = S \,,$$

hence $S = 0$ since $\chi(h) \neq 1$.

In the special case $K = \mathbb{C}$, if g is not the unit element of G, Corollary 2.1.18 shows that there exists $\psi \in \hat{G}$ such that $\psi(g) \neq 1$. The reasoning we have just presented is thus applicable (we set $S = \sum_{\chi} \chi(g)$ and show that $\psi(g)S = S$). □

Note that if $K \neq \mathbb{C}$ (more precisely if K is not an algebraically closed field of characteristic 0 or of characteristic not dividing $|G|$), then (2) is not necessarily true, see Exercise 4.

2.1.4 The Groups $(\mathbb{Z}/m\mathbb{Z})^*$

We first recall that for any commutative ring R, R^* denotes the group of *units* of R, i.e., invertible elements in R. This is equal to $R \setminus \{0\}$ if and only if R is a field.

Lemma 2.1.21. *Let $v \geq 2$ be an integer and $a \in \mathbb{Z}$.*

(1) *If p is an odd prime number, the following statements are equivalent:*
   (a) $a^p \equiv 1 \pmod{p^v}$.
   (b) $a \equiv 1 \pmod{p^{v-1}}$.
   (c) *For any $w \geq v - 1$ there exists $b \in \mathbb{Z}$ coprime to p such that*

$$a \equiv b^{(p-1)p^{v-2}} \pmod{p^w} \,.$$

In particular, $v_p(a^p - 1) = v_p(a - 1) + 1$ when $v_p(a - 1) \geq 1$.
(2) *If $p = 2$ and $a \equiv 1 \pmod 4$ the following statements are equivalent:*
   (a) $a^2 \equiv 1 \pmod{2^v}$.
   (b) $a \equiv 1 \pmod{2^{v-1}}$.

If in addition $v \geqslant 4$, they are also equivalent to the following statement:
c) For any $w \geqslant v - 1$ there exists $b \in \mathbb{Z}$ odd such that

$$a \equiv b^{2^{v-3}} \pmod{2^w}.$$

Proof. (1). To prove that a) is equivalent to b) we set $a = 1 + b$. By Fermat's theorem we have $a^p \equiv a \pmod{p}$; hence we may assume that $p \mid b$, so that

$$a^p = 1 + pb + \sum_{2 \leqslant j \leqslant p-1} \binom{p}{j} b^j + b^p.$$

Since $p \mid \binom{p}{j}$ it follows that all the terms with $2 \leqslant j \leqslant p - 1$ are divisible by pb^2. Furthermore, since $p \geqslant 3$ and $p \mid b$, we also have $pb^2 \mid b^p$. It follows that all the terms after pb have p-adic valuation strictly greater than that of pb; hence $v_p(a^p - 1) = v_p(pb) = 1 + v_p(a - 1)$, proving that a) and b) are equivalent.

By the Euler–Fermat theorem, if b is coprime to p we have $b^{(p-1)p^{v-2}} = b^{\phi(p^{v-1})} \equiv 1 \pmod{p^{v-1}}$; hence c) implies b). To prove the converse, we consider the map f from $(\mathbb{Z}/p^{w-v+2}\mathbb{Z})^*$ to the subgroup G of elements of $(\mathbb{Z}/p^w\mathbb{Z})^*$ congruent to 1 modulo p^{v-1} induced by $y \mapsto y^{(p-1)p^{v-2}}$. By the equivalence of a) and b), $y \equiv 1 \pmod{p^{w-v+2}}$ implies $y^{p^{v-2}} \equiv 1 \pmod{p^w}$; hence the map f is well defined. Since $y^{(p-1)p^{v-2}} \equiv 1 \pmod{p^{v-1}}$ its image lies in G. Furthermore, $f(\overline{y}) = 1$ if and only if $y^{(p-1)p^{v-2}} \equiv 1 \pmod{p^w}$ if and only if $y^{p-1} \equiv 1 \pmod{p^{w-v+2}}$, again by the equivalence of a) and b). Since we will prove below that $(\mathbb{Z}/p^{w-v+2}\mathbb{Z})^*$ is a cyclic group, the number of $\overline{y}$ of order dividing $p-1$ in that group is equal to $p-1$, so that $|\operatorname{Ker}(f)| = p-1$. It follows that $|\operatorname{Im}(g)| = \phi(p^{w-v+2})/p = p^{w-v+1}$, and since clearly $|G| = p^{w-v+1}$, this means that f is surjective, proving the equivalence of b) and c).

(2). If $a = 1 + b$ with $2^{v-1} \mid b$ then $a^2 = 1 + 2b + b^2 \equiv 1 \pmod{2^v}$ since $v \geqslant 2$. Conversely, if $a^2 \equiv 1 \pmod{2^v}$, then $2^v \mid (a-1)(a+1)$, and since $a \equiv 1 \pmod{4}$, $v_2(a+1) = 1$, and therefore $2^{v-1} \mid a - 1$, proving the equivalence of the first two conditions.

If b is odd we have $b^{2v-3} = (b^2)^{2^{v-4}} \equiv 1 \pmod{2^{v-1}}$ by what we have just shown and $b^2 \equiv 1 \pmod{8}$. Conversely, we consider as above the map f from $(\mathbb{Z}/2^{w-v+3})^*$ to the subgroup G of elements of $(\mathbb{Z}/2^w\mathbb{Z})^*$ congruent to 1 modulo 2^{v-1} induced by $y \mapsto y^{2^{v-3}}$. Since $w - v + 3 \geqslant 2$, by what we have just shown $y \equiv 1 \pmod{2^{w-v+3}}$ implies $y^{2^{v-3}} \equiv 1 \pmod{2^w}$; hence f is well defined, and as above $y^{2^{v-3}} \equiv 1 \pmod{2^{v-1}}$ so the image of f lies in G. We have $f(\overline{y}) = 1$ if and only if $y^{2^{v-3}} \equiv 1 \pmod{2^w}$ if and only if $y^2 \equiv 1 \pmod{2^{w-v+4}}$ by what we have just shown. Writing $y^2 - 1 = (y+1)(y-1)$ and noting that $w - v + 4 \geqslant 3$ we see that this is equivalent to $y \equiv \pm 1 \pmod{2^{w-v+3}}$; hence $|\operatorname{Ker}(f)| = 2$. It follows that $|\operatorname{Im}(g)| = \phi(2^{w-v+3})/2 = 2^{w-v+1}$, and since clearly $|G| = p^{w-v+1}$, this again means that f is surjective, finishing the proof. $\qquad \square$

Note also the following generalization, which we will need later.

Lemma 2.1.22. *Let p be a prime number, s an integer such that $s \equiv 1$ (mod p), and let $n \in \mathbb{Z}_{>0}$. When $p = 2$, assume that either $s \equiv 1$ (mod 4) or n is odd. Then*

$$v_p(s^n - 1) = v_p(s - 1) + v_p(n) .$$

Proof. Write $n = p^v m$ with $p \nmid m$. We prove the lemma by induction on v. Assume first that $v = 0$, so that $n = m$. By the binomial theorem, we have

$$s^m - 1 = \sum_{1 \leqslant k \leqslant m} \binom{m}{k} (s - 1)^k .$$

Since $p \nmid m$, we have

$$v_p\left(\binom{m}{1} (s - 1)^1 \right) = v_p(m(s - 1)) = v_p(s - 1) ,$$

while for $2 \leqslant k \leqslant m$ we have

$$v_p\left(\binom{m}{k} (s - 1)^k \right) \geqslant k v_p(s - 1) \geqslant 2 v_p(s - 1) .$$

Since $v_p(s - 1) \geqslant 1$ by assumption it follows that $v_p(s^m - 1) = v_p(s - 1)$ as claimed.

When $v \geqslant 1$ we apply Lemma 2.1.21 (1) to $a = s^{p^{v-1}m} \equiv 1$ (mod p) (since $s \equiv 1$ (mod p)), hence $v_p(a^p - 1) = v_p(a - 1)$, so the result for p odd follows by induction on v. Similarly the result for $p = 2$ follows by induction from Lemma 2.1.21 (2). $\square$

Corollary 2.1.23. *If $p = 2$ and $s \equiv 1$ (mod 2) then for all $n \in \mathbb{Z}_{>0}$ we have*

$$v_p(s^n - 1) = \begin{cases} v_p(s - 1) & \text{if } n \text{ is odd,} \\ v_p(s^2 - 1) + v_p(n/2) & \text{if } n \text{ is even.} \end{cases}$$

Proof. The case n odd is given by the lemma. When n is even, since $s^2 \equiv 1$ (mod 4) the lemma gives $v_p(s^n - 1) = v_p((s^2)^{n/2} - 1) = v_p(s^2 - 1) + v_p(n/2)$. $\square$

Proposition 2.1.24. *Let $m \geqslant 2$ be an integer, and let $m = \prod_{1 \leqslant i \leqslant g} p_i^{v_i}$ be its decomposition into a product of powers of distinct primes. The abelian group structure of $(\mathbb{Z}/m\mathbb{Z})^*$ is given as follows:*

(1) *We have*

$$(\mathbb{Z}/m\mathbb{Z})^* \simeq \prod_{1 \leqslant i \leqslant g} (\mathbb{Z}/p_i^{v_i}\mathbb{Z})^* .$$

(2) *If $p \geqslant 3$ and $v \geqslant 1$, we have*

$$(\mathbb{Z}/p^v\mathbb{Z})^* \simeq \mathbb{Z}/(p^{v-1}(p-1))\mathbb{Z} \; ;$$

in other words the group $(\mathbb{Z}/p^v\mathbb{Z})^*$ is cyclic.
(3) *If $p = 2$ and $v \geqslant 3$, we have*

$$(\mathbb{Z}/2^v\mathbb{Z})^* \simeq \mathbb{Z}/2^{v-2}\mathbb{Z} \times \mathbb{Z}/2\mathbb{Z} \; .$$

In addition, if desired we can always take the class of 5 as generator of the group $\mathbb{Z}/2^{v-2}\mathbb{Z}$, and -1 as generator of $\mathbb{Z}/2\mathbb{Z}$.
(4) *If $p = 2$ and $v \leqslant 2$, then $(\mathbb{Z}/2\mathbb{Z})^*$ is the trivial group and $(\mathbb{Z}/4\mathbb{Z})^* \simeq \mathbb{Z}/2\mathbb{Z}$.*

Proof. (1). I first claim that if $m = m_1 m_2$ with $\gcd(m_1, m_2) = 1$, then $(\mathbb{Z}/m\mathbb{Z})^* \simeq (\mathbb{Z}/m_1\mathbb{Z})^* \times (\mathbb{Z}/m_2\mathbb{Z})^*$. Indeed, there exist integers u and v such that $um_1 + vm_2 = 1$. Denoting by $a + n\mathbb{Z}$ the class in $\mathbb{Z}/n\mathbb{Z}$ of an integer a modulo any integer n (which is in fact the correct notation), we consider the map f_1 from $(\mathbb{Z}/m\mathbb{Z})^*$ to $(\mathbb{Z}/m_1\mathbb{Z})^* \times (\mathbb{Z}/m_2\mathbb{Z})^*$ defined by $f_1(a + m\mathbb{Z}) = (a + m_1\mathbb{Z}, a + m_2\mathbb{Z})$, and the map f_2 in the other direction defined by $f_2(b + m_1\mathbb{Z}, c + m_2\mathbb{Z}) = cm_1u + bm_2v + m\mathbb{Z}$. Since $m = m_1 m_2$ these maps are clearly well defined, and we immediately check that they are group homomorphisms which are inverse to one another, proving my claim. By induction on g, this proves (1).

(2). For any integers m and a, we will say that a is a *primitive root* modulo m if the class of a modulo m generates $(\mathbb{Z}/m\mathbb{Z})^*$ (so that in particular $(\mathbb{Z}/m\mathbb{Z})^*$ is cyclic and $\gcd(a, m) = 1$). By Corollary 2.4.3 below we know that $(\mathbb{Z}/p\mathbb{Z})^*$ is cyclic; in other words there exists $g \in \mathbb{Z}$ that is a primitive root modulo p. By Fermat's theorem, i.e., the fact that $|(\mathbb{Z}/p\mathbb{Z})^*| = p-1$, we know that $g^{p-1} \equiv 1 \pmod{p}$. Assume first that $g^{p-1} \not\equiv 1 \pmod{p^2}$. I claim that g is a primitive root modulo p^v for any $v \geqslant 1$. Indeed, otherwise there would exist a prime divisor q of $\phi(p^v) = p^{v-1}(p-1)$ such that $g^{\phi(p^v)/q} \equiv 1 \pmod{p^v}$. Since $a^p \equiv a \pmod{p}$ for all a, if $q \mid p-1$ we have $g^{p^{v-1}(p-1)/q} \equiv g^{(p-1)/q} \equiv 1 \pmod{p}$, which is absurd since g is a primitive root modulo p. If $q \nmid p-1$ then $q = p$, and since by Lemma 2.1.21, $a^p \equiv 1 \pmod{p^k}$ for $k \geqslant 2$ implies that $a \equiv 1 \pmod{p^{k-1}}$, the congruence $g^{p^{v-2}(p-1)} \equiv 1 \pmod{p^v}$ implies that $g^{p-1} \equiv 1 \pmod{p^2}$, contrary to our assumption.

Assume now that $g^{p-1} \equiv 1 \pmod{p^2}$. Then $g + p$ is also a primitive root modulo p, and since $p \geqslant 3$,

$$(g+p)^{p-1} \equiv g^{p-1} + (p-1)pg^{p-2} \equiv 1 - pg^{p-2} \not\equiv 1 \pmod{p^2} \; ,$$

so it follows from what we have just proved that $g + p$ is a primitive root modulo p^v for all $v \geqslant 1$, proving (2).

(3). Let H denote the subgroup of $(\mathbb{Z}/2^v\mathbb{Z})^*$ formed by the classes of integers congruent to 1 modulo 4. Since any odd integer is congruent to

± 1 modulo 4, we clearly have $(\mathbb{Z}/2^v\mathbb{Z})^* \simeq H \times \mathbb{Z}/2\mathbb{Z}$, so that in particular $|H| = \phi(2^v)/2 = 2^{v-2}$. I claim that H is cyclic, generated by the class of 5. Since the only prime dividing 2^{v-2} is 5, it is enough to show that $5^{2^{v-3}} \not\equiv 1$ (mod 2^v). If we assume the contrary, then since $5 \equiv 1$ (mod 4), using once again Lemma 2.1.21 we would obtain that $5 \equiv 1$ (mod 2^3), a contradiction. (4) is trivial. $\qquad\square$

Corollary 2.1.25. *For $m \geqslant 2$, the group $(\mathbb{Z}/m\mathbb{Z})^*$ is cyclic if and only if $m = 2$, 4, p^k, or $2p^k$ for p an odd prime and $k \geqslant 1$.*

Proof. Note that in a cyclic group the number of elements of order dividing 2 is less than or equal to 2. From the proposition it follows that the number of such elements is exactly equal to $2^{\omega_o(m)+\omega_2(m)}$, where $\omega_o(m)$ is the number of distinct odd prime divisors of m and $\omega_2(m) = 0, 1$, or 2 according to whether $v_2(m) \leqslant 1$, $v_2(m) = 2$, or $v_2(m) \geqslant 3$ respectively. The corollary follows from the inequality $\omega_o(m) + \omega_2(m) \leqslant 1$. $\qquad\square$

Remark. The proofs made in this subsection sometimes use forward references, so we must be careful to check that we do not use a circular argument. Assume for instance p odd, the remark is the same for $p = 2$. The correct order of proof (which would be less practical for presentation) is as follows: the equivalence of (a) and (b) of Lemma 2.1.21 (1), as well as the cyclicity of $(\mathbb{Z}/p\mathbb{Z})^*$ which follows from Corollary 2.4.3, are proved directly, without any reference to results of this subsection. From these two results we deduce by induction as in the proof of Proposition 2.1.24 (2) that $(\mathbb{Z}/p^k\mathbb{Z})^*$ is cyclic for all $k \geqslant 1$. Using this, we can finally prove the equivalence of (b) and (c) of Lemma 2.1.21 (1).

When working in a group $(\mathbb{Z}/p^v\mathbb{Z})^*$ with p and odd prime and $v \geqslant 2$, it is tempting to use the existence of a primitive root g modulo p^v, since all elements a can simply be written as $a = g^x$ for some x uniquely defined modulo $\phi(p^v)$. However, this is not always a good idea. For future reference, we note the following lemma, which usually gives a better representation.

Lemma 2.1.26. *Let p be an odd prime, let $v \geqslant 2$, and let g be a primitive root modulo p^v. For any a coprime to p there exist x and y such that*

$$a \equiv g^{p^{v-1}x}(1+p)^y \pmod{p^v},$$

and x is unique modulo $p - 1$, y is unique modulo p^{v-1}.

Proof. Since g is a primitive root, we can write $a \equiv g^x \pmod{p^v}$, so that $a^{p^{v-1}} \equiv g^{p^{v-1}x} \pmod{p^v}$, and since g has order $p^{v-1}(p-1)$, it is clear that x is unique modulo $p-1$. Since $a^{p^{v-1}-1} \equiv 1 \pmod{p}$ by Fermat's little theorem, we must simply show that for any $b \equiv 1 \pmod{p}$ there exists y such that $b \equiv (1+p)^y \pmod{p^v}$. Indeed, the map $y \mapsto (1+p)^y$ is clearly a group homomorphism from the additive group $\mathbb{Z}/p^{v-1}\mathbb{Z}$ to the multiplicative subgroup of

$(\mathbb{Z}/p^v\mathbb{Z})^*$ of elements congruent to 1 modulo p. The groups having the same cardinality, and the map being injective by Lemma 2.1.22, it follows that it is bijective, showing the existence of y and its uniqueness modulo p^{v-1}.   □

Remark. This lemma can be better understood in the context of p-adic numbers (see Chapter 4): indeed, $1 + p$ is naturally called a *topological generator*, and the exponent y such that $b \equiv (1 + p)^y \pmod{p^v}$ can be given explicitly in terms of p-adic logarithms as $y = \log_p(b)/\log_p(1 + p) \bmod p^v$. Note also that $1 + p$ is only one possible choice, and that we could just as well choose $1 + kp$ for any $k \not\equiv 0 \pmod{p}$.

2.1.5 Dirichlet Characters

According to Proposition 2.1.16, the group $(\widehat{\mathbb{Z}/m\mathbb{Z}})^*$ of characters of $(\mathbb{Z}/m\mathbb{Z})^*$ is (noncanonically) isomorphic to $(\mathbb{Z}/m\mathbb{Z})^*$. It is convenient to extend such a character to the whole of $\mathbb{Z}/m\mathbb{Z}$ by setting it equal to 0 outside of $(\mathbb{Z}/m\mathbb{Z})^*$, and then to $\mathbb{Z}$ by composing with the natural surjection from $\mathbb{Z}$ to $\mathbb{Z}/m\mathbb{Z}$:

Definition 2.1.27. *A Dirichlet character modulo m is a map χ from $\mathbb{Z}$ to $\mathbb{C}$ such that there exists a character $\psi \in (\widehat{\mathbb{Z}/m\mathbb{Z}})^*$ such that $\chi(n) = 0$ if $\gcd(n, m) > 1$, while $\chi(n) = \psi(\overline{n})$ otherwise, where $\overline{n}$ denotes the class of n modulo m.*

Note that χ is still multiplicative, and that $\chi(m + n) = \chi(n)$ for all n, in other words χ is periodic of period dividing m. Furthermore, the values of χ are either equal to 0 or $\phi(m) = |(\mathbb{Z}/m\mathbb{Z})^*|$th roots of unity in $\mathbb{C}$. By abuse of language we will say that χ has *order n* if the corresponding character $\psi \in (\widehat{\mathbb{Z}/m\mathbb{Z}})^*$ has order n, in other words if n is the positive generator of the group of integers k such that χ^k is equal to the trivial character modulo m (see the following definition). Thus the order of a character modulo m divides $\phi(m)$.

Definition 2.1.28. *Let χ be a character modulo m.*

(1) *If $d \mid m$ we say that χ can be defined modulo d if there exists a Dirichlet character χ_d modulo d such that $\chi(n) = \chi_d(n)$ as soon as $\gcd(n, m) = 1$.*
(2) *The conductor of a Dirichlet character is the smallest (for divisibility) positive integer $f \mid m$ such that χ can be defined modulo f.*
(3) *We will say that χ is primitive if the conductor of χ is equal to m, in other words if χ cannot be defined modulo a proper divisor of m.*
(4) *The trivial character, often denoted by χ_0, is the character defined by $\chi(n) = 1$ when $\gcd(n, m) = 1$ and $\chi(n) = 0$ when $\gcd(n, m) > 1$. It is the unique character modulo m of conductor 1.*

Remarks. (1) It is clear that χ is primitive if and only if χ cannot be defined modulo m/p for every prime $p \mid m$.

(2) If χ can be defined modulo $d \mid m$ and χ_d is the corresponding character modulo d, it is clear that $\chi = \chi_0 \chi_d$.

Proposition 2.1.29. *The number of primitive characters modulo m is equal to $q(m)$, where*

$$q(m) = m \prod_{p \| m} \left(1 - \frac{2}{p}\right) \prod_{p^2 | m} \left(1 - \frac{1}{p}\right)^2 .$$

In particular, there are none if and only if $m \equiv 2 \pmod 4$.

Proof. I refer the reader to Section 10.1 for the elementary techniques used here. For any integer f denote by $q(f)$ the number of primitive characters modulo f. By definition we have

$$\phi(m) = |(\mathbb{Z}/m\mathbb{Z})^*| = |\widehat{(\mathbb{Z}/m\mathbb{Z})^*}| = \sum_{f | m} q(f) .$$

In terms of formal Dirichlet series, this means that

$$\zeta(s) \sum_{m \geq 1} q(m) m^{-s} = \sum_{m \geq 1} \phi(m) m^{-s} = \frac{\zeta(s-1)}{\zeta(s)} ;$$

hence $\sum_{m \geq 1} q(m) m^{-s} = \zeta(s-1)/\zeta(s)^2$, and the proposition follows by looking at the Euler factor at p.   □

Corollary 2.1.30. *Let χ be a primitive character modulo m with m even. Then for all n we have $\chi(n + m/2) = -\chi(n)$.*

Proof. By the proposition we know that $4 \mid m$. The result is thus trivial if n is even since both sides vanish; otherwise, denoting by n^{-1} an inverse of n modulo m we have since n is odd

$$\chi(n + m/2) = \chi(n)\chi(1 + (m/2)n^{-1}) = \chi(n)\chi(1 + m/2) .$$

We have $\chi((1+m/2)^2) = \chi(1+m(m/4+1)) = 1$, hence $\chi(1+m/2) = \pm 1$. If we had $\chi(1+m/2) = 1$ then we would have $\chi(k) = \chi(k+(m/2)k) = \chi(k+m/2)$ for all k odd and evidently for all even k, so χ would be defined modulo $m/2$, a contradiction. Thus $\chi(1 + m/2) = -1$; hence $\chi(n + m/2) = -\chi(n)$.   □

A similar reasoning will lead to a useful characterization of primitive characters. We first need a lemma which is useful in many contexts.

Lemma 2.1.31. *If $\gcd(a, b, c) = 1$ there exists an integer k such that*

$$\gcd(a + kb, c) = 1 .$$

Proof. Note that this lemma would immediately follow from Dirichlet's theorem on primes in arithmetic progression (see Theorem 10.5.30), but it is not necessary to use such a powerful tool. In fact, we can give k explicitly: I claim that

$$k = \prod_{\substack{p \mid c \\ p \nmid (a/\gcd(a,b))}} p$$

is a suitable value. Indeed, let p be a prime dividing c. We must show that it does not divide $a + kb$. Assume first that $p \nmid a/\gcd(a,b)$. Thus $p \mid k$; hence $p \nmid a/\gcd(a,b) + kb/\gcd(a,b)$, and since $\gcd(a,b,c) = 1$, we have $p \nmid a + kb$ as desired. Assume now that $p \mid a/\gcd(a,b)$, hence $p \nmid k$. Since $a/\gcd(a,b)$ is coprime to $b/\gcd(a,b)$ by definition of the GCD, it follows that $p \nmid kb/\gcd(a,b)$; hence $p \nmid a/\gcd(a,b) + kb/\gcd(a,b)$ and once again $p \nmid a+kb$ as desired. □

The characterization of primitive characters is a consequence of the following lemma:

Lemma 2.1.32. *Let χ be a character modulo m and let $d \mid m$. Then χ can be defined modulo d if and only if for all a such that $a \equiv 1 \pmod{d}$ and $\gcd(a,m) = 1$ we have $\chi(a) = 1$.*

Proof. The condition is clearly necessary: if $\chi(a) = \chi_d(a)$ for all a such that $\gcd(a,m) = 1$, then if in addition $a \equiv 1 \pmod{d}$ we have $\chi(a) = 1$. Conversely, assume the condition satisfied, and let a be such that $\gcd(a,d) = 1$. We want to define $\chi_d(a)$. By the preceding lemma, there exists k such that $\gcd(a + kd, m) = 1$. We will set $\chi_d(a) = \chi(a + kd)$. Since $\gcd(a + kd, m) = 1$, this is nonzero, and furthermore if k' is another integer such that $\gcd(a + k'd, m) = 1$, then $b = (a + k'd)(a + kd)^{-1}$ (inverse taken modulo m, which makes sense since $\gcd(a + kd, m) = 1$) is such that $b \equiv 1 \pmod{d}$. By assumption it follows that $\chi(b) = 1$, in other words that $\chi(a+k'd) = \chi(a+kd)$, so our definition of $\chi_d(a)$ does not depend on the choice of k. It is then immediate to check that χ_d is a character modulo d such that $\chi_d(a) = \chi(a)$ when $\gcd(a,m) = 1$. □

Corollary 2.1.33. *Let χ be a character modulo m, let $d \mid m$ with $d < m$, and assume that χ cannot be defined modulo d.*

(1) *For all r we have*

$$\sum_{\substack{a \bmod m \\ a \equiv r \pmod{d}}} \chi(a) = 0 \,.$$

(2) *If f is a periodic function of period dividing d, then*

$$\sum_{0 \leqslant a < m} \chi(a) f(a) = 0 \,.$$

In particular, if χ is a primitive character, these properties are true for all $d \mid m$ such that $d < m$.

Proof. The proof of (1) is identical to that of Proposition 2.1.20: by the lemma, there exists $b \equiv 1 \pmod{d}$ with $\gcd(b, m) = 1$ and such that $\chi(b) \neq 1$. The map $a \mapsto ab$ is clearly a bijection from the set of integers modulo m congruent to r modulo m to itself, so that by multiplicativity we have $\chi(b)S = S$, hence $S = 0$, where S is the sum to be computed. For (2) we write $a = qd + r$ so that

$$\sum_{0 \leqslant a < m} \chi(a)f(a) = \sum_{0 \leqslant r < d} f(r) \sum_{\substack{0 \leqslant a < m \\ a \equiv r \pmod{d}}} \chi(a) = 0$$

by (1). $\square$

Proposition 2.1.34. *Let $m \in \mathbb{Z}_{\geqslant 1}$, let m_1 and m_2 be two coprime positive integers such that $m = m_1 m_2$, and let χ be a Dirichlet character modulo m.*

(1) *There exist unique characters χ_i modulo m_i such that $\chi = \chi_1 \chi_2$, in other words such that $\chi(n) = \chi_1(n)\chi_2(n)$ for all n.*
(2) *The order of χ is equal to the LCM of the orders of χ_1 and χ_2.*
(3) *The character χ is primitive if and only if both χ_1 and χ_2 are primitive.*

Proof. (1). Since the m_i are coprime, there exist integers u and v such that $um_1 + vm_2 = 1$. In view of the map f_2 defined in the proof of Proposition 2.1.24 (1), it is natural to set $\chi_1(x) = \chi(xm_2v + m_1u)$ and $\chi_2(x) = \chi(ym_1u + m_2v)$. Since these maps are obtained by composing the homomorphism f_2 with the natural injections of $(\mathbb{Z}/m_i\mathbb{Z})^*$ into $(\mathbb{Z}/m_1\mathbb{Z})^* \times (\mathbb{Z}/m_2\mathbb{Z})^*$, it follows that they are group homomorphisms hence define Dirichlet characters modulo m_1 and m_2 respectively, and it is also clear that $\chi = \chi_1 \chi_2$.

(2). We have $\chi(n)^k = 1$ for all n coprime to m if and only if $\chi_1(n)^k = \chi_2(n)^{-k}$ for all such n, hence if and only if the primitive character equivalent to χ_1^{-k} is equal to the one equivalent to χ_2^{-k}. However, the conductor of χ_i^{-k} divides m_i, and m_1 and m_2 are coprime, so this is possible if and only if χ_1^k and χ_2^k are trivial characters, hence if and only if k is a multiple of the orders of χ_1 and χ_2, proving (2).

The proof of (3) is immediate and left to the reader (Exercise 10). $\square$

It follows in particular from this proposition that any Dirichlet character χ modulo m can be written in a unique way as a product of Dirichlet characters modulo the coprime prime powers dividing m.

Corollary 2.1.35. *Let $m = \prod_p p^{v_p(m)}$ with $v_p(m) \geqslant 1$ be the decomposition into prime powers of $m \in \mathbb{Z}_{\geqslant 1}$. The order of any primitive character modulo m is divisible by $h(m) = \prod_p p^{v_p(m)-1}$, except if $8 \mid m$ in which case it is only divisible by $h(m)/2$.*

Proof. Let χ be a primitive character modulo m. By the above proposition applied inductively we can write $\chi = \prod_p \chi_p$, where χ_p is a primitive character modulo $p^{v_p(m)}$, and the order of χ will be equal to the LCM of the orders of the χ_p. It is thus sufficient to prove that the order of χ_p is divisible by $p^{v_p(m)-1}$, or by $2^{v_2(m)-2}$ if $v_2(m) \geqslant 3$. For simplicity of notation, write v instead of $v_p(m)$. Assume first that $p \geqslant 3$ and let $a \equiv 1 \pmod{p^{v-1}}$. By Lemma 2.1.21 (1) c), there exists b such that $a \equiv b^{(p-1)p^{v-2}} \pmod{p^v}$, hence $\chi_p(a) = \chi_p(b)^{(p-1)p^{v-2}}$. Since $\phi(p^v) = (p-1)p^{v-1}$, if the order of χ_p was not divisible by p^{v-1} it would divide $(p-1)p^{v-2}$, hence we would have $\chi_p(a) = 1$ for all $a \equiv 1 \pmod{p^{v-1}}$. However, by Lemma 2.1.32 this would imply that χ_p can be defined modulo p^{v-1}, contradicting the fact that it is a primitive character modulo p^v and proving the result for $p \geqslant 3$. Assume now that $p = 2$, hence that $v \geqslant 2$ since there are no primitive characters for $m \equiv 2 \pmod 4$. If $v = 2$ or $v = 3$ the only primitive characters modulo 4 or 8 are the characters $\left(\frac{-4}{\cdot}\right)$, $\left(\frac{-8}{\cdot}\right)$, and $\left(\frac{8}{\cdot}\right)$ which have order $2 = h(4) = h(8)/2$, hence we may assume that $v \geqslant 4$. Let $a \equiv 1 \pmod{2^{v-1}}$. By Lemma 2.1.21 (2), there exists b such that $a \equiv b^{2^{v-3}} \pmod{2^v}$, and as for the case $p \geqslant 3$ we deduce that if the order of χ_2 was not divisible by 2^{v-2} it would divide 2^{v-3}, and we again deduce a contradiction. $\square$

In the context of Dirichlet characters, Proposition 2.1.20 reads as follows:

Proposition 2.1.36. *We have*

$$\sum_{a \bmod m} \chi(a) = \begin{cases} \phi(m) & \text{if } \chi \text{ is the trivial character modulo } m, \\ 0 & \text{otherwise.} \end{cases}$$

Dually, if $a \in \mathbb{Z}$ is such that $\gcd(a, m) = 1$ then

$$\sum_{\chi \bmod m} \chi(a) = \begin{cases} \phi(m) & \text{if } a \equiv 1 \pmod m, \\ 0 & \text{otherwise.} \end{cases}$$

Corollary 2.1.37. *Let χ be a nontrivial character modulo m, let $I = [1, m-1]$, and let*

$$S = \sum_{a \in I, \ a \text{ even}} \chi(a) = - \sum_{a \in I, \ a \text{ odd}} \chi(a) .$$

(1) *If either m is even or χ is an even character then $S = 0$.*
(2) *If m is odd and χ is an odd character, then if in addition χ is primitive and m is not a prime power we have $S \equiv 0 \pmod 2$, in other words $S/2$ is an algebraic integer.*

Proof. (1). The fact that the two sums given in the corollary are opposite follows of course from Proposition 2.1.36, and if m is even then $\chi(a) = 0$ for all even a, hence $S = 0$ trivially. On the other hand, if m is odd and χ is even we have

$$S = \sum_{a \in I,\ a \text{ even}} \chi(a) = \sum_{a \in I,\ a \text{ even}} \chi(m - a) = \sum_{a \in I,\ b \text{ odd}} \chi(b) = -S ,$$

hence $S = 0$ also in this case, proving (1).

(2). Consider the sum $S_1 = \sum_{a \in I} \chi(a)a$. We have

$$S_1 \equiv \sum_{a \in I,\ a \text{ odd}} \chi(a)a \equiv S \ (\text{mod } 2) ,$$

so we must show that $S_1 \equiv 0 \ (\text{mod } 2)$. Since m is not a prime power we can write $m = m_1 m_2$ with $m_i > 1$ coprime, and by the Proposition 2.1.34 there exist primitive, hence nontrivial characters χ_i modulo m_i such that $\chi = \chi_1 \chi_2$. Any $a \in [0, m - 1]$ can be written in a unique way as $a = m_1 a_2 + a_1$ with $0 \leqslant a_2 < m_2$ and $0 \leqslant a_1 < m_1$, hence $S_1 = m_1 T_2 + T_1$ with

$$T_j = \sum_{0 \leqslant a_1 < m_1,\ 0 \leqslant a_2 < m_2} \chi_1(a_1)\chi_2(m_1 a_2 + a_1)a_j .$$

Since m_1 and m_2 are coprime we have

$$\sum_{0 \leqslant a_2 < m_2} \chi_2(m_1 a_2 + a_1) = \chi_2(m_1) \sum_{0 \leqslant a_2 < m_2} \chi_2(a_2 + a_1 m_1^{-1}) = 0$$

by Proposition 2.1.36 since χ_2 is nontrivial, hence $T_1 = 0$. On the other hand, if we write $T_2 = \sum_{a_1, a_2} f(a_1, a_2)$ with $f(a_1, a_2) = \chi_1(a_1)\chi_2(m_1 a_2 + a_1)a_2$, then since $m_2 - 1$ is even, it is clear that $f(m_1 - a_1, m_2 - a_2 - 1) \equiv f(a_1, a_2)$ (mod 2), and since this involution has no fixed points (m_1 being odd), we deduce that $T_2 \equiv 0 \ (\text{mod } 2)$, hence that $S_1 \equiv 0 \ (\text{mod } 2)$, proving (2).   □

Remarks. (1) We will study sums generalizing S_1 in much more detail in Section 9.5.1, and use exactly the same reasoning as above.
(2) The result of (2) is not true in general when m is a prime power, as can easily be seen on examples.

2.1.6 Gauss Sums

We are going to meet Gauss sums in two related, but different, contexts: first in the present section, those related to a Dirichlet character, second in the context of finite fields, those related to additive and multiplicative characters of finite fields. Although the results are similar we must prove them separately.

Definition 2.1.38. *Let χ be a (not necessarily primitive) character modulo m. For $a \in \mathbb{Z}$ we define the* Gauss sum $\tau(\chi, a)$ *by the formula*

$$\tau(\chi, a) = \sum_{x \bmod m} \chi(x) \zeta_m^{ax} \ .$$

It is clear that this makes sense, in other words it does not depend on the representatives chosen for x modulo m, both in $\chi(x)$ and in the exponential.

By abuse of notation we will also set $\tau(\chi) = \tau(\chi, 1)$ and call it the Gauss sum associated with the character χ.

Proposition 2.1.39. *If $\gcd(a, m) = 1$ we have*

$$\tau(\chi, a) = \overline{\chi(a)}\tau(\chi) \ .$$

Proof. Since $\gcd(a, m) = 1$, the map multiplication by a is a bijection of $(\mathbb{Z}/m\mathbb{Z})^*$ to itself; hence setting $y = ax$ we have

$$\tau(\chi, a) = \sum_{y \bmod m} \chi(ya^{-1}) \zeta_m^{y} \ .$$

Since $\chi(ya^{-1}) = \chi(y)\overline{\chi(a)}$ because $\chi(a)$ has modulus 1, the proposition follows. □

Proposition 2.1.40. *Let $d = \gcd(a, m)$ and assume that χ cannot be defined modulo m/d. Then $\tau(\chi, a) = 0$.*

Proof. Since χ cannot be defined modulo m/d, by Lemma 2.1.32 we can find b such that $b \equiv 1 \pmod{m/d}$, $\gcd(b, m) = 1$, and $\chi(b) \neq 1$. Thus

$$\chi(b)\tau(\chi, a) = \sum_{x \bmod m} \chi(bx) \zeta_m^{ax} = \sum_{y \bmod m} \chi(y) \zeta_m^{ayb^{-1}} \ .$$

However, since $b \equiv 1 \pmod{m/d}$, we have

$$ayb^{-1} = (a/d)dyb^{-1} \equiv ay \pmod{m}$$

(and not only modulo m/d, because of the factor d); hence

$$\chi(b)\tau(\chi, a) = \sum_{y \bmod m} \chi(y) \zeta_m^{ay} = \tau(\chi, a) \ ,$$

so that $\tau(\chi, a) = 0$ since $\chi(b) \neq 1$. □

Corollary 2.1.41. *If χ is a nontrivial character modulo m, then*

$$\sum_{x \bmod m} \chi(x) = 0 \ .$$

Proof. Apply the above proposition to $a = 0$. $\qquad\square$

This corollary is of course a special case of Proposition 2.1.20.

Corollary 2.1.42. *Assume that χ is a* primitive *character. For all a (not necessarily prime to m) we have*

$$\tau(\chi, a) = \overline{\chi(a)}\tau(\chi) .$$

Proof. If $\gcd(a, m) = 1$ this is Proposition 2.1.39, and if $d = \gcd(a, m) > 1$, then χ cannot be defined modulo m/d so the result follows from Proposition 2.1.40. $\qquad\square$

The reader will find in Exercise 12 a general formula giving $\tau(\chi, a)$.

Corollary 2.1.43. *Assume that χ is a primitive character modulo m and let $n = km$ be a multiple of m. Then*

$$\sum_{x \bmod n} \chi(x)\zeta_n^{ax} = \begin{cases} 0 & \text{if } k \nmid a \\ k\overline{\chi}(a/k)\tau(\chi) & \text{if } k \mid a. \end{cases}$$

Proof. Immediate by writing $x = mq + r$ with $r \bmod m$ and $q \bmod k$ and left to the reader. $\qquad\square$

The above results have in fact little to do with Gauss sums. Indeed, let f be any function defined on mth roots of unity with values in some field, and set $\tau_f(\chi, a) = \sum_{x \bmod m} \chi(x)f(\zeta_m^{ax})$ and $\tau_f(\chi) = \tau_f(\chi, 1)$. Exactly the same proofs as above show the following (Exercise 13).

Proposition 2.1.44. *Set $d = \gcd(a, m)$. If $d = 1$ we have $\tau_f(\chi, a) = \overline{\chi(a)}\tau_f(\chi)$, if χ cannot be defined modulo m/d we have $\tau_f(\chi, a) = 0$, and if χ is a primitive character then for all a we have $\tau_f(\chi, a) = \overline{\chi(a)}\tau_f(\chi)$.*

A very important result concerning Gauss sums is the following.

Proposition 2.1.45. *If χ is a primitive character modulo m then $|\tau(\chi)| = m^{1/2}$.*

Proof. We have $\overline{\tau(\chi)} = \sum_{a \bmod m} \overline{\chi(a)}\zeta_m^{-a}$; hence multiplying by $\tau(\chi)$ and applying the above corollary we obtain

$$|\tau(\chi)|^2 = \sum_{a \bmod m} \tau(\chi, a)\zeta_m^{-a} = \sum_{a \bmod m}\sum_{x \bmod m} \chi(x)\zeta_m^{ax}\zeta_m^{-a}$$

$$= \sum_{x=1}^{m} \chi(x)\sum_{a=1}^{m} \zeta_m^{a(x-1)} .$$

The inner sum is a geometric series, whose sum is equal to 0 if $m \nmid (x-1)$; in other words $x \ne 1$, and is equal to m otherwise, proving the proposition. $\quad\square$

Corollary 2.1.46. *Let χ be a not necessarily primitive character modulo m, and let f be its conductor. Then $|\tau(\chi)| = f^{1/2}$ if m/f is squarefree and coprime to f; otherwise $\tau(\chi) = 0$.*

Proof. This follows from the above proposition and the formula $\tau(\chi) = \mu(m/f)\chi_f(m/f)\tau(\chi_f)$ proved in Exercise 12. □

Corollary 2.1.47. *If χ is a primitive character we have*

$$\tau(\chi)\tau(\overline{\chi}) = \chi(-1)m .$$

In particular, if χ is real, in other words takes only values ± 1 on integers coprime to m, we have $\tau(\chi)^2 = \chi(-1)m$.

Proof. Indeed, by Proposition 2.1.39 we have

$$\overline{\tau(\chi)} = \sum_{x \bmod m} \overline{\chi}(x)\zeta_m^{-x} = \chi(-1)\tau(\overline{\chi}) ,$$

so multiplying by $\tau(\chi)$ and using Proposition 2.1.45 gives $m = \chi(-1)\tau(\chi)\tau(\overline{\chi})$, proving the corollary. □

2.2 The Quadratic Reciprocity Law

2.2.1 The Basic Quadratic Reciprocity Law

Let p be an odd prime. We set $\left(\frac{a}{p}\right) = 0$ if $p \mid a$, and otherwise $\left(\frac{a}{p}\right) = 1$ if a is congruent to a square modulo p, $\left(\frac{a}{p}\right) = -1$ otherwise. This is called the *Legendre symbol.* We recall the following easy result.

Proposition 2.2.1. (1) *The symbol $\left(\frac{a}{p}\right)$ is a real primitive character modulo p, and in particular $\left(\frac{ab}{p}\right) = \left(\frac{a}{p}\right)\left(\frac{b}{p}\right)$.*
(2)

$$a^{(p-1)/2} \equiv \left(\frac{a}{p}\right) \pmod{p} ,$$

and in particular $\left(\frac{-1}{p}\right) = (-1)^{(p-1)/2}$.
(3) *There are exactly $(p-1)/2$ values of a modulo p (called quadratic residues) such that $\left(\frac{a}{p}\right) = 1$, and $(p-1)/2$ values of a modulo p (called quadratic nonresidues) such that $\left(\frac{a}{p}\right) = -1$.*

Proof. By Corollary 2.4.3 below we know that $(\mathbb{Z}/p\mathbb{Z})^*$ is cyclic (of order $p-1$). Let $g \in \mathbb{Z}$ be such that the class of g modulo p is a generator. Then if $a \in \mathbb{Z}$ is coprime to p, there exists an exponent k uniquely defined modulo $p-1$ such that $a \equiv g^k \pmod{p}$. We will call k a *discrete logarithm* of a modulo p to base g, and write $k = \log_g(a)$. It is clear that a is congruent to a square modulo p if and only if $\log_g(a)$ is even (since $p-1$ is even this does not depend on the chosen representative modulo $p-1$); hence $\left(\frac{a}{p}\right) = (-1)^{\log_g(a)}$. This immediately implies that $\left(\frac{a}{p}\right)$ is a real character, and it is primitive since it is not defined modulo 1, i.e., trivial. Also, since g is a generator, $g^{(p-1)/2} \equiv -1$ (mod p), so that

$$a^{(p-1)/2} \equiv g^{(p-1)/2\log_g(a)} \equiv (-1)^{\log_g(a)} \equiv \left(\frac{a}{p}\right) \pmod{p} .$$

In particular, when $a = -1$ both sides are equal to ± 1 and congruent modulo $p > 2$, so they are in fact equal. Finally, there are exactly $(p-1)/2$ even values and $(p-1)/2$ odd values of $k = \log_g(a)$ modulo $p-1$, proving the proposition. $\square$

Remark. When a is not divisible by p and $a \equiv x^2$ (mod p), it is clear that $a^{(p-1)/2} \equiv x^{p-1} \equiv 1$ (mod p); hence one direction of statement (2) is trivial. The converse statement says that if $a^{(p-1)/2} \equiv 1$ (mod p) then there exists x such that $a \equiv x^2$ (mod p). We have just proved this using the cyclicity of $(\mathbb{Z}/p\mathbb{Z})^*$, and indeed the result is not entirely trivial: the algorithmic computation of x (i.e., the square root of a modulo p) can be done using an algorithm due to Tonelli and Shanks; see [Coh0].

We will now prove a lemma that is basic to two of the results that we need, and in particular to the quadratic reciprocity law.

Lemma 2.2.2. *Let χ be a real primitive character modulo m, and let p be an odd prime. Then*

$$\chi(p) = \left(\frac{\chi(-1)m}{p}\right) .$$

Proof. Let $R = \mathbb{Z}[\zeta_m]$, which is a ring and a finitely generated free $\mathbb{Z}$-module since ζ_m is an algebraic *integer* (in fact of degree $\phi(m)$, but we do not need this). We do not need to know that in fact R is the ring of algebraic integers of $\mathbb{Q}(\zeta_m)$.

Let p be any odd prime such that $p \nmid m$. By the binomial theorem, and using either the fact that all intermediate binomial coefficients $\binom{p}{k}$ are divisible by p or the fact that in a ring of characteristic p (here R/pR) the map $x \mapsto x^p$ is additive, we have

$$\tau(\chi)^p \equiv \sum_{x \bmod m} \chi^p(x)\zeta_m^{px} \pmod{pR} .$$

Since χ is a real character and p is odd, $\chi^p(x) = \chi(x)$, so by Proposition 2.1.39 we have

$$\tau(\chi)^p \equiv \chi(p)\tau(\chi) \pmod{pR} .$$

On the other hand, since χ is primitive, Corollary 2.1.47 tells us that $\tau(\chi)^2 = \chi(-1)m$, so that multiplying the above congruence by $\tau(\chi)$ we obtain

$$\chi(-1)m((\chi(-1)m)^{(p-1)/2} - \chi(p)) \equiv 0 \ (\text{mod } pR) .$$

Since $(\chi(-1)m)^{(p-1)/2} \equiv \left(\frac{\chi(-1)m}{p}\right) \ (\text{mod } p)$ and $p \nmid m$, if we multiply the above congruence by any integer u such that $um \equiv 1 \ (\text{mod } p)$ (in $\mathbb{Z}$), we obtain

$$\chi(p) \equiv \left(\frac{\chi(-1)m}{p}\right) \ (\text{mod } pR) .$$

Both sides of this congruence are in fact in $\mathbb{Z}$, and clearly $\mathbb{Z} \cap pR = p\mathbb{Z}$ since R is a free $\mathbb{Z}$-module with basis 1, ζ_m, $\ldots$, ζ_m^{n-1} for some n. Thus the above congruence holds not only modulo pR but modulo $p\mathbb{Z}$, i.e., modulo p. Since both sides are equal to ± 1 and $p > 2$, it follows finally that we have the *equality*

$$\chi(p) = \left(\frac{\chi(-1)m}{p}\right)$$

for p odd not dividing m. Clearly this is also true (both sides vanish) when p divides m, proving the lemma.   □

Corollary 2.2.3 (The basic quadratic reciprocity law). (1) *If p and q are distinct odd primes, we have*

$$\left(\frac{p}{q}\right)\left(\frac{q}{p}\right) = (-1)^{(p-1)(q-1)/4} .$$

(2) *If p is an odd prime, we have the two so-called complementary laws:*

$$\left(\frac{-1}{p}\right) = (-1)^{(p-1)/2} \quad and \quad \left(\frac{2}{p}\right) = (-1)^{(p^2-1)/8} .$$

Proof. Set $\chi(n) = \left(\frac{n}{q}\right)$, which is a real primitive character modulo q. We have $\chi(-1) = (-1)^{(q-1)/2}$, hence the above lemma gives

$$\left(\frac{p}{q}\right) = \left(\frac{(-1)^{(q-1)/2}q}{p}\right) = (-1)^{(p-1)(q-1)/4}\left(\frac{q}{p}\right) ,$$

proving (1). We have already proved the first equality of (2). For the second, set $\chi(x) = (-1)^{(x^2-1)/8}$ for x odd, 0 otherwise. This simply means that $\chi(x) = 1$ for $x \equiv \pm 1 \ (\text{mod } 8)$, and $\chi(x) = -1$ for $x \equiv \pm 3 \ (\text{mod } 8)$. It follows immediately from this that χ is a real character modulo 8, which is primitive, and is such that $\chi(-1) = 1$. Thus, again by the above lemma we obtain $\chi(p) = \left(\frac{8}{p}\right)$, in other words $\left(\frac{2}{p}\right) = (-1)^{(p^2-1)/8}$ as desired.   □

2.2.2 Consequences of the Basic Quadratic Reciprocity Law

We define the Jacobi symbol $\left(\frac{a}{n}\right)$ for a positive odd integer n by requiring multiplicativity in n: if $n = \prod_i p_i^{v_i}$ then

$$\left(\frac{a}{n}\right) = \prod_i \left(\frac{a}{p_i}\right)^{v_i} .$$

Clearly this symbol is still multiplicative in a. Furthermore, since each $\left(\frac{a}{p_i}\right)$ is periodic of period dividing p_i, hence a fortiori dividing n, $\left(\frac{a}{n}\right)$ is periodic of period dividing n.

Proposition 2.2.4. *If m and n are coprime positive odd integers, the same quadratic reciprocity formula holds:*

$$\left(\frac{m}{n}\right)\left(\frac{n}{m}\right) = (-1)^{(n-1)(m-1)/4} .$$

In addition, if n is an odd positive integer, the same complementary laws hold:

$$\left(\frac{-1}{n}\right) = (-1)^{(n-1)/2} \quad and \quad \left(\frac{2}{n}\right) = (-1)^{(n^2-1)/8} .$$

Proof. We note that if n_1 and n_2 are odd, then

$$n_1 n_2 - 1 = (n_1 - 1) + (n_2 - 1) + (n_1 - 1)(n_2 - 1) \equiv (n_1 - 1) + (n_2 - 1) \pmod 4 .$$

It follows that the right-hand side of the first formula is multiplicative in n, and by symmetry it is also multiplicative in m. Since the left-hand side of the formula is also multiplicative in m and in n, to prove equality it suffices to prove it when m and n are coprime odd primes, and then it is simply the basic quadratic reciprocity law.

The first complementary law follows from the same congruence for $n_1 n_2 - 1$, and the second from

$$n_1 n_2^2 - 1 = n_1^2 - 1 + n_2^2 - 1 + (n_1^2 - 1)(n_2^2 - 1) \equiv (n_1^2 - 1)(n_2^2 - 1) \pmod{16} .$$

$\square$

Definition 2.2.5. *The Kronecker symbol (still denoted by $\left(\frac{a}{b}\right)$) is the exten-sion of the Jacobi symbol to $(\mathbb{Z} \setminus \{0\})^2$ obtained by setting $\left(\frac{a}{-1}\right) = \text{sign}(a)$ and $\left(\frac{a}{2}\right) = \left(\frac{2}{a}\right)$ for a odd ($\left(\frac{a}{2}\right) = 0$ for a even), and extending by multiplicativity.*

Note that $\text{sign}(a) = 1$ if $a > 0$ and $\text{sign}(a) < 0$ if $a < 0$.

Proposition 2.2.6. (1) *For two nonzero integers m and n write $m = 2^{v_2(m)} m_1$ and $n = 2^{v_2(n)} n_1$ with m_1 and n_1 odd. Then*

$$\left(\frac{n}{m}\right) = (-1)^{((m_1-1)(n_1-1) + (\text{sign}(m)-1)(\text{sign}(n)-1))/4} \left(\frac{m}{n}\right) .$$

(2) *If D_1 and D_2 are nonzero integers congruent to 0 or 1 modulo 4, we have*

$$\left(\frac{D_2}{D_1}\right) = (-1)^{((\mathrm{sign}(D_1)-1)(\mathrm{sign}(D_2)-1))/4}\left(\frac{D_1}{D_2}\right).$$

Proof. We may assume that either m or n is odd; otherwise they are not coprime and the result is trivial. Since $\left(\frac{2}{a}\right) = \left(\frac{a}{2}\right)$, it is clear that statement (1) follows from Proposition 2.2.4 for m, n both positive. Using the definition of $\left(\frac{m}{-1}\right)$ and considering separately the cases $m > 0$, $n < 0$ (or the reverse), we obtain (1) after a short computation. Statement (2) immediately follows from (1) since either D_1 and D_2 are not coprime, or either D_1 or D_2 is odd, hence congruent to 1 modulo 4. □

Our aim is now to study the periodicity of the Kronecker symbol, especially when the numerator is fixed congruent to 0 or 1 modulo 4.

Lemma 2.2.7. *If m is odd, for any k we have*

$$\left(\frac{a+km}{m}\right) = (-1)^{(\mathrm{sign}(m)-1)(\mathrm{sign}(a+km)-\mathrm{sign}(a))/4}\left(\frac{a}{m}\right).$$

Proof. When $m > 0$, we have periodicity because of the periodicity of the Legendre symbol. If $m < 0$, by definition of the Kronecker symbol we have

$$\left(\frac{a+km}{m}\right) = \mathrm{sign}(a+km)\left(\frac{a+km}{|m|}\right)$$

$$= \mathrm{sign}(a+km)\left(\frac{a}{|m|}\right)\mathrm{sign}(a+km)\,\mathrm{sign}(a)\left(\frac{a}{m}\right),$$

and clearly $\mathrm{sign}(a)\,\mathrm{sign}(b) = (-1)^{(\mathrm{sign}(a)-\mathrm{sign}(b))/2}$. □

Remark. The symbol $\left(\frac{a}{2}\right)$ being periodic of period 8 and not 2, it is in general not useful to consider periodicity of the symbol $\left(\frac{a}{m}\right)$ when m is a fixed even integer.

Lemma 2.2.8. *If m is odd, then writing $n = 2^{v_2(n)}n_1$ and $n + km = 2^{v_2(n+km)}(n+km)_1$, we have*

$$\left(\frac{m}{n+km}\right) = (-1)^{(m-1)((n+km)_1-n_1)/4}\left(\frac{m}{n}\right).$$

Proof. By Proposition 2.2.6 and the above lemma, we have

$$\left(\frac{m}{n+km}\right) = (-1)^{(m-1)((n+km)_1-1)/4+(\mathrm{sign}(m)-1)(\mathrm{sign}(n+km)-1)/4}\left(\frac{n+km}{m}\right)$$

$$= (-1)^{(m-1)((n+km)_1-1)/4+(\mathrm{sign}(m)-1)(\mathrm{sign}(n)-1)/4}\left(\frac{n}{m}\right)$$

$$= (-1)^{(m-1)((n+km)_1-n_1)/4}\left(\frac{m}{n}\right),$$

proving the lemma.   □

We can finally state the main result concerning the Kronecker symbol.

Theorem 2.2.9. *If $m \equiv 0$ or 1 modulo 4 is fixed, the Kronecker symbol $\left(\frac{m}{n}\right)$ is periodic of period dividing $|m|$; in other words for all k, n we have*

$$\left(\frac{m}{n+km}\right) = \left(\frac{m}{n}\right) .$$

Proof. If $m \equiv 1 \pmod 4$ the result follows from the above lemma since in that case $(m-1)((n+km)_1 - n_1) \equiv 0 \pmod 8$. So assume $m \equiv 0 \pmod 4$, and write as usual $m = 2^{v_2(m)} m_1$. We may of course assume that n is odd, so that $n + km$ is also odd. We have therefore $(n+km)_1 - n_1 = km \equiv 0 \pmod 4$, so by the above lemma

$$\left(\frac{m}{n+km}\right) = \left(\frac{2}{n+km}\right)^u \left(\frac{m_1}{n+km}\right) = \left(\frac{2}{n+km}\right)^u \left(\frac{m_1}{n}\right)$$

$$= \left(\frac{2}{n(n+km)}\right)^u \left(\frac{m}{n}\right) .$$

If $u = 2$, $\left(\frac{2}{n(n+km)}\right)^u = 1$ trivially, and if $u \geqslant 3$, then $m \equiv 0 \pmod 8$; hence $n(n+km) \equiv n^2 \equiv 1 \pmod 8$, so $\left(\frac{2}{n(n+km)}\right) = 1$, proving the theorem.   □

Proposition 2.2.10. *There are exactly two extensions $\left(\frac{a}{b}\right)$ of the Jacobi symbol to $(\mathbb{Z} \setminus \{0\})^2$ that are equal to 0 if and only if $\gcd(a,b) \neq 1$, and are multiplicative in a and b and periodic in b of period dividing $|a|$ when $a \equiv 0$ or 1 modulo 4. One is the Kronecker symbol $\left(\frac{a}{b}\right)$ defined above, the other is $\left(\frac{a}{b}\right)\left(\frac{-4}{a}\right)^{v_2(b)}$, where it is understood that $\left(\frac{-4}{a}\right)^{v_2(b)} = 1$ when b is odd.*

Proof. Assume first that $a \equiv 0 \pmod 4$ with $a \neq 0$, and write $a = \text{sign}(a) 2^u m$ with m odd and positive. Then we have

$$\left(\frac{a}{-1}\right) = \left(\frac{a}{|a|-1}\right) = \left(\frac{\text{sign}(a)|a|}{|a|-1}\right) = \text{sign}(a)^{(|a|-2)/2} \left(\frac{|a|}{|a|-1}\right)$$

$$= \text{sign}(a) \left(\frac{2}{2^u m - 1}\right)^u \left(\frac{m}{2^u m - 1}\right) .$$

Now since $u \geqslant 2$, we note that for $u = 2$, $\left(\frac{2}{2^u m - 1}\right)^u = 1$ trivially, while for $u \geqslant 3$, $2^u m - 1 \equiv -1 \pmod 8$, hence $\left(\frac{2}{2^u m - 1}\right)^u = 1$ once again. Thus, applying Proposition 2.2.4 and the fact that $2^u m - 1 \equiv 3 \pmod 4$, we obtain

$$\left(\frac{a}{-1}\right) = \text{sign}(a)(-1)^{(m-1)/2} \left(\frac{2^u m - 1}{m}\right)$$

$$= \text{sign}(a)(-1)^{(m-1)/2} \left(\frac{-1}{m}\right) = \text{sign}(a) ,$$

once again by periodicity in the upper argument since m is odd and positive.

If now $a \neq 0$ is arbitrary, we must have

$$\left(\frac{a}{-1}\right) = \left(\frac{a}{-1}\right)\left(\frac{2}{-1}\right)^2 = \left(\frac{4a}{-1}\right) = \mathrm{sign}(4a) = \mathrm{sign}(a)$$

as desired.

Let us now compute $\left(\frac{a}{2}\right)$. If a is even, we must have $\left(\frac{a}{2}\right) = 0$; hence we may assume that a is odd. Consider first the case $a \equiv 1 \pmod 4$. Then by periodicity and Proposition 2.2.4, we have

$$\left(\frac{a}{2}\right) = \left(\frac{a}{2 + |a|}\right) = \left(\frac{\mathrm{sign}(a)}{2 + |a|}\right)\left(\frac{|a|}{2 + |a|}\right)$$

$$= \mathrm{sign}(a)^{(|a|+1)/2}(-1)^{(|a|-1)(|a|+1)/4}\left(\frac{2 + |a|}{|a|}\right)$$

$$= \mathrm{sign}(a)^{(|a|+1)/2}\left(\frac{2}{|a|}\right) = \mathrm{sign}(a)^{(|a|+1)/2}\left(\frac{2}{a}\right).$$

Thus, when $a \equiv 1 \pmod 4$, if $a > 0$ then $\left(\frac{a}{2}\right) = \left(\frac{2}{a}\right)$, while if $a < 0$, then $|a| \equiv 3 \pmod 4$; hence $\left(\frac{a}{2}\right) = \left(\frac{2}{a}\right)$ once again.

Consider now the case $a \equiv 3 \pmod 4$. By what we have just proved,

$$\left(\frac{a}{2}\right) = \left(\frac{-1}{2}\right)\left(\frac{-a}{2}\right) = \left(\frac{-1}{2}\right)\left(\frac{2}{-a}\right) = \left(\frac{-1}{2}\right)\left(\frac{2}{a}\right).$$

Thus, if we choose $\left(\frac{-1}{2}\right) = 1$ we obtain the first extension, which is the Kronecker symbol, and if we choose $\left(\frac{-1}{2}\right) = -1$ we obtain the second extension, and we have thus shown that only these two possible extensions can exist. Conversely, by definition all the necessary conditions are satisfied, and periodicity in the lower variable for the first extension is the statement of Theorem 2.2.9. Note that if $a \equiv 1 \pmod 4$ then $\left(\frac{-4}{a}\right) = 1$, while if $a \equiv 0 \pmod 4$, then $\left(\frac{a}{b}\right) \neq 0$ implies b odd hence $\left(\frac{-4}{a}\right)^{v_2(b)} = 1$, so periodicity for the second extension follows from that of the first. $\square$

2.2.3 Gauss's Lemma and Quadratic Reciprocity

Another approach to quadratic reciprocity, due to Gauss, deserves to be studied in detail. Although there exist nearly two hundred proofs of the quadratic reciprocity law, it is not just for the sake of it that we will give another proof here, but because we need the results that we will prove elsewhere (see Theorem 11.6.14) and because it sheds some additional light on Kronecker–Jacobi's generalization of the Legendre symbol to composite denominators.

In this subsection, r (or r') will always denote a positive odd integer. By a convenient and common abuse of notation, when r is implicit we will identify integers with their class modulo r in $\mathbb{Z}/r\mathbb{Z}$.

Definition 2.2.11. *A half-system H modulo r is a subset of $\mathbb{Z}/r\mathbb{Z}$ such that $\mathbb{Z}/r\mathbb{Z} = H \cup (-H) \cup \{0\}$, the unions being disjoint, with the evident notation $-H = \{-h/\ h \in H\}$.*

If H is a half-system modulo r we evidently have $|H| = (r-1)/2$.

Now let a be an integer coprime to r. For any $j \in H$ we have $aj \not\equiv 0$ (mod r); hence we can write uniquely

$$aj \equiv \varepsilon_H(j)\sigma_H(j) \ (\text{mod } r) \ ,$$

where $\varepsilon_H(j) = \pm 1$ and $\sigma_H(j) \in H$. It is clear that σ_H is bijective: indeed, since it is a map from the finite set H to itself, it is enough to show that it is injective. But $\sigma_H(j_1) = \sigma_H(j_2)$ implies that $aj_1 \equiv \pm aj_2$ (mod r) hence that $j_1 \equiv \pm j_2$ (mod r) since $\gcd(a, r) = 1$, and since $j_1 \equiv -j_2$ (mod r) is excluded by definition of a half-system, we have therefore $j_1 = j_2$ as claimed, so that σ is indeed a permutation of H.

Define

$$f_H(a, r) = \prod_{j \in H} \varepsilon_H(j) \in \{\pm 1\} \ .$$

Proposition 2.2.12. *The quantity $f_H(a, r)$ does not depend on the half-system H.*

Proof. Let H' be another half-system. Then for any $j \in H$ there exists $\eta(j) = \pm 1$ such that $j = \eta(j)\pi(j)$, where π is a (necessarily bijective) map from H to H'. For simplicity of notation, set $\sigma_1 = \pi^{-1} \circ \sigma_{H'} \circ \pi$, a permutation of H. Then

$$\varepsilon_H(j)\sigma_H(j) \equiv aj = \eta(j)a\pi(j) = \eta(j)\varepsilon_{H'}(\pi(j))\sigma_{H'}(\pi(j))$$
$$= \eta(j)\varepsilon_{H'}(\pi(j))\pi(\sigma_1(j)) = \eta(j)\varepsilon_{H'}(\pi(j))\eta(\sigma_1(j))\sigma_1(j) \ .$$

Since $\sigma_H(j)$ and $\sigma_1(j)$ are both in H, by definition of a half-system we must have $\sigma_1 = \sigma_H$ and moreover

$$\varepsilon_H(j) = \eta(j)\varepsilon_{H'}(\pi(j))\eta(\sigma_1(j)) \ .$$

Since π and σ_1 are bijections, taking the product on $j \in H$ gives

$$f_H(a, r) = f_{H'}(a, r)\left(\prod_{j \in H} \eta(j)\right)^2 = f_{H'}(a, r) \ ,$$

finishing the proof.   □

Since $f_H(a, r)$ does not depend on H, we will of course drop the index H. This proposition will thus allow us to choose the half-system as we please. The main result of this section is the following.

Theorem 2.2.13. *As above let r be odd and positive and let a be coprime to r. Then*

$$f(a, r) = \left(\frac{a}{r}\right) .$$

Proof. Since the Kronecker–Jacobi symbol is defined by complete multiplicativity in the denominator, we will prove the theorem in the same way. Note that it gives an additional justification for this generalization of the Legendre symbol. I claim first that the theorem is true for r an odd prime. Indeed, in that case for any half-system H we have (as usual in $\mathbb{Z}/r\mathbb{Z}$)

$$a^{(r-1)/2} \prod_{j \in H} j = \prod_{j \in H} aj = \prod_{j \in H} \varepsilon_H(j)\sigma_H(j) = f(a, r) \prod_{j \in H} j$$

since σ_H is a permutation. On the other hand, since r is a prime, $\prod_{j \in H} j$ is coprime to r and $a^{(r-1)/2} \equiv \left(\frac{a}{r}\right)$ by definition, proving my claim and the theorem in the prime case.

We must now show that when a is fixed, $f(a, r)$ is a completely multiplicative function of r (restricted to odd r), which will finish the proof. Thus let r and r' be odd integers. We must show that $f(a, rr') = f(a, r)f(a, r')$. Let H and H' be half-systems modulo r and r' respectively. It is immediate to check that

$$J = \{j + rk / \ j \in H, \ k \bmod r'\} \cup \{rj' / \ j' \in H'\}$$

is a half-system modulo rr'. Furthermore, if a is coprime to rr' we have

$$a(j + rk) \equiv \varepsilon_H(j)\sigma_H(j) \pmod{r} ;$$

hence $a(j + rk) = \varepsilon_H(j)(\sigma_H(j) + rk')$, and it follows that $\varepsilon_J(a(j + rk)) = \varepsilon_H(j)$. In addition

$$a(rj') = r(aj') \equiv r(\varepsilon_{H'}(j')\sigma_{H'}(j')) ;$$

hence $\varepsilon_J(rj') = \varepsilon'_H(j')$. Thus

$$f(a, rr') = f_J(a, rr') = \prod_{j \in H, \ k \bmod r'} \varepsilon_H(j) \prod_{j' \in H'} \varepsilon'_H(j')$$

$$= f_H(a, r)^{r'} f_{H'}(a, r') = f(a, r)f(a, r')$$

since r' is odd, proving multiplicativity and finishing the proof of the theorem. $\qquad \square$

Before stating the main corollary of this result, recall that we define $(n - 1)\backslash 2$ to be equal to the integer part of $(n-1)/2$. Furthermore, for two integers n and m with $n > 0$ define

$$S(m, n) = \sum_{1 \leqslant j \leqslant (n-1)\backslash 2} \left\lfloor \frac{jm}{n} \right\rfloor .$$

Corollary 2.2.14. *Let r be an odd positive integer and let a be any integer coprime to r. Then*

(1)

$$(-1)^{S(a,r)} = \begin{cases} \left(\dfrac{a}{r}\right) & \text{when } a \text{ is odd,} \\ \left(\dfrac{2a}{r}\right) & \text{when } a \text{ is even.} \end{cases}$$

(2) *Assume that $a > 0$. Then*

$$(-1)^{S(r,a)} = \begin{cases} (-1)^{(a-1)(r-1)/4}\left(\dfrac{a}{r}\right) & \text{when } a \text{ is odd,} \\ (-1)^{(a-2)(r-1)/4}\left(\dfrac{2a}{r}\right) & \text{when } a \text{ is even.} \end{cases}$$

(3) *If n and m are positive odd coprime integers we have the quadratic reciprocity law*

$$\left(\frac{m}{n}\right)\left(\frac{n}{m}\right) = (-1)^{(m-1)(n-1)/4} .$$

Proof. (1). Choose as half-system H modulo r the integers from 1 to $(r-1)/2$ and keep the above notation. In particular, multiplication by a defines a function $\varepsilon_H(j)$ with values ± 1 and a permutation σ_H of H. For $\varepsilon = 1$ and $\varepsilon = -1$, set

$$R_\varepsilon = \sum_{j \in H/\ \varepsilon_H(j)=\varepsilon} \sigma_H(j) .$$

We have clearly $R_+ + R_- = \sum_{j \in H} j = (r^2 - 1)/8$. On the other hand

$$ja = r\lfloor ja/r \rfloor + \begin{cases} \sigma_H(j) & \text{if } \varepsilon_H(j) = 1, \\ r - \sigma_H(j) & \text{if } \varepsilon_H(j) = -1. \end{cases}$$

Summing over $j \in H$ we obtain

$$a(r^2 - 1)/8 = rS(a,r) + R_+ - R_- + r\ell ,$$

where ℓ is the number of $j \in H$ such that $\varepsilon_H(j) = -1$. Subtracting from this the expression for $R_+ + R_-$, we obtain

$$(a - 1)(r^2 - 1)/8 = rS(a,r) - 2R_- + r\ell .$$

Now we have $(-1)^\ell = \prod_{j \in H} \varepsilon_H(j) = f_H(a,r) = \left(\frac{a}{r}\right)$ by the above theorem. Since r is odd, we therefore obtain

$$(-1)^{S(a,r)} = \varepsilon(r)^{(a-1)}\left(\frac{a}{r}\right) ,$$

where $\varepsilon(r) = (-1)^{(r^2-1)/8}$ depends only on r. We will show below that it is indeed equal to $\left(\frac{2}{r}\right)$ (we are not allowed to use our other proof of quadratic reciprocity), and up to this assumption this proves (1) upon separating the cases a odd and a even.

(2). For simplicity write $a' = (a-1)\backslash 2$, in other words $a' = (a-1)/2$ if a is odd and $a' = (a-2)/2$ if a is even. Consider the lattice points $(i,j) \in \mathbb{Z}^2$ such that $1 \leqslant i \leqslant a'$ and $1 \leqslant j \leqslant (r-1)/2$, which are evidently $a'(r-1)/2$ in number. Since r and a are coprime, the line $y = (r/a)x$ does not go through any lattice point for $0 < x < a$, in particular for $1 \leqslant x \leqslant a'$. It is easy to see, for instance by drawing a picture, that the number of lattice points under that line is exactly $S(r,a)$, while the number of lattice points above that line is $S(a,r)$. It follows that

$$(-1)^{S(r,a)} = (-1)^{a'(r-1)/2}(-1)^{S(a,r)} \ .$$

Using (1), this gives the desired formula when a is odd. When a is even, (1) gives

$$(-1)^{S(r,a)} = (-1)^{(a-2)(r-1)/4}\varepsilon(r)\left(\frac{a}{r}\right) \ .$$

Choosing $a = 2$, we see that $S(r,a) = 0$, and hence $\varepsilon(r) = \left(\frac{2}{r}\right)$, proving the claim made in (1) above and finishing the proof of (2).

(3) Follows immediately from (1) and (2) using the variables m and n instead of a and r. $\qquad\square$

2.2.4 Real Primitive Characters

Real primitive characters are easy to characterize. Recall that a *fundamental discriminant* is 1 or the discriminant of a quadratic field, in other words either a squarefree integer congruent to 1 modulo 4, or 4 times a squarefree integer congruent to 2 or 3 modulo 4.

Theorem 2.2.15. *If D is a fundamental discriminant, the Kronecker symbol $\left(\frac{D}{n}\right)$ defines a real primitive character modulo $m = |D|$. Conversely, if χ is a real primitive character modulo m then $D = \chi(-1)m$ is a fundamental discriminant D and $\chi(n) = \left(\frac{D}{n}\right)$.*

Proof. The definition of the Kronecker symbol and Theorem 2.2.9 show that $\left(\frac{D}{n}\right)$ is a character modulo $|D|$. To show that it is primitive, it is sufficient to show that for any prime $p \mid D$ it cannot be defined modulo D/p. Assume first that $p \neq 2$, and let a be a quadratic nonresidue modulo p. Since D is fundamental and p is odd we have $\gcd(p, 4|D|/p) = 1$; hence by the Chinese remainder theorem there exists $n > 0$ such that $n \equiv a \pmod{p}$ and $n \equiv 1 \pmod{4|D|/p}$, and in particular $n \equiv 1 \pmod 4$. Thus by Theorem 2.2.9 and the quadratic reciprocity law for positive odd numbers we have

$$\left(\frac{D}{n}\right) = \left(\frac{p}{n}\right)\left(\frac{D/p}{n}\right) = \left(\frac{p}{n}\right)\left(\frac{4D/p}{n}\right) = \left(\frac{p}{n}\right) = \left(\frac{n}{p}\right) = -1 \ ,$$

proving that $\left(\frac{D}{n}\right)$ cannot be defined modulo D/p. Assume now that $p = 2$, so that $D \equiv 8$ or 12 modulo 16, and choose $n = 1 + |D|/2$. If $D \equiv 8 \pmod{16}$ we have $n \equiv 5 \pmod 8$ and $n \equiv 1 \pmod{|D|/2}$ and so

$$\left(\frac{D}{n}\right) = \left(\frac{2}{n}\right)\left(\frac{D/2}{n}\right) = \left(\frac{2}{n}\right) = -1$$

since $D/2 \equiv 0 \pmod 4$. If $D \equiv 12 \pmod{16}$ we have $n \equiv 7 \pmod 8$ and $n \equiv 1 \pmod{D/4}$ hence

$$\left(\frac{D}{n}\right) = \left(\frac{-4}{n}\right)\left(\frac{-D/4}{n}\right) = \left(\frac{-4}{n}\right) = -1$$

since $-D/4 \equiv 1 \pmod 4$, proving in both cases that $\left(\frac{D}{n}\right)$ cannot be defined modulo $D/2$ hence that it is a primitive character.

Conversely, let χ be a real primitive character modulo m and let p be any odd prime such that $p \nmid m$. By Lemma 2.2.2 we have $\chi(p) = \left(\frac{D}{p}\right)$ with $D = \chi(-1)m$. Since both sides are multiplicative in p, we deduce that for any *odd* positive n we have $\chi(n) = \left(\frac{D}{n}\right)$. In addition, by definition of the Kronecker symbol we have $\left(\frac{D}{-1}\right) = \operatorname{sign}(D) = \chi(-1)$; hence the equality $\chi(n) = \left(\frac{D}{n}\right)$ is valid for any odd $n \in \mathbb{Z}$.

I now claim that $D \equiv 0$ or $1 \bmod 4$. Indeed, since χ is periodic of period $m = |D|$, by what we have just proved and the properties of the Kronecker symbol, we have

$$1 = \chi(1 + 2D) = \left(\frac{D}{1 + 2D}\right).$$

Thus, if we had $D \equiv 3 \pmod 4$ we would have

$$1 = \left(\frac{-1}{1 + 2D}\right)\left(\frac{-D}{1 + 2D}\right) = (-1)^D\left(\frac{-D}{1}\right) = -1\,,$$

a contradiction, and if we had $D \equiv 2 \pmod 4$ we would have

$$1 = \left(\frac{2}{1 + 2D}\right)\left(\frac{2D}{1 + 2D}\right) = \left(\frac{2}{1 + 2D}\right) = \left(\frac{2}{5}\right) = -1\,,$$

also a contradiction.

We must now prove that $\chi(2) = \left(\frac{D}{2}\right)$. We may of course assume D (or m) odd, otherwise both sides vanish. Thus $(D + 1)/2$ is odd; hence

$$1 = \chi(D + 1) = \chi(2)\chi((D + 1)/2)$$
$$= \chi(2)\left(\frac{D}{(D + 1)/2}\right) = \chi(2)\left(\frac{D}{D + 1}\right)\left(\frac{D}{2}\right) = \chi(2)\left(\frac{D}{2}\right),$$

showing that $\chi(2) = \left(\frac{D}{2}\right)$. By multiplicativity it follows that $\chi(n) = \left(\frac{D}{n}\right)$ for all n.

Finally, since $D \equiv 0$ or 1 modulo 4, we can write (uniquely) $D = D_0 f^2$, where D_0 is a fundamental discriminant. It is clear that the character $\left(\frac{D_0}{n}\right)$ takes the same values as the character $\left(\frac{D}{n}\right)$ on integers n coprime to D; hence $\left(\frac{D}{n}\right)$ is primitive if and only if $D = D_0$, i.e., D is a fundamental discriminant, finishing the proof of the theorem. □

Remark. If χ is a nonprimitive real character modulo m there still exists D such that $\chi(n) = \left(\frac{D}{n}\right)$ (if f is the conductor of χ we can for instance take $D = \chi(-1)fm^2$ by the above theorem), but we cannot in general choose D equal to $\chi(-1)m$: as an example, choose $m = 12$ and let $\chi(n) = \left(\frac{-4}{n}\right)$ for $\gcd(n, 12) = 1$.

2.2.5 The Sign of the Quadratic Gauss Sum

Corollary 2.1.47 gives the square of $\tau(\chi)$ when χ is a real character, in other words by the preceding section, when χ is the Legendre–Kronecker symbol. A more difficult result due to Gauss is that one can give the value of $\tau(\chi)$ itself (Proposition 2.2.24). Before proving it, we need some results of independent interest.

Proposition 2.2.16 (Poisson summation formula). *Let f be a continuous function and locally of bounded variation on some not necessarily bounded interval $[A, B]$. Then*

$$\sideset{}{'}\sum_{A \leqslant n \leqslant B} f(n) = \sum_{m \in \mathbb{Z}} \int_A^B f(t) \exp(2i\pi mt)\, dt\,,$$

where $\sum'$ means that the terms for $n = A$ and $n = B$, if present, must be counted with coefficient $1/2$.

Proof. Let f_1 be a piecewise continuous function locally of bounded variation, that tends to zero sufficiently rapidly (we will in fact have f_1 with compact support, so this is no problem). Set $g(x) = \sum_{n \in \mathbb{Z}} f_1(n + x)$. Then $g(x)$ is an absolutely convergent series that converges normally in any compact subset of $\mathbb{R}$, and clearly $g(x)$ is periodic of period dividing 1. Thus we may apply the standard theorem on Fourier series that tells us that for all x we have

$$\frac{g(x^+) + g(x^-)}{2} = \sum_{m \in \mathbb{Z}} c_m \exp(2i\pi mx)\,,$$

where as usual

$$g(x^\pm) = \lim_{\varepsilon \to 0,\ \mathrm{sign}(\varepsilon) = \pm} g(x + \varepsilon)\,,$$

and the Fourier coefficients c_m are given by

$$c_m = \int_0^1 g(t) \exp(-2i\pi mt)\, dt = \sum_{n \in \mathbb{Z}} \int_0^1 f_1(n+t) \exp(-2i\pi mt)\, dt$$

$$= \sum_{n \in \mathbb{Z}} \int_n^{n+1} f_1(t) \exp(-2i\pi mt)\, dt = \widehat{f_1}(m) \;,$$

where the Fourier transform $\widehat{f_1}(y)$ is defined as usual by

$$\widehat{f_1}(y) = \int_{-\infty}^{+\infty} f_1(t) \exp(-2i\pi yt)\, dt \;.$$

Setting in particular $y = 0$, we obtain

$$\sum_{n \in \mathbb{Z}} \frac{f_1(n^+) + f_1(n^-)}{2} = \sum_{m \in \mathbb{Z}} \widehat{f_1}(m) \;.$$

Choose now $f_1(t) = f(t)$ for $t \in [A, B]$ and $f_1(t) = 0$ elsewhere. Then

$$\widehat{f_1}(y) = \int_A^B f(t) \exp(-2i\pi yt)\, dt \;.$$

Furthermore, since f is continuous on $]A, B[$, when $A < n < B$ we have $(f_1(n^+) + f_1(n^-))/2 = f(n)$, while if $n = A$ (of course only when $A \in \mathbb{Z}$) then $(f_1(n^+) + f_1(n^-))/2 = f(n^+)/2 = f(n)/2$, and similarly if $n = B$ (when $B \in \mathbb{Z}$), then $(f_1(n^+) + f_1(n^-))/2 = f(n^-)/2 = f(n)/2$, proving the proposition after changing m into $-m$. □

Corollary 2.2.17. *Let f be a continuous function and locally of bounded variation on $\mathbb{R}$. Then for all $x \in \mathbb{R}$ we have*

$$\sum_{n \in \mathbb{Z}} f(x+n) = \sum_{m \in \mathbb{Z}} \widehat{f}(m) \exp(2i\pi mx) \;,$$

where as above $\widehat{f}(m)$ is the Fourier transform of f. In particular $\sum_{n \in \mathbb{Z}} f(n) = \sum_{m \in \mathbb{Z}} \widehat{f}(m)$.

Proof. Apply the proposition to $[A, B] = \mathbb{R}$, and note that by an evident change of variable the Fourier transform of $f(x+t)$ at y is $\widehat{f}(y) e^{2i\pi yx}$. □

Lemma 2.2.18. *Let p be an odd prime number, and let $\chi(n) = \left(\frac{n}{p}\right)$ be the Legendre symbol. Then*

$$\tau(\chi) = \sum_{x \bmod p} \zeta_p^{x^2} \;.$$

Proof. This immediately follows from the trivial observation that the number of solutions modulo p to $x^2 \equiv n \pmod{p}$ is equal to $1 + \chi(n)$ and the fact that $\sum_{n \bmod p} \chi(n) = 0$. □

We can now obtain the fundamental result on the sign of the Gauss sum.

Theorem 2.2.19. *Let p be an odd prime number, and let $\chi(n) = \left(\frac{n}{p}\right)$ be the Legendre symbol. Then*

$$\tau(\chi) = \begin{cases} p^{1/2} & \text{if } p \equiv 1 \pmod 4 , \\ p^{1/2}i & \text{if } p \equiv 3 \pmod 4 . \end{cases}$$

Proof. By the above lemma, we have $\tau(\chi) = \sum_{0 \leqslant x \leqslant p-1} \exp(2i\pi x^2/p)$. We apply the Poisson summation formula proved above to $[A, B] = [0, p]$ and $f(x) = \exp(2i\pi x^2/p)$. Since $f(0) = f(p)$, we have

$$\sideset{}{'}\sum_{0 \leqslant n \leqslant p} f(n) = \sum_{0 \leqslant n \leqslant p-1} f(n) = \tau(\chi) .$$

On the other hand,

$$\int_0^p f(t) \exp(2i\pi mt)\, dt = \int_0^p \exp(2i\pi(t^2 + pmt)/p)\, dt$$

$$= \exp(-2i\pi pm^2/4) \int_0^p \exp(2i\pi(t + pm/2)^2/p)\, dt$$

$$= \exp(-2i\pi pm^2/4) \int_{pm/2}^{p(m+2)/2} \exp(2i\pi t^2/p)\, dt .$$

Changing t into $p^{1/2}t$ it follows that

$$\sum_{m \in \mathbb{Z},\ 2|m} \int_0^p f(t) \exp(2i\pi mt)\, dt = \int_{-\infty}^{+\infty} \exp(2i\pi t^2/p)\, dt = p^{1/2}I ,$$

where

$$I = \int_{-\infty}^{+\infty} \exp(2i\pi t^2)\, dt .$$

The value of this integral is well known, but we do not need it since it will follow from the proof. Note that we know in advance that it converges, but this can be checked directly for example by setting $t^2 = x$ and integrating by parts.

Similarly we find that

$$\sum_{m \in \mathbb{Z},\ 2\nmid m} \int_0^p f(t) \exp(2i\pi mt)\, dt = \exp(-2i\pi p/4)p^{1/2}I .$$

Putting everything together, we thus obtain

$$\tau(\chi) = (1 + i^{-p})p^{1/2}I .$$

We can first deduce from this the value of I: indeed, we simply choose a small value of p, for example $p = 3$. Then

$$\tau(\chi) = \exp(2i\pi/3) - \exp(4i\pi/3) = i3^{1/2}$$

hence $I = i/(1+i) = (1+i)/2$.

Thus

$$\tau(\chi) = \frac{(1+i)(1+i^{-p})}{2} p^{1/2} \;,$$

proving the theorem after separation of cases. □

For simplicity of notation, when D is a fundamental discriminant we denote by χ_D the character such that $\chi_D(n) = \left(\frac{D}{n}\right)$. By quadratic reciprocity, we note that the above theorem can be reformulated as $\tau(\chi_D) = D^{1/2}$ for $D = (-1)^{(p-1)/2}p$, where p is an odd prime, choosing the principal branch of the square root, i.e., such that $-\pi/2 < \mathrm{Arg}(z^{1/2}) \leqslant \pi/2$. We are now going to show that this is true for any fundamental discriminant D by proving a few lemmas.

Lemma 2.2.20. *We have $\tau(\chi_D) = D^{1/2}$ for $D = -4$, $D = -8$, and $D = 8$.*

Proof. Clear by direct computation. □

Lemma 2.2.21. *Let D_1 and D_2 be two coprime fundamental discriminants. If $\tau(\chi_{D_1}) = D_1^{1/2}$ and $\tau(\chi_{D_2}) = D_2^{1/2}$, then $\tau(\chi_{D_1 D_2}) = (D_1 D_2)^{1/2}$.*

Proof. First note the important fact that it is *not* true that $(D_1 D_2)^{1/2} = D_1^{1/2} D_2^{1/2}$ (example $D_1 = -3$, $D_2 = -7$).

Since D_1 and D_2 are coprime, by the Chinese remainder theorem a residue modulo $D_1 D_2$ can be written uniquely in the form $n_2 D_1 + n_1 D_2$, where n_2 is modulo D_2 and n_1 is modulo D_1. Thus,

$$
\begin{aligned}
\tau(\chi_{D_1 D_2}) &= \sum_{n \bmod D_1 D_2} \left(\frac{D_1 D_2}{n}\right) \zeta_{|D_1 D_2|}^{n} \\
&= \sum_{n_1 \bmod D_1} \sum_{n_2 \bmod D_2} \left(\frac{D_1 D_2}{n_2 D_1 + n_1 D_2}\right) \zeta_{|D_1 D_2|}^{n_2 D_1 + n_1 D_2} \\
&= \left(\frac{D_1}{D_2}\right)\left(\frac{D_2}{D_1}\right) \sum_{n_1 \bmod D_1} \left(\frac{D_1}{n_1}\right) \zeta_{|D_1|}^{n_1} \sum_{n_2 \bmod D_2} \left(\frac{D_2}{n_2}\right) \zeta_{|D_2|}^{n_2} \\
&= (-1)^{(\mathrm{sign}(D_1)-1)(\mathrm{sign}(D_2)-1)/4} \tau(\chi_{D_1}) \tau(\chi_{D_2})
\end{aligned}
$$

by Proposition 2.2.6. It is clear that $(D_1 D_2)^{1/2} = D_1^{1/2} D_2^{1/2}$ except if both D_1 and D_2 are negative, in which case $(D_1 D_2)^{1/2} = -D_1^{1/2} D_2^{1/2}$, and this is exactly compensated by $(-1)^{(\mathrm{sign}(D_1)-1)(\mathrm{sign}(D_2)-1)/4}$, proving the lemma.

□

Definition 2.2.22. *A fundamental discriminant D is said to be a prime discriminant if it is either equal to -4, -8, or 8, or equal to $(-1)^{(p-1)/2}p$ for p an odd prime.*

Note that all these expressions are indeed fundamental discriminants.

Lemma 2.2.23. *Any fundamental discriminant D can be written in a unique way as a product of prime fundamental discriminants.*

Proof. Since D is fundamental, no odd prime can divide D to a power larger than 1. Thus, we may write $D = 2^u \prod_{p \in S} p$, where S is a finite set of odd primes. It follows that $D = \varepsilon 2^u \prod_{p \in S}(-1)^{(p-1)/2}p$ for some $\varepsilon = \pm 1$. Note that the product over $p \in S$ is congruent to 1 modulo 4. Thus, either $u = 0$, in which case we must have $\varepsilon = 1$ (since $D \equiv 1 \pmod 4$); or $u = 2$, in which case we must have $\varepsilon = -1$ (otherwise $D/4$ is also a discriminant), so the factor in front of the product is -4; or finally $u = 3$, in which case ε can be ± 1, giving the two factors ± 8. Uniqueness of the decomposition is clear.   □

The proof of the result that we are after is now immediate.

Proposition 2.2.24. *Let χ be a real primitive character modulo m, so that $\chi(n) = \left(\frac{D}{n}\right)$ for $D = \chi(-1)m$ a fundamental discriminant. Then*

$$\tau(\chi) = \begin{cases} m^{1/2} & \text{if } \chi(-1) = 1\,, \\ m^{1/2}i & \text{if } \chi(-1) = -1\,. \end{cases}$$

Proof. By Theorem 2.2.15, we know that $\chi = \chi_D$ with $D = \chi(-1)m$ a fundamental discriminant. By Lemma 2.2.23, D is equal to a product of prime fundamental discriminants that are necessarily coprime. By Lemma 2.2.21, it is thus sufficient to prove the proposition for prime fundamental discriminants, and this is exactly the content of Theorem 2.2.19 and Lemma 2.2.20.   □

In view of the functional equation for Dirichlet L-functions that we will study in Chapter 10 we make the following definition:

Definition 2.2.25. *Let χ be any primitive character modulo m. We define the root number $W(\chi)$ by the formula*

$$W(\chi) = \begin{cases} \dfrac{\tau(\chi)}{m^{1/2}} & \text{if } \chi(-1) = 1\,, \\[2mm] \dfrac{\tau(\chi)}{m^{1/2}i} & \text{if } \chi(-1) = -1\,. \end{cases}$$

Thus a restatement of Proposition 2.2.24 is that when χ is real we have $W(\chi) = 1$. In the general case, since $|\tau(\chi)| = m^{1/2}$ we have $|W(\chi)| = 1$, and one can show that $W(\chi)$ is a root of unity if and only if χ is real, in which case $W(\chi) = 1$ (see Exercise 17).

2.3 Lattices and the Geometry of Numbers

2.3.1 Definitions

In this section, we let V be an $\mathbb{R}$-vector space of dimension n.

Proposition 2.3.1. *Let Λ be a sub-$\mathbb{Z}$-module of V. Consider the following three conditions:*

(1) Λ *generates V as an $\mathbb{R}$-vector space.*
(2) Λ *is discrete for the natural topology of V.*
(3) Λ *is a free $\mathbb{Z}$-module of rank n.*

Then any two of these conditions imply the third.

Note that (3) alone does not imply (1) since Λ may be a free $\mathbb{Z}$-module without being a free $\mathbb{R}$-module.

Proof. Assume (1) and (2). Since Λ generates V, by linear algebra there exists a set of n elements $\mathbf{b}_1, \ldots, \mathbf{b}_n$ in Λ that are $\mathbb{R}$-linearly independent, hence that form an $\mathbb{R}$-basis of V, and let Λ_0 be the $\mathbb{Z}$-module generated by the $\mathbf{b}_i$. Since Λ is discrete in V there exists an integer $M > 0$ such that the only element $\sum x_i \mathbf{b}_i$ of V with $|x_i| < 1/M$ for all i and that belongs to Λ is the zero vector. It is clear that the M^n small cubes of the form $m_i/M \leqslant x_i < (m_i+1)/M$ for all i, where m_i are integers such that $0 \leqslant m_i < M$, form a partition of the big cube C defined by $0 \leqslant x_i < 1$ for all i. Let $\beta_1, \ldots, \beta_N$ be some (not necessarily all) representatives of Λ/Λ_0. Translating them if necessary by elements of Λ_0, we may assume that $\beta_j \in C$ for all j. It is then clear that two distinct β_j cannot belong to the same small cube: indeed, if β_j and β_k both belong to the same cube, then $\beta_k - \beta_j$ would be an element of Λ with coordinates $|x_i| < 1/M$ for all i, a contradiction since by assumption the only element of Λ lying in this cube is the origin. Thus the number of β_j is less than or equal to the number of small cubes, in other words $N \leqslant M^n$. It follows that Λ/Λ_0 is finite (since N is uniformly bounded), and since Λ_0 is finitely generated, Λ is also finitely generated.

Thus Λ is a finitely generated $\mathbb{Z}$-module, and is of course torsion-free since $\Lambda \subset V$; hence by the standard theorem on finitely generated torsion-free modules (see Corollary 2.1.2 for the case of $\mathbb{Z}$) we deduce that Λ is a free $\mathbb{Z}$-module. In addition, since Λ/Λ_0 is finite, Theorem 2.1.3 implies that the rank of Λ is equal to the rank of Λ_0, which is equal to n, proving (3).

Assume (1) and (3); hence let $\mathbf{b}_1, \ldots, \mathbf{b}_n$ be a $\mathbb{Z}$-basis of Λ. Thus they also form an $\mathbb{R}$-basis of V. If we consider the neighborhood Ω of 0 consisting of $x = \sum_{1 \leqslant i \leqslant n} x_i \mathbf{b}_i$ with $|x_i| < 1$ for all i, it is clear that the only element of Λ belonging to Ω is 0 itself, proving that Λ is discrete.

Finally, assume (2) and (3), and let W be the $\mathbb{R}$-vector space generated by Λ. Then (1) and (2) hold with V replaced by W; hence by what we have proved, Λ is a free $\mathbb{Z}$-module on $\dim(W)$ generators. It follows that $\dim(W) = n$, hence that $W = V$, proving (1). $\square$

A $\mathbb{Z}$-module Λ satisfying the above three conditions (or any two of them, by the proposition) will be called a *lattice* in V.

From now on, we will assume that V is a Euclidean vector space, in other words equipped with a Euclidean inner product $x \cdot y$. For instance, the most common case $V = \mathbb{R}^n$ will be considered as a Euclidean vector space with the inner product $x \cdot y = \sum_{1 \leqslant i \leqslant n} x_i y_i$ with evident notation. We also let $\|x\| = (x \cdot x)^{1/2}$ be the Euclidean norm.

Definition and Proposition 2.3.2. *Let $(\mathbf{b}_j)_{1 \leqslant j \leqslant n}$ be a family of n vectors in V.*

(1) *The absolute value of the determinant of the matrix of the $\mathbf{b}_j$ on some orthonormal basis of V is independent of that basis. It will be called (with a slight abuse) the* determinant *of the family and denoted by $\det(\mathbf{b}_1, \ldots, \mathbf{b}_n)$.*

(2) *The* Gram matrix *associated with the $\mathbf{b}_j$ is by definition the matrix of scalar products $G = (\mathbf{b}_i \cdot \mathbf{b}_j)_{1 \leqslant i,j \leqslant n}$, and we have $\det(G) = \det(\mathbf{b}_1, \ldots, \mathbf{b}_n)^2$.*

Proof. (1) follows from the fact that two orthonormal bases of V differ by a transition matrix P that is an *orthogonal* matrix, in other words such that $P^t P = I$, hence with determinant equal to ± 1. For (2) we note that if $\mathcal{B}$ is the matrix of the $(\mathbf{b}_j)$ on some orthonormal basis then $G = \mathcal{B}^t \mathcal{B}$; hence $\det(G) = \det(\mathcal{B})^2$. $\qquad\square$

Remark. This terminology is the one used by Cassels and by all the literature dealing with the LLL algorithm, which is the main reason for which we study lattices. It is to be noted however that most modern experts in the geometry of numbers such as Conway–Sloane [Con-Slo] and Martinet [Mar] use a notation that is more adapted to the number-theoretic aspects of lattices: to avoid square roots, they call the determinant the determinant of the Gram matrix, hence the square of what we call the determinant.

Proposition 2.3.3. *Let Λ be a lattice in V and let $(\mathbf{b}_j)_{1 \leqslant j \leqslant n}$ be a $\mathbb{Z}$-basis of Λ.*

(1) *The quantity $\det(\mathbf{b}_1, \ldots, \mathbf{b}_n)$ is independent of the choice of the $\mathbb{Z}$-basis $\mathbf{b}_j$. It is called the* determinant *of the lattice and will be denoted by $\det(\Lambda)$.*

(2) *The determinant of the Gram matrix of the $\mathbf{b}_j$ is equal to $\det(\Lambda)^2$.*

(3) *If $V = \mathbb{R}^n$ the volume of the set $\left\{\sum_{1 \leqslant i \leqslant n} x_j \mathbf{b}_j / 0 \leqslant x_j < 1\right\}$ (called a* fundamental parallelotope *for the lattice Λ) is equal to $\det(\Lambda)$, hence in particular is independent of the basis.*

Thus $\det(\Lambda)$ can also be called the *covolume* of Λ.

Proof. (1). If $\mathbf{b}'_j$ is another $\mathbb{Z}$-basis of Λ the transition matrix from the $\mathbf{b}_i$ to the $\mathbf{b}'_j$ is a matrix P with integral entries whose inverse also has integral entries, hence is such that $\det(P) = \pm 1$, so it follows that the absolute value of the determinant of the matrix $\mathcal{B}$ of the $\mathbf{b}_j$ on some orthonormal basis of V is equal to that of the $\mathbf{b}'_j$.

(2). Clear from the preceding proposition.

(3). This immediately follows from the Jacobian formula for changing variables in multiple integrals. In fact, it is the very reason for the existence of this formula. □

Corollary 2.3.4. *Let* $\mathbf{b}_1, \ldots, \mathbf{b}_n$ *belong to a lattice* Λ, *and let* $\mathcal{B}$ *be the matrix of the* $\mathbf{b}_j$ *on some orthonormal matrix of* V. *The* $(\mathbf{b}_i)$ *form a* $\mathbb{Z}$-*basis of* Λ *if and only if* $|\det(\mathcal{B})| = \det(\Lambda)$.

Proof. Clear. □

Finally, we recall the standard Gram–Schmidt construction.

Proposition 2.3.5. *Let* $(\mathbf{b}_j)_{1 \leqslant j \leqslant n}$ *be an* $\mathbb{R}$-*basis of* V. *There exists a unique orthogonal (but not necessarily orthonormal) basis* $(\mathbf{b}_j^*)_{1 \leqslant j \leqslant n}$ *of* V *whose matrix on the* $\mathbf{b}_i$ *is upper triangular with 1's on the diagonal. It is obtained by the inductive formulas*

$$\mathbf{b}_i^* = \mathbf{b}_i - \sum_{1 \leqslant j < i} \mu_{i,j} \mathbf{b}_j^* \quad \text{with} \quad \mu_{i,j} = \frac{\mathbf{b}_i \cdot \mathbf{b}_j^*}{\mathbf{b}_j^* \cdot \mathbf{b}_j^*} .$$

Proof. The transition matrix is upper triangular with 1 on the diagonal if and only if its inverse is also of this form, hence if and only if $\mathbf{b}_i^* = \mathbf{b}_i - \sum_{1 \leqslant j < i} \mu_{i,j} \mathbf{b}_j^*$ for some $\mu_{i,j} \in \mathbb{R}$. The conditions $\mathbf{b}_i^* \cdot \mathbf{b}_j^* = 0$ for $j < i$ give the formulas for the $\mu_{i,j}$, proving both existence and uniqueness. □

Remark. The coefficient $\mu_{i,j}$ is the coefficient of *column* i and *row* j of the transition matrix, which is the opposite of the usual convention, but which is almost always used when one is dealing with Gram–Schmidt orthogonalization.

Definition 2.3.6. *The Gram–Schmidt basis associated with the* $(\mathbf{b}_j)$ *is the* $\mathbb{R}$-*basis* $(\mathbf{b}_j^*)$ *of* V *constructed above.*

Corollary 2.3.7 (Hadamard's inequality). *Let* $(\mathbf{b}_j)$ *be an* $\mathbb{R}$-*basis of* V *and let* $(\mathbf{b}_j^*)$ *be the associated Gram–Schmidt basis of* V. *We have*

$$\det(\mathbf{b}_1, \ldots, \mathbf{b}_n) = \det(\mathbf{b}_1^*, \ldots, \mathbf{b}_n^*) = \prod_{j=1}^{n} \|\mathbf{b}_j^*\| \leqslant \prod_{j=1}^{n} \|\mathbf{b}_j\| .$$

In particular, if $(\mathbf{b}_j)$ *is a* $\mathbb{Z}$-*basis of a lattice* Λ *we have*

$$\det(\Lambda) = \prod_{j=1}^{n} \|\mathbf{b}_j^*\| \leqslant \prod_{j=1}^{n} \|\mathbf{b}_j\| .$$

Proof. Since the transition matrix from the $(\mathbf{b}_j)$ to the $(\mathbf{b}_j^*)$ has determinant 1, we have $\det(\mathbf{b}_1, \ldots, \mathbf{b}_n) = \det(\mathbf{b}_1^*, \ldots, \mathbf{b}_n^*)$. Furthermore, since the $\mathbf{b}_j^*$ are orthogonal, the Gram matrix of the $\mathbf{b}_j^*$ is the diagonal matrix whose diagonal entries are the $\|\mathbf{b}_j^*\|^2$; hence $\det(\mathbf{b}_1^*, \ldots, \mathbf{b}_n^*)^2 = \prod_{1 \leqslant j \leqslant n} \|\mathbf{b}_j^*\|^2$, proving the first two equalities. On the other hand, the formula $\mathbf{b}_i = \mathbf{b}_i^* + \sum_{1 \leqslant j < i} \mu_{i,j} \mathbf{b}_j^*$ and the orthogonality of the $\mathbf{b}_j^*$ implies that

$$\|\mathbf{b}_i\|^2 = \|\mathbf{b}_i^*\|^2 + \sum_{1 \leqslant j < i} \mu_{i,j}^2 \|\mathbf{b}_j^*\|^2 \geqslant \|\mathbf{b}_i^*\|^2 \;,$$

so that $\prod_{1 \leqslant j \leqslant n} \|\mathbf{b}_j^*\| \leqslant \prod_{1 \leqslant j \leqslant n} \|\mathbf{b}_j\|$, proving the inequality, and the last statement is a trivial rephrasing. $\qquad \square$

Remark. This classical inequality can of course be rephrased purely in matrix terms: the absolute value of the determinant of a matrix is bounded from above by the product of the L^2 norm of its columns.

2.3.2 Hermite's Inequality

We begin with a few preliminary results on orthogonal projections.

Lemma 2.3.8. *Let* $(\mathbf{b}_1, \ldots, \mathbf{b}_n)$ *be an* $\mathbb{R}$*-basis of* V*, let* $W = \mathbf{b}_1^\perp$ *be the orthogonal supplement of* $\mathbf{b}_1$*, and let* $\mathbf{b}_2', \ldots, \mathbf{b}_n'$ *be the orthogonal projection on* W *of* $\mathbf{b}_2, \ldots, \mathbf{b}_n$ *respectively. Then* $\mathbf{b}_2', \ldots, \mathbf{b}_n'$ *is a basis of* W *and we have*

$$\det(\mathbf{b}_1, \ldots, \mathbf{b}_n) = \|\mathbf{b}_1\| \det(\mathbf{b}_2', \ldots, \mathbf{b}_n') \;.$$

Proof. Let $(e_2, \ldots, e_n)$ be an orthonormal basis of W, so that if we set $e_1 = \mathbf{b}_1/\|\mathbf{b}_1\|$, $(e_1, \ldots, e_n)$ is an orthonormal basis of V. For $j \geqslant 2$ we thus have $\mathbf{b}_j = \mathbf{b}_j' + \alpha_j e_1$ for some $\alpha_j \in \mathbb{R}$; hence if $\mathcal{B}$ (respectively $\mathcal{B}'$) denotes the matrix of the $\mathbf{b}_j$ on $(e_1, \ldots, e_n)$ (respectively of the $\mathbf{b}_j'$ on $(e_2, \ldots, e_n)$) we have

$$\mathcal{B} = \begin{pmatrix} \|\mathbf{b}_1\| & \alpha_2 & \cdots & \alpha_n \\ 0 & & & \\ \vdots & & \mathcal{B}' & \\ 0 & & & \end{pmatrix} .$$

We thus have $\det(\mathcal{B}) = \|\mathbf{b}_1\| \det(\mathcal{B}')$, proving the formula and the fact that $\det(\mathcal{B}') \neq 0$, hence that the $\mathbf{b}_j'$ form a basis of W. $\qquad \square$

Corollary 2.3.9. *Let* Λ *be a lattice in* V*, let* $\mathbf{b}_1$ *be an element of a* $\mathbb{Z}$*-basis of* Λ*, let* $W = \mathbf{b}_1^\perp$ *be its orthogonal supplement, and let* Λ' *be the projection of* Λ *on* W*. Then* Λ' *is a lattice in* W *and* $\det(\Lambda) = \|\mathbf{b}_1\| \det(\Lambda')$.

Proof. Applying the above lemma to a $\mathbb{Z}$-basis $(\mathbf{b}_1, \ldots, \mathbf{b}_n)$ of Λ, it is clear that $(\mathbf{b}_2', \ldots, \mathbf{b}_n')$ satisfy conditions (1) and (3) of Proposition 2.3.1; hence Λ'

is a lattice, and the formula for its determinant also comes from the above lemma. $\qquad\Box$

Since a lattice Λ is discrete there exists an element of Λ that has the minimal nonzero Euclidean norm. We can thus set the following definition.

Definition 2.3.10. *We define the* minimum $\min(\Lambda)$ *of a lattice* Λ *to be the minimal norm of a nonzero element of* Λ.

Once again, if we want to do number theory (which is not our purpose in the present context), it would be nicer to define $\min(\Lambda)$ to be the square of the minimal norm so as to avoid square roots.

Lemma 2.3.11. *Keep the notation of the above corollary, and assume that* $\mathbf{b}_1$ *is a nonzero vector of* Λ *with minimal norm. Then every* $x' \in \Lambda'$ *is the orthogonal projection of some* $x \in \Lambda$ *such that* $\|x\|^2 \leqslant (4/3)\|x'\|^2$.

Proof. We may of course assume that $x' \neq 0$. Let x_0 be any element of Λ that projects on x', so that $x_0 = x' - \alpha \mathbf{b}_1$ for some $\alpha \in \mathbb{R}$. The elements of Λ that project on x' are the vectors $x = x_0 + m\mathbf{b}_1 = x' + (m - \alpha)\mathbf{b}_1$, and since $x' \in W = \mathbf{b}_1^\perp$ we have

$$\|x\|^2 = \|x'\|^2 + (m - \alpha)^2 \|\mathbf{b}_1\|^2 .$$

If we choose $m = \lfloor \alpha \rceil$ to be the nearest integer to α we have $(m - \alpha)^2 \leqslant 1/4$, and since $\mathbf{b}_1$ has minimal nonzero norm we have $\|\mathbf{b}_1\|^2 \leqslant \|x\|^2$, hence

$$\|x\|^2 \leqslant \|x'\|^2 + \frac{1}{4}\|x\|^2 ,$$

proving the lemma. $\qquad\Box$

We are now ready to prove Hermite's theorem, which gives an upper bound for $\min(\Lambda)$ in terms of $\det(\Lambda)$.

Theorem 2.3.12 (Hermite's inequality). *Let* Λ *be a lattice in* V. *There exists a* $\mathbb{Z}$-*basis* $(\mathbf{b}_1, \dots, \mathbf{b}_n)$ *of* Λ *such that*

$$\det(\Lambda) \leqslant \prod_{j=1}^{n} \|\mathbf{b}_j\| \leqslant \left(\frac{4}{3}\right)^{n(n-1)/4} \det(\Lambda) .$$

In particular, we have

$$\min(\Lambda) \leqslant \left(\frac{4}{3}\right)^{(n-1)/4} \det(\Lambda)^{1/n} .$$

Proof. The first inequality is simply Hadamard's inequality (Corollary 2.3.7). We prove the second one by induction on n, the case $n = 1$ being trivial. Let $n \geqslant 2$, assume the result true up to $n - 1$, let $\mathbf{b}_1$ be a nonzero vector of Λ with minimal norm, and keep the notation of the above lemmas and corollary. By induction there exists a basis $(\mathbf{b}_2', \ldots, \mathbf{b}_n')$ of Λ' such that $\prod_{2 \leqslant j \leqslant n} \|\mathbf{b}_j'\| \leqslant (4/3)^{(n-1)(n-2)/4} \det(\Lambda')$. Using the above lemma, for $j \geqslant 2$ each $\mathbf{b}_j'$ is the orthogonal projection of some $\mathbf{b}_j \in \Lambda$ such that $\|\mathbf{b}_j\|^2 \leqslant (4/3)\|\mathbf{b}_j'\|^2$. I claim that $(\mathbf{b}_1, \ldots, \mathbf{b}_n)$ is a $\mathbb{Z}$-basis of Λ. Indeed, let $x \in \Lambda$. By definition its projection x' on W is such that $x' = \sum_{2 \leqslant j \leqslant n} x_j \mathbf{b}_j'$ for some $x_j \in \mathbb{Z}$. It follows that the projection of $y = \sum_{2 \leqslant j \leqslant n} x_j \mathbf{b}_j$ is also equal to x', hence $x - y \in \Lambda \cap \mathbb{R}\mathbf{b}_1$. But since $\mathbf{b}_1$ is a vector of minimal norm in Λ it generates $\Lambda \cap \mathbb{R}\mathbf{b}_1$, so that x is indeed a $\mathbb{Z}$-linear combination of the $\mathbf{b}_i$, proving my claim.

By Corollary 2.3.9 we have $\det(\Lambda) = \|\mathbf{b}_1\| \det(\Lambda')$, hence

$$\prod_{j=1}^{n} \|\mathbf{b}_j\|^2 \leqslant \|\mathbf{b}_1\|^2 \left(\frac{4}{3}\right)^{n-1} \prod_{j=2}^{n} \|\mathbf{b}_j'\|^2$$

$$\leqslant \|\mathbf{b}_1\|^2 \left(\frac{4}{3}\right)^{n-1} \left(\frac{4}{3}\right)^{(n-1)(n-2)/2} \det(\Lambda')^2 \leqslant \left(\frac{4}{3}\right)^{n(n-1)/2} \det(\Lambda)^2 ,$$

proving the first inequality by induction. The second inequality follows since by definition $\min(\Lambda) \leqslant \|\mathbf{b}_j\|$ for all j. $\qquad \square$

It is easy to see that the inequality for $\min(\Lambda)/\det(\Lambda)^{1/n}$ given by this theorem is best possible for $n = 2$ (see Exercise 18), and Corollary 2.3.25 below shows that it is not best possible for $n \geqslant 9$ (it can be shown that it is not best possible for all $n \geqslant 3$. For the best known bounds see [Con-Slo] and [Mar]. Note that the best possible bound is known only for $1 \leqslant n \leqslant 8$ and $n = 24$, this latter result having been proved by Elkies et al. in 2004).

An amusing very simple corollary of the above theorem is the following important result due to Fermat.

Corollary 2.3.13 (Fermat). *Every prime $p \equiv 1 \pmod 4$ is the sum of two squares of integers.*

Proof. See Exercise 41. $\qquad \square$

2.3.3 LLL-Reduced Bases

Hermite's theorem clearly shows that there are *good* bases of Λ, in other words bases that are reasonably sized as a function of $\det(\Lambda)$, and we would like to find these bases. In principle the proof of the theorem is completely constructive. Unfortunately the main step in the induction proof is to find a vector of minimal nonzero norm in Λ. Since Λ is discrete this problem can be solved by straightforward enumeration in a suitable compact set, but the time required will be very large. In fact it has been shown that the problem is very close to being NP-complete (whatever that means; just consider that it is probably impossible to solve it in polynomial time). A crucial discovery made in the early 1980s by H. W. Lenstra, A. Lenstra, and L. Lovasz is that even though in general it is not possible to find rapidly a basis satisfying Hermite's conditions, and in particular a minimal vector, it is possible to find a very good *approximation* to it in a very precise sense in *polynomial time*. This LLL algorithm has become the cornerstone of many algorithms in several parts of mathematics, computer science, and operations research.

Definition 2.3.14. *Let γ be a fixed real number such that $\gamma > 4/3$. We say that the basis $(\mathbf{b}_j)$ of Λ is γ-LLL-reduced if the corresponding Gram–Schmidt basis $(\mathbf{b}_j^*)$ (see Proposition 2.3.5) satisfies the following two conditions:*

(1) *For all $j < i$ we have $|\mu_{i,j}| \leqslant 1/2$.*
(2) *For all $i \geqslant 2$ we have*

$$\|\mathbf{b}_i^* + \mu_{i,i-1}\mathbf{b}_{i-1}^*\|^2 \geqslant \left(\frac{1}{\gamma} + \frac{1}{4}\right)\|\mathbf{b}_{i-1}^*\|^2 \ .$$

Note that the second condition is equivalent to

$$\|\mathbf{b}_i^*\|^2 \geqslant \left(\frac{1}{\gamma} + \frac{1}{4} - \mu_{i,i-1}^2\right)\|\mathbf{b}_{i-1}^*\|^2 \ .$$

Proposition 2.3.15. *Let $(\mathbf{b}_j)$ be a γ-LLL-reduced basis of Λ, and let $(\mathbf{b}_j^*)$ be the corresponding Gram–Schmidt basis of $\mathbb{R}^n$.*

(1) *For $1 \leqslant j \leqslant i \leqslant n$ we have $\|\mathbf{b}_j\|^2 \leqslant \gamma^{i-1}\|\mathbf{b}_i^*\|^2$.*
(2) *We have*

$$\det(\Lambda) \leqslant \prod_{j=1}^{n}\|\mathbf{b}_j\| \leqslant \gamma^{n(n-1)/4}\det(\Lambda) \ .$$

(3) *We have $\|\mathbf{b}_1\| \leqslant \gamma^{(n-1)/4}\det(\Lambda)^{1/n}$.*

Proof. (1). Since $|\mu_{i,i-1}| \leqslant 1/2$ we have $\|\mathbf{b}_i^*\|^2 \geqslant \|\mathbf{b}_{i-1}^*\|^2/\gamma$; hence by induction, for $j \leqslant i$ we have $\|\mathbf{b}_j^*\|^2 \leqslant \gamma^{i-j}\|\mathbf{b}_i^*\|^2$. By definition and the inequalities for the $|\mu_{i,j}|$ we thus have

$$\|\mathbf{b}_j\|^2 = \|\mathbf{b}_j^*\|^2 + \sum_{1\leqslant k<j}\mu_{j,k}^2\|\mathbf{b}_k^*\|^2 \leqslant \left(1 + \frac{1}{4}\sum_{1\leqslant k<j}\gamma^{j-k}\right)\|\mathbf{b}_j^*\|^2$$

$$\leqslant \left(1 + \frac{\gamma-1}{\gamma}\left(\frac{\gamma^j-\gamma}{\gamma-1}\right)\right)\gamma^{i-j}\|\mathbf{b}_i^*\|^2 \leqslant \gamma^{i-1}\|\mathbf{b}_i^*\|^2 \ ,$$

using the fact that $\gamma > 4/3$ implies $1/4 < (\gamma-1)/\gamma$, proving (1).

(2). Corollary 2.3.7 (3) implies the first inequality and also that $\det(\Lambda) = \prod_{1 \leqslant j \leqslant n} \|\mathbf{b}_j^*\|$; hence applying (1) above to $i = j$ and multiplying over all values of j we obtain

$$\prod_{j=1}^{n} \|\mathbf{b}_j\|^2 \leqslant \gamma^{n(n-1)/2} \prod_{j=1}^{n} \|\mathbf{b}_j^*\|^2 \leqslant \gamma^{n(n-1)/2} \det(\Lambda)^2 ,$$

proving (2).

(3). Choosing $j = 1$ in (1) and multiplying over all values of i we obtain

$$\|\mathbf{b}_1\|^{2n} \leqslant \gamma^{n(n-1)/2} \prod_{i=1}^{n} \|\mathbf{b}_i^*\|^2 \leqslant \gamma^{n(n-1)/2} \det(\Lambda)^2 ,$$

proving (3). □

We thus see from (2) and (3) that an LLL-reduced basis (whose existence we shall prove in the next subsection) satisfies similar inequalities to those of Hermite's theorem with the number $4/3$ replaced by $\gamma > 4/3$. In addition, we deduce the following information on $\min(\Lambda)$:

Corollary 2.3.16. *Let* $(\mathbf{b}_i)$ *be a* γ-*LLL-reduced basis of* Λ *and let* $(\mathbf{b}_i^*)$ *be the corresponding Gram–Schmidt basis. Set*

$$c_1 = \max_{1 \leqslant i \leqslant n} \frac{\|\mathbf{b}_1\|}{\|\mathbf{b}_i^*\|} .$$

Then:

(1) *We have* $1 \leqslant c_1 \leqslant \gamma^{(n-1)/2}$.
(2) *For any nonzero vector* $\mathbf{x} \in \Lambda$ *we have*

$$\min(\Lambda) \geqslant \|\mathbf{x}\| \geqslant \|\mathbf{b}_1\|/c_1 = \min_{1 \leqslant i \leqslant n} \|\mathbf{b}_i^*\| .$$

Proof. Since $\|\mathbf{b}_1^*\|^2 \leqslant \|\mathbf{b}_1\|^2$ we have $c_1 \geqslant 1$, while by (1) of the proposition we have $\|\mathbf{b}_1\|^2 \leqslant \gamma^{i-1}\|\mathbf{b}_i^*\|^2 \leqslant \gamma^{n-1}\|\mathbf{b}_i^*\|^2$, so (1) is clear. For (2), write

$$\mathbf{x} = \sum_{i=1}^{n} x_i \mathbf{b}_i = \sum_{i=1}^{n} x_i^* \mathbf{b}_i^* ,$$

where $x_i \in \mathbb{Z}$ and $x_i^* \in \mathbb{R}$. If i_0 is the largest index such that $x_i \neq 0$ then by definition of the Gram–Schmidt basis we have $x_{i_0}^* = x_{i_0}$; hence since it is a nonzero integer we have $|x_{i_0}^*| \geqslant 1$, and so

$$\|\mathbf{x}\|^2 = \sum_{i=1}^{n} x_i^{*2} \|\mathbf{b}_i^*\|^2 \geqslant x_{i_0}^{*\,2} \|\mathbf{b}_{i_0}^*\|^2 \geqslant \|\mathbf{b}_{i_0}^*\|^2 \geqslant \|\mathbf{b}_1\|^2/c_1^2$$

by definition of c_1.   □

The final result of this subsection gives an estimate of the distance of a vector $\mathbf{y} \notin \Lambda$ to the vectors of the lattice. If $u \in \mathbb{R}$ we will let $\langle u \rangle = |u - \lfloor u \rfloor|$ be the distance of u to the nearest integer, so that $0 \leqslant \langle u \rangle \leqslant 1/2$.

Corollary 2.3.17. *Let $(\mathbf{b}_i)$ be a γ-LLL-reduced basis of Λ, let $\mathbf{y} \notin \Lambda$, let $Y = (y_i)$ be the vector of coordinates of $\mathbf{y}$ on the basis of the $(\mathbf{b}_i)$, and let i_0 be the largest index such that $\langle y_i \rangle \neq 0$. Then for all $\mathbf{x} \in \Lambda$ we have*

$$\|\mathbf{x} - \mathbf{y}\| \geqslant \langle y_{i_0} \rangle \|\mathbf{b}_1\|/c_1 \ ,$$

where c_1 is as above.

Proof. We use essentially the same proof as the preceding corollary. We write as above

$$\mathbf{x} = \sum_{i=1}^{n} x_i \mathbf{b}_i = \sum_{i=1}^{n} x_i^* \mathbf{b}_i^* \ ,$$

$$\mathbf{y} = \sum_{i=1}^{n} y_i \mathbf{b}_i = \sum_{i=1}^{n} y_i^* \mathbf{b}_i^* \ ,$$

where $x_i \in \mathbb{Z}$ and $y_i, x_i^*, y_i^* \in \mathbb{R}$. Let i_1 be the largest index such that $y_i \neq x_i$, so that as above (applied to the vector $\mathbf{y} - \mathbf{x}$) we have $y_{i_1}^* - x_{i_1}^* = y_{i_1} - x_{i_1}$, hence

$$\|\mathbf{y} - \mathbf{x}\|^2 \geqslant (y_{i_1}^* - x_{i_1}^*)^2 \|\mathbf{b}_{i_1}^*\|^2 \geqslant (y_{i_1} - x_{i_1})^2 \|\mathbf{b}_1\|^2 / c_1^2$$

by definition of c_1. Now if $i_1 < i_0$ we would have $y_{i_0} = x_{i_0} \in \mathbb{Z}$ by definition of $i + 1$, hence $\langle y_{i_0} \rangle = 0$, contradicting the definition of i_0. If $i_1 = i_0$ then $|y_{i_1} - x_{i_1}| = |y_{i_0} - x_{i_0}| \geqslant \langle y_{i_0} \rangle$ by definition of $\langle u \rangle$, giving the desired inequality. Finally, if $i_1 > i_0$ then $y_{i_1} \in \mathbb{Z}$ by definition of i_0, and since $y_{i_1} \neq x_{i_1}$ we have $|y_{i_1} - x_{i_1}| \geqslant 1 \geqslant \langle y_{i_0} \rangle$, proving the inequality also in this case.   □

Thus the two corollaries above enable us to give an explicit lower bound on the quantity $d(\Lambda, \mathbf{y})$ defined to be the minimal distance from $\mathbf{y}$ to a vector of Λ distinct from $\mathbf{y}$ (when $\mathbf{y} \in \Lambda$ this is clearly the same as the norm of the smallest nonzero vector of Λ).

2.3.4 The LLL Algorithms

I refer to my book [Coh0] for a comprehensive treatment of the LLL algorithm and its variants, and many of its applications. In this short subsection we mention only what the reader needs to know about it.

The basic idea is quite simple: we begin with a $\mathbb{Z}$-basis and compute its associated Gram–Schmidt basis. It is then easy to see that by simple $\mathbb{Z}$-linear transformations we can modify the initial lattice basis so that the

Gram–Schmidt coefficients $\mu_{i,j}$ become such that $|\mu_{i,j}| \leqslant 1/2$. We now look at the size condition on the $\|\mathbf{b}_i^*\|^2$. If it is not satisfied, we *exchange* the corresponding vectors of the lattice, backtrack, then start again. We give the algorithm in more detail.

Algorithm 2.3.18 (LLL Algorithm) Given a basis $\mathbf{b}_1, \mathbf{b}_2, \ldots, \mathbf{b}_n$ of a lattice $\Lambda \subset \mathbb{R}^n$, this algorithm transforms the vectors $\mathbf{b}_j$ so that when the algorithm terminates, the $\mathbf{b}_j$ form a γ-LLL-reduced basis of Λ.

1. [Initial Gram–Schmidt] Using the formulas given above, compute the Gram–Schmidt basis of $\mathbb{R}^n$ associated with the $\mathbf{b}_j$, and set $k \leftarrow 2$.

2. [Reduce $\mu_{k,k-1}$] Set $q \leftarrow \lfloor \mu_{k,k-1} \rceil$, $\mathbf{b}_k \leftarrow \mathbf{b}_k - q\mathbf{b}_{k-1}$, $\mu_{k,k-1} \leftarrow \mu_{k,k-1} - q$, and for all i such that $1 \leqslant i \leqslant k-2$, set $\mu_{k,i} \leftarrow \mu_{k,i} - q\mu_{k-1,i}$.

3. [Test LLL condition] If $\|\mathbf{b}_k^* + \mu_{k,k-1}\mathbf{b}_{k-1}^*\|^2 < (1/\gamma + 1/4)\|\mathbf{b}_{k-1}^*\|^2$, exchange $\mathbf{b}_k$ and $\mathbf{b}_{k-1}$ and update the corresponding Gram–Schmidt coefficients $\mu_{i,j}$ and basis vectors $\mathbf{b}_j^*$, set $k \leftarrow \max(2, k-1)$ and go to Step 2. Otherwise, for $l = k-2, k-3, \ldots, 1$, set $q \leftarrow \lfloor \mu_{k,l} \rceil$, $\mathbf{b}_k \leftarrow \mathbf{b}_k - q\mathbf{b}_l$, $\mu_{k,l} \leftarrow \mu_{k,l} - q$, for all $i \leqslant l-1$ set $\mu_{k,i} \leftarrow \mu_{k,i} - q\mu_{l,i}$, and finally set $k \leftarrow k+1$.

4. [Finished?] If $k \leqslant n$, then go to Step 2. Otherwise, output the LLL-reduced basis $(\mathbf{b}_j)$ and terminate the algorithm.

We have not given the detailed formulas for updating the Gram–Schmidt basis in Step 3, but the reader can easily work them out (or see [Coh0]).

An easy examination of this algorithm shows that if it terminates, the output is indeed a γ-LLL-reduced basis of Λ. What must be shown is that it does terminate, in a polynomial number of steps. This can easily be done and is left as an exercise for the reader (Exercise 20).

Of course we have simply given the basic LLL algorithm, and many practical improvements are possible. However, one of the most important, due to B. de Weger, is that if the $\mathbf{b}_j$ have *integral* coordinates, or more generally if the Gram matrix of the $\mathbf{b}_j$ is integral, all the computations in the algorithm (which a priori must be done with rational numbers of possibly very large size) can in fact be done entirely in integers of polynomially bounded size; see Exercise 21. We thus have the following theorem:

Theorem 2.3.19. *There exists a polynomial-time algorithm that, given a basis of a lattice Λ outputs an LLL-reduced basis of Λ. Furthermore, if Λ is a sublattice of $\mathbb{Z}^n$ (or more generally if the Gram matrix of a basis of Λ has integral entries) all the computations can be done in integers of polynomially bounded size.*

Since we always assume that the reader has a number theory package at his disposal, we mention that in GP the commands are `qflll(B)` for the general LLL algorithm on a matrix B, and `qflll(B,1)` for the integral version, which is the one which must be used in the context of Diophantine applications. The output H is the transition matrix from the initial basis to the

LLL-reduced one, so that the matrix of the LLL-reduced basis is in fact the matrix product BH.

Remark. We would of course like to choose the constant $\gamma > 4/3$ as close as possible to $4/3$ to improve the quality of the basis. There are however two good reasons *not* to do this. The first is that the analysis done in Exercise 20 shows that the algorithm will become much slower. Second, in applications to Diophantine equations the quality of the basis is not important, as long as it is γ-LLL reduced for some reasonable value of γ. As a compromise, we will choose $\gamma = 2$ and simply talk of LLL-reduced bases instead of 2-LLL-reduced bases. This is the default in the GP function `qflll`.

2.3.5 Approximation of Linear Forms

One of the most spectacular applications of the LLL algorithm is to linear forms in real or complex numbers. We can either use the algorithm to *find* $\mathbb{Z}$-linear (or more generally algebraic) relations, or we can use it to show that a $\mathbb{Z}$-linear form cannot be too small unless the coefficients of the form are very large. This is explained in rough terms in [Coh0], but here we need precise quantitative statements, which will follow from the corollaries proved above.

We begin with the case where the α_i are all real, and then explain the simple modifications to be made for the general case. Let $\alpha_0, \ldots, \alpha_n$ be real numbers, and fix a (large) positive constant C. If (e_i) is the canonical basis of $\mathbb{R}^n$, for $j \leqslant n - 1$ we set $\mathbf{b}_j = e_j + \lfloor C\alpha_j \rceil e_n$ and $\mathbf{b}_n = \lfloor C\alpha_n \rceil e_n$, so that the matrix $\mathcal{B}$ of the $\mathbf{b}_j$ is the $n \times n$ integer matrix obtained by replacing the last row $(0, 0, \ldots, 1)$ of the identity matrix by $(\lfloor C\alpha_1 \rceil, \ldots, \lfloor C\alpha_n \rceil)$, and let Λ be the lattice generated by the $\mathbf{b}_j$. Finally, set $\mathbf{y} = -\lfloor C\alpha_0 \rceil e_n$. Recall that we have defined $d(\Lambda, \mathbf{y})$ as the distance from $\mathbf{y}$ to the nearest element of Λ distinct from $\mathbf{y}$, and that Corollaries 2.3.16 and 2.3.17 give us lower bounds for $d(\Lambda, \mathbf{y})$.

Proposition 2.3.20. *Keep the above notation, and in particular assume that the α_i are all real. Let $X_1, \ldots, X_n$ be strictly positive integers, set $Q = \sum_{1 \leqslant i \leqslant n-1} X_i^2$, $T = (1 + \sum_{1 \leqslant i \leqslant n} X_i)/2$, and assume that $d(\Lambda, \mathbf{y})^2 \geqslant T^2 + Q$. If the x_i are any integers such that $|x_i| \leqslant X_i$ for all i, then either we have*

$$\left| \alpha_0 + \sum_{1 \leqslant i \leqslant n} x_i \alpha_i \right| \geqslant \frac{\sqrt{d(\Lambda, \mathbf{y})^2 - Q} - T}{C} \, ,$$

or we have $x_1 = \cdots = x_{n-1} = 0$ and $x_n = -\lfloor C\alpha_0 \rceil / \lfloor C\alpha_n \rceil$.

Proof. If we set

$$S = \alpha_0 + \sum_{1 \leqslant i \leqslant n} x_i \alpha_i \quad \text{and} \quad K = \lfloor C\alpha_0 \rceil + \sum_{1 \leqslant i \leqslant n} x_i \lfloor C\alpha_i \rceil$$

then by definition $|K - CS| \leqslant 1/2 + \sum_{1 \leqslant i \leqslant n} X_i/2 = T$, hence

$$|K| = |K - CS + CS| \leqslant |K - CS| + C|S| \leqslant T + C|S| \,.$$

On the other hand, if we set $\mathbf{x} = \sum_{1 \leqslant j \leqslant n} x_j \mathbf{b}_j \in \Lambda$ then by definition of the $\mathbf{b}_j$ we have

$$\mathbf{x} = \sum_{j=1}^{n-1} x_j e_j + \left(\sum_{j=1}^{n} x_j \lfloor C\alpha_j \rfloor \right) e_n \,,$$

so that $\mathbf{x} - \mathbf{y} = \sum_{1 \leqslant j \leqslant n-1} x_j e_j + K e_n$. Thus either $\mathbf{x} = \mathbf{y}$ or

$$d(\Lambda, \mathbf{y})^2 \leqslant \|\mathbf{x} - \mathbf{y}\|^2 = \sum_{1 \leqslant j \leqslant n-1} x_j^2 + K^2 \leqslant Q + (T + C|S|)^2 \,.$$

Since by assumption $d(\Lambda, \mathbf{y})^2 \geqslant T^2 + Q \geqslant Q$ we deduce that $|S| \geqslant (\sqrt{d(\Lambda, \mathbf{y})^2 - Q} - T)/C$ as claimed. When $\mathbf{x} = \mathbf{y}$ we deduce from the formula for $\mathbf{x} - \mathbf{y}$ that $x_j = 0$ for $1 \leqslant j \leqslant n - 1$ and that $K = 0$, hence that $\lfloor C\alpha_0 \rfloor + x_n \lfloor C\alpha_n \rfloor = 0$. $\qquad \square$

Remarks. (1) It is usually impossible to apply the proposition directly since $d(\Lambda, \mathbf{y})$ is unknown. On the other hand, it is clear that in the proposition we may replace $d(\Lambda, \mathbf{y})$ by any lower bound c_2 such as the one given by Corollary 2.3.17 when $\alpha_0 \neq 0$ or by Corollary 2.3.16 when $\alpha_0 = 0$, as long as $c_2^2 \geqslant T^2 + Q$.

(2) This proposition is sufficient for applications to Diophantine problems. However, it is easy to see that the bounds can be improved; see Exercise 24.

(3) To apply Corollary 2.3.17, we should choose C larger than X^n, where $X = \max_{1 \leqslant i \leqslant n} X_i$, for instance C of the order of $10 \cdot X^n$. Indeed, for such a choice of C we have $\det(\Lambda)$ of the order of X^n, hence by Proposition 2.3.15 $\|\mathbf{b}_1\|$ will be of the order of X, hence by Corollary 2.3.17, if c_1 is not too large $d(\Lambda, \mathbf{y})^2$ will have a lower bound also of the order of X^2, which has the same order of magnitude as Q.

Example. To illustrate the above results and remarks we give an example presented in two different ways. We would first like to compute a lower bound for $|x_1 \log(2) + x_2 \pi + x_3 \gamma|$ (where $\gamma = 0.577\ldots$ is Euler's constant, not the constant used in the LLL algorithm), where the x_i are integers such that $|x_i| \leqslant 10^{30}$ and not all equal to 0. We have $X_1 = X_2 = X_3 = 10^{30}$, so we choose $C > 10^{90}$, for instance $C = 10^{100}$, and we form the 3×3 matrix

$$\mathcal{B} = \begin{pmatrix} 1 & 0 & 0 \\ 0 & 1 & 0 \\ \lfloor C \log(2) \rfloor & \lfloor C\pi \rfloor & \lfloor C\gamma \rfloor \end{pmatrix} \,.$$

An application of the (integral) LLL algorithm shows that the first vector of an LLL-reduced basis of the lattice generated by the columns of $\mathcal{B}$ is an explicit vector whose entries have 34 decimal digits, and which we need not write explicitly. We easily compute that $\|\mathbf{b}_1\|/\|\mathbf{b}_2^*\| = 0.969 < 1$ and that $\|\mathbf{b}_1\|/\|\mathbf{b}_3^*\| = 0.704 < 1$, so that $c_1 = 1$. Thus by Corollary 2.3.17 we deduce that

$$d(\Lambda, 0) \geqslant \|\mathbf{b}_1\|/c_1 \geqslant 1.57\,10^{33} .$$

Replacing this lower bound and the values of Q and T in the proposition gives $|S| \geqslant 10^{-67}$. We have thus *proved* that if $|x_i| \leqslant 10^{30}$ for $1 \leqslant i \leqslant 3$ then $|x_1 \log(2) + x_2\pi + x_3\gamma| \geqslant 10^{-67}$.

We give the same example posed differently in a way that is much closer to the type of applications that we have in mind. Assume that we know that $|x_1 \log(2) + x_2\pi + x_3\gamma| \leqslant e^{-X}$ with $X = \max(|x_i|) \leqslant 10^{30}$. We want to compute all possible values of the x_i (they are now finite in number). We perform exactly the same computations as above, but now we conclude that $10^{-67} \leqslant |S| \leqslant e^{-X}$, hence that $X \leqslant 153$. As announced at the beginning, we have thus drastically reduced the bound on X. Now we can start again the whole process, using this much smaller value of X. We choose for instance $C = 10^8 > 153^3$, and apply the (integral) LLL algorithm. We obtain $\mathbf{b}_1 = (-148, -243, 129)^t$, and once again we compute that $c_1 = 1$, hence that $d(\Lambda, 0)^2 \geqslant 118634$, so that replacing this lower bound in the proposition gives $|S| \geqslant 3.5\,10^{-7}$, in other words $X \leqslant 14$. This is again substantially lower than the preceding bound of 154. By choosing $C = 10^6$ the reader can check that we could again reduce the bound to $X \leqslant 9$. However, this is not really necessary since we only need to search for $0 \leqslant x_1 \leqslant 14$ and $-14 \leqslant x_2, x_3 \leqslant 14$, which is very fast, and we find that the only values of (x_1, x_2, x_3) satisfying the given inequality are $(x_1, x_2, x_3) = (0, 0, 0)$, $\pm(1, 0, -1)$, $\pm(2, -1, 3)$, and $\pm(5, 0, -6)$.

It is very easy to modify Proposition 2.3.20 when the α_i are not real. If the $\mathbb{R}$-vector space generated by the α_i for $1 \leqslant i \leqslant n$ has dimension 1, generated by some nonzero complex number z, say, we can apply the proposition to the real numbers α_i/z for $1 \leqslant i \leqslant n$ together with $\Re(\alpha_0/z)$, and we can obtain an even better lower bound if $\Im(\alpha_0/z) \neq 0$ (Exercise 22).

We may therefore assume that at least two of the α_i for $i \geqslant 1$ are $\mathbb{R}$-linearly independent, and by reordering the α_i we may assume that α_{n-1} and α_n are $\mathbb{R}$-linearly independent. The modifications to be done to the above procedure are as follows. For $1 \leqslant j \leqslant n - 2$ we set

$$\mathbf{b}_j = e_j + \lfloor C\Re(\alpha_j) \rceil e_{n-1} + \lfloor C\Im(\alpha_j) \rceil e_n$$

and for $n - 1 \leqslant j \leqslant n$ we set

$$\mathbf{b}_j = \lfloor C\Re(\alpha_j) \rceil e_{n-1} + \lfloor C\Im(\alpha_j) \rceil e_n ,$$

so that the matrix $\mathcal{B}$ of the $\mathbf{b}_j$ is the $n \times n$ integer matrix obtained by replacing the last two rows of the identity matrix by

$$([\lfloor C\Re(\alpha_1)\rceil], \dots, \lfloor C\Re(\alpha_n)\rceil]) \quad \text{and} \quad ([\lfloor C\Im(\alpha_1)\rceil], \dots, \lfloor C\Im(\alpha_n)\rceil]),$$

and let Λ be the lattice generated by the $\mathbf{b}_j$. Finally, set

$$\mathbf{y} = -\lfloor C\Re(\alpha_0)\rceil e_{n-1} - \lfloor C\Im(\alpha_0)\rceil e_n.$$

Then the conclusion of Proposition 2.3.20 is valid almost verbatim; in other words, either

$$\left| \alpha_0 + \sum_{1 \leqslant i \leqslant n} x_i \alpha_i \right| \geqslant \frac{\sqrt{d(\Lambda, \mathbf{y})^2 - Q} - T}{C},$$

or we have $x_1 = \dots = x_{n-2} = 0$ and

$$x_{n-1}\lfloor C\Re(\alpha_{n-1})\rceil + x_n \lfloor C\Re(\alpha_n)\rceil + \lfloor C\Re(\alpha_0)\rceil = 0$$
$$\text{and} \quad x_{n-1}\lfloor C\Im(\alpha_{n-1})\rceil + x_n \lfloor C\Im(\alpha_n)\rceil + \lfloor C\Im(\alpha_0)\rceil = 0.$$

The proof is essentially identical to that of Proposition 2.3.20 and is left to the reader (Exercise 23).

2.3.6 Minkowski's Convex Body Theorem

The aim of this subsection is to prove Minkowski's convex body theorem and a number of corollaries. We assume that $V = \mathbb{R}^n$ and that the subsets of $\mathbb{R}^n$ that we consider are measurable for Lebesgue measure. In actual applications, they will in fact be much nicer than that.

Theorem 2.3.21 (Blichfeldt). *Let S be a (measurable) subset of $\mathbb{R}^n$ with volume $\mathrm{Vol}(S)$, and let Λ be a lattice of $\mathbb{R}^n$. If $\mathrm{Vol}(S) > \det(\Lambda)$ there exist distinct elements a and b in S such that $a - b \in \Lambda$.*

Proof. Let $\mathbf{b}_1, \dots, \mathbf{b}_n$ be a $\mathbb{Z}$-basis of Λ, let as above $U = \{x = \sum_{1 \leqslant j \leqslant n} x_j \mathbf{b}_j / \ 0 \leqslant x_j < 1\}$ be a fundamental parallelotope of Λ, and let $\chi(x)$ be the characteristic function of S, equal to 1 on S and to 0 elsewhere. We thus have

$$\mathrm{Vol}(S) = \int_{\mathbb{R}^n} \chi(x) \, dx = \int_U \left(\sum_{g \in \Lambda} \chi(x + g) \right) dx.$$

Since by Proposition 2.3.3 we have $\mathrm{Vol}(S) > \det(\Lambda) = \int_U dx$, there exists $x_0 \in U$ such that $\sum_{g \in \Lambda} \chi(x_0 + g) > 1$. It follows that there exist distinct elements g_0 and g_1 of Λ such that $a = x_0 + g_0 \in S$ and $b = x_0 + g_1 \in S$, hence $a - b = g_0 - g_1 \in \Lambda$. $\qquad \square$

We will say that a measurable set $C \subset \mathbb{R}^n$ is *symmetric* if $a \in C$ if and only if $-a \in C$. It is *convex* if whenever a, b are in C the line segment $ta + (1 - t)b$ for $0 \leqslant t \leqslant 1$ is in C.

Theorem 2.3.22 (Minkowski). *Let $C \subset \mathbb{R}^n$ be symmetric and convex, let Λ be a lattice in $\mathbb{R}^n$, and assume that $\mathrm{Vol}(C) > 2^n \det(\Lambda)$. Then there exists $c \neq 0$ such that $c \in \Lambda \cap C$.*

Proof. Let $S = C/2 = \{x/2, \; x \in C\}$ be the homothetic of C by a factor $1/2$, so that $\mathrm{Vol}(S) > \det(\Lambda)$. By Blichfeldt's theorem there exist a and b in S such that $c = a - b \in \Lambda$ with $c \neq 0$. Thus $2a$ and $2b$ belong to C; hence $-2b \in C$ by symmetry, so that $c = (1/2)(2a) + (1/2)(-2b) \in C$ by convexity. $\qquad\square$

Corollary 2.3.23. *With the same assumptions, if in addition C is compact, the conclusion of the theorem still holds if we only have $\mathrm{Vol}(C) \geqslant 2^n \det(\Lambda)$.*

Proof. Applying Minkowski's theorem to the homothetic set $(1+\varepsilon)C$ for any $\varepsilon > 0$, we see that there exists $c_\varepsilon \in \Lambda \setminus \{0\}$ such that $(1+\varepsilon)^{-1}c_\varepsilon \in C$. By compactness, the $(1+\varepsilon)^{-1}c_\varepsilon$ have a limit point $c \in C$ when $\varepsilon \to 0^+$, and c is also a limit point of c_ε, hence belongs to $\Lambda \setminus \{0\}$ since it is discrete. $\qquad\square$

Corollary 2.3.24. *For $1 \leqslant j \leqslant n$, let $L_j(y) = \sum_{1 \leqslant i \leqslant n} a_{j,i} y_i$ be a linear form in the n variables y_i with real coefficients, and set $\Delta = |\det(a_{j,i})|$. Let C be symmetric and convex, and assume that $\mathrm{Vol}(C) > 2^n \Delta$. There exists a nonzero element $c \in \mathbb{Z}^n$ such that $(L_1(c), \ldots, L_n(c)) \in C$. If, in addition, $\Delta \neq 0$ and C is compact, the result still holds if we have $\mathrm{Vol}(C) \geqslant 2^n \Delta$.*

Proof. Set $D = \{y \in \mathbb{R}^n \,/\, (L_1(y), \ldots, L_n(y)) \in C\}$. Clearly D is symmetric and convex $(tL(y) + (1-t)L(z) = L(ty + (1-t)z)$ if L is a linear form), and $\mathrm{Vol}(D) = \mathrm{Vol}(C)/\Delta$ (since Δ is the absolute value of the determinant of the Jacobian of the change of variables from the y_i to the $L_i(y)$). Furthermore, D is compact when C is compact and $\Delta \neq 0$. We can thus apply Minkowski's theorem and the preceding corollary to $\Lambda = \mathbb{Z}^n$ and to D, proving the result. $\qquad\square$

As an application of Minkowski's theorem we now show that Hermite's inequality (Theorem 2.3.12) on the minimum of a lattice can be considerably improved.

Corollary 2.3.25 (Minkowski). *If Λ is a lattice in $\mathbb{R}^n$ we have*

$$\min(\Lambda) \leqslant \frac{2}{\pi^{1/2}} \Gamma\left(\frac{n}{2} + 1\right)^{1/n} \det(\Lambda)^{1/n},$$

where $\Gamma(x)$ is the gamma function (see Chapter 9).

Note that

$$\Gamma\left(\frac{n}{2} + 1\right) = \begin{cases} (n/2)! & \text{if } n \text{ is even,} \\ \dfrac{n!}{2^n((n-1)/2)!}\pi^{1/2} & \text{if } n \text{ is odd.} \end{cases}$$

Proof. We choose for $C = C_\lambda$ the closed ball centered at the origin with radius λ, where λ will be chosen presently. It is clear that C_λ is convex, symmetric, and compact; hence if $\mathrm{Vol}(C_\lambda) \geqslant 2^n \det(\Lambda)$ there exists a nonzero vector $c \in \Lambda$ such that $c \in C_\lambda$; in other words, $\|c\| \leqslant \lambda$, so that $\min(\Lambda) \leqslant \lambda$. Clearly $\mathrm{Vol}(C_\lambda) = \lambda^n \mathrm{Vol}(C_1)$, so if we choose $\lambda = 2(\det(\Lambda)/\mathrm{Vol}(C_1))^{1/n}$ we have $\mathrm{Vol}(C_\lambda) \geqslant 2^n \det(\Lambda)$ hence $\min(\Lambda) \leqslant 2(\det(\Lambda)/\mathrm{Vol}(C_1))^{1/n}$. It is a well-known calculus exercise that the volume of the unit ball C_1 is given by $\mathrm{Vol}(C_1) = \pi^{n/2}/\Gamma(n/2+1)$, proving the corollary. $\qquad\square$

By Stirling's formula (see Chapter 9 once again if you do not know it), as $n \to \infty$ we have $\Gamma(n/2+1)^{1/n} \sim (n/(2e))^{1/2}$, so that the upper bound for $\min(\Lambda)/\det(\Lambda)^{1/n}$ is asymptotic to $(2n/(\pi e))^{1/2}$, which is considerably smaller than $(4/3)^{(n-1)/4}$ given by Hermite's inequality. However, for $2 \leqslant n \leqslant 8$, Hermite's bound is better, although not optimal for $n \geqslant 3$.

2.4 Basic Properties of Finite Fields

2.4.1 General Properties of Finite Fields

Let K be a not necessarily commutative finite division algebra (a skew field if you prefer).[1] Consider the natural map s from $\mathbb{Z}$ to K defined by $s(1) = 1$ (where the "1" on the right denotes the identity of K), and extended by additivity. By definition it is a group homomorphism, and it is easily seen that it is in fact a ring homomorphism. Its kernel I is therefore an ideal of $\mathbb{Z}$, i.e., has the form $p\mathbb{Z}$ for a certain $p \in \mathbb{Z}_{\geqslant 0}$, which cannot be equal to 1 since s is not the zero map (otherwise $1 = 0$ in K). It follows that s induces an *injective* map from $\mathbb{Z}/p\mathbb{Z}$ to K. Since K is finite, p is nonzero. Furthermore, since K is a skew field, hence in particular has no zero divisors, $\mathbb{Z}/p\mathbb{Z}$ is an integral domain; hence p is a prime number, called the *characteristic* of K. The image of s in K is thus a subfield k of K isomorphic to $\mathbb{Z}/p\mathbb{Z}$, which we will call the prime subfield of K. Clearly any subfield of K, hence also any field *containing* K, also has characteristic p.

It is trivially checked that the field (or skew field) axioms imply that when we have a field extension such as K/k, then K is naturally a k-vector space. In our case, this implies that *as a vector space K is isomorphic to k^n* for some integer $n = \dim_k(K)$. Of course K is not isomorphic to k^n as a ring, since the latter is not even a field for $n \geqslant 2$. We have thus shown the following:

Proposition 2.4.1. *Let K be a finite skew field. The cardinality of K has the form p^n, where $n \in \mathbb{Z}_{\geqslant 1}$ and p is a prime number equal to the characteristic of K. In addition, the additive group of K is isomorphic to $(\mathbb{Z}/p\mathbb{Z})^n$.*

---

[1] Refer to the introduction for discussion on this terminology. We will see below that K is indeed commutative.

Theorem 2.4.2. *Any finite skew field is commutative, i.e., is a field.*

Proof. Let K be a finite skew field, let C be the center of the multiplicative group K (i.e., the set of elements of K^* that commute with all elements of K) together with 0. Note that C is a subfield of K. We let $q = |C|$ (which is therefore a power of the characteristic p of K) and $n = \dim_C(K)$. If we let K^* act on itself by conjugation, the class equation for groups gives

$$|K^*| = |C^*| + \sum_x [K^* : K_x^*] \,,$$

where the summation is over a system of representatives of the orbits that are not reduced to a single point, K_x^* denotes the stabilizer subgroup of x, and $[K^* : K_x^*]$ denotes the finite group index. Note that the set of elements commuting with x in K form a subfield $K_x = K_x^* \cup \{0\}$ of K distinct from K since $x \notin C$. Thus $|K_x| = q^{n_x}$ with $n_x = \dim_C(K_x)$, and since $\dim_C(K) = \dim_C(K_x)\dim_{K_x}(K)$ we have $n_x \mid n$ and $n_x < n$. We thus obtain an equality of the form

$$q^n - 1 = q - 1 + \sum_x \frac{q^n - 1}{q^{n_x} - 1} \,.$$

We now use some easy properties of cyclotomic polynomials which we will prove in Section 3.5.1. From Definition 3.5.1 and Proposition 3.5.2, since $n_x < n$ we know that $\Phi_n(q)$ divides each quotient $(q^n - 1)/(q^{n_x} - 1)$ and divides also $q^n - 1$. Thus, by the above formula, it divides $q - 1$. However, for $n > 1$ there exists a primitive nth root of unity different from 1, hence if, as in Section 3.5.1, we denote by U_n' the set of primitive nth roots of unity, we have

$$|\Phi_n(q)| = \prod_{\zeta \in U_n'} |q - \zeta| > \prod_{\zeta \in U_n'} |q - 1| = (q - 1)^{\phi(n)} \geqslant (q - 1) \,,$$

contradicting $\Phi_n(q) \mid (q - 1)$ (note that it is crucial to have a strict inequality above). Thus we must have $n = 1$, in other words $K = C$, so that K is commutative. $\qquad\square$

Corollary 2.4.3. *Any finite subgroup of the multiplicative group of a commutative field K is cyclic. In particular, the multiplicative group of a finite field is cyclic; in other words, if K is a finite field with p^n elements then*

$$(K^*, \times) \simeq (\mathbb{Z}/(p^n - 1)\mathbb{Z}, +) \,.$$

Proof. Let G be such a finite subgroup, say of order n. For every $d \mid n$, let $\rho(d)$ be the number of $x \in G$ of order exactly equal to d in G. We clearly have $n = \sum_{d|n} \rho(d)$. On the other hand, since in a commutative field an equation of degree d has at most d roots (trivial, and *not necessarily true* in a noncommutative skew field), the equation $x^d - 1 = 0$ has at most d

solutions in K. If G has at least an element x of order exactly d, then the x^k for $0 \leqslant k < d$ are all the roots of the equation $x^d - 1 = 0$, and among those, $\phi(d)$ are of order exactly equal to d. Thus for every $d \mid n$, either $\rho(d) = 0$ or $\rho(d) = \phi(d)$, so that

$$\rho(n) = n - \sum_{d|n,\ d\neq n} \rho(d) \geqslant n - \sum_{d|n,\ d\neq n} \phi(d) = \phi(n) \ ,$$

the last equality coming from the identity $\sum_{d|n} \phi(d) = n$ (which can trivially be proved directly, or obtained by taking degrees in Definition 3.5.1). In particular, $\rho(n) > 0$, proving that the group G is cyclic, and has in fact $\phi(n)$ generators. □

Remark. The result would be false without the commutativity assumption. For instance, in the field of quaternions over $\mathbb{R}$, the set $\{\pm 1, \pm i, \pm j, \pm k\}$ (with the usual notation) is evidently a noncommutative subgroup of order 8.

Corollary 2.4.4. *Let $y \in \mathbb{F}_q$ and $m \in \mathbb{Z}_{\geqslant 1}$.*

(1) *The number of solutions in $\mathbb{F}_q$ of the equation $x^m = y$ is equal to the number of solutions of $x^d = y$, where $d = \gcd(m, q - 1)$.*
(2) *If $y \neq 0$ and $d \mid (q - 1)$, the number of solutions of $x^d = y$ is equal either to 0 or to d.*

Proof. If $y = 0$ there is the unique solution $x = 0$, so we may assume that $y \neq 0$. Since the group $\mathbb{F}_q^*$ is cyclic, the image of the map $x \mapsto x^m$ is $\mathbb{F}_q^{*d}$, the subgroup of dth powers, and for each $y \in \mathbb{F}_q^{*d}$ it is clear that there are exactly d preimages. □

We will now see that finite fields are characterized by their cardinality. Set $\mathbb{F}_p = \mathbb{Z}/p\mathbb{Z}$, and denote by $\overline{\mathbb{F}_p}$ an algebraic closure of $\mathbb{F}_p$.

Theorem 2.4.5. *For any integer $n \geqslant 1$ there exists a finite subfield of $\overline{\mathbb{F}_p}$ with $q = p^n$ elements. This subfield is unique and is equal to the set of roots in $\overline{\mathbb{F}_p}$ of the equation $X^q - X = 0$. Up to isomorphism, there exists a unique finite field of cardinality $q = p^n$.*

Proof. Assume first that a subfield F of $\overline{\mathbb{F}_p}$ with q elements exists. Since $|F^*| = q - 1$, any element $x \in F^*$ satisfies the equation $x^{q-1} = 1$, hence any element $x \in F$ satisfies the equation $x^q - x = 0$. Conversely, set $\Omega_q(X) = X^q - X$. Note that $\Omega'_q(X) = -1$ (since we are in characteristic p), hence the polynomial Ω_q is separable. Thus, denote by F the set of its q distinct roots in the algebraically closed field $\overline{\mathbb{F}_p}$. For any $x \in F \setminus \{0\}$ we have $(x^{-1})^{q-1} = (x^{q-1})^{-1} = 1$, so that $x^{-1} \in F$, and for any x and y in F, we have $(xy)^q = x^q y^q = (xy)$, hence $xy \in F$, and also since we are in characteristic p, $(x+y)^q = x^q + y^q = x + y$, so that $x + y \in F$, proving the first statement. The last follows immediately from the uniqueness of algebraic closure up to isomorphism. □

Definition 2.4.6. *When $q = p^n$ for $n \geqslant 1$, we denote by $\mathbb{F}_q$ the unique subfield of cardinality q of an algebraic closure $\overline{\mathbb{F}}_p$ of $\mathbb{F}_p$, fixed once and for all.*

Remark. It is important to distinguish between "unique up to isomorphism" and "unique." Here, we *fix* an algebraic closure $\overline{\mathbb{F}}_p$ of $\mathbb{F}_p$, so that $\mathbb{F}_q$ is unique. Similarly, we will see later that although number fields can be considered up to isomorphism, they are better seen as subfields of a *fixed* algebraic closure $\overline{\mathbb{Q}}$ of $\mathbb{Q}$.

We end this subsection with an important remark concerning the lattice of extensions of $\mathbb{F}_p$.

Proposition 2.4.7. *If n and m are in $\mathbb{Z}_{\geqslant 1}$ then*

$$\mathbb{F}_{p^n} \subset \mathbb{F}_{p^m} \iff n \mid m .$$

In particular, $\mathbb{F}_{p^n} \cap \mathbb{F}_{p^m} = \mathbb{F}_{p^{\gcd(n,m)}}$ and $\mathbb{F}_{p^n} \mathbb{F}_{p^m} = \mathbb{F}_{p^{\operatorname{lcm}(n,m)}}$.

Proof. Left as an easy exercise to the reader.   □

Thus, note for instance that $\mathbb{F}_{p^3}$ is *not* an extension of $\mathbb{F}_{p^2}$. We give for example the lattice of subextensions of $\mathbb{F}_{4096}/\mathbb{F}_2$:

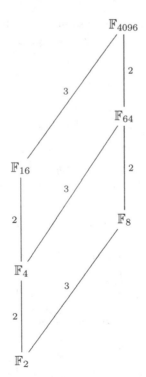

2.4.2 Galois Theory of Finite Fields

We start by studying the automorphism group of a finite field.

Theorem 2.4.8. *Let $\mathbb{E}/\mathbb{F}$ be an extension of finite fields. Then $\mathbb{E}/\mathbb{F}$ is a Galois (i.e., normal and separable) extension and the Galois group $\mathrm{Gal}(\mathbb{E}/\mathbb{F})$ of $\mathbb{F}$-automorphisms of $\mathbb{E}$ is the cyclic group of order $[\mathbb{E} : \mathbb{F}]$ generated by the Frobenius automorphism $\sigma_q : x \mapsto x^q$, where $q = |\mathbb{F}|$.*

Proof. Up to isomorphism, we may assume that we are in a fixed algebraic closure $\overline{\mathbb{F}_p}$ of $\mathbb{F}_p$, and that $\mathbb{F} = \mathbb{F}_q$ (with $q = p^f$ for some f) and $\mathbb{E} = \mathbb{F}_{q^s}$ for $s = [\mathbb{E} : \mathbb{F}]$. Thus, by Theorem 2.4.5, $\mathbb{E}$ is equal to the set of roots in $\overline{\mathbb{F}_p}$ of the separable polynomial $\Omega_{q^s}(X) = X^{q^s} - X$.

Let x be a primitive element of $\mathbb{E}/\mathbb{F}$, i.e., such that $\mathbb{E} = \mathbb{F}(x)$. Note that the primitive element theorem states that such an element exists in any finite separable extension, which is the case here, but in fact it is clear directly that any generator x of the cyclic group $\mathbb{E}^*$ must be a primitive element. Denote by $P_x(X)$ the minimal polynomial of x over $\mathbb{F}$. Since $x \in \mathbb{E}$, by definition of the minimal polynomial, $P_x(X)$ divides $\Omega_{q^s}(X)$; hence all its roots belong to $\mathbb{E}$ and are distinct. Since $\mathbb{E} = \mathbb{F}(x)$ we may define $s = [\mathbb{E} : \mathbb{F}] = \deg(P_x(X))$ automorphisms of $\mathbb{E}/\mathbb{F}$ by sending x to any of the roots of $P_x(X)$. It follows that $\mathbb{E}/\mathbb{F}$ is Galois with $|\mathrm{Gal}(\mathbb{E}/\mathbb{F})| = s$.

Since we are in characteristic p, and since $\mathbb{F}$ is the set of roots of $\Omega_q(X)$ in $\overline{\mathbb{F}_p}$, it is immediately checked that the Frobenius automorphism $\sigma_q : x \mapsto x^q$ is an $\mathbb{F}$-automorphism of $\mathbb{E}$, i.e., it belongs to $\mathrm{Gal}(\mathbb{E}/\mathbb{F})$. To complete the proof, we must show that it has order exactly s. Indeed, if $1 \leqslant k \leqslant s$, then the fixed field of σ_q^k in $\mathbb{E}$ is equal to $\mathbb{F}_{q^s} \cap \mathbb{F}_{q^k} = \mathbb{F}_{q^{\gcd(s,k)}}$ by Proposition 2.4.7, and this is equal to the fixed field of the identity, i.e., to $\mathbb{F}_{q^s}$, if and only if $k = s$, finishing the proof. $\qquad\square$

Corollary 2.4.9. *Let $\mathbb{E}/\mathbb{F}$ be an extension of finite fields, $q = |\mathbb{F}|$, and $s = [\mathbb{E} : \mathbb{F}]$. The trace and norm from $\mathbb{E}$ to $\mathbb{F}$ are given by the formulas*

$$\mathrm{Tr}_{\mathbb{E}/\mathbb{F}}(x) = \sum_{0 \leqslant i < s} x^{q^i} ,$$

$$\mathcal{N}_{\mathbb{E}/\mathbb{F}}(x) = x^{(q^s - 1)/(q-1)} .$$

Corollary 2.4.10. *For any $q = p^f$, the subfield $\mathbb{F}_q$ of $\overline{\mathbb{F}_p}$ is the fixed field of σ_p^f, and*

$$\mathrm{Gal}(\mathbb{F}_q/\mathbb{F}_p) = \sigma_p^{\mathbb{Z}}/\sigma_p^{f\mathbb{Z}} \simeq \mathbb{Z}/f\mathbb{Z} .$$

The statements are clear. Please note the subtle but essential distinction between the first *equality* and the second *isomorphism*.

Proposition 2.4.11. *Let $\mathbb{E}/\mathbb{F}$ be an extension of finite fields with $|\mathbb{F}| = q$.*

(1) The trace map $\mathrm{Tr}_{\mathbb{E}/\mathbb{F}}$ is a surjective homomorphism from $\mathbb{E}$ to $\mathbb{F}$.

(2) *The kernel of* $\mathrm{Tr}_{\mathbb{E}/\mathbb{F}}$ *is the* $\mathbb{F}$-*vector space of elements* $a \in \mathbb{E}$ *of the form* $a = x^q - x$ *for some* $x \in \mathbb{E}$.

(3) *The map* $(x, y) \mapsto \mathrm{Tr}_{\mathbb{E}/\mathbb{F}}(xy)$ *defines a nondegenerate bilinear pairing from* $\mathbb{E} \times \mathbb{E}$ *to* $\mathbb{F}$.

Proof. (1). By Corollary 2.4.9 we have

$$\mathrm{Tr}_{\mathbb{E}/\mathbb{F}}(x) = \sum_{0 \leqslant i < s} x^{q^i}.$$

The right-hand side is a polynomial of degree q^{s-1}, hence has at most $q^{s-1} < q^s = |\mathbb{E}|$ roots in $\mathbb{E}$. Therefore there exists $u \in \mathbb{E}$ such that $\mathrm{Tr}_{\mathbb{E}/\mathbb{F}}(u) = c \neq 0$. By $\mathbb{F}$-linearity it follows that for any $a \in \mathbb{F}$ we have $\mathrm{Tr}_{\mathbb{E}/\mathbb{F}}(ua/c) = (a/c)\mathrm{Tr}_{\mathbb{E}/\mathbb{F}}(u) = a$, proving surjectivity.

(2). For simplicity, write $V = \mathrm{Ker}(\mathrm{Tr}_{\mathbb{E}/\mathbb{F}})$. Since $\mathrm{Tr}_{\mathbb{E}/\mathbb{F}}$ is surjective by (1), it follows that $\dim_{\mathbb{F}}(V) = s-1$, where as usual we set $s = [\mathbb{E} : \mathbb{F}]$. On the other hand, since $x^{q^s} = x$ it follows from Corollary 2.4.9 that $\mathrm{Tr}_{\mathbb{E}/\mathbb{F}}(x^q - x) = 0$, so that the map f defined by $f(x) = x^q - x$ is a map from $\mathbb{E}$ to V, which is evidently linear. Its kernel is the set of x such that $x^q - x = 0$, so that $\dim_{\mathbb{F}}(\mathrm{Ker}(f)) = 1$, from which it follows that $\dim_{\mathbb{F}}(\mathrm{Im}(f)) = s-1 = \dim_{\mathbb{F}}(V)$, so that f is surjective, as claimed.

(3). This is a general property of separable extensions: it is clear that the pairing is bilinear, so the only thing that we need to prove is that it is nondegenerate, but this is clear since if $\mathrm{Tr}_{\mathbb{E}/\mathbb{F}}(ax) = 0$ for all $x \in \mathbb{E}$ with $a \neq 0$ then $\mathrm{Tr}_{\mathbb{E}/\mathbb{F}}(y) = 0$ for all $y \in \mathbb{E}$; in other words, the trace map would be identically zero, contradicting its surjectivity.   □

The $\mathbb{F}$-subspace of $\mathbb{E}$ of elements of the form $x^q - x$ is called the *Artin–Schreier subspace* (or subgroup) of $\mathbb{E}$, and will be used several times; see for instance Section 3.1.8 and Exercise 2 of Chapter 7.

Proposition 2.4.12. *Let* $\mathbb{E}/\mathbb{F}$ *be an extension of finite fields. The norm map* $\mathcal{N}_{\mathbb{E}/\mathbb{F}}$ *is a surjective homomorphism from* $\mathbb{E}^*$ *to* $\mathbb{F}^*$.

Proof. Let g be a generator of the cyclic group $\mathbb{E}^*$. The subgroup $\mathbb{F}^*$ is the unique subgroup of cardinality $q - 1$; hence it is the group generated by $g^{(q^s-1)/(q-1)}$. Thus if $a \in \mathbb{F}^*$ we can write $a = g^{k(q^s-1)/(q-1)}$ for a unique k defined modulo $q - 1$, and thus $\mathcal{N}_{\mathbb{E}/\mathbb{F}}(g^k) = a$ by Corollary 2.4.9.   □

Although the study of infinite topological Galois extensions goes slightly beyond our purpose, we mention here that given the appropriate definitions, it is immediate to deduce from the above that the infinite Galois group $\mathrm{Gal}(\overline{\mathbb{F}}_p/\mathbb{F}_p)$ is a profinite group isomorphic to the profinite completion $\widehat{\mathbb{Z}}$ of $\mathbb{Z}$, and topologically generated by the Frobenius automorphism σ_p.

2.4.3 Polynomials over Finite Fields

We begin with the following proposition.

Proposition 2.4.13. *Let* $\mathbb{F} = \mathbb{F}_q$ *be some finite field and let* $n \geqslant 1$.

(1) *In* $\mathbb{F}[X]$ *we have the decomposition into irreducibles*

$$\Omega_{q^n}(X) = X^{q^n} - X = \prod_{\substack{P \text{ monic irreducible} \\ \deg(P)|n}} P(X) \, ,$$

where the product is over all monic irreducible polynomials of degree dividing n.

(2) *A polynomial* P *of degree* n *is irreducible in* $\mathbb{F}[X]$ *if and only if* $P(X) \mid X^{q^n} - X$ *and* $\gcd(P(X), X^{q^{n/d}} - X) = 1$ *for all* $d \mid n$, $d > 1$.

Proof. (1). Since $\mathbb{F}_q$ is perfect, irreducible polynomials have only simple roots in $\overline{\mathbb{F}_q}$ (see Proposition 3.1.1). Monic irreducible polynomials being pairwise coprime, it follows that both sides of the equation are polynomials with only simple roots in $\overline{\mathbb{F}_q}$. To prove equality, it is thus sufficient to show that they have the same roots. If x is a root of $X^{q^n} - X$, then $x \in \mathbb{F}_{q^n}$, and the minimal polynomial of x defines a subextension of $\mathbb{F}_{q^n}/\mathbb{F}_q$, hence has degree $d \mid n$, and is of course irreducible. Thus, x is a root of the right-hand side. Conversely, if x is a root of an irreducible polynomial P of degree d dividing n, then x belongs to the unique extension $\mathbb{F}_{q^d}/\mathbb{F}_q$ of $\mathbb{F}_q$ of degree d; hence $x \in \mathbb{F}_{q^n}$, so x is a root of $X^{q^n} - X = 0$.

(2). Assume first that P is irreducible. Then by (1), $P(X) \mid X^{q^n} - X$, and if x is a root of P then $x \in \mathbb{F}_{q^n}$, but $x \notin \mathbb{F}_{q^{n/d}}$ for $d > 1$ since otherwise the minimal polynomial of x would be of degree strictly less than n and would divide P, which is absurd. Thus $x^{q^{n/d}} - x \neq 0$, hence $\gcd(P(X), X^{q^{n/d}} - X) = 1$. Conversely, if these conditions are satisfied and x is a root of $P(X)$, then x belongs to $\mathbb{F}_{q^n}$ and to no smaller subextension; hence the minimal polynomial of x has degree n, and since it divides P it is equal to P, so P is irreducible. $\square$

We leave to the reader to show that in fact in (2), it is sufficient to test the GCD condition for d a *prime* divisor of n.

Corollary 2.4.14. *Let* $p(n)$ *be the number of monic irreducible polynomials of degree* n *in* $\mathbb{F}_q[X]$. *Then* $p(n) \geqslant 1$ *and* $p(n)$ *is given by the explicit formula*

$$p(n) = \frac{1}{n} \sum_{d|n} \mu(n/d)q^d \, .$$

Proof. If $x \in \mathbb{F}_{q^n}$ generates $\mathbb{F}_{q^n}$ over $\mathbb{F}_q$, then the degree of the minimal polynomial of x is equal to n; hence $p(n) \geqslant 1$. Taking degrees in the decomposition of $X^{q^n} - X$ given in the above proposition, we obtain the equality

$$q^n = \sum_{d|n} dp(d) .$$

The Möbius inversion formula (Proposition 10.1.5) gives the required equality.
□

If desired, it is immediate to show once again from this equality that $p(n) \geqslant 1$, and in fact much stronger results.

2.5 Bounds for the Number of Solutions in Finite Fields

It is essential to be able to give bounds for the number of solutions (necessarily finite) of a system of algebraic equations over finite fields, in particular with reference to Diophantine equations. This is a very old, important, and extremely difficult subject, which we consider here. In this section, we let $q = p^f$ be a prime power, and we will work in the field $\mathbb{F}_q$ with q elements.

2.5.1 The Chevalley–Warning Theorem

We begin with a simple lemma.

Lemma 2.5.1. (1) *For* $0 \leqslant m < q - 1$ *we have* $\sum_{a \in \mathbb{F}_q} a^m = 0$.
(2) *For any* $m \in \mathbb{Z}$ *we have*

$$\sum_{a \in \mathbb{F}_q^*} a^m = \begin{cases} 0 & when \ (q-1) \nmid m , \\ -1 & when \ (q-1) \mid m . \end{cases}$$

Proof. The result is clear for $m = 0$ since $p \mid q$, so assume $0 < m < q - 1$. It is clear that the map $a \mapsto a^m$ is a group homomorphism χ from $\mathbb{F}_q^*$ to $\mathbb{F}_q^*$. Furthermore, by Corollary 2.4.3 we know that $\mathbb{F}_q^*$ is cyclic, so let g be a generator. Since $0 < m < q - 1$ we have $\chi(g) \neq 1$, so χ is not always equal to 1; hence (1) follows from the orthogonality of characters (Proposition 2.1.20 (1)), and (2) is an immediate consequence since $a^{q-1} = 1$ for all $a \in \mathbb{F}_q^*$ and since $p \mid q$.
□

See also Exercise 29.

Theorem 2.5.2 (Chevalley–Warning). (1) *Let* $(P_i(\underline{X}))_{1 \leqslant i \leqslant r}$ *be a family of* r *polynomials in* $\mathbb{F}_q[X_1, \ldots, X_n]$ *of respective total degrees* d_i, *and let*

$$V = \{(a_1, \ldots, a_n) \in \mathbb{F}_q^n, \ P_i(a_1, \ldots, a_n) = 0 \quad \text{for all } i\}$$

be the set of their common zeros. If $n > \sum_{1 \leqslant i \leqslant r} d_i$ then $|V|$ is divisible by p.

(2) In particular, if $P(\underline{X}) \in \mathbb{F}_q[X_1, \ldots, X_n]$ is a polynomial in n variables of total degree d and if $n > d$, the number of solutions to $P(a_1, \ldots, a_n) = 0$ in $\mathbb{F}_q^n$ is divisible by p.

(3) If, in addition, $P(\underline{X})$ is a nonconstant homogeneous polynomial, there exists $(a_1, \ldots, a_n) \neq (0, \ldots, 0)$ such that $P(a_1, \ldots, a_n) = 0$.

Proof. Define $P(\underline{X}) = \prod_{1 \leqslant i \leqslant r}(1 - P_i(\underline{X})^{q-1})$. Since $a^{q-1} = 1$ when $a \in \mathbb{F}_q^*$, it is clear that if $\underline{A} = (a_1, \ldots, a_n) \in V$ then $P(\underline{A}) = 1$ and if $\underline{A} \notin V$ then $P(\underline{A}) = 0$. It follows that $|V| \equiv \sum_{\underline{A} \in \mathbb{F}_q^n} P(\underline{A}) \pmod{p}$. Note that this makes sense since there is a natural map from $\mathbb{Z}$ to $\mathbb{F}_q$ obtained by composition of the canonical surjection from $\mathbb{Z}$ to $\mathbb{Z}/p\mathbb{Z} = \mathbb{F}_p$ with the canonical injection from $\mathbb{F}_p$ to $\mathbb{F}_q$.

Let d be the total degree of P, so that $d \leqslant (q-1)\sum_{1 \leqslant i \leqslant r} d_i$, and write

$$P(X_1, \ldots, X_n) = \sum_{m_1, \ldots, m_n} c(m_1, \ldots, m_n)X_1^{m_1} \cdots X_n^{m_n}$$

for some coefficients $c(m_1, \ldots, m_n)$, the sum being over certain n-tuples $m_1, \ldots, m_n$ such that $m_1 + \cdots + m_n \leqslant d(q-1) < n(q-1)$ by assumption. It follows that for any such n-tuple there exists a j such that $m_j < q-1$, and by the lemma above, that $\sum_{a_j \in \mathbb{F}_q} a_j^{m_j} = 0$. Thus every term in the expression $\sum_{\underline{A} \in \mathbb{F}_q^n} P(\underline{A})$ is zero, proving (1), and (2) is a special case. For (3) we note that if P is homogeneous then trivially $P(0, \ldots, 0) = 0$; hence by (1) there exist at least $p - 1$ nonzero solutions to $P(a_1, \ldots, a_n) = 0$. $\qquad\square$

Remarks. (1) If we want to apply Hensel's Lemma 4.1.37, the simple existence of a solution in $\mathbb{F}_p$ is not sufficient since we also need a condition on the derivative. The important fact is that the solution that we find is *nonsingular*, in other words that it is not also a root of all the partial derivatives of P, see Corollary 4.1.39.

(2) It has been shown by Ax and Katz that we have the following stronger and essentially optimal statement: with the same notation, if we set

$$k = \left\lceil \frac{n - \sum_{1 \leqslant i \leqslant r} d_i}{\max_{1 \leqslant i \leqslant r} d_i} \right\rceil,$$

then $p^k \mid |V|$; see [Ax] and [Kat1].

2.5.2 Gauss Sums for Finite Fields

We have studied above characters associated with the multiplicative group $(\mathbb{Z}/n\mathbb{Z})^*$ and their corresponding Gauss sums. In the present section we study

Gauss sums associated with characters on finite fields: while the notions are closely related (and in fact almost identical if the finite field is $\mathbb{F}_p$) they are different on $\mathbb{F}_q$ when q is not a prime. In particular the notion of primitive character is irrelevant. I refer to [Ber-Eva-Wil] for an extensive compilation of results on Gauss and Jacobi sums over finite fields.

Definition 2.5.3. (1) *An* additive character ψ *on* $\mathbb{F}_q$ *is a group homomorphism from the additive group of* $\mathbb{F}_q$ *to the multiplicative group* $\overline{\mathbb{Q}}^*$, *and* a *multiplicative character* χ *is a group homomorphism from the multiplicative group* $\mathbb{F}_q^*$ *to* $\overline{\mathbb{Q}}^*$.

(2) *The trivial additive character* ψ_0 *is such that* $\psi_0(x) = 1$ *for all* $x \in \mathbb{F}_q$, *and the trivial multiplicative character* ε *is such that* $\varepsilon(x) = 1$ *for all* $x \in \mathbb{F}_q^*$.

(3) *If* χ *is a multiplicative character we extend* χ *to* $\mathbb{F}_q$ *by setting by convention* $\chi(0) = 0$ *if* $\chi \neq \varepsilon$, *and* $\varepsilon(0) = 1$.

Remarks. (1) Setting $\chi(0) = 0$ for a nontrivial multiplicative character preserves the multiplicative property of χ. In addition, we have $(\chi^{-1})(x) = \overline{\chi}(x)$, even for $x = 0$, denoting by χ^{-1} the inverse of χ in the group of characters of $\mathbb{F}_q^*$.

(2) We will always keep the notation of the definition and reserve the letter χ for multiplicative characters.

(3) If χ is a multiplicative character, it is clear that the order n of χ divides $q - 1$. Furthermore, we note for future reference that $\chi(-1) = 1$ if n is odd, while $\chi(-1) = (-1)^{(q-1)/n}$ if n is even (see Exercise 31).

(4) We will usually assume, either implicitly or explicitly, that the additive characters that we consider are nontrivial. On the other hand, it is necessary to consider the trivial multiplicative character together with the others.

(5) Beware that we define $\chi(0) = 0$ only for $\chi \neq \varepsilon$, while $\varepsilon(0) = 1$. We will see that this convention is quite useful (Lemma 2.5.21). This is the main reason why we denote the trivial character by ε and not by χ_0 as we have done for Dirichlet characters, since in the latter case $\chi_0(0) = 0$ if χ_0 is the trivial character modulo m.

(6) Although very useful, the convention at $x = 0$ has pitfalls: for instance it is *not* true that $(\chi_1\chi_2)(0) = \chi_1(0)\chi_2(0)$ when χ_1 is a nontrivial character and $\chi_2 = \chi_1^{-1}$ in the group of characters.

Let $\zeta_p \in \overline{\mathbb{Q}}$ be a primitive pth root of unity. If $e_1, \ldots, e_f$ is an $\mathbb{F}_p$-basis of $\mathbb{F}_q$, it is easy to see that the additive characters are given (with a slight abuse of notation) by

$$\psi\left(\sum_j x_j e_j\right) = \zeta_p^{\sum_j a_j x_j}$$

for some fixed $a_j \in \mathbb{F}_p$, which determine the character ψ (so that there are $q = p^f$ additive characters, as expected). However, a better description is as follows.

Proposition 2.5.4. *Let $b \in \mathbb{F}_q$ be fixed. The map*

$$\psi_b: \quad x \mapsto \zeta_p^{\mathrm{Tr}_{\mathbb{F}_q/\mathbb{F}_p}(bx)}$$

is an additive character of $\mathbb{F}_q$. Furthermore, the map $b \mapsto \psi_b$ is a group isomorphism from the additive group $\mathbb{F}_q$ to the multiplicative group of additive characters.

Proof. The first statement is clear by linearity of the trace and the fact that we are in characteristic p. For the second statement, we note that the map $b \mapsto \psi_b$ is clearly a group homomorphism from $\mathbb{F}_q$ to $\widehat{\mathbb{F}_q}$, which are two groups with the same cardinality q. Thus to prove that it is an isomorphism it is sufficient to show that its kernel is 0. To say that ψ_b is the trivial additive character means that $\mathrm{Tr}_{\mathbb{F}_q/\mathbb{F}_p}(bx) = 0$ in $\mathbb{F}_p$ for all $x \in \mathbb{F}_q$, and since the trace is nondegenerate (Proposition 2.4.11) it follows that $b = 0$, as claimed. $\square$

We will keep the notation ψ_b in the sequel. Thus, any additive character has the form ψ_b for a unique $b \in \mathbb{F}_q$, and $\psi_b(x) = \psi_1(bx)$.

Definition 2.5.5. *Let χ be a multiplicative character and ψ a nontrivial additive character of $\mathbb{F}_q$. We define the* Gauss sum *attached to χ and ψ by the formula*

$$\tau(\chi, \psi) = \sum_{x \in \mathbb{F}_q^*} \chi(x)\psi(x) \,.$$

Remarks. (1) Note that we omit $x = 0$ in the definition.

(2) Since ψ has order p and χ has order dividing $q - 1$ we have $\tau(\chi, \psi) \in \mathbb{Q}(\zeta_p, \zeta_{q-1})$.

(3) As the reader will notice as we study Gauss sums in the sequel, almost all of the formulas involving Gauss sums as defined above have annoying *signs* in them, the simplest being Lemma 2.5.9 (2) below, which says that $\tau(\varepsilon, \psi) = -1$ for the trivial character. For this and many other good reasons, following A. Weil it would thus be a good idea to include a minus sign in the definition given above, as well as in the corresponding definition for Jacobi sums that we will give below. However, I have preferred not doing so, at the price of keeping minus signs in many formulas.

(4) Independently of the sign issue, several notation are used for Gauss sums in the literature: essentially τ, g, and G. Since we use the letter J for Jacobi sums it would be reasonable to use the letter G, but unfortunately this letter is too often used in connection with Gauss sums to denote finite abelian groups. Thus we stick to τ, which is one of the traditional notations.

We have the following trivial lemmas.

Lemma 2.5.6. *For $b \neq 0$ and any a we have $\tau(\chi, \psi_{ab}) = \chi(b)^{-1}\tau(\chi, \psi_a)$, and in particular $\tau(\chi, \psi_b) = \chi(b)^{-1}\tau(\chi, \psi_1)$.*

Proof. This is the analogue of Proposition 2.1.39 and is proved in the same way:

$$\tau(\chi, \psi_{ab}) = \sum_{x \in \mathbb{F}_q^*} \chi(x)\psi_a(xb) = \sum_{y \in \mathbb{F}_q^*} \chi(yb^{-1})\psi_a(y) = \chi(b)^{-1}\tau(\chi, \psi_a) \ .$$

$\square$

Because of this lemma and Proposition 2.5.4 it is reasonable to set the following definition.

Definition 2.5.7. *If ψ is an additive character on $\mathbb{F}_q$ and b is the unique element of $\mathbb{F}_q$ such that $\psi = \psi_b$, for any $a \in \mathbb{F}_q$ we set $\psi^a = \psi_{ab}$.*

Note that this agrees with the usual definition when $a \in \mathbb{F}_p$ since the trace is $\mathbb{F}_p$-linear.

Lemma 2.5.8. *Let ψ be a nontrivial additive character.*

(1) *We have $\tau(\varepsilon, \psi) = -1$.*
(2) *For any character χ we have*

$$\tau(\chi^{-1}, \psi) = \chi(-1)\overline{\tau(\chi, \psi)} \ .$$

(3) *If $b \in \mathbb{F}_q$ is such that $\psi = \psi_b$ then*

$$\tau(\chi^p, \psi) = \chi^{1-p}(b)\tau(\chi, \psi) = \tau(\chi, \psi_{b^p-1}) \ .$$

Proof. We have $\tau(\varepsilon, \psi) = -\psi(0) + \sum_{a \in \mathbb{F}_q} \psi(a) = -1$ by Proposition 2.1.20, proving (1). For (2), note that $1 = \psi(0) = \psi(x)\psi(-x)$; hence $\psi(-x) = \overline{\psi(x)}$ since it is a root of unity. Since $\chi^{-1}(x) = \overline{\chi(x)}$ we have

$$\overline{\tau(\chi^{-1}, \psi)} = \sum_{x \in \mathbb{F}_q^*} \chi(x)\overline{\psi(x)} = \sum_{y \in \mathbb{F}_q^*} \chi(-y)\psi(y) = \chi(-1)\tau(\chi, \psi) \ ,$$

proving (2). Let us first prove (3) for $\psi = \psi_1$. We note that the map $x \mapsto x^p$ is an automorphism of $\mathbb{F}_q$ (the Frobenius automorphism, which is a canonical generator of $\mathrm{Gal}(\mathbb{F}_q/\mathbb{F}_p)$). Furthermore, for the same reason $\mathrm{Tr}_{\mathbb{F}_q/\mathbb{F}_p}(x) = \sum_{0 \leqslant i < f} x^{p^i}$; hence

$$\mathrm{Tr}_{\mathbb{F}_q/\mathbb{F}_p}(x^p) = \sum_{1 \leqslant i \leqslant f} x^{p^i} = \sum_{0 \leqslant i < f} x^{p^i} + x^q - x = \mathrm{Tr}_{\mathbb{F}_q/\mathbb{F}_p}(x) \ .$$

It follows that

$$\psi_1(x^p) = \zeta_p^{\mathrm{Tr}_{\mathbb{F}_q/\mathbb{F}_p}(x^p)} = \zeta_p^{\mathrm{Tr}_{\mathbb{F}_q/\mathbb{F}_p}(x)} = \psi_1(x) \ ,$$

hence that

$$\tau(\chi, \psi_1) = \sum_{x \in \mathbb{F}_q^*} \chi(x)\psi_1(x) = \sum_{x \in \mathbb{F}_q^*} \chi(x^p)\psi_1(x^p) = \sum_{x \in \mathbb{F}_q^*} \chi^p(x)\psi_1(x) = \tau(\chi^p, \psi_1) \ ,$$

proving (3) for $\psi = \psi_1$. The general case immediately follows from Lemma 2.5.6. $\hfill\square$

One of the main elementary results concerning Gauss sums is the following proposition, which gives their modulus.

Proposition 2.5.9. *Let ψ be a nontrivial additive character.*

(1) *If χ is a nontrivial multiplicative character then $|\tau(\chi, \psi)| = q^{1/2}$.*
(2) *If $\chi = \varepsilon$ then $\tau(\chi, \psi) = -1$.*

Proof. Setting $z = xy^{-1}$, we have

$$|\tau(\chi, \psi)|^2 = \tau(\chi, \psi)\overline{\tau(\chi, \psi)} = \sum_{x,y \in \mathbb{F}_q^*} \chi(x)\overline{\chi}(y)\psi(x)\overline{\psi}(y)$$

$$= \sum_{z \in \mathbb{F}_q^*} \chi(z) \sum_{y \in \mathbb{F}_q^*} \psi(y(z-1)) \ .$$

Now it is clear that $y \mapsto \psi(y(z-1))$ is an additive character, and it is nontrivial if and only if $(z-1) \in \mathbb{F}_q^*$ (since in that case the map $y \mapsto y(z-1)$ is a bijection of $\mathbb{F}_q^*$ onto itself). Thus by orthogonality of characters (Proposition 2.1.20), we have $\sum_{y \in \mathbb{F}_q} \psi(y(z-1)) = 0$ when $z \neq 1$, hence

$$|\tau(\chi, \psi)|^2 = \chi(1)(q-1) + \sum_{z \in \mathbb{F}_q^*, \ z \neq 1} (-1)\chi(z) = q - \sum_{z \in \mathbb{F}_q^*} \chi(z) = q \ ,$$

once again by orthogonality, this time applied to χ, proving (1), and (2) has already been proved above. $\hfill\square$

Corollary 2.5.10. *Let ψ be a nontrivial additive character and let $b \in \mathbb{F}_q^*$. Then*

$$\left| \sum_{x \in \mathbb{F}_q} \psi(bx^m) \right| \leqslant (\gcd(m, q-1) - 1)q^{1/2} \leqslant (m-1)q^{1/2} \ .$$

Proof. Set $d = \gcd(m, q-1)$. By Corollary 2.4.4 we have

$$\sum_{x \in \mathbb{F}_q} \psi(bx^m) = 1 + d \sum_{y \in \mathbb{F}_q^{*d}} \psi(by) \ .$$

By the orthogonality relations in the group $\mathbb{F}_q^*/(\mathbb{F}_q^*)^d$, we clearly have for $c \in \mathbb{F}_q^*$,

$$\sum_{\chi^d = \varepsilon} \overline{\chi}(b)\chi(c) = \begin{cases} d & \text{if } c \in b\mathbb{F}_q^{*d}, \\ 0 & \text{otherwise.} \end{cases}$$

Furthermore, since $d \mid q - 1$, if $c = bx^d$ has one solution, it has exactly d. It follows that

$$\sum_{x \in \mathbb{F}_q} \psi(bx^m) = 1 + \sum_{c \in \mathbb{F}_q^*} \psi(c) \sum_{\chi^d = \varepsilon} \overline{\chi}(b)\chi(c) = 1 + \sum_{\chi^d = \varepsilon} \overline{\chi}(b)\tau(\chi, \psi) .$$

Since

$$\tau(\varepsilon, \psi) = \sum_{x \in \mathbb{F}_q^*} \psi(x) = -\psi(0) = -1$$

by orthogonality, using the proposition and the fact that $|\mathbb{F}_q^*/(\mathbb{F}_q^*)^d| = d$, we obtain the corollary. $\qquad\square$

For instance, this implies that when $p \nmid b$ and $p \equiv 1 \pmod 3$ (otherwise the sum vanishes) we have $|\sum_{0 \leqslant x < p} e^{2i\pi bx^3/p}| \leqslant 2p^{1/2}$, see Exercise 33.

We give the following result as example of an application.

Proposition 2.5.11. *Let a, b, and c be nonzero elements of $\mathbb{F}_q$, let $m \geqslant \mathbb{Z}_{\geqslant 1}$, and set $d = \gcd(m, q - 1)$.*

(1) *The number N of solutions $(x, y, z) \in \mathbb{F}_q^3$ of the equation $ax^m + by^m + cz^m = 0$ satisfies $|N - q^2| \leqslant (d - 1)^3(q - 1)q^{1/2}$.*

(2) *The number M of projective solutions $(x, y, z) \in \mathbb{P}^2(\mathbb{F}_q)$ to the equation $ax^m + by^m + cz^m = 0$ satisfies*

$$|M - (q + 1)| \leqslant (d - 1)^3 q^{1/2} .$$

Proof. By orthogonality, denoting as usual by ψ_0 the trivial additive character, we have

$$qN = \sum_{(x,y,z) \in \mathbb{F}_q^3} \sum_{\psi} \psi(ax^m + by^m + cz^m)$$

$$= q^3 + \sum_{\psi \neq \psi_0} \sum_{(x,y,z) \in \mathbb{F}_q^3} \psi(ax^m + by^m + cz^m) .$$

The inner sum splits into a product of three simple sums; hence using the corollary we obtain $|qN - q^3| \leqslant (q - 1)((d - 1)q^{1/2})^3$, so dividing by q proves (1).

For (2) we simply note that by definition of projective space we have $N = 1 + (q - 1)M$, so (2) immediately follows from (1). $\qquad\square$

Note that although not difficult, the above result is already not entirely trivial. The bounds that we obtain below using Jacobi sums show that we can replace $(d - 1)^3$ by $(d - 1)(d - 2)$, and this is optimal.

2.5.3 Jacobi Sums for Finite Fields

Let ψ be a nontrivial additive character and χ a multiplicative character on $\mathbb{F}_q$. We have already mentioned that $\tau(\chi, \psi) \in \mathbb{Q}(\zeta_p, \zeta_{q-1})$, where ζ_m denotes a primitive mth root of unity. This number field is quite large, and this is one of the reasons why Gauss sums are often unsatisfactory. A second more important reason is that Jacobi sums, which we now introduce, are an essential tool in counting the number of solutions of diagonal forms in finite fields, as we shall see.

Definition 2.5.12. *Let $\chi_1, \ldots, \chi_k$ be multiplicative characters on $\mathbb{F}_q$.*

(1) *We define the* Jacobi sum *with parameter $a \in \mathbb{F}_q$ associated with these characters by the formula*

$$J_k(\chi_1, \ldots, \chi_k; a) = \sum_{\substack{x_i \in \mathbb{F}_q \\ x_1 + \cdots + x_k = a}} \chi_1(x_1) \cdots \chi_k(x_k) \, .$$

(2) *We simply write $J_k(\chi_1, \ldots, \chi_k)$ instead of $J_k(\chi_1, \ldots, \chi_k; 1)$, and call it the Jacobi sum associated with the χ_i's.*

(3) *For notational simplicity, by abuse of notation we will often write $J_k(a)$ instead of $J_k(\chi_1, \ldots, \chi_k; a)$, the characters χ_i being implicit.*

Remarks. (1) We have $J_1(\chi_1) = 1$ for any character χ_1, and more generally $J_1(\chi_1; a) = \chi_1(a)$.

(2) It is clear that the value of $J_k(\chi_1, \ldots, \chi_k; a)$ does not depend on the ordering of the characters χ_i.

(3) As desired we have $J_k(\chi_1, \ldots, \chi_k; a) \in \mathbb{Q}(\zeta_{q-1})$, which is a much smaller number field.

(4) The introduction of a parameter a is analogous to that of $\tau(\chi, a)$ for Gauss sums associated with a Dirichlet character. In fact, as for Gauss sums, the following lemma shows that there is a close link between $J_k(\chi_1, \ldots, \chi_k; a)$ and $J_k(\chi_1, \ldots, \chi_k)$.

Lemma 2.5.13. *For $a \neq 0$ we have*

$$J_k(\chi_1, \ldots, \chi_k; a) = (\chi_1 \cdots \chi_k)(a) J_k(\chi_1, \ldots, \chi_k) \, ,$$

while (abbreviating as above $J_k(\chi_1, \ldots, \chi_k; 0)$ to $J_k(0)$) we have

$$J_k(0) = \begin{cases} q^{k-1} & \text{if } \chi_j = \varepsilon \text{ for all } j \, , \\ 0 & \text{if } \chi_1 \cdots \chi_k \neq \varepsilon \, , \\ \chi_k(-1)(q-1) J_{k-1}(\chi_1, \ldots, \chi_{k-1}) & \text{if } \chi_1 \cdots \chi_k = \varepsilon \text{ and } \chi_k \neq \varepsilon \, . \end{cases}$$

Proof. The formula for $a \neq 0$ is clear by setting $y_k = x_k/a$, so assume $a = 0$. If all the χ_j are equal to ε then $J_k(0)$ is equal to the number of $(x_1, \ldots, x_k) \in \mathbb{F}_q^k$ such that $x_1 + \cdots + x_k = 0$, hence to q^{k-1}, which is the first formula, so we may assume that not all the χ_j are equal to ε, and since $J_k(a)$ is invariant under permutation of the indices we assume that $\chi_k \neq \varepsilon$. We thus have $\chi_k(0) = 0$, hence

$$J_k(0) = \chi_k(0)J_{k-1}(0) + \sum_{\substack{x_i \in \mathbb{F}_q,\ x_k \neq 0 \\ (-x_1/x_k)+\cdots+(-x_{k-1}/x_k)=1}} \chi_1(x_1)\cdots\chi_k(x_k)$$

$$= \chi_k(-1)\sum_{x_k \in \mathbb{F}_q^*}(\chi_1\cdots\chi_k)(-x_k) \sum_{\substack{y_i \in \mathbb{F}_q \\ y_1+\cdots+y_{k-1}=1}} \chi_1(y_1)\cdots\chi_{k-1}(y_{k-1})$$

$$= \chi_k(-1)J_{k-1}(\chi_1,\ldots,\chi_{k-1})\sum_{y \in \mathbb{F}_q^*}(\chi_1\cdots\chi_k)(y) \ ,$$

and since $|\mathbb{F}_q^*| = q - 1$ the result follows from Proposition 2.1.20.

The main result concerning Jacobi sums is the following close link with Gauss sums:

Proposition 2.5.14. *Let ψ be a nontrivial additive character and let χ_1, $\ldots$, χ_k be multiplicative characters of $\mathbb{F}_q$. Denote by t the number of such χ_i equal to the trivial character ε.*

(1) *If $t = k$ then $J_k(\chi_1,\ldots,\chi_k) = q^{k-1}$.*
(2) *If $1 \leqslant t \leqslant k - 1$ then $J_k(\chi_1,\ldots,\chi_k) = 0$.*
(3) *If $t = 0$ and $\chi_1\cdots\chi_k \neq \varepsilon$ then*

$$J_k(\chi_1,\ldots,\chi_k) = \frac{\tau(\chi_1,\psi)\cdots\tau(\chi_k,\psi)}{\tau(\chi_1\cdots\chi_k,\psi)} \ .$$

(4) *If $t = 0$ and $\chi_1\cdots\chi_k = \varepsilon$ then*

$$J_k(\chi_1,\ldots,\chi_k) = -\frac{\tau(\chi_1,\psi)\cdots\tau(\chi_k,\psi)}{q}$$

$$= -\chi_k(-1)\frac{\tau(\chi_1,\psi)\cdots\tau(\chi_{k-1},\psi)}{\tau(\chi_1\cdots\chi_{k-1},\psi)}$$

$$= -\chi_k(-1)J_{k-1}(\chi_1,\ldots,\chi_{k-1}) \ .$$

In particular, in this case we have

$$\frac{\tau(\chi_1,\psi)\cdots\tau(\chi_k,\psi)}{q} = \chi_k(-1)J_{k-1}(\chi_1,\ldots,\chi_{k-1}) \ .$$

Proof. (1) and (2). For $k = 1$ the result is trivial, so assume that $k \geqslant 2$. We can write

$$J_k(\chi_1,\ldots,\chi_k) = \sum_{x_1,\ldots,x_{k-1}\in\mathbb{F}_q} \chi_1(x_1)\cdots\chi_{k-1}(x_{k-1})\chi_k(1-(x_1+\cdots+x_{k-1})) \ .$$

If $t = k$, in other words if all the χ_i are equal to ε, this is evidently equal to q^{k-1}, which is (1). If $1 \leqslant t \leqslant k-1$, then since J_k is invariant by permutation of the characters we may assume that $\chi_k = \varepsilon$. Thus the above sum decomposes into a product:

$$J_k(\chi_1, \ldots, \chi_k) = \prod_{1 \leqslant i \leqslant k-1} \sum_{x_i \in \mathbb{F}_q} \chi_i(x_i) \, .$$

Since $t \leqslant k - 1$, at least one of the χ_i for $i \leqslant k - 1$ is a nontrivial character, hence $\sum_{x_i \in \mathbb{F}_q} \chi_i(x_i) = 0$, proving (2).

(3) and (4). We therefore assume that all the χ_i are nontrivial characters. For simplicity, write $\tau(\chi)$ instead of $\tau(\chi, \psi)$, since in the present proof the additive character ψ does not change. Thus $\tau(\chi_i) = \sum_{x_i \in \mathbb{F}_q} \chi_i(x_i)\psi(x_i)$, where we may include $x_i = 0$, since $\chi_i(0) = 0$, χ_i being nontrivial. Hence grouping terms with given $x_1 + \cdots + x_k = a$ and using the above lemma we have

$$\tau(\chi_1) \cdots \tau(\chi_k) = \sum_{x_i \in \mathbb{F}_q} \chi_1(x_1) \cdots \chi_k(x_k)\psi(x_1 + \cdots + x_k)$$

$$= \sum_{a \in \mathbb{F}_q} \psi(a) \sum_{\substack{x_i \in \mathbb{F}_q \\ x_1 + \cdots + x_k = a}} \chi_1(x_1) \cdots \chi_k(x_k)$$

$$= J_k(0) + \sum_{a \in \mathbb{F}_q^*} \psi(a)(\chi_1 \cdots \chi_k)(a) J_k(\chi_1, \ldots, \chi_k)$$

$$= J_k(0) + \tau(\chi_1 \cdots \chi_k) J_k(\chi_1, \ldots, \chi_k) \, .$$

If $\chi_1 \cdots \chi_k \neq \varepsilon$ the lemma tells us that $J_k(0) = 0$, proving (3). If $\chi_1 \cdots \chi_k = \varepsilon$ then on the one hand by the lemma we have $J_k(0) = \chi_k(-1)(q - 1)J_{k-1}(\chi_1, \ldots, \chi_{k-1})$. However, since $\chi_k \neq \varepsilon$ while $\chi_1 \cdots \chi_k = \varepsilon$ we have $\chi_1 \cdots \chi_{k-1} \neq \varepsilon$ (and all the χ_i still different from ε), so by (3) and Proposition 2.5.8 we have

$$J_{k-1}(\chi_1, \ldots, \chi_{k-1}) = \frac{\tau(\chi_1) \cdots \tau(\chi_{k-1})}{\tau(\chi_k^{-1})} = \chi_k(-1)\frac{\tau(\chi_1) \cdots \tau(\chi_{k-1})}{\tau(\chi_k)}$$

$$= \chi_k(-1)\frac{\tau(\chi_1) \cdots \tau(\chi_{k-1})\tau(\chi_k)}{q}$$

by Proposition 2.5.9, so $J_k(0) = (1 - 1/q)\tau(\chi_1) \cdots \tau(\chi_k)$. On the other hand, again by Proposition 2.5.9 since $\chi_1 \cdots \chi_k = \varepsilon$ we have $\tau(\chi_1 \cdots \chi_k) = -1$, so putting this in the above formula gives the first formula of (4), and the others follows from Lemma 2.5.8 and from (3). $\square$

Corollary 2.5.15. *If χ is a character of order dividing m then $\tau(\chi, \psi)^m \in \mathbb{Q}(\zeta_m)$.*

Proof. This is trivial if χ or ψ is a trivial character. Otherwise we apply the last formula of the above proposition to $k = m$ and $\chi_i = \chi$ for all i, and we deduce the result since evidently $J_k(\chi, \ldots, \chi) \in \mathbb{Q}(\zeta_m)$. $\square$

Corollary 2.5.16. *As above, denote by t the number of indices i such that $\chi_i = \varepsilon$. Then*

$$J_k(0) = \begin{cases} q^{k-1} & \text{if } t = k \text{ ,} \\ 0 & \text{if } 1 \leqslant t \leqslant k-1 \text{ or if } \chi_1 \cdots \chi_k \neq \varepsilon \text{ ,} \\ \left(1 - \dfrac{1}{q}\right) \tau(\chi_1, \psi) \cdots \tau(\chi_k, \psi) & \text{if } t = 0 \text{ and } \chi_1 \cdots \chi_k = \varepsilon \text{ .} \end{cases}$$

Proof. By Lemma 2.5.13, the result is clear for $t = k$ and when $\chi_1 \cdots \chi_k \neq \varepsilon$. Otherwise, we may assume by symmetry that $\chi_k \neq \chi_0$, and the lemma gives $J_k(0) = \chi_k(-1)(q-1)J_{k-1}(\chi_1, \ldots, \chi_{k-1})$. Since $\chi_1 \cdots \chi_{k-1} = \chi_k^{-1} \neq \varepsilon$, the proposition shows that $J_k(0) = 0$ for $1 \leqslant t \leqslant k-1$, and for $t = 0$, the desired result follows from Lemma 2.5.8. $\square$

Remark. Gauss sums are the finite analogue of the gamma function defined in Section 9.6.2. This can be better seen in the context of Gauss sums attached to Dirichlet characters by comparing Proposition 9.6.35 with Definition 2.1.38. The Jacobi sums for $k = 2$ are then the finite analogues of the beta function defined in Proposition 9.6.39, and the above corollary is the exact analogue of that proposition. We will see in Section 3.7 that there also exists an exact analogue of the distribution formula for the gamma function (Proposition 9.6.33) called the Hasse–Davenport product relation.

2.5.4 The Jacobi Sums $J(\chi_1, \chi_2)$

Because of their importance, for simplicity we will usually write $J(\chi_1, \chi_2)$ instead of $J_2(\chi_1, \chi_2)$. We first note that the special case $k = 2$ of the proposition is the following:

Corollary 2.5.17. *Let ψ be a nontrivial additive character of $\mathbb{F}_q$.*

(1) *If χ_1 and χ_2 are two multiplicative characters we have*

$$J(\chi_1, \chi_2) = \begin{cases} \dfrac{\tau(\chi_1, \psi)\tau(\chi_2, \psi)}{\tau(\chi_1\chi_2, \psi)} & \text{if } \chi_1\chi_2 \neq \varepsilon \text{ ,} \\ -\chi_1(-1) & \text{if } \chi_1\chi_2 = \varepsilon \text{ but } \chi_1 \neq \varepsilon \text{ ,} \\ q & \text{if } \chi_1 = \chi_2 = \varepsilon \text{ .} \end{cases}$$

(2) *In particular, if none of χ_1, χ_2, and $\chi_1\chi_2$ is the trivial character we have $|J(\chi_1, \chi_2)| = q^{1/2}$, and if χ_2 is nontrivial we have $J(\varepsilon, \chi_2) = -1$.*

(3) *If q is odd and ρ is a multiplicative character of order 2 then*

$$\tau(\rho,\psi)^2 = \rho(-1)q = (-1)^{(q-1)/2}q \ .$$

We will also need to identify more precisely the Jacobi sums in certain cases. For this we give some equalities and weak congruences in the following sense: if α, β, and γ are algebraic *integers* we will write $\alpha \equiv \beta \pmod{\gamma}$ if $(\alpha - \beta)/\gamma$ is an algebraic integer. In the following, all the characters that occur are nontrivial multiplicative characters.

Proposition 2.5.18. *For any $n \geqslant 2$ denote as usual by ζ_n a primitive nth root of unity.*

(1) *Assume that q is odd. If χ is a character of order $n > 2$ and ρ is the character of order 2 we have the identity*

$$\chi(4)J(\chi,\chi) = J(\chi,\rho) \ .$$

(2) *For $i = 1$, 2, let χ_i be characters of order $n_i > 1$. We have*

$$J(\chi_1,\chi_2) \equiv -q \pmod{(1 - \zeta_{n_1})(1 - \zeta_{n_2})} \ .$$

(3) *For $i = 1$, 2, let χ_i be characters of odd order $n_i > 1$, and let n_3 be the order of the character $\chi_1\chi_2$. We have the more precise congruence*

$$J(\chi_1,\chi_2) \equiv -1 \pmod{(1 - \zeta_{n_1})(1 - \zeta_{n_2})(1 - \zeta_{n_3})} \ ,$$

except if $n_1 = n_2 = n_3 = 3$ (hence $q \equiv 1 \pmod{3}$), in which case

$$J(\chi_1,\chi_2) \equiv q - 2 \pmod{(1 - \zeta_{n_1})(1 - \zeta_{n_2})(1 - \zeta_{n_3})} \ .$$

Proof. (1) is an easy exercise left to the reader (Exercise 40). For (2) we note that since the χ_i are nontrivial we have by orthogonality

$$\sum_{x\in\mathbb{F}_q\backslash\{0,1\}} (1-\chi_1(x))(1-\chi_2(1-x)) = q-2+1+1+J(\chi_1,\chi_2) = q+J(\chi_1,\chi_2) \ .$$

On the other hand, $\chi_i(x)$ is an n_ith root of unity for $x \neq 0$; hence $(1 - \zeta_{n_i}) \mid (1 - \chi_i(x))$, proving (2). Generalizing this method, set

$$T_k(\chi_1,\ldots,\chi_k) = \sum_{x_1+\cdots+x_k=1} \prod_{1\leqslant i\leqslant k} (1 - \chi_i(x_i))$$

$$\text{and} \quad S_k(\chi_1,\ldots,\chi_k) = \sum_{\substack{x_1+\cdots+x_k=1 \\ \forall i\, x_i \neq 0}} \prod_{1\leqslant i\leqslant k} (1 - \chi_i(x_i)) \ .$$

As in (2), we easily find by expanding that

$$T_k(\chi_1,\ldots,\chi_k) = q^{k-1} + (-1)^k J_k(\chi_1,\ldots,\chi_k) \ ,$$

and on the other hand, if we denote by n_i the order of χ_i we have

$$S_k(\chi_1, \ldots, \chi_k) \equiv 0 \ (\text{mod} \ \prod_{1 \leqslant i \leqslant k} (1 - \zeta_{n_i})) \ .$$

If χ is nontrivial it is clear that $S_1(\chi) = 1 - \chi(1) = 0$; hence finally the inclusion–exclusion principle shows that if all the χ_j are nontrivial and if we set $K = \{1, 2, \ldots, k\}$ then

$$T_k(\chi_1, \ldots, \chi_k) = \sum_{J \subset K, \ |J| \geqslant 2} S_{|J|}((\chi_j)_{j \in J})$$

with evident notation.

Specializing to $k = 2$ and $k = 3$ we obtain that

$$S_2(\chi_1, \chi_2) = T_2(\chi_1, \chi_2) = q + J(\chi_1, \chi_2)$$

(which was in fact used in (2)), and

$$q^2 - J_3(\chi_1, \chi_2, \chi_3) = T_3(\chi_1, \chi_2, \chi_3)$$
$$= S_3(\chi_1, \chi_2, \chi_3) + S_2(\chi_1, \chi_2) + S_2(\chi_1, \chi_3) + S_2(\chi_2, \chi_3) \ ;$$

hence

$$J_3(\chi_1, \chi_2, \chi_3) \equiv q^2 - 3q - J(\chi_1, \chi_2) - J(\chi_1, \chi_3) - J(\chi_2, \chi_3)$$
$$(\text{mod} \ (1 - \zeta_{n_1})(1 - \zeta_{n_2})(1 - \zeta_{n_3})) \ .$$

The proof of (3) is now immediate. First note that since the n_i are odd then $\chi_i(-1) = 1$. Next, if $\chi_1 \chi_2$ is the trivial character, then by Corollary 2.5.17 we have $J(\chi_1, \chi_2) = -1$, so the result is trivial. In the above formula we choose $\chi_3 = (\chi_1 \chi_2)^{-1}$, which we may therefore assume to be nontrivial of order n_3. By Proposition 2.5.14 (4) and the symmetry of Jacobi sums we have

$$J_3(\chi_1, \chi_2, \chi_3) = -J(\chi_1, \chi_2) = -J(\chi_2, \chi_3) = -J(\chi_1, \chi_3) \ .$$

Thus, if we set $x = J(\chi_1, \chi_2)$ the above congruence reads

$$2x \equiv q^2 - 3q \ (\text{mod} \ (1 - \zeta_{n_1})(1 - \zeta_{n_2})(1 - \zeta_{n_3})) \ .$$

Furthermore, it is well known (and will be proved in Chapter 3) that $1 - \zeta_n$ is a unit if n is not a prime power, and otherwise $(1 - \zeta_n)^{\phi(n)}$ divides n; hence using again the fact that the n_i are odd we may divide the congruence by 2. Finally, again since $n_i > 1$ is odd, we have $\phi(n_i) \leqslant 2$ if and only if $n_i = 3$, in which case $\phi(n_i) = 2$. Thus $(1 - \zeta_{n_1})(1 - \zeta_{n_2})(1 - \zeta_{n_3})$ divides $q - 1$, hence $(q - 1)/2$, unless $(n_1, n_2, n_3) = (3, 3, 3)$. Since $(q^2 - 3q)/2 \equiv -1$ $(\text{mod} \ (q-1)/2)$, the result follows except in that special case. In that case we have $q \equiv 1 \ (\text{mod} \ 3)$ and $x \equiv q(q-3)/2 \ (\text{mod} \ (1-\zeta_3)^3)$, and since $(1-\zeta_3)^3 \mid 9$ and $q(q - 3)/2 \equiv q - 2 \ (\text{mod} \ 9)$ when $q \equiv 1 \ (\text{mod} \ 3)$ the result follows.   □

Corollary 2.5.19. (1) *Assume that q is odd. If χ is a character of order $n > 2$ then*

$$J(\chi, \chi) \equiv -\chi^{-1}(4)q \ (\mathrm{mod}\ 2(1 - \zeta_n)) \ .$$

(2) *Let ℓ^k be an odd prime power such that $\ell^k \mid (q-1)$, let χ be a character of order ℓ^k, and let a and b be integers such that $\ell \nmid ab(a+b)$. We have*

$$J(\chi^a, \chi^b) \equiv -1 \ (\mathrm{mod}\ (1 - \zeta_{\ell^k})^3) \ ,$$

except for $\ell^k = 3$, in which case

$$J(\chi^a, \chi^b) \equiv q - 2 \ (\mathrm{mod}\ (1 - \zeta_3)^3) \ .$$

Proof. By (1) and (2) of the proposition we have

$$\chi(4)J(\chi, \chi) = J(\chi, \rho) \equiv -q \ (\mathrm{mod}\ 2(1 - i))$$

since $\zeta_2 = -1$. Since $\ell \nmid ab(a+b)$, the characters χ^a, χ^b, and χ^{a+b} have order ℓ^k; hence (2) is a special case of statement (3) of the proposition. □

An important application of Jacobi sums is in the explicit representation by binary quadratic forms. For instance, we have the following:

Proposition 2.5.20. (1) *Let q be a prime power such that $q \equiv 1 \ (\mathrm{mod}\ 4)$, let χ be one of the two characters of order 4 on $\mathbb{F}_q^*$, and write $J(\chi, \chi) = a + bi$. Then $a^2 + b^2 = q$, $2 \mid b$, and $a \equiv -1 \ (\mathrm{mod}\ 4)$. In particular, every prime $p \equiv 1 \ (\mathrm{mod}\ 4)$ is a sum of two squares.*

(2) *Let q be a prime power such that $q \equiv 1 \ (\mathrm{mod}\ 3)$, let χ be one of the two characters of order 3 on $\mathbb{F}_q^*$, and write $J(\chi, \chi) = a + b\rho$, where ρ is a primitive cube root of unity. Then $a^2 - ab + b^2 = q$, $3 \mid b$, $a \equiv -1 \ (\mathrm{mod}\ 3)$, and $a + b \equiv p - 2 \ (\mathrm{mod}\ 9)$. In particular, every prime $p \equiv 1 \ (\mathrm{mod}\ 3)$ has the form $a^2 - ab + b^2$.*

(3) *If p is a prime such that $p \equiv 1 \ (\mathrm{mod}\ 3)$, there exist c and d in $\mathbb{Z}$ such that $4p = c^2 + 27d^2$.*

(4) *If p is a prime such that $p \equiv 1 \ (\mathrm{mod}\ 3)$ then p itself has the form $p = u^2 + 27v^2$ if and only if 2 is a cube in $\mathbb{F}_p^*$, in other words if and only if 2 is a cubic residue modulo p.*

Proof. (1). Since the group of characters of $\mathbb{F}_q^*$ is isomorphic to $\mathbb{F}_q^*$, hence is cyclic of order $q - 1$, for any $n \mid q - 1$ there exists a character of order exactly equal to n. Thus, when $q \equiv 1 \ (\mathrm{mod}\ 4)$ let χ be a character of order 4. By uniqueness of the character ρ of order 2 (coming from the fact that $\mathbb{F}_q^*$ is cyclic) we have $\chi^2 = \rho$; hence $\chi(4) = \rho(2) = \left(\frac{2}{q}\right)$, where $\left(\frac{2}{q}\right)$ is equal to 1 if 2 is a square in $\mathbb{F}_q^*$ and to -1 otherwise. Since $J(\chi, \chi) \in \mathbb{Z}[\zeta_4] = \mathbb{Z}[i]$ we can write $J(\chi, \chi) = a + bi$, and the first congruence of the corollary means that b is even and that $b - a \equiv \left(\frac{2}{q}\right) \ (\mathrm{mod}\ 4)$. On the other hand, we know that $|J(\chi, \chi)|^2 = q = a^2 + b^2$, hence since a is odd that $b^2 \equiv q - 1 \ (\mathrm{mod}\ 8)$.

By separating the cases $q \equiv 1 \pmod 8$ and $q \equiv 5 \pmod 8$ and using the quadratic character of 2 (which is easily seen to be valid also in this more general context), we see that we always have $a \equiv -1 \pmod 4$.

For (2) the proof is simpler: if χ is a character of 3 we have $J(\chi, \chi) \in \mathbb{Z}[\rho]$ and $|J(\chi, \chi)|^2 = |a + b\rho|^2 = a^2 - ab + b^2 = p$, and by Corollary 2.5.19 we have $J(\chi, \chi) \equiv q - 2 \pmod{(1 - \rho)^3}$, in other words $J(\chi, \chi) \equiv q - 2 \pmod{3(1 - \rho)}$ since $3/(1 - \rho)^2 = -\rho^2$ is a unit. Since $q \equiv 1 \pmod 3$ this last congruence considered modulo 3 gives $3 \mid b$ and $a \equiv -1 \pmod 3$, and the full congruence is easily seen to give the congruence modulo 9. Thus, if we set $b = 3d$ we have $4p = 4a^2 - 12ad + 36d^2 = (2a - 3d)^2 + 27d^2$, proving (3).

For (4), we note that in (3) we have simply shown in an explicit manner that a prime $p \equiv 1 \pmod 3$ is split in the field $K = \mathbb{Q}(\rho) = \mathbb{Q}(\sqrt{-3})$ as $p = \pi\overline{\pi}$, with $\pi = (c + 3d\sqrt{-3})/2$. It is clear that π is unique up to conjugation and multiplication by a unit, in other words by a power of $-\rho$, and it is immediate to check that only conjugation, or multiplication by -1, or both, preserve the fact that the coefficient of $\sqrt{-3}$ is divisible by 3. Thus p has the form $u^2 + 27v^2$ if and only if $J(\chi, \chi) = a + b\rho$ with $b = 3d$ *even*, and since $a^2 - ab + b^2 = p$ is odd, if and only if $J(\chi, \chi) \equiv 1 \pmod 2$. Now by Corollary 2.5.19 (1) we have in particular

$$J(\chi, \chi) \equiv -\chi^{-2}(2) \equiv -\chi(2) \pmod 2 \ .$$

Thus p has the given form if and only if $\chi(2) \equiv 1 \pmod 2$, and since neither $(\rho - 1)/2$ nor $(\rho^2 - 1)/2$ is an algebraic integer, this is equivalent to $\chi(2) = 1$. As in the case of *quadratic* residues, by writing $2 \equiv g^x \pmod p$ for some primitive root g modulo p, it is immediate to see that this is equivalent to 2 being a cubic residue modulo p. $\square$

Remarks. (1) Although Jacobi sums give nice explicit formulas for the integers a and b such that $p = a^2 + b^2$ or $p = a^2 - ab + b^2$, they are not at all efficient for computing a and b in practice, since their execution time is linear in p. For this one uses instead Cornacchia's algorithm (see Exercise 41 and Algorithms 1.5.2 and 1.5.3 of [Coh0]), whose execution time is the same as that of the Euclidean algorithm, hence polynomial in $\log(p)$.

(2) All the identities that we have given for Gauss and Jacobi sums, such as $|\tau(\chi, \psi)| = q^{1/2}$ and $J(\chi_1, \chi_2) = \tau(\chi_1, \psi)\tau(\chi_2, \psi)/\tau(\chi_1\chi_2, \psi)$ for nontrivial characters, are quite elementary in nature. Gauss's computation of quadratic Gauss sums (Proposition 2.2.24) lies slightly deeper, but is still reasonably simple. On the other hand, there exist further relations between Gauss sums associated with finite fields, due to Hasse–Davenport, which are considerably more difficult (see Theorems 3.7.3 and 3.7.4). In fact, although they deal only with explicit finite sums, nobody knows of an "elementary" proof except in special cases, so we will delay till Chapter 3 the statement and proofs of these relations. Note, however,

that they are not exotic identities but fundamental results that are used for instance in Weil's proof of the rationality of the zeta function of certain varieties that we will mention below. We will in fact give such an application in Section 3.7.3.

2.5.5 The Number of Solutions of Diagonal Equations

We will now use Jacobi sums to count the number of solutions in $\mathbb{F}_q$ of diagonal equations. Let k and m_i for $1 \leqslant i \leqslant k$ be strictly positive integers, and let $a_1, \ldots, a_k$ be nonzero elements of $\mathbb{F}_q$. We want to count the number $N(q)$ of solutions $(x_1, \ldots, x_k) \in \mathbb{F}_q^k$ to the equation

$$a_1 x_1^{m_1} + \cdots + a_k x_k^{m_k} = 0 .$$

When all the m_i are equal, we will immediately deduce from the result the number of *projective* solutions. We first note that by Corollary 2.4.4 the number of solutions to $x^m = y$ is the same as that of $x^d = y$ for $d = \gcd(m, q-1)$. Thus we set $d_i = \gcd(m_i, q-1)$, and $N(q)$ is equal to the number of solutions to $a_1 x_1^{d_1} + \cdots + a_k x_k^{d_k} = 0$. We may thus replace m_i by d_i, so that $d_i \mid (q-1)$.

Lemma 2.5.21. *Let $d \mid (q-1)$, and denote by G_d the group of multiplicative characters χ on $\mathbb{F}_q$ such that $\chi^d = \varepsilon$. Then $|G_d| = d$, and if $y \in \mathbb{F}_q$ we have*

$$\sum_{\chi \in G_d} \chi(y) = \begin{cases} 0 & \text{if } x^d = y \text{ has no solution,} \\ 1 & \text{if } y = 0, \\ d & \text{if } y \neq 0 \text{ and } x^d = y \text{ has a solution.} \end{cases} .$$

In particular, the number of solutions in $\mathbb{F}_q$ of $x^d = y$ is equal to $\sum_{\chi \in G_d} \chi(y)$.

Proof. The group G_d is canonically isomorphic to the group of characters of the abelian group $\mathbb{F}_q^* / \mathbb{F}_q^{*d}$; hence the lemma is an immediate consequence of the orthogonality of characters (Proposition 2.1.20). $\square$

Note that the result for $y = 0$ comes from the convention $\varepsilon(0) = 1$, and it is the main reason for this choice.

Theorem 2.5.22. (1) *For any nontrivial additive character ψ we have*

$$N(q) = q^{k-1} + \left(1 - \frac{1}{q}\right) \sum_{\substack{\chi_i \in G_{d_i} \setminus \{\varepsilon\} \\ \chi_1 \cdots \chi_k = \varepsilon}} \tau(\chi_1, \psi^{a_1}) \cdots \tau(\chi_k, \psi^{a_k}) .$$

(2) *We have the inequality*

$$|N(q) - q^{k-1}| \leqslant (1 - 1/q) q^{k/2} \sum_{\substack{1 \leqslant y_i \leqslant d_i - 1 \\ \sum_{1 \leqslant i \leqslant k} ((q-1)/d_i) y_i \equiv 0 \pmod{q-1}}} 1 .$$

(3) *In particular, if all the d_i are equal to d then*

$$|N(q) - q^{k-1}| \leqslant \frac{(d-1)^k + (-1)^k(d-1)}{d}\left(1 - \frac{1}{q}\right)q^{k/2} .$$

(4) *If all the m_i are equal to m and $d = \gcd(m, q-1)$, the number $M(q)$ of projective solutions in $\mathbb{P}^{k-1}(\mathbb{F}_q)$ to the equation $a_1 x_1^m + \cdots + a_k x_k^m = 0$ satisfies*

$$|M(q) - |\mathbb{P}^{k-2}(\mathbb{F}_q)|| \leqslant \frac{(d-1)^k + (-1)^k(d-1)}{d}q^{(k-2)/2} .$$

Proof. (1). By the above remark we have

$$N(q) = \sum_{\substack{y_i \in \mathbb{F}_q \\ y_1 + \cdots + y_k = 0}} \prod_{1 \leqslant i \leqslant k} \sum_{\chi \in G_{d_i}} \chi(y_i/a_i)$$

$$= \sum_{\chi_i \in G_{d_i}} (\chi_1(a_1) \cdots \chi_k(a_k))^{-1} J_k(\chi_1, \ldots, \chi_k; 0) .$$

By Corollary 2.5.16 we know that $J_k(0) = 0$ if either $\chi_1 \cdots \chi_k \neq \varepsilon$ or if at least one (but not all) of the χ_i is equal to ε. In addition, $J_k(0) = q^{k-1}$ when all the χ_i are equal to ε. Thus again by the corollary we have

$$N(q) = q^{k-1} + \sum_{\substack{\chi_i \in G_{d_i} \setminus \{\varepsilon\} \\ \chi_1 \cdots \chi_k = \varepsilon}} (\chi_1(a_1) \cdots \chi_k(a_k))^{-1} J_k(\chi_1, \ldots, \chi_k; 0)$$

$$= q^{k-1} + \left(1 - \frac{1}{q}\right) \sum_{\substack{\chi_i \in G_{d_i} \setminus \{\varepsilon\} \\ \chi_1 \cdots \chi_k = \varepsilon}} (\chi_1(a_1) \cdots \chi_k(a_k))^{-1} \tau(\chi_1, \psi) \cdots \tau(\chi_k, \psi)$$

$$= q^{k-1} + \left(1 - \frac{1}{q}\right) \sum_{\substack{\chi_i \in G_{d_i} \setminus \{\varepsilon\} \\ \chi_1 \cdots \chi_k = \varepsilon}} \tau(\chi_1, \psi^{a_1}) \cdots \tau(\chi_k, \psi^{a_k}) ,$$

by Lemma 2.5.8, proving (1).

(2). By Proposition 2.5.9, since all the additive characters ψ^{a_i} are still nontrivial, we have $|\tau(\chi_i, \psi^{a_i})| = q^{1/2}$, hence

$$|N(q) - q^{k-1}| \leqslant (1 - 1/q)q^{k/2}S(d_1, \ldots, d_k) ,$$

where $S(d_1, \ldots, d_k)$ is the number of k-tuples of characters $\chi_i \in G_{d_i}$ different from ε whose product is equal to ε. By duality, this is equal to the number of k-tuples of nonzero elements x_i in $((q-1)/d_i)\mathbb{Z}/(q-1)\mathbb{Z}$ that sum to 0, so (2) follows by setting $x_i = ((q-1)/d_i)y_i$.

(3). The proof of (3) is of course purely combinatorial. We must compute the number $S(d, \ldots, d) = S(d)$ defined above, in other words the number of

k-tuples $(y_1, \ldots, y_k) \in \mathbb{Z}/d\mathbb{Z}$ such that $y_i \neq 0$ for all i while $\sum_{1 \leqslant i \leqslant k} y_i = 0$ in $\mathbb{Z}/d\mathbb{Z}$. We proceed by induction. For $0 \leqslant j \leqslant k$, define $f_k(j)$ to be the number of k-tuples summing to 0 in $\mathbb{Z}/d\mathbb{Z}$ such that $y_i \neq 0$ for $1 \leqslant i \leqslant j$. We want to compute $f_k(k)$. We start the induction by noting that evidently $f_k(0) = d^{k-1}$ for $k \geqslant 1$, since y_k is uniquely determined by y_i for $1 \leqslant i \leqslant k - 1$. Now for $0 \leqslant j \leqslant k - 1$ and $k \geqslant 2$ we have

$$f_k(j) = \sum_{\substack{\sum_i y_i = 0 \\ y_i \neq 0 \text{ for } i \leqslant j+1}} 1 + \sum_{\substack{\sum_i y_i = 0 \\ y_i \neq 0 \text{ for } i \leqslant j \\ y_{j+1} = 0}} 1 = f_k(j+1) + f_{k-1}(j) ,$$

while for $k = 1$ and $j = 0$ we evidently have $f_1(1) = 0$, leading to the convention that $f_0(0) = 1$ if we want the above recurrence to be valid.

Since $f_k(0) = d^{k-1}$ for $k \geqslant 1$, it is now clear by induction that *for $k \geqslant j + 1$ the solution to the above recurrence is $f_k(j) = d^{k-j-1}(d-1)^j$.* On the other hand, set $u_j = (-1)^j f_j(j)$, so that $u_1 = 0$. The above recurrence applied to $k = j + 1$ gives for $j \geqslant 2$, $f_{j+1}(j) = f_{j+1}(j+1) + f_j(j)$, in other words, since $f_{j+1}(j) = (d-1)^j$, $(-1)^{j+1} u_{j+1} = (d-1)^j + (-1)^{j+1} u_j$; hence $u_{j+1} = u_j - (1-d)^j$. Since $u_1 = 0$, this gives

$$u_j = -\sum_{n=1}^{j-1} (1-d)^n = \frac{(1-d)^j - (1-d)}{d} = \frac{(-1)^j (d-1)^j + d - 1}{d} ,$$

so that $f_j(j) = (-1)^j u_j = ((d-1)^j + (-1)^j (d-1))/d$, proving (3).

(4). This is immediate since by the definition of projective space we have $N(q) = 1 + (q-1)M(q)$ and in particular $q^{k-1} = 1 + (q-1)|\mathbb{P}^{k-2}(\mathbb{F}_q)|$. $\square$

Note that when the d_i are not all equal it is still possible to obtain an explicit formula for $S(d_1, \ldots, d_k)$. For instance, if $k = 3$ we have shown above that $S(d, d, d) = (d-1)(d-2) = d^2 - 3d + 2$, but the general formula is

$$S(d_1, d_2, d_3) = \gcd(d_1 d_2, d_1 d_3, d_2 d_3)$$
$$- (\gcd(d_1, d_2) + \gcd(d_1, d_3) + \gcd(d_2, d_3)) + 2 ;$$

see Exercise 42.

Corollary 2.5.23. *Let a, b, and c be nonzero elements of $\mathbb{F}_q$, let $m \geqslant \mathbb{Z}_{\geqslant 1}$, and set $d = \gcd(m, q-1)$. The number $M(q)$ of projective solutions in $\mathbb{P}^2(\mathbb{F}_q)$ to the equation $ax^m + by^m + cz^m = 0$ satisfies*

$$|M(q) - (q+1)| \leqslant (d-1)(d-2)q^{1/2} .$$

This is a stronger result than Proposition 2.5.11, and it can be shown that the bound is optimal.

Corollary 2.5.24. *Assume that $m_i = m$ for all i and that the a_i are all nonzero, and as usual set $d = \gcd(m, q - 1)$. The number $M(q)$ of projective solutions in $\mathbb{P}^{k-1}(\mathbb{F}_q)$ to the equation $a_1 x_1^m + \cdots + a_k x_k^m = 0$ is given by the exact formula*

$$M(q) = \frac{q^{k-1} - 1}{q - 1} + \sum_{\substack{\chi_1, \ldots, \chi_{k-1} \in G_d \setminus \{\varepsilon\} \\ \prod_{1 \leqslant i \leqslant k-1} \chi_i \neq \varepsilon}} \prod_{1 \leqslant i \leqslant k-1} \chi_i(-a_k/a_i) J_{k-1}(\chi_1, \ldots, \chi_{k-1}) \cdot$$

Proof. This immediately follows from Theorem 2.5.22 and Proposition 2.5.14, using as above the fact that $M(q) = (N(q) - 1)/(q - 1)$, and $\chi_k^{-1}(-a_k) = \prod_{1 \leqslant i \leqslant k-1} \chi_i(-a_k)$ when $\prod_{1 \leqslant i \leqslant k} \chi_i = \varepsilon$. □

Corollary 2.5.25. *Assume that $q \equiv 1 \pmod 6$, and let χ be a character of order 3 on $\mathbb{F}_q$. The number $M(q)$ of projective solutions in $\mathbb{P}^2(\mathbb{F}_q)$ to the equation $x^3 + y^3 + z^3 = 0$ is equal to $q + 1 + c$, where c is the unique integer such that $4q = c^2 + 27d^2$ with $c \equiv 1 \pmod 3$.*

Proof. This immediately follows from the above corollary and Proposition 2.5.20 and is left to the reader (Exercise 43 (1)). □

2.5.6 The Weil Bounds

We will place ourselves in the context of *plane projective geometry*. In this context, a (plane projective) curve is an equation $P(X, Y, Z) = 0$, where P is a homogeneous polynomial, and a *point* on the curve is any projective point $(x : y : z)$ such that $P(x, y, z) = 0$ (this makes sense since P is homogeneous). The *affine equation* corresponding to P is the equation $P(x, y, 1) = 0$.

A curve has a genus g, which is a nonnegative integer. Although not difficult to define, we simply give a few examples.

– If the total degree of P is equal to d, then $g \leqslant (d - 1)(d - 2)/2$. In fact, if the plane curve is *nonsingular*, that is if the partial derivatives with respect to x, y, and z never vanish simultaneously, then in fact we have $g = (d - 1)(d - 2)/2$.
– If $P(x, y, 1) = y^2 - f(x)$, where f has no multiple root and is of degree $d \geqslant 2$, then $g = \lfloor (d - 1)/2 \rfloor$.

The fundamental results of Weil, which we will not prove, are the following:

Theorem 2.5.26 (Weil). *Let C be a nonsingular absolutely irreducible projective curve defined on a finite field $\mathbb{F}_q$. Denote by $N_C(q^n)$ the number of projective points on C that are defined over $\mathbb{F}_{q^n}$, and set*

$$\zeta_C(T) = \exp\left(\sum_{n \geqslant 1} N_C(q^n) \frac{T^n}{n}\right)$$

(*this is called the Hasse–Weil zeta function of the curve C*). Then

(1) $\zeta_C(T)$ *is a rational function of* T *of the form*

$$\zeta_C(T) = \frac{P_{2g}(T)}{(1-T)(1-qT)} \, ,$$

where $P_{2g}(T)$ *is a polynomial of degree* $2g$ *with integral coefficients, constant term* 1, *leading term* $q^g T^{2g}$, *and such that* $P_{2g}(1/(qT)) = q^{-g}T^{-2g}P_{2g}(T)$.

(2) *The roots of* $P_{2g}(1/T)$ *all have modulus exactly equal to* $q^{1/2}$; *in other words,*

$$P_{2g}(T) = \prod_{1 \leqslant j \leqslant 2g} (1 - \alpha_j T) \quad \text{with} \quad |\alpha_j| = q^{1/2} \, .$$

Remarks. (1) Terminology: recall that a plane curve defined by some homogeneous equation $F = 0$ is nonsingular if the partial derivatives of F do not all vanish simultaneously on the curve. A similar definition applies for nonplane curves. A curve is *absolutely irreducible* if it is irreducible as a curve over the algebraic closure of the field of definition. For a plane curve, it means that the homogeneous polynomial F defining it is irreducible over the algebraic closure.

(2) Note that since the constant term of $P_{2g}(T)$ is known and since it satisfies a functional equation, the polynomial $P_{2g}(T)$ is entirely determined by the knowledge of $N_C(q^n)$ for $1 \leqslant n \leqslant g$. For instance, if $g = 1$, the number of points over $\mathbb{F}_q$ determines the number of points over $\mathbb{F}_{q^n}$ for all n (see Exercise 45).

(3) Statement (2) (known as the Riemann hypothesis for curves) is the most difficult, and also the most useful part of the theorem.

Example. If C is the projective line, then trivially $N_C(q^n)$ is the number of points of the projective line over $\mathbb{F}_{q^n}$, in other words $q^n + 1$. An immediate computation thus gives in this case

$$\zeta_C(T) = \frac{1}{(1-T)(1-qT)}$$

in accordance with Weil's theorem, since the projective line has genus 0.

Corollary 2.5.27. *With the above notation and assumptions, we have the formula*

$$N_C(q^n) = q^n + 1 - \sum_{1 \leqslant j \leqslant 2g} \alpha_j^n \, .$$

In particular, the number $N_C(q)$ *of projective points on a curve of genus* g *defined over* $\mathbb{F}_q$ *satisfies*

$$|N_C(q) - (q+1)| \leqslant 2g q^{1/2} \, .$$

Proof. Expanding into power series $\log(\zeta_C(T))$ and replacing by the formula given by Weil's theorem immediately gives the first result. The second follows from $|\alpha_j| = q^{1/2}$.   □

It is to be noted that although the above bound is not always optimal, it is usually very close to the truth. As mentioned above, the genus of the curve $ax^d + by^d + cz^d = 0$ is equal to $(d-1)(d-2)/2$, and this is still true over $\mathbb{F}_q$ if q is coprime to $abcd$, so the Weil bound gives $|N_C(q) - (q+1)| \leqslant (d-1)(d-2)q^{1/2}$, which is exactly the bound of Corollary 2.5.23. In fact, it was the study of this example that led Weil to the general case.

2.5.7 The Weil Conjectures (Deligne's Theorem)

The Weil bounds are limited to curves. It is natural to ask for the same kind of results for higher-dimensional algebraic varieties. These results were put forward as conjectures by Weil himself, together with a vast plan for solving them, but it took twenty years to achieve that goal. The Weil conjectures were stated around 1949, and the plan to solve them was put into extremely detailed form by A. Grothendieck and his school between 1956 and 1968. It culminated with Deligne's proof of the Weil conjectures in 1969, using all the tools developed for that goal (but of course his added genius was necessary!).

Although they have been Deligne's theorem for more than thirty years, they are still called the Weil conjectures. This is the case for other famous results in mathematics, in particular, in number theory Ramanujan's conjecture about bounds for the Ramanujan τ function (see Section 10.1.3), also proved by Deligne, and Mordell's conjecture on the finiteness of the number of rational points on a curve of genus $g \geqslant 2$, proved by G. Faltings in the 1980s.

Essentially, what the Weil conjectures assert is that the results of Theorem 2.5.26 and its corollary can be generalized to higher dimensions. It would carry us too far to state them (although it is not difficult). We will prove them in Section 3.7.3 in the special case of diagonal hypersurfaces. For now we simply note one of the corollaries, analogous to Corollary 2.5.27 above.

Theorem 2.5.28. *There exist constants $c(n, d, r)$ such that if V is an algebraic variety defined over $\mathbb{F}_q$ of dimension r and degree d in n-dimensional projective space, then the number N of points of V defined over $\mathbb{F}_q$ satisfies*

$$|N - q^r| \leqslant (d-1)(d-2)q^{r-1/2} + c(n, d, r)q^{r-1} .$$

Since we have seen that the genus of a general curve of degree d is bounded by $(d-1)(d-2)/2$, we see that the above theorem exactly generalizes (without specifying $c(n, d, r)$) Corollary 2.5.27.

2.6 Exercises for Chapter 2

1. Give a direct proof of Corollary 2.1.6 as follows. Let m be the minimum absolute value of the nonzero entries of V. Show first that by exchanging rows we may assume that this minimum is attained for the first entry. Then prove the result by induction on m, directly for $m = 1$, and using Euclidean division by the first entry for $m > 1$.

2. Fill in the details of the proof of Theorem 2.1.10.

3. Prove the existence and uniqueness of the HNF of an integral matrix of maximal rank (see [Coh0] Algorithm 2.4.4 for help).

4. Let G be a finite abelian group of cardinality m, and let K be a commutative field.

(a) Show that Proposition 2.1.20 (2) is still true if we assume only that the equation $x^m - 1 = 0$ has m distinct roots in K, and in particular if K is an algebraically closed field of characteristic 0 or not dividing m, where $\widehat{G}$ is to be interpreted as the group of morphisms from G to K^*.

(b) Give two counterexamples to the above result, the first with K not algebraically closed but of characteristic 0, the second with K algebraically closed with nonzero characteristic.

5. Let p be a prime number, $v \in \mathbb{Z}_{\geqslant 1}$, and $a \in \mathbb{Z}$ coprime to p.

(a) Assume first that $p \geqslant 3$, and let g be a primitive root modulo p^2, hence modulo p^v for all v by the proof of Proposition 2.1.24. Set $u = p/(g^{p-1} - 1) \bmod p$, $b = a^{-1} \bmod p^v$, and let x_1 be a discrete logarithm of a to base g modulo p, in other words such that $g^{x_1} \equiv a \pmod{p}$. Show that if we define x_k for $k \geqslant 2$ by the inductive formula

$$x_k = x_{k-1} - (p-1)u\frac{g^{x_{k-1}}b - 1}{p} \bmod p^{k-1}(p-1)$$

then x_v is a discrete logarithm of a to base g modulo p^v.

(b) Assume now that $p = 2$ and that $v \geqslant 3$ (otherwise the problem is trivial). By Proposition 2.1.24, there exist unique $\varepsilon = \pm 1$ and x defined modulo 2^{v-2} such that $a \equiv \varepsilon 5^x \pmod{2^v}$, and $\varepsilon = \left(\frac{-4}{a}\right) = (-1)^{(a-1)/2}$. Show that if we define $b = \varepsilon a^{-1} \bmod 2^v$, $x_3 = 0$ if $b \equiv 1 \pmod 8$, and $x_3 = 1$ if $b \equiv 5 \pmod 8$, and define x_k for $k \geqslant 4$ by the inductive formula

$$x_k = x_{k-1} - \frac{5^{x_{k-1}}b - 1}{4} \bmod 2^{k-2},$$

then $x_v = x$ is such that $a \equiv \varepsilon 5^x \pmod{2^v}$.

6. The aim of this exercise is to study *Carmichael numbers*. A Carmichael number is an integer $N \geqslant 2$ that is not a prime number and is such that $a^{N-1} \equiv 1 \pmod{N}$ for all a such that $\gcd(a, N) = 1$.

(a) Recall that the *exponent* of a group G is the least integer $e \geqslant 1$ such that $g^e = 1$ for all $g \in G$. Show that for any integer $N \geqslant 1$ (Carmichael number or not), the exponent of the group $(\mathbb{Z}/N\mathbb{Z})^*$ is equal to $\lambda(N)$, where

$$\lambda(N) = \mathrm{lcm}(2^{f(v_2(N))}, \mathrm{lcm}_{p_i \mid N, \ p_i \geqslant 3}(p_i^{v_i - 1}(p_i - 1))),$$

with $f(v) = v - 2$ if $v \geqslant 3$, $f(v) = v - 1$ if $1 \leqslant v \leqslant 2$, and $f(0) = 0$. The function $\lambda(N)$ is called *Carmichael's function*.

(b) Show that $N \geqslant 2$ is a Carmichael number if and only if N is not prime and $\lambda(N) \mid N - 1$.

(c) Show that a Carmichael number N is odd, and then that it is squarefree, i.e., that $v_i = 1$ for all i.

(d) Prove that a Carmichael number has at least three distinct prime factors.

(e) Prove that if N is a Carmichael number, then for all a (not necessarily coprime with N) we have $a^N \equiv a \pmod{N}$.

Note: The Carmichael numbers under 10000 are 561, 1105, 1729, 2465, 2821, 6601, and 8911. In 1994 R. Alford, A. Granville, and C. Pomerance [AGP] succeeded in proving that there are an infinite number of them, in fact asymptotically more than $c \cdot X^{2/7}$ up to X for some strictly positive constant c.

7. Let $p \geqslant 5$ be a prime number.

(a) Prove that when $p \equiv 1 \pmod 6$ then $7^p - 6^p - 1$ is divisible by $77658\,p$, while when $p \equiv 5 \pmod 6$ then $7^p - 6^p - 1 - 1806p$ is divisible by $77658\,p$.

(b) Prove further that when $p \equiv 1 \pmod 6$ then $7^p - 6^p - 1 - 12943p(p-1)$ is divisible by $3339294\,p$, and find the corresponding property for $p \equiv 5 \pmod 6$.

8. (Apostol–Saias.) Let f be a nonzero arithmetic function satisfying $f(mn) = f(m)f(n)$ for all m, n and such that there exists $m \in \mathbb{Z}_{\geqslant 1}$ such that $f(x+m) = f(x)$. Prove that f is a Dirichlet character modulo some divisor d of m.

9. For $r \in \mathbb{Z}$ and $m \geqslant 1$ let $S_r(m) = \sum_{\chi \bmod m,\ \chi \text{ primitive}} \chi(r)$, so that for instance $S_1(m)$ is the number of primitive characters modulo m.

(a) It is clear that $S_r(m) = 0$ when r is not coprime to m. Generalizing Proposition 2.1.29 show that when r is coprime to m we have

$$S_r(m) = \mu\left(\frac{m}{\gcd(m,(r-1)^\infty)}\right) q\left(\gcd(m,(r-1)^\infty)\right),$$

where $(a, b^\infty) = \prod_{p \mid b} p^{v_p(a)}$ (see the notation introduced in the preface).

(b) Let $D(m)$ be the difference between the number of even and odd primitive characters modulo m. Deduce that $D(m) = \mu(m)$ if m is odd, and otherwise $D(m) = 2^{v-2}\mu(m/2^v)$, with $v = v_2(m) \geqslant 2$.

10. Prove Proposition 2.1.34 (3).

11. Recall that for any commutative ring R, $\mathrm{SL}_2(R)$ denotes the group of 2×2 matrices with determinant 1 and coefficients in R. Using Lemma 2.1.31, show that for any $N \geqslant 2$, the natural projection map from $\mathrm{SL}_2(\mathbb{Z})$ to $\mathrm{SL}_2(\mathbb{Z}/N\mathbb{Z})$ is surjective. (Hint: first show that if $\gcd(a, b, c, d) = 1$, there exist u and v such that $\gcd(au+bv, cu+dv) = 1$.) See Lemma 6.3.10 for the proof of the analogous statement for $\mathrm{SL}_n(\mathbb{Z}/N\mathbb{Z})$.

12. (Study the Möbius inversion formula (Proposition 10.1.5) and its uses before solving this exercise.) Let χ be a Dirichlet character modulo m, let f be its conductor, and let χ_f be the character modulo f equal to χ on numbers prime to m. Prove the following general formula:

$$\tau(\chi, a) = \tau(\chi_f) \sum_{d \mid \gcd\left(\frac{m}{f}, a\right)} d\mu\left(\frac{m}{fd}\right) \chi_f\left(\frac{m}{fd}\right) \overline{\chi_f}\left(\frac{a}{d}\right).$$

For this, note that

$$\sum_{x \bmod m} \chi(x) e^{2i\pi ax/m} = \sum_{\substack{\gcd(x,m/f)=1 \\ x \bmod m}} \chi_f(x) e^{2i\pi ax/m} \;,$$

replace the condition $\gcd(x, m/f) = 1$ by a sum of Möbius functions (see Proposition 10.1.5), and finally write $x/d = fx_1 + x_2$. In particular, we have $\tau(\chi) = \mu(m/f)\chi_f(m/f)\tau(\chi_f)$.

13. Prove Proposition 2.1.44.

14. Let p be an odd prime, and set $S = (1+i)^p$. Noting that $S \equiv 1 + i^p \pmod{p\mathbb{Z}[i]}$ and $S \equiv (1+i)i^{(p-1)/2}\left(\frac{2}{p}\right) \pmod{p\mathbb{Z}[i]}$, give another proof of the formula for $\left(\frac{2}{p}\right)$.

15. Generalizing the computation made in Theorem 2.2.19, show that if $m \geqslant 1$ and $a \neq 0$, then

$$\sum_{x \bmod m} e^{2i\pi(ax^2+bx)/m} = \frac{1+i}{2}\sqrt{\frac{m}{a}} e^{-2i\pi b^2/(4am)} \sum_{x \bmod 2a} e^{-2i\pi(mx^2+2bx)/(4a)} \;.$$

16. (H. Stark.) Let χ be a real primitive character modulo $m > 0$, and let Q be a positive definite quadratic form of discriminant $d < 0$. Assuming that $\gcd(m, d) = 1$, prove the following reciprocity formula: for all y and z in $\mathbb{Z}$ we have

$$\sum_{x=1}^{m} \chi(Q(x,y)) e^{2i\pi xz/m} = \sum_{x=1}^{m} \chi(Q(x,z)) e^{2i\pi xy/m} \;.$$

(Hint: use Corollary 2.1.42.)

17. Let χ be a primitive character modulo m, and let $W(\chi)$ be as in Definition 2.2.25, so that $|W(\chi)| = 1$ by Proposition 2.1.45. If χ is real, we know by Proposition 2.2.24 that $W(\chi) = 1$. Prove that if $W(\chi)$ is a root of unity then χ is real, hence $W(\chi) = 1$.

18. Show that there exists a 2-dimensional lattice L in $\mathbb{R}^2$ which reaches the upper bound for $\min(L)/\det(L)^{1/2}$ given by Theorem 2.3.12.

19.

(a) Give the detailed formulas for updating the Gram–Schmidt basis in Step 3 of the LLL algorithm.

(b) Show that after this update the LLL condition is satisfied for the new vectors $\mathbf{b}_{k-1}^*$ and $\mathbf{b}_k^*$.

20. With the notation of the LLL algorithm, let d_k be the determinant of the Gram matrix of the $\mathbf{b}_j$ for $j \leqslant k$ and $D = \prod_{1 \leqslant k \leqslant n-1} d_k$.

(a) Show that $d_k = \prod_{1 \leqslant j \leqslant k} \|\mathbf{b}_j^*\|^2$ and that D stays fixed in the algorithm except in Step 3, where it is multiplied by a factor at most equal to $1/\gamma + 1/4$.

(b) Using Hermite's Theorem 2.3.12, show that d_k is bounded from below by a strictly positive constant depending only on L.

(c) Deduce from this that Step 3 will be executed only a finite number of times, hence that the LLL algorithm will terminate after a number of steps polynomial in n and $\max(\|\mathbf{b}_j\|^2)$. How does this number of steps vary in terms of the constant $\gamma > 4/3$?

Note that it is unknown whether the LLL algorithm is still polynomial time for $\gamma = 4/3$.

21. Assume that the Gram matrix of the $\mathbf{b}_j$ is integral, and as in the preceding exercise let d_k be the determinant of the Gram matrix of the $\mathbf{b}_j$ for $j \leqslant k$.

(a) Prove that for all k we have $d_{k-1}\|\mathbf{b}_k^*\|^2 \in \mathbb{Z}$, for all $j < i$ we have $d_j \mu_{i,j} \in \mathbb{Z}$, and for all m such that $j < m \leqslant i$ we have $d_j \sum_{1 \leqslant k \leqslant j} \mu_{i,k} \mu_{m,k} \|\mathbf{b}_k^*\|^2 \in \mathbb{Z}$.

(b) Deduce from this a modification of the LLL algorithm that works entirely with integers.

(c) Show that these integers will be bounded by a polynomial in 2^n and $\max(\|\mathbf{b}_j\|^2)$.

22. Let $\alpha_0, \ldots, \alpha_n$ be complex numbers, and assume that there exists $z \in \mathbb{C}^*$ such that $\alpha_i/z \in \mathbb{R}$ for $1 \leqslant i \leqslant n$. Using Proposition 2.3.20 give a lower bound for $|\alpha_0 + \sum_{1 \leqslant i \leqslant n} x_i \alpha_i|$ in terms of the same quantities occurring in the proposition together with $\Im(\alpha_0/z)$.

23. Prove the generalization to complex numbers of Proposition 2.3.20 given in the text.

24.

(a) In Proposition 2.3.20 show that one can replace the value of T by $T = \langle C\alpha_0 \rangle + \sum X_i \langle C\alpha_i \rangle$ (where we recall that $\langle u \rangle$ is the distance of u to the nearest integer), which may be considerably smaller especially if we make a little search on C.

(b) Using this improved value and making C vary in a reasonable range, compute a lower bound for $|x_1 \log(2) + x_2 \pi + x_3 \gamma|$, where the x_i are integers not all equal to 0 such that $|x_i| \leqslant 100$.

(c) Make a systematic search for $0 \leqslant x_1 \leqslant 100$ and $-100 \leqslant x_2, x_3 \leqslant 100$ to find the exact minimum of the above expression, and compare with the bound that you have found.

25. As in Proposition 3.5.2, denote by U_n (respectively U_n') the set of nth roots of unity (respectively primitive nth roots of unity) in $\mathbb{C}$. Using cyclotomic polynomials, compute $\sum_{\zeta \in U_n} 1/(x - \zeta)$, $\sum_{\zeta \in U_n'} 1/(x - \zeta)$, $\sum_{\zeta \in U_n} \zeta/(1 - \zeta)^2$, and $\sum_{\zeta \in U_n'} \zeta/(1 - \zeta)^2$.

26. Let $p \geqslant 3$ be a prime number. Prove that there exists a unique value of k modulo $2p$ such that $\sum_{\zeta \in U_p'} (-\zeta)^k/(1 - \zeta) = 0$, and compute k explicitly.

27. Let $\mathbb{E}/\mathbb{F}$ be an extension of finite fields, and set $q = |\mathbb{F}|$ and $s = [\mathbb{E} : \mathbb{F}]$.

(a) Generalizing Proposition 2.4.11 (2), show that if $k \in \mathbb{Z}_{\geqslant 1}$ with $\gcd(k, s) = 1$, the kernel of $\mathrm{Tr}_{\mathbb{E}/\mathbb{F}}$ is equal to the set of $a \in \mathbb{E}$ of the form $a = x^{q^k} - x$ for some $x \in \mathbb{E}$.

(b) Assume that $\gcd(qk, s) = 1$. Show that for any $a \in \mathbb{E}$ there exist $u \in \mathbb{F}$ and $x \in \mathbb{E}$ such that $x^{q^k} - x = a + u$. For instance, if $a \in \mathbb{E} = \mathbb{F}_{2^s}$ with s odd, one of the two equations $x^2 - x = a$ or $x^2 - x = a + 1$ has a solution $x \in \mathbb{E}$.

28. Prove that the polynomial $X^n + X^{n-1} + \cdots + X + 1$ is irreducible in $\mathbb{F}_p[X]$ if and only if $n + 1$ is a prime different from p and the order of p modulo $n + 1$ is equal to n.

29. Give another proof of Lemma 2.5.1 using the identity

$$X^q - X = \prod_{a \in \mathbb{F}_q} (X - a)$$

and the Newton relations between elementary symmetric functions and power sums.

30. Let p be a fixed prime number and define integers a_k by the polynomial identity

$$\prod_{1 \leqslant j \leqslant p-1} (X - j) = \sum_{0 \leqslant k \leqslant p-1} a_k X^k .$$

(a) Using the field $\mathbb{F}_p$ show that $p \mid a_k$ for $1 \leqslant k \leqslant p-2$.

(b) Compute explicitly a_0 and a_1.

(c) Assume that $p \geqslant 5$. Setting $X = p$ in the polynomial identity defining the a_k, prove that $p^2 \mid a_1$.

(d) Deduce Wolstenholme's congruence: if $p \geqslant 5$ is prime then

$$\sum_{1 \leqslant j \leqslant p-1} \frac{1}{j} \equiv 0 \pmod{p^2} \; ;$$

in other words, the numerator of the left-hand side is divisible by p^2.

(e) Prove that if $p \geqslant 5$ is prime then

$$\sum_{1 \leqslant j \leqslant p-1} \frac{1}{j^2} \equiv 0 \pmod{p} \; .$$

(f) More generally, prove that $p^2 \mid a_k$ for every odd k such that $1 \leqslant k \leqslant p-3$, and show that if $p \geqslant k+3$ then

$$\sum_{1 \leqslant j \leqslant p-1} \frac{1}{j^k} \equiv 0 \pmod{p^v} \; ,$$

where $v = 2$ if k is odd and $v = 1$ if k is even.

(g) Deduce from (d) and (e) that if $p \geqslant 5$ and $n \geqslant 1$ we have

$$\binom{(n+1)p-1}{np} = \frac{1}{n+1}\binom{(n+1)p}{np} \equiv 1 \pmod{n(n+1)p^3} \; .$$

Note that this is a special case of a general congruence that we will prove in Chapter 11 (Theorem 11.6.22).

See Exercise 50 of Chapter 11 for a sequel to this exercise giving stronger congruences.

31. Let χ be a multiplicative character of order n on $\mathbb{F}_q$. Prove that $\chi(-1) = 1$ if n is odd and $\chi(-1) = (-1)^{(q-1)/n}$ if n is even.

32. This exercise follows closely the exposition of Section 2.1.6 of [Coh1]. Let $\mathbb{F}_q$ be a finite field of *odd* order, let $P \in \mathbb{F}_q[X]$ be a *monic* irreducible polynomial of degree n, let α be a root of P in some algebraic closure of $\mathbb{F}_q$, let σ denote the Frobenius automorphism $x \mapsto x^q$ of $L = \mathbb{F}_q(\alpha)$, and finally for $k \in \mathbb{Z}$ set

$$D_k = (-1)^k \prod_{0 \leqslant j < n} (\sigma^{j+k}(\alpha) - \sigma^j(\alpha)) \; .$$

(a) Show that $\mathrm{disc}(P) = \prod_{1 \leqslant k \leqslant n-1} D_k$.

(b) Show that D_k is invariant by $\mathrm{Gal}(L/\mathbb{F}_q)$, hence that $D_k \in \mathbb{F}_q$.

(c) Show that $D_{n-k} = D_k$.

(d) Assume that n is even and set

$$E = \prod_{0 \leqslant j < n/2} (\sigma^{j+n/2}(\alpha) - \sigma^j(\alpha)) \; .$$

Prove that $D_{n/2} = E^2$.

(e) Let a be a nonsquare in $\mathbb{F}_q$. Show that $E/\sqrt{a} \in \mathbb{F}_q$.

(f) By abuse of notation write $\left(\frac{x}{q}\right)$ for the value at x of the unique nontrivial character of order 2 on $\mathbb{F}_q$. Conclude from the above

$$\left(\frac{\operatorname{disc}(P)}{q}\right) = (-1)^{n-1} .$$

See Exercise 2 of Chapter 10 for a sequel to this exercise.

33. Let $P(x)$ be an integer-valued polynomial of degree m whose values are not always divisible by a prime p. Generalizing Corollary 2.5.10, it seems that experimentally we always have $|\sum_{x \in \mathbb{F}_p} \psi(P(x))| \leqslant (m-1)p^{1/2}$. Prove or disprove, and if true, generalize to all finite fields (the Weil bounds give $Kp^{1/2}$ with a constant K larger than $m-1$ in general).

34. Using the method of finite Fourier series, show the following.

(a) The number M of projective solutions x, y, z in $\mathbb{F}_q$ of the equation $y^2 z^2 = ax^4 + bz^4$ with a, b in $\mathbb{F}_q^*$ satisfies $|M - (q+1)| \leqslant Cq^{1/2}$, for an absolute constant C.

(b) The number M of projective solutions x, y, z in $\mathbb{F}_q$ of the equation $y^2 z = x^3 + b^2 z^3$ with b in $\mathbb{F}_q^*$ satisfies $|M - (q+1)| \leqslant Cq^{1/2}$, for an absolute constant C.

(c) Using Jacobi sums instead of finite Fourier series, show that in both cases one can choose $C = 2$ (this is a special case of a theorem of Hasse that we will prove in Chapter 7).

35. In the text we have defined and studied Gauss sums attached both to Dirichlet characters and to characters on finite fields, and we have only studied Jacobi sums attached to finite fields. It is of course immediate to define Jacobi sums attached to Dirichlet characters, for instance in the case of two characters modulo m by $J(\chi_1, \chi_2) = \sum_{x \bmod m} \chi_1(x)\chi_2(1-x)$. Prove that if $\chi_1\chi_2$ is a *primitive* character modulo m then, as in the finite field case, we have $J(\chi_1, \chi_2) = \tau(\chi_1)\tau(\chi_2)/\tau(\chi_1\chi_2)$.

36. Let p be a prime such that $p \equiv 1 \pmod 4$, let g be a primitive root modulo p, let χ be a character of order 4 on $\mathbb{F}_p^*$, and write $J(\chi, \chi) = a + ib$. By Proposition 2.5.20 we know that $a^2 + b^2 = p$, $2 \mid b$, and $a \equiv -1 \pmod 4$, so that a and b are uniquely determined apart from a possible sign change of b, which is natural since there are two possible choices for χ. If g is a primitive root modulo p such that $\chi(g) = i$, show that $a + bg^{(p-1)/4} \equiv 0 \pmod p$, which thus determines b uniquely. Similarly, show that if $p \equiv 1 \pmod 3$, χ is a character of order 3 such that $\chi(g) = \rho$, and $J(\chi, \chi) = a + b\rho$, then $a + bg^{(p-1)/3} \equiv 0 \pmod p$.

37. (Jacobstahl.) Let p be a prime such that $p \equiv 1 \pmod 4$, and let r be a quadratic residue and s be a quadratic nonresidue modulo p. Prove that $p = a^2 + b^2$ with

$$a = \frac{1}{2}\sum_{i=1}^{p-1}\left(\frac{i(i^2 - r)}{p}\right) \quad \text{and} \quad b = \frac{1}{2}\sum_{i=1}^{p-1}\left(\frac{i(i^2 - s)}{p}\right) .$$

38. Let q be such that $q \equiv 1 \pmod 6$ and let χ be a character of order exactly equal to 6 on $\mathbb{F}_q$. By Proposition 2.5.20 we can write $J(\chi^2, \chi^2) = a + b\rho$ with $a^2 - ab + b^2 = q$, $3 \mid b$, and $a \equiv -1 \pmod 3$.

(a) Separating cases, compute explicitly $J(\chi^n, \chi^m)$ for $0 \leqslant n, m \leqslant 5$ both even.

(b) Generalizing the results of Section 2.5.4, compute these sums for all n, m.

39. Let p be a prime such that $p \equiv 1 \pmod 3$. Proposition 2.5.20 tells us among other things that if we write $4p = c^2 + 27d^2$ then 2 is a cubic residue modulo p if and only if $2 \mid d$. Show similarly that 3 is a cubic residue modulo p if and only if $3 \mid d$.

40. Let χ be a nontrivial character on $\mathbb{F}_q$ and let ρ be a character of order exactly equal to 2 (so that q is odd).

(a) Prove that the number of solutions of $x^2 = a$ in $\mathbb{F}_q$ is equal to $1 + \rho(a)$, and deduce that $\sum_{x \in \mathbb{F}_q} \chi(q - x^2) = J(\chi, \rho)$.

(b) Prove that for any $m \in \mathbb{F}_q^*$ we have $\sum_{x \in \mathbb{F}_q} \chi(x(m - x)) = \chi(m^2/4) J(\chi, \rho)$.

(c) If $\chi^2 \neq \varepsilon$, deduce that for any nontrivial additive character ψ we have the following formulas:

$$\tau(\chi, \psi)^2 = \chi^{-2}(2) J(\chi, \rho) \tau(\chi^2, \psi) \,,$$
$$J(\chi, \chi) = \chi^{-2}(2) J(\chi, \rho) \,,$$
$$\tau(\chi, \psi) \tau(\chi\rho, \psi) = \chi^{-2}(2) \tau(\chi^2, \psi) \tau(\rho, \psi) \,.$$

Note that this last formula is the special case $m = 2$ of the Hasse–Davenport product relation, which we will prove in Section 3.7.

41. Let p be a prime such that $p \equiv 1 \pmod 4$.

(a) Prove that there exists an integer i such that $i^2 \equiv -1 \pmod p$, and explain how to find one efficiently using a probabilistic algorithm.

(b) Using Hermite's inequality on the lattice Λ in $\mathbb{R}^2$ generated over $\mathbb{Z}$ by the columns of the matrix $\left(\begin{smallmatrix} p & i \\ 0 & 1 \end{smallmatrix} \right)$, prove that p is a sum of two squares of integers.

(c) Using the LLL algorithm, give a fast algorithm for finding these two squares.

(d) Show that in this special case the LLL algorithm reduces to a partial Euclidean algorithm, and write this algorithm explicitly (the resulting algorithm is called Cornacchia's algorithm; see Section 1.5.2 of [Coh0]).

42. Prove the formula for $S(d_1, d_2, d_3)$ given in the text after the proof of Theorem 2.5.22. More generally, find an expression for $S(d_1, \ldots, d_k)$.

43. Let q be such that $q \equiv 1 \pmod 6$ and let χ be a character of order 3 on $\mathbb{F}_q$.

(a) Using Corollary 2.5.24 and Proposition 2.5.20, prove Corollary 2.5.25, in other words that the number of projective solutions to $x^3 + y^3 + z^3 = 0$ is equal to $q + 1 + c$, where $4q = c^2 + 27d^2$ and $c \equiv 1 \pmod 3$.

(b) Prove that $J(\chi, \chi, \chi^2) = q$.

(c) Using once again Corollary 2.5.24, prove that the number of (affine) solutions to the equation $x^3 + y^3 + z^3 = 1$ in $\mathbb{F}_q$ is equal to $q^2 + 6q - c$, where c is as above.

44. Let $K = \mathbb{F}_{2^k}$ be a finite field of characteristic 2.

(a) Show that when k is odd the equation $x^2 + x + 1 = 0$ has no solutions in K.

(b) Using the fact that, when k is even K^* contains a subgroup of order 3, count the number $N(2^k)$ of projective solutions of $x^3 + y^3 + z^3 = 0$ in $\mathbb{P}_2(K)$.

(c) Compute the Hasse–Weil zeta function of the projective curve $x^3 + y^3 + z^3 = 0$ over $\mathbb{F}_2$.

45. Let E be a curve of genus 1 defined over a field $\mathbb{F}_q$, and let $N = N(q)$ be the number of points of E defined over $\mathbb{F}_q$. Give an explicit formula for the number $N(q^n)$ of points of E defined over $\mathbb{F}_{q^n}$ (see Theorem 2.5.26 and remarks following).

46. The goal of this long exercise is to study some elementary properties of *Kloost-erman sums*. Let p be an odd prime, a and b in $\mathbb{Z}$, and set $S(a,b;p) = \sum_{x \bmod p}^{*} \exp(2i\pi(ax + bx^{-1})/p)$, where from now on $\sum^{*}$ means that the sum is restricted to elements coprime to p.

(a) Show that $S(a,b;p) = S(b,a;p)$ and that if $p \nmid n$ we have $S(an,b;p) = S(a,bn;p)$.

(b) For $k \geqslant 1$, set $V_k(p) = \sum_{a \bmod p}^{*} S(a,1;p)^k$. Show that if $p \nmid b$ we have $V_k(p) = \sum_{a \bmod p}^{*} S(a,b;p)^k$, and deduce that

$$(p-1)V_k(p) = \sum_{a,b \bmod p} S(a,b;p)^k - (p-1)^k - 2(-1)^k(p-1).$$

(c) Compute $(p-1)V_k(p)$ in terms of the number $N_k(p)$ of solutions $(x_1, \ldots, x_k) \in ((\mathbb{Z}/p\mathbb{Z})^*)^k$ of the system $\sum_{1 \leqslant i \leqslant k} x_i = \sum_{1 \leqslant i \leqslant k} x_i^{-1} = 0$.

(d) Deduce from this the values of $V_1(p)$ and $V_2(p)$.

(e) Let $M(u,v)$ be the number of solutions $(x,y) \in ((\mathbb{Z}/p\mathbb{Z})^*)^2$ of the system $x + y = u$, $x^{-1} + y^{-1} = v$. Compute $M(0,0)$, $M(0,v)$ for $p \nmid v$, and show that for $p \nmid uv$ we have $M(u,v) = 1 + \left(\frac{1-4(uv)^{-1}}{p}\right)$.

(f) Deduce from this and the nontriviality of the Legendre symbol the value of $N_4(p)$, hence of $V_4(p)$.

(g) Deduce from the values obtained that if $p \nmid ab$ we have $|S(a,b;p)| < 2p^{3/4}$, and that there exists $a \not\equiv 0 \pmod p$ such that $|S(a,1;p)| > \sqrt{2p - 2}$.

Note that it has been shown by Weil that if $p \nmid ab$ we have in fact $S(a,b;p) \leqslant 2p^{1/2}$.

3. Basic Algebraic Number Theory

In this chapter, we recall (sometimes without proof) the main definitions and results that we need from basic algebraic number theory. These can be found in many books, for example [Sam], [Bor-Sha], [Coh0], [Marc], [Frö-Tay], [Ire-Ros].

3.1 Field-Theoretic Algebraic Number Theory

3.1.1 Galois Theory

We begin by recalling (with proof) the main results concerning Galois theory of finite extensions of perfect fields.

Let K be a commutative field. We fix an algebraic closure $\overline{K}$ of K and we assume implicitly that all algebraic extensions and all elements are chosen in this algebraic closure. For simplicity, we will assume that our base field K is *perfect*: this means that either K has characteristic 0, or that K has characteristic $p > 0$ and the map $x \mapsto x^p$ from K to K is surjective. The reason for this hypothesis is the following proposition.

Proposition 3.1.1. *Let K be a perfect field and α an element that is algebraic over K. Then the minimal polynomial of α in $K[X]$ is separable; in other words, it is coprime to its derivative, or equivalently, it has no multiple roots in $\overline{K}$.*

Proof. First, it is easy to check and left to the reader that over any field K a polynomial is coprime to its derivative if and only if it has no multiple roots in $\overline{K}$. Now assume that K is perfect, and let $A \in K[X]$ be the minimal polynomial of α. Since K is a field, it is clear that A is irreducible in $K[X]$ (otherwise one of the factors would have α as a root, contradicting the minimality of A). Since the GCD of A and A' is in particular a divisor of A, it must thus be equal to 1 or A. Assume that it is equal to A. Then A divides A', and since A' has degree strictly less than that of A this means that $A' = 0$. If the characteristic of K is 0, this means that A is constant, which is impossible. If the characteristic of K is equal to $p > 0$, looking at the coefficients, we see that this means that $A(X) = \sum_{0 \leqslant k \leqslant n} a_k X^{pk}$ for some

$a_k \in K$. But since K is perfect, there exist b_k with $a_k = b_k^p$, and since K has characteristic p we have $A(X) = B(X)^p$ with $B(X) = \sum_{0 \leqslant k \leqslant n} b_k X^k$, contradicting the irreducibility of A. We thus have a contradiction, showing that $\gcd(A, A') = 1$, hence the proposition.   □

Fields of characteristic 0 are by definition perfect, as are all finite fields (Exercise 1). An important example of a field that is *not* perfect is $K = \mathbb{F}_p(T)$: it has characteristic p, but $T \in K$ is not a pth power. In fact, if $\alpha = T^{1/p}$, the minimal polynomial of α is $A(X) = X^p - T$, and $A'(X) = 0$, so A is not separable.

Definition 3.1.2. *Let K be a perfect field and let L/K be a finite field extension (in other words, L is a field containing K as a subfield, and L is finite-dimensional as a K-vector space). A map σ from L to L is called a K-automorphism of L if σ is a field isomorphism that leaves K pointwise fixed. A map σ from L to $\overline{K}$ is called a K-embedding of L into $\overline{K}$ if σ is a field homomorphism (necessarily injective) that leaves K pointwise fixed.*

In other words, a K-automorphism σ must be a bijection that preserves the field structure and such that $\sigma(a) = a$ for all $a \in K$. In particular, σ is a K-endomorphism of L. Since L is finite-dimensional over K it follows that the bijectivity of σ is equivalent to its injectivity or its surjectivity.

Proposition 3.1.3. *Let K be a perfect field, and let L/K be an extension of degree n.*

(1) *Any embedding of K into $\overline{K}$ extends to exactly n K-embeddings of L into $\overline{K}$.*

(2) *There exist at most n K-automorphisms of L, which are the K-embeddings σ of L into $\overline{K}$ such that $\sigma(L) \subset L$.*

Proof. By the primitive element theorem (which is true because we have assumed K to be perfect) we can write $L = K(\alpha)$ for some $\alpha \in L$. Let $A(X) \in K[X]$ be the minimal polynomial of α, which is therefore of degree n (a K-basis of L is given by $1, \alpha, \alpha^2, \ldots, \alpha^{n-1}$). Any element of L is therefore of the form $U(\alpha)$ with $U(X) \in K[X]$, and $U(X)$ is unique modulo $A(X)$.

For (1), let τ be an embedding of K into $\overline{K}$. To extend τ to an embedding of L, for any $U \in K[X]$ we must define $\tau(U(\alpha)) = U^\tau(\tau(\alpha))$, where U^τ is the polynomial obtained from U by applying τ to all the coefficients. For this to make sense we must have $0 = \tau(A(\alpha)) = A^\tau(\tau(\alpha))$; hence $\tau(\alpha)$ must be one of the roots of the polynomial A^τ, which has degree n, and since $\overline{K}$ is algebraically closed, A^τ has exactly n roots in $\overline{K}$; this proves (1).

Furthermore, it is clear that if σ is a K-automorphism of L then σ is in particular a K-embedding of L into $\overline{K}$, and conversely, such an embedding is an automorphism if and only if $\sigma(L) = L$ if and only if $\sigma(L) \subset L$ (since $[L : K] < \infty$), proving (2).   □

Definition 3.1.4. *Let $\alpha \in \overline{K}$ and let $A(X) \in K[X]$ be its minimal monic polynomial over K. The roots of $A(X) = 0$ in $\overline{K}$ are called the* conjugates *of α in $\overline{K}$. In other words, two elements α and β of $\overline{K}$ are conjugate if they have the same minimal monic polynomial over K.*

Thus, if $\deg(A) = d$, α has exactly d conjugates. The elements $\sigma(\alpha)$ in the above proof are the conjugates of α.

Proposition 3.1.5. *Let K be a perfect field and L/K a finite extension. The following three properties are equivalent:*

(1) *L is closed under conjugation over K (i.e., if $\alpha \in L$, every conjugate of α also belongs to L).*
(2) *There exist exactly $n = [L : K]$ K-automorphisms of L.*
(3) *If σ is a K-embedding of L into $\overline{K}$, then $\sigma(L) \subset L$.*

Proof. By the proof of the above proposition, if L is closed under conjugation then all the roots of the minimal polynomial $A(X)$ of α belong to L; hence there are indeed n K-automorphisms, so (1) implies (2). The equivalence of (2) and (3) is clear from Proposition 3.1.3. Assume (3), let $\alpha \in L$, let β be a conjugate of α over K, and denote by $A(X) \in K[X]$ their common minimal monic polynomial. As above, we can define a field isomorphism σ from $K(\alpha)$ to $K(\beta)$ by the formula $\sigma(U(\alpha)) = U(\beta)$, and this makes sense only because β is a conjugate of α. In particular, σ is an embedding of $K(\alpha)$ into $\overline{K(\alpha)} = \overline{K}$. By Proposition 3.1.3 (1), σ can be extended to an embedding of L into $\overline{K}$ (in fact, to $[L : K(\alpha)]$ such embeddings, but we do not need more than one). By (3) it follows that $\sigma(\alpha) = \beta \in L$; hence (3) implies (1). $\square$

Definition 3.1.6. *Let K be a perfect field. We say that an extension L/K is* normal *or* Galois *if one of the three equivalent conditions of the proposition is satisfied. The set of K-automorphisms of L forms a group under composition, called the* Galois group *of L/K and denoted by $\mathrm{Gal}(L/K)$.*

Note that this is the definition of a normal extension. A Galois extension is one that is normal and separable. Since we have assumed that K is perfect, this last condition is unnecessary, so the two notions coincide.

We will say that an extension is Abelian (respectively cyclic) if it is Galois and its Galois group is abelian (respectively cyclic). Since the simplest finite groups are the groups $\mathbb{Z}/\ell\mathbb{Z}$ with ℓ prime, clearly the simplest Galois extensions are the cyclic extensions of prime degree. In that case we use the letter ℓ for the cardinality of the Galois group so that the letter p (and $\mathfrak{p}$, etc.) is still available for prime numbers or places.

Proposition 3.1.7. *If $L = K(\alpha_1, \ldots, \alpha_k)$ and L contains the conjugates of all the α_i, then L/K is Galois.*

Proof. Any element $x \in L$ has the form $x = U(\alpha_1, \ldots, \alpha_k)$, where U has coefficients in K. If σ is a K-embedding of L into $\overline{K}$, then the coefficients of U are fixed by σ; hence

$$\sigma(x) = U(\sigma(\alpha_1), \ldots, \sigma(\alpha_k)),$$

and since the $\sigma(\alpha_i)$ are conjugates of α_i, by assumption they belong to L, and hence $\sigma(x) \in L$. It follows from Proposition 3.1.5 that L is Galois over K. □

Corollary 3.1.8. *If L/K is a finite extension, there exists a finite extension N of L that is Galois over K, and any such N will also be Galois over L.*

Proof. Write $L = K(\alpha)$, and let $\alpha_1, \ldots, \alpha_k$ be the conjugates of α in $\overline{K}$. Then by the above proposition, $N = K(\alpha_1, \ldots, \alpha_k)$ is a finite Galois extension of K. Furthermore, if σ is an L-embedding of N into $\overline{L} = \overline{K}$, then it is also a K-embedding; hence it is a K-automorphism of N, hence an L-automorphism of N. □

From now on, we use the following standard notation. If L is a field and H is a group of automorphisms of L, then L^H denotes the *fixed field* of L under H, in other words, the set of elements of L that are fixed by all the elements of H. It is clear that L^H is a subfield of L.

The following proposition is the key result that we need before proving the main theorem of Galois theory.

Proposition 3.1.9. *Let L/K be a Galois extension with Galois group G and let H be a subgroup of G. Then $L^H = K$ if and only if $H = G$.*

Proof. We have clearly $L^H \supset K$ for all H. Choose first $H = G$, assume that $x \in L^G$, and set $K_1 = K(x)$. Then L/K_1 is a field extension, and by assumption every $\sigma \in G$ is a K_1-automorphism of L. It follows from Proposition 3.1.3 that $n = |G| \leqslant [L : K_1] \leqslant [L : K] = n$, so that $[L : K_1] = n$. In other words, $K_1 = K$, proving that $L^G = K$. Now let H be any subgroup, and assume that $L^H = K$. Write $L = K(\alpha)$, and consider

$$A(X) = \prod_{\sigma \in H} (X - \sigma(\alpha)).$$

The coefficients of the polynomial A are the elementary symmetric functions of the $\sigma(\alpha)$, hence are fixed by H, so $A(X) \in K[X]$. Since α is a root of A, it follows that $|H| = \deg(A) \geqslant [L : K] = |G|$, so that $H = G$ as claimed. □

We can now state and prove the fundamental theorem.

Theorem 3.1.10 (Fundamental theorem of Galois theory). *Let K be a perfect field, let L/K be a finite Galois extension, and set $G = \mathrm{Gal}(L/K)$.*

There exists a one-to-one reverse-ordering correspondence between on the one hand subfields L_1 of L containing K and on the other hand, subgroups H of G. The correspondence is as follows: if H is a subgroup of G, the corresponding subfield is L^H. Conversely, if L_1 is a subfield of L containing K, the corresponding subgroup is $\operatorname{Gal}(L/L_1)$. In other words $\operatorname{Gal}(L/L^H) = H$ and $L^{\operatorname{Gal}(L/L_1)} = L_1$.

Furthermore, the extension L^H/K is Galois if and only if H is a normal subgroup of G, and in this case we have a natural isomorphism $G/H \simeq \operatorname{Gal}(L^H/K)$.

Proof. Let L_1 be a subfield of L containing K. Since L/K is Galois, any K-embedding of L into $\overline{K}$ is a K-automorphism; hence any L_1-embedding of L into $\overline{L_1} = \overline{K}$ is an automorphism, hence an L_1-automorphism, so L/L_1 is Galois by Proposition 3.1.5.

Thus, for each subextension L_1 of L/K the group $\operatorname{Gal}(L/L_1)$ exists, and we denote by L_1' the fixed field of L by $\operatorname{Gal}(L/L_1)$. Applying Proposition 3.1.9 to the Galois extension L/L_1, we obtain $L^{\operatorname{Gal}(L/L_1)} = L_1$. Now let H be a subgroup of G and $L_1 = L^H$. By Proposition 3.1.9 once again, $L_1 = L^H$ if and only if $H = \operatorname{Gal}(L/L_1)$. Thus the two maps $H \mapsto L^H$ and $L_1 \mapsto \operatorname{Gal}(L/L_1)$ are indeed inverse maps, proving the first part of the theorem.

For the second part, let H be a subgroup of G and $L_1 = L^H$ the extension corresponding to H under the above correspondence. Clearly, for each $\sigma \in G$ the field corresponding to $\sigma H \sigma^{-1}$ is $\sigma(L_1)$. Now, L_1/K is Galois if and only if $\sigma(L_1) = L_1$ for each K-embedding σ of L_1 into $\overline{K}$, and such an embedding extends to a K-embedding of L, hence to an element of G since L/K is Galois. Thus L_1/K is Galois if and only if $\sigma(L_1) = L_1$ for all $\sigma \in G$, hence by the correspondence if and only if $\sigma H \sigma^{-1} = H$ for all $\sigma \in G$, in other words if and only if H is a normal subgroup of G. Finally, if this is the case, we have a natural group homomorphism from G to $\operatorname{Gal}(L_1/K)$ whose kernel is equal to H. We therefore obtain an injective group homomorphism from G/H to $\operatorname{Gal}(L_1/K)$, and since

$$|G/H| = |G|/|H| = [L:K]/[L:L_1] = [L_1:K] = \operatorname{Gal}(L_1/K) \,,$$

both groups have the same order; hence the homomorphism is an isomorphism. □

Another important result is the following.

Theorem 3.1.11. *Assume that L/K is Galois, and let M/K be any finite extension. Then the extension LM/M is also Galois, and $\operatorname{Gal}(LM/M)$ can be considered as a subgroup of $\operatorname{Gal}(L/K)$ by restriction of automorphisms. Furthermore, we have $\operatorname{Gal}(LM/M) \simeq \operatorname{Gal}(L/K)$ if and only if M and L are linearly disjoint over K, or equivalently in the present case, $M \cap L = K$.*

Proof. Write $L = K(\alpha)$ for some $\alpha \in L$. Clearly $LM = M(\alpha)$, and since the conjugates of α belong to L, hence to LM, Proposition 3.1.7 implies that LM/M is Galois.

The restriction of an M-automorphism of ML to L gives an embedding of L into $\overline{K}$, hence a K-automorphism of L since L/K is Galois, so we have a natural map from $\mathrm{Gal}(ML/M)$ to $\mathrm{Gal}(L/K)$. Furthermore, if $\sigma \in \mathrm{Gal}(ML/M)$ is the identity on L, then since it is an M-automorphism it is also the identity on M, hence on LM, so our map is injective, showing that $\mathrm{Gal}(LM/M)$ can be considered as a subgroup of $\mathrm{Gal}(L/K)$. Finally, let H be the image of $\mathrm{Gal}(LM/M)$ in $\mathrm{Gal}(L/K)$. Since the fixed field of LM under $\mathrm{Gal}(LM/M)$ is M, the fixed field of L under H is $M \cap L$ (in detail: $x \in L^H$ iff $x \in L$ and $\sigma(x) = x$ for all $\sigma \in H$ iff $x \in L$ and $x \in M$). The fundamental theorem of Galois theory that we have just proved shows that $H = \mathrm{Gal}(L/(M \cap L))$, so that $H = \mathrm{Gal}(L/K)$ iff $M \cap L = K$.   □

3.1.2 Number Fields

Recall that a *number field* is a finite extension of the field $\mathbb{Q}$ of rational numbers, i.e., a commutative field of characteristic 0 that is a finite-dimensional $\mathbb{Q}$-vector space (note that *any* field of characteristic 0 is a $\mathbb{Q}$-vector space).

If K is a number field and $x \in K$, then x is an *algebraic number*; in other words, it is a root of a nonzero polynomial equation with rational coefficients. The monic polynomial $P_x(X)$ of lowest degree of which x is a root is called the minimal (monic) polynomial of x. If $Q(X)$ is any polynomial with rational coefficients such that $Q(x) = 0$, then $P_x \mid Q$; in other words, P_x is a generator of the principal ideal of polynomials in $\mathbb{Q}[X]$ that vanish at x (note that $\mathbb{Q}[X]$ is a principal ideal domain (PID) since $\mathbb{Q}$ is a commutative field).

We can thus view K as a subfield of an algebraic closure of $\mathbb{Q}$, which will often be chosen to be the algebraic closure $\overline{\mathbb{Q}}$ of $\mathbb{Q}$ in $\mathbb{C}$.

The $\mathbb{Q}$-dimension of the vector space K is denoted by $[K : \mathbb{Q}]$ and called the *degree* of K.

Proposition 3.1.12. *Let K be a number field and let Ω be an algebraically closed field containing K. If L/K is an extension of degree n, there exist exactly n K-embeddings of L into Ω, i.e., injective field homomorphisms from L to Ω that leave K pointwise fixed. In particular, for $K = \mathbb{Q}$, any number field of degree n has exactly n embeddings into an algebraically closed field of characteristic zero.*

Proof. The proof is essentially the same as that of Proposition 3.1.3 and is left to the reader (Exercise 2).   □

Remark. The proofs of the above proposition and of Proposition 3.1.3 are naturally based on the primitive element theorem, in other words, on the representation of K as $\mathbb{Q}(\alpha)$. It is easily shown (see Exercise 3) that any rational function in α with rational coefficients is in fact a polynomial of degree less than or equal to $n - 1$ with rational coefficients. Thus, $1, \alpha, \ldots, \alpha^{n-1}$ is a $\mathbb{Q}$-basis for K, and of course this basis is not at all canonical since any $\alpha \in K$ of degree n satisfies the conditions. Although this is rarely mentioned in texts, note that this kind of basis, called a *power basis*, is not always the most convenient type of basis. For instance, in M. Bhargava's work on the enumeration of quartic number fields ([Bha1] and especially [Bha2]), such fields are viewed as defined by two *quadratic* equations is two unknowns, instead of one quartic equation in one unknown.

Proposition 3.1.13. *Let K be a number field of degree n. The number of embeddings of K into $\mathbb{C}$ whose image is not contained in $\mathbb{R}$ (we will simply say nonreal embeddings, or sometimes complex embeddings) is even.*

Proof. If σ is such an embedding and if c denotes complex conjugation, it is clear that $c \circ \sigma$ is also such an embedding, and we have $c \circ \sigma \neq \sigma$ since otherwise the image of σ would be contained in $\mathbb{R}$. $\square$

Definition 3.1.14. *Let K be a number field of degree n. The* signature *of K is the pair (r_1, r_2), where r_1 is the number of real embeddings and $2r_2$ is the number of complex embeddings, so that $n = r_1 + 2r_2$.*

Remark. although the notation (r_1, r_2) is the most common, note that some authors also use (r, c) (for real and complex), (r, i) (for real and imaginary), or (r, s) (for real and the letter after r).

When $r_1 = n$ (hence $r_2 = 0$) we say that K is totally real, and when $r_2 = n/2$ (hence $r_1 = 0$) we say that K is totally complex (of course in this case n is even).

The signature of a number field K can easily be found by looking at the proof of Proposition 3.1.12: Using the primitive element theorem, we write $K = \mathbb{Q}(\alpha)$, and we let $A \in \mathbb{Q}[X]$ be the minimal monic polynomial of α (which will in fact then be also the characteristic polynomial of α in K). Then r_1 (respectively $2r_2$) is equal to the number of real (respectively nonreal) roots of A in $\mathbb{C}$. This number can be found using Sturm's algorithm if desired; see for example [Coh0].

An important remark concerning signatures is that Galois theory often forbids certain signatures. The following are two examples, but of course we could give as many as we like.

- If $K/\mathbb{Q}$ is a Galois extension of degree n, then either $(r_1, r_2) = (n, 0)$ or, if n is even, $(r_1, r_2) = (0, n/2)$. Indeed, in that case the images of all the embeddings of K into $\mathbb{C}$ are the same, so either they are all real, or they are all nonreal.

– If $K/\mathbb{Q}$ is an extension of degree n such that the Galois group of the Galois closure of $K/\mathbb{Q}$ is isomorphic to a transitive subgroup of the alternating group A_n, then r_2 must be even. Indeed, the discriminant of such an extension is a square, and it is easily shown that the sign of the discriminant is $(-1)^{r_2}$ (see Definition 3.3.12). For example, if $n = 4$ the only possible signatures for A_4 are $(4, 0)$ and $(0, 2)$.

A purely topological property of embeddings that we will need is the following.

Proposition 3.1.15. *Let σ be a continuous field homomorphism from $\mathbb{R}$ or $\mathbb{C}$ into $\mathbb{C}$. Then either σ is the identity or it is complex conjugation.*

Proof. We have $\sigma(1) = 1$; hence for any positive integer n, by induction we have $\sigma(n) = n$, hence $\sigma(-n) = -n$, so that $\sigma(n) = n$ for all $n \in \mathbb{Z}$, and $\sigma(p/q) = \sigma(p)/\sigma(q) = p/q$ hence $\sigma(x) = x$ for all $x \in \mathbb{Q}$. If $x \in \mathbb{R}$, we can find a sequence of rational numbers x_n tending to x, hence by continuity $\sigma(x)$ is the limit of $\sigma(x_n) = x_n$, hence $\sigma(x) = x$ for all $x \in \mathbb{R}$. Of course continuity is essential here. Finally, if $i = \sqrt{-1}$ we have $\sigma(i)^2 = \sigma(-1) = -1$, hence $\sigma(i) = \varepsilon i$ for some fixed $\varepsilon = \pm 1$. Thus, for any $z = x + iy \in \mathbb{C}$ we have

$$\sigma(z) = \sigma(x) + \sigma(i)\sigma(y) = x + \varepsilon i y ,$$

proving the proposition.   □

Note that if σ is any field homomorphism from $\mathbb{R}$ to $\mathbb{R}$, then σ is necessarily continuous, hence equal to the identity by the above result; see Exercise 4.

3.1.3 Examples

The simplest number field is of course $\mathbb{Q}$. Apart from $\mathbb{Q}$, the simplest number fields are *quadratic fields*, i.e., number fields K that are of degree $n = 2$ over $\mathbb{Q}$. Such a number field has the form $K = \mathbb{Q}(\sqrt{d})$ for some $d \in \mathbb{Q}$, d not a square (see Exercise 7). Because $\mathbb{Q}(\sqrt{d_1})$ is isomorphic to $\mathbb{Q}(\sqrt{d_2})$ if and only if d_2/d_1 is a square, it follows that we can always assume that d is a squarefree integer different from 1. Thus quadratic number fields are exactly the fields $\mathbb{Q}(\sqrt{d})$ for d squarefree integers different from 1.

When $d > 0$, then $\mathbb{Q}(\sqrt{d})$ has 2 real embeddings and 0 nonreal ones, so that $r_1 = 2$, $r_2 = 0$. We then say that it is a real quadratic field. When $d < 0$, then $\mathbb{Q}(\sqrt{d})$ has no real embeddings and 2 complex conjugate nonreal embeddings, so that $r_1 = 0$, $r_2 = 1$. We then say that it is an imaginary quadratic field.

In degree three, new phenomena appear. First of all, since $r_1 + 2r_2 = 3$, we have either $r_1 = 3$, $r_2 = 0$ (so that K is totally real), or $r_1 = 1$, $r_2 = 1$ (then K is neither totally real nor totally complex, and is simply called a complex cubic field). However, there are important subclasses of fields among cubic fields. First of all the *pure cubic fields*, which are the analogues of quadratic fields: they are the fields of the form $K = \mathbb{Q}(\sqrt[3]{d})$ for some $d \in \mathbb{Q}$ not a cube. We may once again assume that d is a cubefree positive integer different from 1. Contrary to the case of quadratic fields, this does not guarantee uniqueness. More precisely, two such fields corresponding to cubefree positive integral d_1 and d_2 are isomorphic if and only if $d_1 = d_2$ or $d_1 d_2$ is a cube.

Second, there are the cyclic cubic fields, i.e., cubic fields K such that the extension $K/\mathbb{Q}$ is a Galois extension, necessarily cyclic. For example, consider the field $K = \mathbb{Q}(\theta)$, where θ is a root of the polynomial

$$T(X) = X^3 + X^2 - 2X - 1 .$$

Then one easily checks that the three roots of this polynomial are θ, $\theta^2 - 2$, and $-\theta^2 - \theta + 1$. The fact that they are polynomials in θ is exceptional, and characterizes Galois extensions. In contrast, if $K = \mathbb{Q}(\theta)$ with $\theta = \sqrt[3]{2}$ a root of $T(X) = X^3 - 2$, the other roots of T are $\theta(-1 \pm \sqrt{-3})/2$, which cannot be expressed as polynomials in θ since they are not real (it is thus clear that a Galois extension is either totally real or totally complex). Note that Galois extensions are rare: any degree-2 extension is of course Galois (change $\sqrt{d}$ into $-\sqrt{d}$), but in higher degrees, Galois extensions can be shown to have density zero in a suitable sense.

3.1.4 Characteristic Polynomial, Norm, Trace

Let α be an algebraic number and let T be its monic minimal polynomial. The roots α_j of T in some algebraic closure of $\mathbb{Q}$ are by definition the *conjugates* of α. By definition, the absolute trace of α is equal to the sum of the α_j, and the absolute norm is equal to the product of the α_j. If we write $T(X) = X^n + a_{n-1}X^{n-1} + \cdots + a_0$, the trace is thus equal to $-a_{n-1}$ and the norm is equal to $(-1)^n a_0$.

However, these notions are not very useful as such: indeed, we would naturally like the trace to be additive, and the norm to be multiplicative, but this is in general not the case. For example, if $\alpha = \sqrt{2}$ and $\beta = \sqrt{3}$, the norm of α is equal to -2 and that of β is equal to -3, while that of $\alpha\beta = \sqrt{6}$ is equal to -6, different from $(-2) \cdot (-3)$.

The reason for this lack of additivity or multiplicativity is that we must stay in a fixed number field. This leads to the following definitions.

Definition 3.1.16. *Let L/K be an extension of number fields of degree n, and let $\alpha \in L$. The characteristic polynomial $C_{L/K,\alpha}(X)$ of α with respect to this extension is the characteristic polynomial of the K-linear map multiplication by α from L to itself. If we write $C_{L/K,\alpha}(X) = X^n + c_{n-1}X^{n-1} + \cdots + c_0$, we set $\mathrm{Tr}_{L/K}(\alpha) = -c_{n-1}$ (the relative trace of α) and $\mathcal{N}_{L/K}(\alpha) = (-1)^n c_0$ (the relative norm of α).*

The following proposition is then immediate and left to the reader.

Proposition 3.1.17. *Let L/K be an extension of number fields of degree n.*

(1) *If α and β are in L, we have $\mathrm{Tr}_{L/K}(\alpha + \beta) = \mathrm{Tr}_{L/K}(\alpha) + \mathrm{Tr}_{L/K}(\beta)$ and $\mathcal{N}_{L/K}(\alpha\beta) = \mathcal{N}_{L/K}(\alpha)\,\mathcal{N}_{L/K}(\beta)$.*

(2) *If $\alpha \in L$ and $a \in K$, then $\mathrm{Tr}_{L/K}(a\alpha) = a\,\mathrm{Tr}_{L/K}(\alpha)$ and $\mathcal{N}_{L/K}(a\alpha) = a^n\,\mathcal{N}_{L/K}(\alpha)$.*

(3) *If σ_i for $1 \leqslant i \leqslant n$ are the n K-embeddings of L into an algebraic closure of K, then $\mathrm{Tr}_{L/K}(\alpha) = \sum_{1 \leqslant i \leqslant n} \sigma_i(\alpha)$ and $\mathcal{N}_{L/K}(\alpha) = \prod_{1 \leqslant i \leqslant n} \sigma_i(\alpha)$.*

(4) *If $\alpha \in L$ and $T_\alpha(X)$ is the minimal monic polynomial of α, of degree m, say, then m divides n, $C_{L/K,\alpha}(X) = T_\alpha(X)^{n/m}$, and hence $\mathrm{Tr}_{L/K}(\alpha) = (n/m)\,\mathrm{Tr}(\alpha)$ and $\mathcal{N}_{L/K}(\alpha) = \mathcal{N}(\alpha)^{n/m}$, where Tr and $\mathcal{N}$ denote the absolute trace and norm.*

One of the most important properties of the trace is the following proposition. It is a general property of separable extensions, and we have already seen it in the context of finite fields (Proposition 2.4.11). In characteristic zero the proof is even simpler.

Proposition 3.1.18. *Let K be a number field of degree n. The map $(x, y) \mapsto \mathrm{Tr}_{K/\mathbb{Q}}(xy)$ defines a nondegenerate $\mathbb{Q}$-bilinear form on $K \times K$ with values in $\mathbb{Q}$.*

Proof. The above proposition shows immediately that this map is a $\mathbb{Q}$-bilinear form. If x is such that $\mathrm{Tr}_{K/\mathbb{Q}}(xy) = 0$ for all $y \in K$, then if $x \neq 0$ we can choose $y = 1/x$, so that $0 = \mathrm{Tr}_{K/\mathbb{Q}}(1) = n$, a contradiction because we are in characteristic zero, proving that the map is nondegenerate. $\qquad\square$

3.1.5 Noether's Lemma

Lemma 3.1.19 (Noether). *Let L/K be a Galois extension with Galois group G, and let ϕ be a map from G to L^*. We will say that ϕ satisfies the cocycle condition if for all g, h in G we have*

$$\phi(gh) = \phi(g) \cdot g(\phi(h)) \ .$$

Then ϕ satisfies the cocycle condition if and only if there exists $\alpha \in L^*$ such that

$$\forall g \in G, \ \phi(g) = \frac{\alpha}{g(\alpha)} \ .$$

Proof. If $\phi(g) = \alpha/g(\alpha)$, we have

$$\phi(g) \cdot g(\phi(h)) = \frac{\alpha}{g(\alpha)} g\left(\frac{\alpha}{h(\alpha)}\right) = \frac{\alpha}{g(h(\alpha))} = \phi(gh) \ ,$$

so ϕ satisfies the cocycle condition. Conversely, assume that ϕ satisfies the cocycle condition. For $x \in L$, set

$$\sigma(x) = \sum_{h \in G} \phi(h)h(x) \ .$$

Then σ is an additive map from L to L. Applying Corollary 3.2.2, which we will prove below, to the distinct homomorphisms $h \in G$, we deduce that σ is not identically zero (recall that $\phi(h) \neq 0$ for all h by assumption). Hence let $x \in L$ be such that $\alpha = \sigma(x) \neq 0$. We have

$$g(\alpha) = g\left(\sum_{h \in G} \phi(h)h(x)\right) = \sum_{h \in G} g(\phi(h))gh(x) \ ;$$

hence by the cocycle condition

$$g(\alpha) = \phi(g)^{-1} \sum_{h \in G} \phi(gh)gh(x) = \phi(g)^{-1} \sum_{h \in G} \phi(h)h(x) = \phi(g)^{-1}\alpha \ ,$$

proving the lemma. □

3.1.6 The Basic Theorem of Kummer Theory

Let K be a commutative field, $\overline{K}$ a fixed algebraic closure of K. We will assume as usual that all algebraic extensions of K are in $\overline{K}$. Let $n \geqslant 1$ be an integer, and denote by ζ_n a primitive nth root of unity. In this section, we make the fundamental assumptions that n is not divisible by the characteristic of K (or that K has characteristic 0) and that $\zeta_n \in K$. What this last statement means is that the equation $X^n - 1 = 0$ (which has no multiple roots in $\overline{K}$ since the characteristic of K does not divide n) has exactly n roots, which are then powers of a single one, which we denote by ζ_n.

Theorem 3.1.20. *Let $n \geqslant 1$ be an integer, and let K be a commutative field of characteristic not dividing n and such that $\zeta_n \in K$. There is a natural bijection between finite subgroups of K^*/K^{*n} and finite Abelian extensions of K whose Galois group has exponent dividing n. This bijection is obtained as follows. If B is a finite subgroup of K^*/K^{*n}, the corresponding Abelian extension is obtained by adjoining to K all nth roots of lifts of elements of B. If L is a finite Abelian extension of K whose Galois group has exponent dividing n, then $B = (L^{*n} \cap K^*)/K^{*n}$. In addition, under this correspondence the Galois group $\mathrm{Gal}(L/K)$ is isomorphic to B.*

Proof. Let B be a finite subgroup of K^*/K^{*n}, and let $S = \{s_1, \ldots, s_k\}$ be a set such that the classes of elements of S in K^*/K^{*n} generate the group B (for example, representatives of all the elements of B). Note that conversely, if S is a finite set, the subgroup of K^*/K^{*n} generated by the classes of the elements of S is also finite, since it has at most $n^{|S|}$ elements. We let

$$K_B = K\left(\sqrt[n]{s_1}, \ldots, \sqrt[n]{s_k}\right) \ .$$

This makes sense since $\zeta_n \in K$ (it could also be made to make sense otherwise). Note for future reference that for all $b \in K^*$ such that $\overline{b} \in B$, the equation $x^n - b = 0$ has a solution in K_B^* (and, in fact, n solutions since $\zeta_n \in K$). Indeed, if $b = \omega^n \prod_i \overline{s_i}^{a_i}$ for some integers a_i and some $\omega \in K^*$, then $x = \omega \prod_i (\sqrt[n]{s_i})^{a_i}$ is a solution.

We are going to prove that the map $B \mapsto K_B$ is the desired bijection. Note first that K_B/K is a finite Abelian extension. Indeed, it is the compositum of the extensions $K(\sqrt[n]{s_i})/K$, and these are Abelian extensions since $\zeta_n \in K$, and so all the roots of the polynomial $X^n - s_i = 0$ belong to $K(\sqrt[n]{s_i})$ for any choice of the root. In fact, all these extensions are cyclic extensions of degree dividing n; hence the Galois group of their compositum is isomorphic to a subgroup of $(\mathbb{Z}/n\mathbb{Z})^k$, hence in particular has an exponent dividing n.

Let G be the Galois group of K_B/K, and denote by $\boldsymbol{\mu}_n = \boldsymbol{\mu}_n(K)$ the subgroup of K^* of nth roots of unity. We define the following pairing $\langle \, , \, \rangle$ from $G \times B$ to $\boldsymbol{\mu}_n$ as follows. Let $\sigma \in G$ and $\overline{b} \in B$. As we have seen, there exists $\beta \in K_B$ such that $\beta^n = b$. We will set

$$\langle \sigma, \overline{b} \rangle = \frac{\sigma(\beta)}{\beta} \, .$$

First, note that this is indeed an nth root of unity. In fact, $(\sigma(\beta)/\beta)^n = \sigma(b)/b = 1$ since $b \in K^*$. Second, the definition does not depend on the choice of β. Indeed, if β' is such that $\beta'^n = b\gamma^n$ for some $\gamma \in K^*$, then for some j we have $\beta'/\beta = \zeta_n^j \gamma \in K^*$, and so $\sigma(\beta')/\beta' = \sigma(\beta)/\beta$.

Furthermore, we evidently have

$$\langle \sigma, \overline{bb'} \rangle = \langle \sigma, \overline{b} \rangle \langle \sigma, \overline{b'} \rangle \quad \text{and}$$

$$\langle \sigma\tau, \overline{b} \rangle = \frac{\sigma\tau(\beta)}{\beta} = \frac{\sigma(\tau(\beta))}{\tau(\beta)} \frac{\tau(\beta)}{\beta} = \tau(\langle \sigma, \overline{b} \rangle) \langle \tau, \overline{b} \rangle \, ,$$

and since τ acts trivially on K, hence on $\boldsymbol{\mu}_n$, we have

$$\langle \sigma\tau, \overline{b} \rangle = \langle \sigma, \overline{b} \rangle \langle \tau, \overline{b} \rangle \, .$$

This means that $\langle \, , \, \rangle$ is a $\mathbb{Z}$-bilinear pairing. In other words, the map $\sigma \mapsto \langle \sigma, \cdot \rangle$ is a group homomorphism from G to $\mathrm{Hom}(B, \boldsymbol{\mu}_n)$, and the map $\overline{b} \mapsto \langle \cdot, \overline{b} \rangle$ is a group homomorphism from B to $\mathrm{Hom}(G, \boldsymbol{\mu}_n)$. We are going to compute the kernels of these two homomorphisms.

First, fix $\sigma \in G$, and assume that $\langle \sigma, \overline{b} \rangle = 1$ for all $b \in B$. Thus, if $\beta^n = b$, then $\sigma(\beta) = \beta$. This implies that for all our generators s_i we have $\sigma(\sqrt[n]{s_i}) = \sqrt[n]{s_i}$, and so $\sigma(x) = x$ for all $x \in K_B$; hence $\sigma = 1$, so the left kernel is trivial.

Second, fix $\overline{b} \in B$, and assume that $\langle \sigma, \overline{b} \rangle = 1$ for all $\sigma \in G$. If $\beta^n = b$, we thus have $\sigma(\beta) = \beta$ for all $\sigma \in G$, and hence by Galois theory, $\beta \in K^*$. Thus, $b \in K^{*n}$, so $\overline{b} = \overline{1}$ in B, and the kernel is again trivial. Therefore, we obtain what is called a *perfect pairing* between G and B.

Thus, the two maps we deduce from the pairing are injective, and in particular we obtain

$$|G| \leqslant |\text{Hom}(B, \boldsymbol{\mu}_n)| \quad \text{and} \quad |B| \leqslant |\text{Hom}(G, \boldsymbol{\mu}_n)| \ .$$

On the other hand, if A is a finite abelian group of exponent dividing n then $\text{Hom}(A, \boldsymbol{\mu}_n) \simeq A$ noncanonically (see Exercise 5). Hence $|\text{Hom}(G, \boldsymbol{\mu}_n)| = |G|$ and $|\text{Hom}(B, \boldsymbol{\mu}_n)| = |B|$, so both our injective homomorphisms are also surjective, from which we deduce that

$$B \simeq \text{Hom}(G, \boldsymbol{\mu}_n) \simeq G \ .$$

Thus, with each finite subgroup B of K^*/K^{*n} we have associated a finite Abelian extension K_B of K whose Galois group G has exponent n and isomorphic to B.

Conversely, let L be such an Abelian extension. We must show that $L = K_B$ for a suitable B. Let G be the Galois group of L/K. We are going to show that $B = (L^{*n} \cap K^*)/K^{*n}$ is such that $L = K_B$. Clearly, B is a subgroup of K^* of exponent dividing n. Let us show that B is finite. Using the same pairing $\langle \ , \ \rangle$ as before, we see that the proof of the injectivity of the map $B \to \text{Hom}(G, \boldsymbol{\mu}_n)$ did not use the finiteness of B. Thus this map is still injective, and since G is a finite group, we deduce that B is finite.

Lemma 3.1.21. *Any homomorphism from G to $\boldsymbol{\mu}_n$ has the form*

$$\sigma \longmapsto \langle \sigma, \overline{b} \rangle$$

for some $\overline{b} \in B$.

Assuming this lemma, it follows that the map $B \to \text{Hom}(G, \boldsymbol{\mu}_n)$ is a bijection and hence that $|B| = |G|$. By definition of B we have $K \subset K_B \subset L$. Since $\text{Gal}(K_B/K) \simeq B$, $\text{Gal}(L/K) = G$, and $|B| = |G|$, it follows that $[K_B : K] = [L : K]$ and so $L = K_B$, as claimed.

To prove the lemma, let ϕ be a homomorphism from G to $\boldsymbol{\mu}_n$. Recall that $\boldsymbol{\mu}_n \subset K$, hence that any element of $G = \text{Gal}(L/K)$ fixes $\boldsymbol{\mu}_n$ pointwise. Thus, for all σ and τ in G we have

$$\phi(\sigma\tau) = \phi(\sigma)\phi(\tau) = \phi(\sigma)\sigma(\phi(\tau)) \ .$$

Thus the map ϕ considered as a map from G to L^* satisfies the conditions of Noether's theorem (Lemma 3.1.19); therefore there exists $\alpha \in L^*$ such that $\phi(\sigma) = \sigma(\alpha)/\alpha$ for all $\sigma \in G$. Since we also have $\phi(\sigma)^n = 1$, we obtain $\sigma(\alpha)^n = \alpha^n$ for all $\sigma \in G$. Hence by Galois theory $\alpha^n \in K^*$, and so $\alpha^n \in L^{*n} \cap K^*$. It is clear that $\overline{b} = \overline{\alpha^n}$ is such that $\phi(\sigma) = \langle \sigma, \overline{b} \rangle$, proving the lemma and hence the theorem. $\qquad\square$

Corollary 3.1.22. *Let K be a commutative field and $n \geqslant 1$ an integer not divisible by the characteristic of K such that $\zeta_n \in K$.*

(1) *An extension L/K is a cyclic extension of degree n if and only if there exists $\alpha \in K^*$ such that $\overline{\alpha}$ is exactly of order n in K^*/K^{*n} and such that $L = K(\sqrt[n]{\alpha})$.*

(2) *The cyclic extensions $L_1 = K(\sqrt[n]{\alpha_1})$ and $L_2 = K(\sqrt[n]{\alpha_2})$ are K-isomorphic if and only if there exists an integer j coprime to n and $\gamma \in K^*$ such that $\alpha_2 = \alpha_1^j \gamma^n$.*

Proof. (1) Let L/K be a cyclic extension of degree n. By Theorem 3.1.20, there exists a subgroup B of K^*/K^{*n} such that $L = K_B$ and $B \simeq \mathrm{Gal}(L/K) \simeq \mathbb{Z}/n\mathbb{Z}$. If $\overline{\alpha}$ is a generator of B, it is clear that $K_B = K(\sqrt[n]{\alpha})$. Conversely, if $L = K(\sqrt[n]{\alpha})$ with $\alpha \in K^*$, then L/K is a cyclic extension of degree n if and only if $\overline{\alpha}$ generates a subgroup of order n of K^*/K^{*n}.

(2) Since L_1 and L_2 are cyclic extensions contained in $\overline{K}$, L_1 and L_2 are isomorphic if and only if they are equal, hence if and only if $B_1 = B_2$, where B_i is the cyclic subgroup of K^*/K^{*n} corresponding to L_i. Let $\overline{\alpha_i}$ be a generator of B_i, so that $L_i = K(\sqrt[n]{\alpha_i})$. Then $B_1 = B_2$ if and only if there exist integers j and k such that $\overline{\alpha_2} = \overline{\alpha_1}^j$ and $\overline{\alpha_1} = \overline{\alpha_2}^k$, hence $\overline{\alpha_1}^{kj-1} = 1$. Since $\overline{\alpha_1}$ is a generator of B_1 it follows that $kj \equiv 1 \pmod{n}$, hence that j is coprime to n, as claimed. $\square$

Definition 3.1.23. *Let K be a commutative field and $n \geqslant 1$ an integer not divisible by the characteristic of K such that $\zeta_n \in K$. Let α_1 and α_2 be elements of K^* of order exactly equal to n in K^*/K^{*n}. We will say that α_1 and α_2 are n-Kummer equivalent (or simply Kummer equivalent if n is understood) if $K(\sqrt[n]{\alpha_1})$ is K-isomorphic to $K(\sqrt[n]{\alpha_2})$, hence by the above corollary, if there exists an integer j coprime to n and $\gamma \in K^*$ such that $\alpha_2 = \alpha_1^j \gamma^n$.*

Since any finite Abelian extension of K can be obtained as a compositum of cyclic extensions of prime-power degree, to build finite Abelian extensions it suffices to build cyclic extensions of prime-power degree. In turn, these extensions can be built as towers of extensions of prime degree (although this is not a nice way to look at such extensions).

3.1.7 Examples of the Use of Kummer Theory

Let K be a commutative field with characteristic 0 or strictly larger than the degrees of the extensions that we will consider. Using Kummer theory, we want to construct explicitly up to K-isomorphisms all finite Abelian extensions L of K of small degree n (we can even do this more generally for non-Galois extensions, but it is beyond the scope of this book; see [Coh1]). We are going to see that this can be done quite explicitly in small cases. We denote by C_n the cyclic group of order n.

– For $n = 2$, we must have $G = C_2$. The condition $\zeta_2 \in K$ is always satisfied, and hence the above results imply that the general C_2-extension is $L = K(\alpha^{1/2})$ for some $\alpha \in K^* \setminus K^{*2}$, Kummer equivalence being simply $\alpha'/\alpha \in K^{*2}$.

– For $n = 3$ and $G = C_3$, we have two cases. If $\zeta_3 \in K$ the situation is as in the C_2 case: $L = K(\alpha^{1/3})$ for some $\alpha \in K^* \setminus K^{*3}$, Kummer equivalence being either α'/α or $\alpha'\alpha$ in K^{*3}.

If $\zeta_3 \notin K$, we may still apply Kummer theory, but we first have to *adjoin* ζ_3 to K: We set $K_z = K(\zeta_3)$, which is an extension of degree 2, and we denote by τ the generator of $\mathrm{Gal}(K_z/K)$. The compositum L_z of K_z and L is an Abelian extension of K with Galois group C_6, and this is easily seen to imply that $L_z = K_z(\theta)$ with $\theta = \alpha^{1/3}$ for some $\alpha \in K_z^* \setminus K_z^{*3}$, and that $\tau(\theta) = \theta^{-1}$ (otherwise we do not have $\theta\sigma = \sigma\theta$, where σ is a generator of $\mathrm{Gal}(L/K) \simeq \mathrm{Gal}(L_z/K_z)$); hence $\alpha\tau(\alpha) = \mathcal{N}_{K_z/K}(\alpha) = 1$, and hence by Hilbert's Theorem 90 (which is trivial for quadratic extensions) $\alpha = \tau(\beta)/\beta$ for some $\beta \in K_z$, which is 3-Kummer equivalent to $\alpha = \beta^2\tau(\beta)$. It can then be shown that $L = K(\theta + \theta^{-1})$. To find an equation for L/K, we write $\beta = (u + v\sqrt{-3})/2$, $e = (u^2 + 3v^2)/4$, and the equation satisfied by $\theta + \theta^{-1}$ is

$$X^3 - 3eX - eu = 0 .$$

The reader can fill in the details in the above construction, and also consider the cases C_4, C_5, and $C_2 \times C_2$.

3.1.8 Artin–Schreier Theory

In the case that the characteristic of K divides n, we must replace Kummer theory by another one called Artin–Schreier theory. Although not more difficult than Kummer theory, and in some sense easier, we will not consider it in detail in this book, but look only at the special case $n = p$, where the prime p is the characteristic of K. We begin with $n = p = 2$, which is especially simple.

Proposition 3.1.24. *Let K be a perfect field of characteristic 2. Quadratic extensions L of K have the form $K(\theta)$, where θ is a root of an Artin–Schreier polynomial $X^2 - X - a$ for some $a \in K$ not of the form $x^2 - x$ for $x \in K$. Furthermore, a and a' define K-isomorphic extensions if and only if $a' - a$ has the form $x^2 - x$ for some $x \in K$.*

Proof. Since K is perfect, the extension is separable, and hence by the primitive element theorem, $L = K(\theta)$, where θ is a root of $X^2 - bX - c = 0$ for some b and c in K. Since K is perfect we cannot have $b = 0$, otherwise $X^2 - c$ would be the square of a polynomial in $K[X]$. Thus if we set $Y = X/b$, we obtain the equation $(bY)^2 - b(bY) - c = 0$; in other words, $Y^2 - Y - a = 0$ with $a = c/b^2$, as desired. Such an equation defines a quadratic extension if and only if it is irreducible, if and only if it has no roots, meaning that

$a \notin G$, where G is the additive subgroup of K formed by the $x^2 - x$ (note that $x^2 - x + y^2 - y = (x + y)^2 - (x + y)$). For the last statement, let ϕ be an isomorphism from K to K' with evident notation. Then ϕ is entirely determined by $\phi(\theta)$, which must be of the form $u\theta' + v$ with u, v in K and $u \neq 0$. Therefore

$$0 = \phi(\theta^2 - \theta - a) = u^2\theta'^2 + v^2 - u\theta' - v - a = (u^2 - u)\theta' + u^2 a' + v^2 - v - a \ .$$

Since 1 and θ' are K-linearly independent and $u \neq 0$, we must have $u = 1$, hence $a - a' = v^2 - v$ as claimed.   □

The general result for $n = p > 2$ is essentially the same (compare it with Corollary 3.1.22), but we must work a little more.

Proposition 3.1.25. *Let K be a perfect field of characteristic p. An extension L/K is a cyclic extension of prime degree p if and only if $L = K(\alpha)$, where α is a root of an Artin–Schreier polynomial $X^p - X - a$ for some $a \in K$ not in the $\mathbb{F}_p$-vector space G of elements of the form $x^p - x$ for $x \in K$. The conjugates of α are then the $\alpha + k$ for $k \in \mathbb{F}_p$. Furthermore, a and a' define K-isomorphic extensions if and only if there exists $j \in \mathbb{F}_p^*$ such that $a' - ja \in G$.*

Proof. Let σ be a generator of $\mathrm{Gal}(L/K)$. By the normal basis Theorem 3.2.12, which we will prove below, there exists $\theta \in L$ such that the $\sigma^i(\theta)$ for $0 \leqslant i \leqslant p - 1$ form a K-basis of L. We can thus write in particular $1 = \sum_{0 \leqslant i \leqslant p-1} a_i \sigma^i(\theta)$ for some $a_i \in K$. Since 1 is stable by σ, applying σ we deduce that $a_i = a_{i-1}$ for all i, in other words that all the a_i are equal to some nonzero $u \in K$, say. Since $u\theta$ is still a normal basis, replacing θ by $u\theta$, we may assume that $u = 1$, in other words that $1 = \sum_{0 \leqslant i \leqslant p-1} \sigma^i(\theta)$.

Now choose $\alpha = -\sum_{0 \leqslant i \leqslant p-1} i\sigma^i(\theta)$. Then

$$\sigma(\alpha) - \alpha = -(p-1)\theta - \sum_{1 \leqslant i \leqslant p-1} (i-1)\sigma^i(\theta) + \sum_{0 \leqslant i \leqslant p-1} i\sigma^i(\theta) = \sum_{0 \leqslant i \leqslant p-1} \sigma^i(\theta) = 1 \ ,$$

so that $\sigma(\alpha) = \alpha + 1$, and hence $\sigma^k(\alpha) = \alpha + k$ for all k. It follows that the characteristic (and minimal) polynomial of α is

$$\prod_{0 \leqslant i \leqslant p-1} (X - \sigma^i(\alpha)) = \prod_{0 \leqslant i \leqslant p-1} (X - (\alpha + i)) = \prod_{i \in \mathbb{F}_p} (X - \alpha - i)$$

$$= (X - \alpha)^p - (X - \alpha) = X^p - X - (\alpha^p - \alpha) \ ,$$

hence is indeed an Artin–Schreier polynomial. Note that, although this is automatic from the proof, $a = \alpha^p - \alpha \in K$ by Galois theory, since $\sigma(a) = (\alpha + 1)^p - (\alpha + 1) = \alpha^p + 1 - (\alpha + 1) = a$.

A necessary condition for $X^p - X - a$ to be irreducible in $K[X]$ is that it have no roots in K, in other words that $a \notin G$, where G is the additive subgroup of the elements of K of the form $x^p - x$ for $x \in K$. This condition is in fact sufficient. Indeed, let $P(X)$ be a nonconstant irreducible factor of $X^p - X - a$ of degree $d \leqslant p$, say. Fix some root α of P in $\overline{K}$. Then all the $\alpha + k$ for $k \in \mathbb{F}_p$ are roots of $X^p - X - a$, and exactly d among them are roots of P. Furthermore, the set of $k \in \mathbb{F}_p$ such that $\alpha + k$ is a root of P forms an additive subgroup of $\mathbb{F}_p$ since if $P(\alpha + k) = 0$ for some α root of P, then $P(X + k)$ is divisible by, hence equal to, $P(X)$, so the set in question is the set of $k \in \mathbb{F}_p$ such that $P(X + k) = P(X)$, which is clearly an additive group. Since the cardinality of this group is d, we have $d \mid p$, hence $d = 1$ or $d = p$. The case $d = 1$ is excluded since we assume that $a \notin G$, so that $d = p$. In other words $X^p - X - a$ is irreducible, as claimed.

Finally, let $L = K(\alpha)$ and $L' = K(\alpha')$ be K-isomorphic cyclic extensions defined by roots of Artin–Schreier polynomials $X^p - X - a$ and $X^p - X - a'$ respectively, and let ϕ be a K-isomorphism from L' to L. Thus $\phi(\alpha')$ is a polynomial in α of degree d such that $1 \leqslant d \leqslant p - 1$, so write $\phi(\alpha') = \sum_{0 \leqslant k \leqslant d} a_k \alpha^k$ for some $a_k \in K$ with $a_d \neq 0$. Since we are in characteristic p and $\alpha^p = \alpha + a$ we have

$$0 = \phi(\alpha'^p - \alpha' - a') = \sum_{0 \leqslant k \leqslant d} a_k^p \alpha^{kp} - \sum_{0 \leqslant k \leqslant d} a_k \alpha^k - a'$$

$$= \sum_{0 \leqslant k \leqslant d} a_k^p (\alpha + a)^k - \sum_{0 \leqslant k \leqslant d} a_k \alpha^k - a' \,.$$

The right-hand side is now a polynomial of degree less than or equal to $d \leqslant p - 1$ in α, hence must be the zero polynomial since α has degree p. Identifying the coefficients of degree d, we see that $a_d^p - a_d = 0$. Identifying now the coefficients of degree $d - 1$ and assuming that $d > 1$, we obtain $a_{d-1}^p - a_{d-1} + daa_d^p = 0$. Since by assumption $a_d \neq 0$, and since $d \neq 0$ also satisfies $d^p - d = 0$ in characteristic p, it follows that $a = x^p - x$, where $x = -a_{d-1}/(da_d)$, which is absurd since by assumption a is not of this form. It follows that we must have $d = 1$. In that case since $a_1^p = a_1$ we can set $j = a_1$ as an element of $\mathbb{F}_p^*$, and the identification of the coefficients of degree 0 gives $a_0^p - a_0 + ja - a' = 0$, in other words $a' - ja \in G$, as claimed. $\square$

3.2 The Normal Basis Theorem

3.2.1 Linear Independence and Hilbert's Theorem 90

In this section, all fields that are considered are commutative (thus corresponding to the usual English meaning). We begin with the Dedekind independence theorem.

Lemma 3.2.1 (Dedekind independence). *Let G be a group, L a field, and let $\sigma_1, \ldots, \sigma_m$ be distinct group homomorphisms from G to L^*. Then* *they are L-linearly independent. In other words, if there exist $a_i \in L$ and a relation $\sum_{1 \leqslant i \leqslant m} a_i \sigma_i(h) = 0$ for all $h \in G$, then $a_i = 0$ for all i.*

Proof. Assume that there exists a nontrivial relation. Choose such a relation of *minimal length*, so that, up to reordering of the σ_i,

$$\forall h \in G \quad \sum_{1 \leqslant i \leqslant k} a_i \sigma_i(h) = 0 \tag{1}$$

with k minimal. For any $g \in G$, we have for all h, $\sum_{1 \leqslant i \leqslant k} a_i \sigma_i(gh) = 0$. Multiplying relation (1) by $\sigma_1(g)$ and subtracting, we obtain that for all g and h in G we have

$$\sum_{1 \leqslant i \leqslant k} a_i(\sigma_i(g) - \sigma_1(g))\sigma_i(h) \;,$$

and since the first coefficient vanishes, this is a relation of length $k-1$ between the characters. By the minimality of k, this must be the trivial relation, and again by minimality the a_i are nonzero; hence $\sigma_i(g) = \sigma_1(g)$ for all $i \leqslant n$ and all $g \in G$. Since the characters are distinct, this implies $n = 1$, hence $\sigma_1 = 0$, which is absurd.   □

Corollary 3.2.2. *Let K and L be fields, and let $\sigma_1, \ldots, \sigma_m$ be distinct field homomorphisms from K to L. Then the σ_i are L-linearly independent in the vector space of linear maps from K to L.*

Proof. Clear by applying the above lemma to $G = K^*$.   □

Corollary 3.2.3. *Let E/F be a finite extension of commutative fields of degree n. The elements of $\mathrm{Gal}(E/F)$ form an E-basis of the space $\mathcal{L}_F(E)$ of F-linear maps from E to E. In other words, we have the direct sum decomposition*

$$\mathcal{L}_F(E) = \bigoplus_{\sigma \in \mathrm{Gal}(E/F)} E\sigma \;.$$

Proof. Indeed, $\mathcal{L}_F(E)$ is an F-vector space of dimension n^2, hence an E-vector space of dimension n, so any family of n E-linearly independent elements form an E-basis.   □

The next result is valid for finite *cyclic* extensions of commutative fields, hence it applies in particular to extensions of finite fields, since finite fields are commutative and extensions of finite fields are cyclic generated by the Frobenius automorphism.

Proposition 3.2.4 (Hilbert's Theorem 90). *Let E/F be a finite cyclic extension of commutative fields of degree n and let σ be a generator of the*

Galois group $G = \mathrm{Gal}(E/F)$. *If* $\alpha \in E^*$ *is such that* $\prod_{0 \leqslant j < n} \sigma^j(\alpha) = 1$, *there exists* $\beta \in E^*$ *such that* $\alpha = \beta/\sigma(\beta)$. *In other words, for the action of the group algebra* $\mathbb{Z}[G]$ *on* E^*, *we have* $\mathrm{Ker}(\sum_{0 \leqslant j < n} \sigma^j) = \mathrm{Im}(1 - \sigma)$.

Proof. Consider the map ϕ from G to E^* defined for $k \geqslant 0$ by $\phi(\sigma^k) = \prod_{0 \leqslant j < k} \sigma^j(\alpha)$. Since $\prod_{0 \leqslant j < n} \sigma^j(\alpha)$ and σ has order n, the map ϕ is well defined, in other words depends only on k modulo n. It is immediately checked that ϕ satisfies the cocycle condition of Noether's lemma:

$$
\phi(\sigma^k)\sigma^k(\phi(\sigma^\ell)) = \prod_{0 \leqslant j < k} \sigma^j(\alpha)\sigma^k\left(\prod_{0 \leqslant i < \ell} \sigma^i(\alpha)\right)
$$
$$
= \prod_{0 \leqslant j < k} \sigma^j(\alpha) \prod_{k \leqslant j < k+\ell} \sigma^j(\alpha)
$$
$$
= \prod_{0 \leqslant j < k+\ell} \sigma^j(\alpha) = \phi(\sigma^{k+\ell}) \,.
$$

It thus follows from Noether's Lemma 3.1.19 that there exists $\beta \in E^*$ such that $\phi(\sigma^k) = \beta/\sigma^k(\beta)$, and the proposition follows by taking $k = 1$ since $\phi(\sigma) = \alpha$. $\qquad\square$

Remarks. (1) There is an *additive* version of Hilbert's Theorem 90, as well as a version for ideals. The modern way of looking at this theorem is to say that a certain 1-cohomology group vanishes; see Section 4.4.4.
(2) If $\beta \in L^*$ is such that $\alpha = \beta/\sigma(\beta)$, then by Galois theory all other possible β have the form $\gamma\beta$ for $\gamma \in K^*$.
(3) Even though Hilbert's Theorem 90 is not true as written for an arbitrary Abelian extension E/F, there exist suitable generalizations to this case. Note that Noether's lemma has no cyclicity assumption.

3.2.2 The Normal Basis Theorem in the Cyclic Case

The following theorem can be called the fundamental theorem of linear algebra.

Theorem 3.2.5. *Let* F *be a commutative field, let* σ *be an endomorphism of a finite-dimensional* F-*vector space* E, *let* P *be its minimal polynomial over* F, *and let* s *be the degree of* P. *There exists* $\theta \in E$ *such that the minimal polynomial of the restriction of* σ *to the vector space generated by* $\theta, \sigma(\theta), \dots, \sigma^{s-1}(\theta)$ *is still equal to* P.

Proof. First, let A and B be coprime polynomials in $F[X]$. We leave as an easy exercise for the reader (Exercise 10) to show that

$$
\mathrm{Ker}((AB)(\sigma)) = \mathrm{Ker}(A(\sigma)) \oplus \mathrm{Ker}(B(\sigma)) \,.
$$

Thus, let $P = \prod_i P_i^{n_i}$ be the factorization into irreducibles in $F[X]$ of P, and set $E_i = \mathrm{Ker}(P_i^{n_i}(\sigma))$. Using the above remark, since the $P_i^{n_i}$ are pairwise coprime, we obtain by induction $E = \mathrm{Ker}(P(\sigma)) = \bigoplus_i E_i$. The subspaces E_i are clearly stable under σ, and the minimal polynomial of the restriction σ_i of σ to E_i must be equal to $P_i^{n_i}$ since otherwise P would not be the minimal polynomial of σ. Thus there exists $\theta_i \in E_i$ such that $P_i^{n_i-1}(\sigma)(\theta_i) \neq 0$. Thus the minimal polynomial of the restriction of σ_i to the vector space generated by $\theta_i, \sigma_i(\theta_i), \ldots$ divides $P_i^{n_i}$ and does not divide $P_i^{n_i-1}$, hence is equal to $P_i^{n_i}$. It is now clear that $\theta = \sum_i \theta_i$ satisfies the required conditions.   □

We can now state and prove the normal basis theorem in the cyclic case. Since extensions of finite fields are always cyclic, this proves in particular the normal basis theorem for finite fields.

Theorem 3.2.6. *Let E/F be a finite cyclic extension of commutative fields of degree n and let $G = \mathrm{Gal}(E/F)$ be its Galois group. There exists an element $\theta \in E$ whose conjugates under the action of G form an F-basis of E, in other words E is a free $F[G]$-module of dimension 1.*

Proof. Let σ be a generator of G (in the case of finite fields, we have seen that we can choose for σ the Frobenius automorphism corresponding to F). Since the n elements σ^i for $0 \leqslant i \leqslant n-1$ are distinct, they are F-linearly independent by Corollary 3.2.2; hence the minimal polynomial of σ over F is equal to $X^n - 1$ (and since it has degree n this is also its characteristic polynomial). By the above theorem, it follows that there exists $\theta \in E$ such that the minimal polynomial of the restriction of σ to the subspace generated by $\theta, \sigma(\theta), \ldots, \sigma^{n-1}(\theta)$ is still equal to $X^n - 1$. In particular, they are F-linearly independent, proving the theorem.   □

The goal of the next subsections is to prove that this theorem is still valid in the noncyclic case. In particular, we will assume that the field F is infinite. This more general result will not be used in the rest of this book, so can be skipped at first. We follow closely [Lan0].

3.2.3 Additive Polynomials

From now on, let F be an *infinite* commutative field. In this case, we note the important fact that if $P \in F[X_1, \ldots, X_n]$ is a polynomial in n variables, then the formal polynomial P can be identified with the function that it induces from F^n to F. In other words, if $P(x_1, \ldots, x_n) = 0$ for all $(x_1, \ldots, x_n) \in F^n$, then P is the zero polynomial (evidently this is not true when F is finite; consider $X^p - X$ over $\mathbb{F}_p$).

Definition 3.2.7. *We say that a polynomial P is an* additive *polynomial if it satisfies one of the following two equivalent conditions:*

(1) *The function corresponding to P from F^n to F is an additive homomorphism.*
(2) *If $X = (X_1, \ldots, X_n)$ and $Y = (Y_1, \ldots, Y_n)$ then $P(X + Y) = P(X) + P(Y)$ as formal polynomials.*

The equivalence of the above two conditions comes from the above remark.

Proposition 3.2.8. *A polynomial $P \in F[X_1, \ldots, X_n]$ is additive if and only if it has the form $\sum_{1 \leqslant i \leqslant n} P_i(X_i)$, where the $P_i(X_i)$ are additive polynomials in one variable. Furthermore,*

(1) *When F has characteristic 0 the additive polynomials in one variable are the polynomials aX with $a \in F$.*
(2) *When F has characteristic $p > 0$, the additive polynomials in one variable are the polynomials*

$$P(X) = \sum_{0 \leqslant k \leqslant m} a_k X^{p^k}$$

with the $a_k \in F$.

Proof. We can write

$$(X_1, \ldots, X_n) = \sum_{1 \leqslant i \leqslant n} (0, \ldots, X_i, \ldots, 0),$$

and so by definition of additive polynomials we have

$$P(X_1, \ldots, X_n) = \sum_{1 \leqslant i \leqslant n} P_i(X_i),$$

where $P_i(X_i) = P(0, \ldots, X_i, \ldots, 0)$ is clearly an additive polynomial in one variable. Thus let now $P(X)$ be an additive polynomial in one variable, and let $a_r X^r$ with $a_r \in F^*$ be a nonzero monomial. The monomials of total degree r in $P(X + Y) - P(X) - P(Y)$ are thus given by $a_r((X + Y)^r - X^r - Y^r)$, which must therefore be identically zero. This is indeed the case if $r = 1$. If $r > 1$ it contains the term $rX^{r-1}Y$, so that $r = 0$. In other words, F has positive characteristic p dividing r. Furthermore, if we write $r = p^k s$ with $p \nmid s$, then

$$(X + Y)^r - X^r - Y^r = (X^{p^k} + Y^{p^k})^s - (X^{p^k})^s - (Y^{p^k})^s,$$

and the same reasoning shows that $s = 1$, so that the nonzero monomials have degree $r = p^k$, as claimed. $\square$

3.2.4 Algebraic Independence of Homomorphisms

Definition 3.2.9. *Let A be an abelian group (written additively), and let $\sigma_1, \ldots, \sigma_n$ be additive homomorphisms from A to F. We say that the σ_i are algebraically dependent over F if there exists a nonzero polynomial $P \in F[X_1, \ldots, X_n]$ such that $P(\sigma_1(x), \ldots, \sigma_n(x)) = 0$ for all $x \in A$. If such a P does not exist, we say that the σ_i are algebraically independent.*

The main result that we need is the following.

Theorem 3.2.10 (Artin). *If the σ_i are algebraically dependent as above, we can choose P to be a (nonzero) additive polynomial.*

Proof. For simplicity, we will write $\Sigma(x)$ instead of $(\sigma_1(x), \ldots, \sigma_n(x))$. Let $P(X_1, \ldots, X_n)$ be the polynomial of lowest possible total degree that is nonzero and is such that $P(\Sigma(x)) = 0$ for all $x \in A$. We will prove that P is additive. Set $Q(X, Y) = P(X + Y) - P(X) - P(Y)$, so that for all x, y in A we have

$$Q(\Sigma(x), \Sigma(y)) = P(\Sigma(x + y)) - P(\Sigma(x)) - P(\Sigma(y)) = 0$$

since the σ_i are additive. Assume by contradiction that Q is not the zero polynomial, hence (since F is infinite) that Q is not identically zero on $F^n \times F^n$. We consider two cases.

Case 1: We have $Q(v, \Sigma(y)) = 0$ for all $v \in F^n$ and all $y \in A$. By assumption there exists $v' \in F^n$ such that $P_1(Y) = Q(v', Y)$ is not the zero polynomial. By definition of Q the degree of P_1 in Y is strictly less than that of P. On the other hand, $P_1(\Sigma(y)) = Q(v', \Sigma(y)) = 0$, so we obtain a contradiction with the minimality of the degree of P.

Case 2: There exist $v \in F^n$ and $y \in A$ such that $Q(v, \Sigma(y)) \neq 0$. Here we set $P_1(X) = Q(X, \Sigma(y))$. Then P_1 is not the zero polynomial, $P_1(\Sigma(x)) = 0$ for all $x \in A$ by the defining property of Q, and the degree of P_1 is strictly less than that of P, again a contradiction.

We have thus shown that Q is identically 0, hence is the zero polynomial, so P is additive. $\qquad\qquad\square$

Theorem 3.2.11. *As above, let F be an infinite field, and let $\sigma_1, \ldots, \sigma_n$ be distinct elements of a finite group of automorphisms of F. Then the σ_i are algebraically independent over F.*

Proof. By Theorem 3.2.10, there exists a nonzero additive polynomial P such that $P(\Sigma(x)) = 0$ for all x. If F has characteristic 0, by Theorem 3.2.8 such a polynomial is simply a linear form, and hence the result follows from linear independence (Corollary 3.2.2). Thus assume now that F has characteristic $p > 0$. By Theorem 3.2.8 we can thus write

$$\sum_{1 \leqslant i \leqslant n} \sum_{1 \leqslant k \leqslant m} a_{i,k} \sigma_i(x)^{p^k} = 0$$

for all $x \in F$, and at least one coefficient $a_{i,k}$ not equal to 0. Denote by ϕ the map $x \mapsto x^p$ from F to itself. Since F has characteristic p, ϕ is a homomorphism of F into F, and $\sigma(x)^{p^k} = \sigma \circ \phi^k(x)$. Note that $\sigma(x)^{p^k}$ is *not* equal to $\sigma^{p^k}(x)$. The above relation is thus a nontrivial linear dependence relation between the homomorphisms $\sigma_i \circ \phi^k$. By Corollary 3.2.2 once again, it follows that these homomorphisms cannot be distinct. In other words, there exist distinct pairs (i, k) and (j, ℓ) such that $\sigma_i \circ \phi^k = \sigma_j \circ \phi^\ell$, so that

$$\sigma_i(x)^{p^k} = \sigma_j(x)^{p^\ell}$$

for all $x \in F$. Now note that in characteristic p we have $(x - y)^p = x^p - y^p$; hence it follows that $x^p = y^p$ implies that $x = y$. Thus if we assume for instance that $k \leqslant \ell$, we obtain

$$\sigma_i(x) = \sigma_j(x^{p^{\ell-k}})$$

for all $x \in F$. If we set $\sigma = \sigma_j^{-1}\sigma_i$, this means that $\sigma(x) = x^{p^{\ell-k}}$ for all $x \in F$. Since σ belongs to a finite group of automorphisms of a certain order r, say, σ^r is the identity. In other words,

$$x^{rp^{\ell-k}} = x$$

for all $x \in F$. But such an equation has only a finite number of roots in the commutative field F, unless $\ell - k = 0$ (and $r = 1$). Since F is infinite, it follows that $\ell = k$, hence $\sigma_i = \sigma_j$, hence $i = j$ (since the σ_i are distinct), in contradiction to the fact that $(i, k) \neq (j, \ell)$. $\qquad\square$

Remark. Do not confuse $\sigma^p(x)$ with $\sigma(x)^p$, for instance. Indeed, $\sigma^p(x)$ means that we compose σ with itself p times, and apply to x, while $\sigma(x)^p = \sigma \circ \phi(x)$ with the above notation. Similarly, the fact that $\sigma^r = \sigma_1$, where σ_1 is the identity automorphism does *not* mean that σ and σ_1 are algebraically dependent, the polynomial P being $P(X, Y) = X^r - Y$. Indeed,

$$P(\sigma(x), \sigma_1(x)) = \sigma(x)^r - x = \sigma(x^r) - x \,,$$

and this has no reason to be equal to 0 for all x since $\sigma(x)^r$ is not in general equal to $\sigma^r(x)$.

3.2.5 The Normal Basis Theorem

We are now in a position to prove the normal basis theorem in complete generality.

Theorem 3.2.12. *Let E/F be a finite Galois extension of commutative fields of degree n, and let $G = \mathrm{Gal}(E/F)$ be its Galois group. There exists an element $\theta \in E$ whose conjugates under the action of G form an F-basis of E, in other words E is a free $F[G]$-module of dimension 1.*

Proof. If F is a finite field, then E/F is a cyclic extension, so that the result is Theorem 3.2.6. We can therefore assume that F is infinite.

Let $\sigma_1, \ldots, \sigma_n$ be the (distinct) elements of G, numbered so that σ_1 is the identity. We can write

$$\sigma_i^{-1}\sigma_j = \sigma_{t(i,j)}$$

for some function t from $[1, n] \times [1, n]$ to $[1, n]$. Set

$$P(X_1, \ldots, X_n) = \det((X_{t(i,j)})_{1 \leqslant i,j \leqslant n}) .$$

Since $X_{t(i,j)} = X_1$ if and only if $\sigma_i = \sigma_j$ if and only if $i = j$, it follows that $P(1, 0, \ldots, 0) = 1$, since it is equal to the determinant of the identity matrix, so that P is a nonzero polynomial. By Theorem 3.2.11, it follows that there exists $\theta \in F$ such that $P(\sigma_1(\theta), \ldots, \sigma_n(\theta)) \neq 0$. Since by definition $\sigma_{t(i,j)}(\theta) = \sigma_i^{-1}(\sigma_j(\theta))$, this can be written

$$\det(\sigma_i^{-1}(\sigma_j(\theta))) \neq 0 .$$

I claim that θ is the desired element. Indeed, assume that there exists a nontrivial linear dependence relation $\sum_{1 \leqslant j \leqslant n} a_j \sigma_j(\theta) = 0$. Applying σ_i^{-1} to this relation for all i shows that

$$\sum_{1 \leqslant j \leqslant n} a_j \sigma_i^{-1}\sigma_j(\theta) = 0 ,$$

which is a nontrivial linear dependence relation between the columns of the matrix $(\sigma_i^{-1}\sigma_j(\theta))_{i,j}$, contradicting the fact that its determinant is nonzero. $\quad\square$

3.3 Ring-Theoretic Algebraic Number Theory

The field-theoretic properties seen in the preceding sections are evidently essential for any further study. However, the most interesting part of algebraic number theory deals with the ring-theoretic properties, which we summarize in this section.

3.3.1 Gauss's Lemma on Polynomials

Definition 3.3.1. *Let $A \in \mathbb{Z}[X]$ be a nonzero polynomial. We define the content of A and denote by $c(A)$ the GCD of all the coefficients of A.*

Proposition 3.3.2 (Gauss's lemma). *If A and B are two nonzero polynomials in $\mathbb{Z}[X]$, we have $c(AB) = c(A)c(B)$.*

Proof. Let us say that a polynomial $A \in \mathbb{Z}[X]$ is *primitive* if its content is equal to 1. Since $A = c(A)A_1$ with A_1 primitive, it is clear that the proposition is equivalent to the statement that the product of two primitive polynomials A and B is primitive. Assume the contrary, so that there exists a prime number p that divides all the coefficients of AB; in other words $\overline{AB} = 0$

where, for any $P \in \mathbb{Z}[X]$, $\overline{P} \in (\mathbb{Z}/p\mathbb{Z})[X]$ denotes the polynomial obtained by reducing the coefficients of P modulo p. Since we evidently have $\overline{AB} = \overline{A}\,\overline{B}$ and since $(\mathbb{Z}/p\mathbb{Z})[X]$ is an integral domain, it follows that $\overline{A} = 0$ or $\overline{B} = 0$, in other words that p divides all the coefficients of A or all the coefficients of B, in contradiction with the fact that A and B are primitive. $\qquad\square$

Corollary 3.3.3. *Let $C \in \mathbb{Z}[X]$ be a monic polynomial and assume that $A \in \mathbb{Q}[X]$ is a monic polynomial such that $A \mid C$ in $\mathbb{Q}[X]$. Then in fact $A \in \mathbb{Z}[X]$.*

Proof. Write $C = AB$ with $B \in \mathbb{Q}[X]$. Let d_A (respectively d_B) be the smallest integer such that $d_A A$ (respectively $d_B B$) is in $\mathbb{Z}[X]$, in other words, the LCM of the denominators of the coefficients of A (respectively B). We can write $d_A d_B C = (d_A A)(d_B B)$. By the minimality assumption, we have $c(d_A A) = c(d_B B) = 1$, hence by Gauss's lemma $c(d_A d_B C) = 1$, and in particular $d_A = 1$, hence $A \in \mathbb{Z}[X]$. $\qquad\square$

3.3.2 Algebraic Integers

We begin with the following basic proposition.

Proposition 3.3.4. *Let α be an algebraic number. The following four properties are equivalent.*

(1) *The number α is a root of a monic polynomial with coefficients in $\mathbb{Z}$.*
(2) *The minimal monic polynomial of α has coefficients in $\mathbb{Z}$.*
(3) *The ring $\mathbb{Z}[\alpha]$ of polynomials in α with integer coefficients is a finitely generated $\mathbb{Z}$-module.*
(4) *There exists a commutative ring with unit R that is a finitely generated $\mathbb{Z}$-module and such that $\alpha \in R$.*

Proof. (1) $\Longrightarrow$ (2): Assume that $P(\alpha) = 0$ with $P \in \mathbb{Z}[X]$ monic, and let T be the minimal monic polynomial of α. By definition, T divides P in $\mathbb{Q}[X]$, and T is monic, so we conclude by Corollary 3.3.3.

(2) $\Longrightarrow$ (3): Let $T(\alpha) = 0$, where T is the minimal monic polynomial of α, hence with integral coefficients, and set $n = \deg(T)$. If L is the $\mathbb{Z}$-module generated by $1, \alpha^1, \ldots, \alpha^{n-1}$, then by assumption $\alpha^n \in L$; hence by induction $\alpha^k \in L$ also for any $k \geqslant n$. Thus $L = \mathbb{Z}[\alpha]$, so that the elements $1, \alpha^1, \ldots, \alpha^{n-1}$ form a generating set of $\mathbb{Z}[\alpha]$; hence $\mathbb{Z}[\alpha]$ is a finitely generated $\mathbb{Z}$-module.

(3) $\Longrightarrow$ (4): Simply choose $R = \mathbb{Z}[\alpha]$.

(4) $\Longrightarrow$ (1): This is the only really amusing part of the proof. Since R is a finitely generated $\mathbb{Z}$-module, there exist $\omega_1, \ldots, \omega_n$ that generate R as a $\mathbb{Z}$-module. Since R is a ring and $\alpha \in R$, there exist $a_{i,j} \in \mathbb{Z}$ such that for $1 \leqslant j \leqslant n$ we have $\alpha \omega_j = \sum_{1 \leqslant i \leqslant n} a_{i,j}\omega_i$. If $A = (a_{i,j})_{1 \leqslant i,j \leqslant n}$ is the

matrix of the $a_{i,j}$, if we set $M = \alpha I_n - A$ with I_n the $n \times n$ identity matrix, and finally if $B = (\omega_1, \ldots, \omega_n)$ is the row vector of the ω_j, then this can be written $BM = 0$. If M was invertible as a matrix with coefficients in the field $\mathbb{Q}(\alpha)$, then multiplying by M^{-1}, we would obtain $B = 0$, hence $R = \{0\}$, contradicting the fact that $1 \in R$ (unless $\alpha = 0$, but in that case the implication is trivial). Thus M is not invertible, so that $\det(M) = 0$. This means that α is a root of $\det(X I_n - A)$, the characteristic polynomial of the matrix A, and this is clearly a monic polynomial with integral coefficients.

$\square$

Note that we could not use directly the Cayley–Hamilton theorem since R is not necessarily a free $\mathbb{Z}$-module, and even so the ω_j are not necessarily $\mathbb{Z}$-linearly independent.

Definition 3.3.5. (1) *An algebraic number satisfying one of the above equivalent properties is called an* algebraic integer.
(2) *A nonzero algebraic integer whose inverse is also an algebraic integer is called a* unit.

By Proposition 3.3.4, when α is not an algebraic integer, $\mathbb{Z}[\alpha]$ is not finitely generated. The simplest example is with $\alpha = 1/2$: the ring $\mathbb{Z}[1/2]$ is the subring of elements of $\mathbb{Q}$ whose denominator is a power of 2. This ring is also not free (although it has no torsion), since two rational numbers are always $\mathbb{Z}$-linearly dependent.

Proposition 3.3.6. *If α and β are algebraic integers, then so are $\alpha + \beta$ and $\alpha\beta$. In other words, algebraic integers belonging to a fixed algebraic closure of $\mathbb{Q}$ form a ring.*

Proof. Consider $R = \mathbb{Z}[\alpha, \beta]$, the ring of polynomials in α and β. Since α and β are algebraic integers, of respective degree m and n, say, it is clear that the $(\alpha^i \beta^j)_{0 \leqslant i < m, \, 0 \leqslant j < n}$ form a finite set that generates R as a $\mathbb{Z}$-module, and since $\alpha + \beta$ and $\alpha\beta$ belong to R we conclude by Proposition 3.3.4. $\square$

It is possible to give a direct (but less elegant) proof of this proposition that directly uses the fact that α and β are roots of monic integral polynomials. This uses the notion of *resultant*, and gives an algorithm for computing the minimal polynomials of $\alpha + \beta$ and of $\alpha\beta$; see Exercise 12.

Proposition 3.3.7. *Let $P(X)$ be a monic polynomial whose coefficients are algebraic integers, and let α be such that $P(\alpha) = 0$. Then α is an algebraic integer.*

Proof. Write $P(X) = X^n + \sum_{1 \leqslant i \leqslant n-1} \beta_i X^i$. Since the β_i are algebraic integers, it follows that $\mathbb{Z}[\beta_1, \ldots, \beta_{n-1}]$ is a finitely generated $\mathbb{Z}$-module. Let $\gamma_1, \ldots, \gamma_N$ be a finite generating set, and let $R = \mathbb{Z}[\alpha, \beta_1, \ldots, \beta_{n-1}]$. As in the proof of the implication (2) $\implies$ (3) of Proposition 3.3.4, it is clear that the $\gamma_i \alpha^j$ for $1 \leqslant i \leqslant N$ and $0 \leqslant j \leqslant n - 1$ form a finite generating set for R. We conclude by Proposition 3.3.4. $\qquad\qquad\qquad\qquad\qquad\qquad\qquad\qquad$ □

When an algebraic number α is not necessarily an algebraic integer, we can still obtain finitely generated free $\mathbb{Z}$-modules as follows.

Proposition 3.3.8 (Dedekind). *Let α be an algebraic number and let $T \in \mathbb{Z}[X]$ be a nonzero polynomial such that $T(\alpha) = 0$. Write $T(X) = a_n X^n + a_{n-1} X^{n-1} + \cdots + a_1 X + a_0$, let $\omega_0 = 1$, and for $1 \leqslant j \leqslant n - 1$, set*

$$\omega_j = a_n \alpha^j + a_{n-1} \alpha^{j-1} + \cdots + a_{n-j} \, .$$

The finitely generated $\mathbb{Z}$-module R generated by the ω_j for $0 \leqslant j \leqslant n - 1$ is a subring of $\mathbb{Z}[\alpha]$. In particular, the ω_j are algebraic integers for all j.

Note that this proposition does not claim that $\alpha \in R$ (otherwise α would be an algebraic integer).

Proof. If we define ω_j for all $j \geqslant 1$ by the formula of the proposition, it is clear by definition that $\omega_j = 0$ for $j \geqslant n$. We can thus consider R as the $\mathbb{Z}$-module generated by all the ω_j for $j \geqslant 0$. Now it is clear that for all $k \geqslant 1$ we have the induction formula $\omega_{k+1} = \alpha \omega_k + a_{n-k-1}$. We are going to prove by induction on k that for all i we have $\omega_i \omega_k \in R$. For $k = 0$ or $i = 0$ this is clear, and for $k = 1$ and $i \geqslant 1$ we have

$$\omega_i \omega_1 = \omega_i (a_n \alpha + a_{n-1}) = a_n (\omega_{i+1} - a_{n-i-1}) + a_{n-1} \omega_i \in R \, .$$

Thus assume $k \geqslant 1$ and that our induction hypothesis is true for k. For $i \geqslant 1$ we have

$$\omega_i \omega_{k+1} = \omega_i (\alpha \omega_k + a_{n-k-1}) = \omega_k (\omega_{i+1} - a_{n-i-1}) + a_{n-k-1} \omega_i$$
$$= \omega_{i+1} \omega_k - a_{n-i-1} \omega_k + a_{n-k-1} \omega_i \in R$$

by our induction hypothesis, proving our claim. It follows that R is a subring of $\mathbb{Z}[\alpha]$, and since all the ω_j belong to the ring R, which is a finitely generated $\mathbb{Z}$-module, it follows by Proposition 3.3.4 that they are algebraic integers. □

Remarks. (1) The result of this proposition is not entirely trivial: for instance try to prove directly that $a_n \alpha^2 + a_{n-1} \alpha$ (which is equal to $\omega_2 - a_{n-2}$) is an algebraic integer.

(2) The ring R is not determined by α alone, but also by the polynomial T, even when α is an algebraic integer. Let us consider an example. Let $\alpha = \sqrt{2}/2$. If we choose $T(X) = 2X^2 - 1$, we have $R = \mathbb{Z}[\sqrt{2}]$. But more generally we can choose $T(X) = d(2X^2 - 1)$ for any integer d, and in that case we have $R = \mathbb{Z}[d\sqrt{2}]$, which is a different ring. We could of course also choose higher-degree polynomials.

(3) If α is an algebraic integer and T is the minimal monic polynomial of α, then since $a_n = 1$ the matrix that sends the α^j to the ω_j is an integral triangular matrix with 1 on the diagonal; hence the ring R is equal to $\mathbb{Z}[\alpha]$ in that case. Otherwise, the best possibility for R is to choose $R = \mathbb{Z}[\alpha] \cap \mathbb{Z}_{\mathbb{Q}(\alpha)}$, which is the largest subring of $\mathbb{Z}[\alpha]$ that is a finitely generated $\mathbb{Z}$-module; see Exercise 14.

We finish this subsection with the following easy but important result due to Kronecker.

Proposition 3.3.9 (Kronecker). *Let $T(X) \in \mathbb{Z}[X]$ be a* monic *polynomial with integer coefficients. Assume that all the roots of T in $\mathbb{C}$ have modulus equal to 1. Then all the roots of T are roots of unity.*

Proof. Without loss of generality we may assume that $T(X)$ is irreducible. Write $T(X) = \prod_{1 \leqslant i \leqslant n}(X - \alpha_i)$ with $\alpha_i \in \mathbb{C}$, and for any $k \geqslant 1$, consider the polynomial $T_k(X) = \prod_{1 \leqslant i \leqslant n}(X - \alpha_i^k)$. The coefficients of $T_k(X)$ are symmetric polynomials with integer coefficients in the α_i, hence are polynomials in the coefficients of T, and in particular are in $\mathbb{Z}$. Furthermore, since $|\alpha_i^k| = 1$ for all i, the coefficient of X^{n-m} of $T_k(X)$ is bounded in absolute value by $\binom{n}{m}$. This implies in particular that the number of possible polynomials $T_k(X)$ is *finite*; hence the number of possible values of α_i^k for $1 \leqslant i \leqslant n$ and $k \geqslant 1$ is finite. Therefore there must exist i and $k_1 \neq k_2$ such that $\alpha_i^{k_1} = \alpha_i^{k_2}$, hence $\alpha_i^{k_1 - k_2} = 1$, so that α_i is a root of unity. Since T has been assumed to be irreducible, all of its roots are conjugate to α_i, hence are also roots of unity. $\qquad\square$

Corollary 3.3.10. *If α is an algebraic* integer *all of whose conjugates have absolute value equal to 1 in $\mathbb{C}$ then α is a root of unity.*

Remarks. (1) The hypothesis of the corollary means that for every embedding σ of $\mathbb{Q}(\alpha)$ in $\mathbb{C}$ we have $|\sigma(\alpha)| = 1$.
(2) The result is trivially false if α is not an algebraic integer: consider for instance $\alpha = (3 + 4\sqrt{-1})/5$.

3.3.3 Ring of Integers and Discriminant

Let K be a number field of degree n. The set of algebraic integers belonging to K is clearly a subring of K, which we denote by $\mathbb{Z}_K$ (many authors denote it by O_K). It is naturally called the ring of algebraic integers of K.

Proposition 3.3.11. *Let K be a number field of degree n. The ring $\mathbb{Z}_K$ is a free $\mathbb{Z}$-module of rank n.*

Proof. Let $\alpha_1, \ldots, \alpha_n$ be a $\mathbb{Q}$-basis of K. Multiplying if necessary each α_i by some nonzero element of $\mathbb{Z}$, we may assume that $\alpha_i \in \mathbb{Z}_K$ for all i. Let Λ be the free $\mathbb{Z}$-module of rank n generated by the α_i, so that $\Lambda \subset$

$\mathbb{Z}_K$. By Proposition 3.1.18, we know that the map $(x, y) \mapsto \mathrm{Tr}_{K/\mathbb{Q}}(xy)$ is a nondegenerate $\mathbb{Q}$-bilinear form from $K \times K$ to $\mathbb{Q}$. We can thus consider the dual basis $\beta_1, \dots, \beta_n$ of $\alpha_1, \dots, \alpha_n$, i.e., the unique $\mathbb{Q}$-basis such that for all i, j we have $\mathrm{Tr}_{K/\mathbb{Q}}(\alpha_i \beta_j) = \delta_{i,j}$, where δ is the Kronecker symbol. Denote by Λ^* the free $\mathbb{Z}$-module of rank n generated by the β_j. Since for any $y \in K$ we have

$$y = \sum_{1 \leqslant j \leqslant n} \mathrm{Tr}_{K/\mathbb{Q}}(\alpha_j y) \beta_j$$

(see why?) it is clear that $y \in \Lambda^*$ if and only if for all j, $\mathrm{Tr}_{K/\mathbb{Q}}(\alpha_j y) \in \mathbb{Z}$. It follows that if $y \in \mathbb{Z}_K$ then $y \in \Lambda^*$. We have thus proved the double inclusion

$$\Lambda \subset \mathbb{Z}_K \subset \Lambda^* .$$

Since Λ and Λ^* are free $\mathbb{Z}$-modules of the same rank n, it follows from Corollary 2.1.5 that $\mathbb{Z}_K$ is also a free $\mathbb{Z}$-module of rank n, proving the proposition.

$\square$

We want to give a measure of the *size* of $\mathbb{Z}_K$. For this we introduce the following definition.

Definition 3.3.12. *Let $\mathcal{O}$ be a free submodule of rank n of K. We define the* discriminant $\mathrm{disc}(\mathcal{O})$ *of $\mathcal{O}$ as the determinant of the $n \times n$ matrix $(\mathrm{Tr}_{K/\mathbb{Q}}(\omega_i \omega_j))_{i,j}$, where $(\omega_i)_{1 \leqslant i \leqslant n}$ is any $\mathbb{Z}$-basis of $\mathcal{O}$. By abuse of language $\mathrm{disc}(\mathbb{Z}_K)$ is simply called the discriminant of K and denoted by $\mathrm{disc}(K)$ or sometimes simply $d(K)$.*

Note that if (ω_i') is another $\mathbb{Z}$-basis of $\mathcal{O}$, the base-change matrix from the ω_i to the ω_i' is an invertible integral matrix P, hence a matrix with determinant equal to ± 1. Since

$$\det(\mathrm{Tr}_{K/\mathbb{Q}}(\omega_i' \omega_j')) = \det(P)^2 \det(\mathrm{Tr}_{K/\mathbb{Q}}(\omega_i \omega_j)) = \det(\mathrm{Tr}_{K/\mathbb{Q}}(\omega_i \omega_j)) ,$$

we see that the definition makes sense. Furthermore, since $\omega_i \omega_j$ is an algebraic integer, its trace is integral, hence $\mathrm{disc}(\mathcal{O}) \in \mathbb{Z}$, and since the ω_i are linearly independent, we have $\mathrm{disc}(\mathcal{O}) \neq 0$. Finally, using the same argument as above it is clear that if $[\mathbb{Z}_K : \mathcal{O}] = f$ then $\mathrm{disc}(\mathcal{O}) = \mathrm{disc}(K) f^2$. For instance, if α is an algebraic *integer* with minimal monic polynomial $A(X)$, then $(1, \alpha, \dots, \alpha^{n-1})$ is a $\mathbb{Z}$-basis of $\mathbb{Z}[\alpha]$, and it is easily checked that $\mathrm{disc}(\mathbb{Z}[\alpha])$ is equal to the discriminant of the polynomial $A(X)$ in the usual sense, so that $\mathrm{disc}(A) = \mathrm{disc}(K) f^2$ with $f = [\mathbb{Z}_K : \mathbb{Z}[\alpha]]$.

It is important to generalize the above notions to the case of a relative extension L/K of number fields. The main difficulty here is that $\mathbb{Z}_K$ is not necessarily a principal ideal domain (PID). It is always, however, a Dedekind domain, and the structure of finitely generated torsion-free modules over Dedekind domains is almost as nice as that over a PID. We state the following only for an extension L/K, but it is true more generally.

Proposition 3.3.13. *Let I be a fractional ideal of L.*

(1) *There exist elements $\omega_i \in I$ and fractional ideals $\mathfrak{a}_i$ of K for $1 \leqslant i \leqslant n$ such that $I = \bigoplus_{1 \leqslant i \leqslant n} \mathfrak{a}_i \omega_i$ in an evident sense.*
(2) *The fractional ideal of K defined by*

$$\mathfrak{d}(I) = \det((\mathrm{Tr}_{L/K}(\omega_i \omega_j))_{1 \leqslant i, j \leqslant n}) \prod_{1 \leqslant i \leqslant n} \mathfrak{a}_i^2$$

is independent of the ω_i and $\mathfrak{a}_i$.

Proof. Left to the reader (Exercise 16).   □

When $I = \mathbb{Z}_L$ is the ring of integers of L, we write $\mathfrak{d}(L/K)$ instead of $\mathfrak{d}(\mathbb{Z}_L)$ and call it the *relative discriminant* of L/K. It is clear that it partially generalizes the usual notion of discriminant, in that $\mathfrak{d}(L/\mathbb{Q}) = \mathrm{disc}(L)\mathbb{Z}$.

3.3.4 Ideals and Units

This is the most important part of the section on ring-theoretic properties of number fields, and was in fact historically the main motivation for the creation of algebraic number theory by Kummer. We recall without proof the following basic definitions and results.

Definition 3.3.14. *Let K be a number field. An* ideal *I of $\mathbb{Z}_K$ is a sub-$\mathbb{Z}_K$-module of $\mathbb{Z}_K$; in other words, it is an additive subgroup of $\mathbb{Z}_K$ such that $\alpha x \in I$ for all $\alpha \in \mathbb{Z}_K$ and $x \in I$. By extension, a* fractional *ideal is a nonzero $\mathbb{Z}_K$-module of the form I/d, where I is an ideal and $d \in K^*$. A nonzero ideal will be called an* integral *ideal.*

If I and J are two ideals we can naturally define their sum (as a sum of $\mathbb{Z}_K$-modules), but also their product: if I and J are ideals then IJ is the smallest ideal containing xy for all $x \in I$ and $y \in J$, in other words, the set of finite $\mathbb{Z}$-linear combinations $\sum x_i y_i$ with $x_i \in I$ and $y_i \in J$.

Proposition 3.3.15. *Let K be a number field such that $[K : \mathbb{Q}] = n$.*

(1) *Any fractional ideal is a free $\mathbb{Z}$-module of rank n, and an integral ideal has finite index in $\mathbb{Z}_K$.*
(2) *The set of fractional ideals forms an abelian group under ideal multiplication.*

Definition 3.3.16. *A* prime ideal *$\mathfrak{p}$ of $\mathbb{Z}_K$ is an integral ideal different from $\mathbb{Z}_K$ such that $\mathbb{Z}_K/\mathfrak{p}$ is an integral domain.*

Since $Z_K/\mathfrak{p}$ is finite when $\mathfrak{p} \neq 0$, and since every finite integral domain is a field, it follows that any nonzero prime ideal $\mathfrak{p}$ is a *maximal ideal*. In other words, $\mathfrak{p} \subset I \subset \mathbb{Z}_K$ for an ideal I implies that $I = \mathfrak{p}$ or $I = \mathbb{Z}_K$. Since the zero ideal is always excluded, when we talk of a prime ideal $\mathfrak{p}$ in the context of number fields, we always implicitly assume that $\mathfrak{p}$ is nonzero, in other words, that $\mathfrak{p}$ is a maximal ideal.

The first important theorem concerning ideals in number fields is the existence and uniqueness of prime ideal factorization.

Theorem 3.3.17. *Let I be a fractional ideal of K. There exists a factorization*

$$I = \prod_{i=1}^{g} \mathfrak{p}_i^{v_i} \ ,$$

where the $\mathfrak{p}_i$ are distinct prime ideals and $v_i \in \mathbb{Z} \setminus \{0\}$, and this factorization is unique up to permutation of the factors.

This theorem is one of the most important consequences of the fact that $\mathbb{Z}_K$ is a *Dedekind domain*.

In Kummer's study of Fermat's last theorem, as in many other Diophantine equations, one side of the equation can be factored algebraically, and the other side is a perfect power. If we are in $\mathbb{Z}$, or more generally in a PID, we can conclude by unique factorization that the algebraic factors are themselves perfect powers, at least up to units (more on units below). Unfortunately, most number rings are not PIDs. However, they are always Dedekind domains, and as such by the above theorem they have unique factorization into prime ideals. Thus each of the algebraic factors is a perfect power of an ideal. Thus assume that we know that an ideal $\mathfrak{a}$ is such that $\mathfrak{a}^n = \gamma \mathbb{Z}_K$ for some *element* γ. This is where the second basic theorem on ideals and units of algebraic number theory comes into play.

Theorem 3.3.18 (Finiteness of the class group). *Define two fractional ideals $\mathfrak{a}$ and $\mathfrak{b}$ to be equivalent if there exists $\alpha \in K^*$ such that $\mathfrak{b} = \alpha \mathfrak{a}$. This equivalence relation is compatible with the multiplicative group structure of ideals, and the quotient group is a finite abelian group.*

The group of ideal classes of K is denoted by $Cl(K)$, and the *class number*, in other words $|Cl(K)|$, is denoted by $h(K)$. Standard group theory implies that for any ideal $\mathfrak{a}$ the ideal $\mathfrak{a}^{h(K)}$ has the form $\beta \mathbb{Z}_K$ for some element β of $\mathbb{Z}_K$. Thus if we know that $\mathfrak{a}^n = \gamma \mathbb{Z}_K$, then if n and $h(K)$ are coprime, the extended Euclidean algorithm implies that $\mathfrak{a}$ itself has the form $\mathfrak{a} = \alpha \mathbb{Z}_K$ for some $\alpha \in \mathbb{Z}_K$. It follows that $\alpha^n \mathbb{Z}_K = \gamma \mathbb{Z}_K$. Thus, even though we are not working in a PID, the conclusion is very similar: the principal ideal generated by γ is indeed equal to the nth power of a principal ideal.

This can be refined further. The above equality can be written $\gamma = \varepsilon \alpha^n$, where ε is a *unit* of $\mathbb{Z}_K$, in other words an element of $\mathbb{Z}_K$ such that $\varepsilon^{-1} \in \mathbb{Z}_K$. The group of units of K will be denoted by $U(K)$. We now need the third basic theorem on ideals and units.

Theorem 3.3.19 (Unit group structure). *There exist units $\varepsilon_0, \varepsilon_1, \ldots, \varepsilon_r$ having the following properties:*

(1) *The group of roots of unity in K is the finite cyclic group (of order $w(K)$, say) generated by ε_0.*

(2) *Any unit ε of K can be written in a* unique *way as*

$$\varepsilon = \prod_{0 \leqslant i \leqslant r} \varepsilon_i^{n_i} \, ,$$

with $n_i \in \mathbb{Z}$ for $1 \leqslant i \leqslant r$ and $0 \leqslant n_0 < w(K)$.

The rank of the unit group is thus equal to r, and we have $r = r_1 + r_2 - 1$, where (r_1, r_2) is the signature of K.

Such a family $(\varepsilon_1, \ldots, \varepsilon_r)$ (with ε_0 not included) is called a basis of *fundamental units* of K.

A last easy but important remark concerning roots of unity: if K is not a totally complex number field, in other words if $r_1 > 0$, then $w(K) = 2$, so that the only roots of unity are ± 1, since all the embeddings of all other roots of unity in $\mathbb{C}$ are nonreal.

3.3.5 Decomposition of Primes and Ramification

Definition and Proposition 3.3.20. *Let L/K be an extension of number fields, let $\mathfrak{p}$ be a prime ideal of K, and let*

$$\mathfrak{p}\mathbb{Z}_L = \prod_{i=1}^{g} \mathfrak{P}_i^{e_i}$$

be its prime ideal decomposition in $\mathbb{Z}_L$, where $e_i \geqslant 1$ and the prime ideals $\mathfrak{P}_i$ are above $\mathfrak{p}$.

(1) *The exponent e_i is denoted by $e(\mathfrak{P}_i/\mathfrak{p})$ and is called the* ramification index *of $\mathfrak{P}_i$.*

(2) *The degree of the finite field extension $[\mathbb{Z}_L/\mathfrak{P}_i : \mathbb{Z}_K/\mathfrak{p}]$ is called the* residual degree, *and denoted by $f(\mathfrak{P}_i/\mathfrak{p})$. If $\mathfrak{p} = p\mathbb{Z}$, we call it simply the* degree of $\mathfrak{P}_i$.

(3) *We have the equality $[L : K] = \sum_{1 \leqslant i \leqslant g} e(\mathfrak{P}_i/\mathfrak{p})f(\mathfrak{P}_i/\mathfrak{p})$.*

(4) *We say that $\mathfrak{P}_i$ is ramified if $e(\mathfrak{P}_i/\mathfrak{p}) \geqslant 2$, and we say that $\mathfrak{p}$ itself is ramified if there exists a ramified $\mathfrak{P}_i$ above $\mathfrak{p}$.*

An easy but fundamental result concerning ramification and residual indices is their *transitivity*.

Proposition 3.3.21. *Let M/L and L/K be extensions of number fields, $\mathfrak{p}$ an ideal of K, $\mathfrak{P}_L$ an ideal of L above $\mathfrak{p}$, and $\mathfrak{P}_M$ an ideal of M above $\mathfrak{P}_L$. We have the transitivity relations:*

$$e(\mathfrak{P}_M/\mathfrak{p}) = e(\mathfrak{P}_M/\mathfrak{P}_L)e(\mathfrak{P}_L/\mathfrak{p}) \quad and \quad f(\mathfrak{P}_M/\mathfrak{p}) = f(\mathfrak{P}_M/\mathfrak{P}_L)f(\mathfrak{P}_L/\mathfrak{p}) \, .$$

Proof. Left to the reader (Exercise 18). □

A simple case in which it is easy to obtain explicitly the prime ideal decomposition is the following, which we state only in the absolute case, although it is immediate to generalize to relative extensions; see [Coh1], Proposition 2.3.9.

Proposition 3.3.22. *Let $K = \mathbb{Q}(\theta)$ be a number field, where θ is an algebraic integer, and denote by $T(X)$ its (monic) minimal polynomial. Let f be the index of θ, i.e., $f = [\mathbb{Z}_K : \mathbb{Z}[\theta]]$. Then for any prime p not dividing f we can obtain the prime decomposition of $p\mathbb{Z}_K$ as follows. Let*

$$T(X) \equiv \prod_{i=1}^{g} \overline{T_i}(X)^{e_i} \pmod{p}$$

be the decomposition of T into monic irreducible factors in $\mathbb{F}_p[X]$. Then

$$p\mathbb{Z}_K = \prod_{i=1}^{g} \mathfrak{p}_i^{e_i} ,$$

where

$$\mathfrak{p}_i = (p, T_i(\theta)) = p\mathbb{Z}_K + T_i(\theta)\mathbb{Z}_K ,$$

with T_i any monic lift of $\overline{T_i}$. Furthermore, the residual index f_i is equal to the degree of T_i.

A basic result concerning ramification is the following.

Proposition 3.3.23. *Let L/K be an extension of number fields, $\mathfrak{p}$ a prime ideal of K, and $\mathfrak{P}$ a prime ideal of L above $\mathfrak{p}$. Then $\mathfrak{p}$ is ramified in the extension L/K if and only if it divides the relative discriminant $\mathfrak{d}(L/K)$.*

Note that there exist much more precise results concerning ramification, which will not be needed.

The next lemma is technical but is needed elsewhere.

Lemma 3.3.24. *Let $L = K(\theta)$, where θ is an algebraic integer, and denote by $f = [\mathbb{Z}_L : \mathbb{Z}_K[\theta]]$ the index of $\mathbb{Z}_K[\theta]$ in $\mathbb{Z}_L$. Let $\mathfrak{p}$ be a prime ideal of $\mathbb{Z}_K$ and let $\mathfrak{P}$ be a prime ideal of $\mathbb{Z}_L$ above $\mathfrak{p}$, and assume that there exist x and y in $\mathbb{Z}_K$ such that the ideal $x\mathbb{Z}_K + y\mathbb{Z}_K$ is coprime to $\mathfrak{p}$ and such that $\mathfrak{P} \mid x + y\theta$. Then either $\mathfrak{p} \mid f$ or $f(\mathfrak{P}/\mathfrak{p}) = 1$; in other words, $\mathfrak{P}$ is a prime ideal of relative degree one.*

Please do not confuse f (traditionally denoted in this way because of the German word Führer) with the residual index $f(\mathfrak{P}/\mathfrak{p})$.

Proof. Clearly $\mathfrak{p} \nmid y$; otherwise, $\mathfrak{P} \mid x$, and hence $\mathfrak{p} \mid x$ and $\mathfrak{p} \mid y$, contradicting the fact that the ideal $x\mathbb{Z}_K + y\mathbb{Z}_K$ is coprime to $\mathfrak{p}$. Assume that $\mathfrak{p} \nmid f$, and let y_1 be an inverse of y modulo $\mathfrak{p}$ and f_1 an inverse of f modulo $\mathfrak{p}$. We have $\theta \equiv -xy_1 \pmod{\mathfrak{P}}$. On the other hand, if $\alpha \in \mathbb{Z}_L$, then $f\alpha \in \mathbb{Z}_K[\theta]$, hence there exists a polynomial $P \in \mathbb{Z}_K[X]$ such that $\alpha \equiv f_1 P(-xy_1) \pmod{\mathfrak{P}}$. Thus any element of $\mathbb{Z}_L$ is congruent modulo $\mathfrak{P}$ to an element of $\mathbb{Z}_K$; hence the natural injection from $\mathbb{Z}_K/\mathfrak{p}$ to $\mathbb{Z}_L/\mathfrak{P}$ is an isomorphism, proving that $f(\mathfrak{P}/\mathfrak{p}) = 1$. $\qquad\square$

Note that it is immediate to replace the condition $\mathfrak{p} \mid f$ by $\mathfrak{p} \mid \mathfrak{f}$, where $\mathfrak{f}$ is the *index-ideal* of $\mathbb{Z}_K[\theta]$ in $\mathbb{Z}_L$; see Definition 1.2.33 of [Coh1].

3.3.6 Galois Properties of Prime Decomposition

If the extension L/K is Galois, the decomposition of a prime ideal in an extension is as regular as we can dream of.

Proposition 3.3.25. *Let L/K be a Galois extension, let $G = \mathrm{Gal}(L/K)$, and let $\mathfrak{p}$ be a prime ideal of $\mathbb{Z}_K$. The decomposition of $\mathfrak{p}$ in L is given by $\mathfrak{p}\mathbb{Z}_L = \prod_{1 \leqslant i \leqslant g} \mathfrak{P}_i^e$, where all the ramification indices $e(\mathfrak{P}_i/\mathfrak{p})$ are equal to a single integer e, all the residual degrees $f(\mathfrak{P}_i/\mathfrak{p})$ are equal to a single integer f, with $efg = n = [L : K]$. Furthermore, the action of G on the ideals $\mathfrak{P}_i$ above $\mathfrak{p}$ is transitive. In other words, for any i and j there exists $\sigma \in G$ such that $\mathfrak{P}_j = \sigma(\mathfrak{P}_i)$.*

Proof. Left to the reader (Exercise 18). $\qquad\square$

Definition 3.3.26. *Keep the same notation. If $\mathfrak{P}$ is a prime ideal of L above $\mathfrak{p}$ the decomposition group $D(\mathfrak{P}/\mathfrak{p})$ is the group of $\sigma \in G$ such that $\sigma(\mathfrak{P}) = \mathfrak{P}$. The inertia group $I(\mathfrak{P}/\mathfrak{p})$ is the group of $\sigma \in G$ such that $\sigma(x) \equiv x \pmod{\mathfrak{P}}$ for all $x \in \mathbb{Z}_L$.*

It is clear that $I(\mathfrak{P}/\mathfrak{p}) \subset D(\mathfrak{P}/\mathfrak{p}) \subset G$.

Proposition 3.3.27. *We have $|D(\mathfrak{P}/p)| = e(\mathfrak{P}/\mathfrak{p})f(\mathfrak{P}/\mathfrak{p})$, $|I(\mathfrak{P}/p)| = e(\mathfrak{P}/\mathfrak{p})$, and $D(\mathfrak{P}/\mathfrak{p})/I(\mathfrak{P}/\mathfrak{p})$ is canonically isomorphic to the (cyclic) Galois group of the finite field extension $\mathrm{Gal}((\mathbb{Z}_L/\mathfrak{P})/(\mathbb{Z}_K/\mathfrak{p}))$.*

Proof. Left to the reader (Exercise 18). $\qquad\square$

Note in particular that $\mathfrak{p}$ is unramified if and only if $I(\mathfrak{P}/\mathfrak{p})$ is the trivial group for one (or for all) prime ideal(s) $\mathfrak{P}$ above $\mathfrak{p}$.

We will also need the following results dealing with prime decomposition in Galois extensions.

Lemma 3.3.28. *Let K_1/K and K_2/K be two extensions of number fields, and let $\mathfrak{p}$ be a prime ideal of $\mathbb{Z}_K$. Denote by $\mathfrak{P}$ a prime ideal of the compositum K_1K_2 above $\mathfrak{p}$, and by $\mathfrak{p}_1$ and $\mathfrak{p}_2$ the prime ideals below $\mathfrak{P}$ in K_1 and K_2 respectively (which are also above $\mathfrak{p}$). Assume that K_1/K is a Galois extension.*

Then K_1K_2/K_2 is a Galois extension, and the restriction map to K_1 induces an injective group homomorphism from $\mathrm{Gal}(K_1K_2/K_2)$ to $\mathrm{Gal}(K_1/K)$, from $D(\mathfrak{P}/\mathfrak{p}_2)$ to $D(\mathfrak{p}_1/\mathfrak{p})$, and from $I(\mathfrak{P}/\mathfrak{p}_2)$ to $I(\mathfrak{p}_1/\mathfrak{p})$. In particular, $e(\mathfrak{P}/\mathfrak{p}_2)$ divides $e(\mathfrak{p}_1/\mathfrak{p})$, so that if $\mathfrak{p}$ is unramified in K_1/K then all the prime ideals above $\mathfrak{p}$ in K_2 are unramified in K_1K_2/K_2.

Proof. Let σ be a K_2-embedding of K_1K_2 into $\mathbb{C}$, so that σ is the identity on K_2, and $\sigma|_{K_1}$ is a K-embedding of K_1 into $\mathbb{C}$. Since K_1/K is Galois, $\sigma(K_1) \subset K_1$, and since $\sigma(K_2) = K_2$ we have $\sigma(K_1K_2) \subset K_1K_2$, so that K_1K_2/K_2 is a Galois extension. The map $\sigma \mapsto \sigma|_{K_1}$ from $\mathrm{Gal}(K_1K_2/K_2)$ is evidently a group homomorphism. If σ belongs to the kernel of this map then σ is the identity on K_2 and on K_1, hence on K_1K_2, proving that the map is injective. We have thus shown that the restriction to K_1 map gives an injection from $\mathrm{Gal}(K_1K_2/K_2)$ to $\mathrm{Gal}(K_1/K)$.

Now let $\sigma \in D(\mathfrak{P}/\mathfrak{p}_2)$ (the decomposition group of $\mathfrak{P}/\mathfrak{p}_2$), so that $\sigma(\mathfrak{P}) \subset \mathfrak{P}$. It follows that

$$\sigma(\mathfrak{p}_1) = \sigma(\mathfrak{P} \cap K_1) \subset \sigma(\mathfrak{P}) \cap \sigma(K_1) \subset \mathfrak{P} \cap K_1 = \mathfrak{p}_1 \,,$$

so that $\sigma|_{K_1} \in D(\mathfrak{p}_1/\mathfrak{p})$. Thus our injective map restricts to an injective map from $D(\mathfrak{P}/\mathfrak{p}_2)$ to $D(\mathfrak{p}_1/\mathfrak{p})$.

Similarly, let $\sigma \in I(\mathfrak{P}/\mathfrak{p}_2)$ (the inertia group of $\mathfrak{P}/\mathfrak{p}_2$), so that $\sigma(x) \equiv x$ (mod $\mathfrak{P}$) for all $x \in \mathbb{Z}_{K_1K_2}$. In particular, if $x \in \mathbb{Z}_{K_1} \subset \mathbb{Z}_{K_1K_2}$ we have $\sigma(x) - x \in \mathfrak{P} \cap K_1 = \mathfrak{p}_1$, so that $\sigma|_{K_1} \in I(\mathfrak{p}_1/\mathfrak{p})$. Thus our injective map restricts also to an injective map from $I(\mathfrak{P}/\mathfrak{p}_2)$ to $I(\mathfrak{p}_1/\mathfrak{p})$. In particular, $e(\mathfrak{P}/\mathfrak{p}_2) = |I(\mathfrak{P}/\mathfrak{p}_2)|$ divides $e(\mathfrak{p}_1/\mathfrak{p}) = |I(\mathfrak{p}_1/\mathfrak{p})|$ as claimed. $\square$

Let $\mathfrak{p}_L$ be a prime ideal above $\mathfrak{p}$. By abuse of language, we will say that $\mathfrak{p}_L/\mathfrak{p}$ is *totally split* if $e(\mathfrak{p}_L/\mathfrak{p}) = f(\mathfrak{p}_L/\mathfrak{p}) = 1$. This corresponds to the usual definition when L/K is Galois.

Lemma 3.3.29. *Let N/K be a Galois extension of number fields with $G = \mathrm{Gal}(N/K)$, let H be a (not necessarily normal) subgroup of G, and let $L = N^H$ be the corresponding subfield, so that $\mathrm{Gal}(N/L) = H$ by Galois theory. Let $\mathfrak{p}$ be a prime ideal of K, $\mathfrak{p}_L$ a prime ideal of L above $\mathfrak{p}$, and $\mathfrak{P}$ a prime ideal of N above $\mathfrak{p}_L$. Then*

$$e(\mathfrak{p}_L/\mathfrak{p}) = [I(\mathfrak{P}/\mathfrak{p}) : I(\mathfrak{P}/\mathfrak{p}) \cap H] \quad and$$

$$e(\mathfrak{p}_L/\mathfrak{p})f(\mathfrak{p}_L/\mathfrak{p}) = [D(\mathfrak{P}/\mathfrak{p}) : D(\mathfrak{P}/\mathfrak{p}) \cap H] \,.$$

In particular, $\mathfrak{p}_L/\mathfrak{p}$ is unramified if and only if $I(\mathfrak{P}/\mathfrak{p}) \subset H$, and $\mathfrak{p}_L/\mathfrak{p}$ is totally split if and only if $D(\mathfrak{P}/\mathfrak{p}) \subset H$.

Proof. By transitivity of ramification and residual indices we have

$$|I(\mathfrak{P}/\mathfrak{p})| = e(\mathfrak{P}/\mathfrak{p}) = e(\mathfrak{p}_L/\mathfrak{p})e(\mathfrak{P}/\mathfrak{p}_L) = e(\mathfrak{p}_L/\mathfrak{p})|I(\mathfrak{P}/\mathfrak{p}_L)|$$

and similarly

$$|D(\mathfrak{P}/\mathfrak{p})| = e(\mathfrak{P}/\mathfrak{p})f(\mathfrak{P}/\mathfrak{p}) = e(\mathfrak{p}_L/\mathfrak{p})f(\mathfrak{p}_L/\mathfrak{p})e(\mathfrak{P}/\mathfrak{p}_L)f(\mathfrak{P}/\mathfrak{p}_L)$$
$$= e(\mathfrak{p}_L/\mathfrak{p})f(\mathfrak{p}_L/\mathfrak{p})|D(\mathfrak{P}/\mathfrak{p}_L)| \ .$$

However, we have by definition $I(\mathfrak{P}/\mathfrak{p}_L) = I(\mathfrak{P}/\mathfrak{p}) \cap H$ and $D(\mathfrak{P}/\mathfrak{p}_L) = D(\mathfrak{P}/\mathfrak{p}) \cap H$, so the lemma follows. $\square$

Lemma 3.3.30. *Let K_1/K and K_2/K be two extensions of number fields, and let $\mathfrak{p}$ be a prime ideal of $\mathbb{Z}_K$. Denote by $\mathfrak{P}$ a prime ideal of the compositum K_1K_2 above $\mathfrak{p}$, and by $\mathfrak{p}_1$ and $\mathfrak{p}_2$ the prime ideals below $\mathfrak{P}$ in K_1 and K_2 respectively (which are also above $\mathfrak{p}$).*

(1) *If $\mathfrak{p}_1/\mathfrak{p}$ is unramified then $e(\mathfrak{P}/\mathfrak{p}) = e(\mathfrak{p}_2/\mathfrak{p})$. In particular, if $\mathfrak{p}_1/\mathfrak{p}$ and $\mathfrak{p}_2/\mathfrak{p}$ are unramified then $\mathfrak{P}/\mathfrak{p}$ is unramified.*
(2) *If $\mathfrak{p}_1/\mathfrak{p}$ is totally split then $f(\mathfrak{P}/\mathfrak{p}) = f(\mathfrak{p}_2/\mathfrak{p})$. In particular, if $\mathfrak{p}_1/\mathfrak{p}$ and $\mathfrak{p}_2/\mathfrak{p}$ are totally split then $\mathfrak{P}/\mathfrak{p}$ is totally split.*

Proof. Denote by N the Galois closure of K_1K_2/K, set $H_i = \mathrm{Gal}(N/K_i)$ for $i = 1, 2$, so that $\mathrm{Gal}(N/K_1K_2) = H_1 \cap H_2$, and let $\mathfrak{P}_N$ be a prime ideal of N above $\mathfrak{P}$.

(1). By Lemma 3.3.29 we have $I(\mathfrak{P}_N/\mathfrak{p}) \subset H_1$, so that

$$I(\mathfrak{P}_N/\mathfrak{p}) \cap (H_1 \cap H_2) = (I(\mathfrak{P}_N/\mathfrak{p}) \cap H_1) \cap H_2 = I(\mathfrak{P}_N/\mathfrak{p}) \cap H_2 \ .$$

Thus, once again by Lemma 3.3.29 we have

$$e(\mathfrak{P}/\mathfrak{p}) = [I(\mathfrak{P}_N/\mathfrak{p}) : I(\mathfrak{P}_N/\mathfrak{p}) \cap (H_1 \cap H_2)]$$
$$= [I(\mathfrak{P}_N/\mathfrak{p}) : I(\mathfrak{P}_N/\mathfrak{p}) \cap H_2] = e(\mathfrak{p}_2/\mathfrak{p}) \ ,$$

proving (1).

(2). Similarly, here we have $D(\mathfrak{P}_N/\mathfrak{p}) \subset H_1$, so that the same reasoning gives $e(\mathfrak{P}/\mathfrak{p})f(\mathfrak{P}/\mathfrak{p}) = e(\mathfrak{p}_2/\mathfrak{p})f(\mathfrak{p}_2/\mathfrak{p})$. By (1) we already know that $e(\mathfrak{P}/\mathfrak{p}) = e(\mathfrak{p}_2/\mathfrak{p})$, so the lemma follows. $\square$

3.4 Quadratic Fields

3.4.1 Field-Theoretic and Basic Ring-Theoretic Properties

By definition a quadratic field K is an extension of degree 2 of $\mathbb{Q}$, hence by Kummer theory (or trivially directly), of the form $K = \mathbb{Q}(\sqrt{d})$, where d is a squarefree integer different from 1. The extension $K/\mathbb{Q}$ is clearly Galois with Galois group isomorphic to $\mathbb{Z}/2\mathbb{Z}$, generated by σ such that $\sigma(a + b\sqrt{d}) = a - b\sqrt{d}$ for a, b in $\mathbb{Q}$. When $d < 0$ we say that K is an imaginary quadratic field, while when $d > 0$ we naturally say that K is a real quadratic field. Note that in the imaginary case σ is complex conjugation, but in the real case σ is a highly nontrivial map from a topological standpoint.

The field-theoretic properties of quadratic fields are thus very simple. We now recall the basic ring-theoretic properties.

Proposition 3.4.1. *Let $K = \mathbb{Q}(\sqrt{d})$ be a quadratic field with d squarefree, and let $\mathbb{Z}_K$ be its ring of integers. A $\mathbb{Z}$-basis of $\mathbb{Z}_K$ is given by $(1, \omega)$, where $\omega = \sqrt{d}$ when $d \equiv 2$ or 3 modulo 4, while $\omega = (1+\sqrt{d})/2$ when $d \equiv 1 \pmod 4$. In the first case the discriminant D of K is equal to $4d$; in the second case it is equal to d.*

Proof. Although the proof is simple, it is not completely trivial, and the reader who has never seen it should try it for himself before reading on. Let $\alpha = a + b\sqrt{d} \in \mathbb{Z}_K$ with a, b in $\mathbb{Q}$. The characteristic polynomial of α is equal to $(X - a)^2 - b^2 d = X^2 - 2aX + a^2 - b^2 d$; hence $\alpha \in \mathbb{Z}_K$ if and only if $2a \in \mathbb{Z}$ and $a^2 - b^2 d \in \mathbb{Z}$. If we write $2b = p/q$ with p and q coprime integers we thus have $q^2 \mid q^2((2a)^2 - 4(a^2 - b^2 d)) = p^2 d$, and since q and p are coprime we have $q^2 \mid d$, so that $q = \pm 1$ since d is squarefree, showing that we also have $2b \in \mathbb{Z}$. We can thus write $a = A/2$ and $b = B/2$ with A and B in $\mathbb{Z}$, satisfying automatically the condition $2a \in \mathbb{Z}$, and the condition $a^2 - b^2 d \in \mathbb{Z}$ gives $A^2 - B^2 d \equiv 0 \pmod 4$, which is immediately seen to imply either that A and B are both even, or that they are both odd when $d \equiv 1 \pmod 4$, proving the first statement. The second follows from Definition 3.3.12 applied to the basis $(1, \omega)$. $\qquad\square$

Definition 3.4.2. *An integer D is called a* fundamental discriminant *if D is either equal to 1 or to the discriminant of a quadratic field, in other words, if either $D \equiv 1 \pmod 4$ is squarefree, or $D \equiv 8$ or 12 modulo 16 and $D/4$ is squarefree.*

We will often exclude $D = 1$ from the set of fundamental discriminants, but this will be explicitly mentioned.

Remark. Thanks to the above proposition, it is clear that if D is the discriminant of K we have $K = \mathbb{Q}(\sqrt{D})$ and $\mathbb{Z}_K = \mathbb{Z}[\omega]$ with $\omega = (D + \sqrt{D})/2$, or more generally, $\omega = (\delta + \sqrt{D})/2$ for any integer δ such that $\delta \equiv D \pmod 2$. This notation has the advantage of being completely uniform, and will therefore be systematically used, instead of notation using the squarefree integer d.

Proposition 3.4.3. *Let $K = \mathbb{Q}(\sqrt{D})$ be a quadratic field of discriminant D, set $\omega = (D + \sqrt{D})/2$, and let p be a prime number. The prime ideal decomposition of $p\mathbb{Z}_K$ is given as follows.*

(1) *If $p \mid D$ then p is ramified; in other words, $p\mathbb{Z}_K = \mathfrak{p}^2$, and we have $\mathfrak{p} = p\mathbb{Z}_K + \omega\mathbb{Z}_K$ except when $p = 2$ and $D \equiv 12 \pmod{16}$, in which case $\mathfrak{p} = p\mathbb{Z}_K + (\omega + 1)\mathbb{Z}_K$.*

(2) *If p is odd and $\left(\frac{D}{p}\right) = -1$, or $p = 2$ and $D \equiv 5 \pmod 8$, then p is inert; in other words, $p\mathbb{Z}_K$ is a prime ideal of K.*

(3) *If p is odd and $\left(\frac{D}{p}\right) = 1$, or $p = 2$ and $D \equiv 1 \pmod 8$, then p is split; in other words, $p\mathbb{Z}_K = \mathfrak{p}_+\mathfrak{p}_-$, where $\mathfrak{p}_+$ and $\mathfrak{p}_-$ are distinct prime ideals given by $\mathfrak{p}_\pm = p\mathbb{Z}_K + (\omega - (D \pm b)/2)\mathbb{Z}_K$, where b is any solution to the congruence $b^2 \equiv D \pmod{4p}$.*

Proof. Since $\mathbb{Z}_K = \mathbb{Z}[\omega]$, this follows immediately from Proposition 3.3.22 and is left to the reader (Exercise 21). □

3.4.2 Results and Conjectures on Class and Unit Groups

The unit group of quadratic fields is also easily described as follows.

Proposition 3.4.4. *Let K be a quadratic field of discriminant D, and let U_K be its unit group, in other words, the group of invertible elements of $\mathbb{Z}_K$.*

(1) *If $D < -4$ we have $U_K = \{\pm 1\}$, hence $|U_K| = 2$.*

(2) *If $D = -4$ we have $U_K = \{\pm 1, \pm i\} = \{i^k / \ 0 \leqslant k \leqslant 3\}$ (where $i^2 = -1$), hence $|U_K| = 4$.*

(3) *If $D = -3$ we have $U_K = \{\pm 1, \pm \rho, \pm \rho^2\} = \{(-\rho)^k / \ 0 \leqslant k \leqslant 5\}$ (where $\rho = (-1 + \sqrt{-3})/2$ is a primitive cube root of unity), hence $|U_K| = 6$.*

(4) *If $D > 0$ there exists $\varepsilon \in U_K$ (called a fundamental unit) such that $U_K = \{\pm 1\} \times \varepsilon^{\mathbb{Z}}$, in other words, such that any $\eta \in U_K$ can be written in a unique way as $\eta = \pm \varepsilon^k$ for some sign $\pm$ and some $k \in \mathbb{Z}$. In addition, ε is unique up to change of sign and change of ε into ε^{-1}.*

Let ε_D be a fundamental unit of the real quadratic field $K = \mathbb{Q}(\sqrt{D})$. By definition the *regulator* $R(D)$ of K is $|\log(|\varepsilon|)|$, which does not depend on the choice of ε. Denote by $h(D)$ the class number of K, and finally by $w(D)$ the cardinality of the torsion group of U_K (hence equal to 2, 4, 6, or 2 in the four cases of the proposition). The *Dirichlet class number formula*, which we do not prove, is the following.

Proposition 3.4.5. *We have*

$$\frac{h(D)R(D)}{w(D)} = \frac{|D|^{1/2}}{c_{\text{sign}(D)}} \sum_{n \geqslant 1} \frac{\left(\frac{D}{n}\right)}{n} = \frac{|D|^{1/2}}{c_{\text{sign}(D)}} L\left(\left(\frac{D}{\cdot}\right), 1\right),$$

where $c_+ = 2$ and $c_- = \pi$, where by convention we set $R(D) = 1$ if $D < 0$, and where $L\left(\left(\frac{D}{\cdot}\right), 1\right)$ is the value at 1 of the L-series associated with the Legendre–Kronecker character $\left(\frac{D}{n}\right)$.

We will study in detail L-series in Chapter 10, and in particular we will see that $L\left(\left(\frac{D}{\cdot}\right), 1\right) = O(\log(|D|))$. On the other hand, it is immediate to see from the definition that $R(D) > \log(D)/2 + o(1)$ for $D > 0$. Since $h(D)$ is an integer, these inequalities imply that $h(D) = O(|D|^{1/2}\log(|D|))$ and that

$R(D) = O(D^{1/2} \log(D))$. Apart from the factor $\log(|D|)$, these upper bounds are close to best possible. However, *lower bounds* are much more difficult to obtain, apart from the trivial lower bound $R(D) > \log(D)/2 + o(1)$. There are many conjectures and a few results on these subjects, some of them quite deep.

Concerning the unit group, and in particular the size of the regulator, the main conjecture, which is at present totally out of reach, is as follows.

Conjecture 3.4.6. *For any $\alpha < 1/2$ there exists an infinite sequence of real quadratic fields $\mathbb{Q}(\sqrt{D_n})$ such that $R(D_n) > cD_n^\alpha$.*

In fact, even $\alpha = 1/2$ may be possible. As far as the author is aware, the best (very weak) result in this direction is due to Yamamoto; see [Yam]:

Theorem 3.4.7. *There exists an infinite sequence of real quadratic fields $\mathbb{Q}(\sqrt{D_n})$ such that $R(D_n) > c\log(D_n)^3$ for some strictly positive constant c.*

Concerning the size of the class group, the situation is evidently quite different for imaginary and real quadratic fields, since in the latter case the size of $R(D)$ comes into play (recall that the quantity that plays a common role is $h(D)R(D)$). In the imaginary quadratic case, the situation is "almost" proved, thanks to the following theorem of Siegel, which is the analogue (and stronger) version of the above conjecture for $R(D)$ for real quadratic fields.

Theorem 3.4.8 (Siegel). *For any $\alpha < 1/2$ there exists a constant $c_\alpha > 0$ such that for all quadratic fields of discriminant D we have $h(D) > c_\alpha|D|^\alpha$ for imaginary quadratic fields and $h(D)R(D) > c_\alpha D^\alpha$ for real quadratic fields.*

This essentially solves the problem for imaginary quadratic fields, since it can be shown that it cannot be true that $h(D) > c_{1/2}|D|^{1/2}$ for some $c_{1/2} > 0$, although this is almost certainly true for an infinite sequence of D.

Unfortunately there is a serious drawback in Siegel's theorem, which is the notion of "ineffectivity." Intrinsically, it is impossible using Siegel's method of proof to find an explicit value for c_α for *any* $\alpha > 0$ or, for that matter, to find any explicit lower bound for $h(D)$ tending to infinity with $|D|$. This latter problem had to wait for much deeper methods of Gross–Zagier, building on previous work of Goldfeld, to indeed obtain an explicit bound $h(D) > f(D)$, where $f(D)$ tends explicitly to infinity (slowly) with D.

Remarks. (1) We will use explicitly the Gross–Zagier formula in Section 8.6 for computing Heegner points.
(2) Although the Goldfeld–Gross–Zagier lower bound is explicit, it is still necessary to do quite a lot of work to find, for instance, all imaginary quadratic fields with $h(D) \leqslant B$ for some given bound B. This has been done for $B = 100$ by M. Watkins.

For real quadratic fields, the most important conjecture, also totally out of reach at present, is certainly the following.

Conjecture 3.4.9. *There exist an infinite number of real quadratic fields of class number* 1. *More precisely, among the real quadratic fields* $\mathbb{Q}(\sqrt{p})$ *with prime discriminant* $p \equiv 1$ (mod 4) *there should exist a strictly positive proportion of class number* 1.

Thanks to Siegel's theorem, this means that the dominating term in the quantity $h(D)R(D)$ should be $R(D)$, in accordance with the above conjecture on the size of $R(D)$.

Note that it is not even known whether there exists an infinite number of nonisomorphic *number fields* of class number 1, with no limitation on the degree.

Essentially nothing is known concerning the above conjecture. However, the so-called Cohen–Lenstra heuristics give conjectural answers to such questions, and in particular, the positive proportion of class-number-1 real quadratic fields of prime discriminant should be equal to an explicit constant close to 0.75446, in excellent agreement with tables; see [Coh-Len] and [Coh-Mar].

3.5 Cyclotomic Fields

Apart from quadratic fields, probably the most important class of number fields is that of *cyclotomic fields*. There are at least two related reasons for this. The first is that the basic algebraic structure of such fields (ring of integers, discriminant, decomposition of primes) can be completely and easily described, as well as a large part of the unit group. On the other hand, as usual the class group is more mysterious. The second reason for their importance is the *Kronecker–Weber* theorem, which states that every Abelian extension of $\mathbb{Q}$ is a subfield of a cyclotomic field, hence inherits most of its nice properties; see Theorem 3.5.13. We begin by studying in some detail the cyclotomic polynomials $\Phi_n(X)$.

3.5.1 Cyclotomic Polynomials

To study most easily cyclotomic polynomials we need some very elementary properties of simple arithmetic functions such as the Möbius function, Euler's ϕ function and the von Mangoldt function $\Lambda(n)$ that the reader can find for instance at the beginning of Chapter 10.

Definition 3.5.1. *We define the cyclotomic polynomials* $\Phi_n(X)$ *as the unique rational functions satisfying*

$$\prod_{d|n} \Phi_d(X) = X^n - 1 \,,$$

in other words by $\Phi_1(X) = X - 1$ *and the induction* $\Phi_n(X) = (X^n - 1)/\prod_{d|n,\ 1 \leqslant d < n} \Phi_d(X)$.

Proposition 3.5.2. (1) *We have*

$$\Phi_n(X) = \prod_{d|n}(X^d - 1)^{\mu(n/d)} \,,$$

where μ is the Möbius function (see Definition 10.1.4).

(2) *Denote by U_n the group of nth roots of unity in $\mathbb{C}$ and by U_n' the subset of primitive nth roots of unity in $\mathbb{C}$, in other words, the numbers ζ_n^a for $0 \leqslant a < n$, $\gcd(a, n) = 1$. We then have*

$$\Phi_n(X) = \prod_{\zeta \in U_n'}(X - \zeta) \,.$$

(3) *$\Phi_n(X)$ is a monic polynomial in $\mathbb{Z}[X]$ of degree $\phi(n)$, where ϕ is Euler's function.*

Proof. (1). Let us show that the right-hand side satisfies the defining property of Φ_n: it is equal to $X - 1$ for $n = 1$, and otherwise

$$\prod_{d|n}\prod_{e|d}(X^e - 1)^{\mu(d/e)} = \prod_{e|n}(X^e - 1)^{S(e)} \,,$$

where

$$S(e) = \sum_{d,\ e|d|n} \mu(d/e) = \sum_{k|n/e} \mu(k) \,,$$

which by definition of μ is equal to 0 unless $n/e = 1$, in which case it is equal to 1. Note that (1) is a special case of the Möbius inversion formula (Proposition 10.1.5).

(2) and (3). Let us define $f_n(X) = \prod_{\zeta \in U_n'}(X - \zeta)$. Since $\zeta \in U_n$ is a primitive dth root of unity for a unique $d \mid n$, it is clear that $\prod_{d|n} f_d(X) = X^n - 1$; hence $f_n(X) = \Phi_n(X)$. It follows that $\Phi_n(X)$ is a polynomial (this was not clear from the inductive definition), that it is monic, and that $\deg(\Phi_n) = |U_n'| = \phi(n)$, since it is the number of $k \in [0, n-1]$ such that $\gcd(k, n) = 1$. Finally, from (1) we see that $\Phi_n(X)$ is the quotient of two monic polynomials with integer coefficients, and since we now know that it is a polynomial, it also has integer coefficients. □

Remark. We will prove later (Proposition 3.5.10 or Corollary 4.1.36) that the polynomials $\Phi_n(X)$ are irreducible in $\mathbb{Q}[X]$.

Lemma 3.5.3. *Let $m \in \mathbb{Z}_{\geqslant 1}$. For all $d \mid m$ we have $\Phi_m(X) \mid \Phi_{m/d}(X^d)$.*

Proof. For any a we have $X^d - a^d = \prod_{k \bmod d}(X - \zeta_d^k a)$, so applying this to $a = \zeta_m^i$, we deduce that

$$X^d - \zeta_{m/d}^i = \prod_{k \bmod d}(X - \zeta_m^{k(m/d)+i}) = \prod_{\substack{j \bmod m \\ j \equiv i \ (\bmod\ m/d)}}(X - \zeta_m^j) \,,$$

so that by definition of $\Phi_{m/d}$ we have

$$\Phi_{m/d}(X^d) = \prod_{\substack{j \bmod m \\ \gcd(j,m/d)=1}} (X - \zeta_m^j) \,.$$

Since, on the other hand,

$$\Phi_m(X) = \prod_{\substack{j \bmod m \\ \gcd(j,m)=1}} (X - \zeta_m^j) \,,$$

the lemma follows.   □

Proposition 3.5.4. *For all $n > 1$ we have*

$$\Phi_n(1) = \prod_{a \in (\mathbb{Z}/n\mathbb{Z})^*} (1 - \zeta_n^a) = \begin{cases} p & \text{if } n = p^k \,, \\ 1 & \text{otherwise.} \end{cases}$$

Proof. By definition of the Möbius function (Definition 10.1.4), for $n > 1$ we have $\sum_{d|n} \mu(d) = \sum_{d|n} \mu(n/d) = 0$. It follows from Proposition 3.5.2 (1) that $\Phi_n(X) = \prod_{d|n}((X^d - 1)/(X - 1))^{\mu(n/d)}$, so that

$$\Phi_n(1) = \prod_{d|n} d^{\mu(n/d)} \,.$$

It follows from Proposition 10.1.14 that $\log(\Phi_n(1)) = \sum_{d|n} \mu(n/d) \log(d) = \Lambda(n)$, where Λ is the von Mangoldt function, so the proposition follows from the explicit formula for $\Lambda(n)$ also given by Proposition 10.1.14.   □

Proposition 3.5.5. (1) *Let a and b be integers coprime to n. The elements $C_{a,b} = (1 - \zeta_n^a)/(1 - \zeta_n^b)$ are units, called* cyclotomic units.
(2) *If $n = p^k$ is a power of a prime p, there exists a unit u such that $(1 - \zeta_n)^{\phi(p^k)} = up$.*
(3) *If n is not a prime power then $1 - \zeta_n$ is a unit.*

Proof. (1). Since a and b are coprime to n there exists c such that $a \equiv bc$ (mod n). It follows that

$$C_{a,b} = \frac{1 - \zeta_n^{bc}}{1 - \zeta_n^b} = \sum_{0 \leqslant j < c} \zeta_n^{bj}$$

is an algebraic integer, and similarly $1/C_{a,b} = C_{b,a}$ is an algebraic integer, proving that $C_{a,b}$ is a unit.

(2). If $n = p^k$, by Proposition 3.5.4 we have

$$p = \prod_{a\in(\mathbb{Z}/n\mathbb{Z})^*} (1 - \zeta_n)C_{a,1} = (1 - \zeta_n)^{\phi(n)}/u ,$$

where $1/u = \prod_{a\in(\mathbb{Z}/n\mathbb{Z})^*} C_{a,1}$, and u is a unit by (1).

(3). By Proposition 3.5.4, if n is not a prime power we have $\prod_{a\in(\mathbb{Z}/n\mathbb{Z})^*}(1-\zeta_n^a) = \Phi_n(1) = 1$; hence all the $1 - \zeta_n^a$ divide 1, hence are units. $\quad\square$

For the next result, we recall the definition of the *resultant* of two polynomials.

Definition 3.5.6. *Let K be a commutative field, let A and B be two polynomials in $K[X]$ with respective leading terms a and b, and let α_i and β_j be the roots of A and B in some algebraic closure of K, repeated with multiplicity. We define the* resultant *$R(A, B)$ of A and B by one of the following equivalent formulas:*

$$R(A, B) = a^{\deg(B)} \prod_i B(\alpha_i) = (-1)^{\deg(A)\deg(B)} b^{\deg(A)} \prod_j A(\beta_j)$$
$$= a^{\deg(B)} b^{\deg(A)} \prod_{i,\,j} (\alpha_i - \beta_j) .$$

It is clear that $R(A, B) = 0$ if and only if A and B have a common root, hence if and only if $\gcd(A, B)$ is not constant. Furthermore, the resultant is clearly multiplicative in A and B, in other words, for instance $R(A_1 A_2, B) = R(A_1, B)R(A_2, B)$. We give without proof the following slightly less trivial proposition.

Proposition 3.5.7. *Let $\mathcal{O}$ be a subring of K, and assume that A and B are in $\mathcal{O}[X]$.*

(1) *We have $R(A, B) \in \mathcal{O}$.*
(2) *There exist polynomials $U(X)$ and $V(X)$ in $\mathcal{O}[X]$ such that*

$$U(X)A(X) + V(X)B(X) = R(A, B) .$$

Note that the second statement of the proposition does not simply follow from the extended Euclidean algorithm.

Proposition 3.5.8. *Let $1 \leqslant m < n$. Then $R(\Phi_m, \Phi_n) = 1$, unless $m \mid n$ and $n/m = p^a$ is a power of a prime p, in which case $R(\Phi_m, \Phi_n) = p^{\phi(m)}$.*

Proof. Assume first that $\gcd(m, n) = 1$ and that $m > 1$. Thus $\Phi_m(X) \mid (X^m - 1)/(X - 1)$, so by multiplicativity of the resultant $R(\Phi_m, \Phi_n) \mid R(X^m - 1, \Phi_n)/R(X - 1, \Phi_n)$. By definition of the resultant we have

$$R(X^m - 1, \Phi_n) = \pm \mathcal{N}_{\mathbb{Q}(\zeta_n)/\mathbb{Q}}(\zeta_n^m - 1) .$$

Since m and n are coprime, $\zeta_n^m - 1$ is a conjugate of $\zeta_n - 1$, hence

$$R(X^m - 1, \Phi_n) = \pm \mathcal{N}_{\mathbb{Q}(\zeta_n)/\mathbb{Q}}(\zeta_n - 1) = \pm R(X - 1, \Phi_n) \,,$$

proving that $R(\Phi_m, \Phi_n) = \pm 1$ when $\gcd(m, n) = 1$ and $m > 1$.

If $\gcd(m, n) = 1$ and $m = 1$, we simply have $R(\Phi_m, \Phi_n) = R(X - 1, \Phi_n) = \Phi_n(1)$, and by Proposition 3.5.4 this is equal to 1 unless $n = p^k$ for some prime p and $k \geqslant 1$, in which case it is equal to p.

Consider now the general case, and set $d = \gcd(m, n)$. By Lemma 3.5.3 we know that $\Phi_m(X) \mid \Phi_{m/d}(X^d)$ and $\Phi_n(X) \mid \Phi_{n/d}(X^d)$, and since the resultant is multiplicative it follows that $R(\Phi_m, \Phi_n) \mid R(\Phi_{m/d}(X^d), \Phi_{n/d}(X^d))$. In addition, it is clear that we also have $R(\Phi_{m/d}(X^d), \Phi_{n/d}(X^d)) \mid R(\Phi_{m/d}, \Phi_{n/d})^d$. Since $\gcd(m/d, n/d) = 1$, by what we have shown in the coprime case we deduce that $R(\Phi_m, \Phi_n) = 1$, except perhaps when $m/d = 1$ (in other words $m \mid n$), and $n/d = n/m = p^k$ is a prime power. We leave to the reader the proof that in this case $R(\Phi_m, \Phi_n) = \pm p^{\phi(m)}$, and that the signs of the resultants are always positive (see Exercise 22). □

3.5.2 Field-Theoretic Properties of $\mathbb{Q}(\zeta_n)$

Definition 3.5.9. *Let $n \geqslant 1$ and let ζ_n be a primitive nth root of unity. The nth cyclotomic field is the field $K_n = \mathbb{Q}(\zeta_n)$.*

Since $\zeta_{4m+2} = -\zeta_{2m+1}$, we have $K_{4m+2} = K_{2m+1}$, so we will in general assume that $n \not\equiv 2 \pmod 4$. The basic field-theoretic properties of cyclotomic fields are summarized in the following proposition.

Proposition 3.5.10. (1) *The polynomial $\Phi_n(X)$ is irreducible in $\mathbb{Q}[X]$; in other words, $\Phi_n(X)$ is the minimal polynomial of ζ_n over $\mathbb{Q}$, and $[\mathbb{Q}(\zeta_n) : \mathbb{Q}] = \phi(n)$, Euler's phi function.*
(2) *The extension $\mathbb{Q}(\zeta_n)/\mathbb{Q}$ is a Galois extension, the Galois group G being canonically isomorphic to $(\mathbb{Z}/n\mathbb{Z})^*$ by the map that sends $a \in (\mathbb{Z}/n\mathbb{Z})^*$ to the automorphism $\sigma_a \in G$ such that $\sigma_a(\zeta_n) = \zeta_n^a$.*

Proof. (I thank J.-F. Jaulent for the following proof of this very classical result.) Let P be an irreducible factor of Φ_n in $\mathbb{Q}[X]$, let ζ be a root of P, and let p be a prime number such that $p \nmid n$. I claim that ζ^p is also a root of P. Indeed, assume otherwise, and let Q be the minimal monic polynomial of ζ^p over $\mathbb{Q}$, so that $Q \in \mathbb{Z}[X]$. Since ζ^p is not a root of P we have $P \neq Q$; hence the irreducible polynomials P and Q are coprime. On the other hand, $P(X) \mid \Phi_n(X) \mid (X^n - 1)$ by assumption, and since $(\zeta^p)^n = 1$ we also have $Q(X) \mid (X^n - 1)$, hence $P(X)Q(X) \mid (X^n - 1)$ since P and Q are coprime. On the other hand, ζ is a root of $Q(X^p)$, so that $P(X) \mid Q(X^p)$, and using $\overline{}$ to denote reduction modulo p we deduce that $\overline{P}(X) \mid \overline{Q}(X^p) = \overline{Q}(X)^p$, since $\overline{Q} \in \mathbb{F}_p[X]$. In particular, if $\overline{R}$ is an irreducible factor of $\overline{P}$ in $\mathbb{F}_p[X]$ then $\overline{R} \mid \overline{Q}$. But since $P(X)Q(X) \mid (X^n - 1)$ it follows that $\overline{R}^2 \mid X^n - \overline{1}$ in $\mathbb{F}_p[X]$, in other words that $X^n - \overline{1}$ is not coprime to its derivative (i.e., is not separable), which is absurd since its derivative is nX^{n-1} and $p \nmid n$, proving my claim.

By induction on the number of prime divisors of k counted with multiplicity, it follows from my claim that for any k coprime with n, ζ^k is a root of P. Since ζ is a root of Φ_n, hence a primitive nth root of unity, it follows that all primitive nth roots of unity are roots of P, hence that $\Phi_n \mid P$, so that $\Phi_n = P$, proving that Φ_n is irreducible. The other statements of (1) have been proved in passing, and (2) is an immediate consequence and left to the reader.   □

Note that the irreducibility of $\Phi_{p^k}(X)$ follows immediately from the Eisenstein criterion (see Corollary 4.1.36).

Corollary 3.5.11. *There is a one-to-one correspondence between subfields of $\mathbb{Q}(\zeta_n)$ and subgroups of $(\mathbb{Z}/n\mathbb{Z})^*$.*

Proof. This is simply Galois theory.   □

Corollary 3.5.12. *The subgroup of roots of unity of $\mathbb{Q}(\zeta_n)$ is the group of $\pm\zeta_n^i$ for $0 \leqslant i < n$, or equivalently, the subgroup of order $2n$ generated by $\zeta_{2n} = -\zeta_n$ when n is odd, or the subgroup of order n generated by ζ_n when $4 \mid n$.*

Proof. Let ζ_m be an mth root of unity in $\mathbb{Q}(\zeta_n)$. We thus have $\mathbb{Q}(\zeta_m) \subset \mathbb{Q}(\zeta_n)$, hence by the proposition $\phi(m) \mid \phi(n)$. Since $\phi(m)$ tends to infinity with m, it follows that m is bounded as a function of n; hence the group of roots of unity in $\mathbb{Q}(\zeta_n)$ is finite (this is of course true in any number field). By Corollary 2.4.3 it follows that it is a cyclic subgroup $\langle \zeta_m \rangle$ generated by ζ_m for some m. We thus have $\zeta_n \in \langle \zeta_m \rangle$, from which we deduce the following two consequences: first $\langle \zeta_n \rangle$ is a subgroup of $\langle \zeta_m \rangle$, so that $n \mid m$. Second $\mathbb{Q}(\zeta_n) \subset \mathbb{Q}(\zeta_m)$; hence $\mathbb{Q}(\zeta_n) = \mathbb{Q}(\zeta_m)$, so that $\phi(n) = \phi(m)$. If we write $m = \prod_p p^{v_p}$ with $v_p \geqslant 1$ and $n = \prod_p p^{w_p}$ we thus have $w_p \leqslant v_p$ for all p, hence $\phi(p^{w_p}) \leqslant \phi(p^{v_p})$ for all p. Since $\phi(n) = \prod_p \phi(p^{w_p}) = \prod_p \phi(p^{v_p})$ we must have $\phi(p^{w_p}) = \phi(p^{v_p})$ for all p. Since $v_p \geqslant 1$, it is clear that this means that $w_p = v_p$ for $p \geqslant 3$, and that either $w_2 = v_2$, or $w_2 = 0$ and $v_2 = 1$. In the first case $m = n$ with n even (hence $4 \mid n$ since $n \not\equiv 2 \pmod 4$), and in the second case $m = 2n$ with n odd, proving the corollary.   □

A special element of $(\mathbb{Z}/n\mathbb{Z})^*$ is the class of -1. Through the bijection with $\mathrm{Gal}(\mathbb{Q}(\zeta_n)/\mathbb{Q})$ it corresponds to σ_{-1}, which sends ζ to ζ^{-1}; hence it corresponds to *complex conjugation*, which we will denote by ι. The fixed field of ι is thus the *maximal totally real subfield* K^+ of $K = \mathbb{Q}(\zeta_n)$, which is such that $[K : K^+] = 2$. It is easy to see that $K^+ = \mathbb{Q}(\zeta + \zeta^{-1})$. We will study it in more detail in Section 3.5.4.

As mentioned above, one of the main reasons for the importance of cyclotomic fields is the following:

Theorem 3.5.13 (Kronecker–Weber). *Any Abelian extension K of $\mathbb{Q}$ is a subfield of some cyclotomic field $\mathbb{Q}(\zeta_n)$.*

Proof. The reader will find a proof of an essential special case of this theorem in Exercises 28 and 29, from which it is not difficult to deduce the general statement.   □

Remarks. (1) The smallest n having this property is called the *conductor* of the field K.

(2) The corresponding theorem for Abelian extensions of an imaginary quadratic field is known, and corresponds to the theory of complex multiplication (the exponential function, which can be used to define cyclotomic fields, is replaced by the Weierstrass $\wp$ function).

(3) On the other hand, very little is known for Abelian extensions of arbitrary number fields. This problem is called Kronecker's Jugendtraum (youthful dream).

An easy special case of the Kronecker–Weber theorem is the following.

Proposition 3.5.14. *Let $k = \mathbb{Q}(\sqrt{D})$ be a quadratic field of discriminant D, and set $m = |D|$. Then k is a subfield of $\mathbb{Q}(\zeta_m)$.*

Proof. Set

$$\tau = \sum_{x \bmod m} \left(\frac{D}{x}\right) \zeta_m^x ,$$

where $\left(\frac{D}{x}\right)$ is the Legendre–Kronecker symbol (see Definition 2.2.5). Since it is a primitive character, Corollary 2.1.47 shows that $\tau^2 = \mathrm{sign}(D)m = D$. Thus $\sqrt{D} = \tau \in \mathbb{Q}(\zeta_m)$.   □

3.5.3 Ring-Theoretic Properties

To simplify notation we write ζ instead of ζ_n and $K = \mathbb{Q}(\zeta)$. We begin with the following.

Proposition 3.5.15. *Assume that $n = p^k$ is a prime power. The ideal $(1 - \zeta)\mathbb{Z}_K$ is the unique prime ideal of K above p, and p is totally ramified in K.*

Proof. By Proposition 3.5.5, there exists a unit $u \in K$ such that $(1 - \zeta)^{\phi(p^k)} = up$. It follows that if we set $\mathfrak{p} = (1-\zeta)\mathbb{Z}_K$ we have the ideal equality $\mathfrak{p}^{\phi(k)} = p\mathbb{Z}_K$, and since $[K : \mathbb{Q}] = \phi(p^k)$, we see that $\mathfrak{p}$ is the unique prime ideal above p in K, that it has degree 1, and that it is totally ramified. This is of course also a consequence of the fact that $\Phi_{p^k}(X + 1)$ is an Eisenstein polynomial, as we shall see in the next chapter.   □

Definition 3.5.16. *The subgroup of the unit group generated by the roots of unity and the cyclotomic units defined in Proposition 3.5.5 is called the group of cyclotomic units.*

For the rest of this section we will assume that $n = p^k$ is a prime power, which simplifies some proofs and is the only case that we need.

Proposition 3.5.17. *Set* $\zeta = \zeta_{p^k}$.

(1) *The ring of integers of* $K = \mathbb{Q}(\zeta)$ *is equal to* $\mathbb{Z}[\zeta]$.
(2) *The discriminant of* K *is equal to* $\pm p^{\phi(p^k)-1}$.
(3) *The only ramified prime in the extension* $K/\mathbb{Q}$ *is* p, *which is totally ramified as* $p\mathbb{Z}_K = (1 - \zeta)^{\phi(p^k)}$.

Proof. Since $\zeta \in \mathbb{Z}_K$ we clearly have $\mathbb{Z}[\zeta] \subset \mathbb{Z}_K$, so we must show the reverse inclusion. Since evidently $K = \mathbb{Q}(\zeta) = \mathbb{Q}(1 - \zeta)$, by Proposition 3.5.10 the $(1 - \zeta)^j$ for $0 \leqslant j < \phi(p^k)$ form a $\mathbb{Q}$-basis of K. Thus, let $\alpha \in \mathbb{Z}_K$, and write

$$\alpha = \sum_{0 \leqslant j < \phi(p^k)} a_j (1 - \zeta)^j \quad \text{with} \quad a_j \in \mathbb{Q}.$$

Denote by $\mathfrak{p}$ the prime ideal $(1 - \zeta)\mathbb{Z}_K$, so that $p\mathbb{Z}_K = \mathfrak{p}^{\phi(p^k)}$. For $a_j \neq 0$ we thus have

$$v_\mathfrak{p}(a_j(1 - \zeta)^j) = \phi(p^k)v_\mathfrak{p}(a_j) + j.$$

Since $0 \leqslant j < \phi(p^k)$ it follows that all these valuations are *distinct* modulo $\phi(p^k)$, and a fortiori they are distinct. Thus $v_\mathfrak{p}(\alpha) = \phi(p^k)v_\mathfrak{p}(a_{j_0}) + j_0$ for the index j_0 realizing the smallest valuation, hence such that for all j

$$\phi(p^k)v_\mathfrak{p}(a_j) + j \geqslant \phi(p^k)v_\mathfrak{p}(a_{j_0}) + j_0 = v_\mathfrak{p}(\alpha) \geqslant 0$$

since α is an algebraic integer. Since $j < \phi(p^k)$ this implies that $v_p(a_j) \geqslant 0$ for all j.

On the other hand, by Proposition 3.5.5 the discriminant of the *polynomial* $\Phi_{p^k}(X)$ is equal to

$$D = \pm \prod_{a,b \in (\mathbb{Z}/p^k\mathbb{Z})^*,\ a \neq b} (\zeta^a - \zeta^b) = u(1 - \zeta)^{\phi(p^k)(\phi(p^k)-1)} = vp^{\phi(p^k)-1},$$

where u and v are units, and since $D \in \mathbb{Z}$ we have $v = \pm 1$. Thus D is a power of p, proving the formula for the discriminant, that p is the only prime number which can ramify in $K/\mathbb{Q}$, and also that $[\mathbb{Z}_K : \mathbb{Z}[\zeta]]$ is a power of p. Thus the a_j in the above reasoning cannot have any other denominators than powers of p, and since we have shown that these cannot occur either it follows that $a_j \in \mathbb{Z}$, as claimed. Thus $\mathbb{Z}_K = \mathbb{Z}[\zeta]$, hence D is also equal to the *field* discriminant of K. $\square$

Although not needed, note that it is easy to determine the sign of D, see Exercise 23.

We know that p is totally ramified in $\mathbb{Q}(\zeta_{p^k})/\mathbb{Q}$. The decomposition of the other primes is given by the following proposition which is not difficult but that we do not prove, and which we state for a general cyclotomic field.

Proposition 3.5.18. *Let $K = \mathbb{Q}(\zeta_m)$ be a cyclotomic field, let q be a prime number not dividing m, and let f be the smallest strictly positive integer such that $q^f \equiv 1 \pmod{m}$. Then $q\mathbb{Z}_K = \prod_{1 \leqslant j \leqslant g} \mathfrak{q}_j$, where the distinct prime ideals $\mathfrak{q}_j$ have degree $f(\mathfrak{q}_j/q) = f$ and $g = \phi(m)/f$, and the decomposition group $D(\mathfrak{q}_j/q)$ is the cyclic subgroup of $\mathrm{Gal}(K/\mathbb{Q})$ generated by the map σ_q sending ζ_m to ζ_m^q.*

3.5.4 The Totally Real Subfield of $\mathbb{Q}(\zeta_{p^k})$

A particularly important subfield of the cyclotomic field $K = \mathbb{Q}(\zeta_{p^k})$ is the field K^+ fixed by complex conjugation ι, which is evidently a totally real field of index 2, hence called *the* maximal totally real subfield of K. As already mentioned it is immediate to check that $K^+ = K(\zeta_{p^k} + \zeta_{p^k}^{-1})$, and it is not difficult to show that $\mathbb{Z}_{K^+} = \mathbb{Z}[\zeta_{p^k} + \zeta_{p^k}^{-1}]$. We denote by $U(K)$ and $U(K^+)$ the corresponding unit groups and by $Cl(K)$ and $Cl(K^+)$ the corresponding class groups. Finally, we set $h_{p^k} = |Cl(K)|$ and $h_{p^k}^+ = |Cl(K^+)|$. We are going to give several results linking the unit and class groups of K and K^+.

Lemma 3.5.19. *Let $\alpha \in K$ be such that $\alpha/\iota(\alpha) \in \mathbb{Z}_K$. Then $\alpha/\iota(\alpha)$ is in fact a root of unity. In particular, if $u \in U(K)$ then $u/\iota(u)$ is a root of unity.*

Proof. Since complex conjugation belongs to $\mathrm{Gal}(K/\mathbb{Q})$, which is abelian, it commutes with all elements of $\mathrm{Gal}(K/\mathbb{Q})$. Thus if $\sigma \in \mathrm{Gal}(K/\mathbb{Q})$ we have $\sigma(\alpha/\iota(\alpha)) = \sigma(\alpha)/\iota(\sigma(\alpha))$, hence $|\sigma(\alpha/\iota(\alpha))| = 1$. Thus $\alpha/\iota(\alpha)$ is an algebraic integer all of whose conjugates have absolute value equal to 1; hence by Kronecker's Corollary 3.3.10 we deduce that it is a root of unity. $\square$

Proposition 3.5.20. *If $p \neq 2$ we have $U(K) = \langle \zeta_{p^k} \rangle U(K^+)$.*

Proof. The right-hand side is clearly a subgroup of the left-hand side. Thus let $u \in U(K)$, and write for simplicity ζ instead of ζ_{p^k}. By the above lemma $u/\iota(u)$ is a root of unity. The group of roots of unity in $\mathbb{Q}(\zeta)$ is equal to $\{\pm 1\} \cdot \langle \zeta \rangle$, hence $u = \varepsilon\iota(u)$ with $\varepsilon = \pm\zeta^j$ for some j and some sign $\pm$. Since p is odd, changing if necessary j into $j + p^k$ we may assume that $j = 2i$ is even. Thus if we set $v = u\zeta^{-i}$ we have $v\zeta^i = \pm\zeta^{2i}\iota(v)\zeta^{-i}$, in other words $v = \pm\iota(v)$. I claim that the $+$ sign holds, so that $v = \iota(v)$, and hence $v \in U(K^+)$ and $u = \zeta^i v \in \langle\zeta\rangle U(K^+)$. Indeed, since $\mathbb{Z}_K = \mathbb{Z}[\zeta]$ we can write $v = \sum_{0 \leqslant j < \phi(p^k)} v_i\zeta^j$ with $v_i \in \mathbb{Z}$, hence $v \equiv \sum_{0 \leqslant j < \phi(p^k)} v_i$ $\pmod{(1-\zeta)\mathbb{Z}_K}$. Since the ideal $(1-\zeta)\mathbb{Z}_K$ and the integers v_i are stable under complex conjugation, it follows that $\iota(v) \equiv v \pmod{(1-\zeta)\mathbb{Z}_K}$. Thus if we had $v = -\iota(v)$ we would have $2v \equiv 0 \pmod{(1-\zeta)\mathbb{Z}_K}$, which is absurd since v is a unit and $1 - \zeta$ divides p, hence is prime to 2, proving my claim and the proposition. $\square$

Note that the above proposition is also true for $p = 2$ (Exercise 30).

Proposition 3.5.21. *If $p \geqslant 3$ is prime, the map sending an ideal $\mathfrak{a}$ of $\mathbb{Z}_{K^+}$ to $\mathfrak{a}\mathbb{Z}_K$ induces an* injective *homomorphism from $Cl(K^+)$ to $Cl(K)$. In particular, $h_{p^k}^+$ divides h_{p^k}.*

Proof. This map evidently induces a group homomorphism from $Cl(K^+)$ to $Cl(K)$, and we must show that it is injective. Thus assume that $\mathfrak{a}$ is an ideal of $\mathbb{Z}_{K^+}$ (which we may assume integral) such that $\mathfrak{a}\mathbb{Z}_K = \alpha\mathbb{Z}_K$ is a principal ideal of K. Since $\iota(\mathfrak{a}) = \mathfrak{a}$ we deduce that $\iota(\alpha)\mathbb{Z}_K = \alpha\mathbb{Z}_K$, hence that $\alpha/\iota(\alpha)$ is a unit of K. It follows from Lemma 3.5.19 that $\alpha/\iota(\alpha)$ is a root of unity; hence since p is odd, $\alpha/\iota(\alpha) = (-\zeta)^j$ for some integer j. I claim that j is even. Indeed, set $\pi = 1 - \zeta$, $\mathfrak{p} = \pi\mathbb{Z}_K$, and let $\mathfrak{p}^+ = \mathfrak{p} \cap \mathbb{Z}_{K^+}$ be the prime ideal below $\mathfrak{p}$. Since p totally ramifies in $\mathbb{Z}_K$ we have $e(\mathfrak{p}/\mathfrak{p}^+) = 2$, hence $v_{\mathfrak{p}}(\mathfrak{a}\mathbb{Z}_K) = 2v_{\mathfrak{p}^+}(\mathfrak{a}) \equiv 0 \pmod 2$, so that $v_{\mathfrak{p}}(\alpha) \equiv 0 \pmod 2$. Since $\pi/\iota(\pi) = (1 - \zeta)/(1 - \zeta^{-1}) = -\zeta$, we have $\alpha/\iota(\alpha) = (\pi/\iota(\pi))^j$, so that if we set $\beta = \alpha\iota(\pi)^j$ we have $\beta = \iota(\beta)$, in other words $\beta \in K^+$. Thus once again $v_{\mathfrak{p}}(\beta) \equiv 0 \pmod 2$; hence $j = v_{\mathfrak{p}}(\iota(\pi)^j) = v_{\mathfrak{p}}(\beta) - v_{\mathfrak{p}}(\alpha) \equiv 0 \pmod 2$ as claimed. Setting $j = 2i$ and $\gamma = \alpha(-\zeta)^{-i}$, we see as in the proof of Proposition 3.5.20 that $\gamma = \iota(\gamma)$, in other words that $\gamma \in K^+$. Since α and γ differ only by a unit, we have $\mathfrak{a}\mathbb{Z}_K = \alpha\mathbb{Z}_K = \gamma\mathbb{Z}_K$. Since $\mathfrak{a}$ and $\gamma\mathbb{Z}_{K^+}$ are ideals of $\mathbb{Z}_{K^+}$, it follows by intersecting with K^+ that $\mathfrak{a} = \gamma\mathbb{Z}_{K^+}$ (see Exercise 17), in other words that $\mathfrak{a}$ is a principal ideal of K^+, proving the proposition. $\qquad\square$

The above result is in fact valid much more generally for so-called *complex-multiplication fields* as follows:

Proposition 3.5.22. *Let K be a totally complex field, K^+ its maximal totally real subfield, and assume that $[K : K^+] = 2$. If we denote by h (respectively h^+) the class number of K (respectively of K^+), then $h^+ \mid h$.*

Proof. We give a proof that assumes the basic definitions and properties of the Hilbert class field. Denote by H the Hilbert class field of K^+, so that H/K^+ is an everywhere unramified Abelian extension, and $[H : K^+] = h^+$. Consider the field compositum HK. Clearly $K^+ \subset H \cap K \subset K$, and since $[K : K^+] = 2$, $H \cap K$ can be equal either to K or to K^+. But $H \cap K = K$ would mean that $H \supset K$, and in particular that K/K^+ is everywhere unramified, which is not the case since all the infinite places of K^+, which are all real, are ramified. Thus $H \cap K = K^+$. This means that H and K are linearly disjoint over K^+, hence that $[HK : K] = [H : K^+] = h^+$ and $[HK : H] = 2$. By Lemma 3.3.28, HK/K is an unramified Abelian extension; hence H is a subfield of the Hilbert class field of K. In particular, $[H : K^+]$ divides h as claimed. $\qquad\square$

Definition 3.5.23. *With the above notation and assumptions, we set $h^- = h/h^+ \in \mathbb{Z}$ and call it the* minus class number. *In particular, for cyclotomic fields we set $h_{p^k}^- = h_{p^k}/h_{p^k}^+$.*

Note that $h_{p^k}^-$ is *easy* to compute for p^k of reasonable size, and we will give a formula for it in Corollary 10.5.27. On the other hand, it is much more difficult to compute $h_{p^k}^+$. For instance, the value of h_p^+ is known rigorously only for $p \leqslant 61$ (for which it is equal to 1), under the generalized Riemann hypothesis (GRH) for $p \leqslant 163$ (for which it is also equal to 1, except for $h_{163}^+ = 4$), although thanks in particular to work of R. Schoof, a conjectural value can be given for much larger p; see pages 420–423 in [Was].

Finally, we give without proof the following fundamental result concerning the unit group and the class group.

Theorem 3.5.24. *Let $K = \mathbb{Q}(\zeta_{p^k})$ and let $U(K)$ be the unit group of K. The subgroup of cyclotomic units of $U(K)$ (see Definition 3.5.16) is a subgroup of finite index of $U(K)$, and this index is equal to $h_{p^k}^+$.*

3.6 Stickelberger's Theorem

3.6.1 Introduction and Algebraic Setting

In this section, let K be a number field, for the moment arbitrary, and let $h(K)$ be its class number. One of the most important applications of the class group is that for any ideal $\mathfrak{a}$ of K the ideal $\mathfrak{a}^{h(K)}$ is a principal ideal. This is in fact the reason for which (in another language) it was introduced by Kummer in his work on Fermat's last theorem. Together with results on the unit group, which is inseparable from the class group, this allows us to treat K almost as if it had unique factorization (as foremost example, see the proof of FLT I for regular primes, i.e., Proposition 6.9.8).

In the case of a cyclotomic field (and more generally of an Abelian extension of $\mathbb{Q}$, which, by the Kronecker–Weber Theorem 3.5.13, is simply a subfield of a cyclotomic field), there is a more powerful theorem, due to Stickelberger, which gives more precise information about the Galois structure of the class group, hence also tells us how to construct principal ideals from arbitrary ideals. This theorem has a large number of important applications, of which we will see three in this book: the Eisenstein reciprocity law, the Hasse–Davenport relations, and Catalan's conjecture. Although perhaps less "basic" than the rest of this chapter, it is important to study it at this point, and since it is slightly less classical than class and unit groups (although it can be found in many excellent textbooks such as [Was]), we give completely detailed proofs. In fact, we will give *two* proofs. In this chapter we give the classical one. As the reader will see, it is quite lengthy and seems roundabout. It uses in a fundamental way the prime ideal decomposition of Gauss sums, and some additional technical machinery. In Chapter 11, however, we will see that thanks to the Gross–Koblitz formula, Gauss sums can be reinterpreted via Morita's p-adic gamma function $\Gamma_p(x)$. Thanks to this interpretation, in

Section 11.7.3 we will give a much shorter and natural proof of the Stickelberger congruence.

We begin by explaining the setting in which we work. Let p be a prime number, let $\zeta = \zeta_p$ be a primitive pth root of unity; set $K = \mathbb{Q}(\zeta)$, so that $\mathbb{Z}_K = \mathbb{Z}[\zeta]$; let $\pi = 1 - \zeta$, and let $\mathfrak{p} = \pi\mathbb{Z}_K$ be the unique ideal above p, hence such that $p\mathbb{Z}_K = \mathfrak{p}^{p-1}$. Finally, let $q = p^f$ be a power of p with $f \geqslant 1$, and let $L = \mathbb{Q}(\zeta_{q-1}, \zeta_p)$. Since p and $q-1$ are coprime, the fields $\mathbb{Q}(\zeta_{q-1})$ and $K = \mathbb{Q}(\zeta_p)$ are linearly disjoint, and in fact it is clear that $L = \mathbb{Q}(\zeta_{p(q-1)})$, although we will never use this. Later we will also have an integer m dividing $q - 1$, and we will consider the fields $\mathbb{Q}(\zeta_m) \subset \mathbb{Q}(\zeta_{q-1})$ and $\mathbb{Q}(\zeta_m, \zeta_p) \subset L$. Since so many different cyclotomic fields are involved, we will use the following notation, which is probably the least confusing. For any integer n we will denote by K_n the cyclotomic field $\mathbb{Q}(\zeta_n)$ and for $m \mid (q-1)$ we will denote by L_m the field $\mathbb{Q}(\zeta_m, \zeta_p)$. Thus for example $K = K_p$, $\mathbb{Q}(\zeta_{q-1}) = K_{q-1}$, and $L = L_{q-1}$.

Let $\mathfrak{P}$ be a *fixed* prime ideal of L above $\mathfrak{p}$. The prime ideal of L_m below $\mathfrak{P}$ will be denoted by $\mathfrak{P}_m$, so that $\mathfrak{P} = \mathfrak{P}_{q-1}$, and if $K_m \subset K_{q-1}$ the prime ideal below $\mathfrak{P}$ will be denoted by $\mathfrak{p}_m$. Since $p \nmid q-1$, we have $e(\mathfrak{p}_{q-1}/p) = 1$; hence by Proposition 3.5.18, $f(\mathfrak{p}_{q-1}/p) = f$ since $q = p^f$ is evidently the smallest power of p congruent to 1 modulo $q-1$. It follows that $\mathfrak{P}$ is unramified in the extension L/K and that we also have $f(\mathfrak{P}/\mathfrak{p}) = f$. The following diagram summarizes all of the above notation.

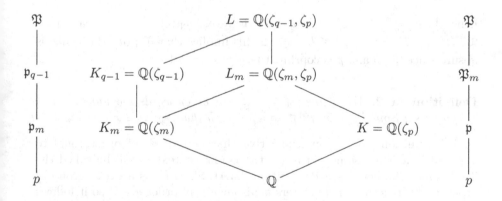

3.6.2 Instantiation of Gauss Sums

In Chapter 2 we introduced the Gauss sum $\tau(\chi, \psi)$ associated with a multiplicative and an additive character on a finite field $\mathbb{F}_q$, and we have seen for instance that $|\tau(\chi, \psi)|^2 = q$ when the characters are nontrivial. Since the values of ψ are in $K = \mathbb{Q}(\zeta)$ and the values of χ are in $K_{q-1} = \mathbb{Q}(\zeta_{q-1})$, it follows that $\tau(\chi, \psi) \in L$. In the present subsection we will see that we can give much more precise information. In particular, we will give the exact description of the principal ideal generated by $\tau(\chi, \psi)$, and congruences satisfied by the Gauss sum. As mentioned above, in Section 11.7.3 we will see how to compute $\tau(\chi, \psi)$ in terms of the p-adic gamma function, leading to a simpler proof of the Stickelberger formula.

To be able to compute $\tau(\chi, \psi)$ we must *instantiate* all the mathematical objects that occur (otherwise, the computation would not make sense), in other words, $\mathbb{F}_q$ and the characters χ and ψ. To instantiate $\mathbb{F}_q$ we use the fixed prime ideal $\mathfrak{P}$ of L chosen above, and set $\mathbb{F}_q = \mathbb{Z}_L/\mathfrak{P}$, which is an extension of $\mathbb{F}_p = \mathbb{Z}_K/\mathfrak{p}$. By Proposition 2.5.4, any additive character ψ has the form $\psi = \psi_b$ for a unique $b \in \mathbb{F}_q$, where $\psi_b(x) = \zeta^{\mathrm{Tr}_{\mathbb{F}_q/\mathbb{F}_p}(bx)}$. Since we will almost always use $\psi = \psi_1$, we will simply write $\tau(\chi)$ instead of $\tau(\chi, \psi_1)$, although it will occasionally be useful to use ψ_b because of the formula $\tau(\chi, \psi_b) = \chi(b)^{-1}\tau(\chi, \psi_1)$ of Lemma 2.5.6. To instantiate multiplicative characters, we need a lemma.

Lemma 3.6.1. *Let* $\mu_{q-1} = \{\zeta_{q-1}^j, \ 0 \leqslant j \leqslant q - 2\} \subset L$ *be the group of* $(q - 1)$*st roots of unity in* L. *The map* $u_{\mathfrak{P}}$ *sending* $x \in \mu_{q-1}$ *to its class modulo* $\mathfrak{P}$ *in* $\mathbb{F}_q^* = (\mathbb{Z}_L/\mathfrak{P})^*$ *is a group isomorphism.*

Proof. It is clear that $u_{\mathfrak{P}}$ is a group homomorphism, and since μ_{q-1} and $\mathbb{F}_q^*$ both have $q - 1$ elements it is sufficient to show that the kernel of $u_{\mathfrak{P}}$ is equal to 1. Now

$$\prod_{1 \leqslant k \leqslant q-2} (x - \zeta_{q-1}^k) = \frac{x^{q-1} - 1}{x - 1} \ ;$$

hence $\prod_{1 \leqslant k \leqslant q-2}(1 - \zeta_{q-1}^k) = q - 1$. However, $u_{\mathfrak{P}}(\zeta_{q-1}^k) = 1$ means that $\mathfrak{P} \mid (\zeta_{q-1}^k - 1)$, hence if $1 \leqslant k \leqslant q - 2$ this implies that $\mathfrak{P} \mid (q - 1)$, which is absurd since $\mathfrak{P} \mid p$ and p is coprime to $q - 1 = p^f - 1$. $\qquad\square$

Definition 3.6.2. *We define* $\omega_{\mathfrak{P}} = u_{\mathfrak{P}}^{-1}$, *in other words the unique group isomorphism from* $\mathbb{F}_q^* = (\mathbb{Z}_L/\mathfrak{P})^*$ *to* μ_{q-1} *such that* $\omega_{\mathfrak{P}}(x) \equiv x \pmod{\mathfrak{P}}$.

By definition $\omega_{\mathfrak{P}}$ is a multiplicative character of $\mathbb{F}_q$ of order equal to $q - 1$ since it is an isomorphism (in the p-adic context it will be called the Teichmüller character; see Proposition 4.3.4). Since $\mathbb{F}_q^*$ is a cyclic group of order $q - 1$, its group of characters is also cyclic of order $q - 1$, so it follows that any multiplicative character χ on $\mathbb{F}_q$ has the form $\omega_{\mathfrak{P}}^{-r}$ for a unique r such that $0 \leqslant r < q - 1$ (equivalently, by transitivity of the Galois action on prime ideals above $\mathfrak{p}$, any multiplicative character χ on $\mathbb{F}_q$ is one of the $\omega_{\mathfrak{Q}}$ for a suitable prime ideal $\mathfrak{Q}$ above p). Thus with an evident abuse of notation, any Gauss sum over $\mathbb{F}_q$ has the form $\tau(\omega_{\mathfrak{P}}^{-r}, \psi_b)$ for a unique $r \in \mathbb{Z}/(q - 1)\mathbb{Z}$ and a unique $b \in \mathbb{F}_q$. This now depends only on the triple $(\mathfrak{P}, r, b)$; hence we will compute its properties only in terms of this triple. In fact, since $\tau(\omega_{\mathfrak{P}}^{-r}, \psi_b) = \omega_{\mathfrak{P}}^r(b)\tau(\omega_{\mathfrak{P}}^{-r})$, we may assume that $b = 1$. Finally, since in this section $\mathfrak{P}$ is fixed, for notational convenience we will simply write ω instead of $\omega_{\mathfrak{P}}$, but we have to keep in mind for later that the definition of ω depends on the choice of the ideal $\mathfrak{P}$ of L above $\mathfrak{p}$. In fact, since by Galois theory all other ideals have the form $\sigma(\mathfrak{P})$ for $\sigma \in \mathrm{Gal}(L/K)$, the following lemma makes explicit the dependence on $\mathfrak{P}$:

Lemma 3.6.3. (1) *For all* $\sigma \in \mathrm{Gal}(L/\mathbb{Q})$ *we have*

$$\omega_{\sigma(\mathfrak{P})} = \sigma \circ \omega_{\mathfrak{P}} \circ \sigma^{-1} .$$

(2) *For all* $\sigma \in \mathrm{Gal}(L/\mathbb{Q})$ *and* $r \in \mathbb{Z}$ *we have*

$$\tau(\omega_{\sigma(\mathfrak{P})}^{-r}) = \sigma(\tau(\omega_{\mathfrak{P}}^{-r})) .$$

(3) *If* $m \mid (q-1)$ *and* $x \in K_m$ *then* $\omega_{\mathfrak{P}}^{-(q-1)/m}(x)$ *depends only on the ideal* $\mathfrak{p}_m$ *of* K_m *below* $\mathfrak{P}$ *and not on* $\mathfrak{P}$ *itself.*

(4) *If* $m \mid (q-1)$ *then* $\tau(\omega_{\mathfrak{P}}^{-(q-1)/m})^m \in K_m$, *and it depends only on the ideal* $\mathfrak{p}_m$ *of* K_m *below* $\mathfrak{P}$ *and not on* $\mathfrak{P}$ *itself.*

Proof. Since $\omega_{\mathfrak{P}}(\sigma^{-1}(x))$ is a $(q-1)$st root of unity, so is $\sigma(\omega_{\mathfrak{P}}(\sigma^{-1}(x)))$. Furthermore, $\omega_{\mathfrak{P}}(\sigma^{-1}(x)) \equiv \sigma^{-1}(x) \pmod{\mathfrak{P}}$, hence applying σ we have $\sigma(\omega_{\mathfrak{P}}(\sigma^{-1}(x))) \equiv x \pmod{\sigma(\mathfrak{P})}$. Since these two properties characterize $\omega_{\sigma(\mathfrak{P})}$, (1) follows. Set $\mathfrak{P}_1 = \sigma(\mathfrak{P})$. With evident notation we have

$$\tau(\omega_{\mathfrak{P}_1}^{-r}) = \sum_{x \in (\mathbb{Z}_L/\mathfrak{P}_1)^*} \omega_{\mathfrak{P}_1}(x)^{-r} \zeta_p^{\mathrm{Tr}_{(\mathbb{Z}_L/\mathfrak{P}_1)/(\mathbb{Z}/p\mathbb{Z})}(x)} ,$$

so that setting $x = \sigma(y)$ and using (1) we obtain

$$\sigma^{-1}(\tau(\chi_{\mathfrak{P}_1})) = \sum_{y \in (\mathbb{Z}_L/\mathfrak{P})^*} \omega_{\mathfrak{P}}(y)^{-r} \zeta_p^{\mathrm{Tr}_{(\mathbb{Z}_L/\mathfrak{P})/(\mathbb{Z}/p\mathbb{Z})}(y)} = \tau(\omega_{\mathfrak{P}}^{-r}) ,$$

proving (2). For (3) and (4), set $d = (q-1)/m$, a notation that we shall in fact use in the rest of this chapter. By definition, $\omega_{\mathfrak{P}}^d(x) \equiv x^d \pmod{\mathfrak{P}}$, and $\omega_{\mathfrak{P}}^d(x)$ is an mth root of unity, and in particular belongs to K_m. If in addition $x \in K_m$, then $\omega_{\mathfrak{P}}^d(x) \equiv x^d \pmod{\mathfrak{P} \cap K_m}$, proving (3) since $\omega_{\mathfrak{P}}^d(x)$ is characterized by the above two properties. For (4), Corollary 2.5.15 tells us that $\tau(\omega_{\mathfrak{P}}^{-d})^m \in K_m$. If $\mathfrak{P}_1$ is some other prime ideal of L above the same prime ideal $\mathfrak{p}_m$ of K_m, by transitivity of the Galois action on prime ideals there exists $\sigma \in \mathrm{Gal}(L/K_m)$ such that $\mathfrak{P}_1 = \sigma(\mathfrak{P})$. Thus by (2) we have $\tau(\omega_{\mathfrak{P}_1}^{-d})^m = \sigma(\tau(\omega_{\mathfrak{P}}^{-d})^m)$. Since σ leaves K_m fixed it does not act on the values of the multiplicative characters, but only on the additive characters; hence by Lemma 2.5.6 we have for some integer a coprime to p,

$$\sigma(\tau(\omega_{\mathfrak{P}}^{-d})^m) = \tau(\omega_{\mathfrak{P}}^{-d}, \psi_a)^m = (\omega_{\mathfrak{P}}(a)^d \tau(\omega_{\mathfrak{P}}^{-d}))^m = \tau(\omega_{\mathfrak{P}}^{-d})^m ,$$

since $\omega_{\mathfrak{P}}(a)$ is a $(q-1)$st root of unity, proving (4). $\qquad\square$

3.6.3 Prime Ideal Decomposition of Gauss Sums

We begin by proving the following congruence for Jacobi sums.

Proposition 3.6.4. *If a and b are integers such that $1 \leqslant a, b \leqslant q - 2$ we have $J(\omega^{-a}, \omega^{-b}) \equiv 0 \pmod{\mathfrak{P}}$ if $a + b \geqslant q$, and otherwise*

$$J(\omega^{-a}, \omega^{-b}) \equiv -\binom{a+b}{a} \equiv -\frac{(a+b)!}{a!b!} \pmod{\mathfrak{P}} .$$

Proof. For any $x \in \mathbb{F}_q^* = (\mathbb{Z}_L/\mathfrak{P})^*$ we let $x' \in \mu_{q-1} \subset \mathbb{Z}_L$ be such that $x' \equiv x \pmod{\mathfrak{P}}$, so that $\omega(x) = x'$. Setting $c = q - 1 - b$ and noting that $\omega^{q-1} = \varepsilon$ we see that $\omega^{-b} = \omega^c$. In particular, for $x \neq 1$ we have $\omega^{-b}(1-x) = \omega^c(1-x) \equiv (1-x')^c \pmod{\mathfrak{P}}$, and we note that this congruence is trivially also true for $x = 1$ since $b < q - 1$, and hence $c > 0$. Thus

$$J(\omega^{-a}, \omega^{-b}) = \sum_{x \in \mathbb{F}_q^*} \omega^{-a}(x)\omega^{-b}(1-x) \equiv \sum_{x \in \mathbb{F}_q^*} x'^{-a}(1-x')^c$$

$$\equiv \sum_{x \in \mathbb{F}_q^*} x^{-a} \sum_{0 \leqslant j \leqslant b} \binom{c}{j}(-1)^j x^j$$

$$\equiv \sum_{0 \leqslant j \leqslant c}(-1)^j \binom{c}{j} \sum_{x \in \mathbb{F}_q^*} x^{j-a} \pmod{\mathfrak{P}} .$$

Thanks to Lemma 2.5.1 we see that the inner sum vanishes except if $j \equiv a \pmod{q-1}$, in other words if $j = a$ since $1 \leqslant a \leqslant q-2$. Thus if $a > c$, in other words $a + b > q - 1$, all the inner sums vanish, so we get $J(\omega^{-a}, \omega^{-b}) \equiv 0 \pmod{\mathfrak{P}}$. On the other hand, if $a \leqslant c$ we have a single nonzero inner sum for $j = a$, which is congruent to -1 modulo p, so that

$$J(\omega^{-a}, \omega^{-b}) \equiv (-1)^{a-1}\binom{c}{a} \pmod{\mathfrak{P}} ,$$

and the proposition follows since

$$\binom{c}{a} = \frac{(q-1-b)(q-2-b)\cdots(q-a-b)}{a!}$$

$$\equiv (-1)^a \frac{(b+1)(b+2)\cdots(b+a)}{a!} \equiv (-1)^a\binom{a+b}{a} \pmod{p} .$$

$\square$

Remarks. (1) It is easy to show that when $a + b \geqslant q$ we have $\binom{a+b}{a} \equiv 0 \pmod{p}$ (see Exercise 25), so the second congruence of the proposition is in fact valid in all generality.

(2) This proposition again shows the analogy between Jacobi sums and the beta function.

Definition 3.6.5. *Let r be an integer such that $0 \leqslant r < q - 1$, and let $r = \sum_{0 \leqslant i < f} r_i p^i$ be the decomposition of r in base p with $0 \leqslant r_i \leqslant p - 1$. We set $s_p(r) = \sum_{0 \leqslant i < f} r_i$ and $t_p(r) = \prod_{0 \leqslant i < f} (r_i)!$. If $r \in \mathbb{Z}$ is arbitrary, we define the $(q-1)$-periodic functions $s(r)$ and $t(r)$ by setting $s(r) = s_p(r \bmod q - 1)$ and $t(r) = t_p(r \bmod q - 1)$, where $r \bmod q - 1$ is the unique integer in $[0, q - 1[$ congruent to r modulo $q - 1$.*

Stickelberger's basic result is the following.

Theorem 3.6.6. *For all $r \in \mathbb{Z}$ we have*

$$\frac{\tau(\omega^{-r})}{(\zeta - 1)^{s(r)}} \equiv -\frac{1}{t(r)} \ (\mathrm{mod} \ \mathfrak{P}) \ .$$

Proof. By periodicity we may assume that $0 \leqslant r < q - 1$. We prove the theorem by induction on $s(r) = s_p(r)$. If $s(r) = 0$ we have $r = 0$, hence $t(0) = 1$ and $\tau(\omega^0) = \tau(\varepsilon) = -1$ by Lemma 2.5.8. The crucial case to be proved is the case $s(r) = 1$, in other words, $r = p^k$. In that case $t(r) = 1$, and since by Lemma 2.5.8 (3) we have $\tau(\omega^{-pa}) = \tau(\omega^{-a})$, it follows that we may assume that $r = 1$. Since ω is a nontrivial character we have

$$\tau(\omega^{-1}) = \sum_{x \in \mathbb{F}_q^*} \omega^{-1}(x) \left(\zeta_p^{\mathrm{Tr}_{\mathbb{F}_q/\mathbb{F}_p}(x)} - 1 \right) \ .$$

Since $\mathfrak{p} = (\zeta_p - 1)\mathbb{Z}_K$ we have

$$\frac{\zeta_p^m - 1}{\zeta_p - 1} = \sum_{0 \leqslant j < m} \zeta_p^j \equiv \sum_{0 \leqslant j < m} 1 \equiv m \ (\mathrm{mod} \ \mathfrak{p}) \ ,$$

hence

$$\frac{\tau(\omega^{-1})}{\zeta_p - 1} \equiv \sum_{x \in \mathbb{F}_q^*} \omega^{-1}(x) \, \mathrm{Tr}_{\mathbb{F}_q/\mathbb{F}_p}(x) \ (\mathrm{mod} \ \mathfrak{p}) \ .$$

Now $\mathrm{Tr}_{\mathbb{F}_q/\mathbb{F}_p}(x) = \sum_{0 \leqslant i < f} x^{p^i} \in \mathbb{F}_p$, and on the other hand, by definition, $\omega^{-1}(x) \equiv x^{-1} \ (\mathrm{mod} \ \mathfrak{P})$. It follows that

$$\frac{\tau(\omega^{-1})}{\zeta_p - 1} \equiv \sum_{0 \leqslant i < f} \sum_{x \in \mathbb{F}_q^*} x^{p^i - 1} \ (\mathrm{mod} \ \mathfrak{P}) \ .$$

Now again by Lemma 2.5.1 the inner sum vanishes if $1 \leqslant i < f$, and it is congruent to -1 modulo p if $i = 0$. It follows that $\tau(\omega^{-1})/(\zeta_p - 1) \equiv -1$ $(\mathrm{mod} \ \mathfrak{P})$, proving the theorem when $s(r) = 1$.

Now let r be such that $0 \leqslant r < q - 1$ with $s(r) > 1$, assume by induction that the theorem is true for all $r' < q - 1$ such that $s(r') < s(r)$, and let $r = \sum_{0 \leqslant i < f} r_i p^i$ with $0 \leqslant r_i \leqslant p - 1$. Thanks once again to Lemma 2.5.8 (3) we may assume that $r_0 \geqslant 1$. It follows in particular that $s(r - 1) = s(r) - 1$

and $r - 1 \geqslant 1$. Since all the characters involved are nontrivial, by Corollary 2.5.17 we have

$$\tau(\omega^{-1})\tau(\omega^{-(r-1)}) = J(\omega^{-1}, \omega^{-(r-1)})\tau(\omega^{-r}) .$$

Using Proposition 3.6.4 we know that

$$J(\omega^{-1}, \omega^{-(r-1)}) \equiv -\binom{r}{1} \equiv -r \equiv -r_0 \pmod{\mathfrak{P}} .$$

Since $1 \leqslant r_0 \leqslant p - 1$, r_0 is invertible modulo $\mathfrak{P}$, so by induction and the case $r = 1$ we see that

$$\frac{\tau(\omega^{-r})}{(\zeta_p - 1)^{s(r)}} \equiv -\frac{1}{r_0}\frac{\tau(\omega^{-1})}{\zeta_p - 1}\frac{\tau(\omega^{-(r-1)})}{(\zeta_p - 1)^{s(r-1)}}$$

$$\equiv -\frac{1}{r_0}\frac{1}{t(r-1)} \equiv -\frac{1}{t(r)} \pmod{\mathfrak{P}} ,$$

since $t(r) = r_0 t(r - 1)$ when $r_0 \neq 0$, proving our induction hypothesis and hence the theorem.   □

To use this theorem, we will also need the following lemma.

Lemma 3.6.7. *As usual denote by $\{z\}$ the fractional part of a real number z.*

(1) *For all $r \in \mathbb{Z}$ we have*

$$s(r) = (p - 1) \sum_{0 \leqslant i < f} \left\{\frac{p^i r}{q - 1}\right\} .$$

(2) *For $0 \leqslant r < q - 1$ we have*

$$t(r) \equiv (-p)^{-v_p(r!)} r! \pmod{p} .$$

Proof. (1). Both sides of the formula are periodic of period dividing $q - 1$; hence we may assume that $0 \leqslant r < q - 1$, so that $r = \sum_{0 \leqslant j < f} r_j p^j$ with $0 \leqslant r_j \leqslant p - 1$. For $0 \leqslant i \leqslant f - 1$ we have

$$p^i r = \sum_{0 \leqslant j < f-i-1} r_j p^{j+i} + \sum_{f-i \leqslant j < f} r_j p^{j+i}$$

$$\equiv \sum_{0 \leqslant j < f-i-1} r_j p^{j+i} + \sum_{f-i \leqslant j < f} r_j p^{j+i-f} \pmod{q - 1} ,$$

hence

$$\left\{\frac{p^i r}{q - 1}\right\} = \frac{1}{q - 1}\left(\sum_{0 \leqslant j < f-i-1} r_j p^{j+i} + \sum_{f-i \leqslant j < f} r_j p^{j+i-f}\right) .$$

It follows that

$$\sum_{0 \leqslant i < f} \left\{ \frac{p^i r}{q-1} \right\} = \frac{1}{q-1} \sum_{0 \leqslant j < f} r_j S_j \ ,$$

where

$$S_j = \sum_{0 \leqslant i < f-j-1} p^{j+i} + \sum_{f-j \leqslant i < f} p^{j+i-f} = \sum_{0 \leqslant i < f} p^i = \frac{p^f - 1}{p - 1} = \frac{q-1}{p-1} \ ,$$

proving (1).

(2) is easily proved by induction on r: it is trivially true for $r \leqslant 1$. Assume $r \geqslant 2$ and that the formula is true for $r - 1$, and let $r = \sum_{k \leqslant j \leqslant f-1} r_j p^j$ be the base-p decomposition of r, with $0 \leqslant r_j \leqslant p-1$ and $r_k \neq 0$. Since

$$r - 1 = \sum_{0 \leqslant j \leqslant k-1} (p-1)p^j + (r_k - 1)p^k + \sum_{k+1 \leqslant j \leqslant f-1} r_j p^j$$

it follows from Wilson's theorem that

$$t(r-1) \equiv (-1)^k (r_k - 1)! \prod_{k+1 \leqslant j \leqslant f-1} (r_j)! \ (\mathrm{mod} \ p) \ ,$$

hence that

$$t(r) \equiv (-1)^k r_k t(r-1) \equiv \frac{r}{(-p)^{v_p(r)}} t(r-1) \quad \mathrm{mod} \ p \ .$$

The result follows by induction. □

Corollary 3.6.8. *We have* $v_{\mathfrak{P}}(\tau(\omega^{-r})) = s(r)$.

Proof. By definition $t(r)$ is coprime to p hence is invertible modulo $\mathfrak{P}$, so by the theorem $v_{\mathfrak{P}}(\tau(\omega^{-r})) = s(r)v_{\mathfrak{P}}(\zeta - 1) = s(r)e(\mathfrak{P}/\mathfrak{p}) = s(r)$. □

Definition 3.6.9. *Let* $m \mid (q-1)$ *be such that* $m > 1$.

(1) *For notational simplicity we will set* $d = (q-1)/m$.
(2) *If* t *is coprime to* m *we denote by* σ_t *the element of* $\mathrm{Gal}(L_m/K)$ *such that* $\sigma_t(\zeta_m) = \zeta_m^t$ *(and of course leaving fixed* ζ_p*).*

Although strictly speaking σ_t depends on m, since these maps are compatible under restriction, there is no possibility of confusion.

Proposition 3.6.10. *Let* $m \mid (q-1)$, *set* $d = (q-1)/m$, *and recall that* $L_m = \mathbb{Q}(\zeta_m, \zeta_p)$ *and that* $\mathfrak{P}_m$ *is the prime ideal of* L_m *below* $\mathfrak{P}$*. Then*

$$\tau(\omega^{-rd})\mathbb{Z}_{L_m} = \prod_{t \in (\mathbb{Z}/m\mathbb{Z})^* / \langle p \rangle} \sigma_t^{-1}(\mathfrak{P}_m)^{s(rtd)} \ .$$

Proof. First note that the values of ω^{-rd} are in K_m, so that $\tau(\omega^{-rd}) \in L_m$. Since $\mathfrak{P}_m$ is a prime ideal of L_m above $\mathfrak{p}$, by Galois theory all the prime ideals of L_m above $\mathfrak{p}$ have the form $\sigma(\mathfrak{P}_m)$ for $\sigma \in \mathrm{Gal}(L_m/K) \simeq \mathrm{Gal}(K_m/\mathbb{Q}) \simeq (\mathbb{Z}/m\mathbb{Z})^*$. By definition of the Gauss sum we have $\sigma_t(\tau(\omega^{-rd})) = \tau(\omega^{-rtd})$. Thus by the above corollary

$$v_{\sigma_t^{-1}(\mathfrak{P}_m)}(\tau(\omega^{-rd})) = v_{\mathfrak{P}_m}(\sigma_t(\tau(\omega^{-rd}))) = v_{\mathfrak{P}}(\tau(\omega^{-rtd}))) = s(rtd) \;,$$

since $\mathfrak{P}$ is unramified over $\mathfrak{P}_m$. Furthermore, the decomposition group $D(\mathfrak{P}_m/\mathfrak{p})$ is isomorphic to $D(\mathfrak{p}_m/p)$; hence by Proposition 3.5.18 it is isomorphic to the cyclic subgroup of $\mathrm{Gal}(L_m/K)$ generated by σ_p, using the same notation as above. This means that the ideals of L_m above $\mathfrak{p}$ are obtained once and only once as $\sigma_t^{-1}(\mathfrak{P}_m)$ for $\sigma_t \in \mathrm{Gal}(L_m/K)/D(\mathfrak{P}_m/\mathfrak{p})$, in other words for $t \in (\mathbb{Z}/m\mathbb{Z})^*/\langle p \rangle x$. Finally, since $\tau(\omega^{-rd})\overline{\tau(\omega^{-rd})} = q = p^f$ for $r \neq 0$, it follows that the only ideals of L_m dividing $\tau(\omega^{-rd})$ are the ideals above p, hence the ideals above $\mathfrak{p}$, in other words the $\sigma_t^{-1}(\mathfrak{P}_m)$ for $t \in (\mathbb{Z}/m\mathbb{Z})^*/\langle p \rangle$, so the proposition follows. $\qquad\square$

Corollary 3.6.11. *Let* $m \mid (q-1)$, *and recall that* $K_m = \mathbb{Q}(\zeta_m)$ *and that* $\mathfrak{p}_m$ *is the prime ideal of* K_m *below* $\mathfrak{P}$. *Then*

$$\tau(\omega^{-d})^m \mathbb{Z}_{K_m} = \prod_{t \in (\mathbb{Z}/m\mathbb{Z})^*/\langle p \rangle} \sigma_t^{-1}(\mathfrak{p}_m)^{v_t} \;,$$

where

$$v_t = \frac{m}{p-1} s\left(t\frac{q-1}{m}\right) = m \sum_{0 \leqslant i < f} \left\{\frac{p^i t}{m}\right\} \;.$$

Proof. By Corollary 2.5.15 we know that $\tau(\omega^{-d})^m \in K_m$. Since $e(\mathfrak{P}_m/\mathfrak{p}_m) = e(\mathfrak{p}/p) = p - 1$ we have

$$v_{\mathfrak{p}_m}(\tau(\omega^{-d})^m) = (m/(p-1))v_{\mathfrak{P}_m}(\tau(\omega^{-d})) \;,$$

so the corollary immediately follows from the proposition, the second formula for v_t coming from Lemma 3.6.7. $\qquad\square$

The above results are expressed much more nicely in the language of group rings, which we recall for the convenience of the reader.

Definition 3.6.12. *Let* A *be a commutative ring and let* G *be a finite abelian group. We define the* group algebra $A[G]$ *by*

$$A[G] = \left\{\sum_{g \in G} a_g g, \; a_g \in A\right\} \;,$$

in other words the set of formal linear combinations with coefficients in A of the elements of G, the addition law coming from that of A and the multiplication law from that of A and of the group law of G, in other words,

$$\left(\sum_{g \in G} a_g g \right) \left(\sum_{h \in G} b_h h \right) = \sum_{g,h \in G} a_g b_h gh = \sum_{g \in G} \left(\sum_{h \in G} a_h b_{h^{-1}g} \right) g \ .$$

It is immediately checked that $A[G]$ is a commutative ring. If E is a group (written multiplicatively, say) on which the group G acts, then E has a natural structure of a (left) $\mathbb{Z}[G]$-module through the formula

$$\left(\sum_{g \in G} a_g g \right) \cdot v = \prod_{g \in G} (g \cdot v)^{a_g} \ .$$

If $\theta = \sum_{g \in G} a_g g$, it is customary to write v^θ instead of $\theta \cdot v$, so that the formula $\theta_1 \cdot (\theta_2 \cdot v) = (\theta_1 \theta_2) \cdot v$ translates into the identity $v^{\theta_1 \theta_2} = (v^{\theta_1})^{\theta_2}$, where it is essential that G be an *abelian* group.

The situation considered in practice will be the following. We will be given an Abelian extension $K/\mathbb{Q}$ with commutative Galois group G. The group G acts naturally on all natural objects linked to K, for instance on elements, units, ideals, ideal classes, etc., so by the above we have an action of $\mathbb{Z}[G]$, written in an exponential manner.

As a first application of this notation we have the following proposition, which is a restatement of Corollary 3.6.11.

Proposition 3.6.13. *Set*

$$\Theta = \sum_{t \in (\mathbb{Z}/m\mathbb{Z})^*} \left\{ \frac{t}{m} \right\} \sigma_t^{-1} = \frac{1}{m} \sum_{\substack{1 \leqslant t \leqslant m-1 \\ \gcd(t,m)=1}} t \sigma_t^{-1} \ .$$

With the same notation as above we have $\tau(\omega^{-d})^m \mathbb{Z}_{K_m} = \mathfrak{p}_m^{m\Theta}$.

Proof. Let T be a system of representatives of $(\mathbb{Z}/m\mathbb{Z})^*$ modulo the cyclic subgroup $\langle p \rangle$ generated by p. Proposition 3.6.11 can be restated by saying that $\tau(\omega^{-d})^m \mathbb{Z}_{K_m} = \mathfrak{p}_m^{m\theta}$ with

$$\theta = \sum_{t \in T} \sum_{0 \leqslant i < f} \{ p^i t / m \} \sigma_t^{-1}$$

(note that it would not make sense to replace T by $(\mathbb{Z}/m\mathbb{Z})^* / \langle p \rangle$ in the above expression since σ_t would not be defined). By definition, as t ranges in T and i ranges from 0 to $f - 1$ the elements $p^i t$ modulo m range through $(\mathbb{Z}/m\mathbb{Z})^*$, so that

$$\theta = \sum_{t\in(\mathbb{Z}/m\mathbb{Z})^*} \{t/m\}\sigma_{t_1}^{-1} ,$$

where t_1 is the representative in T of the class of t modulo $\langle p \rangle$. Since the decomposition group of $\mathfrak{p}_m/p$ is the group generated by σ_p, it follows that $\mathfrak{p}_m^{m\theta} = \mathfrak{p}_m^{m\Theta}$, where Θ is as in the proposition. $\square$

The following corollary is one of the most important consequences.

Corollary 3.6.14. *Let $\mathbb{Q}(\zeta_m)$ be a cyclotomic field and let Θ be defined as above. For all fractional ideals $\mathfrak{a}$ of $\mathbb{Q}(\zeta_m)$ the ideal $\mathfrak{a}^{m\Theta}$ is a principal ideal.*

Proof. We know that in any ideal class there exists an integral ideal coprime to any fixed ideal, in particular to m. In other words, there exists an element α such that $\mathfrak{b} = \alpha\mathfrak{a}$ is an integral ideal coprime to m. If $\mathfrak{p}_m$ is a prime ideal dividing $\mathfrak{b}$ above some prime p we have $p \nmid m$; hence the proposition implies that $\mathfrak{p}_m^{m\Theta}$ is a principal ideal. Since this is true for every prime ideal dividing $\mathfrak{b}$, by multiplicativity it is true for $\mathfrak{b}$ itself, and hence for $\mathfrak{a}$. $\square$

3.6.4 The Stickelberger Ideal

Although the above results and in particular Corollary 3.6.14 are remarkable results, they suffer from the presence of m in the exponent. We are now going to show how to get rid of this exponent m. We keep all the above notation, and we set $G = \mathrm{Gal}(K_m/\mathbb{Q})$, which is canonically isomorphic to $(\mathbb{Z}/m\mathbb{Z})^*$.

Definition 3.6.15. (1) *We define the* Stickelberger ideal $I_s(m)$ *by* $I_s(m) = \mathbb{Z}[G] \cap \Theta\mathbb{Z}[G]$.

(2) *If b is an integer coprime to m we define $\Theta_b \in \mathbb{Q}[G]$ by the formula* $\Theta_b = (\sigma_b - b)\Theta$.

It is clear that $I_s(m)$ is an ideal of the commutative ring $\mathbb{Z}[G]$.

Lemma 3.6.16. *We have $\Theta_b \in I_s(m)$. More precisely, we have*

$$\Theta_b = - \sum_{1\leqslant t\leqslant m-1,\ \gcd(t,m)=1} \left\lfloor \frac{bt}{m} \right\rfloor \sigma_t^{-1} .$$

Proof. Since evidently $\Theta_b \in \mathbb{Z}[G]\Theta$, we must show that $\Theta_b \in \mathbb{Z}[G]$. Setting $u = b^{-1}t$ we have

$$\sigma_b\Theta = \sum_{t\in(\mathbb{Z}/m\mathbb{Z})^*} \{t/m\}\sigma_{b^{-1}t}^{-1} = \sum_{u\in(\mathbb{Z}/m\mathbb{Z})^*} \{bu/m\}\sigma_u^{-1} ,$$

hence $\Theta_b = \sum_{u\in(\mathbb{Z}/m\mathbb{Z})^*} c_u\sigma_u^{-1}$ with

$$c_u = \{bu/m\} - b\{u/m\} \equiv bu/m - b(u/m) \equiv 0 \ (\mathrm{mod}\ \mathbb{Z}) ,$$

proving the first part of the lemma. The formula follows by summing over $u \in [1, m-1]$ coprime to m, since in that case $\{u/m\} = u/m$. $\square$

Example. If $m = p$ is an odd prime and $b = 2$ we have

$$\Theta_2 = - \sum_{(p+1)/2 \leqslant t \leqslant p-1} \sigma_t^{-1} \ .$$

Lemma 3.6.17. *The ideal $I_s(m)$ is generated by the Θ_b as a $\mathbb{Z}$-module (hence also as an ideal). More precisely, it is generated over $\mathbb{Z}$ by Θ_{m+1} and the Θ_b for $1 \leqslant b \leqslant m$ with $\gcd(b, m) = 1$.*

Proof. By definition an element $\delta \in I_s(m)$ has the form $\delta = \gamma\Theta$, where $\gamma\Theta \in \mathbb{Z}[G]$ and $\gamma \in \mathbb{Z}[G]$. If we write $\gamma = \sum_{t\in(\mathbb{Z}/m\mathbb{Z})^*} c_t\sigma_t$ with $c_t \in \mathbb{Z}$ we have

$$\gamma\Theta = \sum_{t,u\in(\mathbb{Z}/m\mathbb{Z})^*} c_t\{u/m\}\sigma_{t^{-1}u}^{-1} = \sum_{v\in(\mathbb{Z}/m\mathbb{Z})^*} d_v\sigma_v^{-1}$$

with

$$d_v = \sum_{t\in(\mathbb{Z}/m\mathbb{Z})^*} c_t\{tv/m\} \ .$$

Since $\gamma\Theta \in \mathbb{Z}[G]$ we have $d_v \in \mathbb{Z}$ for all v, and in particular

$$d_1 = \sum_{t\in(\mathbb{Z}/m\mathbb{Z})^*} c_t\{t/m\} = \sum_{\substack{0\leqslant t\leqslant m-1 \\ \gcd(t,m)=1}} c_t t/m \in \mathbb{Z} \ .$$

Thus

$$\delta = \gamma\Theta = \sum_{t\in(\mathbb{Z}/m\mathbb{Z})^*} c_t(\sigma_t - t)\Theta + \sum_{\substack{0\leqslant t\leqslant m-1 \\ \gcd(t,m)=1}} c_t t\Theta = \sum_{t\in(\mathbb{Z}/m\mathbb{Z})^*} c_t\Theta_t + md_1\Theta \ ,$$

and the lemma follows since $c_t \in \mathbb{Z}$, $d_1 \in \mathbb{Z}$, and $m\Theta = (m + 1 - \sigma_{m+1})\Theta = -\Theta_{m+1}$ (note that $\sigma_{m+1} = \sigma_1 = 1$). $\qquad\square$

Lemma 3.6.18. *Recall that $d = (q - 1)/m$. For all b coprime to m we have $\tau(\omega^{-d})^{\sigma_b-b} \in K_m$.*

Proof. Since m is coprime to p, $\mathrm{Gal}(L_m/K_m)$ is the cyclic group of order $p - 1$ formed by the maps α_k such that $\alpha_k(\zeta_p) = \zeta_p^k$ and $\alpha_k(\zeta_m) = \zeta_m$ for $k \in (\mathbb{Z}/p\mathbb{Z})^{*}$[1]. By Galois theory, to prove the lemma we must show that the left-hand side is invariant by each α_k. Since ω^{-d} has order m and the implicit additive character has order p, we have by Definition 2.5.7 and Lemma 2.5.6,

$$\alpha_k(\tau(\omega^{-d})) = \tau(\omega^{-d}, \psi_k) = \omega(k)^d\tau(\omega^{-d}) \ ,$$

hence since α_k and σ_b commute,

---

[1] It is not possible to use here the notation σ_k since this is reserved for elements of $\mathrm{Gal}(L/K)$, hence those that leave ζ_p invariant.

$$\alpha_k(\tau(\omega^{-d})^{\sigma_b - b}) = \omega(k)^{d(b - \sigma_b)} \tau(\omega^{-d})^{\sigma_b - b} ,$$

and the lemma follows since $\omega(k)^d$ is a power of ζ_m, whence $\omega(k)^{d\sigma_b} = \omega(k)^{db}$.
$\square$

It is now easy to deduce the main theorem on the Stickelberger ideal.

Theorem 3.6.19 (Stickelberger). *The Stickelberger ideal $I_s(m)$ annihilates the class group of $K_m = \mathbb{Q}(\zeta_m)$; in other words, for any $\gamma \in I_s(m)$ and any fractional ideal $\mathfrak{a}$ of K_m the ideal $\mathfrak{a}^\gamma$ is a principal ideal.*

Proof. As in the proof of Corollary 3.6.14, it is sufficient to prove that $\mathfrak{p}_m^\gamma$ is a principal ideal for any $\gamma \in I_s(m)$ and any prime ideal $\mathfrak{p}_m$ coprime to m. Raising the equality of Proposition 3.6.13 to the power $\sigma_b - b$ for b prime to m we obtain

$$\mathfrak{p}_m^{m\Theta_b} = \tau(\omega^{-d})^{m(\sigma_b - b)} \mathbb{Z}_{K_m} = \alpha^m \mathbb{Z}_{K_m} ,$$

where $\alpha = \tau(\omega^{-d})^{\sigma_b - b} \in K_m$ by the above lemma. Since $\mathfrak{p}_m^{\Theta_b}$ and $\alpha \mathbb{Z}_{K_m}$ are ideals of K_m whose mth powers are equal, by uniqueness of the prime ideal decomposition in the Dedekind domain $\mathbb{Z}_{K_m}$ we deduce that they are equal, and in particular that $\mathfrak{p}_m^{\Theta_b}$ is a principal ideal. Since by Lemma 3.6.17 the Θ_b generate $I_s(m)$, it follows that $\mathfrak{p}_m^\gamma$ is a principal ideal for all $\gamma \in I_s(m)$, as was to be proved.
$\square$

It is important to note that the above theorem does not bring any new information on the "plus" part of K_m, but only on the "minus" part. More precisely, we have the following:

Proposition 3.6.20. *Let $m \not\equiv 2 \pmod 4$ and denote by $K_m^+ = \mathbb{Q}(\zeta_m + \zeta_m^{-1})$ the maximal totally real subfield of K_m. Then for any ideal $\mathfrak{a}$ of K_m^+ and any b coprime to m we have*

$$(\mathfrak{a}\mathbb{Z}_{K_m})^{\Theta_b} = \mathcal{N}_{K_m^+/\mathbb{Q}}(\mathfrak{a})^{b-1} \mathbb{Z}_{K_m} .$$

Proof. Recall that we denote complex conjugation by ι, and that $\iota\sigma_t = \sigma_{m-t}$, so that when $\mathfrak{a} \subset K_m^+$ we have $\sigma_{m-t}(\mathfrak{a}\mathbb{Z}_{K_m}) = \sigma_t(\mathfrak{a}\mathbb{Z}_{K_m})$. Thus

$$(\mathfrak{a}\mathbb{Z}_{K_m})^{\Theta_b} = \prod_{1 \leqslant t \leqslant m/2, \ \gcd(t,m)=1} \sigma_t(\mathfrak{a}\mathbb{Z}_{K_m})^{n_t}$$

with $n_t = \lfloor bt/m \rfloor + \lfloor b(m-t)/m \rfloor = b - 1$ as is easily seen, since b and t are coprime to m. Since the σ_t for $1 \leqslant t \leqslant m/2$ with $\gcd(t, m) = 1$ restrict to $\mathrm{Gal}(K_m^+/\mathbb{Q})$ and since $\mathcal{N}_{K_m^+/\mathbb{Q}}(\mathfrak{a})\mathbb{Z}_{K_m} = \prod_{\sigma \in \mathrm{Gal}(K_m^+/\mathbb{Q})} \sigma(\mathfrak{a})$, the result follows.
$\square$

Since $I_s(m)$ is generated by the Θ_b as a $\mathbb{Z}$-module, it follows from the above proposition that for any ideal $\mathfrak{a}$ of K_m^+ and any $\gamma \in I_s(m)$ the ideal

$(\mathfrak{a}\mathbb{Z}_{K_m})^\gamma$ is the principal ideal generated by a power of the norm of $\mathfrak{a}$, so that Stickelberger's ideal theorem does not bring any new information on the class group of K_m^+. We will see in Chapter 16 devoted to Catalan's conjecture that there is a beautiful theorem of F. Thaine that partially replaces Stickelberger's theorem for K_m^+.

3.6.5 Diagonalization of the Stickelberger Element

In this section, we will use some simple analytic results that will be proved in Chapters 9 and 10, but which can easily be proved directly except for the nonvanishing of $L(\chi, 1)$. Thus, in contrast to almost all places in this book, there are a couple of forward references.

Keep the above notation, and in particular $G = \mathrm{Gal}(K_m/\mathbb{Q}) \simeq (\mathbb{Z}/m\mathbb{Z})^*$. Denote as usual by $\widehat{G}$ the group of characters of G, which is noncanonically isomorphic to G by Proposition 2.1.16. For $\chi \in \widehat{G}$ we define

$$e_\chi = \frac{1}{\phi(m)} \sum_{\sigma \in G} \chi(\sigma)\sigma^{-1} \in \mathbb{C}[G] .$$

It is immediately checked that the e_χ form a complete set of orthogonal idempotents, in other words that $e_\chi^2 = e_\chi$, that $e_{\chi_1} e_{\chi_2} = 0$ if $\chi_1 \neq \chi_2$, and that $\sum_{\chi \in \widehat{G}} e_\chi = 1$. Indeed, these properties are immediate consequences of the orthogonality of characters (and in fact are equivalent to them). As an immediate consequence we deduce that

$$\mathbb{C}[G] = \bigoplus_{\chi \in \widehat{G}} e_\chi \mathbb{C}[G] .$$

Since $e_\chi \neq 0$ all the $e_\chi \mathbb{C}[G]$ are nonzero, and on the other hand the number of terms in the direct sum is equal to $|\widehat{G}| = |G| = \dim_{\mathbb{C}} \mathbb{C}[G]$. It follows that all the terms are 1-dimensional, hence that $e_\chi \mathbb{C}[G] = \mathbb{C}e_\chi$, so that the e_χ form a $\mathbb{C}$-basis of $\mathbb{C}[G]$. Since $e_\chi \mathbb{C}[G]$ is the principal ideal of $\mathbb{C}[G]$ generated by e_χ, it follows that $\mathbb{C}e_\chi$ is an ideal of $\mathbb{C}[G]$.

Lemma 3.6.21. *Denote by j the canonical isomorphism from $(\mathbb{Z}/m\mathbb{Z})^*$ to G such that $j(t) = \sigma_t$, and complex conjugation σ_{-1} by ι. We have $\Theta e_\chi = \lambda_\chi e_\chi$ with*

$$\lambda_\chi = \begin{cases} \phi(m)/2 & \text{if } \chi \text{ is the trivial character,} \\ 0 & \text{if } \chi(\iota) = 1 \text{ and } \chi \text{ is nontrivial,} \\ -L(\overline{\chi \circ j}, 0) & \text{if } \chi(\iota) = -1. \end{cases}$$

Proof. By definition of e_χ we have

$$\sigma_t e_\chi = \frac{1}{\phi(m)} \sum_{\sigma \in G} \chi(\sigma)\sigma_t \sigma^{-1} = \frac{1}{\phi(m)} \sum_{\sigma \in G} \chi(\sigma\sigma_t)\sigma^{-1} = \chi(\sigma_t) e_\chi .$$

Thus

$$\Theta e_\chi = \frac{1}{m} \sum_{\substack{1 \leqslant t \leqslant m-1 \\ \gcd(t,m)=1}} t\overline{\chi(\sigma_t)} e_\chi = \lambda_\chi e_\chi \,,$$

say. By definition of the χ-Bernoulli numbers (see Section 9.4.1) we thus have $\lambda_\chi = B_1(\overline{\chi \circ j}) + (m/2)B_0(\overline{\chi \circ j})$. We have $B_0(\overline{\chi \circ j}) = \phi(m)/m$ when χ is trivial, and $B_0(\overline{\chi \circ j}) = 0$ otherwise. Furthermore, by Proposition 9.4.9 we have $B_1(\overline{\chi \circ j}) = 0$ when $\chi \circ j$ is an even character, in other words when $\chi(\iota) = 1$ (even when χ is trivial, since $m > 1$). Finally, using Corollary 10.2.3 we have $B_1(\overline{\chi \circ j}) = -L(\overline{\chi \circ j}, 0)$ when χ is an odd character, proving the lemma. $\square$

Note that although for brevity we have referred to definitions and results of Section 9.4.1, the proof can be done completely independently, apart from the equality $B_1(\overline{\chi \circ j}) = -L(\overline{\chi \circ j}, 0)$.

Remark. The fact that $\lambda_\chi = 0$ when χ is a nontrivial even character is another way of saying that the Stickelberger ideal gives information only on the minus part of the ideal class group, as we have already seen in Proposition 3.6.20. In fact, if we define

$$\mathbb{C}[G]^- = \{x \in \mathbb{C}[G], \, \iota x = -x\} = (1 - \iota)\mathbb{C}[G]$$

(the last equality being immediate), we have the following.

Lemma 3.6.22. (1) *A $\mathbb{C}$-basis of $\mathbb{C}[G]^-$ is given by the e_χ for $\chi(\iota) = -1$, in other words such that $\chi \circ j$ is an odd character.*
(2) *If $m = p^k$ is a power of a prime number, multiplication by Θ induces a bijective $\mathbb{C}$-linear map from $\mathbb{C}[G]^-$ to itself.*

Proof. (1). Since $\sigma_t e_\chi = \chi(\sigma_t) e_\chi$ we have $\iota e_\chi = \chi(\iota) e_\chi = -e_\chi$ if $\chi(\iota) = -1$, so such e_χ are in $\mathbb{C}[G]^-$. When $m \geqslant 2$, $m \not\equiv 2 \pmod 4$ (which is the hypothesis that we always make in the cyclotomic case) the number of odd characters modulo m is equal to $\phi(m)/2$. On the other hand, since $\iota \sigma_t = \sigma_{-t}$, $\sum_{t \in (\mathbb{Z}/m\mathbb{Z})^*} a_t \sigma_t \in \mathbb{C}[G]^-$ if and only if $a_{-t} = -a_t$, so that the σ_t for $1 \leqslant t \leqslant m/2$ such that $\gcd(t,m) = 1$ form a basis of $\mathbb{C}[G]^-$; hence $\dim_\mathbb{C} \mathbb{C}[G]^- = \phi(m)/2$, proving (1) since the e_χ are $\mathbb{C}$-linearly independent.

(2). By the preceding lemma, on the $\mathbb{C}$-basis $(e_\chi)_{\chi(\iota)=-1}$ of $\mathbb{C}[G]^-$ the matrix of multiplication by Θ is diagonal, the diagonal elements being $-L(\overline{\chi \circ j}, 0)$ for $\chi(\iota) = -1$. However, since $m = p^k$, by Lemma 10.2.1 if χ is nontrivial its L-series is *equal* to that of the corresponding primitive character. It follows therefore from the functional equation (or Theorem 10.3.1) that

$$L(\overline{\chi \circ j}, 0) = \frac{\sqrt{p}}{W(\chi \circ j)\pi} L(\chi \circ j, 1) \,,$$

and by the nonvanishing of $L(\chi,1)$ (Theorem 10.5.29) we have $L(\overline{\chi \circ j},0) \neq 0$. It follows that the matrix of multiplication by Θ on the basis of the e_χ is a diagonal matrix with nonzero diagonal entries, hence is invertible, proving (2).

□

Remarks. (1) Since $L(\chi \circ j,0)$ may be equal to 0 when $\chi \circ j$ is a nontrivial character, the above result may not be true when m is not a prime power.

(2) We will see in Proposition 10.5.26 that if $m = p^k$ with $p \geqslant 3$, the determinant of the map multiplication by Θ from $\mathbb{C}[G]^-$ to itself is equal to $-(-2)^{\phi(m)/2-1}h_{p^k}^-/p^k$, where we recall that $h_{p^k}^-$ is the minus class number of the p^kth cyclotomic field (see Definition 3.5.23).

3.6.6 The Eisenstein Reciprocity Law

In this subsection I follow quite closely the exposition of [Ire-Ros]. We restate the theorem on the Stickelberger ideal, but now emphasizing the dependence on p. Thus, instead of letting p be a prime, $q = p^f$, and m a divisor of $q-1$, we fix an integer $m \geqslant 2$ such that $m \not\equiv 2 \pmod 4$, we let Θ be as in Proposition 3.6.13, and we set $\gamma = m\Theta$, in other words,

$$\gamma = \sum_{1 \leqslant t \leqslant m-1,\ \gcd(t,m)=1} t\sigma_t^{-1}\ .$$

For all primes p not dividing m we let f be the order of p modulo m, we set $q = p^f \equiv 1 \pmod m$, so that $m \mid (q-1)$, and as above we set $d = (q-1)/m$.

Important Warning. Since m is fixed, we denote simply by $\mathfrak{p}$ instead of $\mathfrak{p}_m$ a prime ideal of $K_m = \mathbb{Q}(\zeta_m)$ above p. Note that previously, $\mathfrak{p}$ denoted the unique prime ideal of $K = \mathbb{Q}(\zeta_p)$ above p, but here the context is different and p is *not* given but will vary, only m is fixed. Note also that since f is the order of p modulo m, by Proposition 3.5.18 we have $f(\mathfrak{p}/p) = f$, hence $\mathcal{N}(\mathfrak{p}) = p^f = q$.

For any prime ideal $\mathfrak{P}$ of $L = \mathbb{Q}(\zeta_{q-1},\zeta_p)$ above $\mathfrak{p}$ (hence above p) recall that we have defined $\omega_\mathfrak{P}$ and $\tau(\omega_\mathfrak{P}^{-d})$. In addition, in the sequel we will make the following common abuse of terminology: we will say that two (possibly fractional) ideals $\mathfrak{a}$ and $\mathfrak{b}$ (or elements) are coprime if for every prime ideal $\mathfrak{q}$ we have either $v_\mathfrak{q}(\mathfrak{a}) = 0$ or $v_\mathfrak{q}(\mathfrak{b}) = 0$.

Definition 3.6.23. *For $x \in K_m$ coprime to $\mathfrak{p}$ we set*

$$\left(\frac{x}{\mathfrak{p}}\right)_m = \omega_\mathfrak{P}^d(x) \quad and \quad G(\mathfrak{p}) = \tau(\omega_\mathfrak{P}^{-d})^m\ .$$

In addition, to simplify notation we will often write $\chi_\mathfrak{p}(x)$ instead of $\left(\frac{x}{\mathfrak{p}}\right)_m^{-1}$, so that $G(\mathfrak{p}) = \tau(\chi_\mathfrak{p})^m$.

This notation is justified by Lemma 3.6.3 (3) and (4), which tells us that $\omega_{\mathfrak{P}}(x)^d$ and $\tau(\omega_{\mathfrak{P}}^{-d})^m \in K_m$ depend only on $\mathfrak{p} = \mathfrak{P} \cap K_m$. The above definition generalizes the well-known quadratic reciprocity symbol $\left(\frac{x}{p}\right)$ studied in Section 2.2, hence is naturally called the mth-power reciprocity symbol. The following series of definitions and formulas is completely analogous to what is done in the classical case of quadratic reciprocity.

Lemma 3.6.24. (1) $\left(\frac{x}{\mathfrak{p}}\right)_m$ *is characterized by the fact that it is a character of order m such that*

$$\left(\frac{x}{\mathfrak{p}}\right)_m \equiv x^{(q-1)/m} \pmod{\mathfrak{p}} .$$

(2) *We have*

$$\tau(\chi_{\mathfrak{p}}) = \sum_{t \in (\mathbb{Z}_{K_m}/\mathfrak{p})^*} \chi_{\mathfrak{p}}(t)\psi_1(t) ,$$

where as usual $\psi_1(t) = \zeta_p^{\mathrm{Tr}_{(\mathbb{Z}_{K_m}/\mathfrak{p})/(\mathbb{Z}/p\mathbb{Z})}(t)}$.

Proof. (1) is the translation of the corresponding properties of the character $\omega_{\mathfrak{P}}$, and (2) from the fact that the natural inclusion map from K_m to L induces a canonical isomorphism between $\mathbb{Z}_{K_m}/\mathfrak{p}$ and $\mathbb{Z}_L/\mathfrak{P}$. $\qquad\square$

Definition 3.6.25. *Let $\mathfrak{a}$ be an integral ideal of K_m coprime to m, and x be coprime to $\mathfrak{a}$. We define $\left(\frac{x}{\mathfrak{a}}\right)_m$ and $G(\mathfrak{a})$ by the formulas*

$$\left(\frac{x}{\mathfrak{a}}\right)_m = \prod_{\mathfrak{p}|\mathfrak{a}} \left(\frac{x}{\mathfrak{p}}\right)_m^{v_\mathfrak{p}(\mathfrak{a})} \quad and \quad G(\mathfrak{a}) = \prod_{\mathfrak{p}|\mathfrak{a}} G(\mathfrak{p})^{v_\mathfrak{p}(\mathfrak{a})} .$$

If $\mathfrak{a} = \alpha\mathbb{Z}_{K_m}$ is a principal ideal, we will write $\left(\frac{x}{\alpha}\right)_m$ and $G(\alpha)$ instead of $\left(\frac{x}{\alpha\mathbb{Z}_{K_m}}\right)_m$ and $G(\alpha\mathbb{Z}_{K_m})$.

Thus by definition we have $\left(\frac{x}{\mathfrak{ab}}\right)_m = \left(\frac{x}{\mathfrak{a}}\right)_m\left(\frac{x}{\mathfrak{b}}\right)_m$ and $G(\mathfrak{ab}) = G(\mathfrak{a})G(\mathfrak{b})$.

Proposition 3.6.26. *We have:*

(1) $|G(\mathfrak{a})|^2 = \mathcal{N}(\mathfrak{a})^m$.
(2) $G(\mathfrak{a})\mathbb{Z}_{K_m} = \mathfrak{a}^\gamma$, *where* $\gamma = m\Theta$ *is as above.*
(3) *If* $\alpha \in \mathbb{Z}_{K_m}$ *there exists a unit $\varepsilon(\alpha)$ of $\mathbb{Z}_{K_m}$ such that $G(\alpha) = \varepsilon(\alpha)\alpha^\gamma$.*

Proof. If $\mathfrak{p}$ is a prime, by Proposition 2.5.9 we have $|G(\mathfrak{p})|^2 = |\tau(\omega_{\mathfrak{P}}^{-d})|^{2m} = q^m = \mathcal{N}(\mathfrak{p})^m$, so (1) follows by multiplicativity. Statement (2) for a prime ideal is exactly Proposition 3.6.13, and the general result follows by multiplicativity. By (2) we have $G(\alpha)\mathbb{Z}_{K_m} = \alpha^\gamma\mathbb{Z}_{K_m}$, hence $G(\alpha) = \varepsilon(\alpha)\alpha^\gamma$ for some unit $\varepsilon(\alpha)$. $\qquad\square$

We now want to show that $\varepsilon(\alpha)$ is a root of unity. For this we need two lemmas.

Lemma 3.6.27. *For any integral ideal* $\mathfrak{a}$ *prime to* m *and any* $\sigma \in \mathrm{Gal}(K_m/\mathbb{Q})$ *we have* $G(\mathfrak{a})^\sigma = G(\mathfrak{a}^\sigma)$.

Proof. By definition, if $\mathfrak{p}$ is a prime ideal coprime to m we have $G(\mathfrak{p}) = \tau(\omega_{\mathfrak{P}}^{-d})^m$. Thus by Lemma 3.6.3 we have $G(\mathfrak{p})^\sigma = \tau(\omega_{\sigma(\mathfrak{P})}^{-d})^m = G(\sigma(\mathfrak{p}))$ since $\sigma(\mathfrak{p})$ is the prime ideal of K_m below $\sigma(\mathfrak{P})$. As usual, the lemma follows by multiplicativity. □

Lemma 3.6.28. *If* $\alpha \in \mathbb{Z}_{K_m}$ *then* $|\alpha^\gamma|^2 = \mathcal{N}(\alpha)^m$.

Proof. Since σ_{-1} sends ζ_m to ζ_m^{-1} it is complex conjugation. Thus

$$|\alpha^\gamma|^2 = \alpha^\gamma \sigma_{-1}(\alpha^\gamma) = \alpha^{(1+\sigma_{-1})\gamma} .$$

Denoting by $\sum_t^*$ a sum for $1 \leqslant t \leqslant m-1$ such that $\gcd(t,m) = 1$ we have

$$(1 + \sigma_{-1})\gamma = \sum_t^* t\sigma_t^{-1} + \sum_t^* t\sigma_{-t}^{-1} = \sum_t^* t\sigma_t^{-1} + \sum_t^* (m-t)\sigma_t^{-1}$$

$$= m \sum_t^* \sigma_t^{-1} = m \sum_{\sigma \in \mathrm{Gal}(K_m/\mathbb{Q})} \sigma .$$

It follows that

$$\alpha^{(1+\sigma_{-1})\gamma} = \prod_{\sigma \in \mathrm{Gal}(K_m/\mathbb{Q})} \sigma(\alpha)^m = \mathcal{N}_{K_m/\mathbb{Q}}(\alpha)^m .$$

Proposition 3.6.29. *The element* $\varepsilon(\alpha)$ *is a root of unity. In other words, there exist* $i \in \mathbb{Z}$ *and a sign* $\pm$ *such that* $G(\alpha) = \pm\zeta_m^i \alpha^\gamma$.

Proof. By Proposition 3.6.26 (1) and (3) we have

$$|G(\alpha)|^2 = |\mathcal{N}(\alpha)|^m = |\varepsilon(\alpha)|^2 |\alpha^\gamma|^2 = |\varepsilon(\alpha)|^2 \mathcal{N}(\alpha)^m ,$$

hence $|\varepsilon(\alpha)| = 1$ (note that $\mathcal{N}(\alpha)$ is automatically positive). On the other hand, applying Lemma 3.6.27 to $\mathfrak{a} = \alpha \mathbb{Z}_{K_m}$ and using Proposition 3.6.26 (3) we deduce that for all $\sigma \in \mathrm{Gal}(K_m/\mathbb{Q})$,

$$\varepsilon(\alpha)^\sigma \alpha^{\gamma\sigma} = G(\alpha)^\sigma = G(\alpha^\sigma) = \varepsilon(\alpha^\sigma)\alpha^{\gamma\sigma} ,$$

so that $\varepsilon(\alpha)^\sigma = \varepsilon(\alpha^\sigma)$. Since we have shown that $|\varepsilon(\alpha)| = 1$ for all $\alpha \in \mathbb{Z}_{K_m}$, and in particular for α^σ, it follows that $|\varepsilon(\alpha)^\sigma| = 1$ for all σ. We conclude by Kronecker's theorem (Corollary 3.3.10) that $\varepsilon(\alpha)$ is a root of unity, and the proposition follows from Corollary 3.5.12. □

We can now proceed to the statement and proof of Eisenstein's reciprocity law. As the reader will notice, the proof of the following proposition is essentially identical to that of the classical proof of the quadratic reciprocity law (see the proof of Lemma 2.2.2).

Proposition 3.6.30. *Let p_1 and p_2 be two distinct prime numbers not dividing m and let $\mathfrak{p}_1$ and $\mathfrak{p}_2$ be prime ideals of K_m above p_1 and p_2 respectively. Then*

$$\left(\frac{G(\mathfrak{p}_1)}{\mathfrak{p}_2}\right)_m = \left(\frac{\mathcal{N}(\mathfrak{p}_2)}{\mathfrak{p}_1}\right)_m .$$

Proof. Let f_1 and f_2 respectively be the orders of p_1 and p_2 modulo m, so that by Proposition 3.5.18 we have $f(\mathfrak{p}_i/p) = f_i$. Set $q_i = \mathcal{N}(\mathfrak{p}_i) = p_i^{f_i} \equiv 1 \pmod{m}$ and recall the notation of Definition 3.6.23. Since $\chi_{\mathfrak{p}_1}(t)$ is an mth root of unity and $m \mid (q_2 - 1)$, by Lemma 3.6.24 we have

$$\tau(\chi_{\mathfrak{p}_1})^{q_2} \equiv \sum_{t \in (\mathbb{Z}_{K_m}/\mathfrak{p}_1)^*} \chi_{\mathfrak{p}_1}(t)^{q_2} \psi_1(t)^{q_2}$$

$$\equiv \sum_{t \in (\mathbb{Z}_{K_m}/\mathfrak{p}_1)^*} \chi_{\mathfrak{p}_1}(t) \psi_1(q_2 t) \equiv \chi_{\mathfrak{p}_1}(q_2)^{-1} \tau(\chi_{\mathfrak{p}_1}) \pmod{p_2 \mathbb{Z}_{K_m}} .$$

On the other hand, again by Lemma 3.6.24 we have

$$\tau(\chi_{\mathfrak{p}_1})^{q_2-1} = G(\mathfrak{p}_1)^{(q_2-1)/m} \equiv \left(\frac{G(\mathfrak{p}_1)}{\mathfrak{p}_2}\right)_m \equiv \chi_{\mathfrak{p}_2}(G(\mathfrak{p}_1))^{-1} \pmod{\mathfrak{p}_2} .$$

Since $|\tau(\chi_{\mathfrak{p}_1})|^2 = q_1$ is coprime to p_2, it follows from these two congruences that

$$\chi_{\mathfrak{p}_2}(G(\mathfrak{p}_1))^{-1} \equiv \chi_{\mathfrak{p}_1}(q_2)^{-1} \equiv \chi_{\mathfrak{p}_1}(\mathcal{N}(\mathfrak{p}_2))^{-1} \pmod{\mathfrak{p}_2} .$$

Since both sides are mth roots of unity and since p_2 is coprime to m, it follows that they are equal, proving the proposition. □

Corollary 3.6.31. *For all ideals $\mathfrak{a}$ and $\mathfrak{b}$ coprime to m such that $\mathcal{N}(\mathfrak{a})$ and $\mathcal{N}(\mathfrak{b})$ are coprime we have $\left(\frac{G(\mathfrak{a})}{\mathfrak{b}}\right)_m = \left(\frac{\mathcal{N}(\mathfrak{b})}{\mathfrak{a}}\right)_m$.*

Proof. Clear since by definition $\left(\frac{\cdot}{\mathfrak{a}}\right)_m$ is a character and is multiplicative in $\mathfrak{a}$, and $G(\mathfrak{a})$ and $\mathcal{N}(\mathfrak{a})$ are also multiplicative. □

Corollary 3.6.32. *In addition to the assumptions of the preceding corollary, assume that $\mathfrak{a} = \alpha \mathbb{Z}_{K_m}$ is a principal ideal. Then*

$$\left(\frac{\mathcal{N}(\mathfrak{b})}{\alpha}\right)_m = \left(\frac{\varepsilon(\alpha)}{\mathfrak{b}}\right)_m \left(\frac{\alpha}{\mathcal{N}(\mathfrak{b})}\right)_m .$$

Proof. By multiplicativity we have

$$\left(\frac{G(\alpha)}{\mathfrak{b}}\right)_m = \left(\frac{\varepsilon(\alpha)}{\mathfrak{b}}\right)_m \left(\frac{\alpha^\gamma}{\mathfrak{b}}\right)_m = \left(\frac{\varepsilon(\alpha)}{\mathfrak{b}}\right)_m \prod_{1 \leqslant t \leqslant m-1,\ \gcd(t,m)=1} A(t) ,$$

where

$$A(t) = \left(\frac{\sigma_t^{-1}(\alpha)}{\mathfrak{b}}\right)^t = \sigma_t\left(\left(\frac{\sigma_t^{-1}(\alpha)}{\mathfrak{b}}\right)_m\right) = \left(\frac{\alpha}{\sigma_t(\mathfrak{b})}\right)_m$$

by Lemma 3.6.3. The result follows since $\mathcal{N}(\mathfrak{b}) = \prod_{1\leqslant t\leqslant m-1,\ \gcd(t,m)=1} \sigma_t(\mathfrak{b})$.
□

From now on we will assume that $m = \ell$ is a prime number, we will let $R = \mathbb{Z}[\zeta_\ell]$ be the ring of integers of $\mathbb{Q}(\zeta_\ell)$, and we denote by $\mathcal{L} = (1 - \zeta_\ell)R$ the unique prime ideal of R above ℓ.

Lemma 3.6.33. *If $\mathfrak{a}$ is an ideal coprime to ℓ then $G(\mathfrak{a}) \equiv \pm 1 \pmod{\ell R}$.*

Proof. If $\mathfrak{p}$ is a prime ideal coprime to ℓ we have as above

$$G(\mathfrak{p}) = \tau(\chi_\mathfrak{p})^\ell \equiv \tau(\chi_\mathfrak{p}^\ell, \psi_\ell) \equiv \sum_{t\in(\mathbb{Z}/\ell\mathbb{Z})^*} \psi_1(\ell t) \equiv -1 \pmod{\ell R}$$

since ψ_1 is a nontrivial additive character, so the lemma follows by multiplicativity.
□

Definition 3.6.34. *Let $\alpha \in R$. We say that α is primary if it is not a unit and if there exists $x \in \mathbb{Z}$ with $\ell \nmid x$ such that $\alpha \equiv x \pmod{\mathcal{L}^2}$.*

Lemma 3.6.35. *If $\alpha \in R$ is coprime to ℓ there exists $i \in \mathbb{Z}$, unique modulo ℓ, such that $\zeta_\ell^i \alpha$ is primary.*

Proof. Since $f(\mathcal{L}/\ell) = 1$ we have $R/\mathcal{L} \simeq \mathbb{Z}/\ell\mathbb{Z}$, so there exists $a \in \mathbb{Z}$, evidently unique modulo ℓ, such that $\alpha \equiv a \pmod{\mathcal{L}}$. Thus $\beta = (\alpha - a)/(1 - \zeta_\ell) \in R$, so in the same way there exists $b \in \mathbb{Z}$, unique modulo ℓ, such that $\beta \equiv b \pmod{\mathcal{L}}$. It follows that $\alpha \equiv a + b(1 - \zeta_\ell) \pmod{\mathcal{L}^2}$. By the binomial theorem we have $\zeta_\ell^i = (1 - (1 - \zeta_\ell))^i \equiv 1 - i(1 - \zeta_\ell) \pmod{\mathcal{L}^2}$, hence $\zeta_\ell^i \alpha \equiv a + (b - ai)(1 - \zeta_\ell) \pmod{\mathcal{L}^2}$, so we choose $i = ba^{-1} \bmod \ell$, which is possible since $\ell \nmid a$, otherwise α would not be coprime to ℓ.
□

Lemma 3.6.36. *If α is primary we have $\varepsilon(\alpha) = \pm 1$.*

Proof. Since $\mathcal{L}$ is the unique prime ideal above ℓ, for all $\sigma \in \mathrm{Gal}(\mathbb{Q}(\zeta_\ell)/\mathbb{Q})$ we have $\sigma(\mathcal{L}) = \mathcal{L}$. By definition of γ it follows that $\mathcal{L}^\gamma \subset \mathcal{L}$. Thus

$$\alpha^\gamma \equiv x^\gamma \equiv x^{\sum_{1\leqslant t\leqslant \ell-1} t} \equiv x^{\ell(\ell-1)/2} \pmod{\mathcal{L}^2}.$$

By Fermat's theorem we have $x^{(\ell-1)/2} \equiv \pm 1 \pmod{\ell}$, hence $\alpha^\gamma \equiv \pm 1 \pmod{\mathcal{L}^2}$. On the other hand, by the preceding lemma we have $G(\alpha) = \varepsilon(\alpha)\alpha^\gamma \equiv \pm 1 \pmod{\ell R}$, hence $\varepsilon(\alpha) \equiv \pm 1 \pmod{\mathcal{L}^2}$. Now by Proposition 3.6.29 we have $\varepsilon(\alpha) = \pm\zeta_\ell^i$ for some integer i, hence $\zeta_\ell^i \equiv \pm 1 \pmod{\mathcal{L}^2}$. I claim that $\ell \mid i$ (this follows in fact from the uniqueness statement of Lemma 3.6.35). Indeed, as we have already seen, by the binomial theorem

$$\zeta_\ell^i = (1 - (1 - \zeta_\ell))^i \equiv 1 - i(1 - \zeta_\ell) \ (\mathrm{mod}\ \mathcal{L}^2) \,.$$

Thus the $+$ sign must hold in the congruence (otherwise $\mathcal{L} \mid 2$); hence $i(1 - \zeta_\ell) \equiv 0 \ (\mathrm{mod}\ \mathcal{L}^2)$, so that $\mathcal{L} \mid i$, and hence $\ell \mid i$ as claimed. It follows that $\varepsilon(\alpha) = \pm 1$. $\qquad\qquad\square$

Proposition 3.6.37. *If α is primary and $\mathfrak{b}$ is an ideal coprime to ℓ such that $\mathcal{N}(\mathfrak{b})$ is coprime to αR then*

$$\left(\frac{\alpha}{\mathcal{N}(\mathfrak{b})}\right)_\ell = \left(\frac{\mathcal{N}(\mathfrak{b})}{\alpha}\right)_\ell \,.$$

Proof. Since α is primary the above lemma tells us that $\varepsilon(\alpha) = \pm 1$. On the other hand, for all $\mathfrak{p}$ coprime to ℓ, $\chi_\mathfrak{p}$ is a character of order ℓ, we have $\chi_\mathfrak{p}(\pm 1)^\ell = 1$ and $\chi_\mathfrak{p}(\pm 1)^2 = \chi_\mathfrak{p}(1) = 1$, and since ℓ is odd it follows by multiplicativity that for all $\mathfrak{b}$ prime to ℓ we have $\chi_\mathfrak{b}(\pm 1) = 1$, and in particular $\chi_\mathfrak{b}(\varepsilon(\alpha)) = 1$. The result thus follows from Corollary 3.6.32. $\qquad\square$

The Eisenstein reciprocity law immediately follows:

Theorem 3.6.38 (Eisenstein). *Let ℓ be an odd prime, $a \in \mathbb{Z}$ an integer not divisible by ℓ, $\alpha \in R$ a primary element, and assume that αR and aR are coprime. Then*

$$\left(\frac{\alpha}{a}\right)_\ell = \left(\frac{a}{\alpha}\right)_\ell \,.$$

Proof. As usual, by multiplicativity it is enough to prove this when $a = p$ is a prime number different from ℓ and prime to αR. Let $\mathfrak{p}$ be a prime ideal of R above p, $f = f(\mathfrak{p}/p)$, so that $\mathcal{N}(\mathfrak{p}) = p^f$. By the above proposition with $\mathfrak{b} = \mathfrak{p}$ we have $\chi_\mathfrak{p}(\alpha)^f = \chi_\alpha(p)^f$. Since $f \mid (\ell - 1) = [\mathbb{Q}(\zeta_\ell)/\mathbb{Q}]$, f is coprime to ℓ, and since both χ-values are ℓth roots of unity it follows that they are equal. $\qquad\square$

3.7 The Hasse–Davenport Relations

Thanks to Stickelberger's congruence (Theorem 3.6.6), we can return to Gauss sums over finite fields considered in Chapter 2 and prove the important Hasse–Davenport (HD for short) relations. The first one is called the HD product relation; the second is called the HD lifting relation. As mentioned in that chapter, although these relations are explicit finite identities between algebraic numbers, of the same type as those giving the modulus of Gauss sums or the links between Gauss and Jacobi sums, no really "elementary" proof has been found of the product relation, except in special cases (the case $m = 2$ of the product relation is proved in Exercise 40 of Chapter 2). Because of their importance I have thus decided to include *two* proofs of these relations. In the present section, I will show how Stickelberger's congruence combined with some basic algebraic number theory and suitable *distribution relations* leads to the proof of both relations (the product relation only for $p \geqslant 3$). In Exercise 32 I give a direct elementary proof of the HD lifting relation. Finally, in Section 11.7.4 I will show how the Gross–Koblitz formula together with the distribution relation for the p-adic gamma function leads to a painless proof of the HD product relation for all p.

3.7.1 Distribution Formulas

To prove the Hasse–Davenport relations we need *distribution formulas* for the functions $s(r)$ and $t(r)$ given by Definition 3.6.5, analogous to several such relations that we have already seen. The result for $s(r)$ is as follows.

Proposition 3.7.1. *Recall that $\{z\}$ denotes the fractional part of $z \in \mathbb{R}$.*

(1) *If k and m are coprime integers and $z \in \mathbb{R}$ we have*

$$\sum_{0 \leqslant a < m} \left\{ \frac{ka+z}{m} \right\} = \frac{m-1}{2} + \{z\} \ .$$

(2) *If $m \mid (q-1)$ and $d = (q-1)/m$, then for all $b \in \mathbb{Z}$ we have*

$$\sum_{0 \leqslant a < m} s(da+b) - \sum_{0 \leqslant a < m} s(da) = s(mb) \ .$$

In addition,

$$\sum_{0 \leqslant a < m} s(da) = (p-1)(m-1)f/2 \ .$$

Proof. (1). The expression $\{(ka+z)/m\}$ depends only on a modulo m; hence we may replace the summation by $\sum_{a \in \mathbb{Z}/m\mathbb{Z}}$. Since k is coprime to m, multiplication by k is an automorphism of $\mathbb{Z}/m\mathbb{Z}$, and the map $a \mapsto a + \lfloor z \rfloor$ is also an automorphism of $\mathbb{Z}/m\mathbb{Z}$. It follows that if we set $y = \{z\}$ we have

$$\sum_{0 \leqslant a < m} \left\{ \frac{ka+z}{m} \right\} = \sum_{0 \leqslant a < m} \left\{ \frac{a+y}{m} \right\}$$

$$= \sum_{0 \leqslant a < m} \frac{a+y}{m} - \sum_{0 \leqslant a < m} \left\lfloor \frac{a+y}{m} \right\rfloor = \frac{m-1}{2} + y$$

since $0 \leqslant a + y < m$ for $0 \leqslant a < m$, proving (1).

(2). This is an immediate consequence of (1) and of Lemma 3.6.7: by that lemma we have

$$\sum_{0\leqslant a<m} s(da+b) = (p-1)\sum_{0\leqslant i<f}\sum_{0\leqslant a<m}\left\{\frac{p^i(da+b)}{q-1}\right\}$$

$$= (p-1)\sum_{0\leqslant i<f}\sum_{0\leqslant a<m}\left\{\frac{p^i a + p^i b/d)}{m}\right\}$$

$$= (p-1)\sum_{0\leqslant i<f}\left(\frac{m-1}{2}+\{p^i b/d\}\right)$$

$$= (p-1)(m-1)f/2 + (p-1)\sum_{0\leqslant i<f}\{p^i b/d\}$$

since $m \mid (q-1)$ is coprime to p^i, which already gives the second formula of (2). It follows that

$$\sum_{0\leqslant a<m} s(da+b) - \sum_{0\leqslant a<m} s(da) = (p-1)\sum_{0\leqslant i<f}\left\{\frac{p^i mb}{q-1}\right\} = s(mb)$$

by Lemma 3.6.7, proving the proposition.   □

Similarly, we have the following distribution relation for the function $t(r)$, completely analogous to the one for $s(r)$, except that it is multiplicative instead of additive.

Proposition 3.7.2. *If $m \mid (q-1)$ and $d = (q-1)/m$, then for all $b \in \mathbb{Z}$ we have*

$$\frac{m^{mb}\prod_{0\leqslant a<m} t(da+b)}{t(mb)\prod_{0\leqslant a<m} t(da)} \equiv 1 \pmod{p} .$$

Proof. Consider the left-hand side as a function $F(b)$ of b. We have evidently $F(0) = 1$. Furthermore, $t(m(b+d)) = t(mb+q-1) = t(mb)$, $\prod_{0\leqslant a<m} t(da+b+d) = \prod_{1\leqslant a\leqslant m} t(da+b) = \prod_{0\leqslant a<m} t(da+b)$ since $t(dm+b) = t(b)$, and $m^{m(b+d)} = m^{mb}m^{q-1} \equiv m^{mb} \pmod{p}$. It follows that $F(b)$ modulo p is periodic of period d, hence we may assume that $0 \leqslant b < d$. We prove the proposition by induction on $b \in [0, d-1]$. For $b = 0$ the result is clear, so assume that $1 \leqslant b < d$. Since $0 \leqslant da \leqslant da+b < q-1$ for $0 \leqslant a < m$ and using Lemma 3.6.7, for some integer x we have

$$\frac{F(b)}{F(b-1)} \equiv m^m(-p)^x\frac{\prod_{0\leqslant a<m}(da+b)}{(mb)!/(m(b-1))!} \equiv (-p)^x\frac{\prod_{0\leqslant a<m}(mb+(q-1)a)}{\prod_{0\leqslant a<m}(mb-a)}$$

$$\equiv (-p)^x\prod_{0\leqslant a<m}\frac{mb-a+p^f a}{mb-a} \pmod{p} .$$

Since $0 < mb-a < q-1$ we have $v_p(mb-a) < f$, hence $v_p(mb-a+p^f a) = \min(v_p(mb-a), f+v_p(a)) = v_p(mb-a)$, so that $v_p((mb-a+p^f a)/(mb-a)) = 0$. Since we also have $v_p(F(b)) = 0$ it follows that $x = 0$. Furthermore, for the same reason $v_p(p^f a/(mb - a)) \geqslant 1$; hence $(mb - a + p^f a)/(mb - a) \equiv 1$ (mod p), so that $F(b)/F(b - 1) \equiv 1$ (mod p), proving the proposition by induction. $\square$

3.7.2 The Hasse–Davenport Relations

We now have all the tools necessary to prove the Hasse–Davenport relations, at least when $p \geqslant 3$.

Theorem 3.7.3 (Hasse–Davenport product relation). *Let ρ be a multiplicative character of exact order $m \mid (q-1)$, and let χ be any multiplicative character on the finite field $\mathbb{F}_q$. For any nontrivial additive character ψ we have*

$$\prod_{0 \leqslant a < m} \tau(\chi\rho^a, \psi) = -\chi^{-m}(m)\tau(\chi^m, \psi) \prod_{0 \leqslant a < m} \tau(\rho^a, \psi) \, .$$

Proof. Since we know the valuations of Gauss sums for all primes and also their moduli, the proof of this theorem amounts to computing a specific root of unity, and for $p \geqslant 3$ this will follow from the distribution relation for the function t proved above. There is a similar proof for $p = 2$, which will be given as Exercise 54 of Chapter 11.

First note that if χ is equal to a power of ρ then $\chi^m = \varepsilon$, and the identity is trivial since $\tau(\varepsilon, \psi) = -1$. We may therefore assume that this is not the case, hence that none of the characters occurring in the identity is the trivial character.

We use the notation of the proof of Stickelberger's Theorem 3.6.6, and especially that of Section 3.6.2. Let $\mathfrak{P}$ be a prime ideal of $L = \mathbb{Q}(\zeta_{q-1}, \zeta_p)$ above p, and let $\omega = \omega_{\mathfrak{P}}$ be the corresponding character of order $q - 1$. Since ρ has exact order m we have $\rho = \omega^{-dk}$ with $d = (q - 1)/m$ for some k coprime to m, and replacing ρ by ω^{-k} (which does not change the set of ρ^a for $0 \leqslant a < m$), we may assume that $k = 1$. On the other hand, let b be such that $\chi = \omega^{-b}$. We thus have $\tau(\chi\rho^a, \psi) = \tau(\omega^{-(da+b)}, \psi)$. Set

$$\zeta = -\frac{\chi^m(m) \prod_{0 \leqslant a < m} \tau(\chi\rho^a, \psi)}{\tau(\chi^m, \psi) \prod_{0 \leqslant a < m} \tau(\rho^a, \psi)} = -\frac{\omega^{-mb}(m) \prod_{0 \leqslant a < m} \tau(\omega^{-(da+b)}, \psi)}{\tau(\omega^{-mb}, \psi) \prod_{0 \leqslant a < m} \tau(\omega^{-da}, \psi)} \, .$$

It follows from Stickelberger's theorem (more precisely from Corollary 3.6.8) that

$$v_{\mathfrak{P}}(\zeta) = \sum_{0 \leqslant a < m} s(da + b) - \sum_{0 \leqslant a < m} s(da) - s(mb) = 0$$

by the distribution relation for the function $s(a)$. Since this is true for every prime ideal $\mathfrak{P}$ above p and since these are the only ones that can occur in the prime ideal decomposition of Gauss sums over the field $\mathbb{F}_q$, it follows that ζ is a unit in $\overline{\mathbb{Q}}$. Furthermore, from Proposition 2.5.14 we have

$$\zeta = \chi^m(m) \frac{J_m(\chi, \chi\rho, \dots, \chi\rho^{m-1})}{J_m(\chi^m, \rho, \dots, \rho^{m-1})} \in \mathbb{Q}(\zeta_{q-1}) .$$

Since χ is not a power of ρ, each Gauss sum has modulus equal to $q^{1/2}$ except $\tau(\rho^0, \psi) = -1$, so ζ has modulus 1. Since $\sigma_k \in \mathrm{Gal}(\mathbb{Q}(\zeta_{q-1})/\mathbb{Q})$ sends ζ_{q-1} to ζ_{q-1}^k, hence χ to χ^k, it follows for the same reason that all the conjugates of ζ have modulus equal to 1. Thus by Kronecker's theorem (Corollary 3.3.10) it follows that ζ is a root of unity belonging to $\mathbb{Q}(\zeta_{q-1})$; hence $\zeta = \pm\zeta_{q-1}^k$ for some $k \in \mathbb{Z}$. From the above expression of ζ in terms of Jacobi sums, it is clear that ζ does not depend on the choice of the nontrivial additive character ψ, so as usual we will choose $\psi = \psi_1$.

Let again $\mathfrak{P}$ be any prime ideal of L. By Stickelberger's Theorem 3.6.8 we have

$$\tau(\omega^{-r}, \psi_1) \equiv -\frac{(\zeta_p - 1)^{s(r)}}{t(r)} \pmod{\mathfrak{P}^{s(r)+1}} .$$

Since the powers of $\zeta_p - 1$ cancel by the distribution formula for s, it follows from the distribution formula for t that

$$\zeta \equiv \frac{t(mb) \prod_{0 \leqslant a < m} t(da)}{\omega^{mb}(m) \prod_{0 \leqslant a < m} t(da + b)} \equiv (m/\omega(m))^{mb} \equiv 1 \pmod{\mathfrak{P}} ,$$

since by definition $\omega(m) \equiv m \pmod{\mathfrak{P}}$. Now $\prod_{1 \leqslant k < q-1}(1 - \zeta_{q-1}^k) = q - 1$; hence if $\zeta = \zeta_{q-1}^k$ for $1 \leqslant k < q-1$ we would have $(1-\zeta) \mid (q-1)$, so that $1-\zeta$ would be prime to p, in contradiction with the above congruence. Similarly, it is immediate that for $p \geqslant 3$ we have $\prod_{0 \leqslant k < q-1, \, k \neq (q-1)/2}(1 + \zeta_{q-1}^k) = q - 1$, giving again a contradiction if $\zeta = -\zeta_{q-1}^k$. Thus $\zeta = \zeta_{q-1}^0 = 1$, proving the theorem for $p \geqslant 3$. $\qquad\square$

Remarks. (1) For $p = 2$ we have $\prod_{0 \leqslant k < q-1, \, k \neq (q-1)/2}(1 + \zeta_{q-1}^k) = 2$, so that the possibility that $\zeta = -1$ is not excluded by the last argument of the proof. To be able to exclude it we must compute ζ modulo $\mathfrak{P}^2$, thus essentially modulo 4, and this will be done in Exercise 54 of Chapter 11.

(2) Since $m^{p-1} \equiv 1 \pmod{p}$, it follows that as an element of $\mathbb{F}_q$ we have $m^{p-1} = 1$, hence $\chi(m^{p-1}) = 1$, so that $\chi(m)$ is a $(p - 1)$st root of unity, and not simply a $(q - 1)$st root of unity.

(3) By Proposition 2.5.9 the modulus of the product on the right-hand side is equal to $q^{(m-1)/2}$. We will see in Theorem 11.7.16 that it is in fact equal to $q^{(m-1)/2}$ times an explicit fourth root of unity.

Theorem 3.7.4 (Hasse–Davenport lifting relation). *Let $\mathbb{F}_{q^n}/\mathbb{F}_q$ be an extension of finite fields of degree n, let χ (respectively ψ) be a nontrivial multiplicative (respectively additive) character on $\mathbb{F}_q$. Define $\chi^{(n)} = \chi \circ N_{\mathbb{F}_{q^n}/\mathbb{F}_q}$ and $\psi^{(n)} = \psi \circ \mathrm{Tr}_{\mathbb{F}_{q^n}/\mathbb{F}_q}$. Then*

$$\tau(\chi^{(n)}, \psi^{(n)}) = (-1)^{n-1}\tau(\chi, \psi)^n .$$

Despite its simplicity, this theorem is not so easy to prove, although in contrast to the HD product relation there does exist a relatively simple direct proof, which we give in Exercise 32.

Proof. If $x \in \mathbb{F}_{q^n}$, the conjugates of x are the x^{q^i} for $0 \leqslant i < n$. Thus

$$\mathcal{N}_{\mathbb{F}_{q^n}/\mathbb{F}_q}(x) = \prod_{0 \leqslant i < n} x^{q^i} = x^{(q^n-1)/(q-1)} .$$

On the other hand, by transitivity of the trace we have

$$\mathrm{Tr}_{\mathbb{F}_q/\mathbb{F}_p}(\mathrm{Tr}_{\mathbb{F}_{q^n}/\mathbb{F}_q}(x)) = \mathrm{Tr}_{\mathbb{F}_{q^n}/\mathbb{F}_p}(x) .$$

Keeping the notation of the preceding sections, we have $\psi = \psi_b$ and $\chi = \omega^{-a}$ for a unique $b \in \mathbb{F}_q^*$ and integer a modulo $q-1$. It follows that

$$\psi^{(n)}(x) = \psi \circ \mathrm{Tr}_{\mathbb{F}_{q^n}/\mathbb{F}_q}(x) = \mathrm{Tr}_{\mathbb{F}_{q^n}/\mathbb{F}_p}(bx) = \psi_b^{(n)}(x)$$
$$\chi^{(n)}(x) = \omega^{-a}(x^{(q^n-1)/(q-1)}) = (\omega^{(n)})^{-a(q^n-1)/(q-1)} ,$$

using the additional upper index (n) on ψ_b and ω to indicate that they are with respect to the larger finite field $\mathbb{F}_{q^n}$. This implies in particular that $\chi^{(n)}$ is also a nontrivial character since $(q^n - 1) \mid a(q^n-1)/(q-1)$ is equivalent to $(q-1) \mid a$, which is excluded since χ is nontrivial. Equivalently, this also immediately follows from the surjectivity of the norm from $\mathbb{F}_{q^n}$ to $\mathbb{F}_q$ (Proposition 2.4.12), and the surjectivity of the trace (Proposition 2.4.11) implies that $\psi^{(n)}$ is also nontrivial. These immediate preliminaries out of the way, we can now mimic the proof of the HD product relation.

As above, set $L = \mathbb{Q}(\zeta_{q-1}, \zeta_p)$, and set $\zeta = (-1)^{n-1}\tau(\chi^{(n)}, \psi^{(n)})/\tau(\chi, \psi)^n$. Since the values of χ and ψ are in L, we have $\zeta \in L$. Furthermore, $\mathrm{Gal}(L/\mathbb{Q}(\zeta_{q-1}))$ is the group of maps α_k defined by $\sigma_k(\zeta_p) = \zeta_p^k$ for k coprime to p (see the footnote in the proof of Lemma 3.6.18). Clearly

$$\alpha_k(\tau(\chi, \psi)) = \tau(\chi, \psi^k) = \chi^{-1}(k)\tau(\chi, \psi)$$

by Lemma 2.5.6, and similarly

$$\alpha_k(\tau(\chi^{(n)}, \psi^{(n)})) = (\chi^{(n)})^{-1}(k)\tau(\chi^{(n)}, \psi^{(n)})$$
$$= \chi^{-1}(\mathcal{N}_{\mathbb{F}_{q^n}/\mathbb{F}_q}(k))\tau(\chi^{(n)}, \psi^{(n)}) = \chi^{-n}(k)\tau(\chi^{(n)}, \psi^{(n)}) .$$

It follows that $\alpha_k(\zeta) = \zeta$ for all k coprime to p, hence that $\zeta \in \mathbb{Q}(\zeta_{q-1})$ (this part of the proof is the analogue of the use of Jacobi sums to prove the same result).

Let $\mathfrak{P}$ be a prime ideal of L above p. By Corollary 3.6.8 we have

$$v_{\mathfrak{P}}(\tau(\chi^{(n)}, \psi^{(n)})) = v_{\mathfrak{P}}\left(\tau(\omega^{-a(q^n-1)/(q-1)}, \psi^{(n)})\right) = s(a(q^n-1)/(q-1)) .$$

Since $a(q^n - 1)/(q-1) = \sum_{0 \leqslant i < n} aq^i$ and q is a power of p, it follows from the definition of s that when $0 \leqslant a < q - 1$ we have

$$s(a(q^n - 1)/(q - 1)) = \sum_{0 \leqslant i < n} s(aq^i) = \sum_{0 \leqslant i < n} s(a) = ns(a) = nv_{\mathfrak{P}}(\tau(\chi, \psi)) ,$$

proving that $v_{\mathfrak{P}}(\zeta) = 0$ for all $\mathfrak{P}$ above p. Since the moduli of all Gauss sums are powers of p, it follows that ζ is a unit. More precisely, since $\psi^{(n)}$ and $\psi^{(n)}$ are nontrivial we have $|\tau(\chi^{(n)}, \psi^{(n)})| = q^{n/2}$ and $|\tau(\chi, \psi)| = q^{1/2}$, hence $|\zeta| = 1$. Changing ζ_{q-1} into ζ_{q-1}^k for some k coprime to $q - 1$ is equivalent to changing χ into χ^k hence $\chi^{(n)}$ into $(\chi^{(n)})^k$. Thus all the conjugates of ζ also have modulus equal to 1. Since ζ is a unit it follows from Kronecker's theorem that ζ is a root of unity, and since $\zeta \in \mathbb{Q}(\zeta_{q-1})$, that ζ is a $(q-1)$st root of unity. Finally, by Theorem 3.6.6 we have

$$\tau(\chi^{(n)}, \psi^{(n)}) \equiv -\frac{(\zeta_p - 1)^{ns(a)}}{t(a(q^n - 1)/(q - 1))} \pmod{\mathfrak{P}^{ns(a)+1}} .$$

As we have seen above, the digits in base p of $a(q^n - 1)/(q - 1)$ are the same as those of a repeated n times; hence by Lemma 3.6.7 for $0 \leqslant a < q - 1$ we have $t(a(q^n - 1)/(q - 1)) \equiv t(a)^n \pmod{p}$. Thus $\zeta \equiv 1 \pmod{\mathfrak{P}}$, and we conclude as for the HD product relation that $\zeta = 1$, finishing the proof. □

Corollary 3.7.5. *If $\chi_1, \ldots, \chi_k$ are multiplicative characters on $\mathbb{F}_q$ that are not all trivial, and ψ is a nontrivial additive character, then*

$$J_k(\chi_1^{(n)}, \ldots, \chi_k^{(n)}) = (-1)^{(n-1)(k-1)}(J_k(\chi_1, \ldots, \chi_k))^n .$$

Proof. Immediate by inspection of the cases of Proposition 2.5.14. □

Corollary 3.7.6. *Assume that p is odd, let $q = p^f$, and let χ_q be the unique multiplicative character of order 2 of $\mathbb{F}_q$. We have*

$$\tau(\chi_q, \psi_1) = \begin{cases} (-1)^{f-1}q^{1/2} & \text{if } p \equiv 1 \pmod 4 , \\ (-1)^{f-1}i^f q^{1/2} & \text{if } p \equiv 3 \pmod 4 . \end{cases}$$

Proof. Denote by χ_p the Legendre symbol modulo p. We have $\mathcal{N}_{\mathbb{F}_q/\mathbb{F}_p}(x) = \prod_{0 \leqslant i < f} x^{p^i} = x^{(q-1)/(p-1)}$, hence

$$\chi_p \circ \mathcal{N}_{\mathbb{F}_q/\mathbb{F}_p}(x) \equiv x^{(q-1)/2} \equiv \chi_q(x) \pmod{p} ,$$

since evidently $\chi_q(x) = x^{(q-1)/2}$ in the field $\mathbb{F}_q$. It follows from Theorem 3.7.4 that (with an evident abuse of notation)

$$\tau(\chi_q, \psi_1) = \tau(\chi_p^{(f)}, \psi_1^{(f)}) = (-1)^{f-1}\tau(\chi_p, \psi_1)^f .$$

This last Gauss sum over the finite field $\mathbb{F}_p$ is the same as the Gauss sum corresponding to the Dirichlet character χ_p, so thanks to the determination of the sign of the quadratic Gauss sum (Theorem 2.2.19) we know its exact value: if $p \equiv 1 \pmod 4$ it is equal to $p^{1/2}$, so that $\tau(\chi_q, \psi_1) = (-1)^{f-1}q^{1/2}$, and if $p \equiv 3 \pmod 4$ it is equal to $p^{1/2}i$, so that $\tau(\chi_q, \psi_1) = (-1)^{f-1}i^f q^{1/2}$.

□

3.7.3 The Zeta Function of a Diagonal Hypersurface

We have already mentioned (without making the formulas explicit) the Weil conjectures (now proved by Deligne) in Section 2.5.7. These theorems imply very sharp bounds for the number of solutions of systems of polynomial equations over finite fields. In the present subsection we show how the HD lifting relation leads to an easy proof of the Weil conjectures in the special case of a diagonal hypersurface, in other words the set of projective solutions to $\sum_{1 \leqslant i \leqslant k} a_i x_i^d = 0$, where the a_i are nonzero constants and $d \geqslant 1$. Recall from Section 2.5.5 that this depends only on $\gcd(d, q-1)$; hence without loss of generality we may assume that $d \mid (q-1)$.

Theorem 3.7.7. *Assume that $d \mid (q-1)$, and for $1 \leqslant i \leqslant k$ let $a_i \in \mathbb{F}_q^*$. For $n \geqslant 1$ denote by $M_n(q)$ the number of projective solutions in $\mathbb{F}_{q^n}$ of $\sum_{1 \leqslant i \leqslant k} a_i x_i^d = 0$. There exist algebraic numbers α_i (independent of n) indexed by a finite set I, and having the following properties:*

(1)
$$M_n(q) = \frac{q^{n(k-1)} - 1}{q^n - 1} + (-1)^k \sum_{i \in I} \alpha_i^n .$$

(2) *We have $|I| = ((d-1)^k + (-1)^k(d-1))/d$.*
(3) *We have $\alpha_i \in \mathbb{Z}[\zeta_d]$, and in particular the α_i are algebraic integers.*
(4) *All the conjugates of α_i have modulus equal to $q^{(k-2)/2}$.*

Proof. We use the notation of Section 2.5.5 and also keep the notation used in the proof of the HD lifting relation. By Corollary 2.5.24 we have

$$M_n(q) = \frac{q^{n(k-1)} - 1}{q^n - 1} + R_n(q) \quad \text{with}$$

$$R_n(q) = \sum_{\substack{\chi_1', \ldots, \chi_{k-1}' \in G_{n,d} \setminus \{\varepsilon\} \\ \prod_{1 \leqslant i \leqslant k-1} \chi_i' \neq \varepsilon}} \prod_{1 \leqslant i \leqslant k-1} (\chi_i')^{-1}(-a_i) J_{k-1}(\chi_1', \ldots, \chi_{k-1}') ,$$

where $G_{n,d}$ denotes the group of characters of $\mathbb{F}_{q^n}$ of order dividing d. I claim that the map $\chi \mapsto \chi^{(n)}$ from $G_{1,d} = G_d$ to $G_{n,d}$ is a group isomorphism: indeed it clearly maps $G_{1,d}$ to $G_{n,d}$, and is a group homomorphism. Furthermore, since by Lemma 2.5.21, $|G_{n,d}| = |G_{1,d}| = d$, both groups have the same cardinality. Finally, if $\chi^{(n)}$ is the trivial character, we have seen in the proof of the HD lifting relation as a consequence of the surjectivity of the norm on finite fields that χ is also trivial;x hence the kernel of our map is trivial, proving my claim. Using the HD lifting relation, more precisely Corollary 3.7.5, we thus obtain

$$R_n(q) = \sum_{\substack{\chi_1,\ldots,\chi_{k-1}\in G_d\setminus\{\varepsilon\} \\ \prod_{1\leqslant i\leqslant k-1}\chi_i\neq\varepsilon}} \prod_{1\leqslant i\leqslant k-1} (\chi_i^{(n)})^{-1}(-a_i)J_{k-1}(\chi_1^{(n)},\ldots,\chi_{k-1}^{(n)})$$

$$= (-1)^{k(n-1)} \sum_{\substack{\chi_1,\ldots,\chi_{k-1}\in G_d\setminus\{\varepsilon\} \\ \prod_{1\leqslant i\leqslant k-1}\chi_i\neq\varepsilon}} \prod_{1\leqslant i\leqslant k-1} \chi_i^{-1}((-a_i)^n)J_{k-1}(\chi_1,\ldots,\chi_{k-1})^n$$

$$= (-1)^k \sum_{\substack{\chi_1,\ldots,\chi_{k-1}\in G_d\setminus\{\varepsilon\} \\ \prod_{1\leqslant i\leqslant k-1}\chi_i\neq\varepsilon}} S(\chi_1,\ldots,\chi_{k-1})^n \;,$$

where we have set

$$S(\chi_1,\ldots,\chi_{k-1}) = (-1)^k \prod_{1\leqslant i\leqslant k-1} \chi_i^{-1}(-a_i)J_{k-1}(\chi_1,\ldots,\chi_{k-1}) \;.$$

In the above derivation we have of course used in an essential way the fact that a_i belongs to $\mathbb{F}_q$ and not only to $\mathbb{F}_{q^n}$. This proves the first formula of the theorem. The set I is the set of $(k-1)$-tuples $(\chi_1,\ldots,\chi_{k-1})\in(G_d\setminus\{\varepsilon\})^{k-1}$ such that $\prod_{1\leqslant i\leqslant k-1}\chi_i\neq\varepsilon$, whose cardinality has been computed in the proof of part (3) of Theorem 2.5.22. Since $\chi_i^d=1$, the values of χ_i are powers of ζ_d, and by definition, $J_{k-1}(\chi_1,\ldots,\chi_{k-1})\in\mathbb{Z}[\zeta_d]$. For the final statement concerning the moduli of the conjugates of α_i, we first note that the moduli of the conjugates of the values of χ_i are equal to 1. Furthermore, since the χ_i are nontrivial as well as $\prod_{1\leqslant i\leqslant k-1}\chi_i$, we have

$$J_{k-1}(\chi_1,\ldots,\chi_{k-1}) = \frac{\prod_{1\leqslant i\leqslant k-1}\tau(\chi_i,\psi)}{\tau(\prod_{1\leqslant i\leqslant k-1}\chi_i,\psi)} \;;$$

hence for all $\sigma\in\mathrm{Gal}(\mathbb{Q}(\zeta_{q-1},\zeta_d)/\mathbb{Q})$ we have $|\sigma(J_{k-1}(\chi_1,\ldots,\chi_{k-1}))| = q^{(k-2)/2}$, proving the theorem. $\qquad\square$

It follows trivially from this theorem that

$$|M_n(q) - |\mathbb{P}^{k-2}(\mathbb{F}_{q^n})|| \leqslant \frac{(d-1)^k + (-1)^k(d-1)}{d} q^{n(k-2)/2} \;,$$

a result that we have already proved in Theorem 2.5.22. What is important in the present theorem is that thanks to the HD lifting relation we have been able to relate exactly the $M_n(q)$ for different values of n. In fact, it is now immediate to prove Weil's conjecture in the present situation.

Corollary 3.7.8. *Define the zeta function of our hypersurface by* $Z_q(T) = \exp(\sum_{n\geqslant 1} M_n(q)T^n/n)$. *Then* $Z_q(T)$ *is a rational function of* T *of the form*

$$Z_q(T) = \frac{P(T)^{(-1)^{k-1}}}{\prod_{0\leqslant i\leqslant k-2}(1-q^iT)} = \frac{P(T)^{(-1)^{k-1}}}{(1-T)(1-qT)\cdots(1-q^{k-2}T)} \;,$$

where $P(T)$ is a polynomial of degree $((d-1)^k + (-1)^k(d-1))/d$ with constant term 1 whose reciprocal roots (i.e., the inverses of its roots) have modulus equal to $q^{(k-2)/2}$.

Proof. Using the formal expansion of the logarithm and the explicit formula for $M_n(q)$ (where we replace $(q^{n(k-1)} - 1)/(q^n - 1)$ by $\sum_{0 \leqslant i \leqslant k-2} q^{in}$) we find that

$$\sum_{n \geqslant 1} M_n(q) \frac{T^n}{n} = -\sum_{0 \leqslant i \leqslant k-2} \log(1 - q^i T) + (-1)^{k-1} \sum_{i \in I} \log(1 - \alpha_i T) \; ;$$

hence $Z_q(T)$ has the given form with $P(T) = \prod_{i \in I}(1 - \alpha_i T)$, so the other statements of the corollary follow from the theorem. $\square$

3.8 Exercises for Chapter 3

1. Show that every finite field is perfect.
2. Fill in the details of the proof of Proposition 3.1.12.
3. Show that, as claimed in the text, if θ is an algebraic number of degree n then any rational function in θ with rational coefficients is equal to a polynomial in θ with rational coefficients of degree less than or equal to $n - 1$.
4. Let σ be a field homomorphism from $\mathbb{R}$ to $\mathbb{R}$. Using the identity $f(x + a^2) = f(x) + f(a)^2$ and the positivity of a square in $\mathbb{R}$, prove that σ is continuous, hence that $\sigma(x) = x$ for all x.
5. Let A be a finite abelian group of exponent dividing n. Show that $\mathrm{Hom}(A, \mu_n) \simeq A$ noncanonically by first proving it for cyclic groups and then using the elementary divisor theorem giving the structure of finite abelian groups (Theorem 2.1.9).
6. Let p be a prime number, let K be a commutative field of characteristic different from p, set $L = K(\zeta_p)$, $G = \mathrm{Gal}(L/K) \simeq (\mathbb{Z}/p\mathbb{Z})^*$, and $M = L(a^{1/p})$ for some $a \in L^* \setminus L^{*p}$. Using Corollary 3.1.22 prove that the extension M/K is Abelian if and only for all k there exists $c_k \in L$ such that $\sigma_k(a) = c_k^p a^k$, where as usual $\sigma_k(\zeta_p) = \zeta_p^k$.
7. (Kummer extensions of degree 2.) Let k be a commutative field of characteristic different from 2. Show that any extension K of degree 2 of k has the form $K = k(\sqrt{d})$ for some $d \in k$ that is not a square, and that $k(\sqrt{d_1})$ and $k(\sqrt{d_2})$ are k-isomorphic (i.e., there exists a field isomorphism from one to the other which leaves k pointwise fixed) if and only if d_2/d_1 is a square in k.
8. Generalize Proposition 3.1.25 by proving the following. Let K be a perfect field of characteristic p, let $q = p^m$ be a power of p, and assume that the equation $X^q - X = 0$ has all its q roots in K, in other words that K contains an isomorphic copy of $\mathbb{F}_q$ (which we denote by $\mathbb{F}_q$ by abuse of notation). An extension L/K is an Abelian extension with Galois group isomorphic to $\mathbb{F}_q \simeq (\mathbb{Z}/p\mathbb{Z})^m$ if and only if $L = K(\alpha)$, where α is a root of an Artin–Schreier polynomial $X^q - X - a$ for some $a \in K$ not in the $\mathbb{F}_q$-vector space G of elements of the form $x^q - x$ for $x \in K$. The conjugates of α are then the $\alpha + k$ for $k \in \mathbb{F}_q$. Furthermore, a and a' define K-isomorphic extensions if and only if there exists $j \in \mathbb{F}_q^*$ such that $a' - ja \in G$.

9. Generalize Proposition 3.1.25 to cyclic extensions of prime-power order p^m (use induction on m).

10. Prove that, as stated in the text, if A and B are coprime polynomials then with the notation of Theorem 3.2.5 we have

$$\mathrm{Ker}((AB)(\sigma)) = \mathrm{Ker}(A(\sigma)) \oplus \mathrm{Ker}(B(\sigma)) \ .$$

11. In the proof of Proposition 3.3.4, we showed that α is an eigenvalue of the multiplication by α map. Give a corresponding eigenvector.

12. Let α and β be algebraic numbers, and let A and B in $\mathbb{Q}[X]$ be their respective minimal monic polynomials. Denoting by R_Y the resultant with respect to the variable Y (see Definition 3.5.6), show that $\alpha + \beta$ is a root of $R_Y(A(Y), B(X - Y))$ and $\alpha\beta$ is a root of $X^d R_Y(A(Y), B(X/Y))$ for a suitable $d \geqslant 1$. Deduce from this an alternative proof of Proposition 3.3.6.

13. With the notation of Proposition 3.3.8, prove directly (i.e., without using that proposition) that $a_n \alpha^2 + a_{n-1}\alpha$ is an algebraic integer.

14. Let α be an algebraic number. Prove that $R = \mathbb{Z}[\alpha] \cap \mathbb{Z}_{\mathbb{Q}(\alpha)}$ is the largest subring of $\mathbb{Z}[\alpha]$ that is a finitely generated $\mathbb{Z}$-module.

15. (H. W. Lenstra.) As an application of Kronecker's Proposition 3.3.9, prove the following amusing result. Let P be a (closed) polygon in the plane having the following two properties:

(a) All of its sides have the same length.
(b) All the angles between two consecutive sides are rational multiples of 2π, *except* perhaps for two *consecutive* angles.

   Prove that in fact all angles are rational multiples of 2π.

16. Prove Proposition 3.3.13. If you need help, see Section 1.4.1 of [Coh1].

17. Let L/K be an extension of number fields, and let $\mathfrak{a}$ and $\mathfrak{b}$ be ideals of K. Prove that $\mathfrak{a}\mathbb{Z}_L = \mathfrak{b}\mathbb{Z}_L$ if and only if $\mathfrak{a} = \mathfrak{b}$. (Hint: prove that $\mathfrak{a}\mathfrak{b}^{-1}$ and $\mathfrak{b}\mathfrak{a}^{-1}$ are integral ideals.)

18. Prove Propositions 3.3.21, 3.3.25, and 3.3.27. In particular, define the *Frobenius homomorphism* on L when $\mathfrak{p}$ is unramified. For help see any text on algebraic number theory.

19. Let L be a cyclic extension of $\mathbb{Q}$ of degree p, let $M = L(\zeta_p)$, so that by Kummer theory $M = K(a^{1/p})$ for some $a \in K$, and let $\mathfrak{q}$ be a prime ideal of K. If $L/\mathbb{Q}$ is unramified outside p and $p \nmid v_{\mathfrak{q}}(a)$, prove that the prime number q below $\mathfrak{q}$ splits completely in $K/\mathbb{Q}$. (Hint: show that the decomposition group $D(\mathfrak{q}/q)$ is trivial.)

20. Let K be a number field, $\mathfrak{p}$ a prime ideal of K, and p the prime number below $\mathfrak{p}$. For α and β in $\mathbb{Z}_K$ show that the following three properties are equivalent:

(a) $\alpha \equiv \beta \pmod{\mathfrak{p}}$.
(b) $\alpha^p \equiv \beta^p \pmod{\mathfrak{p}}$.
(c) $\alpha^p \equiv \beta^p \pmod{\mathfrak{p}^2}$.

   (Hint: prove that (b) implies (a) implies (c) implies (b). See also Lemma 2.1.21.) In particular, deduce that if p is unramified in K then the same properties are equivalent by replacing $\mathfrak{p}$ by $p\mathbb{Z}_K$.

21. Prove Proposition 3.4.3 using the quadratic reciprocity law.

22. Let m and n be distinct strictly positive integers.

(a) Show that if $m \mid n$ and $n/m = p^k$ is a prime power then $R(\Phi_m, \Phi_n) = \pm p^{\phi(m)}$ (using Proposition 3.5.5 show that $R(\Phi_m, \Phi_n) = u p^{\phi(m)}$ for some unit u, and conclude from the fact that $R(\Phi_m, \Phi_n) \in \mathbb{Z}$).

(b) Show that $R(\Phi_m, \Phi_n) > 0$ except when $m = 2$ and $n = 1$, finishing the proof of Proposition 3.5.8.

23.

(a) Let K be a number field having r_1 real and r_2 pairs of nonreal complex embeddings. Using directly the definition of the discriminant, show that its sign is equal to $(-1)^{r_2}$.

(b) Deduce that the discriminant of the cyclotomic field $K = \mathbb{Q}(\zeta_{p^k})$ is equal to $\varepsilon p^{\phi(p^k)-1}$ with $\varepsilon = 1$ if $p \equiv 1 \pmod 4$ or if $p = 2$ and $k \geqslant 3$, and $\varepsilon = -1$ if $p \equiv 3 \pmod 4$ or if $p^k = 4$.

24. Let $K = \mathbb{Q}(\zeta_n)$ be a cyclotomic field and $\alpha \in \mathbb{Z}_K$ an algebraic integer of K. Assume that $|\alpha|^2 \in \mathbb{Z}$, where $|\alpha|$ denotes the complex modulus of α for some chosen embedding of K in $\mathbb{C}$. Prove that $|\sigma(\alpha)|^2 = |\alpha|^2$ for all $\sigma \in \mathrm{Gal}(K/\mathbb{Q})$. Show that this is not necessarily true if $|\alpha|^2 \notin \mathbb{Z}$.

25. Prove that if $1 \leqslant a, b \leqslant q - 2$ and $a + b \geqslant q$ then $\binom{a+b}{a}$ is divisible by p.

26. Generalize Proposition 3.6.4 by showing that if $1 \leqslant a_i \leqslant q - 2$ we have

$$J_k(\omega^{-a_1}, \dots, \omega^{-a_k}) \equiv (-1)^{k-1} \frac{(a_1 + \dots + a_k)!}{a_1! \cdots a_k!} \pmod{\mathfrak{P}}.$$

27. Let $s(a)$ be as in Definition 3.6.5. Prove that for any $m \mid (q - 1)$ and any t coprime to m the quantity $(m/(p-1))s(t(q-1)/m)$ is an integer (this is proved indirectly in Proposition 3.6.11).

28. The goal of this exercise and the next is to prove the theorem of Kronecker–Weber, using an idea of F. Lemmermeyer in [Lem]. Assume that $p \geqslant 3$, let L be a cyclic extension of $\mathbb{Q}$ of degree p that is unramified outside p, let $K = \mathbb{Q}(\zeta_p)$, let $M = L(\zeta_p) = K(a^{1/p})$ for some $a \in K$, and let $\mathfrak{p} = (1 - \zeta_p)\mathbb{Z}_K$ be the unique ideal of K above p.

(a) Prove that if $\mathfrak{q}$ is a prime ideal of K different from $\mathfrak{p}$ then $p \mid v_{\mathfrak{q}}(a)$.

(b) Using the preceding exercise, prove that $p \mid v_{\mathfrak{p}}(a)$, hence that $a\mathbb{Z}_K = \mathfrak{a}^p$ for some ideal $\mathfrak{a}$ of K.

(c) Using Exercise 6, show that $\sigma_k(\mathfrak{a}) = c_k \mathfrak{a}^k$ for some $c_k \in K^*$.

(d) Using Stickelberger's Corollary 3.6.14 show that $\mathfrak{a}^{p-1}$ is a principal ideal, hence that $\mathfrak{a}$ itself is principal.

(e) Deduce first that $M = K(u^{1/p})$ for some *unit* u of K, then by Proposition 3.5.20 and Exercise 6 applied to complex conjugation σ_{-1}, prove that $M = \mathbb{Q}(\zeta_{p^2})$.

(f) Using Galois theory, show that this implies that if $\mathrm{Gal}(L/\mathbb{Q}) \simeq (\mathbb{Z}/p\mathbb{Z})^n$ for $n \geqslant 1$ and $L/\mathbb{Q}$ unramified outside p, then $n = 1$ and L is the subfield of degree p of $\mathbb{Q}(\zeta_{p^2})$.

(g) Still with $p \geqslant 3$, let now L be a cyclic extension of $\mathbb{Q}$ of degree p^n unramified outside p. Show that L is the subfield of degree p^n of $\mathbb{Q}(\zeta_{p^{n+1}})$. (Hint: if L' is the subfield in question and if $L'L/\mathbb{Q}$ is not cyclic, prove that there exists a subextension whose Galois group is isomorphic to $(\mathbb{Z}/p\mathbb{Z})^2$, thus contradicting the preceding result.)

29. Assume now that $p = 2$.

(a) Show that the only extensions of degree 2 of $\mathbb{Q}$ that are unramified outside 2 are $\mathbb{Q}(\sqrt{D})$ for $D = -8, -4$, and 8.

(b) As in the case p odd, show that this implies that the only *real* extension $L/\mathbb{Q}$ unramified outside 2 such that $\mathrm{Gal}(L/\mathbb{Q}) \simeq (\mathbb{Z}/2\mathbb{Z})^n$ for some $n \geq 1$ is $\mathbb{Q}(\sqrt{8})$.

(c) Now let L be a cyclic extension of degree 2^n for $n \geq 2$ unramified outside 2. If L is not real, prove that L is a subfield of $\mathbb{Q}(\zeta_{2^n+2})$ if and only if so is the maximal real subfield of $L(\zeta_4)$, hence that we can assume that L is real.

(d) If L is a real extension of degree 2^n unramified outside 2 prove that L is the maximal real subfield L' of $\mathbb{Q}(\zeta_{2^n+2})$. (Hint: as in the preceding exercise, consider $L'L/\mathbb{Q}$.)

(e) Deduce from this exercise and the preceding one that if $L/\mathbb{Q}$ is a cyclic extension of degree p^n and unramified outside p, then L is a subfield of a cyclotomic field.

It is then not difficult to show that this implies the full Kronecker–Weber theorem in two steps (see for instance [Marc] for details): if $L/\mathbb{Q}$ is an arbitrary cyclic extension of degree p^n, possibly ramified at other primes than p, one first shows that one can reduce to a similar cyclic extension but ramified at one prime fewer outside p, so that ultimately one can reduce to one that is unramified outside p. As a second, easier, step, this implies that an arbitrary Abelian extension is a subfield of a cyclotomic field by decomposing the Galois group as a product of cyclic groups of prime-power order.

30. Show that Proposition 3.5.20 is also true for $p = 2$.

31. Generalize the Eisenstein reciprocity law (Theorem 3.6.38) to the case that $m = \ell^r$ is a prime power, separating the cases $\ell > 2$ and $\ell = 2$.

32. The goal of this exercise is to give a direct proof of the Hasse–Davenport lifting relation. For hints, the reader is referred to Section 11.4 of [Ire-Ros]. Keep the notation of that relation. For any monic polynomial $f(x) = \sum_{0 \leq i \leq n} a_i x^i$ (thus with $a_n = 1$) we define as usual the trace of f by $\mathrm{Tr}(f) = -a_{n-1}$ and the norm of f by $\mathcal{N}(f) = (-1)^n a_0$. For a nonconstant f we set $\lambda(f) = \chi(\mathcal{N}(f))\psi(\mathrm{Tr}(f))$, and by convention we set $\lambda(1) = 1$.

(a) Prove that $\lambda(fg) = \lambda(f)\lambda(g)$.

(b) Let $x \in \mathbb{F}_{q^n}$, let f be the monic minimal polynomial of x over $\mathbb{F}_q$, and let $d = \deg(f) \mid n$. Prove that $\chi^{(n)}(x)\psi^{(n)}(x) = \lambda(f)^{n/d}$.

(c) Prove the identity

$$\tau(\chi^{(n)}, \psi^{(n)}) = \sum_f \deg(f)\lambda(f)^{n/\deg(f)} ,$$

where the sum is over all monic irreducible polynomials over $\mathbb{F}_q$ with degree dividing n.

(d) Using existence and uniqueness of the decomposition into irreducibles, prove the following formal identity, analogous to the Euler product for the zeta function:

$$\sum_f \lambda(f) X^{\deg(f)} = \prod_f (1 - \lambda(f) X^{\deg(f)})^{-1} ,$$

where the sum is over all monic polynomials, and the product over all monic irreducible polynomials.

(e) Prove that

$$\sum_{\deg(f)=n} \lambda(f) = \begin{cases} \tau(\chi, \psi) & \text{if } n = 1 , \\ 0 & \text{if } n > 1. \end{cases}$$

(f) Finally, deduce the HD lifting relation.

4. p-adic Fields

There are many books dealing with p-adic numbers and local fields. My preference is for the books by Cassels [Cas1] and Serre [Ser2]. The analytic aspects are treated in detail in [Kob1] and [Rob1]. On a more elementary level, we can cite [Ami], [Bac], [Bor-Sha], and [Gou].

In this chapter, all fields are assumed to be commutative (see the warnings at the beginning of this book).

4.1 Absolute Values and Completions

We begin with some general observations valid over any field.

4.1.1 Absolute Values

Definition 4.1.1. *Let K be any field. An absolute value is a map $\| \ \|$ from K to $\mathbb{R}$ satisfying the following properties:*

(1) *(Definiteness.) For all $x \in K$ we have $\|x\| \geq 0$, and $\|x\| = 0$ if and only if $x = 0$.*
(2) *(Multiplicativity.) For all x and y in K we have $\|xy\| = \|x\|\|y\|$.*
(3) *(Generalized triangle inequality.) There exists $a > 0$ such that for all x and y in K we have $\|x + y\|^a \leq \|x\|^a + \|y\|^a$.*

Before going any further, it is important that the reader keep in mind the following three examples:

– If K is a subfield of $\mathbb{R}$, the ordinary absolute value $|x|$ is clearly an absolute value in the above sense, with $a = 1$. More generally, if K is a subfield of $\mathbb{C}$ the modulus $|x|$ is also an absolute value with $a = 1$.
– If K is a subfield of $\mathbb{C}$, the square of the modulus $|x|^2$ is clearly an absolute value in the above sense, with $a = 1/2$.
– If $K = \mathbb{Q}$ and p is a prime number, the map defined by $|0|_p = 0$ and $|x|_p = C^{-v_p(x)}$, where $v_p(x)$ is the exponent of p in the decomposition of x into a product of prime powers and $C > 1$ is a fixed constant, is an absolute value called a p-adic absolute value.

It is clear that an absolute value $\|\ \|$ makes K into a metric space, the distance being defined by $d(x, y) = \|x - y\|^a$. This metric depends on a, but it is immediate that the induced topology does not.

Definition 4.1.2. *Two absolute values on a field K are called* equivalent *if they induce the same topology on K.*

Lemma 4.1.3. *Two absolute values $\|\ \|_1$ and $\|\ \|_2$ are equivalent if and only if there exists $c > 0$ such that $\|x\|_2 = \|x\|_1^c$ for all $x \in K$.*

Proof. If $\|x\|_2 = \|x\|_1^c$, for each $R > 0$ we have $B_1(0, R) = B_2(0, R^c)$, where $B_i(0, R)$ denotes the open ball of radius R for the metric induced by $\|\ \|_i$; hence the topologies are equivalent. Conversely, assume that the topologies are equivalent. Since for any absolute value, $\|x\| < 1$ if and only if $\|x^n\| \to 0$ as $n \to \infty$ if and only if $x^n \to 0$ as $n \to \infty$, it follows that $\|x\|_1 < 1$ if and only if $\|x\|_2 < 1$. We may assume that there exists $x_0 \neq 0$ satisfying this; otherwise, we would have $\|x\|_1 = \|x\|_2 = 1$ for all $x \neq 0$. We define c by the equality $\|x_0\|_2 = \|x_0\|_1^c$, so that by what we have said we have $c > 0$. If $x \in K^*$ is such that $\|x\|_1 < 1$, we can define a real number $\lambda > 0$ by the equality $\|x\|_1 = \|x_0\|_1^\lambda$. If m and n are positive integers such that $m/n > \lambda$, then $\|x_0^m/x^n\|_1 = \|x_0\|_1^{m-n\lambda} < 1$, whence $\|x_0^m/x^n\|_2 < 1$; in other words, $\|x\|_2 > \|x_0\|_2^{m/n}$. If we let m/n tend to λ keeping $m/n \geqslant \lambda$, then since $\|x_0\| < 1$, by continuity we have $\|x\|_2 \geqslant \|x_0\|_2^\lambda$. Using a sequence of rational numbers m/n such that $m/n < \lambda$ and tending to λ, we obtain in the same way $\|x\|_2 \leqslant \|x_0\|_2^\lambda$, so that finally

$$\|x\|_2 = \|x_0\|_2^\lambda = \|x_0\|_1^{c\lambda} = \|x\|_1^c \ ,$$

proving the lemma. $\qquad\qquad\qquad\qquad\qquad\qquad\qquad\qquad\qquad\qquad\qquad\quad \square$

As we have done in the above proof, from now on when we speak of absolute values we will implicitly assume that we exclude the trivial absolute value $\|x\| = 1$ for $x \neq 0$, which induces the discrete topology on K.

Definition 4.1.4. *We say that an absolute value $\|\ \|$ on a field K is* Archimedean *if K has characteristic 0 and if there exists $m \in \mathbb{Z}$ such that $\|m\| > 1$. Otherwise, it is said to be* non-Archimedean.

The important result of this section, called Ostrowski's theorem, gives a complete description of all absolute values on a number field. We begin by describing the Archimedean ones.

4.1.2 Archimedean Absolute Values

Lemma 4.1.5. *Let K be a number field. The Archimedean absolute values are given by $\|\alpha\| = |\sigma(\alpha)|^c$, where $c > 0$ and σ is any fixed embedding from K to $\mathbb{C}$.*

Proof. It is clear that the maps defined in this way are Archimedean absolute values; hence conversely, let $\|\ \|$ be an Archimedean absolute value. Since any element of K may be written as x/y with x and y in $\mathbb{Z}_K$, it is enough to prove the lemma for elements of $\mathbb{Z}_K$. In addition, replacing if necessary $\|\ \|$ by $\|\ \|^a$, we may assume that $a = 1$ in Definition 4.1.1, in other words that we have the ordinary triangle inequality $\|x + y\| \leqslant \|x\| + \|y\|$. We first prove the following claim:

Claim. There exists $c > 0$ such that $\|m\| = |m|^c$ for all $m \in \mathbb{Z}$.

Indeed, by multiplicativity we know that $\|1\| = 1$ and $\|\pm 1\|^2 = \|1\| = 1$; hence again by multiplicativity we may assume that $m > 1$. For the moment fix some $m_0 > 1$, and for any positive m and N write m^N in base m_0, i.e., in the form

$$m^N = \sum_{0 \leqslant n \leqslant N \log(m)/\log(m_0)} a_n m_0^n \text{ with } 0 \leqslant a_n < m_0 .$$

Let A be an upper bound for all the $\|a\|$ with $0 \leqslant a < m_0$. By the triangle inequality, if $\|m_0\| \leqslant 1$ we would have for all positive m and N

$$\|m\|^N \leqslant A(1 + N \log(m)/\log(m_0)) ;$$

hence by taking Nth roots and letting $N \to \infty$ we would obtain $\|m\| \leqslant 1$ for all m, in contradiction with the Archimedean assumption. Since this is true for all $m_0 > 1$, it follows that for all $m > 1$ we have $\|m\| > 1$. Applying again the triangle inequality, but using $\|m_0\| > 1$, we obtain

$$\|m\|^N \leqslant A(1 + N \log(m)/\log(m_0))\|m_0\|^{N \log(m)/\log(m_0)} ,$$

hence again by taking Nth roots and letting $N \to \infty$ we now obtain $\|m\| \leqslant \|m_0\|^{\log(m)/\log(m_0)}$, in other words $\|m\|^{1/\log(m)} \leqslant \|m_0\|^{1/\log(m_0)}$. Exchanging m and m_0, we deduce that we have *equality*, in other words that $C = \|m\|^{1/\log(m)}$ is independent of $m > 1$. Thus

$$\|m\| = C^{\log(m)} = \exp(\log(C)\log(m)) = m^{\log(C)} = m^c$$

with $c = \log(C) > 0$, proving our claim.

We now fix this value of c. For any nonzero $\alpha \in \mathbb{Z}_K$, order the complex embeddings σ_i so that $|\sigma_i(\alpha)| \geqslant |\sigma_{i+1}(\alpha)|$ (ordinary modulus in $\mathbb{C}$). This ordering of course depends on α. Let N be a positive integer, and write

$$\prod_\sigma (X - \sigma(\alpha^N)) = X^n + a_{n-1}X^{n-1} + \cdots + a_0 ,$$

and for all $m \leqslant n$ define $P_m = \prod_{1 \leqslant i \leqslant m} |\sigma_i(\alpha^N)|$. The $\pm a_{n-m}$ are the elementary symmetric functions of the $\sigma(\alpha^N)$, and are equal to a sum of $\binom{n}{m}$ monomials in the $\sigma(\alpha^N)$. Because of our ordering, the largest modulus of these monomials is equal to $\pm P_m$; hence $|a_{n-m}| \leqslant \binom{n}{m} P_m \leqslant 2^n P_m$. In addition, if $|\sigma_{m+1}(\alpha)| < |\sigma_m(\alpha)|$, then when N is large this largest monomial is much larger than any other occurring in a_{n-m}, so that for instance $|a_{n-m}| > P_m/2$. Finally, since the a_{n-m} are symmetric functions they are in $\mathbb{Z}$, so $\|a_{n-m}\| = |a_{n-m}|^c$.

We first prove that $\|\alpha\| = |\sigma_k(\alpha)|^c$ for some k. Indeed, assume first that for some k we have

$$|\sigma_{k+1}(\alpha)|^c < \|\alpha\| < |\sigma_k(\alpha)|^c .$$

If we replace X by α^N in the formula for $\prod_\sigma (X - \sigma(\alpha^N))$, we obtain (with $a_n = 1$)

$$0 = \sum_{0 \leqslant j \leqslant n} a_j \alpha^{Nj} .$$

However, we have

$$\left\| \frac{a_{n-j} \alpha^{N(n-j)}}{a_{n-k} \alpha^{N(n-k)}} \right\| \leqslant \left(\frac{(2^n P_j)^c}{(P_k/2)^c} \right) \|\alpha^{N(k-j)}\| .$$

Thus, if $k > j$ we have

$$\left\| \frac{a_{n-j} \alpha^{N(n-j)}}{a_{n-k} \alpha^{N(n-k)}} \right\| \leqslant \left(\frac{2^{n+1} P_j}{P_k} |\sigma_k(\alpha)|^{N(k-j)} \right)^c \left(\frac{\|\alpha\|}{|\sigma_k(\alpha)|^c} \right)^{N(k-j)} .$$

Since $|\sigma_i(\alpha)| \geqslant |\sigma_k(\alpha)|$ for $i \leqslant k$, it follows that

$$\frac{P_j}{P_k} = \frac{1}{\prod_{j < i \leqslant k} |\sigma_i(\alpha^N)|} \leqslant \frac{1}{|\sigma_k(\alpha^N)|^{k-j}} .$$

Thus,

$$\left\| \frac{a_{n-j} \alpha^{N(n-j)}}{a_{n-k} \alpha^{N(n-k)}} \right\| \leqslant 2^{(n+1)c} \left(\frac{\|\alpha\|}{|\sigma_k(\alpha)|^c} \right)^{N(k-j)} .$$

Since $\|\alpha\| < |\sigma_k(\alpha)|^c$, as N tends to ∞ the right-hand side tends to 0, showing that $a_{n-j} \alpha^{N(n-j)}$ is negligible compared to $a_{n-k} \alpha^{N(n-k)}$. A similar result holds if $j > k$. Since $0 = \sum_{0 \leqslant j \leqslant n} a_{n-j} \alpha^{N(n-j)}$, dividing by $a_{n-k} \alpha^{N(n-k)}$ and letting N tend to infinity gives $0 = 1$, a contradiction. Note that α has been chosen nonzero and that $|a_{n-k}| > P_k/2$ is also nonzero.

A similar proof shows that the inequalities $\|\alpha\| > |\sigma_1(\alpha)|^c$ and $\|\alpha\| < |\sigma_n(\alpha)|^c$ are impossible. It follows that there exists k such that $\|\alpha\| = |\sigma_k(\alpha)|^c$.

This construction of k depends a priori on α, so temporarily we write $k = k(\alpha)$. I claim that $k(\alpha)$ is independent of $\alpha \in \mathbb{Z}_K$. Indeed, if α and β are nonzero elements of $\mathbb{Z}_K$ we have

$$|\sigma_{k(\alpha\beta^N)}(\alpha\beta^N)|^c = \|\alpha\beta^N\| = \|\alpha\|\|\beta\|^N = \|\alpha\||\sigma_{k(\beta)}(\beta)|^{cN} ,$$

hence

$$\frac{|\sigma_{k(\alpha\beta^N)}(\beta)|}{|\sigma_{k(\beta)}(\beta)|} = \left(\frac{\|\alpha\|}{|\sigma_{k(\alpha\beta^N)}(\alpha)|^c}\right)^{1/(cN)}.$$

Now note that when γ is nonzero and fixed, $|\sigma_i(\gamma)|/|\sigma_j(\gamma)|$ can take only a finite number of (nonzero) values. It follows in particular that as N tends to infinity, the right-hand side of the above equality tends to 1. Thus, for N sufficiently large, $|\sigma_{k(\alpha\beta^N)}(\beta)|/|\sigma_{k(\beta)}(\beta)|$ is very close to 1, hence is equal to 1 since there are only a finite number of possible values. It follows that we may choose $k(\alpha\beta^N) = k(\beta)$ (there may be other possibilities for k but we do not care, our only purpose being to show that the same k is valid for all α). Replacing in the above formula, we obtain for N sufficiently large $\|\alpha\| = |\sigma_{k(\beta)}(\alpha)|^c$, so that we may also take $k(\alpha) = k(\beta)$. Since this is true for any nonzero α and β, this shows that $k(\alpha)$ can indeed be chosen independent of α, finishing the proof of the lemma. $\qquad\square$

Corollary 4.1.6. *Let K be a number field and (r_1, r_2) its signature. There exist exactly $r_1 + r_2$ inequivalent Archimedean absolute values on K, corresponding to the r_1 real embeddings and to the r_2 pairs of complex conjugate embeddings.*

Proof. By the above lemma, an Archimedean absolute value is equivalent to one defined by $\|\alpha\| = |\sigma(\alpha)|$ for some fixed embedding σ of K into $\mathbb{C}$, and conversely, any such embedding gives rise to an Archimedean absolute value. Thus, there are at most $n = r_1 + 2r_2$ Archimedean absolute values. The embeddings σ_1 and σ_2 define equivalent absolute values if and only if for all $\alpha \in K$ we have $|\sigma_1(\alpha)| = |\sigma_2(\alpha)|^c$ for some $c \in \mathbb{R}_{>0}$, and choosing for instance $\alpha = 2$, it is clear that $c = 1$. The map $\lambda = \sigma_2\sigma_1^{-1}$ is clearly a field homomorphism from $\sigma_1(K)$ to $\sigma_2(K)$. For $i = 1$ and 2, denote by K_i the completion of $\sigma_i(K)$ in $\mathbb{C}$. If $y \in K_1$, we can write $y = \lim_{n\to\infty} \sigma_1(x_n)$ for $x_n \in K$, where $\sigma_1(x_n)$ is a Cauchy sequence for the absolute value on $\mathbb{C}$. I claim that we can extend λ to K_1 by defining $\lambda(y) = \lim_{n\to\infty} \sigma_2(x_n)$. Indeed, since

$$|\sigma_2(x_n) - \sigma_2(x_m)| = |\sigma_2(x_n - x_m)| = |\sigma_1(x_n - x_m)| = |\sigma_1(x_n) - \sigma_1(x_m)|,$$

it is clear that $\sigma_2(x_n)$ is also a Cauchy sequence. Furthermore, for the same reason the definition of $\lambda(y)$ is independent of the chosen sequence x_n, so λ is a field homomorphism. Since $\lambda \circ \sigma_1 = \sigma_2$, it is clear that this is indeed an extension of the initial λ, and since we can apply the construction to $\lambda^{-1} = \sigma_1\sigma_2^{-1}$, λ is a field isomorphism from K_1 to K_2. By construction, λ is continuous. Thus, applying Proposition 3.1.15 we deduce that λ is either the identity or complex conjugation, proving the corollary. $\qquad\square$

4.1.3 Non-Archimedean and Ultrametric Absolute Values

We now do a similar (but simpler) study for non-Archimedean absolute values. We start with the following simple but essential result:

Lemma 4.1.7. *An absolute value $\| \ \|$ on a field K is non-Archimedean if and only if it satisfies the so-called ultrametric inequality*

$$\|x + y\| \leqslant \max(\|x\|, \|y\|) \ .$$

Thanks to this lemma, from now on we will use indifferently the word ultrametric or the word non-Archimedean, and such an absolute value will be called an ultrametric absolute value.

Proof. Assume first that $\| \ \|$ satisfies the ordinary triangle inequality, i.e., that we can choose $a = 1$ in Definition 4.1.1. Then by the binomial theorem, for any positive integer N we have

$$\|x + y\|^N = \left\| \sum_{0 \leqslant n \leqslant N} \binom{N}{n} x^{N-n} y^n \right\| \leqslant \sum_{0 \leqslant n \leqslant N} \left\| \binom{N}{n} \right\| \|x\|^{N-n} \|y\|^n$$
$$\leqslant (N + 1) \max(\|x\|, \|y\|)^N$$

since $\|\binom{N}{n}\| \leqslant 1$ by the non-Archimedean property. Taking Nth roots and letting $N \to \infty$ gives the result in this case. For a general absolute value $\| \ \|$, we apply the above result to $\| \ \|^a$, where a is as in Definition 4.1.1.

Conversely, since $\|1\| = \| - 1\| = 1$ by definition of an absolute value, it follows that an ultrametric absolute value is non-Archimedean.   □

The p-adic absolute value introduced in one of the above examples is a foremost example of an ultrametric absolute value. The ultrametric property has several interesting consequences, which seem at first surprising:

Corollary 4.1.8. *Let $\| \ \|$ be an ultrametric absolute value.*

(1) *When $\|x\| \neq \|y\|$ we have equality in the ultrametric inequality:*

$$\|x + y\| = \max(\|x\|, \|y\|) \ .$$

(2) *Every "triangle" is an isosceles triangle.*
(3) *Every point inside an open ball of radius R can be taken as the center of the ball, and with the same radius.*
(4) *When two open balls have nonempty intersection one is a subset of the other.*

Proof. (1) Assume for instance that $\|y\| < \|x\|$. Writing $x = (x+y)+(-y)$, we see that $\|x\| \leqslant \max(\|x + y\|, \|y\|)$. The maximum cannot be equal to $\|y\|$, for that would give a contradiction. Thus $\|x\| \leqslant \|x + y\|$. On the other hand, the ultrametric inequality gives directly $\|x+y\| \leqslant \|x\|$, showing equality. (2) is a reformulation of (1) in geometric terms. For (3), let $B(x, R)$ denote the open ball of radius R centered at x, and let $y \in B(x, R)$. Then for any $z \in B(x, R)$ we have $d(z, y) = \|z - y\| = \|(z - x) + (x - y)\| < R$ by the ultrametric inequality, so that $z \in B(y, R)$, whence $B(x, R) \subset B(y, R)$. Exchanging x and y gives the reverse inclusion, proving (3). (4) is an immediate consequence of (3). $\qquad\square$

Although the above results may seem at first amusing, the fourth one is in fact quite annoying. Recall that when using the process of analytic continuation in the complex numbers, we use power series expansions around points near the circle of convergence to enlarge the domain of definition of our functions. Here, nothing of the sort is possible since the new circle of convergence will be the same however close we are to the initial circle of convergence. Thus, to perform analytic continuation we need new tools, which are more sophisticated than in the complex case, for instance "analytic elements," introduced by Krasner (see Section 4.5.2 for a special case).

Let us return to the characterization of non-Archimedean absolute values. We first note the following easy but crucial result.

Proposition 4.1.9. *Let* $\| \ \|$ *be an ultrametric absolute value on a field* K. *Let* $A = \{x \in K/ \ \|x\| \leqslant 1\}$ *and* $\mathfrak{p} = \{x \in K/ \ \|x\| < 1\}$. *Then* A *is a local ring with maximal ideal* $\mathfrak{p}$ *and field of fractions* K.

Proof. The fact that A is a ring immediately follows from the ultrametric inequality. It is clear that if $x \in K^*$ then either x or $1/x$ belongs to A, so K is the field of fractions of A. Finally, $x \in A$ is invertible in A if and only if $\|x\| = 1$, i.e., $x \notin \mathfrak{p}$, showing that A is a local ring with maximal ideal $\mathfrak{p}$, in other words that $\mathfrak{p}$ is a maximal ideal and is the only maximal ideal of A. $\qquad\square$

Definition 4.1.10. *Under the above assumptions and notation, the field* $A/\mathfrak{p}$ *is called the* residue field *of* K *for the absolute value* $\| \ \|$.

Similarly to the case of the p-adic absolute value introduced above, if K is a number field and $\mathfrak{p}$ is a (nonzero) prime ideal of $\mathbb{Z}_K$, we can introduce the $\mathfrak{p}$-adic absolute value in a similar way: for $x \in K^*$ we let $v_{\mathfrak{p}}(x)$ be the exponent of $\mathfrak{p}$ in the decomposition of the principal ideal $x\mathbb{Z}_K$ into a product of powers of prime ideals, and we let $|x|_{\mathfrak{p}} = C^{-v_{\mathfrak{p}}(x)}$ for some $C > 1$ (and $|0|_{\mathfrak{p}} = 0$). As for $\mathbb{Q}$, it is immediate to check that it is a non-Archimedean absolute value. These are in fact the only ones:

Lemma 4.1.11. *Let* $\| \ \|$ *be a non-Archimedean absolute value on a number field* K. *There exists a (nonzero) prime ideal* $\mathfrak{p}$ *of* $\mathbb{Z}_K$ *and a constant* $C > 1$ *such that for all* $\alpha \in K^*$ *we have* $\|\alpha\| = C^{-v_{\mathfrak{p}}(\alpha)}$.

Proof. Let $\alpha \in \mathbb{Z}_K$, $\alpha \neq 0$. Then α is a root of a monic polynomial with integral coefficients, say

$$\alpha^m + a_{m-1}\alpha^{m-1} + \cdots + a_0 = 0$$

with $a_i \in \mathbb{Z}$. If $\|\alpha\| > 1$, the ultrametric inequality implies that

$$\|a_{m-1}\alpha^{m-1} + \cdots + a_0\| \leqslant \|\alpha\|^{m-1} ,$$

while $\|\alpha^m\| > \|\alpha\|^{m-1}$, a contradiction. Thus, for any $\alpha \in \mathbb{Z}_K$ we have $\|\alpha\| \leqslant 1$. Since the absolute value is nontrivial, there exists a nonzero $\alpha \in \mathbb{Z}_K$ such that $\|\alpha\| \neq 1$, so that $\|\alpha\| < 1$. Let $\mathfrak{p}$ be the set of $\alpha \in \mathbb{Z}_K$ such that $\|\alpha\| < 1$. This set is nonzero and strictly included in $\mathbb{Z}_K$, and the ultrametric inequality combined with the just-proved fact that $\|\alpha\| \leqslant 1$ for all $\alpha \in \mathbb{Z}_K$ immediately implies that $\mathfrak{p}$ is a nonzero ideal of $\mathbb{Z}_K$. Since $\|\alpha\beta\| < 1$ implies that $\|\alpha\| < 1$ or $\|\beta\| < 1$, it follows that $\mathfrak{p}$ is a prime ideal.

Let π be a uniformizer of $\mathfrak{p}$, in other words an element of $\mathfrak{p}$ that does not belong to $\mathfrak{p}^2$ (recall that $\mathfrak{p} \neq \mathfrak{p}^2$ since otherwise $\mathfrak{p} = \mathbb{Z}_K$). If $\alpha \in K^*$ we can thus write $\alpha/\pi^{v_{\mathfrak{p}}(\alpha)}\mathbb{Z}_K = \mathfrak{a}_1/\mathfrak{a}_2$, where $\mathfrak{a}_1$ and $\mathfrak{a}_2$ are integral ideals coprime to $\mathfrak{p}$. Thus $\mathfrak{p} \nmid \mathfrak{a}_2$, in other words $\mathfrak{a}_2 \not\subset \mathfrak{p}$, hence we can find $\beta_2 \in \mathfrak{a}_2$ such that $\beta_2 \notin \mathfrak{p}$; hence by definition of $\mathfrak{p}$ we have $\|\beta_2\| = 1$. If we set $\beta_1 = \beta_2\alpha/\pi^{v_{\mathfrak{p}}(\alpha)}$, we have $\beta_1 \in \beta_2\mathfrak{a}_1/\mathfrak{a}_2 \subset \mathfrak{a}_1$; hence $\beta_1 \in \mathbb{Z}_K$, and $\beta_1 \notin \mathfrak{p}$ as well since $v_{\mathfrak{p}}(\beta_1) = v_{\mathfrak{p}}(\beta_2)$. It follows that we also have $\|\beta_1\| = 1$; hence finally, by multiplicativity

$$\|\alpha\| = C^{-v_{\mathfrak{p}}(\alpha)} \text{ with } C = \|\pi\|^{-1} > 1 ,$$

proving the lemma. □

Remark. It is easy to check that the ring A given by Proposition 4.1.9 is the set of $x \in K$ that can be written in the form $x = a/b$ with a, b in $\mathbb{Z}_K$ and $b \notin \mathfrak{p}$. This set is usually denoted by $S^{-1}\mathbb{Z}_K$, with $S = \mathbb{Z}_K \setminus \mathfrak{p}$. It is thus strictly larger than $\mathbb{Z}_K$, and should not be confused with $\mathbb{Z}_{\mathfrak{p}}$ introduced below, which is its *completion*.

It is trivial to show that different prime ideals $\mathfrak{p}$ give inequivalent absolute values: indeed, if $\mathfrak{p}$ and $\mathfrak{q}$ are distinct (nonzero) prime ideals, we can find $\alpha \in \mathfrak{p}$ such that $\alpha \notin \mathfrak{q}$ (otherwise $\mathfrak{q} \mid \mathfrak{p}$), so that $|\alpha|_{\mathfrak{p}} < 1$ while $|\alpha|_{\mathfrak{q}} = 1$.

4.1.4 Ostrowski's Theorem and the Product Formula

The two lemmas above describe completely all the absolute values on a number field. We restate them slightly using the following definition.

Definition 4.1.12. *A place of a number field K is an equivalence class of nontrivial absolute values, where equivalence is characterized either by Definition 4.1.2 or by Lemma 4.1.3.*

Theorem 4.1.13 (Ostrowski). *The places of a number field K are in one-to-one correspondence with on the one hand the r_1 real embeddings and r_2 pairs of nonreal complex conjugate embeddings of K into $\mathbb{C}$, corresponding to the Archimedean absolute values (called the* infinite *places of K), and on the other hand, the (nonzero) prime ideals of $\mathbb{Z}_K$ corresponding to the non-Archimedean absolute values (called the* finite *places of K).*

Although places are determined up to equivalence, there is a canonical way to define them uniquely. This comes from the action of multiplication by an element on a Haar measure, but we simply give the formulas.

If σ is one of the r_1 real embeddings of K, we choose

$$\|\alpha\|_\sigma = |\sigma(\alpha)|\,,$$

in other words $c = 1$ in the above notation.

If $(\sigma, \overline{\sigma})$ is one of the r_2 pairs of complex conjugate nonreal embeddings of K, we choose

$$\|\alpha\|_\sigma = |\sigma(\alpha)|^2 = |\overline{\sigma}(\alpha)|^2 = \sigma(\alpha)\overline{\sigma}(\alpha)\,,$$

in other words $c = 2$ in the above notation.

If $\mathfrak{p}$ is a (nonzero) prime ideal of $\mathbb{Z}_K$, we choose

$$\|\alpha\|_\mathfrak{p} = \mathcal{N}\mathfrak{p}^{-v_\mathfrak{p}(\alpha)}\,,$$

where $\mathcal{N}\mathfrak{p} = |\mathbb{Z}_K/\mathfrak{p}|$ is the norm of the prime ideal $\mathfrak{p}$, in other words $C = \mathcal{N}\mathfrak{p}$ in the above notation.

The main point of these normalizations is the *product formula*:

Proposition 4.1.14 (Product formula). *With the above normalizations, for any $\alpha \in K^*$ the set of places v of K for which $\|\alpha\| \neq 1$ is finite, and we have*

$$\prod_v \|\alpha\|_v = 1\,,$$

where v runs over all the places of K (so that the product is in fact a finite product).

Proof. Let $\alpha\mathbb{Z}_K = \prod_\mathfrak{p} \mathfrak{p}^{v_\mathfrak{p}(\alpha)}$ be the decomposition of the ideal $\alpha\mathbb{Z}_K$ into a product of powers of prime ideals. Thus, $\|\alpha\|_v = 1$ if v is neither an infinite place nor a prime ideal $\mathfrak{p}$ such that $v_\mathfrak{p}(\alpha) \neq 0$, proving the first statement. Then

$$\prod_{v \text{ finite}} \|\alpha\|_v = \prod_\mathfrak{p} \mathcal{N}\mathfrak{p}^{-v_\mathfrak{p}(\alpha)} = 1/\mathcal{N}(\alpha\mathbb{Z}_K) = 1/|\mathcal{N}_{K/\mathbb{Q}}(\alpha)|$$

by multiplicativity of the norm and the fact that $\mathcal{N}(\alpha\mathbb{Z}_K) = |\mathcal{N}_{K/\mathbb{Q}}(\alpha)|$ (note the absolute value). On the other hand,

$$\prod_{v \text{ infinite}} \|\alpha\|_v = \prod_{\sigma \text{ real}} |\sigma(\alpha)| \prod_{(\sigma,\bar{\sigma}) \text{ nonreal}} |\sigma(\alpha)\bar{\sigma}(\alpha)| = \prod_{\sigma} |\sigma(\alpha)| = |\mathcal{N}_{K/\mathbb{Q}}(\alpha)| \,.$$

Taking the product of the finite and infinite primes proves the proposition.

$\square$

4.1.5 Completions

Let K be a field with an absolute value $\| \ \|$ and a corresponding distance $d(x, y) = \|x - y\|$. We recall the following definitions.

Definition 4.1.15. *A Cauchy sequence in K is a sequence (x_n) of elements of K satisfying the following property: for all $\varepsilon > 0$ there exists a positive integer N such that for any n, m both greater than or equal to N, we have $d(x_n, x_m) < \varepsilon$.*

If (x_n) is a convergent sequence, say with limit x, then for N sufficiently large, for all $n \geqslant N$ we have $d(x_n, x) < \varepsilon/2$; hence by the triangle inequality, for all n and m both greater than or equal to N we have $d(x_n, x_m) \leqslant d(x_n, x) + d(x, x_m) < \varepsilon$. It follows that any convergent sequence is a Cauchy sequence.

Definition 4.1.16. *A field K is* complete *if every Cauchy sequence converges (a convergent sequence being always a Cauchy sequence by what we just said).*

If the reader is not very familiar with Cauchy sequences and complete spaces, he should note that the definition is quite remarkable: using Cauchy sequences one can test the convergence of a sequence without knowing or guessing its limit. In many cases including the most fundamental ones, it is the *only* way to prove the convergence of a sequence. For this method to work, one needs of course the space to be complete.

It is a crucial fact that there exists a canonical smallest complete field containing a given field endowed with an absolute value. We recall how this is done.

Proposition 4.1.17. *Let (K, d) be as above. There exists a complete field $(\widehat{K}, \widehat{d})$ and a* uniformly continuous *injective map i from K to $\widehat{K}$ such that $i(K)$ is dense in $\widehat{K}$ and such that $\widehat{d}$ extends d. Any two such complete fields are canonically isomorphic. Furthermore, if (L, e) is any complete field and f is a uniformly continuous map from K to L, there exists a unique uniformly continuous map $\widehat{f}$ from $\widehat{K}$ to L that extends f, i.e., such that $\widehat{f} \circ i = f$.*

Proof. Assume first the existence of $(\widehat{K}, \widehat{d})$. Since i is injective we identify K with $i(K)$. If (L, e) is any complete field and f a continuous map from K to L, then there is only one way to extend f by continuity: since K is dense in $\widehat{K}$, for any $x \in \widehat{K}$ there exists a sequence (x_n) of elements of K

converging to x in $\widehat{K}$, and we must set $f(x) = \lim f(x_n)$. Since f is not only continuous but uniformly continuous, for any $\varepsilon > 0$ there exists $\eta(\varepsilon) > 0$ such that $d(x, y) < \eta(\varepsilon)$ implies $e(f(x), f(y)) < \varepsilon$. Since (x_n) converges to x it is a Cauchy sequence; hence there exists N such that for n, m greater than or equal to N we have $d(x_n, x_m) < \eta(\varepsilon)$, hence $e(f(x_n), f(x_m)) < \varepsilon$, so that $f(x_n)$ is a Cauchy sequence; hence $f(x) = \lim f(x_n)$ exists, and it is clear that it is independent of the sequence x_n of elements of K converging to x. Finally, by writing

$$e(f(x), f(y)) \leqslant e(f(x), f(x_n)) + e(f(x_n), y_m)) + e(f(y_m), f(y))$$

it is immediately checked that f is uniformly continuous on $\widehat{K}$, proving the last assertion. Furthermore, if $(\widehat{K}', \widehat{d}')$ has the same properties as $(\widehat{K}, \widehat{d})$, so that i' is a uniformly continuous injective map from K to $\widehat{K}'$ with dense image, the above reasoning shows that i extends uniquely to a map $\widehat{i}$ from $\widehat{K}'$ to $\widehat{K}$, and i' extends uniquely to a map $\widehat{i}'$ from $\widehat{K}$ to $\widehat{K}'$; hence these maps are canonical inverse isomorphisms, proving the uniqueness statements.

It remains to prove the existence of $(\widehat{K}, \widehat{d})$. This is a standard construction, which we recall here. Denote by $\mathcal{C}$ the set of all Cauchy sequences of elements of K, and by $\mathcal{M}$ the subset consisting of Cauchy sequences of elements of K converging to 0. It is clear that $\mathcal{C}$ is a commutative ring for termwise addition and multiplication, and that $\mathcal{M}$ is an ideal of $\mathcal{C}$. We will set $\widehat{K} = \mathcal{C}/\mathcal{M}$ and show that there exists a distance $\widehat{d}$ on $\widehat{K}$ such that $(\widehat{K}, \widehat{d})$ has the required properties.

By construction, $\widehat{K}$ is a ring. Let us show that it is a field, so let (x_n) be a representative in $\mathcal{C}$ of an element of $\widehat{K}$ that does not belong to $\mathcal{M}$. This means that there exists $\varepsilon_0 > 0$ such that for all N there exists $n(N) \geqslant N$ with $d(0, x_{n(N)}) \geqslant \varepsilon_0$. Since (x_n) is a Cauchy sequence there exists N_0 such that $d(x_n, x_m) < \varepsilon_0/2$ for all n, m greater than or equal to N_0. Thus if $m > N_0$ we have

$$d(x_m, 0) \geqslant d(x_{n(N_0)}, 0) - d(x_m, x_{n(N_0)}) \geqslant \varepsilon_0 - \varepsilon_0/2 = \varepsilon_0/2 \; ;$$

hence $d(x_m, 0)$ is bounded from below. Thus, after changing a finite number of terms in the sequence (x_n), which does not change the class modulo $\mathcal{M}$, we may assume that $x_n \neq 0$ for all n. Furthermore, since

$$d(1/x_n, 1/x_m) = \left\| \frac{1}{x_n} - \frac{1}{x_m} \right\| = \frac{\|x_m - x_n\|}{\|x_n\| \|x_m\|} \leqslant \frac{4}{\varepsilon_0^2} d(x_n, x_m) \; ,$$

it immediately follows that $1/x_n$ is a Cauchy sequence, so $\widehat{K}$ is a field.

Let $x \in \widehat{K}$ and let (x_n) be a representative in $\mathcal{C}$ of x. The triangle inequality gives

$$\|x_n - x_m\| \geqslant |\|x_n\| - \|x_m\|| \; ,$$

from which it immediately follows that $\|x_n\|$ is a Cauchy sequence in $\mathbb{R}$. We set $\|x\|' = \lim_{n \to \infty} \|x_n\|$, which exists since $\mathbb{R}$ is complete. By definition of $\mathcal{M}$, this is independent of the chosen representative (x_n) of x. If fact, we embed K in $\widehat{K}$ by the "diagonal" map i that sends $x \in K$ to the constant sequence $x_n = x$. It is clear that this map is an injective field homomorphism, and that $\| \ \|'$ extends $\| \ \|$, so that i is uniformly continuous. If (x_n) is a Cauchy sequence of elements of K representing an element $x \in \widehat{K}$, it is trivially checked that this sequence converges to x for the topology of $\widehat{K}$, so that $i(K)$ is dense in $\widehat{K}$. Finally, let (x_n) be a Cauchy sequence of elements of $\widehat{K}$. By density, for each n we can find an element $y_n \in K$ such that $\|x_n - y_n\| < 1/n$. Thus

$$\|y_n - y_m\| \leqslant \|y_n - x_n\| + \|x_n - x_m\| + \|x_m - y_m\| \leqslant \|x_n - x_m\| + 1/n + 1/m \ ,$$

and since (x_n) is a Cauchy sequence it follows that (y_n) is a Cauchy sequence of elements of K. If y is the class in $\widehat{K}$ of (y_n), it is clear that y is the limit of the sequence (x_n) in $\widehat{K}$, showing that $\widehat{K}$ is complete and finishing the proof of the proposition.   □

Note that if the map f from K to a complete field L is assumed to be only continuous, it does *not* necessarily extend to a continuous map from $\widehat{K}$ to L; see Exercise 1.

The field $(\widehat{K}, \widehat{d})$ whose existence and uniqueness is ensured by the above proposition is called the *completion* of K for the absolute value $\| \ \|$ (or for the distance d). It is clear that $\widehat{d}$ is ultrametric if and only if d is ultrametric. Also, as for the field $\mathbb{R}$, the whole point of taking completions is that one can do *analysis* in complete fields. For instance, it will be very useful to use *series*:

Lemma 4.1.18. *Let K be a field with an ultrametric absolute value. A sequence (a_n) in K is a Cauchy sequence if and only if $\|a_{n+1} - a_n\|$ tends to 0 as n tends to infinity. In particular, if K is complete, a series $\sum_{k \geqslant 0} u_k$ of elements of K converges if and only if u_k tends to zero as k tends to infinity.*

Proof. The condition is of course necessary. Conversely, by the ultrametric inequality, for $m \geqslant n$ we have

$$\|a_m - a_n\| = \left\| \sum_{n \leqslant j < m} (a_{j+1} - a_j) \right\| \leqslant \max_{n \leqslant j < m} \|a_{j+1} - a_j\| \ ,$$

so that if N is such that $\|a_{n+1} - a_n\| < \varepsilon$ for all $n \geqslant N$, then we have also $\|a_m - a_n\| < \varepsilon$ for all $m \geqslant n \geqslant N$, so that (a_n) is indeed a Cauchy sequence. The last statement follows from this applied to the sequence of partial sums $a_n = \sum_{0 \leqslant k \leqslant n} u_k$.   □

It is convenient to identify K with its isomorphic image $i(K)$ in $\widehat{K}$, so that we can consider K as a subfield of $\widehat{K}$. From now on we make this identification. We will also usually speak of *the* completion of K for the absolute value $\| \ \|$.

In the special case that K is a number field, we also need the following trivial lemma.

Lemma 4.1.19. *Let K be a number field and $\mathfrak{p}$ a (nonzero) prime ideal of $\mathbb{Z}_K$. If we denote by $K_{\mathfrak{p}}$ the completion of K for the $\mathfrak{p}$-adic absolute value, then for all $x \in K_{\mathfrak{p}}^*$ there exists $k \in \mathbb{Z}$ such that $|x|_{\mathfrak{p}} = \mathcal{N}\mathfrak{p}^k$.*

Proof. Let (x_n) be a sequence of elements of K converging to x. Since $x \neq 0$, without loss of generality we may assume that $x_n \neq 0$ for all n. Set $k_n = \log(|x_n|_{\mathfrak{p}})/\log(\mathcal{N}\mathfrak{p})$. It is clear that k_n is a Cauchy sequence in $\mathbb{R}$. However, we have $k_n \in \mathbb{Z}$, which is discrete in $\mathbb{R}$. It follows that k_n is an ultimately constant sequence, so that its limit (which by definition is $\log(|x|_{\mathfrak{p}})/\log(\mathcal{N}\mathfrak{p})$) belongs to $\mathbb{Z}$, proving the lemma. $\square$

It follows from this lemma that it is reasonable to let $v_{\mathfrak{p}}(x)$ be the integer such that $|x|_{\mathfrak{p}} = \mathcal{N}\mathfrak{p}^{-v_{\mathfrak{p}}(x)}$, extending the usual valuation to the completion $K_{\mathfrak{p}}$. Also, we will set $v_{\mathfrak{p}}(0) = +\infty$, so that the formula $|x|_{\mathfrak{p}} = \mathcal{N}\mathfrak{p}^{-v_{\mathfrak{p}}(x)}$ still holds.

4.1.6 Completions of a Number Field

Let K be a number field. Since we know all the absolute values of K, we can describe explicitly all its completions. If v is a place of K (i.e., an equivalence class of nontrivial absolute values), we will denote by K_v a completion of K at v.

- If v is a place at infinity, i.e., corresponds to an Archimedean absolute value induced by an embedding σ of K into $\mathbb{C}$, then K_v is isomorphic to $\mathbb{R}$ if σ is a real embedding, and K_v is isomorphic to $\mathbb{C}$ if not. Since the basic properties of these fields should be well known to the reader, we will not expand more on them.
- Every prime ideal $\mathfrak{p}$ of $\mathbb{Z}_K$, corresponding to a finite place, gives rise to a non-Archimedean (hence ultrametric) absolute value. The completion of K will be denoted by $K_{\mathfrak{p}}$, and called the field of $\mathfrak{p}$-adic numbers. In the case $K = \mathbb{Q}$ and $\mathfrak{p} = p\mathbb{Z}$, we will write $\mathbb{Q}_p$ instead of $\mathbb{Q}_{p\mathbb{Z}}$, and call it the field of p-adic numbers. We describe here some basic properties of these fields.
  We fix a prime ideal $\mathfrak{p}$ of $\mathbb{Z}_K$, and by an abuse of notation that we will always use, we denote by $|\ |_{\mathfrak{p}}$ the extension to $K_{\mathfrak{p}}$ of the $\mathfrak{p}$-adic absolute value on K. By the above lemma, its nonzero values are also of the form $\mathcal{N}\mathfrak{p}^k$ for $k \in \mathbb{Z}$.

Definition 4.1.20. *We denote by $\mathbb{Z}_{\mathfrak{p}}$ the subset of elements $x \in K_{\mathfrak{p}}$ such that $|x|_{\mathfrak{p}} \leqslant 1$ (or equivalently, $v_{\mathfrak{p}}(x) \geqslant 0$). Such an element will be called a $\mathfrak{p}$-adic integer.*

Proposition 4.1.21. *The set $\mathbb{Z}_{\mathfrak{p}}$ is a* discrete valuation ring: *more precisely, the only nonzero ideals of $\mathbb{Z}_{\mathfrak{p}}$ are the $\mathfrak{p}^k\mathbb{Z}_{\mathfrak{p}}$ for $k \geqslant 0$, the only maximal ideal is $\mathfrak{p}\mathbb{Z}_{\mathfrak{p}}$, and $x \in \mathbb{Z}_{\mathfrak{p}}$ is invertible if and only if $x \notin \mathfrak{p}\mathbb{Z}_{\mathfrak{p}}$; in other words if $|x|_{\mathfrak{p}} = 1$.*

Proof. Since $|\ |_{\mathfrak{p}}$ is an ultrametric absolute value, it is clear that $\mathbb{Z}_{\mathfrak{p}}$ is a subring of $K_{\mathfrak{p}}$. Moreover, $x \in \mathbb{Z}_{\mathfrak{p}}$ is invertible in $\mathbb{Z}_{\mathfrak{p}}$ if and only if $|x|_{\mathfrak{p}} = 1$. Thus, let I be a nonzero ideal of $\mathbb{Z}_{\mathfrak{p}}$, and let k be the integer such that $\mathcal{N}\mathfrak{p}^{-k}$ is the largest absolute value of a nonzero element x_0 of I. Denote by π a uniformizer of $\mathfrak{p}$, i.e., an element of $\mathfrak{p} \setminus \mathfrak{p}^2$, so that $|\pi| = 1/\mathcal{N}\mathfrak{p}$. It follows that $x \in I$ implies that $|x/\pi^k|_{\mathfrak{p}} \leqslant 1$, so that $x \in \pi^k\mathbb{Z}_{\mathfrak{p}}$; hence $I \subset \mathfrak{p}^k\mathbb{Z}_{\mathfrak{p}}$. Conversely, $|x_0/\pi^k|_{\mathfrak{p}} = 1$; hence x_0/π^k is invertible in $\mathbb{Z}_{\mathfrak{p}}$, so that $\pi^k \in x_0\mathbb{Z}_{\mathfrak{p}} \subset I$; hence $\mathfrak{p}^k\mathbb{Z}_{\mathfrak{p}} \subset I$, proving the equality $I = \mathfrak{p}^k\mathbb{Z}_{\mathfrak{p}}$. $\square$

Proposition 4.1.22. *Let A and B be two monic polynomials of $K_{\mathfrak{p}}[X]$ such that $B \mid A$. If $A \in \mathbb{Z}_{\mathfrak{p}}[X]$, then also $B \in \mathbb{Z}_{\mathfrak{p}}[X]$.*

Proof. Write $A = BC$, and let π be a uniformizer of $\mathfrak{p}$. Let v_B and v_C be the smallest exponents such that $\pi^{v_B}B(X) \in \mathbb{Z}_{\mathfrak{p}}[X]$ and $\pi^{v_C}C(X) \in \mathbb{Z}_{\mathfrak{p}}[X]$, necessarily nonnegative since B and C are monic. Denoting by $\overline{}$ reduction modulo $\mathfrak{p}$, by minimality of the exponents, $b(X) = \overline{\pi^{v_B}B(X)}$ and $c(X) = \overline{\pi^{v_C}C(X)}$ are nonzero in the integral domain $(\mathbb{Z}_{\mathfrak{p}}/\mathfrak{p}\mathbb{Z}_{\mathfrak{p}})[X]$. It follows that $\overline{\pi^{v_B+v_C}A(X)} = b(X)c(X)$ is also nonzero, hence that $v_B + v_C = 0$, so $v_B = v_C = 0$ as claimed. $\square$

Recall that we identify K with a subfield of $K_{\mathfrak{p}}$.

Lemma 4.1.23. *Under the above identification $\mathbb{Z}_K$ is a subring of $\mathbb{Z}_{\mathfrak{p}}$ and is dense in $\mathbb{Z}_{\mathfrak{p}}$.*

Proof. If $x \in \mathbb{Z}_K$ then x is a root of $x^n + \sum_{0 \leqslant i \leqslant n-1} a_i x^i = 0$ with $a_i \in \mathbb{Z}$. Applying the ultrametric inequality we see that

$$|x|_{\mathfrak{p}}^n = |x^n|_{\mathfrak{p}} = \left| \sum_{0 \leqslant i \leqslant n-1} a_i x^i \right|_{\mathfrak{p}} \leqslant \max_{0 \leqslant i \leqslant n-1} |x|_{\mathfrak{p}}^i \ ,$$

which evidently leads to a contradiction if $|x|_{\mathfrak{p}} > 1$, so that $\mathbb{Z}_K \subset \mathbb{Z}_{\mathfrak{p}}$. Now let $x \in \mathbb{Z}_{\mathfrak{p}}$, and let (x_n) be any sequence of elements of K that converges to x in $K_{\mathfrak{p}}$. In particular, for n sufficiently large we have $|x_n - x|_{\mathfrak{p}} < 1$. Since $|x|_{\mathfrak{p}} \leqslant 1$, the ultrametric inequality implies that $|x_n|_{\mathfrak{p}} = |x_n - x + x|_{\mathfrak{p}} \leqslant 1$ for n sufficiently large, say $n \geqslant N$. Modifying the first N values of x_n, we may therefore assume that this is true for all n.

Write $x_n\mathbb{Z}_K = \mathfrak{a}_n\mathfrak{b}_n^{-1}$, where $\mathfrak{a}_n$ and $\mathfrak{b}_n$ are two coprime integral ideals of $\mathbb{Z}_K$, hence such that $\mathfrak{p} \nmid \mathfrak{b}_n$ for $n \geqslant N$. Since the ideals $\mathfrak{p}^n$ and $\mathfrak{b}_n$ are coprime, we have $\mathfrak{p}^n + \mathfrak{b}_n = \mathbb{Z}_K$, so there exist $u_n \in \mathfrak{p}^n$ and $v_n \in \mathfrak{b}_n$ such that $u_n + v_n = 1$; in other words, $v_n \in \mathfrak{b}_n$ and $v_n \equiv 1 \pmod{\mathfrak{p}^n}$. Thus $\mathfrak{b}_n \mid v_n$, so that $v_n x_n \mathbb{Z}_K = \mathfrak{a}_n(v_n/\mathfrak{b}_n) \subset \mathbb{Z}_K$; hence $v_n x_n \in \mathbb{Z}_K$. On the other hand, v_n tends to 1 in $K_\mathfrak{p}$, so that $v_n x_n$ tends to x, proving the lemma. $\square$

Proposition 4.1.24. *For any $k \geqslant 1$, the natural map from $\mathbb{Z}_K$ to $\mathbb{Z}_\mathfrak{p}$ induces a ring isomorphism from $\mathbb{Z}_K/\mathfrak{p}^k$ to $\mathbb{Z}_\mathfrak{p}/\mathfrak{p}^k \mathbb{Z}_\mathfrak{p}$.*

Proof. It is clear that the map induces a ring homomorphism ϕ from $\mathbb{Z}_K/\mathfrak{p}^k$ to $\mathbb{Z}_\mathfrak{p}/\mathfrak{p}^k \mathbb{Z}_\mathfrak{p}$. Furthermore, $\overline{x} \in \mathbb{Z}_K/\mathfrak{p}^k$ is such that $\phi(x) = \overline{0}$ if and only if for some (or any) representative x of $\overline{x}$ in $\mathbb{Z}_K$ we have $x \in \mathfrak{p}^k \mathbb{Z}_\mathfrak{p}$, hence $v_\mathfrak{p}(x) \geqslant k$, hence $x \in \mathfrak{p}^k$, so that $\overline{x} = \overline{0}$, showing that ϕ is injective. Finally, let $x \in \mathbb{Z}_\mathfrak{p}$. By the above lemma, there exists a sequence (x_n) of elements of $\mathbb{Z}_K$ that converges $\mathfrak{p}$-adically to x. This is a Cauchy sequence, so for N sufficiently large and for all n and m greater than or equal to N we have $|x_n - x_m|_\mathfrak{p} \leqslant \mathcal{N}\mathfrak{p}^{-k}$, hence in particular $|x_n - x_N|_\mathfrak{p} \leqslant \mathcal{N}\mathfrak{p}^{-k}$. By definition, this means that $v_\mathfrak{p}(x_n - x_N) \geqslant k$, in other words that $x_n \equiv x_N \pmod{\mathfrak{p}^k}$. Thus the sequence x_n modulo $\mathfrak{p}^k$ is a constant y for n sufficiently large, and it is clear that the image by ϕ of the class of y modulo $\mathfrak{p}^k$ is equal to the class of x modulo $\mathfrak{p}^k \mathbb{Z}_\mathfrak{p}$. $\square$

In particular, it follows from this proposition that $\mathbb{Z}_\mathfrak{p}/\mathfrak{p}^k \mathbb{Z}_\mathfrak{p}$ is a finite ring. This proposition also allows us to describe in a very concrete way the elements of $K_\mathfrak{p}$.

Proposition 4.1.25. *Let R be a set of representatives in $\mathbb{Z}_K$ of the elements of $\mathbb{Z}_K/\mathfrak{p}$, and for each $m \in \mathbb{Z}$ let $\pi_m \in \mathfrak{p}^m \setminus \mathfrak{p}^{m+1}$ be an element of exact $\mathfrak{p}$-adic valuation m. An element $x \in K_\mathfrak{p}$ can be written in a unique way as the sum of an infinite series of the form*

$$x = \sum_{m \geqslant v_\mathfrak{p}(x)} a_m \pi_m \text{ with } a_m \in R \text{ and } a_{v_\mathfrak{p}(x)} \not\equiv 0 \pmod{\mathfrak{p}} .$$

Note in particular that elements of $\mathbb{Q}_p$ can be written in a unique way in the form $x = \sum_{m \geqslant v_p(x)} a_m p^m$ with $0 \leqslant a_m \leqslant p - 1$.

Proof. Since $a_m \pi_m$ tends to 0 for the $\mathfrak{p}$-adic topology, any series of this form converges. Dividing by $\pi^{v_\mathfrak{p}(x)}$, we may assume that $v_\mathfrak{p}(x) = 0$, i.e., that $x_0 = x$ is an invertible element of $\mathbb{Z}_\mathfrak{p}$.

By the above proposition, there exists a surjective map ϕ from $\mathbb{Z}_\mathfrak{p}$ to R obtained by composing the projection map from $\mathbb{Z}_\mathfrak{p}$ to $\mathbb{Z}_\mathfrak{p}/\mathfrak{p}\mathbb{Z}_\mathfrak{p}$ with the inverse isomorphism of the one given in the proposition for $k = 1$ and finally with the natural lifting map from $\mathbb{Z}_K/\mathfrak{p}$ to R. This map satisfies $\phi(x) \equiv x \pmod{\mathfrak{p}}$. By induction, it is clear that the sequence a_m is determined uniquely by the double recurrence $a_m = \phi(x_m)$, $x_{m+1} = (x_m - a_m)\pi_m/\pi_{m+1}$, and this proves both uniqueness and existence. $\square$

In practice we usually choose $\pi_m = \pi_1^m$, but we need the above generality below.

Corollary 4.1.26. *The field $K_\mathfrak{p}$ is locally compact and totally discontinuous. The ring of $\mathfrak{p}$-adic integers $\mathbb{Z}_\mathfrak{p}$ is compact.*

Proof. We first check that the open unit ball $B = \{x \in K_\mathfrak{p}/\ |x|_\mathfrak{p} < 1\}$ is compact. By the proposition, $x \in B$ if and only if $x = \sum_{m \geqslant 1} a_m \pi_m$ for $a_m \in R$. Thus, if (x_n) is any sequence of elements of B we can write $x_n = \sum_{m \geqslant 1} a_{n,m} \pi_m$, and since R is finite it follows that an infinite number of $a_{n,1}$ are equal, to a_1, say. Among those, an infinite number of $a_{n,2}$ are equal, to a_2, say. By induction, we thus construct an element $x = \sum_{m \geqslant 1} a_m \pi_m \in B$ that is a limit point of (x_n), proving that B is compact. A similar reasoning shows that the closed unit ball is compact. In fact, $|x|_\mathfrak{p} < 1$ being equivalent to $|x|_\mathfrak{p} \leqslant 1/\mathcal{N}\mathfrak{p}$, B is in fact the closed ball of radius $1/\mathcal{N}\mathfrak{p}$.

Finally, recall that a topological space is totally discontinuous if and only if its connected components are reduced to points. By what we have just shown, all balls are both open and closed, so the corollary follows.   □

See also Exercise 2.

Corollary 4.1.27. *Let K be a number field, let L be a finite extension of K, let $\mathfrak{p}$ be a (nonzero) prime ideal of $\mathbb{Z}_K$, and finally let $\mathfrak{P}$ be a prime ideal of $\mathbb{Z}_L$ above $\mathfrak{p}$. Denote as usual by $e(\mathfrak{P}/\mathfrak{p})$ and $f(\mathfrak{P}/\mathfrak{p})$ the ramification index and residual index of $\mathfrak{P}$ over $\mathfrak{p}$, so that $e(\mathfrak{P}/\mathfrak{p}) = v_\mathfrak{P}(\mathfrak{p}\mathbb{Z}_L)$ and $f(\mathfrak{P}/\mathfrak{p}) = [\mathbb{Z}_L/\mathfrak{P} : \mathbb{Z}_K/\mathfrak{p}]$. Then $L_\mathfrak{P}$ is a finite extension of $K_\mathfrak{p}$ of degree $e(\mathfrak{P}/\mathfrak{p})f(\mathfrak{P}/\mathfrak{p})$. Furthermore, $\mathbb{Z}_\mathfrak{P}$ is a free $\mathbb{Z}_\mathfrak{p}$-module of the same rank.*

Proof. Let $\Pi \in \mathfrak{P} \setminus \mathfrak{P}^2$ be a uniformizer of $\mathfrak{P}$ in L, let $\pi \in \mathfrak{p} \setminus \mathfrak{p}^2$ be a uniformizer of $\mathfrak{p}$ in K, and for simplicity write $e = e(\mathfrak{P}/\mathfrak{p})$ and $f = f(\mathfrak{P}/\mathfrak{p})$. It is clear that if we set

$$\Pi_m = \pi^{\lfloor m/e \rfloor} \Pi^{m - e\lfloor m/e \rfloor}$$

we have $v_\mathfrak{P}(\Pi_m) = m$; hence in Proposition 4.1.25 we can choose the Π_m as elements of exact $\mathfrak{P}$-adic valuation m. Furthermore, since $[\mathbb{Z}_L/\mathfrak{P} : \mathbb{Z}_K/\mathfrak{p}] = f$, we can choose a basis $e_1, \ldots, e_f$ of $\mathbb{Z}_L/\mathfrak{P}$ as a $\mathbb{Z}_K/\mathfrak{p}$-vector space, so that a system R of representatives of $\mathbb{Z}_L/\mathfrak{P}$ can be chosen to be the $\sum_{1 \leqslant j \leqslant f} a_j e_j$ with the a_i being chosen in any system of representatives R of $\mathbb{Z}_K/\mathfrak{p}$. Thus, by Proposition 4.1.25, any element $x \in L_\mathfrak{P}$ can be written in a unique way in the form

$$x = \sum_{m \geqslant v_\mathfrak{P}(x)} \sum_{1 \leqslant j \leqslant f} a_{j,m} e_j \Pi_m$$

$$= \sum_{m \geqslant v_\mathfrak{P}(x)} \sum_{1 \leqslant j \leqslant f} a_{j,m} e_j \pi^{\lfloor m/e \rfloor} \Pi^{m - e\lfloor m/e \rfloor}$$

$$= \sum_{0 \leqslant i \leqslant e-1} \Pi^i \sum_{1 \leqslant j \leqslant f} e_j \left(\sum_{k \geqslant k_0(i,j)} a_{i,j,k} \pi^k \right)$$

for some $k_0(i,j)$ and $a_{i,j,k} \in R$. Once again by Proposition 4.1.25, now applied to $\mathfrak{p}$, the expression in parentheses is the expansion of an arbitrary element of $K_{\mathfrak{p}}$ with respect to $\pi_m = \pi^m$. Thus we see that the $\Pi^i e_j$ for $0 \leqslant i \leqslant e-1$ and $1 \leqslant j \leqslant f$ form a $K_{\mathfrak{p}}$-basis of $L_{\mathfrak{p}}$. Furthermore, we clearly have $x \in \mathbb{Z}_{\mathfrak{P}}$ if and only if $k_0(i,j) \geqslant 0$ for all i and j, so that the $\Pi^i e_j$ also form a $\mathbb{Z}_{\mathfrak{p}}$-basis of $\mathbb{Z}_{\mathfrak{P}}$, proving the corollary. $\qquad\square$

Corollary 4.1.28. *With the same notation, if in addition L/K is a Galois extension then $L_{\mathfrak{P}}/K_{\mathfrak{p}}$ is also a Galois extension with Galois group isomorphic to the decomposition group $D(\mathfrak{P}/\mathfrak{p})$ of $\mathfrak{P}$ over $\mathfrak{p}$.*

Proof. Immediate and left to the reader. $\qquad\square$

4.1.7 Hensel's Lemmas

The crucial tool that allows us to work in $\mathfrak{p}$-adic fields is a simple result, called (in many different contexts) Hensel's lemma, and which is nothing else than a non-Archimedean version of Newton's root-finding method. We begin with the following result, where for any object x, we denote by $\overline{x}$ the reduction of x modulo $\mathfrak{p}$. Thanks to Proposition 4.1.24, we will identify $\mathbb{Z}_{\mathfrak{p}}/\mathfrak{p}\mathbb{Z}_{\mathfrak{p}}$ with $\mathbb{Z}_K/\mathfrak{p}$.

Proposition 4.1.29 (Hensel). *Let $f \in \mathbb{Z}_{\mathfrak{p}}[X]$, and assume that $\overline{f}(X) = \phi_1(X)\phi_2(X)$ with $\phi_i \in (\mathbb{Z}_K/\mathfrak{p})[X]$ coprime. There exist polynomials f_1 and f_2 in $\mathbb{Z}_{\mathfrak{p}}[X]$ such that $f(X) = f_1(X)f_2(X)$, $\overline{f_i}(X) = \phi_i(X)$, and $\deg(f_1) = \deg(\phi_1)$.*

Proof. We prove by induction on k that there exist polynomials A_k, B_k, U_k, V_k in $\mathbb{Z}_{\mathfrak{p}}[X]$ such that $f(X) \equiv A_k(X)B_k(X) \pmod{\mathfrak{p}^k}$, $U_k(X)A_k(X) + V_k(X)B_k(X) \equiv 1 \pmod{\mathfrak{p}}$, $A_k(X) - A_{k-1}(X) \equiv B_k(X) - B_{k-1}(X) \equiv 0 \pmod{\mathfrak{p}^{k-1}}$, $\deg(A_k(X)) = \deg(\phi_1(X))$, and $\deg(B_k(X)) \leqslant \deg(f) - \deg(\phi_1(X))$. It is true for $k = 1$ by choosing $A_1(X) = \phi_1(X)$, $A_2(X) = \phi_2(X)$ and using the fact that ϕ_1 and ϕ_2 are coprime in $(\mathbb{Z}_K/\mathfrak{p})[X]$. Assume this assertion true for k, and set $g_k = (f - A_k B_k)/\pi^k \in \mathbb{Z}_{\mathfrak{p}}[X]$ by assumption. We must set $A_{k+1} = A_k + \pi^k S_k$, $B_{k+1} = B_k + \pi^k T_k$, and the main condition $f(X) \equiv A_{k+1}(X)B_{k+1}(X) \pmod{\mathfrak{p}^{k+1}}$ is equivalent to $A_k T_k + B_k S_k \equiv (f - A_k B_k)/\pi^k = g_k \pmod{\mathfrak{p}}$. Since $U_k A_k + V_k B_k \equiv 1 \pmod{\mathfrak{p}}$ the general solution is $S_k \equiv V_k g_k + W A_k \pmod{\mathfrak{p}}$ and $T_k \equiv U_k g_k - W B_k \pmod{\mathfrak{p}}$ for some polynomial W. The degree condition implies that S_k and T_k exist and are unique modulo $\mathfrak{p}$; hence A_{k+1} and B_{k+1} exist and are unique modulo $\mathfrak{p}^{k+1}$. Passing to the limit as k tends to infinity, we obtain the proposition. $\qquad\square$

Note that the degree condition given in the proposition is essential to ensure uniqueness, and that the proof is completely constructive.

Lemma 4.1.30. *Let L/K be an extension of number fields, let $\mathfrak{p}$ be a prime ideal of $\mathbb{Z}_K$, and let $\mathfrak{P}$ be a prime ideal of $\mathbb{Z}_L$ above $\mathfrak{p}$. Finally, let $\alpha \in L_{\mathfrak{P}}$ and let $f(X) = X^m + f_{m-1}X^{m-1} + \cdots + f_0 \in \mathbb{Z}_{\mathfrak{p}}[X]$ be any monic polynomial such that $f(\alpha) = 0$. Then $\alpha \in \mathbb{Z}_{\mathfrak{P}}$.*

Proof. The proof is identical to that of the first part of Lemma 4.1.23 and is left to the reader. $\square$

Proposition 4.1.31. *Let L/K be an extension of number fields, let $\mathfrak{p}$ be a prime ideal of $\mathbb{Z}_K$, and let $\mathfrak{P}$ be a prime ideal of $\mathbb{Z}_L$ above $\mathfrak{p}$. Finally, let $\alpha \in L_{\mathfrak{P}}$ and let $f(X) = f_m X^m + f_{m-1}X^{m-1} + \cdots + f_0$ be its minimal monic polynomial over $K_{\mathfrak{p}}$ (so that $f_m = 1$). Then $\alpha \in \mathbb{Z}_{\mathfrak{P}}$ if and only if $f_j \in \mathbb{Z}_{\mathfrak{p}}$ for all j, and α is a $\mathfrak{P}$-adic unit if and only if in addition f_0 is a $\mathfrak{p}$-adic unit.*

Proof. Note first that the fact that $L_{\mathfrak{P}}$ is a finite extension of $K_{\mathfrak{p}}$ follows from Corollary 4.1.27. Assume first that $\alpha \in \mathbb{Z}_{\mathfrak{P}}$, and by contradiction that not all f_k belong to $\mathbb{Z}_{\mathfrak{p}}$. Let $-v$ with $v > 0$ be the smallest valuation of $v_{\mathfrak{p}}(f_k)$. If the constant term f_0 is of valuation $-v$ and is the only one of minimal valuation, the ultrametric inequality and $f(\alpha) = 0$ with $v_{\mathfrak{P}}(\alpha) \geq 0$ lead to

$$v_{\mathfrak{P}}(f_0) = v_{\mathfrak{P}}\left(-\sum_{1 \leq i \leq m} f_i \alpha^i\right) \geq \min_{1 \leq i \leq m} v_{\mathfrak{P}}(f_i) > v_{\mathfrak{P}}(f_0) ,$$

a contradiction (note that for any $c \in K_{\mathfrak{p}}$ we have $v_{\mathfrak{P}}(c) = e(\mathfrak{P}/\mathfrak{p})v_{\mathfrak{p}}(c)$).

Now consider the polynomial $g(X) = \pi^v f(X) \in \mathbb{Z}_{\mathfrak{p}}[X]$ and its reduction $\overline{g}(X) \in (\mathbb{Z}_{\mathfrak{p}}/\mathfrak{p}\mathbb{Z}_{\mathfrak{p}})[X]$. Since $v > 0$ and $f_m = 1$, it follows that $\deg(\overline{g}) < m$, and by what we have just shown, that $\deg(\overline{g}) > 0$. If we apply Proposition 4.1.29 to the trivial factorization into coprime polynomials $\overline{g}(X) = \overline{g}(X) \cdot 1$, it follows that there exist polynomials g_1 and g_2 in $\mathbb{Z}_{\mathfrak{p}}[X]$ with $g(X) = g_1(X)g_2(X)$, $\overline{g_1}(X) = \overline{g}(X)$, $\overline{g_2}(X) = 1$, and $\deg(g_1) = \deg(\overline{g})$, and since $0 < \deg(\overline{g}) < m$, this gives a nontrivial factorization of $g(X)$ in $\mathbb{Z}_{\mathfrak{p}}[X]$, contradicting the fact that f, being a minimal polynomial, is irreducible.

The converse is Lemma 4.1.30. Finally, if $\alpha \in \mathbb{Z}_{\mathfrak{P}}$, α is a $\mathfrak{P}$-adic unit if and only if $1/\alpha \in \mathbb{Z}_{\mathfrak{P}}$. We then apply the first part of the proposition to the minimal monic polynomial of $1/\alpha$, equal to $X^m + (f_1/f_0)X^{m-1} + \cdots + (f_m/f_0)$. $\square$

Corollary 4.1.32. *Let L/K be an extension of number fields, let $\mathfrak{p}$ be a prime ideal of $\mathbb{Z}_K$, and let $\mathfrak{P}$ be a prime ideal of $\mathbb{Z}_L$ above $\mathfrak{p}$. Then any element $\alpha \in \mathbb{Z}_{\mathfrak{P}}$ is integral over $\mathbb{Z}_{\mathfrak{p}}$. In particular, $\mathrm{Tr}_{L_{\mathfrak{P}}/K_{\mathfrak{p}}}(\alpha)$ and $\mathcal{N}_{L_{\mathfrak{P}}/K_{\mathfrak{p}}}(\alpha)$ are in $\mathbb{Z}_{\mathfrak{p}}$, where the trace and norm are defined as usual in terms of the coefficients of the characteristic polynomial of α in $K_{\mathfrak{p}}[X]$.*

Proof. Clear from the above proposition. $\square$

Another result of the same kind is the following.

Proposition 4.1.33. *Let* $f(X) = f_m X^m + f_{m-1} X^{m-1} + \cdots + f_0 \in K_{\mathfrak{p}}[X]$ *be a monic polynomial (so that* $f_m = 1$*). If* f *is irreducible in* $K_{\mathfrak{p}}[X]$*, then either* $v_{\mathfrak{p}}(f_j) \geqslant v_{\mathfrak{p}}(f_0)$ *for all* j *(and in particular* $v_{\mathfrak{p}}(f_0) \leqslant v_{\mathfrak{p}}(f_m) = 0$*), or* $v_{\mathfrak{p}}(f_j) \geqslant 1$ *(i.e.,* $f_j \in \mathfrak{p}\mathbb{Z}_{\mathfrak{p}}$*) for all* $j < m$*.*

Proof. Let $v = \min_k v_{\mathfrak{p}}(f_k) \leqslant 0$ since $f_m = 1$, and let j_0 be the smallest index j such that $v_{\mathfrak{p}}(f_{j_0}) = v$, so that in particular $v_{\mathfrak{p}}(f_j) > v_{\mathfrak{p}}(f_{j_0})$ for $j < j_0$. If $g(X) = f(X)/f_{j_0} = g_m X^m + \cdots + g_0$, we have therefore $g(X) \in \mathbb{Z}_{\mathfrak{p}}[X]$, $g_{j_0} = 1$, hence $\overline{g(X)} = X^{j_0} \overline{g_2}(X)$ with $\overline{g_2}(0) = 1$. The polynomials $\overline{g_1}(X) = X^{j_0}$ and $\overline{g_2}(X)$ are therefore coprime in $(\mathbb{Z}_K/\mathfrak{p})[X]$; hence by Proposition 4.1.29 there exist g_1 and g_2 in $\mathbb{Z}_{\mathfrak{p}}[X]$ such that $g(X) = g_1(X) g_2(X)$ and $\deg(g_1) = \deg(\overline{g_1}) = j_0$. Since f, hence also g, is irreducible in $K_{\mathfrak{p}}[X]$ it follows that the above factorization must be trivial, so that $j_0 = 0$ or $j_0 = m$. If $j_0 = 0$ then $v_{\mathfrak{p}}(f_j) \geqslant v_{\mathfrak{p}}(f_0)$ for all j, and if $j_0 = m$, then for all $j < m$ we have $v_{\mathfrak{p}}(f_j) > v_{\mathfrak{p}}(f_{j_0}) = 0$, as claimed. $\square$

Corollary 4.1.34. *With the same assumptions as in the proposition, if* $f_0 \in \mathbb{Z}_{\mathfrak{p}}$ *then* $f(X) \in \mathbb{Z}_{\mathfrak{p}}[X]$*, and if* $f_0 \in \mathfrak{p}\mathbb{Z}_{\mathfrak{p}}$ *then* $f_j \in \mathfrak{p}\mathbb{Z}_{\mathfrak{p}}$ *for all* $j < m$*.*

Proof. Clear. $\square$

Still another result of the same kind is the famous "Eisenstein criterion."

Proposition 4.1.35 (Eisenstein). *Let* $f(X) = X^n + f_{n-1} X^{n-1} + \cdots + f_0 \in \mathbb{Z}_{\mathfrak{p}}[X]$ *be a monic polynomial with* $\mathfrak{p}$*-integral coefficients. Assume that* $v_{\mathfrak{p}}(f_j) \geqslant 1$ *for all* j *such that* $0 \leqslant j \leqslant n - 1$ *and that* $v_{\mathfrak{p}}(f_0) = 1$*. Then* f *is irreducible in* $K_{\mathfrak{p}}[X]$*.*

Proof. Note first that if α is a root of $f(X) = 0$ in some finite extension of $\mathbb{Z}_{\mathfrak{p}}$, then by the same reasoning as that used in the "converse" part of the proof of Proposition 4.1.31 (which does not use the irreducibility of f) we deduce that α is a $\mathfrak{P}$-adic integer for a prime ideal $\mathfrak{P}$ above $\mathfrak{p}$. Thus by Proposition 4.1.31 itself, the minimal monic polynomial of α belongs to $\mathbb{Z}_{\mathfrak{p}}[X]$. It follows that the monic irreducible factors of $f(X)$ belong to $\mathbb{Z}_{\mathfrak{p}}[X]$; hence it is enough to prove irreducibility in $\mathbb{Z}_{\mathfrak{p}}[X]$. Assume by contradiction that $f = gh$ with g, h nonconstant monic polynomials in $\mathbb{Z}_{\mathfrak{p}}[X]$ (we may clearly assume g and h monic since their leading terms are $\mathfrak{p}$-adic units). Denote as usual by $\tilde{}$ reduction modulo $\mathfrak{p}$ of elements or polynomials. By the first assumption on f, we have $\overline{f}(X) = X^n$, hence $\overline{g}(X) = X^r$ and $\overline{h}(X) = X^s$, where $r = \deg(f)$ and $r = \deg(g)$. It follows that the constant terms of g and h have $\mathfrak{p}$-adic valuation greater than or equal to 1, so the constant term of $f = gh$, which is the product of the constant terms, has $\mathfrak{p}$-adic valuation greater than or equal to 2, a contradiction. $\square$

A polynomial satisfying the conditions of the above proposition will be called an *Eisenstein polynomial*. See also Exercise 5.

Corollary 4.1.36. *The polynomial*

$$\Phi_{p^n}(X) = \frac{X^{p^n} - 1}{X^{p^{n-1}} - 1} = X^{p^{n-1}(p-1)} + X^{p^{n-1}(p-2)} + \cdots + 1$$

is irreducible in $\mathbb{Q}_p[X]$.

Proof. It suffices to show that $\Phi_{p^n}(X + 1)$ is an Eisenstein polynomial. For $n = 1$, by the binomial theorem we have directly

$$\Phi_p(X + 1) = X^{p-1} + \sum_{2 \leqslant j \leqslant p-2} \binom{p}{j} X^{p-1-j} + p ,$$

so that $\Phi_p(X + 1)$ is indeed an Eisenstein polynomial. For $n > 1$, again by the binomial theorem we have $(X + 1)^p \equiv X^p + 1 \pmod{p\mathbb{Z}[X]}$, so that by induction we have $(X + 1)^{p^{n-1}} \equiv X^{p^{n-1}} + 1 \pmod{p\mathbb{Z}[X]}$. Thus

$$\Phi_{p^n}(X + 1) = \Phi_p((X + 1)^{p^{n-1}}) \equiv \Phi_p(X^{p^{n-1}} + 1) \equiv X^{p^{n-1}(p-1)} \pmod{p\mathbb{Z}[X]}$$

since $\Phi_p(X + 1) \equiv X^{p-1} \pmod{p\mathbb{Z}[X]}$, so that the first condition for an Eisenstein polynomial is satisfied. Furthermore, the constant term of $\Phi_{p^n}(X + 1)$ is $\Phi_{p^n}(1) = \Phi_p(1^{p^{n-1}}) = p$, so the second condition is also satisfied. $\square$

Another version of Hensel's lemma is very useful to show the existence of roots in p-adic fields:

Proposition 4.1.37 (Hensel). *Let $f(X) \in \mathbb{Z}_\mathfrak{p}[X]$ be a monic polynomial and let $\alpha \in \mathbb{Z}_\mathfrak{p}$ be such that $|f(\alpha)|_\mathfrak{p} < |f'(\alpha)|_\mathfrak{p}^2$, where $f'(X)$ is the formal derivative of $f(X)$. There exists a unique root α^* of $f(X) = 0$ in $\mathbb{Z}_\mathfrak{p}$ such that*

$$|\alpha^* - \alpha|_\mathfrak{p} \leqslant \frac{|f(\alpha)|_\mathfrak{p}}{|f'(\alpha)|_\mathfrak{p}} < |f'(\alpha)|_\mathfrak{p} .$$

Proof. We prove by induction that there exists a sequence α_k of elements of $\mathbb{Z}_\mathfrak{p}$ such that

$$|f(\alpha_{k+1})|_\mathfrak{p} \leqslant (|f(\alpha)|_\mathfrak{p}/|f'(\alpha)|_\mathfrak{p}^2)|f(\alpha_k)|_\mathfrak{p} < |f(\alpha_k)|_\mathfrak{p} ,$$

$$|f'(\alpha_{k+1})|_\mathfrak{p} = |f'(\alpha)|_\mathfrak{p}, \quad |\alpha_{k+1} - \alpha_k|_\mathfrak{p} \leqslant |f(\alpha_k)|_\mathfrak{p}/|f'(\alpha)|_\mathfrak{p} .$$

It is clear that these relations will imply that $|f(\alpha_k)|_\mathfrak{p}$ tends to zero (since it is a strictly decreasing sequence of numbers of the form $\mathcal{N}\mathfrak{p}^{-n}$ with $n \in \mathbb{Z}$), hence that $\alpha_{k+1} - \alpha_k$ tends to 0, hence that the sequence α_k converges to some limit α^*, as a sum of a series of terms tending to 0. Since $|f(\alpha_k)|_\mathfrak{p}$ tends to zero, we have $f(\alpha^*) = 0$, and the other inequality also follows by passing to the limit.

To prove the existence of α_k, we use Newton's method. Assuming the existence of α_k, we will set $\alpha_{k+1} = \alpha_k + \beta_k$ with $\beta_k = -f(\alpha_k)/f'(\alpha_k)$, which makes sense since by induction $|f'(\alpha_k)|_{\mathfrak{p}} = |f'(\alpha)|_{\mathfrak{p}} \neq 0$. In fact, by induction, since $|f(\alpha_k)|_{\mathfrak{p}}$ is a strictly decreasing sequence we have

$$|\beta_k|_{\mathfrak{p}} = |f(\alpha_k)|_{\mathfrak{p}}/|f'(\alpha)|_{\mathfrak{p}} \leqslant |f(\alpha)|_{\mathfrak{p}}/|f'(\alpha)|_{\mathfrak{p}}$$

and also

$$|\beta_k|_{\mathfrak{p}}^2 \leqslant (|f(\alpha)|_{\mathfrak{p}}/|f'(\alpha)|_{\mathfrak{p}}^2)|f(\alpha_k)|_{\mathfrak{p}} \ .$$

By Taylor's formula for polynomials, there exists $\gamma_k \in \mathbb{Z}_{\mathfrak{p}}$ such that

$$f(\alpha_{k+1}) = f(\alpha_k) + \beta_k f'(\alpha_k) + \beta_k^2 \gamma_k = \beta_k^2 \gamma_k$$

by our choice of β_k, so that

$$|f(\alpha_{k+1})|_{\mathfrak{p}} \leqslant (|f(\alpha)|_{\mathfrak{p}}/|f'(\alpha)|_{\mathfrak{p}}^2)|f(\alpha_k)|_{\mathfrak{p}} < |f(\alpha_k)|_{\mathfrak{p}} \ .$$

Similarly, there exists $\delta_k \in \mathbb{Z}_{\mathfrak{p}}$ such that

$$f'(\alpha_{k+1}) = f'(\alpha_k) + \beta_k \delta_k \ .$$

Since

$$|\beta_k|_{\mathfrak{p}} \leqslant |f(\alpha)|_{\mathfrak{p}}/|f'(\alpha)|_{\mathfrak{p}} < |f'(\alpha)|_{\mathfrak{p}} = |f'(\alpha_k)|_{\mathfrak{p}} \ ,$$

it follows from the ultrametric inequality that $|f'(\alpha_{k+1})|_{\mathfrak{p}} = |f'(\alpha_k)|_{\mathfrak{p}} = |f'(\alpha)|_{\mathfrak{p}}$, finishing the proof of the existence of the α_k.

To prove uniqueness, assume that α_1^* and α_2^* are distinct roots of $f(X) = 0$ satisfying the inequality of the proposition. We can write $f(X) = (X - \alpha_1^*)(X - \alpha_2^*)g(X)$ for some $g \in \mathbb{Z}_{\mathfrak{p}}[X]$, so that

$$|f'(\alpha_1^*)|_{\mathfrak{p}} = |\alpha_1^* - \alpha_2^*|_{\mathfrak{p}}|g(\alpha_1^*)|_{\mathfrak{p}} \leqslant |\alpha_1^* - \alpha_2^*|_{\mathfrak{p}} < |f'(\alpha)|_{\mathfrak{p}} \ ,$$

a contradiction. $\square$

Note that the above proposition in not true in general if f does not have $\mathfrak{p}$-integral coefficients.

When $f(\alpha) \in \mathfrak{p}\mathbb{Z}_{\mathfrak{p}}$ and $f'(\alpha)$ is a $\mathfrak{p}$-adic unit, the condition of the proposition is satisfied. Here is an important consequence. First recall the following definition.

Definition 4.1.38. *Let $P(X_1, \ldots, X_n) = 0$ be a (not necessarily polynomial) homogeneous equation. We say that a nontrivial solution $(x_1, \ldots, x_n)$ of $P(x_1, \ldots, x_n) = 0$ is* nonsingular *if $\frac{\partial P}{\partial X_j}(x_1, \ldots, x_n) \neq 0$ for at least one index j. We say that the equation itself is* nonsingular *if it has no nontrivial singular solutions.*

Corollary 4.1.39. *Let $P(\underline{X}) \in \mathbb{Z}_p[X_1, \ldots, X_n]$ be a homogeneous polynomial in n variables, and let $(x_1, \ldots, x_n) \in \mathbb{F}_p^n$ be a nontrivial nonsingular solution of $\overline{P}(\underline{X}) = 0$, where $\overline{P}$ is obtained by reducing the coefficients of P modulo p in $\mathbb{Z}_p/p\mathbb{Z}_p \simeq \mathbb{F}_p$. Then there exist $(\alpha_1, \ldots, \alpha_n) \in \mathbb{Z}_p^n$ satisfying $P(\alpha_1, \ldots, \alpha_n) = 0$ such that $\overline{\alpha_i} = x_i$ for all i.*

Proof. By definition, there exists j such that $\frac{\partial P}{\partial X_j}(x_1, \ldots, x_n) \not\equiv 0 \pmod{p\mathbb{Z}_p}$; hence it is a p-adic unit, while $P(x_1, \ldots, x_n) \in p\mathbb{Z}_p$. We can thus apply the above proposition to the single-variable polynomial $P(x_1, \ldots, X, \ldots, x_n)$, where X is at the place of the variable X_j, and we obtain the corollary.   □

Corollary 4.1.40. (1) *Let* $P(\underline{X}) \in \mathbb{Z}_p[X_1, \ldots, X_n]$ *be a homogeneous polynomial in* n *variables of degree* d, *and assume that the reduction* $\overline{P}$ *modulo* $p\mathbb{Z}_p$ *is nonsingular over* $\mathbb{F}_p$ *and that* $d < n$. *Then there exists a nontrivial solution of* $P(x_1, \ldots, x_n) = 0$ *in* $\mathbb{Z}_p$.

(2) *Let* $P(\underline{X}) = \sum_{1 \leqslant j \leqslant n} a_j X_j^k$ *be a diagonal form of degree* k *with coefficients in* $\mathbb{Z}$ *such that* $n > k$. *Then for all primes* $p \nmid k \prod_{1 \leqslant j \leqslant n} a_j$ *there exists a nontrivial solution of* $P = 0$ *in* $\mathbb{Q}_p$.

Proof. (1). When we reduce the polynomial P modulo $p\mathbb{Z}_p$, the degree may decrease; hence the condition $d < n$ is still satisfied for the reduced polynomial. Thus we may apply Chevalley–Warning's Theorem 2.5.2 to deduce the existence of a nontrivial solution in $\mathbb{F}_p$. Since $\overline{P}$ is nonsingular, this solution is nonsingular and we conclude by the preceding corollary.

(2). We have $\frac{\partial P}{\partial X_j} = k a_j X_j^{k-1}$, so that if $p \nmid k \prod_{1 \leqslant j \leqslant n} a_j$, this vanishes modulo p for some $(x_1, \ldots, x_n)$ if and only if $x_j = 0$; hence $\overline{P}$ is nonsingular for these p. We conclude by (1).   □

An example that we shall study in detail in the next chapter is that of ternary quadratic forms $P = aX^2 + bY^2 + cZ^2$: we indeed have $3 = n > k = 2$; hence if $p \nmid 2abc$ we know that $P = 0$ has a nontrivial solution in $\mathbb{Q}_p$.

It is sometimes possible to start the process of successive approximations even when one has a worse starting estimate α. An example which we will use later is the following.

Lemma 4.1.41. *Let* p *be the prime number below* $\mathfrak{p}$, *denote by* $e = e(\mathfrak{p}/p)$ *the absolute ramification index, and assume given* $\mathfrak{p}$-*adic units* α_0 *and* β *such that* $\beta \equiv \alpha_0^p \pmod{\mathfrak{p}^{e+r}}$ *for some integer* r *such that* $r > e/(p-1)$. *Then there exists a* $\mathfrak{p}$-*adic unit* α *such that* $\beta = \alpha^p$.

Proof. We again construct a sequence α_n of $\mathfrak{p}$-adic units such that $\beta \equiv \alpha_n^p$ $\pmod{\mathfrak{p}^{e+r+n}}$ and $\alpha_{n+1} \equiv \alpha_n \pmod{\mathfrak{p}^{n+r}}$. Then α_n converges $\mathfrak{p}$-adically to some $\mathfrak{p}$-adic unit α that clearly satisfies $\beta = \alpha^p$.

The first congruence is satisfied for $n = 0$ by assumption. Assume α_n constructed. To satisfy the second congruence we must set $\alpha_{n+1} = \alpha_n + \pi^{n+r}\gamma_n$ with $\gamma_n \in \mathbb{Z}_{\mathfrak{p}}$. Thus by the binomial theorem

$$\alpha_{n+1}^p \equiv \alpha_n^p + p\pi^{n+r}\alpha_n^{p-1}\gamma_n \pmod{\mathfrak{p}^k}$$

with $k = \min(e + 2(n+r), p(n+r)) \geqslant e + r + n + 1$ since $r(p-1) \geqslant e+1$. Hence to satisfy the first congruence for $n+1$ we must choose γ_n such that

$$\beta - (\alpha_n^p + p\pi^{n+r}\alpha_n^{p-1}\gamma_n) \equiv 0 \pmod{\mathfrak{p}^{e+r+n}},$$

which is possible since the $\mathfrak{p}$-adic valuation of $p\pi^{n+r}\alpha_n^{p-1}$ is exactly equal to $e + n + r$. □

As mentioned above, it would not have been possible to apply Proposition 4.1.37 directly. Indeed, if $f(X) = X^p - \beta$ the condition of that proposition is $v_{\mathfrak{p}}(f(\alpha_0)) > 2v_{\mathfrak{p}}(f'(\alpha_0))$, in other words $e + r > 2e$, i.e., $r > e$, while the above lemma shows that the weaker condition $r > e/(p-1)$ is sufficient. This is due to the special nature of the polynomial $f(X)$.

On the other hand, we could directly apply Corollary 4.2.15 below to obtain $\alpha = \alpha_0(1 + x)^{1/p}$ with $x = (\beta - \alpha_0^p)/\alpha_0^p$.

4.2 Analytic Functions in $\mathfrak{p}$-adic Fields

4.2.1 Elementary Properties

In this section, we let $\mathcal{K}$ be a complete field containing $\mathbb{Q}_p$ whose absolute value $|\ |$ extends that of $\mathbb{Q}_p$, so that it is necessarily ultrametric. If $x \in \mathcal{K}^*$ we define $v_p(x) \in \mathbb{R}$ by the formula $|x| = p^{-v_p(x)}$, which extends the usual definition on $\mathbb{Q}_p$. It will be frequently nicer to work with $v_p(x)$ than with $|x|$ although the two notions are of course equivalent. When $\mathcal{K}$ is the completion of a number field then $v_p(x) \in \mathbb{Q}$; more precisely, when $\mathcal{K} = K_{\mathfrak{p}}$ and $e = e(\mathfrak{p}/p)$, then $v_p(x) = v_{\mathfrak{p}}(x)/e$. It is also true that $v_p(x) \in \mathbb{Q}$ when $x \in \mathbb{C}_p^*$ (which we will define below), but not necessarily in other fields.

The aim of this section is to study functions defined by series expansions. We have already mentioned that a series converges if and only if its general term tends to 0. This already shows that analysis of convergence will be much simpler than over the complex numbers. For instance, interchange of summation usually becomes easy, as the following lemma shows (see, however, Proposition 4.2.7 for a situation in which one must be careful).

Lemma 4.2.1. *Let* $(b_{i,j})$ *be a double sequence of elements of* $\mathcal{K}$. *Assume that for all* $\varepsilon > 0$ *there exists* $N(\varepsilon)$ *such that* $|b_{i,j}| < \varepsilon$ *when* $\max(i,j) \geqslant N(\varepsilon)$. *Then the double series* $\sum_i(\sum_j b_{i,j})$ *and* $\sum_j(\sum_i b_{i,j})$ *both converge and their sums are equal.*

Proof. Immediate from the ultrametric inequality and left to the reader. □

Now let
$$f(X) = f_0 + f_1 X + \cdots + f_n X^n + \cdots$$
be a power series. As for complex numbers, we set

$$R = \frac{1}{\limsup |f_n|^{1/n}},$$

so that $0 \leqslant R \leqslant +\infty$, and call it the radius of convergence of the series.

Lemma 4.2.2. *Denote by $\mathcal{D}$ the set of elements $x \in \mathcal{K}$ for which the series $f(x)$ converges. Then:*

(1) *When $R = 0$ then $\mathcal{D} = \{0\}$.*
(2) *When $R = +\infty$ then $\mathcal{D} = \mathcal{K}$.*
(3) *When $0 < R < +\infty$ and $|f_n|R^n$ tends to 0 then $\mathcal{D} = \{x \in \mathcal{K}/ |x| \leqslant R\}$ is the "closed" ball of radius R.*
(4) *When $0 < R < +\infty$ and $|f_n|R^n$ does not tend to 0 then $\mathcal{D} = \{x \in \mathcal{K}/ |x| < R\}$ is the "open" ball of radius R.*

Proof. This immediately follows from the fact that a series converges if and only if its general term tends to zero. □

Remark. When $\mathcal{K} = \mathcal{K}_p$ and R is not of the form $\mathcal{N}p^v$ for $v \in \mathbb{Z}$, then the "closed" and "open" balls coincide (this is not in general true for an arbitrary ultrametric field). Note however that in any case the "closed" and "open" balls are both open and closed for the ultrametric topology.

Lemma 4.2.3. *Keep the notation of the above lemma, and let $y \in \mathcal{D}$. For any $m \geqslant 0$, set*

$$g_m = \sum_{n \geqslant m} \binom{n}{m} f_n y^{n-m}.$$

Then the series $g(X) = \sum_{m \geqslant 0} g_m X^m$ has again $\mathcal{D}$ as domain of convergence and $g(x) = f(x+y)$ for all $x \in \mathcal{D}$.

Proof. Since $f_n y^n$ tends to 0, we note that the series defining g_m converges. The above lemma shows that the series $g(x)$ converges and the interchange of summation justified by the same lemma shows that $g(x) = f(x+y)$ by the binomial theorem. In particular, the domain of convergence of g contains that of f, but reversing the roles of f and g, we see that they are the same. □

As we have already mentioned several times, this lemma prevents us from using the usual tools of analytic continuation, since we cannot extend the domain of definition of a function in this way.

Corollary 4.2.4. *A function defined by a power series is infinitely differentiable in its domain of convergence. In particular, if two power series with strictly positive radius of convergence coincide on some open ball with nonzero radius, their coefficients coincide.*

Proof. Indeed, the kth derivative of $g(x)$ is equal to $\sum_{m \geqslant k} m(m - 1) \cdots (m - k + 1) g_m x^{m-k}$, and since $|m(m - 1) \cdots (m - k + 1)| \leqslant 1$, the radius of convergence of this series is at least equal (in fact is equal) to that of g, proving the first statement of the corollary, and the second follows. $\square$

As in usual complex analysis, it is immediate to show that the radius of convergence of the sum, difference, or product of two power series is greater than or equal to the smallest one, and that the *value* of the result is respectively equal to the sum, difference, or product of the values of the operands. For the product we sometimes need slightly finer information than the radius of convergence. First note the following:

Lemma 4.2.5. *For all $n \in \mathbb{Z}_{\geqslant 0}$ denote by $s_p(n)$ the sum of the digits of n in base p, and let $k \in \mathbb{Z}_{\geqslant 0}$.*

(1) *We have $s_p(k) \leqslant k$.*
(2) *For all i such that $0 \leqslant i \leqslant k$ we have $s_p(i) + s_p(k - i) \geqslant s_p(k)$.*

Proof. (1) follows from the formula $0 \leqslant v_p(k!) = (k - s_p(k))/(p-1)$ that we shall prove below (Lemma 4.2.8), and (2) from

$$0 \leqslant v_p\left(\binom{k}{i}\right) = \frac{s_p(i) + s_p(k - i) - s_p(k)}{p - 1}.$$

$\square$

Proposition 4.2.6. *Let $A(X) = \sum_{k \geqslant 0} a_k X^k$ and $B(X) = \sum_{k \geqslant 0} b_k X^k$ be two power series with coefficients in $\mathcal{K}$, and set $C(X) = A(X)B(X) = \sum_{k \geqslant 0} c_k X^k$. Assume that there exist constants $\alpha > 0$, $\alpha' \geqslant 0$, $\beta > 0$, and $\beta' \geqslant 0$ such that $v_p(a_k) \geqslant -\alpha k + \alpha' s_p(k)$ and $v_p(b_k) \geqslant -\beta k + \beta' s_p(k)$ for all $k \geqslant 0$. Without loss of generality assume that $\alpha \leqslant \beta$. Then*

$$v_p(c_k) \geqslant -\beta k + \min(\beta - \alpha + \alpha', \beta') s_p(k).$$

Proof. We have $c_k = \sum_{0 \leqslant i \leqslant k} a_i b_{k-i}$, hence

$$v_p(c_k) \geqslant \min_{0 \leqslant i \leqslant k} (v_p(a_i) + v_p(b_{k-i})) \geqslant \min_{0 \leqslant i \leqslant k} (-\alpha i - \beta(k-i) + \alpha' s_p(i) + \beta' s_p(k-i)).$$

Set $\beta_1 = \min(\beta - \alpha + \alpha', \beta')$. By the lemma we have $s_p(i) \geqslant s_p(k) - s_p(k - i)$ and $k - i \geqslant s_p(k - i)$. Since $\beta \geqslant \alpha$ and $0 \leqslant \beta_1 \leqslant \beta'$ we obtain

$$-\alpha i - \beta(k - i) + \alpha' s_p(i) + \beta' s_p(k - i)$$
$$\geqslant -\beta k + \beta_1 s_p(k) + (\beta - \alpha)(k - i) + (\alpha' - \beta_1) s_p(k - i)$$
$$\geqslant -\beta k + \beta_1 s_p(k) + (\beta - \alpha + \alpha' - \beta_1) s_p(k - i)$$
$$\geqslant -\beta k + \beta_1 s_p(k),$$

giving the result. $\square$

It is also important to give estimates for the *composite* of two power series. The situation is as follows. Let $f(X) = \sum_{n \geqslant 0} a_n X^n$ and $g(X) =$

$\sum_{m \geqslant 1} b_m X^m$ be two formal power series (hence with $g(0) = b_0 = 0$). We let $F(X)$ be the composite power series of f and g, defined in the following way. If we set $a_n g(X)^n = \sum_{k \geqslant n} c_{n,k} X^k$, then by definition we have

$$F(X) = f \circ g(X) = \sum_{k \geqslant 0} \left(\sum_{0 \leqslant n \leqslant k} c_{n,k} \right) X^k .$$

The important point to note is that even if all the power series converge, if $x \in K$ we do *not* necessarily have $F(x) = f(g(x))$ (see for instance Proposition 4.2.11). The following proposition gives a sufficient condition for this to be true.

Proposition 4.2.7. *Let $f(X) = \sum_{n \geqslant 0} a_n X^n$ and $g(X) = \sum_{m \geqslant 1} b_m X^m$ be two formal power series, let $F = f \circ g$ be defined as above, and let R be the radius of convergence of f. If $x \in K$ is such that the power series $g(x)$ converges, and if $|b_m x^m| < R$ for all $m \geqslant 1$, then the power series $F(x)$ converges and we have $F(x) = f(g(x))$.*

Proof. We use the above notation. Consider the double series $\sum_{i,j} c_{i,j} x^j$. By definition of the $c_{i,j}$ we have

$$c_{n,m} x^m = \sum_{\substack{k_1, \ldots, k_n \geqslant 1 \\ k_1 + \cdots + k_n = m}} a_n b_{k_1} x^{k_1} \cdots b_{k_n} x^{k_n} .$$

Set $r = \max_{m \geqslant 1} |b_m x^m|$, which exists since $g(x)$ converges. Then

$$|c_{n,m} x^m| \leqslant \max_{k_i} |a_n b_{k_1} x^{k_1} \cdots b_{k_n} x^{k_n}| \leqslant |a_n| r^n .$$

By assumption $r < R$, hence $f(r)$ converges, so that $|a_n| r^n$ tends to 0 as $n \to \infty$; hence $|c_{n,m} x^m|$ tends to 0 as $n \to \infty$ uniformly in m. Furthermore, for n fixed the series $a_n g(x)^n$ converges as a finite product of convergent power series. It follows that $|c_{n,m} x^m|$ tends to 0 as $m \to \infty$, and applying Lemma 4.2.1 proves the proposition.   □

Note that the condition of the proposition is that $|b_m x^m| < R$ for *all* m, and not only for all sufficiently large m.

4.2.2 Examples of Analytic Functions

We begin with an important lemma on the valuation of $n!$ and of binomial coefficients. Recall that for any commutative field K of characteristic 0 we define for $x \in K$ and $n \in \mathbb{Z}_{\geqslant 0}$

$$\binom{x}{n} = \frac{x(x-1) \cdots (x-n+1)}{n!} .$$

When $x \in \mathbb{Z}_{\geqslant 0}$ we recover the usual binomial coefficients, whence the notation.

Lemma 4.2.8. (1) *We have*

$$v_p(n!) = \sum_{k=1}^{\lfloor \log(n)/\log(p) \rfloor} \left\lfloor \frac{n}{p^k} \right\rfloor = \frac{n - s_p(n)}{p - 1} .$$

In particular, for $n \geqslant 1$ we have $v_p(n!) \leqslant (n-1)/(p-1)$, with equality if and only if n is a power of p, and as $n \to \infty$ we have $v_p(n!) = n/(p-1) + O(\log(n))$, and more precisely $p^{-n/(p-1)} \leqslant |n!|_p \leqslant np^{-n/(p-1)}$.

(2) *Let $a \in K$. Then*

(a) *When $v_p(a) < 0$ we have*

$$v_p\left(\binom{a}{n}\right) = -(n|v_p(a)| + v_p(n!)) \sim -n\left(|v_p(a)| + \frac{1}{p-1}\right) .$$

(b) *When $v_p(a) \geqslant 0$, writing $v_p(a) = q - \theta$ with $0 \leqslant \theta < 1$, we have*

$$v_p\left(\binom{a}{n}\right) \geqslant -\frac{n}{p^q}\left(\frac{1}{p-1} + \theta\right) ,$$

and in particular $v_p(\binom{a}{n}) \geqslant -n/(p-1)$.

(c) *When $v_p(a) \geqslant 0$ and $K = \mathbb{Q}_p$ we have*

$$v_p\left(\binom{a}{n}\right) \geqslant 0 ,$$

in other words $\binom{a}{n} \in \mathbb{Z}_p$. If, in addition, $a \neq 0$ and $n \geqslant 1$ then

$$v_p\left(\binom{a}{n}\right) \geqslant \max(v_p(a) - v_p(n), 0) .$$

Proof. (1). We have

$$v_p(n!) = \sum_{1 \leqslant i \leqslant n} v_p(i) = \sum_{k \geqslant 0} k \sum_{1 \leqslant i \leqslant n,\ v_p(i)=k} 1$$

$$= \sum_{k \geqslant 0} k(\lfloor n/p^k \rfloor - \lfloor n/p^{k+1} \rfloor) = \sum_{k \geqslant 1} \lfloor n/p^k \rfloor$$

by Abel summation, proving the first formula, and the others are immediate consequences and left to the reader (Exercise 7).

(2). If $v_p(a) < 0$ then $v_p(a - i) = v_p(a)$ for all $i \in \mathbb{Z}$, so the given formulas are immediate. If $v_p(a) \geqslant 0$ then $v_p(a-i) \geqslant \min(v_p(a), v_p(i))$, so that (setting by convention $v_p(0) = +\infty$)

$$v_p\left(n!\binom{a}{n}\right) = \sum_{0 \leqslant i < n} v_p(a - i) \geqslant \sum_{0 \leqslant i < n} \min(v_p(a), v_p(i))$$

$$\geqslant \sum_{\substack{0 \leqslant i < n \\ v_p(i) < v_p(a)}} v_p(i) + v_p(a) \sum_{\substack{0 \leqslant i < n \\ v_p(i) \geqslant v_p(a)}} 1 .$$

Thus, since $v_p(a) = q - \theta$ with $0 \leqslant \theta < 1$ we have $q = \lceil v_p(a) \rceil$ and $q \in \mathbb{Z}_{\geqslant 0}$, so that $v_p(i) < v_p(a)$ is equivalent to $v_p(i) < q$; hence

$$
v_p\left(n!\binom{a}{n}\right) \geqslant \sum_{0 \leqslant k < q} k \sum_{\substack{0 \leqslant i < n \\ v_p(i) = k}} 1 + v_p(a) \sum_{\substack{0 \leqslant i < n \\ v_p(i) \geqslant q}} 1
$$

$$
= \sum_{0 \leqslant k < q} k(\lfloor n/p^k \rfloor - \lfloor n/p^{k+1} \rfloor) + v_p(a)\lfloor n/p^q \rfloor
$$

$$
= \sum_{1 \leqslant k \leqslant q} \lfloor n/p^k \rfloor - r\lfloor n/p^q \rfloor
$$

again by Abel summation. Since by (1) we have $v_p(n!) = \sum_{k \geqslant 1} \lfloor n/p^k \rfloor$, it follows that

$$
v_p\left(\binom{a}{n}\right) \geqslant -\theta\lfloor n/p^q \rfloor - \sum_{k \geqslant q+1} \lfloor n/p^k \rfloor \,.
$$

In addition

$$
\sum_{k \geqslant q+1} \lfloor n/p^k \rfloor \leqslant \sum_{k \geqslant q+1} n/p^k \leqslant n/(p^q(p-1)) \,,
$$

so that

$$
v_p\left(\binom{a}{n}\right) \geqslant -\frac{n}{p^q}\left(\frac{1}{p-1} + \theta\right) \,,
$$

as claimed.

In the special case $K = \mathbb{Q}_p$, we reason as follows. Since $\binom{a}{n}$ is a polynomial in a it is a continuous function on $\mathcal{K}$. Furthermore, $\binom{a}{n} \in \mathbb{Z}$ when $a \in \mathbb{Z}$, and since $\mathbb{Z}$ is dense in $\mathbb{Z}_p$ it follows that $\binom{a}{n} \in \mathbb{Z}_p$ when $a \in \mathbb{Z}_p$. Since for $n \geqslant 1$ we have

$$
\binom{a}{n} = \frac{a}{n}\binom{a-1}{n-1} \,,
$$

it follows that

$$
v_p\left(\binom{a}{n}\right) = v_p(a) - v_p(n) + v_p\left(\binom{a-1}{n-1}\right) \geqslant v_p(a) - v_p(n) \,,
$$

finishing the proof. $\square$

See also Exercises 9 and 10.

Remark. It is essential to note that when $v_p(a) \geqslant 0$ but $a \notin \mathbb{Z}_p$, we not *not* have in general $v_p(\binom{a}{n}) \geqslant 0$; see for example Exercise 6.

We will also need the following lemma, closely related to (2) above:

Lemma 4.2.9. *As in the preceding lemma, when $v_p(a) \geqslant 0$, write $v_p(a) = q - \theta$ with $0 \leqslant \theta < 1$. Then*

$$\max_{m \geqslant 1} \left(\frac{1/(p-1) - v_p(a) + v_p(m)}{m} \right) = \begin{cases} 1/(p-1) - v_p(a) & \text{when } v_p(a) < 0, \\ (1/(p-1) + \theta)/p^q & \text{when } v_p(a) \geqslant 0. \end{cases}$$

Proof. For simplicity, set $A = 1/(p-1) - v_p(a)$ and $m = p^u n$ with $p \nmid n$. We must compute $\max_{u \geqslant 0, n \geqslant 1} (A + u)/(p^u n)$. Consider for the moment n fixed, and set $f(u) = (A + u)/p^u$. We have

$$f(u) - f(u+1) = \frac{p-1}{p^{u+1}}(u - v_p(a)).$$

When $v_p(a) < 0$ this is strictly positive; hence the maximum of $f(u)$ is attained for $u = 0$, and is equal to A, and since $A > 0$ in that case, the maximum over n is obtained for $n = 1$, proving the lemma for $v_p(a) < 0$. Thus assume now that $v_p(a) \geqslant 0$. By the above equality we have $f(u+1) \leqslant f(u)$ if $u \geqslant \lceil v_p(a) \rceil$ and $f(u+1) > f(u)$ if $u < \lceil v_p(a) \rceil$, so the maximum of f is attained for $u = \lceil v_p(a) \rceil = q$ (and also for $u = q+1$ if $v_p(a) \in \mathbb{Z}$, but we do not need this), and we have

$$\max_{u \geqslant 0} f(u) = \frac{1/(p-1) - v_p(a) + q}{p^q} = \frac{1/(p-1) + \theta}{p^q}.$$

Since this quantity is positive, the maximum over $n \geqslant 1$ is again attained for $n = 1$, proving the lemma.   □

As in the case of $\mathbb{C}$, we can define the exponential and logarithm functions. The main difference is that the radius of convergence of the exponential function is finite.

Proposition 4.2.10. *Let $\mathcal{K}$ be as above, and set $r_p = p^{-1/(p-1)}$. Consider as usual the formal power series*

$$\exp(X) = \sum_{n \geqslant 0} \frac{X^n}{n!} \quad \text{and} \quad \log(1+X) = \sum_{n \geqslant 1} (-1)^{n-1} \frac{X^n}{n}.$$

(1) *The series obtained by replacing the formal variable X by $x \in \mathcal{K}$ in $\exp(X)$ converges if and only if $|x| < r_p = p^{-1/(p-1)}$ (or equivalently, $v_p(x) > 1/(p-1)$), and its sum is denoted by $\exp_p(x)$.*
(2) *The series obtained by replacing the formal variable X by $x \in \mathcal{K}$ in $\log(1+X)$ converges if and only if $|x| < 1$ (or equivalently, $v_p(x) > 0$), and its sum is denoted by $\log_p(1+x)$.*
(3) *If $|x| < r_p$ we have $|\log_p(1+x)| = |x|$ and $|\exp_p(x) - 1| = |x|$, and if $|x| = r_p$ we have $|\log_p(1+x)| \leqslant |x|$.*
(4) *If $\max(|x|, |y|) < r_p$ we have $\exp_p(x+y) = \exp_p(x)\exp_p(y)$, and if $\max(|x|, |y|) < 1$ we have $\log_p((1+x)(1+y)) = \log_p(1+x) + \log_p(1+y)$.*
(5) *If $|x| < r_p$ we have $\log_p(\exp_p(x)) = x$ and $\exp_p(\log_p(1+x)) = 1 + x$.*

Note that it is useful to use a notation that distinguishes the formal power series for $\exp(X)$ and for $\log(1+X)$ (which have nothing to do with $\mathfrak{p}$-adic fields, and can be defined in any field of characteristic 0) from their sum when X is replaced by $x \in \mathcal{K}$, which now depends on the base field, whence the above notation, which we will consistently use.

Proof. (1) and (2). By the above lemma, we have $\limsup(v_p(n!)/n) = 1/(p-1)$, so the radius of convergence of $\exp(X)$ is $1/\limsup(|1/n!|^{1/n}) = p^{-1/(p-1)} = r_p$. Since $v_p(n!) - n/(p-1)$ does not tend to zero as n tends to infinity, the series does not converge on the circle of convergence, proving the first statement. For $\log(1+X)$, we clearly have $1/\limsup(|1/n|)^{1/n} = 1$, so that the radius of convergence is equal to 1, and once again since $|1/n|$ does not tend to 0, the series does not converge on the circle of convergence. Furthermore, if $|x| < r_p$ and $n \geq 1$ we have

$$|x^{n-1}/n!| < r_p^{n-1} p^{v_p(n!)} \leq p^{-(n-1)/(p-1)} p^{(n-1)/(p-1)} \leq 1 .$$

It follows that $|x^n/n!| < |x|$ and a fortiori that $|x^n/n| = |(n-1)! x^n/n!| < |x|$ for all $n \geq 2$, hence that $|\exp(x) - 1| = |x|$ and $|\log(1+x)| = |x|$ by ultrametricity, proving (3) if $|x| < r_p$. If $|x| = r_p$ we only have $|x^n/n| \leq |x^n/n!| \leq |x|$, hence $|\log(1+x)| \leq |x|$. The equalities of (4) are formal consequences of the power series definition since thanks to Lemma 4.2.1 we can rearrange the power series product as we like inside the disks of convergence.

For (5) we must be more careful. We are going to show that the conditions of Proposition 4.2.7 are satisfied. For the first formula we choose $f(X) = \log(1+X)$ and $g(X) = \exp(X) - 1$, so that the radius of convergence R of f is equal to 1. If $|x| < r_p$ then $g(x)$ converges by (1), and for $m \geq 1$ we have

$$|b_m x^m| = |x^m/m!| < |x| r_p^{m-1}/p^{-(m-1)/(p-1)} \leq r_p < 1 ,$$

so Proposition 4.2.7 implies the result. For the second formula we choose $f(X) = \exp(X)$ and $g(X) = \log(1+X)$, so that $R = r_p$. If $|x| < r_p$ then $g(x)$ converges since $r_p < 1$, and

$$|b_m x^m| = |x^m/m| = |(m-1)!||x^m/m!| < r_p$$

by the above inequality, and we again conclude thanks to Proposition 4.2.7. $\qquad \square$

Important Remark. Since $\exp_p(x)$ converges only for $|x| < r_p$, this condition is evidently necessary for the first statement of (5) to make sense. On the other hand, the second statement of (5) may be *false* if we assume only that $|x| < 1$ and $|\log_p(1+x)| < r_p$, which are necessary conditions for the statement to make sense. For instance, we have the following easy result:

Proposition 4.2.11. *Assume that $x \in 2\mathbb{Z}_2$. Then $|\log_2(1+x)| < r_2$, but $\exp_2(\log_2(1+x)) = \varepsilon(1+x)$ with $\varepsilon = 1$ if $x \in 4\mathbb{Z}_2$ and $\varepsilon = -1$ if $x \in 2 + 4\mathbb{Z}_2$.*

Proof. Left to the reader (Exercise 11). □

Lemma 4.2.12. *Let* $f(x) = \sum_{n \geqslant 0} a_n x^n / n!$, *and let* $k \geqslant 1$. *Assume that there exists* $z \in \mathbb{R}$ *such that* $v_p(a_n) \geqslant z$ *for all* $n \geqslant k$, *and that* $v_p(x) > 1/(p-1)$. *Then the series* $f(x)$ *converges and we have*

$$v_p\left(f(x) - \sum_{0 \leqslant n < k} a_n x^n / n! \right) \geqslant k v_p(x) + z - (k-1)/(p-1) > v_p(x) + z \;.$$

If, in addition, $p \geqslant k+1$ *and* $v_p(x) \geqslant 1/(p-k)$ *we have*

$$v_p\left(f(x) - \sum_{0 \leqslant n < k} a_n x^n / n! \right) \geqslant k v_p(x) + z \;.$$

Proof. By Lemma 4.2.8 for $n \geqslant k$ we have by assumption

$$v_p(a_n x^n / n!) \geqslant k v_p(x) + z + (n-k) v_p(x) - (n - s_p(n))/(p-1) \;.$$

Furthermore, if $s_p(n) \geqslant k$ we have

$$(n-k) v_p(x) - (n - s_p(n))/(p-1) \geqslant (n-k)(v_p(x) - 1/(p-1)) > 0 \;,$$

proving that $v_p(a_n x^n / n!) > k v_p(x) + z$, which is what we need for both statements. If $1 \leqslant s_p(n) < k$ then since $(n-k) v_p(x) - (n-1)/(p-1)$ is an increasing function of n and $n \geqslant k$, the minimum is attained for $n = k$, so that

$$(n-k) v_p(x) - (n - s_p(n))/(p-1) \geqslant -(k-1)/(p-1) > -(k-1) v_p(x) \;,$$

proving the first result. If, in addition, $v_p(x) \geqslant 1/(p-k)$, we have instead

$$(n-k) v_p(x) - (n - s_p(n))/(p-1) \geqslant (n-k)/(p-k) - (n-1)/(p-1)$$
$$= (k-1)(n-p)/((p-k)(p-1)) \geqslant 0 \;,$$

since $n \geqslant p$; otherwise, $s_p(n) = n \geqslant k$, proving the second result. □

Corollary 4.2.13. *If* $v_p(x) > 1/(p-1)$ *then*

$$v_p(\log_p(1+x) - x) > 2 v_p(x) - 1/(p-1) \quad \text{and}$$
$$v_p(\exp_p(x) - 1 - x) > 2 v_p(x) - 1/(p-1) \;.$$

If $p \geqslant 3$ *and* $v_p(x) \geqslant 1/(p-2)$ *then*

$$v_p(\log_p(1+x) - x) \geqslant 2 v_p(x) \quad \text{and} \quad v_p(\exp_p(x) - 1 - x) \geqslant 2 v_p(x) \;;$$

in other words, $\log_p(1+x) \equiv x \pmod{x^2 \mathbb{Z}_p}$ *and* $\exp_p(x) \equiv 1 + x \pmod{x^2 \mathbb{Z}_p}$.

Proof. Clear.   □

When $v_p(x) \leqslant 1/(p-1)$ we can also give a rather precise estimate of the valuation of $\log_p(1+x)$ as follows.

Proposition 4.2.14. *Let x be such that $v = v_p(x) > 0$, assume that $v \leqslant 1/(p-1)$, and set*

$$k_0 = \left\lceil \frac{-\log((p-1)v)}{\log p} \right\rceil .$$

(1) *When $v \leqslant 1/(p-1)$ then*

$$v_p(\log_p(1+x)) \geqslant p^{k_0} v - k_0$$

with equality when $1/((p-1)v)$ is not an integral power of p (in particular if it is not an integer).

(2) *In the special case $K = \mathbb{Q}_p$, if $v \leqslant 1/(p-1)$ (which can happen only if $p = 2$ and $v = 1$) we have $v_p(\log_p(1+x)) = v_p(x+2) \geqslant 2$, and more precisely $v_p(\log_p(1+x) - (x+2)) \geqslant 3$.*

Proof. Let $u_n = (-1)^{n-1}x^n/n$ be the nth summand of the series for $\log_p(1+x)$. If we write $n = p^k m$ with $p \nmid m$, then $v_p(u_n) = p^k m v - k$. This is a function of the two integer variables m and k. Since $v > 0$, for fixed k it attains its minimal value for $m = 1$. If $w_k = p^k v - k$ we have $w_{k+1} - w_k = p^k(p-1)v - 1$, and since $v \leqslant 1/(p-1)$ we have $w_{k+1} - w_k \geqslant 0$ if and only if $k \geqslant k_0$ as given in the proposition. It follows that $n = p^{k_0}$ is such that $v_p(u_n)$ has minimum valuation, and it is unique if and only if $1/((p-1)v)$ is not a power of p, proving (1). In the case $K = \mathbb{Q}_2$ and $v = 1$ we can write $x + 2 = 4y$ with $y \in \mathbb{Z}_2$; hence since $2\log_2(-1) = \log_2(1) = 0$ we have $\log_2(1+x) = \log_2(1-4y)$. Now it is immediate that $v_2((4y)^n/n) \geqslant 2n - v_2(n) \geqslant 3$ for $n \geqslant 3$; hence $\log_2(1+x) \equiv -(4y + 8y^2) \equiv 4y \equiv x + 2 \pmod{8\mathbb{Z}_2}$, proving (2) since $v_2(x+2) \geqslant 2$.   □

In the next results we consider $(1+x)^y$ from two points of view, one in which the main variable is x, the other in which it is y. We begin with the most natural one, where we assume that it is x.

Corollary 4.2.15. *Let $a \in K$ be fixed and consider the power series*

$$(1+X)^a = \sum_{n \geqslant 0} \binom{a}{n} X^n .$$

Define

$$V(a) = \begin{cases} 1/(p-1) - v_p(a) & \text{when } v_p(a) < 0 , \\ (1/(p-1) + \theta)/p^q & \text{when } v_p(a) \geqslant 0 , \end{cases}$$

where in this last case we write $v_p(a) = q - \theta$ with $0 \leqslant \theta < 1$.

(1) *The power series for $(1+x)^a$ converges when $v_p(x) > V(a)$, and in particular when $v_p(x) > 1/(p-1) - \min(v_p(a), 0)$, and we have $v_p((1+x)^a) = 0$.*

(2) *When $v_p(a) < 0$ or when $v_p(a) \geqslant 0$ and $v_p(a) \notin \mathbb{Z}$, it converges if and only if $v_p(x) > V(a)$.*

(3) *In the special case in which $a \in \mathbb{Z}_p$ (in particular for $K = \mathbb{Q}_p$), it converges for $v_p(x) > 0$.*

(4) *If $v_p(x) > V(a)$ we also have*

$$(1+x)^a = \exp_p(a \log_p(1+x)) .$$

(5) *In the special case $K = \mathbb{Q}_p$, $a \in \mathbb{Z}_p$, and $v_p(x) > 0$, we have*

$$(1+x)^a = \varepsilon \exp_p(a \log_p(1+x)) ,$$

where $\varepsilon = 1$ except when $p = 2$, $v_p(a) = 0$, and $v_p(x) = 1$, in which case $\varepsilon = -1$.

(6) *If $v_p(x) > 0$ we have*

$$\log_p(1+x) = \lim_{a \to 0} \frac{(1+x)^a - 1}{a} .$$

Proof. (3) and the first and last statement of (1) immediately follow from Lemma 4.2.8. In addition, for $v_p(a) \geqslant 0$ we have $V(a) \leqslant 1/(p-1)$, while $V(a) = 1/(p-1) - v_p(a)$ for $v_p(a) < 0$. Thus $V(a) \leqslant 1/(p-1) - \min(v_p(a), 0)$ in all cases, so that the series converges when $v_p(x) > 1/(p-1) - \min(v_p(a), 0)$, proving the second statement of (1). For (2), note that when $v_p(a) < 0$ we have the *equality*

$$v_p\left(\binom{a}{n}\right) = -(n|v_p(a)| + (n-1)/(p-1))$$

for an infinity of n, so the series does not converge on its circle of convergence. The case $v_p(a) \geqslant 0$ and $v_p(a) \notin \mathbb{Z}$ is left to the reader (Exercise 9).

(4). As in the proof of Proposition 4.2.10 (5) we must be careful, and not only check that everything is well defined. Once again we are going to apply Proposition 4.2.7, this time to $f(X) = \exp(X)$ and $g(X) = a\log(1+X)$, so that $R = r_p = p^{-1/(p-1)}$. Since $v_p(x) > 0$ the series $g(x)$ converges. On the other hand, $g(X) = a\sum_{m \geqslant 1}(-1)^{m-1}x^m/m$, so to apply Proposition 4.2.7 we must show that $|ax^m/m| < r_p$ for all $m \geqslant 1$, in other words that $mv_p(x) + v_p(a) - v_p(m) > 1/(p-1)$. But since $v_p(x) > V(a)$, this is exactly what Lemma 4.2.9 tells us, proving (4).

(5) is proved exactly in the same way as Proposition 4.2.11, and is left to the reader (Exercise 11). Finally, for (6) we note that the real number $V(a)$ tends to 0 as $v_p(a)$ tends to $+\infty$; hence for a sufficiently close p-adically to 0 we have $v_p(x) > V(a)$, so that (4) is applicable, and the power series expansion of the exponential immediately gives the result. $\qquad\square$

Examples and Warnings. (1) As an example of (4), already seen in Proposition 4.2.11, note that for instance $\exp_2(\log_2(1 + 2)) = -3$.

(2) Consider the special case $\mathcal{K} = \mathbb{Q}_7$, $a = 1/2$, $x = 7/9$. Since $|x|_7 < 1$ and $v_7(a) = 0$ the series converges, and by Proposition 4.2.10 (or directly) the sum of the series is one of the square roots of $1 + x = 16/9$, in other words $\pm 4/3$. Since the series begins with $1 + O(x) = 1 + O(7)$ the sum of the series is congruent to 1 modulo $7\mathbb{Z}_7$, hence is equal to $-4/3$. On the other hand, in $\mathbb{R}$ the series also converges, but since $|x| < 1$, it converges to a positive value, hence to $4/3$. Thus, beware that even though a given series may converge to elements of $\mathbb{Q}$ in several p-adic or local fields, the sum is not necessarily the same.

Corollary 4.2.16. *Assume that $v_p(x) = v > 1/(p - 1)$. If $v_p(a) > 1/(p - 1) - v$ the series $(1 + x)^a$ converges, and $v_p((1 + x)^a) = 0$.*

Proof. By the above corollary, if $v_p(a) \geqslant 0$ the series converges when $v_p(x) > 1/(p - 1)$, and if $v_p(a) < 0$ the series converges when $v_p(x) > 1/(p - 1) - v_p(a)$, in other words when $v_p(a) > 1/(p - 1) - v_p(x)$, both of which are true by assumption.   □

Corollary 4.2.17. *Let $a \in \mathbb{Z}_p$.*

(1) *If $p \geqslant 3$ then for all $x \in \mathcal{K}$ such that $v_p(x) \geqslant 1/(p - 2)$ we have*

$$v_p((1 + x)^a - 1 - ax) \geqslant 2v_p(x) + v_p(a) ,$$

in other words $(1 + x)^a \equiv 1 + ax \pmod{ax^2 \mathbb{Z}_p}$.

(2) *If $p = 2$ then for all $x \in \mathcal{K}$ such that $v_p(x) > 1$ we have*

$$v_p((1 + x)^a - 1 - ax) \geqslant 2v_p(x) + v_p(a) - 1 .$$

Proof. By the above corollary (or by Lemma 4.2.12) we have the convergent series $(1 + x)^a = \sum_{n \geqslant 0} \binom{a}{n} x^n$, and since $v_p(n!\binom{a}{n}) = v_p(a(a - 1) \cdots (a - n + 1)) \geqslant v_p(a)$ for $n \geqslant 1$, (1) follows from Lemma 4.2.12, and (2) is proved similarly.   □

Corollary 4.2.18. *Let $a \in \mathcal{K}$ be fixed.*

(1) *When $v_p(a) > 1/(p - 1)$ the power series in x*

$$\phi_a(x) = (1 + a)^x = \exp_p(x \log_p(1 + a))$$

converges if and only if $v_p(x) > 1/(p - 1) - v_p(a)$, and in particular has a radius of convergence strictly greater than 1.

(2) *When $1 \leqslant v_p(a) \leqslant 1/(p - 1)$ the power series $\phi_a(x)$ above converges for*

$$v_p(x) > \frac{1}{p - 1} - p^{k_0} v_p(a) + k_0$$

(and possibly for smaller values of $v_p(x)$), where

$$k_0 = \left\lceil \frac{-\log((p-1)v_p(a))}{\log p} \right\rceil .$$

(3) *In the special case in which $a \in \mathbb{Z}_p$, $v_p(a) = 1$, and $p = 2$, the power series $\phi_a(x)$ converges for $v_p(x) > -1$, and in particular has radius of convergence greater than or equal to 2.*

Note that, contrary to Corollary 4.2.15, in the above corollary the equality $(1+a)^x = \exp_p(x \log_p(1+a))$ is the *definition* of $(1+a)^x$.

Proof. This immediately follows from Proposition 4.2.14 since $\exp_p(x)$ converges when $v_p(x) > 1/(p-1)$, and since $v_2(\log_2(1+a)) \geqslant 2$ when $a \in 2\mathbb{Z}_2$. $\square$

4.2.3 Application of the Artin–Hasse Exponential

The fact that the radius of convergence of the p-adic exponential function is strictly less than 1 is annoying. We introduce a modification of the exponential function, which will no longer be a homomorphism, but which has radius of convergence 1.

Definition 4.2.19. *The* Artin–Hasse exponential *is the power series $E_p(X) = \sum_{k \geqslant 0} e_k X^k$ defined by*

$$E_p(X) = \exp\left(X + \frac{X^p}{p} + \cdots + \frac{X^{p^n}}{p^n} + \cdots\right) .$$

To compute the radius of convergence, we need the following lemma, due to B. Dwork. We state it only over $\mathbb{Q}_p$, but it is easily generalized.

Lemma 4.2.20. *Let $S(X) = \sum_{k \geqslant 0} s_k X^k \in 1 + X\mathbb{Q}_p[[X]]$ be a power series with coefficients in $\mathbb{Q}_p$ such that $s_0 = 1$. Then $s_k \in \mathbb{Z}_p$ for all k (in other words $S(X) \in 1 + X\mathbb{Z}_p[[X]]$) if and only if $S(X)^p/S(X^p) \in 1 + pX\mathbb{Z}_p[[X]]$.*

Proof. If $S(X) = \sum_{k \geqslant 0} s_k X^k \in 1 + X\mathbb{Z}_p[[X]]$ then clearly

$$S(X)^p \equiv \sum_{k \geqslant 0} s_k^p X^{pk} \equiv \sum_{k \geqslant 0} s_k X^{pk} \equiv S(X^p) \pmod{p\mathbb{Z}_p[[X]]} ,$$

hence $S(X)^p/S(X^p) \in 1 + pZ\mathbb{Z}_p[[X]]$ (note that we use the fact that $s^p \equiv s \pmod{p\mathbb{Z}_p}$ when $s \in \mathbb{Z}_p$). Conversely, assume that this is the case. Using an obvious recurrence (valid for any commutative ring R, not only $\mathbb{Q}_p$), we can write $S(X) = \prod_{k \geqslant 1}(1 + a_k X^k)$ for some $a_k \in \mathbb{Q}_p$. Let us show by induction that in fact $a_k \in \mathbb{Z}_p$, which clearly implies that $S(X) \in 1 + X\mathbb{Z}_p[[X]]$. Assume that $a_1, \ldots, a_{n-1}$ are in $\mathbb{Z}_p$, initially with $n = 1$, and let $S_n(X) = \prod_{1 \leqslant k \leqslant n-1}(1 + a_k X^k)$, so that by what we have just shown, $S_n(X)^p / S_n(X^p) \in 1 + p\mathbb{Z}_p[[X]]$, whence $R_n(X)^p / R_n(X^p) \in 1 + p\mathbb{Z}_p[[X]]$ with

$$R_n(X) = \frac{S(X)}{S_n(X)} = \prod_{k \geqslant n}(1 + a_k X^k) = 1 + a_n X^n + O(X^{n+1}) .$$

We thus have

$$\frac{R_n(X)^p}{R_n(X^p)} = \frac{1 + p a_n X^n + O(X^{n+1})}{1 + a_n X^{pn} + O(X^{p(n+1)})} = 1 + p a_n X^n + O(X^{n+1}) ,$$

hence by assumption $p a_n \in p\mathbb{Z}_p$, in other words $a_n \in \mathbb{Z}_p$, proving our induction and the lemma.

Corollary 4.2.21. *The Artin–Hasse power series $E_p(X)$ defined above has p-integral coefficients (in other words belongs to $1 + X\mathbb{Z}_p[[X]]$), and in particular has radius of convergence greater than or equal to 1.*

Proof. Indeed, we have $E_p(X)^p = \exp(pX + \sum_{n \geqslant 1} X^{p^n}/p^{n-1})$ and $E_p(X^p) = \exp(\sum_{n \geqslant 0} X^{p^{n+1}}/p^n)$, hence

$$\frac{E_p(X)^p}{E_p(X^p)} = \exp(pX) = 1 + p\sum_{k \geqslant 1} \frac{p^{k-1}}{k!} X^k .$$

Since $v_p(k!) \leqslant (k-1)/(p-1)$ for $k \geqslant 1$, it follows that $v_p(p^{k-1}/k!) \geqslant k - 1 - (k-1)/(p-1) \geqslant 0$, so that $p^{k-1}/k! \in \mathbb{Z}_p$, and by the lemma it follows that $E_p(X) \in 1 + X\mathbb{Z}_p[[X]]$, proving that the radius of convergence is greater than or equal to 1. $\square$

It can be shown that the radius of convergence of $E_p(X)$ is in fact equal to 1, but we will not need this.

Theorem 4.2.22. *Set $\exp(X + X^p/p) = \sum_{k \geqslant 0} u_k X^k$. Then*

$$v_p(u_k) \geqslant -\frac{2p-1}{(p-1)p^2}(k - s_p(k)) .$$

Proof. We trivially have

$$\exp(X + X^p/p) = E_p(X) \prod_{n \geqslant 2} \exp(-X^{p^n}/p^n) .$$

Write $\exp(-X^{p^n}/p^n) = \sum_{k \geqslant 0} c_{n,k} X^k$. We have $c_{n,k} = 0$ if $p^n \nmid k$, and otherwise $c_{n,k} = (-1)^{k/p^n}/((k/p^n)! p^{nk/p^n})$. It follows in particular that

$$v_p(c_{n,k}) \geqslant -\left(\frac{nk}{p^n} + v_p((k/p^n)!)\right)$$

$$\geqslant -\left(\frac{nk}{p^n} + \frac{k}{p^n(p-1)}\right) + \frac{s_p(k/p^n)}{p-1} \geqslant -\frac{k}{p^n}\left(n + \frac{1}{p-1}\right) + \frac{s_p(k)}{p-1} .$$

Now if we set $a_n = n/p^n$ we have $a_{n+1}/a_n = ((n+1)/n)/p < 1$ for $n \geqslant 2$, so when $n \geqslant 2$, for fixed k the expression $(k/p^n)(n + 1/(p-1))$ is maximal for $n = 2$; hence $v_p(c_{n,k}) \geqslant -\beta_p k + s_p(k)/(p-1)$ with $\beta_p = (2 + 1/(p-1))/p^2 = (2p-1)/((p-1)p^2)$, which gives a lower bound independent of n. It follows by induction from Proposition 4.2.6 that if we write

$$B(X) = \prod_{n \geqslant 2} \exp(-X^{p^n}/p^n) = \sum_{k \geqslant 0} b_k X^k$$

then $v_p(b_k) \geqslant -\beta_p k + s_p(k)/(p-1)$ (note that b_k is obtained by computing only a *finite* number of products). Now if we set $E_p(X) = \sum e_k X^k$ the above corollary tells us that $v_p(e_k) \geqslant 0$; hence by Proposition 4.2.6 applied with $\alpha = \alpha' = 0$ and $\beta = \beta_p$, $\beta' = 1/(p-1)$ we obtain

$$v_p(u_k) \geqslant -\beta_p k + \min(\beta_p, 1/(p-1)) s_p(k) \geqslant -\beta_p(k - s_p(k))$$

since clearly $\beta_p \leqslant 1/(p-1)$ for all $p \geqslant 2$. $\qquad\square$

Note that although proved by "elementary" methods, this result is not trivial (the trivial bound gives $v_p(u_k) \geqslant -k/(p-1)$). It can be shown that the bound given above is essentially best possible (this is essentially equivalent to showing that the radius of convergence of $E_p(X)$ is equal to 1 and no larger), so it is not an artifact of the proof.

Corollary 4.2.23. (1) *Set*

$$u_k = \sum_{j=0}^{\lfloor k/p \rfloor} \frac{1}{p^j j!(k-pj)!} \,.$$

Then

$$v_p(u_k) \geqslant -\frac{2p-1}{(p-1)p^2}(k - s_p(k)) \,.$$

(2) *For* $0 \leqslant r < p$ *set*

$$a_{r,k} = (-1)^k \sum_{j=0}^{k} \frac{p^j j!}{(pj+r)!}\binom{k}{j} \,.$$

Then $v_p(a_{r,0}) = 0$, *and for* $k \geqslant 1$,

$$v_p(a_{r,k}) \geqslant \left(1 - \frac{1}{p}\right)\left(k - \frac{s_p(k)}{p}\right) = \left(1 - \frac{1}{p}\right)^2 v_p((pk)!) \,.$$

Proof. The result for u_k is simply a reformulation of the theorem since the formula for u_k is obtained by computing the product of the series $\exp(X)$ and $\exp(X^p/p)$. For $a_{r,k}$ it is immediate that $a_{r,k} = (-p)^k k! u_{pk+r}$, so the result follows by a short computation. $\qquad\square$

4.2.4 Mahler Expansions

We begin with the following.

Lemma 4.2.24. *For any sequence a_k such that $a_k \in \mathbb{Z}_p$ and tending to 0 p-adically as $k \to \infty$ the function*

$$f(x) = \sum_{k \geqslant 0} a_k \binom{x}{k}$$

is well defined and is continuous on $\mathbb{Z}_p$.

Proof. Since by Lemma 4.2.8 we have $\binom{x}{k} \in \mathbb{Z}_p$, the term $a_k \binom{x}{k}$ tends to 0 as $k \to \infty$ so that $f(x)$ is well defined. Furthermore, for the same reason (i.e., $|\binom{x}{k}|_p \leqslant 1$) it is uniformly convergent on $\mathbb{Z}_p$, hence defines a continuous function. $\qquad\square$

For $n \in \mathbb{Z}_{\geqslant 0}$ we set $b_n = f(n)$. Since $\binom{n}{k} = 0$ when $k > n$, it is clear that b_n depends linearly on $a_0, \dots, a_n$ and the coefficient of a_n is equal to 1. It follows that the a_k are determined by the b_n, i.e., by the values taken by f on $\mathbb{Z}_{\geqslant 0}$. This is of course not surprising since $\mathbb{Z}_{\geqslant 0}$ is dense in $\mathbb{Z}_p$. In fact we have an explicit formula, which we will need.

Proposition 4.2.25. *If*

$$b_n = \sum_{0 \leqslant k \leqslant n} a_k \binom{n}{k}$$

then

$$a_k = \sum_{0 \leqslant m \leqslant k} (-1)^{k-m} b_m \binom{k}{m} .$$

Proof. One can of course prove this directly. However, the simplest proof is by using generating functions: if $A(T) = \sum_{k \geqslant 0} a_k T^k / k!$ and $B(T) = \sum_{k \geqslant 0} b_k T^k / k!$, then by definition $B(T) = \exp(T) A(T)$, so that $A(T) = \exp(-T) B(T)$, giving the formula for a_k. $\qquad\square$

The main theorem of this subsection, due to Mahler, is that the converse of the above lemma is true. More precisely:

Theorem 4.2.26 (Mahler). *Let f be a function from $\mathbb{Z}_p$ to $\mathbb{Z}_p$ and set*

$$a_k = \sum_{0 \leqslant m \leqslant k} (-1)^{k-m} \binom{k}{m} f(m) .$$

(1) *If $f(x)$ is continuous on $\mathbb{Z}_p$ then $v_p(a_k)$ tends to infinity with k.*

(2) *The function $f(x)$ is continuous if and only if $f(x)$ can be written in the form*

$$f(x) = \sum_{k \geqslant 0} a_k \binom{x}{k}$$

for some sequence $a_k \in \mathbb{Z}_p$ tending to 0 p-adically as $k \to \infty$. If $f(x)$ is continuous the coefficients a_k are uniquely determined and given by the above formula, and called the Mahler coefficients of f.

Proof. (1). Introduce the *forward difference operator* Δ on any function g from $\mathbb{Z}_p$ to $\mathbb{Z}_p$ by setting $(\Delta g)(x) = g(x+1) - g(x)$. It is well known and trivial to prove by induction that for all $k \geqslant 0$ we have

$$(\Delta^k g)(x) = \sum_{0 \leqslant m \leqslant k} (-1)^{k-m} \binom{k}{m} g(m+x).$$

Thus $a_k = (\Delta^k f)(0)$.

We prove (1) by induction. Let $r \geqslant 0$, and assume that there exists $N_r \geqslant 0$ such that $v_p((\Delta^k f)(x)) \geqslant r$ for all $k \geqslant N_r$ and all $x \in \mathbb{Z}_p$. This is trivially true for $r = 0$ with $N_0 = 0$ since $f(x) \in \mathbb{Z}_p$ for all x. Let $g(x) = (\Delta^{N_r} f)(x)/p^r$. This is a continuous function from $\mathbb{Z}_p$ to $\mathbb{Z}_p$. Since $\mathbb{Z}_p$ is compact it is uniformly continuous, so there exists $M \geqslant 1$ such that $x \equiv y \pmod{p^M \mathbb{Z}_p}$ implies $g(x) \equiv g(y) \pmod{p \mathbb{Z}_p}$ for all x and y in $\mathbb{Z}_p$. Since $p \mid \binom{p^M}{k}$ for all k except $k = 0$ and $k = p^M$, this implies that for all $x \in \mathbb{Z}_p$ we have

$$(\Delta^{p^M} g)(x) = \sum_{0 \leqslant m \leqslant p^M} (-1)^{p^M - m} \binom{p^M}{m} g(m+x)$$

$$\equiv g(x + p^M) - g(x) \equiv 0 \pmod{p \mathbb{Z}_p}.$$

It follows that $v_p((\Delta^{N_r + p^M} f)(x)) \geqslant r + 1$ for all $x \in \mathbb{Z}_p$; hence by induction $v_p((\Delta^k f)(x)) \geqslant r + 1$ for all $k \geqslant N_{r+1} = N_r + p^M$, proving our induction hypothesis, hence (1) since $a_k = (\Delta^k f)(0)$.

(2). We have already seen in Lemma 4.2.24 that an expansion of the given form defines a continuous function. Conversely, assume that $f(x)$ is continuous on $\mathbb{Z}_p$. By (1), the a_k tend to 0 p-adically as k tends to infinity. Thus, again by Lemma 4.2.24 the function $g(x) = \sum_{k \geqslant 0} a_k \binom{x}{k}$ is well defined and continuous. However, by definition of a_k and Proposition 4.2.25 we have $f(n) = g(n)$ for all $n \in \mathbb{Z}_{\geqslant 0}$. Since f and g are continuous and since $\mathbb{Z}_{\geqslant 0}$ is dense in $\mathbb{Z}_p$ it follows that $f(x) = g(x)$ for all $x \in \mathbb{Z}_p$, proving (2), and hence the theorem. $\square$

Remark. Assume that f is continuous. Since a_k tends to 0 p-adically it follows that the series $\sum_{k \geqslant 0} a_k \binom{x}{k}$ tends to $f(x)$ uniformly in $x \in \mathbb{Z}_p$. It is also easy to show that $\sup_{x \in \mathbb{Z}_p} |f(x)| = \sup_{k \geqslant 0} |a_k|$; see Exercise 16.

Corollary 4.2.27. *Let b_n be a sequence of elements of $\mathbb{Z}_p$. The following conditions are equivalent.*

(1) *This sequence is p-adically continuous; in other words, for all $k \geqslant 0$ there exists $j \geqslant 0$ such that $v_p(n - m) \geqslant j$ implies that $v_p(b_n - b_m) \geqslant k$.*

(2) *There exists a continuous function $f(x)$ from $\mathbb{Z}_p$ to $\mathbb{Z}_p$ such that $f(n) = b_n$ for $n \geqslant 0$.*

(3) *If a_k is given by the formula of Theorem 4.2.26 then $v_p(a_k) \to \infty$ as $k \to \infty$.*

Proof. Clear.   □

Note that Mahler's theorem, which characterizes only *continuous* functions, implies rather weak results. For instance, in Chapter 9 we will study the p-adic gamma function, which can easily be proved to be continuous. This implies that with the notation of Corollary 4.2.23, $v_p(a_{r,k})$ tends to infinity with k, which is much weaker than the statement of the corollary (see Exercise 44 of Chapter 11). To get the full strength we would need to use the analyticity and not only the continuity of the gamma function.

The following proposition gives a lower bound on the radius of convergence of the power series corresponding to a Mahler expansion.

Proposition 4.2.28. *Let $f(x) = \sum_{k \geqslant 0} a_k \binom{x}{k}$, and assume that for all k we have $v_p(a_k) \geqslant \alpha k + \alpha' s_p(k) + \alpha''$ for some constants α, α', and α'' such that $\alpha > 1/(p-1)$ and $\alpha' \geqslant -\alpha$. Then $f(x)$ is equal to the sum of a power series $\sum_{k \geqslant 0} b_k x^k$ with radius of convergence greater than or equal to $R = p^{\alpha - 1/(p-1)} > 1$. More precisely, we have*

$$v_p(b_k) \geqslant (\alpha - 1/(p-1))k + \beta' s_p(k) + \alpha''$$

for all k, where $\beta' = 0$ if $\alpha' \geqslant -1/(p-1)$ and $\beta' = -(\alpha - 1/(p-1))$ if $-\alpha \leqslant \alpha' < -1/(p-1)$.

Proof. Define integers $s(k, j)$ (called Stirling numbers of the first kind) by the formula

$$X(X - 1) \cdots (X - k + 1) = \sum_{j \geqslant 0} (-1)^{k-j} s(k, j) X^j \;,$$

with $s(k, j) = 0$ for $j > k$, and for all n set $f_n(x) = \sum_{0 \leqslant k \leqslant n} a_k \binom{x}{k}$. This is a polynomial of degree less than or equal to n, and we have $f_n(x) = \sum_{j \geqslant 0} b_{n,j} x^j$ with $b_{n,j} = \sum_{j \leqslant k \leqslant n} (-1)^{k-j} s(k, j) a_k / k!$. It follows from the formula for $v_p(k!)$ that

$$v_p(b_{n,j}) \geqslant \min_{j \leqslant k \leqslant n} (\alpha - 1/(p-1))k + (\alpha' + 1/(p-1)) s_p(k) + \alpha'' \;.$$

Since $0 \leqslant s_p(k) < (p-1)(\log(k)/\log(p) + 1)$ for $k \geqslant 1$, it follows that $|a_k/k!| \leqslant Ak^B R^{-k}$ and $|b_{n,j}| \leqslant Aj^B R^{-j}$ for some constants $A > 0$ and B, where as in the proposition $R = p^{\alpha - 1/(p-1)} > 1$.

Let x be fixed such that $|x| < R$, set $u = \max(|x|, 1)$, and let R_1 be such that $u < R_1 < R$. Since $k^B(R_1/R)^k$ tends to 0 as $k \to \infty$ it is bounded, so increasing A if necessary, we may assume that $|a_k/k!| \leqslant AR_1^{-k}$ and $|b_{n,j}| \leqslant AR_1^{-j}$. We have $|x(x-1) \cdots (x-k+1)| \leqslant u^k$, hence $|a_k\binom{x}{k}| \leqslant u^k |a_k/k!| \leqslant A(u/R_1)^{-k}$, which tends to 0 as $k \to \infty$, so that $f_n(x)$ tends to $f(x)$. On the other hand, since $a_k/k!$ tends to 0 $b_{n,j}$ tends to $b_j = \sum_{k \geqslant j}(-1)^{k-j} s(k,j) a_k/k!$ as $n \to \infty$, and more precisely, since $s(k,j) = 0$ for $k < j$, we have

$$|b_{n,j} - b_j| \leqslant \max_{k > n}|s(k,j)a_k/k!| \leqslant \max_{k \geqslant \max(j,n+1)} AR_1^{-k} = AR_1^{-\max(j,n+1)}.$$

In addition, since $|b_{n,j}| \leqslant AR_1^{-j}$ we have $|b_j| \leqslant AR_1^{-j}$, so that the series $g(x) = \sum_{j \geqslant 0} b_j x^j$ converges since $|x| < R_1$. Now, for any n we have $f_n(x) - g(x) = \sum_{j \geqslant 0}(b_{n,j} - b_j)x^j$; hence

$$|f_n(x) - g(x)| \leqslant \max(\max_{0 \leqslant j \leqslant n}|b_{n,j} - b_j||x|^j, \max_{j > n}|b_{n,j} - b_j||x|^j)$$

$$\leqslant \max(AR_1^{-(n+1)}u^n, \max_{j > n} AR_1^{-k}u^j) \leqslant A(u/R_1)^n,$$

which tends to 0 as $n \to \infty$; hence $f_n(x)$ tends to $g(x)$, and since it also tends to $f(x)$ we have $f(x) = g(x)$.

Since $b_{n,j}$ tends to b_j as $n \to \infty$ we deduce from the inequality for $v_p(b_{n,j})$ that $v_p(b_j) \geqslant \min_{k \geqslant j}(\alpha - 1/(p-1))k + (\alpha' + 1/(p-1))s_p(k) + \alpha''$. If $\alpha' \geqslant -1/(p-1)$ we obtain immediately $v_p(b_j) \geqslant (\alpha - 1/(p-1))j + \alpha''$ since $s_p(k) \geqslant 0$. If $\alpha' < -1/(p-1)$, then since by assumption $\alpha' \geqslant -\alpha$ we have $\alpha' + 1/(p-1) \geqslant -(\alpha - 1/(p-1))$, hence

$$v_p(b_j) \geqslant \min_{k \geqslant j}(\alpha - 1/(p-1))(k - s_p(k)) + \alpha''$$

$$= \min_{k \geqslant j}(\alpha - 1/(p-1))(p-1)v_p(k!) + \alpha''$$

$$= (\alpha - 1/(p-1))(p-1)v_p(j!) + \alpha'' = (\alpha - 1/(p-1))(j - s_p(j)) + \alpha''$$

since $v_p(k!)$ is a nondecreasing function of k, finishing the proof of the proposition. $\qquad\square$

Remarks. (1) The proof is simpler when $\alpha' = 0$, but we will need the more precise statement.

(2) It is clear that this proposition implies the first statement of Corollary 4.2.18.

(3) The converse of this proposition is essentially true; see Exercise 17.

Corollary 4.2.29. Let $(c_k)_{k \geqslant 0}$ be a sequence of elements of $\mathbb{Z}$, set

$$a_k = \sum_{0 \leqslant m \leqslant k} (-1)^{k-m} \binom{k}{m} c_m \,,$$

and assume that as $k \to \infty$ we have $v_p(a_k) \geqslant \alpha k + o(k)$ for some $\alpha > 1/(p-1)$. Then there exists a function f having the following properties:

(1) For any $x \in \mathbb{Z}_p$, $f(x)$ is defined, is a continuous function on $\mathbb{Z}_p$, and is such that $f(k) = c_k$ for all $k \in \mathbb{Z}_{\geqslant 0}$.

(2) For any normed field extension $\mathcal{K}$ of $\mathbb{Q}_p$ and any $x \in \mathcal{K}$ such that $|x| < p^{\alpha - 1/(p-1)}$, $f(x)$ is defined as a power series expansion, and is a p-adic analytic function in that disk.

Proof. Clear from the above results. □

It follows from this corollary that if a p-adic function is defined on $\mathbb{Z}_p$ by simple interpolation, we often get for free an extension to a sufficiently small ball of $\mathcal{K}$. We will see examples of this in Chapter 11 for $\mathcal{K} = \mathbb{C}_p$.

4.3 Additive and Multiplicative Structures

In this section we let K be a number field, $\mathfrak{p}$ a (nonzero) prime ideal of $\mathbb{Z}_K$, and $\mathcal{K} = K_{\mathfrak{p}}$ the completion of K for the $\mathfrak{p}$-adic absolute value. We denote by p the prime number below $\mathfrak{p}$ and we let $\pi \in \mathfrak{p} \setminus \mathfrak{p}^2$ be a uniformizer at $\mathfrak{p}$. We denote as usual by $e = e(\mathfrak{p}/p)$ the ramification index and by $f = f(\mathfrak{p}/p)$ the residual degree. Depending on the context, we will use the (rational-valued) p-adic valuation v_p or the (integer-valued) $\mathfrak{p}$-adic valuation $v_{\mathfrak{p}}$, where by definition $v_p(x) = v_{\mathfrak{p}}(x)/e$.

4.3.1 Concrete Approach

By definition, any $x \in \mathbb{Z}_{\mathfrak{p}}$ is the $\mathfrak{p}$-adic limit of a sequence of x_n that can be chosen to be in $\mathbb{Z}_K$. Furthermore, without loss of generality (see Proposition 4.1.25 for instance), we may assume that $v_{\mathfrak{p}}(x_{n+1} - x_n) \geqslant n$. Thus if we let y_n to be the class of x_n in the quotient ring $\mathbb{Z}_K/\mathfrak{p}^n$, we see that y_n is the image of y_{n+1} by the natural surjection ϕ_n from $\mathbb{Z}_K/\mathfrak{p}^{n+1}$ to $\mathbb{Z}_K/\mathfrak{p}^n$. When one has such a setup, one says that $\mathbb{Z}_{\mathfrak{p}}$ is the *projective limit* (or *inverse limit*) of the $\mathbb{Z}_K/\mathfrak{p}^n$ for the natural maps, and that x is the projective limit of the y_n. The general theory of projective limits is not difficult but is of little use to us. We mention this because it corresponds to the concrete approach to $\mathfrak{p}$-adic numbers (for instance Proposition 4.1.25 is more palatable than completions at finite places). Also, this approach naturally gives both the ring structure and the topology of $\mathbb{Z}_{\mathfrak{p}}$, $K_{\mathfrak{p}}$ being then obtained as the quotient field of $\mathbb{Z}_{\mathfrak{p}}$.

We give a few sample computations in $\mathbb{Q}_p$ using this type of representation, which is of course the one used in practice.

By Proposition 4.1.25, any element $x \in \mathbb{Q}_p^*$ can be written in the form $x = a_k p^k + a_{k+1} p^{k+1} + \cdots$ with $a_k \neq 0$. In particular, $x \in \mathbb{Z}_p$ if and only if $k \geqslant 0$. After appropriate scaling by a power of p, we can evidently concentrate on elements of $\mathbb{Z}_p$. Then addition is done *from left to right*, in contrast to ordinary addition. For example, choose $p = 5$, $x = 3 + 2 \cdot 5 + 4 \cdot 5^2 + 1 \cdot 5^3 + 3 \cdot 5^4 + \cdots$, and $y = 4 + 3 \cdot 5 + 3 \cdot 5^2 + 2 \cdot 5^3 + 3 \cdot 5^4 + \cdots$. We compute successively

$$
\begin{aligned}
x + y = {} & 7 + (2 \cdot 5 + 4 \cdot 5^2 + 1 \cdot 5^3 + 3 \cdot 5^4 + \cdots) \\
& + (3 \cdot 5 + 3 \cdot 5^2 + 2 \cdot 5^3 + 3 \cdot 5^4 + \cdots) \\
= {} & 2 + 6 \cdot 5 + (4 \cdot 5^2 + 1 \cdot 5^3 + 3 \cdot 5^4 + \cdots) \\
& + (3 \cdot 5^2 + 2 \cdot 5^3 + 3 \cdot 5^4 + \cdots) \\
= {} & 2 + 1 \cdot 5 + 8 \cdot 5^2 + (1 \cdot 5^3 + 3 \cdot 5^4 + \cdots) + (2 \cdot 5^3 + 3 \cdot 5^4 + \cdots) \\
= {} & 2 + 1 \cdot 5 + 3 \cdot 5^2 + 4 \cdot 5^3 + (3 \cdot 5^4 + \cdots) + (3 \cdot 5^4 + \cdots) \\
= {} & 2 + 1 \cdot 5 + 3 \cdot 5^2 + 4 \cdot 5^3 + 6 \cdot 5^4 + \cdots \\
= {} & 2 + 1 \cdot 5 + 3 \cdot 5^2 + 4 \cdot 5^3 + 1 \cdot 5^4 + \cdots .
\end{aligned}
$$

An alternative way of doing this computation is to find representatives in $\mathbb{Z}$ of the approximations to the p-adic numbers: here $x = 2113 + O(5^5)$, $y = 2219 + O(5^5)$; hence $x + y = 4332 + O(5^5)$, which of course gives the same expansion in base 5.

Note also that -1 has the representation

$$
-1 = (p-1) + (p-1) \cdot p + (p-1) \cdot p^2 + \cdots .
$$

Similarly, multiplication and division can be done either directly, or by using representatives in $\mathbb{Z}$. For instance, if p is odd, by the above representation of -1 we have

$$
1/2 = 1 + (-1/2) = (p+1)/2 + (p-1)/2 \cdot p + (p-1)/2 \cdot p^2 + (p-1)/2 \cdot p^3 + \cdots .
$$

As we can see from the above examples, there is a fundamental difference between computations on p-adic numbers and computations on formal power series $\sum_{m \geqslant m_0} a_m X^m$, which are also done from left to right. In the latter case, there are no carries, while for p-adic numbers we must take into account the carries, from left to right.

There is nothing much to say concerning the additive structure of $K_{\mathfrak{p}}$ or of $\mathbb{Z}_{\mathfrak{p}}$. We know that all ideals of $\mathbb{Z}_{\mathfrak{p}}$ have the form $\mathfrak{p}^k \mathbb{Z}_{\mathfrak{p}}$; hence if π is a uniformizer, multiplication by π^k gives a noncanonical isomorphism between the additive groups of $\mathbb{Z}_{\mathfrak{p}}$ and $\mathfrak{p}^k \mathbb{Z}_{\mathfrak{p}}$. We thus now look at the multiplicative structure.

4.3.2 Basic Reductions

Proposition 4.3.1. *An element* $x \in \mathbb{Z}_{\mathfrak{p}}$ *is invertible in* $\mathbb{Z}_{\mathfrak{p}}$ *if and only if* $v_p(x) = 0$, *i.e.,* $|x|_{\mathfrak{p}} = 1$.

Proof. Clear since $x \in \mathbb{Z}_p$ if and only if $|x|_p \leqslant 1$.   □

An invertible element of $\mathbb{Z}_p$ will be called a $\mathfrak{p}$-*adic unit*, and when the prime ideal $\mathfrak{p}$ is understood, the multiplicative group of $\mathfrak{p}$-adic units will be denoted by U_0. Clearly $U_0 = \mathbb{Z}_p \setminus \mathfrak{p}\mathbb{Z}_p$. Since any $x \in K_{\mathfrak{p}}^*$ can be written in a unique way as $x = \pi^{v_p(x)}y$ with $y \in U_0$, we have

$$K_{\mathfrak{p}}^* = \pi^{\mathbb{Z}} \times U_0 \simeq \mathbb{Z} \times U_0 \, .$$

Note that this isomorphism is not canonical since it depends on the chosen uniformizer π. We are thus reduced to the study of U_0. The following result will be of constant use.

Proposition 4.3.2. *Let ζ_1 and ζ_2 in $\overline{\mathbb{Q}}$ be distinct roots of unity of order not divisible by p. Then $\zeta_1 - \zeta_2$ is invertible modulo p; in other words, there exists an algebraic integer α such that $\alpha(\zeta_1 - \zeta_2) \equiv 1 \pmod{p\mathbb{Z}_{\overline{\mathbb{Q}}}}$. In particular, if ζ_1 and ζ_2 are in $K_{\mathfrak{p}}$ then $\zeta_1 - \zeta_2 \in U_0$.*

Proof. Let m be the LCM of the orders of ζ_1 and ζ_2, so that $\zeta_1^m = \zeta_2^m = 1$ and $p \nmid m$. From the identity

$$\frac{X^m - 1}{X - \zeta_1} = \prod_{\zeta^m = 1, \, \zeta \neq \zeta_1} (X - \zeta) \, ,$$

setting $X = \zeta_1$ we deduce that

$$m\zeta_1^{m-1} = \prod_{\zeta^m = 1, \, \zeta \neq \zeta_1} (\zeta_1 - \zeta) \, .$$

Since ζ_1 is invertible, and m is invertible modulo p, it follows that $\zeta_1 - \zeta$ is invertible modulo p for every $\zeta \neq \zeta_1$, proving the proposition.   □

Before continuing we need a very simple lemma.

Lemma 4.3.3. *Set*

$$k_0 = \left\lfloor \frac{\log(e/(p-1))}{\log(p)} \right\rfloor + 1 \, .$$

Then if $x \equiv 1 \pmod{\mathfrak{p}\mathbb{Z}_p}$, for $k \geqslant k_0$ we have $v_p(x^{p^k} - 1) > 1/(p-1)$.

Proof. We write $x = 1 + y$ with $v_p(y) \geqslant 1$, in other words $v_p(y) \geqslant 1/e$, so that

$$x^{p^k} = 1 + y^{p^k} + \sum_{1 \leqslant m \leqslant p^k - 1} \binom{p^k}{m} y^m \, .$$

For $1 \leqslant m \leqslant p^k - 1$ we have $p \mid \binom{p^k}{m}$; hence $v_p\left(\binom{p^k}{m}y^m\right) \geqslant 1 + m/e \geqslant 1 + 1/e > 1/(p-1)$. Since by definition of k_0 we have $v_p(y^{p^k}) \geqslant p^k/e > 1/(p-1)$, the lemma is proved.   □

Note that $k_0 \geqslant 0$, and that $k_0 = 0$ if and only if $e < p - 1$.

Proposition 4.3.4. *Let $x \in U_0$ be a $\mathfrak{p}$-adic unit.*

(1) *The sequence $x^{\mathcal{N}\mathfrak{p}^n}$ converges to some $\omega(x) \in U_0$ characterized by the two properties $\omega(x) \equiv x \pmod{\mathfrak{p}\mathbb{Z}_\mathfrak{p}}$ and $\omega(x)^{\mathcal{N}\mathfrak{p}-1} = 1$.*

(2) *Furthermore, ω is a group homomorphism; in other words, if x and y are in U_0 we have $\omega(xy) = \omega(x)\omega(y)$.*

(3) *For any m such that $\mathcal{N}\mathfrak{p}^m > e/(p-1)$ we have the explicit formula*

$$\omega(x) = x^{\mathcal{N}\mathfrak{p}^m} \exp_p\left(-\log_p(x^{\mathcal{N}\mathfrak{p}^m\,(\mathcal{N}\mathfrak{p}-1)})/(\mathcal{N}\mathfrak{p}-1)\right)$$

(where $\log_p$ is computed using the power series expansion), so that in the special case of $\mathbb{Q}_p$, for $p > 2$ we have

$$\omega(x) = x\exp_p(-\log_p(x^{p-1})/(p-1))\,.$$

Proof. (1) and (2). Since $|(\mathbb{Z}_\mathfrak{p}/\mathfrak{p}^a\mathbb{Z}_\mathfrak{p})^*| = \mathcal{N}\mathfrak{p}^{a-1}(\mathcal{N}\mathfrak{p}-1)$, when $x \in U_0$ we have $x^{\mathcal{N}\mathfrak{p}^{a-1}(\mathcal{N}\mathfrak{p}-1)} \equiv 1 \pmod{\mathfrak{p}^a}$. It follows that when $n \geqslant m \geqslant N$ we have

$$x^{\mathcal{N}\mathfrak{p}^n} - x^{\mathcal{N}\mathfrak{p}^m} = x^{\mathcal{N}\mathfrak{p}^m}\left(x^{\mathcal{N}\mathfrak{p}^m(\mathcal{N}\mathfrak{p}^{n-m}-1)} - 1\right) \equiv 0 \pmod{\mathfrak{p}^{N+1}}$$

since $(\mathcal{N}\mathfrak{p}-1) \mid (\mathcal{N}\mathfrak{p}^{n-m}-1)$. Thus $x^{\mathcal{N}\mathfrak{p}^n}$ is a Cauchy sequence, hence converges to some $\omega(x)$, and clearly $\omega(x) \in U_0$. Furthermore, again by Fermat's theorem we have $x^{\mathcal{N}\mathfrak{p}^n} - x = x(x^{\mathcal{N}\mathfrak{p}^n-1} - 1) \equiv 0 \pmod{\mathfrak{p}}$, so $\omega(x) \equiv x \pmod{\mathfrak{p}\mathbb{Z}_\mathfrak{p}}$. Since as above $x^{\mathcal{N}\mathfrak{p}^n(\mathcal{N}\mathfrak{p}-1)}$ tends to 1 $\mathfrak{p}$-adically as $n \to \infty$, it follows that $\omega(x)^{\mathcal{N}\mathfrak{p}-1} = 1$. These two conditions characterize $\omega(x)$ since Proposition 4.3.2 tells us that $(\mathcal{N}\mathfrak{p}-1)$st roots of unity are distinct modulo $\mathfrak{p}$. Finally, the fact that ω is a group homomorphism is trivial.

(3). Since $x^{\mathcal{N}\mathfrak{p}-1} \equiv 1 \pmod{\mathfrak{p}}$, $\log_p(x^{\mathcal{N}\mathfrak{p}-1})$ converges. However, Proposition 4.2.14 shows that we do not sufficiently control the valuation of this logarithm unless $v_p(x^{\mathcal{N}\mathfrak{p}-1} - 1) > 1/(p-1)$, which is not true in general, so we need to use the above lemma to increase the valuation. More precisely, if $\mathcal{N}\mathfrak{p}^m > e/(p-1)$ then since $\mathcal{N}\mathfrak{p}^m = p^k$ for $k = fm$, we have $k \geqslant k_0$ with k_0 as in the lemma, so applying that lemma to $x^{\mathcal{N}\mathfrak{p}-1}$ we deduce that $v_p(x^{\mathcal{N}\mathfrak{p}^k(\mathcal{N}\mathfrak{p}-1)} - 1) > 1/(p-1)$, so by Propositions 4.2.14 and 4.2.10 we can write

$$x^{\mathcal{N}\mathfrak{p}^m(\mathcal{N}\mathfrak{p}-1)} = \exp_p\left(\log_p(x^{\mathcal{N}\mathfrak{p}^m(\mathcal{N}\mathfrak{p}-1)})\right)\,,$$

where the logarithm is directly defined by its power series, not by an extension. Thus for $n \geqslant m$ we have

$$x^{\mathcal{N}\mathfrak{p}^n} = x^{\mathcal{N}\mathfrak{p}^m}\,x^{\mathcal{N}\mathfrak{p}^m(\mathcal{N}\mathfrak{p}-1)(\mathcal{N}\mathfrak{p}^{n-m}-1)/(\mathcal{N}\mathfrak{p}-1)}$$

$$= x^{\mathcal{N}\mathfrak{p}^m}\exp_p\left(((\mathcal{N}\mathfrak{p}^{n-m}-1)/(\mathcal{N}\mathfrak{p}-1))\log_p(x^{\mathcal{N}\mathfrak{p}^m(\mathcal{N}\mathfrak{p}-1)})\right)\,,$$

and since $\mathcal{N}\mathfrak{p}^{n-m}$ converges $\mathfrak{p}$-adically to 0 as $n \to \infty$ and $\exp_p$ is a continuous function, we obtain the formula given in the proposition. □

The map ω given by the above lemma is called the *Teichmüller character*. It is essentially the same, in the $\mathfrak{p}$-adic context, as the map $\omega_{\mathfrak{P}}$ given by Definition 3.6.2.

Corollary 4.3.5. *There exists a canonical group isomorphism $\overline{\omega}$ between $(\mathbb{Z}_K/\mathfrak{p})^*$ and the subgroup $\mu_\mathfrak{p}$ of $z \in U_0$ such that $z^{\mathcal{N}\mathfrak{p}-1} = 1$. In particular, $K_\mathfrak{p}$ contains a primitive $(\mathcal{N}\mathfrak{p}-1)$st root of unity.*

Proof. Let ϕ be the canonical isomorphism from $\mathbb{Z}_K/\mathfrak{p}$ to $\mathbb{Z}_\mathfrak{p}/\mathfrak{p}\mathbb{Z}_\mathfrak{p}$ given by Proposition 4.1.24. The map $\overline{\omega} = \omega \circ \phi$ is then a canonical group homomorphism from $(\mathbb{Z}_K/\mathfrak{p})^*$ to $\mu_\mathfrak{p}$. Since $\omega(x) \equiv x \pmod{\mathfrak{p}}$, this map is injective. On the other hand, the equation $X^{\mathcal{N}\mathfrak{p}-1} = 1$ has at most $\mathcal{N}\mathfrak{p}-1$ roots in the field $K_\mathfrak{p}$; hence $\mu_\mathfrak{p}$ has at most $\mathcal{N}\mathfrak{p}-1$ elements, showing that the map is bijective, hence an isomorphism. □

Define

$$U_1 = \{x \in U_0 / \; v_\mathfrak{p}(x-1) \geqslant 1\} = \{x \in U_0 / \; |x-1| < 1\}\,.$$

Note that in the above, and a few times below, we use the $\mathfrak{p}$-adic valuation $v_\mathfrak{p}$ and not the p-adic valuation v_p (with $v_p(x) = v_\mathfrak{p}(x)/e$), which is more natural in this context. The reader should be careful since the symbols are visually very similar.

Corollary 4.3.6. *The Teichmüller character induces a canonical isomorphism from $(\mathbb{Z}_K/\mathfrak{p})^* \times U_1$ to U_0.*

Proof. For $(a,x) \in (\mathbb{Z}_K/\mathfrak{p})^* \times U_1$ define $\psi_1((a,x)) = \overline{\omega}(a) \cdot x \in U_0$, and for $x \in U_0$ define $\psi_2(x) = (\overline{\omega}^{-1}(\omega(x)), x/\omega(x)) \in (\mathbb{Z}_K/\mathfrak{p})^* \times U_1$. These maps are clearly inverse to one another, proving the corollary. □

Corollary 4.3.7. *We have*

$$K_\mathfrak{p}^* = \pi^{\mathbb{Z}} \times \mu_\mathfrak{p} \times U_1 \simeq \mathbb{Z} \times (\mathbb{Z}_K/\mathfrak{p})^* \times U_1\,,$$

where $\mu_\mathfrak{p}$ is the group of $(\mathcal{N}\mathfrak{p}-1)$st roots of unity.

Proof. Clear. □

Please note the difference between the first *equality* and the second *isomorphism*.

We are thus reduced to the study of U_1.

Definition 4.3.8. *For $i \geqslant 1$ we define U_i to be the subgroup of elements $x \in \mathbb{Z}_\mathfrak{p}$ such that $v_\mathfrak{p}(x-1) \geqslant i$.*

Clearly $U_0 \supset U_1 \supset U_2 \supset \cdots U_i \supset \cdots$, and $\bigcap_{i \geqslant 1} U_i = \{1\}$. We have the following easy lemma.

Lemma 4.3.9. *For every $i \geqslant 1$, the multiplicative group U_i/U_{i+1} is non-canonically isomorphic to the additive group $\mathbb{Z}_K/\mathfrak{p}$, hence to $(\mathbb{Z}/p\mathbb{Z})^{f(\mathfrak{p}/p)}$. In particular, for every $i \geqslant 1$ we have $[U_1 : U_i] = \mathcal{N}\mathfrak{p}^{i-1} = p^{(i-1)f(\mathfrak{p}/p)}$.*

Proof. Let π be a uniformizer of $\mathfrak{p}$. If x and y are in U_i we have

$$\frac{xy-1}{\pi^i} - \frac{x-1}{\pi^i} - \frac{y-1}{\pi^i} = \frac{(x-1)(y-1)}{\pi^i} \equiv 0 \pmod{\mathfrak{p}} \; ;$$

hence the map $x \mapsto (x-1)/\pi^i$ induces an isomorphism from U_i/U_{i+1} to $\mathbb{Z}_K/\mathfrak{p}$, and the other isomorphism follows from Proposition 2.4.1.   □

In the special case $p = 2$ and $K = \mathbb{Q}_2$, we have $\omega(x) = 1$, $(\mathbb{Z}_K/\mathfrak{p})^*$ is the trivial group, and $U_0 = U_1$; hence although the above results remain true, the Teichmüller character does not give any interesting information. In this special case we prefer to modify the definition as follows. Since $K = \mathbb{Q}_2$, if $x \in U_0$ then either x or $-x$ is congruent to 1 modulo 4, and we set $\omega(x) = \pm 1$, so that $x \equiv \omega(x) \pmod 4$.

Definition 4.3.10. *For $x \in U_0$, we define $\langle x \rangle = x/\omega(x)$ and call it the diamond of x.*

Proposition 4.3.11. *If we are not in the special case $p = 2$ and $K = \mathbb{Q}_2$ then $\langle x \rangle$ is the unique element of U_1 such that $x/\langle x \rangle$ is an $(\mathcal{N}\mathfrak{p} - 1)$st root of unity. On the other hand, if $p = 2$ and $K = \mathbb{Q}_2$ then $\langle x \rangle$ is the unique element of U_2 such that $x/\langle x \rangle \in \{\pm 1\}$. In particular, for any prime number p, if $K = \mathbb{Q}_p$ then $\exp_p(\log_p(\langle x \rangle)) = \langle x \rangle$.*

Proof. Immediate and left to the reader.   □

4.3.3 Study of the Groups U_i

We begin with the following result.

Proposition 4.3.12. *Set $z(\mathfrak{p}) = \left\lfloor \frac{e(\mathfrak{p}/p)}{p-1} \right\rfloor + 1$. Then for all $i \geqslant z(\mathfrak{p})$ the $\mathfrak{p}$-adic logarithm and exponential give inverse isomorphisms between the multiplicative group U_i and the additive group $\mathfrak{p}^i \mathbb{Z}_\mathfrak{p}$. In particular, if $e(\mathfrak{p}/p) < p-1$ then U_1 is isomorphic to $\mathfrak{p}\mathbb{Z}_\mathfrak{p}$.*

Proof. If $x \in U_{z(\mathfrak{p})}$ then $|x - 1| < 1$, so that the logarithm converges. Furthermore, writing as usual $e = e(\mathfrak{p}/p)$, we have

$$v_p((x-1)^{k-1}/k) = (k-1)v_p(x-1) - v_p(k) \geqslant (k-1)z(\mathfrak{p})/e - v_p(k) \; .$$

Now $v_p(k) \leqslant \log(k)/\log(p)$ (the ordinary logarithm here!) and $z(\mathfrak{p}) \geqslant (e+1)/(p-1)$; hence

$$v_p\left(\frac{(x-1)^{k-1}}{k}\right) \geqslant (k-1)\frac{e+1}{e(p-1)} - \frac{\log(k)}{\log(p)} \,,$$

so that for $k > 1$ we have

$$v_p\left(\frac{(x-1)^{k-1}}{k}\right) > \frac{k-1}{p-1} - \frac{\log(k)}{\log(p)} = \frac{\log(k)}{p-1}\left(\frac{k-1}{\log(k)} - \frac{p-1}{\log(p)}\right) \,.$$

Since the function $(k-1)/\log(k)$ is an increasing function of $k > 1$, it follows that $v_p((x-1)^{k-1}/k) > 0$ for $k \geqslant p$. On the other hand, when $k < p$ we have $v_p(k) = 0$, hence $v_p((x-1)^{k-1}/k) \geqslant (k-1)/e > 0$ for $k > 1$. Thus for each $k \geqslant 2$ the term in $(x-1)^k$ in the power series expansion of $\log_p(x)$ has a p-adic valuation strictly greater than that of the term with $k = 1$. It follows that $v_p(\log_p(x)) = v_p(x-1)$; hence if $x \in U_i$, we have $\log_p(x) \in \mathfrak{p}^i\mathbb{Z}_p$. Conversely, if $y \in \mathfrak{p}^i\mathbb{Z}_p$ for some $i \geqslant z(\mathfrak{p})$, then $v_p(y) \geqslant z(\mathfrak{p})/e > 1/(p-1)$, so that by Proposition 4.2.10 the power series for $\exp_p(y)$ converges, and if $y = \log_p(x)$ we have $\exp_p(y) = x$, and similarly we check that $v_p(\exp_p(y) - 1) = v_p(y) \geqslant i/e$, so that $\log_p(\exp_p(y)) = y$, proving the proposition. $\square$

Corollary 4.3.13. (1) *For every* $i \geqslant 1$, U_i *has a natural* $\mathbb{Z}_p$-*module structure.*

(2) *For* $i \geqslant z(\mathfrak{p})$, U_i *is a free* $\mathbb{Z}_p$-*module of dimension* $[K_{\mathfrak{p}} : \mathbb{Q}_p]$.

(3) *For every* $i \geqslant 1$, U_i *is finitely generated of rank* $[K_{\mathfrak{p}} : \mathbb{Q}_p]$, *and more precisely*

$$U_i \simeq \mu_{\mathfrak{p},i} \times \mathbb{Z}_p^{[K_{\mathfrak{p}}:\mathbb{Q}_p]} \,,$$

where $\mu_{\mathfrak{p},i}$ *is the finite cyclic group of roots of unity in* $K_{\mathfrak{p}}$ *congruent to* 1 *modulo* $\mathfrak{p}^i$, *and* $|\mu_{\mathfrak{p},i}| = p^m$ *for some* m *such that* $0 \leqslant m \leqslant f\lfloor e/(p-1)\rfloor$.

Proof. (1). For $x = 1 + y \in U_i$ with $i \geqslant 1$ and $\alpha \in \mathbb{Z}_p$, we set directly

$$(1+y)^\alpha = \sum_{n \geqslant 0}\binom{\alpha}{n}y^n \,.$$

By Corollary 4.2.15 this series converges and we have $x^\alpha \equiv 1 \pmod{\mathfrak{p}^i\mathbb{Z}_p}$, so thanks again to the above-mentioned corollary this clearly induces a $\mathbb{Z}_p$-module structure on U_i.

(2). If π is a uniformizer of $\mathfrak{p}$, then multiplication by π^i clearly gives a noncanonical isomorphism between the additive groups $\mathbb{Z}_p$ and $\mathfrak{p}^i\mathbb{Z}_p$. By Corollary 4.1.27, $\mathbb{Z}_p$ is a free $\mathbb{Z}_p$-module of dimension $[K_{\mathfrak{p}} : \mathbb{Q}_p]$, proving (2). Note that we have $(1+y)^\alpha = \exp_p(\alpha \log_p(1+y))$, so that the $\mathbb{Z}_p$-structures are the same.

(3). We have already proved this for $i \geqslant z(\mathfrak{p})$. For $i < z(\mathfrak{p})$, we note that by Lemma 4.3.9, $U_{z(\mathfrak{p})}$ has finite index in U_i equal to a power of p, hence is finitely generated with the same rank. Thus by the structure theorem for finitely generated modules over the principal ideal domain $\mathbb{Z}_p$ we deduce that $U_i \simeq T_i \times \mathbb{Z}_p^{[K_\mathfrak{p} : \mathbb{Q}_p]}$, where T_i is the finite torsion subgroup of U_i. In particular, T_i is a finite subgroup of $K_\mathfrak{p}^*$; hence by Corollary 2.4.3, T_i is cyclic, hence contains only roots of unity congruent to 1 modulo $\mathfrak{p}^i$ (and conversely such elements evidently belong to T_i). Finally, since T_i is a finite $\mathbb{Z}_p$-module its order must be a power of p. More precisely, since $U_{z(\mathfrak{p})}$ is torsion-free and $[U_1 : U_{z(\mathfrak{p})}] = \mathcal{N}\mathfrak{p}^{z(\mathfrak{p})-1}$, then a generator y of the cyclic group T_i satisfies $y^{\mathcal{N}\mathfrak{p}^{z(\mathfrak{p})-1}} = 1$, so that $|T_i| \mid \mathcal{N}\mathfrak{p}^{z(\mathfrak{p})-1}$ as claimed. $\square$

Corollary 4.3.14. *As usual set $e = e(\mathfrak{p}/p)$ and $f = f(\mathfrak{p}/p)$. As abelian groups we have the isomorphism*

$$K_\mathfrak{p}^* \simeq \mu_\mathfrak{p}' \times \mathbb{Z} \times \mathbb{Z}_p^{ef} ,$$

where $\mu_\mathfrak{p}'$ is a cyclic group such that

$$|\mu_\mathfrak{p}'| = (\mathcal{N}\mathfrak{p} - 1)p^k \quad \text{for some k such that} \quad 0 \leqslant k \leqslant f\lfloor e/(p-1) \rfloor .$$

If in addition $\mathfrak{p}$ is above 2 then $k \geqslant 1$.

Proof. By Corollary 4.3.7 and the above corollary we have $K_\mathfrak{p}^* \simeq \mu_\mathfrak{p} \times \mu_{\mathfrak{p},1} \times \mathbb{Z} \times \mathbb{Z}_p^{ef}$ and $|\mu_\mathfrak{p}| = \mathcal{N}\mathfrak{p} - 1$, while $|\mu_{\mathfrak{p},1}| \mid \mathcal{N}\mathfrak{p}^{\lfloor e/(p-1) \rfloor}$, so the result follows. If $\mathfrak{p}$ is above 2 we have $-1 \equiv 1 \pmod{\mathfrak{p}}$ hence $-1 \in \mu_{\mathfrak{p},1}$, so $2 \mid |\mu_{\mathfrak{p},1}| \mid |\mu_\mathfrak{p}'|$. $\square$

Examples. If $p \geqslant 3$, then

$$\mathbb{Q}_p^* = \mu_p \times (1 + p\mathbb{Z}_p) \times p^{\mathbb{Z}} \quad \text{and} \quad 1 + p\mathbb{Z}_p = (1+p)^{\mathbb{Z}_p} ,$$

where as above, μ_p is the group of $(p-1)$st roots of unity in $\mathbb{Q}_p$.
   If $p = 2$, then

$$\mathbb{Q}_2^* = \{\pm 1\} \times (1 + 4\mathbb{Z}_2) \times 2^{\mathbb{Z}} \quad \text{and} \quad 1 + 4\mathbb{Z}_2 = 5^{\mathbb{Z}_2} .$$

4.3.4 Study of the Group U_1

We now want to determine explicitly a minimal system of generators for U_1. We begin with a very classical lemma, which is useful in many parts of algebra.

Lemma 4.3.15 (Nakayama). *Let M be a finitely generated $\mathbb{Z}_\mathfrak{p}$-module. The equality $M = \mathfrak{p}M$ implies that $M = 0$. In particular, a set (x_i) of elements of M is a generating set for M if and only if the classes modulo $\mathfrak{p}M$ of the x_i generate $M/\mathfrak{p}M$ as a $\mathbb{Z}_K/\mathfrak{p}$-vector space.*

Proof. Assume that $M = \mathfrak{p}M \neq 0$, and let $(m_i)_{1 \leqslant i \leqslant k}$ be a system of generators of M with k minimal. If π is a uniformizer of $\mathfrak{p}$, we have $\pi M = \mathfrak{p}M$ since any element of $\mathfrak{p}$ can be written as πu with $u \in \mathbb{Z}_\mathfrak{p}$. Thus from $M = \pi M$, we deduce that there exist $\lambda_i \in \mathbb{Z}_\mathfrak{p}$ such that $m_k = \sum_{1 \leqslant i \leqslant k} \pi \lambda_i m_i$, so that $m_k(1 - \pi \lambda_k) = \sum_{1 \leqslant i \leqslant k-1} \pi \lambda_i m_i$. However, since $\lambda_k \in \mathbb{Z}_\mathfrak{p}$, $1 - \pi \lambda_k$ is invertible in $\mathbb{Z}_\mathfrak{p}$; hence m_k is a $\mathbb{Z}_\mathfrak{p}$-linear combination of the other m_i, so it can be suppressed from the generating family, contradicting the minimality of k. The second assertion of the lemma follows by applying the first to the quotient module $M/(\sum_i \mathbb{Z}_\mathfrak{p} x_i)$. $\qquad\square$

Thanks to Nakayama's lemma, to study the $\mathbb{Z}_p$-module structure of U_1 it is enough to study that of the quotient module U_1/U_1^p.

Proposition 4.3.16. *Let π be a uniformizer of $\mathbb{Z}_\mathfrak{p}$, and let $x = 1 + u\pi^i$ with u a $\mathfrak{p}$-adic unit be an element of $U_i \setminus U_{i+1}$. Denote by $e = e(\mathfrak{p}/p)$ the absolute ramification index of $\mathfrak{p}$. Then*

$$x^p \equiv \begin{cases} 1 + u^p \pi^{ip} \pmod{\mathfrak{p}^{ip+1}\mathbb{Z}_\mathfrak{p}} & \text{for } i < e/(p-1)\,, \\ 1 + u^p \pi^{ip} + pu\pi^i \pmod{\mathfrak{p}^{ip+1}\mathbb{Z}_\mathfrak{p}} & \text{for } i = e/(p-1)\,, \\ 1 + pu\pi^i \pmod{\mathfrak{p}^{i+e+1}\mathbb{Z}_\mathfrak{p}} & \text{for } i > e/(p-1)\,. \end{cases}$$

Proof. Simply look at the valuations in the binomial expansion. $\qquad\square$

Definition 4.3.17. *As above, let $K_\mathfrak{p}$ be a $\mathfrak{p}$-adic field, π a uniformizer, and $e = e(\mathfrak{p}/p)$. We set $\varepsilon = -p/\pi^e$, which is a $\mathfrak{p}$-adic unit. We say that the field $K_\mathfrak{p}$ is regular if either $e/(p-1) \notin \mathbb{Z}$, or if $e/(p-1) \in \mathbb{Z}$ and the congruence $u^p \equiv \varepsilon u \pmod{\mathfrak{p}\mathbb{Z}_\mathfrak{p}}$ has no solution in U_0.*

It follows from this definition and the above proposition that if $K_\mathfrak{p}$ is regular then $v_\mathfrak{p}(x^p - 1)$ depends only on $v_\mathfrak{p}(x - 1)$. Indeed, this is clear when $v_\mathfrak{p}(x - 1) = i \neq e/(p-1)$, and when $i = e/(p-1)$ then $e + i = ip$, so that $x^p - 1 \equiv \pi^{ip}(u^p - \varepsilon u) \pmod{\mathfrak{p}^{ip+1}}$, and by definition of a regular field $v_\mathfrak{p}(u^p - \varepsilon u) = 0$.

Theorem 4.3.18. *Let ζ_p be a primitive pth root of unity.*

(1) *We have equality of $\mathfrak{p}$-adic fields*

$$\mathbb{Q}_p((-p)^{1/(p-1)}) = \mathbb{Q}_p(\zeta_p)\,;$$

in other words, the field extensions of $\mathbb{Q}_p$ defined by the irreducible polynomials $X^{p-1} + p$ and $X^{p-1} + \cdots + X + 1$ are equal.

(2) *Let $K_\mathfrak{p}$ be a $\mathfrak{p}$-adic field such that $e/(p-1) \in \mathbb{Z}$. The following conditions are equivalent:*
    (a) *$K_\mathfrak{p}$ is irregular.*
    (b) *$K_\mathfrak{p} \supset \mathbb{Q}_p(\zeta_p)$.*

(c) $K_{\mathfrak{p}} \supset \mathbb{Q}_p((-p)^{1/(p-1)})$.

(d) *The congruence $u^p \equiv \varepsilon u \pmod{\mathfrak{p}\mathbb{Z}_{\mathfrak{p}}}$ has a solution in U_0.*

Proof. (1). The polynomial $X^{p-1} + p$ is an Eisenstein polynomial, and the polynomial $X^{p-1} + \cdots + 1$ is a shifted Eisenstein polynomial (see Corollary 4.1.36), so both are irreducible in $\mathbb{Q}_p[X]$ hence define extensions of the same degree $p-1$. Thus to show equality it is sufficient to show that $(-p)^{1/(p-1)} \in \mathcal{K}$, where $\mathcal{K} = \mathbb{Q}_p(\zeta_p)$ is the pth cyclotomic extension of $\mathbb{Q}_p$. Indeed, if we set $\pi = 1 - \zeta_p$ and let $\mathbb{Z}_{\mathcal{K}}$ be the ring of integers of $\mathcal{K}$ then

$$0 = \frac{1 - \zeta_p^p}{1 - \zeta_p} = \frac{1 - (1-\pi)^p}{\pi} = p + \sum_{2 \leqslant j \leqslant p-1} (-1)^{j-1} \binom{p}{j} \pi^{j-1} + (-1)^{p-1}\pi^{p-1}$$

$$\equiv p + \pi^{p-1} \pmod{p\pi\mathbb{Z}_{\mathcal{K}}} ;$$

hence $-p/\pi^{p-1} \equiv 1 \pmod{\mathfrak{p}\mathbb{Z}_{\mathcal{K}}}$, since π^{p-1}/p is invertible in $\mathcal{K}$. By Corollary 4.2.15 we know that elements of $U_1 = U_1(\mathcal{K})$ are $(p-1)$st powers, so $-p/\pi^{p-1}$; hence $-p$ itself is a $(p-1)$st power, as was to be proved. Note that we do not need to specify *which* $(p-1)$st root of $-p$ we choose since $\mathbb{Q}_p$ contains all $(p-1)$st roots of unity (but see Section 4.4.8 below).

(2). By definition, (a) is equivalent to (d), and (b) is equivalent to (c) by (1). Furthermore, (c) trivially implies (d) by choosing $u = (-p)^{1/(p-1)}/\pi^{e/(p-1)}$. Conversely, if we assume (d) then since $u \in U_0$ is invertible we have $u^{p-1} \equiv \varepsilon$ $\pmod{\mathfrak{p}\mathbb{Z}_{\mathfrak{p}}}$, so we may apply Hensel's Lemma 4.1.37, which tells us that ε is also a $(p-1)$st power in $\mathbb{Z}_{\mathfrak{p}}$; hence $-p = \varepsilon\pi^e = \varepsilon(\pi^{e/(p-1)})^{p-1}$ is a $(p-1)$st power, proving (c). $\qquad\square$

Corollary 4.3.19. *Let $K_{\mathfrak{p}}$ be a regular field, and denote by e and f its ramification and residual indices. Let π be a uniformizer, and let $\zeta_1, \ldots, \zeta_f$ be lifts to $\mu_{\mathfrak{p}}$ of an $\mathbb{F}_p$-basis of the residue field $\mathbb{Z}_{\mathfrak{p}}/\mathfrak{p}\mathbb{Z}_{\mathfrak{p}} \simeq \mathbb{F}_{p^f}$ given by Corollary 4.3.5. The ef elements*

$$\eta_{i,j} = 1 + \zeta_j \pi^i \text{ with } 1 \leqslant j \leqslant f, \ 1 \leqslant i \leqslant \frac{pe}{p-1} \text{ and } p \nmid i$$

constitute a $\mathbb{Z}_p$-basis of the multiplicative module U_1.

Proof. By Lemma 4.3.9, the $\eta_{i,j}$ for fixed i and $1 \leqslant j \leqslant f$ form a generating set of U_i modulo U_{i+1}. Let us call this (for this proof only) a generating set of level i. With the notation of Proposition 4.3.12, when $i = 1, 2, \ldots, \lfloor e/(p-1) \rfloor$, $z(\mathfrak{p})$, $z(\mathfrak{p})+1, \ldots, z(\mathfrak{p})+e-1$, by Proposition 4.3.16, the $\eta_{i,j}^p$ form a generating system of respective levels p, $2p$, $\ldots$, $\lfloor e/(p-1) \rfloor p$, $z(\mathfrak{p}) + e$, $z(\mathfrak{p}) + e + 1, \ldots$. By Nakayama's lemma, all these levels can be removed from the levels of the generating system; hence we obtain levels i for $1 \leqslant i \leqslant e + \lfloor e/(p-1) \rfloor = \lfloor pe/(p-1) \rfloor$, where those divisible by p are suppressed. Since there are ef such elements $\eta_{i,j}$ and we know that the $\mathbb{Z}_p$-rank of $\mathbb{Z}_{\mathfrak{p}}$ is equal to ef, it follows that they form a $\mathbb{Z}_p$-basis. $\qquad\square$

4.3.5 The Group $K_{\mathfrak{p}}^*/K_{\mathfrak{p}}^{*2}$

We begin with the following general result and then specialize to $\mathbb{Q}_p$.

Proposition 4.3.20. *Let $\mathfrak{p}$ be a prime ideal, let p be the prime number below $\mathfrak{p}$, let π be a uniformizer of $\mathfrak{p}$, and set $e = e(\mathfrak{p}/p)$ and $f = f(\mathfrak{p}/p)$.*

(1) *If $\mathfrak{p}$ is not above 2 then $K_{\mathfrak{p}}^*/K_{\mathfrak{p}}^{*2} \simeq (\mathbb{Z}/2\mathbb{Z})^2$, and in particular $|K_{\mathfrak{p}}^*/K_{\mathfrak{p}}^{*2}| = 4$. More precisely, a system of representatives of $K_{\mathfrak{p}}^*/K_{\mathfrak{p}}^{*2}$ is given by $\{1, a, \pi, a\pi\}$, where a is any element of $U_0 \setminus U_0^2$.*

(2) *If $\mathfrak{p}$ is above 2 then $K_{\mathfrak{p}}^*/K_{\mathfrak{p}}^{*2} \simeq (\mathbb{Z}/2\mathbb{Z})^{ef+2}$, and in particular $|K_{\mathfrak{p}}^*/K_{\mathfrak{p}}^{*2}| = 2^{ef+2}$.*

Proof. (1). By Proposition 4.3.7 we have $K_{\mathfrak{p}}^* = \pi^{\mathbb{Z}} \times \mu_{\mathfrak{p}} \times U_1$. By Corollary 4.2.15 for $\mathfrak{p} \nmid 2$ the series $(1+x)^{1/2}$ converges for $|x| < 1$; hence every element of U_1 is a square. Since $|\mu_{\mathfrak{p}}| = \mathcal{N}\mathfrak{p} - 1$ is even, it follows that $K_{\mathfrak{p}}^*/K_{\mathfrak{p}}^{*2} \simeq (\mathbb{Z}/2\mathbb{Z})^2$ as claimed. It is clear that the four elements given in the proposition are not equivalent modulo $K_{\mathfrak{p}}^*$, so they form a system of representatives.

(2). Here we use Corollary 4.3.14, which tells us in particular that when $\mathfrak{p} \mid 2$ we have $K_{\mathfrak{p}}^* \simeq \mu_{\mathfrak{p}}' \times \mathbb{Z} \times \mathbb{Z}_2^{ef}$, where $\mu_{\mathfrak{p}}'$ is a cyclic group of even order. It follows that

$$K_{\mathfrak{p}}^*/K_{\mathfrak{p}}^{*2} \simeq (\mathbb{Z}/2\mathbb{Z})^2 \times (\mathbb{Z}_2/2\mathbb{Z}_2)^{ef} \simeq (\mathbb{Z}/2\mathbb{Z})^{ef+2} .$$

$\square$

Corollary 4.3.21. *Up to isomorphism there are exactly three quadratic extensions of $K_{\mathfrak{p}}$ when $\mathfrak{p} \nmid 2$, and $2^{ef+2} - 1$ when $\mathfrak{p} \mid 2$.*

Proof. Clear since quadratic extensions of a field L of characteristic different from 2 are in one-to-one correspondence with the classes in L^*/L^{*2} other than the unit class. $\square$

We now specialize to $\mathbb{Q}_p$, and note that if $p > 2$ and $a \in \mathbb{Z}_p$, we may define the Legendre symbol $\left(\frac{a}{p}\right)$ as being equal to $\left(\frac{a_0}{p}\right)$ for any $a_0 \in \mathbb{Z}$ such that $a \equiv a_0 \pmod{p}$.

Proposition 4.3.22. *Assume that $p \geqslant 3$, and let $a \in \mathbb{Q}_p^*$. A necessary and sufficient condition for a to be a square in $\mathbb{Q}_p^*$ is that $v(a)$ be even and $\left(\frac{a/p^{v(a)}}{p}\right) = 1$. Furthermore, a system of representatives of $\mathbb{Q}_p^*/\mathbb{Q}_p^{*2}$ is given by 1, a, p, and pa, where a is any integer such that $\left(\frac{a}{p}\right) = -1$.*

Proof. Clear. $\square$

For $p = 2$ we have the following.

Proposition 4.3.23. *A necessary and sufficient condition for some $a \in \mathbb{Q}_2^*$ to be a square in $\mathbb{Q}_2^*$ is that $v(a)$ be even and $a/p^{v(a)} \equiv 1 \pmod 8$. Furthermore, a system of representatives of $\mathbb{Q}_2^*/\mathbb{Q}_2^{*2}$ is given by ± 1, ± 5, ± 2, and ± 6.*

Proof. Since the square of an odd number is congruent to 1 modulo 8, as above the condition is necessary. For the converse, note that by Lemma 4.2.8 we have

$$v_2\left(\binom{1/2}{k}\right) = -(k + v_2(k!)) \geqslant -2k \ ;$$

hence the expansion of $(1+x)^{1/2}$ is 2-adically convergent as soon as $v_2(x) \geqslant 3$. $\qquad\square$

Example. -7 is a square in $\mathbb{Q}_2$. The 2-adic expansion of one of the square roots of -7 is

$$\sqrt{-7} = 1 + 2^2 + 2^4 + 2^5 + 2^7 + 2^{14} + O(2^{15}) \ .$$

Corollary 4.3.24. *If $p \geqslant 3$, up to isomorphism there are three quadratic extensions of $\mathbb{Q}_p$ given by $\mathbb{Q}_p(\sqrt{D})$ for $D = p$, a, and ap, where $\left(\frac{a}{p}\right) = -1$. If $p = 2$, up to isomorphism there are seven quadratic extensions of $\mathbb{Q}_2$ given by $\mathbb{Q}_p(\sqrt{D})$ for $D = -1$, ± 5, ± 2, and ± 6.*

See also Section 4.4.7 for another representation of these seven quadratic extensions of $\mathbb{Q}_2$.

4.4 Extensions of $\mathfrak{p}$-adic Fields

We now consider the situation in which we have several $\mathfrak{p}$-adic fields, all containing $\mathbb{Q}_p$. As above, we will let K be a number field and $\mathfrak{p}$ a (nonzero) prime ideal of K, and we let $\mathcal{K} = K_\mathfrak{p}$.

4.4.1 Preliminaries on Local Field Norms

For the moment, we let $\mathcal{K}$ be a field complete for some absolute value. We will specialize to $\mathcal{K} = K_\mathfrak{p}$ later. In this subsection, we will denote simply by $|\ |$ the absolute value on $\mathcal{K}$. Let V be a $\mathcal{K}$-vector space. Recall that a *norm* on V is a map $\|\ \|$ from V to the nonnegative real numbers, equal to 0 only on the zero vector, satisfying the triangle inequality, and such that for all $x \in V$ and $\lambda \in \mathcal{K}$ we have $\|\lambda x\| = |\lambda| \|x\|$. Clearly a norm makes V into a metric space, and in particular into a topological space.

Recall also that two norms $\|\ \|_1$ and $\|\ \|_2$ are said to be *equivalent* if they induce the same topology on V. This is easily seen to be equivalent to the existence of strictly positive real numbers c_1 and c_2 such that for all

$x \in V$, $c_1\|x\|_1 \leqslant \|x\|_2 \leqslant c_2\|x\|_1$. This is clearly an equivalence relation. The following result is a classical undergraduate result, but considering its importance we prove it again here.

Proposition 4.4.1. *Let $\mathcal{K}$ be a complete field with respect to some absolute value $|\ |$, and let V be a finite-dimensional $\mathcal{K}$-vector space. Any two norms on V are equivalent, and the corresponding topology makes V into a complete metric space.*

Proof. Let $e_1, \ldots, e_n$ be a fixed $\mathcal{K}$-basis of V. We consider the *sup norm*

$$\left\| \sum_{1 \leqslant i \leqslant n} x_i e_i \right\|_\infty = \max_{1 \leqslant i \leqslant n} |x_i| \ .$$

It is indeed clearly a norm, and V is complete for this norm. Thus, it is sufficient to show that any norm $\|\ \|$ is equivalent to it. First, if $x = \sum_{1 \leqslant i \leqslant n} x_i e_i$, by the triangle inequality we have

$$\|x\| \leqslant \sum_{1 \leqslant i \leqslant n} |x_i| \|e_i\| \leqslant c_2 \|x\|_\infty$$

with $c_2 = \sum_{1 \leqslant i \leqslant n} \|e_i\|$, so one inequality is clear. Assume now by contradiction that the other inequality does not hold. This means that for all $\varepsilon > 0$ we can find $b = b(\varepsilon) \in V$, $b \neq 0$, such that $\|b\| \leqslant \varepsilon \|b\|_\infty$. Writing $b = \sum_{1 \leqslant i \leqslant n} b_i e_i$ and permuting the e_i if necessary, we may assume that $\|b\|_\infty = |b_n|$, and replacing b by b/b_n (which does not change the inequality), that $\|b\|_\infty = b_n = 1$. In other words, $b = c + e_n$, where c belongs to the subspace W spanned by the e_i for $1 \leqslant i \leqslant n-1$. Thus, we can find a sequence $c^{(m)}$ of elements of W such that $\|c^{(m)} + e_n\| \to 0$ as $m \to \infty$. By the triangle inequality, this implies that $\|c^{(m_1)} - c^{(m_2)}\| \to 0$ as m_1 and m_2 both tend to infinity; in other words, the sequence $c^{(m)}$ is a Cauchy sequence of elements of W. We can now reason by induction on the dimension n, the result being trivially true for $n = 0$. Thus, by induction we may assume that W, which has dimension $n-1$, is complete. Thus, the Cauchy sequence $c^{(m)}$ converges to some $c^* \in W$. It follows that

$$\|c^* + e_n\| = \lim_{m \to \infty} \|c^{(m)} + e_n\| = 0 \ ,$$

hence $e_n = -c^* \in W$, a contradiction, proving the proposition.   □

Corollary 4.4.2. *Let $\mathcal{K}$ be a complete field with respect to some nontrivial absolute value $|\ |$, and let $\mathcal{L}$ be a finite extension of $\mathcal{K}$. There exists at most one extension $\|\ \|$ of $|\ |$ to $\mathcal{L}$. In addition, $\mathcal{L}$ is complete for this extension, if it exists.*

Proof. We can consider $\mathcal{L}$ as a finite-dimensional $\mathcal{K}$-vector space, and an absolute value $\|\ \|$ on $\mathcal{L}$ clearly satisfies the conditions of a norm. By the above proposition, any two such absolute values $\|\ \|_1$ and $\|\ \|_2$ are equivalent;

in other words, they induce the same topology on $\mathcal{L}$. By Lemma 4.1.3, there exists $c > 0$ such that for all $x \in \mathcal{L}$ we have $\|x\|_2 = \|x\|_1^c$. Since the absolute values coincide on $\mathcal{K}$, choosing any $x \in \mathcal{K}^*$ such that $\|x\|_1 \neq 1$ (which is possible since otherwise we would have a trivial absolute value), we deduce that $c = 1$, hence that the absolute values coincide. $\qquad\square$

We are now ready to prove the main theorem of this subsection.

Theorem 4.4.3. *Let $\mathcal{K}$ be a completion of a number field K with respect to some nontrivial absolute value $|\ |$, and let $\mathcal{L}$ be a finite extension of $\mathcal{K}$ with $[\mathcal{L} : \mathcal{K}] = n$. There exists a unique extension $\|\ \|$ of $|\ |$ to $\mathcal{L}$, given by*

$$\|x\| = |\mathcal{N}_{\mathcal{L}/\mathcal{K}}(x)|^{1/n} ,$$

and $\mathcal{L}$ is complete for this extension.

Proof. Uniqueness and completeness have been proved above. Furthermore, if $x \in \mathcal{K}$ we have $\mathcal{N}_{\mathcal{L}/\mathcal{K}}(x) = x^n$, hence $\|x\| = |x|$, so this is indeed an extension of $|\ |$. It remains to show that it is an absolute value on $\mathcal{L}$. By multiplicativity of the norm we have $\|xy\| = \|x\|\|y\|$, and since $\|1\| = 1$ if $x \in \mathcal{L}^*$ we have $\|x\|\|x^{-1}\| = 1$, hence $\|x\| \neq 0$. It remains to show the triangle inequality. If $\mathcal{K}$ is isomorphic to $\mathbb{R}$ or $\mathbb{C}$, then either $\mathcal{L} = \mathcal{K}$ and there is nothing to prove, or $\mathcal{K}$ is isomorphic to $\mathbb{R}$ and $\mathcal{L}$ is isomorphic to $\mathbb{C}$, in which case the triangle inequality for $\|\ \|$ is nothing else than the triangle inequality for the ordinary modulus on $\mathbb{C}$. We may thus assume that $\mathcal{K}$ is not isomorphic to $\mathbb{R}$ or $\mathbb{C}$, in other words that the absolute value is ultrametric and that $\mathcal{K} = K_{\mathfrak{p}}$ for some number field K and prime ideal $\mathfrak{p}$ of $\mathbb{Z}_K$.

Let x and y be in $\mathcal{L}$, with x and y nonzero (otherwise the inequality is trivial). Exchanging x and y if necessary, we may assume that $\|y\| \leqslant \|x\|$; in other words, $\|a\| \leqslant 1$ with $a = y/x$. Let $C(X) = X^n + C_{n-1}X^{n-1} + \cdots + C_0$ be the characteristic polynomial of a in the extension $\mathcal{L}/\mathcal{K}$, so that $\mathcal{N}_{\mathcal{L}/\mathcal{K}}(a) = (-1)^n C_0$; hence $|C_0| \leqslant 1$ by assumption. Then $C(X) = M(X)^r$ for some $r \geqslant 1$, where $M(X) = X^d + M_{d-1}X^{d-1} + \cdots + M_0$ is the minimal monic polynomial of a. Since $C_0 = M_0^r$ we also have $|M_0| \leqslant 1$. Furthermore, $M(X)$ is irreducible in $\mathcal{K}[X]$. Thus, by Corollary 4.1.34 we have $M \in \mathbb{Z}_{\mathfrak{p}}[X]$, hence also $C \in \mathbb{Z}_{\mathfrak{p}}[X]$. Since $C(X)$ is the characteristic polynomial of a, we have $\mathcal{N}_{\mathcal{L}/\mathcal{K}}(1+a) = (-1)^n C(-1)$, hence $\|1+a\| = |\mathcal{N}_{\mathcal{L}/\mathcal{K}}(1+a)|^{1/n} \leqslant 1$. Replacing a by y/x gives $\|x + y\| \leqslant \|x\| = \max(\|x\|, \|y\|)$, proving the ultrametric inequality. $\qquad\square$

Corollary 4.4.4. *Let $\mathcal{K}$ be a completion of a number field K with respect to some nontrivial absolute value $|\ |$. There exists an extension of $|\ |$ to the algebraic closure $\overline{\mathcal{K}}$ of $\mathcal{K}$, and this extension is unique.*

Proof. Since the algebraic closure is the union of all finite extensions included in it, the result is clear. $\qquad\square$

Remark and Warning. (1) The above theorem is valid for a general field $\mathcal{K}$ complete for some absolute value, and not only for the completion of a number field; see Exercise 26.

(2) The unique extension given by Theorem 4.4.3 is *not* the natural $\mathfrak{p}$-adic absolute value defined in Section 4.1.4 to obtain the product formula. More precisely, we have the following:

Proposition 4.4.5. *Let $\mathcal{K} = K_{\mathfrak{p}}$, let L/K be an extension, let $\mathfrak{P}$ be a prime ideal of L above $\mathfrak{p}$, let $\mathcal{L} = L_{\mathfrak{P}}$, and set $n = [\mathcal{L} : \mathcal{K}] = e(\mathfrak{P}/\mathfrak{p})f(\mathfrak{P}/\mathfrak{p})$. Denote by $\| \ \|$ the extension to $\mathcal{L}$ of the absolute value of $\mathcal{K}$. Then for all $x \in \mathcal{L}$ we have $\|x\| = |x|_{\mathfrak{P}}^{1/n}$.*

Proof. Since both norms define the same topology on $\mathcal{L}$ they are equivalent, so that one is a power of the other. To determine which power, we simply choose $x = \pi$, where π is a uniformizer of $\mathfrak{p}$, and the result immediately follows. $\square$

Remarks. (1) It follows that there are (at least) two possible natural normalizations for the absolute value on a $\mathfrak{p}$-adic field $\mathcal{K}$: when $\mathcal{K}$ is considered as the completion $K_{\mathfrak{p}}$ of a number field K, usually together with other completions, the natural normalization is the one that we chose for the product formula, in other words $\|x\| = \mathcal{N}\mathfrak{p}^{-v_{\mathfrak{p}}(x)}$. On the other hand, if $\mathcal{K}$ is considered as a field with a non-Archimedean absolute value together with extensions of $\mathcal{K}$, then the natural normalization is to choose the one compatible with extensions as given above.

(2) In the rest of this book, unless indicated otherwise, we will use the latter normalization, and since the extension of $|\ |$ to a finite extension or to an algebraic closure exists and is unique, as we have already done in preceding sections we denote it simply by $|\ |$.

4.4.2 Krasner's Lemma

Proposition 4.4.6 (Krasner's lemma). *Let $\mathcal{K}$ be a completion of a number field K with respect to some nontrivial absolute value $|\ |$ and let $\mathcal{L}$ be a finite extension of $\mathcal{K}$.*

(1) *If x and y in $\mathcal{L}$ are conjugate over $\mathcal{K}$ then $|x| = |y|$.*

(2) *If x and y in $\mathcal{L}$ are conjugate over $\mathcal{K}$ then for all $a \in \mathcal{K}$ we have $|x - y| \leqslant |a - x|$.*

(3) *Let $x \in \mathcal{L}$, and assume that $a \in \mathcal{L}$ is such that $|a - x| < |y - x|$ for every conjugate y of x in $\overline{\mathcal{L}}$ different from x. Then $x \in \mathcal{K}(a)$; in other words, $\mathcal{K}(x) \subset \mathcal{K}(a)$.*

Proof. (1) is clear since x and y have the same norm. If (2) were not true, in other words if for some $a \in \mathcal{K}$ we had $|x - y| > |a - x|$, we would have

$$|a - y| = \max(|a - x|, |x - y|) = |x - y| > |a - x| \ ,$$

contradicting (1) since $a - x$ and $a - y$ are conjugate. For (3), let $\mathcal{M}$ be the Galois closure of $\mathcal{L}$ over $\mathcal{K}$. If $x \notin K(a)$, by Galois theory there exists a $\mathcal{K}(a)$-automorphism σ of $\mathcal{M}$ such that $\sigma(x) \neq x$. Applying (2) to the extension $\mathcal{M}/\mathcal{K}(a)$ and to the $\mathcal{K}(a)$-conjugate elements x and $y = \sigma(x)$, we deduce that $|y - x| \leqslant |a - x|$, a contradiction. $\qquad\square$

For the following proposition, if P is a polynomial we denote by $\|P\|$ the maximum of the absolute values of the coefficients of P.

Proposition 4.4.7. *As above let $\mathcal{L}/\mathcal{K}$ be an extension of $\mathfrak{p}$-adic fields of degree n, let $a \in \mathcal{L}$ be such that $\mathcal{L} = \mathcal{K}(a)$, and let $P(X)$ be the minimal monic polynomial of a over $\mathcal{K}$. There exists $\varepsilon > 0$ such that any monic polynomial $Q \in \mathcal{K}[X]$ of degree n such that $\|P - Q\| < \varepsilon$ has a root $b \in \mathcal{L}$ such that $\mathcal{L} = \mathcal{K}(b) = \mathcal{K}(a)$.*

Proof. For any monic $Q \in \mathcal{K}[X]$, let $Q = \prod(X - b_i)$ be the factorization of Q in an algebraic closure of $\mathcal{K}$. We thus have $\prod(a - b_i) = Q(a) = Q(a) - P(a)$, so if we set $M = \max_{0 \leqslant i \leqslant n} |a|^i = \max(1, |a|^n)$ we have

$$\prod |a - b_i| = |Q(a) - P(a)| \leqslant \|Q - P\| M \ .$$

Thus for at least one index i we must have $|a - b_i| \leqslant \|Q - P\|^{1/n} M^{1/n}$, and so by Krasner's lemma (Proposition 4.4.6 (3)) if $\|Q - P\| < \varepsilon$ with ε small enough we will have $\mathcal{K}(b_i) \supset \mathcal{K}(a)$. On the other hand, since b_i is a root of the nth-degree polynomial Q we have $[\mathcal{K}(b_i) : \mathcal{K}] \leqslant n = [\mathcal{K}(a) : \mathcal{K}]$, hence $\mathcal{K}(b_i) = \mathcal{K}(a)$. $\qquad\square$

Corollary 4.4.8. *Let $P \in \mathcal{K}[X]$ be a monic irreducible polynomial, let a be a root of P in an algebraic closure of $\mathcal{K}$, and let (Q_i) be a sequence of monic polynomials in $\mathcal{K}[X]$ of the same degree as P such that Q_i tends to P coefficientwise. Then there exists a sequence (b_i) of roots of the polynomials Q_i such that $b_i \in \mathcal{K}(a)$ for i sufficiently large and such that b_i tends to a as $i \to \infty$.*

Proof. As soon as $\|Q_i - P\| < \varepsilon$ we can apply the above proposition, which shows that $|a - b_i|$ is small for at least one root b_i of Q_i belonging to $\mathcal{K}(a)$, more precisely that $|a - b_i| \leqslant \|Q_i - P\|^{1/n} M^{1/n}$. This inequality shows that $|a - b_i|$ tends to 0, hence that b_i tends to a in $\mathcal{K}(a)$. $\qquad\square$

4.4.3 General Results on Extensions

Proposition 4.4.9. *Any finite extension $\mathcal{L}$ of $K_\mathfrak{p}$ is isomorphic to $L_\mathfrak{P}$ for some finite extension L of K such that $[L : K] = [\mathcal{L} : K_\mathfrak{p}]$ and some prime ideal $\mathfrak{P}$ of L above $\mathfrak{p}$.*

Proof. Set $n = [\mathcal{L} : K_{\mathfrak{p}}]$. By the primitive element theorem, there exists $x \in \mathcal{L}$ such that $\mathcal{L} = K_{\mathfrak{p}}(x)$, and we may of course assume that x is integral, i.e., that $v(x) \geqslant 0$, where v is an extension of the $\mathfrak{p}$-adic valuation on $K_{\mathfrak{p}}$. Thus, let $P(X) = X^n + \sum_{1 \leqslant i < n} a_i X^i$ be the minimal polynomial of x over $K_{\mathfrak{p}}$, with $a_i \in \mathbb{Z}_{\mathfrak{p}}$ for all i. By density, we can find elements $\widetilde{a}_i \in \mathbb{Z}_K$ that are $\mathfrak{p}$-adically as close as we like to a_i. Set

$$\widetilde{P}(X) = X^n + \sum_{1 \leqslant i < n} \widetilde{a}_i X^i ,$$

and let $\widetilde{x}_j$ be the roots of this polynomial in some algebraic closure $\overline{K}$ of K. Since $\overline{K} \subset \overline{K_{\mathfrak{p}}} = \overline{\mathcal{L}}$, we may also consider $\widetilde{x}_j$ as elements of $\overline{\mathcal{L}}$. By Corollary 4.4.8 it follows that when the $\widetilde{a}_i$ are sufficiently close to the a_i one of the roots $\widetilde{x}_j$ of $\widetilde{P}$ tends to x, and also that $K_{\mathfrak{p}}(\widetilde{x}_j) = K_{\mathfrak{p}}(x)$. Thus if we set $L = K(\widetilde{x}_j)$, we have $[L : K] = n$, and

$$L \otimes_K K_{\mathfrak{p}} = K_{\mathfrak{p}}(\widetilde{x}_j) = K_{\mathfrak{p}}(x) = \mathcal{L} .$$

I claim that this implies that there is a *unique* prime ideal $\mathfrak{P}$ of L above $\mathfrak{p}$. Indeed, if there were two distinct such prime ideals $\mathfrak{P}_1$ and $\mathfrak{P}_2$, then $L \otimes K_{\mathfrak{p}}$ would contain $L_{\mathfrak{P}_1} \oplus L_{\mathfrak{P}_2}$, which has zero divisors, in contradiction to the fact that $\mathcal{L}$ is a field, proving my claim. We then clearly have $\mathcal{L} = L \otimes K_{\mathfrak{p}} = L_{\mathfrak{P}}$, finishing the proof of the proposition. This last argument can be replaced by the use of the more general Theorem 4.4.41 (2), which we shall prove below, and whose proof does not depend on the present proposition. $\qquad \square$

By Corollary 4.1.27, we know that $[L_{\mathfrak{P}} : K_{\mathfrak{p}}] = e(\mathfrak{P}/\mathfrak{p})f(\mathfrak{P}/\mathfrak{p})$. If we consider the fields $\mathcal{L}$ and $\mathcal{K} = K_{\mathfrak{p}}$ abstractly, i.e., without reference to the global situation, it is very useful to set $e(\mathcal{L}/\mathcal{K}) = e(\mathfrak{P}/\mathfrak{p})$ and $f(\mathcal{L}/\mathcal{K}) = f(\mathfrak{P}/\mathfrak{p})$. Note that these can be defined directly: f is simply the degree of the residual field extension, and e the index of the value groups of the valuations on $\mathcal{K}$ and $\mathcal{L}$.

As in the absolute case, we set the following definitions.

Definition 4.4.10. *With the above notation, we say that the extension $L_{\mathfrak{P}}/K_{\mathfrak{p}}$ of degree n is:*

- *unramified if $e(\mathfrak{P}/\mathfrak{p}) = 1$ (or equivalently, $f(\mathfrak{P}/\mathfrak{p}) = n$),*
- *totally ramified if $e(\mathfrak{P}/\mathfrak{p}) = n$ (or equivalently, $f(\mathfrak{P}/\mathfrak{p}) = 1$),*
- *tamely ramified if $p \nmid e(\mathfrak{P}/\mathfrak{p})$, where p is the prime number below $\mathfrak{p}$, i.e., the characteristic of $\mathbb{Z}_K/\mathfrak{p}$.*

Clearly any unramified extension is tamely ramified. On the other hand, note that a totally ramified extension may or may not be tamely ramified.

The study of $\mathfrak{p}$-adic fields is very much simplified by the fact that any extension of $\mathfrak{p}$-adic fields can be obtained canonically as a totally ramified extension of an unramified extension, and that these types of extensions can be studied very precisely.

In the sequel, it is useful to let $\mathcal{K} = K_{\mathfrak{p}}$ and $\mathcal{L} = L_{\mathfrak{P}}$, and we let $k = \mathbb{Z}_{\mathfrak{p}}/\mathfrak{p}\mathbb{Z}_{\mathfrak{p}}$ and $l = \mathbb{Z}_{\mathfrak{P}}/\mathfrak{P}\mathbb{Z}_{\mathfrak{P}}$ be the residue fields.

Theorem 4.4.11. *Let $a \in l$ be a class. There exists a representative $\alpha \in \mathbb{Z}_{\mathfrak{P}}$ of $\overline{\alpha}$ such that $[\mathcal{K}(\alpha) : \mathcal{K}] = [k(a) : k]$. Furthermore, the field $\mathcal{K}(\alpha)$ depends only on a and not on the chosen representative, and the extension $\mathcal{K}(\alpha)/\mathcal{K}$ is unramified.*

Proof. Let $\phi(T) \in k[T]$ be the minimal monic polynomial of a over k. Since k is a perfect field, ϕ is separable, so $\phi'(a) \neq 0$. Let Φ be any *monic* lift of ϕ to $\mathbb{Z}_{\mathfrak{p}}[T]$, and let α_0 be any representative of a in $\mathbb{Z}_{\mathfrak{P}}$. We thus have $|\Phi(\alpha_0)| < 1$ and $|\Phi'(\alpha_0)| = 1$. By Hensel's Lemma 4.1.37 applied to the field $\mathcal{K}(\alpha_0)$, there exists $\alpha \in \mathcal{K}(\alpha_0) \subset \mathcal{L}$ such that $\Phi(\alpha) = 0$ and $|\alpha - \alpha_0| < 1$. This means that α is a representative of the class of α_0, i.e., of a, and since the degree of Φ is equal to that of ϕ, we have equality of degrees, as claimed. Furthermore, if we assume that $[\mathcal{K}(\beta) : \mathcal{K}] = [k(a) : k]$ for some other representative β of a, applying the above reasoning with β and α instead of α_0 gives $\mathcal{K}(\beta) = \mathcal{K}(\alpha)$. Finally, since $[\mathcal{K}(\alpha) : \mathcal{K}]$ is equal to the residue field index $[k(a) : k]$, it follows by definition that the extension $\mathcal{K}(\alpha)/\mathcal{K}$ is unramified. $\square$

Corollary 4.4.12. *With the same assumptions, there is a bijection between the unramified subextensions $\mathcal{M}/\mathcal{K}$ of $\mathcal{L}/\mathcal{K}$ and the fields m such that $k \subset m \subset l$. The field m corresponding to $\mathcal{M}$ is $\mathcal{M} \cap \mathbb{Z}_{\mathfrak{P}}$ modulo the prime ideal of $\mathcal{M}$ below $\mathfrak{P}$.*

Proof. Since an extension m/k is an extension of finite fields, it is separable; hence by the primitive element theorem we have $m = k(a)$ for some $a \in m$. The corollary then follows immediately from the theorem. $\square$

Corollary 4.4.13. *There exists a field $\mathcal{M}$ such that $\mathcal{K} \subset \mathcal{M} \subset \mathcal{L}$ such that $\mathcal{M}/\mathcal{K}$ is unramified and such that every $\mathcal{M}' \subset \mathcal{L}$ that is unramified over $\mathcal{K}$ is contained in $\mathcal{M}$. In addition, $\mathcal{L}/\mathcal{M}$ is totally ramified.*

Proof. We simply let $\mathcal{M}$ be the field corresponding to l in the previous corollary. $\square$

Corollary 4.4.14. *Recall that $\mathcal{K} = K_{\mathfrak{p}}$ and $\mathcal{L} = L_{\mathfrak{P}}$. There exists a unique unramified subextension $\mathcal{M}$ of $\mathcal{L}$ that contains all unramified subextensions of $\mathcal{L}/\mathcal{K}$, and that is such that $\mathcal{L}/\mathcal{M}$ is totally ramified. Furthermore, $e(\mathcal{M}/\mathcal{K}) = 1$, $f(\mathcal{M}/\mathcal{K}) = f(\mathfrak{P}/\mathfrak{p})$, $[\mathcal{M} : \mathcal{K}] = f(\mathfrak{P}/\mathfrak{p})$, $e(\mathcal{L}/\mathcal{M}) = e(\mathfrak{P}/\mathfrak{p})$, $f(\mathcal{L}/\mathcal{M}) = 1$, $[\mathcal{L} : \mathcal{M}] = e(\mathfrak{P}/\mathfrak{p})$.*

Proof. The first assertion is a restatement of the above corollary. For the second, note that since $\mathcal{M}/\mathcal{K}$ is unramified, we have $e(\mathcal{M}/\mathcal{K}) = 1$, and since $\mathcal{L}/\mathcal{M}$ is totally ramified we have $e(\mathcal{L}/\mathcal{M}) = [\mathcal{L} : \mathcal{M}]$. Thus by transitivity, $e(\mathfrak{P}/\mathfrak{p}) = e(\mathcal{L}/\mathcal{K}) = e(\mathcal{L}/\mathcal{M})e(\mathcal{M}/\mathcal{K}) = [\mathcal{L} : \mathcal{M}]$, and since $e(\mathcal{L}/\mathcal{M})f(\mathcal{L}/\mathcal{M}) = [\mathcal{L} : \mathcal{M}]$ we have $f(\mathcal{L}/\mathcal{M}) = 1$. The other statements again follow by transitivity.   □

We thus see that, as claimed above, any finite extension of p-adic fields is a totally ramified extension of an unramified extension, and this unramified subextension is maximal and unique.

Corollary 4.4.15. *The residue field of the algebraic closure $\overline{\mathcal{K}}$ of $\mathcal{K}$ is equal to $\overline{k}$, the algebraic closure of k. There exists a (unique) subfield $\mathcal{K}^u$ of $\overline{\mathcal{K}}$ such that a finite extension $\mathcal{L}/\mathcal{K}$ in $\overline{\mathcal{K}}$ is unramified if and only if $\mathcal{L} \subset \mathcal{K}^u$.*

Proof. If $\phi(T) \in k[T]$ is monic and irreducible and $\Phi(T) \in \mathcal{K}[T]$ is any monic lift, then $\overline{\mathcal{K}}$ contains all the roots of $\Phi(T)$; hence its residue field contains all the roots of $\phi(T)$, so that it is equal to the algebraic closure of k. The second statement follows from the above corollary.   □

Obviously, the field $\mathcal{K}^u$ is called the *maximal unramified extension* of $\mathcal{K}$.

Example. Take $\mathcal{K} = \mathbb{Q}_p$, the field of p-adic numbers. It follows from the above results that there is a canonical one-to-one correspondence between the finite fields $\mathbb{F}_{p^n}$ and the unramified extensions of $\mathbb{Q}_p$. If $\mathcal{L}_n$ is the unique (up to isomorphism) degree-n unramified extension of $\mathbb{Q}_p$, then $\mathcal{L}_n$ is called the *canonical lift* of $\mathbb{F}_{p^n}$ in characteristic 0. To obtain it explicitly, it is sufficient to take $\mathcal{L}_n = \mathbb{Q}_p[X]/P_n(X)\mathbb{Q}_p(X)$ with $P_n \in \mathbb{Z}_p[X]$ monic of degree n such that the reduction of P_n modulo p is irreducible in $\mathbb{F}_p[X]$; see examples in Section 4.4.5. Many authors denote by $\mathbb{Q}_q$ this canonical lift, with $q = p^n$, so that $\mathbb{Q}_q/\mathbb{Z}_q = \mathbb{F}_q$.

4.4.4 Applications of the Cohomology of Cyclic Groups

The goal of this (sub)section is the proof of Theorem 4.4.22 and its corollary Proposition 4.4.24, so it can be skipped on first reading, apart from the statement of the results.

First, a friendly word to the reader. When I first heard words like "cohomology," I immediately ran away, and usually stopped listening to talks using this abstract phraseology. Thus, if in turn I use this expression (and some others), I can easily understand the reactions of some of my readers. However, have no fear. In the present book (and in fact only in the present subsection) we will need only the cohomology of finite cyclic groups, and the reader will easily convince himself that it is nothing more than a series of exercises in linear algebra (the only prerequisite is understanding the notion of exact sequences of $\mathbb{Z}$-modules, i.e., of abelian groups, and corresponding diagram chasing), so it does not really need much abstract knowledge. It is simply easier to write H^1 than $\mathrm{Ker}(f)/\mathrm{Im}(g)$. I have found the exposition in the book by Janusz [Jan] to be quite appropriate for my goal, so I follow it very closely.

Let G be a group, for the moment arbitrary, and let A be an abelian group. Recall that we say that G *acts* on A if we are given a group homomorphism from G into the automorphism group of A. The action of $g \in G$ on $a \in A$ can be written indifferently $g \cdot a$, $g(a)$ (or a^g when G is commutative), so that the fundamental homomorphism rule reads $gh \cdot a = g \cdot (h \cdot a)$ or $gh(a) = g(h(a))$. When G acts on A we say that A is a G-*module*. If we let as usual $\mathbb{Z}[G]$ be the group ring associated with G, in other words the set of finite formal $\mathbb{Z}$-linear combinations of elements of G, then A is a $\mathbb{Z}[G]$-module thanks to the action

$$\left(\sum_{g \in G} n_g g\right)(a) = \sum_{g \in G} n_g g(a) \quad \text{or} \quad \left(\sum_{g \in G} n_g g\right)(a) = \prod_{g \in G} g(a)^{n_g},$$

depending on whether the group law on A is written additively or multiplicatively.

Some examples of G-modules that will be important for us are *Galois modules* as follows. Let L/K be a Galois extension of number fields with Galois group G. Then L^*, the group of fractional ideals of L, the ideal class group, and the unit group of L are all G-modules.

From now on let G be a finite cyclic group of order n, and fix a generator σ of G. In particular, if A is a G-module the action of G on A is entirely determined by the action of σ on A. The following are essential elements of $\mathbb{Z}[G]$. First the *norm* $N \in \mathbb{Z}[G]$ defined by $N = \sum_{g \in G} g = \sum_{0 \leqslant i < n} \sigma^i$. The reason for the name is clear, since if $G = \mathrm{Gal}(L/K)$ and $a \in L^*$, then $\mathcal{N}_{L/K}(a) = N(a)$. Second $D = 1 - \sigma$, which by Galois theory has the property that $D(a) = 0$ if and only if $a \in K$. Note that although the element N can be defined for any group, and corresponds to the norm, the element D is specific to cyclic groups: indeed, Galois theory says that $a \in K$ if and only if $g(a) = a$ for any $g \in G$, and this is in general not equivalent to $D(a) = 0$ for a *single* element $D \in \mathbb{Z}[G]$.

In $\mathbb{Z}[G]$ we have $DN = ND = 0$, hence $\mathrm{Im}(D|_A) \subset \mathrm{Ker}(N|_A)$ and $\mathrm{Im}(N|_A) \subset \mathrm{Ker}(D|_A)$, where the symbol $|_A$ is included to emphasize that we consider the action on A. Also, note that $\mathrm{Ker}(D|_A) = A^G$; in other words the group of elements of A that are invariant by G.

Definition 4.4.16. *Let A be a G module, and let D and N be as above. We define the zeroth and first cohomology groups of A by*

$$H^0(A) = \frac{\mathrm{Ker}(D|_A)}{\mathrm{Im}(N|_A)} = \frac{\mathrm{Ker}(D|_A)}{N(A)} \quad and \quad H^1(A) = \frac{\mathrm{Ker}(N|_A)}{\mathrm{Im}(D|_A)} = \frac{\mathrm{Ker}(N|_A)}{D(A)}.$$

Remarks. (1) These groups should more properly be denoted by $H^0(G, A)$ and $H^1(G, A)$, but since in our exposition G will be fixed, we do not include it.

(2) It is easy to define more generally groups $H^i(A)$ for $i \in \mathbb{Z}$. However, for finite cyclic groups it is not difficult to show that $H^{2i}(A) = H^0(A)$ and $H^{2i+1}(A) = H^1(A)$, so we do not need any more than the above two groups.

We leave to the reader to check that if f is a map of G-modules from A to B (in other words a $\mathbb{Z}[G]$-module homomorphism, or again an abelian group homomorphism such that $f(\sigma(a)) = \sigma(f(a))$ for all $a \in A$) then f naturally induces maps f_i from $H^i(A)$ to $H^i(B)$ for $i = 0$ and $i = 1$; in other words, H^i is a *functor*.

When one defines cohomology groups, the first thing is to prove that they satisfy a long cohomology exact sequence. Since here we have only two groups, this is reduced to the following exact hexagon lemma.

Proposition 4.4.17. *Let*

$$0 \longrightarrow A \xrightarrow{f} B \xrightarrow{g} C \longrightarrow 0$$

be an exact sequence of G-modules and G-module homomorphisms. There exist natural homomorphisms δ_i for $i = 0$ and $i = 1$ such that the following hexagon is exact at each group:

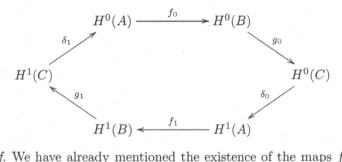

Proof. We have already mentioned the existence of the maps f_i and g_i. Let us define the maps δ_i. First let $\bar{c} \in H^0(C)$, so $c \in \mathrm{Ker}(D|_C)$. There exists $b \in B$ with $g(b) = c$; hence $g(D(b)) = D(g(b)) = D(c) = 0$, so that $D(b) \in \mathrm{Ker}(g) = \mathrm{Im}(f)$. It follows that there exists $a \in A$ such that $f(a) = D(b)$, so that $f(N(a)) = N(f(a)) = N(D(b)) = 0$; hence $N(a) \in \mathrm{Ker}(f) = \{0\}$, so finally $a \in \mathrm{Ker}(N|_A)$. We can thus define δ_0 by sending $\bar{c}$ to the class of a in $H^1(A)$. Let us show that it is well defined. If $c' \in \mathrm{Ker}(D|_C)$ is another representative of $\bar{c}$ and $b' \in B$ and $a' \in A$ are as above, then $c' - c = N(c'')$ for some $c'' \in C$, so there exists $b'' \in B$ with

$$g(b' - b) = c' - c = N(g(b'')) = g(N(b'')) \ .$$

It follows that $b' - b - N(b'') \in \mathrm{Ker}(g) = \mathrm{Im}(f)$, so there exists $a'' \in A$ with $b' - b - N(b'') = f(a'')$; hence

$$f(a' - a) = f(a') - f(a) = D(b') - D(b) = D(f(a'')) = f(D(a'')) \ ,$$

and since f is injective we thus have $a' - a = D(a'')$. We deduce that the classes of a' and a in $H^1(A)$ are equal, as claimed. As usual, this immediately implies that δ_0 is a homomorphism.

In a similar way we define δ_1 by sending the class modulo $D(C)$ of $c \in \mathrm{Ker}(N|_C)$ to the class modulo $N(A)$ of $a \in \mathrm{Ker}(D|_A)$ such that $g(b) = c$ and $f(a) = N(b)$, and show that δ_1 is a homomorphism. The rest of the proof is standard diagram chasing (as in fact was the definition of the δ_i) and is left to the reader. $\square$

Definition 4.4.18. *Let A be a G-module, and assume that the groups $H^i(A)$ are finite. The* Herbrand *quotient $q(A)$ is defined by $q(A) = |H^0(A)|/|H^1(A)|$, and we say that $q(A)$ is defined in this case, otherwise undefined.*

Remark. The usual definition of the Herbrand quotient (for instance that of [Jan]) is $|H^1(A)|/|H^0(A)| = 1/q(A)$, but in view of the results that we shall prove, and also in analogy with the Euler characteristic, the present definition is more natural.

Proposition 4.4.19. *Let $0 \longrightarrow A \longrightarrow B \longrightarrow C \longrightarrow 0$ be an exact sequence of G-modules. If two of the quotients $q(A)$, $q(B)$, $q(C)$ are defined, so is the third, and we have $q(A)q(C) = q(B)$.*

Proof. By the exact hexagon lemma (Proposition 4.4.17) and the standard kernel–image formula we have

$$|H^1(C)| = |\mathrm{Ker}(\delta_1)||\mathrm{Im}(\delta_1)| = |\mathrm{Im}(g_1)||\mathrm{Im}(\delta_1)| \ .$$

If $q(A)$ is defined then $\mathrm{Im}(\delta_1)$ is a subgroup of the finite group $H^0(A)$, hence is finite, and if $q(B)$ is defined then $\mathrm{Im}(g_1)$ is the image of the finite group $H^1(B)$, hence is finite, so if both $q(A)$ and $q(B)$ are defined the group $H^1(C)$ is finite. Similarly the group $H^0(C)$ is finite, so that $q(C)$ is defined. Exactly analogous arguments prove the corresponding result for the two other pairs of groups.

Finally, the exact hexagon lemma and the kernel–image formula give

$$|H^0(A)| \cdot |H^0(C)| \cdot |H^1(B)| = |\mathrm{Im}(\delta_1)||\mathrm{Im}(f_0)| \cdot |\mathrm{Im}(g_0)||\mathrm{Im}(\delta_0)| \cdot |\mathrm{Im}(f_1)||\mathrm{Im}(g_1)| \ ,$$

while

$$|H^1(A)| \cdot |H^1(C)| \cdot |H^0(B)| = |\mathrm{Im}(f_1)||\mathrm{Im}(\delta_0)| \cdot |\mathrm{Im}(\delta_1)|||\mathrm{Im}(g_1)| \cdot |\mathrm{Im}(f_0)||\mathrm{Im}(g_0)|$$

$$= |H^0(A)| \cdot |H^0(C)| \cdot |H^1(B)| \ ,$$

which is equivalent to $q(A)q(C) = q(B)$. $\square$

Corollary 4.4.20. *If $A \subset B$ are G-modules and the index $[B : A]$ is finite then $q(A) = q(B)$ whenever either one is defined.*

Proof. If we set $C = B/A$, by the proposition we must show that $q(C) = 1$ when C is finite. But in that case

$$q(C) = \frac{|\operatorname{Ker}(D|_C)/N(C)|}{|\operatorname{Ker}(N|_C)/D(C)|} = \frac{|\operatorname{Im}(D|_C)||\operatorname{Ker}(D|_C)|}{|\operatorname{Ker}(N|_C)||\operatorname{Im}(N|_C)|} = \frac{|C|}{|C|} = 1 ,$$

proving our claim. □

The fundamental example that we need in the text is the computation of the cohomology of the unit group of a $\mathfrak{P}$-adic extension. We first need the following result of independent interest concerning *permutation modules*, defined as follows. Let R be an integral domain of characteristic 0, let d be a divisor of $n = |G|$, and let $A = R^d$ be the free R-module on d generators. We can define a natural action of G on A by setting $\sigma(e_i) = e_{i+1}$, where $(e_i)_{0 \leqslant i < d}$ is the canonical basis of A, and where here and elsewhere the indices are considered modulo d. Clearly σ^d acts as the identity on A, and we define $G_A = \langle \sigma^d \rangle$, which is the (unique) subgroup of index d in G.

Proposition 4.4.21. *Set $m = n/d$ and assume that R/mR is finite. Then $q(A)$ is defined and equal to $|R/mR|$. In particular, when $R = \mathbb{Z}$ we have $q(A) = m = |G_A|$.*

Proof. We have (again with indices modulo d)

$$\sigma^k \left(\sum_i n_i e_i \right) = \sum_i n_{i-k} e_i ,$$

hence

$$N \left(\sum_i n_i e_i \right) = \sum_{0 \leqslant i < n} n_i \sum_i e_i = m \sum_{0 \leqslant i < d} n_i \sum_i e_i .$$

It follows that

$$\operatorname{Ker}(N) = \left\{ \sum_i n_i e_i / \sum_{0 \leqslant i < d} n_i = 0 \right\}, \quad \operatorname{Im}(N) = mR \sum_i e_i .$$

Furthermore, $\sum_i n_i e_i \in \operatorname{Ker}(D)$ if and only if $n_i = n_{i-1}$ for all i, hence $\operatorname{Ker}(D) = R \sum_i e_i$. Finally, we always have $\operatorname{Im}(D) \subset \operatorname{Ker}(N)$, but conversely if $a = \sum_i n_i e_i \in \operatorname{Ker}(N)$ it is immediately checked that $a = D(a')$ for $a' = \sum_i p_i e_i$ with $p_i = \sum_{1 \leqslant j \leqslant i} n_j$, hence $\operatorname{Im}(D) = \operatorname{Ker}(N)$. Thus $H^0(A) = R/mR$, $H^1(A) = 0$ and the result follows. □

Let us now come to our fundamental situation. Let L/K be an extension of number fields, let $\mathfrak{p}$ be a prime ideal of K and $\mathfrak{P}$ be a prime ideal of L above $\mathfrak{p}$. From Corollary 4.1.27 we know that $L_\mathfrak{P}/K_\mathfrak{p}$ is an extension of degree $e(\mathfrak{P}/\mathfrak{p})f(\mathfrak{P}/\mathfrak{p})$. If this extension is Galois with Galois group G the group $U_\mathfrak{P}$ of $\mathfrak{P}$-adic units is evidently a G-module, and our goal is to compute the cohomology groups of $U_{\mathfrak{P}}$. Note that although we consider a local situation obtained by localizing a global situation, the result is purely local, and we could state it without reference to the global fields K and L. The result is as follows.

Theorem 4.4.22. *If $L_{\mathfrak{P}}/K_{\mathfrak{p}}$ is a cyclic extension we have*

$$[U_{\mathfrak{p}} : \mathcal{N}_{L_{\mathfrak{P}}/K_{\mathfrak{p}}}(U_{\mathfrak{P}})] = |H^0(U_{\mathfrak{P}})| = |H^1(U_{\mathfrak{P}})| = e(\mathfrak{P}/\mathfrak{p}) \,,$$

hence $q(U_{\mathfrak{P}}) = 1$.

Proof. Let $G = \mathrm{Gal}(L_{\mathfrak{P}}/K_{\mathfrak{p}})$ and let σ be a generator of G. We first note that if $A = U_{\mathfrak{P}}$ then $N(A) = \mathcal{N}_{L_{\mathfrak{P}}/K_{\mathfrak{p}}}(U_{\mathfrak{P}})$, and $\mathrm{Ker}(D|_A)$ is the set of $\mathfrak{P}$-adic units of $L_{\mathfrak{P}}$ invariant by σ hence by G, and is thus equal to $U_{\mathfrak{P}} \cap K_{\mathfrak{p}} = U_{\mathfrak{p}}$, so the first equality is clear by definition of H^0.

By definition of the Herbrand quotient the second equality is equivalent to $q(U_{\mathfrak{P}}) = 1$. By Corollary 4.3.13, we know that U_n is a subgroup of finite index of $U_{\mathfrak{P}}$ such that $U_n \simeq \mathfrak{P}^n \mathbb{Z}_{\mathfrak{P}}$ as soon as $n \geqslant z(\mathfrak{P})$. By Corollary 4.4.20 for such an n we thus have

$$q(U_{\mathfrak{P}}) = q(U_n) = q(\mathfrak{P}^n \mathbb{Z}_{\mathfrak{P}}) = q(\mathbb{Z}_{\mathfrak{P}}) \,,$$

since $\mathfrak{P}^n \mathbb{Z}_{\mathfrak{P}}$ has finite index in $\mathbb{Z}_{\mathfrak{P}}$. Now by the normal basis theorem (Theorem 3.2.12) there exists $\alpha \in L_{\mathfrak{P}}$, which after multiplication by a suitable element of $\mathbb{Z}_{\mathfrak{p}}$ we may assume to be in $\mathbb{Z}_{\mathfrak{P}}$, such that the $\sigma^k(\alpha)$ for $0 \leqslant k < |G|$ form a $K_{\mathfrak{p}}$-basis of $L_{\mathfrak{P}}$. If we set

$$B = \sum_{0 \leqslant k < |G|} \sigma^k(\alpha) \mathbb{Z}_{\mathfrak{p}}$$

then B is a free $\mathbb{Z}_{\mathfrak{p}}$-module with the same rank as $\mathbb{Z}_{\mathfrak{P}}$ and is a submodule of $\mathbb{Z}_{\mathfrak{P}}$, hence it has finite index. It follows once again from Corollary 4.4.20 that $q(\mathbb{Z}_{\mathfrak{P}}) = q(B)$, so that $q(U_{\mathfrak{P}}) = q(B)$. However, B is clearly a permutation module in the sense of the above example, so by the above proposition we have $q(B) = |\mathbb{Z}_{\mathfrak{p}}/m\mathbb{Z}_{\mathfrak{p}}|$ with $m = |G|/\dim(B) = |G|/|G| = 1$, hence $q(B) = 1$ as claimed, proving the second equality.

For the third equality we first appeal to Hilbert's Theorem 90 (Proposition 3.2.4) which implies in particular that

$$\mathrm{Ker}(N|_{U_{\mathfrak{P}}}) = U_{\mathfrak{P}} \cap D(L_{\mathfrak{P}}^*) \,.$$

Let Π be a uniformizer of $\mathfrak{P}$. Since σ is a bijection of $\mathfrak{P}$ onto itself $\sigma(\Pi)$ is also a uniformizer, in other words $\sigma(\Pi) = \Pi\omega$ for some $\omega \in U_{\mathfrak{P}}$. It follows that for any $x \in L_{\mathfrak{P}}^*$, if we write $x = \Pi^v y$ with $y \in U_{\mathfrak{P}}$ then

$$D(x) = \frac{x}{\sigma(x)} = \frac{\Pi^v y}{\Pi^v \omega^v \sigma(y)} \in U_{\mathfrak{P}} \,.$$

We thus have $D(L_{\mathfrak{P}}^*) \subset U_{\mathfrak{P}}$, so that $\mathrm{Ker}(N|_{U_{\mathfrak{P}}}) = D(L_{\mathfrak{P}}^*)$. On the other hand, since by Galois theory the kernel of the map D on $L_{\mathfrak{P}}^*$ is $K_{\mathfrak{p}}^*$ we have $D(L_{\mathfrak{P}}^*) \simeq L_{\mathfrak{P}}^*/K_{\mathfrak{p}}^*$ and

$$D(U_{\mathfrak{P}}) \simeq \frac{U_{\mathfrak{P}}}{K_{\mathfrak{p}}^* \cap U_{\mathfrak{P}}} \simeq \frac{K_{\mathfrak{p}}^* U_{\mathfrak{P}}}{K_{\mathfrak{p}}^*} ,$$

hence

$$H^1(U_{\mathfrak{P}}) = \frac{D(L_{\mathfrak{P}}^*)}{D(U_{\mathfrak{P}})} \simeq \frac{L_{\mathfrak{P}}^*}{K_{\mathfrak{p}}^* U_{\mathfrak{P}}} .$$

Finally, if we let π be a uniformizer of $\mathfrak{p}$ then by definition of the ramification index we have $\pi = \Pi^{e(\mathfrak{P}/\mathfrak{p})} u$ for some $u \in U_{\mathfrak{P}}$. It follows that

$$L_{\mathfrak{P}}^* = \pi^{\mathbb{Z}} \times U_{\mathfrak{P}} \qquad \text{and} \qquad K_{\mathfrak{p}}^* U_{\mathfrak{P}} = \pi^{e(\mathfrak{P}/\mathfrak{p})\mathbb{Z}} \times U_{\mathfrak{P}} ,$$

so that $|H^1(U_{\mathfrak{P}})| = e(\mathfrak{P}/\mathfrak{p})$ as claimed, finishing the proof of the theorem. $\square$

Corollary 4.4.23. *If $L_{\mathfrak{P}}/K_{\mathfrak{p}}$ is unramified then the norm is surjective on units.*

Proof. Clear since $L_{\mathfrak{P}}/K_{\mathfrak{p}}$ is cyclic in that case. $\square$

Proposition 4.4.24. *If $L_{\mathfrak{P}}/K_{\mathfrak{p}}$ is a cyclic extension then*

$$[K_{\mathfrak{p}}^* : \mathcal{N}_{L_{\mathfrak{P}}/K_{\mathfrak{p}}}(L_{\mathfrak{P}}^*)] = [L_{\mathfrak{P}} : K_{\mathfrak{p}}] = e(\mathfrak{P}/\mathfrak{p}) f(\mathfrak{P}/\mathfrak{p}) .$$

Proof. With the above notation we have $L_{\mathfrak{P}}^* = \pi^{\mathbb{Z}} \times U_{\mathfrak{P}}$, hence $L_{\mathfrak{P}}^*/U_{\mathfrak{P}}$ is isomorphic to $\mathbb{Z}$ and G acts trivially on it (since $\sigma(\pi)/\pi \in U_{\mathfrak{P}}$). It follows from Proposition 4.4.21 that $q(L_{\mathfrak{P}}^*/U_{\mathfrak{P}}) = |G|$. Since $q(U_{\mathfrak{P}}) = 1$ we have $q(L_{\mathfrak{P}}^*) = |G|$. On the other hand, by definition we have

$$q(L_{\mathfrak{P}}^*) = \frac{|H^0(L_{\mathfrak{P}}^*)|}{|H^1(L_{\mathfrak{P}}^*)|} = |H^0(L_{\mathfrak{P}}^*)|$$

since $H^1(L_{\mathfrak{P}}^*) = 1$ by Hilbert's Theorem 90. It follows that

$$|H^0(L_{\mathfrak{P}}^*)| = |G| = [L_{\mathfrak{P}} : K_{\mathfrak{p}}] = e(\mathfrak{P}/\mathfrak{p}) f(\mathfrak{P}/\mathfrak{p}) ,$$

proving the proposition since by definition $H^0(L_{\mathfrak{P}}^*) = K_{\mathfrak{p}}^*/\mathcal{N}_{L_{\mathfrak{P}}/K_{\mathfrak{p}}}(L_{\mathfrak{P}}^*)$. $\square$

Note that Theorem 4.4.22 is in fact valid more generally for Abelian extensions, but we will omit the proof, which originates in local class field theory:

Theorem 4.4.25. *If $L_{\mathfrak{P}}/K_{\mathfrak{p}}$ is an Abelian extension we have*

$$[U_{\mathfrak{p}} : \mathcal{N}_{L_{\mathfrak{P}}/K_{\mathfrak{p}}}(U_{\mathfrak{P}})] = e(\mathfrak{P}/\mathfrak{p}) .$$

4.4.5 Characterization of Unramified Extensions

Thanks to Corollary 4.4.14, we need only study unramified and totally ramified extensions. In this subsection, we begin with unramified extensions. We have already seen above that unramified extensions are in one-to-one correspondence with extensions of residue fields. Since these are finite fields whose extensions are well known, we can make this more precise.

Lemma 4.4.26. *For every n, up to isomorphism there exists exactly one unramified extension $\mathcal{K}$ of $\mathbb{Q}_p$ of degree n, which is the splitting field of $X^q - X$ over $\mathbb{Q}_p$, where $q = p^n$.*

Proof. The residue field of $\mathbb{Q}_p$ is the finite field $\mathbb{F}_p \simeq \mathbb{Z}/p\mathbb{Z}$. Up to isomorphism, for every n there exists one extension k of $\mathbb{F}_p$ of degree n: it has $q = p^n$ elements and is the splitting field of $X^q - X$ over $\mathbb{F}_p$. By Corollary 4.4.15 there exists exactly one unramified extension $\mathcal{K}$ of $\mathbb{Q}_p$ in the algebraic closure of $\mathbb{Q}_p$ whose residue field is k, and $\mathcal{K} = K_\mathfrak{p}$ for some K and $\mathfrak{p}$ by Proposition 4.4.9. By Theorem 4.4.11, we know that $[\mathcal{K} : \mathbb{Q}_p] = n$. We claim that $\mathcal{K}$ is the splitting field of $f(X) = X^q - X$ over $\mathbb{Q}_p$. Indeed, since $f'(X) = qX^{q-1} - 1$, for all $\alpha \in \mathbb{Z}_p$ we have $|f'(\alpha)| = 1$. Therefore, by Hensel's Lemma 4.1.37, for every $a \in \mathbb{Z}_\mathfrak{p}/\mathfrak{p}\mathbb{Z}_\mathfrak{p}$ there exists a representative α in $\mathbb{Z}_\mathfrak{p}$ such that $f(\alpha) = 0$. It follows that $f(X)$ splits in $\mathcal{K}$. Finally, the splitting field of $f(X)$ over $\mathbb{Q}_p$ cannot be smaller than $\mathcal{K}$ since its residue field must contain at least q elements. $\qquad\square$

Remarks. (1) The element α corresponding to the class a in the above proof is nothing else than the Teichmüller character $\bar{\omega}(a)$ given by Corollary 4.3.5. It is called the *Teichmüller representative* of a.

(2) As mentioned above, if $q = p^n$ it is quite reasonable to use the notation $\mathbb{Q}_q$ to denote the unique (up to isomorphism) unramified extension of degree n of $\mathbb{Q}_p$. Its ring of integers is denoted by $\mathbb{Z}_q$, and its residue field $\mathbb{Z}_q/\mathfrak{m}$ will be isomorphic to $\mathbb{F}_q$, where $\mathfrak{m}$ is the maximal ideal of elements of norm strictly less than 1. We will, however, not make use of this notation in this book.

Now that we have proved the above lemma for $\mathbb{Q}_p$, we can generalize it to any p-adic field.

Corollary 4.4.27. *Let $\mathcal{K} = K_\mathfrak{p}$ be any p-adic field, and let $q = \mathcal{N}\mathfrak{p}$ be the cardinality of its residue field. For every n, up to isomorphism there exists exactly one unramified extension $\mathcal{L}$ of $\mathcal{K}$ of degree n, which is the splitting field of $X^Q - X$ over $\mathcal{K}$, where $Q = q^n$. The extension $\mathcal{L}/\mathcal{K}$ is Galois with cyclic Galois group, and there is a generator σ of $\mathrm{Gal}(\mathcal{L}/\mathcal{K})$ (also called the Frobenius automorphism of $\mathcal{L}/\mathcal{K}$) that induces the Frobenius automorphism $x \mapsto x^q$ on the residue fields.*

Proof. Once again by Corollary 4.4.15 there exists exactly one unramified extension $\mathcal{L}$ of $\mathcal{K}$ in the algebraic closure of $\mathcal{K}$ whose residue field is $\mathbb{F}_Q$, and $\mathcal{L} = L_{\mathfrak{P}}$ for some L and $\mathfrak{P}$ by Proposition 4.4.9. It follows from the above lemma that $\mathcal{L}$ must contain the splitting field $\mathcal{M}$ of $X^Q - X$ over $\mathbb{Q}_p$, so that $\mathcal{L}$ is equal to the compositum of $\mathcal{M}$ with $\mathcal{K}$. It follows that $\mathcal{L}$ is the splitting field of $X^Q - X$ over $\mathcal{K}$.

Every splitting field is Galois. The theory of finite fields tells us that the Galois group of the residue field extensions of $\mathcal{L}$ and $\mathcal{K}$ is cyclic and generated by the Frobenius $x \mapsto x^q$. Since $\mathcal{L}/\mathcal{K}$ is unramified, it is easy to see that its Galois group is canonically isomorphic under the reduction map to that of the corresponding residue fields, so the corollary follows. □

Corollary 4.4.28. *The unramified closure $\mathcal{K}^u$ (see Corollary 4.4.15) is obtained by adjoining to $\mathcal{K}$ the mth roots of unity for all m prime to the residue field characteristic p.*

Proof. By the previous corollary, $\mathcal{K}^u$ is obtained by adjoining the $(q^n - 1)$st roots of unity for all $n \geqslant 1$. Since for any m prime to p, hence to q, there exists n such that $q^n \equiv 1 \pmod{m}$ (the order of the class of q in $(\mathbb{Z}/m\mathbb{Z})^*$), the corollary follows. □

Corollary 4.4.29. *As above let $\mathcal{K} = K_{\mathfrak{p}}$, $q = \mathcal{N}\mathfrak{p}$, $\mathcal{L}$ the unique unramified extension of $\mathcal{K}$ of degree n, and $Q = q^n$, so that $\mathcal{L}$ is the splitting field of $X^Q - X$ over $\mathcal{K}$. If σ is the Frobenius automorphism of $\mathcal{L}/\mathcal{K}$ then for any a such that $a^Q = a$ we have $\sigma(a) = a^q$.*

Proof. Recall first that $\sigma(x) = x^q$ only at the level of the residue field. Let μ_{Q-1} be the group of $(Q-1)$st roots of unity in $\mathcal{L}$. By definition $|\mu_{Q-1}| = Q-1$ and it is a finite subgroup of $\mathcal{L}^*$, so it is cyclic generated by an element a. Thus $\sigma(a)^{Q-1} = \sigma(1) = 1$, so that $\sigma(a) = a^k$ for some $k \in [0, Q-2]$ since a is a generator. On the other hand, $\sigma(a) \equiv a^q \pmod{\mathfrak{P}}$. It follows that a^k and a^q are roots of unity of order dividing $Q-1$, hence prime to p, which are congruent modulo $\mathfrak{P}$. Thus by Proposition 4.3.2 applied to $\mathcal{L}$ we deduce that they must be equal, hence that $\sigma(a) = a^q$. Since this is true for the generator a it is true for all of μ_{Q-1}. □

Remark. The above results are of a theoretical nature, but are also very explicit. If $\mathcal{K} = K_{\mathfrak{p}}$ is a p-adic field and $n \geqslant 1$, to find an equation for the unramified extension of degree n of $\mathcal{K}$, we first search for a monic polynomial $\overline{P}(X) \in (\mathbb{Z}_K/\mathfrak{p})[X]$ of degree n that is irreducible in $(\mathbb{Z}_K/\mathfrak{p})[X]$. A root of $\overline{P}(X)$ defines the finite field of degree n over $\mathbb{Z}_K/\mathfrak{p}$. Therefore, by the above results, if $P(X) \in \mathbb{Z}_K[X]$ is any monic lift of $\overline{P}(X)$, then $\mathcal{L} = \mathcal{K}(\alpha)$ is the unramified extension of degree n of $\mathcal{K}$, where α is a root of $P(X) = 0$.

Examples. Following this procedure, for $p \leqslant 5$ and $n \leqslant 5$, the unramified extension of degree n of $\mathbb{Q}_p$ is equal to $\mathbb{Q}_p(\alpha)$, where α is a root of $P_n(X) = 0$, with:

- For $p = 2$, $P_2(X) = X^2 + X + 1$, $P_3(X) = X^3 + X + 1$, $P_4(X) = X^4 + X + 1$, $P_5(X) = X^5 + X^2 + 1$ (note that $X^5 + X + 1 \equiv (X^2 + X + 1)(X^3 + X^2 + 1)$ (mod 2)).
- For $p = 3$, $P_2(X) = X^2 + 1$, $P_3(X) = X^3 - X + 1$, $P_4(X) = X^4 + X - 1$, $P_5(X) = X^5 - X + 1$.
- For $p = 5$, $P_2(X) = X^2 + 2$, $P_3(X) = X^3 + X + 1$, $P_4(X) = X^4 + 2$, $P_5(X) = X^5 - X - 1$.

See Exercises 28 and 29 for simple methods for explicitly computing the Frobenius automorphism.

4.4.6 Properties of Unramified Extensions

As above, let K be a number field, L a finite extension of K, $\mathfrak{p}$ a (nonzero) prime ideal of K, and $\mathfrak{P}$ a prime ideal of L above $\mathfrak{p}$. We denote by $k = \mathbb{Z}_\mathfrak{p}/\mathfrak{p}\mathbb{Z}_\mathfrak{p}$ and $l = \mathbb{Z}_\mathfrak{P}/\mathfrak{P}\mathbb{Z}_\mathfrak{P}$ the corresponding residue fields. By the normal basis theorem (Theorem 3.2.12), there exists a normal basis $\overline{\alpha}$, i.e., an element of l such that together with its Galois conjugates, $\overline{\alpha}$ generates l as a k-vector space.

Proposition 4.4.30. *Keep the above notation, and assume that $L_\mathfrak{P}/K_\mathfrak{p}$ is an unramified extension of $\mathfrak{p}$-adic fields. Then any lift of a normal basis $\overline{\alpha}$ of l over k gives a normal basis of $\mathbb{Z}_\mathfrak{P}$ over $\mathbb{Z}_\mathfrak{p}$. In particular, such a normal basis exists; in other words, $\mathbb{Z}_\mathfrak{P}$ is a free $\mathbb{Z}_\mathfrak{p}[G]$-module of rank 1, where $G = \mathrm{Gal}(L_\mathfrak{P}/K_\mathfrak{p})$ and $\mathbb{Z}_\mathfrak{p}[G]$ denotes the ring of formal $\mathbb{Z}_\mathfrak{p}$-linear combinations of elements of G.*

Proof. We simply apply Nakayama's Lemma 4.3.15 to the $\mathbb{Z}_\mathfrak{p}$-module $M = \mathbb{Z}_\mathfrak{P}/(\mathbb{Z}_\mathfrak{p}[G]\alpha)$, where α is a lift of a normal basis $\overline{\alpha}$ of l over k. Note that we know in advance the existence of a normal basis by the normal basis theorem itself. The point of the proposition is to show that it can be obtained by lifting the normal basis of the residue field. $\square$

Corollary 4.4.31. *Under the same assumptions, the trace map $\mathrm{Tr}_{L_\mathfrak{P}/K_\mathfrak{p}}$ is a surjective map from $\mathbb{Z}_\mathfrak{P}$ to $\mathbb{Z}_\mathfrak{p}$.*

Proof. Again set $G = \mathrm{Gal}(L_\mathfrak{P}/K_\mathfrak{p})$, and let $\alpha \in \mathbb{Z}_\mathfrak{P}$ be a normal basis of $\mathbb{Z}_\mathfrak{P}$ over $\mathbb{Z}_\mathfrak{p}$. I claim that under the above isomorphism between $\mathbb{Z}_\mathfrak{P}$ and $\mathbb{Z}_\mathfrak{p}[G]$ the set of fixed points of $\mathbb{Z}_\mathfrak{P}$ under G is the image of the trace map. Since by Galois theory this is equal to $\mathbb{Z}_\mathfrak{P} \cap K_\mathfrak{p} = \mathbb{Z}_\mathfrak{p}$, this will prove the corollary. Indeed, $\sum_{\sigma \in G} x_\sigma \sigma(\alpha)$ fixed by G means that for all $\tau \in G$ we have

$$0 = \sum_{\sigma \in G} x_\sigma (\tau(\sigma(\alpha)) - \sigma(\alpha)) = \sum_{\sigma \in G} (x_{\tau^{-1}\sigma} - x_\sigma)\sigma(\alpha) \ .$$

Since the $\sigma(\alpha)$ are linearly independent, this means that $x_\sigma = x_{\tau^{-1}\sigma}$ for all σ and τ, in other words that x_σ is independent of σ, proving our claim.   □

Since an unramified extension is automatically cyclic the above result concerning the trace is also valid for the norm on units, as we have seen above (Theorem 4.4.22). However, considering its importance we restate it here.

Theorem 4.4.32. *Let $L_{\mathfrak{P}}/K_{\mathfrak{p}}$ be an unramified extension of $\mathfrak{p}$-adic fields. Denote by $U_{\mathfrak{p}}$ the group of $\mathfrak{p}$-adic units of $K_{\mathfrak{p}}$, and similarly $U_{\mathfrak{P}}$ for $L_{\mathfrak{P}}$. Then $\mathcal{N}_{L_{\mathfrak{P}}/K_{\mathfrak{p}}}(U_{\mathfrak{P}}) = U_{\mathfrak{p}}$; in other words, the norm is surjective on $\mathfrak{p}$-adic units.*

Since we know that unramified extensions are in bijective correspondence with extensions of the residue fields, these surjectivity results on the trace and norm also follow from Propositions 2.4.11 and 2.4.12.

Corollary 4.4.33. *Let π be a uniformizer of $\mathfrak{p}$. We have*

$$\mathcal{N}_{L_{\mathfrak{P}}/K_{\mathfrak{p}}}(L_{\mathfrak{P}}^*) = \pi^{f(\mathfrak{P}/\mathfrak{p})\mathbb{Z}} U_{\mathfrak{p}} \ .$$

In particular, the map that sends π to the Frobenius automorphism induces an isomorphism

$$K_{\mathfrak{p}}^*/\mathcal{N}_{L_{\mathfrak{P}}/K_{\mathfrak{p}}}(L_{\mathfrak{P}}^*) \simeq \mathrm{Gal}(L_{\mathfrak{P}}/K_{\mathfrak{p}}) \ .$$

Proof. Since $L_{\mathfrak{P}}/K_{\mathfrak{p}}$ is unramified, a uniformizer π of $\mathfrak{p}$ is also one of $\mathfrak{P}$. It follows that $L_{\mathfrak{P}}^* = \pi^{\mathbb{Z}} U_{\mathfrak{P}}$; hence $\mathcal{N}_{L_{\mathfrak{P}}/K_{\mathfrak{p}}}(L_{\mathfrak{P}}^*) = \pi^{f(\mathfrak{P}/\mathfrak{p})\mathbb{Z}} U_{\mathfrak{p}}$ by the above theorem and the fact that $[L_{\mathfrak{P}} : K_{\mathfrak{p}}] = f(\mathfrak{P}/\mathfrak{p})$, proving the first assertion. For the second, consider the map ω sending $\pi^i u$ with $i \in \mathbb{Z}$ and $u \in U_{\mathfrak{p}}$ to $\sigma^i \in \mathrm{Gal}(L_{\mathfrak{P}}/K_{\mathfrak{p}})$, where σ is the Frobenius automorphism, which is clearly a surjective group homomorphism since σ generates the Galois group. We have $\omega(\pi^i u) = 1 \in \mathrm{Gal}(L_{\mathfrak{P}}/K_{\mathfrak{p}})$ if and only if $i \equiv 0 \pmod{f(\mathfrak{P}/\mathfrak{p})}$, which is the order of σ in the Galois group. It follows that

$$\mathrm{Ker}(\omega) = \pi^{f(\mathfrak{P}/\mathfrak{p})\mathbb{Z}} U_{\mathfrak{p}} = \mathcal{N}_{L_{\mathfrak{P}}/K_{\mathfrak{p}}}(L_{\mathfrak{P}}^*)$$

by the first part, proving the corollary.   □

Note that in Proposition 4.4.24 (where we did not assume that the extension was unramified) we obtained only an equality of cardinalities, not an isomorphism.

4.4.7 Totally Ramified Extensions

Since we have seen that any finite extension of a $\mathfrak{p}$-adic field is a totally ramified extension of an unramified extension, it remains to study totally ramified extensions.

Proposition 4.4.34. *An extension $L_{\mathfrak{P}}/K_{\mathfrak{p}}$ is totally ramified if and only if $L_{\mathfrak{P}} = K_{\mathfrak{p}}(\alpha)$, where α is a root of an Eisenstein polynomial (see Proposition 4.1.35).*

Proof. We denote by $|\ |$ the absolute value on $L_{\mathfrak{P}}$ and on $K_{\mathfrak{p}}$ (we know that they are the same). First, if α is a root of an Eisenstein polynomial $f(X) = X^n + f_{n-1}X^{n-1} + \cdots + f_0 \in \mathbb{Z}_{\mathfrak{p}}[X]$, then by the ultrametric inequality, since $|f_j| \leqslant |f_0| < 1$ we have clearly $|\alpha|^n = |f_0|$ (indeed, $|\alpha| \geqslant 1$ together with $|\alpha|^n = |f_{n-1}\alpha^{n-1} + \cdots + f_0| < |\alpha|^{n-1}$ leads to a contradiction, so that $|\alpha| < 1$, and hence $|f_i\alpha^i| < |f_i|$ for all $i > 0$, so that $|\alpha|^n = |f_{n-1}\alpha^{n-1} + \cdots + f_0| = |f_0|$). It follows that $v_{\mathfrak{P}}(f_0) = e(\mathfrak{P}/\mathfrak{p})v_p(f_0) = e(\mathfrak{P}/\mathfrak{p}) = nv_{\mathfrak{P}}(\alpha) \geqslant n$; hence $e(\mathfrak{P}/\mathfrak{p}) = n$ and $K_{\mathfrak{p}}(\alpha)$ is totally ramified.

Conversely, assume that $L_{\mathfrak{P}}/K_{\mathfrak{p}}$ is totally ramified of degree n, and let Π be a uniformizer of $\mathfrak{P}$. Then the Π^j for $0 \leqslant j < n$ are $K_{\mathfrak{p}}$-linearly independent, since each nonzero term $x_j\Pi^j$ in any relation $\sum_{0 \leqslant j < n} x_j\Pi^j = 0$ is unique with $\mathfrak{P}$-adic valuation congruent to j modulo $e(\mathfrak{P}/\mathfrak{p}) = n$. Since the degree of the extension is n, Π^n is a $K_{\mathfrak{p}}$-linear combination of the Π^j for $0 \leqslant j < n$; in other words, there exist $f_j \in K_{\mathfrak{p}}$ such that $f(\Pi) = 0$, with

$$f(X) = X^n + f_{n-1}X^{n-1} + \cdots + f_0 \ .$$

I claim that this is an Eisenstein polynomial. Indeed, in a nontrivial sum equal to 0 the minimal valuation must occur at least twice. Since $\Pi^n + f_{n-1}\Pi^{n-1} + \cdots + f_0 = 0$ and since the valuations of $f_i\Pi^i$ are distinct modulo n, it follows that Π^n and f_0 must be the two terms with minimal and equal valuation. Thus $v_{\mathfrak{P}}(f_0) = nv_p(f_0) = v_{\mathfrak{P}}(\Pi^n) = n$; hence $v_p(f_0) = 1$, and for $1 \leqslant i \leqslant n-1$ we have $v_{\mathfrak{P}}(f_i\Pi^i) = nv_p(f_i) + i \geqslant n$; hence $v_p(f_i) \geqslant 1$, proving my claim and the proposition. $\qquad\square$

Example. In Corollary 4.3.24 we have seen that up to isomorphism there exist seven quadratic extensions of $\mathbb{Q}_2$. Among these we know that there exists a unique unramified extension, and it is easily checked that it is $\mathbb{Q}_2(\sqrt{5}) = \mathbb{Q}_2(\sqrt{-3})$ (these two fields are the same since $5/(-3) \equiv 1 \pmod 8$). The six others are ramified, hence totally ramified. We check that Eisenstein polynomials for $\mathbb{Q}_p(\sqrt{D})$ with $D = -1, -5, \pm 2, \pm 6$ can be taken equal to $X^2 + 2X + 2$, $X^2 + 2X + 6$, $X^2 \mp 2$, and $X^2 \mp 6$ respectively.

It is possible to be much more precise in the case of *tamely ramified* totally ramified extensions.

Theorem 4.4.35. *Let $L_{\mathfrak{P}}/K_{\mathfrak{p}}$ be a totally ramified extension of degree n that is tamely ramified (in other words such that $p \nmid n$). There exists a generator π of $\mathfrak{p}\mathbb{Z}_{K_{\mathfrak{p}}}$ such that $L_{\mathfrak{P}} = K_{\mathfrak{p}}(\alpha)$, where α is a root of $X^n - \pi = 0$.*

Note that since $v_p(\pi) = 1$, $X^n - \pi$ is evidently an Eisenstein polynomial, but of a very special kind.

Proof. Set for simplicity $\mathcal{K} = K_{\mathfrak{p}}$ and $\mathcal{L} = L_{\mathfrak{P}}$, and let π_K and π_L be any generators of $\mathfrak{p}\mathbb{Z}_{\mathcal{K}}$ and $\mathfrak{P}\mathbb{Z}_{\mathcal{L}}$ respectively. Since the extension is totally ramified we have $v_{\mathfrak{P}}(\pi_K) = n$, so $u = \pi_L^n/\pi_K$ is a $\mathfrak{P}$-adic unit of $\mathcal{L}$. Again since the extension is totally ramified we have $f(\mathfrak{P}/\mathfrak{p}) = 1$ so the residue fields are the same. It follows that there exists a $\mathfrak{p}$-adic unit ζ of $\mathcal{K}$ such that $u \equiv \zeta \pmod{\mathfrak{P}\mathbb{Z}_{\mathcal{L}}}$. Thus $\pi_L^n = \pi_K u$ with $u = \zeta + \pi_L v$ for some $v \in \mathbb{Z}_{\mathcal{L}}$; hence if we set $\pi = \zeta\pi_K$ then π is still a generator of $\mathfrak{p}\mathbb{Z}_{\mathcal{K}}$ and $\pi_L^n = \pi + \pi\pi_L w$ with $w = v\zeta^{-1} \in \mathbb{Z}_{\mathcal{L}}$. Let us show that π is suitable. Since $X^n - \pi$ is an Eisenstein polynomial, it is irreducible in $\mathcal{K}[X]$. In addition, $f(\pi_L) = \pi_L^n - \pi = \pi\pi_L w$, so that $|f(\pi_L)| < |\pi|$. Let us factor f in an algebraic closure of $\mathcal{K}$ as $f(X) = X^n - \pi = \prod_{1 \leqslant i \leqslant n}(X - \alpha_i)$, where of course $\prod_{1 \leqslant i \leqslant n} \alpha_i = (-1)^{n-1}\pi$. By Proposition 4.4.6 (1) all the $|\alpha_i|$ are equal, to c say. It follows that $c^n = |\pi|$; hence $|\alpha_i| = c = |\pi|^{1/n} = |\pi_L|$, so that $|\pi_L - \alpha_i| \leqslant \max(|\pi_L|, |\alpha_i|) \leqslant |\pi_L|$. On the other hand,

$$\left| \prod_{1 \leqslant i \leqslant n} (\pi_L - \alpha_i) \right| = |f(\pi_L)| < |\pi| = |\pi_L|^n .$$

It follows that there exists i such that $|\pi_L - \alpha_i| < |\pi_L|$, and we set $\alpha = \alpha_i$. All the roots α_j have the form $\alpha_j = \zeta_j\alpha$, where the ζ_j are the nth roots of unity. Now since the extension is tamely ramified we have $p \nmid n$, so that $\zeta_j - 1$ is a unit when $\zeta_j \neq 1$ by Proposition 4.3.2. Thus when $j \neq i$ we have $|\zeta_j - 1| = 1$; hence $|\alpha_j - \alpha| = |\alpha||\zeta_j - 1| = |\alpha| = c = |\pi_L|$ and $|\pi_L - \alpha| < |\pi_L| = |\alpha_j - \alpha|$. It follows from Krasner's lemma (Proposition 4.4.6 (3)) that $\mathcal{K}(\alpha) \subset \mathcal{K}(\pi_L)$. Since $\mathcal{K}(\pi_L) \subset \mathcal{L}$ and $[\mathcal{L} : \mathcal{K}] = n = [\mathcal{K}(\alpha) : \mathcal{K}]$, this inclusion must be an equality, proving the theorem. $\square$

For instance, if $\mathcal{K} = \mathbb{Q}_p$ and $\mathcal{L} = \mathbb{Q}_p(\zeta_p)$ is the field obtained by adjoining a pth root of unity, which is a totally and tamely ramified extension of degree $p - 1$ of $\mathbb{Q}_p$, we have seen in Proposition 4.3.18 that $\mathcal{L} = \mathbb{Q}_p((-p)^{1/(p-1)})$, so we can choose $\pi = -p$ as a generator of $p\mathbb{Z}_p$ satisfying the conditions of the theorem.

4.4.8 Analytic Representations of pth Roots of Unity

Proposition 4.4.36. *Let $\mathcal{K} = \mathbb{Q}_p(\zeta_p) = \mathbb{Q}_p((-p)^{1/(p-1)})$, and let $\mathfrak{p}$ be the prime ideal of $\mathcal{K}$ above p (generated by $\zeta - 1$ or by $(-p)^{1/(p-1)}$). For each element $\pi \in \mathcal{K}$ such that $\pi^{p-1} + p = 0$ there exists a unique primitive pth root of unity $\zeta_\pi \in \mathcal{K}$ such that $\zeta_\pi \equiv 1 + \pi \pmod{\mathfrak{p}^2}$, and the map $\pi \mapsto \zeta_\pi$ is a bijection between elements π as above and primitive pth roots of unity.*

Proof. In the proof of Theorem 4.3.18 (1) we have seen that if ζ is a primitive pth root of unity there exists π such that $\pi^{p-1} + p = 0$ with $(\zeta - 1)/\pi \equiv 1 \pmod{\mathfrak{p}}$, or equivalently, $\zeta \equiv 1 + \pi \pmod{\mathfrak{p}^2}$. Furthermore, if π_1 and π_2 satisfy this congruence then we have $\pi_2/\pi_1 \equiv 1 \pmod{\mathfrak{p}}$, while π_1/π_2 is a $(p-1)$st root of unity, so by Proposition 4.3.2 we have $\pi_1 = \pi_2$, proving that π is unique. We thus have a well-defined map $\zeta \mapsto \pi$ from primitive pth roots of unity to elements π satisfying $\pi^{p-1} + p = 0$. This map is clearly injective since when $\zeta_1 \neq \zeta_2$ we have $v_p(\zeta_1 - \zeta_2) = v_p(1 - \zeta_2/\zeta_1) = 1$. Since both sets have the same cardinality $p - 1$, it follows that our map is bijective. The map of the proposition is the inverse bijection.   □

Definition 4.4.37. *Let $\pi \in K$ be an element such that $\pi^{p-1} + p = 0$. The Dwork power series $D_\pi(X)$ associated with π is the power series defined formally by*

$$D_\pi(X) = \exp(\pi(X - X^p)) \, .$$

Proposition 4.4.38. (1) *The series $D_\pi(X)$ converges at least for $v_p(x) > -(p-1)/p^2$, and in particular has a radius of convergence strictly greater than 1.*
(2) *More precisely, if we set $D_\pi(X) = \sum_{k \geqslant 0} d_k X^k$ (with d_k depending on π), then $v_p(d_k) \geqslant k(p-1)/p^2$.*
(3) *We have $d_0 = 1$, $d_1 = \pi$, and $v_{\mathfrak{p}}(d_k) \geqslant 2$ for $k \geqslant 2$, or equivalently, $v_p(d_k) \geqslant 2/(p-1)$ for $k \geqslant 2$.*

Proof. (1) and (2). We have

$$D_\pi(X/\pi) = \exp(X + X^p/p) = \sum_{k \geqslant 0} u_k X^k \, ,$$

where $v_p(u_k) \geqslant -k(2p-1)/((p-1)p^2)$ by Theorem 4.2.22. Since $d_k = \pi^k u_k$ and $v_p(\pi) = 1/(p-1)$ we deduce (2), and (1) follows.

(3). The values for d_0 and d_1 are trivial. Assume $k \geqslant 2$. It is preferable to work here with the $\mathfrak{p}$-adic valuation $v_{\mathfrak{p}}$ instead of v_p (which are related by $v_{\mathfrak{p}}(x) = (p-1)v_p(x)$), since it is integer-valued. We already know that $v_{\mathfrak{p}}(d_k) \geqslant k(p-1)^2/p^2$. For $p \geqslant 5$ and $k \geqslant 2$ we have $k(p-1)^2/p^2 \geqslant 2 \cdot 16/25 = 32/25$, so that $v_{\mathfrak{p}}(d_k) \geqslant 2$. If $k \geqslant 5$ we have $k(p-1)^2/p^2 \geqslant 5 \cdot 1/4 = 5/4$, so once again $v_{\mathfrak{p}}(d_k) \geqslant 2$. For $p = 3$ and $k \geqslant 3$ we have $k(p-1)^2/p^2 \geqslant 3 \cdot 4/9 = 12/9$. For $p = 3$ and $k = 2$, we compute directly that $d_2 = \pi^2/2$, so $v_{\mathfrak{p}}(d_2) = 2$. Finally, for $p = 2$ we have $\pi = -2$ and we compute that

$$D_{-2}(X) = \exp(-2X + 2X^2) = 1 - 2X + 4X^2 - \frac{16}{3}X^3 + \frac{20}{3}X^4 + \cdots \, ,$$

so that we also have $v_{\mathfrak{p}}(d_k) \geqslant 2$ for $2 \leqslant k \leqslant 4$.   □

The main result of this section, due to Dwork, is that the value of the function $D_\pi(X)$ at $X = 1$ gives explicitly the pth root of unity ζ_π of Proposition 4.4.36. Note that to compute $D_\pi(x)$ we must compute the power series $\sum_{k \geqslant 0} d_k x^k$, and *not* replace in the formal expression $\exp_p(\pi(x - x^p))$, which would give 1 (note for instance that $\exp_p(\pi)$ does not give a convergent series).

Theorem 4.4.39. *With the above notation we have $D_\pi(1) = \zeta_\pi$; in other words, $D_\pi(1)$ is equal to the unique pth root of unity congruent to $1 + \pi$ modulo $\mathfrak{p}^2$ (and in particular is not equal to 1). Furthermore, for any $a \in \mathbb{Z}_p$ such that $a^p = a$ we have $D_\pi(a) = D_\pi(1)^a = \zeta_\pi^a$.*

Note that since ζ_π is a pth root of unity it makes perfect sense to write ζ_π^a for $a \in \mathbb{Z}_p$.

Proof. Since the radius of convergence of $D_\pi(X)$ is strictly greater than 1, we may compute $D_\pi(a)$ for any $a \in \mathbb{Z}_p$. Furthermore, $D_\pi(X)^p = \exp(p\pi X)\exp(-p\pi X^p)$, and by Proposition 4.2.10, for $a \in \mathbb{Z}_p$ the series $\exp_p(p\pi a)$ and $\exp_p(-p\pi a^p)$ converge (this was not the case before raising to the power p). It follows that if $a^p = a$ we have $D_\pi(a)^p = 1$, so $D_\pi(a)$ is a pth root of unity. In addition, by (2) of the above proposition we have $D_\pi(1) \equiv 1 + \pi \pmod{\mathfrak{p}^2}$, and we conclude by Proposition 4.4.36 that $D_\pi(1) = \zeta_\pi$. Similarly, $D_\pi(a)$ is a pth root of unity such that

$$D_\pi(a) \equiv 1 + a\pi \equiv (1 + \pi)^a \equiv \zeta_\pi^a \pmod{\mathfrak{p}^2},$$

so by the same uniqueness argument used in the proof of Proposition 4.4.36 we deduce that we have the *equality* $D_\pi(a) = \zeta_\pi^a$.   □

See also Exercise 35.

It is useful to generalize the above constructions as follows. Let $f \geqslant 1$ and set $q = p^f$. We define the formal power series

$$D_{\pi,f}(X) = \exp(\pi(X - X^q)),$$

where π is as above such that $\pi^{p-1} = -p$, so that $D_\pi(X) = D_{\pi,1}(X)$. On the other hand, we let $\mathcal{L}$ be the unique unramified extension of degree f of K, which is therefore the splitting field of $X^q - X$ by Lemma 4.4.26.

Proposition 4.4.40. (1) *The series $D_{\pi,f}(X)$ converges at least for $v_p(x) > -(p-1)/p^{f+1}$, and in particular has a radius of convergence strictly greater than 1.*

(2) *More precisely, if we set $D_{\pi,f}(X) = \sum_{k \geqslant 0} d_{k,f} X^k$ then $v_p(d_{k,f}) \geqslant k(p-1)/p^{f+1}$.*

(3) *If $a \in \mathcal{L}$ is such that $a^q = a$ then $D_{\pi,f}(a)$ is a pth root of unity and in fact,*

$$D_{\pi,f}(a) = \zeta_\pi^{\operatorname{Tr}_{\mathcal{L}/K}(a)}.$$

Proof. We can write formally

$$D_{\pi,f}(X) = \prod_{0 \leqslant j < f} \exp(\pi(X^{p^j} - X^{p^{j+1}})) \,.$$

By Proposition 4.4.38, the jth series converges at least for $v_p(x^{p^j}) > -(p-1)/p^2$; in other words, $v_p(x) > -(p-1)/p^{j+2}$. Since $j \leqslant f-1$ this proves (1) since the product of convergent series is convergent.

We prove (2) by induction on f. The case $f = 1$ is Proposition 4.4.38. Assume $f > 1$ and the assertion true for $f - 1$. We can write

$$D_{\pi,f}(X) = D_{\pi,f-1}(X) \exp(\pi(X^{p^{f-1}} - X^{p^f}))$$
$$= D_{\pi,f-1}(X) D_\pi(X^{p^{f-1}}) = \sum_{k \geqslant 0} d_{i,f-1} X^i \sum_{j \geqslant 0} d_j X^{p^{f-1}j} \,.$$

It follows that

$$d_{k,f} = \sum_{0 \leqslant p^{f-1}j \leqslant k} d_j d_{k-p^{f-1}j, f-1} \,;$$

hence by the ultrametric property and the induction hypothesis,

$$v_p(d_{k,f}) \geqslant \min_{0 \leqslant p^{f-1}j \leqslant k} v_p(d_j) + v_p(d_{k-p^{f-1}j, f-1})$$

$$\geqslant (p-1) \min_{0 \leqslant p^{f-1}j \leqslant k} \left(\frac{j}{p^2} + \frac{k - p^{f-1}j}{p^f} \right)$$

$$\geqslant (p-1) \left(\frac{k}{p^f} - \max_{0 \leqslant p^{f-1}j \leqslant k} j \left(\frac{1}{p} - \frac{1}{p^2} \right) \right)$$

$$\geqslant (p-1) \left(\frac{k}{p^f} - \frac{k}{p^{f-1}} \left(\frac{1}{p} - \frac{1}{p^2} \right) \right) \geqslant \frac{k(p-1)}{p^{f+1}} \,,$$

proving the assertion for f.

For (3), since $D_{\pi,f}(X)$ has a radius of convergence strictly greater than 1, $D_{\pi,f}(a)$ makes sense when $v_p(a) \geqslant 0$, and reasoning as in the case $f = 1$, the individual formal power series $\exp(pX)$ and $\exp(pX^q)$ also have a radius of convergence strictly greater than 1, so we may write $D_{\pi,f}(a)^p = \exp_p(pa) \exp_p(-pa^q) = 1$ when $a^q = a$. Since all the series involved have a radius of convergence strictly greater than 1, if $a^q = a$ (or more generally if $v_p(a) \geqslant 0$) we have

$$D_{\pi,f}(a) = \prod_{0 \leqslant j < f} \exp_p(\pi(a^{p^j} - a^{p^{j+1}})) = \prod_{0 \leqslant j < f} D_\pi(a^{p^j}) \,.$$

By Proposition 4.4.38 we have

$$D_\pi(a^{p^j}) \equiv 1 + \pi a^{p^j} \pmod{\mathfrak{p}^2}$$

(note that $D_\pi(a^{p^j})$ is not in general a pth root of unity). It follows that

$$D_{\pi,f}(a) \equiv 1 + \pi \sum_{0 \leqslant j < f} a^{p^j} \pmod{\mathfrak{p}^2} .$$

Since $a^q = a$ and $a \neq 0$, by Corollary 4.4.29 we know that the conjugates of a over K are the a^{p^j}. Thus $\mathrm{Tr}_{L/K}(a) = \sum_{0 \leqslant j < f} a^{p^j}$, so that $D_{\pi,f}(a) \equiv 1 + \pi \, \mathrm{Tr}_{L/K}(a) \pmod{\mathfrak{p}^2}$. Since $\mathrm{Tr}_{L/K}(a) \in K$ and $D_{\pi,f}(a)$ is a pth root of unity, it follows from Dwork's Theorem 4.4.39 that $D_{\pi,f}(a) = \zeta_\pi^{\mathrm{Tr}_{L/K}(a)}$. □

4.4.9 Factorizations in Number Fields

We begin this section with an important warning. Let $L = K(\theta)$ be any finite extension of perfect fields, and let $T(X) \in K[X]$ be the minimal monic polynomial of θ over K. Let $\mathcal{K}$ be *any* extension of K, finite or not. We want to define $\mathcal{K}(\theta)$. There are (at least) two ways to do this.

(1) We can set

$$\mathcal{K}(\theta) = \mathcal{K} \otimes_K K(\theta) = \mathcal{K} \otimes_K L \simeq \frac{\mathcal{K}[X]}{T(X)\mathcal{K}[X]} .$$

In this case, $\mathcal{K}(\theta)$ is not necessarily a field, but only a semisimple algebra, i.e., isomorphic to a finite direct sum of fields. It will be a field if and only if $T(X)$ is irreducible in $\mathcal{K}[X]$.

(2) We can define $\mathcal{K}(\theta)$ to be the set of polynomials in θ with coefficients in $\mathcal{K}$.

To see the difference between these two notions, consider the following example. Let $K = \mathbb{Q}$, $L = K(i)$ (with $i^2 = -1$), and $\mathcal{K} = \mathbb{Q}_5$, chosen because -1 is the square of some element $I \in \mathbb{Q}_5$ by Hensel's lemma. Then according to the first definition, we have

$$\mathbb{Q}_5(i) \simeq \frac{\mathbb{Q}_5[X]}{(X^2 + 1)\mathbb{Q}_5[X]} \simeq \frac{\mathbb{Q}_5[X]}{(X + I)(X - I)\mathbb{Q}_5[X]} \simeq \mathbb{Q}_5 \oplus \mathbb{Q}_5$$

by the Chinese remainder theorem.

On the other hand, according to the second definition we have $\mathbb{Q}_5(i) = \mathbb{Q}_5(I) = \mathbb{Q}_5$. Clearly the first definition is more natural. Therefore in the above context, when we speak of $\mathcal{K}(\theta)$ it is implicitly understood that we use the first definition. This is analogous to the definition of the trace, norm, and characteristic polynomial of θ: we can define them as objects related either to the extension L/K, or only to θ itself. Only the former definition gives reasonable results.

This being said, the aim of this subsection is the proof of the following theorem.

Theorem 4.4.41. *Let* $L = K(\theta)$ *be an extension of number fields, let* $T(X) \in K[X]$ *be the minimal monic polynomial of* θ *over* K, *let* $\mathfrak{p}$ *be a prime ideal of* K, *let* $\mathfrak{p}\mathbb{Z}_L = \prod_{1 \leqslant j \leqslant g} \mathfrak{P}_j^{e_j}$ *be the prime ideal decomposition of* $\mathfrak{p}\mathbb{Z}_L$ *with* $e_j = e(\mathfrak{P}_j/\mathfrak{p})$, *and set* $f_j = f(\mathfrak{P}_j/\mathfrak{p})$. *Then:*

(1) *The only absolute values of* L *extending* $|\ |_\mathfrak{p}$ *are the* $|\ |_{\mathfrak{P}_j}^{1/(e_j f_j)}$.

(2) *There exist canonical isomorphisms*

$$K_\mathfrak{p}(\theta) \simeq K_\mathfrak{p} \otimes_K L \simeq \bigoplus_{1 \leqslant j \leqslant g} L_{\mathfrak{P}_j} \ .$$

(3) *For any algebraic number* θ *the algebra* $K_\mathfrak{p}(\theta)$ *is isomorphic to the direct sum of the completions of* $K(\theta)$ *for the absolute values corresponding to the prime ideals of* $K(\theta)$ *above* $\mathfrak{p}$.

(4) *The decomposition of* $T(X)$ *into monic irreducible factors in* $K_\mathfrak{p}[X]$ *is* $T(X) = \prod_{1 \leqslant j \leqslant g} T_j(X)$, *where* $\deg(T_j) = e_j f_j$.

Proof. (1). It is clear that $|\ |_{\mathfrak{P}_j}^{1/(e_j f_j)}$ is an absolute value and that if $x \in \mathbb{Z}_\mathfrak{p}$, we have

$$|x|_{\mathfrak{P}_j} = \mathcal{N}(\mathfrak{P}_j)^{-v_{\mathfrak{P}_j}(x)} = \mathcal{N}(\mathfrak{p})^{-e_j f_j v_\mathfrak{p}(x)} \ ,$$

so that $|\ |_{\mathfrak{P}_j}^{1/(e_j f_j)}$ extends $|\ |_\mathfrak{p}$. Conversely, if $|\ |$ extends $|\ |_\mathfrak{p}$, then it is non-Archimedean, and by Ostrowski's theorem it has the form $|\ |_\mathfrak{P}^a$ for some prime ideal $\mathfrak{P}$ and some $a > 0$. If $\mathfrak{q}$ is the prime ideal of K below $\mathfrak{P}$ then on K this absolute value has the form $|\ |_\mathfrak{q}^b$ for some $b > 0$, and since distinct prime ideals give inequivalent absolute values, we must have $\mathfrak{q} = \mathfrak{p}$; hence $\mathfrak{P}$ is indeed one of the $\mathfrak{P}_j$, and the exponent must be $1/(e_j f_j)$ as we have just seen, proving (1).

(2). The first isomorphism is the definition of $K_\mathfrak{p}(\theta)$, as we have seen above. Since we can identify K, $K_\mathfrak{p}$, and L with subfields of $L_{\mathfrak{P}_j}$, we have a natural K-bilinear map from $K_\mathfrak{p} \times L$ to $\prod_{1 \leqslant j \leqslant g} L_{\mathfrak{P}_j}$ sending (x, α) to $(x\alpha)_{1 \leqslant j \leqslant g}$. It follows that there exists a natural $K_\mathfrak{p}$-linear map Φ from $K_\mathfrak{p} \otimes_K L$ to $\mathcal{L} = \prod_{1 \leqslant j \leqslant g} L_{\mathfrak{P}_j}$. I claim that the diagonal image of L in $\mathcal{L}$ is dense: indeed, let $(y_1, \ldots, y_g) \in \mathcal{L}$, and let $N > 0$. By definition of the completion, for all k such that $1 \leqslant k \leqslant g$ there exists $x_k \in L$ such that $|y_k - x_k|_k < 2^{-N}$, where $|\ |_k$ denotes the $\mathfrak{P}_k$-adic absolute value. By the Chinese remainder theorem for ideals there exists $x \in L$ such that $|x - x_k|_k < 2^{-N}$; hence $|x - y_k|_k < 2^{-N}$ by the ultrametric inequality, proving my claim.

It follows a fortiori that the image F of Φ in $\mathcal{L}$ is dense. On the other hand, F is a subspace of the $K_\mathfrak{p}$-vector space $\mathcal{L}$, which is finite-dimensional. Since $K_\mathfrak{p}$ is complete, it is well known that such a subspace is *closed*. Since the image of Φ in $\mathcal{L}$ is both dense and closed, it is equal to $\mathcal{L}$; in other words, Φ is surjective.

Finally, $K_\mathfrak{p} \otimes_K L$ is a $K_\mathfrak{p}$-vector space of finite dimension n. On the other hand, by Corollary 4.1.27 the $K_\mathfrak{p}$-dimension of $L_{\mathfrak{P}_j}$ is equal to $e_j f_j$, so that the $K_{\mathfrak{p}}$-dimension of $\mathcal{L}$ is equal to $\sum_{1\leqslant j\leqslant g} e_j f_j = n$. It follows that Φ is also injective, proving (2).

(3). This immediately follows from (1) and (2).

(4). By definition, as already seen above we have canonical isomorphisms

$$K_{\mathfrak{p}} \otimes_K L \simeq K_{\mathfrak{p}}(\theta) \simeq K_{\mathfrak{p}}[X]/T(X)K_{\mathfrak{p}}(X) .$$

Furthermore, $L_{\mathfrak{P}_j}$ is a finite extension of $K_{\mathfrak{p}}$ of degree $e_j f_j$, and since $L = K(\theta)$ it is clear that $L_{\mathfrak{P}_j}$ is generated by θ over $K_{\mathfrak{p}}$ (note that we cannot use the notation $K_{\mathfrak{p}}(\theta)$ since this is reserved for $K_{\mathfrak{p}} \otimes_K L$). Let $T_j(X) \in K_{\mathfrak{p}}[X]$ be the minimal monic polynomial of θ for the extension $L_{\mathfrak{P}_j}/K_{\mathfrak{p}}$. Then $T_j(X)$ is irreducible in $K_{\mathfrak{p}}[X]$, and $\deg(T_j) = [L_{\mathfrak{P}_j} : K_{\mathfrak{p}}] = e_j f_j$ by Corollary 4.1.27. Now, in the above canonical isomorphism the characteristic polynomial of θ is equal to T when θ is considered as an element of the right-hand side, and is equal to $\prod_{1\leqslant j\leqslant g} T_j$ when it is considered as an element of the left-hand side, so we must have equality.   □

4.4.10 Existence of the Field $\mathbb{C}_p$

The first goal of this subsection is the proof of the following result.

Theorem 4.4.42. *The algebraic closure $\overline{\mathbb{Q}_p}$ of $\mathbb{Q}_p$ is not complete for the extension of the natural absolute value given by Corollary 4.4.4. However, the completion $\mathbb{C}_p$ of $\overline{\mathbb{Q}_p}$ is algebraically closed (and evidently complete), and there exists a unique absolute value on $\mathbb{C}_p$ extending that of $\mathbb{Q}_p$.*

Proof. Denote as usual by ζ_m a primitive mth root of unity in $\overline{\mathbb{Q}_p}$, and set

$$\alpha = \sum_{n=1}^{\infty} \zeta_{n'} p^n ,$$

where $n' = n$ if $p \nmid n$ and $n' = 1$ (so $\zeta_{n'} = 1$) if $p \mid n$. Assume by contradiction that $\overline{\mathbb{Q}_p}$ is complete. Since $\zeta_{n'}^{n'} = 1$, we have $|\zeta_{n'}| = 1$, so the series defining α converges. Since $\zeta_{n'} \in \overline{\mathbb{Q}_p}$, it follows that $\alpha \in \overline{\mathbb{Q}_p}$, so the field $\mathcal{K} = \mathbb{Q}_p(\alpha)$ is a finite extension of $\mathbb{Q}_p$. We prove by induction that $\zeta_{n'} \in \mathcal{K}$ for all n, which is trivially true for $n = 1$. Assume this true for all $n < m$, and we want to prove this for $n = m$. Since $n' = 1$ when $p \mid m$, we may assume that $p \nmid m$. Set

$$\alpha_m = p^{-m} \Big(\alpha - \sum_{n=1}^{m-1} \zeta_{n'} p^n \Big) = \sum_{n=m}^{\infty} \zeta_{n'} p^{n-m} .$$

From the first formula we see that $\alpha_m \in \mathcal{K}$, and from the second we have $\alpha_m \equiv \zeta_m \pmod{p}$. Thus α_m is a root modulo p of $X^m - 1 \equiv 0 \pmod{p}$ in $\mathcal{K}$ such that $|\alpha_m| = 1$. By Hensel's lemma (Proposition 4.1.37, since $p \nmid m$) it follows that there exists a root $\zeta \in \mathcal{K}$ of $X^m - 1 = 0$ such that $\zeta \equiv \alpha_m \equiv \zeta_m \pmod{p}$. Since $p \nmid m$, from Proposition 4.3.2 we know that the mth roots of unity are distinct modulo $\mathfrak{p}$ in $\mathcal{K}$, so we have the *equality* $\zeta = \zeta_m$, hence $\zeta_m \in \mathcal{K}$, proving our induction hypothesis. Thus *all* mth roots of unity for $p \nmid m$ are in $\mathcal{K}$, and since they are distinct modulo p (for a fixed m) this means that the set of residue classes modulo $\mathfrak{p}$ (in other words the set of representatives of the ring $\mathbb{Z}_{\mathfrak{p}}/p\mathbb{Z}_{\mathfrak{p}}$, where $\mathfrak{p}$ is the ideal above p in K) is arbitrarily large, which is a contradiction since this set is finite, proving the first statement of the theorem.

We must now show that the completion $\mathbb{C}_p$ of $\overline{\mathbb{Q}_p}$ is algebraically closed. Any polynomial $P(X) \in \mathbb{C}_p[X]$ of degree $d > 0$ is the limit of a sequence of polynomials $P_n(X) \in \overline{\mathbb{Q}_p}[X]$ with the same degree, which by definition of $\overline{\mathbb{Q}_p}$ have d roots in $\overline{\mathbb{Q}_p}$ counted with multiplicity. We must show that we can choose consistently one of these x_n, so that the sequence (x_n) is a Cauchy sequence. Since the limit x of x_n will exist in $\mathbb{C}_p$, and since $P(X)$ is the limit of $P_n(X)$, it will follow that x is a root of $P(X) = 0$ in $\mathbb{C}_p$.

We may of course assume that $P(X)$ and the $P_n(X)$ are monic. If $P(X) = \sum_i a_i X^i$ we set $\|P\| = \max_i |a_i|$. Assume that m is such that $\|P_{n+1} - P_n\| < 1$ for all $n \geqslant m$, and assume by induction on $n \geqslant m$ that we have chosen a root x_j of $P_j(X) = 0$ for $m \leqslant j \leqslant n$ (we can choose any root x_j of $P_j(X)$ for $j \leqslant m$). Denote by $x_{n+1,i}$ for $1 \leqslant i \leqslant d$ the d roots of $P_{n+1}(X) = 0$. Then

$$\prod_{i=1}^{d} |x_n - x_{n+1,i}| = |P_{n+1}(x_n)| = |P_{n+1}(x_n) - P_n(x_n)|$$

$$\leqslant \|P_{n+1} - P_n\| \max(1, |x_n|^d) .$$

It follows that at least one of the roots $x_{n+1,i}$, which we will choose to be x_{n+1}, is such that $|x_n - x_{n+1}| \leqslant \|P_{n+1} - P_n\|^{1/d} \max(1, |x_n|)$. By the ultrametric inequality, this implies that

$$|x_{n+1}| \leqslant \max(|x_n|, \|P_{n+1} - P_n\|^{1/d} \max(1, |x_n|))$$

$$= \max(|x_n|, \|P_{n+1} - P_n\|^{1/d})$$

since $\|P_{n+1} - P_n\| < 1$. It follows first that the sequence $|x_i|$ is bounded by some b (for example $b = \max(\max_{0 \leqslant j \leqslant m} |x_j|, 1)$), and therefore that $|x_n - x_{n+1}| \leqslant \max(1, b)\|P_{n+1} - P_n\|^{1/d}$, so that (x_n) is a Cauchy sequence since (P_n) is a Cauchy sequence. $\qquad\square$

Note that it would be simpler to apply Krasner's lemma (Proposition 4.4.6 (3)) in the second part of this proof, in other words to copy the proof of Proposition 4.4.7, but we cannot do so since we have proved that result only in the context of finite extensions of $\mathbb{Q}_p$, and not in the general context where we would need it here, although it is also true in that case (see Exercise 26).

We denote by $|\ |$ the natural absolute value on $\mathbb{C}_p$. As we have already done in preceding sections, when $x \in \mathbb{C}_p^*$ it is also convenient to denote by $v_p(x)$ the unique exponent such that $|x| = p^{-v_p(x)}$, so that $v_p(x) \in \mathbb{Q}$.

Remark. The image of the absolute value from $\overline{\mathbb{Q}_p}$ or from $\mathbb{C}_p$ to $\mathbb{R}_{\geq 0}$ is the same and is equal to $p^{\mathbb{Q}} \cup \{0\}$ with an evident notation. This is dense in $\mathbb{R}_{\geq 0}$, but is *not* equal to it. Thus we could argue that $\mathbb{C}_p$ is still not large enough, and that we would like to use a larger field such that the image of the absolute value is $\mathbb{R}_{\geq 0}$. This can be done, and leads to the notion of *spherically closed* fields; see [Rob1] for details.

To obtain the multiplicative group $\mathbb{C}_p^*$, we first need a technical but important result, which we give in a special case. This result generalizes Corollary 2.1.17, which corresponds to the case of finite abelian groups.

Proposition 4.4.43. *Let K be an algebraically closed field, let G be an abelian group, and let H be a subgroup of G. Any group homomorphism ϕ from H to K^* can be extended to a homomorphism from G to K^*.*

Proof. Consider the set of subgroups H' of G containing H and on which ϕ can be extended. It is evidently nonempty since H belongs to it, and is ordered by set inclusion. Thus by Zorn's lemma there exists a maximal subgroup H_m in this set (the use of the axiom of choice is necessary in this proof), and let ϕ_m be the extension of ϕ to H_m. Assume by contradiction that $H_m \neq G$, let $g \in G \setminus H_m$, and consider the group H_n generated by H_m and g, in other words the group of $g^k h$ for $k \in \mathbb{Z}$ and $h \in H_m$ (recall that G is abelian). If no power of g except the trivial one belongs to H_m then we can evidently extend ϕ_m to a homomorphism ϕ_n on H_n by setting $\phi_n(g^k h) = \phi_m(h)$, since h is uniquely determined by $g^k h$. Otherwise, let e be the order of the class of g in G/H_m, in other words the $\mathbb{Z}$-generator of all exponents k such that $g^k \in H_m$. Since K is algebraically closed, we may choose $z \in K$ such that $z^e = \phi_m(g^e)$, and we set $\phi_n(g^k h) = z^k \phi_m(h)$. It is immediately checked that this does not depend on the decomposition of an element of H_n as $g^k h$: indeed, if $g^{k_1} h_1 = g^{k_2} h_2$ with $h_i \in H_m$ then on the one hand, $g^{k_1 - k_2} \in H_m$, and hence $e \mid (k_1 - k_2)$, so $\phi_n(g^{k_1 - k_2}) = \phi_m(g^e)^{(k_1 - k_2)/e} = z^{k_1 - k_2}$, and on the other hand,

$$\phi_n(g^{k_1} h_1) \phi_n(g^{k_2} h_2)^{-1} = z^{k_1 - k_2} \phi_m(h_1 h_2^{-1}) = z^{k_1 - k_2} \phi_m(g^{k_2 - k_1}) = 1 \,,$$

as claimed. Furthermore, it is clear that ϕ_n is a group homomorphism from H_n to K^* extending ϕ, contradicting the maximality of H_m and proving the proposition. $\square$

Applying the proposition to $K = \mathbb{C}_p$, $G = \mathbb{Q}$, $H = \mathbb{Z}$, and the map $\phi(a) = p^a$ from $\mathbb{Z}$ to $\mathbb{C}_p$, we see that there exists a group homomorphism ϕ_m from $\mathbb{Q}$ to $\mathbb{C}_p$ extending the map ϕ. By abuse of notation we will still write $\phi_m(a) = p^a \in \mathbb{C}_p$ for $a \in \mathbb{Q}$. Note that we have used Zorn's lemma to define p^a, so the choice is not at all unique. We will denote by $p^{\mathbb{Q}}$ the group of p^a for $a \in \mathbb{Q}$.

Proposition 4.4.44. *Denote by μ the group of roots of unity of order not divisible by p, and by U_1 the group of x such that $|x - 1| < 1$. Then*

$$\mathbb{C}_p^* = p^{\mathbb{Q}} \times \mu \times U_1 .$$

Proof. Let $\alpha \in \mathbb{C}_p^*$. By definition, we can find a sequence of $\alpha_i \in \overline{\mathbb{Q}_p}$ tending to α as $i \to \infty$. Thus for i sufficiently large, $|\alpha_i - \alpha| < |\alpha|$, so that by the ultrametric property, $|\alpha_i| = |\alpha_i - \alpha + \alpha| = |\alpha|$. Since $\alpha_i \in \overline{\mathbb{Q}_p}$, α_i belongs to a finite extension of $\mathbb{Q}_p$, so $|\alpha_i| = p^a$ for some $a \in \mathbb{Q}$ (see Proposition 4.4.5). Conversely, note that since $|p| = p^{-1}$, by multiplicativity we have $|p^{-a}| = p^a$ for all $a \in \mathbb{Q}$, so any p^a is a norm.

It is thus sufficient to restrict to $\alpha \in \mathbb{C}_p^*$ such that $|\alpha| = 1$. Once again, let $\alpha_i \in \overline{\mathbb{Q}_p}$ be sufficiently close to α, and let $\mathcal{L} = \mathbb{Q}_p(\alpha_i)$. We have $|\alpha_i| = |\alpha| = 1$, so α_i is a unit. It follows from Corollary 4.3.7 that $\alpha_i = \omega\beta$, where ω is a root of unity of order prime to p and $|\beta - 1| < 1$. This decomposition is unique since by Proposition 4.3.2, if ω_1 and ω_2 are distinct roots of unity of order prime to p then $|\omega_1 - \omega_2| = 1$, so the proposition follows. $\square$

It follows from this proposition that if $x \in \mathbb{C}_p^*$ is such that $|x| = 1$, we can write uniquely $x = \omega(x)\langle x \rangle$, where $\omega(x)$ is the root of unity of order prime to p such that $|x - \omega(x)| < 1$ and $|\langle x \rangle - 1| < 1$. This generalizes to the p-adic units of $x \in \mathbb{C}_p^*$ the Teichmüller character and the function $\langle x \rangle$ seen above.

4.4.11 Some Analysis in $\mathbb{C}_p$

In Section 4.2 we introduced power series and in particular the p-adic logarithm and exponential functions over an arbitrary complete field containing $\mathbb{Q}_p$, so that this applies in particular to the field $\mathbb{C}_p$. It was necessary to do so at an early stage in this chapter so as to be able to use the logarithm and exponential in the algebraic study of p-adic fields. In this short subsection, we mainly want to study the extension to $\mathbb{C}_p^*$ of the logarithm function $\log_p(x)$, initially defined only for those x such that $|x - 1| < 1$.

Proposition 4.4.45. *There exists a unique extension of $\log_p(x)$ (as defined by the power series for $|x-1| < 1$) to $\mathbb{C}_p^*$ satisfying $\log_p(p) = 0$ and $\log_p(xy) = \log_p(x) + \log_p(y)$ for all x and y in $\mathbb{C}_p^*$. Furthermore, $\log_p(x) = 0$ if and only if x has the form $x = p^a\eta$, where $a \in \mathbb{Q}$ and η is a root of unity of any order.*

Proof. If $\alpha \in \mathbb{C}_p^*$ we can write $\alpha = p^a\omega(x)\langle x \rangle$ with $|\langle x \rangle - 1| < 1$. This decomposition is *not* unique since it depends on the choice of p^a (note that here we do not need the *homomorphism* $a \mapsto p^a$, so we do not need the axiom of choice). However, any other p^a differs multiplicatively from any other by a root of unity η (of order dividing the denominator of a). Since $\log_p(\eta) = 0$, it follows that if we define $\log_p(\alpha) = \log_p(\langle x \rangle)$, this does not depend on the choice of p^a, although $\langle x \rangle$ itself does. It is also clear that this function has the required properties. If $f(\alpha)$ is another extension, then since $\omega^m = 1$ for some m coprime to p, we must have $f(p^a) = af(p) = 0$ (by induction on the numerator and denominator of a), and $mf(\omega) = 0$; hence $f(\omega) = 0$, so $f(\alpha) = f(\langle x \rangle) = \log_p(\langle x \rangle)$, proving uniqueness.

For the second statement, we first note that we clearly have $\log_p(p^a \eta) = 0$. Conversely, assume $\log_p(\alpha) = 0$; in other words, writing $\alpha = p^a \omega(x)\langle x \rangle$, assume that $\log_p(\langle x \rangle) = 0$, where $\langle x \rangle = 1 + y$ with $v_p(y) > 0$. Choose N large enough so that $p^N > 1/((p-1)v_p(y))$, and write

$$\langle x \rangle^{p^N} = (1+y)^{p^N} = 1 + \sum_{1 \leqslant j \leqslant p^N - 1} \binom{p^N}{j} y^j + y^{p^N}.$$

Since $p \mid \binom{p^N}{j}$ for $1 \leqslant j \leqslant p^N - 1$, we have $v_p(\binom{p^N}{j} y^j) \geqslant 1 + v_p(y) > 1/(p-1)$, while by the choice of N we have $v_p(y^{p^N}) > 1/(p-1)$. It follows that $v_p(\langle x \rangle^{p^N} - 1) > 1/(p-1)$; hence by Proposition 4.2.14 we have $|\log_p(\langle x \rangle^{p^N})| = |\langle x \rangle^{p^N} - 1|$, and since by assumption $\log_p(\langle x \rangle^{p^N}) = p^N \log_p(\langle x \rangle) = 0$, we have $\langle x \rangle^{p^N} - 1 = 0$, so $\langle x \rangle$ is a p^Nth root of unity, proving the proposition. $\square$

Remarks. (1) The above definition of the p-adic logarithm is called the Iwasawa logarithm. It would also be possible to set $\log_p(p) = \mathcal{L}$ for a fixed $\mathcal{L} \in \mathbb{C}_p$, but this would lead to slightly more complicated formulas. The choice of the constant $\mathcal{L}$ (here taken to be equal to 0) can be considered as an analogue of the choice of the determination of the logarithm in the complex case.

(2) Using Zorn's lemma it is easy to show that there also exist extensions of the p-adic exponential to the whole of $\mathbb{C}_p$. There are, however, infinitely many, of which none are canonical, but the Iwasawa logarithm is the inverse function to all of them.

For practical computation of $\log_p(x)$, we proceed as follows. Since $\log_p(p) = 0$, after dividing x by p^a for a suitable $a \in \mathbb{Q}$ we may assume that $v_p(x) = 0$, and by choosing a sufficiently close approximation to x (sufficiently close depending on the desired accuracy for $\log_p(x)$) we may assume that x is a $\mathfrak{p}$-adic unit in $K_{\mathfrak{p}}$ for some number field K and some prime ideal $\mathfrak{p}$ of K. Since $|(\mathbb{Z}_{\mathfrak{p}}/\mathfrak{p}\mathbb{Z}_{\mathfrak{p}})^*| = \mathcal{N}\mathfrak{p} - 1$ we know that $x^{\mathcal{N}\mathfrak{p}-1} \equiv 1 \pmod{\mathfrak{p}}$. Thus, writing $x^{\mathcal{N}\mathfrak{p}-1} = 1 + y$, we have $|y| < 1$, so we could compute $\log_p(1+y)$ by the power series expansion and deduce that $\log_p(x) = \log_p(1 + y)/(\mathcal{N}\mathfrak{p} - 1)$. However, the power series usually converges quite slowly, so it is preferable to do one more step before computing the power series. An easy exercise shows that if $z \equiv 1 \pmod{\mathfrak{p}}$ and if $k = \lceil \log(e+1)/\log(p) \rceil$ (where as above $e = e(\mathfrak{p}/p)$ and log is the ordinary real logarithm) then $z^{p^k} \equiv 1 \pmod{\mathfrak{p}^{e+1}}$ (see Exercise 32). Thus $x^{(\mathcal{N}\mathfrak{p}-1)p^k} = 1 + t$ with $v_p(t) \geqslant 1 + 1/e > 1/(p-1)$, so we are in the domain where the $\mathfrak{p}$-adic logarithm and the exponential are inverse functions, and the convergence will be much faster and more controllable. We thus compute $\log_p(x)$ as $\log_p(1+t)/((N\mathfrak{p}-1)p^k)$, where $\log_p(1+t)$ is computed using the power series. This procedure is of course the same as the one that we used to prove Proposition 4.3.4 (3). As an application, we note the following results, which will be used in Chapter 11.

Proposition 4.4.46. *Set $q_p = p$ if $p \geqslant 3$ and $q_2 = 4$, and let $x \in \mathbb{Z}_p^*$.*

(1) *We have*
$$\log_p(x) \equiv \begin{cases} 1 - x^{p-1} \pmod{p^2\mathbb{Z}_p} & \text{if } p \geqslant 3 \\ (x^2-1)/2 \pmod{8\mathbb{Z}_2} & \text{if } p = 2. \end{cases}$$

(2) *We have*
$$\omega(x) \equiv x(1 - \log_p(x)) \pmod{pq_p\mathbb{Z}_p}.$$

Proof. (1). Assume first that $p \geqslant 3$. Note that $1/(p-1) = -1/(1-p) \equiv -(1+p) \pmod{p^2\mathbb{Z}_p}$. Thus by Corollary 4.2.13 we have

$$\log_p(x) = \frac{1}{p-1}\log_p(x^{p-1}) \equiv -(1+p)(x^{p-1}-1) \equiv 1 - x^{p-1} \pmod{p^2\mathbb{Z}_p}.$$

When $p = 2$ we have $x^2 \equiv 1 \pmod{8\mathbb{Z}_2}$; hence by the first statement of Corollary 4.2.13 we have $\log_2(x^2) \equiv x^2 - 1 \pmod{16\mathbb{Z}_2}$, so that

$$\log_p(x) = \frac{1}{2}\log_p(x^2) \equiv \frac{x^2-1}{2} \pmod{8\mathbb{Z}_2},$$

proving (1).

(2). Once again assume first that $p \geqslant 3$. By definition we have $\omega(x) \equiv x \pmod{p\mathbb{Z}_p}$, from which it clearly follows that $\omega(x)^p \equiv x^p \pmod{p^2\mathbb{Z}_p}$ by the binomial expansion. Since $\omega(x)$ is a root of unity of order $p-1$ we deduce that $\omega(x)^p = \omega(x)$, so that $\omega(x) \equiv x^p \equiv x(1 - \log_p(x)) \pmod{p^2\mathbb{Z}_p}$ by (1). When $p = 2$ we can write $x = \varepsilon + 4y$ with $\varepsilon = \omega(x)$. By (1) we have $\log_p(x) \equiv 4\varepsilon y \pmod{8\mathbb{Z}_2}$, hence

$$x(1 - \log_p(x)) \equiv \varepsilon(\varepsilon + 4y)(\varepsilon - 4y) \equiv \varepsilon \equiv \omega(x) \pmod{8\mathbb{Z}_2},$$

proving the proposition. $\square$

Remarks. (1) The notation q_p will be used extensively in Chapter 11 (see Definition 11.2.1).

(2) By Proposition 4.3.4, for $p \geqslant 3$ we have the exact formula

$$\omega(x) = x \sum_{n \geqslant 0} (-1)^n \frac{\log_p(x)^n}{n!} = x\left(1 - \log_p(x) + \frac{\log_p(x)^2}{2!} - \cdots\right),$$

from which (2) is immediate.

Proposition 4.4.47. *Assume that $a \in \mathbb{C}_p$ and $x \in \mathbb{Z}_p^*$. The series $\langle x \rangle^a = \exp_p(a \log_p(\langle x \rangle))$ converges as a power series in $\langle x \rangle - 1$.*

Proof. Immediate from Corollary 4.2.16 since by definition of $\langle x \rangle$ we have $v_p(\langle x \rangle - 1) \geqslant v_p(q_p) > 1/(p-1)$. □

Note that although the conditions are the same, this is not the same statement as Corollary 4.2.18, which gives a result for the convergence of the power series in a with the present notation.

4.5 The Theorems of Strassmann and Weierstrass

From now on, unless explicitly mentioned otherwise, we assume that we work on $\mathbb{C}_p$, with the absolute value $|\ |$, and we let $\mathcal{Z}_p = \{x \in \mathbb{C}_p / |x| \leqslant 1\}$.

4.5.1 Strassmann's Theorem

The following theorem can be useful in applications to Diophantine equations, and is in striking contrast to the case of complex numbers.

Theorem 4.5.1 (Strassmann). *Let $f(X) = f_0 + f_1 X + \cdots + f_n X^n + \cdots$, and assume that f_n tends to 0 (in other words that the domain of convergence contains $\mathcal{Z}_p$, so that the radius of convergence is greater than or equal to 1), but that not all the f_n are equal to 0. Denote by N the largest integer (which exists) such that $|f_N| = \max_n |f_n|$. Then there exist at most N elements $x \in \mathcal{Z}_p$ such that $f(x) = 0$, and in particular there is only a finite number of such x.*

Proof. We use induction on N. First, let $N = 0$ and assume that $f(x) = 0$ for some $x \in \mathcal{Z}_p$, so that $f_0 = -\sum_{n \geqslant 1} f_n x^n$. Since $|f_n| < |f_0|$ for $n \geqslant 1$ and $|x| \leqslant 1$, the ultrametric inequality gives $|\sum_{n \geqslant 1} f_n x^n| \leqslant \max_{n \geqslant 1} |f_n| < f_0$, a contradiction.

Now let $N > 0$, and let $x \in \mathcal{Z}_p$ be such that $f(x) = 0$. For $y \in \mathcal{Z}_p$, we have

$$f(y) = f(y) - f(x) = \sum_{n \geqslant 1} f_n(y^n - x^n) = (y - x) \sum_{n \geqslant 1} f_n \sum_{0 \leqslant j < n} y^j x^{n-1-j} \ .$$

Since by Lemma 4.2.1 we can rearrange terms, we have $f(y) = (y - x)g(y)$ with $g(X) = \sum_j g_j X^j$ and $g_j = \sum_{r \geqslant 0} f_{j+1+r} x^r$. By definition of N we have $|g_j| \leqslant |f_N|$ for all j, $|g_{N-1}| = |f_N|$, and $|g_j| < |f_N|$ for $j > N - 1$. It follows that $N - 1$ is the largest integer such that $|g_{N-1}| = \max_j |g_j|$, so that g satisfies the hypotheses of the theorem with N replaced by $N - 1$. By our induction hypothesis, g has at most $N - 1$ zeros in $\mathcal{Z}_p$, and since $f(y) = 0$ if and only if $y = x$ or $g(y) = 0$, f has at most N zeros in $\mathcal{Z}_p$, giving the result. □

Remark. It is important to note that in fact f has at most N zeros in $\mathcal{Z}_p$, counting multiplicity in a precise sense that we will make clear below. This will follow from Theorem 4.5.12, but can easily be proved directly as above.

As immediate corollaries, the reader can prove the following results:

Corollary 4.5.2. *If $f(X)$ and $g(X)$ converge for all $|x| < R$ with $R > 0$ and if $f(x) = g(x)$ for infinitely many such x, then $f(X)$ and $g(X)$ are identical as power series.*

Corollary 4.5.3. *If $f(X)$ converges in $\mathcal{Z}_p$ and $f(X + a) = f(X)$ for some $a \in \mathcal{Z}_p$ different from 0 then f is constant (both as a function on $\mathcal{Z}_p$ and as a power series).*

4.5.2 Krasner Analytic Functions

Before going on to the Weierstrass preparation theorem, we define and study the basic properties of so-called Krasner analytic functions. As for the next subsection, this is taken with little change from [Cas1].

Definition 4.5.4. *We denote by W the ring of power series that converge in $\mathcal{Z}_p$, in other words whose coefficient of X^n tends to 0, and we set $\|f\| = \max_n |f_n|$.*

It is immediately checked that this is a metric on W that satisfies the ultrametric inequality, and that $\|fg\| = \|f\|\|g\|$.

An example that we will need below is the following.

Lemma 4.5.5. *Let $f(X)$ be a rational function with no poles in $\mathcal{Z}_p$. Then $f \in W$.*

Proof. Decompose f into partial fractions: the polynomial part is of course in W. The polar part is a linear combination of fractions of the form $1/(X - \alpha)^k$ with $k \geqslant 1$. By assumption, $\alpha \notin \mathcal{Z}_p$, so that

$$\frac{1}{(X - \alpha)^k} = \alpha^{-k}(1 - X/\alpha)^{-k} = \alpha^{-k} \sum_{n \geqslant 0} \binom{k + n - 1}{n} \alpha^{-n} X^n \in W$$

since $|\alpha^{-1}| < 1$. $\qquad\square$

Lemma 4.5.6. (1) *The ring W is complete for the topology induced by $\|\ \|$.*
(2) *If $f \in W$ we have $\|f\| = \sup_{x \in \mathcal{Z}_p} |f(x)|$.*
(3) *Let $f^{(n)}$ be a sequence of elements of W. A necessary and sufficient condition that it converge (for the topology induced by $\|\ \|$) is that the sequence $f^{(n)}(x)$ converge uniformly in $\mathcal{Z}_p$.*

Proof. Let $f^j(X) = \sum_{n \geqslant 0} f_n^{(j)} X^n$ be a Cauchy sequence in W. In particular, for each n the sequence $f_n^{(j)}$ is a Cauchy sequence in $\mathbb{C}_p$, hence has a
limit, which we denote by f_n^*. We set $f^*(X) = \sum_{n \geqslant 0} f_n^* X^n$, and we first show
that $f^*(X) \in W$. Thus, let $\varepsilon > 0$. By definition of a Cauchy sequence there
exists $J = J(\varepsilon)$ such that $\|f^{(J)} - f^{(j)}\| < \varepsilon$ for all $j \geqslant J$. Thus by definition
of the metric on W, $|f_n^{(J)} - f_n^*| < \varepsilon$ for all n. Since $f^{(J)} \in W$, there exists
$m = m(\varepsilon)$ such that $|f_n^{(J)}| < \varepsilon$ for all $n \geqslant m$. By the ultrametric inequality it
follows that $|f_n^*| < \varepsilon$ for all $n \geqslant m$, and since ε is arbitrary, that f_n^* tends to
0, hence that $f^* \in W$ as claimed. A similar proof now shows that f^* is the
limit of $f^{(j)}$ in W, proving (1).

(2). By definition we have $|f(x)| \leqslant \|f\|$ for all $x \in \mathcal{Z}_p$, so we must show
that we have equality. Since the result is trivial for $f = 0$, dividing by $\|f\|$
we may assume that $\|f\| = 1$, so that $f_n \in \mathcal{Z}_p$ for all n. Let ϕ be the natural
surjective map from $\mathcal{Z}_p$ to the residue field $\mathbb{F} = \mathcal{Z}_p/\mathfrak{m}$, where $\mathfrak{m} = \{x \in \mathcal{Z}_p, |x| < 1\}$ is the maximal ideal of $\mathcal{Z}_p$. Since f_n tends to 0 it follows that
$\phi(f_n) = 0$ for n sufficiently large, so that $\phi(f)(X) = \sum_{n \geqslant 0} \phi(f_n) X^n$ is a
polynomial in X, in other words belongs to $\mathbb{F}[X]$. However, $\mathbb{F}$ is isomorphic
to the algebraic closure of $\mathbb{F}_p$, and in particular is infinite. It follows that we
may choose some element $\alpha \in \mathbb{F}$ such that $\phi(f)(\alpha) \neq 0$. Since ϕ is surjective,
we thus have $|f(x)| = 1$ for any x such that $\phi(x) = \alpha$.

(3) is an immediate consequence of (1) and (2). $\square$

Lemma 4.5.7. *Let $f \in W$, and assume that f vanishes in some nonempty
open set $S \subset \mathcal{Z}_p$. Then $f = 0$. Equivalently, let f and g be in W and assume
that f and g coincide on some nonempty open subset of $\mathcal{Z}_p$. Then $f = g$.*

Proof. Assume first that $0 \in S$, so that $f(x) = 0$ for all x such that
$|x| < \delta$ for some $\delta > 0$. Let $\zeta \in \mathcal{Z}_p$ be such that $\zeta \neq 0$ and $|z| < \delta$,
and set $F(x) = f(\zeta x)$. Since $\zeta \in \mathcal{Z}_p$ we have $F \in W$, and by assumption
$\sup_{x \in \mathcal{Z}_p} F(x) = 0$. By the above lemma it follows that $\|F\| = 0$, so that
$\zeta^n f_n = 0$ for all n; hence $f_n = 0$ since $\zeta \neq 0$.

If $0 \notin S$, let $x_0 \in S$, and consider the function $f_0(x) = f(x + x_0)$, which
vanishes in a nonempty open set containing 0. It is immediate to check that
we still have $f_0 \in W$, and by (2) of the above lemma we have $\|f_0\| = \|f\|$, so
we are reduced to the previous case. $\square$

Lemma 4.5.8. *Let f be a function from $\mathcal{Z}_p$ to $\mathbb{C}_p$. A necessary and sufficient
condition that f belong to W is that it be a uniform limit of rational functions
all having their poles outside $\mathcal{Z}_p$.*

Proof. Assume first that $f(X) = \sum_{n \geqslant 0} f_n X^n \in W$. Then f is the uniform
limit of the polynomials $h^{(n)}(X) = \sum_{0 \leqslant j \leqslant n} f_j X^j$, which have no poles (or
simply the point at infinity, if we look projectively). Conversely, assume that
f is the uniform limit of rational functions $h^{(n)}$ without poles in $\mathcal{Z}_p$. By

Lemma 4.5.5 these functions are in W, and since W is complete it follows that $f \in W$. □

We can now define Krasner analytic functions. Although they can be defined in much greater generality, we only give two examples.

Definition 4.5.9. *Let S be either the set $\mathcal{Z}_p = \{x \in \mathbb{C}_p, |x| \leqslant 1\}$ or the set $\mathcal{D} = \{x \in \mathbb{C}_p, |x-1| \geqslant 1\}$. We say that a function f defined on S is* Krasner analytic *if f is the uniform limit on S of rational functions having all their poles outside S.*

Thus the above lemma states that f is Krasner analytic on $\mathcal{Z}_p$ if and only if $f \in W$, in other words if and only if its power series coefficients tend to 0. For the set $\mathcal{D}$ the characterization is less simple, but we will need only the analogue of Lemma 4.5.7.

Lemma 4.5.10. *Let f and g be two Krasner analytic functions on the set $\mathcal{D}$ defined above. If f and g coincide on some nonempty open subset of $\mathcal{D}$ then $f = g$.*

Proof. Consider the map from $\mathcal{D}$ to $\mathbb{C}_p$ sending x to $t = 1/(x-1)$. Since $|x-1| \geqslant 1$ for $x \in \mathcal{D}$, this map is a well-defined map from $\mathcal{D}$ to $\mathcal{Z}_p$, and since $x = 1 + 1/t$ for $t \neq 0$, its image is equal to $\mathcal{Z}_p$ minus the origin. A uniformly convergent sequence of rational functions of $x \in \mathcal{D}$ gives a uniformly convergent sequence of rational functions $h^{(n)}$ of $t \in \mathcal{Z}_p \setminus \{0\}$. However, since all the poles of these $h^{(n)}$ are outside $\mathcal{Z}_p$, it follows that the sequence $h^{(n)}(0)$ is well defined and is a Cauchy sequence, hence converges, so that the sequence $h^{(n)}$ is uniformly convergent on the whole of $\mathcal{Z}_p$. The map $t = 1/(x-1)$ thus gives a one-to-one correspondence between Krasner analytic functions on $\mathcal{D}$ and on $\mathcal{Z}_p$, so the lemma follows from Lemma 4.5.7. □

For future reference, we also note the following characterization of elements of $\mathcal{D}$.

Lemma 4.5.11. *Let $z \in \mathbb{C}_p$. The following conditions are equivalent:*

(1) $|z - 1| \geqslant 1$; *in other words, $z \in \mathcal{D}$.*
(2) $|z| \neq 1$ *or* $|z| = |z - 1| = 1$.
(3) $|z/(1 - z)| \leqslant 1$.
(4) *For all $N \geqslant 0$, $|z^{p^N} - 1| \geqslant 1$.*
(5) *For all $N \geqslant 0$, $|z^{p^N}/(1 - z^{p^N})| \leqslant 1$.*

Proof. Left to the reader (Exercise 40). □

4.5.3 The Weierstrass Preparation Theorem

Strassmann's theorem is in fact a consequence of a more precise theorem, which for complex power series is known as the *Weierstrass preparation theorem*. However, due to its importance we have preferred to give Strassmann's theorem separately first.

Theorem 4.5.12 (Weierstrass preparation theorem). *Let* $f(X) = f_0 + f_1 X + \cdots + f_n X^n + \cdots$, *assume that* f_n *tends to 0, and denote by* N *the largest integer such that* $|f_N| = \max_n |f_n|$. *There exists a polynomial* $g(X) = g_0 + g_1 X + \cdots + g_N X^N$ *and a power series* $h(X) = 1 + h_1 X + h_2 X^2 + \cdots$ *satisfying the following conditions:*

(1) $f(X) = g(X)h(X)$.
(2) $|h_n| < 1$ *for* $n \geqslant 1$ *and* h_n *tends to 0 as* $n \to \infty$.
(3) $|g_N| = \max_n |g_n|$.

Since $|h_n| < 1$ for $n \geqslant 1$, the function $h(x)$ cannot have any zeros in $\mathcal{Z}_p$, so that the zeros of $f(X)$ in $\mathcal{Z}_p$ are the zeros of the polynomial $g(X)$ that belong to $\mathcal{Z}_p$ (or to $\mathbb{C}_p$ since $|g_n|/|g_N| \in \mathcal{Z}_p$), hence are at most N in number as claimed by Strassmann's theorem. When there are fewer than N, Strassmann's bound is not attained.

Proof. We need some lemmas.

Lemma 4.5.13. *Let* $R(X) \in \mathbb{C}_p[X]$ *and let* $G(X) = G_0 + G_1 X + \cdots + G_N X^N \in \mathbb{C}_p[X]$ *be a nonzero polynomial such that* $|G_N| = \max_n |G_n|$. *If* $R(X) = L(X)G(X) + M(X)$ *with* $\deg(M) < \deg(G) = N$ *is the Euclidean division of* $R(X)$ *by* $G(X)$, *we have* $\|L\|\|G\| \leqslant \|R\|$ *and* $\|M\| \leqslant \|R\|$.

Proof. Denote by R_m (respectively L_m) the coefficient of X^m in $R(X)$ (respectively in $L(X)$), and let $d = \deg(R(X))$, so that $d - N = \deg(L(X))$. For all j we have

$$R_{d-j} = G_N L_{d-N-j} + G_{N-1} L_{d-N-j+1} + \cdots + G_{N-j} L_{d-N} .$$

Since $|G_N| = \max_n |G_n| = \|G\|$, it follows by induction on j that

$$|L_{d-N-j}|\|G\| \leqslant \|R\| .$$

We thus obtain $\|L\|\|G\| \leqslant \|R\|$, so $\|M\| \leqslant \|R\|$ by the ultrametric inequality. $\qquad\square$

Lemma 4.5.14. *Let* $R(X) \in W$ *and let* $G(X) = G_0 + G_1 X + \cdots + G_N X^N \in \mathbb{C}_p[X]$ *be a nonzero polynomial such that* $|G_N| = \max_n |G_n|$. *There exist* $L(X) \in W$ *and* $M(X) \in \mathbb{C}_p[X]$ *such that*

$$R(X) = L(X)G(X) + M(X) \quad with \quad \deg(M) < N .$$

Furthermore, $\|L\|\|G\| \leqslant \|R\|$ *and* $\|M\| \leqslant \|R\|$.

In other words, we can replace in the preceding lemma a polynomial $R(X)$ by any element of W with the same conclusion.

Proof. Let $R^{(j)}(X) \in \mathbb{C}_p[X]$ be a sequence of polynomials tending to $R(X)$. If we perform the Euclidean division of $R^{(j)}(X)$ by $G(X)$, we can find polynomials $L^{(j)}(X)$ and $M^{(j)}(X)$ such that $R^{(j)}(X) = L^{(j)}(X)G(X) + M^{(j)}(X)$ and $\deg(M^{(j)}) < \deg(G) = N$; hence for all i and j we have

$$R^{(i)}(X) - R^{(j)}(X) = (L^{(i)}(X) - L^{(j)}(X))G(X) + (M^{(i)}(X) - M^{(j)}(X)) .$$

This is the Euclidean division of $R^{(i)}(X) - R^{(j)}(X)$ by $G(X)$, so by the preceding lemma we have $\|L^{(i)} - L^{(j)}\| \leqslant \|R^{(i)} - R^{(j)}\|/\|G\|$ and $\|M^{(i)} - M^{(j)}\| \leqslant \|R^{(i)} - R^{(j)}\|$, so that both $L^{(i)}$ and $M^{(i)}$ are Cauchy sequences. By Lemma 4.5.6, they converge to some limits L and M that clearly satisfy the conditions of the lemma. $\square$

Lemma 4.5.15. *Assume that for some $c < 1$ there exist $G \in \mathbb{C}_p[X]$ and $H \in W$ such that $\deg(G) = N$, $\|f - G\|/\|f\| \leqslant c$, and $\|H - 1\| \leqslant c$. Then there exist G_1 and H_1 satisfying the same conditions and such that $\|f - G_1 H_1\| \leqslant c\|f - GH\|$.*

Proof. We have $f - GH = H(f - G) + f(1 - H)$, hence

$$\|f - GH\| \leqslant \max(\|H\|\|f - G\|, \|f\|\|1 - H\|) \leqslant c\|f\|$$

since $\|H - 1\| \leqslant c < 1$ implies that $\|H\| = 1$. Thus if we set $\delta = \|f - GH\|/\|f\|$, we have $\delta \leqslant c$.

Now since $\|f - G\| < \|f\|$, we have $\|G\| = \|f\| = |f_N|$ and $|f_N - G_N| < |f_N|$, hence $|G_N| = |f_N| = \|G\|$. Thus, we may apply Lemma 4.5.14 to $R = f - GH$. Thus, there exist $L \in W$ and $M \in \mathbb{C}_p[X]$ such that $f - GH = LG + M$, $\deg(M) < N$, $\|L\|\|G\| \leqslant \|f - GH\|$, and $\|M\| \leqslant \|f - GH\|$. It follows in particular that $\|L\| \leqslant \delta\|f\|/\|G\| = \delta$. We set $G_1 = G + M$ and $H_1 = H + L$. Then $\deg(G_1) = N$ and

$$\|f - G_1\| \leqslant \max(\|f - G\|, \|M\|) \leqslant \max(c\|f\|, \delta\|f\|) \leqslant c\|f\| ,$$

since $\delta \leqslant c$ by what we saw above. Furthermore,

$$\|H_1 - 1\| \leqslant \max(\|H - 1\|, \|L\|) \leqslant \max(c, \delta\|f\|/\|f\|)$$

since $\|G\| = \|f\|$. Thus, all the conditions satisfied by G and H are also satisfied by G_1 and H_1. In addition, since $f - GH = LG + M$ we have

$$\|f - G_1 H_1\| = \|(H - 1)M + ML\| \leqslant \max(\|H - 1\|\|M\|, \|M\|\|L\|)$$
$$\leqslant \max(c\delta\|f\|, \delta^2\|f\|) = c\delta\|f\| = c\|f - GH\| ,$$

proving the lemma. Note that this proof has the same nature as the proof of Hensel's lemmas. $\square$

Proof of Theorem 4.5.12. We can of course assume that the coefficients of f are not all 0, since otherwise the result is trivial. By definition of N, we can write $\|f(X) - \sum_{0 \leqslant n \leqslant N} f_n X^n\| = c\|f\|$ for some $c < 1$. By induction, it is clear that we can construct $G^{(j)}(X) \in \mathbb{C}_p[X]$ and $H^{(j)}(X) \in W$ satisfying the conditions of the lemma and such that $\|f - G^{(j)} H^{(j)}\| \leqslant c^{j+1}\|f\|$: for $j = 0$ we set $G^{(0)} = \sum_{0 \leqslant n \leqslant N} f_n X^n$ and $H^{(0)} = 1$, and for $j > 0$ we apply the lemma to $G^{(j-1)}$ and $H^{(j-1)}$. The sequences $G^{(j)}$ and $H^{(j)}$ clearly tend to elements g and h of $\mathbb{C}_p[X]$ and W respectively satisfying the conditions of the lemma, with in addition the equality $f = gh$. Furthermore, since $\|h\| = |h_0| = 1$, replacing g by $h_0 g$ (which preserves $|g_N| = \max_n |g_n|$) and h by h/h_0 we may assume that $h_0 = 1$, as stated. $\qquad\square$

As an application of the Weierstrass preparation theorem, we prove the following result, which is the p-adic analogue of the similar result over $\mathbb{C}$.

Corollary 4.5.16. *Let* $f(X) = \sum_{i \geqslant 0} f_i X^i$ *be an entire function, in other words a power series with infinite radius of convergence. Then either f has a zero in $\mathbb{C}_p$ or $f = f_0$ is constant.*

Proof. Assume that f has no zeros in $\mathbb{C}_p$ (which implies in particular that $f_0 \neq 0$), let $n > 0$, and consider $f_n(X) = f(X/p^n)$. By the preparation theorem there exist a polynomial $g_n(X)$ and a power series $h_n(X) = 1 + \sum_{i \geqslant 1} h_{n,i} X^i$ with $|h_{n,i}| < 1$ for $i \geqslant 1$ such that $f_n(X) = g_n(X)h_n(X)$, in other words $f(X) = g_n(Xp^n)h_n(Xp^n)$. Since f has no zeros in $\mathbb{C}_p$ and $\mathbb{C}_p$ is algebraically closed, the polynomial g_n must be constant, necessarily equal to $f_0 \neq 0$. Thus $f(X) = f_0 h_n(Xp^n)$, in other words $f_i = f_0 h_{n,i} p^{ni}$. Since $|h_{n,i}| < 1$ for $i \geqslant 1$ we thus have $|f_i| < |f_0||p|^{ni} = |f_0|p^{-ni}$ since $|p| = 1/p$. The integer n being arbitrarily large, we deduce that $|f_i| = 0$, in other words $f_i = 0$ for $i \geqslant 1$, proving the corollary. $\qquad\square$

4.5.4 Applications of Strassmann's Theorem

We now come to deeper applications of $\mathfrak{p}$-adic fields to solutions of Diophantine equations. We begin with the following very simple application of Strassmann's theorem.

Proposition 4.5.17. *The only integral solution to the Diophantine equation* $x^3 + 6y^3 = 1$ *is* $(x, y) = (1, 0)$.

Proof. Set $\theta = 6^{1/3}$ and let $K = \mathbb{Q}(\theta)$. Our equation can be rewritten $\mathcal{N}_{K/\mathbb{Q}}(x + y\theta) = 1$, and since x and y are integral this means that $x + y\theta$ is a unit of norm 1 in K. We compute that $\varepsilon = 3\theta^2 - 6\theta + 1$ is a fundamental unit of K, which has norm 1, and since the roots of unity in K are ± 1, and -1 has norm -1, it follows that $x + y\theta = \varepsilon^k$ for some $k \in \mathbb{Z}$. Now we note that $\varepsilon = 1 + 3\alpha$ with $\alpha = \theta^2 - 2\theta \in \mathbb{Z}_K$, so by the binomial theorem we have $x + y\theta = \varepsilon^k = 1 + 3k\alpha + \sum_{j \geqslant 2} 3^j \binom{k}{j} \alpha^j$; hence there exists $\beta \in \mathbb{Z}[\theta]$ such that

$x + y\theta = 1 + 3k(\theta^2 - 2\theta) + 9\beta$. Identifying the coefficients of θ^2 on both sides we deduce that $0 \equiv 3k \pmod 9$, hence that $k \equiv 0 \pmod 3$. This is not at all sufficient for solving our equation, not because we did not work hard enough, but because we used our 3-adic information naïvely.

Indeed, we first note that $X^3 + 6$ is irreducible in $\mathbb{Q}_3[X]$ because it has no roots modulo 9 (or because it is an Eisenstein polynomial). Thus $\mathcal{K} = \mathbb{Q}_3(\theta)$ is a cubic extension of $\mathbb{Q}_3$. Let $\mathfrak{p}$ be the unique ideal of K above 3, so that $e = e(\mathfrak{p}/3) = 3$. We can apply Corollary 4.2.18 to $a = 3\alpha \in K$, since $v_p(a) \geqslant v_p(3) = 1 > 1/2$ (in fact it is clear that $v_p(a) = 4/3$); hence we know that the power series in k for $(1 + 3\alpha)^k$ converges for all $k \in \mathbb{Z}_3$. Now, $\log_3(1 + 3\alpha) \equiv 3\alpha \pmod{3^2 \mathbb{Z}_p}$, so

$$x + y\theta = (1 + 3\alpha)^k = \exp_3(k \log_3(1 + 3\alpha)) = \sum_{j \geqslant 0} c_j k^j$$

with $c_0 = 0$, $c_1 \equiv 3\alpha \pmod{3^2 \mathbb{Z}_p}$, and $c_j \equiv 0 \pmod{3^2 \mathbb{Z}_p}$. Since $1, \theta, \theta^2$ are $\mathbb{Q}_3$-linearly independent, we can identify the coefficients of θ^2 and we obtain $0 = \sum_{j \geqslant 0} f_j k^j$ with $f_0 = 0$, $f_1 \equiv 3 \pmod{3^2 \mathbb{Z}_3}$, and $f_j \equiv 0 \pmod{3^2 \mathbb{Z}_3}$. The hypotheses of Strassmann's theorem are now clearly satisfied with $N = 1$, so there is at most one possible $k \in \mathbb{Z}_3$ satisfying the equation, and since $k = 0$ does satisfy it, $k = 0$ is the only solution, proving that $x + y\theta = \varepsilon^0 = 1$. $\square$

The reader is advised to examine closely the difference between the naïve approach given at the beginning of the proof, which is insufficient to conclude, and the use of Strassmann's theorem, which does not seem to use much more 3-adic information, since we still reason modulo 3^2. The key extra 3-adic information that we use is that the power series for $(1 + 3\alpha)^k$ converges for all k such that $|k|_p \leqslant 1$.

We now give a more sophisticated use of Strassmann's theorem.

Proposition 4.5.18 (Nagell). *Let u_n be the sequence defined by $u_0 = 0$, $u_1 = 1$, and $u_n = u_{n-1} - 2u_{n-2}$ for $n \geqslant 2$. Then $u_n = \pm 1$ if and only if $n = 1, 2, 3, 5,$ and 13.*

Proof. Note that the values of u_n for $0 \leqslant n \leqslant 13$ are respectively $0, 1, 1, -1, -3, -1, 5, 7, -3, -17, -11, 23, 45, -1$, so that the indicated values of n indeed give $u_n = \pm 1$.

If α and β are the roots of the characteristic equation $X^2 - X + 2 = 0$, we obtain the explicit formula $u_n = (\alpha^n - \beta^n)/(\alpha - \beta)$. We have $\alpha = (1 + \sqrt{-7})/2$ and $\beta = (1 - \sqrt{-7})/2$, which considered as complex numbers have the same modulus. Thus we cannot trivially say that $|u_n|$ tends to infinity, for instance. However, we need not work in $\mathbb{C}$; indeed, we can also work in any field of characteristic zero in which $X^2 - X + 2$ splits. Let us choose for example $\mathcal{K} = \mathbb{Q}_{11}$. Since $X^2 - X + 2 \equiv (X - 5)(X - 7) \pmod{11\mathbb{Z}_{11}}$, one of Hensel's lemmas (either Proposition 4.1.29 or Proposition 4.1.37) implies that there exist distinct roots α and β of $X^2 - X + 2$ in $\mathbb{Z}_{11}$, and we check that we may choose $\alpha \equiv 5 + 1 \cdot 11 \equiv 16 \pmod{11^2 \mathbb{Z}_{11}}$, so that $\beta \equiv 1 - \alpha \equiv 106$ $\pmod{11^2 \mathbb{Z}_{11}}$. We would now like to apply Corollary 4.2.18 and Strassmann's theorem. However, $\alpha - 1$ does not satisfy $|\alpha - 1|_{11} < 11^{-1/10}$. Thus, we first apply Fermat's little theorem in $\mathbb{Z}/11\mathbb{Z}$, which implies that $A = \alpha^{10} \equiv 1$ $\pmod{11\mathbb{Z}_{11}}$ and $B = \beta^{10} \equiv 1 \pmod{11\mathbb{Z}_{11}}$. Thus $|A - 1|_{11} \leqslant 1/11$ and $|B - 1|_{11} \leqslant 1/11$, so we can apply Corollary 4.2.18 to $A - 1$ and $B - 1$. Thus, setting $a = A - 1$ and $b = B - 1$, there exist power series ϕ_a and ϕ_b that converge on $\mathbb{Z}_{11}$ and are such that $\phi_a(x) = A^x$ and $\phi_b(x) = B^x$.

For any n, we write $n = 10s + r$ with $0 \leqslant r \leqslant 9$. Thus, $u_{10s+r} = (\alpha^r A^s - \beta^r B^s)/(\alpha - \beta)$. Since $A \equiv B \equiv 1 \pmod{11\mathbb{Z}_{11}}$ and $\alpha - \beta \equiv 9 \not\equiv 0$ $\pmod{11\mathbb{Z}_{11}}$, it follows that $u_{10s+r} \equiv u_r \pmod{11}$, so that by the small table of u_n that we have given, $u_n = \pm 1$ implies that $r = 1, 2, 3$, or 5. For $r = 1$, $2, 3, 5$, and 10, we compute that $\alpha^r \equiv 16, 14, 103, 111$, and 100 modulo 11^2, and $\beta^r \equiv 106, 104, 13, 21$, and 78 modulo 11^2. Thus $a \equiv 99 \pmod{11^2 \mathbb{Z}_{11}}$ and $b \equiv 77 \pmod{11^2 \mathbb{Z}_{11}}$. If we set $\varepsilon = u_r = \pm 1$ for $r = 1, 2, 3$, and 5, we have

$$(\alpha - \beta)(u_{10s+r} - \varepsilon) = \alpha^r \phi_a(s) - \beta^r \phi_b(s) - \varepsilon(\alpha - \beta).$$

Let $\phi(s)$ be the power series on the right-hand side. If we write $\phi(X) = c_0 + c_1 X + c_2 X^2 + \cdots$, then $c_0 = \phi(0) = \alpha^r - \beta^r - u_r(\alpha - \beta) = 0$. Furthermore, by Taylor's theorem, since $\log_{11}(1 + a) \equiv \log_{11}(1 + b) \equiv 0 \pmod{11\mathbb{Z}_{11}}$, it is clear that for $j \geqslant 2$ the coefficients of X^j of $\phi_a(X)$ and $\phi_b(X)$ are divisible by 11^2, so that $c_j \equiv 0 \pmod{11^2 \mathbb{Z}_{11}}$ for $j \geqslant 2$.

Since $\log_{11}(1 + a) \equiv a \equiv 99 \pmod{11^2 \mathbb{Z}_{11}}$ and $\log_{11}(1 + b) \equiv b \equiv 77$ $\pmod{11^2 \mathbb{Z}_{11}}$, it follows that $c_1 \equiv 11(9\alpha^r - 7\beta^r) \pmod{11^2 \mathbb{Z}_{11}}$; hence using the table above, for $r = 1, 2, 3$, and 5 we have $c_1 \equiv 77, 33, 0$, and 55 modulo $11^2 \mathbb{Z}_{11}$. In particular, for $r = 1, 2$, and 5 we have $v_{11}(c_1) = 1$ and $v_{11}(c_j) \geqslant 2$ for all other j, so that we must take $N = 1$ in Strassmann's theorem, proving that $\phi(X)$ has at most 1 zero in $\mathcal{Z}_p$. Since 0 is trivially such a zero, there are no others, so it follows that for $r = 1, 2$, and 5 the only solution to $u_n = \pm 1$ with $n \equiv r \pmod{10}$ is $n = r$.

Since for $r = 3$ we have $c_1 \equiv 0 \pmod{11^2 \mathbb{Z}_{11}}$, we must work a little more. Again by Taylor's theorem we know that $c_j \equiv 0 \pmod{11^3 \mathbb{Z}_{11}}$ when $j \geqslant 3$. Furthermore, a similar computation to that done above shows that

$$\phi_a(X) \equiv 1 + 86 \cdot 11\, X + 2 \cdot 11^2\, X^2 \pmod{11^3 \mathbb{Z}_{11}},$$
$$\phi_b(X) \equiv 1 + 51 \cdot 11\, X + 8 \cdot 11^2\, X^2 \pmod{11^3 \mathbb{Z}_{11}},$$

so

$$\phi(X) \equiv 8 \cdot 11^2\, X + 3 \cdot 11^2\, X^2 \pmod{11^3 \mathbb{Z}_{11}}.$$

In particular, $c_2 \not\equiv 0 \pmod{11^3 \mathbb{Z}_{11}}$; hence we must take $N = 2$ in Strassmann's theorem, proving that $\phi(X)$ has at most two zeros in $\mathcal{Z}_p$. Since we know the trivial zero $s = 0$, but also $s = 1$ (corresponding to $n = 13$), there are no others, finishing the proof of Nagell's result. $\qquad \square$

Remark. There are two important things to note in the above proof. First, we had to choose a prime p such that the polynomial $X^2 - X + 2$ splits in $\mathbb{Q}_p$. This is easy, since one-half of the primes (those that satisfy $\left(\frac{-7}{p}\right) = 1$, in other words $p \equiv 1$, 2, and 4 modulo 7) satisfy this condition. We did not use $p = 2$ for other reasons. On the other hand, we needed to choose a prime p such that the upper bound on the number of solutions of $\phi(X) = 0$ in $\mathbb{Z}_p$ given by Strassmann's theorem is equal to the number of *known* solutions. If this is not the case, we must change the prime p. Note that the failure in using Strassmann's theorem may come from two different reasons. First, of course, Strassmann's bound may not be optimal, as one can see by looking at the Weierstrass preparation Theorem 4.5.12. Second, even if Strassmann's bound *is* optimal, it counts all solutions in $\mathbb{Z}_p$, and there may be some that do not correspond to elements of $\mathbb{Z}$. For example, if we had chosen $p = 23$, we would have found a spurious solution in $\mathbb{Z}_p$, which does not correspond to a solution in $\mathbb{Z}$; see Exercise 41.

Corollary 4.5.19. *The only solutions in $\mathbb{Z}$ of*

$$x^2 + 7 = 2^m$$

are $(x, m) = (\pm 1, 3)$, $(\pm 3, 4)$, $(\pm 5, 5)$, $(\pm 11, 7)$, *and* $(\pm 181, 15)$.

Proof. It is clear that x is odd; hence set $x = 2y - 1$ with $y \in \mathbb{Z}$, so that we obtain $y^2 - y + 2 = 2^{m-2}$. Let $\alpha = (1 + \sqrt{-7})/2$ in some algebraic closure of $\mathbb{Q}$, and $\beta = 1 - \alpha$, so that α and β are the roots of $X^2 - X + 2 = 0$. Then $\mathbb{Z}[\alpha]$ is Euclidean, hence a principal ideal domain. Thus

$$(y - \alpha)(y - \beta) = 2^{m-2} = (\alpha\beta)^{m-2} .$$

Now, α and β have norm 2 hence generate prime ideals, so $(y - \alpha)\mathbb{Z}[\alpha] = \alpha^i \beta^j \mathbb{Z}[\alpha]$ and $(y - \beta)\mathbb{Z}[\alpha] = \alpha^{m-2-i}\beta^{m-2-j}\mathbb{Z}[\alpha]$ for some i and j, and since $y - \beta = \overline{y - \alpha}$ we must have $j = m - 2 - i$. Since the only units in $\mathbb{Z}[\alpha]$ are ± 1, this means that $y - \alpha = \pm\alpha^i\beta^{m-2-i}$ and $y - \beta = \pm\alpha^{m-2-i}\beta^i$.

However, since $\alpha - \beta = \sqrt{-7}$ has odd norm so is coprime to 2, it follows that $y - \alpha$ and $y - \beta$ are coprime in $\mathbb{Z}[\alpha]$ since otherwise a common divisor would divide $\alpha - \beta$ and 2. Thus $\min(i, m - 2 - i) = 0$, in other words $i = 0$ or $i = m - 2$, so that either $y - \alpha = \pm\beta^{m-2}$ and $y - \beta = \pm\alpha^{m-2}$ or $y - \alpha = \pm\alpha^{m-2}$ and $y - \beta = \pm\beta^{m-2}$. In both cases, by subtracting we deduce that $\alpha - \beta = \pm(\alpha^{m-2} - \beta^{m-2})$, which is exactly the problem studied above. Thus the only solutions are for $m - 2 = 1$, 2, 3, 5, and 13, which give the solutions stated in the corollary. $\qquad\square$

4.6 Exercises for Chapter 4

1. In Proposition 4.1.17 choose $K = \mathbb{Q}$ and $\| \ \|$ to be the ordinary absolute value, so that $\widehat{K} = \mathbb{R}$. Let f be the map from $\mathbb{Q}$ to $\mathbb{R}$ such that $f(x) = 0$ for $x^2 < 2$ and $f(x) = 1$ for $x^2 > 2$.

(a) Show that f is continuous on $\mathbb{Q}$.
(b) Show that f cannot be extended by continuity to a map from $\mathbb{R}$ to $\mathbb{R}$.
(c) Let x_n be the sequence in $\mathbb{Q}$ defined by $x_0 = 1$ and $x_{n+1} = x_n + (2 - x_n^2)/2$. Show that (x_n) is a Cauchy sequence, but that $f(x_{2n}) = 0$ and $f(x_{2n+1}) = 1$, so that $f(x_n)$ does not converge.

2. Prove that a locally compact topological space is totally discontinuous if and only if any point has a fundamental system of neighborhoods that are both open and closed.

3. Recall that we set $q_p = p$ for $p \geqslant 3$ and $q_2 = 4$. Let a and b be in $q_p \mathbb{Z}_p$, and assume that $a \equiv b \pmod{p^v \mathbb{Z}_p}$ for some $v \geqslant 2$. Prove that for all $n \geqslant 1$ we have $a^n/n! \in q_p \mathbb{Z}_p$ and that

$$\frac{a^n}{n!} \equiv \frac{b^n}{n!} \pmod{(q_p^{n-1}/n!)p^v \mathbb{Z}_p} .$$

Note that by Lemma 4.2.8 we have $q_p^{n-1}/n! \in \mathbb{Z}_p$, so this implies in particular the congruence modulo $p^v \mathbb{Z}_p$.

4. (N. Elkies.) Let p be a prime number, and let $S_p(k)$ be the series in $\mathbb{Q}_p$ defined by

$$S_p(k) = \sum_{n \geqslant 0} n^k n! .$$

Show that this series converges in $\mathbb{Z}_p$ and that there exist integers $A(k)$ and $B(k)$ in $\mathbb{Z}$, *independent of* p, such that

$$S_p(k) = A(k)S_p(0) + B(k) .$$

Find recurrence formulas and generating functions for $A(k)$ and $B(k)$, and compute the first few values.

5. Generalizing Eisenstein's criterion (Proposition 4.1.35), prove the following. Let $f(X) = X^{2m+1} + f_{2m}X^{2m} + \cdots + f_0 \in \mathbb{Z}_p[X]$ be a polynomial of *odd* degree, and assume that $v_p(f_j) \geqslant 1$ for $m + 1 \leqslant j \leqslant 2m$, $v_p(f_j) \geqslant 2$ for $0 \leqslant j \leqslant m$, and $v_p(f_0) = 2$. Show that f is irreducible in $K_p[X]$.

6. Show that $\binom{\sqrt{2}}{p}$ is not necessarily in $\mathbb{Z}_p[\sqrt{2}]$. More precisely, show that it is in $\mathbb{Z}_p[\sqrt{2}]$ if and only if $p \equiv \pm 1 \pmod 8$.

7.

(a) Show that as claimed in the text we have $\sum_{k \geqslant 1} \lfloor n/p^k \rfloor = (n - s_p(n))/(p - 1)$, where $s_p(n)$ is the sum of the digits of n in base p.
(b) Deduce that the p-adic valuation of a binomial coefficient $\binom{n}{m}$ for $n \geqslant m$ is equal to the number of carries in the base-p subtraction of m from n.

8.

(a) Prove that for all $x \in \mathbb{R}$ the expression

$$\{15x\} + \{10x\} + \{6x\} - \{30x\} - \{x\}$$

takes only the values 0 and 1, where as usual $\{y\}$ denotes the fractional part of y.
(b) (Čebyshev.) Deduce that for all $n \in \mathbb{Z}_{\geqslant 0}$ we have $C_n \in \mathbb{Z}$, where

$$C_n = \frac{(30n)!n!}{(15n)!(10n)!(6n)!} .$$

(c) Deduce that for any prime p we have

$$s_p(15n) + s_p(10n) + s_p(6n) \geqslant s_p(30n) + s_p(n) ,$$

where as usual $s_p(m)$ is the sum of the digits in base p of m.
(d) (More difficult, and due to F. Rodriguez-Villegas.) Prove that the power series $C(T) = \sum_{n \geqslant 0} C_n T^n$ is an *algebraic* function of T, and that the degree of its minimal polynomial over $\mathbb{Q}(T)$ is equal to 483840 (!).

9. Prove that if $v_p(a) \geqslant 0$ but $v_p(a) \notin \mathbb{Z}$, then writing as usual $v_p(a) = q - \theta$ with $0 < \theta < 1$, as $n \to \infty$ we have

$$v_p\left(\binom{a}{n}\right) \sim -n\frac{1}{p^q}\left(\frac{1}{p-1}+\theta\right) ,$$

and deduce that Corollary 4.2.15 (2) is still valid in this case, in other words that the power series $(1 + x)^a$ converges if and only if $v_p(x) > V(a)$ (note that this is trivially false if $v_p(a) \geqslant 0$ and $v_p(a) \in \mathbb{Z}$, as can be seen by choosing $a \in \mathbb{Z}_{>0}$ for instance, since in that case it converges for all x).

10. As a complement to Lemma 4.2.8, prove that if $\mathcal{K} = \mathbb{Q}_p$, $a \in \mathbb{Z}_p$, $a \neq 0$, and $1 \leqslant n < p^{v_p(a)}$, then in fact

$$v_p\left(\binom{a}{n}\right) = v_p(a) - v_p(n) .$$

11. Prove Proposition 4.2.11 and Corollary 4.2.15 (5).

12. Compute the sum of the binomial series expansion for $(1+x)^{1/2}$ at $x = 45/4$ in $\mathbb{Q}_3$ and in $\mathbb{Q}_5$. Are they equal? Find an example for which the series converge to different rational values for two different p-adic fields, and also in $\mathbb{R}$ (necessarily to the same value as in one of the p-adic fields since only two values are possible).

13. (This exercise uses the elementary properties of the Möbius function; see Chapter 10.)

(a) Prove that we have the formal infinite product

$$\exp(X) = \prod_{n \geqslant 1}(1 - X^n)^{-\mu(n)/n} .$$

(b) Show that the Artin–Hasse exponential has the formal infinite product

$$\exp(X + X^p/p + \cdots) = \prod_{n \geqslant 1,\ p \nmid n}(1 - X^n)^{-\mu(n)/n} .$$

(c) Deduce from Corollary 4.2.15 and the above another proof that it has p-integral coefficients, in other words Corollary 4.2.21.

14. By choosing $k = p$ in Corollary 4.2.23, prove Wilson's theorem $(p - 1)! \equiv -1 \pmod{p}$.

15. It follows from Corollary 4.2.23 that for $p \geqslant 3$ we have $1 + 2/(p-1)! + p/(2p-1)! \equiv 0 \pmod{p^2\mathbb{Z}_p}$.

(a) Prove this directly (i.e., without using Dwork's theorem).

(b) Using Exercise 30 of Chapter 2, prove more precisely that $1 + 2/(p-1)! + p/(2p-1)! \equiv (1 + 1/(p-1)!)^2 \pmod{p^3 \mathbb{Z}_p}$ for $p \geqslant 5$.

16. Let f be a continuous function from $\mathbb{Z}_p$ to $\mathbb{Z}_p$, and let a_k be defined as in Mahler's Theorem 4.2.26. Prove that $\sup_{x \in \mathbb{Z}_p} |f(x)| = \sup_{k \geqslant 0} |a_k|$.

17. This exercise gives a converse to Proposition 4.2.28. Let $f(x) = \sum_{k \geqslant 0} b_k x^k$ be a power series, and assume that for all k we have $v_p(b_k) \geqslant \beta k + \beta''$ for some constants β and β'' with $\beta > 0$. Prove that $f(x)$ has a Mahler expansion $f(x) = \sum_{k \geqslant 0} a_k \binom{x}{k}$ with

$$v_p(a_k) \geqslant (\beta + 1/(p-1))k - (1/(p-1))s_p(k) + \beta'' .$$

18. Let $\mu_{\mathfrak{p}}$ as in Corollary 4.3.5 be the group of $z \in \mathbb{Z}_p$ such that $z^{q-1} = 1$, where we set $q = N\mathfrak{p}$. Let g_0 be the image in $\mathbb{Z}_p$ of any lift in $\mathbb{Z}_K$ of a generator of the cyclic group $(\mathbb{Z}_K/\mathfrak{p})^*$. Show that the sequence u_n in $\mathbb{Z}_p$ defined by $u_0 = g_0$ and

$$u_{n+1} = \frac{(q-2)u_n^{q-1} + 1}{(q-1)u_n^{q-2}} \quad \text{for } n \geqslant 0$$

converges to $\chi(g_0)$, which is a generator of $\mu_{\mathfrak{p}}$. Compare the speed of convergence with that of the sequence defining $\chi(g_0)$.

19. Extend the Teichmüller homomorphism ω to all of $\mathbb{Z}_p$ by setting $\omega(x) = 0$ if $x \in p\mathbb{Z}_p$, in other words if x is not a $\mathfrak{p}$-adic unit. Let

$$G_{\mathfrak{p}} = \mu_{\mathfrak{p}} \cap \{0\} = \{z \in \mathbb{Z}_p, \ z^{N\mathfrak{p}} = z\} .$$

Define an addition law $\oplus$ on $G_{\mathfrak{p}}$ by setting $x \oplus y = \omega(x+y)$, and keep the usual multiplication. Prove that $G_{\mathfrak{p}}$ is a (finite) field and that $\overline{\omega}$ is a canonical field isomorphism from $\mathbb{Z}_K/\mathfrak{p}$ to $G_{\mathfrak{p}}$.

20. Let p and q be distinct prime numbers. Prove that when $p \not\equiv 1 \pmod q$ the only qth root of unity in $\mathbb{Q}_p(\zeta_p)$ is 1, while when $p \equiv 1 \pmod q$ there are q such roots of unity, which are all distinct modulo $\mathfrak{p}\mathbb{Z}_p$, where $\mathfrak{p} = (1 - \zeta_p)\mathbb{Z}[\zeta_p]$. (Hint: if η is such a root of unity you may consider the subextension $\mathbb{Q}_p(\eta)/\mathbb{Q}_p$ of $\mathbb{Q}_p(\zeta_p)/\mathbb{Q}_p$.)

21. By Proposition 3.5.5 we know that $(1 - \zeta_p)^{p-1}/p$ is a unit.

(a) Using the fact that $\log_p(1 - (1 - \zeta_p)) = 0$, prove more precisely that

$$\frac{(1 - \zeta_p)^{p-1}}{p} \equiv -1 \pmod{(1 - \zeta_p)\mathbb{Z}[\zeta_p]} .$$

(b) More generally, if $k \leqslant p - 1$ show that

$$k! \frac{(1 - \zeta_p)^{p-1}}{p} \equiv -k! \left(\sum_{1 \leqslant i \leqslant k} \frac{(1 - \zeta_p)^{i-1}}{i} \right) \pmod{(1 - \zeta_p)^k \mathbb{Z}[\zeta_p]} .$$

22. For $k \geqslant 1$, set $\mathrm{Li}_k(X) = \sum_{n \geqslant 1} X^n/n^k$, so that for instance $\mathrm{Li}_1(X) = -\log(1 - X)$. These are the *polylogarithm series*.

(a) Show that we have the formal power series identity

$$\mathrm{Li}_2(X) + \mathrm{Li}_2(-X/(1-X)) = -\frac{1}{2}\log^2(1-X) ,$$

and show that this identity is also valid with X replaced by $x \in p\mathbb{Z}_p$ (use Proposition 4.2.7).

(b) Deduce that for $n \geqslant 1$ we have the (equivalent) combinatorial identities

$$\frac{1}{n^2} + \frac{1}{n}\sum_{k=1}^{n}\frac{(-1)^k}{k}\binom{n}{k} = -\frac{1}{2}\sum_{k=1}^{n-1}\frac{1}{k(n-k)} \quad\text{and}\quad \sum_{k=1}^{n}\frac{(-1)^{k-1}}{k}\binom{n}{k} = \sum_{k=1}^{n}\frac{1}{k} .$$

(c) Show that in the field $\mathcal{K} = \mathbb{Q}_p(\zeta_p)$ we have $\sum_{n\geqslant 1}(1-\zeta_p)^n/n = 0$.

(d) From now on we assume that $p = 2$. Show that in $\mathbb{Q}_2$ we have

$$\mathrm{Li}_1(2) = \sum_{n\geqslant 1}\frac{2^n}{n} = 0 \quad\text{and}\quad \mathrm{Li}_2(2) = \sum_{n\geqslant 1}\frac{2^n}{n^2} = 0 ,$$

but that $v_2(\mathrm{Li}_k(2)) \leqslant -2$ (and in particular $\mathrm{Li}_k(2) \neq 0$) for all $k \geqslant 3$.

(e) Set $S_k(N) = \sum_{1\leqslant n\leqslant N} 2^n/n^k$. Show that in $\mathbb{Q}_2$ we have

$$S_1(N) = -2^N\sum_{j\geqslant 1}\frac{2^j}{j+N} = N2^N\sum_{j\geqslant 1}\frac{2^j}{j(j+N)} = -N^2 2^N\sum_{j\geqslant 1}\frac{2^j}{j^2(j+N)} .$$

(f) Prove that for $m \geqslant 4$ we have $v_2(2^j/(j^2(j+2^m))) \geqslant -1$, except for $j = 4$, for which $v_2(2^j/(j^2(j+2^m))) = -2$.

(g) Deduce that for $m \geqslant 4$ we have $v_2(S_1(2^m)) = 2^m + 2m - 4$.

(h) In a similar way, show that

$$S_2(N) = N2^N\sum_{j\geqslant 1}\frac{2^j(2j+N)}{j^2(j+N)^2} ,$$

and that for $m \geqslant 4$ we have $v_2(S_2(2^m)) = 2^m + m - 1$.

23. Give alternative proofs of Propositions 4.3.22 and 4.3.23 by proving by induction on k that for all $k \geqslant 1$ (when $p \geqslant 3$) or $k \geqslant 3$ (when $p = 2$) there exists b_k such that $a/p^{v_p(a)} \equiv b_k^2 \pmod{p^k}$.

24.

(a) Let $p \geqslant 3$ be a prime. Using Proposition 4.3.22, show that, up to isomorphism, there are exactly 3 quadratic extensions of $\mathbb{Q}_p$ given by $\mathbb{Q}_p(D^{1/2})$ for $D = n$, p, pn, where n is any integer such that $\left(\frac{n}{p}\right) = -1$, and that among those the unique unramified one is for $D = n$.

(b) Similarly, using Proposition 4.3.23, show that, up to isomorphism, there exist exactly seven quadratic extensions of $\mathbb{Q}_2$ given by $\mathbb{Q}_2(D^{1/2})$ for $D = 2, 3, 5, 6, 7, 10, 14$, and that among those the unique unramified one is for $D = 5$.

25. (This is more difficult.) Compute the number of isomorphism classes of *cubic* extensions of $\mathbb{Q}_p$, separating the cases $p \equiv 1 \pmod 3$, $p \equiv 2 \pmod 3$, and $p = 3$. In the first case, show that there exists an integer n such that the congruence $x^3 \equiv n \pmod p$ has no solution and that there exists $\zeta \in \mathbb{Q}_p$ with $\zeta^3 = 1$ and $\zeta \neq 1$.

26.

(a) Prove that Theorem 4.4.3 is still valid if $\mathcal{K}$ is a non-Archimedean complete field of characteristic 0, and not only the completion of a number field.

(b) Under the same conditions show that Krasner's lemma (Proposition 4.4.6 (2) and (3)) is also still valid.

27. Recall that if L/K is any finite extension of commutative fields and $x \in L$, the trace $\mathrm{Tr}_{L/K}(x)$ is the trace of the linear map multiplication by x from L to L. Show that if $K_{\mathfrak{p}}$ is a $\mathfrak{p}$-adic field and if $x \in \mathfrak{p}\mathbb{Z}_{\mathfrak{p}}$ then $\mathrm{Tr}_{K_{\mathfrak{p}}/\mathbb{Q}_p}(x) \in p\mathbb{Z}_p$.

28.

(a) Show that a root θ of the polynomial $X^3 - 3X + 1$ defines an unramified cubic extension of $\mathbb{Q}_2$. Then show that the Frobenius automorphism is given by $\sigma(\theta) = \theta^2 - 2$.

(b) Now consider the unramified cubic extension of $\mathbb{Q}_2$ (necessarily isomorphic to the preceding one) defined by a root α of the polynomial $X^3 + X + 1$. Compute $\sigma(\alpha)$ as a polynomial of degree 2 in α in two different ways: either directly, or by giving explicitly a $\mathbb{Q}_2$-isomorphism with the preceding field. It will be useful to set $v = (-31)^{1/2}/3 \in \mathbb{Q}_2$.

29. Denote by $\mathcal{K}$ the unique unramified cubic extension of $\mathbb{Q}_7$ up to isomorphism. The aim of this exercise is to show how to express the Frobenius automorphism σ of $\mathcal{K}$ depending on the way in which $\mathcal{K}$ is represented.

(a) Show that there exists a unique $u \in \mathbb{Z}_7$ such that $u^2 = -3$ and $u \equiv 2$ (mod $7\mathbb{Z}_7$).

(b) Let θ be a root of $x^3 - 2 = 0$. Show that $\mathbb{Q}_7(\theta)$ is isomorphic to $\mathcal{K}$, give explicitly the three roots of $x^3 - 2 = 0$ in terms of θ and u, and finally compute explicitly $\sigma(\theta)$ as a polynomial of degree less than or equal to 2 in θ.

(c) Let α be a root of $x^3 - 3x + 1 = 0$. Show that $\mathbb{Q}_7(\alpha)$ is also isomorphic to $\mathcal{K}$, check that $\alpha^2 - 2$ is also a root, then as before give explicitly all three roots and compute explicitly $\sigma(\alpha)$.

(d) Let β be a root of $x^3 + x + 1 = 0$. Show that $\mathbb{Q}_7(\beta)$ is also isomorphic to $\mathcal{K}$. Explain how to compute $\sigma(\beta)$ by induction on the p-adic precision, and compute $\sigma(\beta)$ explicitly to precision $O(7^3)$.

30. Let $\mathcal{L}/\mathcal{K}$ be a *ramified* cyclic extension of degree p of $\mathfrak{p}$-adic fields. Prove that $\zeta_p \in \mathcal{K}$, where as usual ζ_p is a primitive pth root of unity.

31. Show that if $\alpha \in \overline{\mathbb{Q}_p}$ then $\log_p(\alpha) \in \mathbb{Q}_p(\alpha)$.

32. With evident notation, set $r = \lceil \log(e(\mathfrak{p}/p) + 1)/\log(p) \rceil$. Show that when $z \equiv 1$ (mod $\mathfrak{p}\mathbb{Z}_{\mathfrak{p}}$) then for $k \geqslant r$ we have $z^{p^k} \equiv 1$ (mod $\mathfrak{p}^{e(k-r)+1}\mathbb{Z}_{\mathfrak{p}}$).

33. Let $n \geqslant 1$.

(a) Prove that for all $x \in \mathbb{C}_p$ such that $|x| \leqslant 1$ we have $|(1 + x)^{p^n} - 1| \leqslant |x| \max(|x|, 1/p)^n$.

(b) Prove that if $|x| \leqslant p^{-1/(p-1)}$ then $|(1+x)^{p^n} - 1| \leqslant |x|/p^n$, which is a stronger inequality when $1/p < |x| \leqslant p^{-1/(p-1)}$.

34. Let p be an odd prime, and let a, b, and d in $\mathbb{Z}$ be such that $d \neq 0$ and $a^2 - db^2 \neq 0$.

(a) Show that

$$\frac{1}{2\sqrt{d}} \log_p \left(\frac{a + b\sqrt{d}}{a - b\sqrt{d}} \right) \in \mathbb{Z}_p .$$

Note that the argument of $\log_p$ is not necessarily in the disk of convergence, and that you must show both that the expression belongs to $\mathbb{Q}_p$ and that it is a p-adic integer.

(b) Assume that $a + b\sqrt{d}$ has norm ± 1, in other words that $a^2 - b^2 d = \pm 1$. Show that $\log_p(a + b\sqrt{d})/(2\sqrt{d}) \in \mathbb{Z}_p$.

(c) Deduce from this that if ζ_5 denotes a primitive 5th root of unity in $\mathbb{C}_5$, then for a suitable choice of $\sqrt{5}$ we have

$$\frac{1}{\sqrt{5}} \log_5(1 - \zeta_5) = \frac{1}{2\sqrt{5}} \log_5\left(\frac{1+\sqrt{5}}{2}\right) \in \mathbb{Z}_5 \ .$$

We will see in Chapter 11 that in fact

$$\frac{1}{\sqrt{5}} \log_5\left(\frac{1+\sqrt{5}}{2}\right) = -\frac{2}{5} \psi_5\left(\frac{1}{5}\right) - \frac{1}{2}\gamma_5 \ ,$$

where ψ_5 is a 5-adic generalization of the logarithmic derivative of the gamma function and γ_5 is the 5-adic Euler constant.

35. With the notation of Definition 4.4.37 and Theorem 4.4.39, show that for $p \geqslant 3$, we have more precisely

$$D_\pi(1) = \zeta_\pi \equiv 1 + \pi + \frac{\pi^2}{2} \pmod{\mathfrak{p}^3} \ ,$$

so that in particular $v_{\mathfrak{p}}(\zeta_\pi - (1 + \pi)) = 2$ (for $p = 2$ we have $\zeta_\pi = 1 + \pi$).

36. Let $f(x) = \sum_{n \geqslant 0} a_n x^{-n}$ be a series that converges in $\mathbb{C}_p$ for all x such that $|x| > 1$. Assume that there exists $a \in \mathbb{Z}_{\geqslant 1}$ such that for all such x we have $f(x + a) = f(x)$. Show that $f(x)$ is a constant equal to a_0 (note that this is not the same result as Corollary 4.5.3).

The following three exercises define and give some properties of the inverse binomial symbol, introduced, I believe, by J. Diamond in [Dia1] and [Dia2]. More properties will be given in the exercises to Chapters 9 and 11.

37. Define the *inverse binomial symbol* $\begin{bmatrix} n \\ x \end{bmatrix}$ by the formula

$$\begin{bmatrix} n \\ x \end{bmatrix} = \frac{n!}{x(x+1)\cdots(x+n)} \ .$$

(a) Prove that the inverse binomial symbol satisfies properties similar to the ordinary binomial symbol, more precisely show that

$$\begin{bmatrix} n \\ x \end{bmatrix} - \begin{bmatrix} n \\ x+1 \end{bmatrix} = \begin{bmatrix} n+1 \\ x \end{bmatrix} \quad \text{and} \quad \begin{bmatrix} n \\ x+1 \end{bmatrix} = \frac{x}{n+1}\begin{bmatrix} n+1 \\ x \end{bmatrix} \ .$$

(b) If $x \in \mathbb{C}_p$ is such that $|x| > 1$, prove that

$$\left|\begin{bmatrix} n \\ x \end{bmatrix}\right| \leqslant \frac{n}{|x|^{n+1} p^{n/(p-1)}} \ ,$$

and if $x \in \mathbb{C}$, using Proposition 9.6.17 prove that $\begin{bmatrix} n \\ x \end{bmatrix} \sim \Gamma(x) n^{-x}$, so that in particular $\left|\begin{bmatrix} n \\ x \end{bmatrix}\right| \leqslant |\Gamma(x)| n^{-\Re(x)}$.

38. Given a sequence $(b_n)_{n \geq 0}$, define its *Stirling transform* as the sequence $(a_n)_{n \geq 0}$ given by the formal identity

$$\sum_{n \geq 0} b_n (1 - e^{-T})^n = \sum_{n \geq 0} a_n T^n / n! \, ,$$

so that the b_n can be recovered from the a_n by the inverse Stirling transform given by the formal identity

$$\sum_{n \geq 0} (a_n / n!)(-\log(1 - T))^n = \sum_{n \geq 0} b_n T^n$$

(so for example $a_0 = b_0$, $a_1 = b_1$, $a_2 = 2b_2 - b_1$, $a_3 = 6b_3 - 6b_2 + b_1$, etc.).

(a) Assume that a function f from $\mathbb{C}_p$ to $\mathbb{C}_p$ has an expansion of the form

$$f(x) = \sum_{n \geq 0} b_n \begin{bmatrix} n \\ x \end{bmatrix}$$

that converges for $|x| > R \geq 1$. Prove that for the same values of x we have the Laurent series expansion

$$f(x) = \sum_{n \geq 0} \frac{a_n}{x^{n+1}} \, ,$$

where a_n is the Stirling transform of b_n (Hint: first show that this is true if $b_n = 0$ for n sufficiently large; then use a suitable uniformity argument to justify passing to the limit.)

(b) Prove that the converse is true.

(c) Prove similar results for a function f from to $\mathbb{C}$ to $\mathbb{C}$.

(d) As special cases, prove that

$$\frac{1}{x^2} = \sum_{n \geq 1} \frac{1}{n} \begin{bmatrix} n \\ x \end{bmatrix} \, ,$$

where $|x| > 1$ if $x \in \mathbb{C}_p$, and $\Re(x) > 0$ if $x \in \mathbb{C}$, and that

$$\frac{1}{x - a} = \frac{1}{x} + \sum_{n \geq 1} \frac{a(a+1) \cdots (a + n - 1)}{x(x+1) \cdots (x+n)} \, ,$$

where $|x| > |a|$ if x and a are in $\mathbb{C}_p$, and $\Re(x) > \Re(a)$ if x and a are in $\mathbb{C}$.

39. Let x be a parameter in $\mathbb{C}$ or in $\mathbb{C}_p$, and let $k \in \mathbb{Z}_{\geq 0}$.

(a) Assuming that $\Re(x) > k + 1$ if $x \in \mathbb{C}$, or $|x| > 1$ if $x \in \mathbb{C}_p$, use the previous exercise to prove the identity

$$\sum_{n \geq 0} \binom{n}{k} \begin{bmatrix} n \\ x \end{bmatrix} = (-1)^{k+1} \begin{bmatrix} k \\ 1 - x \end{bmatrix}$$

(see Exercise 19 of Chapter 11 for another proof).

(b) Prove that if x and y are in $\mathbb{C}$ with $\Re(x) < 1$ and $\Re(x+y) > 1$ or if x and y are in $\mathbb{C}_p$ with $v_p(x) < 0$ and $v_p(y) < 0$, with the notation used above we have

$$\sum_{n \geqslant 0} \begin{bmatrix} n \\ x \end{bmatrix} \begin{bmatrix} n \\ y \end{bmatrix} = -\sum_{n \geqslant 0} \frac{1}{n+y} \begin{bmatrix} n \\ 1-x \end{bmatrix}$$

(note that the symmetry in x and y of the right-hand side of this expression is not clear directly).

40. Prove Lemma 4.5.11.

41. Instead of choosing $p = 11$ in the proof of Proposition 4.5.18, choose $p = 23$. Show that $\alpha \equiv 33 \pmod{23^2}$, $\beta \equiv 497 \pmod{23^2}$, and that if $\phi(s) = (\alpha - \beta)u_{22s+12}$ has the power series expansion $\phi(X) = c_0 + c_1 X + c_2 X^2 + \cdots$, then $v_{23}(c_0) = v_{23}(c_1) = 1$, and $c_j \equiv 0 \pmod{23^2}$ for $j \geqslant 2$, so that Strassmann's theorem says that there exists at most one zero of $\phi(X)$ in $\mathcal{Z}_p$. Show that there indeed exists such a zero, and using Nagell's result that it does not belong to $\mathbb{Z}$. Deduce that for all v there exists s such that $u_{22s+12} \equiv -1 \pmod{23^v}$.

42. Let $\theta = \tan^{-1}(\sqrt{7})$. Using Proposition 4.5.18, find the positive integers n such that $\sin((2n+1)\theta) = \pm \sin(\theta)/2^n$.

43. Using Strassmann's theorem with $p = 3$, find all the integral solutions to the equation $x^3 + 2y^3 = 1$.

44. Using Strassmann's theorem with $p = 3$, find all the integral solutions to the equation $x^3 + 6xy^2 - y^3 = 1$.

45. Let u_n be defined by $u_0 = u_1 = 0$, $u_2 = 1$, and $u_n = 2u_{n-1} - 4u_{n-2} + 4u_{n-3}$ for $n \geqslant 3$. Prove the following result due to M. Mignotte: $u_n = 0$ if and only if $n = 0, 1, 4, 6, 13$, or 52 (use $p = 47$).

46. Let u_n be defined by $u_0 = 1$, $u_1 = 2$, $u_n = 3u_{n-1} - 5u_{n-2}$ for $n \geqslant 2$. Show that $u_n = 1$ if and only if $n = 0, 2$, or 6 (use $p = 3$).

47. Let u_n be defined by $u_0 = 0$, $u_1 = 1$, $u_n = 3u_{n-1} - 7u_{n-2}$ for $n \geqslant 2$. Show (without using Strassmann's theorem!) that $u_n = 0$ only for $n = 0$. Find the smallest $m > 0$ such that $u_m \equiv 0 \pmod{5^4}$, and show that for every v there exists $n > 0$ such that $u_n \equiv 0 \pmod{5^v}$.

48. Let u_n be defined by $u_0 = u_1 = 1$, $u_n = 5u_{n-1} - 11u_{n-2}$ for $n \geqslant 2$. Show that $u_n = 1$ if and only if $n = 0$ or 1 (use $p = 5$). Can one have $u_n = 0$?

49. Let u_n be defined by $u_0 = 0$, $u_1 = 1$, $u_n = 2u_{n-1} - 3u_{n-2}$ for $n \geqslant 2$. Show that $u_n = \pm 1$ if and only if $n = 1$ or 3 (use $p = 11$). Deduce from this all the solutions for $x \in \mathbb{Z}$ and $m \geqslant 0$ of $x^2 + 2 = 3^m$.

50. (Mahler–Lech–Cassels.) More generally, let u_n be defined by a linear recurrence relation with constant coefficients, and let $c \in \mathbb{Z}$. Show that either $u_n = c$ occurs only for finitely many n, or $u_n = c$ for all n in some arithmetic progression (prove first, for instance using one of Hensel's lemmas, that for any polynomial $P(X) \in \mathbb{Q}[X]$ there exists a prime p for which $P(X)$ splits completely in $\mathbb{Q}_p[X]$).

5. Quadratic Forms and Local–Global Principles

The aim of this chapter is to give the most important examples (and also some counterexamples, but these abound) of local to global principles.

In rough terms, a local–global principle is a statement that asserts that a certain property is true globally (usually in a number field) if and only if it is true everywhere locally (usually in p-adic fields). It is important to know when this is indeed true, and when it may not be true.

We will give three local–global principles. The best known is certainly the Hasse–Minkowski theorem, stating that the principle is valid for a single quadratic form in any number of variables. The second is the Hasse norm principle, stating that the principle is valid for the norm form of a cyclic extension of number fields. The third one is a local–global principle for powers. We will prove the Hasse–Minkowski theorem only for the number field $\mathbb{Q}$, the general case being more difficult, and not really more instructive. We will mention the Hasse norm theorem and a couple of consequences, but we will not prove it. As main reference for the Hasse–Minkowski theorem, we will follow very closely the beautiful little book by Serre [Ser1], with also some ideas from [Bor-Sha] (see also an unorthodox presentation in [Con]). For the norm principle we follow [Jan], and for powers we follow [Gras].

5.1 Basic Results on Quadratic Forms

The first four sections of this chapter will be devoted to quadratic forms, culminating in the proof of the Hasse–Minkowski theorem and some applications. In the first section, we study quadratic forms in complete generality over a commutative field K, which will always be assumed to be of characteristic different from 2 (this will be an implicit assumption that will usually not be repeated in the statements). Although the traditional way of studying quadratic forms is to consider them as homogeneous polynomials in n variables, it is much more intrinsic to consider them in the more abstract context of *quadratic modules*, exactly in the same way as doing linear algebra with general vector spaces and linear maps is much cleaner than doing linear algebra on K^n using row or column vectors and matrices.

We then specialize the field K in the same order of complexity as that explained at the end of Chapter 1, in other words first to finite fields, then to p-adic fields (including $p = \infty$), and finally to number fields, in fact only to $K = \mathbb{Q}$. In each case, we solve completely the problem of representation of an element of the field by a quadratic form.

It is to be noted that a slightly different and shorter approach to the proof of the Hasse–Minkowski theorem has been given by J.-H. Conway in [Con], which gives an unconventional, elementary, and useful approach to quadratic forms over $\mathbb{Q}$.

5.1.1 Basic Properties of Quadratic Modules

Let V be a K-vector space of finite dimension n. Recall that a quadratic form on V is a map q from V to K such that $q(ax) = a^2 q(x)$ for all $x \in V$ and $a \in K$, and $b(x,y) = (q(x+y) - q(x) - q(y))/2$ is a (symmetric) bilinear form (so that $q(x) = b(x,x)$). If $(e_i)_{1 \leqslant i \leqslant n}$ is a basis of V, the matrix Q of q in this basis is the matrix $b(e_i, e_j)$, also called the *Gram matrix* of the vectors e_i with respect to q. Thus, if $x = \sum_{1 \leqslant i \leqslant n} x_i e_i \in V$, and if X is the column vector of the x_i, then $q(x) = X^t Q X$ and $b(x,y) = Y^t Q X = X^t Q Y$ with evident notation. The matrix Q is a symmetric matrix. If (e_i') is another basis of V and if P is the matrix expressing the e_i' in terms of the e_i, then again with evident notation $X = PX'$, so that $q(x) = X^t Q X = X'^t P^t Q P X'$; hence the matrix of q in the new basis is equal to $P^t Q P$. In particular, $\det(P^t Q P) = \det(Q) \det(P)^2$, so the class of $\det(Q)$ modulo nonzero squares of K is independent of the chosen basis and called the *discriminant* of q, denoted by $d(q)$.

A pair (V, q) as above will be called a *quadratic module*, and if x and y are elements of (V, q) we will write $x \cdot y$ (the *dot product* of x and y) instead of $b(x,y)$, and in particular $q(x) = x \cdot x$. If f is a map from a quadratic module (V, q) to another (V', q'), we will say that f is a *morphism* if f is a linear map such that for all x and y we have $f(x) \cdot f(y) = x \cdot y$ for the respective dot products.

We will say that two elements x and y are *orthogonal* if $x \cdot y = 0$. The set of elements orthogonal to a given subset H of V will be denoted by $H^\perp$, and it is clearly a subspace of V. Two subspaces V_1 and V_2 of V are said to be orthogonal if $V_1 \subset V_2^\perp$ (or equivalently, if $V_2 \subset V_1^\perp$), in other words if every element of V_1 is orthogonal to every element of V_2. We will say that V is the *orthogonal direct sum* of V_1 and V_2 if $V = V_1 \oplus V_2$ and if V_1 and V_2 are orthogonal. This of course generalizes to any number of subspaces. To indicate the orthogonality, we will write $V = V_1 \oplus^\perp V_2$.

The space $V^\perp$ of vectors orthogonal to all of the vectors of V is called the *radical* of V and denoted by $\mathrm{rad}(V)$. It is clear that if W is a supplement to $\mathrm{rad}(V)$, in other words if $W \oplus \mathrm{rad}(V) = V$, then automatically $V = W \oplus^\perp \mathrm{rad}(V)$.

Let Q be the matrix of q in some basis. It is clear that $X \in V^\perp$ if and only if $Y^t Q X = 0$ for all Y, hence if and only if $QX = 0$. Thus $V^\perp$ is equal to the kernel of the matrix Q, and in particular $\dim(V^\perp) = \dim(V) - \mathrm{rk}(q)$, where $\mathrm{rk}(q)$ is the rank of the matrix Q (which is clearly independent of the chosen basis). Thus $V^\perp = \{0\}$ if and only if Q has maximal rank, hence if and only if $d(q) \neq 0$.

Definition 5.1.1. *We will say that a quadratic form (or the corresponding quadratic module) is* nondegenerate *if* $\mathrm{rad}(V) = V^\perp = \{0\}$, *or equivalently, if* $d(q) \neq 0$.

Proposition 5.1.2. *Let* (V, q) *be a nondegenerate quadratic module.*

(1) *All morphisms from* (V, q) *into some other quadratic module* (V', q') *are injective.*
(2) *If* U *is a subspace of* V *then*

$$(U^\perp)^\perp = U, \ \dim(U) + \dim(U^\perp) = \dim(V), \ \mathrm{rad}(U) = \mathrm{rad}(U^\perp) = U \cap U^\perp .$$

Furthermore, the quadratic module (U, q) *is nondegenerate if and only if* $(U^\perp, q)$ *is nondegenerate, if and only if* $V = U \oplus^\perp U^\perp$.
(3) *If* $V = V_1 \oplus^\perp V_2$, *then* (V_1, q) *and* (V_2, q) *are nondegenerate.*

Proof. (1). If f is a morphism from (V, q) to (V', q') and x is such that $f(x) = 0$, then for all $y \in V$ we have $x \cdot y = f(x) \cdot f(y) = 0$, hence $x = 0$ since q is nondegenerate.

(2) and (3). For any subspace U of V, define a linear map q_U from V to the dual U^* of U by $q_U(x) = (y \mapsto x \cdot y)$. The kernel of q_U is clearly equal to $U^\perp$, and in particular, since q is nondegenerate, q_V is injective, hence bijective. Furthermore, we clearly have $q_U = s_U \circ q_V$, where s_U is the canonical surjection from V^* to U^*. It follows that q_U is surjective. Thus $\dim(\mathrm{Ker}(q_U)) = \dim(V) - \dim(U^*)$; in other words, $\dim(V) = \dim(U) + \dim(U^\perp)$. In particular, U and $(U^\perp)^\perp$ have the same dimension, and since the first is contained in the second they are equal. The other assertions follow immediately from this. $\qquad\square$

Definition 5.1.3. *We say that* $x \in V$ *is* isotropic *if* $q(x) = 0$. *A subspace* U *of* V *is* isotropic *if all its elements are isotropic.*

It is clear that U is isotropic if and only if $U \subset U^\perp$ if and only if the restriction of q to U is identically 0.

Definition 5.1.4. *A quadratic module* (U, q) *is called a* hyperbolic plane *if it has a basis formed by two isotropic elements* x *and* y *such that* $x \cdot y \neq 0$.

Multiplying one of the vectors by $1/(x \cdot y)$, we may assume if desired that in fact $x \cdot y = 1$, and then the matrix of q in this basis will be $\left(\begin{smallmatrix} 0 & 1 \\ 1 & 0 \end{smallmatrix}\right)$. In particular, $d(q) = -1$ so q is nondegenerate.

Lemma 5.1.5. *Let $x \neq 0$ be an isotropic vector of a nondegenerate quadratic module (V, q). There existsn a subspace U of V containing x that is a hyperbolic plane.*

Proof. Since q is nondegenerate, there exists $z \in V$ such that $x \cdot z \neq 0$. The element $y = 2z - (z \cdot z/x \cdot z)x$ is clearly isotropic and $x \cdot y \neq 0$, so the subspace U generated by x and y is a hyperbolic plane containing x. $\square$

Corollary 5.1.6. *If (V, q) is nondegenerate and contains a nonzero isotropic vector then $q(V) = K$. In another language (which we will soon use systematically), if q represents 0 nontrivially then q represents every element of K.*

Proof. Thanks to the lemma we know that there exists an isotropic vector y such that $x \cdot y = 1$. Thus if $a \in K$ it is clear that $a = q(x + (a/2)y)$, proving the corollary. $\square$

Let $\mathcal{B} = (e_1, \ldots, e_n)$ be a basis of V. Recall that the basis is an orthogonal basis if the e_i are pairwise orthogonal. This is equivalent to saying that the matrix Q of q in the basis $\mathcal{B}$ is a diagonal matrix, so that if $x = \sum_{1 \leqslant i \leqslant n} x_i e_i$ then $q(x) = \sum_{1 \leqslant i \leqslant n} a_i x_i^2$.

Proposition 5.1.7. *Every quadratic module has an orthogonal basis.*

Proof. We prove this by induction on $n = \dim(V)$, the case $n = 0$ being trivial. If V is isotropic, then all bases are orthogonal. Otherwise, choose some nonisotropic element $e_1 \in V$, and let $H = e_1^\perp$. This is a subspace of V of dimension exactly $n - 1$, and since $e_1 \notin H$ by assumption we have $V = Ke_1 \oplus^\perp H$. By our induction hypothesis H has an orthogonal basis $(e_2, \ldots, e_n)$, so $(e_1, \ldots, e_n)$ is an orthogonal basis of V. $\square$

Note that there exist more "computational" proofs of this result, based for instance on Gauss's reduction of quadratic forms.

5.1.2 Contiguous Bases and Witt's Theorem

We will now prove a result that is technical, but is crucial for what follows. We begin with the following definition.

Definition 5.1.8. *We say that two orthogonal bases $\mathcal{B}$ and $\mathcal{B}'$ of V are contiguous if they have a common element.*

Theorem 5.1.9. *Let (V, q) be a nondegenerate quadratic module of dimension at least 3 and let $\mathcal{B}$ and $\mathcal{B}'$ be two orthogonal bases of V. There exists a finite sequence $(\mathcal{B}^{(i)})_{0 \leqslant i \leqslant m}$ of orthogonal bases of V such that $\mathcal{B}^{(0)} = \mathcal{B}$, $\mathcal{B}^{(m)} = \mathcal{B}'$ and such that $\mathcal{B}^{(i)}$ and $\mathcal{B}^{(i+1)}$ are contiguous for $0 \leqslant i < m$.*

Proof. Let $\mathcal{B} = (e_1, \ldots, e_n)$ and $\mathcal{B}' = (e_1', \ldots, e_n')$. We consider three different cases.

Case 1: $(e_1 \cdot e_1')^2 \neq (e_1 \cdot e_1)(e_1' \cdot e_1')$.

Thus e_1 and e_1' are not collinear and the plane $P = Ke_1 + Ke_1'$ is *nondegenerate*, since the determinant of q restricted to P is given by the difference of the above two expressions. It follows that there exist vectors v_2 and v_2' such that (e_1, v_2) and (e_1', v_2') are orthogonal bases of P. Let $H = P^\perp$, so that $V = H \oplus^\perp P$ by Proposition 5.1.2, and let $(v_3, \ldots, v_n)$ be an orthogonal basis of H. We then clearly have the following chain of contiguous orthogonal bases from $\mathcal{B}$ to $\mathcal{B}'$ (recall that $n \geqslant 3$):

$$\mathcal{B} \longrightarrow (e_1, v_2, v_3, \ldots, v_n) \longrightarrow (e_1', v_2', v_3, \ldots, v_n) \longrightarrow \mathcal{B}' ,$$

proving the theorem in this case.

Case 2: $(e_1 \cdot e_2')^2 \neq (e_1 \cdot e_1)(e_2' \cdot e_2')$.

Same proof replacing e_1' by e_2'.

Case 3: $(e_1 \cdot e_i')^2 = (e_1 \cdot e_1)(e_i' \cdot e_i')$ for $i = 1$ and $i = 2$.

This case is slightly more difficult. We begin with the following lemma.

Lemma 5.1.10. *There exists $a \in K$ such that $e_a = e_1' + ae_2'$ is not isotropic and generates with e_1 a nondegenerate plane.*

Proof. We enumerate the forbidden values of a. The vector e_a is not isotropic if and only if $a^2 \neq -(e_1' \cdot e_1')/(e_2' \cdot e_2')$. The subspace $Ke_1 + Ke_a$ is a nondegenerate plane if and only if $(e_1 \cdot e_1)(e_a \cdot e_a) - (e_1 \cdot e_a)^2 \neq 0$, and by the assumption of Case 3 this boils down to $-2a(e_1 \cdot e_1')(e_1 \cdot e_2') \neq 0$. Since Case 3 implies that $e_1 \cdot e_i' \neq 0$ for $i = 1$ and $i = 2$, this is equivalent to $a \neq 0$. To summarize, we have to avoid at most three values of $a \in K$, finishing the proof of the lemma if K has at least four elements. Since the characteristic of K is not equal to 2, there remains the case $K = \mathbb{F}_3$, and then we check immediately that $a = 1$ is a suitable value. □

Resuming the proof of the theorem, let us choose e_a satisfying the conditions of the lemma. Since e_a is not isotropic there exists v_2 such that (e_a, v_2) is an orthogonal basis of $Ke_1' + Ke_2'$. Set

$$\mathcal{B}'' = (e_a, v_2, e_3', \ldots, e_n') ,$$

which is clearly an orthogonal basis of V. Since $Ke_1 + Ke_a$ is a nondegenerate plane by construction, using e_a instead of e_1' in the proof of Case 1, we see that there exists a chain of contiguous bases from $\mathcal{B}$ to $\mathcal{B}''$. Since $n \geqslant 3$, $\mathcal{B}''$ is contiguous to $\mathcal{B}'$, so the result follows also in this case. □

The above technical theorem will be essential for the proof of the invariance of a quantity $\varepsilon(q)$ that we will define later (Theorem 5.2.15).

We now come to Witt's theorem, which is probably the most important theorem on general quadratic forms or modules. Let (V, q) and (V', q') be nondegenerate quadratic modules, let U be a subspace of V, and let s be an *injective* morphism of quadratic modules from U to V'. The problem is to extend s to a morphism defined on a subspace larger than U, and if possible to all of V. We begin with the degenerate case, which we will need.

Lemma 5.1.11. *With the above assumptions, if (U, q) is degenerate we can extend s to an injective morphism s_1 from U_1 to V', where U_1 is a subspace of V containing U as a hyperplane.*

Proof. Let $x \in \operatorname{rad}(U)$ be nonzero. There exists a linear form k on U such that $k(x) = 1$. Since V is nondegenerate, as we have seen in the proof of Proposition 5.1.2 the canonical map q_U from V to U^* is surjective; hence there exists $y \in V$ such that $k(u) = u \cdot y$ for all $u \in U$. If we set $z = y - (y \cdot y)x/2$, we note that $u \cdot z = u \cdot y = k(u)$ for all $u \in U$ (since $x \in \operatorname{rad}(U)$), and $z \cdot z = y \cdot y - (y \cdot y)x \cdot y = 0$ since $x \cdot y = k(x) = 1$. We set $U_1 = U \oplus Kz$, which clearly contains U as a hyperplane since otherwise $z \in U$, hence $k(x) = x \cdot z = 0$ since $x \in \operatorname{rad}(U)$, a contradiction. We can apply the same construction to $U' = s(U)$, $x' = s(x)$, and $k' = k \circ s^{-1}$ and obtain $z' \in V'$ and $U'_1 = U' + Kz'$. It is clear that the linear map s_1 from U_1 to V' that coincides with s on U and sends z to z' is an isomorphism from U_1 to U'_1. $\square$

Theorem 5.1.12 (Witt). *If (V, q) and (V', q') are isomorphic and nondegenerate quadratic modules, every injective morphism s from a subspace U of V to V' can be extended to an isomorphism from V to V'.*

Proof. Since V and V' are isomorphic, we may assume that $V = V'$. Furthermore, thanks to the above lemma, as long as (U, q) is degenerate we can extend s to a subspace having one extra dimension. Thus we may assume that (U, q) is nondegenerate. We prove the theorem by induction on the dimension of U.

Assume first that $\dim(U) = 1$, which is in fact the only case in which we have to do some real work. Since (U, q) is nondegenerate, we have $U = Kx$ for some nonisotropic element x, and if we set $y = s(x)$ we have by definition $y \cdot y = x \cdot x$. It is clear that one can choose $\varepsilon = \pm 1$ such that $x + \varepsilon y$ is not isotropic. Indeed, otherwise we would have $2x \cdot x + 2x \cdot y = 2x \cdot x - 2x \cdot y = 0$, hence $x \cdot x = 0$, absurd. Set $z = x + \varepsilon y$ for such a choice of ε, and let $H = z^\perp$, so that $V = Kz \oplus^\perp H$ by Proposition 5.1.2. Let σ be the symmetry with respect to H, defined by $\sigma(w) = w - 2((w \cdot z)/(z \cdot z))z$, which is clearly an automorphism. Noting that $(x - \varepsilon y) \cdot z = x \cdot x - y \cdot y = 0$, we have therefore $\sigma(x - \varepsilon y) = x - \varepsilon y$, and $\sigma(x + \varepsilon y) = \sigma(z) = -x - \varepsilon y$, hence $\sigma(x) = -\varepsilon y$, so that $-\varepsilon \sigma$ is an automorphism extending s.

Assume now that $\dim(U) > 1$, and decompose U as $U = U_1 \oplus^\perp U_2$ with U_1 and U_2 nonzero. By induction, we know that the restriction s_1 of s to U_1 extends to an automorphism σ_1 of V. If $x \in U_1$ and $z \in U_2$, we have

$$x \cdot \sigma_1^{-1}(s(z)) = \sigma_1(x) \cdot s(z) = s(x) \cdot s(z) = x \cdot z = 0 \; ;$$

hence if we set $V_1 = U_1^{\perp}$, we see that $\sigma_1^{-1} \circ s(U_2) \subset V_1$, and by assumption $U_2 \subset V_1$, so by our inductive hypothesis on the dimension, the injective morphism $\sigma_1^{-1} \circ s$ restricted to U_2 extends to an automorphism σ_2 of V_1, so that in particular $\sigma_1 \circ \sigma_2(x) = s(x)$ for all $x \in U_2$. It is now clear that if we define σ to be equal to s on U_1 and to $\sigma_1 \circ \sigma_2$ on V_1, then σ is an automorphism of V extending s.   □

Corollary 5.1.13. *Two isomorphic subspaces of a nondegenerate quadratic module have isomorphic orthogonal supplements.*

Proof. Indeed, we simply extend the isomorphism between the subspaces to an automorphism of the quadratic module, and then restrict to the orthogonal supplements.   □

5.1.3 Translations into Results on Quadratic Forms

If $V = K^n$ and (e_i) is the canonical basis of V, we can identify a quadratic form on V with a homogeneous polynomial of degree 2 in n variables over K by the formula

$$q(x_1, \ldots, x_n) = \sum_{1 \leqslant i, j \leqslant n} q_{i,j} x_i x_j \; ,$$

where the $q_{i,j}$ are the entries of the symmetric matrix Q. We will then simply speak of a quadratic form with coefficients in K. Note that we may also write

$$q(x_1, \ldots, x_n) = \sum_{1 \leqslant i \leqslant n} q_{i,i} x_i^2 + 2 \sum_{1 \leqslant i < j \leqslant n} q_{i,j} x_i x_j \; .$$

Conversely, if q is given as a homogeneous polynomial of degree 2 as above, we can of course consider q as a quadratic form on K^n.

Definition 5.1.14. *We will say that two quadratic forms q and q' are* equivalent *if the quadratic modules (K^n, q) and (K^n, q') are isomorphic, and we write $q \sim q'$.*

Thus q and q' are equivalent if and only if there exists $\gamma \in \mathrm{GL}_n(K)$ such that $q'(x) = q(\gamma(x))$. If P is the matrix of γ in the canonical basis (e_i) of K^n, then $q'(x) = (PX)^t Q(PX)$, so the matrix of q' is $P^t Q P$.

From now on, we will denote by $\oplus$ the *orthogonal sum* of quadratic forms. More precisely, if $q(x_1, \ldots, x_n)$ and $q'(x_1, \ldots, x_m)$ are two quadratic forms in n and m variables respectively, the form $q \oplus q'$ is the quadratic form in $n + m$ variables defined by

$$(q \oplus q')(x_1, \ldots, x_{n+m}) = q(x_1, \ldots, x_n) + q'(x_{n+1}, \ldots, x_{n+m}) \; .$$

We also set $q \ominus q' = q \oplus (-q')$.

In this subsection, we essentially give translations into the more classical language of quadratic forms of the results we have proved on quadratic modules.

Definition 5.1.15. *A form $q(x_1, x_2)$ in two variables is hyperbolic if*

$$q(x_1, x_2) \sim x_1 x_2 \sim x_1^2 - x_2^2 .$$

This clearly means that the quadratic module (K^2, q) is a hyperbolic plane.

Recall that we say that a form q *represents* $a \in K$ if there exists $x \in K^n$ such that $q(x) = a$, with the added condition that $x \neq 0$ when $a = 0$. Thus q represents 0 if and only if (K^n, q) contains a nonzero isotropic element.

Proposition 5.1.16. *If q is nondegenerate and represents 0, then there exists a hyperbolic form h and a nondegenerate form q' such that $q \sim h \oplus q'$. Furthermore, q represents all elements of K.*

Proof. This is Lemma 5.1.5 and Corollary 5.1.6. The fact that q' is nondegenerate follows from Proposition 5.1.2.   □

Corollary 5.1.17. *Let q be a nondegenerate quadratic form in n variables and let $c \in K^*$. The following three conditions are equivalent:*

(1) *The form q represents c.*
(2) *There exists a quadratic form q_1 in $n-1$ variables such that $q \sim cx_0^2 \oplus q_1$.*
(3) *The quadratic form $q \ominus cx_0^2$ represents 0 in K.*

Proof. (2) implies (1) is trivial by taking $x_0 = 1$ and the other variables equal to 0. Conversely, if q represents c, there exists $x \in K^n$ such that $q(x) = x \cdot x = c$. Since q is nondegenerate, if $H = x^\perp$ we have $K^n = H \oplus^\perp Kx$, hence $q \sim q_1 \oplus cx_0^2$, where q_1 is the quadratic form corresponding to a choice of basis of H, proving (2). (1) implies (3) is trivial by choosing $x_0 = 1$ and $(x_1, \ldots, x_n)$ a representation of c. Finally, if $-c\alpha_0^2 + q(\alpha_1, \ldots, \alpha_n) = 0$, then either $\alpha_0 \neq 0$, in which case $c = q(\alpha_1/\alpha_0, \ldots, \alpha_n/\alpha_0)$, or $\alpha_0 = 0$, and we conclude by Proposition 5.1.16.   □

Corollary 5.1.18. *Let q_1 and q_2 be two nonzero nondegenerate quadratic forms and let $q = q_1 \ominus q_2$ as defined above. The following properties are equivalent:*

(1) *The form q represents 0.*
(2) *There exists $c \in K^*$ represented by both q_1 and q_2.*
(3) *There exists $c \in K^*$ such that both $q_1 \ominus cx_0^2$ and $q_2 \ominus cx_0^2$ represent 0.*

Proof. The equivalence of (2) and (3) follows from the above corollary, and (2) implies (1) is trivial. Let us prove that (1) implies (2). If $q = q_1 \ominus q_2$ represents 0 there exist x and y in the corresponding quadratic modules such that $q_1(x) = q_2(y)$, with x and y not both zero. Consider the element $c = q_1(x) = q_2(y)$. If $c \neq 0$, then (2) is proved. Otherwise, since x and y are not both zero, at least one of the forms, say q_1, represents 0; hence by Proposition 5.1.16 it represents all elements of K, and in particular all nonzero values taken by q_2. □

Theorem 5.1.19. *Let q be a quadratic form in n variables. There exists an equivalent form that is a* diagonal *quadratic form; in other words, there exist $a_i \in K$ such that $q \sim \sum_{1 \leqslant i \leqslant n} a_i x_i^2$.*

Proof. This is a translation of Proposition 5.1.7. As already mentioned, this can be proved computationally using Gauss's reduction of quadratic forms into sums of squares, which gives an explicit algorithm for finding the a_i and the linear equivalence from q to the diagonal form. □

Note that things are much easier to read in diagonal form: the rank of q is equal to the number of nonzero a_i, and q is nondegenerate if and only if the rank is equal to n, in which case $d(q) = \prod_{1 \leqslant i \leqslant n} a_i$ up to squares as usual.

However, note that the above theorem is valid only over a *field*. For instance, the quadratic form $x_1^2 + x_1 x_2 + x_2^2$ is not equivalent to a diagonal form over $\mathbb{Z}$, but only over $\mathbb{Q}$ (Exercise 2).

Theorem 5.1.20 (Witt). *Let $q = q_1 \oplus q_2$ and $q' = q_1' \oplus q_2'$ be two nondegenerate quadratic forms. If $q \sim q'$ and $q_1 \sim q_1'$, then $q_2 \sim q_2'$.*

Proof. This is the translation of Corollary 5.1.13, the corollary to Witt's theorem. It is this theorem to which one usually refers when talking about Witt's theorem. □

Corollary 5.1.21. *If q is nondegenerate there exist hyperbolic forms h_i for $1 \leqslant i \leqslant m$ and a form q' that does not represent 0 such that*

$$q \sim h_1 \oplus \cdots \oplus h_m \oplus q',$$

and this decomposition is unique up to equivalence.

Proof. Existence follows from Proposition 5.1.16. Let us prove uniqueness. With evident notation let

$$q \sim \sum_{1 \leqslant i \leqslant m} h_i \oplus q_1 \sim \sum_{1 \leqslant i \leqslant m'} h_i' \oplus q_2,$$

and assume for instance that $m' \leqslant m$. Since all hyperbolic forms are equivalent the above theorem implies that $\sum_{1 \leqslant i \leqslant m-m'} h_i \oplus q_1 \sim q_2$, which is a contradiction if $m \neq m'$, since q_2 does not represent 0, while a hyperbolic form does. □

5.2 Quadratic Forms over Finite and Local Fields

After having studied general properties of quadratic forms, we now specialize the base field K.

5.2.1 Quadratic Forms over Finite Fields

We begin with the simplest possible fields, the finite fields. Recall that what makes them simple is *not* the fact that they are finite, but the fact that their structure and hierarchy are very simple; see Chapter 2. For instance, finite *groups* are extremely complicated and only partly understood objects.

Thus let $q = p^k$ be a prime power, with $p \neq 2$.

Proposition 5.2.1. *A quadratic form over $\mathbb{F}_q$ of rank $n \geqslant 2$ represents all elements of $\mathbb{F}_q^*$, and a quadratic form of rank $n \geqslant 3$ represents all elements of $\mathbb{F}_q$.*

Proof. Corollary 5.1.17 tells us that both statements are equivalent. To prove the first we could apply the Chevalley–Warning Theorem 2.5.2, but we give a direct proof that is a direct application of the so-called pigeonhole principle ("principe des tiroirs" in French). Indeed, by Theorem 5.1.19 we may assume that our quadratic form is $\sum_{1 \leqslant i \leqslant m} a_i x_i^2$ with $a_1 a_2 \neq 0$. Let $a \in \mathbb{F}_q$. We choose $x_i = 0$ for $i \geqslant 3$. Since q is odd the map $x \mapsto x^2$ is a group homomorphism of $\mathbb{F}_q^*$ onto itself, and its kernel has two elements since $\mathbb{F}_q$ is a field. It follows that its image has $(q-1)/2$ elements, so adding the element 0 there are $(q+1)/2$ squares in $\mathbb{F}_q$. Since $a_1 a_2 \neq 0$ it follows that the subsets $\{a_1 x_1^2\}$ and $\{a - a_2 x_2^2\}$ of $\mathbb{F}_q$ also have $(q+1)/2$ elements hence have a nonempty intersection, proving the proposition. $\square$

Proposition 5.2.2. *Let $c \in \mathbb{F}_q^*$ that is not a square in $\mathbb{F}_q^*$. A nondegenerate quadratic form over $\mathbb{F}_q$ is equivalent to $x_1^2 + \cdots + x_{n-1}^2 + ax_n^2$ with $a = 1$ if its discriminant is a square, and with $a = c$ otherwise.*

Proof. Since q is odd, the map $x \mapsto x^2$ is a group homomorphism from $\mathbb{F}_q^*$ to itself with kernel $\{\pm 1\}$; hence $\mathbb{F}_q^*/\mathbb{F}_q^{*2}$ has order 2, generated by the class modulo squares of c. Thus if $n = 1$ the result is true. Assume $n \geqslant 2$ and the result true by induction up to $n - 1$. Thanks to the preceding proposition the form represents any nonzero element, hence 1, so thanks to Corollary 5.1.17 there exists a quadratic form g in $n - 1$ variables such that our form is equivalent to $x_0^2 + g$, and the result follows by induction. $\square$

Corollary 5.2.3. *Two nondegenerate quadratic forms over $\mathbb{F}_q$ are equivalent if and only if they have the same rank and the same discriminant in $\mathbb{F}_q^*/\mathbb{F}_q^{*2}$.*

Proof. Clear from the above proposition. $\square$

5.2.2 Definition of the Local Hilbert Symbol

The crucial case in the study of quadratic forms over p-adic fields is the case of three variables, so it is necessary to study in detail this case first. For this, we introduce the *Hilbert symbol*, which will be sufficient for the local study of quadratic forms.

In this section, we let $\mathcal{K}$ be a completion of $\mathbb{Q}$, in other words either $\mathbb{Q}_p$ or $\mathbb{R}$.

Definition 5.2.4. *If a and b are in $\mathcal{K}^*$, we set $(a, b) = 1$ if the equation $ax^2 + by^2 = z^2$ has a nontrivial solution (in other words with $(x, y, z) \neq (0, 0, 0)$), and $(a, b) = -1$ otherwise. The number (a, b) is called the (local) Hilbert symbol of a and b.*

When using several completions as we will do in the global situation, we will write $(a, b)_p$ or $(a, b)_\infty$ to specify the prime, or simply $(a, b)_v$ to indicate the place v.

It is clear that (a, b) does not change when a or b is multiplied by a nonzero square. Thus (a, b) can be considered as defining a map from $(\mathcal{K}^*/\mathcal{K}^{*2}) \times (\mathcal{K}^*/\mathcal{K}^{*2})$ to $\{\pm 1\}$.

Proposition 5.2.5. *Let a and b be in $\mathcal{K}^*$. We have $(a, b) = 1$ if and only if $a \in \mathcal{N}(\mathcal{K}(\sqrt{b})^*)$, i.e., if and only if a is the norm of an element of $\mathcal{K}(\sqrt{b})^*$.*

Proof. If b is a square, then $(0, 1, \sqrt{b})$ is evidently a solution to our equation, and $\mathcal{K}(\sqrt{b}) = \mathcal{K}$, so the proposition is clear in this case. Otherwise the elements of the quadratic extension $\mathcal{K}(\sqrt{b})$ have the form $u + v\sqrt{b}$, hence a is a norm if and only if it has the form $a = u^2 - bv^2$. If this is the case $(1, v, u)$ is a solution to our equation, and conversely if $ax^2 + by^2 = z^2$, then $x \neq 0$ (otherwise b is a square), hence a is the norm of $(z/x) + (y/x)\sqrt{b}$. $\square$

Proposition 5.2.6. *We have the following formulas, where all the elements that occur are assumed to be nonzero:*

(1) *$(a, b) = (b, a)$ and $(a, c^2) = 1$.*
(2) *$(a, -a) = 1$ and $(a, 1 - a) = 1$.*
(3) *$(a, b) = 1$ implies $(aa', b) = (a', b)$.*
(4) *$(a, b) = (a, -ab) = (a, (1 - a)b)$.*

Proof. (1) is clear. When $b = -a$, (respectively $b = 1 - a$), then $(1, 1, 0)$ (respectively $(1, 1, 1)$) is a nontrivial solution to $ax^2 + by^2 = z^2$, proving (2). For (3), by the preceding proposition if $(a, b) = 1$ then $a \in \mathcal{N}(\mathcal{K}(\sqrt{b})^*)$ hence by multiplicativity of the norm $a' \in \mathcal{N}(\mathcal{K}(\sqrt{b})^*)$ if and only if $aa' \in \mathcal{N}(\mathcal{K}(\sqrt{b})^*)$. Note that this formula is a special case of the bilinearity of the Hilbert symbol $(aa', b) = (a, b)(a', b)$, which we will prove below. Finally, (4) follows immediately from (1), (2), and (3): for instance, since $(-a, a) = 1$, we have $(a, -ab) = (-ab, a) = (b, a) = (a, b)$. $\square$

5.2.3 Main Properties of the Local Hilbert Symbol

The two main results on the local Hilbert symbol are first an explicit formula in terms of the Legendre symbol, and as a corollary the fact that the Hilbert symbol is a nondegenerate bilinear form. We begin with the explicit computation of the Hilbert symbol. When $\mathcal{K} = \mathbb{Q}_p$, we denote as usual by U_p the group of p-adic units.

Theorem 5.2.7. (1) *For $\mathcal{K} = \mathbb{R}$, we have $(a, b) = -1$ if $a < 0$ and $b < 0$, and $(a, b) = 1$ if a or b is positive.*

(2) *For $\mathcal{K} = \mathbb{Q}_p$ with $p \neq 2$, write $a = p^\alpha a_1$, $b = p^\beta b_1$ with a_1 and b_1 in U_p. Then*

$$(a, b) = (-1)^{\alpha\beta(p-1)/2} \left(\frac{a_1}{p}\right)^\beta \left(\frac{b_1}{p}\right)^\alpha .$$

(3) *For $\mathcal{K} = \mathbb{Q}_2$, with the same notation we have*

$$(a, b) = (-1)^{(a_1-1)(b_1-1)/4} \left(\frac{a_1}{2}\right)^\beta \left(\frac{b_1}{2}\right)^\alpha$$

(recall that $\left(\frac{a}{2}\right) = (-1)^{(a^2-1)/8}$ when a is odd).

Proof. (1) is trivial, so we assume that $\mathcal{K} = \mathbb{Q}_p$. Note first the following lemma.

Lemma 5.2.8. *Assume that $b \in U_p$. Then if the equation $px^2 + by^2 = z^2$ has a nontrivial solution in $\mathbb{Q}_p$, it has one such that $x \in \mathbb{Z}_p$ and y and z are in U_p.*

Proof. Let (x, y, z) be a nontrivial solution. Dividing by p^v with $v = \min(v_p(x), v_p(y), v_p(z))$, we may assume that x, y, and z are in $\mathbb{Z}_p$ with at least one of them in U_p. If we had $y \notin U_p$, then $v_p(y) \geqslant 1$, hence $v_p(z) \geqslant 1$, hence $v_p(px^2) \geqslant 2$, so $v_p(x) \geqslant 1$, contradicting the fact that one of x, y, and z is in U_p. Thus $y \in U_p$, hence $z \in U_p$ also. $\qquad\square$

(2). Assume that $p \neq 2$. The Hilbert symbol (a, b) depends only on a and b modulo squares, hence on the parity of α and β. We thus consider three cases.

Case 1: $\alpha = \beta = 0$.

By Proposition 5.2.1, we know that the equation $a_1 x^2 + b_1 y^2 = z^2$ has a nontrivial solution modulo p, and by Hensel's lemma since a_1 and b_1 are p-adic units and $p \neq 2$, this solution lifts to a p-adic solution, so that $(a, b) = 1$ as claimed.

Case 2: $\alpha = 1$, $\beta = 0$.

By Case 1 we have $(a_1, b_1) = 1$. Thus by Proposition 5.2.6 (3) we have $(a, b) = (pa_1, b_1) = (p, b_1)$. If b_1 is a square in $\mathbb{Q}_p$ then $(p, b_1) = 1$ and

$\left(\frac{b_1}{p}\right) = 1$, so we have $(a, b) = \left(\frac{b_1}{p}\right) = 1$ in this case. If b_1 is not a square in $\mathbb{Q}_p$, then $\left(\frac{b_1}{p}\right) = -1$, and the above lemma implies that $px^2 + b_1 y^2 = z^2$ does not have a nontrivial solution; otherwise, since $y \in U_p$, b_1 would be congruent to $(z/y)^2$ modulo p; hence $(a, b) = \left(\frac{b_1}{p}\right) = -1$ in this case.

Case 3: $\alpha = \beta = 1$.

By Proposition 5.2.6 (4) and Case 2, we have

$$(a, b) = (pa_1, pb_1) = (pa_1, -p^2 a_1 b_1) = (pa_1, -a_1 b_1)$$
$$= \left(\frac{-a_1 b_1}{p}\right) = (-1)^{(p-1)/2} \left(\frac{a_1}{p}\right)\left(\frac{b_1}{p}\right) ,$$

giving the desired formula.

(3). Assume now that $p = 2$. The proof will be very similar, but as usual with the added complications coming from the prime 2. We consider the same three cases.

Case 1: $\alpha = \beta = 0$.

We must show that $(a_1, b_1) = 1$ if either a_1 or b_1 is congruent to 1 modulo 4, and to -1 otherwise. Thus assume first that $a_1 \equiv 1 \pmod{4\mathbb{Z}_2}$. Then either $a_1 \equiv 1 \pmod{8\mathbb{Z}_2}$, in which case a_1 is a square by Proposition 4.3.23, hence $(a, b) = 1$ in that case. Or $a_1 \equiv 5 \pmod{8\mathbb{Z}_2}$. But then $a_1 + 4b_1 \equiv 1 \pmod{8\mathbb{Z}_2}$ is the square of some $w \in \mathbb{Z}_2$; hence $(1, 2, w)$ is a nontrivial solution to $a_1 x^2 + b_1 y^2 = z^2$, so once again $(a, b) = 1$.

Assume now that $a_1 \equiv b_1 \equiv -1 \pmod{4\mathbb{Z}_2}$, and by contradiction assume that there exists a nontrivial solution (x, y, z) to the equation $a_1 x^2 + b_1 y^2 = z^2$, where as above we may assume that x, y, and z are in $\mathbb{Z}_2$, with at least one in U_2. Then $x^2 + y^2 + z^2 \equiv 0 \pmod{4\mathbb{Z}_2}$, and since the squares modulo 4 are 0 or 1, this implies that x, y, and z are all three nonunits, a contradiction, so that $(a, b) = -1$ in this case.

Case 2: $\alpha = 1$, $\beta = 0$.

We begin by proving the result when $a_1 = 1$, i.e., we must show that $(2, b_1) = \left(\frac{b_1}{2}\right)$, in other words that $(2, b_1) = 1$ if and only if $b_1 \equiv \pm 1 \pmod{8\mathbb{Z}_2}$. If $(2, b_1) = 1$, the above lemma implies that there exist x, y, and z in $\mathbb{Z}_2$ with y and z in U_2 such that $2x^2 + b_1 y^2 = z^2$. Thus $y^2 \equiv z^2 \equiv 1 \pmod{8\mathbb{Z}_2}$, hence $b_1 \equiv 1 - 2x^2 \pmod{8\mathbb{Z}_2}$, so that $b_1 \equiv \pm 1 \pmod{8\mathbb{Z}_2}$. Conversely, if $b_1 \equiv 1 \pmod{8\mathbb{Z}_2}$ then b_1 is a square, so $(2, b_1) = 1$, and if $b_1 \equiv -1 \pmod{8\mathbb{Z}_2}$ then $-b_1$ is a square; hence $(2, b_1) = (2, -1)$, and $(1, 1, 1)$ is a nontrivial solution to $2x^2 - y^2 = z^2$, so that $(2, b_1) = 1$ also in this case.

We now prove the general result, in other words the fact that $(2a_1, b_1) = (a_1, b_1)(2, b_1)$. By Proposition 5.2.6 (3), this is true when either $(2, b_1) = 1$ or $(a_1, b_1) = 1$, so assume that $(2, b_1) = (a_1, b_1) = -1$. By what we have proved above, this means that $a_1 \equiv b_1 \equiv -1 \pmod{4\mathbb{Z}_2}$ and $b_1 \equiv \pm 3 \pmod{8\mathbb{Z}_2}$, so $b_1 \equiv 3 \pmod{8\mathbb{Z}_2}$. After multiplication by elements that are congruent to 1 modulo $8\mathbb{Z}_2$, hence squares, we may assume that $a_1 = -1$ and $b_1 = 3$ or $a_1 = 3$

and $b_1 = -5$. The equations $-2x^2 + 3y^2 = z^2$ and $6x^2 - 5y^2 = z^2$ having $(1, 1, 1)$ as nontrivial solution, we have therefore $(2a_1, b_1) = 1$ as claimed, finishing the proof of Case 2.

Case 3: $\alpha = \beta = 1$.

As in the case $p > 2$, Proposition 5.2.6 (4) and Case 2 show that

$$(2a_1, 2b_1) = (2a_1, -4a_1b_1) = (2a_1, -a_1b_1)$$

$$= (-1)^{(a_1-1)(b_1-1)/4}\left(\frac{-a_1b_1}{2}\right) = (-1)^{(a_1-1)(b_1-1)/4}\left(\frac{a_1}{2}\right)\left(\frac{b_1}{2}\right),$$

finishing the proof of the theorem.   □

From this theorem it is now easy to deduce the following, of which Proposition 5.2.6 (3) is a special case.

Corollary 5.2.9. *The Hilbert symbol is a nondegenerate bilinear form on the $\mathbb{F}_2$-vector space $\mathcal{K}^*/\mathcal{K}^{*2}$.*

Proof. When $\mathcal{K} = \mathbb{R}$, $\mathcal{K}^*/\mathcal{K}^{*2}$ is a vector space of dimension 1 over $\mathbb{F}_2$ having $\{1, -1\}$ as representatives, and the result is trivial. When $\mathcal{K} = \mathbb{Q}_p$, the bilinearity comes from the multiplicativity of the Legendre–Kronecker symbol. To show that it is nondegenerate, let $\bar{a} \in \mathcal{K}^*/\mathcal{K}^{*2}$ not the identity class. We consider separately the cases $p \neq 2$ and $p = 2$. If $p \neq 2$, by Proposition 4.3.22 we can take as representative in $\mathcal{K}^*$ either $a = n$, p, or np, where n is a quadratic nonresidue modulo p, i.e., an integer such that $\left(\frac{n}{p}\right) = -1$. Then clearly $(n, p) = -1$, so that if we choose b equal respectively to the class of p, n, and n, we have $(a, b) = -1$, showing that the form is nondegenerate. If $p = 2$, by Proposition 4.3.23 we can take as representatives in $\mathcal{K}^*$ the numbers $5, -1, -5, 2, 10, -2, -10$, and we check that $(2a_1, 5) = -1$, while $(5, 2) = (-1, -1) = (-5, -1) = -1$, proving nondegeneracy also in this case.   □

Corollary 5.2.10. *If b is not a square in $\mathcal{K}^*$, then $\mathcal{N}(K(\sqrt{b})^*)$ is a subgroup of index 2 in $\mathcal{K}^*$.*

Proof. Clear since the map $a \mapsto (a, b)$ from $\mathcal{K}^*$ to $\{\pm 1\}$ has kernel $\mathcal{N}(K(\sqrt{b})^*)$ by Proposition 5.2.5, and is surjective since the Hilbert symbol is nondegenerate.   □

Remark. More generally, local class field theory asserts that if $\mathcal{L}/\mathcal{K}$ is an Abelian extension of local fields with Galois group G, then $\mathcal{K}^*/\mathcal{N}(\mathcal{L}^*)$ is isomorphic to G and the extension $\mathcal{L}/\mathcal{K}$ is determined by $\mathcal{N}(\mathcal{L}^*)$.

It is clear that the Hilbert symbol allows us to treat completely the problem of representation of 0 by quadratic forms in three variables. More precisely, we have the following result.

Proposition 5.2.11. *Let* $q(x, y, z) = ax^2 + by^2 + cz^2$ *be a nondegenerate quadratic form in three variables with coefficients in* $\mathbb{Q}_p$ *(including* $p = \infty$*). Set* $\varepsilon = \varepsilon(q) = (a, b)(a, c)(b, c)$, *and let* $d = d(q) = abc$ *be the discriminant of* q. *Then* q *represents* 0 *in* $\mathbb{Q}_p$ *if and only if* $(-1, -d) = \varepsilon$.

Proof. The form q represents 0 if and only if the form $-cq$ does, hence if and only if $-acx^2 - bcy^2 = z^2$ has a nontrivial solution, in other words by definition $(-ac, -bc) = 1$. By bilinearity this condition is

$$1 = (-ac, -bc) = (-1, -1)(-1, a)(-1, b)(a, b)(a, c)(b, c)(c, c) ,$$

and since $(c, c) = (-1, c)$, this can be written $(-1, -abc) = (a, b)(a, c)(b, c)$, proving the proposition. $\qquad\square$

Corollary 5.2.12. *Let* $c \in \mathbb{Q}_p^*$, *and let* $q(x, y) = ax^2 + by^2$ *be a nondegenerate quadratic form in two variables. Then* q *represents* c *in* $\mathbb{Q}_p$ *if and only if* $(c, -ab) = (a, b)$.

Proof. By Corollary 5.1.17, q represents c if and only if the form $q'(x, y, z) = ax^2 + by^2 - cz^2$ represents 0 nontrivially. By the above proposition, this is true if and only if $(-1, abc) = (a, b)(a, -c)(b, -c)$, and since $(c, -c) = 1$, if and only if

$$(a, b) = (-1, abc)(ab, -c) = (-1, abc)(-c, abc) = (c, abc)$$
$$= (c, -c)(c, -ab) = (c, -ab) .$$

$\qquad\square$

Corollary 5.2.13. *Let* $c \in \mathbb{Q}_p^*$. *Then* c *is a sum of two squares of elements of* $\mathbb{Q}_p$ *if and only if one of the following holds:*

(1) $p \equiv 1 \pmod 4$.
(2) $p \equiv 3 \pmod 4$ *and* $v_p(c)$ *is even.*
(3) $p = 2$ *and* $c/2^{v_2(c)} \equiv 1 \pmod 4$.

Proof. By the preceding corollary c is a sum of two squares if and only if $(c, -1) = 1$. By Theorem 5.2.7 we have $(c, -1) = \left(\frac{-1}{p}\right)^{v_p(c)}$ for $p \neq 2$ and $(c, -1) = (-1)^{(c/v_2(c)-1)/2}$ for $p = 2$, giving the result. $\qquad\square$

To finish this subsection, we prove the following lemma, which we will need later.

Lemma 5.2.14. *Let* $\mathcal{K} = \mathbb{Q}_p$ *with* $p \neq \infty$.

(1) *We have* $|\mathcal{K}^*/\mathcal{K}^{*2}| = 2^r$ *with* $r = 2$ *for* $p \neq 2$ *and* $r = 3$ *for* $p = 2$.
(2) *If* $a \in \mathcal{K}^*/\mathcal{K}^{*2}$ *and* $\varepsilon = \pm 1$, *define* $H_\varepsilon(a)$ *to be the set of* $x \in \mathcal{K}^*/\mathcal{K}^{*2}$ *such that* $(x, a) = \varepsilon$. *Then* $|H_1(1)| = 2^r$, $H_{-1}(1) = \emptyset$ *and* $|H_\varepsilon(a)| = 2^{r-1}$ *if* $a \neq 1$.

(3) *Let a and a' in K^*/K^{*2}, and ε and ε' equal to ± 1, and assume that $H_\varepsilon(a)$ and $H_{\varepsilon'}(a')$ are nonempty. Then $H_\varepsilon(a) \cap H_{\varepsilon'}(a') = \emptyset$ if and only if $a = a'$ and $\varepsilon = -\varepsilon'$.*

Proof. (1) has been proved in the preceding chapter (Proposition 4.3.20). For (2), the case $a = 1$ is trivial. When $a \neq 1$, since the Hilbert symbol is nondegenerate and bilinear, $H_1(a)$ is the kernel of the surjective map $x \mapsto (a, x)$, hence has 2^{r-1} elements, and so the same is true for its complement $H_{-1}(a)$. Finally, for (3), if $E = H_\varepsilon(a)$ and $F = H_{\varepsilon'}(a')$ are nonempty and disjoint, by (2) they both have 2^{r-1} elements hence one is the complement of the other. Thus $H_1(a)$ is either E (if $\varepsilon = 1$) or F (if $\varepsilon = -1$), and similarly $H_1(a')$ is either E or F. Since $1 \in H_1(a)$ for all a, $H_1(a)$ and $H_1(a')$ are not disjoint, so $H_1(a) = H_1(a')$. This means that for all x we have $(x, a) = (x, a')$, and since the Hilbert symbol is nondegenerate we thus have $a = a'$, hence necessarily $\varepsilon = -\varepsilon'$, as claimed. $\qquad\square$

5.2.4 Quadratic Forms over $\mathbb{Q}_p$

Thanks to the study of the Hilbert symbol, in other words of quadratic forms of rank 3 over $\mathbb{Q}_p$, we are now in a position to study general quadratic forms over $\mathbb{Q}_p$. Let q be a quadratic form in n variables. Our first goal will be to find necessary and sufficient conditions under which q represents 0 nontrivially in $\mathbb{Q}_p$. Since degenerate forms represent 0 nontrivially, we will always assume that q is nondegenerate. As always, the discriminant $d(q)$ of q, considered as an element of $\mathbb{Q}_p^*/\mathbb{Q}_p^{*2}$, is an invariant of the class of q modulo equivalence. By abuse of notation, we will also denote by $d(q)$ any representative in $\mathbb{Q}_p^*$. We now define a second invariant. Up to equivalence, we can assume that q is in diagonal form as $q(x) = \sum_{1 \leq i \leq n} a_i x_i^2$, and we set

$$\varepsilon((a_1, \ldots, a_n)) = \prod_{1 \leq i < j \leq n} (a_i, a_j),$$

where (a_i, a_j) is the Hilbert symbol, so that $\varepsilon((a_1, \ldots, a_n)) = \pm 1$. We have the following theorem.

Theorem 5.2.15. *The value of $\varepsilon((a_1, \ldots, a_n))$ is independent of the linear change of variables that transforms q into diagonal form, hence is an invariant of the quadratic form itself, which we will denote by $\varepsilon(q)$.*

Proof. We use once again the language of quadratic modules, which is more suitable for this kind of proof. Let $V = K^n$ and let (V, q) be the quadratic module associated with q. The theorem states that ε does not depend on the orthogonal basis $\mathcal{B}$ of V, and we will write $\varepsilon(\mathcal{B})$ instead of $\varepsilon((a_1, \ldots, a_n))$. Consider first the the case $n \leq 2$. If $n = 1$ we have $\varepsilon(\mathcal{B}) = 1$. If $n = 2$, by definition of the Hilbert symbol $\varepsilon(\mathcal{B}) = 1$ if and only if the form $x_0^2 - a_1 x_1^2 - a_2 x_2^2$ represents 0, hence by Corollary 5.1.17 if and only if $a_1x_1^2 + a_2x_2^2$ represents 1, in other words if and only if q represents 1, a condition that is independent of the basis.

Assume now that $n \geqslant 3$, so that we can use Theorem 5.1.9. We use induction on n. Let $\mathcal{B}$ and $\mathcal{B}'$ be two orthogonal bases. By Theorem 5.1.9, it is enough to prove the above theorem when $\mathcal{B}$ and $\mathcal{B}'$ are contiguous, and by definition of ε and the symmetry properties of the Hilbert symbol we may assume that $\mathcal{B} = (e_1, \ldots, e_n)$ and $\mathcal{B}' = (e_1', \ldots, e_n')$ with $e_1' = e_1$, and $a_i = e_i \cdot e_i$, $a_i' = e_i' \cdot e_i'$, hence in particular $a_1' = a_1$. Thus by multiplicativity of the Hilbert symbol and the formula $d(q) = a_1 a_2 \cdots a_n$, we have

$$\varepsilon(\mathcal{B}) = (a_1, a_2 \cdots a_n) \prod_{2 \leqslant i < j \leqslant n} (a_i, a_j) = (a_1, d(q)a_1) \prod_{2 \leqslant i < j \leqslant n} (a_i, a_j) .$$

Similarly

$$\varepsilon(\mathcal{B}') = (a_1, d(q)a_1) \prod_{2 \leqslant i < j \leqslant n} (a_i', a_j') .$$

Applying our induction hypothesis to the orthogonal complement of e_1, we deduce that

$$\prod_{2 \leqslant i < j \leqslant n} (a_i, a_j) = \prod_{2 \leqslant i < j \leqslant n} (a_i', a_j') ,$$

and the result follows.   □

It follows from this theorem that just as for the discriminant $d(q)$, $\varepsilon(q)$ is an invariant of the equivalence class of q.

Theorem 5.2.16. *Let q be a nondegenerate quadratic form in n variables, and set $d = d(q)$ and $\varepsilon = \varepsilon(q)$. Then q represents 0 nontrivially in $\mathbb{Q}_p$ if and only if one of the following holds:*

(1) *$n = 2$ and $d = -1$.*
(2) *$n = 3$ and $(-1, -d) = \varepsilon$.*
(3) *$n = 4$ and either $d \neq 1$, or $d = 1$ and $(-1, -d) = \varepsilon$.*
(4) *$n \geqslant 5$.*

In particular, every quadratic form in at least five variables represents 0 nontrivially in $\mathbb{Q}_p$.

Proof. Since d and ε are invariants, we may assume that q is in diagonal form. Clearly q cannot represent 0 nontrivially when $n = 1$. When $n = 2$, $a_1x_1^2 + a_2x_2^2$ represents 0 nontrivially if and only if $-a_1/a_2 \in \mathbb{Q}_p^{*2}$, hence if and only if $-d = -a_1a_2 \in \mathbb{Q}_p^{*2}$, which means that $d = -1 \in \mathbb{Q}_p^*/\mathbb{Q}_p^{*2}$. The case $n = 3$ is exactly Proposition 5.2.11. It is thus sufficient to prove the cases $n = 4$ and $n \geqslant 5$.

Case $n = 4$.

By the equivalence of (1) and (2) in Corollary 5.1.18, $q(x) = \sum_{1 \leqslant i \leqslant 4} a_i x_i^2$ represents 0 nontrivially if and only if there exists $c \in \mathbb{Q}_p^*$ that is represented simultaneously by the forms $a_1 x_1^2 + a_2 x_2^2$ and $-a_3 x_3^2 - a_4 x_4^2$.

Thus by Corollary 5.2.12, q represents 0 if and only if there exists $c \in \mathbb{Q}_p^*$ such that $(c, -a_1 a_2) = (a_1, a_2)$ and $(c, -a_3 a_4) = (-a_3, -a_4)$. Let A be the set of classes of c modulo $\mathbb{Q}_p^*$ satisfying the first condition, and B the second. We thus know that q does *not* represent 0 if and only if $A \cap B = \emptyset$. On the other hand, A and B are clearly nonempty (for instance $a_1 \in A$ and $-a_3 \in B$). Applying Lemma 5.2.14 (3), we deduce that $A \cap B = \emptyset$ is equivalent to $a_1 a_2 = a_3 a_4$ and $(a_1, a_2) = -(-a_3, -a_4)$. The first relation means that $d = 1 \in \mathbb{Q}_p^*/\mathbb{Q}_p^{*2}$, and if it is satisfied we have

$$\varepsilon = (a_1, a_2)(a_1 a_2, a_3 a_4)(a_3, a_4) = (a_1, a_2)(a_3, a_4)(a_3 a_4, a_3 a_4) \ .$$

By the elementary properties of the Hilbert symbol, we know that for all $x \in \mathbb{Q}_p^*$ we have $(x, x) = (-1, x)(-x, x) = (-1, x)$, hence

$$\varepsilon = (a_1, a_2)(a_3, a_4)(-1, a_3 a_4) = (a_1, a_2)(a_3, a_4)(-1, a_3)(-1, a_4)$$
$$= (a_1, a_2)(-a_4, a_3)(-a_4, -1)(-1, -1) = (a_1, a_2)(-a_3, -a_4)(-1, -1) \ ,$$

so the second condition is equivalent to $\varepsilon = -(-1, -1)$, proving the theorem in the case $n = 4$.

Case $n \geqslant 5$.

It is clearly sufficient to prove the result for $n = 5$. By Corollary 5.2.12, a nondegenerate form $q_1 = a_1 x_1^2 + a_2 x_2^2$ of rank 2 represents $c \in \mathbb{Q}_p^*$ if and only if $(c, -a_1 a_2) = (a_1, a_2)$; hence by Lemma 5.2.14 (2) the number of $c \in \mathbb{Q}_p^*/\mathbb{Q}_p^{*2}$ represented by q_1 is not zero (for instance a_1 is represented), hence is at least 2^{r-1}, hence at least 2 since $r \geqslant 2$. A fortiori this is true for a form in more than two variables, in particular for q. Thus let $c \in \mathbb{Q}_p^*/\mathbb{Q}_p^{*2}$ different from $d(q)$ and represented by q. By Lemma 5.1.17, there exists a form q_1 in four variables x_i for $2 \leqslant i \leqslant 5$ such that up to equivalence $q = c x_1^2 + q_1$, and we have clearly $d(q) = c d(q_1)$. Since $c \neq d(q)$, $d(q_1) \neq 1 \in \mathbb{Q}_p^*/\mathbb{Q}_p^{*2}$, so by the case $n = 4$ we deduce that q_1 represents 0 nontrivially, hence also q, finishing the proof of the theorem. $\qquad\square$

Corollary 5.2.17. *Let $c \in \mathbb{Q}_p^*/\mathbb{Q}_p^{*2}$. A nondegenerate form q in n variables with invariants d and ε represents c if and only if one of the following holds:*

(1) $n = 1$ and $c = d$.
(2) $n = 2$ and $(c, -d) = \varepsilon$.
(3) $n = 3$ and either $c \neq -d$ or $c = -d$ and $(-1, -d) = \varepsilon$.
(4) $n \geqslant 4$.

Proof. By Corollary 5.1.17, q represents c if and only if the nondegenerate form in $n + 1$ variables $q_1 = -c x_0^2 + q$ represents 0 nontrivially. We have

$d(q_1) = -cd(q)$ and $\varepsilon(q_1) = (-c, d(q))\varepsilon(q)$; hence applying the theorem to q_1 we deduce the corollary using the properties of the Hilbert symbol (the case $n = 2$ is Corollary 5.2.12).   □

The classification of quadratic forms over $\mathbb{Q}_p$ is easily performed.

Proposition 5.2.18. *Two quadratic forms over $\mathbb{Q}_p$ are equivalent if and only if they have the same rank, discriminant, and invariant $\varepsilon(q)$.*

Proof. Left to the reader (Exercise 5). See Exercises 6, 7, and 8 for further properties of quadratic forms over $\mathbb{Q}_p$.   □

5.3 Quadratic Forms over $\mathbb{Q}$

After studying in great detail quadratic forms over p-adic fields, we are now in a position to study quadratic forms over $\mathbb{Q}$, and in particular to prove the Hasse–Minkowski theorem. As usual, the crucial case is the case $n = 3$, and for this we need global properties of the Hilbert symbol, which we now study.

5.3.1 Global Properties of the Hilbert Symbol

Let P be the set of places of $\mathbb{Q}$, which can be identified with the prime numbers together with the symbol ∞. For convenience, we will set $\mathbb{Q}_\infty = \mathbb{R}$. As mentioned above, we will now write $(a, b)_v$ or $(a, b)_p$ to denote the Hilbert symbol corresponding to the place v or the prime p. There are two important global properties of the Hilbert symbol: one is the product formula, which is essentially a restatement of the quadratic reciprocity law, and the second is the existence of rational numbers having prescribed Hilbert symbols.

Theorem 5.3.1 (Product formula). *If a and b are in $\mathbb{Q}^*$ then $(a, b)_v = 1$ for almost all $v \in P$ (in other words for all but a finite number), and we have the product formula*

$$\prod_{v \in P} (a, b)_v = 1 \ .$$

Proof. By bilinearity, it is sufficient to prove the theorem when a and b are equal to -1 or to a prime number. In these cases, Theorem 5.2.7 gives the answer:

- If $a = -1$ and $b = -1$, then $(-1, -1)_\infty = (-1, -1)_2 = -1$, and $(-1, -1)_v = 1$ for $v \neq 2$ and ∞, so the product is equal to 1.
- If $a = -1$ and $b = \ell$ with ℓ prime, then if $\ell = 2$ we have $(-1, 2)_v = (-1, (1 - (-1)))_v = 1$ for all $v \in P$, while if $\ell \neq 2$ then $(-1, \ell)_v = 1$ if $v \neq 2$ and ℓ, and

$$(-1, \ell)_2 = (-1, \ell)_\ell = (-1)^{(\ell-1)/2} \ ,$$

so the product is again equal to 1.

– If $a = \ell$ and $b = \ell$, then by Proposition 5.2.6 (4) we have $(\ell, \ell)_v = (-1, \ell)_v$, so we are reduced to the preceding case.

– If $a = 2$ and $b = \ell$ with ℓ an odd prime, then $(2, \ell)_v = 1$ for $v \neq 2$ and ℓ, and

$$(2, \ell)_2 = \left(\frac{\ell}{2}\right) = \left(\frac{2}{\ell}\right) = (2, \ell)_\ell \, ,$$

so here the product is equal to 1 by the complementary law to quadratic reciprocity.

– If $a = \ell$ and $b = \ell'$ with ℓ and ℓ' distinct odd primes, then $(\ell, \ell')_v = 1$ for $v \neq 2$, ℓ, and ℓ', and

$$(\ell, \ell')_2 = (-1)^{(\ell-1)(\ell'-1)/4}, \quad (\ell, \ell')_\ell = \left(\frac{\ell'}{\ell}\right), \quad (\ell, \ell')_\ell = \left(\frac{\ell}{\ell'}\right) \, ,$$

so here the product is equal to 1 by the quadratic reciprocity law.   □

Theorem 5.3.2. *Let $(a_i)_{i \in I}$ be a finite set of elements of $\mathbb{Q}^*$ and let $(\varepsilon_{i,v})_{i \in I, \, v \in P}$ be a set of numbers equal to ± 1. There exists $x \in \mathbb{Q}^*$ such that $(a_i, x) = \varepsilon_{i,v}$ for all $i \in I$ and all $v \in P$ if and only if the following three conditions are satisfied:*

(1) *Almost all of the $\varepsilon_{i,v}$ are equal to 1.*
(2) *For all $i \in I$ we have $\prod_{v \in P} \varepsilon_{i,v} = 1$.*
(3) *For all $v \in P$ there exists $x_v \in \mathbb{Q}_v^*$ such that $(a_i, x_v)_v = \varepsilon_{i,v}$ for all $i \in I$.*

Proof. The necessity of conditions (1) and (2) follows from the above theorem, and that of (3) is trivial by taking $x_v = x$. Thus let us show that these conditions are sufficient. After multiplying the a_i by nonzero squares, we may assume that $a_i \in \mathbb{Z}$ for all i. Denote by S the (finite) subset of P containing ∞, 2, and the prime factors of all the a_i, and by T the (finite) set of $v \in P$ such that there exists $i \in I$ with $\varepsilon_{i,v} = -1$.

Assume first that $S \cap T = \emptyset$, and set

$$a = \prod_{\ell \in T, \ell \neq 2, \infty} \ell \quad \text{and} \quad m = 8 \prod_{\ell \in S, \ell \neq 2, \infty} \ell \, .$$

Since $S \cap T = \emptyset$ the integers a and m are coprime; hence by Dirichlet's theorem on primes in arithmetic progression (Theorem 10.5.30) there exists a prime number $p \equiv a \pmod{m}$ such that $p \notin S \cup T$. I claim that $x = ap$ has the required property. We consider two cases.

Case 1: $v \in S$.

Since $S \cap T = \emptyset$ we have $\varepsilon_{i,v} = 1$ for all i, so we must check that $(a_i, x)_v = 1$. Since $x > 0$, this is clear for $v = \infty$. If $v = \ell$ is a prime number, then $x \equiv a^2 \pmod{m}$, so $x \equiv a^2 \pmod 8$ for $\ell = 2$ and $x \equiv a^2 \pmod{\ell}$. Since a is an ℓ-adic unit, Hensel's lemma implies that x is a square in $\mathbb{Q}_\ell^*$, so $(a_i, x)_v = 1$ for all i.

Case 2: $v = \ell \notin S$.

In this case a_i is an ℓ-adic unit, and since $\ell \neq 2$ by Theorem 5.2.7 we have for all $b \in \mathbb{Q}_\ell^*$,

$$(a_i, b)_\ell = \left(\frac{a_i}{\ell}\right)^{v_\ell(b)}.$$

– If $\ell \notin T \cup \{p\}$, then x is an ℓ-adic unit; hence $(a_i, x)_\ell = 1 = \varepsilon_{i,\ell}$ since $\ell \notin T$.
– If $\ell \in T$ then $v_\ell(x) = 1$, and condition (3) implies that there exists $x_\ell \in \mathbb{Q}_\ell^*$ such that $(a_i, x_\ell) = \varepsilon_{i,\ell}$ for all $i \in I$. Since $\ell \in T$ at least one of the $\varepsilon_{i,\ell}$ is equal to -1; hence $v_\ell(x_\ell) \equiv 1 \pmod{2}$ by Theorem 5.2.7, so for all $i \in I$ we have

$$(a_i, x)_\ell = \left(\frac{a_i}{\ell}\right) = (a_i, x_\ell)_\ell = \varepsilon_{i,\ell}.$$

– Finally, if $\ell = p$ the product formula implies that

$$(a_i, x)_p = \prod_{v \neq p}(a_i, x)_p = \prod_{v \neq p} \varepsilon_{i,v} = \varepsilon_{i,p}$$

by Condition (2), proving the theorem in the special case $S \cap T = \emptyset$.

Consider now the general case. By the approximation theorem, we know that there exists $x' \in \mathbb{Q}^*$ such that $x'/x_v \in \mathbb{Q}_v^{*2}$ for all $v \in S$ (we can for instance ask that $x'/x_v \equiv 1 \pmod{p\mathbb{Z}_p}$ for $v = p \neq 2, \infty$, $x'/x_v \equiv 1 \pmod{8\mathbb{Z}_2}$ for $v = 2$ and $x'/x_v > 0$ for $v = \infty$). Thus $(a_i, x')_v = (a_i, x_v)_v = \varepsilon_{i,v}$ for all $v \in S$. If we set $\eta_{i,v} = (a_i, x')_v \varepsilon_{i,v}$, then clearly the family $\eta_{i,v}$ satisfies conditions (1), (2), and (3), and by definition $\eta_{i,v} = 1$ if $v \in S$. Thus, by the special case that we have treated above there exists $y \in \mathbb{Q}^*$ such that $(a_i, y)_v = \eta_{i,v}$ for all $i \in I$ and $v \in P$, and it is clear that $x = yx'$ has the required properties. $\square$

5.3.2 Statement of the Hasse–Minkowski Theorem

The first important local–global principle is the Hasse–Minkowski theorem, which says that the principle is valid for a single quadratic form.

Theorem 5.3.3 (Hasse–Minkowski). *Let K be a number field and let q be a quadratic form in n variables with coefficients in K. Then q represents 0 in K if and only if it represents 0 in every completion of K.*

The proof of this theorem for a general number field is outside the scope of this book. Even for $K = \mathbb{Q}$, the proof is not short, and we will prove the theorem only in this case, as well as stronger statements.

This theorem is in itself very satisfying. It must be understood, however, that it is quite specific to a single quadratic form: it is in general (not always of course) false for several simultaneous quadratic equations, or for forms of higher degree, as we will see below.

In the modern way of looking at this kind of problem, a local–global principle does not hold if there is some sort of *obstruction* to it, which can usually be realized as a cohomology group. Thus, in the case of a quadratic form there is no obstruction. In the case of an elliptic curve, for instance, the obstruction is essentially given by a group (believed but not known to be always finite) called the Tate–Shafarevich group of the curve.

We will prove the Hasse–Minkowski theorem in the case $K = \mathbb{Q}$, which is due to Minkowski. Of course, the condition that the quadratic form have a nontrivial solution in every completion is *necessary*. Thus we need only prove the converse, i.e., we assume that the quadratic form has a nontrivial solution in every completion, and we must prove that it has a nontrivial solution in $\mathbb{Q}$.

5.3.3 The Hasse–Minkowski Theorem for $n \leqslant 2$

For $n = 1$ the result is trivial since $ax^2 = 0$ has a nonzero solution if and only if $a = 0$. For $n = 2$, we note the following lemma.

Lemma 5.3.4. *Over any field K of characteristic different from 2 the form $ax^2 + bxy + cy^2$ represents 0 nontrivially if and only if $b^2 - 4ac$ is a square in K.*

Proof. We note the formal identity

$$(2ax + by)^2 - y^2(b^2 - 4ac) = 4a(ax^2 + bxy + cy^2) \,.$$

We consider two cases. If $a \neq 0$, then if $(x, y) \neq (0, 0)$ is such that $ax^2 + bxy + cy^2 = 0$ we must have $y \neq 0$ (otherwise $ax^2 = 0$, hence $a = 0$), so $b^2 - 4ac = ((2ax + by)/y)^2$ is a square. Conversely, if $b^2 - 4ac = u^2$, then from the above identity it is clear (since $a \neq 0$) that $(x, y) = (u - b, 2a)$ satisfies $ax^2 + bxy + cy^2 = 0$ and is different from $(0, 0)$. On the other hand, if $a = 0$ then $b^2 - 4ac = b^2$ is a square, and $ax^2 + bxy + cy^2 = y(bx + cy)$ represents 0 nontrivially, for example with $(x, y) = (1, 0)$.   □

Thus, let $q(x, y) = ax^2 + bxy + cy^2$ be a binary quadratic form. Since it represents 0 nontrivially in $\mathbb{R}$, its discriminant $d = b^2 - 4ac$ must be nonnegative (trivially, but also by the above lemma!). If $d = 0$, then q is a square of a linear form hence clearly represents 0 nontrivially in $\mathbb{Q}$. Otherwise, $d > 0$, and let $d = \prod_i p_i^{v_i}$ be the prime power decomposition of d. Since q represents 0 nontrivially in every $\mathbb{Q}_{p_i}$, by the above lemma d is a square in $\mathbb{Q}_{p_i}$. This implies in particular that $v_{p_i}(d) = v_i$ is *even* for all i, hence that d is a square. Thus $q(x, y)$ is in fact the product of two linear forms with coefficients in $\mathbb{Q}$ hence represents 0 nontrivially in $\mathbb{Q}$.   □

5.3.4 The Hasse–Minkowski Theorem for $n = 3$

The most important part of the proof is the case $n = 3$, which we now consider. Up to equivalence (which does not change the problem), we may assume that our quadratic form is a diagonal form $q(x, y, z) = ax^2 + by^2 + cz^2$. If one of the coefficients is equal to 0 then q has clearly a nontrivial zero in $\mathbb{Q}$. Thus we may assume $abc \neq 0$. Furthermore, changing q by a rational multiple, and the variables by rational multiples, we may assume first that a, b, and c are in $\mathbb{Z}$, then (multiplying by a and changing x into x/a) that $a = 1$, and finally that b and c are squarefree. Changing notation, we thus write $q(x, y, z) = x^2 - ay^2 - bz^2$ with a and b squarefree integers, where we assume $|a| \leqslant |b|$. We prove the theorem by induction on $m = |a| + |b|$. If $m = 2$ then $q(x, y, z) = x^2 \pm y^2 \pm z^2$, and since the case $x^2 + y^2 + z^2$ is excluded since q represents 0 in $\mathbb{R}$, in the other cases the form represents 0.

Thus assume now that $m > 2$, in other words $|b| \geqslant 2$, and let $b = \pm \prod_{1 \leqslant i \leqslant k} p_i$ be the prime factorization of the squarefree number b. Let $p = p_i$ for some i. I claim that a is a square modulo p. This is trivial if $a \equiv 0 \pmod{p}$. Otherwise, a is a p-adic unit, and by assumption there exists a nontrivial p-adic solution to $ay^2 + bz^2 = x^2$, where as usual we may assume that x, y, and z are in $\mathbb{Z}_p$ with at least one in U_p. Thus $x^2 \equiv ay^2 \pmod{p\mathbb{Z}_p}$. Now, y is a p-adic unit, since otherwise $v_p(x) \geqslant 1$, so that $v_p(bz^2) \geqslant 2$, and hence $v_p(z) \geqslant 1$ (b being squarefree), contradicting the fact that one of x, y, and z is in U_p. It follows that $a \equiv (x/y)^2 \pmod{p\mathbb{Z}_p}$, so a is a square modulo p, proving my claim. Since this is true for all $p \mid b$, by the Chinese remainder theorem this implies that a is a square modulo b, in other words that there exist b' and k such that $k^2 = a + bb'$, where k may be chosen such that $|k| \leqslant |b|/2$. Since $bb' = k^2 - a$, bb' is a norm in the extension $K(\sqrt{a})/K$, where $K = \mathbb{Q}$ or any $\mathbb{Q}_v$. Thus, as in the proof of Proposition 5.2.5 we deduce that q represents 0 in K if and only if the same is true for q', with $q'(x, y, z) = x^2 - ay^2 - b'z^2$. In particular, by assumption q' represents 0 in all the $\mathbb{Q}_v$. But since $|b| \geqslant 2$ and $|a| \leqslant |b|$, we have

$$|b'| = \left| \frac{t^2 - a}{b} \right| \leqslant \frac{|b|}{4} + 1 < |b| .$$

Thus we may apply our induction hypothesis to the form q' (more precisely to the form q'', where b' is replaced by its squarefree part); hence q' represents 0 in $\mathbb{Q}$, and so the same is true for the form q. $\qquad\square$

To be able to prove the Hasse–Minkowski theorem for $n \geqslant 4$ variables, we need a strengthening of the result for $n = 3$, asserting that we can omit a single place in the assumption of local solubility as follows.

Proposition 5.3.5. *Let $q(x, y, z)$ be a quadratic form in three variables, and assume that $q(x, y, z) = 0$ has a nontrivial solution in every completion of $\mathbb{Q}$ except perhaps in one. Then it has a nontrivial solution in $\mathbb{Q}$, hence in all places.*

Proof. As usual we may assume that q is nondegenerate, since otherwise the result is trivial, and by changing q into an equivalent form, that $q(x, y, z) = ax^2 + by^2 + cz^2$ is a diagonal form. By Proposition 5.2.11, q represents 0 in $\mathbb{Q}_v$ if and only if

$$(-1, -abc)_v = (a, b)_v (a, c)_v (b, c)_v .$$

By assumption this is true for all v except perhaps one. Since both sides satisfy the product formula, it follows that this equality is true for all v; hence once again by the above proposition q represents 0 in $\mathbb{Q}_v$ for all v, hence in $\mathbb{Q}$ by the proof of the Hasse–Minkowski theorem for $n = 3$ given above. $\square$

Remark. This result implies for instance that the existence of solutions in each $\mathbb{Q}_p$ implies the existence of a solution in $\mathbb{R}$, in other words that the quadratic form is indefinite or singular. This has an interesting relationship with quadratic reciprocity. For example, let p and q be distinct odd primes such that $q \equiv 3 \pmod 4$ and $\left(\frac{-p}{q}\right) = 1$. The equation $x^2 + py^2 + qz^2 = 0$ clearly has a nontrivial solution in $\mathbb{Q}_2$ and in $\mathbb{Q}_q$, and since there is no nontrivial solution in $\mathbb{Q}$, there cannot be any nontrivial solution in $\mathbb{Q}_p$; in other words, $\left(\frac{-q}{p}\right) = -1$ (Exercise 9). This is of course not surprising since the proof of the above theorem uses in an essential way the product formula for the Hilbert symbol, which is equivalent to the quadratic reciprocity law.

5.3.5 The Hasse–Minkowski Theorem for $n = 4$

Theorem 5.3.6. *Let q be a quadratic form in four variables such that $q = 0$ has a nontrivial solution in $\mathbb{R}$ and every $\mathbb{Q}_p$. Then $q = 0$ has a nontrivial solution in $\mathbb{Q}$.*

Proof. We may assume that $q = a_1 x_1^2 + a_2 x_2^2 - a_3 x_3^2 - a_4 x_4^2$. Let v be a place of $\mathbb{Q}$. Since q represents 0 in $\mathbb{Q}_v$, Corollary 5.1.18 tells us that there exists $c_v \in \mathbb{Q}_v^*$ that is represented both by $a_1 x_1^2 + a_2 x_2^2$ and by $a_3 x_3^2 + a_4 x_4^2$, and Corollary 5.2.17 (2) (which is trivially true also for $\mathbb{R}$) implies that for all v we have

$$(c_v, -a_1 a_2)_v = (a_1, a_2)_v \quad \text{and} \quad (c_v, -a_3 a_4)_v = (a_3, a_4)_v .$$

By the product formula for the Hilbert symbol, we deduce from Theorem 5.3.2 that there exists $c \in \mathbb{Q}^*$ such that for all places v,

$$(c, -a_1 a_2)_v = (a_1, a_2)_v \quad \text{and} \quad (c, -a_3 a_4)_v = (a_3, a_4)_v .$$

The form in three variables $a_1 x_1^2 + a_2 x_2^2 - cx_0^2$ thus represents 0 in each $\mathbb{Q}_v$, hence by the proof of the Hasse–Minkowski theorem for $n = 3$ also in $\mathbb{Q}$, so c is represented by $a_1 x_1^2 + a_2 x_2^2$. Similarly c is represented by $a_3 x_3^2 + a_4 x_4^2$, so q represents 0. $\square$

Remark. In the above proof it was essential to use Theorem 5.3.2, which needs for its proof Dirichlet's Theorem 10.5.30 on primes in arithmetic progression, which is not a completely elementary result. In [Cas1] it is shown how the use of Dirichlet's theorem can be avoided by introducing an auxiliary quadratic form in *six* variables. See also [Sim1] for an algorithmic application.

5.3.6 The Hasse–Minkowski Theorem for $n \geqslant 5$

Theorem 5.3.7. *Let q be a quadratic form in five variables, and assume that $q = 0$ has a real solution, in other words that q is indefinite or singular. Then $q = 0$ has a solution in $\mathbb{Q}$.*

Note that we no longer need to assume the existence of p-adic solutions since by Theorem 5.2.16 we know that any quadratic form in $n \geqslant 5$ variables has a nontrivial solution in $\mathbb{Q}_p$ for every p.

Proof. The proof is very similar to the case $n = 4$, so we only give a sketch. We may assume q nonsingular and diagonal, and with evident notation we write $g = a_1 x_1^2 + a_2 x_2^2$, $h = -a_3 x_3^2 - a_4 x_4^2 - a_5 x_5^2$, and we assume $a_1 > 0$ and $a_5 < 0$. Thanks to Dirichlet's theorem, we find an integer $a > 0$ representable both by g and h in $\mathbb{R}$ and all the $\mathbb{Q}_p$ except perhaps in $\mathbb{Q}_q$ for a unique odd q not dividing $a_1 a_2 a_3 a_4 a_5$. I claim that g and h also represent a in $\mathbb{Q}_q$. Indeed, for g this follows as above using Proposition 5.3.5 and the auxiliary form in three variables $g_1 = -ax_0^2 + g$. For h, we note that it has a nontrivial zero in $\mathbb{Q}_q$ by Corollary 4.1.40; hence it represents all elements of $\mathbb{Q}_q$ by Lemma 5.1.16, proving my claim.

Using Corollary 5.1.17, we see that the forms $g_1 = -ax_0^2 + g$ and $h_1 = -ax_0^2 + h$ in three and four variables respectively have a nontrivial solution in every completion of $\mathbb{Q}$; hence by the Minkowski–Hasse theorem proved for three and four variables, they have a solution in $\mathbb{Q}$, so that a is representable by g and h in $\mathbb{Q}$, proving as before that $q = 0$ has a nontrivial rational solution. $\quad\square$

Corollary 5.3.8. *Let q be a quadratic form in $n \geqslant 5$ variables, and assume that q represents 0 in $\mathbb{R}$, in other words that q is indefinite or singular. Then q represents 0 in $\mathbb{Q}$.*

Proof. We may assume q nonsingular and diagonal, and that $a_1 > 0$, $a_5 < 0$ for instance. Then $q = q_1 + q_2$ with $q_1 = \sum_{1 \leqslant i \leqslant 5} a_i x_i^2$. By the above theorem $q_1 = 0$ has a nontrivial solution, and we choose $x_i = 0$ for $5 < i \leqslant n$, proving the corollary. $\quad\square$

This finishes the proof of the Hasse–Minkowski theorem for $K = \mathbb{Q}$.

5.4 Consequences of the Hasse–Minkowski Theorem

5.4.1 General Results

Proposition 5.4.1. *A nondegenerate quadratic form with coefficients in $\mathbb{Q}$ represents $c \in \mathbb{Q}^*$ if and only if it represents c in $\mathbb{R}$ and in every $\mathbb{Q}_p$.*

Proof. Clear from Corollary 5.1.17 and the Hasse–Minkowski theorem.
□

Let q be a nondegenerate quadratic form defined over $\mathbb{Q}$. The Hasse–Minkowski theorem gives us local necessary and sufficient conditions for q to represent 0 over $\mathbb{Q}$, a priori infinite in number. We can easily make this more precise as follows.

Proposition 5.4.2. *Let q be a nondegenerate quadratic form in n variables with coefficients in $\mathbb{Q}$, and let $d(q) \in \mathbb{Q}^*/\mathbb{Q}^{*2}$ be its discriminant. Then q represents 0 in $\mathbb{Q}$ if and only one of the following holds:*

(1) $n = 2$ and $-d(q) = 1 \in \mathbb{Q}^*/\mathbb{Q}^{*2}$.
(2) $3 \leqslant n \leqslant 4$ and q represents 0 in $\mathbb{R}$, $\mathbb{Q}_2$, and the $\mathbb{Q}_p$ for the primes p such that $v_p(d(q))$ is odd (note that this makes sense).
(3) $n \geqslant 5$ and q represents 0 in $\mathbb{R}$.

Proof. Clearly the given conditions do not change by an invertible linear change of variables, since the discriminant is well defined modulo squares. We may thus assume that q is in diagonal form. For $n = 2$, $a_1 x_1^2 + a_2 x_2^2$ represents 0 in $\mathbb{Q}$ if and only if $-a_1/a_2$ is a square, so $-d(q)$ is a square. Note that this corresponds to an infinite number of local conditions. For $3 \leqslant n \leqslant 4$, recall that by the explicit computation of the Hilbert symbol, when $p \neq 2$ and $v_p(a) = v_p(b) = 0$ we have $(a, b)_p = 1$. Thus if $p \nmid 2d$, we have $(-1, -d) = \varepsilon = 1$, so that q represents 0 by the theorem in both cases.   □

Remark. Thus to test these conditions for forms in three or four variables we must find those p such that $v_p(d(q))$ is odd. This is essentially equivalent to factoring $d(q)$, which of course may be difficult when the coefficients are large. Once these p have been found, we must look at the form modulo a sufficiently large p^k for which Hensel's lemma is applicable, and we then can check whether the local conditions are satisfied.

Corollary 5.4.3. *Let q be a nondegenerate quadratic form in n variables. A number $c \in \mathbb{Q}^*$ is represented by q in $\mathbb{Q}$ if and only if one of the following holds:*

(1) $n = 1$ and $c/d(q) = 1 \in \mathbb{Q}^*/\mathbb{Q}^{*2}$.
(2) $2 \leqslant n \leqslant 3$ and q represents c in $\mathbb{R}$, $\mathbb{Q}_2$, and the primes p such that $v_p(d(q)) \not\equiv v_p(c) \pmod 2$.

(3) $n \geqslant 4$ and q represents c in $\mathbb{R}$.

Proof. Clear from Corollary 5.1.17 and the above proposition.   □

5.4.2 A Result of Davenport and Cassels

Thanks to the Hasse–Minkowski theorem the problem of representing *rational numbers* by quadratic forms is completely solved in theory, and also in practice thanks to the work of Gauss, Legendre, and very recently D. Simon (see Section 6.3.3). On the other hand, the problem of representing *integers* is much more difficult, and in fact many unsolved problems remain. In this subsection we give an easy but interesting result due to Davenport–Cassels that often allows us to go from the existence of *rational* representations to *integral* representations by quadratic forms. In the next subsection we give some results and conjectures on this question.

First, it is important to distinguish between two types of integral quadratic forms. A quadratic form q is represented over a suitable basis by a symmetric matrix Q, so that $q(x) = X^t Q X$ if X is the (column) vector of coordinates of x in the given basis; in other words, $q(x) = \sum_{i,j} q_{i,j} x_i x_j$ with $q_{i,j} = q_{j,i}$. We will say that q is an *integer-valued* (or simply *integral*) quadratic form if for any $x \in \mathbb{Z}^k$, $q(x) \in \mathbb{Z}$. This is clearly equivalent to saying that $q_{i,i} \in \mathbb{Z}$ and $2q_{i,j} \in \mathbb{Z}$ for $i \neq j$. We will say that q is *matrix-integral* if the bilinear form $b(x,y) = (q(x+y) - q(x) - q(y))/2$ associated with q takes only integral values when x and y are in $\mathbb{Z}^k$. This is equivalent to saying that $q_{i,j} \in \mathbb{Z}$ for all i, j, whence the name. For instance, $q(x_1, x_2) = x_1^2 + x_1 x_2 + x_2^2$ is an integer-valued quadratic form but is not matrix-integral.

Proposition 5.4.4. *Let q be a positive definite matrix-integral quadratic form in k variables. Assume that for every $x \in \mathbb{Q}^k$ there exists $y \in \mathbb{Z}^k$ such that $q(x - y) < 1$. Then if $n \in \mathbb{Z}$ is represented by q in $\mathbb{Q}$, it is also represented by q in $\mathbb{Z}$.*

Proof. If x and y in $\mathbb{Q}^k$ are represented by column vectors X and Y, and if Q is the matrix of q with respect to the canonical basis of $\mathbb{Q}^k$, we denote by $x \cdot y$ their scalar product with respect to the quadratic form q; in other words, $x \cdot y = Y^t Q X = X^t Q Y$, so that in particular $q(x) = x \cdot x$. Let $n \in \mathbb{Z}$ be represented by q in $\mathbb{Q}$. Thus there exist $x \in \mathbb{Z}^k$ and an integer $d > 0$ such that $(x/d) \cdot (x/d) = n$; in other words, $x \cdot x = d^2 n$. Choose d and x such that d is minimal. We are going to show that $d = 1$, which will prove the proposition.

By assumption there exists $y \in \mathbb{Z}^k$ such that $x/d = y + z$ with $q(z) = z \cdot z < 1$. Since q is positive definite, if $z \cdot z = 0$ we have $z = 0$, so $x/d = y \in \mathbb{Z}^k$, and since d is minimal we have $d = 1$. Thus assume that $z \cdot z \neq 0$. We set

$$a = y \cdot y - n, \quad b = 2(nd - x \cdot y), \quad d' = ad + b, \quad x' = ax + by .$$

Clearly a, b, and d' are in $\mathbb{Z}$, $x' \in \mathbb{Z}^k$, and

$$x' \cdot x' = a^2 x \cdot x + 2abx \cdot y + b^2 y \cdot y = a^2 d^2 n + ab(2nd - b) + b^2(n + a)$$
$$= n(a^2 d^2 + 2abd + b^2) = d'^2 n \ .$$

We have thus found a new square multiple of n represented by q in $\mathbb{Z}$. Let us compare d' with d. We have

$$dd' = ad^2 + bd = d^2 y \cdot y - d^2 n + 2d^2 n - 2dx \cdot y$$
$$= d^2 y \cdot y - 2dx \cdot y + x \cdot x = (dy - x) \cdot (dy - x) = d^2 z \cdot z \ .$$

Since by assumption $0 < z \cdot z < 1$, we thus have $0 < d' < d$, contradicting the minimality of d hence proving the proposition. $\qquad\square$

The number of equivalence classes of quadratic forms satisfying the assumptions of the above proposition is in fact finite. Define a matrix-integral quadratic form to be *strongly Euclidean* if the assumptions of the proposition are satisfied, and to be Euclidean if we have the weaker inequality $q(x - y) \leqslant 1$. J. Houriet has kindly computed for me the short list of such forms. The strongly Euclidean forms exist only for $k \leqslant 3$ and are ax_1^2 for $1 \leqslant a \leqslant 3$, $x_1^2 + x_2^2$, $x_1^2 + 2x_2^2$, $2x_1^2 + 2x_1 x_2 + 2x_2^2$, $x_1^2 + x_2^2 + x_3^2$, and $x_1^2 + 2x_2^2 + 2x_2 x_3 + 2x_3^2$. The additional simply Euclidean forms (of which we will meet the fundamental example $\sum_{1 \leqslant i \leqslant 4} x_i^2$ below) are $4x_1^2$, $x_1^2 + 3x_2^2$, $2(x_1^2 + x_2^2)$, $x_1^2 + x_2^2 + 2x_3^2$, $2(x_1^2 + x_2^2 + x_3^2 + x_1 x_2 + x_2 x_3)$, $x_1^2 + x_2^2 + x_3^2 + x_4^2$, $2(x_1^2 + x_2^2 + x_3^2 + x_4^2 - x_1 x_3 - x_2 x_3 - x_3 x_4)$, and

$$2 \left(\sum_{1 \leqslant i \leqslant 8} x_i^2 - \sum_{1 \leqslant i \leqslant 7,\ i \neq 4} x_i x_{i+1} - x_3 x_5 \right) \ .$$

Of these simply Euclidean forms, only $x_1^2 + 3x_2^2$, $x_1^2 + x_2^2 + 2x_3^2$, and $x_1^2 + x_2^2 + x_3^2 + x_4^2$ satisfy the conclusion of the above proposition (Exercise 13).

5.4.3 Universal Quadratic Forms

We begin with the following definitions.

Definition 5.4.5. *Let q be a positive definite integral quadratic form in n variables and S a subset of the nonnegative integers. We say that q is universal for S if for every $k \in S$ there exists $x \in \mathbb{Z}^n$ such that $k = q(x)$.*

For instance, Lagrange's four-square theorem (Corollary 5.4.14 below) states that $x_1^2 + x_2^2 + x_3^2 + x_4^2$ is universal for $S = \mathbb{Z}_{\geqslant 0}$.

Definition 5.4.6. *Let $S \subset \mathbb{Z}_{\geqslant 0}$ and $T \subset S$. We say that T is a witness for S if for any positive definite integral quadratic form q that is not universal* *for S there exists $k \in T$ that is not represented integrally by q. Equivalently, T is a witness for S if a positive definite integral quadratic form q is universal for S if and only if it represents integrally all the elements of T. Similarly, we define a* strong witness *for S by replacing integral quadratic forms by matrix-integral quadratic forms.*

Theorem 5.4.7 (Bhargava). *For any $S \subset \mathbb{Z}_{\geqslant 0}$ there exists a minimal witness $T = \Phi(S)$ for S, in other words such that T' is a witness for S if and only if $T \subset T' \subset S$. This minimal witness is clearly unique and is* finite. *Similarly there exists a minimal strong witness $T = \Phi_s(S)$, and we evidently have $\Phi_s(S) \subset \Phi(S)$.*

Note that we also have the trivial equality $\Phi(S) = \Phi_s(2S)/2$, with evident notation, so if desired we can consider only the map Φ, or only the map Φ_s.

It is sometimes but not always possible to compute $\Phi(S)$ or $\Phi_s(S)$ explicitly. For instance, we have the following results:

$$\Phi_s(\mathbb{Z}_{\geqslant 0}) = \{1, 2, 3, 5, 6, 7, 10, 14, 15\} \,,$$
$$\Phi(\mathbb{Z}_{\geqslant 0}) = \{1, 2, 3, 5, 6, 7, 10, 13, 14, 15, 17, 19, 21, 22, 23, 26, 29, 30, 31,$$
$$34, 35, 37, 42, 58, 93, 110, 145, 203, 290\} \,,$$
$$\Phi_s(2\mathbb{Z}_{\geqslant 0} + 1) = \{1, 3, 5, 7, 11, 15, 33\} \,,$$
$$\Phi_s(P) = \{2, 3, 5, 7, 11, 13, 17, 19, 23, 29, 31, 37, 41, 43, 47, 67, 73\} \,,$$

where P is the set of primes.

The result for $\Phi(\mathbb{Z}_{\geqslant 0})$ (the "290-theorem") is due to Bhargava and Hanke, see [Bha-Han], and involves a considerable amount of computer calculation and a number of tricks. In the same paper, the authors also show that there are exactly 6436 quaternary quadratic forms that are universal for $\mathbb{Z}_{\geqslant 0}$, and of these, 204 are matrix-integral.

A remarkable aspect of Bhargava's theorem stated above is that it is *ineffective*; in other words, there exist sets S for which the computation of $\Phi_s(T)$ or of $\Phi(S)$ is impossible (with present knowledge), even in theory. For instance, although $\Phi_s(2\mathbb{Z}_{\geqslant 0} + 1)$ is given above, $\Phi(2\mathbb{Z}_{\geqslant 0} + 1)$ is unknown. The reason for this is that there exist forms for which it is unknown whether they are universal for S. For instance:

Conjecture 5.4.8. *The positive definite integral quadratic forms*

$$q_1(x, y, z) = x^2 + 2y^2 + 5z^2 + xz \,,$$
$$q_2(x, y, z) = x^2 + 3y^2 + 6z^2 + xy + 2yz \,,$$
$$q_3(x, y, z) = x^2 + 3y^2 + 7z^2 + xy + xz \,,$$

represent integrally all *odd positive integers, in other words are universal for $S = 2\mathbb{Z}_{\geqslant 0} + 1$.*

With all the knowledge on quadratic forms, it is quite incredible that such a simple conjecture has not yet been resolved. It is known that all *sufficiently large* odd positive integers are represented by these forms, but the corresponding upper bounds are ineffective. See [Kap] for more information on this subject.

5.4.4 Sums of Squares

Proposition 5.4.9 (Fermat). *Let n be a positive integer. The following three conditions are equivalent:*

(1) *The integer n is a sum of two squares of elements of $\mathbb{Z}$.*
(2) *The integer n is a sum of two squares of elements of $\mathbb{Q}$.*
(3) *For every prime $p \mid n$ such that $p \equiv 3 \pmod 4$ we have $2 \mid v_p(n)$.*

Proof. By Corollary 5.2.13, n is a sum of two squares in every $\mathbb{Q}_p$ if and only if $n > 0$, $2 \mid v_p(n)$ for every $p \equiv 3 \pmod 4$, and $n/2^{v_2(n)} \equiv 1 \pmod 4$. However, this condition is evidently a consequence of the other, so can be omitted. Thus by the Hasse–Minkowski theorem, (2) and (3) are equivalent, and the equivalence of (1) and (2) immediately follows from Proposition 5.4.4, which is clearly applicable. □

A special case of this theorem is Fermat's important result saying that a prime $p \equiv 1 \pmod 4$ is a sum of two squares, which we have already proved in Corollary 2.3.13.

Using the above proposition and methods from analytic number theory, one can prove the following:

Proposition 5.4.10 (Landau). *As $X \to \infty$ the number $N_2(X)$ of $n \leqslant X$ that are sums of two squares satisfies*

$$N_2(X) \sim C \frac{X}{\sqrt{\log(X)}} ,$$

with

$$C = \frac{1}{\sqrt{2}} \prod_{p \equiv 3 \ (\mathrm{mod}\ 4)} \left(1 - \frac{1}{p^2}\right)^{-1/2} = \frac{\pi}{4} \prod_{p \equiv 1 \ (\mathrm{mod}\ 4)} \left(1 - \frac{1}{p^2}\right)^{1/2}$$
$$= 0.76422365358922066299069873125009232811679054139340951472\ldots .$$

The value of C is computed using the methods explained in Section 10.3.6; see Exercise 53 of Chapter 10.

For three squares we begin with the following lemma.

Lemma 5.4.11. *A number $c \in \mathbb{Q}^*$ is represented over $\mathbb{Q}$ by the quadratic form $x^2 + y^2 + z^2$ if and only if $c > 0$ and $-c$ is not a square in $\mathbb{Q}_2$.*

See Proposition 4.3.23 for a description of the squares of $\mathbb{Q}_2$.

Proof. Let q be the quadratic form $x^2 + y^2 + z^2$. We have trivially $d(q) = \varepsilon(q) = 1$. Thus by Corollary 5.2.17, for $p \neq \infty$, q represents c in $\mathbb{Q}_p$ if and only if either $-c$ is not a square in $\mathbb{Q}_p^*$ or $-c$ is a square and $(-1, -1)_p = 1$. However, by the explicit computation of the Hilbert symbol, we have $(-1, -1)_p = 1$ for $p \neq 2$ and $(-1, -1)_2 = -1$. Thus when $p \neq 2$ the condition is automatically satisfied, so that q represents c in $\mathbb{Q}_p^*$, while when $p = 2$ the condition is that $-c$ is not a square in $\mathbb{Q}_2$, proving the lemma. $\square$

Theorem 5.4.12 (Gauss). *Let n be a positive integer. The following three conditions are equivalent:*

(1) *The integer n is a sum of three squares of elements of $\mathbb{Z}$.*
(2) *The integer n is a sum of three squares of elements of $\mathbb{Q}$.*
(3) *The integer n is not of the form $4^a(8k - 1)$ for $a \geqslant 0$ and $k \geqslant 1$.*

Proof. By Proposition 4.3.23, the condition that n has the form $4^a(8k-1)$ is equivalent to $-n$ being a square in $\mathbb{Q}_2$. Thus by the above lemma the positive integer n is a sum of three squares in $\mathbb{Q}$ if and only if n is not of the form $4^a(8k - 1)$; hence (2) is equivalent to (3), and (1) trivially implies (2). Finally, to show that (2) implies (1) we apply Proposition 5.4.4: indeed if $x = (x_1, x_2, x_3) \in \mathbb{Q}^3$ and if y_i is the closest integer to x_i, then $q(x - y) \leqslant 3/4 < 1$, so the condition of the proposition is satisfied; hence n is indeed a sum of three squares. $\square$

Corollary 5.4.13. *As $X \to \infty$ the number of integers $n \leqslant X$ that are sums of three squares is asymptotic to $5X/6$.*

Proof. Immediate from the proposition and left to the reader (Exercise 15). $\square$

Corollary 5.4.14 (Lagrange). *Every positive integer n is a sum of four squares of integers.*

Proof. We write $n = 4^a(8k + m)$ with $a \geqslant 0$, $4 \nmid m$, and $1 \leqslant m \leqslant 7$. By the above theorem, if $m \neq 7$, then n is a sum of three squares. On the other hand, if $m = 7$, $n - 4^a = 4^a(8k + m - 1)$ is a sum of three squares, so that n is a sum of four squares. $\square$

There exist simpler and more direct proofs of this result. A classical one is as follows.

Proof. We first note the important and easy identity coming from the multiplicativity of the norm on quaternions:

$$(x_1^2 + x_2^2 + x_3^2 + x_4^2)(y_1^2 + y_2^2 + y_3^2 + y_4^2) = z_1^2 + z_2^2 + z_3^2 + z_4^2$$

with

$$z_1 = x_1y_1 - x_2y_2 - x_3y_3 - x_4y_4, \quad z_2 = x_1y_2 + x_2y_1 + x_3y_4 - x_4y_3 ,$$
$$z_3 = x_1y_3 - x_2y_4 + x_3y_1 + x_4y_2, \quad z_4 = x_1y_4 + x_2y_3 - x_3y_2 + x_4y_1 .$$

Assume by contradiction that the result is not true, and let p be the smallest integer that is not a sum of four squares. Thanks to this identity it is clear that p is a prime; otherwise, p would be a product of two strictly smaller integers, hence sums of four squares, so p itself would be one. Thus by Proposition 5.2.1 there exist x_1 and x_2 in $\mathbb{Z}$ such that $x_1^2 + x_2^2 \equiv -1 \pmod{p}$; hence setting $x_3 = 1$ and $x_4 = 0$, there exist $x_i \in \mathbb{Z}$ not all divisible by p such that $x_1^2 + x_2^2 + x_3^2 + x_4^2 = kp$ for some integer k. Note that p is odd (2 is a sum of four squares!), so changing x_i by some multiple of p, we may assume that $|x_i| < p/2$ for $1 \leqslant i \leqslant 4$, hence that $k < p$ (note the strict inequality). Since p is a minimal counterexample, it follows that k is a sum of four squares; hence using again the above identity, $k^2p = k(kp)$ is a sum of four squares. Dividing by k^2, we see that we have shown that p is a sum of four squares in $\mathbb{Q}$. We would now like to apply Proposition 5.4.4. In principle, this proposition is not applicable since if $q(x) = x_1^2 + x_2^2 + x_3^2 + x_4^2$ then for $x \in \mathbb{Q}^4$ there exists $y \in \mathbb{Z}^4$ such that $q(x - y) \leqslant 1$, the inequality being not necessarily strict. However, we have $q(x - y) = 1$ if and only if $x_i \in \frac{1}{2} + \mathbb{Z}$ for all i. Let us follow the proof of the proposition. We choose $d > 0$ minimal such that $q(x) = d^2p$ for some $x \in \mathbb{Z}^4$. There exists $y \in \mathbb{Z}^4$ such that $x/d = y + z$ with $q(z) \leqslant 1$. If $q(z) < 1$, the reasoning of the proposition goes through without change; in other words, we prove that $d = 1$. Thus assume that $q(z) = 1$. As mentioned, this means that $x_i/d \in \frac{1}{2} + \mathbb{Z}$ for $1 \leqslant i \leqslant 4$. In particular, d is even and $x_i/(d/2) \in 1 + 2\mathbb{Z}$. We can thus divide all the x_i (hence d) by $d/2$, and this contradicts the minimality of d except if $d = 2$. We thus have four odd integers x_i such that $\sum_{1 \leqslant i \leqslant 4} x_i^2 = 4p$. Changing x_i into $-x_i$ if necessary, we may assume that $x_i \equiv 1 \pmod 4$. Now if we use our identity above with $y_1 = 1/4$ and $y_2 = y_3 = y_4 = -1/4$, we see that

$$p = \left(\sum_{1 \leqslant i \leqslant 4} x_i^2 \right) / 4 = \sum_{1 \leqslant i \leqslant 4} z_i^2$$

with z_i as given, and the condition $x_i \equiv 1 \pmod 4$ is immediately seen to imply that the z_i are integers, contradicting the minimality of d. $\square$

Remarks. (1) Instead of reasoning by contradiction, we can of course say that we have shown that every prime is a sum of four squares, and this implies the same for all integers thanks to the identity.

(2) The necessity of the last step, where we had to divide by 4, is due to the fact that the canonical ring to consider in the rational quaternions is not the ring with $x_i \in \mathbb{Z}$ but the larger ring where $2x_i$ are integers with the same parity.

Using the theory of modular forms, it is in fact not difficult to give an explicit formula for the number $r_k(n)$ of representations of an integer n as a sum of k squares for $k \leqslant 8$, where all permutations and changes of signs of the variables are considered to give different representations. For the sake of completeness, and because the formulas are not easily found in the literature, we give the result here. The results for k even and for k odd have a completely different nature because in the first case we are dealing with modular forms of integral weight, and in the second case with modular forms of half-integral weight, so we separate the cases. We denote by $\sigma_k(n)$ the sum of the kth powers of the (positive) divisors of n. By convention we set $\sigma_k(n) = 0$ if $n \notin \mathbb{Z}$, and for any nonzero $D \equiv 0$ or 1 modulo 4 we denote by $\chi_D(n) = \left(\frac{D}{n}\right)$ the Kronecker–Legendre symbol and by $L(\chi_D, s)$ the corresponding L-function (see Chapter 10).

Theorem 5.4.15. *Let $n \geqslant 1$ be an integer. We have*

$$r_2(n) = 4 \sum_{d|n} \left(\frac{-4}{d}\right),$$

$$r_4(n) = 8(\sigma_1(n) - 4\sigma_1(n/4)),$$

$$r_6(n) = 4 \sum_{d|n} d^2 \left(4\left(\frac{-4}{n/d}\right) - \left(\frac{-4}{d}\right)\right),$$

$$r_8(n) = 16(\sigma_3(n) - 2\sigma_3(n/2) + 16\sigma_3(n/4)) = (-1)^{n-1}16(\sigma_3(n) - 16\sigma_3(n/2)).$$

Remarks. (1) The formula for $r_2(n)$ can easily be obtained from Proposition 5.4.9; see Exercise 18.

(2) It immediately follows from this theorem that $r_4(n) = 8\sum_{d|n,\ 4\nmid d} d > 0$, giving another proof that any positive integer is a sum of four squares.

For the case k odd we note that any nonzero $m \in \mathbb{Z}$ can be written in a unique way as $m = D(2^v f)^2$, where D is a fundamental discriminant, f is an odd integer, and $v \geqslant -1$.

Theorem 5.4.16. *Let $n \geqslant 1$ be an integer, for k odd write as above $(-1)^{(k-1)/2}n = D(2^v f)^2$, and for $j \geqslant 0$ set*

$$S_j(D, f) = \sum_{d|f} d^j \mu(d) \left(\frac{D}{d}\right) \sigma_{2j+1}(f/d).$$

We have

$$r_3(n) = \quad 12L(\chi_D,\ 0) \left(1 - \left(\frac{D}{2}\right)\right) S_0(D, f),$$

$$r_5(n) = -\frac{40}{7}L(\chi_D, -1) \left(2^{3v+5} + 3 - 2\left(\frac{D}{2}\right)(2^{3v+2} + 3)\right) S_1(D, f),$$

$$r_7(n) = -\frac{28}{31}L(\chi_D, -2) \left(5 \cdot 2^{5v+8} - 9 - 4\left(\frac{D}{2}\right)(5 \cdot 2^{5v+3} - 9)\right) S_2(D, f).$$

Remarks. (1) From the formula for $r_3(n)$ we immediately recover the characterization of integers n that are not sums of three squares given by Theorem 5.4.12. In fact, using Dirichlet's class number formula, the formula for $r_3(n)$ implies that if $n > 4$ and $-n$ is the discriminant of an imaginary quadratic field K then

$$r_3(n) = 12 \left(1 - \left(\frac{-n}{2} \right) \right) h(-n) ,$$

where $h(-n)$ denotes the class number of K.

(2) The above formulas are essentially the same as those that we will give in Corollary 10.3.13.

(3) The quantities $L(\chi_D, -j)$ are in fact rational numbers that can be computed exactly (although not efficiently) using Theorem 10.3.1. More precisely, Corollary 11.4.3 says that they are in $\mathbb{Z}$ with the exception of $D = -3$ and $D = -4$ for $j = 0$, of $D = 5$ for $j = 1$, and of $D = -3$, -4, and -7 for $j = 2$, and they even belong to $2\mathbb{Z}$ with the additional exceptions of $D = -8$ and $D = -p$ with $p \geqslant 7$ prime congruent to 3 modulo 4 for $j = 0$, of $D = 8$ for $j = 1$, and of $D = -8$ for $j = 2$.

5.5 The Hasse Norm Principle

This section should be skipped on first reading. We assume that the reader has some knowledge of local and global class field theory, so we do not define many of the notions that are mentioned.

The second local–global principle that we will consider very briefly is the so-called Hasse norm principle. We state it together with an interesting corollary in a single theorem as follows.

Theorem 5.5.1. *Let L/K be a finite cyclic extension of number fields, denote by $\mathfrak{f}(L/K)$ its conductor, and let $\alpha \in K^*$.*

(1) *The element α is the norm of some element $\beta \in L^*$ if and only if this is the case everywhere locally, i.e., for every place v of K and place w of L above v, α considered as an element of K_v is a norm of some element of L_w.*

(2) *If α is such that $\alpha \equiv 1 \pmod{{}^*\mathfrak{f}(L/K)}$ and is such that there exists an ideal I of L with $\alpha \mathbb{Z}_K = \mathcal{N}_{L/K}(I)$, then α is the norm of some element $\beta \in L^*$.*

Note that it is essential that the extension be cyclic for the theorem to be true, and that we do not need any local conditions at the infinite places.

Proof. The proof of (1) is based on the first basic inequality of global class field theory, and can be found for instance in [Lan1]. It is not especially difficult, but since the present book is not about class field theory we will not reproduce it. We will, however, show that (1) and (2) are equivalent.

(1) implies (2). By (1), to show that α is a norm we must show that α is everywhere a local norm, or equivalently, that it belongs to the kernel of all local Artin maps. But at primes $\mathfrak{p} \mid \mathfrak{f}(L/K)$ this is clear since $\alpha \equiv 1$ (mod *$\mathfrak{p}$). On the other hand, let $\mathfrak{p} \nmid \mathfrak{f}(L/K)$, so that $\mathfrak{p}$ is unramified in L/K, and let v be the corresponding place of K. If $\mathfrak{P}$ is a prime ideal above $\mathfrak{p}$ corresponding to the place w of L, denote by $\text{Art}_\mathfrak{p}$ the local Artin map and by $\sigma_\mathfrak{p}$ the Frobenius automorphism, which generates the local Galois group $\text{Gal}(L_w/K_v)$. For every prime ideal $\mathfrak{P}$ of L above $\mathfrak{p}$ we have $\mathcal{N}_{L/K}(\mathfrak{P}) = \mathfrak{p}^{f(\mathfrak{p})}$, where $f(\mathfrak{p}) = f(\mathfrak{P}/\mathfrak{p})$ is independent of $\mathfrak{P}$ above $\mathfrak{p}$; hence $v_\mathfrak{p}(\alpha) = v_\mathfrak{p}(\mathcal{N}_{L/K}(I))$ is a multiple of $f(\mathfrak{p})$. It follows that $\text{Art}_\mathfrak{p}(\alpha)$ is a power of $\sigma_\mathfrak{p}^{f(\mathfrak{p})}$, so is the identity; hence α is indeed in the kernel of all the local Artin maps, as claimed, proving that (1) implies (2).

(2) implies (1). Set $G = \text{Gal}(L/K)$. Assume first that $\alpha = \mathcal{N}_{L/K}(\beta)$ for some $\beta \in L^*$, denote by $D(w/v)$ the decomposition group of w, in other words the group of elements of G fixing w, and let R be a system of representatives in G of $G/D(w/v)$. We set

$$\beta_w = \prod_{\tau \in R} \tau(\beta) .$$

Since $\tau(\beta) \in L \subset L_w$ we have $\beta_w \in L_w$, and by definition of $D(w/v)$ we clearly have $\mathcal{N}_{L_w/K_v}(\beta_w) = \mathcal{N}_{L/K}(\beta) = \alpha$, so that α is indeed a local norm.

Conversely, assume that α is a local norm everywhere. Let $\mathfrak{p}$ be some prime ideal of K, let $\mathfrak{P}$ be a prime ideal of L above $\mathfrak{p}$, and as above let $f(\mathfrak{p}) = f(\mathfrak{P}/\mathfrak{p})$. To simplify notation we will write $\mathcal{N}_\mathfrak{P}$ instead of $\mathcal{N}_{L_\mathfrak{P}/K_\mathfrak{p}}$. Since α is a local norm there exists $\beta_\mathfrak{P} \in L_\mathfrak{P}$ such that $\alpha = \mathcal{N}_\mathfrak{P}(\beta_\mathfrak{P})$. Let $b = v_\mathfrak{P}(\beta_\mathfrak{P})$, so that $\beta_\mathfrak{P}\mathbb{Z}_\mathfrak{P} = \mathfrak{P}^b\mathbb{Z}_\mathfrak{P}$. Taking norms it follows that $\alpha\mathbb{Z}_\mathfrak{p} = \mathfrak{p}^{f(\mathfrak{p})b}\mathbb{Z}_\mathfrak{p}$, hence that $f(\mathfrak{p}) \mid v_\mathfrak{p}(\alpha)$. We have thus already proved that the principal ideal $\alpha\mathbb{Z}_K$ is the norm of an ideal of L, more precisely

$$\alpha\mathbb{Z}_K = \mathcal{N}_{L/K}\left(\prod_\mathfrak{p} \mathfrak{P}^{v_\mathfrak{p}(\alpha)/f(\mathfrak{p})} \right) ,$$

where as usual $\mathfrak{P}$ denotes any ideal of L above $\mathfrak{p}$.

Write $\mathfrak{f}(L/K) = \prod_i \mathfrak{p}_i^{b_i}$, where we include the infinite places with exponent 0 or 1, for each i let $\mathfrak{P}_i$ be some ideal of L above $\mathfrak{p}_i$ (or some infinite place if $\mathfrak{p}_i$ is infinite), and finally let $e_i = e(\mathfrak{P}_i/\mathfrak{p}_i)$ be the ramification index of $\mathfrak{P}_i$.

By assumption there exist elements $\beta_i \in L_{\mathfrak{P}_i}$ such that $\alpha = \mathcal{N}_{\mathfrak{P}_i}(\beta_i)$. Since L is dense in $L_{\mathfrak{P}_i}$, we can find an element of L as close $\mathfrak{P}_i$-adically as we want to β_i, so we choose $\beta_i' \in L$ such that $\beta_i' - \beta_i \in \mathfrak{P}_i^{b_i e_i}$. By the approximation theorem in L we can find $\gamma \in L$ such that for all i we have

$$\begin{cases} \gamma \equiv \beta_i' \ (\text{mod } *\mathfrak{P}_i^{b_i e_i}) , \\ \gamma \equiv 1 \ (\text{mod } *\mathfrak{P}^{b_i e_i}) \qquad \text{for all } \mathfrak{P} \mid \mathfrak{p}, \ \mathfrak{P} \neq \mathfrak{P}_i . \end{cases}$$

Fix an index i and write $G = \bigcup_j \tau_j D(\mathfrak{P}_i/\mathfrak{p}_i)$ (disjoint union), where as above, the τ_j form a set of representatives for the cosets of G modulo the decomposition group, where we may choose $\tau_1 = 1$. For $j \neq 1$ we have $\tau_j^{-1}(\mathfrak{P}_i) \mid \mathfrak{p}$ but $\tau_j^{-1}(\mathfrak{P}_i) \neq \mathfrak{P}_{i'}$ for any i', hence

$$\gamma \equiv 1 \ (\mathrm{mod}\ {}^*\tau_j^{-1}(\mathfrak{P}_i^{b_i\,e_i}))\quad \text{hence}\quad \tau_j(\gamma) \equiv 1\ (\mathrm{mod}\ {}^*\mathfrak{P}_i^{b_i\,e_i})\ .$$

It follows that

$$N_{L/K}(\gamma) = \prod_j \prod_{\sigma \in D(\mathfrak{P}_i/\mathfrak{p}_i)} \tau_j \sigma(\gamma) \equiv \prod_{\sigma \in D(\mathfrak{P}_i/\mathfrak{p}_i)} \sigma(\gamma)$$
$$= \mathcal{N}_{\mathfrak{P}_i}(\gamma) \equiv \mathcal{N}_{\mathfrak{P}_i}(\beta_i') \ (\mathrm{mod}\ {}^*\mathfrak{P}_i^{b_i\,e_i})\ .$$

On the other hand, since $\alpha = \mathcal{N}_{\mathfrak{P}_i}(\beta_i)$ and $\beta_i' \equiv \beta_i\ (\mathrm{mod}\ {}^*\mathfrak{P}_i^{b_i\,e_i})$ it follows that

$$\alpha \equiv \mathcal{N}_{\mathfrak{P}_i}(\beta_i') \equiv \mathcal{N}_{L/K}(\gamma)\ (\mathrm{mod}\ {}^*\mathfrak{P}_i^{b_i\,e_i})\ .$$

Since both sides are in K and $e_i = e(\mathfrak{P}_i/\mathfrak{p}_i)$, it follows that

$$\alpha \equiv \mathcal{N}_{L/K}(\gamma)\ (\mathrm{mod}\ {}^*\mathfrak{p}_i^{b_i})\ .$$

Since this is true for all i, we finally deduce that

$$\alpha_1 = \alpha/\mathcal{N}_{L/K}(\gamma) \equiv 1\ (\mathrm{mod}\ {}^*\mathfrak{f}(L/K))\ .$$

We have thus found an element $\alpha_1 \in K$ such that $\alpha_1 \equiv 1\ (\mathrm{mod}\ {}^*\mathfrak{f}(L/K))$ and such that the principal ideal $\alpha_1\mathbb{Z}_K$ is the norm of an ideal of L (this is clearly a consequence of the corresponding property of α). This element α_1 satisfies the conditions of (2); hence by (2), $\alpha_1 = \mathcal{N}_{L/K}(\beta_1)$ for some $\beta_1 \in L^*$, so that $\alpha = \mathcal{N}_{L/K}(\beta)$ with $\beta = \beta_1\gamma$, proving that (2) implies (1).   □

The following corollary is an immediate consequence of the proof, but is trivial to prove directly (Exercise 22).

Corollary 5.5.2. *In the above theorem, it is sufficient to consider finite and infinite places v, w such that w is ramified over v, together with finite places v corresponding to prime ideals $\mathfrak{p}$ of K such that $v_{\mathfrak{p}}(\alpha) \neq 0$, and all places w above these v.*

(Recall that an infinite place w is ramified above v if and only if v is real and w is complex.)

Example. Let $K = \mathbb{Q}$ and $L = \mathbb{Q}(i)$ with $i^2 = -1$. The only ramified places v in L/K are the place at infinity and the place corresponding to the prime 2. An element $a \in \mathbb{Q}^*$ is a norm from L to K if and only if a is a sum of two squares of elements of $\mathbb{Q}$. By the above corollary and Corollary 5.2.13, this is true if and only if the following conditions are satisfied:

(1) Local condition at infinity: $a > 0$.
(2) Local condition at 2: $a/2^{v_2(a)} \equiv 1 \pmod 4$.
(3) Other local conditions: for any prime $p \equiv 3 \pmod 4$, $v_p(a)$ is even.

Clearly the second condition is a consequence of the other two, hence can be removed.

5.6 The Hasse Principle for Powers

In this section we follow closely the book by G. Gras [Gras].

5.6.1 A General Theorem on Powers

For any integer n, denote as usual by ζ_n a primitive nth root of unity.

Proposition 5.6.1 (Hilbert's Theorem 90 for roots of unity). *Let K be a field of characteristic zero, let p be a prime number, let $e \geqslant 1$, set $L = K(\zeta_{p^e})$, let μ_{p^e} be the group of p^eth roots of unity in L, and let $G = \mathrm{Gal}(L/K)$. Assume that G is cyclic (which is always the case if $p \neq 2$) generated by σ, say. An element $\zeta \in \mu_{p^e}$ satisfies $\mathcal{N}_{L/K}(\zeta) = 1$ if and only if there exists $\eta \in \mu_{p^e}$ such that $\zeta = \sigma(\eta)/\eta$, except in the following so-called exceptional case: $p = 2$, $e \geqslant 2$, $[L : K] = 2$, $\sigma(\zeta_{2^e}) = \zeta_{2^e}^{-1}$, and ζ is a primitive 2^eth root of unity, i.e., an odd power of ζ_{2^e}.*

Proof. The condition is clearly sufficient. Thus assume that $\mathcal{N}_{L/K}(\zeta) = 1$. We have $\sigma(\zeta_{p^e}) = \zeta_{p^e}^s$ for some $s \in \mathbb{Z}$ coprime to p and defined uniquely modulo p^e. If $s \equiv 1 \pmod{p^e}$ then by Galois theory $L = K$; hence $\mathcal{N}_{L/K}(\zeta) = \zeta$, so there is nothing to prove. Thus we may assume that $s \not\equiv 1 \pmod{p^e}$. We have $\sigma^i(\zeta_{p^e}) = \zeta_{p^e}^{s^i}$, hence $\mathcal{N}_{L/K}(\zeta_{p^e}) = \zeta_{p^e}^S$ with $S = (s^n - 1)/(s - 1)$, where $n = [L : K] = |\mathrm{Gal}(L/K)|$ is the order of σ (or equivalently of s modulo p^e). Thus if ζ has exact order p^ℓ with $0 \leqslant \ell \leqslant e$ (i.e., if $\zeta = \zeta_{p^e}^{p^{e-\ell}m}$ for some m such that $p \nmid m$), then $\mathcal{N}_{L/K}(\zeta) = 1$ if and only if $v_p((s^n - 1)/(s - 1)) \geqslant \ell$.

We want to find an integer k such that $\sigma(\eta)/\eta = \zeta$ for $\eta = \zeta_{p^e}^k$. This can be written

$$(s - 1)k \equiv p^{e-\ell}m \pmod{p^e}.$$

Now, a congruence $ax \equiv b \pmod c$ is soluble in x if and only if the equation $ax - cy = b$ is soluble, hence if and only if the GCD of a and c divides b. In our case, since $p \nmid m$ this means that $\gcd(s - 1, p^e) \mid p^{e-\ell}$, hence that $v_p(s - 1) \leqslant e - \ell$. Assume by contradiction that $v_p(s - 1) > e - \ell$. Since we have excluded the case $s \equiv 1 \pmod{p^e}$, this implies in particular that $\ell \geqslant 2$. Since $v_p((s^n - 1)/(s - 1)) \geqslant \ell$, it follows that $v_p(s^n - 1) > e$.

Consider first the case $p \neq 2$. By Lemma 2.1.22 we thus have $v_p(s - 1) + v_p(n) > e$, and since we have excluded the case $s \equiv 1 \pmod{p^e}$, it follows in particular that $v_p(n) > 0$, hence that $p \mid n$. But then again by Lemma 2.1.22,

$$v_p(s^{n/p} - 1) = v_p(s - 1) + v_p(n) - 1 \geqslant e \,;$$

hence $s^{n/p} \equiv 1 \pmod{p^e}$, so that $\sigma^{n/p}$ is the identity on L, contradicting the fact that σ has exact order n.

Consider now the case $p = 2$. In this case n divides $\phi(2^e) = 2^{e-1}$, so it is a power of 2. If $n = 1$ we have $L = K$ and there is nothing to prove. If $4 \mid n$, Corollary 2.1.23 would give as in the case p odd

$$v_p(s^{n/2} - 1) = v_p(s^2 - 1) + v_p(n/4) = v_p(s^2 - 1) + v_p(n/2) - 1 \geqslant e \,,$$

hence $s^{n/2} \equiv 1 \pmod{p^e}$, again contradicting the fact that σ has exact order n. Thus the proposition can fail only if $p = 2$ and $n = 2$. In this case $v_p((s^n - 1)/(s - 1)) = v_p(s + 1)$. Since $\ell \geqslant 2$, the inequality $v_p((s^n - 1)/(s - 1)) \geqslant \ell$ thus implies that $s \equiv -1 \pmod 4$, hence that $v_p(s - 1) = 1$, and this is strictly greater than $e - \ell$ if and only if $\ell = e$, which is the exceptional case that has been excluded.

We have thus shown that $v_p(s - 1) \leqslant e - \ell$, hence that our congruence is soluble, finishing the proof of the proposition, except when the conditions of the exceptional case are satisfied.

This exceptional case can occur only if $p = 2$, $n = 2$, and $\zeta = \zeta_{2^e}^m$ for some odd m, and for $s \equiv 3 \pmod 4$ and $s^2 \equiv 1 \pmod{2^e}$. Thus $s \equiv -1 \pmod{2^{e-1}}$, so that $\sigma(\zeta_{2^e}) = \pm\zeta_{2^e}^{-1}$. If the sign is $-$, then $\mathcal{N}_{L/K}(\zeta) = \zeta\sigma(\zeta) = -1$, so there are no primitive p^eth roots of unity of norm 1. On the other hand, if the sign is $+$ then all p^eth roots of unity have norm 1, so this is the only case in which there exist counterexamples to the assertion of the proposition. In fact it is clear that

$$\sigma(\zeta_{2^e})/\zeta_{2^e} = \zeta_{2^e}^{-2} = \zeta_{2^{e-1}}^{-1} \,;$$

hence a primitive 2^eth root of unity cannot be of the form $\sigma(\eta)/\eta$ for $\eta \in \mu_{2^e}$.

$\square$

We can now state the following general theorem on powers, which will be essential in the proof of the local–global principle.

Theorem 5.6.2. *Let K be any field of characteristic zero, let p be a prime number, let $e \geqslant 1$ be an integer, and finally let $L = K(\zeta_{p^e})$.*

If $x \in K^*$, then x is a p^eth power in L^* if and only if it is a p^eth power in K^*, except if the following three conditions (comprising the so-called exceptional cases) are all satisfied:

(1) *$p = 2$ and $e \geqslant 2$.*
(2) *There exists n with $0 \leqslant n \leqslant e - 2$ such that*

$$K \cap \mathbb{Q}(\zeta_{2^e}) = \mathbb{Q}(\zeta_{2^{n+2}} + \zeta_{2^{n+2}}^{-1}) \,.$$

(3) *The element x has the form*

$$x = (-1)^{2^{e-n-2}} x_0 y^{2^e}, \quad \text{with } x_0 = (2 + \zeta_{2^{n+2}} + \zeta_{2^{n+2}}^{-1})^{2^{e-1}} \text{ and } y \in K^* \,.$$

In this exceptional case we have $x = ((1 + \zeta_{2^{n+2}})y)^{2^e}$, *which is a* p^e *th power only in the quadratic extension* $K(\zeta_{2^{n+2}})$ *of* K *and not in* K.

Proof. If x is a p^eth power in K (or for the exceptional cases in a quadratic extension of K contained in L), then evidently x is a p^eth power in L. Conversely, let $x \in K^*$ and assume that $x = z^{p^e}$ for some $z \in L^*$. We first assume that G is cyclic, use the same notation as in the proposition, and set $\zeta = \sigma(z)/z$. Since $x \in K^*$ we have $\zeta^{p^e} = \sigma(x)/x = 1$, hence $\zeta \in \mu_{p^e}$. On the other hand, we have $\mathcal{N}_{L/K}(\zeta) = 1$. Thus if we are not in the exceptional case of the proposition there exists $\eta \in \mu_{p^e}$ such that $\zeta = \sigma(\eta)/\eta$. If we set $y = z/\eta$, we thus have $\sigma(y)/y = \zeta/\zeta = 1$, hence $y \in K^*$, and $y^{p^e} = z^{p^e} = x$ since $\eta \in \mu_{p^e}$, proving the theorem in this case. It thus remains to consider the exceptional case of the proposition, and the case in which G is noncyclic, both of these cases occurring only for $p = 2$, which we now consider.

We let $L = K(\zeta_{2^e})$, where we may assume that $e \geqslant 2$, $G = \mathrm{Gal}(L/K)$, and $k = K \cap \mathbb{Q}(\zeta_{2^e})$. By Theorem 3.1.11, we know that G is canonically isomorphic to $\mathrm{Gal}(\mathbb{Q}(\zeta_{2^e})/k)$. Furthermore, $\mathrm{Gal}(\mathbb{Q}(\zeta_{2^e})/k)$ is the subgroup of $\mathrm{Gal}(\mathbb{Q}(\zeta_{2^e})/\mathbb{Q})$ of elements fixing k, hence in particular is a subgroup of the group $(\mathbb{Z}/2^e\mathbb{Z})^*$ generated by the class of -1 (of order 2) and the class of 5 (of order 2^{e-2}).

Now, for instance by applying Theorem 2.1.14, it is easily seen that the subgroups of $(\mathbb{Z}/2^e\mathbb{Z})^*$ (which necessarily have cardinality 2^{e-1-n} for some n such that $0 \leqslant n \leqslant e - 1$) have the following three types (see Exercise 23):

- Case A_n: the cyclic group generated by $\overline{5^{2^{n-1}}}$ for $1 \leqslant n \leqslant e - 1$, when $e \geqslant 3$.
- Case B_n: the cyclic group generated by $\overline{-5^{2^{n-1}}}$ for $1 \leqslant n \leqslant e - 1$, when $e \geqslant 3$
- Case C_n: the noncyclic group generated by $\overline{5^{2^n}}$ and $\overline{-1}$ for $0 \leqslant n \leqslant e - 3$ when $e \geqslant 3$, or the cyclic group generated by $\overline{-1}$ when $e = 2$ and $n = 0$.

Thus G is canonically isomorphic to one of these groups. In cases A_n and B_n, G is cyclic; hence we can apply the proposition, so the theorem is valid as long as we are not in the exceptional case. But this case occurs only if $|G| = 2$ and $\sigma(\zeta_{2^e}) = \zeta_{2^e}^{-1}$, hence if $n = e - 2$ (hence $e \geqslant 3$ since $n \geqslant 1$), and $5^{2^{e-3}} \equiv -1 \pmod{2^e}$ in case A_n, $-5^{2^{e-3}} \equiv -1 \pmod{2^e}$ in case B_n. The case A_n is impossible since $e \geqslant 3$ and $5 \equiv 1 \pmod 4$, and the case B_n is impossible since the class of 5 has exact order 2^{e-2} in $(\mathbb{Z}/2^e\mathbb{Z})^*$. Thus we cannot have any exceptional cases in cases A_n and B_n; hence it remains to consider the case $G \simeq C_n$. In this case, as stated we will set $n = 0$ if $e = 2$, which indeed corresponds to a group of cardinality equal to $2^{e-1-n} = 2$.

In this case $\mathrm{Gal}(\mathbb{Q}(\zeta_{2^e})/k) \simeq G \simeq C_n$, so that k is the fixed field of $\mathbb{Q}(\zeta_{2^e})$ by C_n, which is clearly equal to $\mathbb{Q}(\zeta_{2^{n+2}} + \zeta_{2^{n+2}}^{-1})$, which is the second condition of the exceptional cases given in the theorem. Since this field occurs frequently we set

$$k_n = \mathbb{Q}(\zeta_{2^{n+2}} + \zeta_{2^{n+2}}^{-1}) .$$

Let K_1 be the fixed field of L by the subgroup generated by the class of 5^{2^n} in C_n. By Galois theory K_1 is a quadratic extension of K, and in fact $K_1 = K(\zeta_{2^n+2})$, and $\zeta_{2^n+2} + \zeta_{2^n+2}^{-1} \in k \subset K$. Since $L = K_1(\zeta_{2^e})$ and $\mathrm{Gal}(L/K_1)$ is cyclic, the case A_n implies that if $x \in K^*$ is a 2^eth power in L, it will be a 2^eth power in K_1. Hence write $x = z^{2^e}$ with $z \in K_1$, and let τ (complex conjugation) be the generator of $\mathrm{Gal}(K_1/K)$. Then as usual $(\tau(z)/z)^{2^e} = 1$, so $\tau(z)/z \in \mu_{2^e} \cap K_1 = \mu_{2^n+2}$. Thus if $\tau(z)/z$ is not a primitive 2^{n+2}th root of unity, we are not in the exceptional case of the proposition, so that as usual we deduce that x is a 2^eth power in K. Assume now that $\tau(z)/z$ is a primitive 2^{n+2}th root of unity. Note the identity

$$\tau(1 + \zeta_{2^n+2})/(1 + \zeta_{2^n+2}) = \zeta_{2^n+2}^{-1} .$$

It follows that if we set $z_1 = z/(1 + \zeta_{2^n+2})$, then $\tau(z_1)/z_1$ will indeed be a 2^{n+2}th root of unity of norm 1 in the extension K_1/K, but will not be a *primitive* 2^{n+2}th root. Thus we are not in the exceptional case of the proposition for z_1, so there exists $\eta \in \mu_{2^n+2}$ such that $\tau(z_1)/z_1 = \tau(\eta)/\eta$; hence $z_1 = y\eta$ with $y \in K^*$, so that

$$x = z^{2^e} = (1 + \zeta_{2^n+2})^{2^e} y^{2^e} = \zeta_{2^n+2}^{2^e}(2 + \zeta_{2^n+2} + \zeta_{2^n+2}^{-1})^{2^{e-1}} y^{2^e}$$
$$= (-1)^{2^{e-n-2}}(2 + \zeta_{2^n+2} + \zeta_{2^n+2}^{-1})^{2^{e-1}} y^{2^e} ,$$

showing that up to 2^eth powers in K there exists a single counterexample to the theorem given in the exceptional cases, and finishing the proof.   □

5.6.2 The Hasse Principle for Powers

We can now state and prove the local–global principle for powers.

Theorem 5.6.3. *Let K be a number field, let p be a prime number, let $e \geqslant 1$ be an integer, and let S be a finite set of places of K. Let $x \in K^*$, and assume that for all places $v \notin S$ the element x considered as an element of the completion K_v is a p^eth power. Then x is a p^eth power in K except if the following four conditions (comprising the so-called S-special cases) are all satisfied, where we recall that we have set $k_n = \mathbb{Q}(\zeta_{2^n+2} + \zeta_{2^n+2}^{-1})$.*

(1) *$p = 2$ and $e \geqslant 3$.*
(2) *There exists n with $0 \leqslant n \leqslant e - 3$ such that $K \cap \mathbb{Q}(\zeta_{2^e}) = k_n$.*
(3) *For all places v above 2 and not belonging to S the field K_v contains one of the numbers*

$$1 + \zeta_{2^n+2}, \quad \zeta_{2^n+3} + \zeta_{2^n+3}^{-1}, \quad \zeta_4(\zeta_{2^n+3} + \zeta_{2^n+3}^{-1}) .$$

(4) *The element x has the form*

$$x = x_0 y^{2^e}, \quad \text{with } x_0 = (2 + \zeta_{2^n+2} + \zeta_{2^n+2}^{-1})^{2^{e-1}} \text{ and } y \in K^* .$$

Proof. We set $L = K(\zeta_{p^e})$, and consider the Kummer extension M of L defined by $M = L(x^{1/p^e})$. This extension is evidently a cyclic extension with Galois group canonically isomorphic to a subgroup of $\mathbb{Z}/p^e\mathbb{Z}$. By the Čebotarev density theorem there exists an infinity of prime ideals of L whose corresponding Frobenius automorphism is a generator of $\mathrm{Gal}(M/L)$. In particular, since M/L can be ramified only at prime ideals dividing p and x, if we avoid those prime ideals and those corresponding to the places above S, which are finite in number, we see that there exists a finite place v_0 of L not above a place of S, unramified in M/L, and such that the Frobenius at v_0 is a generator of $\mathrm{Gal}(M/L)$.

Let v be the place of K below v_0. Since $v \notin S$, the element x is a p^eth power in K_v, hence a fortiori in L_{v_0}, which is an extension of K_v. It follows that for any place w_0 of M above v_0 we have $M_{w_0} = L_{v_0}(x^{1/p^e}) = L_{v_0}$; hence, since v_0 is unramified, the place v_0 is totally split in M/L; in other words, the decomposition group of v_0 in the extension M/L, which is generated by the Frobenius at v_0, is trivial. Since the Frobenius at v_0 generates $\mathrm{Gal}(M/L)$, this means that $\mathrm{Gal}(M/L)$ is trivial, hence that $M = L$. We have thus shown that the hypothesis of the theorem implies that x is a p^eth power in L.

By Theorem 5.6.2, this implies that x is indeed a p^eth power in K except in the exceptional cases given by that theorem, finishing the proof of our theorem outside of these exceptional cases.

Let us now consider in detail the exceptional cases. Here we have $p = 2$, $K \cap \mathbb{Q}(\zeta_{2^e}) = k_n$ for some n such that $0 \leqslant n \leqslant e - 2$, and x is of the form $x = (-1)^{2^{e-n-2}} x_0 y^{2^e}$ with $x_0 = (2 + \zeta_{2^{n+2}} + \zeta_{2^{n+2}}^{-1})^{2^{e-1}}$ and $y \in K^*$. I claim first that the theorem is still true if $n = e - 2$. Indeed, in that case x has the form $-z^2$ for some $z \in K$; hence the fact that x is locally a 2^eth power for all $v \notin S$ implies in particular that -1 is locally a square for all $v \notin S$. Reasoning as in the beginning of the proof, we see that if we had $\sqrt{-1} \notin K$ we could find a place v_0 of K totally split in $K(\sqrt{-1})/K$ and not belonging to S, which is absurd since -1 is locally a square at v_0. It follows that $\sqrt{-1} \in K$, but this contradicts the second condition of the exceptional cases stating that $K \cap \mathbb{Q}(\zeta_{2^e}) = \mathbb{Q}(\zeta_{2^e} + \zeta_{2^e}^{-1})$, proving my claim.

Thus we may assume that $0 \leqslant n \leqslant e - 3$, and in particular $e \geqslant 3$. We have $K \cap \mathbb{Q}(\zeta_{2^e}) = k_n$ and $x = x_0 y^{2^e}$ with $x_0 = (2 + \zeta_{2^{n+2}} + \zeta_{2^{n+2}}^{-1})^{2^{e-1}}$. Set $K_1 = K(\zeta_{2^{n+3}})$. Since $n \leqslant e - 3$, K_1/K is a subextension of L/K. In fact, using the notation of the proof of Theorem 5.6.2, K is the fixed field of $L = K(\zeta_{2^e})$ by the group C_n generated by the classes of 5^{2^n} and -1, and K_1 is the fixed field by the group A_{n+2} generated by the class of $5^{2^{n+1}}$. Since the three subgroups of index 2 of A_{n+2} containing C_n are A_{n+1}, B_{n+1}, and C_{n+1}, it follows that K_1 is a biquadratic extension of K, the three quadratic subextensions being generated by $1 + \zeta_{2^{n+2}}$ (fixed by A_{n+1}), $\zeta_{2^{n+3}} + \zeta_{2^{n+3}}^{-1}$ (fixed by C_{n+1}), and $\zeta_4(\zeta_{2^{n+3}} + \zeta_{2^{n+3}}^{-1})$ (fixed by B_{n+1}); see Exercise 24. Furthermore, in K_1 we can write

$$x_0 = (1 + \zeta_{2^{n+2}})^{2^e} = (\zeta_{2^{n+3}} + \zeta_{2^{n+3}}^{-1})^{2^e} = (\zeta_4(\zeta_{2^{n+3}} + \zeta_{2^{n+3}}^{-1}))^{2^e} .$$

Now let v be a place above 2 and not belonging to S, and set $k = K_v \cap \mathbb{Q}(\zeta_{2^e}) \supset k_n$. Assume first that $k = k_n$. By assumption there exists $z \in K_v$ such that $x_0 = z^{2^e}$. It follows (for example) that $1 + \zeta_{2^n+2} = z\zeta_{2^e}^m$ for some integer m; hence $z \in K_v \cap \mathbb{Q}(\zeta_{2^e}) = k = K \cap \mathbb{Q}(\zeta_{2^e})$, so that $z \in K$, and hence x_0 is a 2^eth power in K. Thus we may assume that k is strictly larger than k_n. Since $k \subset \mathbb{Q}(\zeta_{2^e})$, by Galois theory this means that k contains one of the three quadratic extensions of k_n contained in $\mathbb{Q}(\zeta_{2^e})$ (the ones mentioned above), which is exactly the third condition of the S-special case, finishing the proof of the theorem. $\square$

Example. We let $K = \mathbb{Q}$, $e = 3$, $n = 2$, $S = \{2\}$, and $x = 16$. Then x is not an 8th power in K. On the other hand, x is an 8th power in $\mathbb{R}$, and since the four 4th roots of x are ± 2 and $\pm(1 + i)^2$, and since for p odd we have either $\left(\frac{2}{p}\right) = 1$, $\left(\frac{-2}{p}\right) = 1$, or $\left(\frac{-1}{p}\right) = 1$, it follows by Hensel's lemma that 16 is an 8th power in $\mathbb{Q}_p$ for all $p \notin S$. This is a minimal example of failure for the local–global principle for powers.

5.7 Some Counterexamples to the Hasse Principle

A *Hasse principle* is a statement asserting that a given property is true globally if and only if it is true everywhere locally. In the preceding sections we have seen some of the most important examples where it is valid. It is however not at all true in general, and in this short section we give several examples in which it is not applicable (we will see many more in the next chapter). We assume implicitly that our base field is always $\mathbb{Q}$.

Historically, perhaps the most famous counterexample to the Hasse principle, due to Selmer, is the homogeneous cubic equation $3x^3 + 4y^3 + 5z^3 = 0$, which has a nontrivial solution everywhere locally, but not in $\mathbb{Q}$; see Corollary 6.4.12. More generally, the Hasse principle is not valid outside a very narrow range of equations (the most important being the ones seen above), and we will see several examples of failure in the sequel. It is however a very interesting subject to study quantitatively the *obstructions* to this principle, but this is another, much deeper, matter.

We begin with an example that could be considered artificial, but gives a first idea.

Proposition 5.7.1. *The equation*

$$(x^2 - 2)(x^2 - 17)(x^2 - 34) = 0$$

has solutions everywhere locally but not globally.

Proof. It is clear that the equation has a solution in $\mathbb{R}$. It has a solution in $\mathbb{Q}_2$ since 17 is a square in $\mathbb{Q}_2$ by Proposition 4.3.23, and 2 is a square in

$\mathbb{Q}_{17}$ by Proposition 4.3.22, so it has a solution in $\mathbb{Q}_{17}$. If p is not equal to 2 or 17, the product of the three Legendre symbols $\left(\frac{2}{p}\right)$, $\left(\frac{17}{p}\right)$, and $\left(\frac{34}{p}\right)$ is equal to 1; hence at least one of them is equal to 1, so by Proposition 4.3.22 the equation has a solution in $\mathbb{Q}_p$. On the other hand, it is clear that there are no solutions in $\mathbb{Q}$. $\qquad\square$

It is of course immediate to generalize the above example (Exercise 26).

The following example (for which I am indebted to J.-L. Colliot-Thélène) is much less artificial and also quite instructive.

Proposition 5.7.2. *The equation*

$$y^2 + z^2 = (3 - x^2)(x^2 - 2)$$

has solutions everywhere locally but not globally.

Proof. It is clear that the equation has a solution in $\mathbb{R}$. For $p = 3$, we take $x = 1$, and by Corollary 5.2.13, -2 is a sum of two squares in $\mathbb{Q}_3$. For all other p, we choose $x = 0$. Since $v_p(-6) = 0$ for $p \neq 2$ and 3, Corollary 5.2.13 once again tells us that -6 is a sum of two squares for all $p \neq 2$. Finally, since $v_2(-6) = 1$ is odd and $-6/2 = -3 \equiv 1 \pmod 4$, it also tells us that -6 is a sum of two squares in $\mathbb{Q}_2$.

We must now show that this equation does not have solutions in $\mathbb{Q}$. Usually in nontrivial counterexamples to the Hasse principle, this is the most difficult part. Here, luckily we can get away with simple congruence arguments (which in a sense is surprising since we have just shown that the equation is everywhere locally soluble). After clearing out denominators, we can write the equation as

$$Y^2 + Z^2 = (3D^2 - X^2)(X^2 - 2D^2),$$

where now all the variables are integers with $D \neq 0$. Assume that this equation has a solution, and choose one with D minimal. I first claim that the GCD of X and D cannot be divisible by $p = 2$ or by a prime $p \equiv 3 \pmod 4$. Indeed, if that were the case then $Y^2 + Z^2 \equiv 0 \pmod{p^4}$, and since -1 is not a square modulo p^2, we would have $Y \equiv Z \equiv 0 \pmod p$, hence $(Y/p)^2 + (Z/p)^2 \equiv 0 \pmod{p^2}$, hence $Y \equiv Z \equiv 0 \pmod{p^2}$, so we could divide our equation by p^4, contradicting the minimality of D.

Note by positivity that we must have $2D^2 < X^2 < 3D^2$. We consider two cases. If D or X is even (but not both, by what we have just proved) then $3D^2 - X^2 \equiv 3 \pmod 4$ and is positive, hence is divisible by a prime $p \equiv 3$ to an odd power. Since p cannot divide $X^2 - 2D^2$ (otherwise it would divide the GCD of X and D, contradicting what we have proved), it follows that p divides $Y^2 + Z^2$ to the same odd power, which is absurd. In the second case, we assume that D and X are both odd. We apply the same reasoning to $X^2 - 2D^2 \equiv 3 \pmod 4$, which thus also leads to a contradiction, finishing the proof of the proposition. $\qquad\square$

I am indebted to S. Siksek for the following example, initially due to Lind.

Proposition 5.7.3 (Lind). *The equation* $2y^2 = x^4 - 17z^4$ *has nontrivial solutions everywhere locally, but not globally.*

Proof. For the local solubility, we prove the stronger statement that the equation $2y^2 = x^4 - 17$ is everywhere locally soluble. The advantage of this equation is that like any hyperelliptic quartic, it represents a curve of genus 1, which is immediately seen to be nonsingular outside 2, 17, and ∞ (our initial equation is of genus 3). Thus, if p is a prime different from 2 and 17, the Weil bounds tell us that the number of affine (hence nonsingular) solutions modulo p is greater than or equal to $p - 2\sqrt{p}$, which is strictly positive as soon as $p \geqslant 5$, so by Hensel lifting our equation is locally soluble at least for $p \neq 2, 3, 17$, and ∞. We leave as an exercise for the reader that it is also locally soluble for these values of p (Exercise 27).

As usual, the amusing part is to show that our equation has no nontrivial global solution. Assume that it does. Let d be the unique strictly positive rational number such that x/d and z/d are coprime integers (which we can reasonably call the GCD of x and z, even when x and z are in $\mathbb{Q}$). Replacing (x, y, z) by $(x/d, y/d^2, z/d)$ we may therefore assume that $\gcd(x, z) = 1$, and also that $y \in \mathbb{Z}_{>0}$. Let q be an odd prime divisor of y. We cannot have $q \mid z$, so that $(x/z)^4 \equiv 17 \pmod{q}$; hence in particular $\left(\frac{17}{q}\right) = 1$, so by quadratic reciprocity $\left(\frac{q}{17}\right) = 1$. Since $\left(\frac{2}{17}\right) = 1$ and $y > 0$ it follows that $\left(\frac{y}{17}\right) = 1$, so let y_0 be such that $y \equiv y_0^2 \pmod{17}$. We thus have $2y_0^4 \equiv x^4 \pmod{17}$, and since $17 \nmid y_0$ it follows that $2 \equiv (x/y_0)^4 \pmod{17}$. But this is a contradiction since it is immediately checked that 2 is not a fourth power modulo 17. $\square$

This type of reasoning can be generalized to more general equations of the type $y^2 = f(x, z)$, where f is a homogeneous polynomial of even degree; see Exercise 28.

We finish this section with the following example, which is also very instructive for other reasons. Usually, proving local solubility everywhere is not difficult, although for the preceding example we had to use the nontrivial Weil bounds (it is possible to avoid them, however). On the other hand, proving that the equation is not globally soluble is often much harder. The following result, communicated to me by M. Stoll, shows other ways of dealing with both problems.

Proposition 5.7.4. *The equation* $y^2 = -(x^2 + x - 1)(x^4 + x^3 + x^2 + x + 2)$ *has solutions everywhere locally, but not globally.*

Proof. Denote by $f(x)$ the sixth-degree polynomial on the right-hand side. We note that $f(0) = 2$, $f(1) = -2 \cdot 3$, and $f(-2) = -3 \cdot 2^2$; hence in particular $f(0) > 0$, so the equation is soluble in $\mathbb{R}$. If p is a prime different from 2 and 3 the product of the three Legendre symbols $\left(\frac{2}{p}\right)$, $\left(\frac{-2 \cdot 3}{p}\right)$, and $\left(\frac{-3 \cdot 2^2}{p}\right)$ is equal to 1, so that they cannot all be equal to -1. Thus for $x = 0, 1$, or -2, $f(x)$ is a nonzero quadratic residue modulo p, and since $p \neq 2$ it follows that $f(x)$ is a square in $\mathbb{Q}_p$, so the equation is locally soluble for $p \neq 2$ and 3. Note that here it has not been necessary to appeal to the Weil bounds to prove this. For $p = 3$ we note that $f(4) = -6498 = 3^2 + O(3^3)$, so that $f(4) \in (\mathbb{Q}_3^*)^2$, and for $p = 2$ we note that $f(14) = -8646748 = 2^2 + O(2^5)$; hence $f(14) \in (\mathbb{Q}_2^*)^2$, so the equation is also locally soluble for $p = 2$ and 3.

Let us now show that it is not globally soluble. Let $(x, y) \in \mathbb{Q}^2$ be a solution and set $f_1(X) = -(X^2 + X - 1)$ and $f_2(X) = X^4 + X^3 + X^2 + X + 2$, so that $y^2 = f(x) = f_1(x)f_2(x)$. We easily compute that the resultant of f_1 and f_2 is equal to 19, and an easy extended GCD computation shows that in fact $(X^3 + 5X^2 + 2X + 9)f_1(X) + (X + 5)f_2(X) = 19$ (you of course do not need to know anything about resultants to check this!). Writing $x = n/d$ with n and d in $\mathbb{Z}$ coprime, it is clear that this implies that if p is a prime number such that $v_p(d^2 f_1(x)) > 0$ and $v_p(d^4 f_2(x)) > 0$ then $p = 19$. Since $v_p(d^2 f_1(x) d^4 f_2(x)) = v_p(d^6 y^2) \equiv 0 \pmod 2$, this implies that for $p \neq 19$ we have $v_p(d^2 f_1(x)) \equiv v_p(d^4 f_2(x)) \equiv 0 \pmod 2$, hence that $f_1(x) = mu^2$ and $f_2(x) = mv^2$, where $m \in \{\pm 1, \pm 19\}$ and u and v are in $\mathbb{Q}$. It is now easy to obtain a contradiction: since $f_2(x)$ does not have any real roots, we have $f_2(x) > 0$ for all x, so that $m < 0$ is impossible. On the other hand, it is immediately checked by looking at the six possible cases that either $d^2 f_1(x)$ or $d^4 f_2(x)$ is congruent to 2 modulo 3; hence since $19 \equiv 1 \pmod 3$ they cannot both be congruent to squares modulo 3, so that $m = 1$ and $m = 19$ are also impossible. □

5.8 Exercises for Chapter 5

1. Let q be a nondegenerate quadratic form in n variables that represents 0, and let $c \in K^*$. Proposition 5.1.16 being only an existence result, find explicitly an $x \in K^n$ such that $q(x) = c$. It will be useful to first reduce to diagonal form, although algorithmically this may not be a good idea.

2. Show that $x^2 + xy + y^2$ is not equivalent over $\mathbb{Z}$ to a diagonal form.

3. (A. Pfister.) Let K be a field of characteristic different from 2.

(a) Using a simple identity, show that if $a \in K^*$ is represented by the form $bx^2 + cy^2$ then $bx^2 + cy^2 \sim ax^2 + abcy^2$, where as usual $\sim$ denotes equivalence of quadratic forms.

(b) Deduce that for any quadratic forms q_1 and q_2 we have $bq_1 \oplus cq_2 \sim aq_1 \oplus abcq_2$. Let q be a nondegenerate quadratic form. Define q to be *multiplicative* if for every $c \in K^*$ represented by q we have $cq \sim q$. Let q be a multiplicative quadratic form, let b and c (possibly equal to 0) be represented by q, let $a \in K^*$ be arbitrary (not necessarily represented by q), and set $d = b + ac$. We assume that $d \neq 0$.

(c) Show that if $b = 0$ or if $c = 0$, $d(q \oplus aq) \sim q \oplus aq$.

(d) Show that if b and c are nonzero then $d(q \oplus aq) \sim dq \oplus abcdq$.

(e) Deduce from (b) that also in that case we have $d(q \oplus aq) \sim q \oplus aq$, hence that if q is multiplicative then $q \oplus aq$ is also multiplicative for any $a \in K^*$.

(f) For instance, show that for any $k \geqslant 1$ the form in 2^k variables $\sum_{1 \leqslant i \leqslant 2^k} x_i^2$ is multiplicative.

4. Continuing the previous exercise, assume that -1 is a sum of squares in K (it is known that this is equivalent to the fact that there is no ordering on K compatible with the field structure), and let $m(K)$ be the minimum number of squares necessary to represent -1. Let k be the unique integer such that $2^k \leqslant m(K) < 2^{k+1}$.

(a) Show that there is a representation $-1 = (b+1)/a$, where a is a nonzero sum of 2^k squares and b is a sum of $2^k - 1$ squares.
(b) Deduce from the preceding exercise that -1 is a sum of 2^k squares, hence that $m(K) = 2^k$. In other words, the minimal number of squares necessary to represent -1 is always a power of 2.
(c) Show that any element of K is a sum of squares.
(d) Let $s(K)$ be the smallest integer, if it exists, such that every element of K is a sum of $s(K)$ squares. Using Corollary 5.1.17 show that $s(K)$ exists and that $s(K) = m(K)$ or $s(K) = m(K) + 1$.
(e) Compute $m(K)$ and $s(K)$ for $K = \mathbb{F}_q(X_1, \ldots, X_n)$ and for $K = \mathbb{Q}_p(X_1, \ldots, X_n)$ for $n \geqslant 0$ (distinguish between p or q equal to 2 and congruent to 1 or 3 modulo 4).

5. Show that two quadratic forms over $\mathbb{Q}_p$ are equivalent if and only if they have the same rank, discriminant $d \in \mathbb{Q}_p^*/\mathbb{Q}_p^{*2}$, and ε invariant (use induction on the rank and Corollary 5.2.17).

6. Show that up to equivalence there exists a unique form of rank 4 that does not represent 0 over $\mathbb{Q}_p$, the form $x^2 - ay^2 - bz^2 + abt^2$, where a and b are such that $(a, b) = -1$.

7. Let $d \in \mathbb{Q}_p^*/\mathbb{Q}_p^{*2}$ and $\varepsilon = \pm 1$. Show that there exists a form q of rank n, discriminant d, and invariant ε if and only if $n = 1$ and $d = \varepsilon = 1$, $n = 2$ and $d \neq -1$, $n = 2$ and $\varepsilon = 1$, or $n \geqslant 3$.

8. Compute the number of classes of quadratic forms of rank n over $\mathbb{Q}_p$.

9. Fill in the details of the remark following the proof of Proposition 5.3.5.

10. Let K be a field with at least seven elements and characteristic different from 2.

(a) If a and δ are in K^*, show that for any $b \in K^*$ there exist α and β in K^* (i.e., nonzero) such that $a\delta^2 = a\alpha^2 + b\beta^2$.
(b) Let $q(x) = \sum_{1 \leqslant i \leqslant n} a_i x_i^2$ be a diagonal quadratic form on K with $a_i \neq 0$ for all i, and assume that q represents 0. Then there exist $(\alpha_i)_{1 \leqslant i \leqslant n}$ with $\sum_{1 \leqslant i \leqslant n} a_i \alpha_i^2 = 0$ with all the α_i not equal to 0.
(c) Deduce from this an alternative proof of the Hasse–Minkowski theorem for $n = 4$ (see [Bor-Sha] for hints).

11. (I. Cassels.) Reasoning as in the proof of Proposition 5.4.4, show the following. Let q be a nondegenerate quadratic form over a field K of characteristic different from 2. If q represents $P \in K[X]$ in $K(X)$, then q already represents P in $K[X]$.

12. This exercise requires some knowledge of the geometry of numbers as can be found in [Con-Slo] and [Mar]. Prove the results of Houriet mentioned after Proposition 5.4.4; in other words:

(a) Show that, up to equivalence, there is only a finite number of quadratic forms q satisfying the assumption of Proposition 5.4.4, and compute the list of all such forms.

(b) Compute the list of forms satisfying the weaker assumption that for every $x \in \mathbb{Q}^k$ there exists $y \in \mathbb{Z}^k$ such that $q(x - y) \leqslant 1$.

(c) Find among the above forms those that satisfy the conclusion of Proposition 5.4.4, in other words such that if $n \in \mathbb{Z}$ is represented by q in $\mathbb{Q}$ it is also represented by q in $\mathbb{Z}$ (we have shown for example in the second proof of Corollary 5.4.14 that this is the case for $x_1^2 + x_2^2 + x_3^2 + x_4^2$).

13. Show that the simply Euclidean forms found by Houriet that satisfy the conclusion of Proposition 5.4.4 are exactly those given in the text. (Hint: look at the second proof of Corollary 5.4.14.)

14. Call an integer *triangular* if it has the form $m(m+1)/2$. Using Theorem 5.4.12, prove that every positive integer is the sum of 3 triangular numbers.

15. Prove Corollary 5.4.13.

16. Prove the formula for $r_4(n)$ given by Theorem 5.4.15 in the special case $n = 2^k$.

17. Show that the two formulas for $r_8(n)$ given by Theorem 5.4.15 are equivalent.

18. Prove directly that if $\gcd(m, n) = 1$ we have $r_2(m)r_2(n) = 4r_2(mn)$ (in other words that $r_2(n)/4$ is *multiplicative*; see Section 10.1.2), and deduce from Fermat's Proposition 5.4.9 the formula for $r_2(n)$ given in Theorem 5.4.15.

19.

(a) Using Theorem 5.4.15, prove that for $k = 1, 2, 4$, and 8 the functions $f_k(n) = r_k(n)/(2k)$ are *multiplicative*, in other words that $\gcd(n, m) = 1$ implies that $f_k(nm) = f_k(n)f_k(m)$ (see also Section 10.1.2).

(b) Compute $f_k(6) - f_k(2)f_k(3)$ as a polynomial in k and deduce that the above values of k are the only ones for which $f_k(n)$ is multiplicative.

20. Let D be the discriminant of a quadratic field and p a prime number.

(a) Generalizing Corollary 5.2.13, find a necessary and sufficient condition for an element $a \in \mathbb{Q}^*$ to be of the form $a = x^2 - Dy^2$ in $\mathbb{Q}_p$. It will of course be necessary to distinguish the cases $p \mid D$ and $p \nmid D$.

(b) Generalizing the example given in Section 5.5, deduce from the above necessary and sufficient conditions for a to be of the form $x^2 - Dy^2$ with x and y in $\mathbb{Q}$.

21. Show that the Hasse norm principle is in general not valid for Abelian but noncyclic extensions by studying the following example. Let $K = \mathbb{Q}$ and $L = \mathbb{Q}(\sqrt{13}, \sqrt{17})$. Show that any element of $\mathbb{Q}^{*2}$ is everywhere a local norm, but show that for instance 25 is not a global norm.

22. Prove directly Corollary 5.5.2.

23. Using Theorem 2.1.14, prove that, as stated in the proof of Theorem 5.6.2, the subgroups of $(\mathbb{Z}/2^e\mathbb{Z})^*$ having cardinality 2^{e-1-n} are given by cases A_n, B_n, and C_n.

24. With the notation of the proof of Theorem 5.6.3, show that the quadratic subextensions of K_1/K are indeed generated by the given elements.

25. Using the preceding exercise, prove that for all $n \geqslant 0$ the element $2 - \zeta_{2^n} - \zeta_{2^n}^{-1}$ is a square in $\mathbb{Q}(\zeta_{2^{n+1}} + \zeta_{2^{n+1}}^{-1})$.

26. Let p_1 and p_2 be distinct prime numbers. Assume that $\left(\frac{p_1}{p_2}\right) = \left(\frac{p_2}{p_1}\right) = 1$, where (contrary to the usual definition) $\left(\frac{p}{2}\right)$ is to be understood as equal to 1 if and only if $p \equiv 1 \pmod 8$. Generalizing Proposition 5.7.1, show that the equation $(x^2 - p_1)(x^2 - p_2)(x^2 - p_1p_2) = 0$ is everywhere locally soluble but not globally soluble. Generalize to the case that the p_i can be of the form $\pm q_i$ for some primes q_i.

27. Prove that the equation $2y^2 = x^4 - 17$ has a nontrivial solution in $\mathbb{R}$ and in $\mathbb{Q}_p$ for $p = 2$, 3, and 17.

28. (S. Siksek.) Generalizing the method used for Proposition 5.7.3, consider the Diophantine equation $y^2 = f(x, z)$, where f is a homogeneous polynomial of even degree. Assume that there exists a nontrivial solution (x, y, z). Without loss of generality we may assume that $\gcd(x, z) = 1$ and $z > 0$. For any u and v such that $\gcd(u, v) = 1$, write $F(u, v) = ab^2$ with a squarefree.

   (a) Prove that a is a quadratic residue modulo $vx - uz$, so that we can obtain congruences for $vx - uz$, and repeating with several pairs (u, v), we can hope to reach a contradiction.

   (b) As an application, consider the equation

$$y^2 = F(x, z) = -4x^4 + 4x^3 z + 92x^2 z^2 - 104xz^3 - 727z^4 ,$$

   which occurs naturally in the search for rational points on an elliptic curve of conductor 571 using the 2-descent method that we will explain in Section 8.3.4, where as usual we may assume that $\gcd(x, z) = 1$. By looking modulo powers of 2 show that $4 \nmid z$, and that if q is an odd prime dividing z then $q \equiv 1 \pmod 4$. Then noting that $F(-53, 16) = 2^2$ show that the equation has no nontrivial solutions.

Part II

Diophantine Equations

6. Some Diophantine Equations

6.1 Introduction

This chapter can be considered as the culmination of the tools that we have introduced in the first part of this book. We have already solved a number of Diophantine problems, but here we are going to solve many more. Although we have already mentioned that each Diophantine equation poses a new problem, there does exist a large number of general techniques, and in this introduction we will briefly describe these techniques and give a simple example of each (where "simple" is relative to the technique: for instance FLT is the "simplest" example (!) of the use of Ribet's level-lowering theorem, which we will study in Chapter 15).

Whatever method is used, a general principle is that it is usually easier to show that a Diophantine equation has no solutions at all than it is to show that it has only a specific nonempty set of solutions. A case in point is FLT, since it unfortunately has the solution $1^p + (-1)^p = 0$, which in a certain sense is nontrivial.

6.1.1 The Use of Finite Fields

We can use finite fields in two opposite ways. The first is when we want to prove that an equation does *not* have a solution. In that case the finite field that is used is $\mathbb{F}_p$ for a suitable prime p. We have seen in Chapter 1 the toy example $x^2 + y^2 = 3z^2$, which is seen to have no nonzero solutions by working in $\mathbb{F}_3$.

The second way is on the contrary to prove that an equation does have a solution in a finite field. If both the equation and the finite field are given, this is at least in theory very easy, since we simply make the variables of the equation range over the finite number of elements of our field. The situation changes completely when we are studying either a fixed equation, but over all finite fields at once, or a family of equations over a finite field, or both. In that case we must use general theorems such as those given in Chapter 2, and in particular the powerful Weil bounds (Corollary 2.5.27), either due to Weil himself in the case of curves, or from Deligne's proof of the Weil conjectures in the general case.

As an example we prove the following proposition.

Proposition 6.1.1. *Let $\ell \geqslant 3$ be prime and let C be the affine equation $y^2 = x^\ell + t$ for some fixed $t \in \mathbb{Z}$. This equation has a solution in $\mathbb{Z}_p$ for all p if and only if it has one for primes p of the form $p = 2k\ell + 1$ with $1 \leqslant k \leqslant (\ell - 3)/2$.*

Proof. Consider first the corresponding projective curve over $\mathbb{F}_p$. Even though the equation is singular at infinity when $\ell \geqslant 5$, if $t \neq 0$ in $\mathbb{F}_p$ and the characteristic of $\mathbb{F}_p$ is different from 2 and ℓ the Weil bounds apply, and since it is a hyperelliptic curve its genus is equal to $(\ell - 1)/2$, so we have

$$-(\ell - 1)p^{1/2} \leqslant |C(\mathbb{F}_p)| - (p + 1) \leqslant (\ell - 1)p^{1/2} .$$

In fact it is easy to compute directly $|C(\mathbb{F}_p)|$ in the cases $t = 0$ or characteristic 2 or ℓ and to see that these bounds are still valid; see Exercise 3.

In particular, $|C(\mathbb{F}_p)| \geqslant p + 1 - (\ell - 1)p^{1/2}$, and this is strictly greater than 1 for $p > (\ell - 1)^2$, so $|C(\mathbb{F}_p)| \geqslant 2$ for such p. In addition, if $p \not\equiv 1 \pmod{\ell}$ the map $x \mapsto x^\ell$ is a bijection from $\mathbb{F}_p$ to itself, so that for a given y, $x^\ell = y^2 - t$ has one solution, so $|C(\mathbb{F}_p)| = p + 1 \geqslant 2$ also for such p.

Since we must exclude the point at infinity, it follows in particular that there exists an affine nonsingular point in $C(\mathbb{F}_p)$ for all $p > (\ell - 1)^2$ and for $p \not\equiv 1 \pmod{\ell}$. Then a standard Hensel-type argument shows that we can lift this solution to $\mathbb{Z}_p$. On the other hand, if $p \leqslant (\ell - 1)^2$ and $p \equiv 1 \pmod{\ell}$ we can clearly write $p = 2k\ell + 1$ with $1 \leqslant k \leqslant (\ell - 3)/2$.   □

Remark. The equation $y^2 = x^\ell + t$ is always locally soluble; in other words, it always has a solution in every $\mathbb{Q}_p$, as opposed to $\mathbb{Z}_p$: simply choose $x = 1/p^2$ and $y = (1/p^\ell)(1 + p^{2\ell}t)^{1/2}$, which is p-adically convergent.

Corollary 6.1.2. *Let ℓ be a prime such that $3 \leqslant \ell \leqslant 31$. The equation $y^2 = x^\ell + t$ has a solution in $\mathbb{Z}_p$ for all p if and only if the following conditions are satisfied:*

(1) *For $\ell = 3$, 7, 13, 17, 19, and 31, no condition.*

(2) *For $\ell = 5$, $t \not\equiv 7 \pmod{11}$.*

(3) *For $\ell = 11$, $t \not\equiv 21$ or 22 modulo 23.*

(4) *For $\ell = 23$, $t \not\equiv 30$, 39, 40, 44, or 45 modulo 47, and $t \not\equiv 18$, 60, or 61 modulo 139.*

(5) *For $\ell = 29$, $t \not\equiv 31$, 32, 33, 38, 39, 43, 55 modulo 59.*

Proof. Since the above proposition reduces the problem to a reasonably small finite computation, this corollary is proved by a simple computer search, and can be extended at will.   □

We will come back to this equation in Section 6.7.

6.1.2 Local Methods

Since we have performed Hensel lifts, we have of course already used local methods. They can be used in several ways. One of the most common, as above, is to show that a Diophantine equation does not have any solutions in $\mathbb{Z}$, or at least to specify as much as possible which congruence classes a possible solution can belong to. We have seen in Chapter 5 the important example of *quadratic forms*, for which the Hasse–Minkowski theorem asserts that everywhere local solubility is a necessary and sufficient condition for global solubility, and we have also given methods to check local solubility at the finite number of places where this must be done.

As an additional isolated example taken almost at random from Chapter 14, consider the Diophantine equation to be solved in integers

$$y^2 = -2x^4 - 12x^2z^2 + 6z^4 \ .$$

We of course exclude the trivial solution $(x, y, z) = (0, 0, 0)$. Otherwise, if $d = \gcd(x, z)$ then $d^4 \mid y^2$, hence $d^2 \mid y$, so replacing (x, y, z) by $(x/d, y/d^2, z/d)$ we may assume that $\gcd(x, z) = 1$. We must have $2 \mid y$, so setting $y = 2y_1$ we obtain $2y_1^2 = -x^4 - 6x^2z^2 + 3z^4$. Thus x and z have the same parity, and since they are coprime they are both odd. Since the square of an odd number is congruent to 1 modulo 8 it follows that $2y_1^2 \equiv -1 - 6 + 3 = -4 \pmod 8$, so $y_1^2 \equiv -2 \pmod 4$, which is clearly impossible. We have thus shown the impossibility of our Diophantine equation by working in $\mathbb{Z}_2$, and not only modulo powers of 2. More precisely, if we do not remove the GCD then modulo 2^m we always have the nonzero solution $x = y = 0$ and $z = 2^{\lceil m/4 \rceil}$, which tends 2-adically to the trivial solution. On the other hand, after removal of the GCD, our proof shows that the equation does not have any solution modulo 16, but $(1, 0, 1)$ is clearly a solution modulo 8.

A legitimate question is to ask how one checks the local solubility of an equation. As we have seen above, the usual way is to prove solubility in the residue field, and then apply a Hensel lift. It is sometimes necessary to work modulo higher powers of the prime before performing the lift. Generally speaking, checking local solubility for a given prime is easy, and usually to check everywhere local solubility one needs to consider only a finite number of primes.

A more sophisticated use of local methods is through p-adic analysis, for instance Strassmann's theorem; see Section 4.5.4 for examples. Here the fact that we work fnot only modulo p^k for all k but in the *characteristic-zero topological field* $\mathbb{Q}_p$ gives us new tools of analytic nature.

6.1.3 Global Methods

Given a Diophantine problem, the first thing to do is always to see whether the problem has a solution locally, using one of the methods mentioned above.

If it has none, the problem is solved since we know that our equation has
no solution. Evidently the only interesting problems are those for which the
equations are everywhere locally soluble. The local information that we obtain
may already completely solve the problem, or may give useful information on
the global problem through local–global principles. However, such principles
are rather rare (see Chapter 5), so it is necessary to study the equation
globally, either by simply factoring over $\mathbb{Z}$, or by working in some appropriate
number field K. There are many methods for doing this, which we mention
briefly in turn, since we will come back to them in much more detail in the
rest of this book.

– The first and most classical method, originating with Fermat, Euler, Gauss,
  and especially Kummer, applies when it is possible to *factor* the Diophantine equation in K, a typical example being FLT, where one factors the
  equation in the cyclotomic field $K = \mathbb{Q}(\zeta_p)$. It is then essential to know
  explicitly the structure of the class group of K and of the unit group (i.e., a
  system of fundamental units), and to be able to find explicitly a generator
  of a principal ideal.

  It is important to note that in the twenty-first century, these computational
  problems can (for reasonable K) be solved at the click of a computer mouse
  button using computer algebra systems specializing in such tasks, such as
  Kant/Kash, magma, and Pari/GP. We will therefore always assume that we
  have available the basic data concerning the number fields that occur. This
  method will be used at length in the present chapter, as well as in Chapter
  14.

– A second global method for solving Diophantine equations is based on
  Diophantine approximation techniques, and on Baker-type results on linear forms in logarithms of algebraic numbers, and I refer to Chapter 12 for
  a survey of the method. It is used in particular to solve *Thue equations*, in
  other words equations of the form $f(x, y) = m$, where f is a homogeneous
  polynomial in two variables. This is now in complete but quite technical algorithmic form; see Algorithm 12.10.3. In Section 8.7 we will study
  in detail a variant that involves linear forms in *elliptic logarithms*, which
  paradoxically is easier to explain. This will enable us to find in reasonable
  cases all integral points on an elliptic curve.

– A third global method for solving certain types of Diophantine equations,
  mostly those that can be reduced to a cubic, is the use of elliptic curves,
  either via the method of *infinite descent* (initiated by Fermat, and which
  does not necessarily involve elliptic curves explicitly), or via the Birch and
  Swinnerton-Dyer conjecture (BSD for short), which we will state and study
  in detail in Chapter 8. As mentioned in the introduction, this remarkable
  conjecture enables us to predict the $\mathbb{Z}$-rank of the group of points of an
  elliptic curve over $\mathbb{Q}$ by computing a purely analytic quantity, and in particular tells us whether this group is finite or infinite. The fact that this
  method is based on a *conjecture* is not important, since either the analytic

result says that the group is finite, and in that case BSD is proved, or it says that the group is infinite, and we can then search for generators of the group using other techniques. All this will become much clearer in the numerous examples that we will give.

Although the theory of elliptic curves, and in particular of elliptic curves over $\mathbb{Q}$ and the BSD conjecture, is explained only in Chapters 7 and 8, I will assume in some places in the present chapter that the reader is familiar with this theory. Thus, to be able to fully understand the corresponding sections, the reader not conversant with the theory of elliptic curves is advised to read first the corresponding chapters.

– The most modern and sophisticated method for solving Diophantine equations is that used by Ribet, Wiles, and Taylor–Wiles for solving completely Fermat's last theorem, using modular forms and Galois representations. The kind of Diophantine equations that it is able to solve is usually of the form $a + b + c = 0$, where a, b, and c are highly divisible by certain integers (FLT being a typical example). This is linked to the famous abc conjecture (Conjecture 14.6.4). This method is based on a combination of a theorem of Ribet on "level lowering" of modular forms with the theorem of Wiles and Taylor–Wiles saying that the L-function attached to an elliptic curve defined over $\mathbb{Q}$ is in fact the L-function of a modular form. The proof of these theorems is very difficult, and Wiles's theorem has justly been celebrated as one of the great mathematical achievements of the end of the twentieth century. However, it is not necessary to understand the proof to *use* the theorems, if one understands the underlying concepts. Thanks to S. Siksek, I have included as Chapter 15 a detailed black-box explanation of the method. I advise the reader to look also at the expository paper by M. Bennett [Ben2] (see also [Ben-Ski]), which has a similar purpose.

6.2 Diophantine Equations of Degree 1

The simplest of all Diophantine equations are equations of degree 1. The two-variable case is well known: the equation $ax + by = c$ has a solution in integers x, y if and only if $\gcd(a, b)$ divides c, and in that case if (x_0, y_0) is a particular solution, the general solution is given by $x = x_0 + kb/\gcd(a, b)$, $y = y_0 - ka/\gcd(a, b)$ for any integer k. Furthermore a particular solution can easily be found with the extended Euclidean algorithm.

The case of more than two variables is slightly more difficult, because of the necessity of writing down explicitly the solution to the homogeneous equation (once again it is easy to find a particular solution with the extended Euclidean algorithm). For example, in the case of three variables, the equation $ax + by + cz = d$ has a solution if and only if $\gcd(a, b, c)$ divides d, and in that case if (x_0, y_0, z_0) is a particular solution, the general solution is given by $x = x_0 + mb/\gcd(a, b) - \ell c/\gcd(a, c)$, $y = y_0 + kc/\gcd(b, c) - ma/\gcd(a, b)$,

$z = z_0 + \ell a / \gcd(a, c) - kb / \gcd(b, c)$ for any integers k, ℓ, and m; see Exercise 4.

To state the solution in the case of n variables, we must use the notion of *Hermite normal form* (HNF) of a general (not necessarily square) integer matrix. However, it is not simpler to state it in that case than in the general case of a system of m linear Diophantine equations in n variables. The definition and result are as follows.

Definition 6.2.1. *Let $H = (h_{i,j})$ be an $m \times n$ matrix with $m \geqslant n$. We will say that H is in Hermite normal form (HNF) if there exists a strictly increasing function f from $[1, n]$ to $[1, m]$ such that for all $j \leqslant n$ we have $m_{f(j),j} \geqslant 1$, $m_{i,j} = 0$ for $i > f(j)$, and $0 \leqslant m_{f(j),k} < m_{f(j),j}$ for $k > j$.*

For instance, if $m = n$ we have necessarily $f(j) = j$, so in that very common and important case an integer matrix H is in HNF if and only if it is upper triangular with strictly positive diagonal elements, and its off-diagonal elements are nonnegative and strictly less than the diagonal element in the same row, which is exactly Definition 2.1.13 that we gave for square matrices.

Proposition 6.2.2. *Let A be an $m \times n$ integer matrix, and let B be an m-component integer column vector. There exists a matrix $U \in \mathrm{GL}_n(\mathbb{Z})$ and a matrix H in HNF such that $AU = (0|H)$. If k is the number of zero columns in the right-hand side, write $U = (U_1|U_2)$, where U_1 and U_2 are $n \times k$ and $n \times (n - k)$ matrices respectively. Then the Diophantine system $AX = B$ has a solution if and only if there exists an inverse image Z_2 of B by H (which can easily be checked), and in that case the general solution is given by $U_2 Z_2 + U_1 Y$, for any k-component integer vector Y.*

Proof. Recall that $\mathrm{GL}_n(\mathbb{Z})$ denotes the group of integer matrices that are invertible, i.e., of determinant ± 1. The first statement is proved in a manner very similar to the existence of the column echelon form (proved using Gaussian elimination). Here, we must perform all operations using only integer matrices of determinant ± 1. This is done by using as elementary operations either column exchanges or operations transforming a matrix $\left(\begin{smallmatrix} x & y \\ a & b \end{smallmatrix}\right)$ into a matrix of the form $\left(\begin{smallmatrix} A & B \\ 0 & D \end{smallmatrix}\right)$ by right multiplication by $\left(\begin{smallmatrix} -b/\gcd(a,b) & u \\ a/\gcd(a,b) & v \end{smallmatrix}\right)$, where u and v are such that $au + bv = \gcd(a, b)$. We leave the (well-known) details to the reader (see [Coh0], Section 2.4.2).

Once this basic statement proved, the rest is immediate: the equation $AX = B$ is equivalent to $AUX_1 = B$ (with $X_1 = U^{-1}X$), hence to $(0|H)X_1 = B$, which is soluble if and only if $HZ_2 = B$ is soluble. This last equation is in echelon form, so its solubility can be checked immediately one component after the other. If such a vector Z_2 exists, we can choose for X_1 the vector $\left(\begin{smallmatrix} 0 \\ Z_2 \end{smallmatrix}\right)$ with evident notation. We then have $X = UX_1 = (U_1|U_2)X_1 = U_2 Z_2$ as claimed. Finally, if X_1 is a general solution

to $AUX_1 = B$, then $AU\left(X_1 - \left(\frac{0}{Z_2}\right)\right) = 0$, hence $(0|H)\left(X_1 - \left(\frac{0}{Z_2}\right)\right) = 0$.
If we write $X_1 - \left(\frac{0}{Z_2}\right) = \left(\frac{Y_1}{Y_2}\right)$, then we obtain $HY_2 = 0$, and since H is in HNF and in particular in column echelon form, the columns of H are linearly independent, so that $Y_2 = 0$. Thus $X_1 = \left(\frac{Y_1}{Z_2}\right)$, so that $X = (U_1|U_2)X_1 = U_1Y_1 + U_2Z_2$, finishing the proof of the proposition.   □

In the special case in which $A = (a_j)$ has a single row, thus corresponding to a single linear Diophantine equation, the equation $AX = b$ has a solution if and only if the GCD of the a_j divides b, but the general solution must be written as explained in the proposition.

6.3 Diophantine Equations of Degree 2

We have already studied Diophantine equations of degree 2 in Section 5.3.2 in the context of the Hasse–Minkowski theorem. We will study them in more detail in this section.

6.3.1 The General Homogeneous Equation

Let $f(x_1, \ldots, x_n)$ be a quadratic form in n variables with integer coefficients, represented by a symmetric matrix Q with integral diagonal entries and half-integral off-diagonal entries. If $X = (x_1, \ldots, x_n)^t$ is a *column* vector, then $f(x_1, \ldots, x_n) = X^tQX$. The discriminant D of the form f is by definition $D = (-1)^{n(n-1)/2}\det(2Q)$. We will always assume that f is nondegenerate, in other words that $D \neq 0$ (a degenerate quadratic form being equivalent to a quadratic form with a strictly smaller number of variables, we do not lose any generality in doing so). Since we are looking for rational solutions, we will also always assume that f is not positive definite or negative definite, so that the condition at the place at infinity of the Hasse–Minkowski theorem is satisfied. In this case, we will simply say that f is *indefinite*.

By the Hasse–Minkowski theorem, we can determine whether a nontrivial rational solution to $f = 0$ exists by looking at the equation locally. More precisely, we have the following proposition.

Proposition 6.3.1. *The Diophantine equation $f(x_1, \ldots, x_n) = 0$ has a nontrivial rational solution if and only if it has a nontrivial solution in every $\mathbb{Q}_p$ for every p such that $p \mid 2D$.*

Proof. The necessity of the conditions is clear. Conversely, assume that they are satisfied. By the Chevalley–Warning Theorem 2.5.2, if $n \geqslant 3$ the equation has a nontrivial solution $X_0 = (x_{0,1}, \ldots, x_{0,n})^t \in \mathbb{F}_p^n$ for all p. Now if $p \nmid 2D$, the partial derivatives of f cannot all vanish modulo p at X_0. If for instance $\frac{\partial f}{\partial x_i}(X_0) \not\equiv 0 \pmod{p}$, the simple form of Hensel's Lemma 4.1.37 tells us that there exists $\alpha_i \in \mathbb{Z}_p$ such that $f(x_{0,1}, \ldots, \alpha_i, \ldots, x_{0,n}) = 0$, so that there exists a local solution for all these p. We conclude by the Hasse–Minkowski theorem.

For $n = 2$ we reason differently. Write $f(x_1, x_2) = ax_1^2 + bx_1x_2 + cx_2^2$, so that $D = b^2 - 4ac$. For any field K of characteristic zero it is clear that $f(x_1, x_2) = 0$ has a nontrivial solution in K if and only if D is a square in K. Thus by assumption D is a square in every $\mathbb{Q}_p$ such that $p \mid 2D$. But in particular this means that $v_p(D)$ is even for every $p \mid D$, in other words that D is a square in $\mathbb{Q}$, so that $f(x_1, x_2) = 0$ has a nontrivial rational solution. Finally, for $n = 1$ the equation evidently has no nontrivial solutions.   □

The test of local solubility at the "bad" primes p dividing $2D$ is done by looking modulo p^k for a suitable k for which Hensel's lemma can be used, hence is easy. The only difficult part is thus factoring the discriminant D.

Finding explicitly a nontrivial rational solution can be done using an efficient algorithm. We will see below such an algorithm in the important special case $n = 3$.

Once a solution $X_0 = (x_{1,0}, \ldots, x_{n,0})^t$ is found, we can ask for the general solution in *rational* numbers (the general solution in *integers* is a more difficult task, which we will consider only in special cases). The result is as follows.

Proposition 6.3.2. *Let $X_0 = (x_{1,0}, \ldots, x_{n,0})^t$ be a nontrivial solution to the Diophantine equation $f(x_1, \ldots, x_n) = X^t Q X = 0$.*

(1) *The general solution X to the equation in rational numbers such that $X^t Q X_0 \neq 0$ is given by $X = d((R^t Q R)X_0 - 2(R^t Q X_0)R)$, where $R \in \mathbb{Q}^n$ is a vector of parameters such that $R^t Q X_0 \neq 0$, and $d \in \mathbb{Q}^*$.*
(2) *In addition, if we choose a matrix $M \in \mathrm{GL}_n(\mathbb{Q})$ whose last column M_n is equal to X_0, we may assume that R is a $\mathbb{Z}$-linear combination of the first $n - 1$ columns of M, with the GCD of the coefficients equal to 1.*
(3) *In (2) the matrix R is unique up to changing R into $-R$, and the coefficient $d \in \mathbb{Q}$ is unique.*

Proof. (1). Since f is homogeneous, we consider nontrivial solutions of $f = 0$ as elements of the projective space $\mathbb{P}_n(\mathbb{Q})$. The parametric equation of a general line passing through X_0 is $X = uX_0 + vR$ with $(u, v) \in \mathbb{P}_2(\mathbb{Q})$, for some fixed $R \in \mathbb{P}_n(\mathbb{Q})$ not equal (projectively) to X_0. Let us find the values of (u, v) for which such an X is a solution of our equation. We write

$$0 = X^t Q X = u^2 X_0^t Q X_0 + 2uv R^t Q X_0 + v^2 R^t Q R = v(2u R^t Q X_0 + v R^t Q R) .$$

Solutions with $X^t Q X_0 \neq 0$ correspond to $v R^t Q X_0 \neq 0$, hence to $v \neq 0$ and $R^t Q X_0 \neq 0$, so that we can choose $(u, v) = (R^t Q R, -2R^t Q X_0) \in \mathbb{P}_2(\mathbb{Q})$. Since we have considered X as an element of the projective space, to obtain all solutions we must multiply by an arbitrary $d \in \mathbb{Q}^*$, proving (1).

(2). Since X_0 is nonzero, there exists a matrix M such that $M \in \mathrm{GL}_n(\mathbb{Q})$ whose last column M_n is equal to X_0. If we set $S = M^{-1}R = (y_1, \ldots, y_n)^t$, then $R = MS = \sum_{1 \leqslant j \leqslant n} y_j M_j = T + y_n X_0$, so that

$$(R^t Q R)X_0 - 2(R^t Q X_0)R = (T^t Q T + 2y_n T^t Q X_0)X_0 - 2(T^t Q X_0)(T + y_n X_0)$$
$$= (T^t Q T)X_0 - 2(T^t Q X_0)T ,$$

proving that we can replace R by T, in other words take only a linear combination of the first $n - 1$ columns of M. Furthermore, if u is the unique positive rational number such that the y_i/u for $1 \leqslant i \leqslant n - 1$ are integers with global GCD equal to 1, we may replace T by T/u and d by du^2, proving (2).

(3). For simplicity, let us say that a $\mathbb{Z}$-linear combination is primitive if the GCD of the coefficients is equal to 1. If we set $M^{-1}X = (y_1, \ldots, y_n)^t$ for some $y_i \in \mathbb{Q}$, we have $X = \sum_{1 \leqslant i \leqslant n} y_i M_i$. Since $X^t Q X_0 \neq 0$, the vector R has the form $R = aX + bX_0$ for some a, b in $\mathbb{Q}$. It follows that

$$R = \sum_{1 \leqslant i \leqslant n-1} (ay_i)M_i + (ay_n + b)M_n ,$$

and since the columns of M are linearly independent, if R is a primitive $\mathbb{Z}$-linear combination of the first $n - 1$ columns this means that $ay_n + b = 0$ and that for $1 \leqslant i \leqslant n - 1$ the ay_i are integers with global GCD equal to 1. Thus, if as in (2) we denote by u the unique positive rational number such that the y_i/u for $1 \leqslant i \leqslant n - 1$ are integers with global GCD equal to 1, we have necessarily $a = \pm 1/u$, hence $b = \pm y_n/u$, so that R is indeed determined up to sign, as claimed. In addition, we have

$$(R^t Q R)X_0 - 2(R^t Q X_0)R = 2ab(X^t Q X_0)X_0 - 2a(X^t Q X_0)(aX + bX_0)$$
$$= -2a^2(X^t Q X_0)X ,$$

and this is equal to X/d with $d = -1/(2a^2 X^t Q X_0)$, so that d is also uniquely determined.   □

Note that clearly the solutions of our Diophantine equation such that $X^t Q X_0 = 0$ cannot be attained by this parametrization, since $X^t Q X_0 = 0$ is equivalent to $R^t Q X_0 = 0$, which is excluded.

6.3.2 The Homogeneous Ternary Quadratic Equation

As we have seen during the proof of the Hasse–Minkowski theorem, the case $n = 3$ is the most important. In that case, the above proposition can be refined. We begin with a lemma.

Lemma 6.3.3. *Let Q be a nonsingular 3×3 real symmetric matrix, and let X_0 be a nonzero real vector such that $X_0^t Q X_0 = 0$. For $X \in \mathbb{R}^3$, $X^t Q X = X^t Q X_0 = 0$ is equivalent to $X = \lambda X_0$ for some $\lambda \in \mathbb{R}$.*

Proof. By diagonalizing Q, we may assume that Q is a diagonal matrix with real nonzero diagonal entries a, b, c. Thus $X^t Q X = X^t Q X_0 = 0$ is equivalent to $ax^2 + by^2 + cz^2 = axx_0 + byy_0 + czz_0 = 0$. Since $(x_0, y_0, z_0) \neq (0, 0, 0)$, we may assume for instance that $z_0 \neq 0$. Thus $z = -(axx_0 + byy_0)/(cz_0)$; hence

$$0 = X^t Q X = (ax^2 + by^2)cz_0^2 + (axx_0 + byy_0)^2$$
$$= (axx_0 + byy_0)^2 - (ax^2 + by^2)(ax_0^2 + by_0^2) = -ab(xy_0 - yx_0)^2 ,$$

so that $xy_0 - yx_0 = 0$, and hence

$$cz_0(zx_0 - xz_0) = -x_0(axx_0 + byy_0) + x(ax_0^2 + by_0^2) = by_0(xy_0 - yx_0) = 0 ,$$

so that X and X_0 are proportional, as claimed. $\square$

Proposition 6.3.4. *Let $X_0 = (x_0, y_0, z_0)$ be a nontrivial solution of the Diophantine equation $f(x, y, z) = X^t Q X = 0$ and let M be a matrix in $\mathrm{GL}_3(\mathbb{Q})$ whose last column M_3 is equal to X_0. The general rational solution X to the equation is given by $X = d((R^t Q R)X_0 - 2(R^t Q X_0)R)$, where $R = sM_1 + tM_2$, s and t are coprime integers, and $d \in \mathbb{Q}$.*

Proof. By Proposition 6.3.2, the above parametrization with $R^t Q X_0 \neq 0$ gives all solutions such that $X^t Q X_0 \neq 0$. Furthermore, if $R = s_1 M_1 + s_2 M_2$ for some s_1 and s_2 in $\mathbb{Q}$ not both 0, setting $u = \gcd(s_1, s_2)$, $s = s_1/u$, $t = s_2/u$, hence changing R into R/u, and finally changing d into du^2, it is clear that we may assume that s and t are coprime integers. In addition, by the above lemma, the solutions such that $X^t Q X_0 = 0$ are the multiples of X_0. Now since Q is nonsingular and $X_0 \neq 0$, we have $QX_0 \neq 0$. It follows that the subspace $V \subset \mathbb{Q}^3$ of R such that $R^t Q X_0 = 0$ is exactly 2-dimensional. I claim that there exists R equal to a linear combination of M_1 and M_2 that belongs to V, and that is not proportional to X_0. Indeed, since $M \in \mathrm{GL}_3(\mathbb{Q})$, no nonzero linear combination of M_1 and M_2 is proportional to $M_3 = X_0$. Furthermore, for $R = sM_1 + tM_2$ the equation $R^t Q X_0 = 0$ reads $sM_1^t Q X_0 + tM_2^t Q X_0 = 0$. If, for instance, $M_1^t Q X_0 = 0$, we can choose $R = M_1$. Otherwise, we choose $s_1 = -M_2^t Q X_0$, $t_1 = M_1^t Q X_0$, and set $s = s_1/\gcd(s_1, t_1)$, $t = t_1/\gcd(s_1, t_1)$, proving my claim. Thus, using once again the above lemma, it follows that for this R we have $R^t Q R \neq 0$, hence $X = d((R^t Q R)X_0 - 2(R^t Q X_0)R) = d(R^t Q R)X_0$. Since $R^t Q R \neq 0$, by choosing a suitable value of $d \in \mathbb{Q}$ we can thus obtain any multiple of X_0. $\square$

Remark. The above construction is of course explicit: to obtain X_0 itself for instance, either $M_1^t Q X_0 = 0$, in which case we choose $R = M_1$ and $d = 1/(M_1^t Q M_1)$, which exists by the lemma, or $M_1^t Q X_0 \neq 0$, and we choose s and t as explained, $R = sM_1 + tM_2$, and then $d = 1/(R^t Q R)$.

Corollary 6.3.5. *Let $f(x, y, z)$ be a nonsingular rational quadratic form in three variables. There exist three polynomials P_x, P_y, P_z with integer coefficients that are homogeneous of degree 2 in two variables (i.e., integral binary quadratic forms) such that the general rational solution of the Diophantine equation $f(x, y, z) = 0$ is given by the parametrization $x = dP_x(s, t)$, $y = dP_y(s, t)$, and $z = dP_z(s, t)$, where s and t are coprime integers and $d \in \mathbb{Q}$, uniquely determined (up to simultaneous change of sign of s and t) by x, y, z.*

Proof. Clear from the above proposition. Note that by multiplying d by a suitable rational number we may indeed assume that the polynomials P_x, P_y, and P_z have integral coefficients. □

Remark. Although we have proved the above corollary in the context of quadratic forms defined over $\mathbb{Q}$, it is clear that the proofs remain valid over any field of characteristic different from 2; hence the corollary is true if we remove all mention of integrality, coprimeness, or uniqueness.

Corollary 6.3.6. *Assume that $ABC \neq 0$, let (x_0, y_0, z_0) be a particular nontrivial solution of $Ax^2 + By^2 = Cz^2$, and assume that $z_0 \neq 0$. The general solution in rational numbers to the equation is given by*

$$x = d(x_0(As^2 - Bt^2) + 2y_0 Bst),$$
$$y = d(2x_0 Ast - y_0(As^2 - Bt^2)),$$
$$z = dz_0(As^2 + Bt^2),$$

where s and t are coprime integers and d is any rational number. Moreover, s, t, and d are uniquely determined, up to a simultaneous change of sign of s and t.

Proof. We apply the above proposition to the diagonal quadratic form with diagonal $(A, B, -C)$, and to the particular solution $(-x_0, -y_0, z_0)$ (so as to obtain a parametrization with fewer minus signs). Since $z_0 \neq 0$, we may choose

$$M = \begin{pmatrix} 1 & 0 & -x_0 \\ 0 & 1 & -y_0 \\ 0 & 0 & z_0 \end{pmatrix} \in \mathrm{GL}_3(\mathbb{Q}),$$

so that $R = (s, t, 0)^t$ with s and t coprime integers, and set $E = (R^t Q R)X_0 - 2(R^t Q X_0)R$. We compute that

$$E = (As^2 + Bt^2)(-x_0, -y_0, z_0)^t + 2(Asx_0 + Bty_0)(s, t, 0)^t$$
$$= (x_0(As^2 - Bt^2) + 2y_0 Bst, -y_0(As^2 - Bt^2) + 2x_0 Ast, (As^2 + Bt^2)z_0)^t,$$

giving the above parametrization. The uniqueness statement (up to sign) has been proved in complete generality above. □

Remark. Although evidently $(-x, y, z)$ (for instance) is also a solution if (x, y, z) is one, it is not clear from the formulas how to obtain it. We leave this as an exercise for the industrious reader; see Exercise 5.

Assume now that $f(x, y, z)$ has integral coefficients, and that we want to parametrize all *integral* solutions. Writing $d = u/v$ with $\gcd(u, v) = 1$, we see that we want $v \mid P_x(s, t)$, $v \mid P_y(s, t)$, and $v \mid P_z(s, t)$. By a well-known property of resultants, there exist polynomials U and V say with *integer* coefficients such that

$$U(S, T)P_x(S, T) + V(S, T)P_y(S, T) = R_S(P_x(S, T), P_y(S, T)) \,,$$

where R_S denotes the resultant with respect to the variable S. Clearly $R_S(P_x(S, T), P_y(S, T)) = r(x, y)T^4$ for some $r(x, y) \in \mathbb{Z}$, and by homogeneity we also have $R_T(P_x(S, T), P_y(S, T)) = r(x, y)S^4$ for the same constant $r(x, y)$, where R_T denotes the resultant with respect to T. By abuse of language we will call $r(x, y)$ the resultant of the polynomials P_x and P_y. It follows that $v \mid r(x, y)s^4$ and $v \mid r(x, y)t^4$, and since $\gcd(s, t) = 1$ we have $v \mid r(x, y)$. We have thus proved the following.

Proposition 6.3.7. *In the parametrization of Corollary 6.3.5, if x, y, and z are integers, and if we write $d = u/v$ with $\gcd(u, v) = 1$, then $v \mid \gcd(r(x, y), r(x, z), r(y, z))$.*

Note that the converse is not necessarily true.

In the context of Corollary 6.3.6, assume that A, B, and C are integers and that (x_0, y_0, z_0) is chosen to be an integral solution. We then have the following:

Corollary 6.3.8. *In the parametrization given by Corollary 6.3.6, if x, y, and z are integers then $d = u/v$ with $v \mid 2BCz_0^2$.*

Proof. If we simply used the above proposition, we would obtain $v \mid 4ABCz_0^4$. However, we can do better (also in the general case) by using *reduced resultants*. Without entering into this theory, we simply note that if we set $P_x(S, T) = x_0(AS^2 - BT^2) + 2y_0BST$, $P_y(S, T) = 2x_0AST - y_0(AS^2 - BT^2)$, and $P_z(S, T) = z_0(AS^2 + BT^2)$ then

$$U(S, T)P_x(S, T) + V(S, T)P_y(S, T) = 2BCz_0^2T^2 \,,$$

where

$$U(S, T) = A(y_0S - 2x_0T) \quad \text{and} \quad V(S, T) = Ax_0S + 2By_0T \,.$$

As above, we deduce that $v \mid 2BCz_0^2$, proving the corollary. Note that if we also consider P_x and P_z or P_y and P_z, we would obtain a right-hand side equal to $2BCz_0^3T^3$, hence a multiple of the above, so we would not obtain any additional information. $\qquad\square$

6.3.3 Computing a Particular Solution

We see from the above results that the main task for finding a parametriza-tion of a homogeneous ternary equation $X^t Q X = 0$ is to find a particular solution X_0. Note that the proof of the Hasse–Minkowski theorem is com-pletely effective, but it leads to a rather inefficient algorithm for finding X_0. Although this book is not mainly algorithmic in nature, we give a very elegant and efficient algorithm for doing so, initially due to Gauss and Legendre, but streamlined in the present nice form by D. Simon; see [Sim1]. We begin with two lemmas of independent interest, where as usual we denote by $\mathcal{M}_n(\mathbb{Z})$ the ring of $n \times n$ matrices with integral entries.

Lemma 6.3.9. *Let $M \in \mathcal{M}_n(\mathbb{Z})$, let p be a prime number, and let $d = \dim_{\mathbb{F}_p}(\mathrm{Ker}(\overline{M}))$, where $\overline{M}$ denotes the reduction of M modulo p. Then $p^d \mid \det(M)$; in other words, $d \leqslant v_p(\det(M))$.*

Proof. Let $\overline{U}_1, \dots, \overline{U}_d$ be an $\mathbb{F}_p$-basis of $\mathrm{Ker}(\overline{M})$, let $\overline{U}_1, \dots, \overline{U}_n$ be a completion to an $\mathbb{F}_p$-basis of $\mathbb{F}_p^n$, let $\overline{U} \in \mathrm{GL}_n(\mathbb{F}_p)$ be the matrix whose columns are the $\overline{U}_j$, and finally let U be any lift of $\overline{U}$ to $\mathcal{M}_n(\mathbb{Z})$. By assumption the first d columns of the matrix MU are divisible by p, so $p^d \mid \det(MU) = \det(M) \det(U)$. On the other hand, $p \nmid \det(U)$ since $\overline{U} \in \mathrm{GL}_n(\mathbb{F}_p)$, so $p^d \mid \det(M)$ as claimed. $\qquad \square$

Lemma 6.3.10. *For any $\overline{M} \in \mathrm{SL}_n(\mathbb{Z}/p\mathbb{Z})$ there exists a lift M such that $M \in \mathrm{SL}_n(\mathbb{Z})$; in other words, the natural reduction map from $\mathrm{SL}_n(\mathbb{Z})$ to $\mathrm{SL}_n(\mathbb{Z}/p\mathbb{Z})$ is surjective (here p is not necessarily a prime number).*

Proof. The following proof is taken from [Shi]. We prove this by induction on the size n of the matrix, the result being trivial for $n = 1$. Assume $n > 1$ and the result true for $n - 1$, and let N be any lift to $\mathcal{M}_n(\mathbb{Z})$ of the matrix $\overline{M}$. By the elementary divisor theorem (i.e., the Smith normal form in algorithmic terms) we can find two matrices U and V in $\mathrm{SL}_n(\mathbb{Z})$ such that $UNV = D = \mathrm{diag}(d_1, \dots, d_n)$ is a diagonal matrix with diagonal elements d_i such that $d_n \mid d_{n-1} \mid \cdots \mid d_1$, and we have $\det(D) = d_1 \cdots d_n \equiv 1 \pmod{p}$. If we can find a matrix $E \in \mathrm{SL}_n(\mathbb{Z})$ such that $E \equiv D \pmod{p}$ then $U^{-1} E V^{-1} \in \mathrm{SL}_n(\mathbb{Z})$ will be such that $U^{-1} E V^{-1} \equiv N \equiv \overline{M} \pmod{p}$. Thus we may assume that $N = D$ and forget the matrices U and V. Set $b = d_2 \cdots d_n$, $a = \det(D) = d_1 b$, and define

$$W = \begin{pmatrix} 1 & -1 \\ 1-b & b \end{pmatrix} \quad \text{and} \quad X = \begin{pmatrix} 1 & d_2 \\ 0 & 1 \end{pmatrix},$$

both of determinant 1. We check that

$$W \begin{pmatrix} 1 & 0 \\ 1-d_1 & d_1 d_2 \end{pmatrix} X = \begin{pmatrix} d_1 & 0 \\ 1-a & d_2 \end{pmatrix} \equiv \begin{pmatrix} d_1 & 0 \\ 0 & d_2 \end{pmatrix} \pmod{p}$$

since $a \equiv 1 \pmod{p}$ by assumption.

By our induction hypothesis there exists a matrix $C \in \mathrm{SL}_{n-1}(\mathbb{Z})$ such that $C \equiv \mathrm{diag}(d_1 d_2, d_3, \dots, d_n) \pmod{p}$. It follows that if we let W_1, X_1 in $\mathrm{SL}_n(\mathbb{Z})$ be defined as block matrices by

$$W_1 = \begin{pmatrix} W & 0 \\ 0 & I_{n-2} \end{pmatrix} \quad \text{and} \quad X_1 = \begin{pmatrix} X & 0 \\ 0 & I_{n-2} \end{pmatrix},$$

where as usual I_{n-2} is the identity matrix of order $n-2$, and if we set

$$C_1 = \begin{pmatrix} 1 & 0 & \cdots & 0 \\ 1 - d_1 & & & \\ 0 & & & \\ \vdots & & C & \\ 0 & & & \end{pmatrix},$$

which is also clearly in $\mathrm{SL}_n(\mathbb{Z})$, then $W_1 C_1 X_1 \equiv D \pmod{p}$ and $W_1 C_1 X_1 \in \mathrm{SL}_n(\mathbb{Z})$, proving the lemma by induction.   □

The algorithm for finding a particular solution to a homogeneous ternary quadratic equation is based on the following theorem.

Theorem 6.3.11. *Let $Q \in \mathcal{M}_3(\mathbb{Z})$ be the (symmetric) matrix of a nondegenerate ternary quadratic form that has a nontrivial solution in $\mathbb{Q}_p$ for every $p \mid \det(Q)$. There exists a matrix $V \in \mathcal{M}_3(\mathbb{Z})$ such that*

$$\det(V) = |\det(Q)| \quad \text{and} \quad Q_1 = \frac{1}{\det(Q)} V^t Q V \in \mathcal{M}_3(\mathbb{Z}),$$

and in particular $\det(Q_1) = 1$.

Furthermore, if the prime factorization of $\det(Q)$ is known, V can be found by a polynomial-time algorithm, and the entries of V are bounded by a polynomial in $\det(Q)$.

Proof. We prove the theorem by induction on $|\det(Q)| \geqslant 1$. If $\det(Q) = \pm 1$ there is nothing to prove. Thus let p be a prime number dividing $\det(Q)$, so that in particular $\mathrm{Ker}(\overline{Q})$ is nontrivial, and let $d = \dim_{\mathbb{F}_p}(\mathrm{Ker}(\overline{Q})) \geqslant 1$, so that $p^d \mid \det(Q)$ by the above lemma. We consider three cases.

Case 1: $v_p(\det(Q)) = 1$. By Lemma 6.3.9 we must have $d = \dim_{\mathbb{F}_p}(\mathrm{Ker}(\overline{Q})) = 1$, and as in the proof of that lemma let $\overline{U} \in \mathrm{GL}_3(\mathbb{F}_p)$ such that the first column of $\overline{U}$ forms a basis of $\mathrm{Ker}(\overline{Q})$. Multiplying a column of $\overline{U}$ by a suitable element of $\mathbb{F}_p^*$ we may assume that $\overline{U} \in \mathrm{SL}_3(\mathbb{F}_p)$. From Lemma 6.3.10 it follows that we can lift $\overline{U}$ to a matrix $U \in \mathrm{SL}_3(\mathbb{Z})$, whose columns we denote by U_i. By assumption we know that $p \mid Q U_1$, so $p \mid U_i^t Q U_1$ for all i. Thus if we set $R = U^t Q U = (r_{i,j})$, the first column (hence the first row) of R is divisible by p. Clearly $p^2 \nmid r_{1,1}$; otherwise, $p^2 \mid \det(R) = \det(Q)$, contrary to our assumption, as can be seen by dividing by p the first row and then the first column. By assumption, we know that $X^t R X = 0$ has a nontrivial p-adic solution $X = (a_1, a_2, a_3)^t$, where after suitable rescaling we may assume that the a_i are p-integral with one of them a p-adic unit. I claim that either a_2 or a_3 is a p-adic unit. Indeed, otherwise we would have $v_p(a_2) \geqslant 1$, $v_p(a_3) \geqslant 1$, hence $v_p(a_1) = 0$, so setting $Y = (0, a_2, a_3)^t$ and $e_1 = (1, 0, 0)^t$ we would have

$$0 = X^t R X = a_1^2 e_1^t R e_1 + 2 a_1 Y^t R e_1 + Y^t R Y ,$$

and we would have

$$v_p(a_1^2 e_1^t R e_1) = 2 v_p(a_1) + v_p(r_{1,1}) = 1 ,$$

$v_p(2 a_1 Y^t R e_1) \geqslant 2$ (since $p \mid Y$ and $p \mid R e_1$, which is the first column of R), and $v_p(Y^t R Y) \geqslant 2$ since $p \mid Y$, leading to a contradiction and proving my claim.

Exchanging the indexes 2 and 3 if necessary we may assume that $v_p(a_2) = 0$, and let $x \in \mathbb{Z}$ be such that $x \equiv a_3 a_2^{-1} \pmod{p}$. Set

$$N = \begin{pmatrix} 1 & 0 & 0 \\ 0 & p & x \\ 0 & 0 & 1 \end{pmatrix}$$

and $V = U N$. It is clear that $V \in \mathcal{M}_3(\mathbb{Z})$ with $\det(V) = p$. Furthermore, the above computation of $X^t R X$ shows that $Y^t R Y \equiv 0 \pmod{p}$; hence $N_3^t R N_3 \equiv 0 \pmod{p}$, where as usual N_j denotes the jth column of N. Since $p \mid N_2$ and $N_1 = e_1$, it immediately follows that the matrix $N^t R N = V^t Q V$ is divisible by p. Thus we can replace Q by $V^t Q V / p$, whose determinant is equal to $\det(Q)/p$, hence is strictly smaller than that of Q in absolute value.

Algorithmic Remarks.

(1) When $p \mid r_{2,2}$ it is not necessary to use this construction since it is immediate that $N^t R N$ is divisible by p for

$$N = \begin{pmatrix} 1 & 0 & 0 \\ 0 & 1 & 0 \\ 0 & 0 & p \end{pmatrix} .$$

(2) Although we have used p-adic solubility we do not need an explicit p-adic solution. We only want $N_3^t R N_3 \equiv 0 \pmod{p}$ with $N_3 = (0, x, 1)^t$, in other words a solution to the quadratic equation

$$r_{2,2} x^2 + 2 r_{2,3} x + r_{3,3} \equiv 0 \pmod{p} .$$

Thus we take $x \in \mathbb{Z}$ such that

$$x \equiv \left(-r_{2,3} + \sqrt{r_{2,3}^2 - r_{2,2} r_{3,3}} \right) r_{2,2}^{-1} \pmod{p} ,$$

and the existence of V is equivalent to the existence of the square root.

(3) We have $p \nmid (r_{2,3}^2 - r_{2,2}r_{3,3})$, otherwise $p^2 \mid \det(Q)$. Thus for $p = 2$ the square root always exists, so we do not need to assume local solvability at 2 in this case.

Case 2: $v_p(\det(Q)) \geqslant 2$ **and** $d = \dim_{\mathbb{F}_p}(\mathrm{Ker}(\overline{Q})) = 1$. Let $U \in \mathrm{SL}_3(\mathbb{Z})$ be defined as in Case 1 and set $R = U^t Q U = (r_{i,j})$. We know that the first row and column of R are divisible by p, so expanding $\det(R)$ this implies that

$$\det(Q) = \det(R) \equiv r_{1,1}\det(S) \pmod{p^2}, \quad \text{where} \quad S = \begin{pmatrix} r_{2,2} & r_{2,3} \\ r_{3,2} & r_{3,3} \end{pmatrix}.$$

Since $p \mid r_{1,j}$ we cannot have $p \mid \det(S)$; otherwise, the last two columns of R would be linearly dependent over $\mathbb{F}_p$, so that $d \geqslant 2$, contrary to our assumption. Thus $p \nmid \det(S)$ and since $p^2 \mid \det(Q)$ we deduce that $p^2 \mid r_{1,1}$. If we set $V = UN$ with

$$N = \begin{pmatrix} 1 & 0 & 0 \\ 0 & p & 0 \\ 0 & 0 & p \end{pmatrix}$$

it is clear that $\det(V) = p^2$ and $V^t Q V$ is divisible by p^2. We thus replace Q by $V^t Q V / p^2$, whose determinant is equal to $\det(Q)/p^2$ hence strictly smaller than that of Q in absolute value.

Case 3: $v_p(\det(Q)) \geqslant 2$ **and** $d = \dim_{\mathbb{F}_p}(\mathrm{Ker}(\overline{Q})) \geqslant 2$. Here we take for U a matrix in $\mathrm{SL}_3(\mathbb{Z})$ whose first two columns reduced modulo p are linearly independent elements of $\mathrm{Ker}(\overline{Q})$ and we set $R = U^t Q U$. The first two rows and columns of R are divisible by p, so it is clear that if we set $V = UN$ with

$$N = \begin{pmatrix} 1 & 0 & 0 \\ 0 & 1 & 0 \\ 0 & 0 & p \end{pmatrix}$$

we have $\det(V) = p$ and $V^t Q V$ is divisible by p. We thus replace Q by $V^t Q V / p$, whose determinant is equal to $\det(Q)/p$, hence strictly smaller than that of Q in absolute value.

Since in all three cases we have obtained a new symmetric matrix with strictly smaller determinant in absolute value, the first statement of the theorem is proved by induction. The other statements immediately follow from the proof.   □

Thanks to Theorem 6.3.11 the search for a particular solution to $X^t Q X = 0$ can be algorithmically reduced to the search for a particular solution to such an equation for which $\det(Q) = \pm 1$. The second result that enables us to find such a solution is based on a natural modification of the LLL algorithm (Algorithm 2.3.18) due to D. Simon.

Denote by $\cdot$ the bilinear form associated with the quadratic form Q. Since Q is indefinite, this is never a scalar product, but the notation is useful nonetheless. We will write $(x)^2$ instead of $Q(x) = x \cdot x$, since for the same reason we cannot write $Q(x) = \|x\|^2$. Thus $(x)^2$ may be negative, so we will have to include absolute values in all the necessary inequalities.

Now let $(\mathbf{b}_j)_{1 \leqslant j \leqslant n}$ be a $\mathbb{Z}$-basis of the lattice $\Lambda = \mathbb{Z}^n$ (which is more than what we ask in the usual LLL algorithm, where the $\mathbf{b}_j$ are required only to be linearly independent) and let $(\mathbf{b}_i^*)$ be the corresponding Gram–Schmidt vectors obtained using the standard formulas of Proposition 2.3.5. First note that the induction used to define these vectors may *fail* since some vector $\mathbf{b}_j^*$ may be such that $\mathbf{b}_j^* \cdot \mathbf{b}_j^* = 0$. If this happens either here or in the rest of the algorithm we are in fact happy since we have a nonzero vector $\mathbf{b}_j^*$ such that $Q(\mathbf{b}_j^*) = 0$, and since $\mathbf{b}_j^* \in \mathbb{Q}^n$ throughout the algorithm, it is not necessary to search any further because this is a particular solution. We will thus implicitly assume that this never happens, so that we indeed have a Gram–Schmidt basis.

Let $\gamma > 4/3$ be fixed. We define the notion of γ-LLL reduced basis in the same way as in Definition 2.3.14, except that we must add absolute value signs to the norms: in other words we must have $|\mu_{i,j}| \leqslant 1/2$ for all $j < i$ and

$$\left| (\mathbf{b}_i^* + \mu_{i,i-1} \mathbf{b}_{i-1}^*)^2 \right| \geqslant \left(\frac{1}{\gamma} + \frac{1}{4} \right) \left| (\mathbf{b}_{i-1}^*)^2 \right| ,$$

the absolute value signs being necessary since the form Q is indefinite (we will see that in our special case we must take $4/3 < \gamma < 2$). Note that contrary to the positive definite case this is not equivalent to $|(\mathbf{b}_i^*)^2| \geqslant (1/\gamma + 1/4 - \mu_{i,i-1}^2)|(\mathbf{b}_{i-1}^*)^2|$; see Exercise 8.

Given any $\mathbb{Z}$-basis $(\mathbf{b}_j)$ of $\mathbb{Z}^n$ we apply a straightforward modification of the LLL algorithm to obtain a γ-LLL-reduced basis of $\mathbb{Z}^n$ by adding suitable absolute values in Step 3 of Algorithm 2.3.18.

The proof used in the positive definite case shows that the algorithm terminates in polynomial time and that the final basis that we obtain is in particular such that

$$1 \leqslant |(\mathbf{b}_1)^2| \leqslant \gamma^{(n-1)/2} |\det(Q)|^{1/n} .$$

An easy modification of the proof shows that if Q is indefinite the above inequality can be slightly improved (see [Sim1]), but this is not important although useful in practice.

The second result, although very easy, is the key to finding a particular solution.

Proposition 6.3.12. *Assume that $n \leqslant 5$, that Q is a quadratic form with integral entries such that $\det(Q) = \pm 1$, and choose γ such that $4/3 < \gamma < 2^{2/(n-1)}$. Then either we find a $\mathbf{b}_j^*$ such that $Q(\mathbf{b}_j^*) = 0$ during the algorithm (hence a particular solution), or the Gram matrix of Q on the final LLL-reduced basis is diagonal with diagonal entries equal to ± 1.*

Proof. It follows from the inequality for $|(\mathbf{b}_1)^2|$ that $1 \leqslant |(\mathbf{b}_1)^2| < 2$, so that $(\mathbf{b}_1)^2 = \pm 1$ since it is an integer. Since $\mathbf{b}_1^* = \mathbf{b}_1$, for $1 < i \leqslant n$ we have

$$\mu_{i,1} = \frac{\mathbf{b}_i \cdot \mathbf{b}_1}{(\mathbf{b}_1)^2} = \pm \mathbf{b}_i \cdot \mathbf{b}_1 \,,$$

and since the Gram matrix has integral entries and $|\mu_{i,1}| \leqslant 1/2$ we have $\mu_{i,1} = \mathbf{b}_i \cdot \mathbf{b}_1 = 0$. It follows that $\mathbf{b}_2^* = \mathbf{b}_2$, so we can continue the same reasoning by induction. Note that the double inequality for γ is possible only if $n \leqslant 5$, whence the restriction.   □

In our case we have $n = 3$ and the inequality for γ is $4/3 < \gamma < 2$. Summarizing, to find a particular solution of $X^t Q X$ for an indefinite ternary quadratic form Q that is p-adically soluble for all $p \mid \det(Q)$ we proceed as follows. After factoring $\det(Q)$ (which is theoretically the longest part of the algorithm) we apply the algorithm implicit in the proof of Theorem 6.3.11 to find a matrix $V \in \mathcal{M}_3(\mathbb{Z})$ such that $Q_1 = V^t Q V / \det(Q) \in \mathcal{M}_3(\mathbb{Z})$ and such that $\det(Q_1) = 1$. We then use the modified LLL algorithm explained above applied to Q_1 and the canonical basis of $\mathbb{Z}^3$. Then either we directly find a vector $X_1 = \mathbf{b}_j^*$ such that $Q_1(\mathbf{b}_j^*) = 0$, in which case $X = V X_1$ is a particular nontrivial solution to $X^t Q X = 0$, or we find a matrix $W \in \mathrm{GL}_3(\mathbb{Z})$ such that $Q_2 = W^t Q_1 W$ is a diagonal matrix with diagonal entries equal to ± 1. Since Q_2 is indefinite the signs of two diagonal entries must be opposite, so we can trivially find a solution X_2 to $X_2^t Q_2 X_2 = 0$ of the form $(1, 1, 0)^t$ up to permutation. It follows that $X = V W X_2$ is a particular nontrivial solution to $X^t Q X = 0$.

6.3.4 Examples of Homogeneous Ternary Equations

We now apply the above results to a number of important special cases.

Corollary 6.3.13. *Up to exchange of x and y the general* integral *solution to the Pythagorean equation $x^2 + y^2 = z^2$ is given by $x = d(s^2 - t^2)$, $y = 2dst$, $z = d(s^2 + t^2)$, where s, t are coprime integers of opposite parity and $d \in \mathbb{Z}$. In addition, we have $|d| = \gcd(x, y) = \gcd(x, z) = \gcd(y, z)$. The general solution with x and y coprime is (up to exchange of x and y) $x = s^2 - t^2$, $y = 2st$, $z = \pm(s^2 + t^2)$.*

This is the well-known parametrization of Pythagorean triples, which we will use several times.

Proof. Using Corollary 6.3.6 with the particular solution $(1, 0, 1)$, we obtain the formulas of the corollary. If $s \equiv t \pmod{2}$, we set $s_1 = (s+t)/2$ and $t_1 = (s-t)/2$, so that $s = s_1 + t_1$, $t = s_1 - t_1$. Then $x = 4dst$, $y = 2d(s_1^2 - t_1^2)$, $z = 2d(s_1^2 + t_1^2)$, which is the same parametrization with x and y exchanged and d replaced by $2d$. Since $\gcd(s, t) = 1$, s and t are odd, so s_1 and t_1 have opposite parity, showing that we can always reduce to this case. Finally, if this is the case then $\gcd(s^2 - t^2, 2st) = 1$, so x and y are in $\mathbb{Z}$ if and only if $d \in \mathbb{Z}$ (and then $|d| = \gcd(x, y)$ and the other statement follows). Finally, if we want $\gcd(x, y) = 1$, we must have $d = \pm 1$, but two of the signs in the formulas can be absorbed by exchanging s and t, and by changing s into $-s$, proving the corollary.   □

Corollary 6.3.14. (1) *Let* $p = \pm 2$. *The general integral solution of* $x^2 + py^2 = z^2$ *with* x *and* y *coprime is given by* $x = \pm(s^2 - pt^2)$, $y = 2st$, $z = \pm(s^2 + pt^2)$, *where* s *and* t *are coprime integers with* s *odd and the* $\pm$ *signs are independent.*

(2) *The general integral solution of* $x^2 - 2y^2 = -z^2$ (*in other words of* $x^2 + z^2 = 2y^2$) *with* x *and* y *coprime is given by* $x = \pm(s^2 - 2st - t^2)$, $y = \pm(s^2 + t^2)$, $z = \pm(s^2 + 2st - t^2)$, *where* s *and* t *are coprime integers of opposite parity and the* $\pm$ *signs are independent.*

Proof. Using Corollary 6.3.6 with the particular solution $(1, 0, 1)$ we obtain the formulas $x = d(s^2 - pt^2)$, $y = 2dst$, $z = d(s^2 + pt^2)$, where s and t are coprime integers and $d \in \mathbb{Q}$. We consider two cases.

If $2 \nmid s$, then $\gcd(s^2 - pt^2, 2st) = 1$, so for x and y to be coprime integers we must have $d = \pm 1$, giving the first parametrization.

If $2 \mid s$ then t is odd, so that $\gcd(s^2 - pt^2, 2st) = 2$. Thus for x and y to be coprime integers we must have $d = \pm 1/2$, and this leads to another parametrization, which is the same as the first with s and t exchanged and some signs changed, proving (1).

In a similar manner for (2), using the particular solution $(1, 1, 1)$ we obtain the parametrization $x = \pm(s^2 - 4st + 2t^2)$, $y = \pm(s^2 - 2st + 2t^2)$, $z = \pm(s^2 - 2t^2)$ with s and t coprime and s odd, and (2) follows by replacing s with $s + t$, giving the more symmetrical parametrization of the corollary.   □

Corollary 6.3.15. *Let* p *be a positive or negative prime number with* $p \neq 2$. *The general integral solution of* $x^2 + py^2 = z^2$ *with* x *and* y *coprime is given by one of the following two disjoint parametrizations.*

(1) $x = \pm(s^2 - pt^2)$, $y = 2st$, $z = \pm(s^2 + pt^2)$, *where* s *and* t *are coprime integers of opposite parity such that* $p \nmid s$.

(2) $x = \pm(((p-1)/2)(s^2 + t^2) + (p+1)st)$, $y = s^2 - t^2$, $z = \pm(((p+1)/2)(s^2 + t^2) + (p-1)st)$, *where* s *and* t *are coprime integers of opposite parity such that* $s \not\equiv t \pmod{p}$.

In the above, the $\pm$ *signs are independent.*

Proof. Using Corollary 6.3.6 with the particular solution $(1, 0, 1)$ we again obtain the formulas $x = d(s^2 - pt^2)$, $y = 2dst$, $z = d(s^2 + pt^2)$, whnere s and t are coprime integers and $d \in \mathbb{Q}$. We consider two cases.

If $p \nmid s$, we easily check that $\gcd(s^2 - pt^2, 2st) = \gcd(s^2 - pt^2, 2)$, and since p is odd, this is equal to 1 if s and t have opposite parity, and to 2 otherwise. If it is equal to 1, for x and y to be coprime integers we must have $d = \pm1$, giving the first parametrization. If it is equal to 2, then s and t have the same parity, so they are both odd. We set $s_1 = (s + t)/2$, $t_1 = (t - s)/2$, which are coprime of opposite parity such that $p \nmid (s_1 - t_1)$, and we obtain $x = -d((p - 1)(s_1^2 + t_1^2) + 2(p + 1)s_1t_1)$, $y = 2d(s_1^2 - t_1^2)$, $z = d((p + 1)(s_1^2 + t_1^2)^2 + 2(p - 1)s_1t_1)$. For x and y to be coprime integers we must have $d = \pm1/2$, giving the second parametrization.

If $p \mid s$, then $p \nmid t$, so if we exchange s/p and t and change d into d/p we reduce to the preceding case, up to the sign of x, which plays no role, so that we do not obtain any extra parametrizations.   □

6.3.5 The Pell–Fermat Equation $x^2 - Dy^2 = N$

Introduction and Reductions.

A degree-2 equation of another kind that deserves special mention is the Pell–Fermat equation, often simply called the Pell equation.

This equation is $x^2 - Dy^2 = N$ for given integers D and N, to be solved in integers x and y. It is evidently closely linked to the arithmetic properties of $\mathbb{Q}(\sqrt{D})$. Its nature is very different from that of the degree-2 equations that we have studied above for two reasons. First, it is not a homogeneous equation. But most importantly, we want the solutions in *integers*, and not in rational numbers. Indeed, finding for instance the *rational* solutions to $x^2 - Dy^2 = 1$ is very easy; see Exercise 10.

We make a number of reductions.

– We may assume that $D \geqslant 0$, since otherwise there is only a finite number of pairs (x, y) to be checked. This can be done either in a naïve manner, or more intelligently by working in the imaginary quadratic field $\mathbb{Q}(\sqrt{D})$ and using a computer algebra system (CAS) to check whether N is the norm of an element.

– We may assume that D is not a square, since otherwise if $D = d^2$ the equation can be written $(x - dy)(x + dy) = N$, so it is only a matter of listing all (positive or negative) divisors g of N such that $g + N/g$ is even and $d \mid (N/g - g)$.

– Hence we are reduced to working in the real quadratic field $\mathbb{Q}(\sqrt{D})$, where D is not necessarily the discriminant of the field. As in the case $D < 0$, the *existence* of a solution can be proved intelligently using a CAS to check whether N is the norm of an element in the real quadratic *order* $\mathbb{Z}[\sqrt{D}]$. If a solution does exist, multiplicativity of the norm (equivalent to the simple identity $(x^2 - Dy^2)(z^2 - Dt^2) = (xz + Dyt)^2 - D(xt + yz)^2$) implies that the general solution is obtained by multiplying the corresponding element $x + y\sqrt{D}$ of the quadratic field by a *unit* of norm 1, i.e., a solution to $x^2 - Dy^2 = 1$.

Once these reductions are made, it is natural to consider the following three special cases:

(1) $x^2 - Dy^2 = 1$.

(2) $x^2 - Dy^2 = \pm 1$, where ± 1 means that we accept both signs as solutions.

(3) $x^2 - Dy^2 = \pm 4$, when $D \equiv 0$ or 1 modulo 4.

We note that each equation is a special case of the next one. Indeed, if (x, y) is a solution of (1), it is also a solution of (2). But conversely, we will see that the set of solutions of (2) has the form $x + y\sqrt{D} = \pm(x_0 + y_0\sqrt{D})^k$ for any $k \in \mathbb{Z}$, and then either (x_0, y_0) is already a solution of (1), in which case all solutions of (2) are also solutions of (1), or it is not, in which case it is immediately seen that the solutions of (1) are the solutions of (2) with k even. In both cases the solutions of (1) are also given by $x + y\sqrt{D} = \pm(x_0 + y_0\sqrt{D})^k$ for any $k \in \mathbb{Z}$ for suitable (possibly different) (x_0, y_0).

Finally, if (x, y) is a solution of (2), then (X, y) is a solution of (3) with $X = 2x$ and D replaced by $4D$. Conversely, any solution to $X^2 - DY^2 = \pm 4$ with $X \equiv 0 \pmod 2$ gives a solution to $x^2 - (D/4)y^2 = \pm 1$ with $x = X/2$. This is automatic if $D \equiv 0 \pmod 4$. If $D \equiv 1 \pmod 4$ and X (hence Y) is odd, then we obtain a solution to $x^2 - Dy^2 = \pm 1$ by setting $x = X(X^2 + 3DY^2)/8$ and $y = Y(3X^2 + DY^2)/8$, which are easily seen to be integral, and correspond to the identity $x + y\sqrt{D} = ((X + Y\sqrt{D})/2)^3$.

To summarize, what we will call *the* Pell equation is an equation of the form $x^2 - Dy^2 = \pm 4$ with $D > 0$ nonsquare and congruent to 0 or 1 modulo 4. There are two main results concerning this equation. One deals with the structure of the set of solutions, the other with the algorithmic construction of that set.

The Structure Theorem.

Proposition 6.3.16. *If $D > 0$ is not a square and is congruent to 0 or 1 modulo 4 the Pell equation $x^2 - Dy^2 = \pm 4$ has an infinity of solutions given in the following way. If (x_0, y_0) is a solution with the least strictly positive y_0 (and $x_0 > 0$, say), the general solution is given by*

$$\frac{x + y\sqrt{D}}{2} = \pm \left(\frac{x_0 + y_0\sqrt{D}}{2} \right)^k$$

for any $k \in \mathbb{Z}$.

Proof. The equation can be written $\mathcal{N}(\varepsilon) = \pm 1$, with $\varepsilon = (x + y\sqrt{D})/2$, and since x and y are integers and $x \equiv Dy \pmod 2$, ε is an algebraic integer of norm equal to ± 1, hence a unit. Since the set of elements of the form $(x + y\sqrt{D})/2$ with $x \equiv yD \pmod 2$ is the quadratic order of discriminant D, we must thus find the structure of the group of units of this order.

Let D_0 be the discriminant of the quadratic field $\mathbb{Q}(\sqrt{D})$, so that $D = D_0 f^2$ for some positive integer f. We prove the result by induction on the number of prime factors of f. If $f = 1$, by an easy special case of Dirichlet's unit theorem we know that $\varepsilon = \pm \varepsilon_0^k$ for some sign $\pm$ (having nothing to do with the sign of the equation), and $k \in \mathbb{Z}$, where ε_0 is the fundamental unit of $\mathbb{Q}(\sqrt{D_0})$, i.e., the solution of the equation with the smallest strictly positive y.

Assume now that f is arbitrary, and by induction that the result has been proved for all $g \mid f$ having a strictly smaller number of prime factors. We can thus write $D = D_1 p^2$, where p is a prime, and by induction we may assume that the result has already been proved for D_1. Thus, our Pell equation can be written $x^2 - D_1(py)^2 = \pm 4$, so that by induction the equation is equivalent to

$$\frac{x + py\sqrt{D_1}}{2} = \pm \left(\frac{x_1 + y_1\sqrt{D_1}}{2} \right)^k ,$$

where $\varepsilon_1 = (x_1 + y_1\sqrt{D_1})/2$ is the fundamental unit of the order of discriminant D_1. We have

$$\pm py = \frac{((x_1 + y_1\sqrt{D_1})/2)^k - ((x_1 - y_1\sqrt{D_1})/2)^k}{\sqrt{D_1}} .$$

It will thus be sufficient to prove the following lemma.

Lemma 6.3.17. *The set of $k \in \mathbb{Z}$ such that $(\varepsilon_1^k - \overline{\varepsilon_1}^k)/\sqrt{D_1} \in p\mathbb{Z}$ has the form $k_0\mathbb{Z}$, where $k_0 \mid p - \left(\frac{D_1}{p} \right)$.*

Proof. We may clearly assume that $p \nmid y_1$, since otherwise we can choose $k_0 = 1$. Assume first that $p \neq 2$. If $p \mid D_1$, expanding the right-hand side gives $\pm 2^{k-1} py \equiv k x_1^{k-1} y_1 \pmod{p}$, so that the set of suitable k is $p\mathbb{Z}$ if $p \nmid x_1 y_1$, and $\mathbb{Z}$ otherwise, so that $k_0 = p$ or 1 respectively. Otherwise, $\varepsilon_1, \overline{\varepsilon_1}$ can be considered as *distinct* elements e_1 and $\overline{e_1}$ of $\mathbb{F}_{p^2}$ (since $(e_1 - \overline{e_1})^2 = y_1^2 D_1 \neq 0$ since $p \nmid y_1$), so our equation is equivalent to $(\overline{e_1}/e_1)^k = 1$. The first claim of the lemma is thus proved, with k_0 being equal to the order of $(\overline{e_1}/e_1)$ in $\mathbb{F}_{p^2}$. For the second, we note that if $\left(\frac{D_1}{p} \right) = 1$ then in fact $e_1 \in \mathbb{F}_p$, so $k_0 \mid (p - 1)$. On the other hand, if $\left(\frac{D_1}{p} \right) = -1$ then by the theory of finite fields we have $\overline{e_1} = e_1^p$ (the action of the Frobenius automorphism). This can also be seen directly by applying the Frobenius automorphism to the equation $e_1^2 - x_1 e_1 \pm 1 = 0$. Thus the order of $\overline{e_1}/e_1 = e_1^{p-1}$ divides $(p^2 - 1)/(p - 1) = p + 1$, proving the lemma, hence the proposition. $\qquad\square$

The Algorithmic Method.

Now that we know the structure of the solution set of $x^2 - Dy^2 = \pm 4$, it remains to find in an efficient manner the fundamental solution (x_0, y_0), or equivalently, the fundamental unit of the quadratic order of discriminant D. There are essentially four methods for doing this, which differ by their complexity. The first is the naïve method, consisting in trying $y = 1$, 2 etc., until $Dy^2 \mp 4$ is a square. This method is absolutely correct but highly inefficient, since one can prove that the number of *binary digits* of y_0 may often be larger than $\sqrt{D}$. Thus, the running time can be of order $O(\exp(D^{1/2}))$. In fact, since the result we are computing is so large, the fundamental unit must be *represented* in a nontrivial manner (the so-called *compact representation*, see [Coh0]), which we will not discuss here. We will always assume that this representation is used.

The second method uses *continued fractions*, and will be described more precisely in a moment. Its running time has order $O(D^{1/2})$.

The third method is a remarkable improvement of the continued fraction method due to D. Shanks, using a combination of two of his most important algorithmic ideas: the *baby-step giant-step* algorithm and the *infrastructure* of the continued fraction cycle. We refer the reader to Chapter 5 of [Coh0] for a detailed description of these ideas and algorithms. The running time of this method has order $O(D^{1/4})$.

The fourth method is a combination of the third method with a method coming from the theory of factoring, the use of *factor bases*, introduced in this context by J. Buchmann. Its heuristic running time, well supported by practical evidence, is *subexponential*, of order $O(\exp(c\sqrt{\log(D)\log(\log(D))}))$ for a small strictly positive constant c. However, its correctness depends on the truth of the Generalized Riemann Hypothesis (GRH) (although a slower variant has recently been found that runs in time $O(D^{1/6+\varepsilon})$, and which does not need the GRH). Using this record-breaking method, it is possible to compute the fundamental unit in a reasonable amount of time for discriminants up to 10^{80}, say.

The continued fraction method is based on the following result, which is not difficult but which we will not prove.

Proposition 6.3.18. *Let $D > 0$ be a nonsquare integer congruent to 0 or 1 modulo 4. Denote by r the largest integer such that $r^2 < D$ and $r \equiv D$ (mod 2). The continued fraction expansion of the quadratic number $\alpha = (r + \sqrt{D})/2$ is purely periodic. Furthermore, if $(a_0, a_1, \ldots, a_{n-1})$ is the period of that expansion, then the rational number p_{n-1}/q_{n-1} whose continued fraction expansion is given by that period is such that $\varepsilon = p_{n-1} - q_{n-1}(r - \sqrt{D})/2$ is a fundamental unit of the quadratic order of discriminant D.*

To apply this proposition we simply note that to compute the continued fraction expansion of a quadratic number we must *not* compute any decimal or other approximation to the number, but work formally directly only on quadratic numbers. We leave the (easy) details to the reader, or refer once again to [Coh0].

6.4 Diophantine Equations of Degree 3

From now on, as mentioned at the beginning of this chapter, it will be necessary to have some knowledge of elliptic curves over $\mathbb{Q}$ as explained in Chapters 7 and 8.

6.4.1 Introduction

In the case of a Diophantine equation of degree 3 or higher (or of several equations of degree 2), a new and very annoying phenomenon occurs: the failure of the Hasse principle. In other words our equations may be everywhere locally soluble without having a global solution. We have already seen examples of this in the preceding chapter, and we will see more here. Note that it is usually very easy to check for local solubility everywhere. On the other hand, to prove that a Diophantine equation has no global solutions is often very difficult, and often gives rise to unsolved problems. Even in the simplest case of a homogeneous diagonal equation of degree 3 in three variables, no algorithm is known, and it would be a remarkable advance in number theory to find one (we will see that this problem can be reduced to the computation of the Mordell–Weil group of an elliptic curve; see Proposition 6.4.15).

We thus consider Diophantine equations of the form $f(x_1, \ldots, x_n) = 0$, where f is a homogeneous polynomial of degree 3 with integral coefficients. We will always assume that f is nonsingular (in other words that the partial derivatives of f do not simultaneously vanish except at the origin); otherwise, the problem is much easier and essentially reduces to Diophantine equations of degree 1 or 2. When $n = 2$ there is nothing much to say since by dehomogenizing the equation the problem boils down to the determination of rational roots of a polynomial in one variable, which can easily be done (see Exercise 11). When $n = 3$, then, *if we know a nontrivial solution* we are by definition dealing with an elliptic curve, and we will see in Section 7.2.4 how to transform our equation into an equivalent one in *Weierstrass form* $y^2 z = x^3 + axz^2 + bz^3$ for suitable a and b. We will devote the whole of Chapter 8 to the study of the global solubility of such equations. In particular, we will prove the Mordell–Weil theorem, which states that the set of rational points on the corresponding projective curve is a finitely generated abelian group.

When $n \geqslant 4$, contrary to the case of quadratic forms, there is no really simple reduction to a canonical form. For $n = 4$ we are dealing with a *cubic surface* S. If P and Q are distinct rational points on S, the line through P and Q either intersects S in a single third point, which must be rational, or is entirely contained in S, so that we may obtain new points by this secant process. Starting from a single point P we can consider the tangent plane of S at P. It will intersect S along a singular cubic curve, and any tangent at P with rational slope will intersect this curve, hence S, at a third point, which will also be rational, or again be entirely contained in S. This will be called a tangent process. We will see that for elliptic curves this does lead to the whole set of rational points starting from a finite number (the Mordell–Weil theorem). There is a conjecture of Yu. Manin that states that the same should be true here:

Conjecture 6.4.1 (Manin). *Let S be a cubic surface defined by a projective equation $f(x_1, x_2, x_3, x_4) = 0$, where f has integer coefficients and is* *nonsingular. There exists a* finite *number of rational points* $P_1, \ldots, P_r$ *on* S *such that any rational point of* S *can be obtained from them by a succession of secant and tangent processes.*

In view of all the above, in this section we will consider *diagonal equations*, in other words equations of the form $\sum_{1 \leqslant i \leqslant n} a_i x_i^3 = 0$, or inhomogeneous versions.

6.4.2 The Equation $ax^p + by^p + cz^p = 0$: Local Solubility

Some explanations about the title and subject matter of this subsection and the next two are in order. There are essentially two distinct methods to study equations of the form $ax^3 + by^3 + cz^3 = 0$. The first one, mainly used in the nineteenth century, but expanded later by Selmer and Cassels in the twentieth, is the use of tools coming from algebraic number theory: factorization in suitable number fields together with the use of class and unit groups, and the use of reciprocity laws (such as the Eisenstein reciprocity law that we will use in Section 6.9.6 to prove Wieferich's criterion for the first case of Fermat's last theorem). The second one, more recent, uses elliptic curves and in some sense is more complete; we will study it in detail in Section 6.4.4; see in particular Proposition 6.4.15 for the definitive result.

The methods using algebraic number theory apply with little change to the more general equations $ax^p + by^p + cz^p = 0$ where p is an odd prime, so it is natural to consider the case of general p together with the case $p = 3$. It is important to note, however, that p will be assumed to be quite *small*, for instance $3 \leqslant p \leqslant 31$, and suitable assumptions will have to be made to obtain interesting results. In addition, our study will *not* include the study of the equation $x^p + y^p + z^p = 0$, in other words the initial Fermat equation for two reasons: first because that equation *does* have the nontrivial solution $(x, y, z) = (1, -1, 0)$ for instance, and second because in the context of Fermat's theorem p can be large. We will study FLT itself in Section 6.9, where we will of course be looking for solutions with $xyz \neq 0$. Finally note that, to simplify the presentation, we will not use reciprocity laws such as Eisenstein's, although they do give important additional information, as the proof of Wieferich's criterion shows (see Corollary 6.9.10); see [Sel1] for more details on the use of reciprocity laws for the equation $ax^3 + by^3 + cz^3 = 0$, and [Den] for the equation $x^p + y^p + cz^p = 0$.

The method using elliptic curves is specific to the equation $ax^3 + by^3 + cz^3 = 0$ since this represents a curve of genus 1. However, it can be partially generalized to the equation $ax^p + by^p + cz^p = 0$ by using Chabauty-type techniques on a suitable *hyperelliptic* curve, and we will also briefly mention this; see Proposition 6.4.13 and the comments following it.

Thus, until further notice, we let p be an odd prime, and consider the equation $ax^p + by^p + cz^p = 0$ with $abc \neq 0$. We begin of course with local solubility.

First note that even if a, b, and c are initially in $\mathbb{Q}$, after multiplying by a suitable denominator we may assume that they are integers such that $\gcd(a, b, c) = 1$. Furthermore, we may also assume that a, b, and c are pth power-free (in other words not divisible by pth power of a prime), since if for instance $\ell^p \mid a$, we can rewrite our equation as $(a/\ell^p)(\ell x)^p + by^p + cz^p = 0$ (see Exercise 13, however). We will always implicitly or explicitly make these two reductions in the sequel.

The question of local solubility is answered by the following theorem (note that since p is odd the equation $ax^p + by^p + cz^p = 0$ has a trivial solution in $\mathbb{R}$).

Theorem 6.4.2. *Let a, b, and c be nonzero pth power-free integers such that $\gcd(a, b, c) = 1$, let ℓ be a prime number, and denote by S_ℓ the statement that the equation $ax^p + by^p + cz^p = 0$ has a nontrivial solution in $\mathbb{Q}_\ell$. Set $L = \ell$ if $\ell \neq p$, and $L = \ell^2 = p^2$ otherwise. Reorder a, b, and c so that $0 = v_\ell(a) \leqslant v_\ell(b) \leqslant v_\ell(c)$, and set $v_b = v_\ell(b)$ and $v_c = v_\ell(c)$.*

(1) If $1 \leqslant v_b < v_c$ then S_ℓ is false.
(2) If $1 \leqslant v_b = v_c$, then S_ℓ is true if and only if b/c is a pth power in $(\mathbb{Z}/L\mathbb{Z})^*$ (in other words, if and only if either $p \nmid \phi(L)$, or $p \mid \phi(L)$ and $(b/c)^{\phi(L)/p} \equiv 1 \pmod{L}$), or if $v_b = v_c = p - 1$ and $\ell = p$.
(3) If $v_b = 0$ and $L \mid c$, then S_ℓ is true if and only if a/b is a pth power in $(\mathbb{Z}/L\mathbb{Z})^*$.
(4) If $v_b = 0$, $v_c = 1$, and $\ell = p$, then S_ℓ is true.
(5) Assume finally that $v_b = v_c = 0$, and denote by G_ℓ the group of pth powers in $(\mathbb{Z}/L\mathbb{Z})^*$. Then S_ℓ is true if and only if at least one of the following conditions is satisfied:
   (a) $p \nmid \phi(L)$.
   (b) $\ell \geqslant ((p-1)(p-2))^2$;
   (c) $-b/a \in G_\ell$, $-c/a \in G_\ell$, $-c/b \in G_\ell$, or there exists $m \in G_\ell$ such that $-(c + am)/b \in G_\ell$.

Note that since $v_\ell(a) = 0$, the condition $v_b = v_c = 0$ means that $\ell \nmid abc$.

Proof. Let (x, y, z) be a nontrivial solution in $\mathbb{Q}_\ell$, where we may clearly assume that x, y, and z are in $\mathbb{Z}_\ell$ and that at least one of them is an ℓ-adic unit, so that in fact at least *two* are ℓ-adic units since the coefficients are pth power-free.

(1). It is clear that if $1 \leqslant v_b < v_c$ then $ax^p + by^p \equiv 0 \pmod{\ell^2}$; hence $\ell \mid x$, so $by^p \equiv 0 \pmod{\ell^2}$; hence $\ell \mid y$ since $v_b = 1$, which is absurd since x or y is an ℓ-adic unit.

(3). If $v_b = 0$ and $L \mid c$, then a necessary condition for solubility in $\mathbb{Q}_\ell$ is that $ax^p + by^p \equiv 0 \pmod{L}$ be nontrivially soluble, in other words that a/b be a pth power in $(\mathbb{Z}/L\mathbb{Z})^*$ (since both x and y must be ℓ-adic units). But conversely, if $a/b \equiv u^p \pmod{L}$ then a/b is a pth power of $y \in \mathbb{Z}_\ell$ given by the series $y = u(1 + (a/(bu^p) - 1))^{1/p}$, which is convergent in $\mathbb{Z}_\ell$ since either $\ell \neq p$, or $\ell = p$ and $v_\ell(L) = 2$, so we can choose $z = 0$ and $x = -1$.

(4). If $v_b = 0$, $v_c = 1$, and $\ell = p$, consider the set E of integers of the form $bu^p + cv^p$, with $1 \leqslant u < p$ and $0 \leqslant v < p$. I claim that E is a set of representatives of $(\mathbb{Z}/p^2\mathbb{Z})^*$. Indeed, there are $p(p-1) = \phi(p^2)$ pairs (u, v), and since $p \mid c$ and $p \nmid bx$, it is clear that if $z \in E$ then $p \nmid z$; finally, the elements of E are distinct modulo p^2: indeed, if $b(u_1^p - u_2^p) + c(v_1^p - v_2^p) \equiv 0$ (mod p^2) then $u_1 \equiv u_1^p \equiv u_2^p \equiv u_2$ (mod p), so that $u_1 = u_2$, and hence $v_1 \equiv v_1^p \equiv v_2^p \equiv v_2$ (mod p) since $v_p(c) = 1$, so that we also have $v_1 = v_2$, proving my claim. In particular, there exist x and y such that $bx^p + cy^p \equiv -a$ (mod p^2), and by Hensel's lemma we conclude that there exists a nontrivial solution to our equation in $\mathbb{Q}_p$, proving (4).

(2). If $1 \leqslant v_b = v_c < p$, then necessarily $\ell \mid x$, so our equation is equivalent to $(a\ell^{p-v_b})(x/\ell)^p + (b/\ell^{v_b})y^p + (c/\ell^{v_c})z^p = 0$, which has one of the forms that we have studied, hence is soluble in $\mathbb{Q}_\ell$ if and only if either $-c/b$ is a pth power in $(\mathbb{Z}/L\mathbb{Z})^*$ (when $v_\ell(L) \leqslant p - v_b$), or if $v_b = p - 1$ and $\ell = p$.

(5). The only remaining case is $v_b = v_c = 0$. If $p \nmid \phi(L)$, the map $x \mapsto x^p$ is a bijection of $(\mathbb{Z}/L\mathbb{Z})^*$ onto itself, so it is clear that our equation has a nontrivial solution modulo L, hence in $\mathbb{Z}_\ell$ by Hensel's lemma, proving (a). Since $v_b = v_c = 0$ we have $\ell \nmid abc$, so that if $N(\mathbb{F}_\ell)$ denotes the number of projective points on the curve with coordinates in $\mathbb{F}_\ell$, Corollary 2.5.23 implies that when $\ell \neq p$ we have $N(\mathbb{F}_\ell) > \ell + 1 - (p-1)(p-2)\ell^{1/2}$; hence if $\ell \geqslant ((p-1)(p-2))^2$ (which implies that $\ell \neq p$) we have $N(\mathbb{F}_\ell) > 0$, proving (b). For (c), assume for instance that $-b/a \in G_\ell$. Then $-b/a \equiv u^p$ (mod L), so that $au^p + b(1)^p + c(0)^p \equiv 0$ (mod L), and we conclude as usual by Hensel's lemma that our solution has a solution in $\mathbb{Q}_\ell$, and the same conclusion holds if $-c/a \in G_\ell$ or if $-c/b \in G_\ell$. If this is not the case we have necessarily $v_\ell(z) = 0$, otherwise $v_\ell(x) = v_\ell(y) = 0$, so that $-b/a \equiv (x/y)^p$ (mod ℓ^p), and in particular $-b/a \in G_\ell$. Thus, setting $m = (x/z)^p$ we evidently have $m \in G_\ell$, and $-(c + am)/b = (y/z)^p \in G_\ell$, proving (c). $\qquad\square$

Remarks. (1) It is easy to see that for $p = 3$ condition (c) of (5) can be removed, and the proof shows that condition (a) can be replaced by $\ell \neq p$; see Exercise 32. For $p = 5$, condition (a) can be weakened to $\ell > 11$, and for $p = 7$ it can be weakened to $\ell > 71$, these values being optimal; see Exercise 33. More generally, it seems experimentally that it can be weakened perhaps to something like $\ell > p^3/3$, but I do not know if this is true.

(2) If $\ell \nmid u$, then $u \in G_\ell$ if and only if $u^{\phi(L)/p} \equiv 1$ (mod L), which is easier to test, although not any faster if the elements of G_ℓ have been sorted.

Lemma 6.4.3. *Let a, b, and c be pth power-free integers. If the equation $ax^p + by^p + cz^p = 0$ has a nontrivial solution with x, y, z in $\mathbb{Q}$, it has a nontrivial solution with x, y, z in $\mathbb{Z}$ pairwise coprime. Furthermore, if in* *addition a, b, and c are pairwise coprime, then ax^p, by^p, and cz^p are pairwise coprime.*

Proof. Easy and left to the reader; see Exercise 13.   □

6.4.3 The Equation $ax^p + by^p + cz^p = 0$: Number Fields

We first note that over an integral domain of characteristic zero the equation $ax^p + by^p + cz^p = 0$ is equivalent to the equation $X^p + BY^p + Cz^p = 0$ with for instance $B = a^{p-1}b$, $C = a^{p-1}c$, and $(X, Y, Z) = (ax, y, z)$. It is thus sufficient to study such equations with $a = 1$. As already mentioned, in the case $p = 3$ we can use methods using elliptic curves, and we will do this is the next subsection; in particular, 3-descent will give us a complete theoretical answer to the problem of the existence of a global solution, which unfortunately relies on the explicit computation of the Mordell–Weil group of the associated elliptic curve, which is feasible in practice for small cases, but which is one of the major unsolved algorithmic problems on elliptic curves. In the present section, however, we will apply classical tools of algebraic number theory (class numbers, units, etc.) to give sufficient conditions that imply that these equations are *not* soluble in $\mathbb{Q}$. This has two advantages: first, it gives an alternative (although not complete) approach in the case $p = 3$, but second and most importantly it is the *only* method that enables us to study equations $ax^p + by^p + cz^p = 0$ with $p > 3$ prime. Note also that, in contrast to some equations that we will study later (for instance, in Section 6.4.5), there are no "evident" solutions. We begin with a few lemmas.

Lemma 6.4.4. *Let c be a pth power-free integer different from ± 1, set $\theta = c^{1/p}$, let $K = \mathbb{Q}(\theta)$, let $f = [\mathbb{Z}_K : \mathbb{Z}[\theta]]$ be the index of $\mathbb{Z}[\theta]$ in $\mathbb{Z}_K$, let m be a nonzero integer, and let $u_0, u_1, \ldots, u_{p-1}$ be integers.*

(1) *If $m \mid (u_0 + u_1\theta)\mathbb{Z}_K$ then $m \mid \gcd(u_0, u_1)$.*
(2) *If $m \mid \left(\sum_{0 \leqslant i \leqslant p-1} u_i\theta^i\right)\mathbb{Z}_K$ then $m \mid f\gcd(u_0, u_1, \ldots, u_{p-1})$.*

Proof. (1). Set $\alpha = (u_0 + u_1\theta)/m$, so that $(m\alpha - u_0)^p - cu_1^p = 0$. The characteristic polynomial of α is thus $\sum_{1 \leqslant k \leqslant p} \binom{p}{k}(-u_0/m)^{p-k}X^k - (u_0^p + cu_1^p)/m^p$. Since α is an algebraic integer the coefficients of this polynomial are in $\mathbb{Z}$, and in particular for $k = 1$ we have $m^{p-1} \mid pu_0^{p-1}$, hence $m \mid u_0$ (look at the valuations), so that $m^p \mid cu_1^p$, and since c is pth power-free $m \mid u_1$ (again look at valuations), as claimed.

(2). Set $\alpha = \left(\sum_{0 \leqslant i \leqslant p-1} u_i\theta^i\right)/m$. Since by assumption $\alpha \in \mathbb{Z}_K$, by definition of f we have $f\alpha \in \mathbb{Z}[\theta]$, in other words $m \mid fu_i$ for all i.   □

Lemma 6.4.5. *Let K be a number field, and let $\mathfrak{a}$, $\mathfrak{b}$, $\mathfrak{c}_1$, and $\mathfrak{c}_2$ be integral ideals of K. Assume the following:*

(1) *We have an ideal equality $\mathfrak{c}_1\mathfrak{c}_2 = \mathfrak{b}\mathfrak{a}^p$.*

(2) *We have* $\gcd(\mathfrak{c}_1, \mathfrak{c}_2, \mathfrak{a}) = 1$.

Then there exist integral ideals $\mathfrak{a}_i$ and $\mathfrak{b}_i$ such that $\mathfrak{c}_i = \mathfrak{b}_i \mathfrak{a}_i^p$ for $i = 1, 2$, $\mathfrak{b}_1 \mathfrak{b}_2 = \mathfrak{b}$, $\mathfrak{a}_1 \mathfrak{a}_2 = \mathfrak{a}$, $\gcd(\mathfrak{a}_1, \mathfrak{a}_2) = 1$, and $\gcd(\mathfrak{c}_1, \mathfrak{c}_2) = \gcd(\mathfrak{b}_1, \mathfrak{b}_2)$.

Proof. If we set $\mathfrak{d} = \gcd(\mathfrak{c}_1, \mathfrak{c}_2)$ then by assumption $\mathfrak{d}$ and $\mathfrak{a}$ are coprime; hence $\mathfrak{d}^2 \mid \mathfrak{b}$, so that

$$(\mathfrak{c}_1 \mathfrak{d}^{-1})(\mathfrak{c}_2 \mathfrak{d}^{-1}) = (\mathfrak{b}\mathfrak{d}^{-2}) \mathfrak{a}^p .$$

Since by definition of $\mathfrak{d}$ the two factors on the left are coprime, for any ideal I we have

$$\gcd((\mathfrak{c}_1 \mathfrak{d}^{-1})(\mathfrak{c}_2 \mathfrak{d}^{-1}), I) = \gcd(\mathfrak{c}_1 \mathfrak{d}^{-1}, I) \gcd(\mathfrak{c}_2 \mathfrak{d}^{-1}, I) ;$$

hence if we set $\mathfrak{d}_i = \gcd(\mathfrak{c}_i \mathfrak{d}^{-1}, \mathfrak{b}\mathfrak{d}^{-2})$ and $\mathfrak{e}_i = \mathfrak{c}_i \mathfrak{d}^{-1} \mathfrak{d}_i^{-1}$ we have $\mathfrak{d}_1 \mathfrak{d}_2 = \mathfrak{b}\mathfrak{d}^{-2}$, hence $\mathfrak{e}_1 \mathfrak{e}_2 = \mathfrak{a}^p$. Now, since $\mathfrak{c}_1 \mathfrak{d}^{-1}$ and $\mathfrak{c}_2 \mathfrak{d}^{-1}$ are coprime, a fortiori so are $\mathfrak{e}_1$ and $\mathfrak{e}_2$; hence there exist coprime integral ideals $\mathfrak{a}_i$ such that $\mathfrak{e}_i = \mathfrak{a}_i^p$ and $\mathfrak{a}_1 \mathfrak{a}_2 = \mathfrak{a}$. Thus if we set $\mathfrak{b}_i = \mathfrak{d}\mathfrak{d}_i$ we have $\mathfrak{c}_i = \mathfrak{b}_i \mathfrak{a}_i^p$ and $\mathfrak{b}_1 \mathfrak{b}_2 = \mathfrak{d}^2 \mathfrak{d}_1 \mathfrak{d}_2 = \mathfrak{b}$. By construction we have $\mathfrak{d} = \gcd(\mathfrak{c}_1, \mathfrak{c}_2) \mid \gcd(\mathfrak{b}_1, \mathfrak{b}_2)$, and conversely it is clear that $\gcd(\mathfrak{b}_1, \mathfrak{b}_2) \mid \gcd(\mathfrak{c}_1, \mathfrak{c}_2)$, proving the lemma.   □

Recall that an integral ideal $\mathfrak{a}$ is said to be *primitive* if $m = 1$ is the only $m \in \mathbb{Z}_{>0}$ such that $\mathfrak{a}/m$ is an integral ideal. For simplicity we introduce the following definition, which will only be used in this section.

Definition 6.4.6. *Let θ be an algebraic integer, $K = \mathbb{Q}(\theta)$, and $f = [\mathbb{Z}_K : \mathbb{Z}[\theta]]$. We say that an integral ideal $\mathfrak{b}$ of K dividing $b\mathbb{Z}_K$ is a* suitable divisor *of $\mathfrak{b}$ (relative to θ) if it satisfies the following three conditions:*

(1) *$\mathfrak{b}$ is primitive and $\mathfrak{b}/\mathcal{N}(\mathfrak{b})$ is the pth power of a rational number.*
(2) *If $m \in \mathbb{Z}$ divides $b\mathbb{Z}_K/\mathfrak{b}$ then $m \mid \gcd(b, f)$.*
(3) *Every prime ideal dividing $\mathfrak{b}$ and not dividing $f\mathbb{Z}_K$ has degree 1.*

For instance, if b and f are coprime then this means that $\mathfrak{b}$ and $b\mathbb{Z}_K/\mathfrak{b}$ are primitive, that $\mathfrak{b}/\mathcal{N}(\mathfrak{b})$ is the pth power of a rational number, and that all prime ideals dividing $\mathfrak{b}$ have degree 1. The following lemma is the key to the theorems that we are going to prove.

Lemma 6.4.7. *Let c be an integer not equal to a pth power and such that $c^{p-1} \not\equiv 1 \pmod{p^2}$, set $\theta = c^{1/p}$, and let $K = \mathbb{Q}(\theta)$. Let b be a nonzero integer, let x, y, and z be pairwise coprime integers such that $x^p + by^p + cz^p = 0$, and set $L = x + z\theta$ and $Q = \sum_{0 \leqslant k < p} (-1)^k x^{p-1-k} z^k \theta^k$.*

(1) *There exist integral ideals $\mathfrak{a}_i$ and $\mathfrak{b}_i$ of K such that $L\mathbb{Z}_K = \mathfrak{b}_1 \mathfrak{a}_1^p$, $Q\mathbb{Z}_K = \mathfrak{b}_2 \mathfrak{a}_2^p$, $\mathfrak{b}_1 \mathfrak{b}_2 = b\mathbb{Z}_K$, $\mathfrak{a}_1 \mathfrak{a}_2 = y\mathbb{Z}_K$, and the $\mathfrak{a}_i$ are coprime to $p\mathbb{Z}_K$.*
(2) *If c is pth power-free the ideal $\mathfrak{b}_1$ is a suitable divisor of b.*

Proof. (1). We note first that $p \nmid y$: indeed, if $p \mid y$ then $p \nmid xz$ by coprimality, so $c \equiv (-x/z)^p \pmod{p^2}$, and hence $c^{p-1} \equiv (-x/z)^{\phi(p^2)} \equiv 1$ $\pmod{p^2}$, contrary to our assumption. Now with the notation of the lemma, in the field K our equation can be written $LQ = -by^p$. Let $\mathfrak{q}$, if it exists, be a prime ideal of K dividing both L and Q. Then $\mathfrak{q}$ divides

$$px^{p-1} = Q + \sum_{2 \leqslant k \leqslant p} (-1)^k \binom{p}{k} L^{k-1} x^{p-k}$$

(this identity is an easy combinatorial argument), so that either $\mathfrak{q} \mid p\mathbb{Z}_K$ or $\mathfrak{q} \mid x$, and since y is coprime to x and is not divisible by p it follows that $\mathfrak{q} \nmid y\mathbb{Z}_K$, so $\gcd(L\mathbb{Z}_K, Q\mathbb{Z}_K)$ is coprime to $y\mathbb{Z}_K$. By Lemma 6.4.5 it follows that there exist integral ideals $\mathfrak{a}_i$ and $\mathfrak{b}_i$ such that $L\mathbb{Z}_K = \mathfrak{b}_1 \mathfrak{a}_1^p$, $Q\mathbb{Z}_K = \mathfrak{b}_2 \mathfrak{a}_2^p$, $\mathfrak{b}_1 \mathfrak{b}_2 = b\mathbb{Z}_K$, and $\mathfrak{a}_1 \mathfrak{a}_2 = y\mathbb{Z}_K$, proving (1).

(2). If $m \in \mathbb{Z}_{>0}$ divides $\mathfrak{b}_1$ then $m \mid (x + z\theta)\mathbb{Z}_K$; hence by Lemma 6.4.4 $m \mid \gcd(x, z) = 1$, so $\mathfrak{b}_1$ is primitive. By the same lemma if $m \in \mathbb{Z}_{>0}$ divides $b\mathbb{Z}_K/\mathfrak{b}_1 = \mathfrak{b}_2$ then $m \mid (\sum_{0 \leqslant k < p}(-1)^k x^{p-1-k} z^k \theta^k)\mathbb{Z}_K$; hence $m \mid f\gcd(x^{p-1}, z^{p-1}) = f$ since $\gcd(x, z) = 1$, and since $\mathfrak{b}_2 \mid b\mathbb{Z}_K$ we also have $m \mid b$, so $m \mid \gcd(b, f)$. Since $L\mathbb{Z}_K = \mathfrak{b}_1 \mathfrak{a}_1^p$ and $\mathcal{N}(L) = x^p + cz^p = -by^p$, it follows that $\mathcal{N}(\mathfrak{b}_1)\mathcal{N}(\mathfrak{a}_1)^p = |by^p|$, so that $b/\mathcal{N}(\mathfrak{b}_1) = (\pm \mathcal{N}(\mathfrak{a}_1)/y)^p$ is the pth power of a rational number. Finally, let $\mathfrak{q}$ be a prime ideal divisor of $\mathfrak{b}_1$ and q the prime number below $\mathfrak{q}$. Since by assumption $\mathfrak{q} \nmid f\mathbb{Z}_K$, we have $q \nmid f$, and since x and z are coprime, by Lemma 3.3.24 we deduce that $\mathfrak{q}$ has degree 1, proving that $\mathfrak{b}_1$ is a suitable ideal divisor of b.   □

Thanks to the above lemma we can easily give some sufficient conditions for the insolubility of the equation $x^p + by^p + cz^p = 0$. We give two results, corresponding to the cases in which the class number of $K = \mathbb{Q}(c^{1/p})$ is divisible by p or not.

Theorem 6.4.8. *Let b and c be pth power-free integers not equal to ± 1, set $\theta = c^{1/p}$, and let $K = \mathbb{Q}(\theta)$. Assume that the following conditions are satisfied:*

(1) *We have $c^{p-1} \not\equiv 1 \pmod{p^2}$.*
(2) *The class number $h(K)$ of K is divisible by p.*
(3) *For every suitable divisor $\mathfrak{b}$ of b the ideal $\mathfrak{b}^{e/p}$ is not principal, where $e \mid h(K)$ is the exponent of the class group of K.*

Then the equation $x^p + by^p + cz^p = 0$ has no nontrivial rational solutions.

Proof. By Lemma 6.4.3 it is sufficient to prove that the equation has no solution with x, y, z in $\mathbb{Z}$ pairwise coprime. By Lemma 6.4.7 there exist integral ideals $\mathfrak{a}_i$ and $\mathfrak{b}_i$ of K such that $L\mathbb{Z}_K = \mathfrak{b}_1 \mathfrak{a}_1^p$, $Q\mathbb{Z}_K = \mathfrak{b}_2 \mathfrak{a}_2^p$, $\mathfrak{b}_1 \mathfrak{b}_2 = b\mathbb{Z}_K$, $\mathfrak{a}_1 \mathfrak{a}_2 = y\mathbb{Z}_K$, and since c is pth power-free $\mathfrak{b}_1$ is a suitable divisor of b. Since $L^{e/p}\mathbb{Z}_K = \mathfrak{b}_1^{e/p} \mathfrak{a}_1^e$ and by definition of the exponent the ideal $\mathfrak{a}_1^e$ is principal, it follows that $\mathfrak{b}_1^{e/p}$ is also principal, a contradiction.   □

Example. Among the 84 fields $\mathbb{Q}(c^{1/3})$ with c cubefree such that $2 \leqslant c \leqslant 100$, 41 are such that $c \not\equiv \pm 1 \pmod 9$ and class number divisible by 3.

Before stating the next theorem we need an easy lemma.

Lemma 6.4.9. *Keep the above notation and assume that $p \nmid f = [\mathbb{Z}_K : \mathbb{Z}[\theta]]$. Then if $\alpha \in \mathbb{Z}_K$ we have $\alpha^p \equiv \mathcal{N}(\alpha) \pmod{p\mathbb{Z}_K}$. In particular, if α is coprime to p there exists $a \in \mathbb{Z}$ with $p \nmid a$ and such that $\alpha^p \equiv a \pmod{p\mathbb{Z}_K}$.*

Proof. By definition of f we have $f\alpha \in \mathbb{Z}[\theta]$, so that $f\alpha = \sum_{0 \leqslant j < p} a_j \theta^j$ with $a_j \in \mathbb{Z}$. Thus $(f\alpha)^p \equiv \sum_{0 \leqslant j < p} a_j^p c^j \pmod{p\mathbb{Z}[\theta]}$. Denote by ζ_p a primitive pth root of unity and set $L = K(\zeta_p)$. We thus have

$$\mathcal{N}(f\alpha) = \prod_{0 \leqslant i < p} \sum_{0 \leqslant j < p} a_j \zeta_p^{ij} \theta^j \equiv \prod_{0 \leqslant i < p} \sum_{0 \leqslant j < p} a_j \theta^j = (f\alpha)^p$$

$$\equiv \sum_{0 \leqslant j < p} a_j^p c^j \pmod{(1 - \zeta_p)\mathbb{Z}_L} .$$

Since both ends of this congruence are integers and since $(1 - \zeta_p)\mathbb{Z}_L \cap \mathbb{Z} = p\mathbb{Z}$, it follows that $(f\alpha)^p \equiv \mathcal{N}(f\alpha) \pmod{p\mathbb{Z}[\theta]}$, and since $p \nmid f$ and α is coprime to p if and only if $p \nmid \mathcal{N}(\alpha)$, the lemma follows. $\qquad \square$

We can now prove a result which is also valid when the class number is not necessarily divisible by p.

Theorem 6.4.10. *Let b and c be pth power-free integers, set $\theta = c^{1/p}$, let $K = \mathbb{Q}(\theta)$, $f = [\mathbb{Z}_K : \mathbb{Z}[\theta]]$, denote by $e = e(K)$ the exponent of the class group of K, set $r = e \bmod p$, so that $r \equiv e \pmod p$ and $0 \leqslant r < p$, and as usual let $U(K)$ be the unit group of K. Assume that the following conditions are satisfied:*

(1) *We have $p^2 \nmid c$ and $c^{p-1} \not\equiv 1 \pmod{p^2}$; equivalently, we have $p \nmid f$.*
(2) *For every suitable divisor $\mathfrak{b}$ of b, let β be a fixed generator of $\mathfrak{b}^e$. For any $\varepsilon \in U(K)$ modulo pth powers, set $\varepsilon\beta = \sum_{0 \leqslant j < p} b_j \theta^j$ with $b_j \in \mathbb{Q}$ (of course depending on ε), and let $P(X)$ be the polynomial $P(X) = \sum_{0 \leqslant j \leqslant r} f b_j X^j$. Assume that for every pair $(\mathfrak{b}, \varepsilon)$, either there exists j such that $r < j < p$ with $v_p(b_j) = 0$, or there exists k such that $0 \leqslant k \leqslant r - 2$ with $v_p(\text{disc}(P^{(k)}(X))) = 0$ (where $P^{(k)}(X)$ denotes the kth derivative of $P(X)$ and disc the discriminant).*

Then the equation $x^p + by^p + cz^p = 0$ has no nontrivial rational solutions.

Proof. First note that it follows from the Dedekind criterion (see Theorem 6.1.4 of [Coh0]) that the condition on c in (1) is indeed equivalent to $p \nmid f$. Thus, by elementary algebraic number theory the prime ideal decomposition of p in the extension $K/\mathbb{Q}$ is the same as the decomposition of $X^p - c$ in $\mathbb{F}_p[X]$. Since $X^p - c \equiv (X - c)^p \pmod{p\mathbb{Z}[X]}$ it follows that p is totally ramified in $K/\mathbb{Q}$; in other words there exists a unique prime ideal $\mathfrak{p}$ of K above p, and we have $p\mathbb{Z}_K = \mathfrak{p}^p$. Now I claim that if $e \geqslant p$ and if $\mathfrak{b}$ is a suitable divisor of b then $\mathfrak{b}$ is coprime to $p\mathbb{Z}_K$ (equivalently, to $\mathfrak{p}$). Indeed, otherwise $p\mathbb{Z}_K = \mathfrak{p}^p \mid \mathfrak{b}^p \mid \mathfrak{b}^e$, so if β is a generator of $\mathfrak{b}^e$ then $\beta = p\gamma$ for some $\gamma \in \mathbb{Z}_K$. With the notation of the theorem, it follows that for any $\varepsilon \in U(K)$ we have $p \mid fb_j$ for all j, so that $v_p(b_j) \geqslant 1$ for all j and $v_p(\mathrm{disc}(P^{(k)})) \geqslant 1$ for all $k \leqslant p - 2$, and in particular for all $k \leqslant r - 2$, in contradiction with our assumption, proving my claim. We can now begin the proof proper.

As before, by Lemma 6.4.3 we may assume that x, y, z are pairwise coprime integers. Writing our equation $LQ = -by^p$ as above, by Lemma 6.4.7 there exist integral ideals $\mathfrak{a}_i$ and $\mathfrak{b}_i$ of K such that $L\mathbb{Z}_K = \mathfrak{b}_1\mathfrak{a}_1^p$, $Q\mathbb{Z}_K = \mathfrak{b}_2\mathfrak{a}_2^p$, $\mathfrak{b}_1\mathfrak{b}_2 = b\mathbb{Z}_K$, $\mathfrak{a}_1\mathfrak{a}_2 = y\mathbb{Z}_K$, where $\mathfrak{b}_1$ is a suitable divisor of b and the $\mathfrak{a}_i$ are coprime to $p\mathbb{Z}_K$. By assumption we have $\mathfrak{b}_1^e = \beta\mathbb{Z}_K$ with $\beta \in \mathbb{Z}_K$, so that $(x + z\theta)^e\mathbb{Z}_K = \beta\alpha^p\mathbb{Z}_K$, where α is a generator of $\mathfrak{a}_1^e$. It follows that there exists a unit ε such that $(x+z\theta)^e = \varepsilon\beta\alpha^p$, so that if we set $\varepsilon\beta = \sum_{0\leqslant j<p} b_j\theta^j$ with $b_j \in \mathbb{Q}$ we have

$$(x + z\theta)^e = \left(\sum_{0\leqslant j<p} b_j\theta^j\right)\alpha^p .$$

Now note that $\mathfrak{a}_1^e = \alpha\mathbb{Z}_K$ is coprime to $p\mathbb{Z}_K$, and since $p \nmid f$ by (1), it follows from Lemma 6.4.9 that there exists $m \in \mathbb{Z}$ with $p \nmid m$ and such that

$$(fx + fz\theta)^e \equiv m \sum_{0\leqslant j<p} fb_j\theta^j \pmod{p\mathbb{Z}[\theta]} ,$$

where all the coefficients are in $\mathbb{Z}$. I claim that in fact

$$(fx + fz\theta)^r \equiv n \sum_{0\leqslant j<p} fb_j\theta^j \pmod{p\mathbb{Z}[\theta]}$$

for some integer n such that $p \nmid n$. This is trivial if $e < p$ since in that case $r = e$. Otherwise, by what we have shown above $\mathfrak{b}_1$ is coprime to $p\mathbb{Z}_K$; hence $(fx + fz\theta)\mathbb{Z}_K = f\mathfrak{b}_1\mathfrak{a}_1^e$ is also coprime to $p\mathbb{Z}_K$, so again by Lemma 6.4.9 we have $(fx + fz\theta)^e \equiv m'(fx + fz\theta)^r \pmod{p\mathbb{Z}[\theta]}$ for some $m' \in \mathbb{Z}$ such that $p \nmid m'$, proving my claim. Note also that by the same lemma, the congruence depends only on ε modulo pth powers of units.

Identifying the coefficients of θ^j for $r < j < p$ we deduce that $0 \equiv nfb_j$ $\pmod{p}$, and since $p \nmid nf$, this implies that $v_p(b_j) \geqslant 1$ for all such j. Thus, if we set $P(X) = \sum_{0\leqslant j\leqslant r}(fb_j)X^j$, using the fact that the θ^j are $\mathbb{Q}$-linearly independent, we deduce that in $\mathbb{F}_p[X]$ we have $\overline{nP}(X) = (u+vX)^r$ for some u and v in $\mathbb{F}_p$. Thus $\overline{nP^{(k)}}(X) = k!\binom{r}{k}(u+vX)^{r-k}$, whose discriminant is equal to 0 for $0 \leqslant k \leqslant r - 2$, so that $p \mid \mathrm{disc}(P^{(k)})$ for all such k since $p \nmid n$. Since by assumption either one of these discriminants or one of the b_j for $r < j < p$ is coprime to p we obtain a contradiction, proving the theorem.   □

Remarks. (1) When $p \geqslant 7$, it becomes costly to test the existence of a congruence of the type $(fx + fz\theta)^r \equiv nf\beta\varepsilon \pmod{p\mathbb{Z}[\theta]}$ for all $\varepsilon \in U(K)/U(K)^p$, since there are $p^{(p-1)/2}$ such classes of units. The problem can in fact easily be *linearized*, so that the number of tests is only polynomial-time instead of exponential-time in p; see Exercise 17.

(2) Recall that we replaced the equation $ax^p + by^p + cz^p = 0$ by the equation $x^p + B(y/a)^p + C(z/a)^p = 0$ with $B = ba^{p-1}$ and $C = ca^{p-1}$. Evidently the initial equation has a solution if and only if $a'x^p + b'y^p + c'z^p = 0$ has one, where (a', b', c') is any permutation of (a, b, c). Since we completely broke the symmetry by dividing specifically by a and by considering the field $\mathbb{Q}(C^{1/p})$, it follows that to apply Theorems 6.4.8 and 6.4.10, one should try the six permutations. On the equation $x^p + by^p + cz^p = 0$ this corresponds to the six pairs $(b, c) = (b, c)$, (c, b), $(b^{p-1}, b^{p-1}c)$, $(b^{p-1}c, b^{p-1})$, $(c^{p-1}, c^{p-1}b)$, and $(c^{p-1}b, c^{p-1})$, which should all be tested to use the theorems to their maximum extent.

Corollary 6.4.11. *The equations $x^p + by^p + cz^p = 0$ have a nontrivial solution in every completion of $\mathbb{Q}$ but no nontrivial solutions in $\mathbb{Q}$:*

(1) *For $p = 3$ and $b \leqslant c \leqslant 22$, if and only if $(b, c) = (3, 20)$, $(3, 22)$, $(4, 15)$, $(5, 12)$, $(6, 10)$, $(6, 11)$, $(6, 17)$, $(10, 15)$, $(10, 22)$, $(11, 15)$, $(11, 20)$, $(12, 17)$, $(15, 17)$, $(15, 20)$, $(15, 22)$, $(17, 20)$, or $(17, 22)$.*

(2) *For $p = 5$ and $b \leqslant c \leqslant 12$, if and only if $(b, c) = (2, 9)$, $(2, 10)$, $(3, 7)$, $(3, 10)$, $(4, 7)$, $(5, 7)$, $(5, 12)$, $(6, 10)$, $(7, 10)$, or $(7, 12)$.*

(3) *For $p = 7$ and $b \leqslant c \leqslant 12$, if and only if $(b, c) = (2, 5)$, $(2, 7)$, $(3, 6)$, $(3, 10)$, $(4, 9)$, $(5, 7)$, $(6, 11)$, $(7, 9)$, $(7, 10)$, or $(7, 12)$.*

(4) *For $p = 11$ and $b \leqslant c \leqslant 22$, if and only if $(b, c) = (2, 21)$, $(2, 22)$, $(3, 19)$, $(3, 21)$, $(3, 22)$, $(5, 17)$, $(9, 22)$, $(10, 22)$, $(15, 22)$, or $(19, 22)$.*

Proof. Since we have at our disposal a CAS that can say immediately whether the conditions of the theorems are satisfied, a simple computer program proves the corollary, which can of course be extended at will.   □

Remark. For $p = 3$ and $(b, c) = (19, 30)$, or for $p = 3$ and $(b, c) = (25, 29)$, or for $p = 5$ and $(b, c) = (19, 24)$ (and many other examples), the equation is everywhere locally soluble, none of the above theorems apply, and a search does not seem to give any global solutions. For $p = 3$ and $(b, c) = (19, 30)$ we can prove that the equation does not have any global solution by using elliptic curves as described in the next section. For $p = 3$ and $(b, c) = (25, 29)$ the same method shows that the equation does have global solutions. For $p = 5$ and $(b, c) = (19, 24)$, using coverings by *hyperelliptic* genus 2 curves (see Proposition 6.4.13), at my request the problem has been solved by M. Stoll. It can also probably be solved by using modular methods, in particular Theorem 15.8.1.

Remarks. (1) We will come back to the equation $x^3 + 15y^3 + 22z^3 = 0$ at the end of Section 7.2.4.

(2) For $p = 3$ and if the equation is everywhere locally soluble, Theorem 6.4.10 is applicable only when the fundamental unit ε is such that $\varepsilon \equiv \pm 1$ (mod $3\mathbb{Z}_K$), in which case in condition (3) it is sufficient to test the condition for a single generator α of $\mathfrak{b}^e$; see Exercise 16. However, this is not true when $p \geqslant 5$, as can easily be seen on examples.

(3) Again for $p = 3$, it seems experimentally that when $3 \mid e$, then if Theorem 6.4.10 is applicable, so is Theorem 6.4.8; I do not know how to prove this, although it may be easy.

Example. Among the 84 fields $\mathbb{Q}(c^{1/3})$ with c cubefree such that $2 \leqslant c \leqslant 100$, 19 are such that $c \not\equiv \pm 1$ (mod 9), $c \not\equiv 0$ (mod 9), and class number not divisible by 3, and among those, 7 satisfy $\varepsilon \equiv \pm 1$ (mod $3\mathbb{Z}_K$).

Corollary 6.4.12 (Selmer). *The equation $3x^3 + 4y^3 + 5z^3 = 0$ has a nontrivial solution in every completion of $\mathbb{Q}$ but no nontrivial solutions in $\mathbb{Q}$.*

Proof. Multiplying the equation by 2 and setting $(X, Y, Z) = (2y, x, z)$, it is clear that its solubility in any field of characteristic 0 is equivalent to that of the equation $X^3 + 6Y^3 + 10Z^3 = 0$, and it is easy to check that the conditions of Theorem 6.4.10 are satisfied for $c = 6$ and $b = 10$ (the pair $(6, 10)$ is among those given above). $\qquad\square$

This equation has historical value since it was the first example of an equation of that type violating the Hasse principle.

To conclude this subsection, let me repeat once again that there are a number of other sufficient conditions that are also based on algebraic number theory that imply the nonglobal solubility of our equations, in particular conditions based on reciprocity laws; see for instance the first paper by Selmer [Sel1] on the subject.

6.4.4 The Equation $ax^p + by^p + cz^p = 0$: Hyperelliptic Curves

It is also natural to study the global solubility of the equation $ax^p + by^p + cz^p = 0$ using algebraic geometry. This is particularly efficient for $p = 3$, where the problem reduces to computing the Mordell–Weil group of an elliptic curve, but we first briefly mention the general case.

By Faltings's theorem on Mordell's conjecture we know that for $p \geqslant 5$ and $abc \neq 0$ the projective curve $ax^p + by^p + cz^p = 0$ has only a finite number of rational points. The proof is *ineffective*, and although several methods are known which give effective results in special cases, they are almost impossible to apply directly, in particular because the genus of the curve is equal to $(p-1)(p-2)/2$, which is large when $p \geqslant 5$. However, the following immediate proposition shows that we can reduce to a curve of much lower genus.

Proposition 6.4.13. *Assume that p is odd. If $ax^p + by^p + cz^p = 0$ with $x \neq 0$ then*

$$Y^2 = X^p + a^2(bc)^{p-1}/4 , \quad \text{with}$$

$$X = -bcyz/x^2 \quad \text{and} \quad Y = (-bc)^{(p-1)/2}(by^p - cz^p)/(2x^p) .$$

Proof. The proof is of course a simple verification: using the given formulas for X and Y, since p is odd we have

$$4\,x^{2p}(Y^2 - X^p) = (bc)^{p-1}(by^p - cz^p)^2 + 4(bc)^p y^p z^p$$
$$= (bc)^{p-1}(by^p + cz^p)^2 = a^2(bc)^{p-1}x^{2p} ,$$

proving the result.   □

The main advantage of this proposition is that the hyperelliptic curve $Y^2 = X^p + a^2(bc)^{p-1}/4$ has genus $(p-1)/2$, which is much smaller than the genus of our initial curve, so that Chabauty-type methods may be applicable; see Chapter 13 (these methods would not be practical for the initial equation). In addition, note that by permuting a, b, and c, for $p > 3$ we have in fact three hyperelliptic curves available, thus improving the chances that Chabauty methods be applicable. If we succeed in proving that the hyperelliptic curve has no rational point, this of course implies that our initial equation has no nontrivial global solution. On the other hand, if the hyperelliptic curve has rational points, then it is necessary to find all of them, and it is then an immediate matter to check whether they correspond to global solutions to our equation.

In the case $p = 3$, the hyperelliptic curve is an elliptic curve, so we can say much more. First note the following easy result.

Proposition 6.4.14. *Assume that neither b/a, c/a, nor c/b is the cube of a rational number. If the elliptic curve E with affine equation $Y^2 = X^3 + (4abc)^2$ has zero rank then the equation $ax^3 + by^3 + cz^3$ has no nontrivial rational solutions.*

Proof. This is essentially a restatement of Corollary 8.1.15, and also immediately follows from the above proposition after rescaling. Note that Proposition 8.4.3 tells us that the elliptic curve $Y^2 = X^3 - 432(abc)^2$ is 3-isogenous with the elliptic curve $Y^2 = X^3 + (4abc)^2$.   □

The general result on the equation $ax^3 + by^3 + cz^3 = 0$ is a very nice application of 3-descent (see Section 8.4), and I thank T. Fisher for explanations. The relation is the following. Let a, b, and c be three nonzero rational numbers, and as above, let $E = E_{abc}$ be the elliptic curve with projective equation $Y^2 Z = X^3 + (4abc)^2 Z^3$. This curve has the point $T = (0, 4abc, 1)$ as rational point of order 3. In Section 8.4 we will define a 3-descent map α from $E(\mathbb{Q})$ to $\mathbb{Q}^*/\mathbb{Q}^{*3}$ by setting $\alpha(\mathcal{O}) = 1$, $\alpha(T) = (abc)^2$, and for $P = (X, Y, Z)$

with $P \neq \mathcal{O}$ and $P \neq T$ by setting $\alpha(P) = Y/Z - 4abc$, modulo multiplication by cubes of $\mathbb{Q}^*$. The fundamental property of α is that it induces a group homomorphism from $E(\mathbb{Q})/3E(\mathbb{Q})$ to $\mathbb{Q}^*/\mathbb{Q}^{*3}$ (Proposition 8.4.8).

Proposition 6.4.15. *Let $\mathcal{C}$ be the cubic curve with projective equation $ax^3 + by^3 + cz^3 = 0$, let E be the elliptic curve with projective equation $Y^2Z = X^3 + (4abc)^2Z^3$, and define*

$$\phi(x, y, z) = (-4abcxyz, \ -4abc(by^3 - cz^3), \ ax^3) .$$

(1) *The map ϕ sends $\mathcal{C}(\mathbb{Q})$ into $E(\mathbb{Q})$.*

(2) *The image $\phi(\mathcal{C}(\mathbb{Q}))$ is equal to the set of $P \in E(\mathbb{Q})$ such that $\alpha(P) = b/c \in \mathbb{Q}^*/\mathbb{Q}^{*3}$. More precisely, if $P = (X, Y, Z) \in \phi(\mathcal{C}(\mathbb{Q}))$ with P different from $\mathcal{O}$ and T, and if $\lambda \in \mathbb{Q}^*$ is such that $Y/Z - 4abc = (b/c)\lambda^3$ then*

$$(x, y, z) = (2bc\lambda Z, \ -cX, \ b\lambda^2 Z) \in \mathcal{C}(\mathbb{Q})$$

is a preimage of P; in addition, $\mathcal{O} \in \phi(\mathcal{C}(\mathbb{Q}))$ if and only if $c/b = \lambda^3$ for some $\lambda \in \mathbb{Q}^*$, and in that case $(0, -\lambda, 1)$ is a preimage of $\mathcal{O}$; finally, $T \in \phi(\mathcal{C}(\mathbb{Q}))$ if and only if $b/a = \lambda^3$ for some $\lambda \in \mathbb{Q}^*$, and in that case $(-\lambda, 1, 0)$ is a preimage of T.

(3) *The set $\mathcal{C}(\mathbb{Q})$ is nonempty if and only if the class of b/c modulo cubes belongs to the image of the 3-descent map α from $E(\mathbb{Q})$ to $\mathbb{Q}^*/\mathbb{Q}^{*3}$. In addition, $\mathcal{C}(\mathbb{Q})$ is infinite if and only if the class of b/c modulo cubes is equal to $\alpha(G)$ for some nontorsion point $G \in E(\mathbb{Q})$.*

Proof. (1). As for Proposition 6.4.13, of which up to rescaling this is a special case, the proof is a simple verification: setting $\phi(x, y, z) = (X, Y, Z)$, then if $ax^3 + by^3 + cz^3 = 0$ we check that $Y^2Z - (4abc)^2Z^3 = X^3$. Furthermore, we cannot have $X = Y = Z = 0$ since otherwise $x = 0$, so that $by^3 + cz^3 = 0$, and $by^3 - cz^3 = 0$, and hence $x = y = z = 0$, which is excluded. Thus $\phi(x, y, z) \in E(\mathbb{Q})$, proving (1).

(2). Note that if $P = \phi(x, y, z) = (X, Y, Z)$, then

$$c(Y - 4abcZ) = -4abc^2(by^3 - cz^3 + ax^3) = 8abc^3z^3 = bZ\lambda^3$$

with $\lambda = 2cz/x \in \mathbb{Q}^*$ when x and z are nonzero, so that $\alpha(P) = b/c \in \mathbb{Q}^*/\mathbb{Q}^{*3}$. Now, x can be equal to 0 if and only if $b/c = (-z/y)^3 \in \mathbb{Q}^{*3}$, and in that case we have $\phi(0, y, z) = \mathcal{O}$ and $\alpha(\mathcal{O}) = b/c \in \mathbb{Q}^*/\mathbb{Q}^{*3}$. Similarly z can be equal to 0 if and only if $b/a = (-x/y)^3 \in \mathbb{Q}^{*3}$, and in that case it is clear that $\phi(x, y, 0) = T$ and $\alpha(T) = (abc)^2 = (b/c)(ac)^3(b/a) = b/c \in \mathbb{Q}^*/\mathbb{Q}^{*3}$, proving (2).

(3). Evidently $\mathcal{C}(\mathbb{Q})$ is nonempty if and only if $\phi(\mathcal{C}(\mathbb{Q}))$ is nonempty, and since clearly the number of preimages of an element of $\phi(\mathcal{C}(\mathbb{Q}))$ is finite, $\mathcal{C}(\mathbb{Q})$ is infinite if and only if $\phi(\mathcal{C}(\mathbb{Q}))$ is infinite. Thus, it follows from (2) that $\mathcal{C}(\mathbb{Q})$ is nonempty if and only if b/c modulo cubes belongs to the image of

α, and it is infinite if and only if $b/c = \alpha(G)$ for a point of infinite order, proving (3). □

Thus, to check whether an equation $ax^3 + by^3 + cz^3 = 0$ has a nontrivial solution we proceed as follows. Using either `mwrank` or the 2-descent methods, we compute if possible the complete Mordell–Weil group $E(\mathbb{Q})$ of the curve with affine equation $Y^2 = X^3 + (4abc)^2$, and we also compute the torsion subgroup (which we know will contain the subgroup of order 3 generated by T). If $P_1, \ldots, P_r$ form a basis of the free part of $E(\mathbb{Q})$ then $P_0 = T, P_1, \ldots,$ P_r form an $\mathbb{F}_3$-basis of $E(\mathbb{Q})/3E(\mathbb{Q})$. We then check whether the class of b/c modulo cubes belongs to the group generated by the $\alpha(P_i) \in \mathbb{Q}^*/\mathbb{Q}^{*3}$, which is done using simple linear algebra over $\mathbb{F}_3$.

Example. To illustrate, consider the equation $x^3 + 55y^3 + 66z^3$. Using Theorem 6.4.2 we check that it is everywhere locally soluble, but none of the results given in Section 6.4.3 allow us to determine whether the equation is globally soluble.

Thus we consider the curve E with affine equation $Y^2 = X^3 + (4 \cdot 55 \cdot 66)^2$. We find that the torsion subgroup has order 3 and is generated by $P_0 = T = (0, 14520)$. In less than a second the `mwrank` program tells us that the curve has rank 1, a generator being $P_1 = (504, 18408)$ (so that Proposition 6.4.14 is not applicable). We have (modulo cubes) $\alpha(P_0) = 2^2 \cdot 3^2 \cdot 5^2 \cdot 11$, $\alpha(P_1) = 2 \cdot 3^2$, and $b/c = 2^2 \cdot 3^2 \cdot 5$. Here the linear algebra can be done naïvely: if $b/c = \alpha(uP_0 + vP_1) = \alpha(P_0)^u \alpha(P_1)^v$, where u and v are defined modulo 3, we see that $u = 0$ because of the 11 factor, which is impossible since there is a factor of 5 in b/c and none in $\alpha(P_1)$. This shows that our equation has no nontrivial solutions in $\mathbb{Q}$, although our curve has nonzero rank.

For the convenience of the reader, in the following table we give in a very compact form detailed information on the solubility of the equation $x^3 + by^3 + cz^3 = 0$ for $1 \leqslant b, c \leqslant 64$. The entry in line b and column c of the table means the following: – means locally insoluble, L, M, and E mean everywhere locally soluble but not globally soluble (hence a failure of the Hasse principle) obtained using Theorems 6.4.8 and 6.4.10, and Proposition 6.4.15, used in that order (note that in the range of our table it is sufficient to use Proposition 6.4.14 instead of Proposition 6.4.15, but this is of course not true in general). In every other case, the equation is globally soluble, hence the curve $x^3 + by^3 + cz^3 = 0$ is the projective equation of an elliptic curve, and the entry (0, 1, 2, or 3 in the limits of the table) gives the *rank* of the curve (see Chapters 7 and 8 for all these notions). In particular, if the entry is equal to 0 this means that the equation has only a finite nonzero number of (projective) solutions, which can all easily be found. By Corollary 8.1.15 this implies that either b, c, or b/c is a cube. The same corollary also gives the torsion subgroup. Note that some solutions are rather large, for instance $149105^3 + 17 \cdot (-140161)^3 + 41 \cdot 101988^3 = 0$ for $(b, c) = (17, 41)$ and $147267^3 + 41 \cdot (-6040)^3 + 59 \cdot (-37793)^3 = 0$ for $(b, c) = (41, 59)$.

Note that the visual rows, columns, and diagonals that can easily be seen in the table reflect the existence of simple global solutions. For instance, on the diagonal $c = 64 - b$ we see that the equation is always globally soluble, and this is clear since there exists the solution $(x, y, z) = (-4, 1, 1)$.

```
 c |            1111111111122222222222333333333334444444444555555555566666
 b | 123456789012345678901234567890123456789012345678901234567890123456789012345678901234
----+----------------------------------------------------------------------------------
  1| 000001101001101010210100010102101110200001100001111010010101001110
  2| 001--1-0-11---201-------11-0-1L--L1---------2--111--L-102-------2-0
  3| 01111--0-21----11--L-L111-0-11-1-L------11--2-E1--L2-11E--L1-2--0
  4| 0-101--0---1--L-1------11--0---202------11--1----L-1-L---12---0
  5| 0-1102-0-1-L2-1-1----211-10--L-1L------101--L-L-2--L-1-1--L1L---0
  6| 11--2011-ML--1-1M---22L-L-1-L---1L2-1--22----M201L--M1L1-2L----1
  7| 1----111------1----1------1-2---1------1-----1------112------1
  8| 000001101001101010210100010102101110200001100001111010010101001110
  9| 1------101-------21------------1------111----2-1------1
 10| 012-1M-0102---M111---LL2--0-LE--LL--2--1L---11LM--L-111--L2----0
 11| 011--L-0-202--L11-2L----11-0-1L--1L--21----E1-21L--L-111-1--L2--0
 12| 1--1L--1--201-1-M-12--L-L-1-2--1------2LL--M--L----1M-L---L2---1
 13| 1---2--1---112------2----1-----------2--1-----1-E-------L---1
 14| 0----1-0----202------1---0--------1---1-----1-1----L-----0
 15| 12-L1-11-ML1-202M--L-L2---122-1L-M1----121-L-LL-1---M221-L2--2-1
 16| 001--1-0-11---201-------11-0-1L--L1---------2--111--L-102-------2-0
 17| 11111M-1211M--M111--L-L111--1-1L-1M1-1---11--1111M-L1-211--L13---1
 18| 0------011---112------20---------------21----1-L------0
 19| 2------2--21-----201------2--E-----------22--1-------1-----2
 20| 1-L---11--L2--L-L-112--L--11L---L-1------2--2---3-L-L12--L1--1
 21| 0----2-0----2-----212---011---2------1-2---2-L----2------0
 22| 1-L-22-1-L---1L-L---212L--1--1--L-L----2L2---L-21L3-L--LL----1
 23| 0-111L-0-L-L--2-1----201--0--L21L---2--11--E-1-L-2L-1-2--L1L-3-0
 24| 01111--0-21----11--L-L111-0-11-1-L------11--2-E1--L2-11E--L1-2--0
 25| 011--L-0--1L---11------1010-1----2L----1----L-21L--L111------0
 26| 1---1--1------------2------111----1--211----1--E-2---3------1
 27| 000001101001101010210100010102101110200001100001111010010101001110
 28| 1------1------2----11----111-----2----1---------1------1
 29| 011--L20-L12--211--L1--11-0101--2L1-2----L-L-L1L--L-1122---L-3-0
 30| 2L1-L---2-EL----LL-E--1L1--2-112--3---2-LL----LL----L-LLL-1LL--E-2
 31| 1--2--1------1------2--1--2121-----2-----3-------3-L-E-1
 32| 0-101--0---1--L-1------11--0---202------11--1----L-1-L---12---0
 33| 1L-2L1-1-L1----LM--L-LL-2-1-2-1203-----L2--M-L21-L--ML2--L22L2-1
 34| 11L--L11-LL--M11----2--LL11-L3--312---L--2---LLL--L-L1L1----1--1
 35| 1----2-1-----11----1-L-----1-1---211---1-2----2------2------221
 36| 0------01-------1---------02------112------1M-----1-2-------10
 37| 2----1-2-22-----------2---2-2------202-----2-21--1--------L---2
 38| 0------0--1-----------20--2-----203------1-------1------0
 39| 0---1--0---2---------110--2--L--311--L--1---1-----------0
 40| 0-1102-0-1-L2-1-1----211-10--L-1L------101--L-L-2--L-1-1--L1L---0
 41| 0-1112-0-L-L-12-1---1L11--01-L-12-1----102-2-E-2222-1-2--L1L---0
 42| 1-----11------1------2----1-L---2----201-----2----31-------1
 43| 12-----1--E----2----2----1--------2-----113--2---1--2-------1
 44| 0-21L--01-1M1-L-1--2--E2L-0-L--1M--1--LL2-302-L---L11-1---LL---0
 45| 0-----011------122------0------M2----212-----1-1-------0
 46| 01E-LM-0112---L1112--L1E2-0-LL3-LL---1-LE---202M-L3-11E-ELL--E-0
 47| 011--2-0-L1L--L11--2---1110-1L--2L--2-1---2L-202--E1112-----L-3-0
 48| 11--2011-ML--1-1M---22L-L-1-L---1L2-1--22----M201L--M1L1-2L----1
 49| 1----1-1------1------1----1-----------2----111----1-2-----1
 50| 1-L--L-1------1--L-1-LL2L--1--------L------22--L-L111-L----22L----1
 51| 1L2LL--1-LL-1--L1--3-3L2LE1-LL-L-L--1--L2-1L-3E--102LL3--L2----1
 52| 0------0---1-----------1-0----------------1--1---201-----2--0
 53| 11111M-1211ME-M121-L-L11121-1L-1ML-1--111--1111M-LL1111--L1L2--1
 54| 001--1-0-11---201-------11-0-1L--L1---------2--111--L-102-------2-0
 55| 02EL1L10111L-L221L-L2L2E--012L-L2L22---123-11E2L1-3-1201-LLL--30
 56| 1----111------1----1----1-2----1------1------112------1
 57| 0-----20--1-------2-----30--1--------1------E--22-----212-----0
 58| 1-L-L2-1-L----L-L-1--LLL--1--L3-L------LL----L-2-2L-L-L-212--L-1
 59| 0-111L-0-2-L--2-1----L11--0--L-12----11--L-L-L-L2-1-L--201---0
 60| 0--2L--0--L-L---3--L--L---0-L-L22---L--LL--L--L----2L-L---112--0
 61| 1-2----1--2--------1--2--1-----L1-------------2-------212-1
 62| 12-----1-----22-----3---1-3EE-2-2----------E3------2---L-2121
 63| 1------1------------1--------21--------------------3------201
 64| 000001101001101010210100010102101110200001100001111010010101001110
```

Solubility of $x^3 + by^3 + cz^3 = 0$ up to $\max(b, c) = 64$

6.4.5 The Equation $x^3 + y^3 + cz^3 = 0$

This equation has the evident global solution $(x, y, z) = (1, -1, 0)$, hence is everywhere locally soluble, so we are now going to study whether it has global rational solutions with $xyz \neq 0$. Note that proving that the equation has no such solutions for $c = 1$ is Fermat's "last theorem" for exponent 3, which was probably already solved by Fermat using the method of infinite descent, together with a touch of algebraic number theory, although a proof was given only later by Euler. Also for $c = 1$, a (slightly) more complicated proof that generalizes to the more general Fermat equation $x^p + y^p + z^p = 0$ for so-called *regular* prime exponents will be given in Proposition 6.9.14 below.

Remarks. (1) The solubility of the equation $x^3 + y^3 + cz^3 = 0$ with $z \neq 0$ is evidently equivalent to the representability of c as a sum of two rational cubes. This question will be considered in more detail (but with some proofs omitted) in Section 6.4.6.
(2) The algebro-theoretic method used in Section 6.4.3 is not applicable here since the theorems that we have seen only prove the nonexistence of solutions, while the present equation does have a solution.
(3) The general result of Section 6.4.4 is of course applicable, but we need to determine whether the curve $Y^2 = X^3 + 16c^2$ has nonzero rank. This is often done using 2-descent (or 3-descent in this case since there exists a rational 3-torsion point) as we will see in Chapter 8, but it is instructive and more elementary to use 2-descent directly on the equation without explicitly mentioning elliptic curves, and this is what we are going to do in this section.

We begin with the following proposition, which is typical of the type of reasoning which one uses to solve Diophantine equations by factoring and algebraic number theory. We will see many other such examples in this chapter and in Chapter 14.

Proposition 6.4.16. *The equation* $x^2 - 3xy + 3y^2 = z^3$ *with x and y coprime integers has the three disjoint parametrizations*

$$(x, y, z) = (s^3 + 3s^2t - 6st^2 + t^3, 3st(s - t), s^2 - st + t^2) ,$$
$$(x, y, z) = (s^3 + 3s^2t - 6st^2 + t^3, s^3 - 3st^2 + t^3, s^2 - st + t^2) ,$$
$$(x, y, z) = ((s + t)(s - 2t)(2s - t), s^3 - 3st^2 + t^3, s^2 - st + t^2) ,$$

where in all three s and t are coprime integers such that $3 \nmid s + t$, and the parametrizations correspond to the solutions for which $6 \mid y$, $6 \mid x - y$, and $6 \mid x - 2y$ respectively.

Proof. Let $\rho = (-1 + \sqrt{-3})/2$ be a primitive cube root of unity. In the principal ideal domain $\mathbb{Z}[\rho]$ our equation factors as $(x - (1 - \rho)y)(x - (1 - \rho^2)y) = z^3$. If $\mathfrak{p}$ is a prime ideal of $\mathbb{Z}[\rho]$ that divides the two factors on the left then $\mathfrak{p}$ divides their difference and $(1 - \rho)$ times the second minus $(1 - \rho^2)$ times the first, so $\mathfrak{p}$ divides $(1 - \rho)\gcd(x, y)$; hence $\mathfrak{p} \mid (1 - \rho)$ since x and y are coprime, so that $3 \mid z$. However, this would imply $3 \mid x$, so $9 \mid 3y^2$, and hence $3 \mid y$, contradicting $\gcd(x, y) = 1$. Thus $3 \nmid z$ and the two factors on the left are coprime. It follows that there exist $\alpha \in \mathbb{Z}[\rho]$ and an integer k with $0 \leqslant k \leqslant 2$ such that $x - (1 - \rho)y = \rho^k \alpha^3$, so $z = \alpha\overline{\alpha}$. Writing $\alpha = s + t\rho$, the condition $3 \nmid z$ translates into $3 \nmid s + t$, and choosing successively $k = 0$, 1, and 2 gives the three parametrizations, where in the second we have exchanged s and t. The divisibilities by 6 are trivially checked, and show that the parametrizations are disjoint. $\square$

Thanks to this proposition we can now prove that many of our equations $x^3 + y^3 + cz^3 = 0$ do not have any global solutions with $xyz \neq 0$:

Theorem 6.4.17. (1) *Let p be a prime number such that $p \equiv 2 \pmod{3}$. The equation $x^3 + y^3 + cz^3 = 0$ has no solutions in nonzero integers x, y, and z when $c = 1$, 3, p, or p^2 with $p \equiv 2$ or 5 modulo 9, except for $c = 2$, where it has the unique solution $(x, y, z) = (1, 1, -1)$ (up to multiplication by a constant).*

(2) *If p is a prime number such that $p \equiv 8 \pmod{9}$, the equation $x^3 + y^3 + cz^3 = 0$ has no nontrivial solutions with $3 \mid z$ when $c = p$ or p^2.*

(3) *If p and q are prime numbers such that $p \equiv 2 \pmod{9}$ and $q \equiv 5 \pmod{9}$, the equation $x^3 + y^3 + cz^3 = 0$ has no nontrivial solutions with $3 \mid z$ when $c = pq$.*

Proof. We may clearly assume that x, y, and z are pairwise coprime integers. We prove all the results simultaneously, and consider two cases.

Case 1: $3 \mid cz$. Then $3 \mid x + y$; hence $3 \mid (x^2 - xy + y^2) = ((x + y)^2 - 3xy)$, so that $9 \mid cz^3 = -(x^3 + y^3)$. Since $9 \nmid c$ it follows that $3 \mid z$, so we set $y_1 = (x + y)/3$, $z_1 = -z/3$ and our equation is $y_1(x^2 - 3xy_1 + 3y_1^2) = 3cz_1^3$. Since y_1 and x are coprime the two factors on the left are coprime. Since $3 \mid z$ we have $3 \nmid x$ hence $3 \mid y_1$. Furthermore, if a prime number ℓ divides $x^2 - 3xy_1 + 3y_1^2$ then $\ell \nmid y_1$, and we check that $(2x/y_1) - 3$ is a square root of -3 modulo ℓ, so $\ell \equiv 1 \pmod{3}$. Since in all cases considered in the theorem c has no such prime divisors, it follows that $x^2 - 3xy_1 + 3y_1^2$ is coprime to $3c$. Therefore our equation implies that there exist coprime integers a and b such that $y_1 = 3ca^3$ and $x^2 - 3xy_1 + 3y_1^2 = b^3$. By Proposition 6.4.16 this last equation has three disjoint parametrizations, but we keep only the first since we know that $3 \mid y_1$. Thus there exist coprime integers s and t with $3 \nmid s + t$ such that in particular $y_1 = 3st(s - t)$, so $ca^3 = st(s - t)$. To symmetrize we write $u = -s$, $v = t$, $w = s - t$, which are pairwise coprime and satisfy $u + v + w = 0$ and $uvw = c(-a)^3$.

In statements (1) and (2) of the theorem, c is a power of a prime, and since s and t are coprime it follows that c divides one and only one of u, v, and w, and without loss of generality we may assume that $c \mid w$. Then c is coprime to u and v, so that $u = e^3$, $v = f^3$, $w = cg^3$, and hence $e^3 + f^3 + cg^3 = 0$, so we have found a new solution to our initial equation. Clearly $efg \neq 0$, and it is easily checked that $|efg| < |xyz|$, so that the magic of Fermat's descent method applies: if we had started with a nontrivial solution with minimal $|xyz|$, we would have thus obtained a smaller one, a contradiction that proves the impossibility of the initial equation.

In (3), we have $c = pq$. Then up to permutation of u, v, and w, either $pq \mid w$, or $p \mid v$ and $q \mid w$. But in this last case, we would have $u = e^3$, $v = pf^3$, $w = qg^3$, hence $e^3 + pf^3 + qg^3 = 0$. Since a cube is congruent to 0 or ± 1 modulo 9 and $p \equiv 2 \pmod 9$ and $q \equiv 5 \pmod 9$, it is easily checked that the congruence $e^3 + pf^3 + qg^3 \equiv 0 \pmod 9$ implies that e, f, and g are all divisible by 3, which is absurd since u, v, and w are pairwise coprime. Thus this case is impossible, so that $pq = c \mid w$ and we can descend as above.

Case 2: $3 \nmid cz$. Thus here $c = 1$, p, or p^2 with $p \equiv 2$ or 5 modulo 9. In the special case $c = 1$ our equation is completely symmetrical in x, y, and z, so we may assume $3 \nmid xyz$, and we deduce an immediate contradiction modulo 9 since we have x^3, y^3, and z^3 all congruent to ± 1 modulo 9. Thus we assume $c = p$ or p^2.

We set here $y_1 = x + y$, so our equation is $y_1(y_1^2 - 3xy_1 + 3x^2) = -cz^3$, and since $3 \nmid z$, we have $3 \nmid y_1$ so the factors on the left are coprime. As above, a prime $p \equiv 2 \pmod 3$ cannot divide $y_1^2 - 3xy_1 + 3x^2$ when x and y_1 are coprime, so that there exist integers a and b such that $y_1 = ca^3$, $y_1^2 - 3xy_1 + 3x^2 = b^3$. By Proposition 6.4.16 once again this last equation has three disjoint parametrizations. However, we note that $3 \nmid x$ and $3 \nmid y$; otherwise, $x^3 + y^3 \equiv \pm 1 \pmod 9$, so $c \equiv \pm 1 \pmod 9$, which is impossible when $c \equiv 2$ or 5 modulo 9 (this is where we must exclude 8 modulo 9, for which the theorem would be false). Thus we can keep only the third parametrization, for which $x \equiv y \pmod 3$, and we deduce that there exist coprime s and t with $3 \nmid s+t$ such that in particular $y_1 = (s+t)(s-2t)(2s-t)$. To symmetrize we set $u = s + t$, $v = s - 2t$, $w = t - 2s$, which are pairwise coprime since $3 \nmid s+t$, and satisfy $u + v + w = 0$ and $uvw = c(-a)^3$. Exactly the same reasoning as in the first case allows us to conclude that the descent method works, with one exception: if $c = 2$ and $x = y = -z = 1$, we obtain the *same* solution. Thus the descent also works in this case and shows that $(x, y, z) = (1, 1, -1)$ is the only solution.   □

Remarks. (1) As already mentioned, this theorem includes in particular Fermat's last theorem for exponent 3.

(2) It is clear that the theorem is still valid for $c = p^k$ for all k if $p \equiv 2$ or 5 modulo 9, since it is then a special case of $x^3 + y^3 + p^m z^3 = 0$ with $m = 0$, 1, or 2.

(3) When $c = p \equiv 8 \pmod 9$, solutions not only may exist, but we will see below that as a consequence of the BSD conjecture solutions should exist for every p. For instance, we have $18^3 + (-1)^3 + 17(-7)^3 = 0$.

(4) When $c = 3^2$ there is the trivial solution $(x, y, z) = (1, 2, -1)$, and it is immediate to check using the methods of Chapter 8 that this gives a point of infinite order on the corresponding elliptic curve, so that there exists an infinity of distinct coprime solutions. On the other hand, for $c = 2$ the above descent method shows that $(x, y, z) = (1, 1, -1)$ is the only coprime solution (up to sign), and this means that the point is a *torsion point* on the corresponding elliptic curve.

(5) When $c = pq$ with p and q primes such that $p \equiv 2 \pmod 9$ and $q \equiv 5 \pmod 9$, there are in fact no solutions also when $3 \nmid z$; see Exercise 15.

6.4.6 Sums of Two or More Cubes

The problem of representing integers or rational numbers by sums of *squares* is completely understood thanks to the Hasse–Minkowski theorem and Proposition 5.4.4 (see Section 5.4.4). On the other hand the problem for cubes is much more difficult and far from being understood. First, as we have seen above in several analogous situations (Section 6.4.3) the local–global principle fails. The analogue of Proposition 5.4.4 also fails: representations as sums of cubes of rational numbers and as sums of cubes of integers are two quite different problems. For instance, we will see below that every integer is the sum of three cubes of rational numbers, while it is trivial to see that the integer 4 cannot be equal to the sum of three cubes of integers. Hence in this subsection we give an assortment of results and conjectures on those subjects, including the proof of a beautiful result of Dem'yanenko.

Finally, note that the natural setting for the problems that we consider is $\mathbb{Q}$ or $\mathbb{Z}$, and *not* the positive rationals or the positive integers, so we do not consider the problem of representations as sums of *positive* cubes, in other words Waring's problem (in brief: every positive integer is a sum of 9 nonnegative cubes and 9 is optimal; every sufficiently large integer is a sum of 7 nonnegative cubes, and it is conjectured that every sufficiently large integer is a sum of 4 nonnegative cubes, and 4 is evidently optimal). Thus when we speak of an *integer*, we always mean an element of $\mathbb{Z}$, not necessarily of $\mathbb{Z}_{\geq 0}$. In addition, since the exponent 3 is odd, in contrast to the quadratic case we do not need to look at signs. Note however the following result.

Proposition 6.4.18. *An integer $n \geq 1$ is a sum of two rational cubes if and only if it is a sum of two* nonnegative *rational cubes.*

Proof. We may evidently assume that n is not a cube, so let $n = x_0^3 + y_0^3$ be a decomposition of n as a sum of two cubes. If x_0 and y_0 are nonnegative there is nothing to prove, so we may assume without loss of generality that $y_0 > 0$ and $x_0 < 0$. It is easily checked that if $x_k^3 + y_k^3 = n$ then $x_{k+1}^3 + y_{k+1}^3 = n$ with

$$x_{k+1} = \frac{y_k^4 + 2x_k^3 y_k}{y_k^3 - x_k^3} \quad \text{and} \quad y_{k+1} = -\frac{x_k^4 + 2x_k y_k^3}{y_k^3 - x_k^3}.$$

The existence of such an identity comes from Fermat's tangent method and will be explained in detail in Chapter 7 (see also Exercise 10 of Chapter 7), but the direct verification is immediate. We thus define a sequence of points on the curve $x^3 + y^3 = n$. I claim that there exists $k \in \mathbb{Z}_{\geqslant 0}$ such that $x_k > 0$ and $y_k > 0$. Indeed, if we set $u_k = y_k/x_k$ this is equivalent to $u_k > 0$ (since $n > 0$), and u_k satisfies the recurrence $u_{k+1} = f(u_k)$ with

$$f(x) = -\frac{2x^3 + 1}{x(x^3 + 2)} \ .$$

Furthermore, since $y_0 > 0$, $x_0 < 0$, and $x_0^3 + y_0^3 = n$, we have $u_0^3 + 1 < 0$, so that $u_0 < -1$. Now we have

$$f(x) - 8x - 7 = -\frac{(x+1)^2(8x^3 - 9x^2 + 12x + 1)}{x(x^3 + 2)} \ ,$$

and since for $x < -1$ we have $8x^3 - 9x^2 + 12x + 1 < -28$ it follows in particular that for $-\sqrt[3]{2} < x < -1$ we have $f(x) - 8x - 7 < 0$. Thus if $-\sqrt[3]{2} < u_k < -1$ we have $u_{k+1} + 1 < 8(u_k + 1)$. Since $u_0 + 1 < 0$ it follows that there exists $k \geqslant 0$ such that $u_k < -\sqrt[3]{2}$. But then clearly $u_{k+1} = f(u_k) > 0$, as was to be proved. $\qquad\square$

Proposition 6.4.19. *There are infinitely many integers that are not the sum of two cubes of rational numbers.*

Proof. Indeed, Theorem 6.4.17 tells us for instance that odd primes congruent to 2 or 5 modulo 9 cannot be the sum of two cubes of rational numbers, and by Dirichlet's theorem on primes in arithmetic progression (Theorem 10.5.30) there are infinitely many such primes. $\qquad\square$

It is reasonable to ask whether, in a manner analogous to Proposition 5.4.9, we can characterize the integers that are sums of two cubes, either of integers or of rational numbers. The answer for integers is trivial. For rational numbers there is a conjecture and a theorem for prime numbers that is a combination of work of N. Elkies with a theorem of F. Rodriguez-Villegas and D. Zagier [Rod-Zag].

Proposition 6.4.20. *An integer n is a sum of two cubes of (positive or negative) integers if and only if there exists a positive divisor d of n such that $(4n/d - d^2)/3$ is the square of an integer. For example, a (positive) prime p is a sum of two cubes of integers if and only if $p = 2$ or p has the form $p = 3x^2 - 3x + 1$ for $x \geqslant 2$.*

Proof. Left to the reader (Exercise 18). $\qquad\square$

The conjecture for sums of two cubes of rational numbers is the following.

Conjecture 6.4.21. (1) *Any squarefree integer congruent to 4, 6, 7, or 8 modulo 9 is a sum of two rational cubes.*

(2) *Denote by $S(X)$ the set of squarefree integers less than or equal to X that are congruent to 1, 2, 3, or 5 modulo 9 and that are the sum of two rational cubes. Then $S(X)$ has density 0, and more precisely there exists a strictly positive constant c such that*

$$S(X) \sim cX^{5/6} \log(X)^{\sqrt{3}/2 - 1/8} .$$

The first conjecture immediately follows from the BSD conjecture, hence is almost certainly correct. The second conjecture has been obtained using methods coming from random matrix theory, and is more speculative although supported by numerical evidence [Kea]. Note that it is possible to state a conjecture for *cubefree* integers, which is a more natural condition, but the statement is more complicated.

To state the theorems of Elkies and Rodriguez-Villegas–Zagier we first need to define a sequence of polynomials.

Definition 6.4.22. *We define the Villegas–Zagier polynomials $V_n(t)$ by the initial conditions $V_{-1}(t) = 0$, $V_0(t) = 1$ and the recurrence*

$$V_{n+1}(t) = (8t^3 - 1)V_n'(t) - (16n + 3)t^2 V_n(t) - 4n(2n - 1)tV_{n-1}(t)$$

for $n \geqslant 1$.

Theorem 6.4.23. *Assume the Birch and Swinnerton-Dyer conjecture (Conjecture 8.1.7), and let p be a prime number. Then p is the sum of two cubes of rational numbers if and only if one of the following conditions is satisfied:*

(1) $p = 2$.
(2) $p \equiv 4$, 7, *or* 8 *modulo* 9.
(3) $p \equiv 1 \pmod 9$ *and* $p \mid V_{(p-1)/3}(0)$.

The proof of this theorem uses the classical theory of modular forms, complex multiplication, and central values of L-series. For example, it implies that the only $p \leqslant 100$ such that $p \equiv 1 \pmod 9$ that are sums of two cubes of rational numbers are $p = 19$ and $p = 37$.

Note that in Theorem 6.4.17 we have proved by descent arguments that if $p = 3$ or $p \geqslant 5$ and $p \equiv 2$ or 5 modulo 9 then p is not a sum of two cubes. When $p \equiv 4$ or 7 modulo 9, N. Elkies has shown that p is a sum of two cubes of rational numbers without assuming BSD. (Note that there does not even exist a preprint of this proof, and that no one else has been able to reconstruct it.) The result for $p \equiv 1 \pmod 9$ is also independent of the correctness of BSD. The only case for which the BSD conjecture is needed is for $p \equiv 8 \pmod 9$.

It is possible that this theorem can be extended to a complete characterization of all integers that are sums of two cubes of rational numbers.

Note, however, that if we are willing to reason in a heuristic manner and in particular to believe the BSD conjecture, it is easy to determine whether a *given* integer (or rational number) c is a sum of two cubes. Without loss of generality assume that c is a cubefree integer and that $c > 2$. It follows from Proposition 7.2.3 and the remarks following it, combined with Corollary 8.1.15 and BSD, that c is a sum of two cubes if and only if the elliptic curve E_c with affine equation $y^2 = x^3 + 16c^2$ has rank greater than or equal to 1. To check this, we first compute the root number (in the case of squarefree c it will be equal to -1 for $c \equiv 4$, 6, 7, or 8 modulo 9, and to 1 otherwise, explaining Conjecture 6.4.21 (1)). If it is equal to -1, then by BSD c is a sum of two cubes (and BSD is not necessary if the rank is equal to 1, which can be checked by computing $L'(E_c, 1)$ using Corollary 8.5.14). If the root number is equal to 1, we compute $L(E_c, 1)$ using Corollary 8.5.11, which gives a rapidly convergent series for $L(E_c, 1)$. If the result is different from 0 (which can always be proved if true), then by the proved results on the BSD conjecture we know that the rank of E_c is equal to 0, hence that c is not a sum of two cubes. On the other hand, if the result seems to be very close to 0, then it is highly plausible (but not proved, even assuming BSD) that c is a sum of two cubes since the rank of E_c will probably be greater than or equal to 2.

```
       1111111111112222222222233333333333444444444455555555555566666
     123456789012345678901234567890123456789012345678901234567890123
   0 TT00011T1001101T1021010001T10210111020000110000111101T010100111
  63 T2011111100100110000112101121010111000011111002000110100100121T2
 126 2T01021101001111100000001220011011112101001111000111111022001101
 189 000101011002120110122111117222001110001101010010102101102111T10
 252 0200011000011010102121101111102111110100001120000101011012100110
 315 1001011111000011010211010127121021010000111110001100100110010100
 378 210101111110211111110201011101021101011000102110110111T202201101
 441 0011211110121001111102111112200001002000010111000100111122101111
 504 0200011T200110121121110001001111110001011011021210121010010010
 567 1001110101020210000112101101010111111010111022001100001001211
 630 2101201120001111100120001103201101010010121110100111100T0201100
 693 00110111100001001112211111120001012T2011011110101001010101111111
 756 0000011010011010101101001101021111022000110000110101201210 0111
 819 10011211120001010100100101111002013210011110021001101211101210
 882 21210011110010111202021100001101211210102010021111112002011 01
 945 201111011012101110102101110201001101001101210020120112T20111112
1008 0000011200031217002101001111021111111000000100221101121010 0110
1071 12001111000000111000110101020020010100111100000103101001121211
1134 010102111100111210002101100101101011210110101001011100200011 01
1197 0001210120120001102001111121002011000011010300010101101102111111
1260 200021300020110001101010011010211111000101210011011001210 0111
1323 1101113T2000000130002010101101210211112021111102000010122110101
1386 210100112100101110020101100111010100011302111100111111002011100
1449 100121217210122111001011010200031000011210101021011002011111 1
1512 202001101001001010212100110010110011110010020 03100001101011202100010
1575 1201111111001201000211010112121021121001111022021110100001010
1638 0101021111011111102000011120011012111101020011201111110101011 111
1701 02110101101001001101011011T200000120212121001000212121001 0110
1764 00000110201110101000121001110121113102002011010011020110101 01010
1827 1001111110020011010011110320021001111102111100012011000011011 11
1890 2301201110001111100000011010021010210100110000011111022000101
1953 02112101001012211010001111220210112000112121 00T2102111120111130
```

Sums of Two Rational Cubes up to 2016

In this manner, it is easy to construct the above table, whose validity does not depend on BSD since in the range of the table we have curves only of analytic rank 0 or 1, or curves with proved rank 2 or 3. The entry in row numbered R and column C (with R going from 0 to 1953 and C going from 1 to 63) gives the rank of the curve E_c with $c = R + C$, except when the rank is 0 and the curve has nontrivial torsion (which occurs if and only if c is equal to a cube or twice a cube), in which case the letter T occurs, corresponding to the trivial solutions $0^3 + 1^3 = 1$ and $1^3 + 1^3 = 2$, and their multiples.

To find explicitly the decomposition of a cubefree integer $c \leqslant 2016$ as a sum of two cubes the reader should proceed as follows. If the above table indicates 0, there are no solutions. If it indicates T, there is only one solution, either $c = x^3$ if c is a cube, or $c = x^3 + x^3$ if c is twice a cube. If it indicates 2 or 3, then for all $c \leqslant 2016$ the mwrank program or the 2-descent methods of Chapter 8 will succeed in finding a nontrivial point on the elliptic curve $y^2 = x^3 + 16c^2$, which can then be transformed into a solution of $x^3 + y^3 = c$ thanks to Proposition 7.2.3 and the remarks following it. Finally, if it indicates 1, then either mwrank or 2-descent will find a nontrivial rational point, or we apply the Heegner point method described in Section 8.6.

We now consider the case of three cubes. The main conjecture, now widely believed to be true, is the following.

Conjecture 6.4.24. *An integer n is a sum of three cubes of integers if and only if $n \not\equiv \pm 4 \pmod 9$.*

Note that since the cube of an integer is congruent to 0 or ± 1 modulo 9, an integer $n \equiv \pm 4 \pmod 9$ cannot be the sum of three cubes. The conjecture claims that this is the only restriction.

It is also conjectured that any integer not congruent to ± 4 modulo 9 is a sum of three cubes of integers in an infinite number of ways. However, to the author's knowledge the only known representations of the integer 3 are $3 = 1^3 + 1^3 + 1^3 = 4^3 + 4^3 + (-5)^3$ (and permutations of the latter).

A large amount of computer work has been done on these conjectures. For a long time the smallest positive integer $n \not\equiv \pm 4 \pmod 9$ not known to be a sum of three cubes was $n = 30$, until the discovery by M. Beck, E. Pine, W. Tarrant, and K. Yarbrough of the decomposition

$$30 = (-283059965)^3 + (-2218888517)^3 + (2220422932)^3 .$$

As of 2007 the only integers n such that $0 \leqslant n \leqslant 100$ and $n \not\equiv \pm 4 \pmod 9$ that are not known to be a sum of three cubes are $n = 33, 42, 52$, and 74. The size of the solutions found suggests that they are at least exponential in n.

Note that, contrary to similar results for squares, we must not assume that n is cubefree. To take a simple example, 5 is evidently not a sum of three cubes, but $135 = 3^3 \cdot 5 = 2^3 + (-6)^3 + 7^3$ is one.

Indeed, as for two cubes, the situation changes dramatically for the representation with *rational* numbers:

Proposition 6.4.25. *Every integer (and in fact every rational number) is the sum of three cubes of rational numbers. For instance, we have $n = x^3 + y^3 + z^3$, with*

$$x = m - 1, \ y = \frac{3(m^2 + m)}{m^2 + m + 1}, \ z = \frac{-m^3 + 3m + 1}{m^2 + m + 1},$$

where we have set $m = n/9$.

Proof. Just check. Of course this does not explain how to obtain such identities or why they exist. □

We now consider the case of four or more cubes. Because of the above proposition we no longer need to consider representations as sums of cubes of rational numbers. For integers there is a conjecture (a weak version and a strong version) and two results.

Conjecture 6.4.26. (1) (*Weak version.*) *Every integer is a sum of four cubes of integers.*
(2) (*Strong version.*) *Every integer has the form $2x^3 + y^3 + z^3$, hence is a sum of four cubes of integers of which two at least are equal.*

The two known results on this subject are as follows. The first is very easy, and the second is due to Dem'yanenko [Dem1].

Proposition 6.4.27. *Every integer is a sum of five cubes of integers, where we can in fact assume that at least two are equal. In other words, every integer has the form $2x^3 + y^3 + z^3 + t^3$.*

Proof. The identity $6x = (x - 1)^3 + (-x)^3 + (-x)^3 + (x + 1)^3$ shows that every multiple of 6 is a sum of four cubes of which two are equal. If n is an integer, $n - n^3$ is divisible by 6 hence is a sum of four cubes, so $n = (n - n^3) + n^3$ is a sum of five cubes of which two are equal. □

Theorem 6.4.28 (Dem'yanenko). *Every integer n such that $n \not\equiv \pm 4$ (mod 9) is a sum of four cubes of integers.*

Note that this theorem was certainly conjectured as far back as the end of the nineteenth century, but was proved only in 1966 by Dem'yanenko; see [Dem1].

Proof. First, the polynomial identities

$$6x = (x - 1)^3 + (-x)^3 + (-x)^3 + (x + 1)^3 \text{ and}$$
$$6x + 3 = x^3 + (-x + 4)^3 + (2x - 5)^3 + (-2x + 4)^3$$

show that every multiple of 3 is a sum of four cubes. Next, the identities

$$18x + 1 = (2x + 14)^3 + (-2x - 23)^3 + (-3x - 26)^3 + (3x + 30)^3 \,,$$
$$18x + 7 = (x + 2)^3 + (6x - 1)^3 + (8x - 2)^3 + (-9x + 2)^3 \,,$$
$$18x + 8 = (x - 5)^3 + (-x + 14)^3 + (-3x + 29)^3 + (3x - 30)^3 \,,$$

together with the complementary identities obtained by changing x into $-x$ and multiplying by -1 show that every $n \equiv \pm 1, \pm 7$, or ± 8 modulo 18 is a sum of four cubes. The only remaining congruence classes are $n \equiv \pm 2 \pmod{18}$. Finally, the polynomial identities

$$54x + 2 = (29484x^2 + 2211x + 43)^3 + (-29484x^2 - 2157x - 41)^3$$
$$+ (9828x^2 + 485x + 4)^3 + (-9828x^2 - 971x - 22)^3$$
$$54x + 20 = (3x - 11)^3 + (-3x + 10)^3 + (x + 2)^3 + (-x + 7)^3$$
$$216x - 16 = (-27x + 13)^3 + (24x - 12)^3 + (18x - 8)^3 + (3x - 3)^2 \text{ and}$$
$$216x + 92 = (3x - 164)^3 + (-3x + 160)^3 + (x - 35)^3 + (-x + 71)^3$$

together with their complementary identities leave only the congruence classes $n \equiv \pm 38 \pmod{108}$, so let $n \in \mathbb{Z}$ be of this form. Changing if necessary n into $-n$, we may assume that $n \equiv 38 \pmod{108}$.

In the sequel, denote by p the prime number $p = 83$. Assume first that $p \mid n$. Then $n/p \equiv 38p^{-1} \equiv 46 \pmod{108}$, and the identity

$$83(108x + 46) = (29484x^2 + 25143x + 5371)^3 + (-29484x^2 - 25089x - 5348)^3$$
$$+ (9828x^2 + 8129x + 1682)^3 + (-9828x^2 - 8615x - 1889)^3$$

shows that n is a sum of four cubes. We may thus assume that $n \equiv 38 \pmod{108}$ with $p \nmid n$.

Let a, b, m be integers and set

$$w = -(24m - 25a + 2937b), \quad x = 27m - 19a + 2746b,$$
$$y = -(19m + 9a + 602b), \quad z = 10m + 27a - 928b \,.$$

We check that $w^3 + x^3 + y^3 + z^3 = 18p(a^2 - 3420b^2)m + P(a, b)$, with

$$P(a, b) = (25a - 2937b)^3 + (-19a + 2746b)^3 + (-9a - 602b)^3 + (27a - 928b)^3 \,.$$

Thus, if a and b are chosen as solutions of the Pell equation $a^2 - 3420b^2 = 1$ we have $w^3 + x^3 + y^3 + z^3 = 18pm + P(a, b)$; hence, given such a pair (a, b), the equation $w^3 + x^3 + y^3 + z^3 = n$ is solvable in m if and only if $P(a, b) \equiv n$ is solvable modulo 2, 9, and p. Since $n \equiv 2 \pmod{18}$, the condition modulo 2 is $2 \mid b$, and the condition modulo 9 is easily seen to be $a \equiv 1 \pmod 3$ and $b \equiv 2 \mod 3$. Thus the condition modulo 18 is $a \equiv 1 \pmod 3$ and $b \equiv 2 \pmod 6$. There remains the condition modulo p.

The fundamental unit of the order $\mathbb{Z}[\sqrt{3420}]$ is easily computed to be $\varepsilon = 3041 + 52\sqrt{3420}$, which has norm 1. Thus $a + b\sqrt{3420} = s\varepsilon^k$ for any $k \in \mathbb{Z}$ and $s = \pm 1$, in other words

$$a = s\frac{\varepsilon^k + \bar{\varepsilon}^k}{2} \quad \text{and} \quad b = s\frac{\varepsilon^k - \bar{\varepsilon}^k}{2\sqrt{3420}} \, .$$

Since a and b satisfy a second-order linear recurrence, it is immediately seen that the congruence conditions modulo 3 and 6 on a and b are equivalent to $a + b\sqrt{3420} = (-\varepsilon)^k$ with $k \equiv 1 \pmod 3$. Thus if we set $\eta = -\varepsilon^3$ and $j = (k-1)/3$ we have

$$a = \frac{(-\varepsilon)\eta^j + \overline{(-\varepsilon)}\bar{\eta}^j}{2} \quad \text{and} \quad b = \frac{(-\varepsilon)\eta^j - \overline{(-\varepsilon)}\bar{\eta}^j}{2\sqrt{3420}} \, .$$

We note that $3420 \equiv 10^2 \pmod p$, hence that $\varepsilon \equiv 75 \pmod p$, $\bar{\varepsilon} \equiv 31 \pmod p$, $\eta \equiv 14 \pmod p$, and $\bar{\eta} \equiv 6 \pmod p$, so that

$$a \equiv \frac{8 \cdot 14^j - 31 \cdot 6^j}{2} \pmod p \quad \text{and} \quad b \equiv \frac{8 \cdot 14^j + 31 \cdot 6^j}{2 \cdot 10} \pmod p \, ,$$

and replacing gives finally

$$P(a,b) \equiv 71 \cdot 50^j \pmod p \, .$$

Now it is immediately checked that 50 is a primitive root modulo p. Since $\gcd(71, p) = 1$ it follows that for any n such that $p \nmid n$ there exists $j \in \mathbb{Z}$ with $71 \cdot 50^j \equiv n \pmod p$; hence there exist a and b such that $P(a,b) \equiv n \pmod p$, proving that the condition modulo p can be satisfied and finishing the proof of the theorem. $\square$

Remarks. (1) The above proof is translated essentially verbatim from Dem'yanenko's paper, but with a few minor improvements. First it is not at all clear from this proof how one obtains the second-degree polynomial identities and why we define w, x, y, and z as above (for instance, why use the identity $(-24)^3 + 27^3 + (-19)^3 + 10^3 = 0$ among so many similar ones?). This has been explained by M. Watkins in an unpublished manuscript. Furthermore, Dem'yanenko chooses the prime $p = 3323$, but Watkins shows that one can use the smaller prime $p = 83$, so I have used this p instead. Finally, the identity for $216x - 16$ was sent to me by D. Alpern; it can be obtained very simply from the identity for $18x + 7$ by a linear change of variable. It replaces a more complicated identity involving quadratic polynomials analogous to the one for $54x + 2$ found by Dem'yanenko.

(2) I would like to emphasize that the above proof gives a covering set of 83 identities for all integers $n \equiv 38 \pmod{108}$, one for $n = 83(108x + 46) = 108(83x + 35) + 38$ involving quadratic polynomials, and the 82 others involving linear polynomials. For instance, the choice $a + b\sqrt{3420} = -\varepsilon$ and a linear change of variable leads to the identity

$$108(83x - 2) + 38 = (-144x - 213614285)^3 + (162x + 240317344)^3$$
$$+ (-114x - 169113356)^3 + (60x + 89004059)^3 \,,$$

and the choice $a + b\sqrt{3420} = (-\varepsilon)^{-2}$ leads to a similar identity for $108(83x + 16) + 38$.

(3) The use of polynomial identities is essential in the above proof. It is natural to ask whether it is possible to use such identities also for integers congruent to ± 4 modulo 9 (with a left-hand side that is *linear* in x). It has been proved by Schinzel, Mordell, and successors that such an identity does not exist with polynomials of degree less than or equal to 7. It is reasonable to conjecture that no such identity exists.

To conclude this section, we note the following related result.

Proposition 6.4.29. *Up to permutation of the variables the equation* $w^3 + x^3 + y^3 + z^3 = 0$ *in* $\mathbb{Q}$ *has the trivial parametrization* $x = -w$, $z = -y$, *and the parametrization*

$$w = -d((s - 3t)(s^2 + 3t^2) + 1) \,,$$
$$x = d((s + 3t)(s^2 + 3t^2) + 1) \,,$$
$$y = -d((s^2 + 3t^2)^2 + (s + 3t)) \,,$$
$$z = d((s^2 + 3t^2)^2 + (s - 3t)) \,,$$

with d, s, *and* t *in* $\mathbb{Q}$.

Proof. If we set $W = (w + x)/2$, $X = (x - w)/2$, $Y = (y + z)/2$, and $Z = (z - y)/2$ the equation is equivalent to $W(W^2 + 3X^2) = -Y(Y^2 + 3Z^2)$. Excluding the trivial parametrization we have $W \neq 0$ and $Y \neq 0$, so that if we define s and t by $(Y + Z\sqrt{-3})/(W + X\sqrt{-3}) = s + t\sqrt{-3}$, we have by definition $sW - 3tX = Y$ and $sX + tW = Z$, and the equation is equivalent to $W = -Y(s^2 + 3t^2)$. Thus $-3tX = Y(1 + s(s^2 + 3t^2))$, or equivalently, $X = d(1 + s(s^2 + 3t^2))$ and $Y = -3dt$; hence $W = 3dt(s^2 + 3t^2)$ and $Z = d(s + (s^2 + 3t^2)^2)$, so we obtain the given parametrization.   □

When s, t, and d are integers the solution is trivially integral, but the converse is not true (choose for instance $d = -361/42$, $s = -10/19$, $t = -7/19$, which gives one of the smallest nontrivial integral solutions $(w, x, y, z) = (12, 1, -10, -9)$). No complete parametrization of the equation in integers is known, but nontrivial partial ones are easy to find; see Exercise 20. Note that Elkies gives the following homogeneous rational parametrization:

$$w = d(-(r+s)t^2 + (s^2 + 2r^2)t - s^3 + rs^2 - 2r^2s - r^3) \,,$$
$$x = d(t^3 - (r+s)t^2 + (s^2 + 2r^2)t + rs^2 - 2r^2s + r^3) \,,$$
$$y = d(-t^3 + (r+s)t^2 - (s^2 + 2r^2)t + 2rs^2 - r^2s + 2r^3) \,,$$
$$z = d((s - 2r)t^2 + (r^2 - s^2)t + s^3 - rs^2 + 2r^2s - 2r^3) \,.$$

Here $d = 1/7$, $r = 1$, $s = -4$, and $t = -2$ give the solution $(9, -1, 10, -12)$.

6.4.7 Skolem's Equations $x^3 + dy^3 = 1$

The aim of this section is to prove the following theorem.

Theorem 6.4.30 (Skolem). *If $d \in \mathbb{Z}$ is given with $d \neq 0$, there exists at most one pair $(x, y) \in \mathbb{Z} \times \mathbb{Z}$ with $y \neq 0$ such that $x^3 + dy^3 = 1$.*

Proof. If $d = c^3$ with $c \in \mathbb{Z}$, then $(x+cy) \mid (x^3+dy^3) = 1$, so $x+cy = \pm 1$. Replacing in the equation gives $\pm 1 - 3cy \pm 3c^2y^2 = 1$, so looking modulo 3 we have $\pm = +$, hence $cy(cy-1) = 0$. Thus since we assume $d \neq 0$ and $y \neq 0$, we obtain $y = 1/c$ as the only possible solution (if $d = \pm 1$), otherwise none. Thus, assume that d is not a perfect cube, and let $K = \mathbb{Q}(\theta)$ with $\theta = d^{1/3}$ be the corresponding pure cubic field. In particular, its signature is $(1, 1)$, so by Dirichlet's unit theorem there exists a fundamental unit ε such that any unit has the form $\pm\varepsilon^k$ for $k \in \mathbb{Z}$. Changing if necessary ε into $-\varepsilon$ we may assume that ε has norm $+1$.

Assume now that there exist two solutions (x_1, y_1) and (x_2, y_2) to our equation with $y_1 \neq 0$ and $y_2 \neq 0$, and set $\varepsilon_i = x_i + y_i\theta$ for $i = 1$ and 2. Our equation is equivalent to $\mathcal{N}_{K/\mathbb{Q}}(\varepsilon_i) = 1$; hence since the ε_i are algebraic integers, they are units in K of norm 1, so that $\varepsilon_i = \varepsilon^{p_i}$ for some $p_i \in \mathbb{Z}$. Writing $p_1/p_2 = n_1/n_2$ with $\gcd(n_1, n_2) = 1$, we thus have $(x_1 + y_1\theta)^{n_2} = (x_2+y_2\theta)^{n_1}$, and if necessary exchanging (x_1, y_1) and (x_2, y_2) we may assume that $3 \nmid n_1$. Thus $N = n_2/n_1$ can be considered as an element of $\mathbb{Z}_3$. To simplify, write $x = x_1$ and $y = y_1$. By definition of θ, we have

$$(x + y\theta)^3 = x^3 + 3x^2y\theta + 3xy^2\theta^2 + dy^3 = 1 + 3xyG \quad \text{with} \quad G = x\theta + y\theta^2 \,.$$

We will work in $\mathbb{Q}_3(\theta)$, which by Proposition 4.4.41 is isomorphic to the direct sum of the completions of $\mathbb{Q}(\theta)$ for the absolute values corresponding to the prime ideals of $\mathbb{Z}_K$ dividing 3. Write $N = 3M + r$ with $0 \leqslant r \leqslant 2$. Since $N \in \mathbb{Z}_3$, by Corollary 4.2.15 we can thus write $x_2 + y_2\theta = (1 + 3xyG)^M(x + y\theta)^r$, where $(1 + 3xyG)^M$ is *defined* by a convergent binomial series. Thus, using Corollary 4.2.17 (which was proved for a $\mathfrak{p}$-adic field but is clearly still true for a product of such), we have

$$x_2+y_2\theta = (1+3xyG)^M(x+y\theta)^r = (1+3Mxy(x\theta+y\theta^2)+9Mx^2y^2B)(x+y\theta)^r$$

for some $B \in \mathbb{Z}_3[\theta]$. Note that even though $\mathbb{Q}_3(\theta)$ is not in general a field, 1, θ, and θ^2 are still $\mathbb{Q}_3$-linearly independent, so that we may identify the coefficients of powers of θ in the above formula. In particular, if we write $B = B_0 + B_1\theta + B_2\theta^2$ with $B_i \in \mathbb{Z}_3$ and identify the coefficients of θ^2 we obtain

$$0 = \begin{cases} 3Mxy^2(1 + 3xB_2) & \text{for } r = 0 , \\ 3Mx^2y^2(2 + 3(yB_1 + xB_2)) & \text{for } r = 1 , \\ y^2(1 + 9Mx^2(x + B_2x^2 + 2B_1xy + B_0y^2)) & \text{for } r = 2 . \end{cases}$$

Since x and y are nonzero, and N is not equal to 0 or 1, we can divide respectively by $3Mxy^2$, $3Mx^2y^2$, and y^2, and we immediately obtain a contradiction modulo 3.   □

6.4.8 Special Cases of Skolem's Equations

Since we can replace y^3 by c^3y^3 for any $c \in \mathbb{Z}$, in Skolem's theorem we may assume that d is a positive cubefree integer.

Corollary 6.4.31. *For $d = 1, 2, 7, 9, 17, 19, 20, 26, 28, 37, 43, 63, 65$, and 91 the equation $x^3 + dy^3 = 1$ has a (necessarily unique) integral solution with $y \neq 0$, given respectively by $(x, y) = (0, 1), (-1, 1), (2, -1), (-2, 1), (18, -7),$ $(-8, 3), (-19, 7), (3, -1), (-3, 1), (10, -3), (-7, 2), (4, -1), (-4, 1)$, and $(9, -2)$.*

Proof. Clear by direct check, the uniqueness coming from Skolem's theorem.   □

Corollary 6.4.32. *The only integral solutions to the equation $y^2 = x^3 + 1$ are $(x, y) = (-1, 0), (0, \pm 1)$, and $(2, \pm 3)$.*

Proof. We rewrite the equation as $(y - 1)(y + 1) = x^3$. If y is even, $y - 1$ and $y + 1$ are coprime, hence are both cubes, and since the only two cubes that differ by 2 are -1 and 1 we deduce that $(x, y) = (-1, 0)$. Otherwise y is odd, so x is even. Changing if necessary y into $-y$ we may assume that $y \equiv 1$ (mod 4). Thus $((y - 1)/4)((y + 1)/2) = (x/2)^3$; hence there exist integers a and b such that $y - 1 = 4a^3$ and $y + 1 = 2b^3$, so that $b^3 - 2a^3 = 1$. By Skolem's theorem above, the only solutions to this equation are $(a, b) = (0, 1)$ and $(-1, -1)$, giving the solutions $(x, y) = (0, \pm 1)$ and $(2, \pm 3)$.   □

The following result can be shown using only slightly more complicated methods.

Theorem 6.4.33 (Delone). *Let d be a positive cubefree integer. The equation $x^3 + dy^3 = 1$ has a nontrivial integral solution if and only if the fundamental unit ε of the ring $\mathbb{Z}[d^{1/3}]$ (which is not necessarily equal to the full ring of integers of $\mathbb{Q}(d^{1/3})$) such that $0 < \varepsilon < 1$ has the form $x + yd^{1/3}$ with x and y in $\mathbb{Z}$.*

Using this theorem, it is immediate to check that the values of d given in Corollary 6.4.31 are the *only* positive cubefree values of d less than or equal to 100 for which Skolem's equation has a solution with $y \neq 0$. The only additional such values for $d \leqslant 1000$ are $d = 124, 126, 182, 215, 217, 254, 342, 422, 511, 614, 635, 651, 730,$ and 813. Let us show how we can prove that there are no nontrivial solutions in the particular case $d = 11$.

Proposition 6.4.34. *The only integral solution to $x^3 + 11y^3 = 1$ is the trivial solution $(x, y) = (1, 0)$.*

Proof. We work in the number field $\mathbb{Q}(\theta)$ with $\theta = 11^{1/3}$. A fundamental unit is $\varepsilon = 1 + 4\theta - 2\theta^2$, with $\mathcal{N}_{K/\mathbb{Q}}(\varepsilon) = 1$. Since our Diophantine equation is equivalent to $\mathcal{N}_{K/\mathbb{Q}}(x + y\theta) = 1$, Dirichlet's unit theorem tells us that $x + y\theta = \varepsilon^n$ for some $n \in \mathbb{Z}$.

The smallest prime p in which $X^3 - 11$ splits completely is $p = 19$, sox we work in $\mathbb{Q}_{19}$, in which the three roots are $c_1 \equiv -3 + 5 \cdot 19 \pmod{19^2}$, $c_2 \equiv -2 + 8 \cdot 19 \pmod{19^2}$, and $c_3 \equiv 5 + 6 \cdot 19 \pmod{19^2}$. The corresponding values of the embeddings of ε are $e_1 \equiv 9 + 2 \cdot 19 \pmod{19^2}$, $e_2 \equiv 4 \pmod{19^2}$, $e_3 \equiv 9 + 16 \cdot 19 \pmod{19^2}$, and for $j = 1, 2,$ and 3 we have $x + yc_j = e_j^n$. Since $\mathrm{Tr}_{K/\mathbb{Q}}(\theta) = \mathrm{Tr}_{K/\mathbb{Q}}(\theta^2) = 0$, we have $\sum_j c_j = \sum_j c_j^2 = 0$, so that $c_1 e_1^n + c_2 e_2^n + c_3 e_3^n = 0$. On the other hand, since $\mathcal{N}_{K/\mathbb{Q}}(\varepsilon) = 1$ we have $e_1 e_2 e_3 = 1$, so replacing e_1 by $(e_2 e_3)^{-1}$ and multiplying by $(e_2 e_3)^n$ we obtain

$$c_1 + c_2 e_2^{2n} e_3^n + c_3 e_2^n e_3^{2n} = 0.$$

We first consider this equation modulo 19. We obtain $16 + 17 \cdot 11^n + 5 \equiv 0 \pmod{19}$, in other words $11^n \equiv 1 \pmod{19}$, or equivalently, since the order of 11 modulo 19 is equal to 3, $n \equiv 0 \pmod 3$. Thus, we must have $n = 3m$ for some $m \in \mathbb{Z}$. But then we have $(e_2^2 e_3)^3 \equiv 1 + 7 \cdot 19 \pmod{19^2}$ and $(e_2 e_3^2)^3 \equiv 1 + 11 \cdot 19 \pmod{19^2}$. Thus, with the notation of Corollary 4.2.18, we have $(e_2^2 e_3)^n = \phi_a(m)$ and $(e_2 e_3^2)^n = \phi_b(m)$ for $a = e_2^2 e_3 - 1$ and $b = e_2 e_3^2 - 1$. We immediately see that $\phi_a(X) = 1 + 7 \cdot 19 X \pmod{19^2}$ and $\phi_b(X) = 1 + 11 \cdot 19 X \pmod{19^2}$, and since $c_1 + c_2 + c_3 = 0$, our equation has the form $\phi(m) = 0$ with $\phi(X) \equiv 3 \cdot 19 X \pmod{19^2}$. In the notation of Strassmann's theorem we thus have $N = 1$, so there exists only one solution $m = 0$ corresponding to $(x, y) = (1, 0)$, as claimed. $\qquad\square$

6.4.9 The Equations $y^2 = x^3 \pm 1$ in Rational Numbers

In Corollary 6.4.32 we have found all *integral* solutions to the equation $y^2 = x^3 + 1$. It is instructive to see how to find all *rational* solutions to this equation. The method that we use is not related to Skolem's, but is an example of a *descent* method that we will explore in more detail in Section 8.2. This proof is essentially due to L. Euler. I would like to thank B. de Weger and R. Schoof for showing me their versions, and the one below is a (slight) blend of the two. We slightly simplify Euler's argument by using the following lemma.

Lemma 6.4.35. *Let $K = \mathbb{Q}(\sqrt{-3})$ and $\alpha \in \mathbb{Z}_K$. Then $\alpha\bar{\alpha}$ is a square in $\mathbb{Z}$ if and only if there exist $n \in \mathbb{Z}$ and $\beta \in \mathbb{Z}_K$ such that $\alpha = n\beta^2$.*

Proof. Since $\mathbb{Z}_K$ is a principal ideal domain, simply decompose α into a product of a root of unity and a product of powers of prime elements of $\mathbb{Z}_K$. The details are left to the reader (Exercise 22). □

The key descent result of Euler is the following.

Proposition 6.4.36. *Let $\varepsilon = \pm 1$. The only nonzero integral solutions to the Diophantine equations $Y^2 = XZ(X^2 - 3\varepsilon XZ + 3Z^2)$ with $\gcd(X, Z) = 1$ are for $\varepsilon = 1$, with $(X, Z) = \pm(1, 1)$ (hence $Y = \pm 1$) or $\pm(3, 1)$ (hence $Y = \pm 3$).*

Proof. Since the discriminant of $X^2 - 3\varepsilon X + 3$ is negative it follows that $XZ > 0$, so X and Z have the same sign. Thus if necessary changing (X, Z) into $(-X, -Z)$ we may assume they are both positive.

Assume first that $3 \nmid X$, and consider a solution to our equation where $|Y| > 1$ is *minimal*. As always in descent arguments we are going to construct another solution with a strictly smaller value of $|Y|$, hence giving a contradiction. Thus X, Z, and $X^2 - 3\varepsilon XZ + 3Z^2$ are pairwise coprime, and since they are all positive they are all three squares, so we write $X = x^2$, $Z = z^2$, and $X^2 - 3\varepsilon XZ + 3Z^2 = a^2$, say. If we set $\alpha = X + Z(-3\varepsilon + \sqrt{-3})/2 \in \mathbb{Z}_K$, we see that $\alpha\bar{\alpha} = a^2$, so that by the above lemma we have $\alpha = n\beta^2$ with $n \in \mathbb{Z}$ and $\beta \in \mathbb{Z}_K$. Since $(1, (-3\varepsilon + \sqrt{-3})/2)$ is a $\mathbb{Z}$-basis of $\mathbb{Z}_K$, we write $\beta = u + v(-3\varepsilon + \sqrt{-3})/2$, and equating coefficients we obtain $X = n(u^2 - 3v^2)$, $Z = n(2uv - 3\varepsilon v^2)$. Since X and Z are coprime, it follows that $n = \pm 1$, that u and v are coprime, and $3 \nmid u$. Since X is a square and $3 \nmid u$ we have $X \equiv nu^2 \equiv n \pmod 3$, hence $n \equiv 1 \pmod 3$, so in fact $n = 1$. We thus obtain the system of equations $u^2 = x^2 + 3v^2$, $z^2 = v(2u - 3\varepsilon v)$. If $\alpha_1 = x + v\sqrt{-3}$ we have $\alpha_1\bar{\alpha_1} = u^2$, so again by the above lemma there exists $\beta_1 = (s + t\sqrt{-3})/2 \in \mathbb{Z}_K$ (thus, with $s \equiv t \pmod 2$) and $n_1 \in \mathbb{Z}$ such that $\alpha_1 = n_1\beta_1^2$, which gives by equating coefficients $x = n_1(s^2 - 3t^2)/4$, $v = n_1 st/2$, hence $u = n_1(s^2 + 3t^2)/4$. Replacing in the formula for z^2, we obtain $z^2 = (n_1/2)^2 st(s^2 - 3\varepsilon st + 3t^2)$. It follows that $Y_1 = z/(n_1/2)$ is an integer such that $Y_1^2 = st(s^2 - 3\varepsilon st + 3t^2)$, so we have obtained a new solution to our Diophantine equation. Evidently s and t are nonzero (otherwise v, hence Z, is zero). If $g = \gcd(s, t)$ (equal in fact to 1 or 2), replacing (s, t, Y_1) by $(s/g, t/g, Y_1/g^2)$ we may assume that s and t are coprime. Let us show that $|Y_1| < |Y|$. Indeed, we have

$$\frac{Y^2}{Y_1^2} = \frac{XZ(X^2 - 3\varepsilon XZ + 3Z^2)}{4z^2/n_1^2} \geqslant \frac{X(X^2 - 3\varepsilon XZ + 3Z^2)}{4}.$$

Now $X^2 - 3\varepsilon XZ + 3Z^2 \geqslant 7$ for $\varepsilon = -1$. For $\varepsilon = 1$, since $Z = z^2$ we have $X^2 - 3XZ + 3Z^2 = (X - 3Z/2)^2 + 3Z^2/4 \geqslant 3z^4/4 > 1$ for $|z| \geqslant 2$. For $\varepsilon = Z = 1$, since $X = x^2$ we have $X(X^2 - 3\varepsilon X + 3) \geqslant x^2(x^4 - 3x^2 + 3) > 4$

for $|x| > 1$. It follows from all this that $|Y| > |Y_1|$ unless $\varepsilon = X = Z = 1$. But in that case we have $|Y| = 1$, and since we have initially assumed that $|Y| > 1$, this gives the desired contradiction showing that when $3 \nmid X$ the only possible solution has $|Y| = 1$, which is indeed possible with $\varepsilon = 1$ and $X = Z = 1$, but not possible if $\varepsilon = -1$.

If $3 \mid X$ then $3 \mid Y$, so $(Y/3)^2 = Z(X/3)(Z^2 - 3\varepsilon Z(X/3) + 3(X/3)^2)$, and since $\gcd(X, Z) = 1$ we have $3 \nmid Z$, so by what we have just proved we have $\varepsilon = 1$, $(Z, X/3) = \pm(1, 1)$, hence $(X, Z) = \pm(3, 1)$.   □

Corollary 6.4.37. *The only* rational *solution to the equation* $y^2 = x^3 - 1$ *is* $(x, y) = (1, 0)$, *and the only rational solutions to the equation* $y^2 = x^3 + 1$ *are* $(x, y) = (-1, 0)$, $(0, \pm 1)$, *and* $(2, \pm 3)$.

As already mentioned we will later give a similar proof of this result (Proposition 8.2.14), this time using 2-descent explicitly.

Proof. Write $x = m/n$ with $\gcd(m, n) = 1$. Multiplying the equation $y^2 = x^3 + \varepsilon$ by n^4 we see that $n(m^3 + \varepsilon n^3)$ is a square, and if we set $c = m + \varepsilon n$ this means that $nc(m^2 - \varepsilon mn + n^2) = nc(c^2 - 3\varepsilon nc + 3n^2)$ is a square. Clearly $\gcd(c, n) = \gcd(m, n) = 1$, $n \neq 0$, and $c \neq 0$ except if $m = -\varepsilon n$, i.e., $x = -\varepsilon$. Thus by the above proposition we deduce that otherwise we have $\varepsilon = 1$, and $(c, n) = \pm(1, 1)$ or $\pm(3, 1)$, giving $x = 0$ or $x = 2$ respectively, and proving the corollary.   □

6.5 The Equations $ax^4 + by^4 + cz^2 = 0$ and $ax^6 + by^3 + cz^2 = 0$

Equations of the type $ax^p + by^q + cz^r = 0$ are called super-Fermat equations, and we will devote a special chapter to them (Chapter 14). Simple heuristic reasoning shows that if $1/p + 1/q + 1/r < 1$, we expect only a finite number of solutions up to a reasonable notion of equivalence, and if $1/p + 1/q + 1/r > 1$ we expect infinitely many solutions (see Chapter 14 for details). The intermediate case $1/p + 1/q + 1/r = 1$ reduces to the study of elliptic curves, and the existence or not of solutions essentially depends on the *rank* of the curve. It is clear that up to permutation of p, q, and r we have $(p, q, r) = (3, 3, 3)$, $(4, 4, 2)$, or $(6, 3, 2)$. We have studied in great detail the case $(p, q, r) = (3, 3, 3)$ in Section 6.4. It is thus natural to study the other two cases here, and in fact we are going to see that the $(4, 4, 2)$ case is very similar to the $(3, 3, 3)$ case, although the equation is not homogeneous.

6.5.1 The Equation $ax^4 + by^4 + cz^2 = 0$: Local Solubility

The question of local solubility is answered by the following proposition.

Proposition 6.5.1. *Let* a, b, *and* c *be nonzero integers such that* a *and* b *are 4th power-free and* c *is squarefree, and such that* $\gcd(a, b, c) = 1$.

(1) *The equation $ax^4 + by^4 + cz^2 = 0$ has a nontrivial solution in every $\mathbb{Q}_p$ for which $p \nmid 2abc$, and it has a nontrivial solution in $\mathbb{R}$ if and only if a, b, and c do not all have the same sign.*

(2) *Let $p \mid abc$, $p \neq 2$, reorder a and b so that $v_p(a) \leqslant v_p(b)$, and set $\mathbf{v} = (v_p(a), v_p(b), v_p(c))$.*

    (a) *If $\mathbf{v} = (2, 2, 0)$ the equation has a nontrivial solution in $\mathbb{Q}_p$.*

    (b) *If $\mathbf{v} = (0, 2, 1)$ or $\mathbf{v} = (1, 3, 0)$ the equation does not have any nontrivial solutions in $\mathbb{Q}_p$.*

    (c) *If $\mathbf{v} = (0, 0, 1)$, $\mathbf{v} = (1, 1, 0)$, or $\mathbf{v} = (3, 3, 0)$ the equation has a nontrivial solution in $\mathbb{Q}_p$ if and only if $-a/b$ is a fourth power in $\mathbb{F}_p^*$.*

    (d) *If $\mathbf{v} = (0, 2, 0)$ the equation has a nontrivial solution in $\mathbb{Q}_p$ if and only if either $-a/c$ or $-b/(p^2c)$ is a square in $\mathbb{F}_p^*$.*

    (e) *Otherwise, if $\mathbf{v} = (0, 1, 0)$ or $\mathbf{v} = (0, 3, 0)$, set $\alpha = -a/c$, if $\mathbf{v} = (0, 1, 1)$ set $\alpha = -b/c$, if $\mathbf{v} = (0, 3, 1)$ or $\mathbf{v} = (1, 2, 0)$ set $\alpha = -b/(p^2c)$, and if $\mathbf{v} = (2, 3, 0)$ set $\alpha = -a/(p^2c)$. The equation has a nontrivial solution in $\mathbb{Q}_p$ if and only if α is a square in $\mathbb{F}_p^*$.*

(3) *Assume that $p = 2$, reorder a and b so that $v_p(a) \leqslant v_p(b)$, and set $\mathbf{w} = (v_p(a), v_p(b), v_p(c))$. For $\mathbf{w} = (0, 3, 0)$, $(0, 3, 1)$, $(1, 1, 0)$, $(1, 2, 0)$, $(2, 2, 0)$, $(2, 3, 0)$, or $(3, 3, 0)$ replace (a, b, c) by $(b/8, 2a, 2c)$, $(b/8, 2a, c/2)$, $(a/2, b/2, 2c)$, $(a/2, b/2, 2c)$, $(a/4, b/4, c)$, $(a/4, b/4, c)$, or $(a/8, b/8, 2c)$ respectively, otherwise keep a, b, and c unchanged. Finally, set $\mathbf{v} = (v_p(a), v_p(b), v_p(c))$.*

    (a) *If $\mathbf{v} = (0, 2, 1)$ or $\mathbf{v} = (1, 3, 0)$ the equation does not have any nontrivial solutions in $\mathbb{Q}_2$.*

    (b) *If $\mathbf{v} = (0, 0, 0)$ the equation has a nontrivial solution in $\mathbb{Q}_2$ if and only if $8 \mid (a + c)$, $8 \mid (b + c)$, $16 \mid (a + b)$, or $16 \mid (a + b + 4c)$.*

    (c) *If $\mathbf{v} = (0, 0, 1)$ the equation has a nontrivial solution in $\mathbb{Q}_2$ if and only if $8 \mid (a + b)$ or $16 \mid (a + b + c)$.*

    (d) *If $\mathbf{v} = (0, 1, 0)$ the equation has a nontrivial solution in $\mathbb{Q}_2$ if and only if $8 \mid (a + c)$ or $8 \mid (a + b + c)$.*

    (e) *If $\mathbf{v} = (0, 2, 0)$ the equation has a nontrivial solution in $\mathbb{Q}_2$ if and only if $8 \mid (a + c)$, $8 \mid (a + b + c)$, or $16 \mid (b + 4c)$.*

    (f) *If $\mathbf{v} = (0, 1, 1)$ the equation has a nontrivial solution in $\mathbb{Q}_2$ if and only if $16 \mid (b + c)$.*

Proof. This follows from an easy but tedious case-by-case examination analogous to the proof of Theorem 6.4.2 and is left to the reader (Exercise 28). $\qquad\square$

In the special case of the equation $x^4 + y^4 = cz^2$ the above proposition simplifies considerably.

Corollary 6.5.2. *Assume that c is squarefree. The equation $x^4 + y^4 = cz^2$ is everywhere locally soluble if and only if $c > 0$ and the odd prime divisors of c are congruent to 1 modulo 8.*

Proof. Indeed, by the proposition the equation is locally soluble in $\mathbb{Q}_p$ for an odd prime divisor p of c if and only if -1 is a fourth power in $\mathbb{F}_p^*$, hence if $p \equiv 1 \pmod 8$, and it is locally soluble in $\mathbb{Q}_2$ if and only if $c \equiv 1 \pmod 8$ or $c \equiv 2 \pmod{16}$, which is automatically satisfied when the conditions at the odd primes are.   □

To check global solubility we have at least two methods: one using algebraic number theory, the other using elliptic curves. We begin with the first one, but we treat only special cases.

6.5.2 The Equations $x^4 \pm y^4 = z^2$ and $x^4 + 2y^4 = z^2$

The equation $x^4 + y^4 = z^2$ was solved by Fermat using the method of infinite descent, similar but simpler than the one we used in Theorem 6.4.17. The proof is also based on the parametric solution of a simpler equation, here of the Pythagorean equation $x^2 + y^2 = z^2$, which we have given in Corollary 6.3.13.

Proposition 6.5.3 (Fermat). *Let $\varepsilon = \pm 1$. The Diophantine equation $x^4 + \varepsilon y^4 = z^2$ has no solutions with $xyz \neq 0$.*

Proof. We may clearly assume that x, y, and z are pairwise coprime. Assume first that z is even. This can happen only if $\varepsilon = -1$, since otherwise we would get a contradiction modulo 8. Writing the equation as $y^4 + z^2 = x^4$ with y odd, by Corollary 6.3.13 we obtain $y^2 = s^2 - t^2$, $z = 2st$, and $x^2 = s^2 + t^2$ for some coprime s and t of opposite parity. It follows that $s^4 - t^4 = (xy)^2 = u^2$ with u odd, so we have reduced our equation to one in which the right-hand side is odd.

We may thus assume that z is odd. If $\varepsilon = 1$ we exchange x and y if necessary so that x is odd and y is even, while if $\varepsilon = -1$, reasoning modulo 4 we see that these conditions are automatic. By Corollary 6.3.13 there exist coprime s and t of opposite parity such that $x^2 = s^2 - \varepsilon t^2$, $y^2 = 2st$, and $z = \pm(s^2 + \varepsilon t^2)$ (the sign of x^2 for $\varepsilon = -1$ can be removed since $x^2 \geqslant 0$). Since $st \geqslant 0$, changing if necessary (s, t) into $(-s, -t)$ we may assume that $s \geqslant 0$ and $t \geqslant 0$. Exchanging if necessary s and t if $\varepsilon = -1$, we may assume that s is odd and t is even, this being automatic if $\varepsilon = 1$. Using once again Corollary 6.3.13 on the equation $x^2 = s^2 - \varepsilon t^2$, we deduce the existence of coprime u and v of opposite parity such that $x = \pm(u^2 - \varepsilon v^2)$, $s = \pm(u^2 + \varepsilon v^2)$, and $t = 2uv$, and since $t \geqslant 0$ we may assume $u \geqslant 0$ and $v \geqslant 0$. The last remaining equation to be solved is therefore $(y/2)^2 = \pm uv(u^2 + \varepsilon v^2)$, where the $\pm$ sign must be $+$ if $\varepsilon = 1$, and it can be removed by exchanging u and v otherwise. Since $\gcd(u, v) = 1$ the three factors on the right are clearly pairwise coprime and are nonnegative, so each one is a square. Thus, if $u = u_1^2$, $v = v_1^2$, and $u^2 + \varepsilon v^2 = w^2$ we have $u_1^4 + \varepsilon v_1^4 = w^2$. This is exactly our initial equation with new values of the variables. However, following through the reductions it is immediate that $|w| < |z|$ when $z \neq 0$. Thus, if we start with a solution with the smallest nonzero value of $|z|$ we obtain a strictly smaller one, a contradiction that shows that there cannot be any such solution. $\square$

Proposition 6.5.4. *The Diophantine equation* $x^4 + 2y^4 = z^2$ *has no solution with* $y \neq 0$.

Proof. Let (x, y, z) be integers such that $x^4 + 2y^4 = z^2$, where we may assume that x, y, and z are pairwise coprime. By Corollary 6.3.14 there exist coprime integers s and t with s odd such that $x^2 = \pm(s^2 - 2t^2)$ and $y^2 = 2st$. It follows that $s = \pm u^2$ and $t = \pm 2v^2$, so $x^2 = \pm(u^4 - 8v^4)$, and u is odd and coprime to v. If the sign were $-$, we would have $x^2 + u^4 = 8v^4$, and since u and x are odd, $x^2 + u^4 \equiv 2 \pmod 8$, a contradiction. Thus the sign is $+$, so that $x^2 + 8v^4 = u^4$, so again by Corollary 6.3.14 there exist coprime a and b such that $2v^2 = 2ab$ and $u^2 = a^2 + 2b^2$. It follows that $a = \pm c^2$, $b = \pm d^2$, so that $c^4 + 2d^4 = u^2$, which is our initial Diophantine equation, and we conclude by the usual descent argument since clearly $|u| < |z|$. $\square$

We will see below that the equation $x^4 - 2y^4 = z^2$ has nontrivial solutions (in fact infinitely many), for instance $(x, y, z) = (3, 2, 7)$.

6.5.3 The Equation $ax^4 + by^4 + cz^2 = 0$: Elliptic Curves

We will say that two nonzero rational solutions (x, y, z) and (x', y', z') of our equation are the same under *twisted projective equivalence* if there exists $\lambda \in \mathbb{Q}^*$ such that $x' = \lambda x$, $y' = \lambda y$, and $z' = \lambda^2 z$. It is easy to see that in each equivalence class there exists $(x, y, z) \in \mathbb{Z}^3$ such that $\gcd(x, y) = 1$, and (x, y, z) is unique up to sign.

In Definition 8.2.3, we will see that on the elliptic curve $Y^2 Z = X^3 + AX^2 Z + BXZ^2$ with 2-torsion point $T = (0, 0, 1)$, we can define (in projective coordinates) a 2-descent map α from $E(\mathbb{Q})$ to $\mathbb{Q}^*/\mathbb{Q}^{*2}$ by setting $\alpha(\mathcal{O}) = 1$, $\alpha(T) = B$, and for $P = (X, Y, Z)$ with $P \neq \mathcal{O}$ and $P \neq T$ by setting $\alpha(P) = X/Z$, modulo multiplication by squares of $\mathbb{Q}^*$. The fundamental property of α is that it induces a group homomorphism from $E(\mathbb{Q})/2E(\mathbb{Q})$ to $\mathbb{Q}^*/\mathbb{Q}^{*2}$ (see Proposition 8.2.4, which in fact gives a better result). The main result concerning global solvability is the following.

Proposition 6.5.5. *Let* E *be the elliptic curve with projective equation* $Y^2 Z = X^3 + abc^2 X Z^2$, *and define* $\phi(x, y, z) = (-bcxy^2, bc^2 yz, x^3)$.

(1) *The map* ϕ *sends the set of nonzero rational solutions of* $ax^4 + by^4 + cz^2 = 0$ *up to twisted projective equivalence into* $E(\mathbb{Q})$.
(2) *Let* S *be the set of* $(X, Y, Z) \in E(\mathbb{Q}) \setminus \{\mathcal{O}, T\}$ *such that there exists* $\lambda \in \mathbb{Q}^*$ *such that* $-bcX = Z\lambda^2$. *The image of* ϕ *is equal to* S, *together with* $\mathcal{O}$ *if* $-b/c$ *is a square, and with* T *if* $-a/c$ *is a square.*

More precisely, if (X, Y, Z) is such a point different from $\mathcal{O}$ and T then $(x, y, z) = (c\lambda Z, -cX, c\lambda Y Z)$ is a preimage of (X, Y, Z), $(0, 1, \beta)$ is a preimage of $\mathcal{O}$ if $-b/c = \beta^2$ is a square, and $(1, 0, \gamma)$ is a preimage of T if $-a/c = \gamma^2$ is a square.

(3) *The equation $ax^4 + by^4 + cz^2$ has nonzero solutions if and only if the class of $-b/c$ modulo squares belongs to the image of the 2-descent map α from $E(\mathbb{Q})$ to $\mathbb{Q}^*/\mathbb{Q}^{*2}$. In addition, it has an infinity of inequivalent solutions if and only if the class of $-b/c$ modulo squares is equal to $\alpha(G)$ for some nontorsion point $G \in E(\mathbb{Q})$.*

Proof. Setting $X = -bcxy^2$, $Y = bc^2yz$, and $Z = x^3$, it is clear that $P = \phi((x, y, z)) = (X, Y, Z) \in E(\mathbb{Q})$, and P is unchanged as a point in projective space if (x, y, z) is replaced by a twisted projectively equivalent solution. Conversely, if $P = (X, Y, Z) \in E(\mathbb{Q}) \setminus \{\mathcal{O}, T\}$ is such that $-bcX = Z\lambda^2$ for some $\lambda \in \mathbb{Q}^*$, then we check that $(x, y, z) = (c\lambda Z, -cX, c\lambda Y Z)$ is a solution to our equation that is a preimage of (X, Y, Z) by ϕ, and it is nonzero if and only if $X \neq 0$, in other words if and only if $P \neq \mathcal{O}$ and $P \neq T$. On the other hand, $P = \mathcal{O}$ corresponds to the solutions with $x = 0$, which exist if and only if $-b/c$ is a square, and $P = T$ corresponds to the solutions with $y = 0$, which exist if and only if $-a/c$ is a square, proving (1) and (2). Statement (3) is an immediate consequence since by definition $-b/c = \alpha(P)$ with $P = (X, Y, Z)$ if and only if $-bcX = Z\lambda^2$ for some $\lambda \in \mathbb{Q}^*$ when $P \neq \mathcal{O}$ and $P \neq T$, and $-b/c = \alpha(\mathcal{O})$ if and only if $-b/c$ is a square, and $-b/c = \alpha(T)$ if and only if $-a/c$ is a square. $\qquad\square$

Thus, after testing for everywhere local solubility, to check whether an equation $ax^4 + by^4 + cz^2 = 0$ has a nontrivial solution we proceed as follows. Using either `mwrank` or the 2-descent methods, we compute if possible the complete Mordell–Weil group $E(\mathbb{Q})$ of the curve with affine equation $Y^2 = X^3 + abc^2X$, together with its torsion subgroup (given by Proposition 8.1.14). If $G_1, \ldots, G_k$ are representatives of an $\mathbb{F}_2$-basis of $E(\mathbb{Q})/2E(\mathbb{Q})$, we then check whether the class of $-b/c$ modulo squares belongs to the group generated by the $\alpha(G_i) \in \mathbb{Q}^*/\mathbb{Q}^{*2}$, which is done using simple linear algebra over $\mathbb{F}_2$. It is also immediate to check in this way whether the equation has infinitely many solutions.

6.5.4 The Equation $ax^4 + by^4 + cz^2 = 0$: Special Cases

Several special cases of this equation are worth mentioning. In Section 6.12 we will define a *congruent number* as a rational number equal to the area of a Pythagorean triangle, i.e., a right triangle with rational sides, and we will prove the elementary fact that c is congruent if and only if the elliptic curve $y^2 = x(x^2 - c^2)$ has nonzero rank. The most important result concerning congruent numbers is Tunnell's Theorem 6.12.4 which allows us to check very rapidly whether or not c is congruent. Thus any result which can at least partly reduce to testing congruent numbers is very useful.

A first result of this kind is the following, which corresponds to the special case $a = 1$ and $b = -1$ (or, more generally, $b = -a$).

Proposition 6.5.6. *Let c be a nonzero integer. The equation $x^4 - y^4 = cz^2$ has a solution with $xyz \neq 0$ if and only if $|c|$ is a congruent number. More precisely, if $x^4 - y^4 = cz^2$ with $xyz \neq 0$ then $Y^2 = X(X^2 - c^2)$ with $(X, Y) = (-cy^2/x^2, c^2yz/x^3)$, and conversely if $Y^2 = X(X^2 - c^2)$ with $Y \neq 0$ then $x^4 - y^4 = cz^2$, with*

$$x = X^2 + 2cX - c^2, \quad y = X^2 - 2cX - c^2, \quad and \quad z = 4Y(X^2 + c^2).$$

Proof. These formulas can of course be checked directly, but are *not* inverse of each other. More precisely the formula for (X, Y) comes directly from Proposition 6.5.5, and the formula for (x, y, z) is the inverse of the formula giving *twice* the point (X, Y) plus the 2-torsion point $(0, 0)$. □

See Section 6.12.3 for a small table of congruent numbers.

A similar result, now valid in the special case $a = b = 1$ (or, more generally, $a = b$), gives a *necessary* condition for our equation to be globally soluble, which is usually also sufficient.

Corollary 6.5.7. *Assume that $c \in \mathbb{Z}_{\geq 3}$ is squarefree and that the equation $x^4 + y^4 = cz^2$ is everywhere locally soluble, in other words that the odd prime divisors of c are congruent to 1 modulo 8, and let E_c be the elliptic curve with affine equation $Y^2 = X^3 + c^2X$.*

(1) *The equation $x^4 + y^4 = cz^2$ has nonzero solutions if and only if the class of c modulo squares belongs to the image of the 2-descent map α, and in that case it has infinitely many inequivalent solutions.*

(2) *If $x^4 + y^4 = cz^2$ has a nonzero solution then $2c$ is a congruent number. More precisely, if $x^4 + y^4 = cz^2$ then $Y^2 = X(X^2 - 4c^2)$ with*

$$X = -\frac{4x^2y^2}{z^2} \quad and \quad Y = \frac{4xy(x^4 - y^4)}{z^3}.$$

Proof. It follows from Proposition 6.5.5 that the torsion points of E_c cannot come from nonzero solutions to our equation and that T is the only nontrivial torsion point, proving (1). Note that for $c = 2$ the torsion subgroup of E_c has order 4, generated by $P = (2, 4)$, which corresponds to the solutions $(\pm 1, \pm 1, \pm 1)$ to our equation, which are the only ones since E_c has rank 0. For (2) we note that by Proposition 8.2.1 the curve E_c is 2-isogenous to the curve $Y^2 = X^3 - 4c^2X$, which is the curve corresponding to the congruent number $n = 2c$ by Proposition 6.12.1, and the explicit formulas follow from Proposition 8.2.1. Using Proposition 6.12.1 one can also easily give the formulas for the sides of the corresponding right triangle; see Exercise 29, and see also Exercise 30. □

Remarks. (1) Since $\alpha(T) = c^2 = 1$ modulo squares in this case, it is not
necessary to include T in an $\mathbb{F}_2$-basis of $E(\mathbb{Q})/2E(\mathbb{Q})$ to see whether $-b/c$
belongs to the image of α.

(2) Since we assume that c is squarefree and that its odd divisors are congruent to 1 modulo 8, it is not difficult to show that the root number of
the curve E_c is $+1$, so that assuming the BSD conjecture, the rank of E_c
is always even.

(3) It is *not* true that when the local conditions are satisfied, the equation
$x^4 + y^4 = cz^2$ has nonzero solutions if and only if $2c$ is a congruent
number. The smallest counterexample is $c = 1513 = 17 \cdot 89$, which satisfies
the local conditions and is such that $2c$ is a congruent number, but for
which the above corollary enables us to prove that the equation $x^4 +
y^4 = cz^2$ does not have any nonzero solution. Nonetheless, thanks to
Tunnell's Theorem 6.12.4 we can very rapidly check that our equation
has no nonzero solution when $2c$ is not congruent, and if $2c$ is indeed
congruent, there is a good chance that it does have a solution.

In fact, thanks to these remarks and the above corollary, it is easy to
build the following table, which lists the squarefree values of $c \leqslant 10001$ such
that $x^4 + y^4 = cz^2$ has a nonzero solution (and these solutions can be given
explicitly). For $c = 1513, 2993, 4658, 4777, 7361, 8633, 9266,$ and 9881, which
are the other squarefree values of $c \leqslant 10001$ satisfying the local conditions
and such that $2c$ is a congruent number, the above corollary shows that
the equation $x^4 + y^4 = cz^2$ does *not* have any nonzero solution (I thank
J. Cremona for computing for me the Mordell–Weil group for $c = 2801$,
which I could not do with `mwrank` alone, but which *can* be done using 4-
descent, as programmed in the recent version of `magma`; see Exercise 31). For
completeness we include $c = 0, 1,$ and 2, but for $3 \leqslant c \leqslant 10001$ the rank of
the corresponding elliptic curve is always equal to 2.

0	1	2	17	82	97	113	193	257	274
337	433	514	577	593	626	641	673	706	881
914	929	1153	1217	1297	1409	1522	1777	1873	1889
1921	2129	2402	2417	2434	2482	2498	2642	2657	2753
2801	2833	2897	3026	3121	3137	3298	3329	3457	3649
3697	4001	4097	4129	4177	4226	4289	4481	4546	4561
4721	4817	4993	5281	5554	5617	5666	5729	5906	6002
6353	6449	6481	6497	6562	6577	6673	6817	6866	7057
7186	7489	7522	7537	7633	7762	8017	8081	8737	8753
8882	8962	9281	9298	9553	9586	9649	9778	9857	10001

Solubility of $x^4 + y^4 = cz^2$, c **squarefree up to** 10001

A final interesting special case is the following:

Corollary 6.5.8. (1) *For any $a \in \mathbb{Z} \setminus \{0\}$ the equation $x^4 + ay^4 = z^2$ has
solutions with $xy \neq 0$ if and only if the elliptic curve $Y^2 = X^3 + aX$*

has nonzero rank, and in that case it has infinitely many inequivalent solutions.

(2) Assume the BSD conjecture and that a is squarefree. Then if $a > 0$ and $a \equiv 3, 5, 13$, or 15 modulo 16, or if $a < 0$ and $a \equiv 1, 2, 6, 7, 9, 10, 11$, or 14 modulo 16, the equation $x^4 + ay^4 = z^2$ has infinitely many coprime solutions with $xy \neq 0$.

Note that this equation is everywhere locally soluble since it has the non-trivial global solution $(x, y, z) = (0, 1, 1)$.

Proof. (1). The solutions with $xy = 0$ correspond to the points $\mathcal{O}$ and T on the elliptic curve E. Thus by Proposition 6.5.5 our equation has solutions with $xy \neq 0$ if and only if there exists a nontorsion point G such that $\alpha(G) = 1$ modulo squares. If such solutions exist then E has nonzero rank, but conversely if H is a nontorsion point of $E(\mathbb{Q})$ then $G = 2H$ is a nontorsion point such that $\alpha(G) = \alpha(H)^2 = 1$ modulo squares, so the condition of the proposition is satisfied.

(2). Using an algorithm to compute root numbers, which we have not given, but which is included in packages such as Pari/GP and magma, it is easy to show that the root number of the elliptic curve E is -1 for the given values of a, and $+1$ for the others, so that assuming the BSD conjecture the rank is odd, hence nonzero when the sign is -1. □

6.5.5 The Equation $ax^6 + by^3 + cz^2 = 0$

We first note that this equation has the nontrivial global solution $(x, y, z) = (0, -bc, b^2c)$, so in particular it is everywhere locally soluble; see also Exercise 54. Here we will say that two nonzero rational solutions (x, y, z) and (x', y', z') are the same under twisted projective equivalence if there exists $\lambda \in \mathbb{Q}^*$ such that $x' = \lambda x$, $y' = \lambda^2 y$, and $z' = \lambda^3 z$. The result is as follows.

Proposition 6.5.9. Let E be the elliptic curve with projective equation $Y^2Z = X^3 - ab^2c^3Z^3$.

(1) The map $(x, y, z) \mapsto (X, Y, Z) = (-bcxy, bc^2z, x^3)$ is a one-to-one correspondence between the nonzero rational solutions of $ax^6 + by^3 + cz^2 = 0$ up to twisted projective equivalence and the points of $E(\mathbb{Q})$. More precisely $(x, y, z) = (bcZ, -bcXZ, b^2cYZ^2)$ is a preimage of $(X, Y, Z) \in E(\mathbb{Q}) \setminus \{\mathcal{O}\}$, and $(x, y, z) = (0, -bc, b^2c)$ is a preimage of $\mathcal{O}$.

(2) The equation $ax^6 + by^3 + cz^2 = 0$ has infinitely many solutions up to twisted projective equivalence if and only if E has nonzero rank.

(3) The equation $ax^6 + by^3 + cz^2 = 0$ always has solutions with $x = 0$, for instance $(0, -bc, b^2c)$. Otherwise, if E has rank zero it has solutions with $x \neq 0$ only in the following cases:

  (a) If $-a/b = m^3$ is a cube and not a sixth power the only solutions have $z = 0$, for instance $(1, m, 0)$.

(b) If $-a/c = m^2$ is a square and not a sixth power, the only solutions have $y = 0$, for instance $(1, 0, m)$.

(c) If $-ab^2c^3 = m^6$ is a sixth power, in addition to the solutions with y or z equal to 0, we have the two solutions $(bc, -2bcm^2, \pm 3b^2cm^3)$, which are unique up to twisted projective equivalence.

(d) If $ab^2c^3/432 = m^6$ is a sixth power, we have the two solutions $(bc, -12bcm^2, \pm 36b^2cm^3)$, which are unique up to twisted projective equivalence.

Proof. (1). Setting $X = -bcxy$, $Y = bc^2z$, and $Z = x^3$, it is clear that $P = (X, Y, Z) \in E(\mathbb{Q})$, and P is unchanged as a point in projective space if (x, y, z) is replaced by a twisted projectively equivalent solution. Conversely, if $P = (X, Y, Z) \in E(\mathbb{Q})$ then $(x, y, z) = (bcZ, -bcXZ, b^2cYZ^2)$ is a solution to our equation that is nonzero if and only if $Z \neq 0$, in other words if and only if $P \neq \mathcal{O}$, while $P = \mathcal{O}$ corresponds to the solutions with $x = 0$, for instance to $(x, y, z) = (0, -bc, b^2c)$.

(2) and (3). It follows that our equation has a solution with $x \neq 0$ if and only if either the curve E has nonzero rank, in which case it has infinitely many inequivalent solutions, or if E has nontrivial torsion. By Proposition 8.1.14, we have four cases:

- If $-a/b$ is a cube and not a sixth power we have $E_t(\mathbb{Q}) \simeq (\mathbb{Z}/2\mathbb{Z})$, so the only nontrivial torsion point corresponds to $Y = 0$, hence to $z = 0$.
- If a/c is a square and not a sixth power, we have $E_t(\mathbb{Q}) \simeq (\mathbb{Z}/3\mathbb{Z})$, and the nontrivial torsion points correspond to $X = 0$, hence to $xy = 0$.
- If $-ab^2c^3 = m^6$ is a sixth power, we have $E_t(\mathbb{Q}) \simeq (\mathbb{Z}/6\mathbb{Z})$, and the nontrivial torsion points are, in addition to the above three points, the two points $(X, Y, Z) = (2m^2, \pm 3m^3, 1)$, corresponding to the solutions given in the proposition.
- If $-ab^3c^3 = -432m^6$ we have $E_t(\mathbb{Q}) \simeq (\mathbb{Z}/3\mathbb{Z})$, and the nontrivial torsion points are the two points $(X, Y, Z) = (12m^2, \pm 36m^3, 1)$, corresponding to the solutions given in the proposition.

Otherwise E has no torsion, so our equation does not have any solutions with $x \neq 0$ if the rank of E is equal to 0. □

6.6 The Fermat Quartics $x^4 + y^4 = cz^4$

For a more detailed study of these Diophantine equations, in particular over number fields, I refer to [Cal].

An evident *necessary* condition for the existence of solutions to the equation of the title is the existence of solutions to the simpler equation $x^4 + y^4 = cZ^2$, which we have studied in detail in Section 6.5.3, so for a given value of c one should first consider the simpler equation. In particular, we know that $2c$ must be a congruent number. However, it is interesting to study the equation $x^4 + y^4 = cz^4$ independently.

We will denote by $\mathcal{C}_c$ the projective curve defined by the equation $x^4 + y^4 = cz^4$. We may clearly assume that c is 4th power-free, in other words that it is not divisible by a fourth power strictly larger than 1.

6.6.1 Local Solubility

We begin by studying local solubility over $\mathbb{Q}$, in order to give a necessary and sufficient condition for the equation to be everywhere locally soluble.

Proposition 6.6.1. *Assume that c is 4th power-free.*

(1) *We have $\mathcal{C}_c(\mathbb{Q}_2) \neq \emptyset$ if and only if $c \equiv 1$ or 2 modulo 16.*
(2) *If p is an odd prime divisor of c, then $\mathcal{C}_c(\mathbb{Q}_p) \neq \emptyset$ if and only if $p \equiv 1$ (mod 8).*
(3) *If $p \equiv 3$ (mod 4) is a prime not dividing c then $\mathcal{C}_c(\mathbb{Q}_p) \neq \emptyset$.*
(4) *If $p \geqslant 37$ is a prime not dividing c then $\mathcal{C}_c(\mathbb{Q}_p) \neq \emptyset$.*
(5) *We always have $\mathcal{C}_c(\mathbb{Q}_{17}) \neq \emptyset$.*
(6) *Let $p \in \{5, 13, 29\}$ be a prime not dividing c. Then*
    (a) *$\mathcal{C}_c(\mathbb{Q}_5) \neq \emptyset$ if and only if $c \not\equiv 3$ or 4 modulo 5.*
    (b) *$\mathcal{C}_c(\mathbb{Q}_{13}) \neq \emptyset$ if and only if $c \not\equiv 7$, 8, or 11 modulo 13.*
    (c) *$\mathcal{C}_c(\mathbb{Q}_{29}) \neq \emptyset$ if and only if $c \not\equiv 4$, 5, 6, 9, 13, 22, or 28 modulo 29.*

Proof. If $(x : y : z) \in \mathcal{C}_c(\mathbb{Q}_p)$, we may clearly assume that x, y, and z are p-adic integers and that at least one is a p-adic unit. If $p \nmid c$, reduction modulo p gives a projective curve $\overline{\mathcal{C}_c}$ over $\mathbb{F}_p$, which is smooth (nonsingular) if $p \neq 2$.

(1). Let u be a 2-adic unit. I claim that $u \in \mathbb{Q}_2^4$ if and only if $u \equiv 1$ (mod $16\mathbb{Z}_2$). Indeed, if v is a 2-adic unit we can write $v = 1 + 2t$ with $t \in \mathbb{Z}_2$, and

$$v^4 = 1 + 8t + 24t^2 + 32t^3 + 16t^4 \equiv 1 + 8(t(3t + 1)) \equiv 1 \pmod{16} \ .$$

Conversely, if $u \equiv 1$ (mod $16\mathbb{Z}_2$) we write $u = 1 + x$ with $v_2(x) \geqslant 4$, and it is easy to check that the binomial expansion for $(1 + x)^{1/4}$ converges for $v_2(x) \geqslant 4$. Alternatively, we set $f(X) = X^4 - u$ and use Hensel's lemma (Proposition 4.1.37): if $u \equiv 1$ (mod 32) we have $|f'(1)|_2 = 1/4$ and $|f(1)|_2 \leqslant 1/32 < |f'(1)|_2^2$, while if $u \equiv 17$ (mod 32) we have $|f'(5)|_2 = 1/4$ and $|f(5)|_2 \leqslant 1/32 < |f'(5)|_2^2$, proving my claim.

Now assume that $x^4 + y^4 = cz^4$. Since $v_2(c) \leqslant 3$, either x or y is a 2-adic unit. It follows that $x^4 + y^4 \equiv 1$ or 2 modulo 16; hence z is a 2-adic unit, so that $c \equiv 1$ or 2 modulo 16 as claimed. Conversely, if $c \equiv 1$ (mod 16) then $c = t^4$ by my claim above, so that $(t : 0 : 1) \in \mathcal{C}_c(\mathbb{Q}_2)$, while if $c \equiv 2$ (mod 16), then $c - 1 = t^4$ for some t, so that $(t : 1 : 1) \in \mathcal{C}_c(\mathbb{Q}_2)$, proving (1).

(2). Assume that $p \mid c$ is odd. Since $v_p(c) \leqslant 3$, x and y are p-adic units, so that -1 is a fourth power in $\mathbb{F}_p$. If g is a generator of the cyclic group $\mathbb{F}_p^*$, then $-1 = g^{(p-1)/2}$, so that -1 is a fourth power in $\mathbb{F}_p$ if and only if $p \equiv 1 \pmod 8$. If this is the case, let $x_0 \in \mathbb{Z}$ be such that $x_0^4 \equiv -1 \pmod p$. By Hensel's lemma (which is trivial here since the derivative of $X^4 + 1$ at x_0 is a p-adic unit), there exists $x \in \mathbb{Z}_p$ such that $x^4 = -1$, so that $(x : 1 : 0) \in \mathcal{C}_c(\mathbb{Q}_p)$, proving (2).

The following lemma shows that for the remaining p it is sufficient to consider the equation in $\mathbb{F}_p$.

Lemma 6.6.2. *Let $p \nmid 2c$ be a prime number. Then $\mathcal{C}_c(\mathbb{Q}_p) \neq \emptyset$ if and only if $\overline{\mathcal{C}_c}(\mathbb{F}_p) \neq \emptyset$. In particular, if $p \not\equiv 1 \pmod 8$ then*

$$\mathcal{C}_c(\mathbb{Q}_p) \neq \emptyset \quad \text{if and only if} \quad c \bmod p \in \mathbb{F}_p^4 + \mathbb{F}_p^4 .$$

Proof. One direction is clear. Conversely, assume that $\overline{\mathcal{C}_c}(\mathbb{F}_p) \neq \emptyset$, and let $(x_0 : y_0 : z_0)$ with x_0, y_0, and z_0 not all divisible by p such that $x_0^4 + y_0^4 \equiv cz_0^4$ $\pmod p$. Since $p \nmid c$, either $p \nmid x_0$ or $p \nmid y_0$. Assume for instance that $p \nmid x_0$, and set $f(X) = X^4 + y_0^4 - cz_0^4$. Clearly $|f'(x_0)|_p = 1$ and $|f(x_0)|_p < 1$, so that by Hensel's lemma there exists $t \in \mathbb{Q}_p$ such that $f(t) = 0$; hence $(t : y_0 : z_0) \in \mathcal{C}_c(\mathbb{Q}_p)$, proving the converse.

Finally, assume that $p \not\equiv 1 \pmod 8$. If $x^4 + y^4 \equiv cz^4 \pmod p$ with x or y not divisible by p, we cannot have $p \mid z$; otherwise, $x^4 \equiv -y^4 \pmod p$, so that -1 is a fourth power modulo p, a contradiction. Thus $p \nmid z$, so $(xz^{-1})^4 + (yz^{-1})^4 \equiv c \pmod p$, finishing the proof of the lemma. $\square$

(3). Let $p \nmid c$, $p \equiv 3 \pmod 4$. I claim that there exist x and y in $\mathbb{Z}$ such that $x^4 + y^4 \equiv c \pmod p$. Indeed, in a finite field $\mathbb{F}$ any element is a sum of two squares (in characteristic 2 any element is a square so the result is trivial; otherwise, if $q = |\mathbb{F}|$ then there are $(q+1)/2$ squares hence $(q+1)/2$ elements of the form $c - y^2$, so the two sets have a nonempty intersection; see Proposition 5.2.1). Thus there exist u and v such that $c \equiv u^2 + v^2 \pmod p$. However, when $p \equiv 3 \pmod 4$ we have $\mathbb{F}_p^{*2} = \mathbb{F}_p^{*4}$: indeed, we have a trivial inclusion, and the kernel of the map $x \mapsto x^4$ from $\mathbb{F}_p^*$ into itself is ± 1, so that $|\mathbb{F}_p^{*4}| = (p-1)/2 = |\mathbb{F}_p^{*2}|$, proving the equality. Thus $c = x^4 + y^4$, as claimed, and the above lemma proves (3).

(4). If $p \nmid 2c$ we may apply Corollary 2.5.23, which tells us in particular that $|\overline{\mathcal{C}_c}(\mathbb{F}_p)| \geqslant p + 1 - 6p^{1/2}$. This is strictly positive (for p prime) if and only if $p \geqslant 37$, so that (4) follows from the above lemma. Note that Corollary 2.5.23 is a special case of the Weil bounds, but in the present (diagonal) case we do not need these general bounds but only the case that we have proved.

(5) and (6). Thanks to the above cases, it remains to consider the primes p not dividing c such that $3 \leqslant p \leqslant 31$ and $p \equiv 1 \pmod 4$, in other words $p \in \{5, 13, 17, 29\}$. For such a p, -1 is a fourth power modulo p only for $p = 17$,

in which case as usual Hensel's lemma shows that there exists $t \in \mathbb{Q}_{17}$ such that $-1 = t^4$, proving (5). Otherwise, we compute that

$$\mathbb{F}_5^4 = \{0, 1\}, \quad \mathbb{F}_{13}^4 = \{0, 1, 3, 9\}, \quad \mathbb{F}_{29}^4 = \{0, 1, 7, 16, 20, 23, 24, 25\},$$

and we deduce the list of nonzero elements of $\mathbb{F}_p^4 + \mathbb{F}_p^4$, proving (6).   □

Note that it is easy to generalize this proposition to the more general equation $ax^4 + by^4 + cz^4 = 0$; see Exercise 52.

Corollary 6.6.3. *The curve $\mathcal{C}_c$ is everywhere locally soluble (i.e., has points in every completion of $\mathbb{Q}$) if and only if $c > 0$ and the following conditions are satisfied:*

(1) $c \equiv 1$ *or 2 modulo* 16.
(2) $p \mid c$, $p \neq 2$ *implies* $p \equiv 1 \pmod{8}$.
(3) $c \not\equiv 3$ *or 4 modulo* 5.
(4) $c \not\equiv 7, 8$, *or 11 modulo* 13.
(5) $c \not\equiv 4, 5, 6, 9, 13, 22$, *or 28 modulo* 29.

Proof. Clear.   □

Corollary 6.6.4. *For all primes p such that $p \equiv 1 \pmod{1160}$ the curve $\mathcal{C}_{p^2}$ is everywhere locally soluble, but is not globally soluble, so is a counterexample to the Hasse principle.*

Proof. It is clear that the above conditions are satisfied modulo 16, 5, and 29, and also modulo 13 since 7, 8, and 11 are quadratic nonresidues modulo 13. On the other hand, by Fermat's Proposition 6.5.3, the equation $x^4 + y^4 = Z^2$ does not have any nontrivial solutions, so this is in particular the case for our equation $x^4 + y^4 = (pz^2)^2$.   □

Since by Dirichlet's theorem on primes in arithmetic progressions (Theorem 10.5.30) there exist infinitely many primes $p \equiv 1 \pmod{1160}$ this corollary gives infinitely many counterexamples to the Hasse principle.

6.6.2 Global Solubility: Factoring over Number Fields

Now that we know necessary and sufficient conditions for our equation to be everywhere locally soluble, we begin the study of *sufficient* conditions for our equation to have no global solutions, since is does not seem reasonable to hope for necessary and sufficient ones. We will give two types of conditions. The first uses classical techniques of algebraic number theory and the other uses the theory of elliptic curves. Thus, for an integer $c \geqslant 1$ we consider the equation

$$x^4 + y^4 = cz^4 .$$

Without loss of generality we may assume that c is 4th power-free, so we may also assume that x, y, and z are pairwise coprime. The cases $c = 1$ and $c = 2$ (which are in fact the easiest) are treated in Section 6.5.2 and Exercise 24; hence we will assume that $c \geqslant 3$, so that in particular $xyz \neq 0$. Finally, we assume that our equation is everywhere locally soluble (otherwise there is nothing more to be done), in other words that the conditions of Corollary 6.6.3 are satisfied.

Since x and y are coprime one of them at least is odd, so by exchanging x and y if necessary, we can assume that x is odd. If necessary, by changing the signs of x and y we may assume that $x \equiv 1 \pmod 4$, and that either y is even or $y \equiv 1 \pmod 4$. In addition, since $c \equiv 1$ or 2 modulo 16 it follows that z is necessarily odd.

In this section, to study our equation we will *factor* it over the natural number field that occurs, which is here the field $K = \mathbb{Q}(\zeta)$, where $\zeta = \zeta_8$ is a primitive 8th root of unity (whose minimal polynomial is $P(X) = X^4 + 1$). Luckily, the ring of integers $\mathbb{Z}_K$ of K is as simple as can be desired: we have $\mathbb{Z}_K = \mathbb{Z}[\zeta]$, $\mathbb{Z}_K$ has class number 1, in other words is a principal ideal domain, and the group of units of $\mathbb{Z}_K$ is the group of elements of the form $\zeta^j \varepsilon^k$ with $0 \leqslant j \leqslant 7$, $k \in \mathbb{Z}$, and for instance $\varepsilon = 1 + \zeta - \zeta^3$, which is equal to $1 + \sqrt{2}$ if we choose $\zeta = (1 + i)/\sqrt{2}$ as primitive 8th root of unity. The prime 2 is totally ramified in K as $2\mathbb{Z}_K = \mathfrak{p}^4$, and we have $\mathfrak{p} = (1+\zeta)\mathbb{Z}_K$. It is also clear that $G = \mathrm{Gal}(K/\mathbb{Q}) = \{\sigma_1, \sigma_3, \sigma_5, \sigma_7\}$ is a Klein 4-group, where σ_j sends ζ to ζ^j.

All these facts are obtained immediately using a CAS, but are also very easy to show directly. Finally, to simplify notation we will denote by $\mathcal{N}$ the absolute norm $\mathcal{N}_{K/\mathbb{Q}}$ from K to $\mathbb{Q}$.

Definition 6.6.5. *Let $\gamma = A + B\zeta + C\zeta^2 + D\zeta^3 \in \mathbb{Z}[\zeta]$ be such that $v_{\mathfrak{p}}(\gamma) \leqslant 1$.*

(1) *If $v_{\mathfrak{p}}(\gamma) = 0$ we say that γ is* normalized *if*

$$A \equiv 1 \pmod 4, \quad B \equiv 0 \pmod 2, \quad C \equiv 0 \pmod 2, \quad and \ D \equiv 0 \pmod 4 \ .$$

(2) *If $v_{\mathfrak{p}}(\gamma) = 1$ we say that γ is* normalized *if*

$$A \equiv 1 \pmod 4, \quad B \equiv 1 \pmod 4, \quad C \equiv 0 \pmod 2, \quad and \ D \equiv 0 \pmod 4 \ .$$

Lemma 6.6.6. *Let γ be such that $v_{\mathfrak{p}}(\gamma) \leqslant 1$.*

(1) *There exists a unit u of K such that $u\gamma$ is normalized; in other words, there exists an associate of γ that is normalized.*
(2) *If γ and $u\gamma$ are both normalized, with u a unit, then u is equal to ε^{4k} for some $k \in \mathbb{Z}$.*

Proof. (1). Assume first that $v_{\mathfrak{p}}(\gamma) = 0$. Multiplication by ζ maps the coefficients (A, B, C, D) of γ to $(-D, A, B, C)$; in other words, it is a circular permutation up to sign changes. Since γ is coprime to 2 we have $A + B +$

$C + D \equiv 1 \pmod{2}$; hence either a single coefficient is odd or a single one is even, so that with a suitable circular permutation we may put the single odd coefficient as the constant coefficient, or the single even coefficient as the coefficient of ζ^2. It follows that there is an associate of γ such that A is odd and C even, and then necessarily $B \equiv D \pmod{2}$.

Now multiplication by the unit $\zeta\varepsilon$ is easily seen to change (A, B, C, D) into $(A+B-D, A+B+C, B+C+D, -A+C+D)$. This transformation preserves the fact that A is odd, C even, and $B \equiv D \pmod{2}$, but changes the parity of B and D. Therefore we may assume that B, C, and D are all even, and changing γ into $-\gamma$ that $A \equiv 1 \pmod{4}$. These congruences being satisfied we check that multiplication by $-\zeta^2\varepsilon^2$ preserves all the congruences, and changes (A, B, C, D) into (A', B', C', D') with $(A', B', C', D') \equiv (A, B + 2, C, D + 2)$ $\pmod{4}$, so that we may assume that $D \equiv 0 \pmod{4}$, and hence the result is normalized.

Assume now that $v_{\mathfrak{p}}(\gamma) = 1$ and set $\gamma_1 = \gamma/(1+\zeta) = A_1 + B_1\zeta + C_1\zeta^2 + D_1\zeta^3 \in \mathbb{Z}_K$, so that $v_{\mathfrak{p}}(\gamma_1) = 0$. By the first part of the proof there exists an associate of γ_1 that is normalized. Since $A = A_1 - D_1$, $B = A_1 + B_1$, $C = B_1 + C_1$, and $D = C_1 + D_1$, we deduce that $A \equiv 1 \pmod{4}$, $B \equiv 1 \pmod{2}$, and $C \equiv D \equiv 0 \pmod{2}$. Multiplication by ε^2 preserves these congruences and changes (C, D) into $(2B + 3C + 2D, -2A + 2C + 3D) \equiv (C + 2, D + 2)$ $\pmod{4}$, so that we may assume that $D \equiv 0 \pmod{4}$. If $B \equiv 1 \pmod{4}$ then γ is already normalized. Thus assume that $B \equiv 3 \pmod{4}$. When $C \equiv 0$ $\pmod{4}$, multiplication by $\zeta^2\varepsilon$ preserves existing congruences and leads to $B \equiv 1 \pmod{4}$. When $C \equiv 2 \pmod{4}$, the same is true with $\zeta^6\varepsilon^3$ instead of $\zeta^2\varepsilon$, thus proving the existence of a normalized associate of γ in all cases.

(2). If $v_{\mathfrak{p}}(\gamma) = 0$ and γ and $u\gamma$ are normalized, then $u \equiv 1 \pmod{2\mathbb{Z}[\zeta]}$. An immediate computation among units of the form $\zeta^j\varepsilon^k$ for $0 \leqslant j \leqslant 7$ and $0 \leqslant k \leqslant 3$ shows that the only such units up to powers of ε^{4k} are ± 1 and $\pm\varepsilon^2$, and it is immediate that the only one of these that respects the additional congruences modulo 4 is $u = 1$. Since $\varepsilon^4 \equiv 1 \pmod{4\mathbb{Z}[\zeta]}$ the result follows in this case. If $v_{\mathfrak{p}}(\gamma) = 1$ and $C \equiv 0 \pmod{4}$, we must now have $u\gamma \equiv \gamma$ $\pmod{4}$ (since it is easily checked that the coefficient of ζ^3 of $u\gamma$ is congruent to C modulo 4); hence $u \equiv 1 \pmod{4/\mathfrak{p}}$, and this implies as above that u is a power of ε^4. Finally, if $v_{\mathfrak{p}}(\gamma) = 1$ and $C \equiv 2 \pmod{4}$, we have eight units $u \equiv 1 \pmod{2/\mathfrak{p}}$ of the form $\zeta^j\varepsilon^k$ for $0 \leqslant j \leqslant 7$, and it is immediate that the congruences modulo 4 imply $u = 1$, proving the lemma. $\qquad\square$

Definition 6.6.7. *Let p be a prime number such that $p \equiv 1 \pmod{8}$. For each of the four values of $r \in \mathbb{F}_p$ such that $r^4 = -1$ we denote by $\phi_{p,r}$ the ring homomorphism from $\mathbb{Z}[\zeta]$ to $\mathbb{F}_p$ that sends 1 to 1 and ζ to r.*

Note that it is clear that $\phi_{p,r}$ is well defined, and since $8 \mid (p - 1)$ that there exist eight distinct 8th roots of unity in $\mathbb{F}_p$, of which four are such that $r^4 = -1$.

The main result that we are going to prove is the following, due in essence to Bremner and Morton [Bre-Mor].

Proposition 6.6.8. *Let $c \geqslant 3$ be 4th power-free. A necessary condition for the global solubility in $\mathbb{Q}$ of $x^4 + y^4 = cz^4$ is the existence of a normalized divisor γ of c in $\mathbb{Z}[\zeta]$ of the form $\gamma = A + B\zeta + C\zeta^2 + D\zeta^3$ satisfying the following properties:*

(1) *There exists $\alpha \in \mathbb{Z}[\zeta]$ such that $x + y\zeta = \gamma\alpha^4$.*
(2) *We have $v_{\mathfrak{p}}(\gamma) = 0$ when $c \equiv 1 \pmod{16}$ and $v_{\mathfrak{p}}(\gamma) = 1$ when $c \equiv 2 \pmod{16}$.*
(3) *When $c \equiv 1 \pmod{16}$ the conjugates of γ are pairwise coprime, and when $c \equiv 2 \pmod{16}$ the conjugates of $\gamma/(1 + \zeta)$ are pairwise coprime.*
(4) *We have $\mathcal{N}(\gamma) = c$.*
(5) *The coefficients of γ satisfy the congruences*

$$C \equiv D \equiv 0 \pmod 4 \,,$$
$$AC \equiv BD \pmod 8 \,,$$
$$C(A + C) \equiv D(B - D) \pmod 3 \,.$$

(6) *For each odd prime number p dividing c (hence such that $p \equiv 1 \pmod 8$), there exists a unique fourth root r of -1 in $\mathbb{F}_p$ such that $\phi_{p,r}(\gamma) = 0$. Then both*

$$\frac{1 - r^2}{2} \frac{\phi_{p,r}(\sigma_5(\gamma))}{\phi_{p,r}(\sigma_3(\gamma))} \quad and \quad \frac{1 + r^2}{2} \frac{\phi_{p,r}(\sigma_5(\gamma))}{\phi_{p,r}(\sigma_7(\gamma))}$$

are fourth powers in $\mathbb{F}_p^*$.

Proof. (1) and (2). Factoring our equation in $\mathbb{Z}_K$ gives

$$x^4 + y^4 = (x + \zeta y)(x + \zeta^3 y)(x + \zeta^5 y)(x + \zeta^7 y) = cz^4 \,.$$

Assume that π is a prime element such that $\pi \mid \gcd(x + \zeta^m y, x + \zeta^n y)$ for two distinct odd exponents m and n such that $1 \leqslant m, n \leqslant 7$. Then $\pi \mid (\zeta^m - \zeta^n)y$ and $\pi \mid \zeta^m(x + \zeta^n y) - \zeta^n(x + \zeta^m y) = (\zeta^m - \zeta^n)x$, so that $\pi \mid (\zeta^m - \zeta^n)$ since x and y are coprime. Since the norm of $\zeta^m - \zeta^n$ is a power of 2, it follows that π is a prime dividing 2; in other words (up to multiplication by a unit), $\pi = 1 + \zeta$. From our factored equality and the fact that z is odd, we deduce that when $c \equiv 1 \pmod{16}$ the four factors on the left are pairwise coprime, and when $c \equiv 2 \pmod{16}$ (hence with x and y odd) the factors divided by $1 + \zeta$ are (algebraic integers and) also pairwise coprime. Assume for instance that $c \equiv 1 \pmod{16}$ and set $\gamma = \gcd(x + \zeta y, c)$, defined for the moment only up to multiplication by a unit. Since $(x + \zeta y)/\gamma$ is coprime to the other three factors and to c/γ, it follows that it must be a fourth power of an ideal in $\mathbb{Z}_K$. Since $\mathbb{Z}_K$ is a principal ideal domain, this means that there exist α and γ (equal to a unit multiple of the initially chosen one) in $\mathbb{Z}_K$, coprime to 2, and such that $x + \zeta y = \gamma\alpha^4$. Since we may change simultaneously γ into γv^4

and α into α/v for any unit v, we note for future reference that if some γ is fixed, it is necessary to consider only associates of γ modulo fourth powers of units. Similarly, if $c \equiv 2 \pmod{16}$ we deduce that $x + \zeta y = \gamma \alpha^4$, where α is coprime to 2 and $v_{\mathfrak{p}}(\gamma) = 1$, proving (1) and (2).

(3) and (4). We take the norm down to $\mathbb{Q}$ of the relation obtained in (1). Setting $m = \mathcal{N}(\alpha)$ we thus obtain $cz^4 = x^4 + y^4 = \mathcal{N}(\gamma)m^4$; hence $m^4 \mid cz^4$, and since c is 4th power-free we have $m \mid z$, so that $\mathcal{N}(\gamma) = c(z/m)^4$, and hence $c \mid \mathcal{N}(\gamma)$. Conversely, we have $\gamma \mid x + \zeta y$, so for any $\sigma \in G = \mathrm{Gal}(K/\mathbb{Q})$ we have $\sigma(\gamma) \mid x + \sigma(\zeta)y$. When $c \equiv 1 \pmod{16}$ the numbers $x + \sigma(\zeta)y$ are pairwise coprime, so the conjugates $\sigma(\gamma)$ are pairwise coprime. Since $\sigma(\gamma) \mid \sigma(c) = c$ it follows that $\mathcal{N}(\gamma) = \prod_{\sigma \in G} \sigma(\gamma) \mid c$, and combining this with $c \mid \mathcal{N}(\gamma)$ we deduce that $c = \mathcal{N}(\gamma)$. When $c \equiv 2 \pmod{16}$ the same reasoning shows that the $\sigma(\gamma/(1 + \zeta))$ are pairwise coprime hence that $\mathcal{N}(\gamma)/2 \mid c/2$, and we conclude in the same way, proving (3) and (4).

(5). Possibly after changing γ into $-\gamma$, we may multiply α by any power of ζ without changing (1). If we write $\alpha = a + b\zeta + c\zeta^2 + d\zeta^3$, the condition α coprime to 2 means that $a + b + c + d \equiv 1 \pmod{2}$; in other words, either one or three of the coefficients are odd, and the others are even. Since multiplication by ζ sends (a, b, c, d) to $(-d, a, b, c)$, in other words is a circular permutation up to changes of sign, it is clear that we may assume that a is odd and c even, and then $b \equiv d \pmod{2}$.

Write $\alpha^4 = U + V\zeta + W\zeta^2 + X\zeta^3$. An easy calculation shows that

$$U \equiv 1 \pmod{8}, \quad V \equiv X \equiv 4b \pmod{8}, \quad \text{and} \quad W \equiv 0 \pmod{8} \ .$$

On the other hand, since $\gamma = A + B\zeta + C\zeta^2 + D\zeta^3$ we have

$$\begin{aligned}
x + \zeta y &= \gamma \alpha^4 \\
&= (AU - BX - CW - DV) + (AV + BU - CX - DW)\zeta \\
&\quad + (AW + BV + CU - DX)\zeta^2 + (AX + BW + CV + DU)\zeta^3 \ .
\end{aligned}$$

It follows in particular that $AW + BV + CU - DX = 0$ and $AX + BW + CV + DU = 0$. Since V, W, and X are divisible by 4 and U is odd, we already deduce that $4 \mid C$ and $4 \mid D$. Also, since we have chosen $x \equiv 1 \pmod{4}$ we have $1 \equiv x \equiv AU - BX - CW - DV \equiv A \pmod{4}$. If $c \equiv 1 \pmod{16}$ we must have $B \equiv 0 \pmod{2}$, so that γ is normalized. If $c \equiv 2 \pmod{16}$ then since we have chosen $y \equiv 1 \pmod{4}$ we have $1 \equiv y \equiv AV + BU - CX - DW \equiv B \pmod{4}$, so that γ is normalized also in this case.

Working now modulo 8, we deduce from the same two equations above that $4bB + C - 4bD \equiv 4bA + 4bC + D \equiv 0 \pmod{8}$, and since $4 \mid C$ and $4 \mid D$, we have $C \equiv 4bB \pmod{8}$ and $D \equiv 4bA \pmod{8}$, from which we evidently obtain $AC \equiv BD \pmod{8}$. For the result modulo 3, another easy calculation using the trivial relations $x^3 \equiv x \pmod{3}$ and $xy(x^2 - y^2) \equiv 0 \pmod{3}$ shows that $U \equiv a^2 - b^2 + c^2 - d^2 \pmod{3}$, $V \equiv X \equiv ab - bc + cd + da \pmod{3}$, and $W \equiv 0 \pmod{3}$. The equalities obtained above thus imply that

$(B - D)V + CU \equiv 0 \pmod 3$ and $(A + C)V + DU \equiv 0 \pmod 3$,

hence that

$$(D(B - D) - C(A + C))V \equiv (D(B - D) - C(A + C))U \equiv 0 \pmod 3.$$

Now we cannot have $U \equiv V \equiv 0 \pmod 3$. Indeed, otherwise

$$(a + b + d)^2 + (c - b + d)^2 \equiv (a - b - d)^2 + (c + b - d)^2 \equiv 0 \pmod 3$$

by the congruences obtained above, and since a sum of two squares is divisible by 3 if and only if both squares are, we would have $a + b + d \equiv c - b + d \equiv a - b - d \equiv c + b - d \equiv 0 \pmod 3$, hence $3 \mid \gcd(a, b, c, d)$, so $3 \mid \gcd(x, y)$ in contradiction with our assumption, proving the congruence modulo 3 of the proposition.

(6). Let $r \in \mathbb{F}_p$ be such that $r^4 = -1$. The four roots of this equation are r^j for $1 \leqslant j \leqslant 7$, j odd, and it is clear that $\phi_{p,r}(\sigma_j(\gamma)) = \phi_{p,r^j}(\gamma)$. It follows that

$$\prod_{1 \leqslant j \leqslant 7,\ j\ \text{odd}} \phi_{p,r^j}(\gamma) = \phi_{p,r}(\mathcal{N}(\gamma)).$$

Since $\mathcal{N}(\gamma) = c$ and $p \mid c$ we have $\phi_{p,r}(\mathcal{N}(\gamma)) = 0$; hence there exists j such that $\phi_{p,r^j}(\gamma) = 0$, proving the existence of r such that $\phi_{p,r}(\gamma) = 0$.

From the equation $x + \zeta y = \gamma \alpha^4$ we deduce by application of the σ_j that $x + \zeta^j y = \sigma_j(\gamma) \sigma_j(\alpha)^4$ for j odd; hence by application of the homomorphism $\phi_{p,r}$ we deduce that there exist f_3, f_5, and f_7 such that in $\mathbb{F}_p$ we have $x + ry = 0$ and

$$x + r^3 y = \phi_{p,r}(\sigma_3(\gamma))f_3^4, \quad x + r^5 y = \phi_{p,r}(\sigma_5(\gamma))f_5^4, \quad x + r^7 y = \phi_{p,r}(\sigma_7(\gamma))f_7^4.$$

From the first equation we deduce that $x = -ry$, so replacing and using $r^4 = -1$ we obtain

$$(r^3 - r)y = \phi_{p,r}(\sigma_3(\gamma))f_3^4, \quad -2ry = \phi_{p,r}(\sigma_5(\gamma))f_5^4, \quad -(r^3 + r)y = \phi_{p,r}(\sigma_7(\gamma))f_7^4.$$

Since $r^3 - r \neq 0$, $2r \neq 0$, and $r^3 + r \neq 0$, this shows in particular that $\phi_{p,r}(\sigma_j(\gamma)) \neq 0$ for $j \neq 1$, since otherwise $y \equiv 0 \pmod p$, so $x \equiv 0 \pmod p$, contradicting the assumption that x and y are coprime, proving the uniqueness of r since $\phi_{p,r}(\sigma_j(\gamma)) = \phi_{p,r^j}(\gamma)$. Finally, by dividing the second relation by the first and the third respectively we obtain the conditions given in (6). $\square$

Remarks. (1) A short computation shows that if $c \equiv 1 \pmod{16}$ and γ is normalized then $\mathcal{N}(\gamma) \equiv 1 + 2C^2 \pmod{16}$, and since $\mathcal{N}(\gamma) = c$ it follows that in this case the condition $C \equiv D \equiv 0 \pmod 4$ is automatic as soon as γ is normalized.

(2) Once γ is known to be normalized it is clear that the congruence modulo 8 is equivalent to $C \equiv 0 \pmod 8$ when $c \equiv 1 \pmod{16}$, and to $C \equiv D \pmod 8$ when $c \equiv 2 \pmod{16}$.

Lemma 6.6.9. *Let $c \equiv 1 \pmod{16}$ (respectively $c \equiv 2 \pmod{16}$). An element γ is normalized and satisfies conditions* (1) *to* (6) *of Proposition 6.6.8 if and only if $\sigma_5(\gamma)$ (respectively $\zeta\sigma_7(\gamma)$) does.*

Proof. Since $\sigma_5(A + B\zeta + C\zeta^2 + D\zeta^3) = A - B\zeta + C\zeta^2 - D\zeta^3$ and $\zeta\sigma_7(A + B\zeta + C\zeta^2 + D\zeta^3) = B + A\zeta - D\zeta^2 - C\zeta^3$, the lemma is clear. $\qquad\square$

Definition 6.6.10. *We will say that γ is a suitable divisor of c if γ is a normalized divisor of c which satisfies properties* (3) *and* (4) *of Proposition 6.6.8, in other words such that $\mathcal{N}(\gamma) = c$, and γ coprime to its conjugates if $c \equiv 1 \pmod{16}$, or $\gamma/(1 + \zeta)$ coprime to its conjugates if $c \equiv 2 \pmod{16}$.*

Corollary 6.6.11. *Assume that the equation $x^4 + y^4 = cz^4$ is everywhere locally soluble, in other words that c satisfies the conditions of Corollary 6.6.3. Let $\mathcal{F}$ be a set of representatives of suitable divisors γ of c modulo multiplication by powers of ε^4, and modulo σ_5 if $c \equiv 1 \pmod{16}$, or modulo the action of $\zeta\sigma_7$ if $c \equiv 2 \pmod{16}$. If for every $\gamma \in \mathcal{F}$ one of the conditions* (5) *and* (6) *of Proposition 6.6.8 is not satisfied, then the equation $x^4 + y^4 = cz^4$ has no nontrivial global solution. Furthermore, if c is not a square or twice a square we have $|\mathcal{F}| = 2^{2k-1}$, where k is the number of distinct odd prime divisors of c.*

Proof. Indeed, it is clear that multiplication of γ by a power of ε^4 does not change the conditions in (6). Furthermore, it is immediate to check that multiplication by ε^4 does not change $AC - BD \bmod 8$ or $C(A + C) - D(B - D) \bmod 3$, so that it is enough to consider γ up to powers of ε^4 (we have already mentioned this fact in the proof of (4)), and by the preceding lemma up to the action of σ_5 or $\zeta\sigma_7$. Finally, it is clear from the uniqueness of the normalization up to powers of ε^4 that the number of suitable γ is equal to 4^k, since every prime congruent to 1 modulo 8 splits completely into 4 factors (since γ is coprime to its conjugates we cannot mix different factors above the same prime even when c is not squarefree). Now we check that if $c \equiv 1 \pmod{16}$ and $\sigma_5(\gamma) = \gamma$ then $c = \mathcal{N}(\gamma) = (A^2 + C^2)^2$, and if $c \equiv 2 \pmod{16}$ and $\zeta\sigma_7(\gamma) = \gamma$ then $c = \mathcal{N}(\gamma) = 2(A^2 - 2AC - C^2)^2$, so that when c is not a square or twice a square we have $|\mathcal{F}| = 2^{2k-1}$. $\qquad\square$

Note that if c is a square our equation has no global solutions by Proposition 6.5.3, and if c is twice a square it has no solutions for $c > 2$ by Exercise 24.

Remark. The conditions of Proposition 6.6.8 are all useful. For instance, to exclude $c = 4801$ the condition modulo 8 is the only one that applies. To exclude $c = 5266$ the condition modulo 3 is the only one that applies.

To exclude $c = 5281$ we need the condition on fourth powers for one of the values of γ. In many examples we can use only one of the two conditions on fourth powers; in others we can use both. We will give a summary of results in a table below.

6.6.3 Global Solubility: Coverings of Elliptic Curves

Although Corollary 6.6.11 is very powerful in proving that a Fermat quartic has no global solution, it is not the whole story. For instance, as will be clear from the table given below, of the 107 suitable values of c such that $3 \leqslant c \leqslant 10000$, 99 can be treated using this corollary, leaving eight indeterminate cases. Another natural (and in fact easier) approach is to use maps from the curve $\mathcal{C}_c$ with affine equation $x^4 + y^4 = c$ to two elliptic curves, and then to use results on elliptic curves to conclude. This will enable us to solve five of the remaining eight cases with $c \leqslant 10000$. This section assumes that the reader is familiar with the theory of elliptic curves over $\mathbb{Q}$, which will be explained in Chapter 8.

Let c be as above, and consider the two elliptic curves with affine Weierstrass equations

$$E_c : Y^2 = X^3 - cX \quad \text{and} \quad F_c : Y^2 = X^3 + c^2 X .$$

It is immediate to check that the maps ϕ and ψ defined in affine coordinates by

$$\phi((x,y)) = (-x^2, xy^2) \quad \text{and} \quad \psi((x,y)) = (cx^2/y^2, c^2 x/y^3)$$

are maps from $\mathcal{C}_c$ to E_c and F_c respectively. Since all rational points of $\mathcal{C}_c$ are affine, it is clear that if $P \in \mathcal{C}_c(\mathbb{Q})$ then $\phi(P) \in E_c(\mathbb{Q})$ and $\psi(P) \in F_c(\mathbb{Q})$. In particular, since the inverse image of a point by ϕ or ψ is finite, if *either* $E_c(\mathbb{Q})$ or $F_c(\mathbb{Q})$ is an explicit finite set, it will be immediate to determine $\mathcal{C}_c(\mathbb{Q})$. We will see in Chapter 8 that this means that $E_c(\mathbb{Q})$ or $F_c(\mathbb{Q})$ has rank 0, hence is equal to its easily determined torsion subgroup. Thus in this favorable case it is very easy to determine $\mathcal{C}_c(\mathbb{Q})$:

Proposition 6.6.12. *Let $c \geqslant 3$ be a 4th power-free integer. If either $E_c(\mathbb{Q})$ or $F_c(\mathbb{Q})$ has rank 0 then $\mathcal{C}_c(\mathbb{Q}) = \emptyset$.*

Proof. Note first that the trivial 2-torsion point $(X, Y) = (0, 0)$ on E_c corresponds to $x = 0$ on $\mathcal{C}_c$, hence to $c = y^4$, which is absurd by assumption. We now use Proposition 8.1.14, which we will prove in Chapter 8. It tells us that there can be other torsion points only if c (for E_c) or $-c^2$ (for F_c) is equal to m^2 or to $-4m^4$. Consider first E_c. Since $c > 0$ (or because c is 4th power-free) we cannot have $c = -4m^4$. On the other hand, $c = m^2$ is a priori possible, and gives as affine torsion points the ones with $Y = 0$. But this implies that either x or y is equal to 0, which is impossible, proving the result for E_c. Consider now F_c. We cannot have $-c^2 = m^2$, so the only possibility is

$c = 2m^2$, and the extra torsion points are clearly $(X, Y) = (2m^2, \pm 4m^3)$. The inverse images (x, y) of these points by the map ψ are easily seen to be such that $y = \pm x$ and $y^3 = \pm x$, and since $x \neq 0$ this gives $m = \pm x^2$. Thus x is an integer and $c = 2m^2 = 2x^4$, which implies that $c = 2$ since c is assumed to be 4th power-free, and this is excluded since we have assumed that $c \geqslant 3$. $\quad\square$

Examples. We first choose $c = 562$, for which C_c is locally soluble by Corollary 6.6.3. The 2-descent method that we will study in Section 8.2 or Cremona's `mwrank` program shows that $E_c(\mathbb{Q})$ has rank 1, but that $F_c(\mathbb{Q})$ has rank 0, so that $C_{562}(\mathbb{Q}) = \emptyset$.

We now choose $c = 226$ or $c = 977$. The 2-descent methods and `mwrank` tell us only that the rank of $F_c(\mathbb{Q})$ is equal to 0 or 2. However, the computation of $L(F_c, 1)$ (see Section 8.5) shows that in both cases we have $L(F_c, 1) \neq 0$; hence since the BSD conjecture is a theorem in this case (in fact here due to Coates–Wiles [Coa-Wil] since F_c has complex multiplication), this proves that the rank of $F_c(\mathbb{Q})$ is equal to 0, and once again this implies that $C_c(\mathbb{Q}) = \emptyset$.

It is important to note that we have the following stronger result due to Dem'yanenko (see [Dem3]). See Section 13.3.1 for an indication of the method of proof.

Theorem 6.6.13 (Dem'yanenko). *If $c \geqslant 3$ is a 4th power-free integer and if the rank of $E_c(\mathbb{Q})$ is less than or equal to 1 then $C_c(\mathbb{Q}) = \emptyset$.*

This theorem sometimes allows us to show that the Fermat quartic is not globally soluble, even when both $E_c(\mathbb{Q})$ and $F_c(\mathbb{Q})$ have nonzero rank. For instance, if $c = 2642$, for which C_c is again everywhere locally soluble, it can be shown that $E_c(\mathbb{Q})$ has rank 1 and $F_c(\mathbb{Q})$ has rank 2, so that thanks to Dem'yanenko's theorem we can conclude that $C_c(\mathbb{Q}) = \emptyset$, which we would not have been able to do using only the rank-0 conditions.

Remarks. (1) It is easy to compute that the *root number* (see Theorem 8.1.4) of E_c is equal to 1 when $c \equiv 1 \pmod{16}$ and to -1 when $c \equiv 2 \pmod{16}$, and that the root number of F_c is always equal to 1. Thus, it follows from a weak form of the BSD conjecture that F_c will always have even rank, and E_c will have even or odd rank according to whether c is odd or even. This shows in particular that Dem'yanenko's theorem is absolutely necessary if we want to use the curve E_c when $c \equiv 2 \pmod{16}$.

(2) The existence of the maps ϕ and ψ, together with the map $\overline{\phi}$ from C_c to E_c defined by $\overline{\phi}((x, y)) = (-y^2, yx^2)$, means that the Jacobian of C_c is isogenous to $E_c \times E_c \times F_c$, the map $(\phi, \overline{\phi}, \psi)$ composed with the isogeny giving the embedding of C_c into its Jacobian (see Section 13.2 for these notions). The proof of Dem'yanenko's theorem amounts to showing that if $(x, y) \in C_c(\mathbb{Q})$, the two points $\phi((x, y))$ and $\overline{\phi}((x, y))$ in E_c are generically independent, so that E_c must have rank greater than or equal to 2.

6.6.4 Conclusion, and a Small Table

When there do exist solutions to our Fermat quartic, for instance when $c = 1$, 2, or 17, we can ask for *all* the solutions (since a Fermat quartic is a curve of genus 3, we know by Faltings's theorem that there are only finitely many). For $c = 1$, Fermat's theorem for $n = 4$ (which follows from Proposition 6.5.3 below) tells us that the points $(\pm 1, 0)$ and $(0, \pm 1)$ are the only rational points on the curve $x^4 + y^4 = 1$. By a similar method of descent, for $c = 2$ it is easy to show that the points $(\pm 1, \pm 1)$ are the only rational points on the curve $x^4 + y^4 = 2$ (Exercise 24). On the other hand, it is much more difficult to prove that for $c = 17$ the only rational points on $x^4 + y^4 = 17$ are $(\pm 1, \pm 2)$ and $(\pm 2, \pm 1)$. This problem was posed by J.-P. Serre as the simplest nontrivial case of the Fermat quartic equations, and also because it was known that the standard methods using Chabauty-type techniques failed on this curve. It was solved only in 1999 by Flynn and Wetherell using covering techniques, and their proof is not easy (see also Section 13.3.4).

A table up to 10000. The following table gives numerical data obtained using all of the preceding results. It first lists the 109 values of $c \geqslant 1$ that are 4th power-free and for which the equation $x^4 + y^4 = c$ is everywhere locally soluble, together with $c = 0$. Below each such value of c is either a pair (x, y) of rational numbers such that $x^4 + y^4 = c$ (except for $c = 5906$, where $*$ means $(x, y) = (25/17, 149/17)$), or a code made with one or more letters. The letter A (for Algebraic) means that we can prove the nonexistence of global solutions using Corollary 6.6.11. The letter D means that $E_c(\mathbb{Q})$ has rank 1, so that Dem'yanenko's Theorem 6.6.13 is applicable (this can occur only for c even, see above), E means that $E_c(\mathbb{Q})$ has rank 0 (this can occur only for c odd), and F means that $F_c(\mathbb{Q})$ has rank 0, so that in both of these cases Proposition 6.6.12 is applicable. Thus if at least one of these letters occurs this implies that $C_c(\mathbb{Q}) = \emptyset$. Finally, the letter U (for undetermined), which occurs three times, means that the results given above, together with a computer search, do not allow us to conclude anything. We refer to [Bre-Mor] for still other methods that can prove nonglobal solubility of our equation in other cases, in particular for the first undetermined case $c = 4481$. Evidently all the values of c for which any of the letters A, D, E, and F occur are counterexamples to the Hasse principle. It can be seen that the purely algebraic method using the factorization of our equation is much more powerful than the method using elliptic curves, although the latter is necessary in five cases. If we push the computation to 10^5, there are 833 suitable values of $c \geqslant 3$ for which the equation is everywhere locally soluble, 90 for which we find a global solution, 692 that can be shown to have no global solutions by the algebraic method, and of the 49 remaining ones, 33 can be shown to have no global solutions using elliptic curves, leaving 16 undetermined cases (the smallest above 10000 being $c = 33377$).

0	1	2	17	82	97	146	226	257	337
(0,0)	(0,1)	(1,1)	(1,2)	(1,3)	(2,3)	ADF	AF	(1,4)	(3,4)
482	562	577	626	641	706	802	881	977	1042
ADF	ADF	A	(1,5)	(2,5)	(3,5)	ADF	(4,5)	AF	AF
1186	1201	1297	1361	1522	1777	1921	2017	2066	2161
DF	AF	(1,6)	AEF	A	A	(5,6)	AF	ADF	EF
2306	2402	2417	2482	2642	2657	2722	2801	2866	3026
ADF	(1,7)	(2,7)	(3,7)	AD	(4,7)	ADF	A	ADF	(5,7)
3041	3106	3121	3202	3217	3442	3506	3617	3697	3761
AF	AF	AE	DF	AF	ADF	ADF	AF	(6,7)	AF
3826	4097	4162	4177	4226	4241	4306	4322	4481	4657
ADF	(1,8)	ADF	(3,8)	AD	AF	ADF	ADF	U	AEF
4721	4786	4801	4946	5186	5266	5281	5297	5426	5521
(5,8)	ADF	AF	ADF	AF	ADF	AE	AF	AF	AF
5617	5906	5986	6242	6337	6497	6562	6577	6626	6722
A	*	AF	AF	AF	(7,8)	(1,9)	(2,9)	ADF	ADF
6817	6961	6977	7121	7186	7297	7361	7537	7666	7762
(4,9)	AF	EF	AF	(5,9)	AEF	A	U	ADF	D
7841	8161	8306	8402	8482	8546	8737	8882	8962	9026
AEF	AF	ADF	ADF	ADF	AF	AE	U	(7,9)	AF
9122	9266	9281	9346	9377	9442	9586	9697	9857	9986
ADF	A	A	ADF	AEF	ADF	A	AF	AE	ADF

Sums of Two Rational Fourth Powers up to 10000

An amusing corollary of this table is the following result, due to Bremner and Morton [Bre-Mor]:

Corollary 6.6.14. *The integer $c = 5906$ is the smallest integer that is the sum of two fourth powers of rational numbers, and not the sum of two fourth powers of integers.*

Proof. Indeed, for all smaller values of c except $c = 4481$ we see that either the equation $x^4 + y^4 = c$ has no *rational* solutions, or it has an integral solution. It is an immediate verification that 5906 is not a sum of two fourth powers of integers, and it is the sum of the two fourth powers of rational numbers given above. There remains to prove that $x^4 + y^4 = 4481z^4$ is not globally soluble, and this is done using the more general factoring methods explained in [Bre-Mor]. □

6.7 The Equation $y^2 = x^n + t$

For general results on this equation, we refer to [Cohn1] and [Cohn2], from which a large part of this section is taken.

In this section, we look for *integral* solutions to the equation $y^2 = x^n + t$, where t and n are given integers with $n \geqslant 3$ (otherwise the equation is trivial). If n is even this equation factors as $(y - x^{n/2})(y + x^{n/2}) = t$, which is trivially solved: if $t < 0$ (which will be the main case that we will consider), we may assume that $x > 0$, so we let $d = x^{n/2} - y$, which will be a positive divisor of $|t|$ less than or equal to $|t|^{1/2}$, and the condition to be satisfied is that $(d + |t|/d)/2 = x^{n/2}$ must be an exact $(n/2)$th power with $n/2 \geqslant 2$. From this, a short calculation shows the following, which we give for future reference:

Proposition 6.7.1. *Let $n \geqslant 4$ be even and let t be a squarefree negative integer not congruent to 1 modulo 8. The only values of n and t with $n \geqslant 4$ even and $-100 < t \leqslant -1$ squarefree and not congruent to 1 modulo 8 for which the Diophantine equation $y^2 = x^n + t$ has solutions are for $t = -1$ with solutions $(x, y) = (\pm 1, 0)$, and for $(n, t) = (4, -17)$, $(6, -53)$, $(4, -65)$, $(4, -77)$, and $(4, -97)$ with respective solutions $(x, y) = (\pm 3, \pm 8)$, $(\pm 3, \pm 26)$, $(\pm 3, \pm 4)$, $(\pm 3, \pm 2)$, and $(\pm 7, \pm 48)$.*

We leave to the reader to state and prove the corresponding statement for $1 \leqslant t \leqslant 100$ squarefree and not congruent to 1 modulo 8 (Exercise 35).

If n is odd and $p \mid n$, we can write $x^n = (x^{n/p})^p$, so we can reduce to the exponent p. Thus in the sequel we will usually assume that n is an odd prime number p.

6.7.1 General Results

For reasons that will soon become clear, we make the following definitions.

Definition 6.7.2. *We will say that condition $H(p, t)$ is satisfied if p is an odd prime, t is a squarefree negative integer not congruent to 1 modulo 8, and p does not divide the class number of the imaginary quadratic field $\mathbb{Q}(\sqrt{t})$. By abuse of notation we will say that $H(t)$ is satisfied if t is a squarefree negative integer not congruent to 1 modulo 8.*

Proposition 6.7.3. *Assume $H(p, t)$, and define $A_p(t)$ to be the (possibly empty) set of nonnegative integers a such that*

$$\sum_{k=0}^{(p-1)/2} \binom{p}{2k} a^{2k} t^{(p-1)/2-k} = \pm 1 .$$

The set of solutions $(x, y) \in \mathbb{Z}^2$ to the Diophantine equation $y^2 = x^p + t$ is given by the pairs

$$(x, y) = \left(a^2 - t, \pm \sum_{k=0}^{(p-1)/2} \binom{p}{2k+1} a^{2k+1} t^{(p-1)/2-k} \right)$$

for each $a \in A_p(t)$, with in addition the so-called special pairs

$$(x, y) = (a^2 + 2\varepsilon, \pm(a^3 + 3\varepsilon a))$$

if $p = 3$ and $t = -(3a^2 + 8\varepsilon)$ for any $\varepsilon = \pm 1$ and odd a such that $a \geqslant 1$ if $\varepsilon = 1$ or $a \geqslant 3$ if $\varepsilon = -1$.

Proof. Let (x, y) be a solution to the equation $y^2 = x^p + t$. In the quadratic field $K = \mathbb{Q}(\sqrt{t})$ we can write $(y - \sqrt{t})(y + \sqrt{t}) = x^p$. I claim that the ideals generated by the two factors on the left are coprime. Indeed, assume otherwise, so let $\mathfrak{q}$ be a prime ideal of $\mathbb{Z}_K$ dividing these factors. It thus divides their sum and difference, so if q is the prime number below $\mathfrak{q}$ we have $q \mid 2y$ and $q \mid 2t$. If x is even, then y is odd since otherwise $4 \mid t$, contradicting the squarefreeness of t, so $t = y^2 - x^p \equiv 1 \pmod 8$, contradicting our assumption on t. Thus x is odd; hence $\mathfrak{q}$ cannot be above 2, so $q \mid \gcd(y, t)$, and hence $q \mid x$, so $q^2 \mid t$, again contradicting the fact that t is squarefree and proving my claim. Since the product of the two coprime ideals $(y - \sqrt{t})\mathbb{Z}_K$ and $(y + \sqrt{t})\mathbb{Z}_K$ is a pth power, it follows that $(y + \sqrt{t})\mathbb{Z}_K = \mathfrak{a}^p$ for some ideal $\mathfrak{a}$ of $\mathbb{Z}_K$. On the other hand, if h denotes the class number of K then essentially by definition the ideal $\mathfrak{a}^h$ is a principal ideal. Since by assumption p and h are coprime, there exist integers v and w such that $vp + wh = 1$, so that $\mathfrak{a} = (\mathfrak{a}^p)^v(\mathfrak{a}^h)^w$ is itself a principal ideal, say $\mathfrak{a} = \alpha\mathbb{Z}_K$ for some $\alpha \in \mathbb{Z}_K$ (this type of reasoning involving the class number is typical, and will be met again, for instance in Fermat's last theorem). We thus deduce that there exists a unit $\varepsilon \in K$ such that $y + \sqrt{t} = \varepsilon\alpha^p$. However, since K is an *imaginary* quadratic field, there are not many units, and more precisely the group of units is $\{\pm 1\}$ except for $t = -1$ and $t = -3$, for which it has order 4 and 6 respectively. Since p is odd, it follows that apart from the special case $(p, t) = (3, -3)$ the order of the group of units is coprime to p; hence any unit is a pth power, so in these cases we are reduced to the equation $y + \sqrt{t} = \alpha^p$ with $\alpha \in \mathbb{Z}_K$. We will see in Proposition 6.7.5 below that there are no solutions for $(p, t) = (3, -3)$. Otherwise, we write $\alpha = (a + b\sqrt{t})/d$ with a and b integral, where either $d = 1$, or, only in the case $t \equiv 5 \pmod 8$, also $d = 2$ and a and b odd. Expanding the relation $y + \sqrt{t} = \alpha^p$ gives the two equations

$$d^p y = \sum_{k=0}^{(p-1)/2} \binom{p}{2k+1} a^{2k+1} b^{p-2k-1} t^{(p-1)/2-k} \text{ and}$$

$$d^p = \sum_{k=0}^{(p-1)/2} \binom{p}{2k} a^{2k} b^{p-2k} t^{(p-1)/2-k} .$$

Note that we may assume $a \geqslant 0$ since changing a into $-a$ does not change the second equation, and changes y into $-y$ in the first. From the second equation we deduce that $b \mid d^p$, and since b is coprime to d this means that $b = \pm 1$. It follows that

$$d^p = \pm \sum_{k=0}^{(p-1)/2} \binom{p}{2k} a^{2k} t^{(p-1)/2-k} .$$

If $d = 1$, we obtain the formula for y by replacing in the first equation, and we have $x = a^2 - b^2 t$. If $d = 2$, then since p is an odd prime we have

$$2 \equiv 2^p \equiv \pm t^{(p-1)/2} \equiv \pm \left(\frac{t}{p} \right) \equiv 0, \pm 1 \pmod{p} ,$$

which is possible only for $p = 3$, giving $t = -(3a^2 \mp 8)$, $y = -a^3 \pm 3a$, $x = a^2 \mp 2$, whence the additional cases of the proposition. $\square$

Remarks. (1) When t is not squarefree, it is easy to obtain similar but more complicated results; see for example Exercise 36.

(2) For *given* t and p, it is trivial to find all possible values of $a \in A_p(t)$. What is considerably more difficult in general is to find the sets $A_p(t)$ when only p is fixed. Thanks to a remarkable theorem of Bilu, Hanrot, and Voutier, this problem is completely solved; see below.

(3) Considering the formula modulo p, it is clear that the $\pm$ sign on the right-hand side of the formula defining $A_p(t)$ is equal to $\left(\frac{t}{p} \right)$.

(4) Much more important is the fact that in the cases that we have not treated ($t \equiv 1 \pmod 8$, $t > 0$, or p not coprime to the class number of $\mathbb{Q}(\sqrt{t})$) the problem is considerably more difficult but can be reduced to a finite number of so-called Thue equations. First, if $t \equiv 1 \pmod 8$ (positive or negative, but squarefree), we see that either x is odd, in which case y is even and the proof goes through as above, or x is even, and hence y is odd, so that $y - \sqrt{t}$ and $y + \sqrt{t}$ are both divisible by 2 in $\mathbb{Z}_K$, and we can easily deduce that $(y + \sqrt{t})/2 = \mathfrak{p}_2^{p-2} \mathfrak{a}^p$ for some ideal $\mathfrak{a}$ and some ideal $\mathfrak{p}_2$ above 2. Thus for *any* squarefree t we can reduce our equation either to $(y + \sqrt{t})/d = \mathfrak{a}^p$, or to $(y + \sqrt{t})/d = \mathfrak{p}_2^{p-2} \mathfrak{a}^p$ for $d = 1$ or 2. Since the number of possibilities for d and $\mathfrak{p}_2$ is finite, since the class group and the unit group modulo pth powers are also finite, an easy argument shows that our Diophantine equation reduces to a *finite* set of equations of the form $y + \sqrt{t} = \beta_i \alpha_i^p$, for a known finite set of elements $\beta_i \in K^*$, and unknowns $\alpha_i \in \mathbb{Z}_K$ (the above proof corresponds to the special case $\beta_i = 1$). When we expand this equation after writing $\alpha_i = (a + b\sqrt{t})/d$, the enormous simplification of having b as a common factor of all the coefficients of $\sqrt{t}$ (implying $b = \pm 1$) no longer occurs, and we have to solve equations of the form $P(a, b) = d^p$, where P is a homogeneous polynomial of degree p in a and b with integral coefficients depending on t. These are called Thue equations, and there are excellent methods for solving them, based on linear forms in logarithms; see Algorithm 12.10.3. The problem is that the equations depend on t, so for a fixed t and p it is "easy" to solve the equation; however, when we fix p, say, and let t vary it is more difficult to give a general solution.

Proposition 6.7.3 can be rephrased in a more positive way as follows.

Corollary 6.7.4. *Let $p \geqslant 3$ be a prime, let x and y be integers, and assume that $t = y^2 - x^p$ satisfies $H(t)$ (so that in particular x and y are coprime and x is odd). Assume in addition that $t + x$ is not a square, and furthermore when $p = 3$ that we do not have $(x, t) = (a^2 + 2\varepsilon, -(3a^2 + 8\varepsilon))$ for some odd a and some $\varepsilon = \pm 1$. Then the class number of the imaginary quadratic field $\mathbb{Q}(\sqrt{t})$ is divisible by p.*

Proof. Clear since $x = a^2 - t$, except in the given special case.   □

6.7.2 The Case $p = 3$

To apply Proposition 6.7.3 (assuming $H(p, t)$), there remains to find the sets $A_p(t)$. As already mentioned, this is trivial if p and t are fixed. The difficulty is to give general results when only one of these two variables is fixed. We will give detailed results below, and in particular complete results for some fixed values of t. In the next subsections we give the complete results for fixed p.

Proposition 6.7.5. *Assume $H(3, t)$.*

(1) *When $t \equiv 2$ or 3 modulo 4 then if t is not of the form $t = -(3a^2 \pm 1)$ the equation $y^2 = x^3 + t$ has no integral solutions. If $t = -(3a^2 + \varepsilon)$ with $\varepsilon = \pm 1$, the integral solutions are $x = 4a^2 + \varepsilon$, $y = \pm(8a^3 + 3\varepsilon a)$.*

(2) *When $t \equiv 5 \pmod 8$ then if t is not of the form $t = -(12a^2 - 1)$ or $-(3a^2 \pm 8)$, both with a odd, the equation $y^2 = x^3 + t$ has no integral solutions. If $t = -(12a^2 - 1)$ with a odd, the integral solutions are $x = 16a^2 - 1$, $y = \pm(64a^3 - 6a)$. If $t = -(3a^2 + 8\varepsilon)$ with $\varepsilon = \pm 1$ and a odd, the integral solutions are $x = a^2 + 2\varepsilon$, $y = \pm(a^3 + 3a\varepsilon)$.*

Proof. This case is especially simple since the equations defining the sets $A_3(t)$ are *linear* in t. We obtain $t = -(3a^2 \pm 1)$, $x = a^2 - t$, and $y = \pm(a^3 + 3at)$, giving the solutions of the proposition, to which we must add the solutions for the special case $t = -(3a^2 \pm 8)$. Recall that we have postponed the case $t = -3$, which we now consider. In the proof of Proposition 6.7.3 we found that $y + \sqrt{t} = ua^3$ for some unit u. Thus either we are led to the equations of the proposition (if $u = \pm 1$), or there exists $\varepsilon = \pm 1$ such that $y + \sqrt{t} = ((a + b\sqrt{t})/2)^3(-1 + \varepsilon\sqrt{t})/2$. Equating coefficients of $\sqrt{t}$ gives

$$16 = \varepsilon(a^3 - 9b^2a) - 3b(a^2 - b^2).$$

If $a \equiv 0 \pmod 3$, the right-hand side is divisible by 3, a contradiction. If $b \equiv 0 \pmod 3$, the right-hand side is congruent to ± 1 modulo 9 since a cube is such, again a contradiction. Thus neither a nor b is divisible by 3; hence $a^2 \equiv b^2 \equiv 1 \pmod 3$, so the right-hand side is still congruent to ± 1 modulo 9, a contradiction once again, so there are no solutions for $t = -3$.   □

Note that the case $t = -2$ of the above equation was already solved by Fermat, who posed it as a challenge problem to his English contemporaries.

Although usually the problem for $t > 0$ or for $t \equiv 1$ (mod 8) is much more difficult, in certain cases it is quite easy to find the set of integral solutions. A classical example is the following, for which the special case $t = 7$ is also due to Fermat.

Proposition 6.7.6. *Let a be an odd integer and let b be an integer such that $3 \nmid b$. Assume that either $t = 8a^3 - b^2$ with b odd, or that $t = a^3 - 4b^2$ with $t \not\equiv 1$ (mod 8). Then if t is squarefree but of any sign, the equation $y^2 = x^3 + t$ has no integral solution.*

Proof. Note that in both cases t is odd. I claim that x must be odd and y even. Indeed, if x were even then $y^2 = x^3 + t$ would be odd, so that $t \equiv y^2 - x^3 \equiv 1$ (mod 8), which contradicts the assumption of the second case and contradicts the congruence $t \equiv -b^2 \equiv -1$ (mod 8) of the first case. We now separate the cases.

We rewrite the first equation $y^2 = x^3 + (8a^3 - b^2)$ as

$$y^2 + b^2 = (x + 2a)((x - a)^2 + 3a^2) \ .$$

Since x and a are odd it follows that $(x - a)^2 + 3a^2 \equiv 3$ (mod 4), and since this is a positive number (why is this needed?) this implies that there exists a prime $p \equiv 3$ (mod 4) dividing it to an *odd* power. Thus $y^2 + b^2 \equiv 0$ (mod p), and since $\left(\frac{-1}{p}\right) = -1$ this implies that p divides b and y. I claim that $p \nmid (x + 2a)$. Indeed, since

$$(x - a)^2 + 3a^2 = (x + 2a)(x - 4a) + 12a^2$$

if we had $p \mid (x + 2a)$ we would have $p \mid 12a^2$, so either $p \mid a$ or $p = 3$ ($p = 2$ is impossible since $p \equiv 3$ (mod 4)). But $p \mid a$ implies $p^2 \mid t = 8a^3 - b^2$, a contradiction since t is squarefree, and $p = 3$ implies $3 \mid b$, which has been excluded, proving my claim. Thus the p-adic valuation of $y^2 + b^2$ is equal to that of $(x - a)^2 + 3a^2$ hence is odd, a contradiction since this would again imply that $\left(\frac{-1}{p}\right) = 1$.

Since y is even we write $y = 2y_1$, so we rewrite the second equation $y^2 = x^3 + (a^3 - 4b^2)$ as

$$4(y_1^2 + b^2) = x^3 + a^3 = (x + a)(x(x - a) + a^2) \ .$$

Since $x - a$ is even and a is odd, it follows that $4 \mid (x + a)$. Writing $x + a = 4x_1$, we obtain

$$y_1^2 + b^2 = x_1((4x_1 - a)(4x_1 - 2a) + a^2) = x_1(16x_1^2 - 12ax_1 + 3a^2) \ .$$

Since a is odd we have $16x_1^2 - 12ax_1 + 3a^2 \equiv 3$ (mod 4), so as in the preceding proof there exists a prime $p \equiv 3$ (mod 4) dividing it to an odd power. As above, this implies that p divides y_1 and b. I claim that $p \nmid x_1$. Indeed, otherwise $p \mid 3a^2$, so either $p \mid a$ or $p = 3$. As above, $p \mid a$ is impossible since it implies $p^2 \mid t$, a contradiction since t is squarefree, and $p = 3$ implies $3 \mid b$, which has been excluded. Thus the p-adic valuation of $y_1^2 + b^2$ is odd, a contradiction since this would imply $\left(\frac{-1}{p}\right) = 1$.   □

Proposition 6.7.5 applies to negative squarefree t not congruent to 1 modulo 8 such that the class number of $\mathbb{Q}(\sqrt{t})$ is not divisible by 3. The above proposition solves our equation for the following additional values of t with $|t| < 250$:

$t = -241, -129$ (class number divisible by 3), 7, 11, 13, 23, 39, 47, 53, 61, 67, 83, 87, 95, 109, 139, 155, 159, 167, 191, 215, 239 ($t > 0$).

Finally, note that we have already solved the case $t = 1$ (Corollary 6.4.32), as an application of Skolem's theorem.

6.7.3 The Case $p = 5$

In this case we can also give the complete answer as follows.

Proposition 6.7.7. *Assume $H(5, t)$. The only values of t for which the equation $y^2 = x^5 + t$ has a solution are $t = -1$ (with only solution $(x, y) = (1, 0)$), $t = -19$ (with only solutions $(x, y) = (55, \pm 22434)$), and $t = -341$ (with only solutions $(x, y) = (377, \pm 2759646)$).*

Proof. The equation defining the set $A_5(t)$ of Proposition 6.7.3 is $t^2 + 10a^2 t + 5a^4 \pm 1 = 0$. This has a rational solution in t if and only if the discriminant is a square, hence if and only if $20a^4 \mp 1 = b^2$ for some integer b. Looking modulo 4 shows that the sign must be $+$, so $b^2 = 20a^4 + 1$, and hence $(2b)^2 = 5(2a)^4 + 4$. This is one of the equations that we will solve in Corollary 6.8.4 as a consequence of our study of squares in Lucas and Fibonacci sequences. We deduce from that corollary that $a = 0$ or $a = \pm 6$. The value $a = 0$ leads to $t = -1$ (giving the universally trivial solution $(x, y) = (1, 0)$), and $a = \pm 6$ leads to $t = -19$, $(x, y) = (55, \pm 22434)$ and $t = -341$, $(x, y) = (377, \pm 2758646)$.   □

The following is a strengthening of Corollary 6.7.4 in the case $p = 5$.

Corollary 6.7.8. *Let x and y be integers such that the pair (x, y) is not equal to $(1, 0)$, $(55, \pm 22434)$, or $(377, \pm 2758646)$. Assume that $t = y^2 - x^5$ satisfies $H(t)$ (so that in particular x and y are coprime and x is odd). Then the class number of the imaginary quadratic field $\mathbb{Q}(\sqrt{t})$ is divisible by 5.*

Proof. Clear.   □

Note that it is necessary to impose *some* conditions on x and y. For instance, if $(x, y) = (2, 5)$ we have $t = -7 \equiv 1 \pmod{8}$, but the class number of $\mathbb{Q}(\sqrt{-7})$ is 1. However, it can be shown that if we assume only that x and y are coprime, but t not necessarily squarefree, then the class number of the quadratic *order* of discriminant t (or $4t$ if $t \equiv 2$ or 3 modulo 4) is divisible by 5.

6.7.4 Application of the Bilu–Hanrot–Voutier Theorem

To treat the case $p \geqslant 7$, we use a remarkable theorem of the above authors. We need a definition.

Definition 6.7.9. *Let α and β be such that $\alpha + \beta$ and $\alpha\beta$ are nonzero coprime integers and such that α/β is not a root of unity.*

(1) *The Lucas sequence associated with α, β is the sequence defined by $u_n = u_n(\alpha, \beta) = (\alpha^n - \beta^n)/(\alpha - \beta)$ for $n \in \mathbb{Z}_{\geqslant 0}$.*
(2) *We say that a prime number p is a primitive divisor of u_n if $p \mid u_n$ but $p \nmid u_i$ for $0 < i < n$ and $p \nmid (\alpha - \beta)^2$.*

Note that I use the original name "Lucas sequence," but it should more properly be called a generalized Fibonacci sequence since the usual Lucas sequence is rather $u_n = \alpha^n + \beta^n$.

A special case of the theorem of Bilu, Hanrot, and Voutier is the following (see [Bil-Han-Vou], and also [Abou] for a slight addition):

Theorem 6.7.10. *Let $u_n = u_n(\alpha, \beta)$ be a Lucas sequence as above. Then*

(1) *If $n > 30$, u_n always has a primitive divisor.*
(2) *If $5 < n < 30$ is prime and $u_n(\alpha, \beta)$ has no primitive divisors, then either $n = 7$ and $(\alpha, \beta) = ((1 + \sqrt{-7})/2, (1 - \sqrt{-7})/2)$, or $n = 7$ and $(\alpha, \beta) = ((1 + \sqrt{-19})/2, (1 - \sqrt{-19})/2)$, or $n = 13$ and $(\alpha, \beta) = ((1 + \sqrt{-7})/2, (1 - \sqrt{-7})/2)$.*

This theorem solves a century-old problem, and its proof involves both very delicate estimates on linear forms in logarithms and new algorithms for solving Thue equations. It is thus a beautiful mixture of difficult mathematics with an extensive rigorous computer computation (as the epigraph of the paper remarkably illustrates).

An immediate application of the above theorem to our problem is the following.

Corollary 6.7.11. *Let $p \geqslant 7$ be prime, and assume $H(p, t)$. The only value of t for which the equation $y^2 = x^p + t$ has a solution is $t = -1$ with only solution $(x, y) = (1, 0)$.*

Proof. Indeed, the equation defining the set $A_p(t)$ is $(\alpha^p - \beta^p)/(\alpha - \beta) = \pm 1$ with $\alpha = a + \sqrt{t}$ and $\beta = a - \sqrt{t}$; hence with the notation of the above definition $u_p(\alpha, \beta) = \pm 1$. We have $0 \in A_p(t)$ if and only if $t = \pm 1$, so that

$t = -1$ since we assume $t < 0$. Otherwise, it is clear that $\alpha + \beta$ and $\alpha\beta$ are integers, and α/β belongs to the imaginary quadratic field $\mathbb{Q}(\sqrt{t})$. Since $\alpha/\beta \neq \pm 1$ for $a \neq 0$ it can be a root of unity only when $t = -1$ or $t = -3$. However, it is easily checked that for $t = -1$ the only nonzero values of a such that α/β is a root of unity are $a = \pm 1$, and for $t = -3$ they are $a = \pm 1$ and $a = \pm 3$; see Exercise 41. In all these cases the same exercise shows that for $p \geqslant 7$ we have $|u_p(\alpha, \beta)| > 1$. Thus these cases do not give any elements of $A_p(t)$, so we may apply the above theorem. Thus for $p > 30$, $A_p(t)$ must have a primitive divisor, and in particular it cannot be equal to ± 1, while for $7 \leqslant p < 30$ all the possibilities listed in the theorem give $\alpha = (u + \sqrt{v})/2$ with u and v odd, which thus cannot be of the form $a + \sqrt{t}$. $\qquad\square$

I would again like to emphasize that the above corollary, which immediately follows from the theorem of Bilu, Hanrot, and Voutier, is very deep.

6.7.5 Special Cases with Fixed t

The above corollary essentially solves the problem when the condition $H(p,t)$ is satisfied. However, it is not completely satisfactory for two reasons. The first is mathematical: we must comment on what we can do when the condition $H(p,t)$ is not satisfied, although we will have to assume $H(t)$. The second is pedagogical: in the study of Catalan's equation we will need the case $t = -1$, and to be entirely self-contained we treat it without using the difficult theorem of Bilu–Hanrot–Voutier. The following result is due to V.-A. Lebesgue (see [Leb]).

Proposition 6.7.12 (V.-A. Lebesgue). *For $p \geqslant 3$ prime the only integral solution to the equation $y^2 = x^p - 1$ is $(x,y) = (1,0)$.*

Proof. First note that since the class number of $\mathbb{Q}(\sqrt{-1})$ is equal to 1 the condition $H(p,-1)$ is satisfied for all p. Furthermore, $t = -1$ does not occur in the special cases, so we must show that 0 is the only element of $A_p(t)$. Thus, let $a \in A_p(t)$, so that by definition we have

$$\sum_{k=0}^{(p-1)/2} \binom{p}{2k} a^{2k} (-1)^{(p-1)/2-k} = \pm 1 \ .$$

Since $p \mid \binom{p}{2k}$ for $1 \leqslant k \leqslant (p-1)/2$ and $p \geqslant 3$ it follows by looking at the equation modulo p that the right-hand side is equal to $(-1)^{(p-1)/2}$. We thus have $L = 0$ with

$$L = \sum_{k=1}^{(p-1)/2} \binom{p}{2k} a^{2k} (-1)^{(p-1)/2-k} \ .$$

I claim that a is even. Indeed, otherwise looking at the equation modulo 2 we would obtain

$$\sum_{k=0}^{(p-1)/2} \binom{p}{2k} \equiv 1 \pmod 2 ,$$

which is absurd since the left-hand side is equal to 2^{p-1}, which is even. Now set

$$u_k = \binom{p}{2k} a^{2k} = \frac{p(p-1)}{2k(2k-1)} \binom{p-2}{2k-2} a^{2k} .$$

Since $u_1 = p(p-1)a^2/2$, we have

$$\frac{u_k}{u_1} = \frac{1}{k(2k-1)} \binom{p-2}{2k-2} a^{2k-2} ,$$

so that for $k > 1$ (hence $p > 3$) we have

$$v_2(u_k) - v_2(u_1) \geqslant (2k-2)v_2(a) - v_2(k) \geqslant (2k-2) - v_2(k)$$

since a is even. It is immediately checked that the rightmost expression is always greater than or equal to 1, hence that $v_2(u_k) > v_2(u_1)$ for $k > 1$. Since $L = \sum_{k=1}^{(p-1)/2}(-1)^{(p-1)/2-k} u_k$ it follows that $v_2(L) = v_2(u_1) = v_2((p(p-1)/2)a^2)$, which is impossible if $a \neq 0$ since $L = 0$. Thus we must have $a = 0$, proving the proposition. $\qquad\square$

It is easy to generalize the above reasoning to other values of t (see Exercises 38 and 39), but thanks to the Bilu–Hanrot–Voutier theorem we do not need to do so. In fact, for small values of t satisfying $H(t)$ we have the following definitive result:

Theorem 6.7.13. *Assume $H(t)$, in other words that t is a squarefree negative integer such that $t \not\equiv 1 \pmod 8$. For $n \geqslant 3$ and $-100 \leqslant t \leqslant -1$ the Diophantine equations $y^2 = x^n + t$ do not have any integral solutions except for the solutions with $y = 0$ when $t = -1$, and for the pairs (t,n) given in the table below, for which the only solutions (x,y) are as indicated.*

Proof. For $t = -1$, we have the trivial solutions $(x,y) = (1,0)$ if n is odd, $(x,y) = (\pm 1, 0)$ if n is even. Thanks to Proposition 6.7.1 we know that for $-100 < t < -1$ the only ones for which there are solutions with n even are the given ones. Otherwise we may restrict to $n = p$ an odd prime and deduce the others from that case. When the condition $H(p,t)$ is satisfied, we obtain the equations and the solutions of the theorem. The only values of t such that $-100 < t < -1$ for which $H(t)$ is true but the condition $H(p,t)$ is not satisfied are $t = -26$, -29, -38, -53, -59, -61, -83, and -89 for $p = 3$, and $t = -74$ and $t = -86$ for $p = 5$. In the case $p = 3$ we must find all integral solutions to $y^2 = x^3 + t$ for the eight given values of t. This is done without difficulty using the techniques of Section 8.7; see Exercise 29 of Chapter 8. On the other hand, in the case $p = 5$ and $t = -74$ or $t = -86$, we must find the integral points on $y^2 = x^5 + t$, which is a hyperelliptic curve of genus 2. This is more difficult, and I refer either to Chapter 13 for the general methods of attack of this kind of problem, or to the original paper by Mignotte and de Weger [Mig-Weg].   □

(t,n)	(x,y)	(t,n)	(x,y)
$(-2,3)$	$(3,\pm5)$	$(-11,3)$	$(3,\pm4)$ and $(15,\pm58)$
$(-13,3)$	$(17,\pm70)$	$(-17,4)$	$(\pm3,\pm8)$
$(-19,3)$	$(7,\pm18)$	$(-19,5)$	$(55,\pm22434)$
$(-26,3)$	$(3,\pm1)$ and $(35,\pm207)$	$(-35,3)$	$(11,\pm36)$
$(-53,3)$	$(9,\pm26)$ and $(29,\pm156)$	$(-53,6)$	$(3,\pm26)$
$(-61,3)$	$(5,\pm8)$	$(-65,4)$	$(\pm3,\pm4)$
$(-67,3)$	$(23,\pm110)$	$(-74,3)$	$(99,\pm985)$
$(-74,5)$	$(3,\pm13)$	$(-77,4)$	$(\pm3,\pm2)$
$(-83,3)$	$(27,\pm140)$	$(-83,9)$	$(3,\pm140)$
$(-89,3)$	$(5,\pm6)$	$(-97,4)$	$(\pm7,\pm48)$

Solubility of $y^2 = x^n + t$ for (t,n) as in the Theorem

6.7.6 The Equations $ty^2 + 1 = 4x^p$ and $y^2 + y + 1 = 3x^p$

We study these equations in this section because the methods are very similar, and because we will need the second one in Chapter 16 in the proof of Theorem 16.1.11.

Proposition 6.7.14. *Let $t \in \mathbb{Z}_{\geq 1}$ and let p be an odd prime not dividing the class number of the imaginary quadratic field $K = \mathbb{Q}(\sqrt{-t})$. The equation $ty^2 + 1 = 4x^p$ has no integer solutions except for $t = 3$, for which it has the solutions $(x,y) = (1,\pm1)$ for any p.*

Proof. If the equation has a solution then $ty^2 \equiv 3 \pmod 4$, so y is odd, and hence $t \equiv 3 \pmod 4$. Furthermore, writing $-t = Df^2$ for a fundamental discriminant D, our equation is equivalent to $-D(fy)^2 + 1 = 4y^p$, so we may assume that $-t$ is a fundamental discriminant. In K our equation can be written $\beta\bar\beta = x^p$ with $\beta = (1 + y\sqrt{-t})/2$. Since $\beta + \bar\beta = 1$ the ideals $\beta\mathbb{Z}_K$ and $\bar\beta\mathbb{Z}_K$ are coprime, so that each is a pth power of an ideal. Since p does not divide the class number of K, as in the proof of Proposition 6.7.3 we deduce that there exist $\alpha \in \mathbb{Z}_K$ and a unit ε of K such that $\beta = \varepsilon\alpha^p$. Since K is imaginary quadratic ε is a pth power, except perhaps in the case $(p,t) = (3,3)$, which we leave to the reader (Exercise 42), so that we can simply write $\beta = \alpha^p$. Setting $\alpha = (a + b\sqrt{-3})/2$ with a and b in $\mathbb{Z}$, since p is odd the equation $\beta + \bar\beta = 1$ translates into $(\gamma + a)^p - \gamma^p = 1$, with $\gamma = -\bar\alpha = (-a + b\sqrt{-t})/2 \in \mathbb{Z}_K$. Expanding the left-hand side of this equation by the binomial theorem we see that it is divisible by a, so that a is a unit, and therefore $a = \pm1$. On the other hand, looking at the equation modulo $p\mathbb{Z}_K$ gives $a \equiv a^p \equiv 1 \pmod p$, so that we must have $a = 1$ since

$p \neq 2$. Since $\gamma \neq 0$ (otherwise $\beta = (-\overline{\gamma})^p = 0$), we deduce that $P(\gamma) = 0$, where

$$P(X) = \frac{(X+1)^p - X^p - 1}{pX}.$$

Since p is prime it is clear from the binomial theorem that $P \in \mathbb{Z}[X]$ and that P is monic with constant term 1, so that the roots of P are *units*. When $t \neq 3$ the units of K are ± 1, so the equation $P(\gamma) = 0$ with $\gamma \in K$ implies that $\gamma = \pm 1$, so that $\beta = (1 + y\sqrt{-t})/2 = (-\overline{\gamma})^p = \pm 1$, which is impossible. When $t = 3$ the units of K are the 6th roots of unity, so similarly $\beta = (1 + y\sqrt{-3})/2 = (-\overline{\gamma})^p$ is a 6th root of unity. The 6th roots of unity of the form $(1 + y\sqrt{-3})/2$ being $(1 \pm \sqrt{-3})/2$, we have $y = \pm 1$, giving the two solutions for $t = -3$, and if $-t = -3f^2$ for $f > 1$ it also shows that there are no solutions for such t.   □

Corollary 6.7.15 (Nagell). *Let $p \geq 3$ be a prime. The only integer solutions to the equation $y^2 + y + 1 = 3x^p$ are $(x, y) = (1, 1)$ and $(x, y) = (1, -2)$.*

Proof. This equation is equivalent to $(2y + 1)^2 + 3 = 12x^p$; it follows that $3 \mid (2y + 1)$, so setting $2y + 1 = 3z$ with z odd we obtain $3z^2 + 1 = 4x^p$, and the result follows from the proposition.   □

6.8 Linear Recurring Sequences

6.8.1 Squares in the Fibonacci and Lucas Sequences

We have already seen in Section 4.5.4 how to apply p-adic methods to find specific values of linear recurring sequences. I emphasize the fact that these methods (for instance using Strassmann's theorem) are really p-adic in nature, and not simply based on simple congruence arguments. In the present section, we will study similar problems that on the other hand can be solved by congruence arguments and quadratic reciprocity.

We let $\alpha = (1 + \sqrt{5})/2$ and $\beta = (1 - \sqrt{5})/2$ be the two roots of the equation $x^2 - x - 1 = 0$. Recall that we define the classical Fibonacci and Lucas sequences F_n and L_n by the formulas

$$F_n = \frac{\alpha^n - \beta^n}{\alpha - \beta} \quad \text{and} \quad L_n = \alpha^n + \beta^n,$$

so that these sequences both satisfy the linear recurrence $u_{n+1} = u_n + u_{n-1}$, with initial terms $F_0 = 0$, $F_1 = 1$ and $L_0 = 2$, $L_1 = 1$.

Before stating the Diophantine theorems, we need some elementary properties of these sequences, summarized in the following proposition.

Proposition 6.8.1. (1) $L_{-n} = (-1)^n L_n$, $F_{-n} = (-1)^{n-1} F_n$, $L_n^2 - 5F_n^2 = 4(-1)^n$, $2L_{m+n} = 5F_m F_n + L_m L_n$, $2F_{m+n} = F_m L_n + F_n L_m$, $L_{2m} = L_m^2 + 2(-1)^{m-1}$, $F_{2m} = F_m L_m$.

(2) $\gcd(L_n, F_n) = 1$ *if* $3 \nmid n$, $\gcd(L_n, F_n) = 2$ *if* $3 \mid n$.
(3) *When* $k \equiv \pm 2 \pmod 6$ *then for all* $t \in \mathbb{Z}$ *we have*

$$L_{n+2kt} \equiv (-1)^t L_n \pmod{L_k} \quad \text{and} \quad F_{n+2kt} \equiv (-1)^t F_n \pmod{L_k} .$$

Proof. The formulas of (1) are proved by direct computation from the definitions in terms of α and β, which can be summarized by the equality $(L_n + F_n\sqrt5)/2 = \alpha^n$. For (2), we note that since $|L_n^2 - 5F_n^2| = 4$, the GCD of L_n and F_n is equal to 1 or 2. Because of that same formula it is equal to 2 if and only if $2 \mid F_n$, and since evidently the sequence F_n modulo 2 is periodic of period 3, we see that $2 \mid F_n$ if and only if $3 \mid n$, proving (2).

For (3), since $2 \mid k$ we have

$$2L_{n+2k} = 5F_n F_{2k} + L_n L_{2k} = 5F_n F_k L_k + L_n(L_k^2 - 2) \equiv -2L_n \pmod{L_k} .$$

Since $3 \nmid k$ we have $2 \nmid L_k$, hence $L_{n+2k} \equiv -L_n \pmod{L_k}$, so the result for L follows by induction on t. Similarly

$$2F_{n+2k} = F_n L_{2k} + F_{2k} L_n = F_n(L_k^2 - 2) + F_k L_k L_n \equiv -2F_n \pmod{L_k} ,$$

and we conclude as before.

Theorem 6.8.2. (1) *For* $n \geq 0$ *we have* $L_n = x^2$ *with* $x \in \mathbb{Z}$ *if and only if* $n = 1$ *or* $n = 3$.
(2) *Similarly* $L_n = 2x^2$ *if and only if* $n = 0$ *or* $n = 6$.

Proof. Since $L_{2m} = L_m^2 + 2(-1)^{m-1}$, $L_{2m} = x^2$ implies that $|x^2 - L_m^2| = 2$, which is impossible. Thus we may assume that n is odd. Clearly $L_1 = 1$ and $L_3 = 4$ are squares, so we may assume that $n > 3$. We can write $n = r + 2 \cdot 3^s k$ with $r = 1$ or 3, $2 \mid k$, and $3 \nmid k$, so $k \equiv \pm 2 \pmod 6$ and $k > 0$. By the above proposition we thus have

$$L_n \equiv (-1)^{3^s} L_r \equiv -L_r \pmod{L_k} .$$

On the other hand, since $L_1 = 1$ and $L_3 = 4$, we have $-L_r = -1$ or -4. Note also that since $k/2 \equiv \pm 1 \pmod 3$, we have

$$L_k = L_{k/2}^2 + 2(-1)^{k/2-1} \equiv 1 \pm 2 \equiv 3 \pmod 4 ,$$

hence

$$\left(\frac{L_n}{L_k} \right) = \left(\frac{-L_r}{L_k} \right) = \left(\frac{-1}{L_k} \right) = -1 .$$

If follows that L_n cannot be a square, proving (1).

To prove (2), let $L_n = 2x^2$ with $x \in \mathbb{Z}$. We consider several cases.

- If n is odd, then $4x^4 = L_n^2 = 5F_n^2 - 4$; hence F_n is even, so that $x^4 = 5(F_n/2)^2 - 1$. But $x^4 \equiv 0$ or 1 modulo 8, so $(F_n/2)^2 \equiv 5$ or 2 modulo 8, a contradiction.

- If $4 \mid n$ and $n \neq 0$, we can write $n = 2 \cdot 3^s k$ with $k \equiv \pm 2 \pmod 6$, so that by the proposition

$$2L_n \equiv -2L_0 \equiv -4 \pmod{L_k} ;$$

hence as above, $\left(\frac{2L_n}{L_k}\right) = -1$, so that $2L_n$ cannot be a square.

- If $n \equiv 6 \pmod 8$ and $n \neq 6$, we can write $n = 6 + 2 \cdot 3^s k$ with $k \equiv \pm 2 \pmod 6$, so that by the proposition

$$2L_n \equiv -2L_6 \equiv -36 \pmod{L_k} .$$

On the other hand, note that $3 \mid L_m$ if and only if $m \equiv 2 \pmod 4$, so that since $4 \mid k$ we have $3 \nmid L_k$. Thus as above $\left(\frac{2L_n}{L_k}\right) = -1$, so that $2L_n$ cannot be a square.

- If $n \equiv 2 \pmod 8$ then $L_{-n} = L_n$ and $-n \equiv 6 \pmod 8$, so the preceding reasoning (which is applicable for $n < 0$) shows that $L_n = 2x^2$ if and only if $-n = 6$, and in particular $n < 0$.   □

Theorem 6.8.3. (1) *We have $F_n = x^2$ with $x \in \mathbb{Z}$ if and only if $n = 0$, ± 1, 2, or 12.*

(2) *Similarly $F_n = 2x^2$ if and only if $n = 0$, ± 3, or 6.*

Proof. The proof of this theorem is similar and left to the reader (see Exercise 43).   □

As an application of the above theorem, we give the following corollary.

Corollary 6.8.4. *Consider the Diophantine equation*

$$y^2 = 5x^4 + a .$$

(1) *For $a = 1$, the only integral solutions are $(x, y) = (0, \pm 1)$ and $(\pm 2, \pm 9)$.*

(2) *For $a = -1$, the only integral solutions are $(x, y) = (\pm 1, \pm 2)$.*

(3) *For $a = 4$, the only integral solutions are $(x, y) = (0, \pm 2)$, $(\pm 1, \pm 3)$, $(\pm 12, \pm 322)$.*

(4) *For $a = -4$, the only integral solutions are $(x, y) = (\pm 1, \pm 1)$.*

Proof. If we write the equations as $y^2 - 5x^4 = \varepsilon b^2$ with $\varepsilon = \pm 1$ and $b = 1$ or $b = 2$, we see that we can apply the solution to the Pell equation $y^2 - 5X^2 = \pm 1$ or ± 4. The fundamental unit of the order $\mathbb{Z}[\sqrt 5]$ is $2 + \sqrt 5 = \alpha^3$, of norm -1, so the general solution with $X \geq 0$ to $y^2 - 5X^2 = \pm 1$ is given by $y + X\sqrt 5 = \alpha^{3n}$, and hence

$$2x^2 = 2X = \frac{\alpha^{3n} - \beta^{3n}}{\alpha - \beta} = F_{3n} .$$

By the above theorem, this implies $n = 0$, ± 1, or 2. The value $n = 0$ gives the solution $x = 0$ to the first equation (hence $y = \pm 1$), and $n = 2$ gives the solution $x = \pm 2$ to the first equation (hence $y = \pm 9$). On the other hand, $n = \pm 1$ gives the solution $x = \pm 1$ to the second equation (hence $y = \pm 2$).

Similarly the general solution to $y^2 - 5X^2 = \pm 4$ is given by $(y + X\sqrt{5})/2 = \alpha^n$, so $x^2 = X = F_n$. By the above theorem this implies $n = 0, \pm 1, 2$, or 12. The values $n = 0, 2$, and 12 give the solutions $x = 0$ (hence $y = \pm 2$), $x = \pm 1$ (hence $y = \pm 3$), and $x = \pm 12$ (hence $y = \pm 322$) to the third equation, while $n = \pm 1$ gives the solution $x = \pm 1$ (hence $y = \pm 1$) to the fourth equation. □

Remark. Using a combination of Baker-type methods giving lower bounds for linear forms in two or three logarithms of algebraic numbers (see Chapter 12), and the Ribet–Taylor–Wiles level lowering method (see Chapter 15), Y. Bugeaud, M. Mignotte, and S. Siksek have recently proved the following remarkable result:

Theorem 6.8.5 (Bugeaud–Mignotte–Siksek). (1) *The only nontrivial perfect powers in the Fibonacci sequence are $F_1 = F_2 = 1$, $F_6 = 8$, and $F_{12} = 144$.*
(2) *The only nontrivial perfect powers in the Lucas sequence are $L_1 = 1$ and $L_3 = 4$.*

We also mention without proof the following results of the same type.

Theorem 6.8.6 (Ljunggren, Ellenberg). *The only integral solutions to the Diophantine equation $y^2 = 2x^4 - 1$ are $(x, y) = (\pm 1, \pm 1)$ and $(x, y) = (\pm 13, \pm 239)$.*

Theorem 6.8.7 (Cohn, Ljunggren). (1) *For fixed D, the Diophantine equation $x^4 - Dy^2 = 1$ has at most one integral solution with $x > 0$ and $y > 0$, except for $D = 1785$, for which it has the two solutions $(x, y) = (13, 4)$ and $(239, 1352)$.*
(2) *If D is prime, the above equation has such a solution if and only if $D = 5$ or $D = 29$, for which it has the respective solutions $(x, y) = (3, 4)$ and $(x, y) = (99, 1820)$.*

6.8.2 The Square Pyramid Problem

A classical problem due to E. Lucas asks for all integral solutions to the equation $y^2 = 1^2 + 2^2 + \cdots + x^2$, in other words to the Diophantine equation $x(x + 1)(2x + 1) = 6y^2$ (this problem is equivalent to finding all possible numbers of "cannonballs" that can be piled up in a square pyramid as well as on one level as a square, so it is also called the cannonball problem). This problem was until relatively recently solved only using rather sophisticated methods, but in the 1980s a completely elementary proof was found, which we paraphrase in this section; see [Ma] and [Ang]. We need some preliminary results.

Lemma 6.8.8. *The only integral solutions to the Diophantine equation* $y^2 = 8x^4 + 1$ *are* $(x, y) = (0, \pm 1)$ *and* $(\pm 1, \pm 3)$.

Proof. We may assume $y > 0$. Since y is odd we can write $y = 2s + 1$, so that $2x^4 = s(s + 1)$. If s is even there exist coprime integers u and v such that $s = 2u^4$ and $s + 1 = v^4$, so that $1 + 2u^4 = v^4$. By Proposition 6.5.4 this implies $u = 0$, hence $s = 0$, $x = 0$, and $y = \pm 1$. So assume that s is odd. In this case there exist coprime integers u and v such that $s = u^4$ and $s + 1 = 2v^4$, so that $u^4 + 1 = 2v^4$. This implies that u is odd, hence (by looking modulo 8) that v is odd. If we set $a = |v^2 - u|$ and $b = |v^2 + u|$, we see that $a^2 + b^2 = 2v^4 + 2u^2 = (u^2 + 1)^2$, so that $(a, b, u^2 + 1)$ are the three sides of a Pythagorean triangle, and its area is equal to $ab/2 = |(v^4 - u^2)/2| = ((u^2 - 1)/2)^2$, a square. Since we will show that 1 is not a congruent number (Proposition 6.12.2), it follows that the triangle must be degenerate, i.e., that $u = \pm 1$, so that $s = 1$, $x = \pm 1$, and $y = 3$, proving the lemma.   □

Remark. It would have been more pleasing to consider the more general equation $y^2 = 8x^4 + z^4$ with x, y, and z pairwise coprime. Unfortunately, as is shown in Exercise 13 of Chapter 8, this equation has an *infinity* of integral solutions (because the corresponding elliptic curve has nonzero rank), so the result would not be suitable for our purposes.

We can now solve Lucas's problem for even values of x.

Proposition 6.8.9. *The only integral solutions to the Diophantine equation* $x(x + 1)(2x + 1) = 6y^2$ *with* $y \neq 0$ *and* x *even are* $(x, y) = (24, \pm 70)$.

Proof. Clearly x is nonnegative. Since x is even and x, $x + 1$, and $2x + 1$ are pairwise coprime, the equation implies that the odd numbers $x + 1$ and $2x + 1$ are either squares or triples of squares. It follows that $x + 1 \not\equiv 2$ (mod 3) and $2x + 1 \not\equiv 2$ (mod 3), which is equivalent to $x \equiv 0$ (mod 3). Thus we can find integers s, t, and u such that $x = 6s^2$, $x + 1 = t^2$, and $2x + 1 = u^2$, and of course u and t are odd and coprime. The equality $6s^2 = (u - t)(u + t)$ implies that $4 \mid 6s^2$ hence that $2 \mid s$. Thus write $s = 2v$, so that $6v^2 = ((u - t)/2)((u + t)/2)$. Since u and t are coprime so are $(u - t)/2$ and $(u + t)/2$. Changing if necessary the signs of u and t, it follows that there exist integers a and b such that either $(u + t)/2 = 6a^2$ and $(u - t)/2 = b^2$, or $(u + t)/2 = 3a^2$ and $(u - t)/2 = 2b^2$. In the first case we have $t = 6a^2 - b^2$ and $s = 2ab$, and since $6s^2 + 1 = t^2$ we obtain the equation $24a^2b^2 + 1 = (6a^2 - b^2)^2$, hence $36a^4 - 36a^2b^2 + b^4 = 1$, which can be rewritten by completing the square $(6a^2 - 3b^2)^2 - 8b^4 = 1$. By the above lemma, since $3 \mid (6a^2 - 3b^2)$ we have $b = \pm 1$ and $a = \pm 1$, giving $s = 2ab = \pm 2$, so that $x = 24$ and hence $y = \pm 70$. In the second case we have $t = 3a^2 - 2b^2$ and $s = 2ab$, giving here the equation $24a^2b^2 + 1 = (3a^2 - 2b^2)^2$, hence $9a^4 - 36a^2b^2 + 4b^4 = 1$, so that $1 + 2(2b)^4 = (3a^2 - 6b^2)^2$. By Proposition 6.5.4 (which is stronger than what we need) we deduce that $b = 0$, hence $9a^4 = 1$, which is impossible, so there are no solutions in the second case.   □

To solve the problem in the odd case we make an analysis similar to that done in Section 6.8.1. We set $\alpha = 2 + \sqrt{3}$, $\beta = 2 - \sqrt{3}$, $M_n = (\alpha^n + \beta^n)/2$, and $G_n = (\alpha^n - \beta^n)/(\alpha - \beta)$. The reason for dividing by 2 in M_n is that $\alpha^n + \beta^n$ is trivially always an even integer. Clearly M_n and G_n both satisfy the linear recurrence $u_{n+1} = 4u_n - u_{n-1}$ with initial terms $M_0 = 1$, $M_1 = 2$, $G_0 = 0$, $G_1 = 1$. We have of course an exact analogue of Proposition 6.8.1, which is in fact slightly simpler since $\alpha\beta = 1$ instead of -1 (i.e., a fundamental unit of norm 1) and since there is no denominator 2 in α and β (the full ring of integers of $\mathbb{Q}(\sqrt{3})$ is $\mathbb{Z}[\sqrt{3}]$).

Proposition 6.8.10. (1) $M_{-n} = M_n$, $G_{-n} = -G_n$, $M_n^2 - 3G_n^2 = 1$, $M_{m+n} = 3G_mG_n + M_mM_n$, $G_{m+n} = G_mM_n + G_nM_m$, $M_{2m} = 2M_m^2 - 1$, $G_{2m} = 2G_mM_m$.
(2) $\gcd(M_n, G_n) = 1$.
(3) *For any integers k and t in $\mathbb{Z}$ we have*

$$M_{n+2kt} \equiv (-1)^t M_n \pmod{M_k} \quad and \quad G_{n+2kt} \equiv (-1)^t G_n \pmod{M_k} .$$

Proof. Essentially identical to the proof of Proposition 6.8.1, this time using $M_n + G_n\sqrt{3} = \alpha^n$.   □

Lemma 6.8.11. *Assume that n is even. Then M_n is odd, $5 \nmid M_n$, $\left(\frac{5}{M_n}\right) = 1$ if and only if $3 \mid n$ and $\left(\frac{-2}{M_n}\right) = 1$ if and only if $4 \mid n$.*

Proof. Write $n = 2m$. Since $M_{2m} = 2M_m^2 - 1$, it is clear that M_n is odd, and since $M_m^2 \equiv 0$ or ± 1 modulo 5, $M_n \equiv -1, 1$, or 2 modulo 5 and in particular $5 \nmid M_n$. Since M_n satisfies a linear recurrence with integral coefficients it is periodic modulo k for any given k, and for $k = 5$ the period is clearly $(1, 2, 2)$ of length 3, and for $k = 8$ the period is $(1, 2, 7, 2)$ of length 4, leading immediately to the desired results.   □

The key proposition in the proof is the following result, first proved by Ma in [Ma].

Proposition 6.8.12 (Ma). *If $n \geqslant 0$, then M_n has the form $4x^2 + 3$ if and only if $n = 2$, so $M_n = 7$.*

Proof. Assume that $M_n = 4x^2 + 3$, so that $M_n \equiv 3$ or 7 modulo 8. Since the period of M_n modulo 8 is $(1, 2, 7, 2)$, this implies that $n \equiv 2 \pmod 4$, or equivalently, that $n \equiv \pm 2 \pmod 8$. Assume that $n > 2$, hence that $M_n > 7$, and write $n = 2t \cdot 2^s \pm 2$ with t odd and $s \geqslant 2$ (for $n = 2$ we could not choose t odd). By Proposition 6.8.10 we have

$$M_n \equiv (-1)^t M_{\pm 2} \equiv -7 \pmod{M_{2^s}} ,$$

so $4x^2 \equiv -10 \pmod{M_{2^s}}$. Since $s \geqslant 2$, M_{2^s} is odd, so it follows that

$$\left(\frac{-2}{M_{2^s}}\right)\left(\frac{5}{M_{2^s}}\right) = \left(\frac{-10}{M_{2^s}}\right) = \left(\frac{4x^2}{M_{2^s}}\right) = 1$$

since everything is nonzero by the above lemma. On the other hand, the above lemma also tells us that $\left(\frac{-2}{M_{2^s}}\right) = 1$ since $s \geqslant 2$, and that $\left(\frac{5}{M_{2^s}}\right) = -1$ since $3 \nmid 2^s$, so we obtain a contradiction, proving the proposition. $\square$

Thanks to this proposition we can now completely solve Lucas's problem.

Theorem 6.8.13. *The only integral solutions with $y \neq 0$ to the Diophantine equation $x(x+1)(2x+1) = 6y^2$ are $(x, y) = (1, \pm 1)$ and $(x, y) = (24, \pm 70)$.*

Proof. The case that x is even has been proved in Proposition 6.8.9. So assume that x is odd. As in the even case, since x, $x+1$, and $2x+1$ are pairwise coprime, x is either a square or three times a square, so $x \not\equiv 2 \pmod 3$. Since $x + 1$ is even and the other two factors are odd, it is either twice a square or six times a square, so $x + 1 \not\equiv 1 \pmod 3$. Thus $x \equiv 1 \pmod 3$, so $x + 1 \equiv 2$ (mod 3) and $2x + 1 \equiv 0 \pmod 3$. It follows that there exist pairwise coprime integers s, t, and u such that $x = s^2$, $x + 1 = 2t^2$, and $2x + 1 = 3u^2$. We thus obtain the equation

$$(6u^2 + 1)^2 - 3(4tu)^2 = (4x + 3)^2 - 8(x+1)(2x+1) = 1 \ ,$$

and since $2 + \sqrt{3}$ is the fundamental unit of $\mathbb{Z}[\sqrt{3}]$, we deduce that there exists $n \in \mathbb{Z}$ such that, using the above notation,

$$6u^2 + 1 + 4tu\sqrt{3} = \pm(M_n + G_n\sqrt{3}) \ .$$

In particular, $M_n = \pm(6u^2 + 1) = 6u^2 + 1$, since $M_n \geqslant 0$. On the other hand, $6u^2 + 1 = 4x + 3 = 4s^2 + 3$, so that M_n has the form $4s^2 + 3$. By Ma's result above this implies that $n = \pm 2$ and $M_n = 7$; hence $s = \pm 1$, so $x = 1$, as claimed, finishing the proof of the theorem. $\square$

6.9 Fermat's "Last Theorem" $x^n + y^n = z^n$

6.9.1 Introduction

This is certainly the most famous of all Diophantine equations. It claims that for all $n \geqslant 3$ there are no integral solutions to $x^n + y^n = z^n$ with $xyz \neq 0$. We may clearly assume that x, y, and z are pairwise coprime. Furthermore, since any integer $n \geqslant 3$ is divisible either by an odd prime number or by 4, it is sufficient to prove the result for $n = 4$ and for $n = p$ an odd prime.

For $n = 2$ Fermat's equation does have solutions, which we have parametrized completely in Corollary 6.3.13, and which we restate as follows.

Proposition 6.9.1. *The general solution in $\mathbb{Z}$ to the equation $x^2 + y^2 = z^2$ is*

$$x = d(s^2 - t^2), \quad y = 2dst, \quad z = d(s^2 + t^2),$$

where s and t are coprime integers such that $s \not\equiv t \pmod{2}$ and $d \in \mathbb{Z}$, or the same with x and y exchanged. Furthermore, we have $\gcd(x, y) = \gcd(x, z) = \gcd(y, z) = |d|$.

Corollary 6.9.2. *The general solution of $x^2 + y^2 = z^2$ with $\gcd(x, y) = 1$ and x odd is*

$$x = s^2 - t^2, \quad y = 2st, \quad z = \pm(s^2 + t^2),$$

and the general solution of $x^2 - y^2 = z^2$ with $\gcd(x, y) = 1$ and x odd is

$$x = \pm(s^2 + t^2), \quad y = 2st, \quad z = s^2 - t^2,$$

where s and t are as above.

Proof. Immediate from the above proposition and the fact that two out of three sign changes can be included in the exchange of s and t or in the exchange of s with $-s$. □

We have shown in Theorem 6.4.17 that Fermat's equation does not have any solution for $n = 3$. We will give below a slightly different proof, which has the advantage of being generalizable to all *regular* prime exponents. In Proposition 6.5.3 we have shown that the equation $x^4 + y^4 = z^2$ has no nontrivial solutions, so a fortiori Fermat's equation does not have any nontrivial solution for $n = 4$. We may thus assume that n is an odd prime.

6.9.2 General Prime n: The First Case

From now on we assume that $n = p$ is an odd prime. In the traditional attacks on Fermat's "Last Theorem" (FLT for short), one distinguishes the so-called *first case* (or FLT I), in which we assume that $p \nmid xyz$, and the *second case* (or FLT II), in which we assume that $p \mid xyz$. In the modern attack that culminated in the work of Wiles and Taylor–Wiles solving FLT completely, the distinction between these two cases is unimportant.

The first case is much easier to treat (although nobody knows a complete proof of the first case using traditional methods). Using Wendt's criterion (Proposition 6.9.6), it is easy to check the validity of FLT I for a *given* p. Furthermore, using generalizations of Wieferich's criterion (Corollary 6.9.10), it is possible to check FLT I for *all* primes p up to a reasonable bound. This has in fact been done on a computer for $p < 10^{18}$ at least. On the other hand, the second case depends on some "luck" that in practice is always true but cannot be proved to be the case. Thus it may be that for some p the traditional method for proving FLT II fails, in that it does not succeed in proving the result (of course since FLT II is true by Wiles, it will not find a counterexample either).

There exist several elementary but nontrivial results on FLT I. We prove two of them. The first is a straightforward p-adic approach (more precisely a congruence approach modulo p^2), and I am indebted to A. Kraus for pointing it out in an unpublished manuscript. A second is a remarkable result due to Sophie Germain, and generalized by Wendt.

6.9.3 Congruence Criteria

We begin with the following.

Proposition 6.9.3. *The following three conditions are equivalent:*

(1) *There exist three p-adic units α, β, and γ such that $\alpha^p + \beta^p = \gamma^p$ (in other words, FLT I is soluble p-adically).*

(2) *There exist three integers a, b, c in $\mathbb{Z}$ such that $p \nmid abc$ with $a^p + b^p \equiv c^p$ (mod p^2).*

(3) *There exists $a \in \mathbb{Z}$ such that a is not congruent to 0 or -1 modulo p with $(a+1)^p \equiv a^p + 1$ (mod p^2).*

Proof. From the binomial theorem it is clear that if $u \equiv 1$ (mod $p\mathbb{Z}_p$) then $u^p \equiv 1$ (mod $p^2\mathbb{Z}_p$). Thus if $u \equiv v$ (mod $p\mathbb{Z}_p$) and u and v are p-adic units, then $u^p \equiv v^p$ (mod $p^2\mathbb{Z}_p$). We will use this several times without further mention. Taking a, b, and c to be residues modulo p of α, β, and γ thus shows that (1) implies (2). Conversely, assume (2). We would like to apply Hensel's lemma. However, the congruence is not quite good enough, so we have to do one step by hand. Let $a^p + b^p = c^p + kp^2$ for some $k \in \mathbb{Z}$, and set $d = c + kp$, so that $p \nmid d$. Then by the binomial theorem $d^p \equiv c^p + kp^2 c^{p-1}$ (mod p^3), so that

$$a^p + b^p - d^p \equiv kp^2(1 - c^{p-1}) \equiv 0 \ (\text{mod } p^3)$$

since $p \nmid c$. We can now apply Hensel's lemma (Proposition 4.1.37) to the polynomial $f(X) = X^p + b^p - d^p$ and to $\alpha = a$: we have $|f'(a)|_p = |pa^{p-1}|_p = 1/p$ since $p \nmid a$, while $|f(a)|_p \leqslant 1/p^3$ by the above, so $|f(a)|_p < |f'(a)|_p^2$, and hence Hensel's lemma is applicable, proving (1).

Clearly (3) implies (2). Conversely, assume (2), i.e., that $c^p \equiv a^p + b^p$ (mod p^2) with $p \nmid abc$. In particular, $c \equiv a + b$ (mod p). Thus, if we set $A = ba^{-1}$ modulo p, then by the above remark $A^p \equiv b^p a^{-p}$ (mod p^2) and $(A+1)^p \equiv c^p a^{-p}$ (mod p^2), so that $(A+1)^p \equiv A^p + 1$ (mod p^2), proving (3) and the proposition. □

Corollary 6.9.4. *FLT I cannot be proved by congruence conditions (i.e., p-adically) if and only if condition (3) of the proposition is satisfied for some a such that $1 \leqslant a \leqslant (p-1)/2$.*

Proof. Indeed, condition (3) is invariant when we change a modulo p, and also under the change $a \mapsto p - 1 - a$, so the result is clear. $\square$

Corollary 6.9.5. *If for all $a \in \mathbb{Z}$ such that $1 \leqslant a \leqslant (p-1)/2$ we have $(a+1)^p - a^p - 1 \not\equiv 0 \pmod{p^2}$, then the first case of FLT is true for p.*

Proof. Indeed, if $a^p + b^p = c^p$ with $p \nmid abc$ then condition (2) of the proposition is satisfied; hence by (3), as above there exists a such that $1 \leqslant a \leqslant (p-1)/2$ with $(a+1)^p - a^p - 1 \equiv 0 \pmod{p^2}$, proving the corollary. $\square$

For instance, thanks to this corollary we can assert that FLT I is true for $p = 3, 5, 11, 17, 23, 29, 41, 47, 53, 71, 89, 101, 107, 113, 131, 137, 149, 167, 173, 191, 197$, which are the prime numbers less than 200 satisfying the condition of the corollary.

We will see below that a theorem of Wieferich (Corollary 6.9.10) says that it is sufficient to take $a = 1$ in the above corollary, in other words that FLT I is true as soon as $2^p - 2 \not\equiv 0 \mod p^2$.

6.9.4 The Criteria of Wendt and Germain

As mentioned there is another elementary and more powerful approach to FLT I, initially due to Sophie Germain, and generalized by Wendt. Recall that we denote by $R(P, Q)$ the resultant of two polynomials P and Q.

Proposition 6.9.6 (Wendt). *Let $p > 2$ be an odd prime. If there exists an integer $k \geqslant 1$ such that $q = kp + 1$ is a prime number satisfying*

$$q \nmid (k^k - 1)R(X^k - 1, (X+1)^k - 1) ,$$

then FLT I is valid; in other words, $x^p + y^p + z^p = 0$ implies $p \mid xyz$.

Proof. Assume that $x^p + y^p + z^p = 0$ with $p \nmid xyz$. We may of course as usual assume that x, y, and z are pairwise coprime. We can write

$$-x^p = y^p + z^p = (y+z)(y^{p-1} - y^{p-2}z + \cdots + z^{p-1}) .$$

Clearly the two factors are relatively prime: we cannot have $p \mid (y+z)$, since otherwise $p \mid x$, and if $r \neq p$ is a prime dividing both factors then $y \equiv -z \pmod{r}$; hence the second factor is congruent to py^{p-1} modulo r, and since $r \neq p$ we have $r \mid y$, hence $r \mid z$, contradicting the fact that y and z are coprime. Since p is odd (otherwise we would have to include signs), it follows that there exist coprime integers a and s such that $y + z = a^p$ and $y^{p-1} - y^{p-2}z + \cdots + z^{p-1} = s^p$. By symmetry, there exist b and c such that $z + x = b^p$ and $x + y = c^p$.

Consider now the prime $q = kp + 1$. The Fermat equation implies that

$$x^{(q-1)/k} + y^{(q-1)/k} + z^{(q-1)/k} \equiv 0 \pmod{q} .$$

I claim that $q \mid xyz$. Indeed, assume by contradiction that $q \nmid xyz$, and let $u = (x/z)^{(q-1)/k} \bmod q$, which makes sense since $q \nmid z$. Since $q \nmid x$ we have $u^k - 1 \equiv 0 \pmod q$. On the other hand, $u + 1 \equiv -(y/z)^{(q-1)/k} \pmod q$, and since k is even and $q \nmid y$ we deduce that $(u+1)^k - 1 \equiv 0 \pmod q$. It follows that the polynomials $X^k - 1$ and $(X+1)^k - 1$ have the common root u modulo q, contradicting the assumption that $q \nmid R(X^k - 1, (X+1)^k - 1)$.

Thus $q \mid xyz$, and by symmetry we may assume for instance that $q \mid x$. Thus

$$0 \equiv 2x = (x+y) + (z+x) - (y+z) = c^p + b^p + (-a)^p$$
$$= c^{(q-1)/k} + b^{(q-1)/k} + (-a)^{(q-1)/k} \pmod q .$$

As above, it follows that $q \mid abc$. Since $q \mid x$ and x, y, and z are pairwise coprime, we cannot have $q \mid b^p = z + x$ or $q \mid c^p = x + y$. Thus $q \mid a$. It follows that $y \equiv -z \pmod q$, so that $s^p \equiv py^{p-1} \pmod q$. On the other hand, $y = (x+y) - x \equiv c^p \pmod q$, so that

$$s^{(q-1)/k} = s^p \equiv pc^{((q-1)/k)(p-1)} \pmod q ,$$

and since $q \nmid c$ we have $p \equiv d^{(q-1)/k} \pmod q$ with $d - s/c^{p-1}$ modulo q. Since a and s are coprime we have $q \nmid s$, and hence $q \nmid d$, so $p^k \equiv 1 \pmod q$. Since k is even it follows that

$$1 = (-1)^k = (kp - q)^k \equiv k^k p^k \equiv k^k \pmod q ,$$

contradicting the assumption that $q \nmid k^k - 1$.   □

A computer search shows that for every prime $p \geqslant 3$ up to very large bounds we can find an integer k satisfying the conditions of the proposition, and it can reasonably be conjectured that such a k always exists, so that in practice FLT I can always be checked thanks to this criterion; see also Exercise 47.

The following is S. Germain's initial criterion:

Corollary 6.9.7. *Let $p > 2$ be an odd prime, and assume that $q = 2p + 1$ is also a prime. Then FLT I is valid; in other words, if $x^p + y^p + z^p = 0$ then $p \mid xyz$.*

Proof. Since for $k = 2$ we have $(k^k - 1)R(X^k - 1, (X+1)^k - 1) = -3^2$, the condition of the proposition is $q \neq 3$, which is always true.   □

6.9.5 Kummer's Criterion: Regular Primes

A less elementary attack on FLT I uses algebraic number theory. It gives a result that is usually weaker than the above proposition since an infinity of p cannot be obtained by this attack. However, it has the great advantage that it can be generalized to the second case FLT II, while none of the above elementary approaches can.

In the sequel, we let $\zeta = \zeta_p$ be a primitive pth root of unity in $\mathbb{C}$, we let $K = \mathbb{Q}(\zeta)$, and we recall that the ring of integers of K is equal to $\mathbb{Z}[\zeta]$. We set $\pi = 1 - \zeta$, and recall that the ideal $\pi\mathbb{Z}_K$ is a prime ideal such that $(\pi\mathbb{Z}_K)^{p-1} = p\mathbb{Z}_K$, and p is the only prime number ramified in K. The first successful attacks on FLT were based on the possibility of unique factorization in $\mathbb{Z}[\zeta]$. Unfortunately this is true for only a limited number of small values of p. With the work of E. Kummer it was realized that one could achieve the same result with the much weaker hypothesis that p does not divide the class number h_p of $\mathbb{Z}_K$. Such a prime is called a *regular* prime. Note that it is known that there are infinitely many irregular (i.e., nonregular) primes, see Proposition 9.5.27, but that it is unknown (although widely believed) whether there are infinitely many regular primes. In fact, there should be a positive density equal to $\exp(-1/2)$ of regular primes among all prime numbers (see Exercise 48). The irregular primes below 100 are $p = 37$, 59, and 67. See Exercise 11 of Chapter 9 for an efficient regularity test.

We thus assume that p is a regular prime, i.e., that $p \nmid h_p$. The usefulness of this assumption comes from the following easy fact: if an ideal $\mathfrak{a}$ of K is such that $\mathfrak{a}^p$ is a principal ideal, then so is $\mathfrak{a}$ itself. Indeed, since p and h_p are coprime, we can find integers u and v such that $up + vh_p = 1$, so that $\mathfrak{a} = (\mathfrak{a}^p)^u(\mathfrak{a}^{h_p})^v$. Now by assumption $\mathfrak{a}^p$ is a principal ideal, and by definition of the class group, so is $\mathfrak{a}^{h_p}$, proving our claim.

Proposition 6.9.8. *If $p \geqslant 3$ is a regular prime then FLT I holds.*

Proof. First note that if $p = 3$ and $p \nmid xyz$ we have x^3, y^3, and z^3 congruent to ± 1 modulo 9, which is impossible if $x^3 + y^3 = z^3$, so we may assume that $p \geqslant 5$. The equation $x^p + y^p = z^p$ can be written

$$(x + y)(x + \zeta y) \cdots (x + \zeta^{p-1}y) = z^p .$$

Since x and y are coprime, as ideals (otherwise it does not make sense since $\mathbb{Z}_K$ is not necessarily a PID) the factors on the left-hand side are pairwise coprime: indeed, if some prime ideal $\mathfrak{p}$ divides $x + \zeta^i y$ and $x + \zeta^j y$ for $i \neq j$, it divides also $(\zeta^i - \zeta^j)y$ and $(\zeta^j - \zeta^i)x$, hence $\zeta^i - \zeta^j$. Thus $\mathfrak{p} = \pi$, so that $\pi \mid z$; hence $p \mid z$, contrary to our hypothesis. We thus have a product of pairwise coprime ideals that is equal to the pth power of an ideal, so that each of them is a pth power. Thus for each j we have $(x + \zeta^j y)\mathbb{Z}_K = \mathfrak{a}_j^p$ for some ideal $\mathfrak{a}_j$. By the above remark, since p is a regular prime this implies that $\mathfrak{a}_j$ itself is a principal ideal, say $\mathfrak{a}_j = \alpha_j\mathbb{Z}_K$. In particular, for $j = 1$ we can write $x + \zeta y = \alpha^p u$ with u a unit of K.

Denote complex conjugation by $\overline{}$. By Lemma 3.5.19, which is an immediate consequence of Kronecker's theorem on roots of unity, $u/\overline{u}$ is a root of unity, hence of the form $\eta = \pm\zeta^m$ for some m. On the other hand, since $\pi \mid (\zeta^j - \zeta^{-j})$ for all j, it is clear that for any $\beta \in \mathbb{Z}[\zeta]$ we have $\overline{\beta} \equiv \beta \pmod{\pi}$, so $\overline{\alpha} \equiv \alpha \pmod{\pi}$. Since $\pi \nmid z$, it follows that $\pi \nmid \alpha$, so that $\overline{\alpha}/\alpha \equiv 1 \pmod{\pi}$. Using the binomial expansion and the fact that $\pi^{(p-1)} \mid p\mathbb{Z}_K$, we deduce that $(\overline{\alpha}/\alpha)^p \equiv 1 \pmod{\pi^p}$. Dividing $x + \zeta y$ by its complex conjugate (and remembering that both are coprime to π), we obtain $(x + \zeta y)/(x + \zeta^{-1}y) \equiv \eta \pmod{\pi^p}$, in other words

$$x + \zeta y - \eta(x + \zeta^{-1}y) \equiv 0 \pmod{\pi^p} .$$

I claim that $m = 1$. Indeed, assume otherwise. If $m = 0$ we multiply the above congruence by ζ, and if $m = p - 1$ we multiply it by ζ^2; otherwise we do nothing. Thus we see that there exists a polynomial $f(T) \in \mathbb{Z}[T]$ of degree less than or equal to $p - 2 \geqslant 3$ (since we have assumed $p \geqslant 5$), not divisible by p, and such that $f(\zeta) \equiv 0 \pmod{\pi^p}$. Set $g(X) = f(1 - X)$. It is also of degree less than or equal to $p - 2$, it not divisible by p, and $g(\pi) \equiv 0 \pmod{\pi^p}$. However, it is clear that different monomials in $g(\pi)$ have valuations which are noncongruent modulo $p - 1$, hence are distinct, a contradiction. It follows that $m = 1$, proving my claim. Thus $\eta = \pm\zeta$, and our congruence reads $x + \zeta y \mp (x\zeta + y) = (x \mp y)(1 \mp \zeta) \equiv 0 \pmod{\pi^p}$, so $x \mp y \equiv 0 \pmod{p}$. We cannot have $x + y \equiv 0 \pmod{p}$, since otherwise $p \mid z$. Thus $y \equiv x \pmod{p}$. We may now apply the same reasoning to the equation $(-x)^p + z^p = y^p$ and deduce that $-z \equiv x \pmod{p}$. It follows that $0 = x^p + y^p - z^p \equiv 3x^p \pmod{p}$, and since $p \nmid x$, we obtain $p = 3$, which has been excluded and treated directly, finishing the proof of FLT I when p is a regular prime.   □

For instance, the irregular primes less than or equal to 200 are $p = 37$, 59, 67, 101, 103, 131, 149, 157, so that FLT I is true up to $p = 200$ for all but those primes.

Remark. It is interesting to note that the prime numbers p for which FLT I can be proved using congruence conditions (i.e., Proposition 6.9.5) and those for which it can be proved using global methods as above (i.e., Proposition 6.9.8) are essentially independent. For instance, by Proposition 6.9.5, FLT I cannot be proved by congruence conditions for $p = 7$ or $p = 13$, but since these are regular primes, FLT I follows from Proposition 6.9.8. On the other hand, since $p = 101$ and $p = 131$ are irregular primes, FLT I does not follow immediately from Proposition 6.9.8, although it does follow by congruence conditions from Proposition 6.9.5. In fact, combining the two approaches, we have thus proved FLT I for all primes $p \leqslant 200$ except for $p = 37, 59, 67, 103, 157$.

If we go up to $p = 5000$, there are 668 odd primes, and among those 279 can be solved by local considerations (Corollary 6.9.5), 407 by global considerations (Proposition 6.9.8), and 522 by one or the other. Of course, using Wendt's criterion given above (Proposition 6.9.6), all cases can be solved.

Asymptotically, it is expected (but not proved) that Corollary 6.9.5 can solve a proportion of $1 - \exp(-1/2) = 0.393\ldots$ of prime numbers, while Proposition 6.9.8 can solve a proportion of $\exp(-1/2) = 0.607\ldots$ of prime numbers, so that if they are independent, one or the other can solve a proportion of $1 - \exp(-1/2) + \exp(-1) = 0.761\ldots$.

6.9.6 The Criteria of Furtwängler and Wieferich

Theorem 6.9.9 (Furtwängler). *Let $p \geqslant 3$ be prime, let x, y, and z be pairwise coprime nonzero integers such that $x^p + y^p = z^p$, and assume that $p \nmid yz$. Then for every $q \mid yz$ we have $q^{p-1} \equiv 1 \pmod{p^2}$.*

Note that since x, y, and z are pairwise coprime, at most one can be divisible by p, so the condition $p \nmid yz$ can always be achieved by permuting x, y, and $-z$.

Proof. By multiplicativity, it is sufficient to prove the result for a prime number q such that $q \mid yz$, and by symmetry we may assume that $q \mid y$. Let $\zeta = \zeta_p$. As in the proof of Kummer's theorem on FLT I (Proposition. 6.9.8), since x and y are coprime and $p \nmid z$ the ideals $(x + \zeta^i y)\mathbb{Z}[\zeta]$ are pth powers of ideals for all i. In particular, if we set $\alpha = (x + y)^{p-2}(x + \zeta y)$, the ideal $\alpha\mathbb{Z}[\zeta]$ is a pth power. Furthermore, we have $x + \zeta y = x + y + (\zeta - 1)y$, hence $\alpha = (x+y)^{p-1} + (\zeta - 1)u$ with $u = y(x+y)^{p-2} \in \mathbb{Z}$. Since $(x+y) \mid x^p + y^p = z^p$ we have $p \nmid (x + y)$, hence $(x + y)^{p-1} \equiv 1 \pmod p$, and in particular $\alpha \equiv 1 + (\zeta - 1)u \pmod{\mathfrak{p}^2}$, where $\mathfrak{p} = (\zeta - 1)\mathbb{Z}[\zeta]$ is the unique prime ideal of $\mathbb{Q}(\zeta)$ above p. On the other hand, $\zeta^{-u} = (1 + (\zeta - 1))^{-u} \equiv 1 - (\zeta - 1)u \pmod{\mathfrak{p}^2}$, so that $\zeta^{-u}\alpha \equiv 1 \pmod{\mathfrak{p}^2}$; hence $\zeta^{-u}\alpha$ is a primary element in the sense of Eisenstein reciprocity (see Definition 3.6.34, where ℓ must be replaced by p and $\mathcal{L}$ by $\mathfrak{p}$).

Since $p \nmid y$ and $q \mid y$, we have $q \neq p$, and it is immediate to check that $q \nmid \mathcal{N}(\alpha)$, since y is coprime to x and z. Thus applying Eisenstein's reciprocity law (Theorem 3.6.38) we have

$$\left(\frac{q}{\zeta^{-u}\alpha}\right)_p = \left(\frac{\zeta^{-u}\alpha}{q}\right)_p = \left(\frac{\zeta}{q}\right)_p^{-u}\left(\frac{\alpha}{q}\right)_p .$$

Since $\zeta^{-u}\alpha\mathbb{Z}[\zeta] = \mathfrak{a}^p$ for some ideal $\mathfrak{a}$, by definition we have $\left(\frac{q}{\zeta^{-u}\alpha}\right)_p = \left(\frac{q}{\mathfrak{a}}\right)_p^p = 1$ since $\left(\frac{\cdot}{\mathfrak{a}}\right)_p$ has order p. Furthermore, since $q \mid y$ we have $x + \zeta y = x + y + y(\zeta - 1) \equiv x + y \pmod{q\mathbb{Z}[\zeta]}$, hence $\alpha \equiv (x + y)^{p-1} \pmod{q\mathbb{Z}[\zeta]}$. Since the value of $\left(\frac{\alpha}{q}\right)_p$ depends only on the class of α in $\mathbb{Z}[\zeta]/q\mathbb{Z}[\zeta]$, and since $(x + y)^{p-1} \equiv 1 \pmod p$ is trivially a primary element, it follows once again from Eisenstein reciprocity that

$$\left(\frac{\alpha}{q}\right)_p = \left(\frac{(x+y)^{p-1}}{q}\right)_p = \left(\frac{q}{(x+y)^{p-1}}\right)_p = 1 ,$$

because the ideal $(x + y)\mathbb{Z}[\zeta]$ is a pth power. Combining all these relations we deduce that $\left(\frac{\zeta}{q}\right)_p^u = 1$.

Let $q\mathbb{Z}[\zeta] = \prod_{1 \leqslant i \leqslant g} \mathfrak{q}_i$ be the prime ideal decomposition of q in $\mathbb{Z}[\zeta]$, so that $\mathcal{N}(\mathfrak{q}_i) = q^f$ and $g = (p-1)/f$ for some $f \mid (p-1)$. By definition, for any $\mathfrak{q} = \mathfrak{q}_i$ we have

$$\left(\frac{\zeta}{\mathfrak{q}}\right)_p \equiv \zeta^{(q^f-1)/p} \pmod{\mathfrak{q}},$$

and since both sides are pth roots of unity and p and q are distinct primes it follows that we have *equality* in the above congruence, hence by multiplicativity that

$$\left(\frac{\zeta}{q}\right)_p = \zeta^{g(q^f-1)/p}.$$

It follows that the identity $\left(\frac{\zeta}{q}\right)_p^u = 1$, which we have proved above, is equivalent to $ug(q^f - 1)/p \equiv 0 \pmod{p}$. Since $g \mid (p-1)$, $p \nmid g$, and since $p \nmid y$ and $p \nmid (x+y)$, we have $p \nmid u$. It follows that $q^f \equiv 1 \pmod{p^2}$, proving the theorem since $f \mid (p-1)$. $\qquad\square$

Corollary 6.9.10 (Wieferich). *If FLT I for a prime exponent $p \geqslant 3$ has a nonzero solution then $2^{p-1} \equiv 1 \pmod{p^2}$.*

Proof. Indeed, if $x^p + y^p = z^p$ with $p \nmid xyz$, then exactly one of x, y, and z is even. We may thus assume that $2 \mid y$, so the result follows from the theorem. $\qquad\square$

Remarks. (1) The only known values of p such that $2^{p-1} \equiv 1 \pmod{p^2}$ are $p = 1093$ and $p = 3511$, and there are no others up to $1.25 \cdot 10^{15}$. On simple probabilistic grounds it is however believed that there exist infinitely many, and that their number up to X should be on the order of $\log(\log(X))$.

(2) Wieferich's criterion has been generalized by many authors, replacing 2 by larger integers, and it is by combining these criteria that FLT I has been proved by "classical" methods up to 10^{18} as has already been mentioned.

6.9.7 General Prime n: The Second Case

The second case of FLT, denoted by FLT II, is more difficult for several reasons. We begin with a p-adic remark.

Proposition 6.9.11. *For every prime number ℓ there exist nonzero elements α, β, and γ in $\mathbb{Z}_\ell$ such that $\alpha\beta\gamma \equiv 0 \pmod{\ell}$ and $\alpha^p + \beta^p = \gamma^p$.*

Proof. Set $F(X) = X^p + \ell^p - 1$. Assume first that $\ell \neq p$. Then $F(X) \equiv (X-1)(X^{p-1} + \cdots + 1) \pmod{\ell}$. Since $\ell \neq p$, it follows that 1 is a simple root of $F(X) \equiv 0 \pmod{\ell}$, so by Hensel's lemma (Proposition 4.1.37) $F(X)$

has a root $\alpha \in \mathbb{Z}_\ell$, thus proving the proposition in this case with $\beta = \ell$ and $\gamma = 1$. Assume now that $\ell = p$. Then $|F(1)|_p = p^{-p}$ and $|F'(1)|_p = p^{-1}$. Since $p \geqslant 3$, we can once again conclude from Hensel's lemma that $F(X)$ has a root in $\mathbb{Z}_p$, proving the proposition. $\square$

We keep the notation of the first case. We begin with two results of Kummer on units.

Lemma 6.9.12. *Let p be a prime number, let $\zeta = \zeta_p$ be a primitive pth root of unity, let $K = \mathbb{Q}(\zeta)$, and let $\pi = 1 - \zeta$ generate the prime ideal $\mathfrak{p}$ of $\mathbb{Z}_K$ such that $p\mathbb{Z}_K = \mathfrak{p}^{p-1}$. Let $\beta \in \mathbb{Z}_K$ be prime to $\mathfrak{p}$ and assume that the congruence $\alpha_0^p \equiv \beta \pmod{\mathfrak{p}^p}$ has a solution in $\mathbb{Z}_K$, or even in $\mathbb{Z}_\mathfrak{p}$. If $L = K(\beta^{1/p})$, then $\mathfrak{p}$ is unramified in the extension L/K.*

Proof. Assume first that $\alpha_0^p \equiv \beta \pmod{\mathfrak{p}^{p+1}}$. Since the absolute ramification index $e = e(\mathfrak{p}/p)$ is equal to $p - 1$, Lemma 4.1.41 with $r = 2$ tells us that there exists a $\mathfrak{p}$-adic unit α such that $\beta = \alpha^p$. Thus the polynomial $X^p - \beta$ is totally split in $K_\mathfrak{p}$, and since by Theorem 4.4.41 the splitting of a prime ideal $\mathfrak{p}$ in L/K mimics the splitting of the defining polynomial of L/K in $K_\mathfrak{p}$, it follows that $\mathfrak{p}$ is totally split in L/K, and in particular is unramified.

Assume now that $v_\mathfrak{p}(\beta - \alpha_0^p) = p$. Since the statement is trivial when $L = K$, we may assume that $L \neq K$. Set $\eta = (\beta^{1/p} - \alpha_0)/\pi$, so that $L = K(\eta)$. The minimal monic polynomial $f(X)$ of η over K is

$$((\pi X + \alpha_0)^p - \beta)/\pi^p \equiv X^p + p\pi\alpha_0^{p-1}/\pi^p X + (\alpha_0^p - \beta)/\pi^p \pmod{\mathfrak{p}} \ .$$

Since this polynomial is monic and all its coefficients are $\mathfrak{p}$-integral (recall that p/π^{p-1} is even a $\mathfrak{p}$-adic unit), it follows that η is $\mathfrak{p}$-integral (more correctly $\mathfrak{P}$-integral for any prime ideal $\mathfrak{P}$ of L above $\mathfrak{p}$, but it is shorter to talk this way), and since the only prime ideals that can divide the denominator of η are divisors of $\mathfrak{p}$, it follows that $\eta \in \mathbb{Z}_L$. Now recall that the discriminant of η is up to sign the resultant of $f(X)$ with $f'(X)$. However, since p/π^{p-1} is a $\mathfrak{p}$-adic unit, the formula above shows that for any $x \in \mathbb{Z}_L$ we have $f'(x) \equiv px^{p-1} + p/\pi^{p-1}\alpha_0^{p-1} \equiv p/\pi^{p-1}\alpha_0^{p-1} \not\equiv 0 \pmod{\mathfrak{p}}$, so that the discriminant of η is coprime to $\mathfrak{p}$. Since $L = K(\eta)$, the relative discriminant ideal of the extension L/K divides that of η, hence is prime to $\mathfrak{p}$, so that $\mathfrak{p}$ is unramified in L/K as claimed. In fact, in this case it is not difficult to show that $\mathfrak{p}$ is inert in L/K. $\square$

Corollary 6.9.13. *Let p be a regular prime, and let ε be a unit of K such that the congruence $\varepsilon \equiv \alpha^p \pmod{\pi^p}$ has a solution in $\mathbb{Z}_K$. Then $\varepsilon = u^p$ for some $u \in \mathbb{Z}_K$ (necessarily a unit).*

Proof. Assume the contrary, and consider the extension $L = K(\varepsilon^{1/p})$. Since ε is not a pth power and $\zeta \in K$ it follows that L/K is a cyclic extension of degree p (the simplest case of a Kummer extension). The relative ideal discriminant of this extension divides the discriminant of the polynomial $X^p - \varepsilon$, which is equal to $(-1)^{(p-1)/2} p^p \varepsilon^{p-1}$. Since ε is a unit, it follows that it divides $p^p \mathbb{Z}_K$, hence that $\mathfrak{p} = \pi \mathbb{Z}_K$ is the only prime ideal that can divide it. However, by the above lemma, under the above conditions we know that even $\mathfrak{p}$ is unramified. Thus no finite prime can ramify in the extension L/K, and since K is totally complex, L/K is an unramified Abelian extension. Applying one of the basic results of class field theory, this tells us that L/K is a subextension of the Hilbert class field H/K. In particular, $p = [L : K]$ divides $h_p = [H : K]$, contrary to the assumption that p is a regular prime.

$\square$

The above proof using quite elementary results of class field theory is very simple. A direct proof without using class field theory would take two pages and be much more painful.

We now begin the proof of FLT II for regular primes. We will use Fermat's method of infinite descent. For this to work we need to study an equation that will descend to itself, so we will prove a stronger result.

Proposition 6.9.14. *Let $p \geqslant 3$ be a regular prime, and recall that $\pi = 1 - \zeta$ and $\mathfrak{p} = \pi \mathbb{Z}_K$. There are no solutions to $x^p + y^p = \varepsilon z^p$ with x, y, z in $\mathbb{Z}_K$, with $\mathfrak{p} \mid z$, $\mathfrak{p} \nmid xy$, and ε a unit of K. In particular, FLT II holds.*

Proof. Assume the contrary. We again write the equation in the form

$$(x + y)(x + \zeta y) \cdots (x + \zeta^{p-1} y) = \varepsilon z^p .$$

At least one of the factors on the left must be divisible by $\mathfrak{p}$, so all of them are. On the other hand, if $\mathfrak{q}$ is any ideal of $\mathbb{Z}_K$, it is clear that if $i \neq j$, $\mathfrak{q}$ divides both $(x + \zeta^i y)\mathbb{Z}_K$ and $(x + \zeta^j y)\mathbb{Z}_K$ if and only if both $x + \zeta^i y$ and $x + \zeta^j y$ belong to $\mathfrak{q}$, which implies that πy and πx belong to $\mathfrak{q}$, hence that $\mathfrak{q}$ divides $\mathfrak{p}\mathfrak{a}$, where $\mathfrak{a}$ is the ideal GCD of $x\mathbb{Z}_K$ and $y\mathbb{Z}_K$. But conversely, $\mathfrak{a}$ clearly divides both $(x + \zeta^i y)\mathbb{Z}_K$ and $(x + \zeta^j y)\mathbb{Z}_K$ and is coprime to $\mathfrak{p}$ since x and y are, so that $\mathfrak{p}\mathfrak{a}$ divides both. We have thus proved that the ideal GCD of any two distinct factors in the above product is equal to $\mathfrak{p}\mathfrak{a}$. In particular, the p residues modulo $\mathfrak{p}$ of the $(x + \zeta^j y)/\pi$ are all distinct, and since $\mathbb{Z}_K/\mathfrak{p}$ has p elements, these residues form a complete system of representatives modulo $\mathfrak{p}$. In particular, exactly one of them is divisible by $\mathfrak{p}$. Changing y into $y\zeta^j$ for some j, we may assume that $\mathfrak{p}^2 \mid (x + y)$. It follows that $v_{\mathfrak{p}}(x + \zeta^j y) = 1$ for $1 \leqslant j \leqslant p - 1$, hence that $v_{\mathfrak{p}}(x + y) = p(n - 1) + 1$, where $n = v_{\mathfrak{p}}(z)$. In particular, we see that $n \geqslant 2$.

Since the product of the ideals $(x + \zeta^j y)\mathbb{Z}_K$ is the pth power of an ideal and since the GCD of any two is equal to $\mathfrak{p}\mathfrak{a}$, it follows that there exist ideals $\mathfrak{b}_j$ such that $(x + \zeta^j y)\mathbb{Z}_K = \mathfrak{p}\mathfrak{a}\mathfrak{b}_j^p$ for $0 \leqslant j \leqslant p - 1$. Now, we know that in any ideal class there exists an integral ideal coprime to any fixed ideal. In particular, we can find an integral ideal $\mathfrak{c}_0$ belonging to the ideal class of $\mathfrak{a}^{-1}\mathfrak{b}_0^{-1}$ and coprime to $\mathfrak{p}$. We set $\mathfrak{c} = \mathfrak{a}\mathfrak{c}_0$, which is still coprime to $\mathfrak{p}$ and belongs to the ideal class of $\mathfrak{b}_0^{-1}$. Since $\mathfrak{a}\mathfrak{b}_0^p = ((x + y)/\pi)\mathbb{Z}_K$ and $\mathfrak{c}\mathfrak{b}_0$ are principal ideals, it follows that $\mathfrak{a}^{-1}\mathfrak{c}^p = (\mathfrak{a}\mathfrak{b}_0^p)^{-1}(\mathfrak{c}\mathfrak{b}_0)^p$ is a principal ideal $\beta\mathbb{Z}_K$, say, and $\beta \in \mathbb{Z}_K$ since $\mathfrak{a} \mid \mathfrak{c}$. Multiplying by β/π our p equations, we obtain

$$((\beta x + \zeta^j \beta y)/\pi)\mathbb{Z}_K = (\mathfrak{c}\mathfrak{b}_j)^p .$$

Thus the pth power of the ideal $\mathfrak{c}\mathfrak{b}_j$ is a principal ideal, and since p is a regular prime, as in FLT I we deduce that $\mathfrak{c}\mathfrak{b}_j$ itself is a principal ideal $\alpha_j\mathbb{Z}_K$, so that for some units ε_j we have

$$\beta x + \zeta^j \beta y = \pi\varepsilon_j\alpha_j^p .$$

Recall that $\mathfrak{a}$ and $\mathfrak{c}$ are prime to $\mathfrak{p}$; hence so is β. Since we know $v_{\mathfrak{p}}(x + \zeta^j y)$ for all j, we deduce that α_j is prime to $\mathfrak{p}$ for $1 \leqslant j \leqslant p - 1$, and that $v_{\mathfrak{p}}(\alpha_0) = n - 1$. Adding to the equation for $j = 1$ ζ times the equation for $p - 1$ we obtain

$$(1 + \zeta)\beta(x + y) = \pi(\varepsilon_1\alpha_1^p + \zeta\varepsilon_{p-1}\alpha_{p-1}^p) ,$$

and since the equation for $j = 0$ gives $\beta(x + y) = \pi\varepsilon_0\alpha_0^p$, we get

$$\varepsilon_1\alpha_1^p + \zeta\varepsilon_{p-1}\alpha_{p-1}^p = (1 + \zeta)\varepsilon_0\alpha_0^p .$$

Since $n \geqslant 2$, $v_{\mathfrak{p}}(\alpha_0) = n - 1$, and $v_{\mathfrak{p}}(\alpha_j) = 0$ for $1 \leqslant j \leqslant p - 1$, it follows that $\zeta\varepsilon_{p-1}\varepsilon_1^{-1} \equiv (-\alpha_1/\alpha_{p-1})^p \pmod{\mathfrak{p}^p}$. Now by the crucial Corollary 6.9.13 proved above, this implies that $\zeta\varepsilon_{p-1}\varepsilon_1^{-1} = \eta^p$ for some unit η. Note that this is really the only difficult step in the proof, the rest being quite standard and automatic. Thus, dividing by ε_1 we obtain

$$\alpha_1^p + (\eta\alpha_{p-1})^p = (1 + \zeta)\varepsilon_0\varepsilon_1^{-1}\alpha_0^p ,$$

where we note that $(1 + \zeta)\varepsilon_0\varepsilon_1^{-1}$ is a unit, for instance because $(1 + \zeta)(\zeta + \zeta^3 + \cdots + \zeta^{p-2}) = -1$. We have thus obtained a new solution to our Diophantine equation $x^p + y^p = \varepsilon z^p$, such that $v_{\mathfrak{p}}(z) = v_{\mathfrak{p}}(\alpha_0) = n - 1$. If we had started with a solution for which $v_{\mathfrak{p}}(z)$ was minimal, we would have obtained a solution with a strictly smaller value of $v_{\mathfrak{p}}(z)$, a contradiction, proving the first statement of the proposition. In addition, if $x^p + y^p = z^p$ with $p \mid x$ for instance, we can write instead $y^p + (-z)^p = (-x)^p$, so we may always assume in FLT II that $p \mid z$, proving the second statement, hence FLT in general for a regular prime. □

Remark. Denote by h_p^+ the class number of the totally real subfield $K^+ = \mathbb{Q}(\zeta + \zeta^{-1})$ of K. By Proposition 3.5.21, we know that $h_p^+ \mid h_p$. It can be shown that if the much weaker condition $p \nmid h_p^+$ is satisfied then it is not difficult to check whether FLT holds. The advantage of this is that in fact we do not know of *any* p such that $p \mid h_p^+$. The hypothesis that such p do not exist is known as *Vandiver's conjecture*. It is however believed among experts that this conjecture is probably false, although the smallest counterexample may be rather large (it has been verified up to several million). The problem with this is that, although as already mentioned, there is an algorithm to verify FLT I, if one finds a p such that $p \mid h_p^+$ (a counterexample to Vandiver's conjecture) it may be that one does not know of any way to prove FLT using classical methods (i.e., not using Wiles) for that p.

6.10 An Example of Runge's Method

Descriptions of this method can be found in [Ten] and [Wals]. We will only give a typical example, and note that we will again use this method in the context of Catalan's equation; see Section 6.11. We begin with the following lemma, which is typical of Diophantine approximation techniques in which we need to bound both the denominator and the absolute value of certain coefficients.

Lemma 6.10.1. *Let $S(X) = \sum_{k \geq 0} s_k X^k$ be a power series with integral coefficients such that $s_0 = 1$ and not identically equal to 1, let $d \geq 2$ be an integer, and write $S(X)^{1/d} = \sum_{k \geq 0} a_k X^k$ with $a_0 = 1$. Then*

(1) *We have $D_k a_k \in \mathbb{Z}$, where $D_k = d^k \prod_{p \mid d} p^{v_p(k!)}$.*
(2) *Let k_0 be the smallest strictly positive index such that $s_{k_0} \neq 0$, and assume that there exists a prime p dividing d such that $v_p(s_{k_0}) = 0$. Then when $k_0 \mid k$ we have $v_p(a_k) = -(k v_p(d) + v_p(k!))$.*
(3) *Assume that S has a nonzero radius of convergence R in $\mathbb{C}$, and let $M = \inf_{|z| < R, \ S(z)=0} |z|$ be the infimum of the zeros of $S(z)$ in the open disk of radius R if there exists such a zero (otherwise, let M be arbitrary such that $0 < M < R$), and finally let $N = \sum_{k \geq 0} |s_k| M^k$. Then $M \leq 1$, and for all k we have the inequality $|a_k| \leq N^{1/d} M^{-k}$.*

Proof. (1). Set $g(X) = \sum_{k \geq 1} s_k X^k$ and write $g(X)^j = \sum_{k \geq j} g_{j,k} X^k$. We have

$$S(X)^{1/d} = (1 + g(X))^{1/d} = 1 + \sum_{j \geq 1} \binom{1/d}{j} \sum_{k \geq j} g_{j,k} X^k$$

$$= 1 + \sum_{k \geq 1} X^k \sum_{1 \leq j \leq k} \binom{1/d}{j} g_{j,k} ,$$

so that $a_k = \sum_{1 \leq j \leq k} \binom{1/d}{j} g_{j,k}$. Since $g_{j,k} \in \mathbb{Z}$, Lemma 4.2.8 implies that $D_k a_k \in \mathbb{Z}$ with $D_k = \prod_{p \mid d} p^{k v_p(d) + v_p(k!)}$, proving (1).

(2). We have $g(X) = \sum_{k \geq k_0} s_k X^k$, hence $g(X)^j = \sum_{k \geq j k_0} g_{j,k} X^k$ with $g_{j,jk_0} = s_{k_0}^j$, and $a_k = \sum_{1 \leq j \leq k/k_0} \binom{1/d}{j} g_{j,k}$. By Lemma 4.2.8, if $p \mid d$ we have $v_p(\binom{1/d}{j}) = -(j v_p(d) + v_p(j!))$, which is a strictly decreasing function of j.

Since $v_p(g_{k/k_0,k}) = (k/k_0)v_p(s_0) = 0$, it follows that $v_p(a_k) = -(kv_p(d) + v_p(k!))$, proving (2).

(3). First note that the series S defines an analytic function for $|z| < R$, so $S(z)$ has only a finite number of zeros in this disk, and since $S(0) = 1$ we deduce that $0 < M < R$. I claim that $M \leqslant 1$. We consider two cases. If $S(X)$ is not a polynomial, then $s_k \neq 0$ for an infinity of k, and since $s_k \in \mathbb{Z}$ it follows that $R \leqslant 1$, so that $M < R \leqslant 1$. On the other hand, if $S(X)$ is a polynomial of degree n, say, then the product of the roots of S is equal to $(-1)^n/s_n$, and since $s_n \in \mathbb{Z}$ we have $|(-1)^n/s_n| \leqslant 1$, so at least one root must have a modulus less than or equal to 1, as claimed.

To obtain an inequality for $|a_k|$ we simply apply Cauchy's formula. If C_r denotes the circle of radius r centered at the origin, then if $r < M$ we have

$$a_k = \frac{1}{2i\pi} \int_{C_r} \frac{S(z)^{1/d}}{z^{k+1}} \, dz \, ,$$

since $S(z)$ has no zeros or poles in $|z| \leqslant r$, so that we can choose a fixed determination of the logarithm to define $S(z)^{1/d} = \exp(\log(S(z))/d)$. Thus

$$|a_k| \leqslant r^{-k} \left(\sup_{|z|=r} |S(z)| \right)^{1/d} \leqslant N^{1/d} r^{-k} \, .$$

Since this is true for all $r < M$ we obtain (3).   □

We also need the following completely elementary lemma.

Lemma 6.10.2. *Assume that for some integer $d \geqslant 1$ and real numbers a, b, and r we have the inequality $(a - r)^d < b < (a + r)^d$. Then we have $|\operatorname{sign}(r)b^{1/d} - a| < |r|$.*

Proof. If d is odd, then $a - r < b^{1/d} < a + r$ (where from now on, $b^{1/d}$ denotes the unique dth root of b when d is odd), so that $r > 0$ and $|b^{1/d} - a| < r$. If d is even then $b > 0$ and $|a - r| < b^{1/d} < |a + r|$ (where from now on $b^{1/d}$ denotes the unique positive dth root of b when d is even and $b > 0$). The first inequality gives $r - b^{1/d} < a < r + b^{1/d}$. The second inequality gives $a > -r + b^{1/d}$ or $a < -r - b^{1/d}$. If $r > 0$ the inequalities $a > r - b^{1/d}$ and $a < -r - b^{1/d}$ are incompatible, so that $-r + b^{1/d} < a < r + b^{1/d}$; in other words, once again $|b^{1/d} - a| < r$. If $r < 0$ the inequalities $a < r + b^{1/d}$ and $a > -r + b^{1/d}$ are incompatible, so that $r - b^{1/d} < a < -r - b^{1/d}$; in other words, $|b^{1/d} + a| < |r|$.   □

Proposition 6.10.3. *Let $f(X) = \sum_{0 \leqslant i \leqslant n} f_i X^i \in \mathbb{Z}[X]$ be a monic polynomial of degree n, let $r \geqslant 2$ be an integer, set $d = \gcd(r, n)$, $m = n/d$, and let $h(X)$ be the polynomial of degree m obtained by truncating the power series expansion in $1/X$ of $f(X)^{1/d}$. We assume that $f(X)$ is not identically equal to $h(X)^d$ (so that in particular $d > 1$). Let U (respectively L) be the largest* *(respectively smallest) real number that is a root of one of the two polynomials* $g_1(X) = f(X) - (h(X) - 1/D_m)^d$ *and* $g_{-1}(X) = f(X) - (h(X) + 1/D_m)^d$.

(1) *If* (x, y) *is an integral solution to the Diophantine equation* $y^r = f(x)$ *then either* $L \leqslant x \leqslant U$ *or* x *is a root of the nonzero polynomial* $f(x) - h(x)^d$.

(2) *Let* k_0 *be the largest index such that* $k_0 < n$ *with* $f_{k_0} \neq 0$ *(which must exist; otherwise,* $f(X) = (X^{n/d})^d$ *is a dth power), and assume that* $(n - k_0) \mid m$. *If in addition there exists a prime* $p \mid d$ *such that* $v_p(f_{k_0}) = 0$ *then when* $x \in \mathbb{Z}$ *we have* $f(x) - h(x)^d \notin \mathbb{Z}$, *and in particular* $f(x) - h(x)^d$ *has no integral roots.*

Proof. By definition of $h(x)$ we formally have $f(X)^{1/d} = h(X) + O(1/X)$; hence for $a = \pm 1$ we obtain $g_a(X) = (da/D_m)X^{n-m} + O(X^{n-m-1})$, so that the degree of $g_a(X)$ is equal to $n - m$ and the sign of its leading term is equal to a.

If $x > U$ we have $g_1(x) > 0$ and $g_{-1}(x) < 0$, in other words $(h(x) - 1/D_m)^d < f(x) < (h(x) + 1/D_m)^d$, so by the above lemma $|f(x)^{1/d} - h(x)| < 1/D_m$. Similarly, if $x < L$ we have $\text{sign}(g_1(x)) = (-1)^{n-m}$ and $\text{sign}(g_{-1}(x)) = (-1)^{n-m-1}$, so if $n - m$ is even we obtain the same conclusion, while if $n - m$ is odd we have $(h(x) + 1/D_m)^d < f(x) < (h(x) - 1/D_m)^d$; hence by the above lemma $|f(x)^{1/d} + h(x)| < 1/D_m$, and this can happen only if d is even. Thus in any case when $x > U$ or $x < L$ there exists $\varepsilon_1 = \pm 1$ such that $|f(x)^{1/d} - \varepsilon_1 h(x)| < 1/D_m$, and we have $\varepsilon_1^d = 1$.

Now let (x, y) be an integral solution to $y^r = f(x)$ with $x > U$ or $x < L$. Writing $Y = y^{r/d}$ we see that $f(x)^{1/d} = \varepsilon_2 Y \in \mathbb{Z}$, where $\varepsilon_2 = \pm 1$ is such that $\varepsilon_2^d = 1$. If we set $I = \varepsilon_2 D_m Y - \varepsilon_1 D_m h(x)$ it follows from the above inequality that $|I| < 1$. On the other hand, by Lemma 6.10.1 applied to the series $S(X) = X^n f(1/X)$ and the fact that $D_k \mid D_m$ if $k \leqslant m$, we know that $D_m h(X) \in \mathbb{Z}[X]$, so that $D_m h(x) \in \mathbb{Z}$, so $I \in \mathbb{Z}$. Since $|I| < 1$ it follows that $I = 0$, in other words $Y = \varepsilon_1 \varepsilon_2 h(x)$. Since $\varepsilon_1^d = \varepsilon_2^d = 1$ we thus have $f(x) = Y^d = h(x)^d$, so that x is a root of the nonzero polynomial $f(x) - h(x)^d$, proving (1).

For (2), let $x \in \mathbb{Z}$ be such that $f(x) - h(x)^d \in \mathbb{Z}$, so that $h(x)^d \in \mathbb{Z}$. Since $h(X) \in \mathbb{Q}[X]$, $h(x) \in \mathbb{Q}$ and $h(x)$ is an algebraic integer, so that $h(x) \in \mathbb{Z}$. However, with the notation of Lemma 6.10.1, we have $h(x) = \sum_{0 \leqslant k \leqslant m} a_k x^{m-k}$, and since $k_0 \mid m$ we have by the lemma $v_p(a_m) = -(m v_p(d) + v_p(m!))$, while for $k < m$ we have

$$v_p(a_k x^{m-k}) \geqslant v_p(a_k) \geqslant -(k v_p(d) + v_p(k!)) > -(m v_p(d) + v_p(m!)) = v_p(a_m).$$

Thus $v_p(h(x)) = v_p(a_m) < 0$, so $h(x) \notin \mathbb{Z}$, a contradiction. $\square$

Remarks. (1) It is easy to see that the type of reasoning used in the proof can be generalized as soon as we are able to compute y (or some integral power of y) as a formal power series in x. This is the case, for instance, for hyperelliptic equations of the form $y^2 = f(x)$ with $f \in \mathbb{Z}[X]$, the leading term of $f(X)$ being of the form $a^2 X^{2k}$ for $a \in \mathbb{Z}$. More generally still, the method can easily be extended to equations of the form $g(y) = f(x)$, where f and g are *monic* polynomials of *noncoprime degree* (see Exercise 50).

(2) More generally, let $E \in \mathbb{Q}$ be some expression involving a possible solution to a Diophantine equation. Then we say that we use Runge's method if on the one hand we find some *analytic bound* of the form $|E| < \varepsilon$ for some small ε, say, and on the other hand, if we can find an *arithmetic bound* D for the denominator of E. Then if $D\varepsilon < 1$ we deduce as above that $E = 0$, leading to very strict restrictions on the possible solution.

(3) It has not been necessary to use the bounds for a_k obtained in Lemma 6.10.1, since we can obtain much better inequalities for x directly as we have done above. In other situations, however, these bounds (or stronger ones obtained by similar methods) are the only available ones.

(4) Clearly this type of method can apply only to the search for *integer* solutions to Diophantine equations, and not rational solutions.

(5) The above method cannot apply to equations $y^r = f(x)$ in which r and n are coprime (for instance, think of the problem of finding integral points on elliptic curves $y^2 = x^3 + ax + b$), or to such equations in which f is nonmonic with a leading term not an exact rth power (for instance, think of the "trivial" Pell equation $y^2 = dx^2 + 1$).

As an example, we have the following:

Corollary 6.10.4. (1) *The only integer solutions to* $y^2 = x^4 + x^3 + x^2 + x + 1$ *are* $(x, y) = (-1, \pm 1)$, $(0, \pm 1)$, *and* $(3, \pm 11)$.

(2) *The only integer solutions to* $y^2 = x^6 - x^4 + 1$ *are* $(x, y) = (\pm 2, \pm 7)$, $(x, y) = (\pm 1, \pm 1)$, *and* $(0, \pm 1)$.

Proof. For (1), we easily find that $L = -1$ and $U = 3$, and the second condition is satisfied with $k_0 = 3$. Thus we need only look at $-1 \leqslant x \leqslant 3$ to prove the corollary, which is immediate. We leave (2) to the reader (Exercise 49). □

6.11 First Results on Catalan's Equation

I am very much indebted to Yu. Bilu and R. Schoof for help in writing this section and showing me their simplifications of the proofs of Cassels's and Ko Chao's theorems. I also invite the reader to read the notes of M. Mischler available on the Web [Boe-Mis].

6.11.1 Introduction

Catalan's conjecture, now a theorem, is the following:

Theorem 6.11.1 (Mihăilescu). *If n and m are greater than or equal to 2 the only nonzero integral solutions to*

$$x^m - y^n = 1$$

are $m = 2$, $n = 3$, $x = \pm 3$, $y = 2$.

This conjecture was formulated by Catalan in 1844 (see [Cat]) and received much attention. As already mentioned in Chapter 1, it was finally solved in 2002 by P. Mihăilescu. Complete proofs are available on the Web and at least two books are being written on the subject. We will prove this conjecture in two parts. First, in this section we prove the classical results of Cassels on the subject (see [Cas3]), which are essential for the final proof. Then in Chapter 16 the reader will find Mihăilescu's complete proof of the conjecture in an essentially self-contained form except that we will have to assume the validity of an important theorem of F. Thaine.

First, as for FLT we may evidently restrict ourselves to the case that m and n are prime numbers (we do not have to treat the special case n or m equal to 4 since the conjecture is enunciated also for $n = 2$ or $m = 2$). In addition, the conjecture is clearly true if $m = n$; see Exercise 55. Thus, Mihăilescu's theorem can be stated as follows:

Theorem 6.11.2 (Mihăilescu). *Let p and q be distinct primes, and let x and y be nonzero integers such that $x^p - y^q = 1$. Then $p = 2$, $q = 3$, $x = \pm 3$, and $y = 2$.*

Note that we have already proved this theorem for $q = 2$ (see Proposition 6.7.12). We will prove it for $p = 2$ below (see Theorem 6.11.8).

Since we can write $y^q = (x - 1)((x^p - 1)/(x - 1))$, we can expect as usual that each factor on the right will be close to a qth power. Indeed, first note the following.

Lemma 6.11.3. *Let p be prime, let $x \in \mathbb{Z}$ be such that $x \neq 1$, and set $r_p(x) = (x^p - 1)/(x - 1)$.*

(1) *If p divides one of the numbers $(x - 1)$ and $r_p(x)$ it divides both.*
(2) *If $d = \gcd(x - 1, r_p(x))$ then $d = 1$ or $d = p$.*
(3) *If $d = p$ and $p > 2$, then $r_p(x) \equiv p \pmod{p^2}$.*

Proof. Expanding $r_p(x) = ((x - 1 + 1)^p - 1)/(x - 1)$ by the binomial theorem we can write

$$r_p(x) = (x - 1)^{p-1} + p + (x - 1) \sum_{k=1}^{p-2} \binom{p}{k+1} (x - 1)^{k-1},$$

and all three results of the lemma immediately follow from this and the fact that $p \mid \binom{p}{k+1}$ for $1 \leqslant k \leqslant p - 2$. Note that (3) is trivially false for $p = 2$. $\quad\square$

Corollary 6.11.4. *Let (x, y, p, q) be such that $x^p - y^q = 1$. Then $\gcd(r_p(x), x - 1) = p$ if $p \mid y$ and $\gcd(r_p(x), x - 1) = 1$ otherwise.*

Proof. Since $y^q = (x - 1)r_p(x)$ it follows that $p \mid y$ if and only if p divides either $x - 1$ or $r_p(x)$, hence by the above lemma, if and only if $\gcd(r_p(x), x - 1) = p$. $\qquad\square$

The fundamental result of Cassels is the following.

Theorem 6.11.5 (Cassels). *Let p and q be primes, and let x and y be nonzero integers such that $x^p - y^q = 1$. Then $p \mid y$ and $q \mid x$.*

Thus the case $\gcd(r_p(x), x - 1) = 1$ of the above corollary does not happen. The proof of this theorem is the object of the next subsections, but we immediately give the most important consequence.

Corollary 6.11.6. *If x and y are nonzero integers and p and q are odd primes such that $x^p - y^q = 1$, there exist nonzero integers a and b and positive integers u and v with $q \nmid u$ and $p \nmid v$ such that*

$$x = qbu, \ x - 1 = p^{q-1}a^q, \ \frac{x^p - 1}{x - 1} = pv^q,$$

$$y = pav, \ y + 1 = q^{p-1}b^p, \ \frac{y^q + 1}{y + 1} = qu^p .$$

Proof. Since $p \mid y$, by the above corollary we have $\gcd(r_p(x), x - 1) = p$, so by Lemma 6.11.3 (3) we have $r_p(x) \equiv p \pmod{p^2}$, and in particular $v_p(r_p(x)) = 1$. Thus the relation $y^q = (x - 1)r_p(x)$ implies that there exist integers a and v with $p \nmid v$ such that $x - 1 = p^{q-1}a^q$, $r_p(x) = pv^q$; hence $y = pav$, and since $r_p(x) > 0$, we also have $v > 0$. This shows half of the relations of the theorem, and the other half follow by symmetry, changing (x, y, p, q) into $(-y, -x, q, p)$ and noting that p and q are odd. $\qquad\square$

6.11.2 The Theorems of Nagell and Ko Chao

Since by Proposition 6.7.12 we know that the equation $x^p - y^q = 1$ has no solutions with $xy \neq 0$ for $q = 2$, we may assume that $q \neq 2$. The first important step, due to Nagell, is to prove Theorem 6.11.5 for $p = 2$. This will enable us to finish the proof of Catalan's conjecture in that case (or equivalently, of the equation $y^2 = x^n + t$ with $t = 1$).

Proposition 6.11.7 (Nagell). *If x and y are nonzero integers and q is a prime such that $x^2 - y^q = 1$ then $2 \mid y$ and $q \mid x$.*

Proof. As already mentioned, we may assume that $q \neq 2$, and since $xy \neq 0$, we have $y > 0$ and we may assume $x > 0$. If y is odd, then x is even, so $x - 1$ and $x + 1$ are coprime, and since $(x - 1)(x + 1) = y^q$ this means that $x - 1$ and

$x + 1$ are both qth powers, which is impossible since two distinct qth powers cannot differ by 2. Thus $2 \mid y$.

Assume by contradiction that $q \nmid x$. From the equality $x^2 = (y + 1)r_q(-y) = (y + 1)((y^q + 1)/(y + 1))$ and Lemma 6.11.3, we deduce that $y + 1$ and $(y^q + 1)/(y + 1)$ (which are positive) are both squares, so we write $y + 1 = a^2$, $(y^q + 1)/(y + 1) = b^2$, so that $x = ab$, with $a > 0$, $b > 0$. In particular, since $y \neq 0$, y is not a square.

On the other hand, if $\alpha = x + y^{(q-1)/2}\sqrt{y} \in \mathbb{Z}[\sqrt{y}]$, then the norm of α is equal to 1, and α is an algebraic integer, so it is a unit of the order $\mathbb{Z}[\sqrt{y}]$. By Proposition 6.3.16 we know that the group of units of a real quadratic order is equal to $\{\pm 1\}$ times an infinite cyclic group. Furthermore, $\varepsilon = a + \sqrt{y}$ is clearly a fundamental unit (i.e., a generator strictly greater than 1 of the infinite cyclic group): indeed, let $\varepsilon_0 = u + v\sqrt{y}$ be the fundamental unit, so that $\varepsilon = \varepsilon_0^k$ for some k. Then $\varepsilon_0 - \overline{\varepsilon_0} = 2v\sqrt{y}$ divides $\varepsilon_0^k - \overline{\varepsilon_0^k} = 2\sqrt{y}$; hence $v \mid 1$, so $v = 1$ and $\varepsilon = \varepsilon_0$ as claimed. It follows that there exists $k > 0$ such that

$$x + y^{(q-1)/2}\sqrt{y} = (a + \sqrt{y})^k .$$

We first reduce this equation modulo y. We obtain $x \equiv a^k + ka^{k-1}\sqrt{y}$ (mod $y\mathbb{Z}[\sqrt{y}]$), in other words $y \mid a^k - x$ and $y \mid ka^{k-1}$, and since y and a are coprime, $y \mid k$. Since y is even, it follows that k is even.

We now reduce the above equality modulo a, using $x = ab \equiv 0$ (mod a) and $y = a^2 - 1 \equiv -1$ (mod a), so we obtain $(-1)^{(q-1)/2}\sqrt{y} \equiv y^{k/2} \equiv (-1)^{k/2}$ (mod $a\mathbb{Z}[\sqrt{y}]$), in other words $a \mid 1$, so $a = 1$, contradicting the assumption $y \neq 0$. $\qquad\square$

We can now easily prove the theorem of Ko Chao (see [Ko]), using a proof due to E. Chein.

Theorem 6.11.8 (Ko Chao). *If q is prime there are no nonzero solutions to the equation $x^2 - y^q = 1$ apart from $(x, y) = (\pm 3, 2)$ for $q = 3$.*

Proof. We may clearly assume $q \neq 2$. Furthermore, we have proved in Corollary 6.4.32 that there are no solutions for $q = 3$ apart from the given ones. We may thus assume that $q \geqslant 5$. By Nagell's result, we know that x is odd, and we may of course assume $x > 0$. Choose $\varepsilon = \pm 1$ such that $x \equiv \varepsilon$ (mod 4). As in the proof of Corollary 6.11.6 the equality $(x - \varepsilon)(x + \varepsilon) = y^q$ with $v_2(x + \varepsilon) = 1$ implies that there exist positive integers a and b such that $x - \varepsilon = 2^{q-1}a^q$ and $x + \varepsilon = 2b^q$. Since $q \geqslant 5$ we have $a^q = (b^q - \varepsilon)/2^{q-2} < b^q$, so that $a < b$. On the other hand, we have

$$(b^2 - 2\varepsilon a)\frac{b^{2q} - (2\varepsilon a)^q}{b^2 - 2\varepsilon a} = b^{2q} - (2\varepsilon a)^q = \left(\frac{x + \varepsilon}{2}\right)^2 - 2\varepsilon(x - \varepsilon) = \left(\frac{x - 3\varepsilon}{2}\right)^2 .$$

By Nagell's proposition above, we know that $q \mid x$. Since $q \geqslant 5$, it follows that $q \nmid (x - 3\varepsilon)/2$; hence by Lemma 6.11.3 the two factors on the left are coprime, hence are both squares. However, since we have seen above that $a < b$, and $a > 0$, we have

$$(b - 1)^2 = b^2 - 2b + 1 < b^2 - 2a < b^2 < b^2 + 2a < b^2 + 2b + 1 = (b + 1)^2 \,,$$

which shows that $b^2 - 2\varepsilon a$ cannot be a square, a contradiction.   □

6.11.3 Some Lemmas on Binomial Series

Before proceeding to the proof of Cassels's theorem below we need an arithmetic result and two analytic results. The arithmetic result is the following.

Lemma 6.11.9. *Set $w(j) = j + v_q(j!)$. Then $q^{w(j)} \binom{p/q}{j}$ is an integer not divisible by q, and $w(j)$ is a strictly increasing function of j.*

Proof. By Lemma 4.2.8 (2) (a) and (c), we know that $\binom{p/q}{j}$ is an ℓ-adic integer for $\ell \neq q$, and that its q-adic valuation is equal to $-w(j)$, proving the first assertion. Since $w(j+1) - w(j) = 1 + v_q(j+1) \geqslant 1$ the second assertion is also clear.   □

The first analytic result that we need is the following.

Lemma 6.11.10. (1) *For all $x > 0$ we have $(x + 1)\log(x + 1) > x\log(x)$.*
(2) *Let $b \in \mathbb{R}_{>1}$. The function $(b^t + 1)^{1/t}$ is a decreasing function of t from $\mathbb{R}_{>0}$ to $\mathbb{R}_{>0}$ and the function $(b^t - 1)^{1/t}$ is an increasing function of t from $\mathbb{R}_{>0}$ to $\mathbb{R}_{>0}$.*
(3) *Assume that $q > p \in \mathbb{R}_{>0}$. If $a \in \mathbb{R}_{\geqslant 1}$ then $(a^q + 1)^p < (a^p + 1)^q$ and if $a \in \mathbb{R}_{>1}$ then $(a^q - 1)^p > (a^p - 1)^q$.*

Proof. Since $\log(x)$ is an increasing function of x and $\log(x + 1) > 0$ we have $(x + 1)\log(x + 1) > x\log(x + 1) > x\log(x)$, so (1) is clear. For (2), we note that for $\varepsilon = \pm 1$ the derivative of the logarithm of $(b^t + \varepsilon)^{1/t}$ is equal to $b^t \log(b^t) - (b^t + \varepsilon)\log(b^t + \varepsilon)$,n which has the sign of $-\varepsilon$ by (1), so (2) follows. Applying the first inequality of (2) to $t = p$ and $t = q$, we deduce that $(a^q + 1)^{1/q} < (a^p + 1)^{1/p}$, giving the first inequality of (3) for $a > 1$, and the second follows similarly. Note that the first inequality of (3) is also trivially true if $a = 1$.   □

The second analytic result that we need is more delicate.

Lemma 6.11.11. *Assume that $p > q$, set $F(t) = ((1 + t)^p - t^p)^{1/q}$, let $m = \lfloor p/q \rfloor + 1$, and denote by $F_m(t)$ the sum of the terms of degree less than or equal to m in the Taylor series expansion of $F(t)$ around $t = 0$. Then for all $t \in \mathbb{R}$ such that $|t| \leqslant 1/2$ we have*

$$|F(t) - F_m(t)| \leqslant \frac{|t|^{m+1}}{(1 - |t|)^2} \,.$$

Proof. Set $G(t) = (1+t)^{p/q}$. It is clear that the Taylor coefficients of $F(t)$ and $G(t)$ around $t = 0$ are the same to order strictly less than p, and in particular to order m since $m \leqslant p/3 + 1 < p$ (since $p \geqslant 5$). In what follows, assume that $|t| < 1$. By the Taylor–Lagrange formula applied to the functions $x^{1/q}$ and $G(x)$ respectively there exist t_1 and t_2 such that

$$|F(t) - F_m(t)| \leqslant |F(t) - G(t)| + |G(t) - F_m(t)|$$

$$\leqslant \frac{|t|^p}{q} t_1^{1/q-1} + |t|^{m+1} \frac{1}{(m+1)!} G^{(m+1)}(t_2)$$

$$\leqslant \frac{|t|^p}{q} t_1^{1/q-1} + |t|^{m+1} \binom{p/q}{m+1} (1+t_2)^{p/q-m-1},$$

with t_1 between $(1+t)^p$ and $(1+t)^p - t^p$, and t_2 between 0 and t. Now note that $p/q < m \leqslant p/q + 1$, so that $-1 \leqslant p/q - m < 0$ and for all $j \geqslant 1$ $0 < p/q - (m-j) = j - (m - p/q) < j$; hence

$$0 < \prod_{1 \leqslant j \leqslant m} (p/q - (m-j)) < \prod_{1 \leqslant j \leqslant m} j = m!.$$

It follows that

$$\left| \binom{p/q}{m+1} \right| = \frac{(m - p/q) \prod_{1 \leqslant j \leqslant m}(p/q - (m-j))}{m!} \leqslant \frac{1}{m+1}.$$

Since $1/q - 1 < 0$ and $p/q - m - 1 < 0$ we must estimate t_1 and $1 + t_2$ from below. If $t > 0$ both $(1+t)^p$ and $(1+t)^p - t^p$ are greater than 1, so $t_1 > 1 > 1 - t^p$. If $t < 0$ then $(1+t)^p = (1 - |t|)^p$ and $(1+t)^p - t^p = (1 - |t|)^p + |t|^p > (1 - |t|)^p$, so that $t_1 > (1 - |t|)^p$ in all cases. On the other hand, we have trivially $|1 + t_2| \geqslant 1 - |t|$. Putting everything together we obtain

$$|F(t) - F_m(t)| \leqslant \frac{|t|^p}{q} (1 - |t|)^{-p+p/q} + \frac{|t|^{m+1}}{m+1} (1 - |t|)^{p/q-m-1}.$$

The above inequality is valid for all t such that $|t| < 1$. If we assume that $|t| \leqslant 1/2$ then $|t|^{p-m-1} \leqslant (1 - |t|)^{p-m-1}$ (since $m \leqslant p - 1$), so that $|t|^p(1 - |t|)^{-p+p/q} \leqslant |t|^{m+1}(1 - |t|)^{p/q-m-1}$. It follows that

$$|F(t) - F_m(t)| \leqslant \left(\frac{1}{q} + \frac{1}{m+1} \right) |t|^{m+1}(1 - |t|)^{p/q-m-1}.$$

Since $p/q - m - 1 \geqslant -2$ and $1/q + 1/(m+1) \leqslant 1$ the lemma follows. $\qquad \square$

6.11.4 Proof of Cassels's Theorem 6.11.5

We now prove Cassels's theorem saying that if p and q are primes and $x^p - y^q = 1$ with $xy \neq 0$ then $q \mid x$ and $p \mid y$. We have already seen that the case $p = q$ is impossible. By Proposition 6.7.12 the case $q = 2$ is impossible, and Nagell's Proposition 6.11.7 is the special case $p = 2$ (in fact in this case Ko Chao's Theorem 6.11.8 shows that the only nontrivial solutions occur for $(x, y) = (\pm 3, 2)$ and $q = 3$). We may thus assume that p and q are distinct odd primes. It is then sufficient to prove that $p \mid y$ since when p and q are odd we can change (p, q, x, y) into $(q, p, -y, -x)$. The proof of Theorem 6.11.5 will be done by considering separately the cases $p < q$ and $p > q$. We begin with the case $p < q$, which is considerably simpler.

Proposition 6.11.12. *Let x and y be nonzero integers and p and q be odd primes such that $x^p - y^q = 1$. Then if $p < q$ we have $p \mid y$.*

Proof. Assume on the contrary that $p \nmid y$. It follows from Corollary 6.11.4 that $x - 1$ and $r_p(x)$ are coprime, and since their product is a qth power, they both are. We can thus write $x - 1 = a^q$ for some integer a, and $a \neq 0$ (otherwise $y = 0$) and $a \neq -1$ (otherwise $x = 0$), so $(a^q + 1)^p - y^q = 1$. Consider the function $f(z) = (a^q + 1)^p - z^q - 1$, which is trivially a decreasing function of z. Assume first that $a \geqslant 1$. Then $f(a^p) = (a^q + 1)^p - a^{pq} - 1 > 0$ by the binomial expansion, while $f(a^p + 1) = (a^q + 1)^p - (a^p + 1)^q - 1 < 0$ by (3) of Lemma 6.11.10. Since f is strictly decreasing it follows that the root of $f(y) = 0$ is not an integer, a contradiction. Similarly, assume that $a < 0$, so that in fact $a \leqslant -2$, and set $b = -a$. Then since p and q are odd, $f(a^p) = (a^q + 1)^p - a^{pq} - 1 = -((b^q - 1)^p - b^{pq} + 1) > 0$ by the binomial expansion, while $f(a^p + 1) = (a^q + 1)^p - (a^p + 1)^q - 1 = -((b^q - 1)^p - (b^p - 1)^q + 1) < 0$ again by (3) of the Lemma 6.11.10 since $b > 1$. Once again we obtain a contradiction, proving the proposition. $\square$

The following corollary, essentially due to S. Hyyrö, will be used for the case $p > q$.

Corollary 6.11.13. *With the same assumptions as above (and in particular $p < q$) we have $|y| \geqslant p^{q-1} + p$.*

Proof. Since by the above proposition we have $p \mid y$, as in Corollary 6.11.6 we deduce that there exist integers a and v with $a \neq 0$ and $v > 0$ such that $x - 1 = p^{q-1} a^p$, $(x^p - 1)/(x - 1) = p v^q$ and $y = pav$. Set $P(X) = X^p - 1 - p(X - 1)$. Since $P(1) = P'(1) = 0$, it follows that $(X - 1)^2 \mid P(X)$, hence that $(x - 1) \mid (x^p - 1)/(x - 1) - p = p(v^q - 1)$. Since $p^{q-1} \mid x - 1$ it follows that $v^q \equiv 1 \pmod{p^{q-2}}$. However, the order of the multiplicative group modulo p^{q-2} is equal to $p^{q-3}(p - 1)$, and since $q > p$ this is coprime to q. As usual this implies that $v \equiv 1 \pmod{p^{q-2}}$.

On the other hand, I claim that $v > 1$. Indeed, assume otherwise that $v = 1$, in other words $x^{p-1} + \cdots + x + 1 = p$. If $x > 1$ then $2^{p-1} > p$, so this is impossible. Since p and q are odd primes and $a \neq 0$ we have $|x - 1| = p^{q-1}|a|^p \geqslant 9$, so that when $x \leqslant 1$ we must have in fact $z = -x \geqslant 8$. But then since $p - 1$ is even we have

$$p = z^{p-1} - z^{p-2} + \cdots + 1 \geqslant z^{p-1}(z-1) \geqslant z^{p-1} \geqslant 2^{p-1} \;,$$

a contradiction that proves my claim. Since $v \equiv 1 \pmod{p^{q-2}}$, it follows that $v \geqslant p^{q-2} + 1$, hence $|y| = pav \geqslant pv \geqslant p^{q-1} + p$, proving the corollary.   □

We now prove the more difficult case $p > q$ of Cassels's theorem.

Proposition 6.11.14. *Let x and y be nonzero integers and p and q be odd primes such that $x^p - y^q = 1$. Then if $p > q$ we have $p \mid y$.*

Proof. We keep all the notation of Lemma 6.11.11 and begin as for the case $p < q$ (Proposition 6.11.12): assuming by contradiction that $p \nmid y$ and using Corollary 6.11.4, we deduce that there exists $a \in \mathbb{Z} \setminus \{0\}$ such that $x - 1 = a^q$; hence $y^q = (a^q + 1)^p - 1$, so that $y = a^p F(1/a^q)$. Thus if we set $z = a^{mq-p} y - a^{mq} F_m(1/a^q)$ we have $z = a^{mq}(F(1/a^q) - F_m(1/a^q))$. Applying Lemma 6.11.11 to $t = 1/a^q$ (which satisfies $|t| \leqslant 1/2$ since $a \neq \pm 1$) we obtain

$$|z| \leqslant \frac{|a|^q}{(|a|^q - 1)^2} \leqslant \frac{1}{|a|^q - 2} \leqslant \frac{1}{|x| - 3} \;.$$

By Taylor's theorem we have $t^m F_m(1/t) = \sum_{0 \leqslant j \leqslant m} \binom{p/q}{j} t^{m-j}$, and by Lemma 6.11.9, $D = q^{m + v_q(m!)}$ is a common denominator of all the $\binom{p/q}{j}$ for $0 \leqslant j \leqslant m$. It follows that $Da^{mq} F_m(1/a^q) \in \mathbb{Z}$, and since $mq \geqslant p$ that $Dz \in \mathbb{Z}$. We now estimate the size of Dz. By Hyyrö's Corollary 6.11.13 (with (p, q, x, y) replaced by $(q, p, -y, -x)$) we have $|x| \geqslant q^{p-1} + q \geqslant q^{p-1} + 3$, so by the above estimate for $|z|$ we have

$$|Dz| \leqslant \frac{D}{|x| - 3} \leqslant q^{m + v_q(m!) - (p-1)} \;.$$

Now for $m \geqslant 1$ we have $v_q(m!) < m/(q-1)$, and since $m < p/q + 1$ we have

$$m + v_q(m!) - (p-1) < m\frac{q}{q-1} - (p-1) = \frac{3 - (p-2)(q-2)}{q-1} \leqslant 0$$

since $q \geqslant 3$ and $p \geqslant 5$ (note that it is essential that the above inequality be strict). Thus $|Dz| < 1$, and since $Dz \in \mathbb{Z}$, it follows that $Dz = 0$. However, note that

$$Dz = Da^{mq-p} y - \sum_{0 \leqslant j \leqslant m} D\binom{p/q}{j} a^{q(m-j)} \;,$$

and by Lemma 6.11.9 we have

$$v_q\left(\binom{p/q}{j}\right) < v_q\left(\binom{p/q}{m}\right) = v_q(D)$$

for $0 \leqslant j \leqslant m - 1$, so that $0 = Dz \equiv D\binom{p/q}{m} \not\equiv 0 \pmod{q}$ by the same lemma. This contradiction finishes the proof of the proposition, hence of Cassels's theorem.   □

Remark. The reasoning that we have just used is a special case of Runge's method seen in Section 6.10 in a slightly different context.

We have seen that Corollary 6.11.6 summarizes the most important consequences of Cassels's theorem. For future reference, we note that Hyrrö's Corollary 6.11.13 is valid without restriction on p and q:

Proposition 6.11.15. *Let p, q be odd primes and x, y be nonzero integers such that $x^p - y^q = 1$. Then $|x| \geqslant q^{p-1} + q$ and $|y| \geqslant p^{q-1} + p$.*

Proof. Since we can change (p, q, x, y) into $(q, p, -y, -x)$, it is enough to prove the statement for y. If $p < q$, this is Hyrrö's result. Otherwise we have $p \geqslant q$, hence $p > q$ since $p \neq q$. By Cassels's Corollary 6.11.6 we have $y+1 = q^{p-1}b^p$, hence $|y| \geqslant q^{p-1}-1$. I claim that when $p > q$ we have $q^{p-1} > p^{q-1}+p$, which will prove the proposition. Indeed, set $f(x) = \log(x)/(x-1)$, so that

$$f(q) - f(p) = \frac{\log(q^{p-1}) - \log(p^{q-1})}{(p-1)(q-1)} .$$

The inequality to be proved is thus equivalent to $f(q) - f(p) > \log(1 + 1/p^{q-2})/((p-1)(q-1))$, and since $\log(1+x) < x$ for $x > 0$, this will follow from the inequality $f(q) - f(p) > 1/(p^{q-2}(p-1)(q-1))$. Now by the mean value theorem we have $f(q) - f(p) = (q-p)f'(c)$ for some $c \in]q, p[$. We have $f'(x) = -(x\log(x) - (x-1))/(x(x-1)^2)$, and this is easily seen to be strictly negative as soon as $x > 1$. Furthermore, we easily check that $f''(x) > 0$ for $x \geqslant 2$; hence it follows that $f'(q) < f'(c) < f'(p) < 0$, so that $(q-p)f'(q) > (q-p)f'(c) > (q-p)f'(p) > 0$ since $p > q$. It is thus sufficient to prove that $(q-p)f'(p) > 1/(p^{q-2}(p-1)(q-1))$, in other words that $(p-q)(q-1)p^{q-2}(p\log(p) - (p-1))/(p(p-1)) > 1$, or

$$(p-q)(q-1)p^{q-2} \left(\frac{\log(p)}{p-1} - \frac{1}{p} \right) > 1 .$$

Now an immediate study shows that for $x \geqslant 5$ we have $\log(x)/(x-1) > 2/x$. Since $p > q \geqslant 3$ are odd we have $p \geqslant 5$, hence

$$(p-q)(q-1)p^{q-2} \left(\frac{\log(p)}{p-1} - \frac{1}{p} \right) > 2(p-q)(q-1)p^{q-3} > 1$$

since $q \geqslant 3$, proving the proposition.   □

6.12 Congruent Numbers

We give a short description of the congruent number problem, and refer to the excellent book by N. Koblitz [Kob2], which is entirely devoted to that problem.

6.12.1 Reduction to an Elliptic Curve

Recall from the introduction that a *congruent number* is an integer n that is the area of a right triangle with rational sides (i.e., a Pythagorean triangle). Since an area is homogeneous of degree 2, it is clear that we can assume without loss of generality that n is squarefree. For instance, from the well-known $(3, 4, 5)$ triangle we deduce that $n = 6$ is a congruent number. Several problems can be asked about congruent numbers, but the most important are the following: give a criterion for determining whether a given number n is congruent; if it is, determine a corresponding Pythagorean triangle. Both problems are difficult, and we will say a little of what is known about both.

Proposition 6.12.1. *A number n is a congruent number if and only if there exists a rational point on the curve $y^2 = x(x^2 - n^2)$ with $y \neq 0$. More precisely, if (a, b, c) is a Pythagorean triangle of area n, then the four points $(a(a \pm c)/2, a^2(a \pm c)/2)$ and $(b(b \pm c)/2, b^2(b \pm c)/2)$ are points on the curve with nonzero y coordinate, and conversely such a point (x, y) gives rise to a Pythagorean triangle (a, b, c) of area n with $a = |y/x|$, $b = 2n|x/y|$, and $c = (x^2 + n^2)/|y|$.*

Proof. The proof consists in simple verifications: if for example $x = a(a + c)/2$, $y = a^2(a + c)/2$, and $n = ab/2$ is the area of the triangle, then

$$x(x^2 - n^2) = a\frac{a + c}{2}\frac{a^2(a + c)^2 - a^2b^2}{4} = \frac{a^3(a + c)}{8}(a^2 + 2ac + c^2 - b^2)$$

$$= \frac{a^4(a + c)^2}{4} = y^2 \,,$$

since $c^2 = a^2 + b^2$. The other cases follow by exchanging a and b and/or changing c into $-c$ (even if this has little geometrical meaning). Conversely, if (x, y) is a rational point on the curve with $y \neq 0$ and if a, b, c are as given in the proposition, then a, b, c are strictly positive, and we have

$$a^2 + b^2 = \frac{y^2}{x^2} + 4n^2\frac{x^2}{y^2} = \frac{x^2 - n^2}{x} + \frac{4n^2x}{x^2 - n^2} = \frac{(x^2 - n^2)^2 + 4n^2x^2}{x(x^2 - n^2)}$$

$$= \frac{(x^2 + n^2)^2}{y^2} = c^2 \,.$$

$\square$

Thanks to this proposition, an easy computer search reveals for instance that the integers $n = 5$, 6, and 7 are congruent numbers. However, the corresponding triangles are not as simple as the one for 6: for $n = 5$ we find (for instance) the point $(x, y) = (-4, 6)$, giving the triangle $(3/2, 20/3, 41/6)$; for $n = 7$ we find the point $(x, y) = (-63/16, 735/64)$, giving the triangle $(35/12, 24/5, 337/60)$, which is already a little more complicated. On the other hand, a more extended computer search does not give any solution for $n = 1$, 2, and 3, and indeed these are not congruent numbers. However, this is more difficult and needs proof. We simply give the example of $n = 1$.

Proposition 6.12.2. *The number $n = 1$ is not congruent.*

Proof. Assume by contradiction that 1 is a congruent number, so that there exists $(x, y) \in \mathbb{Q}^2$ with $y \neq 0$ such that $y^2 = x(x^2 - 1)$. Writing $x = p/q$ and $y = u/v$ with $\gcd(p, q) = \gcd(u, v) = 1$, we obtain $(q^4/v^2)u^2 = pq(p^2 - q^2)$. Since $\gcd(u, v) = 1$, it follows that $v^2 \mid q^4$, i.e., $v \mid q^2$, so that $pq(p^2 - q^2)$ is the square of an integer. Since $\gcd(p, q) = 1$, the three factors are pairwise coprime, so they are all squares. Writing $p = p_1^2$, $q = q_1^2$, and $p^2 - q^2 = w^2$ we obtain the equation $p_1^4 - q_1^4 = w^2$. By Proposition 6.5.3, we know that this equation has no nontrivial solutions. Since $q \neq 0$ hence $q_1 \neq 0$, the only possible solution is thus with $w = 0$, in other words $p = \pm q$, hence $y = 0$, a contradiction. $\square$

6.12.2 The Use of the Birch and Swinnerton-Dyer Conjecture

By Proposition 6.12.1, we know that n is a congruent number if and only if there exists a point (x, y) on the curve $y^2 = x(x^2 - n^2)$ with $y \neq 0$. Such curves are called *elliptic curves*, and are among the most beautiful objects in mathematics, certainly in number theory. Fermat noticed in the seventeenth century (in another language) that such curves, considered in projective coordinates, have an abelian *group law*, obtained simply by taking secants and tangents though known points. This observation was strengthened by Mordell in the beginning of the twentieth century, who proved that this group is finitely generated, in other words isomorphic to $T \times \mathbb{Z}^r$, where T is a finite group. The group T can easily be determined. For instance, for our curves it is independent of n and always isomorphic to $\mathbb{Z}/2\mathbb{Z} \times \mathbb{Z}/2\mathbb{Z}$ (the elements of T are the three points with $y = 0$ together with the point at infinity of projective coordinates $(0, 1, 0)$). On the other hand, the *rank* r is in general very difficult to compute. From Proposition 6.12.1 and our assertion concerning T, it is clear that n is a congruent number if and only if the rank of the corresponding elliptic curve is strictly positive. In particular, if n is congruent, i.e., if there exists a Pythagorean triangle of area n, then there exist infinitely many such triangles, obtained by taking multiples of the given one for the group law on the curve (see Exercise 51).

Luckily, the BSD conjecture predicts that the rank r should be equal to the order of vanishing at $s = 1$ of a certain natural analytic L-function attached to the elliptic curve. Unfortunately, even this conjecture does not answer the problem completely, although on a computer it does give strong indications: the reason is that it is impossible to prove (except of course in certain cases) that a certain analytic function vanishes exactly at a given point.

There is one important special case for which it does give a result. Like most L-functions, the L-function of an elliptic curve satisfies a functional equation, specifically of the form $\Lambda(2 - s) = \varepsilon\Lambda(s)$, where $\Lambda(s)$ is equal to $L(s)$ times a suitable gamma and exponential factor, and $\varepsilon = \pm 1$ is the so-called *sign of the functional equation* (what else?). Thus, when $\varepsilon = -1$, we *know* that $L(1) = 0$, so that assuming the BSD conjecture we have $r > 0$, so n is a congruent number. It is easily shown that when n is integral and squarefree (which we always assume), then $\varepsilon = -1$ if and only if $n \equiv 5$, 6, or 7 modulo 8. It follows that, assuming the conjecture, all of these numbers should be congruent, and indeed, we have seen that 5, 6, and 7 are congruent.

On the other hand, when $\varepsilon = 1$, the order of vanishing of $L(s)$ at $s = 1$ is even, so assuming the conjecture, the rank r should be *even*. It is in fact very often equal to 0, but not always. For instance, we have $r = 0$ for $n = 1$, 2, and 3, so that these numbers are not congruent. On the other hand, it can be shown that we have $r = 2$ for $n = 34$, 41, and 65 for instance (and for no other squarefree $n \leqslant 100$), so that these numbers are indeed congruent.

The most precise conjecture on the distribution of congruent numbers is the following, where the second part comes from random matrix theory as in the case of sums of two cubes, so is quite speculative but well supported by numerical evidence; see [Kea].

Conjecture 6.12.3. (1) *Any squarefree integer congruent to* 5, 6, *or* 7 *modulo* 8 *is a congruent number.*
(2) *Denote by $C(X)$ the set of squarefree integers less than or equal to X that are congruent to* 1, 2, *or* 3 *modulo* 8 *and that are congruent numbers. Then $C(X)$ has density* 0; *more precisely there exists a strictly positive constant c such that*

$$C(X) \sim cX^{3/4}\log(X)^{11/8} \, .$$

6.12.3 Tunnell's Theorem

The congruent number problem was finally completely solved by Tunnell in 1980, up to a weak form of the BSD conjecture. For this, in addition to the standard ingredients in the theory of elliptic curves and the related theory of modular forms, he used modular forms of *half-integral weight*. In fact, we will see in Chapter 10 that the theta function attached to a Dirichlet character is the prototypical example of such a form. It is impossible to enter into the details of Tunnell's proof, but we give his result; see [Tun].

Theorem 6.12.4 (Tunnell). *Let n be a squarefree natural number, and set $u = 1$ if n is odd and $u = 2$ if n is even. If n is a congruent number, the number of solutions in $\mathbb{Z}$ of $n/u = x^2 + 2uy^2 + 8z^2$ with z odd is equal to the number of solutions with z even. Furthermore, if a weak form of the BSD conjecture holds (more precisely if $L(E, 1) = 0$ implies that $r > 0$ for the corresponding elliptic curve), then the converse also holds.*

The enormous advantage of this theorem is that it is very easy to check Tunnell's conditions, since we are dealing with representations by positive definite ternary forms that can easily be enumerated. In particular, it is easy (up to BSD) to make exhaustive tables of congruent numbers up to any desired reasonable limit. Thus the problem is completely solved, except of course that we must wait for the solution to the BSD conjecture to be absolutely sure. Note that this is one of the most beautiful and important conjectures in all of mathematics, and that a one-million-dollar Clay prize has been offered for its solution (see also Section 10.6).

For instance, we see from Tunnell's result what was already expected from the BSD conjecture, i.e., that the (squarefree integral) congruent numbers less than or equal to 100 are the numbers congruent to 5, 6, or 7 modulo 8 together with the three numbers $n = 34$, 41, and 65.

In the following table, which does not depend on BSD since in the range of the table we have curves of analytic rank only 0 or 1 or of proved rank 2 or 3, the entry in row numbered R and column C (with C going from 1 to 64) gives the rank of the elliptic curve corresponding to $n = R + C$, so that n is congruent if and only if the entry is greater than or equal to 1.

```
          1111111111122222222222333333333334444444444455555555555566666
          1234567890123456789012345678901234567890123456789012345678901234
      0 | 0000111100000111000011111000111100200111020001110000111110001110
     64 | 2000111000001110000111110001111000011100001111000011111100011110
    128 | 0000111122200111020001111100201111020002110000011100001111000111 0
    192 | 0200111100000111102011111100211111020011100001111000011110001110
    256 | 2002111102000111000011110001111120201110002011100201111120011111
    320 | 0020111002001110001111100011111200011100001111202111110001111111
    384 | 0200111100020111000011110201110000011002001110020111102101111111
    448 | 0000111102000111102001110001110001110001110000011100011112011110
    512 | 0200111100000111000011111000111202021112000011102001111000111110
    576 | 0002111102000111000011110201111200011002000011100011111000111110
    640 | 0000211100020111202011100011100110020011000001110200111100011110
    704 | 0000111100000111202111110021111100001110000011100011111200111110
    768 | 0000111122200111000011112001110001110000011000001110001111000111
    832 | 0000111120000111102011110001110200110000211100011000112011111111
    896 | 0000111122200111000210011110001110000011000001110000111100011111
    960 | 0000111100000111000111112021111100002011002001110000111100111110
   1024 | 2002111100000111200111110001110200211000011002001110200111201110
   1088 | 0000111100000111200111110021112000110002001110000011111122011110
   1152 | 0220111000021112020011110020111022001110002211020011100111001110
   1216 | 2000111200001111200111112201111200213100001110001110001111000111
   1280 | 0200111000021112000011110201110000011000012200111020011110021111
   1344 | 0200111100021110001111100011110200210002001120011110001111001111
   1408 | 0022111100020111000011110201110000020111020001110000111100011111
   1472 | 0002111100020211100011110001110011002000011100001111100011001111
   1536 | 0000111120000111100011112001110000011000021111000011000211110
   1600 | 0000111100200111000011112001111202011120000111202111110021110
   1664 | 0000111100000111100011110001110000011221000011110000111111001110
   1728 | 0020111200001111220111100011102001120200111000011100111201111
   1792 | 0200111100000111000011111000111100020111000001110000111110001111
   1856 | 0202111100000111100011110201110000011100001110001110011110011111
   1920 | 0000111100000111100021111000111000001100011111000211110001111111
   1984 | 0000111100020111000011112001111000211100000111102211110001110
```

Congruent Numbers up to $n = 2048$

Note that by Proposition 6.5.6, the above also gives a table of integers c for which the equation $x^4 - y^4 = cz^2$ has a solution with $xyz \neq 0$.

6.13 Some Unsolved Diophantine Problems

Evidently there are infinitely more unsolved Diophantine problems than there are solved ones, and we have already mentioned a few. We recall them here, and add a couple more. They all have some aesthetic value, and many people have tried to solve them.

For many other such problems, we refer to the vast literature on the subject, for instance the books by Mordell [Mord] and Guy [Guy]. Note that we do not mention other well-known problems such as Waring's problem, Goldbach's conjecture, or the twin prime conjecture, since these problems are of quite a different nature and are not tackled using the tools developed in this book.

In most cases, the reason for which the Diophantine equations are unsolved is that they reduce to finding rational points on curves or higher-dimensional varieties that are of *general type*, in a suitable sense. For curves, this means curves of genus greater than or equal to 2; for surfaces it really means surfaces of general type. We have no algorithmic way of searching for the complete set of points except in very special situations. On the other hand, if the Diophantine problem reduces to finding rational points on more special kinds of varieties, for instance curves of genus 1 or K3 surfaces, then even though we do not have real algorithms for finding rational points, we do have a large number of available methods.

Here is a small list, including some that we have already mentioned.

(1) Show that the quadratic forms $x^2+2y^2+5z^2+xz$, $x^2+3y^2+6z^2+xy+2yz$, and $x^2+3y^2+7z^2+xy+xz$ represent all odd positive integers (see Section 5.4.3). It is known that they represent all sufficiently large odd integers, but the bound is ineffective.

(2) Show that any squarefree integer congruent to 4, 6, 7, or 8 modulo 9 is a sum of two cubes of elements of $\mathbb{Q}$ (Conjecture 6.4.21).

(3) Show that an integer n is a sum of three cubes of integers if and only if $n \not\equiv 4 \pmod 9$ (it is clear that this latter condition is necessary). In addition, show that there are infinitely many representations, in other words that if $n \not\equiv 4 \pmod 9$ the Diophantine equation $x^3 + y^3 + z^3 = n$ has infinitely many integer solutions (Conjecture 6.4.24).

(4) Prove that every integer is a sum of four cubes of integers, in other words that for all n the Diophantine equation $x^3 + y^3 + z^3 + t^3 = n$ has an integer solution (Dem'yanenko's Theorem 6.4.28 shows that this is true when $n \not\equiv \pm 4 \pmod 9$). In fact, show that it has an integer solution with $t = x$, in other words that the equation $2x^3 + y^3 + z^3 = n$ has an integer solution (Conjecture 6.4.26).

(5) Prove that any squarefree integer n congruent to 5, 6, or 7 modulo 8 is a congruent number, in other words that the equation $y^2 = x^3 - n^2x$ has a solution in rational numbers with $y \neq 0$ (Conjecture 6.12.3). This would

follow from the Birch and Swinnerton-Dyer conjecture, in fact from a weak form of it.

(6) The *rational cuboid* problem: does there exist a rectangular parallelepiped all of whose sides, face diagonals, and main diagonals are rational? In other words, do there exist nonzero rational numbers a, b, and c such that $a^2 + b^2$, $a^2 + c^2$, $b^2 + c^2$, and $a^2 + b^2 + c^2$ are all rational squares? The answer is positive if any one condition is dropped: for instance $(a, b, c) = (44, 117, 240)$ satisfies the first three conditions but not the fourth, and $(a, b, c) = (117, 520, 756)$ satisfies the first, second, and fourth conditions, but not the third.

(7) The $4/n$ problem: is it true that for any integer $n > 1$ there exist positive integers a, b, and c such that $4/n = 1/a + 1/b + 1/c$? Note that it is very easy to find arithmetic progressions of n for which this is true other than the set of multiples of a given integer (for instance $n = 3k + 2$), that the number of counterexamples has asymptotic density zero, that the smallest counterexample, if any, is necessarily a prime number, and that for a given n there seems to be a large number of solutions a, b, c; see Exercises 56 and 57. The problem is that we do not know how to prove that this large number is greater than or equal to 1! See Exercise 58 for a very similar but much easier problem.

6.14 Exercises for Chapter 6

1.

(a) Solve the Diophantine equation $y^2 = (x + 1)^3 - x^3$ in integers.
(b) Solve the Diophantine equation $(x - y)^5 = x^3 - y^3$ by reducing it to the above equation.

2. Prove that for any positive integer n there exist x, y, and z such that $n = x^2 + y^2 + z^3$.

3. Let C be the curve $y^2 = x^\ell + t$ with $\ell \geqslant 3$ prime. Compute $|C(\mathbb{F}_q)|$ in characteristics 2 and ℓ and when $t = 0$ in $\mathbb{F}_q$.

4. Show that, as stated in the text, the general integral solution to $ax + by + cz = 0$ is $x = mb/\gcd(a, b) - \ell c/\gcd(a, c)$, $y = kc/\gcd(b, c) - ma/\gcd(a, b)$, $z = \ell a/\gcd(a, c) - kb/\gcd(b, c)$ for any integers k, ℓ, and m.

5. Consider the parametrization given by Proposition 6.3.6. It is quite trivial to see how to obtain the values of s, t, and d corresponding to the solutions $(x_0, -y_0, z_0)$ and $(-x_0, y_0, z_0)$.

(a) Find the values for s, t and d (which are unique up to a simultaneous change of sign of s and t) corresponding to the point (x_0, y_0, z_0) itself.
(b) More generally, if (x, y, z) is a solution of $Ax^2 + By^2 = Cz^2$ with the parameters s, t, and d, find the corresponding parameters for the solutions $(-x, y, z)$, $(x, -y, z)$, and $(x, y, -z)$.

6. Using the particular solution $(1, 0, 1)$ to the equation $x^2 + Ny^2 = z^2$, give a complete family of disjoint parametric solutions to this equation. It will be

useful to distinguish the cases N odd, $N \equiv 2 \pmod 4$, $4 \mid N$ with $v_2(N)$ even, and finally $4 \mid N$ with $v_2(N)$ odd.

7. Prove that the general integral solution of $x^2 + y^2 = 2z^2$ with x and y coprime is given by $x = \pm(s^2 + 2st - t^2)$, $y = \pm(s^2 - 2st - t^2)$, $z = \pm(s^2 + t^2)$, where s and t are coprime integers of opposite parity and the $\pm$ signs are independent.

8. Let Q be an indefinite quadratic form, and with the usual Gram–Schmidt notation assume that $|\mu_{i,j}| \leqslant 1/2$ for all $j < i$. Show that $|(\mathbf{b}_i^* + \mu_{i,i-1}\mathbf{b}_{i-1}^*)^2| \geqslant (1/\gamma + 1/4)|(\mathbf{b}_{i-1}^*)^2|$ implies $|(\mathbf{b}_i^*)^2| \geqslant (1/\gamma + 1/4 - \mu_{i,i-1}^2)|(\mathbf{b}_{i-1}^*)^2|$, but that the converse is not necessarily true.

9. Let C be a cube of side a in Euclidean three-space $\mathbb{R}^3$. Assume that all the vertices of C have coordinates in $\mathbb{Z}^3$. Translating C, we assume that one of its vertices is at the origin, and we denote by (x_j, y_j, z_j) the coordinates of the three vertices of C adjacent to the origin.

(a) Let M be the 3×3 matrix whose rows are the (x_j, y_j, z_j). Compute explicitly MM^t.

(b) For $1 \leqslant j \leqslant 3$ let α_j be the complex number $\alpha_j = x_j + iy_j$. Deduce from (a) that $\alpha_1^2 + \alpha_2^2 + \alpha_3^2 = 0$.

(c) Find the general solution to the equation $x^2 + y^2 + z^2 = 0$ in the Euclidean domain $\mathbb{Z}[i]$, generalizing Corollary 6.3.13.

(d) Deduce finally a parametrization of triples $((x_1, y_1), (x_2, y_2), (x_3, y_3))$ of points in $\mathbb{Z}^2$ that are the orthogonal projections of cubes C as above.

(e) Give a few numerical examples of such triples, and draw the corresponding pictures.

10. Let D be a nonsquare integer (in fact rational number is sufficient). Prove that the general *rational* solution to the Diophantine equation $x^2 - Dy^2 = 1$ is given by $x = \pm(s^2 + D)/(s^2 - D)$, $y = 2s/(s^2 - D)$ for $s \in \mathbb{Q}$.

11. Let $f(X) = a_n X^n + \cdots + a_0 \in \mathbb{Z}[X]$ be a polynomial with integer coefficients with $a_n \neq 0$ and $a_0 \neq 0$. If $c/d \in \mathbb{Q}$ is a root of $f(X) = 0$ with $\gcd(c, d) = 1$, show that $d \mid a_n$ and $c \mid a_0$.

12. Let S be the cubic surface with affine equation $x^3 + y^3 + z^3 = 10$, on which there is the evident point $P = (1, 1, 2)$.

(a) In view of Manin's Conjecture 6.4.1, using tangents at the point P find a two-parameter family of rational points on S.

(b) Show that none of these points (except the point P) have all three coordinates strictly positive.

(c) By iterating the process starting from one of the new rational points, find a rational point on S with strictly positive coordinates other than $(1, 1, 2)$, $(1, 2, 1)$, and $(2, 1, 1)$.

13.

(a) Prove Lemma 6.4.3.

(b) Show that the equation $27x^3 + 2y^3 + 3z^3 = 0$ has no solutions with x, y, z integers such that $\gcd(y, z) = 1$, although it has an infinity of *rational* solutions (you may need to use the chapters on elliptic curves for this). This shows that the pth power-free condition in Lemma 6.4.3 is necessary.

14. Show that an immediate corollary of Theorem 6.4.17 is the following: the equation $x^3 + cy^3 + cz^3 = 0$ has no solutions in nonzero integers when $c = 1, 9, p$, or p^2 with $p \equiv 2$ or 5 modulo 9, except for $c = 4$, in which case it has the unique solution $(x, y, z) = (-2, 1, 1)$ (up to multiplication by a constant), and it has no solutions with $3 \mid x$ when $c = p$ or p^2 with $p \equiv 8 \pmod 9$.

15. (Taken from [Cas2].) Let p and q be prime numbers such that $p \equiv 2 \pmod 9$ and $q \equiv 5 \pmod 9$, and let $c = pq$. Theorem 6.4.17 (3) asserts that the equation $x^3 + y^3 + cz^3 = 0$ does not have any solutions with $3 \mid z$. The aim of this exercise is to show that it does not have any nontrivial solutions at all. For this, let ρ be a primitive cube root of unity, and set $\lambda = \rho - \rho^{-1} = \sqrt{-3}$. We will show more generally by descent that our equation has no solutions in $\mathbb{Z}[\rho]$. Without loss of generality, let (x, y, z) be a pairwise coprime solution to our equation in $\mathbb{Z}[\rho]$ with $|xyz|$ minimal.

(a) By factoring our equation, show that there exist elements α, β, γ, u, v, and w in $\mathbb{Z}[\rho]$ with u, v, and w pairwise coprime, such that either

$$x + y = \alpha u^3, \quad \rho x + \rho^{-1} y = \beta v^3, \quad \rho^{-1} x + \rho y = \gamma w^3, \quad \alpha\beta\gamma = c, \quad \text{or}$$
$$x + y = \lambda \alpha u^3, \quad \rho x + \rho^{-1} y = \lambda \beta v^3, \quad \rho^{-1} x + \rho y = \lambda \gamma w^3, \quad \alpha\beta\gamma = c,$$

and hence $\alpha u^3 + \beta v^3 + \gamma w^3 = 0$ and $\alpha\beta\gamma = c$ in both cases.

(b) Noting that we may multiply simultaneously x, y, and z by any unit, show that without loss of generality we may assume that (α, β, γ) is a permutation of $(\pm 1, \pm 1, \pm c)$ or of $(\pm 1, \pm p, \pm q)$.

(c) As in the proof of Theorem 6.4.17 (3), prove that $u^3 + pv^3 + qw^3 \equiv 0 \pmod{9\mathbb{Z}[\rho]}$ is impossible with u, v, and w pairwise coprime.

(d) Prove that $|uvw| < |xyz|$, and hence deduce by descent that our equation $x^3 + y^3 + cz^3 = 0$ has no nontrivial solutions.

16. Assume that $p = 3$ and that the equation $x^3 + by^3 + cz^3 = 0$ is everywhere locally soluble.

(a) Show that if conditions (1) and (2) of Theorem 6.4.10 are satisfied, then $\varepsilon_1 \equiv 1 \pmod{3\mathbb{Z}_K}$, where ε_1 is a fundamental unit of $K = \mathbb{Q}(c^{1/3})$, so that in (2) it is sufficient to test a single generator α of $\mathfrak{b}^e$ instead of all the $\varepsilon\alpha$ for ε a unit modulo cubes.

(b) Under the same assumptions, if in addition we assume that $3 \nmid e$, show that in fact $c \equiv \pm 3 \pmod 9$.

   Note that the condition $\varepsilon_1 \equiv 1 \pmod{3\mathbb{Z}_K}$ means that the p-adic regulator $R_3(K) = \log_3(\varepsilon_1)$ is divisible by 3, so that the combination of this condition with the condition $3 \mid h(K)$ of Theorem 6.4.8 means that $3 \mid h(K)R_3(K)$, which is essentially the residue at 1 of the p-adic zeta function of K.

17. With the notation of Theorem 6.4.10, let $\mathfrak{p}$ be the unique prime ideal of K above p, so that $\mathfrak{p}^p = p\mathbb{Z}_K$.

(a) Set $L(x) = \sum_{1 \leqslant k \leqslant p-1} (-1)^{k-1} x^k / k$ and $E(x) = \sum_{0 \leqslant k \leqslant p-1} x^k / k!$, which are truncations of the power series for $\log(1+x)$ and $\exp(x)$ respectively. Show that if the $\alpha_i \in \mathbb{Z}_K$ are such that $v_\mathfrak{p}(\alpha_i) \geqslant 1$ then $E(\alpha_1 + \alpha_2) - E(\alpha_1)E(\alpha_2) \in \mathfrak{p}\mathbb{Z}_K$, and if $\alpha \in \mathbb{Z}_K$ is such that $v_\mathfrak{p}(\alpha) \geqslant 1$ then $E(L(\alpha)) - (1 + \alpha) \in \mathfrak{p}\mathbb{Z}_K$ and $L(E(\alpha) - 1) - \alpha \in \mathfrak{p}\mathbb{Z}_K$.

(b) Assuming that β is coprime to $\mathfrak{p}$, show how to replace the conditions on β and ε of Theorem 6.4.10 by conditions on $L(\beta/b_0 - 1)$ and $L(\varepsilon/e_0 - 1)$ for suitable integers b_0 and e_0, and deduce that the conditions of Theorem 6.4.10 can be easily checked by linear algebra instead of by exhaustive enumeration of the $p^{(p-1)/2}$ elements $\varepsilon \in U(K)/U(K)^p$.

(c) Generalize to the case where β is not coprime to p.

18. Prove Proposition 6.4.20 by writing $x^3 + y^3 = (x + y)(x^2 - xy + y^2)$.

19. Find parameters d, s, and t in Proposition 6.4.29 giving the solution $(w, x, y, z) = (1, 6, 8, -9)$ to $w^3 + x^3 + y^3 + z^3 = 0$.

20. Let (a_1, a_2, a_3, a_4) be four integers satisfying $a_1^3 + a_2^3 + a_3^3 + a_4^3 = 0$ with $a_i \neq -a_j$ for any (i, j).

(a) Show that there exists a (partial) parametrization of $w^3 + x^3 + y^3 + z^3 = 0$ of the form

$$(w, x, y, z) = (a_1 u^2 + buv - a_2 v^2, a_2 u^2 - buv - a_1 v^2,$$
$$a_3 u^2 + cuv - a_4 v^2, a_4 u^2 - cuv - a_3 v^2)$$

with b and c *rational* numbers (not necessarily integers) if and only if $P = -(a_1 + a_2)(a_3 + a_4)$ is a square, and express b and c as a rational function of the a_i and of the square root of P.

(b) Find the parametrizations coming from the integral solutions $(3, 5, 4, -6)$ and $(1, -9, -10, 12)$, and show that $(1, 6, 8, -9)$ (in any order) does not give rise to any such parametrization.

(c) By considering the integral solution $(12, 86, 159, -167)$, show that b is not always an integer (when b or c is not an integer one can multiply the a_i, b, and c by a common denominator of b and c to obtain an integral parametrization).

(d) Give a complete parametrization of (a_1, a_2, a_3, a_4) such that $a_1^3 + a_2^3 + a_3^3 + a_4^3 = 0$, P a square, and $b = -a_1$, $c = -a_3$, and deduce that there exists an infinity of parametrizations as in (a).

21. Find all integral solutions to the Diophantine equation $y^2 = x^3 + 16$.

22. Let K be a quadratic field and let $\alpha \in \mathbb{Z}_K$.

(a) Assume that K is an imaginary quadratic field of odd class number and different from $\mathbb{Q}(i)$. Show that $\alpha \bar{\alpha}$ is a square in $\mathbb{Z}$ if and only if $\alpha = n\beta^2$ for some $n \in \mathbb{Z}$ and $\beta \in \mathbb{Z}_K$, thus in particular proving Lemma 6.4.35.

(b) How must this statement be modified if K is a real quadratic field of odd class number?

(c) Give examples showing that the result is false if $K = \mathbb{Q}(i)$ or if K does not have odd class number.

23.

(a) Find an analogue of Corollary 6.6.3 for everywhere local solubility in every completion of $K = \mathbb{Q}(i)$, with $i^2 = -1$ (show for instance that the local condition at 2 is $c \equiv 1$, 2, or -3 modulo $\mathfrak{p}_2^7$, where $\mathfrak{p}_2 = (1 + i)\mathbb{Z}_K$).

(b) What about the field $K = \mathbb{Q}(\zeta_8)$ generated by a primitive 8th root of unity?

24. Using a descent method, find all coprime integer solutions to the Diophantine equation $x^4 + y^4 = 2z^4$.

25. Using Corollary 6.6.11 prove that the equation $x^4 + y^4 = c$ with $c = 7361$ has no rational solutions, although it is everywhere locally soluble by Corollary 6.6.3, and although the groups $E_c(\mathbb{Q})$ and $F_c(\mathbb{Q})$ have rank 2.

26. It follows from Proposition 5.7.3 and its proof that the equation $2y^2 = x^4 - 17$ is everywhere locally soluble, but not globally soluble. The aim of this exercise is to give an alternative proof of this last fact.

(a) Prove that if x, y is a rational solution, there exist a, b, and c in $\mathbb{Z}$ such that $x = a/c$, $y = b/c^2$, $\gcd(a, b, c) = 1$, and

$$(5a^2 + 17c^2 + 4b)(5a^2 + 17c^2 - 4b) = 17(a^2 + 5c^2)^2 .$$

(b) Show that $p = 2$ is the only prime that can divide both factors on the left.

(c) Deduce that for a suitable choice of signs and $e = 1$ or 2, there exist u and v in $\mathbb{Z}$ such that

$$5a^2 + 17c^2 \pm 4b = 17eu^2, \quad 5a^2 + 17c^2 \mp 4b = ev^2, \quad \text{and} \quad a^2 + 5c^2 = euv \,.$$

(d) Show that there does not exist any solution of these equations in $\mathbb{Q}_{17}$, hence that our initial equation has no global solution.

27. Generalize Exercise 26 as follows. Assume that c is a sum of two squares, say $c = c_0^2 + c_1^2$. Using the identity

$$(ca^2 + c_0 d^2 + c_1 b)(ca^2 + c_0 d^2 - c_1 b) = c(c_0 a^2 + d^2)^2$$

already used in Exercise 26, give sufficient conditions for the equation $y^2 = cx^4 - 1$ to have no rational solutions (hence for the equation $x^4 + y^2 = cz^4$ to have no nontrivial integer solutions).

28. Prove Proposition 6.5.1 (Hint: for (1), reason as in Proposition 6.6.1 for $p \geqslant 37$ and for $p \equiv 3 \pmod 4$, and for the remaining four values of p perform a systematic search).

29. Assume that $c \in \mathbb{Z}_{\geqslant 3}$ is squarefree, and let x, y, and z be integers such that $x^4 + y^4 = cz^2$ with $z \neq 0$. By Corollary 6.5.7 we know that $2c$ is a congruent number. Show that (u, v, w) are the sides of a Pythagorean triangle with area $2c$, where

$$u = \frac{|x^4 - y^4|}{xyz}, \quad v = \frac{4cxyz}{|x^4 - y^4|}, \quad \text{and} \quad w = \frac{x^8 + 6x^4 y^4 + y^8}{xyz|x^4 - y^4|} \,.$$

30. Using the group law on the elliptic curve $Y^2 = X^3 + c^2 X$ (and not by a simple check) show that if (x, y, z) is a solution to $x^4 + y^4 = cz^2$ then so is (x_2, y_2, z_2) with

$$x_2 = x(3c^2 z^4 - 4x^8), \quad y_2 = y(3c^2 z^4 - 4y^8), \quad z_2 = z(c^4 z^8 + 24x^4 y^4(x^8 + y^8)) \,,$$

and for instance find a coprime integrer solution to $x^4 + y^4 = 17z^2$ different from $(\pm 1, \pm 2, \pm 1)$ and $(\pm 2, \pm 1, \pm 1)$.

31. The text states that 2-descent methods are not sufficient to determine the global solubility of $x^4 + y^4 = cz^2$ with $c = 2801$. Check that in fact

$$x = 2703612950089664684650, \quad y = 90462483365506215707,$$
$$z = 138975206507312317343180998848020200625001$$

is a solution.

32. Assume that $p = 3$.

(a) Show that condition (5) (a) of Theorem 6.4.2 can be replaced by $\ell \neq p$; in other words, show that if $\ell \nmid pabc$ condition S_ℓ is satisfied.

(b) Assume that $\ell = p = 3$. Show that if condition (5) (c) of Theorem 6.4.2 is satisfied, then so is condition (5) (b), so that condition (5) (c) can be removed; in other words, show that if $3 \nmid abc$ and if $a \pm b \pm c \equiv 0 \pmod 9$ for some signs $\pm$, then in fact either $a \pm b \equiv 0 \pmod 9$, $a \pm c \equiv 0 \pmod 9$, or $b \pm c \equiv 0 \pmod 9$.

33. Let p and ℓ be distinct odd primes. We will say that condition $C(p, \ell)$ is satisfied if for all integers a, b, and c such that $\ell \nmid abc$ the equation $ax^p + by^p + cz^p = 0$ has a nontrivial solution in $\mathbb{Z}_\ell$.

(a) Show that the equation $x^5 + 2y^5 + 4z^5 \equiv 0 \pmod{11}$ has no nontrivial solutions, so that $C(5, 11)$ is not satisfied.

(b) Show that if $\ell > 11$ then $C(5, \ell)$ is satisfied.

(c) Similarly, show that the equation $x^7 + 2y^7 + 6z^7 \equiv 0 \pmod{71}$ has no nontrivial solutions, so that $C(7, 71)$ is not satisfied, but that if $\ell > 71$ then $C(7, \ell)$ is satisfied.

(d) From now on, let p be fixed and assume that $C(p, \ell)$ is *not* satisfied. Show that $\ell \equiv 1 \pmod{2p}$.

(e) Using the Weil bounds, show that $\ell \leqslant B(p)$, where $B(p)$ is an explicit function depending only on p. It follows that for a given p the set $E(p)$ of primes ℓ such that condition $C(p, \ell)$ is not satisfied is finite.

(f) Using a computer, show that $E(3) = \emptyset$, $E(5) = \{11\}$, $E(7) = \{29, 43, 71\}$, $E(11) = \{23, 67, 89, 199, 419\}$, and $E(13) = \{53, 79, 131, 157, 313, 547\}$. Show also that $B(5) = 11$, $B(7) = 71$, $B(11) = 419$, $B(13) = 547$, $B(17) = 1429$, $B(19) = 1597$, $B(23) = 1979$, $B(29) = 5279$, and $B(31) = 7193$.

(g) For $p = 5, 7, 11$, and 13 give explicit values of a, b, and c for which one can easily prove that $ax^p + by^p + cz^p = 0$ has no nontrivial solution in $\mathbb{Z}$ by congruence arguments.

34. Show that for $b = 8, 9, 10$, and 12 the equations $x^5 + by^5 + 19z^5 = 0$ are everywhere locally soluble but do not have any nontrivial solutions in $\mathbb{Q}$.

35. Prove the analogue of Proposition 6.7.1, but now for squarefree t such that $1 \leqslant t \leqslant 100$.

36. Generalizing Propositions 6.7.3, 6.7.5, and 6.7.7, find the integral solutions to the Diophantine equation $y^2 = x^p + 4t$ for general p, then for $p = 3$ and $p = 5$, with the same assumptions on t as in the above propositions. As an example, find all the integral solutions to $y^2 = x^3 - 4$.

37. Assume $H(p, t)$. Prove that if $a \in A_p(t)$ then $|a| \leqslant p\sqrt{-t}/\pi$, hence that apart from the special solutions, if (x, y) is a solution to $y^2 = x^p + t$ we have $x = a^2 - t$ with $|a| \leqslant p\sqrt{-t}/\pi$. (Hint: write the defining equation for $A_p(t)$ in terms of $\theta = \operatorname{atan}(\sqrt{-t}/a)$.)

38. Assume $H(p, t)$. Looking now also at 3-adic valuations and using a similar reasoning to that of the preceding exercise, prove that if t is congruent to 3, 12, 15, 21, or 24 modulo 27 and not congruent to -1 or 3 modulo 16, or when $t \equiv 14 \pmod{24}$, there are no integer solutions to the Diophantine equation $y^2 = x^p + t$ (for these two exercises, see [Cohn1] if you need help).

39. Assume $H(p, t)$. Using similar reasoning to that of Proposition 6.7.12, prove that if $t \equiv 3 \pmod 4$ with $v_2(t+1)$ odd the equation $y^2 = x^p + t$ has no integral solutions.

40. Give explicitly 6 (respectively 16) integral points $(x, y) \in \mathbb{Z}^2$ satisfying the Diophantine equation $y^2 = x^3 + t$ for $t = -39$ (respectively $t = 17$). Note that in these cases, we have $t \equiv 1 \pmod 8$ (and even $t > 0$ in the second). Using the techniques of Section 8.7, one can show that there are no other solutions.

41.

(a) Prove that the only roots of unity of the form $(a + \sqrt{t})/(a - \sqrt{t})$ with $a \neq 0$ are obtained for $a = \pm 1$ when $t = -1$, and for $a = \pm 1$ or $a = \pm 3$ when $t = -3$.

(b) With the notation of the proof of Corollary 6.7.11, prove that if $p \geqslant 7$ we have $|u_p(a + \sqrt{t}, a - \sqrt{t})| > 1$ for the above values of t and a.

42. Prove the special case $(p, t) = (3, 3)$ of Proposition 6.7.14, which has been omitted from the proof.

43. Prove Theorem 6.8.3 (1) by considering separately the cases $n \equiv 1 \pmod{4}$, $n \equiv 3 \pmod{4}$, $n \equiv 0 \pmod{6}$, and $n \equiv \pm 2 \pmod{6}$. Similarly, prove (2).

44. Using the results of Section 6.8.1, find all integral solutions to the Diophantine equations $5y^2 = x^4 + a$ for $a = \pm 1$ and $a = \pm 4$.

45. Find all integers n such that $F_n = 3x^2$ or $L_n = 3x^2$ for some $x \in \mathbb{Z}$.

46. (Bremner–Tzanakis.) Let P, Q be nonzero integers, and consider the sequence $F_n = F_n(P,Q)$ defined by $F_0 = 0$, $F_1 = 1$, and $F_{n+1} = PF_n - QF_{n-1}$. This generalizes the Fibonacci sequence that corresponds to $(P,Q) = (1,-1)$. A natural problem, which we have solved in the text for the Fibonacci sequence, is to ask for which n is $F_n(P,Q)$ a perfect square. We now ask the converse problem: given $n \geqslant 1$, for which (P,Q) can $F_n(P,Q)$ be a perfect square?

(a) Show that the answer to the direct problem is trivial when $(P,Q) = (\pm 1, 1)$ and $(\pm 2, 1)$.
(b) Show that if we do not assume $\gcd(P,Q) = 1$, there exists a solution to the inverse problem for any even n. Thus from now on we exclude the values found in (a) and we assume $\gcd(P,Q) = 1$.
(c) Give a complete parametrization of the coprime pairs (P,Q) other than that of (a) that solve the converse problem for $1 \leqslant n \leqslant 6$. See Exercise 8 of Chapter 8 for the other values of n.

47. Show that the integers k occurring in Wendt's criterion (Proposition 6.9.6) satisfy $k \equiv \pm 2 \pmod{6}$.

48. Assuming that the numerators of the Bernoulli numbers $B_2, B_4, \ldots, B_{p-3}$ are divisible by p with probability $1/p$, and that these probabilities are independent (a strong assumption), show that the density of regular primes should be equal to $\exp(-1/2)$.

49.

(a) Prove Corollary 6.10.4 (2).
(b) Show that there are 14 integer solutions (x,y) to the Diophantine equation $y^2 = x^4 - 3x^3 - 4x^2 + 2x + 4$.
(c) Show that the only integer solutions (x,y) to the Diophantine equation $y^2 = x^4 - 5x^3 - 5x^2 - 5x - 2$ are $(x,y) = (-129, \pm 16958)$, $(-1, \pm 2)$, and $(6, \pm 2)$.
(d) Show that there are 16 integer solutions (x,y) to the Diophantine equation $y^2 = x^6 - 3x^5 + 3x^4 - x^3 - 4x^2 + 4x + 1$.
(e) Find all the solutions to Diophantus's equation $y^2 = x^6 + x^2 + 1$ for which x is a rational number whose denominator is less than or equal to 4 (see Section 13.3.4).

50. Try to generalize Proposition 6.10.3 to equations of the form $g(y) = f(x)$, where f is a monic polynomial of degree n, g a monic polynomial of degree d such that $d \mid n$, or such that $\gcd(d,n) > 1$ (for help, see for instance [Ten]).

51. Assume that (a,b,c) are the sides of a Pythagorean triangle of area n. Using the group law on the corresponding elliptic curve (more precisely the formula for doubling a point, obtained by computing the coordinates of the third point of intersection of a tangent), find another Pythagorean triangle with the same area. It is easy to show that repeating this process gives an infinite number of them.

52. Let a, b, and c be nonzero 4th power-free integers such that $\gcd(a,b,c) = 1$. It is clear that the equation $ax^4 + by^4 + cz^4 = 0$ has a nontrivial solution in $\mathbb{R}$ if and only if a, b, and c do not have the same sign. Now let ℓ be a prime number, and as in Theorem 6.4.2, denote by S_ℓ the condition that the equation

$ax^4 + by^4 + cz^4 = 0$ has a nontrivial solution in $\mathbb{Q}_\ell$. Generalizing Corollary 6.6.3, prove the following results, which are quite similar to those of Theorem 6.4.2. Reorder a, b, and c so that $0 = v_\ell(a) \leqslant v_\ell(b) \leqslant v_\ell(c) < 4$ and set $v_b = v_\ell(b)$ and $v_c = v_\ell(c)$.

(a) Show that S_2 is true if and only either $a + b$, $a + c$, $a + b + c$, $(b + c)/b$, or $(16a + b + c)/b$ is divisible by 16.

(b) Assume that $\ell \mid abc$ with $\ell \neq 2$. Show that if $1 \leqslant v_b < v_c$ then S_ℓ is false, if $1 \leqslant v_b = v_c$ then S_ℓ is true if and only if $-b/c$ is a fourth power in $(\mathbb{Z}/\ell\mathbb{Z})^*$, and if $v_b = 0$ then S_ℓ is true if and only if $-a/b$ is a fourth power in $(\mathbb{Z}/\ell\mathbb{Z})^*$.

(c) Assume finally that $\ell \nmid 2abc$. Show that if $\ell \geqslant 37$ or $\ell \equiv 3 \pmod 4$ then S_ℓ is true. Otherwise, for $\ell = 5, 13, 17$, and 29, set $G_\ell = \{1\}$, $\{1, 3, 9\}$, $\{1, 4, 13, 16\}$, and $\{1, 7, 16, 20, 23, 24, 25\}$ respectively. show that S_ℓ is true if and only if $\alpha \in G_\ell$ for $\alpha = -b/a$, $-c/a$, $-c/b$, or $-(c + ma)/b$ for some $m \in G_\ell$.

53. In Section 6.6 we have studied in detail the solubility of the equation $x^4 + y^4 = cz^4$. Consider now the equation $x^4 - y^4 = cz^4$, where we do not necessarily assume that c is 4th power-free.

(a) Show that if one of the elliptic curves $y^2 = x^3 - cx$, $y^2 = x^3 + cx$, or $y^2 = x^3 - c^2x$ has rank 0 our equation has no solution with $xyz \neq 0$, and using `mwrank` or 2-descent, give the list of the 76 values of c with $1 \leqslant c \leqslant 100$ for which one can deduce in this way that the equation has no solution with $xyz \neq 0$.

(b) By making a systematic search for $1 \leqslant y < x \leqslant 50$, give a list of 7 values of c with $1 \leqslant c \leqslant 100$ for which there exists a solution to our equation with $xyz \neq 0$.

(c) By factoring the equation, try to solve the remaining 17 cases using algebraic methods (I have not tried to do this; the first five values to be studied are $c = 14, 20, 21, 31$, and 37).

54. Let a, b, and c be nonzero integers, and p, q, and r be in $\mathbb{Z}_{\geqslant 2}$.

(a) Show that if at least two among p, q, and r are coprime, the equation $ax^p + by^q + cz^r$ has a nontrivial integer solution.

(b) Show that the equation $x^6 + y^{10} + 4z^{15} = 0$ has no nontrivial solution in $\mathbb{Q}_2$, although $\gcd(6, 10, 15) = 1$.

55. Let $m \in \mathbb{Z}_{\geqslant 2}$. Show that $x^m - y^m = 1$ is impossible in nonzero integers x and y (when for instance $x > y > 0$ prove that $x^m - y^m \geqslant m + 1$, and proceed similarly otherwise).

56. This exercise and the next give some easy results on the $4/n$ problem. Let us say that a positive integer n is *Egyptian* if there exist positive integers a, b, and c such that $4/n = 1/a + 1/b + 1/c$.

(a) Show that if there exists $k \equiv 3 \pmod 4$ such that $k \mid (n+4)$ then n is Egyptian.

(b) Deduce that if n is *not* Egyptian then $n \equiv 1 \pmod 3$.

(c) Deduce also that if $n + 4$ is *not* the sum of two integer squares, then n is Egyptian (by Proposition 5.4.10, the number of integers $n \leqslant X$ that are sums of two squares is asymptotic to $CX/\sqrt{\log(X)}$ for a suitable constant $C > 0$, so this shows that non-Egyptian numbers, if they exist, have asymptotic density zero).

(d) Find more general criteria than (a).

(e) Assume that $4/n = 1/a + 1/b + 1/c$ with $a \leqslant b \leqslant c$. Prove that $(n + 1)/4 \leqslant a \leqslant 3n/4$, that $an/(4a - n) < b \leqslant 2an/(4a - n)$, and deduce that for given n the number $N(n)$ of solutions is finite. More precisely, show that $N(n) \leqslant n^2 \log(n)/16 + O(n^2)$ (this bound is far from optimal).

57. Let $n \in \mathbb{Z}_{\geqslant 2}$ be an integer.

(a) Prove that there exist positive integers a and b such that $3/n = 1/a + 1/b$ if and only if not all prime divisors of n are congruent to 1 modulo 3.

(b) Deduce that if n is not represented by the quadratic form $x^2 - xy + y^2$, then n is Egyptian (as in the preceding exercise, the number of n that *are* represented by the quadratic form has asymptotic density zero).

58. Show that every positive rational number m/n can be written as $m/n = 1/a_1 + 1/a_2 + \cdots + 1/a_s$ for some s, with $1 \leqslant a_1 < a_2 < \cdots < a_s$. (Hint: reduce to rational numbers less than $1/N$ for some N by using the divergence of the harmonic series $\sum 1/k$, and then use induction.)

7. Elliptic Curves

7.1 Introduction and Definitions

7.1.1 Introduction

To make this book as self-contained as possible we include here most of the important results that we need on elliptic curves, sometimes without proof. We urge the reader to read further on the subject (for the proofs and for a large amount of additional material) in the remarkable books by Cassels [Cas2] and Silverman [Sil1], [Sil2], as well as the more elementary treatment in Silverman–Tate [Sil-Tat]. The book by Darmon [Dar], which is available on his home page, contains a great deal of very useful recent material.

Since we could literally write thousands of pages on elliptic curves, and not as well as the above masters, we must target our needs. Our main purpose is the *Diophantine properties* of elliptic curves, in other words properties of elliptic curves defined over $\mathbb{Q}$, or more generally over a number field or a global function field. This will be done in detail over $\mathbb{Q}$ in the next chapter. Thus, as usual in Diophantine problems it will be useful to study elliptic curves over completions of number fields, in other words over $\mathbb{R}$, $\mathbb{C}$, and $\mathfrak{p}$-adic fields $K_{\mathfrak{p}}$. Again as usual, to have properties over $\mathfrak{p}$-adic fields it is necessary to start by studying curves over the residue fields $\mathbb{F}_{\mathfrak{p}} = \mathbb{F}_q$. For technical reasons, we will also study special kinds of curves over function fields in one variable $k(T)$.

The plan of this chapter is thus as follows. We first spend several sections describing properties of elliptic curves that are independent of the field of definition (or with mild restrictions such as being of characteristic zero), and then study curves over the above-mentioned fields, specifically over $\mathbb{C}$, $\mathbb{R}$, $k(T)$, $\mathbb{F}_q$, and $K_{\mathfrak{p}}$. We will then be equipped with the necessary tools for the Diophantine study of the next chapter.

7.1.2 Weierstrass Equations

We should begin with the following abstract definition: an elliptic curve over a field K is a smooth projective algebraic curve over K (i.e., a projective algebraic variety of dimension 1 with no singular points) of genus 1, together with a point defined over K, which by abuse of language we will call a *rational point*, even when $K \neq \mathbb{Q}$. This may require some effort from the reader, in particular in the precise understanding of the notion of genus. Thanks to the Riemann–Roch theorem, it is however not necessary to use this general definition, although it is quite useful. Indeed, this theorem implies that there exists a plane model of the curve with projective equation (called a general Weierstrass equation)

$$y^2 z + a_1 xyz + a_3 yz^2 = x^3 + a_2 x^2 z + a_4 xz^2 + a_6 z^3 \,,$$

where $a_i \in K$ and (x, y, z) are the projective coordinates, and we can take this as the definition of an elliptic curve. The numbering of the coefficients is accepted worldwide and must not be changed; it is in fact very logical since x can be considered of weight 2, y of weight 3, z of weight 0, and a_i is then of weight i.

Note that with this equation the implicit rational point on the curve is the point $\mathcal{O} = (0, 1, 0)$, which is the only point at infinity (i.e., with $z = 0$) on the curve. Note also that the fact that the curve is nonsingular translates into the nonvanishing of a certain universal polynomial in the a_i called the *discriminant* of E and denoted by disc(E), which we will define below. Since we will need the notation anyway, we recall the additional construction of the standard numbers attached to E.

Assume for the moment that the characteristic is different from 2 and 3. Setting $Y = 2y + a_1 x + a_3 z$ we have

$$zY^2 = 4x^3 + b_2 x^2 z + 2b_4 xz^2 + b_6 z^3$$

with

$$b_2 = a_1^2 + 4a_2, \ b_4 = a_1 a_3 + 2a_4, \text{ and } b_6 = a_3^2 + 4a_6 \,,$$

and setting $X = x + b_2/12$ we have

$$zY^2 = 4X^3 - \frac{c_4}{12} X z^2 - \frac{c_6}{216} z^3$$

with

$$c_4 = b_2^2 - 24b_4 \quad \text{and} \quad c_6 = -b_2^3 + 36b_2 b_4 - 216b_6 \,.$$

This can be rewritten $z(Y/2)^2 = X^3 - (c_4/48)Xz^2 - (c_6/864)z^3$ or $z(108Y)^2 = (36X)^3 - 27c_4(36X)z^2 - 54c_6 z^3$, and an equation of the form $ZY^2 = X^3 + aXZ^2 + bZ^3$ will be called a *simple* Weierstrass equation. We define the discriminant of an elliptic curve given in this form as $-16(4a^3 + 27b^2)$, in other words 16 times the discriminant of the cubic polynomial. We will soon see the reason for this factor 16. Thus when E is given by a general Weierstrass equation it is natural to define disc(E) as the discriminant of the corresponding simple equation, and we find that disc(E) $= (c_4^3 - c_6^2)/1728$. All this is valid only in characteristic different from 2 and 3. However, a short computation shows that we can express disc(E) as

$$\mathrm{disc}(E) = -b_2^2 b_8 - 8b_4^3 - 27b_6^2 + 9b_2 b_4 b_6 \;,$$

$$\text{where} \quad b_8 = a_1^2 a_6 + 4a_2 a_6 - a_1 a_3 a_4 + a_2 a_3^2 - a_4^2 \;,$$

and this expression now makes sense in any characteristic, and is the precise definition of $\mathrm{disc}(E)$. This is the main reason for the inclusion of the factor 16 in the definition of $\mathrm{disc}(E)$. Finally, we define the j-invariant of the curve by $j(E) = c_4^3 / \mathrm{disc}(E)$, which exists since we assume that $\mathrm{disc}(E) \neq 0$. In all discussions dealing with elliptic curves, the quantities a_i, b_i, c_i, $\mathrm{disc}(E)$, and $j(E)$ will have the above meaning. Note that in characteristic different from 2 we have $b_8 = (b_2 b_6 - b_4^2)/4$, but this formula cannot be used in characteristic 2.

The human mind usually prefers to work with *affine* coordinates, so in general we will write a Weierstrass equation as $y^2 + a_1 xy + a_3 y = x^3 + a_2 x^2 + a_4 x + a_6$, and keep in mind the additional point at infinity. It is easy to show that the allowed affine transformations which preserve the Weierstrass form are

$$x = u^2 x' + r, \quad y = u^3 y' + su^2 x' + t \;,$$

where $u \in K^*$ and r, s, and t are in K. The coefficients a_i' of the Weierstrass equation satisfied by x' and y' are given by the following formulas, that we give for completeness:

$$ua_1' = a_1 + 2s$$
$$u^2 a_2' = a_2 - sa_1 + 3r - s^2$$
$$u^3 a_3' = a_3 + ra_1 + 2t$$
$$u^4 a_4' = a_4 - sa_3 + 2ra_2 - (t + rs)a_1 + 3r^2 - 2st$$
$$u^6 a_6' = a_6 + ra_4 + r^2 a_2 + r^3 - ta_3 - t^2 - rta_1 \;.$$

In practice, an elliptic curve can be given in many other forms than by a Weierstrass equation. We will see in Section 7.2 how to deal with the most common cases. Note, however, that it is not always necessary or useful to transform the equation into a Weierstrass equation. For instance, we shall see below that the group law on an elliptic curve in Weierstrass form is obtained by intersecting with a line. However, this is true more generally for any nonsingular plane cubic, so it is worthwhile to use the group law in this more general case without transferring to Weierstrass form. For other models the group law is obtained differently; for instance, in the case of hyperelliptic quartics with rational tangents it is obtained by intersecting with a parabola with vertical axis; see Exercise 10 (c).

7.1.3 Degenerate Elliptic Curves

Let $f(x, y, z) = 0$ be a general Weierstrass equation with

$$f(x, y, z) = y^2 z + a_1 xyz + a_3 yz^2 - (x^3 + a_2 x^2 z + a_4 xz^2 + a_6 z^3) .$$

We begin with the following.

Proposition 7.1.1. *A general Weierstrass equation over any field K defines an absolutely irreducible curve (i.e., irreducible over an algebraic closure of K).*

Proof. Without loss of generality we may assume that K is algebraically closed. Let $Ax + By + Cz = 0$ be the general equation of a line in the projective plane over K with $(A, B, C) \neq (0, 0, 0)$, and set $P(x, z) = x^3 + a_2 x^2 z + a_4 xz^2 + a_6 z^3$. Assume that the line is a component of the curve defined by f, in other words that there exists a homogeneous quadratic polynomial $Q(x, y, z) \in K[x, y, z]$ such that we have

$$y^2 z + a_1 xyz + a_3 yz^2 - P(x, z) = (Ax + By + Cz)Q(x, y, z) .$$

Looking at the coefficient of x^3 we see that $A \neq 0$, so dividing the equation of the line by A we may assume that $A = -1$. We can thus replace x by $By + Cz$ in the above equation, so we obtain

$$y^2 z + a_1(By + Cz)yz + a_3 yz^2 - P(By + Cz, z) = 0 .$$

Looking now at the coefficient of y^3 we see that $B = 0$, so that there remains $y^2 z + (a_1 C + a_3)yz^2 - P(Cz, z) = 0$, which is impossible since the coefficient of $y^2 z$ is equal to 1. $\square$

By abuse of notation, if L is a field containing the field of definition K of E and if $(x, y, z) \in L^3$, we will write from now on $(x, y, z) \in E$ as a shorthand for $(x, y, z) \in E(L)$, in other words when the triple (x, y, z) satisfies the projective equation of the curve.

In the rest of this section we assume that $f(x, y, z) = 0$ does *not* define an elliptic curve, in other words that it has singularities, or equivalently, that $\text{disc}(E) = 0$.

Proposition 7.1.2. *If a general Weierstrass equation has singularities it has exactly one, and it is on the line $2y + a_1 x + a_3 z = 0$. In particular, if the characteristic of K is different from 2 and $a_1 = a_3 = 0$, it is on the x-axis.*

Proof. A point $P = (x_0, y_0, z_0) \in E$ is singular if and only if the three partial derivatives of the equation vanish at (x_0, y_0, z_0) (note that because of Euler's relation $\sum x_i \frac{\partial f}{\partial x_i} = kf$ for a homogeneous function f of degree k we must add the condition $P \in E$ only in characteristic 3). We have $\frac{\partial f}{\partial y} = z(2y + a_1 x + a_3 z)$. Note that the point $\mathcal{O}$ (which is the only point with $z_0 = 0$) is nonsingular since one checks that $\frac{\partial f}{\partial z}(\mathcal{O}) = 1$. Thus we must have $2y_0 + a_1 x_0 + a_3 z_0 = 0$ as claimed. Assume first that this defines a line in projective space, in other words either that the field does not have characteristic 2, or that $(a_1, a_3) \neq (0,0)$. If the curve had more than one singularity the line would intersect the curve in at least four points counting multiplicities, since a singularity counts at least for two, and this is impossible for a cubic curve, except if the line is entirely in the curve, which is impossible by Proposition 7.1.1.

If we are in characteristic 2 and $a_1 = a_3 = 0$ we have the equations $\frac{\partial f}{\partial x} = x^2 + a_4 z^2 = 0$ and $\frac{\partial f}{\partial z} = y^2 + a_2 x^2 + a_6 z^2 = 0$. Thus, given z_0, which we know is different from 0, we have $x_0^2 = a_4 z_0^2$, giving at most one value of x_0 (recall that we are in characteristic 2), and similarly $y_0^2 = a_2 x_0^2 + a_6 z_0^2$, giving again at most one value for y_0, and proving that also in this case there is at most one singular point. $\square$

Remark. It is interesting to note that if K is a perfect field of characteristic 2, in particular if K is finite or algebraically closed, then the equation $y^2 z = x^3 + a_2 x^2 + a_4 x + a_6$ always has a singular point given by the above equations. On the other hand, if K is not perfect (typically $K = \mathbb{F}_2(T)$) then there may not exist a singular point *defined over* K but only on a finite extension; see Exercise 4.

Let us now consider the possible types of singularity. For simplicity, we assume that the characteristic of K is different from 2, so that we can also assume that $a_1 = a_3 = 0$. The reader can check for himself that the situation is similar in characteristic 2. In affine coordinates the equation of the curve is thus $y^2 = P(x)$ with $P(x) = x^3 + a_2 x^2 + a_4 x + a_6$, and the above proposition says that the singularity, if any, is such that $y = 0$; in other words, x is equal to a root α of P. Clearly such a root gives a singularity if and only if it is a multiple root. A line passing through $(\alpha, 0)$ has an equation of the form $ny + m(x - \alpha) = 0$, and it is tangent to the curve if and only if its multiplicity of intersection with the curve is greater than or equal to 3, since the singularity already counts for at least 2. Thus, either $n = 0$, $m = 1$, and we obtain $y^2 = 0$, giving multiplicity only 2, so this can never happen, or $n \neq 0$, so we may assume that $n = -1$; hence replacing we obtain $m^2(x - \alpha)^2 = (x - \alpha)^2(x - \beta)$, where β is the third root of P, possibly equal to α if α is a triple root; in other words, $(x - \alpha)^2(x - \beta - m^2) = 0$. Thus the multiplicity at $x = \alpha$ is greater than or equal to 3 if and only if $m^2 = \alpha - \beta$. We thus have three cases:

A cusp. When $\beta = \alpha$, in other words when α is a triple root of P, there is a single tangent of slope $m = 0$, and the figure over $\mathbb{R}$ explains why this case is called a cusp.

A double point with tangents defined over K. When $\alpha - \beta$ is a nonzero square in K there exist two distinct values of $m \in K^*$ (since we are in characteristic different from 2), so two distinct tangents defined over K.

A double point with tangents not defined over K. When $\alpha - \beta \neq 0$ is not a square in K then there exist two distinct values of m, but defined in a quadratic extension of K.

In all these cases, since the curve is singular we have in fact a curve of genus 0 with a distinguished point, hence a rational curve. More precisely, changing x into $x - \alpha$ we may assume that the singularity is at the origin so that our equation has the shape $y^2 = x^2(x - \beta)$. If we set $t = x/y$ ($t = \infty$ if $y = 0$ and $x \neq 0$, i.e., $x = \beta$), which makes sense on nonsingular points, we have $1/t^2 = y^2/x^2 = x - \beta$; hence $x = \beta + 1/t^2$, $y = \beta/t + 1/t^3$, and this gives a rational parametrization of the nonsingular points of our curve by the projective line $\mathbb{P}_1(K)$, except that we have to exclude the values of t giving the singular point, in other words such that $1/t^2 = -\beta$, if they exist. Note that the point at infinity is parametrized by $t = 0$.

For future reference, note the following easy lemma.

Lemma 7.1.3. *Let $p \geqslant 3$, let $K = \mathbb{F}_p$, let E be a degenerate curve over $\mathbb{F}_p$ as above, and let $c_6 = c_6(E)$ be the invariant defined in Section 7.1.2. Then E has a cusp (respectively a double point with tangents defined over $\mathbb{F}_p$, respectively a double point with tangents not defined over K) if and only if $\left(\frac{-c_6}{p}\right) = 0$ (respectively $\left(\frac{-c_6}{p}\right) = 1$, respectively $\left(\frac{-c_6}{p}\right) = -1$).*

Proof. Since E is degenerate it has a singular point, and changing coordinates we may assume that it is at the origin, so that we can choose the equation of our curve to be $y^2 = x^2(x + a)$ for some $a \in \mathbb{F}_p$. One computes that $c_6(E) = -64a^3$. On the other hand, the general equation of a line through the origin is $y = tx$, so such a line is tangent if and only if $t^2 = a$. Thus, we have a cusp when $a = 0$ (so that $\left(\frac{c_6}{p}\right) = 0$), and the tangents are defined over $\mathbb{F}_p$ if and only if a is a square; in other words, $\left(\frac{-c_6}{p}\right) = 1$ since $-c_6 = (8a)^2 a$. $\qquad\square$

7.1.4 The Group Law

In the sequel we let E be an elliptic curve defined over a field K by a generalized Weierstrass equation given in affine form as $y^2 + a_1 xy + a_3 y = x^3 + a_2 x^2 + a_4 x + a_6$. One of the most important properties of elliptic curves is that there is a natural abelian group law on E that is defined by rational equations. More precisely, we have the following:

Theorem 7.1.4. *Let $y^2 + a_1 xy + a_3 y = x^3 + a_2 x^2 + a_4 x + a_6$ be the generalized Weierstrass equation of an elliptic curve E (hence nonsingular). We define an addition law on E by asking that the point at infinity $\mathcal{O}$ be the identity element, and that $P_1 + P_2 + P_3 = \mathcal{O}$ if and only if the points P_i are (projectively) collinear.*

(1) *This addition law defines an abelian group structure on E.*
(2) *If $P_1 = (x_1, y_1)$ and $P_2 = (x_2, y_2)$ are two points on E different from $\mathcal{O}$, their sum is equal to $\mathcal{O}$ if $x_1 = x_2$ and $y_2 = -y_1 - a_1 x - a_3$, and otherwise is given as follows. Set*

$$m = \begin{cases} \dfrac{y_2 - y_1}{x_2 - x_1} & \text{if } P_1 \neq P_2 \,, \\[2mm] \dfrac{3x_1^2 + 2a_2x_1 + a_4 - a_1y_1}{2y_1 + a_1x_1 + a_3} & \text{if } P_1 = P_2. \end{cases}$$

Then $P_1 + P_2 = P_3 = (x_3, y_3)$ *with* $x_3 = m(m + a_1) - x_1 - x_2 - a_2$ *and* $y_3 = m(x_1 - x_3) - y_1 - a_1x_3 - a_3$.

Proof. The proof of (1) is very classical and not too difficult, although associativity is painful if one tries to prove it directly from the formulas given in (2), but it is immediate in terms of divisors; see Exercise 6. The formulas of (2) follow from an immediate computation; see Exercise 5.   □

Remarks. (1) If we write the equation as $f(x, y) = 0$, the case $P_1 = P_2$ gives $m = -f'_x/f'_y$ as it should.

(2) In the case of a simple Weierstrass equation $y^2 = x^3 + ax + b$, the formulas reduce to $m = (y_2 - y_1)/(x_2 - x_1)$ if $P_1 \neq P_2$, $m = (3x_1^2 + a)/(2y_1)$ if $P_1 = P_2$, and $x_3 = m^2 - x_1 - x_2$, $y_3 = m(x_1 - x_3) - y_1$.

(3) The opposite of (x, y) is $(x, -y - a_1x - a_3)$ (hence $(x, -y)$ in the simple case), and in particular the points of order dividing 2 are $\mathcal{O}$ together with those such that $2y + a_1x + a_3 = 0$ (hence $y = 0$ in the simple case).

When the equation defines a singular curve, we still have a group law, but now on the set of nonsingular points:

Proposition 7.1.5. *Let* $y^2 = x^3 + a_2x^2 + a_4x + a_6$ *be a singular curve* E *with singular point* $P_0 = (\alpha, 0)$. *The set* $G = E(K) \backslash \{P_0\}$ *has a natural group structure, and furthermore* $G \simeq (K, +)$ *when* P_0 *is a cusp,* $G \simeq (K^*, \times)$ *when* P_0 *is a double point with distinct tangents defined over* K, *and finally* G *is isomorphic to the multiplicative group of elements of relative norm 1 in the quadratic extension of* K *generated by the slopes of the tangents at* P_0 *when* P_0 *is a double point with distinct tangents not defined over* K.

Proof. As in the preceding section we may assume that our equation has the form $y^2 = x^2(x - \beta)$ and is parametrized by $(\beta + 1/t^2, \beta/t + 1/t^3)$. Thus if the points P_1 and P_2 (which we may assume to be different from the point at infinity) correspond to the respective parameters t_1 and t_2, a short computation using the formulas given above, which are still valid in the nonsingular case, shows that $P_3 = P_1 + P_2$ is a nonsingular point that corresponds to the parameter

$$t_3 = \frac{t_1 + t_2}{1 - \beta t_1 t_2} \,.$$

Since one also checks that

$$\beta + \frac{1}{t_3^2} = \frac{t_1^2 t_2^2 (\beta + 1/t_1^2)(\beta + 1/t_2^2)}{(t_1 + t_2)^2} \,,$$

it follows that t_3 cannot correspond to the singular point (recall that $t_i \neq 0$ since we assume that the points P_i are not at infinity). The results in the elliptic curve case thus imply that we have an abelian group law on the nonsingular points. However, as stated, we can say more about the group structure G: if $\beta = 0$, we simply have $t_3 = t_1 + t_2$, so evidently $G \simeq (K, +)$ (the excluded value of t being ∞). If $-\beta = \gamma^2$ is a square in K^* (the case in which we have two distinct tangents defined over K), we set $u_i = (1 + \gamma t_i)/(1 - \gamma t_i)$, in other words $t_i = (u_i - 1)/(\gamma(u_i + 1))$. An easy computation shows that $u_3 = u_1 u_2$, so $G \simeq (K^*, \cdot)$ (the excluded values of u being 0 and ∞, corresponding to $1/t^2 = -\beta$). If $-\beta$ is not a square in K, we let $L = K(\gamma)$ with $\gamma^2 = -\beta$. The group law on G still corresponds to the multiplicative group law on L^*, but now the acceptable values of u are such that $u\bar{u} = 1$, where $\bar{u}$ is the Galois conjugate of u over K. Conversely, if $u\bar{u} = 1$ then by Hilbert's Theorem 90 (which is here trivial by setting $\alpha = u + u\bar{u} = u + 1$), there exists $\alpha \in L^*$ such that $u = \alpha/\bar{\alpha}$. Writing $\alpha = s + \gamma t$ with s and t in K, we note that if $s \neq 0$, dividing by s if necessary we may assume that $s = 1$, hence that $u = (1 + \gamma t)/(1 - \gamma t)$ as claimed. But the case $s = 0$ corresponds to $t = \infty$, giving the value $u = -1$. To summarize, in this last case G is isomorphic to the multiplicative group of elements of $K(\gamma)$ of norm 1. $\square$

Because of this proposition, the case that P_0 is a cusp is also called a case of *additive reduction*, while the case that P_0 is a double point with distinct tangents defined over K (respectively not defined over K) is called a case of *split multiplicative reduction* (respectively *nonsplit multiplicative reduction*).

Example. Assume that $K = \mathbb{F}_q$.

(1) In the case of additive reduction we have $G \simeq \mathbb{F}_q$, a cyclic group of order q, so that $|E(\mathbb{F}_q)| = q + 1$.
(2) In the case of split multiplicative reduction we have $G \simeq \mathbb{F}_q^*$, a cyclic group of order $q - 1$, so that $|E(\mathbb{F}_q)| = q$.
(3) In the case of nonsplit multiplicative reduction, up to isomorphism $\mathbb{F}_{q^2}$ is the unique quadratic extension of K, and since $\mathcal{N}_{\mathbb{F}_{q^2}/\mathbb{F}_q}(x) = x^{q+1}$, the group G isomorphic to the unique (cyclic) subgroup of $\mathbb{F}_{q^2}^*$ of order $q + 1$, so that $|E(\mathbb{F}_q)| = q + 2$.

7.1.5 Isogenies

As usual in mathematics, when a class of objects is defined, we must also define and study the natural maps between these objects (this is exactly the definition of a *category*, the maps being called *morphisms*). An elliptic curve E has *three* natural structures. The first one is the definition itself, which says that E is an algebraic curve (of genus 1 with a rational point), whether it is defined by a Weierstrass equation or by more complicated equation(s). The second is the fact that it is an analytic manifold of dimension 1 over $\mathbb{C}$, i.e., a compact Riemann surface. The third is that it has a natural abelian group structure.

Thus to define morphisms between two elliptic curves it is natural to want them to respect all three structures, and this is of course correct. However, there is an amazing simplification, which in fact essentially reduces to the first structure. First we have the beautiful theorem due to Riemann saying that all compact Riemann surfaces are *algebraic*, i.e., can be defined by algebraic equations, and that analytic morphisms between them are also algebraic. Note that this is quite a remarkable theorem since algebraic curves are very rigid, while a priori analytic manifolds can be deformed. Thus the first two structures on an elliptic curve are in fact one and the same. The additional theorem that enables us to almost forget about the abelian group structure is the following.

Theorem 7.1.6. *Let E and E' be two elliptic curves with identity elements $\mathcal{O}$ and $\mathcal{O}'$ respectively, and let ϕ be a morphism of algebraic curves from E to E' (i.e., ϕ is defined by rational functions). Then ϕ is a group homomorphism from E to E' if and only if $\phi(\mathcal{O}) = \mathcal{O}'$.*

The condition is of course necessary, but what is remarkable is that the simple fact that ϕ is a morphism of algebraic curves (together with the very weak condition $\phi(\mathcal{O}) = \mathcal{O}'$) implies that ϕ preserves the group law. Note that when $\phi(\mathcal{O}) \neq \mathcal{O}'$ then $\phi_1(P) = \phi(P) - \phi(\mathcal{O})$ (subtraction on the curve E') is still a morphism of algebraic curves, and is such that $\phi_1(\mathcal{O}) = \mathcal{O}'$, so is a group homomorphism by the theorem. This leads to the following definition.

Definition 7.1.7. *Let E and E' be two elliptic curves with identity elements $\mathcal{O}$ and $\mathcal{O}'$ respectively. An isogeny ϕ from E to E' is a morphism of algebraic curves from E to E' such that $\phi(\mathcal{O}) = \mathcal{O}'$. A nonconstant isogeny is one such that there exists $P \in E$ such that $\phi(P) \neq \mathcal{O}'$. We say that E and E' are isogenous if there exists a nonconstant isogeny from E to E'.*

We will implicitly assume that our isogenies are nonconstant. By the above theorem, an isogeny ϕ preserves the group law, in other words is such that $\phi(P + P') = \phi(P) + \phi(P')$, where addition on the left is on the curve E, and on the right is on the curve E'.

The following results summarize the main properties of isogenies; see [Sil1] for details and proofs.

Theorem 7.1.8. *Let ϕ be a nonconstant isogeny from E to E' defined over an algebraically closed field K. Then*

(1) *The map ϕ is surjective.*
(2) *ϕ is a finite map; in other words, the fiber over any point of E' is constant and finite.*

From these properties it is easy to see that ϕ induces an injective map from the function field of E' to that of E over some algebraic closure of the base field. The degree of the corresponding field extension is finite and called the *degree* of ϕ. If this field extension is *separable*, the degree of ϕ is also equal to the cardinality of a fiber, in other words to $|\operatorname{Ker}(\phi)|$, but this is not true in general. Thus, as algebraic curves, or equivalently, over an algebraically closed field extension of the base field, a nonconstant isogeny induces an isomorphism from $E/\operatorname{Ker}(\phi)$ to E', where $E/\operatorname{Ker}(\phi)$ must be suitably defined as an elliptic curve. If there exists a nonconstant isogeny ϕ from E to E' of degree m, we say that E and E' are m-isogenous. Conversely, we have the following:

Proposition 7.1.9. *If G is a finite subgroup of E there exists a natural elliptic curve E' and an isogeny ϕ from E to E' whose kernel (over some algebraic closure) is equal to G. The elliptic curve E' is well defined up to isomorphism and is denoted by E/G.*

Note that the equation of E' can be given explicitly by formulas due to Vélu [Vel].

Two isogenous elliptic curves are very similar, but are in general not isomorphic. For instance, Theorem 8.1.3 tells us that two elliptic curves defined over $\mathbb{Q}$ that are isogenous over $\mathbb{Q}$ have for instance the same rank and the same L-function. However, they do not necessarily have the same torsion subgroup: for instance, it follows from Proposition 8.4.3 that the elliptic curves $y^2 = x^3 + 1$ and $y^2 = x^3 - 27$ are 3-isogenous, but it is easily shown using for instance the Nagell–Lutz Theorem 8.1.10 that the torsion subgroup of the former has order 6, while the torsion subgroup of the latter has order 2.

Proposition 7.1.10. *Let ϕ be a nonconstant isogeny from an elliptic curve E to E' of degree m. There exists an isogeny ψ from E' to E, called the* dual *isogeny of ϕ, such that*

$$\psi \circ \phi = [m]_E \quad and \quad \phi \circ \psi = [m]_{E'} \,,$$

where $[m]$ denotes the multiplication-by-m map on the corresponding curve.

An isogeny of degree m will also be called an m-isogeny. We define the degree of the constant isogeny to be 0. We will see several examples of isogenies in the next chapter, for instance in Section 8.2 on rational 2-descent, where the basic tools are 2-isogenies.

7.2 Transformations into Weierstrass Form

7.2.1 Statement of the Problem

In this section, we explain how to transform the most commonly encountered equations of elliptic curves into Weierstrass form (simple or not, since it is trivial to transform into simple Weierstrass form by completing the square or the cube, if the characteristic permits). We will usually assume that the characteristic of the base field is different from 2 and 3, although some of the transformations are valid in more general cases. Recall that a *birational transformation* is a rational map with rational inverse, outside of a finite number of poles. Although not entirely trivial, it can be shown that in the case of *curves* (but not of higher-dimensional varieties), two curves are isomorphic if and only if they are birationally equivalent, in other words if there exists a birational transformation from one to the other.

It will be slightly simpler to work in projective coordinates instead of affine ones. Thus whenever projective coordinates (x, y, z) appear, it is always implicit that $(x, y, z) \neq (0, 0, 0)$. Apart from simple or generalized Weierstrass equations, an elliptic curve can be given in the following ways, among others:

(1) $f(x, y, z) = 0$, where f is a homogeneous cubic polynomial whose three partial derivatives do not vanish simultaneously, together with a known rational point (x_0, y_0, z_0).

(2) $y^2 z^2 = f(x, z)$, where $f(x, z)$ is a homogeneous polynomial of degree 4 such that $f(1, 0) \neq 0$ and without multiple roots, together with a known rational point (x_0, y_0, z_0) (this type of equation is called a *hyperelliptic quartic*). Note that in this case the point at infinity $(0, 1, 0)$ is a singular point with distinct tangents, and if the given point is at infinity we ask that the slopes of the tangents be rational. This is equivalent to the fact that $f(x, 1)$ is a fourth-ndegree polynomial whose leading coefficient is a square.

(3) $f_1(x, y, z, t) = f_2(x, y, z, t) = 0$, where f_1 and f_2 are two homogeneous *quadratic* polynomials together with a common projective rational solution (x_0, y_0, z_0, t_0), and additional conditions to ensure that the corresponding curve is nonsingular and of genus 1.

We first explain how to transform each of the above equations into Weierstrass form. More precisely, we will show how (3) and (2) transform into (1), and explain how to transform (1) into Weierstrass form. In fact we will see that (2) can also be directly transformed into Weierstrass form.

7.2.2 Transformation of the Intersection of Two Quadrics

Assume that we are given the homogeneous quadratic equations $f_1(x, y, z, t) = f_2(x, y, z, t) = 0$ with common projective rational solution (x_0, y_0, z_0, t_0), and assume that the intersection of the corresponding quadrics is nonsingular and of genus 1. For $i = 1$ and 2 write

$$f_i(x, y, z, t) = A_i t^2 + L_i(x, y, z)t + Q_i(x, y, z),$$

where A_i is a constant, L_i is linear, and Q_i quadratic. By making a linear coordinate change, we may send the rational solution to the projective point

$(0, 0, 0, 1)$, so that in the new coordinates we have $A_i = 0$; hence the equations take the form $tL_i(x, y, z) + Q_i(x, y, z) = 0$. I claim that the linear forms L_1 and L_2 are linearly independent: indeed, otherwise we could replace one of the equations, f_1 say, by a suitable linear combination of f_1 and f_2 to make the L_1 term disappear, so that the equations would read $Q_1(x, y, z) = 0$ and $tL_2(x, y, z) + Q_2(x, y, z) = 0$. This second equation expresses t rationally in terms of x, y, and z, and the first is a conic, which is of genus 0, a contradiction that proves my claim.

Eliminating t between the two equations $tL_i(x, y, z) + Q_i(x, y, z) = 0$, we thus have a new equation $C(x, y, z) = 0$ with $C = L_1 Q_2 - L_2 Q_1$. This is a homogeneous cubic equation with a projective rational point obtained by solving the homogeneous system of linear equations $L_1(x, y, z) = L_2(x, y, z) = 0$, which has a unique projective solution since the L_i are independent. This shows how (3) can be transformed into (1).

7.2.3 Transformation of a Hyperelliptic Quartic

Assume now that we are given the equation $y^2 z^2 = f(x, z)$ with $f(x, z)$ a homogeneous polynomial of degree 4, and a rational point (x_0, y_0, z_0), assumed to have rational tangents if $z_0 = 0$. If $z_0 = 0$, we do nothing. Otherwise, by a translation $x \mapsto x + kz$ for a suitable $k \in K$, we may assume that $x_0 = 0$, so that the equation is $y^2 z^2 = f(x, z)$ with $f(0, z) = (y_0^2/z_0^2) z^4$. Exchanging x and z gives $y^2 x^2 = g(x, z)$ with $g(x, z) = f(z, x)$, so $g(x, 0) = (y_0^2/z_0^2) x^4$; hence finally, setting $Y = xy/z^2$ we obtain an equation of the form $Y^2 z^2 = g(x, z)$, where the term in $x^4 z^0$ of g is a square. If we had had $z_0 = 0$ from the start, our equation would have already been in this form.

We have thus transformed our equation into one of the form $y^2 z^2 = a^2 x^4 + z P(x, z)$, where P is a homogeneous polynomial of degree 3. We set $y = ax^2/z + y_1$, and we obtain $2ax^2 y_1 + y_1^2 z - P(x, z) = 0$, which is the equation of a cubic having the rational point $(0, 1, 0)$. This shows how (2) can be transformed into (1). In the next subsection we will see how (1) can be transformed into Weierstrass form. Here, however, the equation of the cubic is quite special, so we can do the transformation directly, and I thank J. Cremona for this remark. To simplify we give the formulas in affine coordinates.

Proposition 7.2.1. *Let* $y^2 = a^2 x^4 + bx^3 + cx^2 + dx + e$ *be the equation of a hyperelliptic quartic with rational point $(0, 1, 0)$ and rational tangents at infinity. We set*

$$(a_1, a_2, a_3, a_4, a_6) = \left(\frac{b}{a}, \ -\frac{b^2 - 4a^2 c}{4a^2}, \ 2ad, \ -4a^2 e, \ (b^2 - 4a^2 c)e \right)$$

and

$$X = 2a(ax^2 - y) + bx, \quad Y = x\left(4a^3x^2 + 2abx - 4a^2y - \frac{b^2 - 4a^2c}{2a}\right),$$

$$x = \frac{Y}{2a(X+c) - b^2/(2a)}, \quad y = ax^2 + \frac{b}{2a}x - \frac{X}{2a}.$$

Then the maps $(x,y) \mapsto (X,Y)$ and $(X,Y) \mapsto (x,y)$ are inverse birational transformations from the quartic to the elliptic curve $Y^2 + a_1XY + a_3Y = X^3 + a_2X^2 + a_4X + a_6$.

Proof. This is a simple verification, which is left to the reader (Exercise 8). $\qquad \square$

Corollary 7.2.2. Let $y^2 = x^4 + cx^2 + e$ be the equation of an even hyperelliptic quartic with leading coefficient 1. Set

$$X = 2x^2 - 2y + c, \quad Y = 2x(2x^2 - y + c), \quad x = \frac{Y}{2X}, \quad y = \frac{Y^2}{4X^2} - \frac{X-c}{2}.$$

Then the maps $(x,y) \mapsto (X,Y)$ and $(X,Y) \mapsto (x,y)$ are inverse birational transformations from the quartic to the elliptic curve $Y^2 = X(X^2 - 2cX + c^2 - 4e)$.

Proof. This is a special case of the above proposition, after changing X into $X + c$. $\qquad \square$

7.2.4 Transformation of a General Nonsingular Cubic

Finally, we must show how to transform a general cubic to Weierstrass form, given a known rational point. Let

$$f(x,y,z) = Az^3 + L(x,y)z^2 + Q(x,y)z + C(x,y),$$

where A, L, Q, and C are constant, linear, quadratic, and cubic homogeneous polynomials respectively, and assume known a projective point $P_0 = (x_0, y_0, z_0)$ such that $f(x_0, y_0, z_0) = 0$. Denote by $\mathcal{C}$ the projective curve $f(x,y,z) = 0$. We consider two cases.

Case 1. Assume that P_0 is an inflection point, in other words that the multiplicity of intersection of the tangent to $\mathcal{C}$ at P_0 is equal to 3. Then using simply a linear projective transformation we can send this tangent to the line at infinity $z = 0$ and the point P_0 to the point $(0,1,0)$. The intersection of the line at infinity with the curve is given by $C(x,y) = 0$, and since this must be a triple intersection at the point $(x,y,z) = (0,1,0)$, in the new coordinates we must have $C(x,y) = cx^3$ for some nonzero constant c. It is now immediate to transform the equation $Az^3 + L(x,y)z^2 + Q(x,y)z + cx^3 = 0$ into Weierstrass form.

Case 2. Assume now that the given rational point P_0 is not an inflection point, so that the tangent at P_0 to the curve $\mathcal{C}$ intersects it at another point P, necessarily unique. Using a linear projective transformation, we may assume that the tangent at P_0 becomes the line $x = 0$ and that $P = (0, 0, 1)$ is the origin of the affine coordinate axes. In the new coordinates the term in z^3 must disappear, so that the affine form of our equation is $L(x, y) + Q(x, y) + C(x, y) = 0$. By homogeneity, the intersection of our curve with the line $x = 0$ is given by $yL(0, 1) + y^2 Q(0, 1) + y^3 C(0, 1)$. The solutions to this equation are $y = 0$, corresponding to the point P, and the two others must be equal since $x = 0$ is tangent to the curve at P_0, so that $Q^2(0, 1) - 4L(0, 1)C(0, 1) = 0$.

The final trick is now to consider the intersections of $\mathcal{C}$ with the lines $y = tx$ other than $(x, y) = (0, 0)$. By homogeneity once again we obtain $L(1, t) + xQ(1, t) + x^2 C(1, t) = 0$, which can be rewritten

$$(2xC(1, t) + Q(1, t))^2 = Q(1, t)^2 - 4C(1, t)L(1, t) \ .$$

Thus by a birational transformation we have obtained a hyperelliptic quartic equation. But in fact since we know that $Q^2(0, 1) - 4L(0, 1)C(0, 1) = 0$, it is immediately checked that it is a hyperelliptic cubic, hence essentially already in Weierstrass form.

Remark. In [Coh0] I have given an explicit algorithm that is essentially the translation of the above description. There are, however, a number of misprints in that algorithm, most (but not all) of which have been corrected in the errata sheets available on the author's home page. Since the aim of the present book is not primarily algorithmic, I have thus preferred to give the outline of the algorithm, closely following the exposition in [Cas2], and leave to the reader the writeup of the detailed algorithm (Exercise 9).

Example. Consider the general Fermat cubic equation $x^3 + y^3 = cz^3$ with $c \neq 0$, hence let $f(x, y, z) = x^3 + y^3 - cz^3$ and $P_0 = (-1, 1, 0)$. The equation of the tangent L_0 at P_0 is $x + y = 0$, and the intersection of L_0 with the curve is $z = 0$, so P_0 is an inflection point. We are thus in Case 1. To send L_0 to the line at infinity and P_0 to the point $(0, 1, 0)$ we set for example $(X, Y, Z) = (z, y, x + y)$; hence $(x, y, z) = (-Y + Z, Y, X)$, so replacing in our equation gives $Z^3 - 3YZ^2 + 3Y^2 Z = cX^3$. To transform this into simple Weierstrass form, we first multiply by $1728c^2$, set $Y_1 = 72cY$ and $X_1 = 12cX$, thus giving $Y_1^2 Z - 72cY_1 Z^2 + 1728c^2 Z^3 = X_1^3$, then set $Y_2 = Y_1 - 36cZ$, obtaining finally $Y_2^2 Z = X_1^3 - 432c^2 Z^3$. Putting everything together, we have thus established the following result, expressed in affine coordinates:

Proposition 7.2.3. *The curves* $x^3 + y^3 = c$ *and* $Y^2 = X^3 - 432c^2$ *are birationally equivalent over* $\mathbb{Q}$ *via the transformations*

$$X = \frac{12c}{x + y}, \quad Y = \frac{36c(y - x)}{x + y}, \quad x = -\frac{Y - 36c}{6X}, \quad y = \frac{Y + 36c}{6X} \ .$$

More generally, using the same method we can prove the following.

Proposition 7.2.4. *Let a, b, c be nonzero rational numbers, and assume that there exists a rational point $(x_0, y_0, z_0) \neq (0,0,0)$ on the cubic $ax^3 + by^3 + cz^3 = 0$. Then this cubic is birationally equivalent to the elliptic curve whose affine Weierstrass equation is $Y^2 = X^3 - 432(abc)^2$.*

Proof. See Exercise 11. □

Warning. The reader will notice that the coefficients of the Weierstrass equation do not depend on the given rational point (x_0, y_0, z_0). In geometric terms, the curve $Y^2 = X^3 - 432(abc)^2$ is the *Jacobian* of the cubic $ax^3 + by^3 + cz^3 = 0$, and this does not depend on the existence of a rational point; see Section 13.2 for the general definitions. The Jacobian of an elliptic curve (i.e., of a curve of genus 1 having a rational point) is isomorphic to the curve, but this is of course not the case for the Jacobian of a curve of genus 1 with no rational points. Consider the following examples. The cubic $x^3 + y^3 + 9z^3$ has the rational point $(1, 2, -1)$, hence is an elliptic curve, so is isomorphic to its Jacobian with Weierstrass equation $Y^2 = X^3 - 432 \cdot 9^2$. Cremona's `mwrank` program or the methods that we will study in the next chapter show that this curve has rank 1, so that our initial cubic has an infinity of (projective) rational solutions; see Exercise 9 of Chapter 8.

Consider now the cubic $x^3 + 15y^3 + 22z^3$, which we have already mentioned after Corollary 6.4.11. Although this equation is everywhere locally soluble nontrivially, we know that it does not have any nontrivial solution in $\mathbb{Q}$. However, we can compute that its Jacobian $Y^2 = X^3 - 432 \cdot 330^2$ has rank 1, hence has an infinity of rational points. This phenomenon indicates that the 3-part of the Tate–Shafarevich group Ш of the elliptic curve is nontrivial, and in fact using the method of Chapter 8 together with the BSD conjecture, which is proved in the rank-1 case, one can show that $|Ш| = 36$.

Remarks. (1) We will see in Proposition 8.4.3 that an elliptic curve with equation $y^2 = x^3 - 27d$ is birationally equivalent (therefore isomorphic) to the elliptic curve with equation $Y^2 = X^3 + d$, where the maps are given by

$$X = \frac{x^3 - 108d}{9x^3}, \quad Y = y\frac{x^3 + 216d}{27x^3}, \quad x = \frac{X^3 + 4d}{X^2}, \quad y = Y\frac{X^3 - 8d}{X^3},$$

and the two curves thus have the same rank. It follows that to study the solubility of $x^3 + y^3 = c$, or more generally of $ax^3 + by^3 + cz^3 = 0$, we can replace the Jacobian with equation $y^2 = x^3 - 432(abc)^2$ by the elliptic curve with equation $y^2 = x^3 + (4abc)^2$, whose rank is easier to compute since its coefficients are smaller. This is in fact the curve E_{abc} that we considered in Section 6.4.4.

(2) In Section 6.4.4 we have given the explicit equations for the map from the curve $ax^3 + by^3 + cz^3 = 0$ to the curve E_{abc}. For instance, the map

from $x^3 + y^3 = c$ to the curve $Y^2 = X^3 + 16c^2$ is clearly given by $X = -4xy$, $Y = 4(x^3 - y^3)$. The equation of the reverse map is however more complicated; see Exercise 12.

7.2.5 Example: The Diophantine Equation $x^2 + y^4 = 2z^4$

As an application of the above transformations, we solve the above equation, where we assume that x, y, and z are integers with x and z coprime (which immediately implies that x, y, and z are pairwise coprime). Several methods can be used, and we begin with the one using the intersection of two quadrics (which is not the simplest by far). We will in fact not use the method described above, but yet another one, so as to illustrate the wide variety of available methods.

By Corollary 6.3.14 there exist coprime integers s and t of opposite parity such that $x = \pm(s^2 - 2st - t^2)$, $y^2 = \pm(s^2 + 2st - t^2)$, and $z^2 = s^2 + t^2$. We can consider the first equation as simply giving the value of x in terms of s and t, and the two others as defining the intersection of two quadrics. Exchanging s and t and changing simultaneously the sign of s does not change z^2 and only changes the $\pm$ signs. Making this change if necessary, we may thus assume that s is odd and t even. Looking modulo 4 we thus see that $y^2 = s^2 + 2st - t^2$. Using the solution to the Pythagorean equation (Corollary 6.3.13), we deduce that there exist coprime integers u and v of opposite parity such that $s = u^2 - v^2$, $t = 2uv$, and $z = \pm(u^2 + v^2)$. The values of x and z are thus determined by u and v, hence we obtain the hyperelliptic quartic equation

$$y^2 = (u^2 - v^2)^2 + 2(u^2 - v^2)(2uv) - (2uv)^2 = u^4 + 4vu^3 - 6v^2u^2 - 4v^3u + v^4 \ .$$

Since u and v are coprime, the case $v = 0$ implies $u = \pm 1$, hence leads to $s = 1$, $t = 0$, giving the evident solution $(x, y, z) = (\pm 1, \pm 1, \pm 1)$. If $v \neq 0$, setting $Y = y/v^2$ and $X = u/v$ we obtain the affine equation

$$Y^2 = X^4 + 4X^3 - 6X^2 - 4X + 1 \ .$$

Note that since u and v are coprime, so are y and v; hence y and v^2 are determined by the value of Y.

We have thus transformed the intersection of two quadrics to this hyperelliptic quartic. For a general equation it would also have been necessary to follow the general algorithm, and also to keep track of the coordinate transformations. Here there are evident rational points on the quartic that we can use, including the point at infinity; in other words, the coefficient of X^4 is already a square.

At this point we should directly use Proposition 7.2.1, but to make life more difficult we want to use the general algorithm of Section 7.2.4 instead. As mentioned above, we set $Y = X^2 + Z$, and we obtain the affine cubic equation

$$Z^2 + 2X^2Z - 4X^3 + 6X^2 + 4X - 1 = 0 \,,$$

with for instance $(X, Z) = (0, 1)$ as evident point.

The general algorithm tells us that to transform into Weierstrass form we can make the change of variables

$$X = 4\frac{y - 4x - 64}{(x + 16)(x + 32)}, \quad Z = \frac{x^3 + 16x^2 - 8xy - 2048x - 32768}{(x + 16)(x + 32)^2} \,,$$

whose inverse is given by

$$x = 32\frac{Z + 2}{Z + 2X - 1} \,,$$

$$y = 64\frac{Z^2 + 4ZX^2 - 6ZX + 4Z - 8X^3 + 28X^2 - 18X - 5}{(Z + 2X - 1)^2} \,,$$

leading to the Weierstrass equation $y^2 = x^3 + 48x^2 + 256x - 4096$. If we set $x = 16x_1$, $y = 64y_1$ this gives the simpler (although not simple) equation $y_1^2 = x_1^3 + 3x_1^2 + x_1 - 1$. The $\mathtt{mwrank}$ program or the descent methods that we will explain in the next chapter tell us that the group of rational points on this curve is equal to $(\mathbb{Z}/2\mathbb{Z})(-1, 0) \oplus \mathbb{Z}(1, 2)$. It is easy to check that adding the point $(-1, 0)$ does not change the solution to $x^2 + y^4 = 2z^4$, and that the opposite of a point simply leads to a solution with the sign of y changed. Thus, up to signs all the solutions will be obtained from $k \cdot (1, 2)$ for $k \geqslant 1$. For $k \leqslant 5$ this gives (up to signs) the solutions

$$(x, y, z) = (1, 1, 1),$$
$$(239, 1, 13),$$
$$(2750257, 1343, 1525),$$
$$(3503833734241, 2372159, 2165017),$$
$$(2543305831910011724639, 9788425919, 42422452969).$$

There are no other solutions than those obtained in this way, and in particular the above five solutions are the smallest.

A simpler method for solving the Diophantine equation $x^2 + y^4 = 2z^4$ would have been to set $Y = x/z^2$, $X = y/z$, so that the equation gives $Y^2 = -X^4 + 2$, with the evident point $(X, Y) = (1, 1)$. We set $X = X_1 + 1$, so that $Y^2 = -X_1^4 - 4X_1^3 - 6X_1^2 - 4X_1 + 1$ with the point $(X_1, Y) = (0, 1)$. Setting $Y_1 = Y/X_1^2$ and $X_2 = 1/X_1$ we obtain the equation $Y_1^2 = X_2^4 - 4X_2^3 - 6X_2^2 - 4X_2 - 1$, whose leading term in X_2^4 is a square, and we can then use Proposition 7.2.1 and a linear change of variable so as to obtain the Weierstrass equation $y^2 = x^3 - 6x^2 - 12x - 24$ isogenous to the one obtained above, which of course leads to the same solutions. Note that Proposition 6.5.5 deals with the general equations $ax^4 + by^4 + cz^2 = 0$, of which our equation is a special case.

Remark. Recall that we have solved the similar Diophantine equation $x^2 + y^4 = z^4$ by purely algebraic considerations (factoring and descent) in Proposition 6.5.3. Here no amount of such manipulations would have led to the result. In fact, what we did in Proposition 6.5.3 using descent is to prove that the corresponding elliptic curve had finitely many points, in other words rank 0 instead of strictly positive rank.

As an application of the solution of the above equation, we consider *Fermat triangles*: a Fermat triangle is a right triangle with integer sides (hence a Pythagorean triangle) such that both the hypotenuse and the sum of the two other sides are perfect squares. Writing with evident notation $c^2 = a^2 + b^2$, $c = z^2$, $a + b = y^2$ we have

$$2z^4 - y^4 = 2c^2 - (a+b)^2 = (a-b)^2 = x^2 \,,$$

for $x = a - b$. Thus if (a, b, c) is a Fermat triangle with $a \geqslant b$, say, then $(x, y, z) = (a-b, (a+b)^{1/2}, c^{1/2})$ is a positive integral solution to the equation $x^2 + y^4 = 2z^4$, and conversely if (x, y, z) is such a solution then x and y have the same parity, so $a = (x + y^2)/2$, $b = (y^2 - x)/2$, and $c = z^2$, and such a triangle will exist if and only if $b > 0$, in other words $y^2 > x$. Among the first five solutions given above, only the fourth satisfies this condition, giving the smallest Fermat triangle with sides

$$(a, b, c) = (4565486027761, 1061652293520, 4687298610289) \,,$$

thus having 13-digits. It is easy to show that there exists an infinity of solutions such that $y^2 > x$, hence an infinity of essentially distinct Fermat triangles; see Exercise 13.

7.3 Elliptic Curves over $\mathbb{C}$, $\mathbb{R}$, $k(T)$, $\mathbb{F}_q$, and $K_{\mathfrak{p}}$

The most interesting fields on which to consider elliptic curves are $K = \mathbb{C}$, $K = \mathbb{R}$, $K = \mathbb{Q}_p$ (or a $\mathfrak{p}$-adic field $K_{\mathfrak{p}}$), $K = \mathbb{Q}$ (or a number field), and $K = \mathbb{F}_q$, a finite field. We will also consider special kinds of curves over $K = k(T)$. Each of these types of fields gives rise to an important and different aspect of the subject, all useful for the case $K = \mathbb{Q}$, and we will mention a few. We will devote all of the next chapter to the case $K = \mathbb{Q}$.

7.3.1 Elliptic Curves over $\mathbb{C}$

Here the most important aspect is that an elliptic curve over $\mathbb{C}$ is isomorphic (in a very strong sense, both analytically and algebraically) to the quotient of $\mathbb{C}$ by a 2-dimensional *lattice* $\Lambda = \omega_1 \mathbb{Z} + \omega_2 \mathbb{Z}$ with $\omega_1/\omega_2 \notin \mathbb{R}$. Recall that we define the *Weierstrass $\wp$-function* $\wp_\Lambda(z)$ associated with Λ by the formula

$$\wp_\Lambda(z) = \frac{1}{z^2} + \sum_{w \in \Lambda \setminus \{0\}} \left(\frac{1}{(z-w)^2} - \frac{1}{w^2} \right)$$

$$= \frac{1}{z^2} + \sum_{(m,n) \in \mathbb{Z}^2 \setminus \{(0,0)\}} \left(\frac{1}{(z-(m\omega_1 + n\omega_2))^2} - \frac{1}{(m\omega_1 + n\omega_2)^2} \right).$$

This defines a doubly periodic (or Λ-periodic) meromorphic function on $\mathbb{C}$ with poles only for $z \in \Lambda$. The crucial (and easy) result is that $\wp_\Lambda(z)$ satisfies the algebraic differential equation

$$\wp'_\Lambda(z)^2 = 4\wp_\Lambda(z)^3 - g_2(\Lambda)\wp_\Lambda(z) - g_3(\Lambda) = 0$$

with

$$g_2(\Lambda) = 60 \sum_{w \in \Lambda \setminus \{0\}} \frac{1}{w^4} \quad \text{and} \quad g_3(\Lambda) = 140 \sum_{w \in \Lambda \setminus \{0\}} \frac{1}{w^6}$$

(this is an easy application of Liouville's theorem; see Exercise 36 of Chapter 10). We have therefore an explicit isomorphism between $\mathbb{C}/\Lambda$ and an elliptic curve over $\mathbb{C}$ by sending a representative $z \in \mathbb{C}$ of an element of $\mathbb{C}/\Lambda$ to the pair $(\wp_\Lambda(z), \wp'_\Lambda(z))$, and all elliptic curves are obtained in this way. The group law is simply induced by addition on $\mathbb{C}$ and corresponds to the fact that the $\wp$-function has an *addition formula*; see Exercise 14. The point at infinity is the unique pole of $\wp$ (up to translation by Λ), i.e., $0 \in \mathbb{C}$ as it should, since it is the neutral element for addition. The above isomorphism is reminiscent of the ordinary exponential (which also has an addition formula), and the reverse isomorphism is thus called the *elliptic logarithm*. It is more natural to consider the elliptic logarithm with values in $\mathbb{C}$ and not in $\mathbb{C}/\Lambda$, but then it is evidently defined only up to addition of an element of Λ, exactly as the ordinary complex logarithm is defined up to addition of an element of the 1-dimensional lattice $2i\pi\mathbb{Z}$.

If (ω_1, ω_2) is a given basis of Λ, the *principal determination* of the elliptic logarithm is by definition the unique one belonging to the fundamental parallelogram $\{x_1\omega_1 + x_2\omega_2, \; -1/2 < x_1, x_2 \leqslant 1/2\}$.

The lattice Λ and elliptic logarithms can be expressed in terms of the equation of the elliptic curve thanks to elliptic integrals, themselves computed thanks to the arithmetic–geometric mean (AGM); see for instance [Coh0] Chapter 7 for details.

Another aspect of elliptic curves over $\mathbb{C}$ (which is also important over other fields) is the phenomenon of *complex multiplication*. This is best understood with the help of an example. Consider the two elliptic curves E_4 with affine equation $y^2 = x^3 + ax$ and E_3 with equation $y^2 = x^3 + a$, in both cases with $a \neq 0$. For any elliptic curve E and any integer m we have an endomorphism of E denoted by $[m]$ that sends P to $m \cdot P = P + P + \cdots + P$. However, for the curves E_3 and E_4 we clearly have an additional endomorphism: in the case of E_4 we have the map sending (x, y) to $(-x, iy)$ (with $i^2 = -1$) and in the case of E_3 the map sending (x, y) to $(\rho x, y)$, where $\rho^2 + \rho + 1 = 0$, i.e., ρ is a primitive cube root of unity. It is easy to show that these are indeed endomorphisms of the curve.

The general situation over $\mathbb{C}$ (or more generally over any field of characteristic 0) is as follows. Clearly the set $\mathrm{End}(E)$ of endomorphisms of an elliptic curve E is a ring, where addition is induced by that of E and multiplication is composition of endomorphisms. This ring contains the maps $[m]$ for all $m \in \mathbb{Z}$, and it is easily shown that these maps are distinct, so $\mathbb{Z} \subset \mathrm{End}(E)$ in a canonical way. The main result is that either $\mathrm{End}(E) = \mathbb{Z}$, i.e., there are no additional endomorphisms, or $\mathrm{End}(E)$ is a (not necessarily maximal) order in an imaginary quadratic field, in other words $\mathrm{End}(E) \simeq \mathbb{Z}\tau + \mathbb{Z}$ with $\tau = (D + \sqrt{D})/2$ for some $D < 0$ such that $D \equiv 0$ or 1 modulo 4. In this last case we say that E has *complex multiplication* by the order of discriminant D.

Note that in the above two examples of E_3 and E_4 it was easy to give explicitly the action of τ on E. In more general situations the computations are more complicated, but completely algorithmic; see for instance [Sta2] for a reasonable method.

When the curve E is defined over $\mathbb{Q}$, the theory of complex multiplication leads to remarkable results, in particular to the complete description of Abelian extensions of an imaginary quadratic field.

7.3.2 Elliptic Curves over $\mathbb{R}$

Let E be an elliptic curve defined over $\mathbb{R}$ by a generalized Weierstrass equation. After the same transformations that are done over $\mathbb{C}$ we can transform our curve into an equation of the form $Y^2 = 4X^3 - g_2 X - g_3$ with $g_2 = c_4/12$, $g_3 = c_6/216$, and discriminant

$$\mathrm{disc}(E) = \frac{c_4^3 - c_6^2}{1728} = g_2^3 - 27 g_3^2 \; .$$

We write
$$4X^3 - g_2 X - g_3 = 4(X - e_1)(X - e_2)(X - e_3) \; ,$$

where the e_i are the three complex roots of the cubic polynomial, necessarily distinct since $\mathrm{disc}(E) \neq 0$. Note the important relation $e_1 + e_2 + e_3 = 0$.

Warning. the coordinates that we are considering are evidently the coordinates on the equation $Y^2 = 4X^3 - g_2 X - g_3$, and *not* on the initial equation. More precisely, since the transformation on the x-coordinate has only been $X = x + b_2/12$ (see above), we must make this replacement in all the formulas.

There are now two cases, which must always be distinguished in practice in dealing with elliptic curves defined over $\mathbb{R}$ (and in particular for the Diophantine case in which the curve is defined over $\mathbb{Q}$):

Case 1: $\operatorname{disc}(E) > 0$. In that case all three roots e_i are real, which we order as $e_1 < e_2 < e_3$. It follows that the (affine) graph of the curve has two connected components:

- A compact component called the *egg* because of its shape, formed by the points (X, Y) such that $e_1 \leqslant X \leqslant e_2$, and denoted by E^{gg}.
- A noncompact component, called the component at infinity, and denoted by E^0, formed by the points (X, Y) such that $X \geqslant e_3$.

An easy exercise on the group law shows that the map from $E(\mathbb{R})$ to $\{\pm 1\}$ sending P to 1 if $P \in E^0$ and to -1 if $P \in E^{gg}$ is a *group homomorphism*. Another way of saying this is that E^0 is the exact image of the map $P \mapsto 2P$ from $E(\mathbb{R})$ to itself (Exercise 15).

All this can easily be seen on the lattice Λ associated with E. When $\operatorname{disc}(E) > 0$ it is easy to show that we can choose a basis (ω_1, ω_2) of the lattice such that $\omega_1 \in \mathbb{R}_{>0}$ and $\omega_2/i \in \mathbb{R}_{<0}$ (because we require that $\Im(\omega_1/\omega_2) > 0$). It follows that the associated fundamental domain is a rectangle, so that the case $\operatorname{disc}(E) > 0$ can be called the *rectangular case*. Under the canonical isomorphism between E and $\mathbb{C}/\Lambda$ the points of E^0 correspond to $z \in \mathbb{C}$ such that $z \in \mathbb{R} \pmod{\Lambda}$, in other words $E^0 \simeq \mathbb{R}/(\omega_1\mathbb{Z})$, while the points of E^{gg} correspond to $z \in \mathbb{C}$ such that $z - \omega_2/2 \in \mathbb{R} \pmod{\Lambda}$, in other words $E^{gg} \simeq (\omega_2/2 + \mathbb{R})/(\omega_1\mathbb{Z})$. The above statements then become clear.

It is useful to note that the basis elements ω_i above can be given as integrals. Set $f(X) = 4X^3 - g_2X - g_3 = 4(X - e_1)(X - e_2)(X - e_3)$. We then have the formulas

$$\omega_1 = 2\int_{e_1}^{e_2} \frac{dt}{\sqrt{f(t)}} = 2\int_{e_3}^{\infty} \frac{dt}{\sqrt{f(t)}} \quad \text{and}$$

$$\omega_2 = -2i\int_{e_2}^{e_3} \frac{dt}{\sqrt{-f(t)}} = -2i\int_{-\infty}^{e_1} \frac{dt}{\sqrt{-f(t)}}.$$

The above integral equalities are simple calculus exercises; see Exercise 16. Note that all the square roots are of nonnegative real numbers.

Case 2: $\operatorname{disc}(E) < 0$. In that case there is a single real root, say e_3, and the other two are complex conjugates. It follows that the (affine) graph of the curve has a single connected component, which is noncompact. We again denote it by E^0, and it is again formed by the points (X, Y) such that $X \geqslant e_3$, so that $E^0 = E(\mathbb{R})$. The same exercise as above now shows that the map $P \mapsto 2P$ from $E(\mathbb{R})$ to itself is surjective, in other words that its image is E^0.

Here we can choose a basis for Λ of the form (ω_1, ω_2) with $\omega_1 \in \mathbb{R}_{>0}$, $\Re(\omega_2) = \omega_1/2$, and $\Im(\omega_2) < 0$, so this can be called the *triangular case*. The points of $E^0 = E(\mathbb{R})$ correspond to $z \in \mathbb{C}$ such that $z \in \mathbb{R} \pmod{\Lambda}$, in other words $E^0 \simeq \mathbb{R}/(\omega_1\mathbb{Z})$.

The integral formulas for the ω_i are now

$$\omega_1 = 2 \int_{e_3}^{\infty} \frac{dt}{\sqrt{f(t)}} \quad \text{and}$$

$$\omega_2 = \frac{\omega_1}{2} - i \int_{-\infty}^{e_3} \frac{dt}{\sqrt{-f(t)}} .$$

Independently of the sign of $\mathrm{disc}(E)$, the formula for the elliptic logarithm of a point $P \in E^0$ is the same: since it must be the inverse function of a function satisfying the differential equation $y'^2 = 4y^3 - g_2 y - g_3$, the formula for the derivative of an inverse function implies that for $P = (X, Y) \in E^0$ we can choose

$$\psi(P) = \mathrm{sign}(Y) \int_{\infty}^{X} \frac{dt}{\sqrt{f(t)}} .$$

If $\mathrm{disc}(E) > 0$ and $P \in E^{gg}$ then

$$\psi(P) = \frac{\omega_2}{2} + \mathrm{sign}(Y) \int_{e_1}^{X} \frac{dt}{\sqrt{f(t)}} ,$$

in both cases with the convention $\mathrm{sign}(0) = 1$, and this is exactly the principal determination of the elliptic logarithm (recall that in terms of our initial coordinates we must replace X by $x + b_2/12$).

We have given all these integral formulas mainly for theoretical purposes: indeed, in practice they are computed using the arithmetic–geometric mean (AGM), and are included in many computer algebra systems; see [Coh0] and [Sma]. In fact, with GP the constants b_i, c_i, $\mathrm{disc}(E)$, $j(E)$, and the periods ω_i are among those that are computed with the command e=ellinit([a$_1$,a$_2$,a$_3$,a$_4$,a$_6$]). More precisely, the periods are obtained as the two components e.omega. Note, however, that the second component of e.omega is $\overline{\omega_2}$ (equal to $-\omega_2$ or to $\omega_1 - \omega_2$ depending on the sign of $\mathrm{disc}(E)$) and not ω_2, because the convention chosen in GP is to have $\Im(\omega_2/\omega_1) > 0$. When $z \in \mathbb{C}$, the point of E corresponding to z under the canonical isomorphism (essentially, but not exactly, $(\wp_\Lambda(z), \wp'_\Lambda(z))$ since some transformations have been made, see the above warning) can be obtained using the command ellztopoint(e,z). Finally, the inverse isomorphism, in other words the elliptic logarithm $\psi(P)$ of $P \in E(\mathbb{Q})$ is given by the command ellpointtoz(e,P), and the result is defined only up to addition of a point of Λ. However, as explained above, since the curve is defined over $\mathbb{R}$ we can choose $\psi(P) \in \mathbb{R}$ when $P \in E^0$, and $\psi(P) - \omega_2/2 \in \mathbb{R}$ otherwise (i.e., in the case $\mathrm{disc}(E) > 0$ and $P \in E^{gg}$).

7.3.3 Elliptic Curves over $k(T)$

Before studying this we prove a very elementary but useful result.

Proposition 7.3.1. *Let R be a principal ideal domain with field of fractions K, let E be an elliptic curve given by a generalized Weierstrass equation as*

$Y^2 + a_1 XY + a_3 Y = X^3 + a_2 X^2 + a_4 X + a_6$ with $a_i \in R$, and let $(X, Y) \in E(K)$ be an affine point defined over K. There exist M, N, and D in R such that

$$X = \frac{M}{D^2}, \quad Y = \frac{N}{D^3}, \quad \gcd(M, D) = \gcd(N, D) = 1 .$$

Proof. Write $X = M/A$ and $Y = N/B$ with $\gcd(M, A) = \gcd(N, B) = 1$. Substituting in the equation and clearing denominators gives

$$A^3 N^2 + a_1 A^2 BMN + a_3 A^3 BN = B^2 M^3 + a_2 B^2 AM^2 + a_4 B^2 A^2 M + a_6 B^2 A^3 .$$

Let us call the sides of this equation the LHS and RHS. Since B^2 divides the RHS, it follows that $B \mid A^3 N^2$, and since $\gcd(B, N) = 1$ we have $B \mid A^3$. But B^2 divides A times the LHS, and since $B \mid A^3$ we deduce that $B^2 \mid A^4 N^2$, hence that $B^2 \mid A^4$, so that $B \mid A^2$. But then replacing once again we see that $B^2 \mid A^3 N^2$, so that $B^2 \mid A^3$. By a similar process, let us prove the converse. Since A^2 divides the LHS, it follows that $A \mid B^2 M^3$, so that $A \mid B^2$ since $\gcd(A, M) = 1$. Replacing in the RHS, we see that $A^2 \mid B^2 M^3$; hence $A^2 \mid B^2$, so that $A \mid B$. Thus the LHS is divisible by A^3, and replacing one last time in the RHS we see that $A^3 \mid B^2 M^3$, so that $A^3 \mid B^2$. It follows that there exists a unit $u \in R^*$ such that $uB^2 = A^3$, and since $A \mid B$, if we set $D = B/A$ we see that $D^2 = B^2/A^2 = A/u$ and that $B = dA = uD^3$, proving the proposition after dividing M and N by u. $\square$

In this subsection we let k be a commutative field, and to simplify the exposition we will assume that the characteristic of k is different from 2, although everything goes through in general. We set $R = k[T]$, $K = k(T)$, and consider an elliptic curve E over k, which we may assume without loss of generality to be given by an equation $y^2 = x^3 + ax^2 + bx + c$, where a, b, and c belong to k. We first show that the group $E(K)$ of points of E with values not only in k but in the function field $K = k(T)$ is equal to $E(k)$ (this result is true in much greater generality but we prove it only in this case).

Proposition 7.3.2. *If E is an elliptic curve defined over k and $K = k(T)$ we have $E(K) = E(k)$.*

Proof. It is of course sufficient to prove that $E(K) \subset E(k)$. Thus, let $(X, Y) \in E(K) \setminus \{\mathcal{O}\}$. If the characteristic of k is equal to $p \neq 0$, then it is clear that if $X'(T) = 0$ we have $X = X_1(T^p)$ for some $X_1 \in k(T)$, and $2YY' = f'(X)X' = 0$, and since $p \neq 2$ we have $Y' = 0$, hence $Y = Y_1(T^p)$ for some $Y_1 \in k(T)$. Since E is a constant elliptic curve it is clear that $(X_1, Y_1) \in E(K)$. Continuing this process, we may therefore assume that if $X'(T) = 0$ then X and Y are constant.

By the above proposition there exist polynomials M, N, and D with D coprime to M and N and such that $(X, Y) = (M/D^2, N/D^3)$, and our equation is $N^2 = M^3 + aM^2 D^2 + bMD^4 + cD^6$. Differentiating the equation

$Y^2 = X^3 + aX^2 + bX + c$ with respect to T gives $2YY' = (3X^2 + 2aX + b)X'$, so after replacing and clearing denominators we obtain

$$2N(DN' - 3ND') = (3M^2 + 2aD^2M + bD^4)(DM' - 2MD') .$$

Since E is an elliptic curve the polynomials $x^3 + ax^2 + bx + c$ and $3x^2 + 2ax + b$ are coprime as polynomials in x, and it is immediate to check that this implies that the GCD of the polynomials $M^3 + aM^2D^2 + bMD^4 + cD^6$ and $3M^2 + 2aD^2M + bD^4$ divides a power of D. Since the first of these polynomials is equal to N^2, which is coprime to D, it follows that $3M^2 + 2aD^2M + bD^4$ is coprime to N. We thus deduce from the above equation that $N \mid (DM' - 2MD')$ and $(3M^2 + 2aD^2M + bD^4) \mid (DN' - 3ND')$. For simplicity, call d, m, and n the degrees of D, M, and N, respectively. We consider three cases.

Case 1: $d < m/2$. Then $n = 3m/2$ and if $DM' - 2MD' \neq 0$ we have

$$3m/2 = n \leqslant \deg(DM' - 2MD') \leqslant d + m - 1 < 3m/2 - 1 ,$$

a contradiction showing that $DM' - 2MD' = 0$, in other words that $X' = 0$, hence by the reduction made above, X and Y are constant.

Case 2: $d > m/2$. Since E is an elliptic curve we must have either b or c different from 0. If $c \neq 0$ we have $n = 3d$, while if $c = 0$ and $b \neq 0$ we have $n = m/2 + 2d$, so in both cases $n \geqslant m/2 + 2d$. On the other hand, since $N \mid (DM' - 2MD')$ then if $DM' - 2MD' \neq 0$ we have

$$m/2 + 2d \leqslant n \leqslant \deg(DM' - 2MD') \leqslant d + m - 1 ,$$

hence $d \leqslant m/2 - 1$, a contradiction showing that $DM' - 2MD' = 0$. As in case 1 we deduce that X and Y are constant.

Case 3: $d = m/2$. In this case, call ℓ the quotients of the leading terms of the numerator and denominator of X, in other words the "limit as $T \to \infty$" of $X(T)$. Our equation tells us that $n = 3d$ if and only if ℓ is not a root of the polynomial $f(x) = x^3 + ax^2 + bx + c$. If ℓ is not a root then as in Case 2 we have $N \mid (DM' - 2MD')$, and $DM' - 2MD' \neq 0$ thus implies

$$3d \leqslant \deg(DM' - 2MD') \leqslant d + m - 1 = 3d - 1 ,$$

a contradiction. On the other hand, if ℓ is a root, then since E is an elliptic curve, ℓ is not a root of $f'(x)$, so that $\deg(3M^2 + 2aD^2M + bD^4) = 2m$; hence if $DN' - 3ND' \neq 0$ we have

$$4d = 2m \leqslant \deg(DN' - 3ND') \leqslant d + n - 1 \leqslant 4d - 2 ,$$

a contradiction showing that $DN' - 3ND' = 0$, hence that $Y' = 0$, so that Y, and hence X, is constant thanks to the reduction made above. $\quad\square$

An interesting alternative proof of this proposition has been communicated to me by J. Cremona; see Exercises 19 and 20.

The above proposition shows that constant elliptic curves over $K = k(T)$ are not very interesting. On the other hand, we can consider *quadratic twists* E_g of such curves (see Definition 7.3.15 below), whose equations have the form $g(T)Y^2 = X^3 + aX^2 + bX + c$, where a, b, and c are again in the base field k, for some not necessarily constant $g \in k(T)$, and where the curve E whose equation is $y^2 = x^3 + ax^2 + bx + c$ is assumed to be an elliptic curve over k. When g is a polynomial of degree less than or equal to 2, the same proof as above shows a similar result; see Exercise 18. Thus to be really interesting we will need g to be of degree at least 3, and this will indeed be the case in the application to Manin's proof of Hasse's theorem that we will give in the next subsection.

The main result that we need on quadratic twists E_g is that the naïve notion of "height" of a point is in fact a quadratic form. More precisely, if $P \in E_g(K)$ we define a *height function* $h(P)$ as follows. If P is the point at infinity, we set $h(P) = 0$. Otherwise, for $P = (X_P, Y_P) \in E_g(K)$ we write $X_P = N_P/D_P$ with N_P and D_P coprime polynomials (this should not be confused with the notation used in Proposition 7.3.1 above), and we set $h(P) = \max(\deg(N_P), \deg(D_P))$. Clearly $h(P) \in \mathbb{Z}_{\geq 0}$ for all P, and under a mild assumption $h(P) = 0$ if and only if $P = (u, 0)$ with $u \in k$ (see Lemma 7.3.8 below).

Theorem 7.3.3. *Let* $f(X) = X^3 + aX^2 + bX + c \in k[X]$ *be such that* $\gcd(f, f') = 1$, *and let* E_g *be the elliptic curve defined over* $K = k(T)$ *by an equation* $g(T)Y^2 = f(X)$, *where* $g(T) \in k(T)$ *is nonzero. The map* h *defined above is a quadratic form on* $E_g(K)$; *in other words,* $B(P, Q) = (h(P+Q) - h(P) - h(Q))/2$ *is a bilinear form on the* $\mathbb{Z}$*-module* $E_g(K)$.

Note that this theorem is *false* for a general elliptic curve defined over K; see Exercise 22. As we will do over $\mathbb{Q}$, to obtain a quadratic form it is necessary to modify the definition of h, but we will not consider this question in this book, although the method is essentially the same.

Proof. By Exercise 21, we must show that $h(P+Q) + h(P-Q) = 2(h(P) + h(Q))$ for all P and Q in $E_g(K)$. If P or Q is the point at infinity this is clear, so we assume that this is not the case. We thus write $P = (X_P, Y_P)$, $Q = (X_Q, Y_Q)$, and $X_P = N_P/D_P$, $X_Q = N_Q/D_Q$ with $\gcd(N_P, D_P) = \gcd(N_Q, D_Q) = 1$. In a first part we assume that $Q \neq \pm P$. If we write $P \pm Q = (X_\pm, Y_\pm)$ the addition law gives

$$X_\pm = g(T)\left(\frac{Y_P \mp Y_Q}{X_P - X_Q}\right)^2 - (X_P + X_Q + a)$$

$$= \frac{(X_P X_Q + b)(X_P + X_Q) + 2aX_P X_Q + 2c \mp 2g(T)Y_P Y_Q}{(X_P - X_Q)^2}.$$

After a short computation it follows that

$$X_+ X_- = \frac{(X_P X_Q - b)^2 - 4c(X_P + X_Q + a)}{(X_P - X_Q)^2} \quad \text{and}$$

$$\frac{X_+ + X_-}{2} = \frac{(X_P X_Q + b)(X_P + X_Q) + 2a X_P X_Q + 2c}{(X_P - X_Q)^2} ;$$

hence setting $X_\pm = N_\pm / D_\pm$ with $\gcd(N_\pm, D_\pm) = 1$, this gives

$$\frac{N_+ N_-}{D_+ D_-} = \frac{A_{N,N}}{A_{D,D}} \quad \text{and} \quad \frac{N_+ D_- + N_- D_+}{2 D_+ D_-} = \frac{A_{N,D}}{A_{D,D}} ,$$

where

$$A_{N,N} = (N_P N_Q - b D_P D_Q)^2 - 4c D_P D_Q (N_P D_Q + N_Q D_P + a D_P D_Q),$$
$$A_{D,D} = (N_P D_Q - N_Q D_P)^2, \quad \text{and}$$
$$A_{N,D} = (N_P N_Q + b D_P D_Q)(N_P D_Q + N_Q D_P) + 2a N_P N_Q D_P D_Q + 2c D_P^2 D_Q^2 .$$

To prove Theorem 7.3.3, we need two results: One is of an arithmetic nature, saying that up to constant multiples the numerators and denominators of the above equalities of rational functions match. The second gives estimates for the degrees of $N_\pm$ and $D_\pm$. The proof of this second result will be a slightly painful case distinction and should be skipped at first. Note that there is a "highbrow" way of proving Theorem 7.3.3 in a simple manner using some algebraic geometry, but to stay in the spirit of this book we give an "elementary" but tedious proof.

Lemma 7.3.4. *There exists $u \in k^*$ such that*

$$N_+ N_- = u A_{N,N}, \quad D_+ D_- = u A_{D,D}, \quad \text{and} \quad \frac{N_+ D_- + N_- D_+}{2} = u A_{N,D} .$$

Proof. Define $S_1 = \gcd(N_+ N_-, D_+ D_-)$ and $S_2 = \gcd(A_{N,N}, A_{D,D})$. We thus have

$$S_2 N_+ N_- = S_1 A_{N,N}, \quad S_2 D_+ D_- = S_1 A_{D,D}, \quad S_2(N_+ D_- + N_- D_+) = 2 S_1 A_{N,D} .$$

We are going to show that $S_1 \mid S_2$ and $S_2 \mid S_1$, which is clearly equivalent to the lemma by setting $u = S_2 / S_1$. First let F be an irreducible divisor of S_1. By symmetry we may assume that $F \mid D_+$. By coprimality it follows that $F \nmid N_+$, hence $F \mid N_-$, hence $F \nmid D_-$, so in particular F is coprime to $N_+ D_- + N_- D_+$. Since this is true for every irreducible $F \mid S_1$ it follows that S_1 is coprime to $N_+ D_- + N_- D_+$; hence by the third equation we deduce that $S_1 \mid S_2$. Conversely, let F be an irreducible divisor of S_2 / S_1, hence a divisor of $A_{N,N}$ and $A_{D,D}$. It is clear that if $F \mid D_P$ we would have $F \nmid N_P$; hence by the first two equations $F \mid N_Q$ and $F \mid D_Q$, which is absurd. By symmetry, $F \nmid D_Q$. Using $N_Q D_P \equiv N_P D_Q \pmod{F}$, multiplying the first and third equations above by D_P^2 and D_P respectively, and dividing by D_Q^2, which we can do since $F \nmid D_Q$, we easily obtain

$$(3N_P^2 + 2aN_P D_P + bD_P^2)^2$$
$$- 4(2N_P + aD_P)(N_P^3 + aN_P^2 D_P + bN_P D_P^2 + cD_P^3) \equiv 0 \pmod{F} \quad \text{and}$$
$$N_P^3 + aN_P^2 D_P + bN_P D_P^2 + cD_P^3 \equiv 0 \pmod{F} .$$

Since by assumption $\gcd(f, f') = 1$ and $\gcd(N_P, D_P) = 1$, as we have seen in the proof of Proposition 7.3.2 this implies that F divides a power of D_P, a contradiction that finishes the proof of the lemma. $\qquad\square$

To state the analytic lemma we introduce the following notation. The *degree* $z(X)$ of a rational function $X(T)$ is as usual the difference between the degree of the numerator and that of the denominator, with $z(0) = -\infty$ (we use the notation z instead of the more natural d because we will need d below). When $X \neq 0$, the *leading coefficient* $\ell(X)$ of X is the quotient of the leading coefficients of the numerator and denominator, so that $X(T) = \ell(X)T^{z(X)} + O(T^{z(X)-1})$ as $T \to \infty$. It is clear that the function z satisfies $z(X_1 X_2) = z(X_1) + z(X_2)$ and $z(X_1 + X_2) \leqslant \max(z(X_1), z(X_2))$, with equality if and only if either $z(X_1) \neq z(X_2)$, or $z(X_1) = z(X_2)$ but $\ell(X_1) \neq -\ell(X_2)$. For simplicity we will denote by z_P, z_Q, z_+, and z_- the degrees of X_P, X_Q, X_+, and X_-, and similarly for ℓ.

Lemma 7.3.5. *Recall that we have set $f(x) = x^3 + ax^2 + bx + c$.*

(1) *If z_P and z_Q are not both equal to 0 we have:*
    (a) *If $z_P \geqslant 0$ and $z_Q \geqslant 0$ then $z_+ \geqslant 0$ and $z_- \geqslant 0$.*
    (b) *If $z_P \geqslant 0$ and $z_Q < 0$ or if $z_P < 0$ and $z_Q \geqslant 0$ then $z_+ \leqslant 0$ and $z_- \leqslant 0$.*
    (c) *If $z_P < 0$ and $z_Q < 0$ and $c \neq 0$ then either $z_+ \leqslant 0$ and $z_- > 0$ or $z_+ > 0$ and $z_- \leqslant 0$.*
    (d) *If $z_P < 0$ and $z_Q < 0$ and $c = 0$ then $z_+ > 0$ and $z_- > 0$.*
(2) *If $z_P = z_Q = 0$ we have:*
    (a) *If $\ell_P \neq \ell_Q$ then $z_+ \leqslant 0$ and $z_- \leqslant 0$.*
    (b) *If $\ell_P = \ell_Q$ and $f(\ell_P) \neq 0$ then either $z_+ \leqslant 0$ and $z_- > 0$ or $z_+ > 0$ and $z_- \leqslant 0$.*
    (c) *If $\ell_P = \ell_Q$ and $f(\ell_P) = 0$ then $z_+ > 0$ and $z_- > 0$.*

Proof. From the formulas that we have proved for $X_\pm$ and $X_+ X_-$ it follows that

$$X_\pm = \frac{A_\mp}{C} = \frac{B}{A_\pm} ,$$

where

$$A_\pm = (X_P X_Q + b)(X_P + X_Q) + 2aX_P X_Q + 2c \pm 2\varepsilon(f(X_P)f(X_Q))^{1/2},$$
$$B = (X_P X_Q - b)^2 - 4c(X_P + X_Q + a), \quad \text{and} \quad C = (X_P - X_Q)^2 ,$$

where $\varepsilon = \pm 1$ depends on the choice of the square root. Thanks to the properties of the function $z(X)$ given above, the lemma is easy to prove by inspection of all the cases, using the two formulas for $X_\pm$ just given. To illustrate, let us show a few of the more subtle cases.

Assume that $z_P < 0$ and $z_Q < 0$. If $c \neq 0$ we have $z(f(X_P)) = z(f(X_Q)) = 0$ and $\ell(f(X_P)) = \ell(f(X_Q)) = c$. Choosing the sign of the square root such that $(f(X_P)f(X_Q))^{1/2} = c + O(1/T)$, we have $z(A_\varepsilon) = 0$, $z(B) \leqslant 0$, and $z(C) < 0$, hence $z(X_\varepsilon) = z(B) - z(A_\varepsilon) \leqslant 0$ and $z(X_{-\varepsilon}) = z(A_\varepsilon) - z(C) > 0$, proving case 1 (c). On the other hand, if $z_P < 0$, $z_Q < 0$ and $c = 0$ then necessarily $b \neq 0$, since otherwise E would not be an elliptic curve. It follows that $z(B) = 0$ and $z(A_\pm) < 0$, so that $z(X_\pm) > 0$, proving case 1 (d).

Assume that $z_P = z_Q = 0$ and $\ell_P = \ell_Q = \ell$, say. We have $z(B) \leqslant 0$, $z(C) < 0$, $z(A_\pm) \leqslant 0$, and $z(A_\pm) = 0$ if and only if $f(\ell)(1 \pm \varepsilon) \neq 0$. Thus if $f(\ell) \neq 0$ we have $z(A_\varepsilon) = 0$ and $z(A_{-\varepsilon}) < 0$, so $z(X_\varepsilon) = z(B) - z(A_\varepsilon) \leqslant 0$ and $z(X_{-\varepsilon}) = z(A_\varepsilon) - z(C) > 0$, proving case 2 (b). Finally, if $f(\ell) = 0$ then $z(A_\pm) < 0$, and $f'(\ell) \neq 0$, since otherwise E would not be an elliptic curve. We have $z(B) \leqslant 0$ and $z(B) = 0$ if and only if $(\ell^2 - b)^2 - 4c(2\ell + a) \neq 0$. However, we check that

$$(\ell^2 - b)^2 - 4c(2\ell + a) = (3\ell^2 + 2a\ell + b)^2 - 4(2\ell + a)f(\ell) = f'(\ell)^2 \neq 0 ,$$

so that $z(B) = 0$; hence $z(X_\pm) = z(B) - z(A_\pm) > 0$, proving case 2 (c).

The other cases are easier and left to the reader (Exercise 23).   □

Remark. We have given this lemma in complete generality. However, in many cases it can be considerably simplified. For instance, if $z(g)$ is *odd* (which will be the case in the application to Hasse's theorem in the next section) and if $c \neq 0$, then it is clear that $z(X) < 0$ is impossible, and (whether or not $c \neq 0$) $z(X) = 0$ implies that $f(\ell(X)) = 0$, in both cases because otherwise $z(Y) = -z(g)/2 \notin \mathbb{Z}$.

It is now easy to finish the proof of Theorem 7.3.3 when $P \neq \pm Q$, by combining the inequalities for $z(X_\pm)$ given by Lemma 7.3.5 with the identities of Lemma 7.3.4. For simplicity denote by n_P, n_Q, n_+, d_P, etc., the degrees of the polynomials N_P, N_Q, N_+, D_P, etc., and by h_P, h_Q, and $h_\pm$ the heights of the points P, Q, and $P \pm Q$. Once again we give only a few illustrative examples, leaving the other cases to the reader.

Assume that $z_P < 0$ and $z_Q < 0$, in other words that $d_P > n_P$, $d_Q > n_Q$, hence $h_P = d_P$, $h_Q = d_Q$. If $c \neq 0$ we have seen that $z_+ \leqslant 0$ and $z_- > 0$ (up to exchange of $+$ with $-$), so $n_+ \leqslant d_+$, $n_- > d_-$; hence $h_+ = d_+$ and $h_- = n_-$. In particular, $z(N_+D_-) < z(N_-D_+)$ (note that the strict inequality $z_- > 0$ is essential), so that $z(N_+D_- + N_-D_+) = h_+ + h_-$; hence by the third identity of Lemma 7.3.4 we have, since $c \neq 0$,

$$h_+ + h_- = 2(d_P + d_Q) = 2(h_P + h_Q) ,$$

as desired. On the other hand, if $c = 0$ we have seen that $b \neq 0$ and that $z_+ \geqslant 0$ and $z_- \geqslant 0$; hence $h_+ = n_+$ and $h_- = n_-$, so that by the first identity of Lemma 7.3.4 we have once again $h_+ + h_- = 2(d_P + d_Q) = 2(h_P + h_Q)$.

Assume that $z_P = z_Q = 0$ and $\ell_P = \ell_Q = \ell$. If $f(\ell) \neq 0$ we have seen that $z_+ \leqslant 0$ and $z_- > 0$ (up to exchange of $+$ and $-$), and as above, the strict inequality $z_- > 0$ together with the third identity of Lemma 7.3.4 gives $h_+ + h_- = 2(d_P + d_Q)$, since the coefficient of $T^{d_P + d_Q}$ is equal to $2f(\ell) \neq 0$. If $f(\ell) = 0$ then $z_+ > 0$ and $z_- > 0$, and we conclude again by the first identity of Lemma 7.3.4 since we have seen that the coefficient of $T^{d_P + d_Q}$ is equal to $f'(\ell)^2 - 4(2\ell + a)f(\ell) = f'(\ell)^2 \neq 0$.

As above, the other cases are similar and left to the reader (Exercise 23).

To finish the proof of the theorem, we must now treat the much simpler case $Q = \pm P$, so we must show that $h(2P) = 4h(P)$. The addition law gives

$$X_{2P} = X_+ = \frac{f'(X_P)^2 - 4(2X_P + a)f(X_P)}{4f(X_P)},$$

so that with the same notation as in the general case we have $N_+/D_+ = A_N/A_D$ with

$$
\begin{aligned}
A_N &= (3N_P^2 + 2aN_P D_P + bD_P^2)^2 \\
&\quad - 4(2N_P + aD_P)(N_P^3 + aN_P^2 D_P + bN_P D_P^2 + cD_P^3) \\
&= (N_P^2 - bD_P^2)^2 - 4cD_P^3(2N_P + aD_P) \quad \text{and} \\
A_D &= 4D_P(N_P^3 + aN_P^2 D_P + bN_P D_P^2 + cD_P^3).
\end{aligned}
$$

Lemma 7.3.6. *There exists $u \in k^*$ such that $N_+ = uA_N$ and $D_+ = uA_D$.*

Proof. This immediately follows from the fact that A_N and A_D are coprime and is left to the reader (Exercise 24). □

Lemma 7.3.7. *Keep all the above notation.*

(1) *If $z_P > 0$ then $z_+ > 0$.*
(2) *If $z_P < 0$ and $c \neq 0$ then $z_+ \leqslant 0$.*
(3) *If $z_P < 0$ and $c = 0$ then $z_+ > 0$.*
(4) *If $z_P = 0$ and $f(\ell_P) \neq 0$ then $z_+ \leqslant 0$.*
(5) *If $z_P = 0$ and $f(\ell_P) = 0$ then $z_+ > 0$.*

Proof. Identical to that of Lemma 7.3.5 and left to the reader (Exercise 24). □

As in the general case, the proof of the identity $h(2P) = 4h(P)$ is obtained as a combination of the above two lemmas and is left to the reader, finishing the proof of the theorem. □

To finish, we prove the following lemma (recall that we denote by $\mathcal{O}$ the identity element of E_g, in other words the unique point at infinity).

Lemma 7.3.8. *Assume that $g(T)$ is not equal to a constant times the square of a rational function, and let $P \in E_g(K)$. The following are equivalent:*

(1) $h(P) = 0$.
(2) $P \in E_g(k)$.
(3) *Either $P = \mathcal{O}$ or there exists $u \in k$ such that $P = (u, 0)$.*
(4) *For all $Q \in E_g(K)$ we have $h(P+Q) = h(Q)$; in other words, $B(P,Q) = 0$.*

Proof. Since the result is trivial when $P = \mathcal{O}$, assume that this is not the case and write as usual $P = (X_P, Y_P)$. Clearly $h(P) = 0$ if and only if $X_P \in k$, so we can write $X_P = u$. We thus have $g(T)Y_P^2 = f(u)$, so that if $f(u) \neq 0$, $g(T)$ is equal to a constant times the square of the rational function $1/Y_P$, contrary to our assumption. Thus $f(u) = 0$, so $Y_P = 0$ and $P = (u, 0)$, proving that (1) implies (3), and (3) implies (2) implies (1) is trivial, so that (1), (2), and (3) are equivalent. To prove that (3) implies (4) we note that if $Q = \pm P$ we have $h(2P) = 4h(P) = 0$ and $h(P - P) = 0$, so (4) is clear, and (4) is also clear if $Q = \mathcal{O}$. Otherwise, a simple computation using the addition law gives $X_{P+Q} = u + f'(u)/(X_Q - u)$, and since $f(u) = 0$ we have $f'(u) \neq 0$, and this is easily seen to imply that $h(P + Q) = h(Q)$ as desired. Finally, (4) applied to $Q = kP$ for any $k \in \mathbb{Z}$ gives $(k+1)^2 h(P) = h(P + Q) = h(Q) = k^2 h(P)$; hence $h(P) = 0$, so (4) implies (1). $\square$

7.3.4 Elliptic Curves over $\mathbb{F}_q$

We begin with the following very easy lemma.

Lemma 7.3.9. *Let q be an odd prime power, let ρ be the unique multiplicative character of order 2 on $\mathbb{F}_q$, and let $y^2 = f(x)$ be the equation of an elliptic curve over $\mathbb{F}_q$, where $f(X)$ is a polynomial of degree 3. Then $|E(\mathbb{F}_q)| = q + 1 - a_q$ with $a_q = -\sum_{x \in \mathbb{F}_q} \rho(f(x))$. In particular, if $q = p$ is prime we have $|E(\mathbb{F}_p)| = p + 1 - a_p$ with*

$$a_p = -\sum_{x \in \mathbb{F}_p} \left(\frac{f(x)}{p} \right).$$

Proof. For a given $X \in \mathbb{F}_q$ it is clear that the number of $y \in \mathbb{F}_q$ such that $y^2 = X$ is equal to $1 + \rho(X)$. Counting the point at infinity separately it follows that

$$|E(\mathbb{F}_q)| = 1 + \sum_{x \in \mathbb{F}_q} (1 + \rho(f(x))) = q + 1 + \sum_{x \in \mathbb{F}_q} \rho(f(x)),$$

proving the lemma. $\square$

The main result for curves over finite fields is Hasse's theorem giving precise bounds on the cardinality of $E(\mathbb{F}_q)$. Considering the importance of this result, we give a proof due to Manin that is taken from a book by Knapp [Kna]. We have done most of the necessary groundwork in the preceding subsection.

Theorem 7.3.10 (Hasse). *For all elliptic curves E over a finite field $\mathbb{F}_q$ we have the inequality*

$$q + 1 - 2q^{1/2} \leqslant |E(\mathbb{F}_q)| \leqslant q + 1 + 2q^{1/2} .$$

Proof. We first assume that the characteristic of $\mathbb{F}_q$ is not equal to 2, and we will see below how to modify the proof in characteristic 2. In this case we know that E has a plane model whose affine equation is $y^2 = f(x)$ with $f(x) = x^3 + ax^2 + bx + c$.

We define the curve E_f over $K = \mathbb{F}_q(T)$ whose affine equation over K is

$$Y^2 = \frac{1}{T^3 + aT^2 + bT + c}(X^3 + aX^2 + bX + c) = \frac{1}{f(T)}f(X) .$$

This is therefore a quadratic twist of a constant elliptic curve, hence of the form E_g studied in the preceding subsection in the special case $g = f$. We thus know that the height function $h(P)$ is a quadratic form.

Apart from the point at infinity the point $P = (T, 1)$ is evidently in $E_f(K)$. Furthermore, we have a Frobenius homomorphism σ from $E_f(K)$ to $E_f(K)$ defined by

$$\sigma(X, Y) = (X^q, f(T)^{(q-1)/2}Y^q) .$$

Indeed, this belongs to $E_f(K)$ since

$$f(T)\left(f(T)^{(q-1)/2}Y^q\right)^2 = (f(T)Y^2)^q = f(X)^q = f(X^q) ,$$

since $f \in \mathbb{F}_q[T]$. In particular, the point $Q = \sigma(P) = (T^q, f(T)^{(q-1)/2})$ belongs to $E_f(K)$. We will set $h_n = h(Q + nP)$, where h is the height function.

Proposition 7.3.11. *We have $h_n = n^2 + a_q n + q$, where $a_q = q + 1 - |E(\mathbb{F}_q)|$.*

Proof. By Theorem 7.3.3 we have

$$h_n = n^2 h(P) + 2nB(P, Q) + h(Q) = n^2 + a_q n + q$$

for some constant $a_q = 2B(P, Q)$ to be determined, where $B(P, Q)$ is the bilinear form associated with h. To compute a_q, we note that $h(Q - P) = h_{-1} = q + 1 - a_q$, so that $a_q = q + 1 - h(Q - P)$. On the other hand, we note that $T^q \neq T$; hence $Q \neq \pm P$, so the addition law gives

$$X_{Q-P} = f(T) \left(\frac{f(T)^{(q-1)/2} + 1}{T^q - T} \right)^2 - T^q - T - a$$

$$= \frac{T^{2q+1} + 2f(T)^{(q+1)/2} + T^{q+2} + 2aT^{q+1} + b(T^q + T) + 2c}{(T^q - T)^2}$$

$$= \frac{T^{2q+1} + O(T^{2q})}{(T^q - T)^2}$$

in an evident sense. Thus if we set $X_{Q-P} = A/B$ with A and B coprime we have $z(X_{Q-P}) = \deg(A) - \deg(B) = 1$, and in particular $h(Q - P) = \deg(A)$ (in fact it is easily shown that $z(X_{Q+nP}) = 1$ for all $n \neq 0$; see Exercise 18).

For $\varepsilon = -1, 0, 1$ let N_ε be the number of $u \in \mathbb{F}_q$ such that $f(u)^{(q-1)/2} = \varepsilon$, so that $N_{-1} + N_0 + N_1 = q$. Since the roots of $T^q - T = 0$ are the elements of $\mathbb{F}_q$ and since $f(T)$ has no multiple roots, it is clear that the GCD of $f(T)^{(q-1)/2} + 1$ and $T^q - T$ will be of degree N_{-1} and the GCD of $f(T)$ and $(T^q - T)^2$ will be of degree N_0. It thus follows from the above expression for A/B that $\deg(B) = 2q - 2N_{-1} - N_0$. On the other hand, we have

$$|E(\mathbb{F}_q)| = 2N_1 + N_0 + 1 = 2q + 1 - (2N_{-1} + N_0)$$
$$= \deg(B) + 1 = \deg(A) = h(Q - P) \,,$$

showing that $a_q = q + 1 - |E(\mathbb{F}_q)|$ and proving the proposition.   □

We can now finish the proof of Hasse's theorem.

Proof of Theorem 7.3.10. If the characteristic of $\mathbb{F}_q$ is different from 2, by Theorem 7.3.3 we have

$$h(nP + mQ) = n^2 h(P) + 2nmB(P,Q) + m^2 h(Q) = n^2 + a_q nm + qm^2 \,.$$

Since $h(nP + mQ) \geqslant 0$, it follows that $x^2 + a_q xy + qy^2$ is a positive semi-definite quadratic form, hence that $a_q^2 - 4q \leqslant 0$, proving Hasse's inequality in this case.

In characteristic 2, the process is similar but the formulas rather different. In essence we must use "Artin–Schreier theory" (in other words the solution of a second-degree equation in characteristic 2). Write $q = 2^n$, and let $y^2 + a_1 xy + a_3 y = x^3 + a_2 x^2 + a_4 x + a_6 = f(x)$ be an affine plane model of our elliptic curve with $a_i \in \mathbb{F}_q$. Since this curve is nonsingular, we cannot have $a_1 = a_3 = 0$. Thus we can define E_f by the equation

$$Y^2 + a_1 XY + a_3 Y = f(X) + \left(\frac{a_1 X + a_3}{a_1 T + a_3} \right)^2 f(T) \,,$$

and our points P and Q are

$$P = (T, 0) \quad \text{and} \quad Q = (T^q, (a_1 T + a_3)^q A(f(T)/(a_1 T + a_3)^2)) \,,$$

where $A(Z)$ is the Artin–Schreier polynomial

$$A(Z) = \sum_{0 \leqslant j \leqslant n-1} Z^{2^j} = Z + Z^2 + Z^4 + \cdots + Z^{2^{n-1}}.$$

It is easy to check that the discriminant of E_f is equal to that of E, so that E_f is indeed an elliptic curve, and that the points P and Q are in $E_f(K)$. It is now a routine (but slightly painful) matter left to the reader (Exercise 25) to check that everything that we have done above carries through to this case, proving the theorem in general.   □

In addition to Hasse's theorem, the following result, mainly due to Deuring, tells us which values of $|E(\mathbb{F}_q)|$ are possible; see [Ruc] and the references therein.

Theorem 7.3.12. *Let $q = p^n$ be the cardinality of a finite field, and let $t \in \mathbb{Z}$ be such that $|t| \leqslant 2q^{1/2}$. There exists an elliptic curve E defined over $\mathbb{F}_q$ such that $|E(\mathbb{F}_q)| = q + 1 - t$ if and only if one of the following conditions is satisfied:*

(1) $p \nmid t$.
(2) n *is even and* $t = \pm 2q^{1/2}$.
(3) n *is even,* $p \not\equiv 1 \pmod 3$, *and* $t = \pm q^{1/2}$.
(4) n *is even,* $p \not\equiv 1 \pmod 4$, *and* $t = 0$.
(5) n *is odd,* $p = 2$ *or* 3, *and* $t = \pm p^{(n+1)/2}$.
(6) n *is odd and* $t = 0$.

Thus, there are no restrictions if $q = p$. On the other hand, if for instance $q = 8$ we cannot have $|E(\mathbb{F}_q)| = 9$ or $|E(\mathbb{F}_q)| = 11$.

Note also that it is easy to show that the group $E(\mathbb{F}_q)$ is the product of at most two cyclic groups, and if we write $E(\mathbb{F}_q) \simeq (\mathbb{Z}/d_1\mathbb{Z}) \times (\mathbb{Z}/d_2\mathbb{Z})$ with $d_2 \mid d_1$, then $d_2 \mid q - 1$.

As for any curve, we can define the so-called *Hasse–Weil zeta function* attached to E by

$$\zeta_q(E, T) = \exp\left(\sum_{n \geqslant 1} \frac{|E(\mathbb{F}_{q^n})|}{n} T^n\right).$$

It is not too difficult to show that $\zeta_q(E, T)$ is a rational function, and more precisely that

$$\zeta_q(E, T) = \frac{1 - a_q T + qT^2}{(1 - T)(1 - qT)},$$

where as above, $a_q = q + 1 - |E(\mathbb{F}_q)|$. When $q = p$ is a prime number we will set

$$L_p(E, T) = \frac{1}{1 - a_p T + pT^2}.$$

This will be used as the local Euler factor in the definition of the global L-function of an elliptic curve defined over $\mathbb{Q}$ when p is a prime of good reduction.

The structure of the ring $\mathrm{End}(E)$ of $\overline{\mathbb{F}_q}$-endomorphisms of an elliptic curve defined over $\mathbb{F}_q$ is slightly more complicated than over $\mathbb{C}$ (or over a field of characteristic 0). We have seen that $\mathrm{End}(E)$ always contains a copy of $\mathbb{Z}$ thanks to the multiplication-by-m maps $[m]$. Here $\mathrm{End}(E)$ contains an additional automorphism, the *Frobenius automorphism* ϕ defined (in affine coordinates) by $\phi((x,y)) = (x^q, y^q)$. Indeed, more generally let $P(x_1, \ldots, x_n) = 0$ be some polynomial equation with coefficients in $\mathbb{F}_q$. If we raise to the qth power and use the identity $(X + Y)^q = X^q + Y^q$ together with the definition of $\mathbb{F}_q$, we see that $P(x_1^q, \ldots, x_n^q) = 0$. This shows that in our special case (x^q, y^q) is still on our curve, and it is easily checked that it preserves the group law (this is in fact automatic since any algebraic map from E to E preserving the point $\mathcal{O}$ preserves the whole group law).

Coming back to elliptic curves, it can be shown that ϕ (like any isogeny from E to E) satisfies a quadratic equation with discriminant less than or equal to zero, and it is easy to see that ϕ cannot be equal to $[m]$ for any m, so that the discriminant is in fact strictly negative. It follows that over $\mathbb{F}_q$ the ring $\mathrm{End}(E)$ always contains an order in an imaginary quadratic field; in other words, E always has complex multiplication.

The ring $\mathrm{End}(E)$ may be even larger than this: in that case we say that the curve E is *supersingular*. More precisely, denote by $E[m] \subset E(\overline{K})$ the kernel of the multiplication-by-m map from $E(\overline{K})$ to itself. We then have the following results, that we will not prove:

Proposition 7.3.13. *Let K be a field of characteristic $p > 0$ and let E be an elliptic curve defined over K. Then either $E[p^r] \simeq \mathbb{Z}/p^r\mathbb{Z}$ for all r, or $E[p^r] = \{0\}$ for all $r \geqslant 1$, in which case the curve is said to be* supersingular.

Note that in if K has characteristic 0, we always have $E[m] \simeq (\mathbb{Z}/m\mathbb{Z}) \times (\mathbb{Z}/m\mathbb{Z})$, where we again emphasize that the torsion subgroup $E[m]$ is taken over $\overline{K}$.

Theorem 7.3.14. *Let E be an elliptic curve over $\mathbb{F}_q$ with $q = p^n$, and let $a_q = q + 1 - |E(\mathbb{F}_q)|$.*

(1) *The Frobenius endomorphism ϕ satisfies the equation*

$$\phi^2 - a_q\phi + q = 0 \ .$$

(2) *The curve E is supersingular if and only if $p \mid a_q$, or equivalently, $|E(\mathbb{F}_q)| \equiv 1 \pmod{p}$.*

(3) *If E is not supersingular the ring $\mathrm{End}(E)$ is isomorphic to the imaginary quadratic order of discriminant $a_q^2 - 4q < 0$.*

(4) *If E is supersingular the ring $\mathrm{End}(E)$ is isomorphic to an order in a definite quaternion algebra, and in particular is a $\mathbb{Z}$-module of dimension 4.*

To finish this section, we need a very simple result on *quadratic twists*.

Definition 7.3.15. *Let E be an elliptic curve defined over a field K of characteristic different from 2 by an equation of the form $y^2 = f(x)$ for some cubic polynomial f, and let $D \in K^*$. The* quadratic twist *of E by D is the elliptic curve with equation*

$$Dy^2 = f(x) .$$

If $f(x) = x^3 + ax^2 + bx + c$ and we want to convert this equation to ordinary Weierstrass form we first note that if D is a square in K^* we simply set $Y = y\sqrt{D}$. Otherwise, we multiply the equation by D^3 and set $Y = D^2 y$ and $X = Dx$, giving the equation $Y^2 = X^3 + aDX^2 + bD^2X + cD^3$.

The result that we need is the following.

Proposition 7.3.16. *Let E be an elliptic curve defined over $\mathbb{F}_p$ by a Weierstrass equation of the form $y^2 = f(x)$ for some cubic polynomial f, where p is an odd prime, let $D \in \mathbb{F}_p^*$, let E_D be the quadratic twist of E by D, and finally as above let $a_p(E) = p + 1 - |E(\mathbb{F}_p)|$ and similarly $a_p(E_D)$. Then*

$$a_p(E_D) = \left(\frac{D}{p}\right) a_p(E) ,$$

where $\left(\frac{D}{p}\right)$ is the usual Legendre symbol. More generally this result is true for any finite field $\mathbb{F}_q$ of odd characteristic, replacing $\left(\frac{D}{p}\right)$ by $\rho(D)$, where ρ is the unique multiplicative character of order 2.

Proof. By Lemma 7.3.9 we have

$$a_p(E) = -\sum_{x \in \mathbb{F}_p} \left(\frac{f(x)}{p}\right) .$$

On the other hand, the equation of E_D is $Dy^2 = f(x)$, which is isomorphic to $Y^2 = Df(x)$ since $D \neq 0$ in $\mathbb{F}_p$, so

$$a_p(E_D) = -\sum_{x \in \mathbb{F}_p} \left(\frac{Df(x)}{p}\right) = \left(\frac{D}{p}\right) a_p(E)$$

by multiplicativity of the Legendre symbol. The same proof clearly works for any finite field of odd characteristic. □

This proposition will in particular be used in Section 8.5.2 to compute efficiently $a_p(E)$ when E comes from an elliptic curve defined over $\mathbb{Q}$ with *complex multiplication*.

7.3.5 Constant Elliptic Curves over $R[[T]]$: Formal Groups

In the next section we are going to study elliptic curves over $\mathfrak{p}$-adic fields $K_{\mathfrak{p}}$. For this, we must first study elliptic curves over formal power series rings $R[[T]]$, although if we only want to study curves over the field $\mathbb{Q}_p$ of p-adic numbers, this is not really necessary. In this section we thus let R be an integral domain of characteristic zero (we will later specialize to the case $R = \mathbb{Z}_p$), and we let $R[[T]]$ be the ring of formal power series with coefficients in R. We denote by $\mathcal{L} = R[[T]][1/T]$ the ring of formal power series in T, where a finite number of negative integral powers of T are allowed.

Let E be an elliptic curve over R given by a generalized Weierstrass equation $y^2 + a_1 xy + a_3 y = x^3 + a_2 x^2 + a_4 x + a_6$, where we may assume that the a_i are in R.

Proposition 7.3.17. *There exists a unique formal power series $s(T) = \sum_{k \geqslant 0} u_k T^k \in R[[T]]$ with $u_0 \neq 0$ such that if we set $(x(T), y(T)) = (s(T)/T^2, -s(T)/T^3)$ then $(x(T), y(T)) \in E(\mathcal{L})$.*

Proof. We use a Hensel-type argument with the T-adic valuation instead of a $\mathfrak{p}$-adic one. Set $f(x, y) = y^2 + a_1 xy + a_3 y - (x^3 + a_2 x^2 + a_4 x + a_6)$. Identifying coefficients of T^{-6} in $f(x(T), y(T))$ we find that $u_0^2 = u_0^3$, so that $u_0 = 1$ since we have assumed $u_0 \neq 0$. For all $K \geqslant 0$ let $s_K(T) = \sum_{0 \leqslant i < K} u_k T^k$ be the truncated power series, and define $x_K(T) = s_K(T)/T^2$ and $y_K(T) = -s_K(T)/T^3$. Let $K \geqslant 1$, assume that we have proved existence and uniqueness of u_i for $i < K$ such that $f(x_K(T), y_K(T)) = O(T^{K-6})$, which is indeed the case for $K = 1$, and write $f(x_K(T), y_K(T)) = c_K T^{K-6} + O(T^{K-5})$. By Taylor's formula we have

$$f(x_{K+1}(T), y_{K+1}(T)) = f(x_K(T) + u_K T^{K-2}, y_K(T) - u_K T^{K-3})$$
$$= f(x_K(T), y_K(T)) + u_K T^{K-2} f_x'(x_K(T), y_K(T))$$
$$- u_K T^{K-3} f_y'(x_K(T), y_K(T)) + O(T^{2K-6}) .$$

Since $K \geqslant 1$ we have $2K - 6 \geqslant K - 5$. Furthermore,

$$f_x'(x_K(T), y_K(T)) = a_1 y_K(T) - (3x_K(T)^2 + 2a_2 x_K(T) + a_4)$$
$$= -3T^{-4} + O(T^{-3}) \quad \text{and}$$
$$f_y'(x_K(T), y_K(T)) = 2y_K(T) + a_1 x_K(T) + a_3 = -2T^{-3} + O(T^{-2}) ,$$

so we deduce that

$$f(x_{K+1}(T), y_{K+1}(T)) = (c_K - u_K)T^{K-6} + O(T^{K-5}) .$$

We must thus choose $u_K = c_K$, simultaneously proving existence and uniqueness by induction. $\square$

Remark. It is important to note that in this proof we did not need any division, so all the coefficients u_K are indeed in R. In fact, the following proposition shows that this is also true for the "invariant differential" $dx/(2y + a_1 x + a_3)$:

Proposition 7.3.18. *We have* $x'(T)/(2y(T) + a_1 x(T) + a_3) \in R[[T]]$.

Proof. Since $x'(T) = -2T^{-3} + O(T^{-2})$ we can write

$$g(T) = x'(T)/(2y(T) + a_1 x(T) + a_3) = \sum_{k \geqslant 0} g_k T^k .$$

Because of the factor 2 in the denominator we know only that $2^k g_k \in R$. However, because of the Weierstrass equation we also have

$$g(T) = y'(T)/(3x^2(T) + 2a_2 x(T) + a_4 - a_1 y(T)) ,$$

and the factor 3 implies that $3^k g_k \in R$. Since 2^k and 3^k are coprime (in $\mathbb{Z}$, not in R!) the proposition follows. $\square$

Although we will not need them, for completeness we give the beginning of the expansions of $s(T)$ and of the invariant differential:

$$s(T) = 1 - a_1 T - a_2 T^2 - a_3 T^3 - (a_1 a_3 + a_4) T^4$$
$$- (a_1^2 a_3 + a_1 a_4 + a_2 a_3) T^5 + O(T^6) ,$$

$$\frac{x'(T)}{2y(T) + a_1 x(T) + a_3} = 1 + a_1 T + (a_1^2 + a_2) T^2 + (a_1^3 + 2a_1 a_2 + 2a_3) T^3$$
$$+ (a_1^4 + 3a_1^2 a_2 + 6a_1 a_3 + a_2^2 + 2a_4) T^4 + O(T^5) .$$

For the next result, we will need the ring $R[[T_1, T_2]]$ of formal power series in two variables. Note that in writing an expansion in this ring the "remainder term" cannot simply be written $O(T_1^k)$ or $O(T_2^k)$, since all possible monomials could appear. We will thus use the following practical abuse of notation: if we write $F(T_1, T_2) = G(T_1, T_2) + O(T^k)$, this means that all the monomials occurring in the formal power series $F - G$ have total degree greater than or equal to k.

Proposition 7.3.19. (1) *There exists a unique formal power series in two variables* $F(T_1, T_2) \in R[[T_1, T_2]]$ *such that* $F(T_1, T_2) = T_1 + T_2 + O(T^2)$ *and satisfying*

$$(x(T_1), y(T_1)) + (x(T_2), y(T_2)) = (x(F(T_1, T_2)), y(F(T_1, T_2))) ,$$

where the + sign in the left-hand side denotes of course the addition law on the elliptic curve E.

(2) *We have* $F(T_2, T_1) = F(T_1, T_2)$, $F(F(T_1, T_2), T_3) = F(T_1, F(T_2, T_3))$, $F(0, T) = F(T, 0) = T$, *and* $F(T, i(T)) = F(i(T), T) = 0$, *where* $i(T) = -T/(1 - a_1 T + a_3/y(T)))$, *and* $y(T)$ *is as above.*

Proof. For $i = 1$ and 2 let $P_i = (x(T_i), y(T_i))$, and let $mx + ny = 1$ be the equation of the line through P_1 and P_2 (since T_1 and T_2 are formal variables and $x(T) = T^{-2} + O(T^{-1})$ there are no special cases). By definition the third point of intersection $P_3 = (x_3, Y_3)$ will be such that $P_1 + P_2 + P_3 = \mathcal{O}$, so $P_1 + P_2 = (x_3, y_3)$ with $y_3 = -Y_3 - a_1 x_3 - a_3$. Since we will want to expand in $T_i = -x(T_i)/y(T_i)$ we set $t = -x/y$ and $w = -1/y$, so that the equation of the line is $w = mt - n$, while the equation of E becomes

$$t^3 + (a_1 t + a_2 t^2 - 1)w + (a_3 + a_4 t)w^2 + a_6 w^3 = 0 .$$

Writing the intersection gives

$$t_1 + t_2 + t_3 = \frac{-a_1 m - a_3 m^2 + n(a_2 + 2a_4 m + 3a_6 m^2)}{1 + a_2 m + a_4 m^2 + a_6 m^3} .$$

We must now compute m and n. Since $y(T) = -T^{-3} + \sum_{k \geqslant 1} (-u_k) T^{k-3}$ with $u_k \in R$, we note that $w(T) = -1/y(T) = \sum_{k \geqslant 3} v_k T^k$ with $v_3 = 1$ and again $v_k \in R$ for all k. It follows that

$$m = \frac{w(T_2) - w(T_1)}{T_2 - T_1} = \sum_{k \geqslant 3} v_k \sum_{0 \leqslant i \leqslant k-1} T_1^i T_2^{k-1-i} \in R[[T_1, T_2]] ,$$

so of course $n = mT_1 - w(T_1) \in R[[T_1, T_2]]$. In fact, we have more precisely $m = T_1^2 + T_1 T_2 + T_2^2 + O(T^3)$, so that $n = T_1 T_2 (T_1 + T_2) + O(T^4)$, and $1 + a_2 m + a_4 m^2 + a_6 m^3 = 1 + O(T^2)$ is invertible in $R[[T_1, T_2]]$. Thus, using the formula obtained above we obtain that $t_3 \in R[[T_1, T_2]]$ and that $t_3 = -T_1 - T_2 - a_1(T_1^2 + T_1 T_2 + T_2^2) + O(T^3)$, so that

$$w_3 = mt_3 - n = -(T_1 + T_2)(T_1^2 + T_1 T_2 + T_2)^2 - T_1 T_2(T_1 + T_2) + O(T^4)$$
$$= -(T_1 + T_2)^3 + O(T^4) .$$

We have thus obtained the (t, w)-coordinates of the point $P_3 = -(P_1 + P_2)$. Recall that we have written $P_3 = (x_3, Y_3)$ and $P_1 + P_2 = (x_3, y_3)$. The t-coordinate of $P_1 + P_2$ is thus equal to

$$-\frac{x_3}{y_3} = \frac{x_3}{Y_3 + a_1 x_3 + a_3} = \frac{x_3/Y_3}{1 + a_1(x_3/Y_3) + a_3/Y_3} = \frac{-t_3}{1 - a_1 t_3 - a_3 w_3}$$
$$= (T_1 + T_2 + a_1(T_1^2 + T_1 T_2 + T_2^2))(1 - a_1(T_1 + T_2)) + O(T^3)$$
$$= T_1 + T_2 - a_1 T_1 T_2 + O(T^3) ,$$

and since $a_1 t_3 + a_3 w_3 = O(T)$, we deduce that $-x_3/y_3 \in R[[T_1, T_2]]$. Similarly the w-coordinate of $P_1 + P_2$ is given by

$$-\frac{1}{y_3} = \frac{1}{Y_3 + a_1 x_3 + a_3} = \frac{1/Y_3}{1 + a_1 x_3/Y_3 + a_3/Y_3} = \frac{-w_3}{1 - a_1 t_3 - a_3 w_3}$$
$$= (T_1 + T_2)^3 + O(T^4),$$

also with $-1/y_3 \in R[[T_1, T_2]]$.

Thus if we set $F(T_1, T_2) = -x_3/y_3$ it is clear that $F(T_1, T_2) \in R[[T_1, T_2]]$ with $F(T_1, T_2) = T_1 + T_2 - a_1 T_1 T_2 + O(T^3)$, and by definition $(x(T_1), y(T_1)) + (x(T_2), y(T_2)) = (x(F(T_1, T_2)), y(F(T_1, T_2)))$, proving (1).

For (2), it is clear that the given identities are equivalent to the commutativity, associativity, and the existence of a unit element and of an inverse in the group of points of E with coordinates in the fraction field of $\mathcal{L}$. $\quad\square$

As above, it is immediate to compute that

$$F(T_1, T_2) = T_1 + T_2 - a_1 T_1 T_2 - a_2 T_1 T_2 (T_1 + T_2)$$
$$- T_1 T_2 (2a_3(T_1^2 + T_2^2) + (3a_3 - a_1 a_2)T_1 T_2) + O(T^5).$$

Note that since $F(T, 0) = F(0, T) = T$ all the homogeneous terms starting with degree 2 must be divisible by $T_1 T_2$.

A formal power series F having the above properties is called a *formal group*. We have thus shown that there is a unique formal group associated with an elliptic curve. The essential property of F that we shall use is that $F \in R[[T_1, T_2]]$.

The two other foremost examples of formal groups are those associated with degenerate elliptic curves, in other words to the additive group and to the multiplicative group: for the additive group it is simply $F(T_1, T_2) = T_1 + T_2$, for the multiplicative group it is

$$F(T_1, T_2) = T_1 + T_2 + T_1 T_2 = (1 + T_1)(1 + T_2) - 1.$$

Although we will not need to study general properties of formal groups, we need to define the formal logarithm and exponential. Denote by K the field of fractions of R.

Proposition 7.3.20. *Let* $F(T_1, T_2) \in R[[T_1, T_2]]$ *be a formal group.*

(1) *There exists a unique formal power series* $L \in K[[T]]$ *such that* $L(T) = T + O(T^2)$ *and satisfying* $L(F(T_1, T_2)) = L(T_1) + L(T_2)$. *In addition,* $L'(T) \in R[[T]]$.

(2) *If* $E = T + O(T^2) \in K[[T]]$ *is the unique formal power series such that* $E(L(T)) = L(E(T)) = T$ *then* $F(T_1, T_2) = E(L(T_1) + L(T_2))$.

Proof. Denote by F_1' and F_2' the (formal) partial derivative of F with respect to the first and second variables respectively. Assuming the existence of L, we differentiate with respect to T_1 its defining equation. We obtain $F_1'(T_1, T_2) L'(F(T_1, T_2)) = L'(T_1)$, so that setting $T_1 = 0$ and using $L(T) = $

$T + O(T^2)$ we obtain $F_1'(0, T_2)L'(F(0, T_2)) = 1$, in other words $L'(T) = 1/F_1'(0, T)$ (which exists since $F(T_1, T_2) = T_1 + T_2 + O(T^2)$). Since $L(0) = 0$, this shows uniqueness, and also the fact that $L'(T) \in R[[T]]$. We now prove existence. Differentiating the "associativity law" $F(T_3, F(T_1, T_2)) = F(F(T_3, T_1), T_2)$ with respect to T_3 and setting $T_3 = 0$ we obtain

$$F_1'(0, F(T_1, T_2)) = F_1'(0, T_1)F_1'(T_1, T_2) .$$

Thus, if we *define* $L(T)$ to be the unique formal power series with 0 constant term such that $L'(T) = 1/F_1'(0, T)$ we have $F_1'(T_1, T_2)L'(F(T_1, T_2)) = L'(T_1)$, which is the identity obtained above. Integrating with respect to T_1, it follows that $L(F(T_1, T_2)) = L(T_1) + M(T_2)$, where $M(T_2)$ is some power series in T_2, which is immediately seen to be equal to $L(T_2)$ by setting $T_1 = 0$, proving (1). (2) is an immediate consequence since any formal power series $f(T) = T + O(T^2)$ has a functional inverse $g(T) = T + O(T^2)$ such that $f(g(T)) = g(f(T)) = T$; see Exercise 26.   □

The series L and E are called the *formal logarithm* and the *formal exponential* associated with the formal group F. Note that the formal logarithm and exponential associated with the multiplicative formal group $F(T_1, T_2) = T_1 + T_2 + T_1 T_2$ are $\log(1 + T)$ and $\exp(T) - 1$ respectively. The above proposition implies in particular that the formal group F can easily be computed explicitly by the sole knowledge of the power series in one variable L. For the formal group associated with an elliptic curve we have explicitly

$$L(T) = T + a_1 \frac{T^2}{2} + (a_1^2 + a_2)\frac{T^3}{3} + (a_1^3 + 2a_1 a_2 + 2a_3)\frac{T^4}{4} + O(T^5) ,$$

$$E(T) = T - a_1 \frac{T^2}{2!} + (a_1^2 - 2a_2)\frac{T^3}{3!} - (a_1^3 - 8a_1 a_2 + 12a_3)\frac{T^4}{4!} + O(T^5) .$$

Remarks. (1) It can be shown by an explicit but unilluminating computation that in fact $L'(T) = x'(T)/(2y(T) + a_1 x(T) + a_3)$, which indeed checks with the expansions given above. Since we will not need this, we leave the proof to the reader (Exercise 27).

(2) We could have *defined* $L(T)$ as the antiderivative of $x'(T)/(2y(T) + a_1 x(T) + a_3)$ vanishing at $T = 0$, then $E(T)$ as the functional inverse of $L(T)$, and finally $F(T_1, T_2)$ as $E(L(T_1) + L(T_2))$. But we would then have to prove that F is indeed a formal group (this is essentially equivalent to the above exercise), but above all we would need to prove that $F \in R[[T_1, T_2]]$, which is again a painful computation.

Definition 7.3.21. *For any $m \in \mathbb{Z}$ we define the formal power series $[m]T$ by the formula $[m]T = E(mL(T))$.*

We note that by definition of L and E we have $[0]T = 0$, $[1]T = T$, $[-1]T = i(T) = -T/(1 - a_1 T + a_3/y(T))$, and

$$[m+n]T = E(mL(T) + nL(T)) = E(L([m]T) + L([n]T)) = F([m]T, [n]T) ,$$

so we could also define $[m]T$ by induction using these formulas. Note also that since $L(T) = T + O(T^2)$ and $E(T) = T + O(T^2)$ we have $[m]T = mT + O(T^2)$.

Proposition 7.3.22. (1) *For any $m \in \mathbb{Z}$ we have*

$$([m]T)' = mL'(T)/L'([m]T) .$$

(2) *If p is a prime number there exist formal power series $f(T) = T + O(T^2) \in R[[T]]$ and $g(T) = aT + O(T^2) \in R[[T]]$ such that $[p]T = pf(T) + g(T^p)$. In other words when $p \nmid k$ the coefficient of T^k in $[p]T$ is divisible by p (i.e., its quotient by p belongs to R).*

Proof. If we differentiate the formula $E(L(T)) = T$ with respect to T we obtain $E'(T) = 1/L'(E(T))$, so replacing T by $mL(T) = L([m]T)$ gives $E'(mL(T)) = 1/L'(E(L([m]T))) = 1/L'([m]T)$, and since by the chain rule we have $([m]T)' = mL'(T)E'(mL(T))$ this proves (1). For (2) we note that since $L'(T) = 1 + O(T) \in R[[T]]$ and $[m]T = mT + O(T^2)$ we have $L'(T)/L'([m]T) = 1 + O(T) \in R[[T]]$, hence $(m[T])' = mh_m(T)$ with $h_m \in R[[T]]$. On the other hand, and this is the crucial fact, since $F \in R[[T_1, T_2]]$ it is clear that $[m]T \in R[[T]]$ thanks to the induction $[m]T = F([m-1]T, T)$ seen above, so we can write $[m]T = \sum_{n \geqslant 1} u_n T^n$ with $u_n \in R$. We thus have $h_m(T) = \sum_{n \geqslant 1} (n/m) u_n T^{n-1} \in R[[T]]$. Fix some $n \geqslant 1$, set $d = \gcd(n, m)$ (in $\mathbb{Z}$), and write $(n/m)u_n = v_n \in R$. By definition there exist x and y in $\mathbb{Z}$ such that $1 = x(n/d) + y(m/d)$, so

$$u_n = (m/d)(x(n/m)u_n + yu_n) = (m/d)(xv_n + yu_n) \in (m/d)R .$$

In particular, if m is coprime to n we have $u_n \in mR$, which immediately implies (2). $\square$

7.3.6 Reduction of Elliptic Curves over $K_{\mathfrak{p}}$

In this section we let $K_{\mathfrak{p}}$ be the completion of a number field K at some prime ideal $\mathfrak{p}$ above a prime number p. For simplicity of notation we write $v(\alpha)$ instead of $v_{\mathfrak{p}}(\alpha)$, with the usual convention that $v(0) = +\infty$, and we write $|\ |$ for the $\mathfrak{p}$-adic absolute value $|\ |_{\mathfrak{p}}$. We denote by $\mathbb{Z}_{\mathfrak{p}}$ the ring of $\mathfrak{p}$-adic integers of $K_{\mathfrak{p}}$, i.e., the set of α such that $|\alpha| \leqslant 1$, which is a local ring with maximal ideal $\mathfrak{p}\mathbb{Z}_{\mathfrak{p}}$, the set of α such that $|\alpha| < 1$. We write $\mathbb{F}_{\mathfrak{p}} = \mathbb{Z}_{\mathfrak{p}}/\mathfrak{p}\mathbb{Z}_{\mathfrak{p}} \simeq \mathbb{Z}_K/\mathfrak{p}$ for the residue field of $K_{\mathfrak{p}}$ (and of K), we let $q = |\mathbb{F}_{\mathfrak{p}}| = \mathcal{N}\mathfrak{p} = p^{f(\mathfrak{p}/p)}$ be its cardinality, so that $\mathbb{F}_{\mathfrak{p}} \simeq \mathbb{F}_q$. Finally, we let π be a uniformizer of $\mathfrak{p}$, hence such that $v(\pi) = 1$. Although not really necessary, for the sake of concreteness we normalize the absolute value so that $|\alpha| = q^{-v(\alpha)}$ for all $\alpha \in K_{\mathfrak{p}}$, and in particular $|\pi| = 1/q$.

Although we will only need elliptic curves over $\mathbb{Q}_p$, the proofs are essentially the same in the case of a general $\mathfrak{p}$-adic field. The only added difficulty is that we explicitly need formal groups as introduced in the preceding section, while over $\mathbb{Q}_p$ this can be bypassed; see Exercise 30. If desired, the reader can thus restrict to this case.

As usual with $\mathfrak{p}$-adic questions, everything ultimately boils down to looking at the residue field $\mathbb{F}_\mathfrak{p}$ and performing suitable Hensel liftings. We follow [Cas2], but generalize to the number field case.

Since we work in projective coordinates, it is useful to make the following convention: when (x, y, z) is used to represent projective coordinates for a point P in the projective plane over $K_\mathfrak{p}$, we will always implicitly assume that $\min(v(x), v(y), v(z)) = 0$, in other words that $\max(|x|, |y|, |z|) = 1$. This can always be achieved by dividing (x, y, z) by $\pi^{\min(v(x), v(y), v(z))}$. In particular, one of x, y, and z will be a $\mathfrak{p}$-adic unit, and the projective coordinates of P will now be defined up to multiplication by a $\mathfrak{p}$-adic *unit* and not by an arbitrary element of $K_\mathfrak{p}^*$. These will be called *canonical coordinates* for P, thus defined up to multiplication by a $\mathfrak{p}$-adic unit. If desired we can even assume that one of the coordinates is equal to 1, and that the others are in $\mathbb{Z}_\mathfrak{p}$.

If $x \in \mathbb{Z}_\mathfrak{p}$, we will denote by $\overline{x}$ the class of x in $\mathbb{F}_\mathfrak{p}$. Thanks to the above remark, we can similarly define $\overline{P}$ for a point P in the projective plane over $K_\mathfrak{p}$. Finally, to simplify we will assume that an elliptic curve E is given by a plane homogeneous cubic equation $f(x, y, z) = 0$ (not necessarily in Weierstrass form), the general case being similar. Once again by multiplying by a suitable power of π we may assume that all the coefficients of f are in $\mathbb{Z}_\mathfrak{p}$ and one of them at least is a $\mathfrak{p}$-adic unit (or even equal to 1 if desired). Thus $\overline{f}$ (obtained by reducing modulo $\mathfrak{p}$ all the coefficients of f) is not identically zero, hence defines a curve over $\mathbb{F}_\mathfrak{p}$, which we will denote by $\overline{E}$.

Important Remarks. (1) This is a dangerous although practical abuse of notation: indeed, $\overline{E}$ does not depend only on E, but explicitly on the model chosen to represent E, in other words on the function f (if we chose a plane model).

(2) The curve $\overline{E}$ may be singular, hence not an elliptic curve. In this case, we say that E (or more precisely the model that we have chosen for E) has *bad reduction* (at $\mathfrak{p}$, but here $\mathfrak{p}$ is fixed), otherwise that it has *good reduction*. Note that because of the first remark, E may have bad reduction in a given model, and good reduction in another. To give a specific example, the model $y^2 = x^3 + 16$ of an elliptic curve E has evidently bad reduction at 2, but the change of variable $y = 8Y + 4$, $x = 4X$ gives the model $Y^2 + Y = X^3$, which has good reduction at 2.

(3) Because of the above remarks, it is important to note and is easily shown that there always exists a *minimal model*, which is the one having the lowest possible $\mathfrak{p}$-adic valuation of the discriminant, and there is an explicit algorithm due to Tate for computing it; see Algorithms 7.5.1 and 7.5.2 of [Coh0], and Exercise 28. In particular, if the minimal model has bad reduction, then all models will. Note finally that over a local field min-

imal models always exist, but this is not necessarily the case for *global* minimal models over a global field, which do exist over $\mathbb{Q}$, but not in general over number fields of class number strictly greater than 1.

Clearly if $P \in E(K_{\mathfrak{p}})$, then $\overline{P} \in \overline{E}(\mathbb{F}_{\mathfrak{p}})$, whether E has good reduction or not. The following proposition shows that the converse is partially true.

Proposition 7.3.23. *Let E be an elliptic curve defined over $K_{\mathfrak{p}}$, and let $r \in \overline{E}(\mathbb{F}_{\mathfrak{p}})$ be a nonsingular point. Then there exists $R \in E(K_{\mathfrak{p}})$ such that $r = \overline{R}$.*

Proof. Let E be given by a plane model $f(x, y, z) = 0$ chosen as above, and write $r = \overline{P}$, where $P = (x_0, y_0, z_0)$ is any lift of r to $\mathbb{Z}_{\mathfrak{p}}$. Since r is a nonsingular point of $\overline{E}$ the partial derivatives of $\overline{f}$ do not simultaneously vanish at r, so for instance we may assume that $\frac{\partial \overline{f}}{\partial x}(\overline{x_0}, \overline{y_0}, \overline{z_0}) \neq 0$, in other words that $\frac{\partial f}{\partial x}(x_0, y_0, z_0)$ is a $\mathfrak{p}$-adic unit. Since $|f(x_0, y_0, z_0)| < 1$ by assumption, we may apply Hensel's lemma (Proposition 4.1.37) and conclude that there exists $x \in \mathbb{Z}_{\mathfrak{p}}$ such that $f(x, y_0, z_0) = 0$ and $\overline{x} = \overline{x_0}$, finishing the proof. $\square$

If r is a singular point, then a point R such that $r = \overline{R}$ may or may not exist; see Exercise 29.

If $f(x, y, z) = 0$ is the equation of an irreducible nonsingular cubic curve over $K_{\mathfrak{p}}$, it is possible that $\overline{f}(x, y, z)$ becomes reducible, as the example $f(x, y, z) = x^3 + y^3 + 3xz^2$ and $p = 3$ shows. However, if f is a (possibly generalized) Weierstrass equation, Proposition 7.1.1 tells us that this cannot happen.

7.3.7 The $\mathfrak{p}$-adic Filtration for Elliptic Curves over $K_{\mathfrak{p}}$

In this section we let E be an elliptic curve over $K_{\mathfrak{p}}$ given by a generalized Weierstrass equation $y^2 z + a_1 xyz + a_3 yz^2 = x^3 + a_2 x^2 z + a_4 xz^2 + a_6 z^3$, where without loss of generality we may assume that $a_i \in \mathbb{Z}_{\mathfrak{p}}$. We have seen above that $\overline{E}$, which may be singular, is at least absolutely irreducible over $\mathbb{F}_{\mathfrak{p}}$ by Proposition 7.1.1. We let $G = E(K_{\mathfrak{p}})$ be the set of points of E defined over $K_{\mathfrak{p}}$, as usual represented by projective coordinates (x, y, z) in canonical coordinates. Although the curve $\overline{E}$ may have a singular point, by Proposition 7.1.5 the set $\overline{G}_0$ of nonsingular points of $\overline{E}(\mathbb{F}_{\mathfrak{p}})$ (which may be all of $\overline{E}(\mathbb{F}_{\mathfrak{p}})$ if E has good reduction) does have a natural group structure. If we denote by G_0 the set of elements of G that reduce modulo $\mathfrak{p}$ to a nonsingular point, we have the following easy result:

Proposition 7.3.24. *The group G_0 is a subgroup of G of finite index, and the natural reduction map modulo $\mathfrak{p}$ induces a group homomorphism from G_0 to $\overline{G}_0 \subset \overline{E}(\mathbb{F}_{\mathfrak{p}})$. In particular, if $\overline{E}$ is nonsingular, it induces a group homomorphism from $E(K_{\mathfrak{p}})$ to $\overline{E}(\mathbb{F}_{\mathfrak{p}})$.*

Proof. Left to the reader (Exercise 7).   □

Definition 7.3.25. *The index* $[G : G_0]$ *is called the* Tamagawa number *of* E *(or of* $E/K_{\mathfrak{p}}$*).*

It is clear that $[G : G_0] = 1$ if $\overline{E}$ is nonsingular, and more generally $[G : G_0]$ can easily be computed using Tate's algorithm; see for instance Algorithms 7.5.1 and 7.5.2 of [Coh0].

By Proposition 7.3.23 we know that the natural reduction map $^{-}$ from G_0 to $\overline{G}_0$ is surjective. Since $\overline{\mathcal{O}}$ is not a singular point (otherwise $\overline{G}_0$ could not be a group!), we can consider the kernel G_1 of the reduction map $^{-}$, in other words the group of elements of G_0 (or of G) that map to $\overline{\mathcal{O}}$.

Lemma 7.3.26. *A point* $P = (x, y, z) \neq \mathcal{O}$ *in canonical coordinates is in the kernel of the reduction if and only if there exists an integer* $N \geqslant 1$ *such that* $v(x) = N$, $v(y) = 0$, *and* $v(z) = 3N$, *or equivalently,* $|x| = q^{-N}$, $|y| = 1$, *and* $|z| = q^{-3N}$.

Proof. Since $\overline{O} = (0, 1, 0)$, in canonical coordinates P is in the kernel of reduction modulo $\mathfrak{p}$ if and only if $|x| < 1$, $|z| < 1$, and $|y| = 1$. It follows from the non-Archimedean inequality and $|a_i| \leqslant 1$ that

$$|y^2 z - x^3| = |a_2 x^2 z + a_4 x z^2 + a_6 z^3 - (a_1 x y z + a_3 y z^2)| < |z|,$$

and since $|y^2 z| = |z|$ we have $|x^3| = |y^2 z - (y^2 z - x^3)| = |z|$, proving the lemma since $|x| = q^{-v(x)} < 1$.   □

Definition 7.3.27. *The integer* $N \geqslant 1$ *as above will be called the* level *of the point* $P \neq \mathcal{O}$. *We define the level of* $\mathcal{O}$ *to be* ∞ *(note that* $q^{-\infty} = 0$*), and the level of a point of* G_0 *that is not in the kernel of reduction to be* 0 *(note that in that case we have* $|z| = 1$ *but not necessarily* $|x| = |y| = 1$*). Finally, we define* G_N *to be the set of points of level greater than or equal to* N.

Lemma 7.3.28. (1) *The sets* G_N *are groups such that* $G \supset G_0 \supset G_1 \supset \cdots G_N \supset \cdots$ *and we have* $\bigcap_N G_N = \{\mathcal{O}\}$.
(2) *The quotient* G_0/G_1 *is isomorphic to the group* $\overline{G}_0$ *of nonsingular points of* $\overline{E}(\mathbb{F}_{\mathfrak{p}})$.
(3) *For* $N \geqslant 1$ *the quotient* G_N/G_{N+1} *is isomorphic to the additive group* $\mathbb{F}_{\mathfrak{p}}$, *hence to* $(\mathbb{Z}/p\mathbb{Z})^{f(\mathfrak{p}/p)}$. *In particular, if* $\mathfrak{p}$ *is a prime ideal of degree 1 then* G_N/G_{N+1} *is cyclic of order* p.

Proof. For $N \geqslant 1$ make the change of variables $x = \pi^N X$, $y = Y$, and $z = \pi^{3N} Z$ in the Weierstrass equation. We obtain a new curve E_N with equation

$$Y^2 Z + a_1 \pi^N X Y Z + a_3 \pi^{3N} Y Z^2 = X^3 + a_2 \pi^{2N} X^2 Z + a_4 \pi^{4N} X Z^2 + a_6 \pi^{6N} Z^3 \, ;$$

hence the reduced curve $\overline{E_N}$ has the simple equation $Y^2Z = X^3$ with singular point $(0,0,1)$. A point (X,Y,Z) maps into the singular point of E_N if and only if $\min(v(X), v(Y)) > v(Z)$, hence transferring back into the initial coordinates, if and only if $\min(v(x) - N, v(y)) > v(z) - 3N$. If (x,y,z) is not in the kernel of reduction we have $v(z) = 0$, hence $\min(v(x) - N, v(y)) \geqslant \min(-N, 0) = -N > -3N$. If (x,y,z) is in the kernel of reduction with a certain level $M \geqslant 1$, then by Lemma 7.3.26 the condition $\min(v(x)-N, v(y)) > v(z)-3N$ can be rewritten $\min(M-N, 0) > 3(M-N)$, which is true if and only if $M < N$. To summarize, the points that map into the singular point of $\overline{E_N}$ are exactly those with level strictly less than N.

It follows that G_N is the set of points that map into the nonsingular points of $\overline{E_N}(\mathbb{F}_{\mathfrak{p}})$, and the reduction map is surjective by Proposition 7.3.23. Thus G_N is a group. Furthermore, a point is in the kernel of reduction for E_N if and only if $v(Z) > 0$, in other words $v(z) > 3N$, hence by Lemma 7.3.26 once again if and only if its level is strictly greater than N, i.e., if and only if it belongs to G_{N+1}. It follows that G_N/G_{N+1} is isomorphic to the group of nonsingular points of $\overline{E_N}(\mathbb{F}_{\mathfrak{p}})$, which is isomorphic to $\mathbb{F}_{\mathfrak{p}}$ by Proposition 7.1.5 since $\overline{E_N}$ has a cusp. The other statements of the lemma are clear (note that G_1 is by definition the kernel of reduction on the curve E). $\square$

The above sequence of groups G_i is called the $\mathfrak{p}$-*adic filtration* associated with E. It is reminiscent of the filtration given by upper ramification groups in algebraic number theory, which has similar but slightly more complicated properties.

The aim of this section is to study the torsion subgroup of G_N, and in particular to prove that G_N is torsion-free for N sufficiently large (see Theorem 7.3.31). Note first the following consequence of the above lemma (where we recall that p is the characteristic of $\mathbb{F}_{\mathfrak{p}} = \mathbb{F}_q$).

Corollary 7.3.29. *If $P \in G_1$ is a torsion point, its order is a power of p.*

Proof. Assume that $kP = \mathcal{O}$ with $k \neq 0$, write $k = p^u m$ with $p \nmid m$, and consider $Q = p^u P$. If $Q = \mathcal{O}$ then the order of P divides p^u so is a power of p. Otherwise, by Lemma 7.3.28 there exists a unique $N \geqslant 1$ such that $Q \in G_N$ and $Q \notin G_{N+1}$. But then $mQ = kP = \mathcal{O} \in G_{N+1}$, and since G_N/G_{N+1} is a power of $\mathbb{Z}/p\mathbb{Z}$ we have $pG_N \subset G_{N+1}$, hence $pQ \in G_{N+1}$. Since p and m are coprime, it follows that $Q \in G_{N+1}$, a contradiction that proves the corollary. $\square$

To go further, we will now use the machinery of formal groups that we developed in Section 7.3.5, for the case in which the base ring R is the ring $\mathbb{Z}_{\mathfrak{p}}$ of $\mathfrak{p}$-adic integers. Recall in particular that we proved the existence of a formal power series $s(T) = 1 + O(T) \in \mathbb{Z}_{\mathfrak{p}}[[T]]$ such that $(x(T), y(T))$ is a formal affine point on the curve, where $x(T) = s(T)/T^2$ and $y(T) = -s(T)/T^3$.

Proposition 7.3.30. *For all $t \in \mathfrak{p}\mathbb{Z}_\mathfrak{p}$ set $P(t) = (ts(t), -s(t), t^3) \in \mathbb{P}^2(K_\mathfrak{p})$, where $s(T) \in \mathbb{Z}_\mathfrak{p}[[T]]$ is as above. Then G_N is the group of points $P(t)$ for $t \in \mathfrak{p}^N\mathbb{Z}_\mathfrak{p}$ together with the group law $P(t_1) \oplus P(t_2) = P(F(t_1, t_2))$, identity element $P(0) = (0, -1, 0)$, and inverse $\ominus P(t) = P(i(t))$.*

Proof. Note first that since $|t| < 1$ and $s(T) \in \mathbb{Z}_\mathfrak{p}[[T]]$ the series $s(t)$ converges, and more precisely $s(t) - 1 \in \mathfrak{p}\mathbb{Z}_\mathfrak{p}$, which also shows that $P(t)$ is in canonical coordinates since $s(t)$ is a $\mathfrak{p}$-adic unit. Since $(s(T)/T^2, -s(T)/T^3)$ is a formal point on E, it follows that $P(t) \in E(K_\mathfrak{p})$, and its level is equal to $v(t)$ since $s(t)$ is a unit. Conversely, it is clear that if $(x, y, z) \in G_N$, then if we set $t = -x/y$ we have $t \in \mathfrak{p}^N\mathbb{Z}_\mathfrak{p}$ and $P(t) = (x, y, z)$, proving that G_N is indeed equal to the set of points $P(t)$ for $t \in \mathfrak{p}^N\mathbb{Z}_\mathfrak{p}$. Now thanks to the crucial fact that the formal group F associated with E satisfies $F \in \mathbb{Z}_\mathfrak{p}[[T_1, T_2]]$ the the double series $F(t_1, t_2)$ will converge in $\mathbb{Z}_\mathfrak{p}$ when $|t_1| < 1$ and $|t_2| < 1$. The defining properties of a formal group immediately imply that $\oplus$ is a group law with the given identity and inverse. Since by definition of F we have $P(t_1) + P(t_2) = P(F(t_1, t_2))$, where here the $+$ sign is addition on the elliptic curve, it follows that the group law $\oplus$ and the addition law on the elliptic curve coincide, proving the proposition. $\qquad\square$

Note that this proposition implies that G_N/G_{N+1} is isomorphic to $\mathfrak{p}^N\mathbb{Z}_\mathfrak{p}/\mathfrak{p}^{N+1}\mathbb{Z}_\mathfrak{p}$, which is Lemma 7.3.28 (2). We have preferred proving that lemma directly since it does not need the machinery of formal groups, which we now use to bound the torsion in G_N. Recall from Corollary 7.3.29 that if P is a torsion point in G_1, its order must be a power of p.

Theorem 7.3.31. *Let $N \geqslant 1$. If $P \in G_N$ is a torsion point of order p^k with $k \geqslant 1$ then*

$$p^k \leqslant \frac{pe(\mathfrak{p}/p)}{(p-1)N}, \quad \text{or equivalently,} \quad N \leqslant \frac{e(\mathfrak{p}/p)}{p^k - p^{k-1}}.$$

Proof. We may clearly assume that P has exact order k. We prove the result by induction on $k \geqslant 1$. By the above proposition we let $P = P(t)$ with $t \neq 0$, and assume first that $pP(t) = \mathcal{O}$, or equivalently by the proposition, that $[p]t = 0$. By Proposition 7.3.22 we deduce that there exist $f(T) = T + O(T^2) \in \mathbb{Z}_\mathfrak{p}[[T]]$ and $g(T) = aT + O(T^2) \in \mathbb{Z}_\mathfrak{p}[[T]]$ such that $pf(t) + g(t^p) = 0$. Since $v(f(t)) = v(t)$ and $v(p) = e(\mathfrak{p}/p)$ we have

$$e(\mathfrak{p}/p) + v(t) = v(pf(t)) = v(g(t^p)) \geqslant v(t^p) \geqslant pv(t),$$

hence $v(t) \leqslant e(\mathfrak{p}/p)/(p-1)$, proving the result for $k = 1$ since the level of P is equal to $v(t)$. Assume now the result true for $k \geqslant 1$, and let $P = P(t)$ have exact order p^{k+1}, so that $pP(t) = P([p]t)$ has exact order p^k. By the induction hypothesis and Proposition 7.3.22 once again we have

$$e(\mathfrak{p}/p)/(p^k - p^{k-1}) \geqslant v([p]t) = v(pf(t) + g(t^p)) \geqslant \min(v(pf(t)), g(t^p))$$
$$\geqslant \min(v(pt), v(t^p)),$$

since $f(T) = T + O(T^2)$ and $g(T) = aT + O(T^2)$. However, since $k \geqslant 1$ and $v(t) \geqslant 1$ we have $v(pt) \geqslant e(\mathfrak{p}/p) + 1 > e(\mathfrak{p}/p) \geqslant e(\mathfrak{p}/p)/(p^k - p^{k-1})$, so the minimum cannot be equal to $v(pt)$. It follows that $e(\mathfrak{p}/p)/(p^k - p^{k-1}) \geqslant v(t^p) = pv(t)$, proving our induction hypothesis. $\square$

We now specialize to $\mathbb{Q}$. For this, first note the following trivial lemma:

Lemma 7.3.32. *Let G be an abelian group and G_t its torsion subgroup. If H is a torsion-free subgroup of G the natural map from G_t to G/H is injective.*

Proof. Indeed, if $g \in G_t$ maps to the class of 0 in G/H, in other words to H, we have $g \in G_t \cap H = \{0\}$ since H is torsion-free, so $g = 0$. $\square$

Corollary 7.3.33. *Assume that $K = \mathbb{Q}$, so that $\mathfrak{p} = p\mathbb{Z}$.*

(1) *If $p > 2$ or a_1 is even the group G_1 is torsion-free.*
(2) *If $p = 2$ and a_1 is odd the group G_2 is torsion-free, and the group G_1 is either torsion-free or isomorphic to $G_2 \times (\mathbb{Z}/2\mathbb{Z})$.*
(3) *If (X, Y) are the affine coordinates of a torsion point of $E(\mathbb{Q}_p)$ different from $\mathcal{O}$ then either X and Y are in $\mathbb{Z}_p$, or else $p = 2$, a_1 is odd, $v_2(X) = -2$, and $v_2(Y) = -3$, this last case occurring for at most a single torsion point of order 2.*

Proof. (1) and (2). Since $e(\mathfrak{p}/p) = 1$ for all p we have $e(\mathfrak{p}/p)/(p^k - p^{k-1}) < 1$ for all $p^k \geqslant 3$, and $e(\mathfrak{p}/p)/(p^k - p^{k-1}) = 1$ for $p^k = 2$. Thus G_2 is always torsion-free, and G_1 is also torsion-free for $p > 2$, and for $p = 2$ its only possible torsion is 2-torsion. Let $P = (x, y, z) \in G_1$ be a 2-torsion point in canonical coordinates. We thus have $a_1 x = -2y - a_3 z$, so when $p = 2$ we have $v(-2y) = 1$ and $v(-a_3 z) \geqslant 3$, hence $v(a_1 x) = 1$, so $v(a_1) \leqslant 0$ since $v(x) \geqslant 1$, showing that a 2-torsion point can exist only if a_1 is odd.

By Lemma 7.3.28 we have $G_1/G_2 \simeq \mathbb{Z}/2\mathbb{Z}$. Denote temporarily by G_t the torsion subgroup of G_1. By Lemma 7.3.32 the natural map from G_t to G_1/G_2 is injective, so its image is either equal to the trivial group, in which case $G_t = \{\mathcal{O}\}$ so G_1 is already torsion-free, or there exists a nonzero $P_0 \in G_1$ such that $2P_0 = \mathcal{O}$, and we clearly have $G_1 \simeq G_2 \times \{0, P_0\}$. In particular, we see that P_0 is the only torsion point of G_1 different from $\mathcal{O}$.

(3). By definition elements of $G \setminus G_1$ are characterized in canonical coordinates by $v(z) = 0$, and those of $G_1 \setminus G_2$ by $v(x) = 1$, $v(y) = 0$, and $v(z) = 3$, so the result follows since $X = x/z$ and $Y = y/z$. $\square$

Remark. When $p = 2$ and a_1 is odd the group G_1 may indeed have 2-torsion as the following example shows. Consider the curve with equation $y^2 z + xyz = x^3 + 4x^2 z + xz^2$. It is easily checked that it is nonsingular. Completing the square and passing to affine coordinates we see that the 2-torsion points are such that $X = x/z$ is a root of $X^3 + 17X^2/4 + X = 0$. The three roots of this equation are rational, and it is clear that $(x, y, z) = (-2, 1, 8)$, which corresponds to $X = -1/4$, is a 2-torsion point that belongs to G_1.

In the same way, it is easy to work out the corresponding results for number fields other than $\mathbb{Q}$. For instance, we leave as an exercise to the reader the proof of the following proposition (Exercise 31).

Proposition 7.3.34. *Let K be a quadratic field of discriminant D, let $\mathfrak{p}$ be a prime ideal of K, and let p be the prime number below $\mathfrak{p}$.*

(1) *If $p \nmid 2D$ then G_1 is torsion-free.*

(2) *If $p = 2$ and D is odd then G_2 is torsion-free, and G_1 is either torsion-free or is isomorphic to $G_2 \times \mathbb{Z}/2\mathbb{Z}$.*

(3) *If $p \mid D$ and $p > 3$ then G_1 is torsion-free.*

(4) *If $p = 3$ and $p \mid D$ then G_2 is torsion-free, and G_1 is either torsion-free or is isomorphic to $G_2 \times \mathbb{Z}/3\mathbb{Z}$.*

(5) *If $p = 2$ and $p \mid D$ then G_3 is torsion-free, and G_1 is either torsion-free or isomorphic to $G_2 \times \mathbb{Z}/2\mathbb{Z}$, $G_3 \times \mathbb{Z}/4\mathbb{Z}$, or $G_3 \times \mathbb{Z}/2\mathbb{Z} \times \mathbb{Z}/2\mathbb{Z}$.*

We finish this section with the following result.

Proposition 7.3.35. *Let again $K = \mathbb{Q}$ and $\mathfrak{p} = p\mathbb{Z}$, and assume that $p \nmid \operatorname{disc}(E)$. The torsion subgroup G_t of $E(\mathbb{Q}_p)$ is either isomorphic to a subgroup of $\overline{E}(\mathbb{F}_p)$ or is such that $G_t/\{0, P_0\}$ is isomorphic to a subgroup of $\overline{E}(\mathbb{F}_p)$ for the element P_0 of order 2 of G_1, this last case possibly occurring only when $p = 2$ and a_1 is odd.*

Proof. First note that by definition of the discriminant $p \nmid \operatorname{disc}(E)$ means that the reduced curve $\overline{E}$ is nonsingular (this is true even for $p = 2$), in other words that $G = E(\mathbb{Q}_p) = G_0$. By Proposition 7.3.28, since $\overline{E}$ is nonsingular we have $G_0/G_1 \simeq \overline{E}(\mathbb{F}_p)$; hence when G_1 is torsion-free (in particular when $p > 2$ or a_1 is even) we deduce from Lemma 7.3.32 that the torsion subgroup of $G_0 = E(K_{\mathfrak{p}})$ is isomorphic to a subgroup of $\overline{E}(\mathbb{F}_p)$. Otherwise, the same proof as that lemma shows that $G_t/\{0, P_0\}$ is isomorphic to a subgroup of $\overline{E}(\mathbb{F}_p)$, where P_0 is the unique element of order 2 in G_1. $\square$

7.4 Exercises for Chapter 7

1.

(a) Let E be an elliptic curve over some field K of characteristic 2. Show that there exists an extension L of K of degree 24 such that E is L-isomorphic to a curve with equation $y^2 + axy + y = x^3$ for some $a \in L$.

(b) Show that this curve is supersingular (see Proposition 7.3.13) if and only if $a = 0$, so that there exists exactly one supersingular elliptic curve up to L-isomorphism (but not up to K-isomorphism, see the next exercise).

2. Study Proposition 2.4.11 and solve Exercise 27 of Chapter 2 before solving this exercise.

(a) Using Proposition 7.3.13, show that over a field K of characteristic 2 any nonsupersingular elliptic curve has an equation of the form $y^2 + xy = x^3 + ax^2 + b$ with $b \neq 0$, and any supersingular elliptic curve has an equation of the form $y^2 + a_3 y = x^3 + ax + b$ with $a_3 \neq 0$, in both cases with coefficients in K.

(b) In the nonsupersingular case show that a can be chosen modulo the Artin–Schreier $\mathbb{F}_2$-subspace A of elements of K of the form $c^2 - c$ (see Proposition 2.4.11), and in particular that when K is finite we can choose $a = 0$ or $a = a_0$ a fixed element such that $a_0 \notin A$ (or, equivalently, $\mathrm{Tr}_{K/\mathbb{F}_2}(a_0) = 1$).

(c) In the supersingular case, show that a_3 can be chosen modulo the Kummer multiplicative subgroup K^{*3} of K^*.

(d) Assume that $K = \mathbb{F}_{2^n}$ is a finite field with n odd. Deduce that any elliptic curve over K has a unique equation of the form $y^2 + xy = x^3 + ax^2 + b$ with $a \in \mathbb{F}_2$ and $b \neq 0$, or of the form $y^2 + y = x^3 + ax + b$ with a and $b \in \mathbb{F}_2$ and $(a, b) \neq (0, 1)$. In particular, there are exactly 3 supersingular elliptic curves over K up to K-isomorphism.

(e) Assume now that $K = \mathbb{F}_{2^n}$ with n even, let g be a generator of K^*, let $\rho = g^{(2^n - 1)/3}$ be a primitive cube root of 1 in $\mathbb{F}_4 \subset K$, and finally let $c \in K$ be such that $\mathrm{Tr}_{K/\mathbb{F}_4}(c) = 1$, which exists by Proposition 2.4.11. Show that a supersingular elliptic curve over K has a unique equation of the form $y^2 + g^k y = x^3 + acx + b\rho g^{2k}$ with $0 \leqslant k \leqslant 2$ and $b \in \mathbb{F}_2$, and $a = 0$ when $k \neq 0$, or $a \in \mathbb{F}_2$ when $k = 0$, so that up to K-isomorphism there exist 8 supersingular elliptic curves over K.

3.

(a) Show that over a field K of characteristic 3 any nonsupersingular elliptic curve has an equation of the form $y^2 = x^3 + ax^2 + b$ with $ab \neq 0$, and any supersingular elliptic curve has an equation of the form $y^2 = x^3 + ax + b$ with $a \neq 0$, in both cases with coefficients in K.

(b) Assume that $K = \mathbb{F}_{3^n}$. Similarly to the previous exercise, give unique equations for elliptic curves over K up to K-isomorphism, and in particular compute the number of nonisomorphic supersingular curves.

4. Show that the curve $y^2 z = x^3 + T z^3$ defined over $K = \mathbb{F}_2[T]$ does not have any singular points defined over K, but that it has a singular point over a finite extension of K.

5. Prove the equations given in the text for point addition on an elliptic curve given in (general) Weierstrass form.

6. Prove that three points P_1, P_2, and P_3 on an elliptic curve are collinear (in other words are such that $P_1 + P_2 + P_3 = \mathcal{O}$) if and only if there exists a rational function of x and y that has a simple zero at P_1, P_2, and P_3, and a triple pole at $\mathcal{O}$ (i.e., at infinity), and no zeros or poles elsewhere. Deduce from this that the addition law on an elliptic curve is associative.

7. Using a similar method to that of the previous exercise, prove Proposition 7.3.24.

8. Check the formulas of Proposition 7.2.1.

9. Write the detailed algorithm corresponding to Section 7.2.4 (see Algorithm 7.4.10 of [Coh0], but beware that there are several misprints in that algorithm, essentially in the formulas for the inverse coordinate transformations).

10.

(a) Let a, b, and c be nonzero elements in a field of characteristic 0. If $P_i = (x_i, y_i)$ for $i = 1$ and 2 are points on the cubic $ax^3 + by^3 + c = 0$, compute by the chord and tangent method a third point on the cubic (since there is no point chosen as the origin we cannot give the group law).

(b) As a special case, give the group law on the cubic $x^3 + y^3 + c = 0$, where the origin is taken as the unique point at infinity with projective coordinates $(1 : -1 : 0)$.

(c) Let $u \neq 0$, set $f(x) = u^2 x^4 + a x^3 + b x^2 + c x + d$, and assume that f has no multiple roots. We know that the curve $y^2 = f(x)$ defines an elliptic curve, the origin being the point at infinity $(1 : u : 0)$. Write explicitly the group law on this elliptic curve. (Hint: thanks to Section 7.2.3, show that one must use parabolas $y = u x^2 + m x + n$ going through the two points.)

11. Prove Proposition 7.2.4, and give explicitly the birational transformations from one curve to the other.

12. In the remarks following Proposition 7.2.4 we have seen that there is a very simple birational map from $x^3 + y^3 = c$ to the elliptic curve $Y^2 = X^3 + 16c^2$. Show that the inverse map is as follows (try to do the computation without looking at the result):

$$x = \frac{36cX^3 + 8Y^3 - 9YX^3}{6X(4Y^2 - 3X^3)}, \quad y = \frac{36cX^3 - 8Y^3 + 9YX^3}{6X(4Y^2 - 3X^3)}.$$

13. Using the method explained in the text for constructing Fermat triangles, show that there exists an infinity of essentially distinct Fermat triangles.

14. Using the algebraic differential equation of the $\wp$-function and the addition formulas for elliptic curves, write explicitly the addition formula for the $\wp$-functions giving $\wp(z_1 + z_2)$ in terms of $\wp(z_1)$, $\wp(z_2)$, $\wp'(z_1)$, and $\wp'(z_2)$.

15. Let E be an elliptic curve defined over $\mathbb{R}$ by a reduced Weierstrass equation.

(a) Show algebraically that the map from $E(\mathbb{R})$ to $\{\pm 1\}$ sending P to 1 if $P \in E^0$ and to -1 if $P \in E^{gg}$ is a group homomorphism.

(b) Show algebraically that E^0 is the image of the map $P \mapsto 2P$ from $E(\mathbb{R})$ to itself.

16. Let E be an elliptic curve defined over $\mathbb{R}$ defined by an equation $Y^2 = f(X)$ with $f(X) = 4X^3 - g_2 X - g_3$, and assume that $\operatorname{disc}(E) > 0$. By making a change of variable of the form $t = (e_1 t_1 + b)/(t_1 - e_1)$ for a suitable value of b show that

$$\int_{e_1}^{e_2} \frac{dt}{\sqrt{f(t)}} = \int_{e_3}^{\infty} \frac{dt}{\sqrt{f(t)}},$$

and similarly show that

$$\int_{e_2}^{e_3} \frac{dt}{\sqrt{-f(t)}} = \int_{-\infty}^{e_1} \frac{dt}{\sqrt{-f(t)}}.$$

17. Let R be a principal ideal domain with field of fractions K, let $g \geq 0$ be an integer, and let $f(X) \in R[X]$ and $h(X) \in R[X]$ be two polynomials with coefficients in R such that $\deg(f) = 2g + 1$ and $\deg(h) \leq g$. Consider the *hyperelliptic curve* C given by the affine equation $Y^2 + h(X)Y = f(X)$. Generalizing Proposition 7.3.1, show that if $(X, Y) \in C(K)$ there exist M, N, and D in R such that

$$X = \frac{M}{D^2}, \quad Y = \frac{N}{D^{2g+1}}, \quad \gcd(M, D) = \gcd(N, D) = 1.$$

18. This exercise generalizes Proposition 7.3.2. Let E be an elliptic curve defined over a commutative field k of characteristic different from 2 by an equation $y^2 = x^3 + ax^2 + bx + c$, and consider the quadratic twist E_g with equation $g(T)Y^2 = X^3 + aX^2 + bX + c$, where g is a polynomial with no multiple roots.

(a) Show that if $\deg(g) \leqslant 2$ then $X'(T) = 0$.

(b) Deduce that if $\deg(g) = 1$ or 2 then $X(T) = u \in k$ with $u^3 + au^2 + bu + c = 0$, so that $Y(T) = 0$ if $g(u) \neq 0$.

(c) Assume that $\deg(g) = 3$. Show that either $X'(T) = 0$, or $z(X) = \pm 1$, or $z(X) = 0$ and $f(\ell(X)) = 0$, where $z(X)$ and $\ell(X)$ are as in Lemma 7.3.5.

(d) Show that with the notation of the proof of Theorem 7.3.10, for all $n \neq 0$ we have $z(Q + nP) = 1$ and $\ell(Q + nP) = 1/n^2$.

19. This exercise is given as a preparatory result for the next one. Let k be an algebraically closed field, and let A and B be nonconstant coprime polynomials in $k[T]$. Consider the set of polynomials $S = \{aA + bB, \ a, b \in k\}$. Since A and B are nonconstant and coprime, these polynomials are all distinct.

(a) Assume that S contains at least four polynomials that are squares, and that are not constant multiples of one another. Show that there exist polynomials F and G and nonzero constants c and d such that $F^2 - cG^2$ and $F^2 - dG^2$ are squares of polynomials that are not constant multiples of one another.

(b) By writing $F^2 - cG^2 = (F - c^{1/2}G)(F + c^{1/2}G)$, and similarly for the other, prove that $S_1 = \{aF + bG, \ a, b \in k\}$ has the same property as S, in other words that S_1 contains at least four polynomials that are squares, and that are not constant multiples of one another.

(c) Deduce a contradiction, in other words that such a set S cannot exist.

(d) Show that this is still true when k is not algebraically closed.

20. The goal of this exercise is to give an alternative proof of Proposition 7.3.2. Let k be an algebraically closed field and let E be an elliptic curve given by a Weierstrass equation $Y^2 = f(X) = X^3 + aX^2 + bX + c$. Set $K = k(T)$ and let $P = (X, Y) \in E(K)$, so that $X = M/D^2$, $Y = N/D^3$ with $\gcd(M, D) = \gcd(N, D) = 1$ by Proposition 7.3.1. Let e_i be the roots in k of $f(X) = 0$.

(a) Show that $M(T) - e_i D(T)^2$ is the square of a polynomial, and deduce that $S = \{aM + bD^2\}$ satisfies the properties of the preceding exercise, in other words that it contains at least four polynomials that are squares, and that are not constant multiples of one another.

(b) Deduce from this that X and Y are constant, hence that $E(K) = E(k)$.

(c) Show that this is still true when k is not algebraically closed. (Hint: $k(T) \cap \overline{k} = k$.)

21. Let G be an abelian group, let h be a map from G to a field K of characteristic different from 2, and define the map B from $G \times G$ to K by $B(P, Q) = (h(P + Q) - h(P) - h(Q))/2$. Show that B is a bilinear form (i.e., $B(P + Q, R) = B(P, R) + B(Q, R)$) if and only if for all P, Q in G we have $h(P+Q)+h(P-Q) = 2(h(P) + h(Q))$ (this is not as easy as it looks).

22. Let k be a commutative field of characteristic different from 2 and 3, let E be the elliptic curve defined over $K = k(T)$ by the Weierstrass equation $Y^2 = X^3 - (T^3 - T^2)$, and let $P = (T, T) \in E(K)$ so that $h(P) = 1$.

(a) Show that $h(2P) = 2$, hence that h is not a quadratic form on $E(K)$.

(b) For any $Q = (X_Q, Y_Q) \in E(K)$ different from $\mathcal{O}$ write as usual $X_Q = N_Q/D_Q$ with N_Q and D_Q coprime polynomials. Assume that Q is such that $N_Q(T) = TM_Q(T)$, where $M_Q(T)$ is a polynomial coprime to $T(T - 1)$. Prove that $N_{2Q}(T) = N_Q(T)(TM_Q(T)^3 + 8(T - 1)D_Q(T))$ and $D_{2Q}(T) = 4D_Q(T)(TM_Q(T)^3 - (T-1)D_Q(T)^3$, and deduce that we also have $N_{2Q}(T) = TM_{2Q}(T)$, where $M_{2Q}(T)$ is a polynomial coprime to $T(T - 1)$.

(c) Deduce from the preceding question that $\deg(N_{2^n P}) = (4^n + 2)/3$ and $\deg(D_{2^n P}) = (4^n - 4)/3$, hence that $h(2^n P) = (4^n + 2)/3$, so that the "canonical height" of P, defined as $\lim_{n\to\infty}(h(2^n P)/4^n)$, is equal to $1/3$.

23. Finish the proof of Lemma 7.3.5 and of Theorem 7.3.3 by filling in the easy cases not done in the text.

24. Prove Lemmas 7.3.6 and 7.3.7.

25. Prove Hasse's Theorem 7.3.10 in characteristic 2 using the indications given in the text.

26. Show by induction or by a Hensel-type argument that for any formal power series $f(T) = T + O(T^2)$ there exists a unique formal power series $g(T) = T + O(T^2)$ such that $f(g(T)) = g(f(T)) = T$.

27. Let $g(T)$ be the formal power series defined by the formula $g(T) = x'(T)/(2y(T) + a_1 x(T) + a_3)$. Prove in any order the following two statements:

(a) The derivative of the formal logarithm $L(T)$ associated with an elliptic curve satisfies $L'(T) = f(T)$.
(b) We have $g(F(T_1, T_2)) = g(T_1) + g(T_2)$, where as usual $F(T_1, T_2)$ is the formal group associated with the elliptic curve.

28. Let E be an elliptic curve defined over $\mathbb{Q}_p$ with $p \geqslant 5$, so that we may assume that E is given by the simple Weierstrass equation $y^2 = x^3 + ax + b$ with a and b in $\mathbb{Z}_p$. Show that if $p^4 \mid a$ and $p^6 \mid b$ this model is not minimal, and that conversely if $p^4 \nmid a$ or $p^6 \nmid b$ the model is minimal. Deduce an algorithm for finding a minimal model over $\mathbb{Q}_p$ for $p \geqslant 5$.

29. Give examples of (models of) elliptic curves over $\mathbb{Q}_p$ for which the singular point r of the reduction modulo p has or does not have a lift R to the curve (i.e., such that $r = \overline{R}$), both when the reduction has a cusp and when it has a double point with distinct tangents.

30. The aim of this exercise is to prove Corollary 7.3.33 without using formal groups. Let E be an elliptic curve given by a generalized Weierstrass equation with coefficients in $\mathbb{Z}_p$. We assume either that $p > 2$ or that $p = 2$ and a_i is even. If $(x, y, z) \in G_1$ is not equal to $\mathcal{O}$ define $u(P) = x/(y + (a_1/2)x + (a_3/2)z)$.

(a) Show that $u(P)$ is well defined, and that $v(u(P)) = N$, where N is the level of P.
(b) Show that if P_1 and P_2 are in G_1 then

$$|u(P_1 + P_2) - u(P_1) - u(P_2)| \leqslant \max(|u(P_1)|^3, |u(P_2)|^3)$$

(this is the longest part of the exercise: you will evidently have to use the straight line going through P_1 and P_2).
(c) Deduce by induction that $|u(kP) - ku(P)| \leqslant |u(P)|^3$ and $|u(kP)| \leqslant |u(P)|$ for all $k \in \mathbb{Z}_{\geqslant 0}$. Then deduce by a different induction that in fact $|u(kP)| = |k||u(P)|$.
(d) Deduce that G_1 is torsion-free.
(e) By replacing E by a different curve, deduce without further computation that when $p = 2$ and a_1 is odd, G_2 is torsion-free.

31. Prove Proposition 7.3.34, and find the analogue of Corollary 7.3.33 (3).

8. Diophantine Aspects of Elliptic Curves

8.1 Elliptic Curves over $\mathbb{Q}$

8.1.1 Introduction

The Hasse–Minkowski theorem implies that the existence of rational points on a curve of genus 0 can be decided by local arguments, and then the rational points have a parametrization in terms of rational functions of a single parameter $t \in \mathbb{P}_1(\mathbb{Q})$, or equivalently a pair of coprime integers (see for example Corollary 6.3.6).

On the other hand, the parametrization of the group of rational points on an elliptic curve is of a more difficult kind, and we have already seen several examples in which the local–global principles fail. Since it is the simplest case after curves of genus 0, the Diophantine aspects of elliptic curves have been extensively studied, and even though they are far from being solved, several powerful techniques have been developed. Many proofs are quite difficult and involved, so some of them will be omitted.

There are two main questions, and correspondingly two main theorems about Diophantine aspects of elliptic curves. The first one is the existence and structure of the set of *rational* solutions. The answer to this is that this set is an abelian group (in essence this is due to Fermat), but the more difficult theorem due to Mordell is that this group is *finitely generated*, in other words isomorphic to $E_t \times \mathbb{Z}^r$, where E_t is a finite abelian group consisting of the rational torsion points on the curve. It is very easy to compute E_t effectively. On the other hand, the integer r, called the algebraic *rank* of the curve, is much more difficult to compute, and no general algorithm is known.

The second question concerns the set of *integral* points on the curve. Here the situation is more satisfactory: a theorem of C. L. Siegel says that this set is *finite*, without giving any effective way of computing it. However, recent techniques based on Baker-type bounds due to S. David on elliptic logarithms, combined with the use of the LLL algorithm, make the search for the complete set of integral points almost automatic when (and that is of course a big "when") one knows explicitly the group of rational points; see for example the book by Smart [Sma]. We will give an outline of this method.

8.1.2 Basic Results and Conjectures

There are basically five main results and conjectures on this subject (not counting the existence of the group law, due in essence to Fermat in the seventeenth century). In increasing order of difficulty, these are the Mordell–Weil theorem, proved in the case of $\mathbb{Q}$ by Mordell in the 1920s; Siegel's theorem, proved in the 1930s; the isogeny conjecture, proved by Faltings in the 1980s together with the Mordell conjecture; the Taniyama–Shimura–Weil conjecture, proved by Wiles et al. between 1995 and 2000; and finally the Birch–Swinnerton-Dyer conjecture, still unproved.

We begin with the celebrated theorem of Mordell (or Mordell–Weil in the case of number fields or of Abelian varieties).

Theorem 8.1.1 (Mordell). *Let E be an elliptic curve defined over $\mathbb{Q}$. Then $E(\mathbb{Q})$ is a finitely generated abelian group. In other words, the torsion subgroup $E_t(\mathbb{Q})$ of all $T \in E(\mathbb{Q})$ such that there exists a nonzero integer k such that $k \cdot T = 0$ is finite, and there exists an integer r (called the algebraic rank) and points $P_i \in E(\mathbb{Q})$ for $1 \leqslant i \leqslant r$ such that any point $P \in E(\mathbb{Q})$ can be written uniquely as $P = T + \sum_{1 \leqslant i \leqslant r} x_i \cdot P_i$, with $T \in E_t(\mathbb{Q})$ and the $x_i \in \mathbb{Z}$.*

The proof of this theorem is not too difficult, and we will give it below (see Theorem 8.2.7 and Corollary 8.3.8). However, an important point must be noted: it is easy to compute $E_t(\mathbb{Q})$ algorithmically (more on this below), but it is difficult (and in fact there is no rigorous algorithm known) to compute the rank r and a fortiori the generators P_i. It is conjectured that r is unbounded, and the present record is $r = 28$ due to Elkies [Elk2], using a method first introduced by Mestre; see Exercise 11 for a weak but nontrivial version.

The second theorem, due to Siegel, deals with *integral points*. Here an important remark must be made. The group of rational points does not depend on the particular *model* chosen for the curve E: if we transform the equation(s) of E by a birational transformation, the structure of the group of rational points will be unchanged. This is absolutely not true for the set of integral points, which depends on the chosen model. To give an example in the even simpler case of genus 0, the curve $x^2 + y^2 = 1$ has only $(\pm 1, 0)$ and $(0, \pm 1)$ as integral points, while the $\mathbb{Q}$-isomorphic curve $x^2 + y^2 = 25$ has $(\pm 5, 0)$, $(\pm 4, \pm 3)$, $(\pm 3, \pm 4)$, and $(0, \pm 5)$ as integral points. Thus when one speaks of the set of integral points, it is always with respect to an equation or a family of equations. Furthermore, the notion of projective coordinates loses much of its meaning. (When is the projective point (x, y, z) an integral point? it cannot be when x, y, and z are integral since any rational point has a representative of that form. It could be when x/z and y/z are integral, but why choose z as special coordinate?) Siegel's theorem is as follows.

Theorem 8.1.2 (Siegel). *Let $f(x, y) = 0$ be the affine equation of a nonsingular plane cubic with integer coefficients. There exists only a finite number* *of pairs* $(a, b) \in \mathbb{Z}^2$ *such that* $f(a, b) = 0$*; in other words, the equation has only a finite number of integral points (possibly none).*

This theorem is fundamentally *ineffective*; in other words, it does not give any bound on the size of the solutions, or even on their number. A breakthrough in this and many other types of similar subjects was made by Baker at the end of the 1960s by his results on linear forms in logarithms of algebraic numbers (see Chapter 12 for an overview). Baker's results transformed all of this type of problem into *effective* results, although with huge constants. Soon afterward it was realized that the use of lattice reduction algorithms, and in particular the LLL algorithm when it was invented, could drastically reduce these huge bounds to a point where they can be used for practical computations.

As noted in the introduction, finding integral points on an elliptic curve in practice has become a routine (if not completely trivial) task, and we will devote Section 8.7 to a detailed explanation. We will proceed as follows. We first need a basis $(P_i)_{1 \leqslant i \leqslant n}$ of the torsion-free part of the Mordell–Weil group, and this is of course the hardest part. Second, we use Baker-type bounds on linear forms in *elliptic logarithms* to find a huge but effective upper bound on the integer coefficients x_i of the integral points expressed as a linear combination of the P_i (for the group law of the curve). Third, using the LLL algorithm in a suitable manner, possibly two or three times, we reduce the upper bound to something manageable, often less than 10; see Section 2.3.5. As a fourth and final step, we explore systematically all the possible linear combinations of the P_i with coefficients up to the bound that has been found, and all the integral points will be found during this search. Of course many things must be explained, and many tricks exist to improve on the above method (see [Sma]), but the main thing to understand is that the method is quite straightforward.

To explain in more detail the other results, we now introduce the notion of *minimal model* and of *reduction* modulo a prime. By making a suitable change of variables, we may always assume that our elliptic curve is given by an equation $y^2 = x^3 + ax + b$ with a and b integral. If a prime number p does not divide the discriminant of the curve $\mathrm{disc}(E) = -16(4a^3 + 27b^2)$, it is clear that the curve obtained by taking the reduction modulo p of a and b is still an elliptic curve, i.e., is nonsingular. For those p that divide $\mathrm{disc}(E)$ the curve is singular, but we may hope that by using other changes of variables certain other p may become acceptable. The fact is that the right context in which to consider this problem (which readily follows from the Riemann–Roch theorem) is that of generalized Weierstrass equations of the form

$$y^2 + a_1xy + a_3y = x^3 + a_2x^2 + a_4x + a_6$$

already considered in the preceding chapter.

The main result is that there exists a *minimal model* in generalized Weierstrass form with the $a_i \in \mathbb{Z}$, which among other properties has the smallest possible discriminant. However, this is not a satisfactory definition: the important property is that if p divides this minimal discriminant then whatever birational transformation is applied, the equation of the curve will remain singular at p. It is to be noted that the existence of a minimal model is due in large part (but not only) to the fact that $\mathbb{Z}$ is a principal ideal domain. An elliptic curve over a number field of class number strictly greater than 1 does not always have a global minimal model.

Let E be given by a minimal Weierstrass equation. We now introduce the global L-function of E. When p does not divide the (minimal) discriminant of E, the reduction of E modulo p is nonsingular hence defines an elliptic curve over $\mathbb{F}_p$. We have seen in Section 7.3.4 how to define $L_p(E, T)$ in that case. When p does divide the minimal discriminant, we must proceed a little differently. The singularity of the curve is necessarily unique, and by Proposition 7.1.5 (which can easily be extended to more general equations) it can come in one of three types. Assume for simplicity that the equation is $y^2 = x^3 + a_2 x^2 + a_4 x + a_6$ as in the proposition. The only possible singularity is at a point $P_0 = (x_0, 0)$, where x_0 is a multiple root of the third-degree polynomial. If P_0 is a cusp (respectively a double point with distinct tangents defined over K, respectively a double point with distinct tangents not defined over K), in other words if we have additive (respectively split multiplicative, respectively nonsplit multiplicative) reduction, we set $L_p(E, T) = 1$ (respectively $L_p(E, T) = 1/(1-T)$, respectively $L_p(E, T) = 1/(1+T)$). By the example following Proposition 7.1.5, these three formulas can be unified using the single formula $L_p(E, T) = 1/(1 - a_p T)$, where as usual $a_p = p + 1 - |E(\mathbb{F}_p)|$. Thus for *all* primes p we can write

$$L_p(E, T) = \frac{1}{1 - a_p T + \chi(p)pT^2} ,$$

where $\chi(p) = 1$ if p is a prime of good reduction, and $\chi(p) = 0$ otherwise.

Now that we have all the local L-functions, the global L-function of E is defined as the Euler product

$$L(E, s) = \prod_p L_p(E, p^{-s}) = \prod_p \frac{1}{1 - a_p p^{-s} + \chi(p)p^{1-2s}} ,$$

which gives of course a Dirichlet series $L(E, s) = \sum_{n \geqslant 1} a_n n^{-s}$ by expanding. Hasse's inequality immediately implies that the above Euler product (as well as the Dirichlet series) is absolutely convergent for $\Re(s) > 3/2$. The third important result on elliptic curves is the following:

Theorem 8.1.3 (Faltings). *Let E and E' be two elliptic curves defined over $\mathbb{Q}$. If E and E' are isogenous over $\mathbb{Q}$ then $L(E, s) = L(E', s)$, and conversely, if $L(E, s) = L(E', s)$ then E and E' are isogenous over $\mathbb{Q}$.*

The first part of this theorem is not difficult, but the converse (known previously as the *isogeny conjecture*) is a deep theorem of Faltings, proved in the same paper as the proof of Mordell's conjecture. Note that this theorem is important mainly for theoretical reasons.

The fourth important theorem on elliptic curves over $\mathbb{Q}$ is the extremely difficult and famous result of Wiles et al. (which, by using an older but also highly nontrivial result of Ribet, implies FLT; see Chapter 15 for details) proving the Taniyama–Shimura–Weil conjecture, and which states the following.

Theorem 8.1.4 (Wiles et al.). *The function $L(E, s)$ has an analytic continuation to the whole complex plane into a holomorphic function. Furthermore, there exists a positive integer N (which has the same prime divisors as the discriminant $\mathrm{disc}(E)$ of a minimal model, and which divides it), called the* conductor *of the curve, such that if we set*

$$\Lambda(E, s) = N^{s/2}(2\pi)^{-s}\Gamma(s)L(E, s) \,,$$

then Λ satisfies the functional equation $\Lambda(E, 2 - s) = \varepsilon(E)\Lambda(E, s)$, where $\varepsilon(E) = \pm 1$.

The number $\varepsilon(E)$ is called the (global) *root number*, or simply the sign of the functional equation. It is to be noted that there exists a tedious but easy algorithm due to Tate for computing the minimal model and the conductor (see for example [Coh0]). There exists a more recent and even more tedious algorithm for computing $\varepsilon(E)$, due to Mestre–Henniart and Halberstadt.

Another way to state the above theorem that is useful for computation is the following. For lack of space we cannot give the definition and properties of modular forms, but we will come back to them in Chapter 15.

Theorem 8.1.5. *There exists a modular cusp form f of weight 2 for $\Gamma_0(N)$ that is a Hecke eigenform for all Hecke operators (in other words a newform), such that the L-function $L(f, s)$ is equal to $L(E, s)$.*

We will also need the following property of the conductor, which we of course assume since we have not defined it.

Proposition 8.1.6. *Let E be an elliptic curve defined over $\mathbb{Q}$, let N be its conductor, let p be a prime number, and denote by $\overline{E}$ the reduction modulo p of a minimal model of E, considered as a curve over $\mathbb{F}_p$. Then $p \mid N$ if and only if $\overline{E}$ is singular, $p^2 \mid N$ if and only if the singularity of $\overline{E}$ is a cusp (i.e., additive reduction); hence $p \| N$ if and only if the singularity of $\overline{E}$ is a node (i.e., multiplicative reduction).*

The fifth and last important aspect of the theory of elliptic curves over $\mathbb{Q}$ is unfortunately in a conjectural state: it is the conjecture of Birch and Swinnerton-Dyer (BSD for short). As mentioned elsewhere, in the author's opinion it is the most beautiful and important conjecture in the whole of number theory (together with analogous or more general conjectures of the same type), and probably in the whole of mathematics.

Conjecture 8.1.7 (Birch–Swinnerton-Dyer). *Let E be an elliptic curve defined over $\mathbb{Q}$. The algebraic rank r defined by Mordell's theorem is equal to the order of vanishing of $L(E, s)$ at $s = 1$. More precisely,*

$$\lim_{s \to 1} \frac{L(E, s)}{(s - 1)^r} = \frac{L^{(r)}(E, 1)}{r!} = \omega_1(E) \frac{|\text{III}(E)| R(E) c_\infty(E) \prod_{p \mid N} c_p(E)}{|E_t(\mathbb{Q})|^2} ,$$

where $\omega_1(E)$ is the real period of E, $\text{III}(E)$ is the so-called Tate–Shafarevich group of E, $R(E)$ is the regulator of E, $c_\infty(E)$ is the number of connected components of $E(\mathbb{R})$, and for each $p \mid N$, $c_p(E)$ is the Tamagawa number of $E/\mathbb{Q}_p$ (see Definition 7.3.25), which is a small easily computed integer.

The real period $\omega_1(E)$ has been defined in Section 7.3.2, but we will not give here the definitions of $\text{III}(E)$ and $R(E)$, which will be (partially) introduced when necessary. The main point to note is that all the quantities on the right-hand side are in principle computable (although there is no known *algorithm* to compute $R(E)$), except for $\text{III}(E)$, which is not even known to be finite in general, except when $r \leqslant 1$. See Section 8.5.6 for an example.

Because of this conjecture the order of $L(E, s)$ at $s = 1$ is called the *analytic rank*, and the main statement of the conjecture is that it is equal to the ordinary (or algebraic) rank.

The main results concerning this conjecture are due to Coates–Wiles, Gross–Zagier, Kolyvagin, Rubin, and others. A weak but sufficient form of the known results is the following:

Theorem 8.1.8 (Kolyvagin et al.). *Let E be an elliptic curve defined over $\mathbb{Q}$. Then*

(1) *If the analytic rank is equal to 0, in other words if $L(E, 1) \neq 0$, then $r = 0$.*

(2) *If the analytic rank is equal to 1, in other words if $L(E, 1) = 0$ and $L'(E, 1) \neq 0$, then $r = 1$.*

(3) *In both of these cases, $\text{III}(E)$ is finite and the BSD conjecture is valid up to a controlled rational factor.*

Remark. Note that it is easy to check numerically that a given quantity such as $L(E, 1)$ or $L'(E, 1)$ is *nonzero*, but that it is impossible in general to prove numerically that a certain quantity is *equal* to 0. Thus, when we say that $L(E, 1) = 0$, we mean in fact that the sign of the functional equation $\varepsilon(E)$ is equal to -1 (which can be checked algorithmically), so that indeed $L(E, 1) = 0$. We also have the following:

Corollary 8.1.9. *Let E be an elliptic curve defined over $\mathbb{Q}$, let r be its (algebraic) rank, denote by r_{an} its analytic rank, and let $\varepsilon(E)$ the sign of the functional equation. Then*

(1) *If $r \geqslant 2$, then $r_{an} \geqslant 2$, in other words $L(E, 1) = L'(E, 1) = 0$.*
(2) *If $r = 2$, $\varepsilon(E) = 1$, and $L''(E, 1) \neq 0$, then $r_{an} = 2$.*
(3) *If $r = 3$, $\varepsilon(E) = -1$, and $L'''(E, 1) \neq 0$, then $r_{an} = 3$.*

Proof. Immediate and left to the reader (Exercise 1). $\square$

Notwithstanding this corollary, which is in fact a restatement of the theorem, one can reasonably say that nothing is known on the BSD conjecture when the analytic rank is greater than or equal to 2, even for a single curve.

Let us give two examples. To be able to find explicit lower bounds for the class number of imaginary quadratic fields, Goldfeld had shown long ago that it would be sufficient to find an L-function with suitable properties, and having a zero of order at least 3 at $s = 1$. The L-functions attached to modular elliptic curves over $\mathbb{Q}$ (at the time it was not known that all elliptic curves over $\mathbb{Q}$ are modular by Wiles et al.) do satisfy the necessary properties, but there remained to prove that one has a zero of order at least 3. The above corollary tells us that to find such an L-function it is enough to find an elliptic curve of rank at least 3, which is very easy. For instance, there exists a curve of rank 3 of conductor 5077 (and thanks to the work of Cremona, it is known that this is the smallest conductor); see Section 8.5.6 for more properties of this curve. To prove that the L-function has a zero of order 3 is immediate since we can compute algorithmically the sign of the functional equation, which is equal to -1 (it better be, since otherwise BSD would be false!), and using the methods of Section 8.5.3 it is also easy to compute numerically that $L'''(E, 1) \neq 0$.

As a second example, let E be an elliptic curve of algebraic rank 4 and $\varepsilon(E) = 1$ (infinite families of such curves are known). Then $L'(E, 1) = L'''(E, 1) = 0$ because of the functional equation, by the above corollary we have $r_{an} \geqslant 2$, so that in particular $L(E, 1) = 0$, and a numerical computation easily shows that $L''''(E, 1) \neq 0$. The BSD conjecture implies that $L''(E, 1) = 0$. This can easily be checked numerically to as many decimal places as one likes, but nobody has any idea how to prove this. In fact, if it could be proved in a single instance, it would be an exceedingly important advance on the subject, certainly worth a million dollars from the Clay prize plus a Fields medal.

From now on we assume that the curve E is given by a Weierstrass equation $y^2 z = x^3 + axz^2 + bz^3$ with $4a^3 + 27b^2 \neq 0$, where we may assume without loss of generality that a and b are in $\mathbb{Z}$. Furthermore, we will work in affine coordinates, simply remembering that the point at infinity is the neutral element for the group law, and is the given rational point, so we write our equation as $y^2 = x^3 + ax + b$. We would like to determine the group of rational points on this curve. This is extremely difficult to do in complete generality (no algorithm is known), but we can obtain a great deal of information from different points of view, both rigorous and conjectural. In this section some proofs will be omitted, and we refer to the numerous books on the subject such as [Cas2], [Cre2], [Dar], [Sil1], [Sil2], or [Sil-Tat].

8.1.3 Computing the Torsion Subgroup

There are several algorithms that can be used to compute $E_t(\mathbb{Q})$. The most efficient uses analytic techniques, and will not be described here. We begin with the Nagell–Lutz theorem, which is sufficient for small cases.

Theorem 8.1.10 (Nagell, Lutz). *Let E be given by a Weierstrass equation $y^2 = x^3 + ax^2 + bx + c = f(x)$ with a, b, and c in $\mathbb{Z}$. If $T = (x, y) \in E_t(\mathbb{Q}) \setminus \{\mathcal{O}\}$, then either T has order 2, in other words $y = 0$, or x and y are integers such that y^2 divides $D = -(4a^3c - a^2b^2 - 18abc + 4b^3 + 27c^2) = \mathrm{disc}(f)$. In particular, if the equation is $y^2 = x^3 + ax + b$ then either $y = 0$ or $y^2 \mid D = -(4a^3 + 27b^2)$.*

Proof. The statement concerning points of order 2 is clear, so assume that T is not of order 2. Since the natural map from $E(\mathbb{Q})$ to $E(\mathbb{Q}_p)$ is injective, it follows from Corollary 7.3.33 (3) that $(x, y) \in \mathbb{Z}_p^2$ for all p including $p = 2$, hence that $(x, y) \in \mathbb{Z}^2$. But $2T$ is also a torsion point different from $\mathcal{O}$, so if we write $2T = (x_3, y_3)$ we also have $(x_3, y_3) \in \mathbb{Z}^2$. By the addition formula, $x_3 = m^2 - 2x - a$ with $m = (3x^2 + 2ax + b)/(2y)$. Since $x_3 \in \mathbb{Z}$ it follows that m is a rational number that is a root of a monic second degree equation with integral coefficients, hence that $m \in \mathbb{Z}$, so that in particular $y \mid 3x^2 + 2ax + b = f'(x)$. Now we have the identity (see Exercise 2)

$$(27f(x) - (4a^3 - 18ab + 54c))f(x) - (f'(x) - a^2 + 3b)f'(x)^2 = \mathrm{disc}(f) = D \,,$$

and since $y^2 = f(x)$ and $y \mid f'(x)$ it follows that $y^2 \mid D$. $\qquad\square$

Note that the same proof shows that if E is given by a general Weierstrass equation $y^2 + a_1xy + a_3y = x^3 + a_2x^2 + a_4x + a_6$ with $a_i \in \mathbb{Z}$ and if $T = (x, y)$ is a torsion point of order not dividing 2, then again $(x, y) \in \mathbb{Z}^2$ but with the slightly weaker condition $(2y + a_1x + a_3)^2 \mid 4\,\mathrm{disc}(E)$; see Exercise 3.

The following corollary is important.

Corollary 8.1.11. *If $P = (x, y)$ is a rational point on an elliptic curve given as above (i.e., with integral coefficients), then P is a nontorsion point if and only if there exists k such that $k \cdot P$ has nonintegral coordinates.*

Proof. If $k \cdot P$ has nonintegral coordinates, then it cannot be a torsion point by the above theorem, so P is also nontorsion. Conversely, if $k \cdot P$ has integral coordinates for all k, these points cannot be distinct; otherwise, we would have an infinity of integral points, which is impossible by Siegel's Theorem 8.1.2. Thus two of them coincide for distinct values of k, so that P is a torsion point. $\square$

Note that a point satisfying the hypothesis of the Nagell–Lutz theorem is not necessarily a torsion point. For instance, the point $P = (-1, 1)$ on the curve $y^2 = x^3 - 2x$, which satisfies the conditions, is nontorsion since $2 \cdot P = (9/4, -21/8)$ does not have integral coordinates.

A variant of the Nagell–Lutz theorem which is useful in some cases is the following.

Proposition 8.1.12. *Assume that E is given by an equation of the form $y^2 = x^3 + ax^2 + bx$ with a and b integral, in other words that up to translation of the x-coordinate the curve has a rational 2-torsion point. Then if $(x, y) \in E_t(\mathbb{Q})$ with $y \neq 0$, we have $x \in \mathbb{Z}$, $x \mid b$, and $x + a + b/x$ is a square.*

Proof. Assume that $T = (x, y) \in E_t(\mathbb{Q})$ with $y \neq 0$. Then $2T \in E_t(\mathbb{Q})$ and $2T \neq \mathcal{O}$. The x-coordinate of $2T$ is equal to $(b - x^2)^2/(4x(x^2 + ax + b))$, and by the Nagell–Lutz theorem this must be an integer. Let $d = \gcd(b, x)$, $b_1 = b/d$, $x_1 = x/d$ so that $\gcd(b_1, x_1) = 1$ and $4x_1(dx_1^2 + ax_1 + b_1) \mid (b_1 - dx_1^2)^2$. In particular, $x_1 \mid b_1^2$, and since x_1 and b_1 are coprime we have $x_1 = \pm 1$, in other words $d = \pm x$ so $x \mid b$. Thus $x^2 \mid y^2$, so writing $y = \pm x y_1$ we deduce that $y_1^2 = x + a + b/x$, hence the latter quantity is a square. $\square$

Another consequence of the Nagell–Lutz theorem that is very useful for computing $E_t(\mathbb{Q})$ is the following.

Proposition 8.1.13. *Let E be given by $y^2 = x^3 + ax^2 + bx + c = f(x)$, and let ℓ be a prime number such that $\ell \nmid \mathrm{disc}(E) = -16D$, where $D = \mathrm{disc}(f)$ is as in Theorem 8.1.10. Then $E_t(\mathbb{Q})$ is isomorphic to a subgroup of $\overline{E}(\mathbb{F}_\ell)$, and in particular $|E_t(\mathbb{Q})|$ divides $|\overline{E}(\mathbb{F}_\ell)|$ for all such ℓ.*

Proof. Since $\ell \neq 2$ and $\ell \nmid \mathrm{disc}(f)$ the reduction $\overline{E}$ of the curve E modulo ℓ is again an elliptic curve. By the Nagell–Lutz theorem all the points of $E_t(\mathbb{Q})$ different from the point at infinity $\mathcal{O}$ have integral coordinates. The map reduction modulo ℓ is thus well defined from $E_t(\mathbb{Q})$ to $\overline{E}(\mathbb{F}_\ell)$ by sending the point at infinity of E to that of $\overline{E}$, and sending $(x, y) \in E_t(\mathbb{Q})$ to $(\overline{x}, \overline{y}) \in \overline{E}(\mathbb{F}_\ell)$. By Proposition 7.3.24 this map is a group homomorphism with trivial kernel, hence is injective, proving the proposition. $\square$

Thus for instance if we find two suitable values of ℓ for which the cardinalities $|\overline{E}(\mathbb{F}_\ell)|$ are coprime, we immediately know that $E_t(\mathbb{Q})$ is reduced to the point at infinity. As an application we give the following classical result.

Proposition 8.1.14. *Let d be a nonzero integer.*

(1) *Let E be given by $y^2 = x^3 - dx$. Then $E_t(\mathbb{Q}) \simeq \mathbb{Z}/2\mathbb{Z} \times \mathbb{Z}/2\mathbb{Z}$ if and only if d has the form $d = m^2$, $E_t(\mathbb{Q}) \simeq \mathbb{Z}/4\mathbb{Z}$ if and only if d has the form $d = -4m^4$, and otherwise $E_t(\mathbb{Q}) \simeq \mathbb{Z}/2\mathbb{Z}$.*

(2) *Let E be given by $y^2 = x^3 + d$. We have $E_t(\mathbb{Q}) \simeq \mathbb{Z}/2\mathbb{Z}$ if and only if d is a cube and not a sixth power, $E_t(\mathbb{Q}) \simeq \mathbb{Z}/3\mathbb{Z}$ if and only if d is either of the form $-432m^6$ or a square and not a sixth power, $E_t(\mathbb{Q}) \simeq \mathbb{Z}/6\mathbb{Z}$ if and only if d is a sixth power, and otherwise $E_t(\mathbb{Q})$ is trivial.*

Proof. The proof of both statements relies on the essential fact that the two types of curves under consideration have complex multiplication by $\mathbb{Z}[i]$ and $\mathbb{Z}[\rho]$ respectively, where $i^2 + 1 = 0$ and $\rho^2 + \rho + 1 = 0$, but there is no need to know the theory of CM to understand the very simple proof.

(1). We note that the discriminant of E is equal to $-64d^3$. Let ℓ be a prime not dividing d and congruent to 3 modulo 4. Since $\left(\frac{-1}{\ell}\right) = -1$, it follows that for each $x \in \mathbb{F}_\ell$ either $x^3 - dx = 0$, or exactly one of $x^3 - dx$ and $(-x)^3 - d(-x) = -(x^3 - dx)$ is a quadratic residue modulo ℓ. If k denotes the number of roots of $x^3 - dx = 0$ in $\mathbb{F}_\ell$, it follows that counting the point at infinity we have $|\overline{E}(\mathbb{F}_\ell)| = 1 + k + \ell - k = \ell + 1$. As mentioned above, this reflects the fact that a prime congruent to 3 modulo 4 is inert in the complex multiplication field $\mathbb{Q}(i)$.

When ℓ varies among all primes congruent to 3 modulo 4 and not dividing d, it is easy to see that the GCD of $\ell + 1$ is equal to 4. Indeed, assume the contrary, and let p be a common prime divisor of all such $(\ell + 1)/4$. Assume first that $p \neq 3$. By Dirichlet's theorem on primes in arithmetic progression (Theorem 10.5.30), since $\gcd(4p, 3) = 1$ we can find an infinity of primes ℓ such that $\ell = 4kp + 3$, and in particular one such that $\ell > d$, hence that does not divide d. But then $(\ell + 1)/4 = kp + 1$ must be divisible by p, which is absurd. If now $p = 3$, we consider instead the arithmetic progression $4kp - 5$, and we again obtain a contradiction.

It thus follows from Proposition 8.1.13 that $|E_t(\mathbb{Q})| \mid 4$. Since $(0,0)$ is evidently a point of order 2 in $E(\mathbb{Q})$, we have $E_t(\mathbb{Q}) \simeq \mathbb{Z}/2\mathbb{Z}$, $\mathbb{Z}/2\mathbb{Z} \times \mathbb{Z}/2\mathbb{Z}$, or $\mathbb{Z}/4\mathbb{Z}$. Since points of order 2 other than the point at infinity are those of the form $(x, 0)$, it follows that $E_t(\mathbb{Q}) \simeq \mathbb{Z}/2\mathbb{Z} \times \mathbb{Z}/2\mathbb{Z}$ if and only if $x^2 - d = 0$ has two rational roots, hence if and only if d is a square. On the other hand, P is a point of order 4 if and only if $2P$ has zero y-coordinate. If $P = (x, y)$, a short computation shows that this happens if and only if $x^2 + d = 0$ or $x^4 - 6dx^2 + d^2 = 0$. This last case cannot occur since it would imply that the equation $X^2 - 6X + 1 = 0$ has the rational root x^2/d. Thus $d = -x^2$, so that x and y are in $\mathbb{Z}$, and $y^2 = x^3 - dx = 2x^3$, so that $(2x)^3 = (2y)^2$; therefore $2x$ is a square, so $x = 2m^2$ for some m; hence $d = -4m^4$ as claimed, and in that case the point $(2m^2, 4m^3)$ has order 4.

(2). This case is completely similar. The discriminant of E is equal to $-432d^2$, and we choose primes ℓ not dividing d and congruent to 5 modulo 6. Since $3 \nmid (\ell - 1)$, for such primes the map $x \mapsto x^3$ from $\mathbb{F}_\ell$ to itself is a bijection; hence for each $y \in \mathbb{F}_\ell$ there exists exactly one $x \in \mathbb{F}_\ell$ such that $x^3 = y^2 - d$, so it follows once again that $|\overline{E}(\mathbb{F}_\ell)| = \ell + 1$, and this time it is because a prime congruent to 5 modulo 6 is inert in the complex multiplication field $\mathbb{Q}(\rho)$. Reasoning exactly as for (1) shows that this implies that $|E_t(\mathbb{Q})|$ divides 6.

Clearly $|E_t(\mathbb{Q})|$ is even if and only if there exists a rational point of order 2, hence if and only if d is a cube. On the other hand, there exists a point $P = (x, y)$ that has order 3 if and only if $2P = -P = (x, -y)$. It is immediately checked that this occurs if and only if $x(x^3 + 4d) = 0$, hence either when $x = 0$, so $d = y^2$ is a square, or if $x^3 = -4d$. But then $x = 2x_1$, and hence $d = -2x_1^3$ and $y^2 = x^3 + d = 6x_1^3$, so $(6x_1)^3 = (6y)^2$; hence $6x_1$ is a square, so $x_1 = 6m^2$ for some m; hence $d = -432m^6$, and in that case the point $(12m^2, 36m^3)$ has order 3, proving (2). $\square$

Corollary 8.1.15. *Let a, b, c be nonzero rational numbers, and assume that there exists a (projective) rational point $(x_0 : y_0 : z_0)$ on the cubic $ax^3 + by^3 + cz^3 = 0$.*

(1) *If there are only a finite number of such points then either b/a, c/a, or c/b is the cube of a rational number.*
(2) *Let T be the group of torsion points of the projective cubic, considered as an elliptic curve. We have $T \simeq \mathbb{Z}/3\mathbb{Z}$ if and only if b/a, c/a, and c/b are all cubes, $T \simeq \mathbb{Z}/2\mathbb{Z}$ if and only if up to permutation of a, b, and c we have that b/a and $c/(2a)$ are cubes, or $a/(2b)$ and $a/(2c)$ are cubes, otherwise T is trivial.*

Proof. By Proposition 7.2.4 we know that the cubic is birationally equivalent to the elliptic curve E whose affine Weierstrass equation is $Y^2 = X^3 - 432(abc)^2$. It follows that if there are k projective points on our cubic, the curve E has rank 0 and $|E_t(\mathbb{Q})| = k$. We consider two cases.

Case 1: abc is not a cube or twice a cube. It follows from the proposition that $E_t(\mathbb{Q})$ is trivial; hence $k = 1$, so the cubic must have a single projective rational point. However, if we compute the intersection of the tangent to the cubic at $(x_0 : y_0 : z_0)$ with the cubic we find that

$$(x_1 : y_1 : z_1) = (x_0(by_0^3 - cz_0^3) : y_0(cz_0^3 - ax_0^3) : z_0(ax_0^3 - by_0^3))$$

is another projective point on the cubic. Note that it is well defined (i.e., $(x_1, y_1, z_1) \neq (0, 0, 0)$): indeed otherwise, by symmetry assume that $x_0 \neq 0$. Then $by_0^3 = cz_0^3$, so $y_0 \neq 0$ (otherwise $z_0 = 0$, so $x_0 = 0$ since $ax_0^3 + by_0^3 + cz_0^3 = 0$); hence $ax_0^3 = by_0^3 = cz_0^3$, which implies that $0 = ax_0^3 + by_0^3 + cz_0^3 = 3ax_0^3$, so $x_0 = 0$, again absurd and proving our claim.

Since $k = 1$ we must have $(x_1 : y_1 : z_1) = (x_0 : y_0 : z_0)$. Since we cannot have $ax_0^3 = by_0^3 = cz_0^3$, once again by symmetry we may assume that $by_0^3 \neq cz_0^3$. If we had the equality

$$by_0^3 - cz_0^3 = cz_0^3 - ax_0^3 = ax_0^3 - by_0^3 \, ,$$

then by adding these three quantities we would obtain $0 = 3(by_0^3 - cz_0^3)$, contradicting our assumption that $by_0^3 \neq cz_0^3$. It follows that for instance $cz_0^3 - ax_0^3 \neq by_0^3 - cz_0^3$, and since $(x_1 : y_1 : z_1) = (x_0 : y_0 : z_0)$ this implies that $y_0 = 0$. But then $z_0 \neq 0$ and $ax_0^3 + cz_0^3 = 0$, so $c/a = (-x_0/z_0)^3$ is a cube, as claimed.

Case 2: abc is a cube or twice a cube. By elementary manipulations that we have already explained in Section 6.4.3, without loss of generality we may assume that $a = 1$ and that b and c are cubefree integers, and these manipulations only modify the ratios b/a, c/a, and c/b by cubes. Assume first that $abc = bc = m^3$, and let p be a prime divisor of m. We have $v_p(b) + v_p(c) \equiv 0 \pmod 3$ and $v_p(b) + v_p(c) \neq 0$, hence $v_p(b) \equiv 1 \pmod 3$ and $v_p(c) \equiv 2 \pmod 3$ or the reverse. But this is absurd since then $v_p(ax^3) = v_p(x^3) \equiv 0 \pmod 3$, $v_p(by^3) \equiv 1 \pmod 3$, and $v_p(cz^3) \equiv 2 \pmod 3$, although these three quantities sum to zero. Thus p cannot exist, so that $m = \pm 1$, and since b and c are integers, $b = \pm c = \pm 1$, so for instance b is a cube. If $abc = bc = 2m^3$, the same reasoning shows that the only possible prime divisor of m is $p = 2$, so $b = \pm 2^{j_1}$ and $c = \pm 2^{j_2}$ with $j_1 + j_2 \equiv 1 \pmod 3$. It follows that $(j_1, j_2) \equiv (0, 1)$, $(1, 0)$, or $(2, 2)$ modulo 3, so that either b, c, or c/b is a cube, as claimed. We leave the proof of (2) to the reader (Exercise 5). $\qquad\square$

Although the determination of $E_t(\mathbb{Q})$ is very easy, the following deep theorem of B. Mazur can also be useful.

Theorem 8.1.16 (Mazur). *The group $E_t(\mathbb{Q})$ is isomorphic either to $\mathbb{Z}/N\mathbb{Z}$ with $1 \leqslant N \leqslant 10$ or $N = 12$, or to $(\mathbb{Z}/N\mathbb{Z}) \times (\mathbb{Z}/2\mathbb{Z})$ with $N = 2, 4, 6,$ or 8.*

For instance, if we find a point of order 7 in $E_t(\mathbb{Q})$, it is not necessary to go any further ($E_t(\mathbb{Q})$ has order 7). If we find a point of order 5, then since it is trivial to check whether there exists a point of order 2 (the points with $y = 0$), we can immediately determine that $E_t(\mathbb{Q})$ is cyclic of order 5 or 10. Note that it is easy to show that there is an infinity of nonisomorphic elliptic curves E such that $E_t(\mathbb{Q})$ is isomorphic to one of the 15 groups given above, and they can also be rationally *parametrized*; see Exercise 18.

8.1.4 Computing the Mordell–Weil Group

Now that we have seen that $E_t(\mathbb{Q})$ is easily accessible, we consider the nontorsion part. There are essentially three different methods to attack the problem. The first and historically the oldest (initiated by Fermat) is the method of 2-descent. The second method is that of Heegner points, initiated in principle by Heegner in 1954, but really developed by Stark, Birch, and others starting in 1967. The third method is partly conjectural since it is based on the Birch–Swinnerton-Dyer (BSD) conjecture, which is proved (up to constants) only in rank 0 and 1, but at least says what to expect.

The 2-descent method is most useful when $r = 0$, i.e., when there are no points of infinite order. It is then often (but not always) possible to prove this in an elementary way, as we shall see below. However, even when $r > 0$ it gives very useful information, and in many favorable cases allows the rigorous computation of r and the P_i.

The Heegner point method is applicable if and only if $r = 1$. This may seem like a severe restriction, but tables and heuristics seem to show that the density (in some reasonable sense) of elliptic curves with $r > 1$ is equal to 0, and that the others are equally divided between $r = 0$ and $r = 1$. Since the group of rational points of an elliptic curve of rank 0 is reduced to its torsion part, which is easily found, the Heegner point construction should be applicable to 100% of all elliptic curves for which the Mordell–Weil group needs to be computed.

Finally, the third method (which appears in several guises, for instance Manin's conditional algorithm) is based on the BSD conjecture, stating among other things that the algebraic rank r should be equal to the analytic rank, which is the order of vanishing at $s = 1$ of the Dirichlet series $L(E, s)$ attached to the elliptic curve E. Even that order is not easy to compute rigorously (in fact, as already mentioned, nobody has any idea how to prove that $L(E, s)$ vanishes to order greater than or equal to 4 when it should), but at least we can use numerical approximations to guess its exact value. This then gives strong guidelines on how to use the rigorous methods.

In the next sections we will describe the three methods described above. Since the 2-descent method is the closest in spirit to the rest of this book we will describe it in more detail than the two others.

8.1.5 The Naïve and Canonical Heights

Before studying practical methods for computing the rank and if possible, also generators, an important point must be settled, which in fact is essential for the completion of the proof of the Mordell–Weil theorem. Consider the following problem. Let E be an elliptic curve defined over $\mathbb{Q}$. If we are given a point $P \in E(\mathbb{Q})$, it is easy to determine whether P has infinite order, for instance by using Corollary 8.1.11. But now assume that P and Q are two points in $E(\mathbb{Q})$, and for instance that there is no torsion. How do we check that P and Q are independent points in $E(\mathbb{Q})$, in other words that $mP + nQ \neq \mathcal{O}$ for all $(m, n) \neq (0, 0)$? The answer is not as simple as one mightn think, but luckily there is a very nice answer, given by the notion of $canonical\ height$.

Let us begin by defining the (naïve) height of a nonzero rational number x. Writing $x = n/d$ with $\gcd(n, d) = 1$ we define $h(x) = \max(\log(|n|), \log(|d|))$. This is also natural if we view x as an element of $\mathbb{P}^1(\mathbb{Q})$ with coordinates $(n : d)$. Thus more generally if $P \in \mathbb{P}^n(\mathbb{Q})$ we can write (uniquely up to a sign change) $P = (x_0 : x_1 : \cdots : x_n)$, where $x_i \in \mathbb{Z}$ and $\gcd(x_0, \ldots, x_n) = 1$, and we define $h(P) = \max_i(\log(|x_i|))$, where by convention $\log(0) = -\infty$.

Now assume for simplicity that the elliptic curve E is given by a single equation, hence defined as a curve in $\mathbb{P}^2$, and let $P \in E(\mathbb{Q}) \subset \mathbb{P}^2(\mathbb{Q})$. We could define the height of P as a point in $\mathbb{P}^2(\mathbb{Q})$. However, for several reasons we prefer to define the (naïve) height of P as the height of its x-coordinate. In other words, thanks to Proposition 7.3.1, we can write the affine coordinates of P as $(m/d^2, n/d^3)$ with $\gcd(m,d) = \gcd(n,d) = 1$, and we define the height as $h(P) = \max(\log(|m|), \log(d^2))$, called again the *naïve height* of the point P. Note that because of the equation of the curve, when $h(P)$ is large then $\log(|n|)$ is comparable to $3h(P)$, so $h(P)$ does take into account the y-coordinate.

Please note that the function $h(P)$ is defined only on $E(\mathbb{Q})$ (or more generally on a number field K), but not on $E(\mathbb{C})$. Although it has some nice properties, we also need a more regular function of P, called the *canonical height* of P, and defined as follows. First note that when experimentally computing $h(kP)$ for increasing values of k, we find immediately that it has the appearance of a parabola, which in fact is approximately true. For instance, if E is the curve $y^2 = x^3 - 2x$ and $P = (-1, 1)$, then the first values of the integers $\exp(h(kP))$ are

```
1
9
169
12769
2325625
3263037129
5627138321281
68970122119586689
1799664515907016914961
197970893765498628138595401
58648738806449243564537197828441
113430878631471464907295822495116028129
32398460974000521187196405196075267458328192  1
57163008369984740944839327879387136420685658888848009
2043089963462381155150392740589608447914207325218257654  30625
```

which visually form a parabola (see also Exercise 6). We thus define the canonical height by the formula

$$\widehat{h}(P) = \lim_{k \to \infty} \frac{h(kP)}{k^2} \ .$$

Note that we evidently have $h(P) \geqslant 0$, hence $\widehat{h}(P) \geqslant 0$ for all P. The following theorem summarizes the main properties of the canonical height.

Theorem 8.1.17. *The above limit exists, and defines a nonnegative function $\widehat{h}(P)$ on $E(\mathbb{Q})$ with the following properties:*

(1) (*Quadratic form.*) *The function* $\widehat{h}(P)$ *is a quadratic form on* $E(\mathbb{Q})$; *in other words, if we define* $\langle P, Q \rangle$ *by the formula*

$$\langle P, Q \rangle = (\widehat{h}(P + Q) - \widehat{h}(P) - \widehat{h}(Q))/2$$

then $\langle P, Q \rangle$ *is a symmetric bilinear form on* $E(\mathbb{Q})$ *such that* $\langle P, P \rangle = \widehat{h}(P)$, *so that* $\widehat{h}(kP) = k^2 \widehat{h}(P)$.

(2) (*Nondegeneracy.*) *We have* $\widehat{h}(P) = 0$ *if and only if* $P \in E_t(\mathbb{Q})$, *so that* $\widehat{h}$ *induces a positive definite quadratic form on the finitely generated free abelian group* $E(\mathbb{Q})/E_t(\mathbb{Q})$.

(3) (*Independence.*) *Points* $(P_i)_{1 \leqslant i \leqslant n}$ *in* $E(\mathbb{Q})$ *are linearly independent in* $E(\mathbb{Q})/E_t(\mathbb{Q})$ *if and only if the determinant of the so-called* height pairing *matrix* $M = ((\langle P_i, P_j \rangle)_{1 \leqslant i,j \leqslant n})$ *is not equal to* 0. *More precisely,* $\sum_{1 \leqslant j \leqslant n} b_j P_j$ *is a torsion point if and only if* $\sum_{1 \leqslant j \leqslant n} b_j M_j = 0$, *where* M_j *is the* jth *column of* M.

(4) (*Bound.*) *There exists an explicitly computable constant* $C(E)$ *depending only on* E *such that for all* $P \in E(\mathbb{Q})$ *we have* $|\widehat{h}(P) - h(P)| \leqslant C(E)$ *(see below for a more precise estimate).*

(5) (*Finiteness.*) *For any* $B > 0$ *there exists only a finite number of points* $P \in E(\mathbb{Q})$ *such that* $\widehat{h}(P) \leqslant B$ *(or equivalently,* $h(P) \leqslant B$*).*

We refer to [Sil-Tat] for proofs of the above properties, which are not difficult.

Note that in practice, to check that points are independent (or dependent) modulo the torsion subgroup one must use some care, since the determinant of the matrix M is a real number that cannot be computed exactly. If this determinant seems to be nonzero, then one should give an error bound on the computation of the determinant so as to *prove* rigorously that the determinant is nonzero. On the other hand, if the determinant seems to be equal to 0, one must then find a nonzero element of the kernel of the matrix M, which must exist and have entries very close to an integer after multiplication by a suitable denominator. Although it is usually impossible to prove rigorously that a real number is exactly equal to 0, here it is possible because one simply checks that the (integral) entries of the given element of the kernel produce a linear combination of the generators that is in the torsion subgroup. If this is the case, all is well; we have shown that the points are dependent modulo torsion; otherwise, it shows that the determinant computation has not been accurate enough, and it should be redone with a higher accuracy.

Because of the above theorem and remarks, it is essential to be able to compute heights numerically. The definition can be used, but it is not very well suited to accurate computation. A much better algorithm is given for instance in Chapter 7 of [Coh0]. This is implemented in GP as the function 2*ellheight(P). Note that in versions at least up to 2.3 it is important to multiply by 2 the result given by ellheight(P), since it corresponds to a different normalization (this may change in future releases of the package).

If we had taken a slightly different definition of the naïve height, such as for instance $\max(\log(|md|), \log(|n|), \log(d^3))$, which is the naïve height on the projective plane, using the same definition it can be shown that we would obtain a canonical height *equal to* (up to a constant multiple) the canonical height defined above.

For practical applications, it is essential to give explicit bounds for the difference between the naïve and canonical heights. Such a bound is the following (see [Sil3], and see [Cre-Pri-Sik] for much better bounds).

Theorem 8.1.18. *Let E be an elliptic curve defined over $\mathbb{Q}$ by a generalized Weierstrass equation. With the usual notation, set*

$$\mu(E) = \frac{\log(|\operatorname{disc}(E)|) + \log^+(j(E))}{6} + \log^+(b_2/12) + \log(2^*) \,,$$

where $\log^+(x) = \max(1, \log(|x|))$ and $2^* = 2$ if $b_2 \neq 0$ and $2^* = 1$ otherwise. Then for $P \in E(\mathbb{Q})$ we have

$$-\frac{h(j(E))}{12} - \mu(E) - 1.946 \leqslant \widehat{h}(P) - h(P) \leqslant \mu(E) + 2.14$$

(recall that if $\gcd(n, d) = 1$ then $h(n/d) = \max(\log(|n|), \log(|d|)))$.

As a direct application, we see that in the computation of the torsion subgroup, for instance using Theorem 8.1.10, then if $P = (x, y) \in E_t(\mathbb{Q})$ we have $\widehat{h}(P) = 0$, hence $h(P) = h(x) \leqslant h(j(E))/12 + \mu(E) + 1.946$, and since we know that $x \in \mathbb{Z}$, this gives a (usually small) upper bound for $|x|$.

8.2 Description of 2-Descent with Rational 2-Torsion

I emphasize from the start that my purpose is not to give the most efficient algorithms, which are in fact in constant progress, but to describe a simple version of the method that is already useful in treating many small cases. We closely follow [Sil-Tat].

8.2.1 The Fundamental 2-Isogeny

As above, in this section we fix an elliptic curve E given by a not necessarily reduced Weierstrass equation $y^2 = x^3 + ax^2 + bx + c$ with integers a, b, and c and nonzero discriminant. We denote by $\mathcal{O}$ its point at infinity, which is the neutral element for the group law. In this section we make the crucial simplifying assumption that there exists a rational point of order 2 different from $\mathcal{O}$, i.e., that there exists $x_0 \in \mathbb{Q}$ (hence in $\mathbb{Z}$) such that $x_0^3 + ax_0^2 + bx_0 + c = 0$. We will explain in Section 8.3 what must be done if this assumption is not satisfied.

By setting $x = X + x_0$, we can send the point $(x_0, 0)$ to the origin $T = (0,0)$, which is therefore a point of order 2, and our equation will now have the form $y^2 = x^3 + ax^2 + bx$ for some other integers a and b. We will work with equations of this form. It is easy to see that the discriminant of the third-degree polynomial is given by the formula $D = b^2(a^2 - 4b)$, so that $\mathrm{disc}(E) = 16b^2(a^2 - 4b)$.

In this section we will work with a pair of elliptic curves, one being E and the other denoted by $\widehat{E}$. All quantities and variables relative to $\widehat{E}$ will be denoted with a $\widehat{}$, which should not cause any notational confusion. The curve $\widehat{E}$ is defined by the equation $y^2 = x^3 + \widehat{a}x^2 + \widehat{b}x$ with $\widehat{a} = -2a$ and $\widehat{b} = a^2 - 4b$. Note that $\widehat{\widehat{a}} = 4a$ and $\widehat{\widehat{b}} = 16b$, so the curve $\widehat{\widehat{E}}$ is the curve $y^2 = x^3 + 4ax^2 + 16bx$, which is trivially isomorphic to E by replacing x by $4x$ and y by $8y$.

Proposition 8.2.1. *For any $P = (x, y) \in E$ set*

$$\phi(P) = (\widehat{x}, \widehat{y}) = \left(\frac{y^2}{x^2}, \frac{y(x^2 - b)}{x^2} \right)$$

for P not equal to T or $\mathcal{O}$, and set $\phi(T) = \phi(\mathcal{O}) = \widehat{\mathcal{O}}$. Then ϕ is a group homomorphism from E to $\widehat{E}$, whose kernel is equal to $\{\mathcal{O}, T\}$. Applying the same process to $\widehat{E}$ gives a map ϕ_1 from $\widehat{E}$ to $\widehat{\widehat{E}}$, and $\widehat{\widehat{E}}$ is isomorphic to E via the map $(x, y) \mapsto (x/4, y/8)$. Thus there is a homomorphism $\widehat{\phi}$ from $\widehat{E}$ to E defined for $\widehat{P} = (\widehat{x}, \widehat{y})$ different from $\widehat{T}$ and $\widehat{\mathcal{O}}$ by

$$\widehat{\phi}(\widehat{P}) = (x, y) = \left(\frac{\widehat{y}^2}{4\widehat{x}^2}, \frac{\widehat{y}(\widehat{x}^2 - \widehat{b})}{8\widehat{x}^2} \right)$$

and by $\widehat{\phi}(\widehat{T}) = \widehat{\phi}(\widehat{\mathcal{O}}) = \mathcal{O}$. Furthermore, for all $P \in E$ we have $\widehat{\phi} \circ \phi(P) = 2P$, and for all $\widehat{P} \in \widehat{E}$ we have $\phi \circ \widehat{\phi}(\widehat{P}) = 2\widehat{P}$.

Proof. The proof consists in a series of explicit verifications, where in each case we must separate the points $\mathcal{O}$ and T from the other points. It is done with utmost detail in [Sil-Tat], to which we refer. We will simply show that ϕ maps E to $\widehat{E}$, and that it maps three collinear points of E to three collinear points of $\widehat{E}$. This is the essential part of the proof. Also, to simplify we will assume that all the points that occur are distinct and different from $\mathcal{O}$, T, $\widehat{\mathcal{O}}$, and $\widehat{T}$.

Let (x, y) be a point on E, and $(\widehat{x}, \widehat{y}) = \phi(x, y)$. We compute that

$$\widehat{x}^3 + \widehat{a}\widehat{x}^2 + \widehat{b}\widehat{x} = \frac{y^2}{x^2} \left(\frac{y^4}{x^4} - 2a\frac{y^2}{x^2} + a^2 - 4b \right) = \frac{y^2}{x^6}((y^2 - ax^2)^2 - 4bx^4)$$

$$= \frac{y^2}{x^6}((x^3 + bx)^2 - 4bx^4) = \left(\frac{y(x^2 - b)}{x^2} \right)^2 = \widehat{y}^2,$$

proving that $(\widehat{x}, \widehat{y})$ is on the curve $\widehat{E}$.

Now for $i = 1$, 2, and 3 let $P_i = (x_i, y_i)$ be three collinear points on E (so that $P_1 + P_2 + P_3 = \mathcal{O}$ by definition of the group law). We will show that the points $\phi(P_i) = (\widehat{x}_i, \widehat{y}_i)$ are collinear. Let $y = mx + n$ be the equation of the line through the points P_i. We have $n \neq 0$ since otherwise one of the points would be equal to $T = (0,0)$, which we have excluded. I claim that the points $\phi(P_i)$ are on the line $y = \widehat{m}x + \widehat{n}$, with

$$\widehat{m} = \frac{nm - b}{n} \quad \text{and} \quad \widehat{n} = \frac{n^2 - anm + bm^2}{n}.$$

Using the equation of the curve and the relations $y_i = mx_i + n$ we compute that

$$\begin{aligned}
\widehat{m}\widehat{x}_i + \widehat{n} &= \frac{(nm - b)y_i^2 + (n^2 - anm + bm^2)x_i^2}{nx_i^2} \\
&= \frac{nm(y_i^2 - ax_i^2) - b(y_i - mx_i)(y_i + mx_i) + n^2 x_i^2}{nx_i^2} \\
&= \frac{m(x_i^3 + bx_i) - b(y_i + mx_i) + nx_i^2}{x_i^2} \\
&= \frac{x_i^2(mx_i + n) - by_i}{x_i^2} = \frac{y_i(x_i^2 - b)}{x_i^2} = \widehat{y}_i,
\end{aligned}$$

proving my claim. The rest of the verifications are simpler and left to the reader.

The proofs of the formulas for $\widehat{\phi}$ and that $\widehat{\phi} \circ \phi(P) = 2P$ and $\phi \circ \widehat{\phi}(\widehat{P}) = 2\widehat{P}$ are also verifications left to the reader. $\square$

It follows from Definition 7.1.7 that ϕ is an isogeny from E to $\widehat{E}$, and that $\widehat{\phi}$ is its dual isogeny. Furthermore, since we are in characteristic zero and the kernels (over $\overline{\mathbb{Q}}$) have two elements, these maps are 2-isogenies. This is why this method is called 2-descent via 2-isogenies (we will study general 2-descent in Section 8.3 below).

8.2.2 Description of the Image of ϕ

Although we know by Theorem 7.1.8 that ϕ is surjective over $\overline{\mathbb{Q}}$, we now restrict to *rational* points, and we want to determine the image of ϕ on rational points (since T is assumed to be a rational point, here in fact $(0,0)$, the kernel of ϕ is of course still equal to $\{\mathcal{O}, T\}$). This is given by the following result.

Proposition 8.2.2. *Denote by $I = \phi(E(\mathbb{Q}))$ the image of the rational points of E in $\widehat{E}(\mathbb{Q})$. Then*

(1) $\widehat{O} \in I$, and $\widehat{T} \in I$ if and only if $\mathrm{disc}(E)$ is a square in $\mathbb{Q}^*$, or equivalently, if $\widehat{b} = a^2 - 4b$ is a square in $\mathbb{Q}^*$.

(2) Otherwise, a general point $\widehat{P} = (\widehat{x}, \widehat{y}) \in \widehat{E}(\mathbb{Q})$ with $\widehat{x} \neq 0$ belongs to I if and only if $\widehat{x}$ is a square in $\mathbb{Q}$.

Proof. Since $\phi(\mathcal{O}) = \widehat{O}$ the first statement is trivial. Since $x = 0$ implies $y = 0$, hence $(x, y) = T$ so $\phi((x, y)) = \widehat{O}$, for the other statements we may assume $x \neq 0$. Then $\widehat{T} \in I$ if and only if there exists $x \neq 0$ such that $y^2/x^2 = 0$; hence $y^2 = x(x^2 + ax + b) = 0$, so x is a root of $x^2 + ax + b$. Thus x exists if and only if the discriminant $a^2 - 4b$ of this quadratic is a square, proving (1).

For (2), the definition of ϕ shows that $\widehat{x}$ is a square. Conversely, assume that $(\widehat{x}, \widehat{y}) \in \widehat{E}(\mathbb{Q})$ with $\widehat{x} \neq 0$ and $\widehat{x} = u^2$, and for $\varepsilon = \pm 1$ set

$$x_\varepsilon = \frac{u^2 - a + \varepsilon \widehat{y}/u}{2}, \quad y_\varepsilon = \varepsilon x_\varepsilon u .$$

I claim that both points $(x_\varepsilon, y_\varepsilon)$ are in $E(\mathbb{Q})$ and that $\phi(x_\varepsilon, y_\varepsilon) = (\widehat{x}, \widehat{y})$ (since the kernel of ϕ has order 2, we must indeed have two preimages). To prove that they are in $E(\mathbb{Q})$, using the equation of $\widehat{E}$ we compute that

$$x_1 x_{-1} = \frac{(\widehat{x} - a)^2 - \widehat{y}^2/\widehat{x}}{4} = \frac{\widehat{x}^3 - 2a\widehat{x}^2 + a^2\widehat{x} - \widehat{y}^2}{4\widehat{x}} = b .$$

Thus

$$x_\varepsilon + a + \frac{b}{x_\varepsilon} = x_\varepsilon + x_{-\varepsilon} + a = u^2 ,$$

so that

$$x_\varepsilon^3 + ax_\varepsilon^2 + bx_\varepsilon = (ux_\varepsilon)^2 = y_\varepsilon^2 ,$$

proving that both points are on E, and of course with rational coordinates. Furthermore, we have $\phi(x_\varepsilon, y_\varepsilon) = (x', y')$ with

$$x' = \frac{y_\varepsilon^2}{x_\varepsilon^2} = u^2 = \widehat{x} ,$$

and using once again the equality $b = x_\varepsilon x_{-\varepsilon}$,

$$y' = \frac{y_\varepsilon(x_\varepsilon^2 - b)}{x_\varepsilon^2} = \varepsilon u(x_\varepsilon - x_{-\varepsilon}) = \varepsilon u(\varepsilon \widehat{y}/u) = \widehat{y}$$

as claimed. $\square$

8.2.3 The Fundamental 2-Descent Map

The fact that the image of ϕ consists essentially of points $(\widehat{x}, \widehat{y})$ for which $\widehat{x}$ is a square is quite remarkable and will now be exploited in full.

Definition 8.2.3. *We define the* 2-*descent map* α *from the group* $E(\mathbb{Q})$ *to the multiplicative group* $\mathbb{Q}^*/\mathbb{Q}^{*2}$ *as follows.*

(1) $\alpha(\mathcal{O}) = 1$, $\alpha(T) = b$.
(2) *When* $x \neq 0$ *and* $(x, y) \in E(\mathbb{Q})$ *then* $\alpha((x, y)) = x$.

In the above, all the values are of course understood modulo the multiplicative action of $\mathbb{Q}^{*2}$.

The main result is the following.

Proposition 8.2.4. (1) *The* 2-*descent map* α *is a group homomorphism.*
(2) *The kernel of* α *is equal to* $\widehat{\phi}(\widehat{E}(\mathbb{Q}))$, *so* α *induces an injective group homomorphism from* $E(\mathbb{Q})/\widehat{\phi}(\widehat{E}(\mathbb{Q}))$ *to* $\mathbb{Q}^*/\mathbb{Q}^{*2}$.
(3) *Let* p_i *for* $1 \leqslant i \leqslant t$ *be the distinct primes dividing* b. *The image of* α *is contained in the subgroup of* $\mathbb{Q}^*/\mathbb{Q}^{*2}$ *generated by the classes modulo squares of* -1 *and the* p_i.
(4) *The index* $[E(\mathbb{Q}) : \widehat{\phi}(\widehat{E}(\mathbb{Q}))]$ *divides* 2^{t+1}.

Proof. (1) Clearly if $P = (x, y) \neq T$ then $\alpha(-P) = \alpha((x, -y)) = x$; hence $\alpha(P)\alpha(-P) = x^2 \in \mathbb{Q}^{*2}$, and $\alpha(T)\alpha(-T) = \alpha(T)^2 = b^2 \in \mathbb{Q}^{*2}$, so α sends inverses to inverses. Thus to prove (1) we must prove that if $P_1 + P_2 + P_3 = \mathcal{O}$ then $\alpha(P_1)\alpha(P_2)\alpha(P_3) \in \mathbb{Q}^{*2}$. If one of the P_i is equal to $\mathcal{O}$, we are in the case we have just treated. Let us first assume that none of the P_i is equal to T. As usual, let $y = mx + n$ be the equation of the line passing through the three points (the only other possible lines $x = n$ are excluded since none of the P_i is equal to $\mathcal{O}$). Writing the intersection of the line with the cubic equation, we see that the three abscissas x_i of the points P_i are the three roots of the equation

$$x^3 + (a - m^2)x^2 + (b - 2mn)x - n^2 = 0$$

(this is of course how the algebraic formula for the group law is obtained in the first place). In particular, $x_1 x_2 x_3 = n^2 \in \mathbb{Q}^{*2}$, proving (1) when none of the P_i is equal to T. If one of the P_i is equal to T (and only one since otherwise the third point is equal to $\mathcal{O}$), we may assume for instance that $P_1 = T$. The three abscissas are now $x_1 = 0$, x_2, and x_3, and a line going through $P_1 = T = (0, 0)$ has equation $y = mx$, so $n = 0$. It follows that the x_i are the roots of $x^3 + (a - m^2)x^2 + bx = 0$, so x_2 and x_3 are the two roots of $x^2 + (a - m^2)x + b = 0$. Thus $x_2 x_3 = b$, so $\alpha(P_1)\alpha(P_2)\alpha(P_3) = b^2 \in \mathbb{Q}^{*2}$, finishing the proof of (1) and explaining why we must choose $\alpha(T) = b$.

(2) Applying Proposition 8.2.2 with $\widehat{E}$ instead of E and $\widehat{\phi}$ instead of ϕ, we see that α has in fact been constructed so that its kernel is exactly equal to $\widehat{\phi}(\widehat{E})(\mathbb{Q})$ (note that $\widehat{a}^2 - 4\widehat{b} = 16b \equiv b \pmod{\mathbb{Q}^{*2}}$).

(3) Let $P = (x, y) \in E(\mathbb{Q})$. We want to find conditions on $x = \alpha(P)$ modulo squares. By Proposition 7.3.1 we know that there exist integers m,

n, and d such that $x = m/d^2$, $y = n/d^3$, and $\gcd(m,d) = \gcd(n,d) = 1$. Replacing in the equation of E and clearing denominators gives

$$n^2 = m^3 + am^2d^2 + bmd^4 = m(m^2 + amd^2 + bd^4) \, .$$

This is the key to the proposition: we have a product of two integers equal to a square, so that as we have so often done in the study of Diophantine equations, both are close to squares. To see how close, we must compute the GCD of both factors. Assume first that $x \neq 0$. Since $\gcd(m,d) = 1$, we see that the GCD of the factors is equal to $\gcd(m,b)$, and in particular is a divisor of b. Thus if $p \nmid b$, $v_p(m)$ is even. This means that m, hence x, is up to a multiplicative square in the group generated by ± 1 and the p_i, as claimed. If $x = 0$, then $P = T$ and $\alpha(P) = b$, which of course belongs to the group generated by its prime divisors and by -1.

(4) The subgroup described in (3) is the group of classes of the distinct representatives $\prod_{0 \leqslant i \leqslant t} p_i^{e_i}$ with $p_0 = -1$ and $e_i = 0$ or 1, which has 2^{t+1} elements. Thus (4) follows from (2) and (3).   □

Although the aim of the above results is to describe an explicit method for computing the Mordell–Weil group in practice, it is to be noted that they comprise a large part of the Mordell–Weil theorem itself, at least for the type of curve that we are considering (having a rational torsion point of order 2).

Now note the following purely abelian group-theoretic lemma.

Lemma 8.2.5. *Let A and B be abelian groups written additively, and let ϕ from A to B and $\widehat{\phi}$ from B to A be two group homomorphisms. Assume that the indexes $[B : \phi(A)]$ and $[A : \widehat{\phi}(B)]$ are finite. Then the index $[A : \widehat{\phi} \circ \phi(A)]$ is also finite, and more precisely we have*

$$[A : \widehat{\phi} \circ \phi(A)] \big| [A : \widehat{\phi}(B)][B : \phi(A)] \, .$$

Proof. We have

$$[A : \widehat{\phi} \circ \phi(A)] = [A : \widehat{\phi}(B)][\widehat{\phi}(B) : \widehat{\phi}(\phi(A))] \, .$$

On the other hand, it is clear that the map $\widehat{\phi}$ induces a surjective map from $B/\phi(A)$ to $\widehat{\phi}(B)/\widehat{\phi}(\phi(A))$, so the cardinality of the latter quotient divides that of the former, proving the lemma.   □

We now immediately deduce what is commonly called the *weak Mordell–Weil theorem*, and we will see that it easily implies the full theorem.

Corollary 8.2.6. *The group $E(\mathbb{Q})/2E(\mathbb{Q})$ is finite. More precisely, its cardinality divides 2^{s+t+2}, where t is the number of distinct prime divisors of b, and s is the number of distinct prime divisors of $a^2 - 4b$.*

Proof. By Proposition 8.2.4, we have $[E(\mathbb{Q}) : \widehat{\phi}(\widehat{E}(\mathbb{Q}))] \mid 2^{t+1}$. Applying the proposition to $\widehat{E}$ and ϕ, we have $[\widehat{E}(\mathbb{Q}) : \phi(E(\mathbb{Q}))] \mid 2^{s+1}$. The result thus follows from the lemma, since $\widehat{\phi} \circ \phi$ is the multiplication by 2 map. $\qquad\square$

We can now prove the strong form of Mordell's theorem.

Theorem 8.2.7 (Mordell). *Let E be an elliptic curve defined over $\mathbb{Q}$, and assume known that for some $m \geqslant 2$ we know that $E(\mathbb{Q})/mE(\mathbb{Q})$ is finite (by the above corollary this is true for $m = 2$ when E has a rational 2-torsion point). Then $E(\mathbb{Q})$ is a finitely generated abelian group. More precisely, if B is the largest canonical height of a system of representatives of $E(\mathbb{Q})$ modulo $mE(\mathbb{Q})$, then the (finite) set S of rational points $P \in E(\mathbb{Q})$ such that $\widehat{h}(P) \leqslant B$ generates $E(\mathbb{Q})$.*

Proof. Assume by contradiction that the subgroup H of $E(\mathbb{Q})$ generated by S is not equal to $E(\mathbb{Q})$, and let $Q_1 \in E(\mathbb{Q}) \setminus H$. The set of points in $E(\mathbb{Q}) \setminus H$ of height less than or equal to that of Q_1 is finite, so let $Q \in E(\mathbb{Q}) \setminus H$ be of minimal height. By assumption there exist $P \in S$ (in fact in our chosen system of representatives modulo $mE(\mathbb{Q})$) and $R \in E(\mathbb{Q})$ such that $Q = P + mR$. Since $P \in S \subset H$ and $Q \notin H$ we have $R \notin H$, so $\widehat{h}(R) \geqslant \widehat{h}(Q)$ by our minimality assumption. Thus, since $\widehat{h}$ is a nonnegative quadratic form we obtain

$$\widehat{h}(P) = \frac{1}{2}(\widehat{h}(Q + P) + \widehat{h}(Q - P)) - \widehat{h}(Q) \geqslant \frac{1}{2}\widehat{h}(mR) - \widehat{h}(Q)$$

$$\geqslant \frac{m^2}{2}\widehat{h}(R) - \widehat{h}(Q) \geqslant 2\widehat{h}(R) - \widehat{h}(Q) \geqslant \widehat{h}(Q) > B$$

since $Q \notin H$, and a fortiori $Q \notin S$. This is a contradiction since $P \in S$ and hence $\widehat{h}(P) \leqslant B$. $\qquad\square$

An important consequence of the proof of this theorem is that once $E(\mathbb{Q})/mE(\mathbb{Q})$ is known for some m (for instance for $m = 2$), obtaining a system of generators for $E(\mathbb{Q})$ is completely algorithmic. Thus the only obstruction to the existence of an algorithm to compute $E(\mathbb{Q})$ lies in the computation of the finite group $E(\mathbb{Q})/mE(\mathbb{Q})$ for some m. In practice, however, better algorithms are used than the one implicit in the proof of the theorem.

8.2.4 Practical Use of 2-Descent with 2-Isogenies

Now that we have seen how to use 2-descent for theoretical purposes, we will show how it can be used in practice to bound the rank of an elliptic curve, and sometimes to compute it exactly. For this, we must analyze more precisely the images of the 2-descent maps.

We will denote by r the algebraic rank of the group $E(\mathbb{Q})$. Since E and $\widehat{E}$ are isogenous through the maps ϕ and $\widehat{\phi}$, it is clear that r is also the rank of

$\widehat{E}$. We naturally denote by $\widehat{\alpha}$ the 2-descent map from $\widehat{E}(\mathbb{Q})$ to $\mathbb{Q}^*/\mathbb{Q}^{*2}$. We begin with the following proposition.

Proposition 8.2.8. *We have the equality*

$$|\alpha(E(\mathbb{Q}))||\widehat{\alpha}(\widehat{E}(\mathbb{Q}))| = 2^{r+2} .$$

Proof. As an abstract abelian group, we have $E(\mathbb{Q}) \simeq E_t(\mathbb{Q}) \oplus \mathbb{Z}^r$, hence

$$E(\mathbb{Q})/2E(\mathbb{Q}) \simeq E_t(\mathbb{Q})/2E_t(\mathbb{Q}) \oplus (\mathbb{Z}/2\mathbb{Z})^r .$$

Furthermore, for any finite abelian group A, the exact sequence

$$0 \longrightarrow A[2] \longrightarrow A \longrightarrow A \longrightarrow A/2A \longrightarrow 1 ,$$

where the middle map is multiplication by 2 and $A[2]$ is the kernel of that map, shows that $|A/2A| = |A[2]|$ (in fact $A/2A$ is noncanonically isomorphic to $A[2]$). In our case, with $A = E_t(\mathbb{Q})$ the points of order 2 are exactly $\mathcal{O}$ and those with $y = 0$, hence $x = 0$, plus the two points corresponding to the roots of $x^2 + ax + b = 0$ if $a^2 - 4b$ is a square. Thus

$$|E(\mathbb{Q})/2E(\mathbb{Q})| = 2^{r+1+\delta} ,$$

where $\delta = 1$ or 0 according to whether $a^2 - 4b$ is a square or not.

On the other hand, let us consider our 2-isogenies ϕ and $\widehat{\phi}$. Since $\widehat{\phi} \circ \phi$ is the multiplication by 2 map, we evidently have

$$|E(\mathbb{Q})/2E(\mathbb{Q})| = [E(\mathbb{Q}) : \widehat{\phi}(\widehat{E}(\mathbb{Q}))][\widehat{\phi}(\widehat{E}(\mathbb{Q})) : \widehat{\phi}(\phi(E(\mathbb{Q})))] .$$

Now for any group homomorphism $\widehat{\phi}$ and subgroup B of finite index in an abelian group A we evidently have

$$\frac{\widehat{\phi}(A)}{\widehat{\phi}(B)} \simeq \frac{A}{B + \mathrm{Ker}(\widehat{\phi})} \simeq \frac{A/B}{(B + \mathrm{Ker}(\widehat{\phi}))/B} \simeq \frac{A/B}{\mathrm{Ker}(\widehat{\phi})/(\mathrm{Ker}(\widehat{\phi}) \cap B)} .$$

Thus

$$[\widehat{\phi}(A) : \widehat{\phi}(B)] = \frac{[A : B]}{[\mathrm{Ker}(\widehat{\phi}) : \mathrm{Ker}(\widehat{\phi}) \cap B]} .$$

We are going to use this formula with $A = \widehat{E}(\mathbb{Q})$ and $B = \phi(E(\mathbb{Q}))$. We know that $\mathrm{Ker}(\widehat{\phi})$ has two elements $\widehat{\mathcal{O}}$ and $\widehat{T}$, and we have shown in Proposition 8.2.2 that $\widehat{T} \in \phi(E(\mathbb{Q}))$ if and only if $a^2 - 4b$ is a square. Using the δ-notation above, it follows that

$$[\widehat{\phi}(\widehat{E}(\mathbb{Q})) : \widehat{\phi}(\phi(E(\mathbb{Q})))] = \frac{[\widehat{E}(\mathbb{Q}) : \phi(E(\mathbb{Q}))]}{2^{1-\delta}} .$$

Putting everything together we obtain

$$2^{r+2} = [E(\mathbb{Q}) : \widehat{\phi}(\widehat{E}(\mathbb{Q}))][\widehat{E}(\mathbb{Q}) : \phi(E(\mathbb{Q}))] \, ,$$

proving the proposition thanks to Proposition 8.2.4 (2).   □

It remains to give a reasonably practical method to compute $|\alpha(E(\mathbb{Q}))|$ (which we will of course also use for $\widehat{\alpha}(\widehat{E}(\mathbb{Q}))$). We have seen in Proposition 8.2.4 (3) and (4) how $\alpha(E(\mathbb{Q}))$ can be determined in principle by looking at the factorization of b. We make this more precise in the following theorem.

Theorem 8.2.9. *The group $\alpha(E(\mathbb{Q}))$ is equal to the classes modulo squares of 1, b, and the positive and negative divisors b_1 of b such that the quartic equation*

$$Y^2 = b_1 X^4 + a X^2 Z^2 + (b/b_1) Z^4$$

has a solution with X, Y, and Z pairwise coprime integers such that $XZ \neq 0$. If (X, Y, Z) is such a solution we will have $\gcd(b/b_1, X) = \gcd(b_1, Z) = 1$, and the point $P = (b_1 X^2/Z^2, b_1 XY/Z^3)$ is in $E(\mathbb{Q})$ and such that $\alpha(P) = b_1$.

Proof. Clearly $1 \in \alpha(E(\mathbb{Q}))$, so we can forget the point at infinity. Let $(x, y) \in E(\mathbb{Q})$, and assume for the moment that $y \neq 0$, hence $x \neq 0$. We have seen in the proof of Proposition 8.2.4 that we can write $x = m/d^2$, $y = n/d^3$ with $d \neq 0$, $\gcd(m, d) = \gcd(n, d) = 1$ and the equation $n^2 = m(m^2 + amd^2 + bd^4)$. Let us now go further. Set $b_1 = \text{sign}(m)\gcd(m, b)$. We can thus write $m = b_1 m_1$, $b = b_1 b_2$ with $m_1 > 0$ and $\gcd(m_1, b_2) = 1$. Substituting, we obtain $n^2 = b_1^2 m_1(b_1 m_1^2 + am_1 d^2 + b_2 d^4)$. It follows that $b_1 \mid n$, so we write $n = b_1 n_1$, so that $n_1^2 = m_1(b_1 m_1^2 + am_1 d^2 + b_2 d^4)$. Since $\gcd(m_1, b_2) = 1$ and $\gcd(m_1, d) \mid \gcd(m, d) = 1$ it follows that both factors are relatively prime, so each of them is a square (since $m_1 > 0$). Thus there exist coprime integers X and Y such that $m_1 = X^2$, $b_1 m_1^2 + am_1 d^2 + b_2 d^4 = Y^2$, and $n_1 = XY$. Setting $Z = d$, this gives the desired quartic $Y^2 = b_1 X^4 + a X^2 Z^2 + b_2 Z^4$, and coming back to the initial point we have $x = m/Z^2 = b_1 X^2/Z^2$ and $y = n/Z^3 = b_1 XY/Z^3$. Thus given a point on the quartic we can come back to a point on $E(\mathbb{Q})$, proving that we have exactly described $\alpha(E(\mathbb{Q}))$ outside of the image of the points for which $y = 0$. Since $\gcd(m, d) = \gcd(n, d) = 1$ we deduce that $\gcd(X, Z) = \gcd(Y, Z) = 1$, so that X, Y, and Z are pairwise coprime. Finally, the points with $y = 0$ are either the point $T = (0, 0)$, which is such that $\alpha(T) = b$, which is taken into account, or, when $a^2 - 4b = e^2$ is a square, the points with $x = (-a \pm e)/2$. But in that case $((-a - e)/2)((-a + e)/2) = (a^2 - e^2)/4 = b$, so we can choose $b_1 = (-a \pm e)/2$, and clearly the point $(X, Y, Z) = (1, 0, 1)$ is on the corresponding quartic $Y^2 = b_1 X^4 + a X^2 Z^2 + (b/b_1)Z^4$, so these points will be included in the count. Note that $Z = d \neq 0$, and that in every case $X \neq 0$. Finally, a simple inspection of the quartic equation shows that if X, Y, and Z are pairwise coprime then $\gcd(b/b_1, X) = \gcd(b_1, Z) = 1$.   □

To use this theorem in practice it is useful to have some additional results.

Definition 8.2.10. *For any nonzero integer* $N \in \mathbb{Z}$, *denote by* $s(N)$ *the squarefree part of* N, *i.e., the unique squarefree integer such that there exists an integer* f *with* $N = s(N)f^2$.

Proposition 8.2.11. *A divisor* b_1 *of* b *is such that the quartic* $Y^2 = b_1 X^4 + aZ^2 X^2 + (b/b_1)Z^4$ *is solvable with pairwise coprime* X, Y, *and* Z *with* $XZ \neq 0$ *if and only if the quartic* $Y^2 = s(b_1)X^4 + aZ^2 X^2 + (b/s(b_1))Z^4$ *is solvable with* $\gcd(X, Z) = 1$ *and* $XZ \neq 0$.

Proof. Assume that b_1 is such that the quartic $Y^2 = b_1 X^4 + aZ^2 X^2 + (b/b_1)Z^4$ is solvable with pairwise coprime X, Y, and Z with $XZ \neq 0$ and write $b_1 = s(b_1)f^2$. Then

$$(Yf)^2 = s(b_1)(Xf)^4 + aZ^2(Xf)^2 + (b/s(b_1))Z^4 ,$$

so that if we set $Y_1 = Yf$, $X_1 = Xf$, the coprimality conditions of the theorem imply that X and $f \mid b_1$ are coprime to Z; hence so is X_1.

Conversely, assume that $Y^2 = s(b_1)X^4 + aZ^2 X^2 + (b/s(b_1))Z^4$ is solvable with $XZ \neq 0$ and $\gcd(X, Z) = 1$, and set $f = \gcd(X, Y)$, which is coprime to Z since it divides X. Thus f^2 divides $(b/s(b_1))Z^4$, hence also $b/s(b_1)$, so we can write

$$(Y/f)^2 = s(b_1)f^2(X/f)^4 + aZ^2(X/f)^2 + (b/(s(b_1)f^2))Z^4 .$$

It follows that $(X/f, Y/f, Z)$ is a solution to the quartic with $s(b_1)f^2$ (still dividing b) instead of b_1, but now with $\gcd(X/f, Y/f) = 1$. Next we have evidently $\gcd(X/f, Z) = 1$, and if p is a prime dividing $\gcd(Y, Z)$, then $p^2 \mid s(b_1)X^4$; hence since $s(b_1)$ is squarefree, $p \mid X$, a contradiction since $p \mid Z$ and $\gcd(X, Z) = 1$. Thus $\gcd(Y, Z) = \gcd(Y/f, Z) = 1$. By Theorem 8.2.9, the pairwise coprimality of X/f, Y/f, and Z implies the two other coprimality conditions. $\square$

Corollary 8.2.12. *Let* b_1 *be a divisor of* b *such that both* b_1 *and* b/b_1 *are squarefree (which is in particular the case if* b *is squarefree). If* (X, Y, Z) *satisfies* $Y^2 = b_1 X^4 + aZ^2 X^2 + (b/b_1)Z^4$ *with* $XZ \neq 0$ *and* $\gcd(X, Z) = 1$, *then* X, Y, *and* Z *are pairwise coprime.*

Proof. Note that $b_1 = s(b_1)$. Thus from the proof of the above proposition we see that if we set $f = \gcd(X, Y)$ then $f^2 \mid b/b_1$, so that $f = 1$ since we assume b/b_1 squarefree. As in the above proof we also deduce that $\gcd(Y, Z) = 1$. $\square$

Corollary 8.2.13. *The group* $\alpha(E(\mathbb{Q}))$ *is equal to the set of classes modulo squares of* 1, *of* $s(b)$, *and of* b_1 *and* b/b_1 *for all positive and negative divisors* b_1 *of* b *such that* b_1 *is squarefree,* $|b_1| \leqslant |b|^{1/2}$, *and such that the quartic equation*

$$Y^2 = b_1 X^4 + a Z^2 X^2 + (b/b_1) Z^4$$

has a solution with X, Y, Z integral, $XZ \neq 0$, and $\gcd(X, Z) = 1$.

Proof. Denote by G the set of classes modulo squares of the elements described in this corollary. Clearly the classes of 1 and the squarefree part of b belong to $\alpha(E(\mathbb{Q}))$. If b_1 is squarefree we have $s(b_1) = b_1$, so the proposition and the theorem imply that the class of b_1 is in $\alpha(E(\mathbb{Q}))$; hence so is the class of b/b_1 since $\alpha(E(\mathbb{Q}))$ is a group. We have thus shown that $G \subset \alpha(E(\mathbb{Q}))$. Conversely, let b_1 be an arbitrary divisor of b such that there exist pairwise coprime integers X, Y, and Z with $XZ \neq 0$ such that $Y^2 = b_1 X^4 + a Z^2 X^2 + (b/b_1) Z^4$. By the proposition, the class of $s(b_1)$, which is equal to that of b_1, is such that the corresponding quartic is solvable with $XZ \neq 0$ and $\gcd(X, Z) = 1$. If $|b_1| \leqslant |b|^{1/2}$, hence $|s(b_1)| \leqslant |b|^{1/2}$, this implies that the class of b_1 is in G. If $|b_1| > |b|^{1/2}$, then $|b/b_1| < |b|^{1/2}$, and we have $Y^2 = (b/b_1) Z^4 + a Z^2 X^2 + (b/(b/b_1)) X^4$, so the quartic is solvable with X and Z interchanged, $XZ \neq 0$, and X, Y, Z pairwise coprime. By the proposition we deduce that the class of $s(b/b_1)$, hence of b/b_1, is in G, hence also that of $b_1 = b/(b/b_1)$ by definition of G. It follows that $G = \alpha(E(\mathbb{Q}))$, as claimed. $\square$

Remark. In the above results we have used in part the fact that $\alpha(E(\mathbb{Q}))$ is a group. In practice, this fact must be used to its maximum extent.

8.2.5 Examples of 2-Descent using 2-Isogenies

Let us consider several simple examples of 2-descent using 2-isogenies.

Proposition 8.2.14. (1) *The curve $y^2 = x^3 - 1$ has rank 0 and torsion group of order 2.*
(2) *The curve $y^2 = x^3 + 1$ has rank 0 and torsion group of order 6.*

Note that we have already proved this result in Corollary 6.4.37, also using 2-descent.

Proof. We treat both curves simultaneously. The point $\mathcal{O}$ and either $(1, 0)$ for the first curve or $(-1, 0)$ for the second clearly are the only points of order dividing 2. By the Nagell–Lutz Theorem 8.1.10 any other torsion point is such that y is integral and $y^2 \mid 27$, so that $y^2 = 1$ or 9. For the first curve this does not correspond to rational values of x, and for the second curve it gives the points $(0, \pm 1)$ and $(2, \pm 3)$, which one can check are torsion points, the torsion group being of order 6 generated by the point $(2, 3)$.

Let us now compute the rank using 2-descent. Let $y^2 = x^3 - \varepsilon$ be the equation of our curve, with $\varepsilon = \pm 1$. We first set $x = x_1 + \varepsilon$ to put the curve in the form that we have treated: $y^2 = x_1^3 + 3\varepsilon x_1^2 + 3x_1$. Thus $a = 3\varepsilon$ and $b = 3$. The group $\alpha(E(\mathbb{Q}))$ contains 1 and 3. The divisors of b are ± 1 and ± 3, so it is sufficient to check whether $b_1 = -1$ gives a solvable quartic with $X \neq 0$. The quartic equation is $Y^2 = -X^4 + 3\varepsilon X^2 Z^2 - 3Z^4$. Since the discriminant of the quadratic $-u^2 + 3\varepsilon u - 3$ is negative, the quadratic is always negative, so our quartic does not even have real solutions. Thus $|\alpha(E(\mathbb{Q}))| = 2$.

We must now compute $|\widehat{\alpha}(\widehat{E}(\mathbb{Q}))|$. The equation of $\widehat{E}$ is $y^2 = x_1^3 - 6\varepsilon x_1^2 - 3x_1$. Thus $b = -3$, and the group $\widehat{\alpha}(\widehat{E}(\mathbb{Q}))$ contains 1 and -3. Once again it is sufficient to check whether $b_1 = -1$ gives a solvable quartic with $X \neq 0$. The quartic equation is $Y^2 = -X^4 - 6\varepsilon X^2 Z^2 + 3Z^4$. There exist real solutions to this quartic, so we cannot get away with the local condition in $\mathbb{R}$. On the other hand, there are no solutions modulo 3 since $Y^2 \equiv -(X^2)^2 \pmod{3}$ implies that $3 \mid X$ and $3 \mid Y$ hence $9 \mid 3Z^4$ hence $3 \mid \gcd(X, Z)$, a contradiction since $\gcd(X, Z) = 1$. Thus once again $|\widehat{\alpha}(\widehat{E}(\mathbb{Q}))| = 2$, and we deduce from Proposition 8.2.8 that $r = 0$. $\qquad\square$

Proposition 8.2.15. *Let p be a positive or negative prime number, and let E_p be the elliptic curve with equation $y^2 = x^3 - px$. The torsion group of E_p has order 2, and if r_p is the rank of $E_p(\mathbb{Q})$ we have the following results:*

(1) *When $p > 0$ and $p \equiv 3$, 11, or 13 modulo 16 or if $p < 0$ and $p = -2$ or $p \equiv 5$ or 9 modulo 16 we have $r_p = 0$.*

(2) *When $p > 0$ and $p = 2$ or $p \equiv 5$, 7, 9, or 15 modulo 16, or if $p < 0$ and $p \equiv 1$, 3, 11, or 13 modulo 16, we have $r_p = 0$ or 1 (and $r_2 = 1$).*

(3) *When $p > 0$ and $p \equiv 1 \pmod{16}$ or $p < 0$ and $p \equiv -1 \pmod{8}$ we have $r_p = 0$, 1, or 2.*

Remark. Assuming a very weak form of the Birch–Swinnerton-Dyer conjecture, in case (2) of the proposition we always have $r = 1$ and in case (3) we always have $r = 0$ or 2, and both cases can occur.

Proof. The points $\mathcal{O}$ and $(0,0)$ are clearly the only points of order dividing 2. By the Nagell–Lutz Theorem 8.1.10 and its refinement Proposition 8.1.12, any other torsion point has integral x and y with $x \mid p$ and $y^2 \mid 4p^3$, in other words $y^2 \mid 4p^2$. This clearly implies that $x = -\operatorname{sign}(p)$ with $|p - 1|$ a square dividing $4p^2$, hence dividing 4 since it is coprime to p, or $x = \operatorname{sign}(p)p = |p|$ with $p^2 \mid p - 1|$ a square dividing $4p^2$, hence $|p - 1|$ a square dividing 4, as before. However, Proposition 8.1.12 also implies that $x + p/x$ is a square, so here that $|p + 1|$ is a square. Since two squares cannot differ by 2, this is impossible, proving the statement concerning the torsion subgroup (we could also look at the finite number of remaining possibilities).

Let us now apply 2-descent. With the notation used in that context we have $a = 0$ and $b = -p$. Thus $\alpha(E(\mathbb{Q}))$ contains the classes of 1 and of $-p$. The only other divisors of b are -1 and p, and they will both be in $\alpha(E(\mathbb{Q}))$ if and only if $b_1 = -1$ is, hence if the quartic $Y^2 = -X^4 + pZ^4$ has a solution with $XZ \neq 0$. If $p < 0$ this quartic has no real solution, so in that case $|\alpha(E(\mathbb{Q}))| = 2$. On the other hand, if $p > 0$ there are cases in which there exist solutions, and others in which there are none. For $p = 2$ we have the trivial solution $(X, Y, Z) = (1, 1, 1)$, so $|\alpha(E(\mathbb{Q}))| = 4$; hence we exclude that case, so p is odd. Recall that by Theorem 8.2.9 we have $\gcd(X, Z) = \gcd(Y, Z) = \gcd(p, X) = \gcd(X, Y) = 1$. Since $p \nmid X$, Y/X^2 is a square root of -1 modulo p, so when $p \equiv 3 \pmod 4$ once again we obtain that $|\alpha(E(\mathbb{Q}))| = 2$. We may thus assume that $p > 0$ with $p \equiv 1 \pmod 4$. If Z is even then X must be odd, hence Y also, which is impossible if $Y^2 \equiv -X^4 \pmod{16}$. Thus Z is odd, so that $Z^4 \equiv 1 \pmod{16}$. If X is even, we have $1 \equiv Y^2 \equiv p \pmod 8$. If X is odd we have $X^4 \equiv 1 \pmod{16}$ and Y is even, hence $Y^2 \equiv 0$ or 4 modulo 16, so $p \equiv 1$ or 5 modulo 16. Thus if $p \equiv 13 \pmod{16}$ we see once again that $|\alpha(E(\mathbb{Q}))| = 2$. On the other hand, for $p \equiv 1$, 5, and 9 modulo 16 there are no 2-adic conditions, and a short computer search shows that the quartic is often (but not always) solvable (the only exceptions for $p \leqslant 500$ and a search up to $d = 1000$ are $p = 113$, 193, and 353, out of 33 possible primes; of course this does not imply that the quartic has no solutions for these values of p).

To summarize, if $p < 0$ or $p \equiv 3 \pmod 4$ or $p \equiv 13 \pmod{16}$ we have $|\alpha(E(\mathbb{Q}))| = 2$, if $p = 2$ we have $|\alpha(E(\mathbb{Q}))| = 4$, and otherwise (i.e., if $p > 0$ and $p \equiv 1$, 5, or 9 modulo 16) we have $|\alpha(E(\mathbb{Q}))| = 2$ or 4.

Let us now consider $\widehat{\alpha}(\widehat{E})(\mathbb{Q})$. The equation of $\widehat{E}$ is $y^2 = x^3 + 4px$, so $a = 0$ and $b = 4p$, and we will apply Corollary 8.2.13 since b is not squarefree. The classes of 1 and p belong to $\widehat{\alpha}(\widehat{E})(\mathbb{Q})$, and otherwise the quartic to be considered is $Y^2 = b_1 X^4 + (4p/b_1)Z^4$. The possible squarefree values of b_1 less than $|b|^{1/2}$ in absolute value are ± 1 and ± 2, except if $p = \pm 3$, for which we also have ± 3. When $b_1 = -1$ the quartic is $Y^2 = -X^4 - 4pZ^4$, hence has no real solutions when $p > 0$. When $p < 0$ we cannot have X odd; otherwise, Y is also odd, which is impossible modulo 4. Thus X and Y are both even, and writing $Y = 2Y_1$, $X = 2X_1$ we obtain $Y_1^2 = -4X_1^4 - pZ^4$. Since $\gcd(X, Z) = 1$, Z is odd, and $p \neq -2$ since otherwise Y_1 is even and hence $4 \mid pZ^4$, so p is also odd. Thus Y_1 is odd, so $1 + p \equiv 0 \pmod 4$ hence $p \equiv 3 \pmod 4$. To summarize, if either $p > 0$ or $p < 0$ and $p \not\equiv 3 \mod 4$ then we cannot take $b_1 = -1$ in Corollary 8.2.13.

When $b_1 = \pm 2$ the quartic is $Y^2 = \pm(2X^4 + 2pZ^4)$. If $b_1 = -2$ we must have $p < 0$ since otherwise there are no real solutions. Writing $Y = 2Y_1$, we have $2Y_1^2 = \pm(X^4 + pZ^4)$. If p is odd, X must be odd; otherwise, Z is even and $2 \mid \gcd(X, Z)$. Thus Z is also odd, so we deduce that $p + 1 \equiv 0$, ± 2, or 8 modulo 16, hence $p \equiv 1$, 7, 13, or 15 modulo 16. When $p = 2\varepsilon$ with $\varepsilon = \pm 1$, X must be even, hence Z must be odd, and writing $X = 2X_1$ we have $Y_1^2 = \pm(8X_1^4 + \varepsilon Z^4)$. It follows that we must have $\varepsilon = \pm = \operatorname{sign}(b_1)$, since otherwise we have a contradiction modulo 4, and for $\varepsilon = \pm$ we have the solution $(X_1, Y_1, Z) = (1, 3, 1)$ when $b_1 = p = 2$ and $(X_1, Y_1, Z) = (1, 1, 3)$ when $b_1 = p = -2$. To summarize, if $p \neq b_1$ or p odd and $p \not\equiv \pm 2 - 1$, 7, or 15 modulo 16 we cannot take $b_1 = \pm 2$, while for $p = b_1$ we can, but we already have p in our list.

Assume now that $p = 3\varepsilon$ with $\varepsilon = \pm 1$, so that we must also consider $b_1 = \pm 3$. Since the class of p already belongs to $\widehat{\alpha}(\widehat{E})(\mathbb{Q})$, it is thus sufficient to consider $b_1 = -p$, so that the quartic is $Y^2 = -3\varepsilon X^4 - 4Z^4$. This has no real solution if $\varepsilon = 1$. If $\varepsilon = -1$, i.e., $p = -3$, it has no solution modulo 4 with X odd. Thus X, hence Y, is even, so Z is odd since $\gcd(X, Z) = 1$, and writing $X = 2X_1$ and $Y = 2Y_1$ we obtain $Y_1^2 = 12X_1^4 - Z^4$, which has also no solution modulo 4. Thus in all cases we do not obtain any extra element of our group when $b_1 = \pm 3$.

Using Corollary 8.2.13, we finally have the following cases.

– If $p = \pm 2$, then $\widehat{\alpha}(\widehat{E})(\mathbb{Q})$ is equal to the classes of 1 and p, hence has 2 elements.
– If $p > 0$ and $p \equiv 3, 5, 9, 11$, or 13 modulo 16, then $\widehat{\alpha}(\widehat{E})(\mathbb{Q})$ is equal to the classes of 1 and p, hence has 2 elements.
– If $p > 0$ and $p \equiv 1, 7$, or 15 modulo 16, then $\widehat{\alpha}(\widehat{E})(\mathbb{Q})$ may have 2 or 4 elements (when the class of $b_1 = 2$ belongs to it). Both cases can occur.
– If $p < 0$ and $p \equiv 5$ or 9 modulo 16, then $\widehat{\alpha}(\widehat{E})(\mathbb{Q})$ is equal to the classes of 1 and p, hence has 2 elements.
– If $p < 0$ and $p \equiv 1 \pmod{16}$, then $\widehat{\alpha}(\widehat{E})(\mathbb{Q})$ may have 2 or 4 elements (when the class of $b_1 = 2$ belongs to it). Both cases can occur.
– If $p < 0$ and $p \equiv 13 \pmod{16}$, then $\widehat{\alpha}(\widehat{E})(\mathbb{Q})$ may have 2 or 4 elements (when the class of $b_1 = -2$ belongs to it). Both cases can occur.
– If $p < 0$ and $p \equiv 3$ or 11 modulo 16, then $\widehat{\alpha}(\widehat{E})(\mathbb{Q})$ may have 2 or 4 elements (when the class of $b_1 = -1$ belongs to it). Both cases can occur.
– If $p < 0$ and $p \equiv 7$ or 15 modulo 16, then $\widehat{\alpha}(\widehat{E})(\mathbb{Q})$ may have 2, 4, or 8 elements (depending on the classes of $b_1 = -1$ and $b_1 = \pm 2$).

Putting together the results on both groups, we obtain the results of the proposition. □

As mentioned above, assuming a weak form of BSD, in case (3) we have either $r = 0$ or $r = 2$. We give here an example in which $r = 2$, and in the next section we will give an example with $r = 0$.

Proposition 8.2.16. *For $p = -73$ we have $r_p = 2$, generators of $E(\mathbb{Q})$ modulo torsion being $(9/16, 411/64)$ and $(4/9, 154/27)$.*

Proof. Since $p < 0$ we already know that $|\alpha(E(\mathbb{Q}))| = 2$, so we consider only $\widehat{E}$ whose equation is $y^2 = x^3 - 292x$. The squarefree divisors b_1 of $b = -292 = -2^2 \cdot 73$ less than $|b|^{1/2}$ are $b_1 = \pm 1$ and ± 2. The corresponding quartics are $Y^2 = b_1 X^4 - (292/b_1)Z^4$, and for $b_1 = -1, 2$, and -2 we obtain $(X, Y, Z) = (4, 6, 1), (3, 4, 1)$, and $(1, 12, 1)$ respectively as solutions. It follows that $|\widehat{\alpha}(\widehat{E}(\mathbb{Q}))| = 8$, hence $r_p = 2$, as claimed. To find the corresponding points on E, we proceed as follows. By Theorem 8.2.9, we find the points $(-16, -24), (18, 24)$ on the curve $\widehat{E}$. We do not need the third point corresponding to $b_1 = -2 = -1 \cdot 2$ since it will be the sum or difference of the first two (in fact it is the difference), and we do not need the points corresponding to b/b_1, which will be the opposites. We now apply the map $\widehat{\phi}$ from $\widehat{E}$ to E, thus obtaining the two points $(9/16, -411/64)$ and $(4/9, 154/27)$. These points are necessarily independent, and one can prove that they generate $E(\mathbb{Q})$ modulo torsion. $\hfill\square$

8.2.6 An Example of Second Descent

We now give an example showing how descent can be pushed one step further, and also showing that the case $r_p = 0$ can also occur in case (3) of Proposition 8.2.15.

Proposition 8.2.17. *For $p = -17$, we have $r_p = 0$.*

Proof. As above, we note that since $p < 0$ we already know that $|\alpha(E(\mathbb{Q}))| = 2$, so we consider only $\widehat{E}$ whose equation is $y^2 = x^3 - 68x$. The squarefree divisors b_1 of $b = -68 = -2^2 \cdot 17$ less than $|b|^{1/2}$ are $b_1 = \pm 1$ and ± 2, so we must consider the quartics $Y^2 = -X^4 + 68Z^4$ and $Y^2 = \varepsilon(2X^4 - 34Z^4)$ for $\varepsilon = \pm 1$. It is not difficult to see that these quartics are everywhere locally soluble. On the other hand, a quick search does not produce any solutions. We thus must work some more to show that they indeed have no solutions.

Consider the first quartic. Dividing through by Z^4 gives the conic $y^2 = -x^2 + 68$ with $y = Y/Z^2$ and $x = X^2/Z^2$. Conversely, if we have a rational point (x, y) on that conic with $x \in \mathbb{Q}^{*2}$, we can write $x = X^2/Z^2$ with $\gcd(X, Z) = 1$ and set $Y = yZ^2$, so we will have a suitable integer point on our quartic. Now $(x, y) = (2, 8)$ is an evident point on our conic, so to parametrize it we set $y - 8 = t(x - 2)$ and intersect with the conic. An easy computation gives the parametrization $x = 2(t^2 - 8t - 1)/(t^2 + 1)$, $y = -4(2t^2 + t - 2)/(t^2 + 1)$. Thus, writing $t = u/v$ with $\gcd(u, v) = 1$ we are looking for such pairs (u, v) with $2(u^2 - 8uv - v^2)/(u^2 + v^2) \in \mathbb{Q}^{*2}$.

This is equivalent to the equation $z^2 = 2(u^2 - 8uv - v^2)(u^2 + v^2)$, which is a new quartic, and we could hope to show that this quartic is not locally soluble. However, it is simpler to proceed as follows. Writing $2(u^2 - 8uv - v^2)/(u^2 + v^2) = a^2/b^2$ with $\gcd(a, b) = 1$, we see that there exists $\lambda \in \mathbb{Z}$ (which we may assume squarefree if we drop the condition $\gcd(a, b) = 1$) such that $2(u^2 - 8uv - v^2) = \lambda a^2$ and $u^2 + v^2 = \lambda b^2$. Now note that

$$(-8u + 66v)(u^2 + v^2) + (4u - v)(2(u^2 - 8uv - v^2)) = 68v^3 ,$$

hence (exchanging u and v, and replacing v by $-v$)

$$(8v + 66u)(u^2 + v^2) + (4v + u)(2(u^2 - 8uv - v^2)) = 68u^3 ,$$

so that $\lambda \mid 68$ since $\gcd(u, v) = 1$. Since $\lambda = (u^2 + v^2)/b^2 > 0$ and is squarefree, it follows that $\lambda = 1, 2, 17,$ or 34.

Assume first that λ is odd. Then $u^2 + v^2 = \lambda b^2$, so u and v have opposite parities; otherwise, they are both odd, and hence $\lambda b^2 \equiv 2 \pmod 8$, so $2 \mid b$

which is absurd. But then $u^2 - 8uv - v^2$ is odd, so that $\lambda a^2 \equiv 2 \pmod 4$ which again is impossible since it implies $2 \mid a$. Thus λ must be even, i.e., $\lambda = 2$ or 34, so that $\lambda \equiv 2 \pmod{32}$. Then $u^2 + v^2 = \lambda b^2$ and $\gcd(u, v) = 1$ imply that u and v are both odd. We thus have

$$a^2 \equiv (\lambda/2)a^2 = u^2 - 8uv - v^2 = 2((\lambda/2)b^2 - v^2 - 4uv) \equiv 2(b^2 + 3) \pmod{16} ,$$

hence $a^2 \equiv 6$, 8, or 14 modulo 16, which is impossible. Thus all four values of λ are excluded, showing that our quartic $Y^2 = -X^4 + 68Z^4$ has no solutions.

We proceed similarly for the quartics $Y^2 = \varepsilon(2X^4 - 34Z^4)$ with $\varepsilon = \pm 1$. We want to look for rational points on the conic $y^2 = \varepsilon(2x^2 - 34)$ for which x is a rational square. Clearly $(x, y) = (4 + \varepsilon, 4)$ is on the conic, so we set $y - 4 = t(x - 4 - \varepsilon)$. An easy computation gives the parametrization

$$x = \frac{t^2(4 + \varepsilon) - 8t + 8\varepsilon + 2}{t^2 - 2\varepsilon}, \quad y = \frac{-4t^2 + 4t(4\varepsilon + 1) - 8\varepsilon}{t^2 - 2\varepsilon} .$$

As above, we write $t = u/v$ with $\gcd(u, v) = 1$ and $x = a^2/b^2$ hence we deduce as above that there exists a squarefree integer λ such that

$$u^2(4 + \varepsilon) - 8uv + (8\varepsilon + 2)v^2 = \lambda a^2 \quad \text{and} \quad u^2 - 2\varepsilon v^2 = \lambda b^2 ,$$

and since

$$(-10u - 9v)(u^2 - 2v^2) + (2u + 5v)(5u^2 - 8uv + 10v^2) = 68v^3$$

and

$$(-6u + 25v)(u^2 + 2v^2) + (2u - 3v)(3u^2 - 8uv - 6v^2) = 68v^3$$

and similar identities with $68u^3$ on the right-hand side, we deduce as above that $\lambda \mid 68$ hence that $\lambda = \pm 1$, ± 2, ± 17, or ± 34.

When $\varepsilon = 1$, the quadratic form $5u^2 - 8uv + 10v^2$ has negative discriminant; hence it is always positive, so we must have $\lambda > 0$. When $\varepsilon = -1$, $u^2 + 2v^2 > 0$, so once again $\lambda > 0$. Thus in both cases we must have $\lambda = 1$, 2, 17, or 34.

Assume first that λ is odd, so that $\lambda \equiv 1 \pmod 8$. From $u^2 - 2\varepsilon v^2 = \lambda b^2$ we deduce that b is odd; otherwise, $4 \mid \lambda b^2$ so $4 \mid u^2$ hence $2 \mid v^2$, contradicting $\gcd(u, v) = 1$. Thus u is odd. It follows that

$$u^2(4 + \varepsilon) - 8uv + 8\varepsilon v^2 + 2v^2 \equiv 4 + 2\varepsilon - \varepsilon\lambda \pmod 8$$

is odd; hence a is odd, so that $3 + \varepsilon \equiv 0 \pmod 8$, which is absurd.

Assume now that λ is even, so that $\lambda/2 \equiv 1 \pmod 8$. Then u is even, so v is odd. We thus have

$$2(u/2)^2(4 + \varepsilon) - 8(u/2)v + (4\varepsilon + 1)v^2 = (\lambda/2)a^2 ;$$

hence a is odd, so

$$2\varepsilon(u/2)^2 + 4\varepsilon + 1 \equiv 1 \;(\text{mod } 8)\,,$$

hence $(u/2)^2 \equiv 2 \;(\text{mod } 4)$, a contradiction.

To conclude, we see that all values of λ are excluded, showing that our quartics $Y^2 = \varepsilon(2X^4 - 34Z^4)$ have no solutions. Putting everything together, we obtain that $|\widehat{\alpha}(\widehat{E}(\mathbb{Q}))| = 2$ hence that $r_p = r(E) = 0$.   □

Remark. For all three quartics we have been able to show local insolubility at 2 directly. In general, however, it will be necessary to parametrize one of the conics using the general theory of Diophantine equations of degree 2, and replace it in the other.

8.3 Description of General 2-Descent

From now on, we do *not* assume that our elliptic curve $y^2 = x^3 + ax + b$ has a rational 2-torsion point, and in fact we explicitly assume that it does not, in other words that the polynomial $x^3 + ax + b$ is irreducible. As usual we may always assume that a and b are rational integers.

There are essentially two methods to deal with this case. The first method is algebraic, and consists in imitating the above method by placing ourselves in a larger number field containing a 2-torsion point. This has the advantage of being easy to explain since it is a simple generalization, and also of being useful also for the computation of the group of points of an elliptic curve over an arbitrary number field in addition to $\mathbb{Q}$. It has the disadvantage of not being very efficient for small examples, although for large ones it is competitive. The second method consists in using invariant theory. It is often more efficient than the first, but has the disadvantage of being applicable only over $\mathbb{Q}$. We will describe only the first method, and refer to [Cre2] for complete details on the second method.

8.3.1 The Fundamental 2-Descent Map

Let $K = \mathbb{Q}(\theta)$ be the number field generated over $\mathbb{Q}$ by a root θ of the equation $x^3 + ax + b = 0$. Consider the map α from $E(\mathbb{Q})$ to K^*/K^{*2} defined by $\alpha(\mathcal{O}) = 1 \;(\text{mod } K^{*2})$ and

$$\alpha(P) = x - \theta \;(\text{mod } K^{*2}) \text{ if } P = (x,y) \neq \mathcal{O}\,,$$

where of course modulo is taken in the multiplicative sense. As in the rational 2-torsion case, the main usefulness of this map comes from the following result.

Proposition 8.3.1. (1) *The map α is a group homomorphism from $E(\mathbb{Q})$ to K^*/K^{*2}.*

(2) *The kernel of α is equal to $2E(\mathbb{Q})$.*

Proof. (1). We treat the generic case, leaving the (easy) special cases to the reader. Clearly if $P = (x, y)$ then

$$\alpha(-P)\alpha(P) = \alpha((x, -y))\alpha((x, y)) = (x - \theta)^2 \equiv 1 \ (\mathrm{mod}\ K^{*2}) \ ,$$

so α sends inverses to inverses. Thus we must prove that if $P_1 + P_2 + P_3 = \mathcal{O}$ then $\alpha(P_1)\alpha(P_2)\alpha(P_3) \equiv 1 \ (\mathrm{mod}\ K^{*2})$. Let $y = mx + n$ be the equation of the line passing through the three points. Writing the intersection of the line with the cubic equation, we see that the three abscissas x_i of the points P_i are the three roots of the equation $A(x) = 0$ with

$$A(X) = (X^3 + aX + b) - (mX + n)^2 \ ,$$

so by definition of θ we have

$$(x_1 - \theta)(x_2 - \theta)(x_3 - \theta) = -A(\theta) = (m\theta + n)^2 \equiv 1 \ (\mathrm{mod}\ K^{*2}) \ ,$$

proving (1).

(2). It follows from (1) that

$$\alpha(2P) \equiv \alpha(P)^2 \equiv 1 \ (\mathrm{mod}\ K^{*2}) \ ,$$

so that $2E(\mathbb{Q}) \subset \mathrm{Ker}(\alpha)$. Conversely, assume that $Q = (x, y) \in \mathrm{Ker}(\alpha)$ with $Q \neq \mathcal{O}$, in other words that $x - \theta = u^2$ for some $u = u_2\theta^2 + u_1\theta + u_0 \in K$ with $u_i \in \mathbb{Q}$ for all i. Expanding u^2 and using $\theta^3 = -a\theta - b$, we obtain the three equations

$$\begin{cases} au_2^2 - u_1^2 - 2u_0u_2 = 0 \ , \\ bu_2^2 + 2au_1u_2 - 2u_0u_1 = 1 \ , \\ -2bu_1u_2 + u_0^2 = x \ . \end{cases}$$

Clearly $u_2 \neq 0$, since otherwise $u_1 = 0$ by the first equation, so $0 = 1$ by the second. I claim that the point $P = (u_1/u_2, 1/u_2)$ is in $E(\mathbb{Q})$ and is such that $Q = \pm 2P$ for a suitable sign $\pm$. As in the rational 2-torsion case this is a simple but tedious verification. Indeed, first note that using the above equations we have

$$\left(\frac{u_1}{u_2}\right)^3 + a\left(\frac{u_1}{u_2}\right) + b = \frac{u_1^3 + au_1u_2^2 + bu_2^3}{u_2^3} = \frac{u_1(au_2^2 - 2u_0u_2) + au_1u_2^2 + bu_2^3}{u_2^3}$$

$$= \frac{2au_1u_2 - 2u_0u_1 + bu_2^2}{u_2^2} = \frac{1}{u_2^2} \ ,$$

hence $P \in E(\mathbb{Q})$. Furthermore, multiplying the first of the above equations by $2u_0$ and subtracting u_1 times the second, we obtain the identity

$$u_1 = u_2(bu_1u_2 + 2au_1^2 - 2au_0u_2 + 4u_0^2) \ .$$

Thus if we set $2P = (x_3, y_3)$ we have $x_3 = m^2 - 2(u_1/u_2)$ with

$$m = \frac{3(u_1/u_2)^2 + a}{2/u_2} = \frac{3u_1^2 + au_2^2}{2u_2} = \frac{3(au_2^2 - 2u_0u_2) + au_2^2}{2u_2} = 2au_2 - 3u_0 \ .$$

Using the above identity, it follows that

$$x_3 = 4a^2u_2^2 - 12au_0u_2 + 9u_0^2 - 2(bu_1u_2 + 2au_1^2 - 2au_0u_2 + 4u_0^2)$$
$$= u_0^2 - 8au_0u_2 - 4au_1^2 + 4a^2u_2^2 - 2bu_1u_2$$
$$= u_0^2 - 2bu_1u_2 + 4a(au_2^2 - 2u_0u_2 - u_1^2) = x$$

by the first and third of the basic relations above. Since $P \in E(\mathbb{Q})$, it follows that $y_3 = \pm y$ hence that $Q = \pm 2P$ as claimed, proving the proposition. $\square$

Corollary 8.3.2. *The map α induces an injective group homomorphism from $E(\mathbb{Q})/2E(\mathbb{Q})$ to K^*/K^{*2} (which by abuse of notation we will still denote by α). In addition, if the image of α is finite then $E(\mathbb{Q})$ is finitely generated of rank r equal to the dimension of the image of α as an $\mathbb{F}_2$-vector space.*

Proof. The first statement is clear. For the second we note that by assumption E has no rational 2-torsion; hence if $d = \dim_{\mathbb{F}_2}(\mathrm{Im}(\alpha))$ we have $|E(\mathbb{Q})/2E(\mathbb{Q})| = |\mathrm{Im}(\alpha)| = 2^d$, so Theorem 8.2.7 implies that $E(\mathbb{Q})$ is finitely generated of some rank r such that $2^r = |E(\mathbb{Q})/2E(\mathbb{Q})|$, hence that $r = d$ as claimed. $\square$

To have precise information on $E(\mathbb{Q})$ we must therefore determine the image of α. For this we need the notion of the T-Selmer group of a number field (not to be confused with the Selmer group of the elliptic curve, although the notions are related).

8.3.2 The T-Selmer Group of a Number Field

For the reader's convenience in the following definitions we have included the classical definitions of T-unit group and T-class group.[1]

Definition 8.3.3. *Let T be a finite set of finite places of K.*

(1) *We say that an element $u \in K^*$ is a T-unit if $v_{\mathfrak{p}}(u) = 0$ for every prime ideal $\mathfrak{p}$ such that $\mathfrak{p} \notin T$. The group of T-units is denoted by $U_T(K)$.*

(2) *We define the T-class group $Cl_T(K)$ as the quotient group of the ordinary class group $Cl(K)$ by the subgroup generated by the classes of the elements of T.*

---

[1] We use T instead of the more standard S to avoid notation such as $S_S(K)$.

(3) *We say that an element $u \in K^*$ is a T-virtual square if $v_\mathfrak{p}(u) \equiv 0 \pmod{2}$ for every prime ideal $\mathfrak{p}$ such that $\mathfrak{p} \notin T$.*

(4) *We define the T-Selmer group $S_T(K)$ as the set of classes of virtual squares modulo K^{*2}.*

Remark. Most authors use the notation $K(T, 2)$ instead of $S_T(K)$.

Denote by $\mathcal{A}$ the group of fractional ideals generated by the elements of T. The reader can easily check that u is a T-unit if and only if $u\mathbb{Z}_K \in \mathcal{A}$ and that u is a T-virtual square if and only if $u\mathbb{Z}_K = \mathfrak{q}^2\mathfrak{a}$ for some ideal $\mathfrak{q}$ and some $\mathfrak{a} \in \mathcal{A}$.

The main properties of these notions are summarized in the following proposition.

Proposition 8.3.4. *Let K be a number field of signature (r_1, r_2), let T be a finite set of finite places of K, and denote by t its cardinality.*

(1) *The group $U_T(K)$ is a finitely generated abelian group of rank $r_1 + r_2 + t - 1$, whose torsion subgroup is independent of T and equal to the (cyclic) group of roots of unity of K. In particular,*

$$\left| \frac{U_T(K)}{U_T(K)^2} \right| = 2^{r_1 + r_2 + t} \ .$$

(2) *We have a natural split exact sequence*

$$1 \longrightarrow \frac{U_T(K)}{U_T(K)^2} \longrightarrow S_T(K) \longrightarrow Cl_T(K)[2] \longrightarrow 1 \ ,$$

where as usual for an abelian group G, $G[2]$ denotes the subgroup of G killed by 2. In particular, $S_T(K)$ is finite and its cardinality is equal to $2^{r_1+r_2+t+s'}$, where s' denotes the 2-rank of $Cl_T(K)$, so $|S_T(K)|$ divides $2^{r_1+r_2+t+s}$, where s denotes the 2-rank of $Cl(K)$.

Proof. (1). Although the proof is well known and easy we repeat it here. We have a natural exact sequence

$$1 \longrightarrow U(K) \longrightarrow U_T(K) \longrightarrow \mathcal{A} \longrightarrow Cl(K) \longrightarrow Cl_T(K) \longrightarrow 1 \ ,$$

where the map starting from $U_T(K)$ sends u to the ideal $u\mathbb{Z}_K$, and the map starting from $\mathcal{A}$ sends an ideal to its ideal class. It is immediately checked that the sequence is indeed exact. Since $Cl(K)$ and a fortiori $Cl_T(K)$ are finite groups, it follows that $U_T(K)$ is finitely generated and its rank is equal to that of $U(K)$ $(r_1 + r_2 - 1)$ plus that of $\mathcal{A}$, equal to t. The statement concerning the torsion subgroup is clear.

(2). Let $\bar{u} \in U_T(K)$, so that $u\mathbb{Z}_K = \mathfrak{q}^2\mathfrak{a}$ for some $\mathfrak{a} \in \mathcal{A}$. We send $\bar{u}$ to the class of $\mathfrak{q}$ in $Cl_T(K)$. Clearly this does not depend on the decomposition

$\mathfrak{q}^2\mathfrak{a}$ or on the chosen representative u of $\overline{u}$ in K^*. Since $\mathfrak{q}^2 = u\mathfrak{a}^{-1}$ it is clear that the class of $\mathfrak{q}$ belongs in fact to $Cl_T(K)[2]$. With this map defined, it is then easily checked that the given sequence is exact and split. The statements concerning the cardinality of $S_T(K)$ follow.   □

It is clear that $S_T(K)$ is an $\mathbb{F}_2$-vector space, and from the existence of these two exact sequences it is not difficult to give an $\mathbb{F}_2$-basis for $S_T(K)$. As always in this book we assume that we have at our disposal a CAS such as `Pari/GP` that can efficiently compute class and unit groups of number fields. We first compute explicitly $U_T(K)$ using the algorithm given in [Coh1] Proposition 7.4.7, and therefore also an $\mathbb{F}_2$-basis of $U_T(K)/U_T(K)^2$. We compute $Cl_T(K)$ as a quotient of $Cl(K)$ using the general quotient algorithm for abelian groups ([Coh1] Algorithm 4.1.7), and we can then easily compute $Cl_T(K)[2]$ and use the splitting of the exact sequence to obtain an $\mathbb{F}_2$-basis of $S_T(K)$. In the frequent special case in which the class number $h(K) = |Cl(K)|$ of K is *odd* (in particular when it is equal to 1), then $S_T(K) = U_T(K)/U_T(K)^2$, and as an $\mathbb{F}_2$ basis of $S_T(K)$ we can take the disjoint union of generators of the $h(K)$th power of each prime ideal of T (which are principal ideals) together with a system of fundamental units and a generator of the group of roots of unity of K. We will see several explicit examples below.

8.3.3 Description of the Image of α

With these definitions and properties, it is now easy to determine the image of α. We keep all the above assumptions and notation; in other words, E is an elliptic curve defined over $\mathbb{Q}$ by a Weierstrass equation $y^2 = x^3 + ax + b$ with a and b in $\mathbb{Z}$, we let θ be a root of $x^3 + ax + b = 0$, assumed to be irreducible, and we set $K = \mathbb{Q}(\theta)$. Finally, we set $I(\theta) = [\mathbb{Z}_K : \mathbb{Z}[\theta]]$, the index of $\mathbb{Z}[\theta]$ in the full ring of integers $\mathbb{Z}_K$. Thus, if $d(K)$ is the discriminant of the number field K we have $-(4a^3 + 27b^2) = d(K)I(\theta)^2$.

Proposition 8.3.5. *Let $P = (x, y) \in E(\mathbb{Q}) \setminus \{\mathcal{O}\}$, assume that $\mathfrak{q}$ is a prime ideal of K such that $v_{\mathfrak{q}}(x-\theta)$ is odd, and denote by q the prime number below $\mathfrak{q}$. Then $v_{\mathfrak{q}}(x - \theta) \geqslant 1$, $q \mid (3\theta^2 + a)$, and $q \mid \gcd(y, 3x^2 + a, I(\theta))$ (so that in particular $q^2 \mid (4a^3 + 27b^2)$).*

Proof. Set $\gamma = x - \theta \in K$. We can write $y^2 = \gamma C$ with $C = \gamma^2 + 3\theta\gamma + 3\theta^2 + a$. If $v_{\mathfrak{q}}(\gamma)$ were negative we would have $v_{\mathfrak{q}}(\gamma) = v_{\mathfrak{q}}(y^2) - v_{\mathfrak{q}}(C) = 2(v_{\mathfrak{q}}(y) - v_{\mathfrak{q}}(\gamma)) \equiv 0 \pmod 2$, since a and θ are integral. Thus when $v_{\mathfrak{q}}(\gamma)$ is odd we have $v_{\mathfrak{q}}(\gamma) \geqslant 1$, and we deduce from the expression for C that $v_{\mathfrak{q}}(C) \geqslant 0$. Since $v_{\mathfrak{q}}(C) = 2v_{\mathfrak{q}}(y) - v_{\mathfrak{q}}(\gamma) \equiv 1 \pmod 2$, we have $v_{\mathfrak{q}}(C) \geqslant 1$, hence $v_{\mathfrak{q}}(3\theta^2 + a) = v_{\mathfrak{q}}(C - \gamma(\gamma + 3\theta)) \geqslant 1$, proving the first two results. Since $y^2 = \gamma C$ and $\mathfrak{q} \mid \gamma$ (or C), we have $v_{\mathfrak{q}}(y) \geqslant 1$. Furthermore, $3x^2 + a = 3(\gamma + \theta)^2 = 3\gamma^2 + 6\gamma\theta + 3\theta^2 + a$, and since $\mathfrak{q} \mid \gamma$ and $\mathfrak{q} \mid (3\theta^2 + a)$ it follows that $v_{\mathfrak{q}}(3x^2 + a) \geqslant 1$. There remains to prove that $q \mid I(\theta)$, a result and proof that were communicated to me by D. Simon. Assume for the moment the following lemma.

Lemma 8.3.6. *If $q \nmid I(\theta)$ then $\mathfrak{q}$ is the only ideal $\mathfrak{q}_i$ above q such that $x - \theta \in \mathfrak{q}_i$, and it has residual degree 1.*

If we assume by contradiction that $q \nmid I(\theta)$ we thus have $(x - \theta)\mathbb{Z}_K = \mathfrak{q}^v \mathfrak{a}$, where $v = v_\mathfrak{q}(x - \theta)$, and $\mathfrak{a}$ is an ideal coprime to all ideals above q. Since $f(\mathfrak{q}/q) = 1$ we thus have

$$y^2 = |x^3 + ax + b| = |\mathcal{N}_{K/\mathbb{Q}}(x - \theta)| = \mathcal{N}_{K/\mathbb{Q}}(\mathfrak{q})^v \mathcal{N}_{K/\mathbb{Q}}(\mathfrak{a}) = q^v \mathcal{N}_{K/\mathbb{Q}}(\mathfrak{a}) ,$$

so $v = v_q(x - \theta) = 2v_q(y)$ is even, in contradiction to the assumption of the proposition, and finishing the proof. $\square$

Proof of Lemma 8.3.6. By Proposition 3.3.22, since $q \nmid I(\theta)$ the decomposition of $q\mathbb{Z}_K$ into prime ideals copies the decomposition of the polynomial $R(X) = X^3 + aX + b$ modulo q. Thus, write $R(X) \equiv \prod_i R_i(X)^{e_i} \pmod{q}$, where $R_i(X)$ are monic polynomials in $\mathbb{Z}[X]$. We then have $q\mathbb{Z}_K = \prod_i \mathfrak{q}_i^{e_i}$, where $\mathfrak{q}_i = q\mathbb{Z}_K + R_i(\theta)\mathbb{Z}_K$, and $f(\mathfrak{q}_i/q) = \deg(R_i)$, and we reorder the $\mathfrak{q}_i$ so that $\mathfrak{q}_1 = \mathfrak{q}$. If we write $x = n/d$ with coprime n and d in $\mathbb{Z}$, we see that $v_q(d) = 0$; otherwise, $v_\mathfrak{q}(x - \theta) = v_\mathfrak{q}(n/d - \theta) < 0$ since $v_\mathfrak{q}(\theta) \geqslant 0$. Thus if we set $x_1 = nd^{-1} \bmod q$ we have $v_\mathfrak{q}(x - x_1) \geqslant 1$, hence $x_1 - \theta \in \mathfrak{q}$. Since $(R_1(X) - R_1(x_1))/(X - x_1) \in \mathbb{Z}[X]$, it follows that $R_1(\theta) - R_1(x_1) \in \mathfrak{q}$, hence that $R_1(x_1) \in \mathfrak{q}$, in other words $R_1(x_1) \equiv 0 \pmod{q}$. Since $\overline{R_1}$ is irreducible in $(\mathbb{Z}/q\mathbb{Z})[X]$, this means that $R_1(X) = X - x_1$, and in particular that $\deg(R_1) = 1$, so that $f(\mathfrak{q}/q) = 1$. Furthermore, since the R_i are pairwise coprime modulo q, x_1 cannot be a root of R_i for $i \neq 1$, so $x - \theta$ cannot belong to $\mathfrak{q}_i$ for $i \neq 1$. $\square$

Corollary 8.3.7. *Denote by T the set of prime ideals $\mathfrak{q}$ of K such that $\mathfrak{q} \mid (3\theta^2 + a)$ and $q \mid I(\theta)$, where q is the prime number below $\mathfrak{q}$. The image of α is equal to the group of $\overline{u} \in S_T(K)$ such that $\mathcal{N}_{K/\mathbb{Q}}(u)$ is a square in $\mathbb{Q}$ for some (or every) lift of $\overline{u}$ to K^*, and for which there exists a lift u of the form $x - \theta$.*

Proof. Let $P = (x, y) \in E(\mathbb{Q}) \setminus \{\mathcal{O}\}$, so that $\alpha(P) = x - \theta$. By the proposition, if $v_\mathfrak{q}(x - \theta)$ is odd we have $\mathfrak{q} \mid (3\theta^2 + a)$ and $q \mid I(\theta)$, hence $\mathfrak{q} \in T$; in other words, the class $\overline{u}$ of $\alpha(P)$ belongs to $S_T(K)$. It is evidently the class of an element of the form $x - \theta$, and since $\mathcal{N}_{K/\mathbb{Q}}(u) = x^3 + ax + b$ it follows that $(x, y) \in E(\mathbb{Q})$ if and only if $\mathcal{N}_{K/\mathbb{Q}}(u) = y^2$ is a square in $\mathbb{Q}$. $\square$

Remarks. (1) If E is given by an equation $y^2 = R(x)$ with $R(x) = x^3 + ax^2 + bx + c$, it is clear that the above corollary is still valid if we replace $\mathfrak{q} \mid 3\theta^2 + a$ by $\mathfrak{q} \mid R'(\theta)$.

(2) It can be shown (see [Sch-Sto]) that an additional condition on q is that the Tamagawa number c_q be even (see Definition 7.3.25), and it is easy to see on examples that this condition does not follow from the others; see Exercise 16.

Corollary 8.3.8 (Mordell). *The group $E(\mathbb{Q})/2E(\mathbb{Q})$ is finite and the group $E(\mathbb{Q})$ is finitely generated. More precisely, $|E(\mathbb{Q})/2E(\mathbb{Q})|$ divides $2^{r_1+r_2+t+s}$, using the above notation.*

Proof. The finiteness and the bound for $E(\mathbb{Q})/2E(\mathbb{Q})$ follow from the above proposition and Proposition 8.3.4. The statement for $E(\mathbb{Q})$ follows from Theorem 8.2.7.   □

Although our aim is practical, note that we have finished the proof of the Mordell–Weil theorem over $\mathbb{Q}$.

8.3.4 Practical Use of 2-Descent in the General Case

We now explain how to use the above results in practice, keeping in mind that, as in the rational 2-torsion case, there does not exist any unconditional algorithm for computing the rank.

We begin by computing $\mathrm{disc}(R)$, the number field $K = \mathbb{Q}(\theta)$, its discriminant $d(K)$ as well as the index $I(\theta) = \sqrt{\mathrm{disc}(R)/d(K)}$, and finally the set T of prime ideals $\mathfrak{q}$ of K such that $\mathfrak{q} \mid (3\theta^2 + a)$ and $\mathfrak{q} \mid I(\theta)$. Using the algorithms explained at the end of Section 8.3.2 we then compute an $\mathbb{F}_2$-basis of $S_T(K)$. Using [Coh1], Algorithm 4.1.11, we then compute the kernel $S_T(K,1)$ of the norm map from $S_T(K)$ to $\mathbb{Q}^*/\mathbb{Q}^{*2}$. By Corollary 8.3.7 the image of α is exactly the group of elements $\overline{u} \in S_T(K,1)$ that have a lift $u \in K^*$ of the form $x - \theta$. Up to this point the computation is completely algorithmic. However, the determination of such elements $\overline{u}$ is the nonalgorithmic part of the method since we are going to see that, as in the rational 2-torsion case, it leads to the determination of rational points on hyperelliptic quartics.

Let $u = u_2\theta^2 + u_1\theta + u_0$ be any lift of $\overline{u}$. We must determine whether there exists $\gamma = c_2\theta^2 + c_1\theta + c_0 \in K^*$ such that $u\gamma^2 = x - \theta$. Expanding, we have

$$u\gamma^2 = q_2(c_0, c_1, c_2)\theta^2 - q_1(c_0, c_1, c_2)\theta + q_0(c_0, c_1, c_2) \,,$$

where the q_i are explicit integral quadratic forms in the c_j. Thus we must solve the equations $q_2(c_0, c_1, c_2) = 0$, $q_1(c_0, c_1, c_2) = 1$, and then x is determined thanks to $q_0(c_0, c_1, c_2) = x$. The solubility of the first equation can easily be determined thanks to the Hasse–Minkowski theorem, an explicit solution can then be found using the algorithm explained in Section 6.3.3, and the general solution is given by Proposition 6.3.4 and its corollary. Thus there exist quadratic polynomials $P_i(X, Y)$ such that $q_2(c_0, c_1, c_2) = 0$ if and only if there exist coprime integers s and t and $d \in \mathbb{Q}$ such that $c_i = dP_i(s, t)$ for $0 \leqslant i \leqslant 2$. The equation $q_1(c_0, c_1, c_2) = 1$ can thus be written

$$q_1(P_0(s,t), P_1(s,t), P_2(s,t)) = 1/d^2 ,$$

which is a hyperelliptic quartic equation. The rest of the process is similar
to the case of rational 2-torsion: we must determine whether this quartic
equation is everywhere locally soluble. If it is not, we exclude $\overline{u}$ from the
consideration of the points in the image of α. If it is, we look as intelligently
as possible for rational points on the quartic. If we find one, we include $\overline{u}$ in
the image of α (and $x = d^2 q_0(P_0(s,t), P_1(s,t), P_2(s,t))$ is the explicit abscissa
of the corresponding point in $E(\mathbb{Q})$). If we cannot find one, we are stuck and
cannot determine $E(\mathbb{Q})$ without further work such as a second descent.

Remark. The group of $\overline{u} \in S_T(K,1)$ such that the corresponding quartic
is everywhere locally soluble is the smallest group containing $E(\mathbb{Q})/2E(\mathbb{Q})$
that can be determined algorithmically using only a 2-descent. It is called the
2-*Selmer group* of the elliptic curve E and denoted by $S_2(E)$. The quotient
of $S_2(E)$ by its subgroup $E(\mathbb{Q})/2E(\mathbb{Q})$ is the part of the so-called *Tate–
Shafarevich* group $Ш(E)$ of E killed by 2, so that we have an exact sequence
(analogous to the one for $S_T(K)$, whence the name and the notation)

$$1 \longrightarrow \frac{E(\mathbb{Q})}{2E(\mathbb{Q})} \longrightarrow S_2(E) \longrightarrow Ш(E)[2] \longrightarrow 1 .$$

The group $Ш(E)[2]$ is the the obstruction to performing a 2-descent. In the
rational 2-torsion case, the groups $Ш(E)[2]$ and $Ш(\widehat{E})[2]$ are both obstructions to performing a 2-descent, although if either of them is trivial there is
no obstruction to performing a second descent (I owe this remark to J. Cremona).

8.3.5 Examples of General 2-Descent

As examples of general 2-descent we consider two curves that we will need
for the proof of Corollary 14.6.11 in Chapter 14.

Proposition 8.3.9. (1) *The curve E defined by $y^2 = x^3 - 16$ has rank 0
over $\mathbb{Q}$ and trivial torsion group; in other words, $E(\mathbb{Q}) = \{\mathcal{O}\}$.*
(2) *The curve E defined by $y^2 = x^3 + 16$ has rank 0 over $\mathbb{Q}$ and torsion
group of order 3 generated by $(x,y) = (0,4)$; in other words, $E(\mathbb{Q}) = \{\mathcal{O}, (0, \pm 4)\}$.*

Proof. The statements concerning the torsion subgroups easily follow from
the results of Section 8.1.3, so we only compute the ranks. In both cases
we have $K = \mathbb{Q}(\beta)$ with $\beta^3 = 2$, and $\theta = 2\varepsilon\beta$, where $\varepsilon = 1$ for the first
equation and $\varepsilon = -1$ for the second. We compute that $\operatorname{disc}(R) = -2^8 \cdot 3^3$,
and $\operatorname{disc}(K) = -2^2 3^3$; hence $I(\theta) = 2^6$, so T contains only prime ideals
above 2. Since 2 is totally ramified as $2\mathbb{Z}_K = \mathfrak{p}_2^3$ with $\mathfrak{p}_2 = \beta\mathbb{Z}_K$ and $\mathfrak{p}_2 \mid \theta$,
we take $T = \{\mathfrak{p}_2\}$. Since the class number of K is equal to 1, it follows that

an $\mathbb{F}_2$-basis of $S_T(K)$ is given by the classes modulo squares of the union of a generator of $\mathfrak{p}_2$ with the fundamental units and generator of torsion units, so here by the classes modulo squares of β, $\beta - 1$, and -1 of respective absolute norms 2, 1, and -1. Thus if $u = \beta^a(\beta - 1)^b(-1)^c$ the norm of u is a square in $\mathbb{Q}$ if and only if a and c are even. We deduce that the group $S_T(K, 1)$ of elements of $S_T(K)$ whose norm is a square is an $\mathbb{F}_2$-vector space of dimension 1 generated by $\beta - 1$.

It follows that for both curves the only quartics to consider are those corresponding to $\beta - 1$. Let us compute explicitly the quadratic forms q_0, q_1, and q_2 as above, in other words such that

$$(\beta - 1)(4c_2\beta^2 + 2\varepsilon c_1\beta + c_0)^2 = q_2(c_0, c_1, c_2)\theta^2 - q_1(c_0, c_1, c_2)\theta + q_0(c_0, c_1, c_2) ,$$

where we recall that $\theta = 2\varepsilon\beta$. We obtain

$$q_2(c_0, c_1, c_2) = 8c_2^2 - 2c_0c_2 - c_1^2 + \varepsilon c_0c_1 ,$$
$$q_1(c_0, c_1, c_2) = 16\varepsilon c_2^2 - 16c_1c_2 + 2c_0c_1 - \varepsilon c_0^2/2 ,$$
$$q_0(c_0, c_1, c_2) = -32\varepsilon c_1c_2 + 16c_0c_2 + 8c_1^2 - c_0^2 .$$

The equation $q_2(c_0, c_1, c_2) = 0$ has the evident solution $(c_0, c_1, c_2) = (\varepsilon, 1, 0)$. Thus by Proposition 6.3.4 we can easily parametrize the general solution: we may choose

$$M = \begin{pmatrix} 1 & 0 & \varepsilon \\ 0 & 0 & 1 \\ 0 & 1 & 0 \end{pmatrix} ,$$

so that $R = (s, 0, t)^t$; hence (after replacing d by εd in the formula of the proposition) the general solution is given by

$$c_0 = d(8t^2 - s^2), \quad c_1 = d\varepsilon(8t^2 - 2st), \quad c_2 = d(2t^2 - st) .$$

The condition $q_1 = 1$ gives the hyperelliptic quartic equations

$$\varepsilon(-96t^4 + 96st^3 - 24s^2t^2 + 4s^3t - s^4/2) = 1/d^2 ,$$

and since s and t are (coprime) integers we have $1/d = Y$ with $Y \in \mathbb{Z}$ (we cannot have $1/d = Y/2$ with Y odd; otherwise, the left-hand side would have a denominator 4), so s is even, and hence t is odd, and writing $s = 2s_1$ we see that $8 \mid Y^2$, so that $4 \mid Y$; hence writing $Y = 4Y_1$, this implies that $s_1 = 2s_2$ is even, giving the equation

$$2\varepsilon(-3t^4 + 12s_2t^3 - 12s_2^2t^2 + 8s_2^3t - 4s_2^3) = Y_1^2 ,$$

implying that Y_1 is even, hence that t is also even, a contradiction since s and t are assumed coprime. It follows that the quartic associated with $\beta - 1$ is not 2-adically soluble, so the image of the 2-descent map α is trivial, proving that the rank of the two curves is equal to 0 by Corollary 8.3.2.   □

8.4 Description of 3-Descent with Rational 3-Torsion Subgroup

Although 2-descent (possibly followed by a second descent) often works, there are many cases in which it does not. The obstruction to this is the fact that the Tate–Shafarevich group III of the curve has a nontrivial 2-part, analogous to the fact that the obstruction to nonunique factorization in number fields is a nontrivial class group. It is not necessary to understand precisely the definition of III to grasp the underlying philosophy.

When 2-descent does not work, we can try p-descent for a larger prime p (I will not give the precise definition), hoping that the p-part of III is trivial. In the present section I will give the example of 3-descent when there exists a rational 3-torsion subgroup, and I thank T. Fisher and J. Cremona for many explanations. It is very analogous to 2-descent when there exists a rational 2-torsion point (Section 8.2.3).

8.4.1 Rational 3-Torsion Subgroups

We first have to emphasize that there is a difference between having a rational 3-torsion *point* and having a rational 3-torsion *subgroup*, the latter meaning that there exists a subgroup of order 3 of $E(\mathbb{Q})$ that is stable under the action of Galois conjugation, but not necessarily composed of three rational points. More precisely, we set the following definition.

Definition 8.4.1. *Let E be an elliptic curve defined over a perfect commutative field K, and let $\mathcal{T}$ be a finite subgroup of $E(L)$ for some extension L/K, which without loss of generality we may assume to be finite and Galois. We say that $\mathcal{T}$ is a K-rational subgroup of E if it is globally stable by any $\sigma \in \mathrm{Gal}(L/K)$, in other words if $T \in \mathcal{T}$ implies that $\sigma(T) \in \mathcal{T}$.*

A more elegant (but strictly equivalent way) of expressing this definition is simply to say that $\mathcal{T}$ is stable under $\mathrm{Gal}(\overline{K}/K)$, without introducing the field L.

Proposition 8.4.2. *Let E be an elliptic curve defined over a perfect commutative field K of characteristic different from 2 and having a K-rational subgroup of order 3, necessarily of the form $\mathcal{T} = \{\mathcal{O}, T, -T\}$.*

(1) *The abscissa $x(T)$ of T is in K.*
(2) *Up to a change of x into $x - x_0$ for some $x_0 \in K$ the equation of E is $y^2 = x^3 + d(ax + 1)^2$ for some $d \in K^*$ and $a \in K$, and then $T = (0, \sqrt{d})$.*
(3) *If in addition E has a K-rational point T of order 3, up to the same change the equation of E is $y^2 = x^3 + (ax + b)^2$ for some $a \in K$ and $b \in K^*$, and then $T = (0, b)$.*

Proof. We have necessarily $T = \{\mathcal{O}, (x, y), (x, -y)\}$ with $x = x(T)$ and $y = y(T)$. Let L be as in the definition of a K-rational subgroup. If $\sigma \in \mathrm{Gal}(L/K)$ we must have $(\sigma(x), \sigma(y)) \in T$; therefore since it is an affine point we have $(\sigma(x), \sigma(y)) = (x, \pm y)$ for a suitable sign $\pm$. In particular, $\sigma(x) = x$ for all $\sigma \in \mathrm{Gal}(L/K)$, so by Galois theory $x \in K$, proving (1). For (2), we note that thanks to (1), changing x into $x - x_0$ we may assume that $x(T) = 0$, so that $T = (0, \theta)$ with θ not necessarily in K. If the equation of E in the new coordinates is $y^2 = x^3 + Ax^2 + Bx + C$ then $C = \theta^2$, and $C \neq 0$ since otherwise T is also of order 2, which is impossible since $T \neq \mathcal{O}$. We thus have $x(2T) = (B/(2y(T)))^2 - A = (B^2 - 4AC)/(4C)$, and T has order 3 if and only if $x(2T) = x(-T) = x(T) = 0$, hence if and only if $B^2 - 4AC = 0$. Since $C \neq 0$ we thus have $Ax^2 + Bx + C = d(ax + 1)^2$ for $d = C$ and $a = B/(2C)$, proving (2), and (3) is an immediate consequence.   □

When E has a K-rational subgroup of order 3 it will be more convenient to work with a more general equation of the form

$$y^2 = x^3 + d(ax + b)^2$$

instead of $y^2 = x^3 + d(ax + 1)^2$, which is of course equivalent to the preceding form since $d(ax + b)^2 = db^2((a/b)x + 1)^2$ and $b \neq 0$, keeping in mind that we can change (d, a, b) into $(df^2, a/f, b/f)$. We note for future reference that the discriminant of the elliptic curve E is given by

$$\mathrm{disc}(E) = 16d^2 b^3 (27b - 4a^3 d) ,$$

so that in particular $d \neq 0$, $b \neq 0$, and $27b - 4a^3 d \neq 0$.

8.4.2 The Fundamental 3-Isogeny

From now on we follow what we have done for 2-descent in Section 8.2. The proofs are very similar, the main difference being that we will have to deal with elements of $\mathbb{Q}(\sqrt{d})$ and not only of $\mathbb{Q}$.

Thus from now on let E be an elliptic curve defined over $\mathbb{Q}$ having a rational subgroup of order 3, so that by the above proposition, up to translation of the x-coordinate we may assume that E is given by an equation of the form $y^2 = x^3 + d(ax + b)^2$. We fix the 3-torsion point $T = (0, b\sqrt{d})$, which may not be in $E(\mathbb{Q})$, but the group $\{\mathcal{O}, T, -T\}$ of order 3 is a rational subgroup.

As in Section 8.2 we will work with a pair of elliptic curves E and $\widehat{E}$, defined by a similar equation $y^2 = x^3 + \widehat{d}(\widehat{a}x + \widehat{b})^2$, where

$$\widehat{d} = -3d, \quad \widehat{a} = a, \quad \widehat{b} = \frac{27b - 4a^3 d}{9} .$$

Note that $\widehat{b} = -\mathrm{disc}(E)/(144d^2 b^3) \neq 0$, and since

$$\mathrm{disc}(\widehat{E}) = 16\widehat{d}^2\widehat{b}^3(27\widehat{b} - 4\widehat{a}^3\widehat{d}) = 1296\widehat{d}^2\widehat{b}^3 b ,$$

the curve $\widehat{E}$ is indeed nonsingular, hence is an elliptic curve.

This curve has the same form as E, and it thus has a rational subgroup of order 3 generated by

$$\widehat{T} = \left(0, \frac{27b - 4a^3d}{9}\sqrt{-3d} \right) .$$

Note that $\widehat{\widehat{d}} = 9d$, $\widehat{\widehat{a}} = a$, and $\widehat{\widehat{b}} = 9b$, so that the curve $\widehat{\widehat{E}}$ is the curve $y^2 = x^3 + 9d(ax + 9b)^2$, which is trivially isomorphic to E by replacing x by $9x$ and y by $27y$.

Proposition 8.4.3. *For any $P = (x, y) \in E$ set*

$$\phi(P) = (\widehat{x}, \widehat{y}) = \left(\frac{x^3 + 4d((a^2/3)x^2 + abx + b^2)}{x^2}, \ \frac{y(x^3 - 4db(ax + 2b))}{x^3} \right)$$

for P not equal to $\pm T$ or $\mathcal{O}$, and set $\phi(T) = \phi(-T) = \phi(\mathcal{O}) = \widehat{\mathcal{O}}$. Then ϕ is a group homomorphism from E to $\widehat{E}$, whose kernel is equal to $\{\mathcal{O}, T, -T\}$. Dually, there exists a homomorphism $\widehat{\phi}$ from $\widehat{E}$ to E defined for $\widehat{P} = (\widehat{x}, \widehat{y})$ different from $\pm\widehat{T}$ and $\widehat{\mathcal{O}}$ by

$$\widehat{\phi}(\widehat{P}) = (x, y) = \left(\frac{\widehat{x}^3 + 4\widehat{d}((\widehat{a}^2/3)\widehat{x}^2 + \widehat{a}\widehat{b}\widehat{x} + \widehat{b}^2)}{9\widehat{x}^2}, \ \frac{\widehat{y}(\widehat{x}^3 - 4\widehat{d}\widehat{b}(\widehat{a}\widehat{x} + 2\widehat{b}))}{27\widehat{x}^3} \right)$$

and by $\widehat{\phi}(\widehat{T}) = \widehat{\phi}(-\widehat{T}) = \widehat{\phi}(\widehat{\mathcal{O}}) = \mathcal{O}$. Furthermore, for all $P \in E$ we have $\widehat{\phi} \circ \phi(P) = 3P$, and for all $\widehat{P} \in \widehat{E}$ we have $\phi \circ \widehat{\phi}(\widehat{P}) = 3\widehat{P}$.

Proof. As in the 2-descent case, it is enough to check the given formulas. However, this is not satisfactory and does not explain how they have been obtained. I give here a partial justification. For $P = (x, y) \in E$ we will set with evident notation $\widehat{x} = x(P) + x(P + T) + x(P - T) - x(\widehat{T})$ and $\widehat{y} = y\frac{d}{dx}\widehat{x}$ (in the case of a p-isogeny with a point T of order p we would set $\widehat{x} = \sum_{0 \leqslant i \leqslant p-1} x(P + iT)$ up to some constant translation). A small computation gives the formula of the proposition. In any case, we check that $\phi(P) \in \widehat{E}$ and that its kernel (more precisely the inverse image of $\widehat{\mathcal{O}}$) is our given group of order 3. We must now show that ϕ is a group homomorphism. In fact, since ϕ is a morphism of algebraic curves and sends $\mathcal{O}$ to $\widehat{\mathcal{O}}$ this follows from Theorem 7.1.6, but let us show this directly, as usual putting ourselves in the generic situation. Thus let P_1, P_2, P_3 be three points on E such that $P_1 + P_2 + P_3 = \mathcal{O}$, and let $y = mx + n$ be the line through those three points (which has this form since we are in the generic case). Once again I could give directly the equation of the line passing through the $\widehat{P}_i$, but instead let us find this equation. We thus want to find $\widehat{m}$ and $\widehat{n}$ such that $\widehat{y} = \widehat{m}\widehat{x} + \widehat{n}$

for $(x, y) = (x_i, y_i)$, $1 \leqslant i \leqslant 3$. Since $y_i = mx_i + n$ this implies that the x_i are three roots of the equation

$$(mx + n)(x^3 - 4db(ax + 2b))/x^3 = \widehat{m}(x^3 + 4d((a^2/3)x^2 + abx + b^2))/x^2 + \widehat{n} \,,$$

in other words

$$(m - \widehat{m})x^4 + (n - \widehat{n} - (4/3)a^2 d\widehat{m})x^3$$
$$- 4abd(m + \widehat{m})x^2 - 4bd(b(2m + \widehat{m}) + an) - 8b^2 dn = 0 \,.$$

Since we are in a generic situation this means that this polynomial must be divisible by the third-degree polynomial of which the x_i are roots, in other words by $x^3 + d(ax + b)^2 - (mx + n)^2$. Computing the remainder, we obtain three linear equations in the two unknowns $\widehat{m}$ and $\widehat{n}$, and after some computation we find that they are compatible and that

$$\widehat{m} = \frac{(n^2 + 3db^2)m - 4adbn}{n^2 - db^2} \quad \text{and}$$

$$\widehat{n} = \frac{n^3 - (4/3)a^2 dmn^2 + (4abm^2 + (4/3)a^3 bd - 9b^2)dn - 4db^2 m^3}{n^2 - db^2} \,.$$

As in the 2-descent case, we could now start from these values and check that they satisfy $\widehat{y}_i = \widehat{m}\widehat{x}_i + \widehat{n}$ for $1 \leqslant i \leqslant 3$, but here it is not necessary.

Applying the first part of the proposition to $\widehat{E}$ gives a map $\widehat{\phi}_1$ from $\widehat{E}$ to $\widehat{\widehat{E}}$, and composing with the isomorphism $(x, y) \mapsto (x/9, y/27)$ between $\widehat{\widehat{E}}$ and E gives the map $\widehat{\phi}$ of in the proposition. $\qquad\square$

8.4.3 Description of the Image of ϕ

Although we want to copy almost verbatim what we have done in the 2-descent case, a difficulty arises from the fact that the 3-torsion point $T = (0, b\sqrt{d})$ does not necessarily have rational coordinates, although the group it generates is rational. It will thus be necessary to work in the field $K_d = \mathbb{Q}(\sqrt{d})$, which is equal to $\mathbb{Q}$ if d is a square, and is a quadratic field otherwise. Note however that this field is only a necessary *tool*, but that we will *not* need to consider the whole group $E(K_d)$.

Proposition 8.4.4. *Denote by $\widehat{I} = \phi(E(\mathbb{Q}))$ the image of the rational points of E in $\widehat{E}(\mathbb{Q})$.*

(1) *$\widehat{O} \in \widehat{I}$, and $\pm\widehat{T} \in \widehat{I}$ if and only if $\widehat{d} = -3d$ is a square and $\mathrm{disc}(E) = 144d^2 b^3 \widehat{b}$ is a cube in $\mathbb{Q}^*$ (or equivalently, $\widehat{d}/(2\widehat{b})$ is a cube).*

(2) *Otherwise, a general point $\widehat{P} = (\widehat{x}, \widehat{y}) \in \widehat{E}(\mathbb{Q})$ different from $\pm\widehat{T}$ belongs to $\widehat{I}$ if and only if there exists $\gamma \in K_{\widehat{d}} = \mathbb{Q}(\sqrt{-3d})$ such that*

$$\gamma^3 = \widehat{y} - (\widehat{a}\widehat{x} + \widehat{b})\sqrt{\widehat{d}} \,.$$

Proof. (1). Since $\widehat{I} \subset \widehat{E}(\mathbb{Q})$, it is clear that a necessary condition for $\widehat{T}$ to be in $\widehat{I}$ is that its ordinate be in $\mathbb{Q}$, in other words that $\widehat{d} = \delta^2$ for some $\delta \in \mathbb{Q}$. Thus assume that this is the case. Since the only affine points with zero x-coordinates on $\widehat{E}$ are $\pm\widehat{T}$, the definition of ϕ shows that $\widehat{T} \in \widehat{I}$ if and only if there exists $x \in \mathbb{Q}^*$ such that $x^3 + 4d((a^2/3)x^2 + abx + b^2) = 0$, and this last equation implies $x \neq 0$ since $bd \neq 0$. We compute that the discriminant of this polynomial is equal to $-(16/27)b^2d^2(4a^3d - 27b)^2$, which is always strictly negative, so that our cubic equation has only one real root. An application of Cardano's formula or a direct check shows that this root is given by the simple formula

$$x = -\frac{3b\delta}{a\delta - (a^3\delta^3 + (81/4)b\delta)^{1/3}} \, .$$

It follows that there exists a rational root if and only if $a^3\delta^3 + (81/4)b\delta$ is a cube in $\mathbb{Q}^*$, hence if and only if its square $\widehat{d}(a^3\widehat{d} + (81/4)b)^2 = (27/4)^2\widehat{d}\widehat{b}^2$ is a cube in $\mathbb{Q}^*$, hence if and only if $\widehat{d}/(2\widehat{b})$ is a cube, if and only if $\mathrm{disc}(E) = (\widehat{d}/(2\widehat{b}))^2(4b\widehat{b})^3$ is a cube, proving (1).

(2). Since $x = 0$ implies $(x, y) = \pm T$ hence $\phi((x,y)) = \widehat{\mathcal{O}}$, we may assume $x \neq 0$. Thus let $(x,y) \in E(\mathbb{Q})$ with $x \neq 0$. A short computation shows that

$$\widehat{y} - (\widehat{a}\widehat{x} + \widehat{b})\sqrt{\widehat{d}} = \left(\frac{y - ((a/3)x + b)\sqrt{\widehat{d}}}{x}\right)^3 ,$$

showing that this expression is a cube in $K_{\widehat{d}}$, more precisely that it is equal to γ^3, where $\gamma = (y - ((a/3)x + b)\sqrt{\widehat{d}})/x$. Conversely, assume that $\widehat{P} = (\widehat{x}, \widehat{y}) \in \widehat{E}(\mathbb{Q})$ is such that there exists γ such that $\gamma^3 = \widehat{y} - (\widehat{a}\widehat{x} + \widehat{b})\sqrt{\widehat{d}}$. Note that $\gamma = 0$ implies that
$$0 = \widehat{y}^2 - \widehat{d}(\widehat{a}\widehat{x} + \widehat{b})^2 = \widehat{x}^3 ,$$
hence that $\widehat{x} = 0$, i.e., $\widehat{P} = \pm\widehat{T}$, and conversely. Therefore in the present case we have $\gamma \neq 0$. We have the following lemma.

Lemma 8.4.5. *Set $u = (\gamma + \widehat{x}/\gamma)/2$ and $v = (\gamma - \widehat{x}/\gamma)/(2\sqrt{\widehat{d}})$.*

(1) *u and v are in $\mathbb{Q}$.*

(2) *We have*
$$(\widehat{x}/\gamma)^3 = \widehat{y} + (\widehat{a}\widehat{x} + \widehat{b})\sqrt{\widehat{d}} .$$

(3) *We have $b = -(v + a/3)(u^2 - d(v - 2a/3)^2)$.*

Proof. (1). This is trivial if $\sqrt{\widehat{d}} \in \mathbb{Q}$, so assume that this is not the case. Then denoting by $\sigma(\gamma)$ the conjugate of γ in the quadratic field $K_{\widehat{d}}$ we have

$$(\gamma\sigma(\gamma))^3 = \widehat{y}^2 - \widehat{d}(\widehat{a}\widehat{x} + \widehat{b})^2 = \widehat{x}^3 .$$

Since $\gamma\sigma(\gamma)$ and $\widehat{x}$ are in $\mathbb{Q}$ this implies that $\gamma\sigma(\gamma) = \widehat{x}$, so that $\widehat{x}/\gamma = \sigma(\gamma)$, proving (1).

(2). This is clear since

$$\left(\frac{\widehat{x}}{\gamma}\right)^3 = \frac{\widehat{y}^2 - \widehat{d}(\widehat{a}\widehat{x} + \widehat{b})^2}{\widehat{y} - (\widehat{a}\widehat{x} + \widehat{b})\sqrt{\widehat{d}}} = \widehat{y} + (\widehat{a}\widehat{x} + \widehat{b})\sqrt{\widehat{d}}\,.$$

(3). We may thus write $\gamma = u + v\sqrt{\widehat{d}}$ and $\widehat{x}/\gamma = u - v\sqrt{\widehat{d}}$, hence $\widehat{x} = u^2 - \widehat{d}v^2 = u^2 + 3dv^2$. We compute $\gamma^3 - (\widehat{x}/\gamma)^3$ in two different ways. On the one hand, we have

$$\gamma^3 - (\widehat{x}/\gamma)^3 = 2\sqrt{\widehat{d}}(3u^2v + v^3\widehat{d}) = -2\sqrt{\widehat{d}}(3dv^3 - 3u^2v)\,,$$

while by (2) we have

$$\gamma^3 - (\widehat{x}/\gamma)^3 = -2\sqrt{\widehat{d}}(\widehat{a}\widehat{x} + \widehat{b}) = -2\sqrt{\widehat{d}}(a(u^2 + 3dv^2) + (27b - 4a^3d)/9)\,.$$

Identifying both expressions gives

$$b = -u^2(v + a/3) + d(v^3 - av^2 + 4a^3/27) = (v + a/3)(d(v - 2a/3)^2 - u^2)\,,$$

proving the lemma. □

It is now easy to finish the proof of the proposition. Since $b \neq 0$, the lemma implies that $v + a/3 \neq 0$. Thus we can set

$$x = -b/(v + a/3) = u^2 - d(v - 2a/3)^2 \quad \text{and} \quad y = ux = u^3 - du(v - 2a/3)^2\,.$$

We thus have $y^2 = u^2x^2 = u^2b^2/(v + a/3)^2$, while by the lemma we have

$$\begin{aligned}(v + a/3)^3(x^3 + d(ax + b)^2) &= -b^3 + db^2(v + a/3)(-a + v + a/3)^2 \\ &= b^2(-b + d(v + a/3)(v - 2a/3)^2) \\ &= b^2(v + a/3)u^2\,,\end{aligned}$$

so that we indeed have $y^2 = x^3 + d(ax + b)^2$, hence $(x, y) \in E(\mathbb{Q})$. Furthermore, we have $\phi((x, y)) = (\widehat{x}_1, \widehat{y}_1)$, with

$$\begin{aligned}\widehat{x}_1 &= x + 4d(a^2/3 + a(b/x) + (b/x)^2) \\ &= u^2 - d(v - 2a/3)^2 + d(4a^2/3 - 4a(v + a/3) + 4(v + a/3)^2) \\ &= u^2 + 3dv^2 = \widehat{x}\end{aligned}$$

and

$$\begin{aligned}\widehat{y}_1 &= u(x - 4d(a(b/x) + 2(b/x)^2)) \\ &= u(u^2 - d(v - 2a/3)^2 + d(4a(v + a/3) - 8(v + a/3)^2)) = u(u^2 - 9v^2d)\,.\end{aligned}$$

On the other hand, by the above lemma we have

$$\widehat{y} = (\gamma^3 + (\widehat{x}/\gamma)^3)/2 = u^3 + 3uv^2\widehat{d} = u(u^2 - 9v^2d) = \widehat{y}_1 \ ,$$

finishing the proof.   □

The following corollary is immediate by considering the dual isogeny.

Corollary 8.4.6. *Set* $I = \widehat{\phi}(\widehat{E}(\mathbb{Q}))$.

(1) $\mathcal{O} \in I$, *and* $\pm T \in I$ *if and only if* d *is a square and* $d/(2b)$ *is a cube in* $\mathbb{Q}^*$.

(2) *Otherwise, a general point* $P = (x, y) \in E(\mathbb{Q})$ *different from* $\pm T$ *belongs to* I *if and only if there exists* $\gamma \in K_d = \mathbb{Q}(\sqrt{d})$ *such that* $\gamma^3 = y - (ax + b)\sqrt{d}$.

8.4.4 The Fundamental 3-Descent Map

We continue to imitate what we have done for 2-descent. We keep the assumptions and notation of the preceding sections. In particular, recall that we have set $K_d = \mathbb{Q}(\sqrt{d})$, which is equal to $\mathbb{Q}$ when d is a square.

Definition 8.4.7. *We define the* 3-descent map α *from the group* $E(\mathbb{Q})$ *to the multiplicative group* K_d^*/K_d^{*3} *as follows:*

(1) $\alpha(\mathcal{O}) = 1$, *and if* $T \in E(\mathbb{Q})$ *(in other words if* $\sqrt{d} \in \mathbb{Q}^*$*) then* $\alpha(T) = 4db^2$.

(2) *When* $P = (x, y) \in E(\mathbb{Q})$ *with* $P \neq T$ *then* $\alpha((x, y)) = y - (ax + b)\sqrt{d}$.

In the above, all the values are of course understood modulo the multiplicative action of K_d^{*3}.

The main result is the following.

Proposition 8.4.8. (1) *The* 3-descent map α *is a group homomorphism.*

(2) *The kernel of* α *is equal to* $\widehat{\phi}(\widehat{E}(\mathbb{Q}))$.

(3) *The map* α *induces an injective group homomorphism from* $E(\mathbb{Q})/\widehat{\phi}(\widehat{E}(\mathbb{Q}))$ *to the subgroup of* K_d^*/K_d^{*3} *of elements whose norm is trivial in* $\mathbb{Q}^*/\mathbb{Q}^{*3}$ *when* $\sqrt{d} \notin \mathbb{Q}$, *and to* $\mathbb{Q}^*/\mathbb{Q}^{*3}$ *otherwise.*

Proof. If $P = (x, y) \neq T$, then $\alpha(-P) = \alpha((x, -y)) = -y - (ax + b)\sqrt{d}$, so $\alpha(P)\alpha(-P) = -(y^2 - d(ax + b)^2) = (-x)^3 \in \mathbb{Q}^{*3}$, and if $P = T = (0, b\sqrt{d})$ then by definition

$$\alpha(T)\alpha(-T) = 4db^2\alpha((0, -b\sqrt{d})) = (-2b\sqrt{d})^3 \in K_d^{*3} \ ,$$

so α sends inverses to inverses. Thus we must show that if $P_1 + P_2 + P_3 = \mathcal{O}$ then $\alpha(P_1)\alpha(P_2)\alpha(P_3) \in K_d^{*3}$. If one of the P_i is equal to $\mathcal{O}$ we are in the case that we have just treated, so we exclude that case. If one of the P_i is equal to T then either all three P_i are equal to T, and the result is clear, or only one, say P_1, is equal to T. In that case none of the P_i is equal to $2T$, since otherwise the third one would be equal to $\mathcal{O}$. Thus by what we have proved about inverses we have

$$\alpha(P_1)\alpha(P_2)\alpha(P_3) = (\alpha(-P_1)\alpha(-P_2)\alpha(-P_3))^{-1}$$

(as usual modulo the multiplicative action of K_d^{*3}), and now none of the $-P_i$ is equal to $\mathcal{O}$ or to T, so we are in the generic case, which we now treat.

Let $y = mx + n$ be the equation of the line going through the points $P_i = (x_i, y_i)$. Note that we can indeed choose the equation to be of this form since the excluded equations are $x = k$, which intersect the curve E in two affine points together with $\mathcal{O}$, a case that we have excluded. The x_i are thus the three roots of the polynomial $f(x) = x^3 + d(ax + b)^2 - (mx + n)^2$. An easy computation shows that

$$\prod_i (y_i - (ax_i + b)\sqrt{d}) = \prod_i (mx_i + n - (ax_i + b)\sqrt{d}) = (n - b\sqrt{d})^3 \in K_d^3 ,$$

finishing the proof of (1).

(2). This follows immediately from Corollary 8.4.6: $\mathcal{O}$ is evidently in the kernel of α, and by definition $T \in \mathrm{Ker}(\alpha)$ if and only if $4db^2$ is a cube, if and only if $(d/2b) = 4db^2/(2b)^3$ is a cube, hence by the corollary if and only if $T \in I$. Finally, a point $P = (x, y)$ different from $\mathcal{O}$ and T is in the kernel of α if and only if there exists $\gamma \in K_d^*$ such that $\gamma^3 = y - (ax + b)\sqrt{d}$, hence by the corollary if and only if $P \in I$, proving (2).

(3). From (1) and (2) we deduce that α induces an injection $\overline{\alpha}$ from $E(\mathbb{Q})/\widehat{\phi}(\widehat{E}(\mathbb{Q}))$ to K_d^*/K_d^{*3}. If $\sqrt{d} \in \mathbb{Q}^*$ the image of $\overline{\alpha}$ is in $\mathbb{Q}^*/\mathbb{Q}^{*3}$ and there is nothing else to say. Otherwise, $T \notin E(\mathbb{Q})$ and for $P = (x, y) \neq \mathcal{O}$ we have that $\mathcal{N}_{K_d/\mathbb{Q}}(\alpha((x, y))) = x^3$ is a cube in $\mathbb{Q}^*$, proving the proposition. $\qquad\square$

We leave to the reader to state and prove an analogue of Proposition 8.2.4 (3) and (4), see Exercise 17, and we will not study 3-descent any further. See however Section 6.4.4 for a very nice application of 3-descent to the Diophantine equations $ax^3 + by^3 + cz^3$.

8.5 The Use of $L(E, s)$

8.5.1 Introduction

We have seen in Section 8.1.2 the definition and main properties of the L-function $L(E, s)$ attached to an elliptic curve E defined over $\mathbb{Q}$. Thanks to the work of Wiles et al. (see Theorem 8.1.4), we know that $L(E, s)$ extends to an entire function with a functional equation of the form $\Lambda(E, 2-s) = \varepsilon(E)\Lambda(s)$, where $\Lambda(E, s) = N^{s/2}(2\pi)^{-s}\Gamma(s)L(E, s)$ and $\varepsilon(E) = \pm 1$. Finally, the Birch–Swinnerton-Dyer Conjecture 8.1.7 predicts that the rank $r(E)$ should be equal to the order of vanishing of $L(E, s)$ at $s = 1$. This conjecture is a theorem when the order of vanishing is equal to 0 or 1.

Even though BSD is a conjecture, it suggests many useful approaches. First, it implies the *parity conjecture*, saying that $(-1)^{r(E)} = \varepsilon(E)$. Since there exists an algorithm to compute $\varepsilon(E)$, this gives a conjectural parity for $r(E)$. For instance, this explains the remark made after Proposition 8.2.15.

When the parity does not suffice to determine the rank, we proceed as follows. We search more or less intelligently for rational points on the curve. There are many methods to do this, but we simply mention that even if a 2-descent has not succeeded in giving the rank, it still may help in the search for points.

If after a sufficiently long search we find sufficiently many independent points compared to the upper bound on the rank given by descent arguments, we are happy and can conclude. Unfortunately, in many cases this does not happen, either because the points we are looking for have a very large height, or more simply because the rank is simply not equal to the upper bound given by descent. It is in this case that we must appeal to the computation of $L(E, s)$, hence rely on the BSD conjecture.

The numerical computation of $L(E, s)$ (and of its derivatives) involves two completely different tasks: first the evaluation of some transcendental functions (the exponential function in the simplest case), which will be studied in great detail below, and second the arithmetic computation of the coefficients $a_p(E)$. In the case of a general elliptic curve these coefficients are computed either using Legendre symbol sums (Lemma 7.3.9) for small p, by the Shanks–Mestre baby-step giant-step method (see Algorithm 7.4.12 of [Coh0]) for moderate p, which should be sufficient for the computation of $L(E, s)$, or even by the Schoof–Elkies–Atkin algorithm for very large p (see for instance [Coh-Fre]).

However, in the special case of an elliptic curve E with *complex multiplication* the computation of $a_p(E)$ can be done much more efficiently, and we now treat this case.

8.5.2 The Case of Complex Multiplication

In this section we explain how to compute efficiently $a_p(E)$ when the elliptic curve E defined over $\mathbb{Q}$ has complex multiplication (CM). Apart from the classical cases $j = 1728$ and $j = 0$, this section is a writeup of a C program written by M. Watkins based on a series of published and unpublished papers,[2] and does not contain any proof, except for the special cases $j = 1728$

---

[2] To tell the truth, it is not even sure that the results have been proved, although they are certainly correct

and $j = 0$ mentioned above. Note that the "naïve" method using Lemma 7.3.9 requires time $O(p)$, that Shanks's baby-step giant-step method as modified by Mestre (see for instance [Coh0], Algorithm 7.4.12) requires time $O(p^{1/4})$, while the method that we are going to study in this section requires time $O(\log^2(p))$.

We first recall the important differences between isomorphism classes of curves over $\mathbb{C}$ and over $\mathbb{Q}$. Let $y^2 = x^3 + ax + b$ be a reduced Weierstrass equation of an elliptic curve E defined over $\mathbb{Q}$, where we may assume that a and b are in $\mathbb{Z}$. The curves isomorphic to E *over* $\mathbb{C}$ are the curves having the same j-invariant $j(E) = 12^3(4a^3/(4a^3 + 27b^2))$, and whose reduced Weierstrass equation is $y^2 = x^3 + u^2ax + u^3b$ for some $u \in \mathbb{C}^*$, or equivalently, by the immediate change of variable $(x, y) \mapsto (ux, u^{3/2}y)$, the equation $uy^2 = x^3 + ax + b$. Thus, they are the *quadratic twists* E_u of E by u (see Definition 7.3.15).

Assume first that $j(E) \neq 0$ and $j(E) \neq 12^3 = 1728$, in other words that a and b are nonzero. The curve E_u is defined over $\mathbb{Q}$ if and only if u^2 and u^3 are both in $\mathbb{Q}^*$, hence if $u = u^3/u^2 \in \mathbb{Q}^*$, and the curves E_{u_1} and E_{u_2} are isomorphic *over* $\mathbb{Q}$ if and only if u_1/u_2 is a square in $\mathbb{Q}$. Thus, multiplying if necessary u by a square of $\mathbb{Q}$, we may always assume that u is a fundamental discriminant, in other words either equal to 1 or to the discriminant of a quadratic field.

Assume now that $b = 0$, so that the reduced equation of E is $y^2 = x^3 + ax$ and $j(E) = 1728 = 12^3$. The curve E is in fact a *quartic twist* of any curve in the family, for instance of the curve $y^2 = x^3 - x$, and the curve with equation $y^2 = x^3 + a'x$ is isomorphic to E if and only if a'/a is a fourth power in $\mathbb{Q}$.

Similarly assume that $a = 0$, so that the reduced equation of E is $y^2 = x^3 + b$ and $j(E) = 0$. The curve E is now a *sextic twist* of any curve in the family, for instance of the curve $y^2 = x^3 + 1$, and the curve with equation $y^2 = x^3 + b'$ is isomorphic to E if and only if b'/b is a sixth power in $\mathbb{Q}$.

Computation of $a_p(E)$: the Case $j = 1728$

We first consider the case of complex multiplication by $\mathbb{Z}[i]$, in other words of curves such that $j(E) = 1728$.

Proposition 8.5.1. *Let E_N be the elliptic curve with equation $y^2 = x^3 - Nx$. When $p \mid 2N$ or $p \equiv 3 \pmod 4$ we have $a_p(E_N) = 0$, and when $p \equiv 1 \pmod 4$ and $p \nmid N$ we have*

$$a_p(E_N) = 2\left(\frac{2}{p}\right) \begin{cases} -a & \text{if } N^{(p-1)/4} \equiv 1 \pmod p, \\ a & \text{if } N^{(p-1)/4} \equiv -1 \pmod p, \\ -b & \text{if } N^{(p-1)/4} \equiv -a/b \pmod p, \\ b & \text{if } N^{(p-1)/4} \equiv a/b \pmod p, \end{cases}$$

where $p = a^2 + b^2$ with $a \equiv -1 \pmod 4$.

Proof. Assume that $p \equiv 3 \pmod 4$. Since $\left(\frac{-1}{p}\right) = -1$ it follows that for each x not equal to 0 or ± 1 there is exactly one value in $\{x, -x\}$ such that $x^3 - x$ is a square. It follows that $a_p(E_N) = 0$ in that case, and it is also immediate that $a_2(E_N) = 0$ (this is a special case of a general fact on curves with CM that we will state in Proposition 8.5.5 below). Assume now that $p \equiv 1 \pmod 4$. By Proposition 2.5.20 we know that there exists a character χ of order 4 and that $J(\chi, \chi) = a + bi$ with $a^2 + b^2 = p$, $2 \mid b$, and $a \equiv -1 \pmod 4$. It is slightly more natural to reason backwards. Since χ has exact order 4, When n is not a square we have $\chi(n) = \pm i$, hence $\Re(\chi(n)) = 0$. Since χ^2 is equal to the Legendre symbol we have

$$2\Re(\chi(N)J(\chi,\chi)) = 2\sum_{x\in\mathbb{F}_p} \Re(\chi(Nx(1-x))) = 2 \sum_{\substack{x\in\mathbb{F}_p \\ \exists y\in\mathbb{F}_p,\ Nx(1-x)=y^2}} \left(\frac{y}{p}\right)$$

$$= \sum_{\substack{x,y\in\mathbb{F}_p \\ (2Nx-N)^2=N^2-4Ny^2}} \left(\frac{y}{p}\right) = \left(\frac{2}{p}\right) \sum_{\substack{X,Y\in\mathbb{F}_p \\ X^2=N^2-NY^2}} \left(\frac{Y}{p}\right)$$

$$= \left(\frac{2}{p}\right) \sum_{Y\in\mathbb{F}_p} \left(\frac{Y}{p}\right) \sum_{X\in\mathbb{F}_p,\ X^2=N^2-NY^2} 1$$

$$= \left(\frac{2}{p}\right) \sum_{Y\in\mathbb{F}_p} \left(\frac{Y}{p}\right)\left(1 + \left(\frac{N^2-NY^2}{p}\right)\right)$$

$$= \left(\frac{2}{p}\right) \sum_{Y\in\mathbb{F}_p} \left(\frac{N(Y^3-NY)}{p}\right)$$

since $\sum_{Y\in\mathbb{F}_p} \left(\frac{Y}{p}\right) = 0$ and $\left(\frac{-4}{p}\right) = 1$. Since by Lemma 7.3.9 we have $a_p(E_N) = -\sum_{Y\in\mathbb{F}_p}\left(\frac{Y^3-NY}{p}\right)$, it follows that

$$a_p(E_N) = -\left(\frac{2N}{p}\right)2\Re(\chi(N)(a+bi)) .$$

If $\left(\frac{N}{p}\right) = 1$ we have $\chi(N) = \pm 1$, where the sign is determined by $\chi(N) \equiv N^{(p-1)/4} \pmod p$, which gives the first two cases. If $\left(\frac{N}{p}\right) = -1$, we let g be a primitive root modulo p such that $\chi(g) = i$. By Exercise 36 of Chapter 2 we have $a+bg^{(p-1)/4} \equiv 0 \pmod p$. If $N \equiv g^k \pmod p$ then either $k \equiv 1 \pmod 4$, and then $\chi(N) = i$ and $a + bN^{(p-1)/4} \equiv 0 \pmod p$, or $k \equiv 3 \pmod 4$, and then $\chi(N) = -i$ and $a - bN^{(p-1)/4} \equiv 0 \pmod p$, proving the other two cases. $\square$

Corollary 8.5.2. *For $n \neq 0$ let E_n be the elliptic curve with affine equation $y^2 = x^3 - n^2x$. When $p \mid 2n$ or $p \equiv 3 \pmod 4$ we have $a_p(E_n) = 0$, and when $p \nmid 2n$ and $p \equiv 1 \pmod 4$ we have*

$$a_p(E_n) = -\left(\frac{2n}{p}\right) 2a \ ,$$

where $p = a^2 + b^2$ *with* $a \equiv -1 \mod 4$.

Proof. Clear from the proposition, and also from Proposition 7.3.16 together with the proposition applied to $N = 1$. □

Remark. It follows from this corollary that, even without using Tunnell's Theorem 6.12.4, but still assuming BSD, it is very easy to make large tables of congruent numbers: using Cornacchia's algorithm as explained in the remarks following Proposition 2.5.20 we compute decompositions $p = a^2 + b^2$ with $a \equiv -1 \pmod 4$ for a large number of primes $p \equiv 1 \pmod 4$. Thanks to the above corollary it is then immediate to compute an approximate value of $L(E_n, 1)$, and we conclude thanks to BSD and Proposition 6.12.1.

Computation of $a_p(E)$: the Case $j = 0$

We now consider the case of complex multiplication by $\mathbb{Z}[\rho]$, where $\rho = \zeta_3$ is a primitive cube root of unity, in other words curves such that $j(E) = 0$.

Proposition 8.5.3. *Let E_N be the elliptic curve with affine equation $y^2 = x^3 + N$. When $p \mid 3N$ or $p \equiv 2 \pmod 3$ we have $a_p(E_N) = 0$, and when $p \nmid 3N$ and $p \equiv 1 \pmod 3$ we have*

$$a_p(E_N) = \left(\frac{N}{p}\right) \begin{cases} b - 2a & \text{if} \ \ (4N)^{(p-1)/3} \equiv 1 \pmod p \ , \\ a + b & \text{if} \ \ (4N)^{(p-1)/3} \equiv -a/b \pmod p \ , \\ a - 2b & \text{if} \ \ (4N)^{(p-1)/3} \equiv -b/a \pmod p \ , \end{cases}$$

where $p = a^2 - ab + b^2$ *with* $3 \mid b$ *and* $a \equiv -1 \mod 3$.

Proof. The case $p \mid 3N$ being immediate we assume $p \nmid 3N$. When $p \equiv 2$ (mod 3) the map $x \mapsto x^3$ is a bijection of $\mathbb{F}_p$ onto itself, since $3 \nmid |\mathbb{F}_p^*| = p - 1$. It follows that for every value of y there is exactly one value of x, hence that $|E_N(\mathbb{F}_p)| = p + 1$, so that $a_p(E_N) = 0$. Assume now that $p \equiv 1 \pmod 3$, fix a primitive root g modulo p, and let χ be a character of order 3. Changing if necessary χ into $\overline{\chi} = \chi^2$ we may assume that $\chi(g) = \rho$. By Proposition 2.5.20, we know that $J(\chi, \chi) = a + b\rho$ with $a^2 - ab + b^2 = p$, $3 \mid b$, and $a \equiv -1$ (mod 3). On the other hand, since $1 + \rho + \rho^2 = 0$ it is clear that the expression $1 + \chi(x) + \chi(x^2)$ is equal to 1 if $x = 0$, to 3 if x is a cube in $\mathbb{F}_p^*$, and to 0 otherwise. Since $\chi(x^2) = \overline{\chi}(x)$ it follows that for a given y the number of x such that $x^3 = y^2 - 1$ is equal to $1 + \chi(y^2 - 1) + \overline{\chi}(y^2 - 1)$, so that

$$|E_N(\mathbb{F}_p)| = p + 1 + 2\Re\left(\sum_{y \in \mathbb{F}_p} \chi(y^2 - N)\right) ,$$

in other words

$$a_p(E_N) = -2\Re\left(\sum_{y \in \mathbb{F}_p} \chi(y^2 - N)\right).$$

Assume first that $\left(\frac{N}{p}\right) = 1$, and let $k \in \mathbb{Z}$ be such that $k^2 \equiv N \pmod{p}$. Setting $y = k(2t - 1)$ (which is a bijection since $p \nmid 2N$) and using $\chi(-1) = 1$, we thus have

$$a_p(E_N) = -2\Re\left(\chi(-4k^2)\sum_{t \in \mathbb{F}_p} \chi(t - t^2)\right) = -2\Re(\chi(4N)J(\chi, \chi)).$$

Now by Exercise 36 of Chapter 2 we know that the integers a and b such that $J(\chi, \chi) = a + b\rho$ are completely determined by the congruence $a + bg^{(p-1)/3} \equiv 0 \pmod{p}$, since we have chosen χ such that $\chi(g) = \rho$, so that $g^{(p-1)/3} \equiv -a/b \pmod{p}$. Since g is a primitive root modulo p there exists k such that $4N \equiv g^k \pmod{p}$, so that $\chi(4N) = \chi(g)^k = \rho^k = \rho^{k \bmod 3}$. On the other hand,

$$(4N)^{(p-1)/3} \equiv g^{k(p-1)/3} \equiv (-a/b)^k \pmod{p},$$

and $k \bmod 3$ is determined by this congruence. The proposition follows when N is a quadratic residue modulo p by distinguishing the three possible values of $k \bmod 3$ and noting that $(-a/b)^2 \equiv -b/a \pmod{p}$.

If N is a quadratic nonresidue modulo p, we note that the twist $E_{N,N}$ of E_N by N itself has equation $Ny^2 = x^3 + N$, which is isomorphic to $Y^2 = X^3 + N^4$ by setting $Y = Ny$ and $X = Ny$. Since N^4 is trivially a quadratic residue, on the one hand by what we have just proved we have

$$a_p(E_{N,N}) = -2\Re\left(\chi(4N^4)(a + b\rho)\right) = -2\Re\left(\chi(4N)(a + b\rho)\right),$$

and on the other hand, by Proposition 7.3.16 we have $a_p(E_{N,N}) = \left(\frac{N}{p}\right)a_p(E_N)$, so the result follows in general. $\square$

In the special case $N = 1$, since by Proposition 2.5.20 we know the cubic character of 2, hence of 4, we obtain the following:

Corollary 8.5.4. *When $p = 3$ or $p \equiv 2 \pmod 3$ we have $a_p(E_1) = 0$, and when $p \equiv 1 \pmod 3$ we have*

$$a_p(E_1) = \begin{cases} b - 2a & \text{if } b \text{ is even,} \\ a + b & \text{if } a \text{ and } b \text{ are odd,} \\ a - 2b & \text{if } a \text{ is even and } b \text{ is odd,} \end{cases}$$

where $p = a^2 - ab + b^2$ with $3 \mid b$ and $a \equiv -1 \mod 3$.

Proof. See Exercise 19. $\square$

Remark. As in the case $j = 1728$ it follows from the proposition that, assuming BSD, it is very easy to make large tables of $L(E_n, 1)$, since once again using Cornacchia's algorithm it is immediate to compute the decompositions $p = a^2 - ab + b^2$ with $3 \mid b$ and $a \equiv -1 \pmod 3$. Thanks to Proposition 7.2.3 and the remarks following it, this enables us for instance to compute tables of integers c that are sums of two rational cubes, by computing $L(E_{-432c^2}, 1)$ or $L(E_{16c^2}, 1)$ (which are equal since the two curves are isogenous).

Computation of $a_p(E)$: the General CM Case

From now on we assume that $j \neq 0$ and $j \neq 1728$. The set of $\mathbb{Q}$-isomorphism classes of elliptic curves E such that $j(E) = j$ is thus parametrized by a fundamental discriminant u (or equivalently, by the group $\mathbb{Q}^*/\mathbb{Q}^{*2}$). By Proposition 7.3.16 we know that if we can compute $a_p(E)$ for a *single* curve in this set, then it is immediate to compute $a_p(E')$ for all curves E' having the same j-invariant: let $y^2 = x^3 + a'x + b'$ be a reduced equation for E'. If $p = 2$ or 3 it is trivial to compute $a_p(E')$ directly, and if E' has bad reduction at p then by Lemma 7.1.3 we have $a_p(E') = \left(\frac{-c_6(E')}{p}\right) = \left(\frac{6b'}{p}\right)$ since $c_6(E') = -864b'$. For all other p we compute $u = ab'/(ba')$ since $ba' \neq 0$, so that $E' = E_u$, and we use the formula $a_p(E_u) = \left(\frac{u}{p}\right) a_p(E)$. In fact, since $b' = u^3 b$ we have $\left(\frac{b'}{p}\right) = \left(\frac{u}{p}\right)\left(\frac{b}{p}\right)$, so that we have more simply, without even computing u,

$$a_p(E') = \left(\frac{bb'}{p}\right) a_p(E) .$$

We now come specifically to the CM case. Let E be an elliptic curve defined over $\mathbb{Q}$, and assume that E has CM. It follows that $\text{End}(E)$ is isomorphic to an imaginary quadratic order of class number 1, hence either to one of the 9 maximal orders of discriminant $D = -3, -4, -7, -8, -11, -19, -43, -67$, or -163, or to one of the 4 nonmaximal orders of discriminant $-12, -16, -27$, or -28. Furthermore, since CM is a property of elliptic curves over $\mathbb{C}$, a curve E has CM if and only if its j-invariant is equal to one of the 13 values of the elliptic modular function $j(\tau)$ for $\tau = ((-\delta + \sqrt{D})/2)$. Thanks to the above discussion, to be able to compute efficiently $a_p(E)$ for *all* elliptic curves E with CM and defined over $\mathbb{Q}$, it is sufficient to be able to do this for 13 specific curves, one per discriminant. We have already done so for $D = -3$ and $D = -4$, so we will assume that $D \leqslant -7$.

Note first the following easy result, which we do not prove.

Proposition 8.5.5. *Let E be an elliptic curve defined over $\mathbb{Q}$ and having CM by the imaginary quadratic order $\mathcal{O}_D$ of discriminant D, and let p be a prime of good reduction and not dividing D. If $\left(\frac{D}{p}\right) = -1$ we have $a_p(E) = 0$, while if $\left(\frac{D}{p}\right) = 1$ we have $a_p(E) = \pi + \overline{\pi}$, where $\pi \in \mathcal{O}_D$ is such that $\pi\overline{\pi} = p$.*

We can write $\pi = (a + b\sqrt{D})/2$ with $a \equiv bD \pmod 2$, so we have $a_p(E) = a$, where $a^2 + |D|b^2 = 4p$. An important algorithmic fact is that, given D and

p, by using a variant of Euclid's algorithm called Cornacchia's algorithm that we have already mentioned (see Exercise 41 of Chapter 2 and Algorithms 1.5.2 and 1.5.3 of [Coh0]), it is easy to compute a pair (a, b). The main problem is that it is not *quite* unique. More precisely, if π is a solution to $\pi\overline{\pi} = p$, then since $\mathcal{O}_D$ is a principal ideal domain it is immediate to see that all other solutions are given by $\pi' = u\pi$ or $\pi' = \overline{u\pi}$ for some unit $u \in \mathcal{O}_D$. Since π and $\overline{\pi}$ give the same a, the only problem is the unit u. Since we have assumed that $D \leqslant -7$ we have $u = \pm 1$, so that a is determined up to sign, and this is the difficult problem. One solution, which was used for some time, is to note that since $|E(\mathbb{F}_p)| = p + 1 - a_p(E)$, and if $P \in E(\mathbb{F}_p)$, we have $(p + 1 - a_p(E)) \cdot P = \mathcal{O}$. Choosing a random point P, we can expect that $(p + 1 - a) \cdot P = \mathcal{O}$ for only one of the two possible values of a. This is evidently not very elegant, although quite fast, but still seems like a waste of time since after all we only want to determine a among two possibilities.

There is, however, a much nicer solution. A first result is the following.

Proposition 8.5.6. *Assume that $D \leqslant -7$ is a fundamental discriminant such that $\mathcal{O}_D$ has class number 1, in other words $D = -7, -8, -11, -19, -43, -67,$ or -163, and let $j = j(\tau)$ with $\tau = (-\delta + \sqrt{D})/2$ as above, which by the fundamental theorem of CM belongs to $\mathbb{Z}$.*

(1) *The integer j is always the cube of an integer, and we will denote by j_3 the unique cube root of j in $\mathbb{Z}$.*
(2) *Set $s_8 = -1$ for $D = -8$ and $s_8 = 1$ otherwise. The integer $s_8(j - 1728)D$ is always the square of an integer, and we will denote by j_2 its positive square root.*

Proof. Assuming the fundamental theorem of CM, which tells us that $j \in \mathbb{Z}$, the proof is a simple verification: we compute numerically the seven values of j to sufficient accuracy (this can be done very simply), round to the nearest integer, and check.   □

Of course, this is not the "reason" for this proposition. A much more general result due to Gross–Zagier says among other things that $j(\tau_1) - j(\tau_2)$ is always very highly factorable. A number such as $j(\tau)$ with τ imaginary quadratic is called a *singular modulus*.

For the sake of completeness, for D ranging in order through the seven given values we have $j_3 = -15, 20, -32, -96, -960, -5280, -640320$, and $j_2 = 189, 224, 616, 4104, 195048, 3140424, 6541681608$.

Definition 8.5.7. *Let $D \leqslant -7$ be one of the 11 discriminants of an imaginary quadratic order of class number 1, and let $j = j(\tau)$ be the corresponding j-invariant. We define the* basic CM elliptic curve *with invariant j as follows:*

(1) *If D is one of the seven fundamental discriminants, we set $c_4 = s_8 j_3 D$ and $c_6 = s_7 j_2 D$, where $s_7 = -1$ if $D = -7$ and $s_7 = 1$ otherwise.*

(2) *If D is one of the four nonfundamental discriminants, in other words if $D = -12, -16, -27,$ or $-28,$ we set $(c_4, c_6) = (720, -19008),$ $(528, 12096), (1440, -54648),$ or $(1785, 75411)$ respectively.*

The basic CM elliptic curve E_j is the elliptic curve with invariants c_4 and c_6, in other words with reduced Weierstrass equation $y^2 = x^3 - 27c_4x - 54c_6$ (see Section 7.1.2).

Remark. It is immediate to check that we indeed have $j(E_j) = j$. Furthermore, it is also easy to see that the seemingly random values of c_4 and c_6 given for the four nonfundamental discriminants are in fact also closely related to $j^{1/3}D$ and to $(D(j-1728))^{1/2}D$ respectively.

The point of the above construction is not to construct elliptic curves E_j such that $j(E) = j$, which is immediate, but that a remarkable property of these curves is that their coefficients $a_p(E_j)$ can easily be computed as follows.

Theorem 8.5.8. *Let $D \leqslant -7$ be one of the 11 discriminants considered above, let $E = E_j$ be the corresponding basic CM curve, and let D_0 be the discriminant of the quadratic field $\mathbb{Q}(\sqrt{D})$, in other words the unique fundamental discriminant such that $D = D_0 f^2$ for some $f \in \mathbb{Z}$. For any prime number $p \neq 2$ such that $\left(\frac{D}{p}\right) = 1$ we have the following results:*

(1) *If $D \neq -8$ and $D \neq -16$ we have*

$$\left(\frac{D_0}{a_p(E)}\right) = -s_7 ,$$

where $s_7 = 1$ if $D_0 \neq -7$ and $s_7 = -1$ if $D_0 = -7$.
(2) *If $D = -8$ we have $a_p(E) \equiv 2 \pmod 8$ when $p \equiv 1$ or 11 modulo 16, and $a_p(E) \equiv -2 \pmod 8$ when $p \equiv 3$ or 9 modulo 16.*
(3) *If $D = -16$ we have*

$$a_p(E) \equiv 2\left(\frac{-8}{p}\right) \pmod 8 .$$

Since $D < 0$, the properties of the Legendre–Kronecker symbol tell us that $\left(\frac{D}{-a_p(E)}\right) = -\left(\frac{D}{a_p(E)}\right)$, so in all cases this theorem allows us to distinguish between $a_p(E)$ and its opposite, which was the goal of the construction of the basic CM elliptic curve.

8.5.3 Numerical Computation of $L^{(r)}(E, 1)$

In this section we will explain how to compute numerically the derivatives of $L(E, s)$. Once again we emphasize that using a suitable error analysis it is always easy to prove that a given real number (here $L^{(r)}(E, 1)$) is *not* equal to 0 (of course when it is not), while it is impossible to prove that it *is* equal to 0. The only thing we can do is have a reasonable certainty if the value we obtain is less in absolute value than 10^{-20}, say.

We refer to [Dok] for a detailed analysis and implementation of the general problem of computing special values of L-functions and their derivatives, using a slightly different approach from that given here.

By Wiles et al., we know that $L(E, s)$ satisfies a functional equation of standard form. It is not sufficiently well known that this automatically implies that there exists an exponentially convergent series for computing $L(E, s)$ and its derivatives numerically. The result is as follows (see for instance [Coh1] Section 10.3 for a proof). Recall first the following definition:

Definition 8.5.9. *The incomplete gamma function is defined for $\Re(x) > 0$ and all $s \in \mathbb{C}$ by*

$$\Gamma(s, x) = x^s \int_1^\infty e^{-xt} t^s \, \frac{dt}{t} \,,$$

so that in particular if $x \in \mathbb{R}_{>0}$

$$\Gamma(s, x) = \int_x^\infty t^s e^{-t} \, \frac{dt}{t} \,.$$

Proposition 8.5.10. *Write $L(E, s) = \sum_{n \geqslant 1} a_n(E) n^{-s}$, and let N and $\varepsilon(E)$ be as above. Then for all t_0 such that $\Re(t_0) > 0$ we have*

$$L(E, s) = F(E, s, t_0) + \varepsilon(E) \big(2\pi/\sqrt{N}\big)^{2s-2} F(E, 2 - s, 1/t_0) \,,$$

where

$$F(E, s, t_0) = \sum_{n \geqslant 1} \frac{a_n(E)}{n^s} \Gamma\!\left(s, \frac{2\pi n}{t_0 \sqrt{N}}\right) .$$

If $\varepsilon(E) = -1$ we clearly have $L(E, 1) = 0$. If $\varepsilon(E) = 1$ we obtain the following:

Corollary 8.5.11. *Assume that $\varepsilon(E) = 1$. Then*

$$L(E, 1) = 2 \sum_{n \geqslant 1} \frac{a_n(E)}{n} e^{-2\pi n/\sqrt{N}} \,.$$

Proof. We simply choose $t_0 = 1$ in the proposition, and note that

$$\Gamma(1, x) = \int_x^\infty e^{-t} \, dt = e^{-x} \,.$$

$\square$

We see that we obtain an exceedingly simple and fast formula for $L(E, 1)$. Note however that it is useful only when N is not too large, say $N < 10^{15}$. If N is much larger, it is difficult to estimate $L(E, 1)$ by this method.

Remark. Most number-theory-oriented packages such as `Pari/GP` and `magma` provide built-in functions for computing the conductor N and the sign of the functional equation $\varepsilon(E)$, which are necessary in using the above formulas. However, if these are not available we can easily compute them indirectly using the free parameter t_0 occurring in the formula for $L(E, s)$. Since the result must be independent of t_0, it is not difficult to compute N and $\varepsilon(E)$, aided by the fact that the prime divisors of N are the same as those of $\mathrm{disc}(E)$. In practice it is reasonably easy to compute N using Tate's algorithm, and so the only quantity that we really need to compute if we do not have a suitable CAS available is $\varepsilon(E)$, and this requires only two distinct values of t_0.

We now need to compute derivatives. For this we set the following definition.

Definition 8.5.12. *We define by induction the functions $\Gamma_r(s, x)$ by*

$$\Gamma_{-1}(s, x) = e^{-x} x^s \quad and \quad \Gamma_r(s, x) = \int_x^\infty \frac{\Gamma_{r-1}(s, t)}{t}\, dt \quad for \quad r \geqslant 0\,.$$

For instance, $\Gamma_0(s, x) = \Gamma(s, x)$, the incomplete gamma function; hence for example $\Gamma_0(1, x) = e^{-x}$ and

$$\Gamma_1(1, x) = \int_x^\infty \frac{e^{-t}}{t}\, dt = E_1(x)\,,$$

the *exponential integral.*

The functions Γ_r should not be confused with the higher gamma functions of Barnes.

Proposition 8.5.13. *Set*

$$\omega = \log\left(\frac{2\pi}{t_0 \sqrt{N}}\right)\,.$$

We have the formula

$$\frac{L^{(r)}(E, s)}{r!} = \sum_{n \geqslant 1} \frac{a_n(E)}{n^s} \Gamma_r\left(s, \frac{2\pi n}{t_0 \sqrt{N}}\right)$$

$$+ (-1)^r \varepsilon(E) \left(\frac{2\pi}{\sqrt{N}}\right)^{2s-2} \sum_{n \geqslant 1} \frac{a_n(E)}{n^{2-s}} \Gamma_r\left(2 - s, \frac{2\pi n t_0}{\sqrt{N}}\right)$$

$$+ \sum_{k=1}^r (-1)^{k-1} \frac{L^{(r-k)}(E, s)}{(r-k)!} \frac{\omega^k}{k!}\,.$$

Proof. It immediately follows from the first formula of Proposition 8.5.15 below that

$$\frac{d}{ds}\Gamma_r(s, x) = \Gamma_r(s, x)\log(x) + (r+1)\Gamma_{r+1}(s, x) \ .$$

Using Proposition 8.5.10, the above proposition easily follows by induction on r. □

Generalizing Corollary 8.5.11, we have the following.

Corollary 8.5.14. *Assume that $\varepsilon(E) = (-1)^r$ and in addition that $L^{(k)}(E, 1) = 0$ when $0 \leqslant k \leqslant r - 1$, $k \equiv r \pmod 2$. Then*

$$\frac{L^{(r)}(E, 1)}{r!} = 2\sum_{n\geqslant 1}\frac{a_n(E)}{n}\Gamma_r\left(1, \frac{2\pi n}{\sqrt{N}}\right) \ .$$

In particular, if $\varepsilon(E) = -1$ then

$$L'(E, 1) = 2\sum_{n\geqslant 1}\frac{a_n(E)}{n}E_1\left(\frac{2\pi n}{\sqrt{N}}\right) \ ,$$

where as above,

$$E_1(x) = \int_x^\infty \frac{e^{-t}}{t}\,dt$$

is the exponential integral function.

Proof. Since $\varepsilon(E) = (-1)^r$, the functional equation implies that $L^{(k)}(E, 1) = 0$ for all $k \not\equiv r \pmod 2$. Thus the hypotheses of the corollary and the above proposition applied with $t_0 = 1$ give the first formula, and the second follows from $\Gamma_1(1, x) = E_1(x)$. □

There remains the problem of numerically computing the functions $\Gamma_r(1, x)$ for positive x. This is done in two completely different ways depending on whether x is small or large, and we will treat these cases separately.

8.5.4 Computation of $\Gamma_r(1, x)$ for Small x

We begin with the following essential integral representation, useful whether x is small or large.

Proposition 8.5.15. *For $r \geqslant 0$ we have*

$$\Gamma_r(s, x) = x^s\int_1^\infty \frac{\log(t)^r}{r!}e^{-xt}t^s\,\frac{dt}{t} = \int_x^\infty \frac{\log(t/x)^r}{r!}e^{-t}t^s\,\frac{dt}{t} \ .$$

Proof. We prove this by induction on r, calling $g_r(x)$ the first integral on the right-hand side. It is clear that it is true for $r = 0$. It is also clear that $g_r(x)$ tends exponentially fast to 0 as x tends to infinity, so by definition of $\Gamma_r(s, x)$ we must show that $g_r'(x) = -g_{r-1}(x)/x$. We have

$$g_r'(x) = sx^{s-1} \int_1^\infty \frac{\log(t)^r}{r!} e^{-xt} t^s \frac{dt}{t} - x^s \int_1^\infty \frac{\log(t)^r}{r!} e^{-xt} t^{s+1} \frac{dt}{t} .$$

Now integration by parts shows that

$$\int_1^\infty \frac{\log(t)^r}{r!} e^{-xt} t^{s+1} \frac{dt}{t} = \frac{1}{x} \int_1^\infty e^{-xt} t^s \left(\frac{\log(t)^{(r-1)}}{(r-1)!} + s \frac{\log(t)^r}{r!} \right) \frac{dt}{t} ,$$

so that

$$g_r'(x) = -x^{s-1} \int_1^\infty e^{-xt} t^s \frac{\log(t)^{(r-1)}}{(r-1)!} \frac{dt}{t} = -\frac{g_{r-1}(x)}{x}$$

as claimed.   □

Recall that $\zeta(s) = \sum_{n \geqslant 1} n^{-s}$ for $\Re(s) > 1$ (we will study this function in much more detail in later chapters) and that Euler's constant γ is defined by

$$\gamma = \lim_{N \to \infty} \left(\sum_{n=1}^N \frac{1}{n} - \log(N) \right) = \lim_{s \to 1+} \left(\zeta(s) - \frac{1}{s-1} \right) .$$

Proposition 8.5.16. *Set*

$$G_r(x) = \sum_{n \geqslant 1} (-1)^{n-1} \frac{x^n}{n^r n!} ,$$

and define constants a_k by the formal equality

$$\exp \left(\sum_{k \geqslant 1} \frac{\zeta(k)}{k} x^k \right) = \sum_{k \geqslant 0} a_k x^k ,$$

where by convention we set $\zeta(1) = \gamma$. Then

$$(-1)^r \Gamma_r(1, x) = \sum_{k=0}^r a_k \frac{\log(x)^{r-k}}{(r-k)!} - G_r(x) .$$

Proof. When $r = 0$ we have $\Gamma_0(1, x) = e^{-x}$ and $G_0(x) = 1 - e^{-x}$; hence the formula is true in that case, so we may assume that $r \geqslant 1$. Integrating by parts we have

$$\Gamma_r(1, x) = \int_x^\infty \frac{\log(t/x)^r}{r!} e^{-t} \, dt = \int_x^\infty \frac{\log(t/x)^{r-1}}{(r-1)!} e^{-t} \frac{dt}{t}$$

$$= \int_1^\infty \frac{\log(t/x)^{r-1}}{(r-1)!} e^{-t} \frac{dt}{t} + \int_0^1 \frac{\log(t/x)^{r-1}}{(r-1)!} (e^{-t} - 1) \frac{dt}{t}$$

$$+ \int_x^1 \frac{\log(t/x)^{r-1}}{(r-1)!} \frac{dt}{t} - \int_0^x \frac{\log(t/x)^{r-1}}{(r-1)!} (e^{-t} - 1) \frac{dt}{t} .$$

First, let $I_r(x)$ be the sum of the first two integrals from 1 to ∞ and from 0 to 1. Expanding by the binomial theorem, it is clear that

$$I_r(x) = \sum_{k=0}^{r-1} (-1)^{r-1-k} C_k \frac{\log(x)^{r-1-k}}{(r-1-k)!} \, ,$$

where

$$C_k = \int_1^\infty \frac{\log(t)^k}{k!} e^{-t} \frac{dt}{t} + \int_0^1 \frac{\log(t)^k}{k!} (e^{-t} - 1) \frac{dt}{t} \, .$$

Thus

$$\sum_{k \geqslant 0} C_k s^k = \int_1^\infty e^{-t} t^s \frac{dt}{t} + \int_0^1 (e^{-t} - 1) t^s \frac{dt}{t} \, ,$$

which is clearly valid for $\Re(s) > -1$. Thus, for $\Re(s) > 0$ we have

$$\sum_{k \geqslant 0} C_k s^k = -\frac{1}{s} + \int_0^\infty e^{-t} t^s \frac{dt}{t} = \Gamma(s) - \frac{1}{s} \, ,$$

and by analytic continuation this last equality is also valid for $\Re(s) > -1$. It follows that

$$1 + \sum_{k \geqslant 0} C_k s^{k+1} = \Gamma(s+1) = \exp\left(\sum_{k \geqslant 1} (-1)^k \frac{\zeta(k)}{k} s^k\right) = \sum_{k \geqslant 0} (-1)^k a_k s^k$$

by the well-known power series expansion of $\log(\Gamma(s+1))$ (see Proposition 9.6.15 for a proof). It follows that $C_k = (-1)^{k+1} a_{k+1}$ for $k \geqslant 0$, so that

$$I_r(x) = (-1)^r \sum_{k=1}^r a_k \frac{\log(x)^{r-k}}{(r-k)!} \, ,$$

giving the first term in the formula, apart from the $k = 0$ summand.
     Furthermore, we have

$$\int_x^1 \frac{\log(t/x)^{r-1}}{(r-1)!} \frac{dt}{t} = \left. \frac{\log(t/x)^r}{r!} \right|_x^1 = (-1)^r \frac{\log(x)^r}{r!} \, ,$$

giving the $k = 0$ summand since $a_0 = 1$.
     Finally,

$$\int_0^x \frac{\log(t/x)^{r-1}}{(r-1)!} (e^{-t} - 1) \frac{dt}{t} = \int_0^1 \frac{\log(t)^{r-1}}{(r-1)!} (e^{-tx} - 1) \frac{dt}{t} = \sum_{n \geqslant 1} (-1)^n \frac{x^n}{n!} J_{n, r-1} \, ,$$

where

$$J_{n,k} = \int_0^1 t^n \frac{\log(t)^k}{k!} \frac{dt}{t} = \int_0^\infty e^{-nt} \frac{(-t)^k}{k!} dt = \frac{(-1)^k}{n^{k+1}}$$

by definition of the gamma function, so thatn

$$\int_0^x \frac{\log(t/x)^{r-1}}{(r-1)!}(e^{-t}-1)\,\frac{dt}{t} = (-1)^r \sum_{n \geqslant 1}(-1)^{n-1}\frac{x^n}{n^r\,n!} = (-1)^r G_r(x) \, ,$$

proving the proposition.   □

Example. We have $a_0 = 1$, $a_1 = \gamma$, $a_2 = (\gamma^2 + \zeta(2))/2$, $a_3 = (\gamma^3 + 3\gamma\zeta(2) + 2\zeta(3))/6$.

The above proposition reduces the computation of $\Gamma_r(1,x)$ to that of $G_r(x)$. When x is small (say $x < 10$) this is perfectly fine. When x is larger there are two closely related pitfalls that plague the computation, coming from numerical cancellation. First, since $G_r(x)$ is an alternating series, when x is large we will lose a great deal of accuracy in the computation. By comparison with the series for $\exp(-x)$, it can be shown that to obtain D decimal digits of relative accuracy we need to perform the computations to $D + 2x/\log(10)$ decimal digits in all the intermediate computations. This becomes prohibitive for x large. The second closely related pitfall is that since $\Gamma_r(1,x)$ is exponentially small, the additional polynomial in $\log(x)$ that must be subtracted from $G_r(x)$ will be of comparable size, so we will have an expression of the form $a - b$ with a and b possibly very accurate real numbers, but almost equal, the nightmare of the numerical analyst. Note that no naïve rearrangement of the alternating series can help with this.

The following proposition shows that we can at least dispense with the problem of alternating series, although it does not remove the second pitfall (cancellation with the logarithmic terms), which can be avoided only by the use of methods specific to the case in which x is large, which we shall study in the next section.

Proposition 8.5.17. *Recall that*

$$G_r(x) = \sum_{n \geqslant 1}(-1)^{n-1}\frac{x^n}{n^r\,n!} \, .$$

Set $H_k(n) = \sum_{1 \leqslant j \leqslant n} 1/j^k$, and define arithmetic functions $A_k(n)$ by a formal expansion similar to that giving the constants a_k:

$$\exp\left(\sum_{k \geqslant 1}\frac{H_k(n)}{k}x^k\right) = \sum_{k \geqslant 0} A_k(n)x^k \, .$$

Then

$$G_r(x) = e^{-x}\sum_{n \geqslant 1}\frac{x^n}{n!}A_r(n) \, .$$

Proof. We prove the proposition by induction on r, the case $r = 0$ being clear since

$$G_0(x) = 1 - e^{-x} = e^{-x} \sum_{n \geq 1} \frac{x^n}{n!} \ .$$

Define, by induction on k, $A_0(n) = 1$ for $n \geq 0$, and for $k \geq 1$,

$$A_k(n) = \sum_{j=1}^{n} \frac{A_{k-1}(j)}{j}$$

(we will see below that this is the same definition as in the proposition). Assume the proposition true for $r - 1$, in other words that

$$G_{r-1}(x) = e^{-x} \sum_{n \geq 1} \frac{x^n}{n!} A_{r-1}(n) \ .$$

By our induction hypothesis we have

$$G_r(x) = \int_0^x \sum_{n \geq 1} (-1)^{n-1} \frac{t^{n-1}}{n^{r-1} \, n!} \, dt$$

$$= \int_0^x \frac{G_{r-1}(t)}{t} \, dt = \sum_{k \geq 1} \frac{A_{r-1}(k)}{k!} \int_0^x e^{-t} t^{k-1} \, dt \ .$$

Now by induction we have

$$\int_0^x e^{-t} t^{k-1} = (k-1)! \left(1 - e^{-x} \sum_{n=0}^{k-1} \frac{x^n}{n!} \right) = (k-1)! e^{-x} \sum_{n \geq k} \frac{x^n}{n!} \ ,$$

so that

$$G_r(x) = e^{-x} \sum_{k \geq 1} \frac{A_{r-1}(k)}{k} \sum_{n \geq k} \frac{x^n}{n!} = e^{-x} \sum_{n \geq 1} \frac{x^n}{n!} \sum_{1 \leq k \leq n} \frac{A_{r-1}(k)}{k} \ ,$$

proving the result by induction on r.

It remains to see that the $A_k(n)$ defined above are given as in the proposition. Set $f(x, n) = \sum_{k \geq 0} A_k(n) x^k$. Then by definition

$$\left(1 - \frac{x}{n} \right) f(x, n) = \sum_{r \geq 0} \left(A_r(n) - \frac{A_{r-1}(n)}{n} \right) x^r = f(x, n-1) \ ,$$

and since $f(x, 0) = 1$ we obtain

$$f(x, n) = \prod_{1 \leq j \leq n} 1/(1 - x/j) \ ,$$

so the result follows by taking logarithms and expanding formally. $\square$

Note that all the series and integral manipulations are justified by absolute convergence, all the series having infinite radius of convergence. For the same reason, this enables us to compute $G_r(x)$ even for large x without loss of accuracy. However, we still need $x^n/n!$ to be small, hence n at least of the order of x (plus a small amount to account for the desired accuracy), and also we must keep in mind the cancellation phenomenon with the logarithmic terms in the formula of Proposition 8.5.16. In practice we do not advise to use the above formulas for x larger than 50, say, and even for $x > 10$.

Examples. We have $A_0(n) = 1$, $A_1(n) = H_1(n)$, $A_2(n) = (H_1(n)^2 + H_2(n))/2$, etc., the formulas being formally identical to those for the coefficients a_k (note that $\zeta(k) = \lim_{n \to \infty} H_k(n)$ for $k \geqslant 2$).

Corollary 8.5.18. *With the notation of Proposition 8.5.16 and the above proposition we have*

$$(-1)^r \Gamma_r(1, x) = \sum_{k=0}^{r} a_k \frac{\log(x)^{r-k}}{(r-k)!} - e^{-x} \sum_{n \geqslant 1} \frac{x^n}{n!} A_r(n) .$$

8.5.5 Computation of $\Gamma_r(1, x)$ for Large x

For large x we can use the following proposition.

Proposition 8.5.19. *Define arithmetic functions $C_k(n)$ by the formal expansion*

$$\exp\left(\sum_{k \geqslant 1} (-1)^{k-1} \frac{H_k(n)}{k} x^k\right) = \sum_{k \geqslant 0} C_k(n) x^k .$$

Then for $r \geqslant 1$ we have

$$\Gamma_r(1, x) = e^{-x} \sum_{n \geqslant 0} \frac{(-1)^{n+r+1} n! C_{r-1}(n)}{x^{n+1}} ,$$

where the divergent series is to be interpreted as meaning that $e^x \Gamma_r(1, x)$ is always between two successive partial sums of the series.

Proof. Set

$$E_r(x) = \int_x^\infty \frac{e^{-t}}{t^r} dt .$$

Integrating by parts gives $E_r(x) = e^{-x}/x^r - r E_{r+1}(x)$. It follows that

$$E_r(x) = \frac{e^{-x}}{(r-1)!} \sum_{n=0}^{m-1} \frac{(-1)^n (n+r-1)!}{x^{n+r}} + (-1)^m \frac{(m+r-1)!}{(r-1)!} E_{m+r}(x) .$$

Since $E_{m+r}(x) > 0$ for all x, with the interpretation of the divergent series given in the proposition we can write

$$E_r(x) = \frac{e^{-x}}{(r-1)!} \sum_{n \geqslant 0} \frac{(-1)^n (n+r-1)!}{x^{n+r}}.$$

Thus since $\Gamma_1(1, x) = E_1(x)$ the proposition is proved for $r = 1$.

Define $C_0(n) = 1$ for $n \geqslant 0$ and by induction on $k \geqslant 1$,

$$C_k(n) = \sum_{j=1}^{n} \frac{C_{k-1}(j-1)}{j}$$

(once again we will see below that these are the same as those defined in the proposition). Let $r \geqslant 2$ and assume by induction that for $k \leqslant r - 1$ we have

$$\Gamma_k(1, x) = e^{-x} \sum_{n=0}^{m-1} \frac{(-1)^{n+k-1} n! C_{k-1}(n)}{x^{n+1}} + (-1)^{m+k-1} I_{m,k}(x)$$

with $I_{m,k}(x) > 0$ for all x. We now use the induction formula

$$\Gamma_r(1, x) = \int_x^\infty \frac{\Gamma_{r-1}(1, t)}{t} \, dt$$

the expression obtained for $E_r(x)$ applied to a suitable m, and denoting by the generic letter P a nonnegative quantity, we obtain

$$\Gamma_r(1, x) = \sum_{n=0}^{m-1} (-1)^{n+r} n! C_{r-2}(n) E_{n+2}(x) + (-1)^{m+r} \int_x^\infty \frac{I_{m,r}(t)}{t} \, dt$$

$$= \sum_{n=0}^{m-1} \frac{(-1)^{n+r} n! C_{r-2}(n)}{(n+1)!} \left(e^{-x} \sum_{k=0}^{m-n-1} \frac{(-1)^k (k+n+1)!}{x^{k+n+2}} + (-1)^{m-n} P \right)$$

$$+ (-1)^{m+r} P$$

$$= e^{-x} \sum_{N=0}^{m-1} \frac{(-1)^{N+r}(N+1)!}{x^{N+2}} \sum_{n+k=N} \frac{C_{r-2}(n)}{n+1} + (-1)^{m+r} P$$

$$= e^{-x} \sum_{n=1}^{m} \frac{(-1)^{n+r+1} n!}{x^{n+1}} \sum_{1 \leqslant k \leqslant n} \frac{C_{r-2}(k-1)}{k} + (-1)^{m+r} P,$$

proving the formula of the proposition.

As before, it remains to see that the $C_k(n)$ defined above are given as in the proposition. Set $g(x, n) = \sum_{k \geqslant 0} C_k(n) x^k$. Then by definition

$$\left(1 + \frac{x}{n}\right) g(x, n-1) = \sum_{r \geqslant 0} \left(C_r(n-1) + \frac{C_{r-1}(n-1)}{n} \right) x^r = g(x, n),$$

and since $g(x, 0) = 1$ we obtain

$$g(x, n) = \prod_{1 \leqslant j \leqslant n} (1 + x/j) \,,$$

so the result follows once again by taking logarithms and expanding formally.

$\square$

Examples. We have $C_1(n) = 1$, $C_2(n) = H_1(n)$, $C_3(n) = (H_1(n)^2 - H_2(n))/2$, etc., the formulas being formally the same as for $A_k(n)$, changing $H_k(n)$ into $-H_k(n)$ for even values of k.

Remarks. (1) To compute $\Gamma_r(1, x)$, say to approximately 18 decimal digits, which is almost always sufficient for practical uses, we thus suggest the following method. If $x \leqslant 50$, we use directly the power series expansion given by Corollary 8.5.18, remembering to take into account the cancellation that occurs between the logarithmic terms and the power series. Note that 50 is not just any number but chosen such that $\exp(-50) < 10^{-20}$. If $x > 50$ we use the asymptotic expansion given by Proposition 8.5.19. This gives good results. If it is really necessary to compute to more than 18 decimal digits, simply increase 50 to a larger value.

(2) Note that in Corollary 8.5.14 we need the values of $\Gamma_r(1, x)$ to a given *absolute* accuracy. Since it tends to zero exponentially fast as $x \to \infty$, when x is large it is not necessary to compute it to a large relative accuracy. We leave the details to the reader.

(3) In the special case $r = 1$ we can do better. It can be shown that the asymptotic expansion can be expanded into the following *continued fraction*, which should of course be used instead (see Exercise 21):

$$\Gamma_1(1, x) = E_1(x) = \cfrac{e^{-x}}{x + 1 - \cfrac{1^2}{x + 3 - \cfrac{2^2}{x + 5 - \cfrac{3^2}{x + 7 - \ddots}}}}$$

This continued fraction converges for all $x > 0$, and rapidly for large x, see Exercise 22.

(4) A continued fraction corresponds to linear recurrent sequences of order 2. It can be shown that there exist similar recurrent sequences but of order $r + 1$ for $\Gamma_r(1, x)$, also leading to faster methods to compute them, but this is beyond the scope of this book (and not really essential in practice).

8.5.6 The Famous Curve $y^2 + y = x^3 - 7x + 6$

The above curve (written in minimal form) is famous because it was used to give an explicit solution to an old problem of Gauss on lower bounds for class numbers of imaginary quadratic fields. It is the curve with smallest conductor of rank 3 over $\mathbb{Q}$. Let us prove that it has rank 3. After completing the square and changing x into $x/4$ we obtain the equation $y^2 = x^3 - 112x + 400$. This curve has no torsion, so we will compute its rank using the method explained in Section 8.3.4. We let θ be a root of $P(x) = x^3 - 112x + 400 = 0$ and $K = \mathbb{Q}(\theta)$. The discriminant of the polynomial $P(x)$ is equal to $2^8 \cdot 5077$ with 5077 prime, so the only primes of K to consider are those above 2 (because of the condition $p^2 \mid \operatorname{disc}(P)$, which follows from Corollary 8.3.7). In fact, 2 ramifies completely as $2\mathbb{Z}_K = \mathfrak{p}^3$, so T is reduced to the single prime ideal $\mathfrak{p}$, which of course automatically divides $3\theta^2 + a$. To find $U_T(K)$ we should use Proposition 7.4.7 of [Coh1], but here we are lucky since we find that $\mathfrak{p}$ is a principal ideal generated by $\varepsilon_1 = -\theta/2 + 4$. We compute that the class number of K is equal to 2, that a generator of the square of a generator of the class group of K is equal to $\varepsilon_2 = \theta^2/2 + 4\theta - 25$, and that the fundamental and torsion units are $\varepsilon_3 = \theta^2/4 + 2\theta - 13$, $\varepsilon_4 = \theta^2 - 12\theta + 33$, and $\varepsilon_5 = -1$. It follows that $S_T(K)$ is generated over $\mathbb{F}_2$ by the classes modulo squares of the five elements that we have just listed. The respective norms of these elements being 2, -25, -1, 1, and -1 it follows that the kernel $S_T(K, 1)$ of the norm map modulo squares is generated by $-\varepsilon_2$, $-\varepsilon_3$, and ε_4. We could now proceed with the algorithm, and for each of these generators compute the quadratic forms $q_i(c_0, c_1, c_2)$, find a particular solution to $q_2 = 0$, then the general solution, and replace in $q_1 = 1$ to obtain the quartic, one for each generator. We then have to search for a point on these three quartics. This can be done. However, we will cheat to avoid such tedious computations. Since the dimension of $S_T(K, 1)$ is equal to 3 we know that the rank $r(E)$ of our curve is less than or equal to 3. On the other hand, on the initial equation $y^2 + y = x^3 - 7x + 6$ we readily discover the points $P_1 = (2, 0)$, $P_2 = (-1, 3)$, and $P_3 = (4, 6)$. To show that they are independent we use Theorem 8.1.17. Using for instance the algorithms of [Coh0] we compute that the determinant of the height pairing matrix of the P_i is equal to $0.41714355875838397\ldots$, hence is definitely different from 0, so that the points are independent. It follows that $r(E) \geqslant 3$, and since we have shown that $r(E) \leqslant 3$ it follows that $r(E) = 3$ as claimed. We have *not* shown that the P_i are generators, only that they generate a subgroup of $E(\mathbb{Q})$ of finite index, but this is indeed true.

We now use the method of the preceding sections to compute $L(E, s)$ and its derivatives. A computation shows that the sign of the functional equation is -1. It follows that $L^{(r)}(E, 1) = 0$ for all even $r \geqslant 1$, and in particular $L(E, 1) = L''(E, 1) = 0$. On the other hand, we compute numerically that $L'(E, 1)$ is almost equal to 0 (equal within the limits of accuracy of our computation). Note that however precisely we perform our computation we will never be able to *prove* that $L'(E, 1) = 0$. Here we must apply an important theorem of Gross–Zagier (Theorem 8.6.9 below), which implies in particular that if $L'(E, 1) \neq 0$ then the curve has rank 1. Since we know that it has three independent points this is not possible, *proving* that $L'(E, 1) = 0$. Finally, using the formulas given above we compute that

$$\frac{L'''(E,1)}{3!} = 1.73184990011930068979\ldots.$$

On the other hand, the quantity $\omega_1(E)$ that enters in the BSD conjecture is easily computed to be $2.07584399154346652494\ldots$. Since there is no torsion, the Tamagawa numbers $c_p(E)$ for finite p can be shown to be equal to 1, and $c_\infty(E) = 2$, so we deduce the equality

$$|\text{III}(E)|R(E) = 0.41714355875838397\ldots.$$

This is exactly the determinant of the height pairing matrix that we have found above by completely different methods. Thus the BSD conjecture tells us both that the points P_i given form a basis of $E(\mathbb{Q})$ (this is easy to show directly) and that $\text{III}(E)$ is the trivial group, which at present nobody knows how to prove, even on this specific curve.

8.6 The Heegner Point Method

I would like to thank C. Delaunay for writing a large part of this section.

8.6.1 Introduction and the Modular Parametrization

The Heegner point method is applicable if and only if the elliptic curve E has analytic rank exactly equal to 1. We therefore assume that we know this for a fact by having computed that $\varepsilon(E) = -1$ and that $L'(E,1) \neq 0$, which can be done rigorously as already explained. We then know that $E(\mathbb{Q})$ has a point of infinite order on it, and the purpose of the method is to find it explicitly, and even to find a generator of the torsion-free part of $E(\mathbb{Q})$.

Remarks. (1) If we have done a 2-descent showing that the rank is equal to 1, it is not necessary to use this method since a nontorsion point can easily be computed explicitly from the 2-descent method.
(2) Once a point of infinite order has been found, finding a generator is straightforward; see Exercise 24. It follows that we only want to find a point of infinite order.

Recall that the BSD conjecture involves quantities that we know how to compute, and others that we do not know directly. More precisely, given an elliptic curve E defined over $\mathbb{Q}$, we can algorithmically compute its conductor N, which enters in the functional equation and which is divisible exactly by the primes where E has bad reduction, and we can compute the (finite number of) Euler factors corresponding to these bad primes, and the so-called Tamagawa numbers c_p for $p \mid N$. All these steps are done simultaneously using variants of an algorithm due to Tate. In addition, we can compute the real period $\omega_1(E)$ as an elliptic integral using the arithmetic–geometric mean, and the torsion subgroup $E_t(\mathbb{Q})$ using one of the methods explained above. Next, for any given $p \nmid N$ we can compute $a_p = p + 1 - |E(\mathbb{F}_p)|$ hence the corresponding Euler factor, and so as many terms as we want of the Dirichlet series for the global L-function $L(E, s)$. Finally, using the method of Section 8.5.3, we can compute $L^{(r)}(E, 1)$ to any desired accuracy, and in particular use Corollaries 8.5.11 and 8.5.14.

In addition, we can also compute the *volume* $\text{Vol}(E)$ of E, in other words the volume (or determinant) of the lattice Λ such that $E(\mathbb{C}) = \mathbb{C}/\Lambda$. Although this is not necessary for the BSD formula, it will be needed in the Gross–Zagier formula below.

All the algorithms (and many more) for doing these computations are explained in great detail in [Cre2]; see also [Coh0].

The quantities that we do not know how to compute at first are the regulator $R(E)$, which is known only once a generating set for the Mordell–Weil group has been computed, and the Tate–Shafarevich group order $|\text{III}(E)|$, of which we know little apart from the fact that, if finite, it will be the square of an integer. Note that since our goal is to find a point of infinite order on $E(\mathbb{Q})$, we can assume any conjecture for doing so.

The *Heegner point method*, which we consider in this section is based on a number of facts that are outside the scope of this book, but which lead to an algorithm that is sufficiently simple and important to be explained here. Thus several of the terms used below will not be defined, and we ask the reader to be patient until we come to the actual description of the algorithm. Note however that the theory that we sketch in a few lines is very beautiful and described in many papers and textbooks; see for example [Cre1], [Dar], and [Zag].

First, if $L(E, s) = \sum_{n \geq 1} a_n n^{-s}$ then the Taniyama–Shimura–Weil conjecture asserts that $f_E(\tau) = \sum_{n \geq 1} a_n q^n$ (as usual with $q = e^{2i\pi\tau}$ and $\tau \in \mathcal{H}$, the upper half-plane) is a modular form of weight 2 on $\Gamma_0(N)$. Thanks to Wiles and successors this conjecture is now a theorem, but as already explained, for the work that we are doing we may assume any conjecture that we like.

Since $f_E(\tau)d\tau$ is a holomorphic differential the integral

$$\tilde{\phi}(\tau) = 2i\pi \int_{i\infty}^{\tau} f_E(z)dz$$

(where $i\infty$ is the point at infinity in the upper half-plane) is independent of the chosen path hence defines a map from $\mathcal{H}$ to $\mathbb{C}$. Explicitly, for $\tau \in \mathcal{H}$ we clearly have

$$\tilde{\phi}(\tau) = \sum_{n \geq 1} \frac{a_n}{n} q^n \,,$$

where as usual $q = e^{2i\pi\tau}$. The modularity property of f_E is equivalent to the fact that $\tilde{\phi}$ induces an analytic map ϕ from $X_0(N)$ to $\mathbb{C}/\Lambda$, where $X_0(N) =$

$(\mathcal{H} \cup \mathbb{P}_1(\mathbb{Q}))/\Gamma_0(N)$ is the modular curve associated with $\Gamma_0(N)$, and Λ is the lattice formed by the *periods* of f_E, corresponding to the values of $\tilde{\phi}$ for $\tau = \gamma(\infty)$ and $\gamma \in \Gamma_0(N)$ (which cannot be directly computed from the infinite series but only from the integral definition).

The lattice Λ is very often a sublattice of the lattice Λ_E associated with the minimal model of the elliptic curve E, hence in this case $\tilde{\phi}$ induces an analytic map ϕ from $X_0(N)$ to $\mathbb{C}/\Lambda_E$. To come back to points $(x,y) \in E(\mathbb{C})$ we use the classical isomorphism from $\mathbb{C}/\Lambda_E$ to $E(\mathbb{C})$ given by the Weierstrass $\wp$ function and its derivative.

However, in principle it might happen that Λ is not a sublattice of Λ_E, due to the fact that the so-called "Manin constant" of E may not be equal to 1 (even assuming Manin's conjecture saying that this should be the case for the strong Weil curve in the isogeny class of f). In practice it does not happen, but if it did it would be easy to deal with.

Putting everything together, we see that for any $\tau \in \mathcal{H}$ (and even $\tau \in \mathcal{H} \cup \mathbb{P}_1(\mathbb{Q})$, but we do not need this extra generality) we can associate a point $\varphi(\tau) \in E(\mathbb{C})$, where $\varphi = \wp \circ \phi$. The map φ from $X_0(N)$ to E is called the *modular parametrization* of E, and Wiles's theorem states that such a parametrization exists (and is usually unique up to sign, unless the curve has complex multiplication).

8.6.2 Heegner Points and Complex Multiplication

We begin by defining Heegner points.

Definition 8.6.1. (1) *Let $\tau \in \mathcal{H}$. We say that τ is a* complex multiplication point *(or simply a* CM point*) if it is a root of a quadratic equation $A\tau^2 + B\tau + C = 0$ with A, B, C integral and $B^2 - 4AC < 0$.*

(2) *If in addition we choose A, B, C such that $\gcd(A, B, C) = 1$ and $A > 0$ (which makes them unique), then $(A, B, C) = Ax^2 + Bxy + Cy^2$ is called the (positive definite binary) quadratic form associated with τ, and its discriminant $\Delta(\tau) = B^2 - 4AC$ is called the discriminant of τ.*

(3) *For a given integer $N \geqslant 1$, we will say that τ is a Heegner point of level N if $\Delta(N\tau) = \Delta(\tau)$.*

The above definition of a Heegner point is due to B. Birch. The following proposition shows that this notion depends only on the class of τ in $X_0(N)$.

Proposition 8.6.2. *If $\gamma \in \mathrm{SL}_2(\mathbb{Z})$ (in particular if $\gamma \in \Gamma_0(N)$) then $\Delta(\gamma(\tau)) = \Delta(\tau)$, and if $\gamma \in \Gamma_0(N)$ and τ is a Heegner point of level N, so is $\gamma(\tau)$.*

Proof. This comes from the fundamental group equality

$$\Gamma_0(N) = \Gamma \cap \begin{pmatrix} N & 0 \\ 0 & 1 \end{pmatrix}^{-1} \Gamma \begin{pmatrix} N & 0 \\ 0 & 1 \end{pmatrix}$$

with $\Gamma = \mathrm{SL}_2(\mathbb{Z})$, and is left to the reader. $\square$

Proposition 8.6.3. *Let $\tau \in \mathcal{H}$ be a quadratic irrationality and let (A, B, C) be the quadratic form with discriminant D associated with τ. Then τ is a Heegner point of level N if and only if $N \mid A$ and one of the following equivalent conditions is satisfied:*

(1) $\gcd(A/N, B, CN) = 1$.
(2) $\gcd(N, B, AC/N) = 1$.
(3) *There exists $F \in \mathbb{Z}$ such that $B^2 - 4NF = D$ with $\gcd(N, B, F) = 1$.*

Proof. We have $\tau = (-B + \sqrt{D})/(2A)$, hence $N\tau = (-NB + N\sqrt{D})/(2A)$. For this to have the same discriminant, it must be of the form $(-B' + \sqrt{D})/(2A')$; hence by identification of imaginary parts we have $A = NA'$, and hence $B' = B$, so that $N \mid A$. It follows that $N\tau$ is a root of $(A/N)(N\tau)^2 + B(N\tau) + CN = 0$, and since this equation has discriminant $D = B^2 - 4AC$, this must be the smallest equation satisfied by $N\tau$, so τ is a Heegner point of level N if and only if $N \mid A$ and $\gcd(A/N, B, CN) = 1$. The equivalence with the other two properties is a straightforward exercise left to the reader. $\square$

Corollary 8.6.4. *If τ is a Heegner point of level N and discriminant D, then so is $W(\tau) = -1/(N\tau)$.*

Proof. Indeed, if (A, B, C) is the quadratic form associated with τ then clearly $(CN, -B, A/N)$ is the quadratic form associated with $-1/(N\tau)$, so the result follows from the proposition. Equivalently, since $\Delta(-1/\tau) = \Delta(\tau)$ it is immediate to check from the definition that $\Delta(NW(\tau)) = \Delta(W(\tau))$. $\square$

The operator W is called the *Fricke involution*.

From now on we assume that D is a *fundamental discriminant*, in other words the discriminant of the quadratic field $K = \mathbb{Q}(\sqrt{D})$. Recall that the class group $Cl(K)$ of K is in one-to-one correspondence with classes of positive definite primitive quadratic forms (A, B, C) of discriminant D modulo the action of $SL_2(\mathbb{Z})$. More precisely, to the class of such a form (A, B, C) we associate the class of the ideal $\mathbb{Z} + (-B + \sqrt{D})/(2A)\mathbb{Z}$.

Proposition 8.6.5. *Let τ be a Heegner point of discriminant D and level N. If D is a fundamental discriminant the condition $\gcd(N, B, F) = 1$ of the above proposition is automatically satisfied and for all $p \mid \gcd(D, N)$ we have $p \| N$, in other words $v_p(N) = 1$.*

Proof. Let p be a prime dividing $\gcd(N, B, F)$. Since $B^2 - 4NF = D$ we deduce that $p^2 \mid D$, which implies that $p = 2$ and $D/4 \equiv 2$ or 3 modulo 4 since D is fundamental. But then $(B/2)^2 = (D/4) + NF \equiv (D/4) \equiv 2$ or 3 modulo 4 since $2 \mid N$ and $2 \mid F$, which is absurd, proving the first statement. Now let $p \mid \gcd(D, N)$ and assume that $p^2 \mid N$. Since $B^2 - 4NF = D$ we deduce that $p \mid B$, hence $p^2 \mid D$ hence $p = 2$. But once again since $4 \mid N$ this gives $(B/2)^2 \equiv (D/4) + NF \equiv 2$ or 3 modulo 4, which is absurd. $\square$

We have the following.

Proposition 8.6.6. *There is a one-to-one correspondence between on the one hand classes modulo $\Gamma_0(N)$ of Heegner points of discriminant D and level N, and on the other hand, pairs $(\beta, [\mathfrak{a}])$ where $\beta \in \mathbb{Z}/2N\mathbb{Z}$ is such that $b^2 \equiv D \pmod{4N}$ for any lift b of β to $\mathbb{Z}$, and $[\mathfrak{a}] \in Cl(K)$ is an ideal class. The correspondence is as follows: if $(\beta, [\mathfrak{a}])$ is as above, there exists a primitive quadratic form (A, B, C) whose class is equal to $[\mathfrak{a}]$ and such that $N \mid A$ and $B \equiv \beta \pmod{2N}$, and the corresponding Heegner point is $\tau = (-B + \sqrt{D})/(2A)$. Conversely, if (A, B, C) is the quadratic form associated with a Heegner point τ we take $\beta = B \bmod 2N$ and $\mathfrak{a} = \mathbb{Z} + \tau\mathbb{Z}$.*

Proof. This consists in a series of easy verifications, which are essentially identical to those made in checking that the ideal class group of K is isomorphic to the group of classes of positive definite primitive quadratic forms of discriminant D, and the details are left to the reader. Note that if b is defined modulo $2N$ then b^2 is indeed defined modulo $4N$.   □

Thanks to this proposition, it is often more natural to consider a (class of) Heegner point as a pair $(\beta, [\mathfrak{a}])$ rather than as a complex number.

We need one last ingredient from algebraic number theory. Let K be a number field (in our case $K = \mathbb{Q}(\sqrt{D})$ will be an imaginary quadratic field). There exists a finite extension H of K, called the Hilbert class field, which has many remarkable properties. The most important one for us is that it is an Abelian extension of K whose Galois group is canonically isomorphic to the class group $Cl(K)$ through a completely explicit map Art from $Cl(K)$ to $\mathrm{Gal}(H/K)$. In other words, any element of $\mathrm{Gal}(H/K)$ has the form $\mathrm{Art}([\mathfrak{a}])$ for a unique ideal class $[\mathfrak{a}]$. The action of an element $\sigma \in \mathrm{Gal}(H/K)$ on $h \in H$ will be written h^σ.

The theorem that makes the whole method work is the main theorem of *complex multiplication*, which we will not prove. The results are due to Deuring and Shimura, but in the present context I refer to the paper of Gross [Gro].

Theorem 8.6.7. *Let $\tau = (\beta, [\mathfrak{a}])$ be a Heegner point of level N and discriminant D, let $K = \mathbb{Q}(\sqrt{D})$, and denote by H the Hilbert class field of K. Then $\varphi(\tau) \in E(H)$, and we have the following properties:*

(1) *For any $[\mathfrak{b}] \in Cl(K)$ then*

$$\varphi((\beta, [\mathfrak{a}]))^{\mathrm{Art}([\mathfrak{b}])} = \varphi((\beta, [\mathfrak{a}\mathfrak{b}^{-1}])) .$$

(2)

$$\varphi(W(\beta, [\mathfrak{a}])) = \varphi((-\beta, [\mathfrak{a}\mathfrak{n}^{-1}])) ,$$

where $\mathfrak{n} = N\mathbb{Z} + \dfrac{B + \sqrt{D}}{2}\mathbb{Z}$ and B is any integer whose class modulo $2N$ is equal to β.

(3) *If $\overline{}$ denotes complex conjugation then*

$$\overline{\varphi((\beta, [\mathfrak{a}]))} = \varphi((-\beta, [\mathfrak{a}^{-1}])) \ .$$

Thus we see that using the analytic function φ, we can obtain a point with coordinates in H, hence with algebraic coordinates. This is the "miracle" of complex multiplication, which generalizes the fact that the exponential function evaluated at rational multiples of $2i\pi$ gives algebraic numbers.

The first formula gives all the conjugates over K of $\varphi(\tau)$, and is called Shimura's reciprocity law. In particular, we can compute the trace P of $\varphi(\tau)$ as a point in $E(H)$ as

$$P = \sum_{\sigma \in \mathrm{Gal}(H/K)} \varphi((\beta, [\mathfrak{a}]))^\sigma = \sum_{[\mathfrak{b}] \in Cl(K)} \varphi((\beta, [\mathfrak{a}\mathfrak{b}^{-1}])) = \sum_{[\mathfrak{b}] \in Cl(K)} \varphi((\beta, [\mathfrak{b}])) \ ,$$

the sum being computed with the group law of E. By Galois theory we will have $P \in E(K)$, so we have considerably reduced the field of definition of the algebraic point found on E. Finally, we have the following easy result:

Lemma 8.6.8. *If $\varepsilon = -1$, then in fact $P \in E(\mathbb{Q})$.*

Proof. Indeed, it is easy to see that $\varepsilon = -1$ is equivalent to saying that $\varphi \circ W = \varphi$, so that

$$\overline{\varphi((\beta, [\mathfrak{b}]))} = \overline{\varphi(W(\beta, [\mathfrak{b}]))} = \overline{\varphi((-\beta, [\mathfrak{b}\mathfrak{n}^{-1}]))} = \varphi((\beta, [\mathfrak{b}^{-1}\mathfrak{n}])) \ ,$$

hence

$$\overline{P} = \sum_{[\mathfrak{b}] \in Cl(K)} \varphi((\beta, [\mathfrak{b}^{-1}\mathfrak{n}])) = \sum_{[\mathfrak{b}] \in Cl(K)} \varphi((\beta, [\mathfrak{b}])) = P \ ,$$

so by Galois theory once again we deduce that $P \in E(\mathbb{Q})$. $\square$

We thus see that the Heegner point method does give us a point in $E(\mathbb{Q})$ (which of course may be a torsion point). It immediately follows from the Gross–Zagier Theorem 8.6.9 and the work of Kolyvagin that if the rank (analytic or algebraic) is strictly greater than 1, this point *will* always be a torsion point, so the method is useless. Furthermore, a similar proof to that of the above lemma shows that if $\varepsilon = 1$ then $P + \overline{P}$ is a torsion point, so once again the method is useless. Hence, as claimed from the beginning, the Heegner point method is applicable only in the rank-1 case. Without any exaggeration it can be said that this is the *only* reason for which nothing is known on the BSD conjecture when the rank is strictly greater than 1.

8.6.3 The Use of the Theorem of Gross–Zagier

Although it would already be possible to use the method as explained above, an important additional result due to Gross–Zagier usually simplifies the computations. Recall from Definition 7.3.15 that the *quadratic twist* of an elliptic curve E given by a Weierstrass equation $y^2 = x^3 + ax^2 + bx + c$ by a fundamental discriminant D is the curve E_D with equation $y^2 D = x^3 + ax^2 + bx + c$. If desired this can be put in ordinary Weierstrass form, and extended to curves in generalized Weierstrass form. The important (and easy) point is that the L-function of E_D can easily be obtained from that of E: in our case, since D is a fundamental discriminant that is a square modulo $4N$, Proposition 8.6.5 tells us that if $p \mid \gcd(D, N)$ then $p^2 \nmid N$. This implies that the conductor N_D of E_D is equal to $ND^2/\gcd(D, N)$ and that if $L(E, s) = \sum_{n \geq 1} a_n n^{-s}$, then

$$L(E_D, s) = \sum_{n \geq 1} \left(\frac{D}{n}\right) \frac{a_n}{n^s} \ .$$

Furthermore, it is not difficult to show that the sign of the functional equation for $L(E_D, s)$ is equal to $\left(\frac{D}{-N}\right)$ times that of $L(E, s)$. In particular, in our context that of E is equal to -1, $\left(\frac{D}{N}\right) = 1$, and $D < 0$, so the sign of the functional equation for $L(E_D, s)$ is equal to $+1$. Finally, recall that we have defined a canonical height function $\widehat{h}$ on $E(\mathbb{Q})$; see Theorem 8.1.17.

Theorem 8.6.9 (Gross–Zagier). *If* $\gcd(D, 2N) = 1$ *and* $D \neq -3$ *the point* P *computed above satisfies*

$$\widehat{h}(P) = \frac{\sqrt{|D|}}{4\,\mathrm{Vol}(E)} L'(E, 1) L(E_D, 1)$$

(for $D = -3$ *the right-hand side should be multiplied by 9, see below).*

Since $L(E_D, 1)$ can easily be computed using the exponentially convergent series given above, this allows us to check whether $\widehat{h}(P)$ is close to 0, hence whether we will obtain a torsion point. But it is especially interesting to combine it with the BSD formula: indeed, in the rank-1 case we have $R(E) = \widehat{h}(G)$, where G is a generator of the Mordell–Weil group of E. Since $P \in E(\mathbb{Q})$ it has the form $\ell G + Q$ for some torsion point Q, hence $\widehat{h}(P) = \ell^2 \widehat{h}(G)$. The combination of the two formulas thus reads

$$\frac{\ell^2}{|\mathrm{III}(E)|} = \omega_1(E) \frac{c(E)\sqrt{|D|}}{4\,\mathrm{Vol}(E)|E_t(\mathbb{Q})|^2} L(E_D, 1) \ ,$$

where $c(E)$ is the product of the Tamagawa numbers $c_p(E)$ including c_∞.

Although $|\mathrm{III}(E)|$ is unknown, it is usually small and very often equal to 1 (and in our case is known to be finite hence equal to the square of an integer), so this very often gives the value of ℓ.

It is useful to be able to generalize this formula to the case $\gcd(D, 2N) > 1$ and also to $D = -3$ and $D = -4$. This leads us to formulate the following reasonable conjecture.

Conjecture 8.6.10. *Let E be an elliptic curve of analytic rank 1 (in other words $\varepsilon(E) = -1$ and $L'(E, 1) \neq 0$), and let D be a negative fundamental discriminant that is a square modulo $4N$. Assume that $L(E_D, 1) \neq 0$ and that for any $p \mid \gcd(D, N)$ we have $a_p = -1$. Then*

$$\frac{\ell^2}{|\mathrm{III}(E)|} = \omega_1(E) \frac{c(E)\sqrt{|D|}(w(D)/2)^2}{4\,\mathrm{Vol}(E)|E_t(\mathbb{Q})|^2} 2^{\omega(\gcd(D,N))} L(E_D, 1)\ ,$$

where $w(D)$ is the number of roots of unity in $\mathbb{Q}(\sqrt{D})$ ($w(-3) = 6$, $w(-4) = 4$, and $w(D) = 2$ for $D < -4$), and as usual $\omega(\gcd(D, N))$ is the number of distinct prime factors of $\gcd(D, N)$.

The condition $a_p = -1$ for $p \mid \gcd(D, N)$ is necessary to obtain a non-torsion Heegner point, and for the validity of the formula. Furthermore, the conditions on D in the conjecture imply that $\varepsilon(E_D) = 1$, so $L(E_D, 1)$ can also be computed by the exponentially convergent series given above. The truth of this conjecture is of course mainly supported by the work of Gross–Zagier, but the additional terms are due to work of Y. Hayashi (see [Gro], [Hay]). In addition, it has also been verified in numerous cases. As already mentioned, to compute a rational point we can always assume any reasonable conjecture.

8.6.4 Practical Use of the Heegner Point Method

The most lengthy computations will be the evaluations of $\phi((-B+\sqrt{D})/(2A))$ for the $|Cl(K)|$ classes of quadratic forms (A, B, C). Two remarks must be made.

Remarks. (1) Since the convergence of the series for $\phi(\tau)$ is essentially that of a geometric series with ratio $\exp(-2\pi\Im(\tau)) = \exp(-2\pi\sqrt{|D|}/(2A))$, and since $N \mid A$, it will be particularly slow when N is large. The method will thus be inapplicable for large conductors, say $N > 10^8$.

(2) Thanks to the relation $\overline{\varphi((\beta, [\mathfrak{a}]))} = \varphi((\beta, [\mathfrak{a}^{-1}\mathfrak{n}]))$, which we have used above, we need to compute only approximately half of the necessary values of ϕ: indeed, if $[\mathfrak{a}]$ corresponds to the class of the form (A, B, C), then $[\mathfrak{a}^{-1}\mathfrak{n}]$ corresponds to the class of the form $(CN, B, A/N)$, so the value of ϕ on $(CN, B, A/N)$ is simply the conjugate of $\phi((A, B, C))$ modulo the lattice Λ.

We give the method as an algorithm, and then apply it to a reasonably large example.

Algorithm 8.6.11 (Heegner Point Method) Let E be an elliptic curve defined over $\mathbb{Q}$ with conductor N, and assume that E has analytic rank 1, in other words that $\varepsilon(E) = -1$ and $L'(E, 1) \neq 0$. We assume that E is given by a minimal Weierstrass equation $y^2 + a_1xy + a_3y = x^3 + a_2x^2 + a_4x + a_6$ with integer coefficients. This algorithm outputs a nontorsion rational point in $E(\mathbb{Q})$.

We assume computed the standard values associated with E, and in particular we denote by $\omega_1(E)$ and $\omega_2(E)$ standard $\mathbb{Z}$-generators of the period lattice Λ of E with $\omega_1(E) \in \mathbb{R}_{>0}$, and by $\mathrm{Vol}(E)$ the area of the fundamental parallelogram.

1. [Compute necessary accuracy] Compute the product $|\mathrm{III}(E)|R(E)$ thanks to the BSD formula

$$|\mathrm{III}(E)|R(E) = \frac{|E_t(\mathbb{Q})|^2 L'(E,1)}{c(E)\omega_1(E)} ,$$

where $L'(E,1)$ is computed thanks to Corollary 8.5.14. Compute the height difference bound HB given by Theorem 8.1.18, in other words compute $HB = h(j(E))/12 + \mu(E) + 1.946$, where $\mu(E)$ is given by the above-mentioned theorem, and finally set $d = 2(|\mathrm{III}(E)|R(E) + HB)$. All computations will now be done with a default accuracy of $dd = \lceil d/\log(10)\rceil + 10$ decimal digits, and in particular recompute all the floating-point quantities such as $\omega_i(E)$ and $\mathrm{Vol}(E)$ to that accuracy.

2. [Loop on fundamental discriminants] For each successive negative fundamental discriminant $D = -3$, -4, etc., execute the rest of the algorithm until a nontorsion point of $E(\mathbb{Q})$ is found. Check that D is a square modulo $4N$, that $a_p = -1$ for each $p \mid \gcd(D,N)$, and by computing

$$L(E_D,1) = 2\sum_{n\geqslant 1} \frac{a_n}{n}\left(\frac{D}{n}\right)\exp\left(\frac{-2\pi n}{\sqrt{ND^2/\gcd(D,N)}}\right)$$

check that $L(E_D,1)$ is numerically not equal to 0. If any of these conditions is not satisfied, choose the next fundamental discriminant. Otherwise fix $\beta \in \mathbb{Z}/(2N)\mathbb{Z}$ such that $D \equiv \beta^2 \pmod{4N}$ and compute $m > 0$ such that

$$m^2 = \omega_1(E)\frac{c(E)\sqrt{|D|}(w(D)/2)^2}{4\,\mathrm{Vol}(E)|E_t(\mathbb{Q})|^2}2^{\omega(\gcd(D,N))}L(E_D,1) .$$

This m should be very close to an integer, or at least to a rational number with small denominator.

3. [Find List of Forms] Using Subalgorithm 8.6.12 below, compute a list $\mathcal{L}$ of $|Cl(K)|$ representatives (A,B,C) of classes of positive definite quadratic forms of discriminant D, where A must be chosen divisible by N and minimal, and $B \equiv \beta \pmod{2N}$ (this is always possible). Whenever possible pair elements (A,B,C) and (A',B',C') of this list such that (A',B',C') is equivalent to $(CN,B,A/N)$ by computing the unique canonical reduced form equivalent to each.

4. [Main Computation] Compute the complex number

$$z = \sum_{(A,B,C)\in\mathcal{L}} \phi\left(\frac{-B+\sqrt{D}}{2A}\right) \in \mathbb{C} ,$$

using the formula $\phi(\tau) = \sum_{n \geqslant 1}(a_n/n)q^n$ (with $q = e^{2i\pi\tau}$) given above
and the fact that $\phi((-B' + \sqrt{D})/(2A')) = \overline{\phi((-B + \sqrt{D})/(2A))}$, where
(A', B', C') is paired with (A, B, C) as in Step 3. The number z should be
computed to at least dd decimal digits of accuracy. This means that the
number of terms to be taken in the series for $\phi((-B + \sqrt{D})/(2A))$ should be
a little more than $Ad/(\pi\sqrt{|D|})$.

5. [Find Rational Point] Let e be the exponent of the group $E_t(\mathbb{Q})$, let $\ell = \gcd(e, m^\infty) = \gcd(e, m^3)$, and $m' = m\ell$. For each pair $(u, v) \in [0, m' - 1]^2$, set $z_{u,v} = (\ell z + uw_1(E) + vw_2(E))/m'$. Compute $x = \wp(z_{u,v})$, where $(\wp, \wp')$ is the isomorphism from $\mathbb{C}/\Lambda$ to $E(\mathbb{C})$. For each (u, v) such that the corresponding point $(x, y) \in E(\mathbb{C})$ has real coordinates (in fact, we can know in advance which of these m or $2m$ points are real; see the remarks below), test whether x is close to a rational number with a square denominator f^2. If the computation has been performed with sufficient accuracy, at least one of these points x must be a rational number. Otherwise, we must slightly increase the accuracy used in the computations. Once x is found corresponding to a nontorsion point, compute y using the equation of the curve (which must be a rational number whose denominator is equal to f^3) and terminate the algorithm.

To compute the list of forms necessary in Step 3, we could use a sophisticated method. However, since the time spent on doing this is completely negligible compared to the time spent in the main computation of Step 4, the following naïve algorithm is sufficient.

Subalgorithm 8.6.12 (Compute list of forms) Given a fundamental negative discriminant D and β such that $\beta^2 \equiv D \pmod{4N}$, this subalgorithm computes a list $\mathcal{L}$ of forms as in Step 3 above.

1. [Initialize] Using any method (since D is small), compute the number $h(D)$ of classes of forms of discriminant D, set $\mathcal{L} \leftarrow \emptyset$, $\mathcal{L}_r \leftarrow \emptyset$, and let b be such that $b \equiv \beta \pmod{2N}$.

2. [Fill lists] Set $R \leftarrow (b^2 - D)/(4N)$, and for all positive divisors d of R do as follows: set $f \leftarrow (dN, b, R/d)$ and let f_r be the unique reduced quadratic form (A, B, C) equivalent to f (in other words $|B| \leqslant A \leqslant C$ and $B \geqslant 0$ if either $|B| = A$ or $A = C$). If $f_r \notin \mathcal{L}_r$, set $\mathcal{L}_r \leftarrow \mathcal{L}_r \cup \{f_r\}$, using Subalgorithm 8.6.13 below, find a form $f' = (A', B', C')$ equivalent to f still with $N \mid A'$ and $B' \equiv \beta \pmod{2N}$, but with A' minimal, and set $\mathcal{L} \leftarrow \mathcal{L} \cup \{f'\}$.

3. [Finished?] If $|\mathcal{L}| < h(D)$ set $b \leftarrow b + 2N$ and go to Step 2; otherwise, terminate the subalgorithm.

In Step 2 the reduction from f to f_r is done using a standard algorithm such as Algorithm 5.4.2 of [Coh0]. The reduction of f to f' is done as follows.

Subalgorithm 8.6.13 (Compute minimal A) Given a positive definite form $f = (A, B, C)$ of discriminant $D < 0$ with $N \mid A$, this algorithm finds a form

$f' = (A', B', C')$ equivalent to f with $N \mid A'$ and $B' \equiv B \pmod{2N}$ with A' minimal.

1. [Initialize] Set $u \leftarrow -B/(2A/N)$, $v_2 \leftarrow |D|/(2A/N)^2$. By any reasonable method (see below) find some (c_0, d_0) with $\gcd(c_0 N, d_0) = 1$ and such that $(c_0 u + d_0)^2 + c_0^2 v_2 = m_0$ is as small as possible by the chosen method. Note that we can always set $c_0 \leftarrow 0$, $d_0 \leftarrow 1$, and $m_0 \leftarrow 1$. Set $L \leftarrow \sqrt{m_0/v_2}$. If $L \leqslant 1$, output (A, B, C) and terminate the algorithm; otherwise, set $c \leftarrow 0$.

2. [Loop on c] Set $c \leftarrow c + 1$, and if $c \geqslant L$ go to Step 3. Otherwise, set $d \leftarrow \lfloor -cu \rceil^*$ (the nearest integer to $-cu$ prime to cN) and set $r \leftarrow (cu+d)^2 + c^2 v_2$. If $r < m_0$, set $m_0 \leftarrow r$, $c_0 \leftarrow c$, $d_0 \leftarrow d$, and $L \leftarrow \sqrt{m_0/v_2}$. Go to Step 2.

3. [Find form] If $m_0 = 1$, output (A, B, C). Otherwise, using the extended Euclidean algorithm, compute a_0 and b_0 such that $a_0 d_0 - b_0 N c_0 = 1$, let $f' = (A', B', C')$ be the form $f(a_0 x + b_0 y, N c_0 x + d_0 y)$ and output f'. Terminate the subalgorithm.

Proof. If we write $\tau = (-B + \sqrt{D})/(2A) = x + iy$, then $x = \Re(\tau) = -B/(2A)$ and $y = \Im(\tau) = \sqrt{|D|}/(2A)$. It follows that finding a minimal A' is equivalent to finding the corresponding τ' with the largest imaginary part. It is immediately checked that the relations $N \mid A'$ and $B' \equiv B \pmod{2N}$ are preserved by $\gamma = \left(\begin{smallmatrix} a & b \\ c & d \end{smallmatrix}\right) \in \mathrm{SL}_2(\mathbb{Z})$ (for all forms f) if and only if $\gamma \in \Gamma_0(N)$, so we write Nc instead of c. Finally, it is clear that

$$\Im(\gamma(\tau)) = \frac{\Im(\tau)}{|Nc\tau + d|^2},$$

so we must make $|Nc\tau + d|^2$ minimal, where

$$|Nc\tau + d|^2 = (Ncx + d)^2 + N^2 c^2 y^2 = (cu + d)^2 + c^2 v_2$$

using the notation of the algorithm. This quantity can trivially be made equal to m_0 by choosing $(c, d) = (c_0, d_0)$ of Step 1. Hence if we want it to be strictly less than m_0 we must have $c < \sqrt{m_0/v_2} = L$, and c being fixed, the optimal value of d is the one making $|cu + d|$ minimal, in other words the nearest integer to $-cu$ coprime to cN, since we must apply the extended Euclidean algorithm in Step 3. $\square$

To find a good (c_0, d_0) in Step 1 of the above subalgorithm, we can, for example, either use a continued fraction approximation to $-Nx \approx d_0/c_0$, or use Gaussian reduction of the quadratic form $(c, d) \mapsto (cu + d)^2 + c^2 v_2$ in order to find small vectors. Having a small m_0 in Step 1 is important for the efficiency of Step 3.

Remarks. (1) As already mentioned, the above two subalgorithms are not at all optimal. However, since the time spent in computing the necessary forms is negligible compared to the time spent in computing the values of ϕ, it does not matter at all. What is essential is that we choose a representative (A', B', C') with a minimal A'; otherwise, the computation of ϕ will be much longer.

(2) Since by Proposition 7.3.1 we have $x = n/f^2$ and $y = n'/f^3$, the detection of y as a rational number is much more costly than that of x, so it is preferable to compute x first, then y. This proposition also explains the choice of the accuracy d made in Step 1 of the main algorithm.

(3) The default accuracy d should depend not only on E, but also very slightly on the chosen discriminant D. If the choice of D is reasonable (say $|D| < 10^6$), the added constant 10 more than compensates for this dependence.

(4) We could directly compute $(\wp(z), \wp'(z))$, which will be a nontorsion point of $E(\mathbb{Q})$, and identify its coordinates as rational numbers. However, in general we need very high accuracy for doing this computation, so this can be done only on small examples (see below). The use of the integer m in the above algorithm considerably reduces the necessary accuracy.

(5) If P denotes the point of $E(\mathbb{Q})$ corresponding to $z \in \mathbb{C}/\Lambda$, we have $P = mG+T$, where G is a nontorsion point and $T \in E_t(\mathbb{Q})$. If the order of T is prime to m we can also write $P = mG + mT'$ for a suitable $T' \in E_t(\mathbb{Q})$, so the point $G + T'$ will correspond to $z/m + \omega$ for some $\omega \in \Lambda/m$. Unfortunately, the order of T may not be prime to m. Nevertheless, we have $\ell P = m'G + \ell T$ and the order of ℓT is prime to m', so that we can apply the previous case with ℓz instead of z. Note that Mazur's Theorem 8.1.16 implies that $\gcd(e, m^\infty) = \gcd(e, m^3)$, where we recall that $\gcd(e, m^\infty) = \lim_{n\to\infty} \gcd(e, m^n)$.

(6) A point $z = \lambda_1\omega_1(E) + \lambda_2\omega_2(E)$ will correspond to a real point if and only if $\lambda_2 \in \mathbb{Z}$ if $\Delta < 0$, or $\lambda_2 \in \frac{1}{2}\mathbb{Z}$ if $\Delta > 0$. It is therefore easy to find for which $(u, v) \in [0, m' - 1]^2$ the point $z_{u,v}$ corresponds to a real point.

(7) Even after dividing by m' as in $z_{u,v}$ above, the point that we will obtain may be a large multiple of the Mordell–Weil generator; more precisely, it will be equal to $\sqrt{|\text{III}(E)|}G+T$ for some torsion point T. Thus, this will occur when $\text{III}(E)$ is nontrivial. In this case we can either increase the accuracy of the computations, or choose small multiples of m instead of m itself.

(8) The number of coefficients a_n that must be used in the series for ϕ can well exceed the capacity of a computer. In this case they must be computed inductively; see [Buh-Gro] for one way of doing this.

(9) As noted above, it may happen that the so-called Manin constant of the curve is not equal to 1. In that case we must on the contrary *multiply* the values $z_{u,v}$ by a small constant, which technically is the degree of the isogeny between E and the strong Weil curve in its isogeny class.

(10) Although the main algorithm suggests choosing discriminants D in increasing absolute value, it is clear that the best choice is to choose D for which the smallest value of $\sqrt{|D|}/A$ in Step 4 is as large as possible. The smallest $|D|$ is not always the best for this.

8.6.5 Improvements to the Basic Algorithm, in Brief

Any algorithm is subject to improvement. However, in the case of the Heegner point algorithm, the possible gains both in time and space are so large that even though the present book is not primarily oriented toward algorithms, it is essential to present, at least briefly and without proof, some of the major improvements.

Atkin–Lehner Operators

We have seen that the set of Heegner points of discriminant D and level N is invariant under the action of $\Gamma_0(N)$, and that it is also invariant under the Fricke involution W defined by $W(\tau) = -1/(N\tau)$, or in matrix form, $W = \left(\begin{smallmatrix} 0 & -1 \\ N & 0 \end{smallmatrix}\right)$. Atkin and Lehner remarked that when N is not a prime power, there exist more general operators W_Q, one for each positive *primitive* divisor Q of N, in other words such that $\gcd(Q, N/Q) = 1$, which have very similar properties: up to the action of $\Gamma_0(N)$ they are involutions, and they also preserve the set of Heegner points of discriminant D and level N. The great advantage of having these extra operators is that we can hope to decrease the values of A in the forms (A, B, C) that we have to choose, so as to increase the imaginary parts of the points $\tau \in \mathcal{H}$, making the computation of ϕ faster; see Exercises 27 and 28. However, these operators W_Q do not always preserve $B \bmod 2N$, so we have to choose Q suitably so that $B \bmod 2N$ is preserved. Thus, we must slightly adapt Subalgorithm 8.6.13 and note that $\varphi \circ W_Q = \varepsilon_Q \varphi + T_Q$, where $T_Q \in E_t(\mathbb{Q})$ and $\varepsilon_Q = \pm 1$ can easily be computed. Note also that we do not need to compute T_Q.

The Cremona–Silverman Trick

This is probably the most important improvement to the basic algorithm, since it can be applied in complete generality, while the Atkin–Lehner operators can be applied only when the conductor has several distinct prime factors, and the CM method can be applied only to complex multiplication curves and/or to quadratic twists.

We first remark that the x-coordinate of an element of $E(\mathbb{Q})$ is of the form n/d^2 for some coprime integers n and d. This is a very special form for rational numbers. Nonetheless, in the basic algorithm this is not used at all, and the rational number is recognized via continued fractions as any other rational number with a similar but nonsquare denominator. This seems like a waste, because intuitively one would like to make the computations to approximately one half of the accuracy so as to recognize d, which would both increase considerably the speed (by a factor of 4 approximately) and also decrease the number of necessary coefficients in the sum. Unfortunately, as far as the author is aware, there is no LLL-like recognition algorithm specific to rational numbers of the form n/d^2.

The very clever idea of Cremona, based on a paper of Silverman, is to use as much as possible all available information on the point $P = (x, y) \in E(\mathbb{Q})$ computed by the Heegner point method.

- First of all, thanks to the Heegner point computation, we know a real approximation to x.
- Second, thanks to the Gross–Zagier theorem, we know the canonical height $\widehat{h}(P)$ of P.
- For every prime $p \mid N$ and for $p = \infty$ it is possible to define a *local height* $\widehat{h}_p(P)$, which will be a *rational* multiple of $\log(p)$, and if $x = n/d^2$ as before, we have

$$2\log(d) = \widehat{h}(P) - \widehat{h}_\infty(P) - \sum_{\substack{p \mid N \\ p^2 \mid \mathrm{disc}(E) \\ p \nmid d}} \widehat{h}_p(P)$$

(as we shall see, the fact that the right-hand side depends on d should not worry us). The main points are now the following.

- First of all, the height at infinity $\widehat{h}_\infty(P)$ can easily be computed knowing only a real approximation to x (see for instance [Coh0], Algorithm 7.5.7; note that this algorithm also needs a real approximation to y, which is easily obtained from the equation of the curve).
- Second, the number of possibilities for $\widehat{h}_p(P)$ is *finite* and can easily be given explicitly. Since we do not know in advance whether $p \nmid d$, it is sufficient to include systematically 0 in the list of possible $\widehat{h}_p(P)$ that will be given below.

Thus, by looping through all possibilities for $\widehat{h}_p(P)$, we find a finite number of possibilities for d, and n is recovered thanks to $n = \lfloor d^2 x \rfloor$. In practice, this allows us to work as we wanted, with an accuracy slightly more than half of the accuracy used in the basic algorithm.

The list of possible values for $\widehat{h}_p(P)$ depends on the *Kodaira type* K of the elliptic curve, which is easily computed thanks to Tate's algorithm. It is of the form $V \log(p)$, where V is a list of rational numbers given as follows (see [Cre-Pri-Sik] for proofs and more details):

(1) If $K = I_m$, let $n = v_p(\mathrm{disc}(E))$, then $V = \{j^2/(n-j),\ 0 \leqslant j \leqslant \lfloor m/2 \rfloor\}$.
(2) If $K = I_m^*$, then $V = \{0, -1, -1 - m/4\}$.
(3) If $K = II$, $K = II^*$, or $K = I_0^*$, then $V = \{0\}$.
(4) If $K = III$, then $V = \{0, -1/2\}$, and if $K = III^*$, then $V = \{0, -3/2\}$.
(5) If $K = IV$, then $V = \{0, -2/3\}$, and if $K = IV^*$, then $V = \{0, -4/3\}$.

The Case of CM Curves

An important special type of curves for which the Heegner point method can be considerably speeded up is the case of curves with *complex multiplication*, such as for instance the *congruent number curves* $y^2 = x^3 - n^2 x$ (see Proposition 6.12.1 and the complete example $n = 157$ given below). Since our curves are defined over $\mathbb{Q}$, the ring of endomorphisms of E must be an imaginary quadratic order of class number 1. By the work of Heegner and Stark, the only possibilities are for the nine maximal orders of respective discriminants -3, -4, -7, -8, -11, -19, -43, -67, and -163, together with the four nonmaximal orders of respective discriminants -12, -16, -27, and -28.

A large part of the computation time in the Heegner point method is taken by the computation of the $a_p(E) = p + 1 - |E(\mathbb{F}_p)|$. In the case of CM curves, the computation of $a_p(E)$ can be considerably speeded up thanks to the use of Cornacchia's algorithm. We refer to the detailed explanation and algorithms given in Section 8.5.2.

For other practical improvements on this method, we refer to ongoing work of Delaunay, Watkins, and Cremona.

8.6.6 A Complete Example

We consider the following problem. By Tunnell's theorem or by the BSD conjecture, we know that 157 is a congruent number; in other words, there exists a rational nontorsion point on the elliptic curve $y^2 = x^3 - 157^2 x$. We want to compute such a point (from which it is easy to compute explicitly a Pythagorean triangle with area 157). We give explicitly the GP commands so that the reader can reproduce the computations himself (we use the basic algorithm, and none of the improvements mentioned afterward). We should first choose a sufficiently large stack (200 MB is sufficient). We begin with the command e = ellinit([0, 0, 0, -157^2, 0]), which computes a number of needed constants. In particular, the period lattice is generated by om1 = e.omega[1] and om2 = e.omega[2], and since the discriminant is positive the real period is om = 2*om1. The command et = elltors(e) shows that $|E_t(\mathbb{Q})| = 4$, which we knew already in the congruent number problem, vole = e.area gives the volume, ered = ellglobalred(e) gives the conductor N = ered[1], the Tamagawa product c = ered[3], and the fact that the equation for E is in fact a minimal model. We now compute the necessary accuracy: we find that $|Ш(E)|R(E) \approx 54.6$ and $HB \approx 10.6$; hence $d \approx 130.4$, so we will perform our computations with a default accuracy of 67 decimal digits.

Since we performed the ellinit command with the default accuracy of 28, we must now change the default accuracy with the command \p 67, and then recompute ellinit and the values of the corresponding floating point-numbers given above.

We now search for suitable fundamental discriminants. We find that up to $D = -40$, only $D = -31$ and $D = -39$ are squares modulo 4N. The quantity $m^2 = m^2(D)$ of the algorithm can be computed by defining the function

```
m2(D) = v = ellan(e, 1000000); q = exp(-2*Pi/sqrt(N*D^2)); \
q1 = 1.; s = 0.; for (n = 1, 1000000, q1 *= q; \
s += v[n]/n*kronecker(D,n)*q1); \
          sqrt(-D)*c*om/(4*vole*et[1]^2)*2*s.
```

In the above, 1000000 is overkill, but we are lazy since the computation is very fast. We find that m2(-31) is very close to 0, so -31 is not suitable, but m2(-39) is very close to 16, so we choose $D = -39$ and $m = 4$. We easily find that $b = 1275547$ satisfies $b^2 \equiv D \pmod{4N}$, so we write D = -39 and b = 1275547.

There are four classes of quadratic forms of discriminant -39, and with the notation of the algorithm, it is easy to see that we have $z = 2\Re(\phi(x_1)+\phi(x_2))$ with $x_j = (-b+\sqrt{-39})/(2jN)$. The largest value of A in the quadratic forms (A, B, C) is thus $A = 2N$. It follows that to compute z we will need more than $Ad/(\pi\sqrt{|D|})$ terms, giving here approximately 10.5 million. We thus write v=ellan(e,10500000). This requires only 84 seconds on a 3 Ghz PC. Being lazy, we then write the function

ph(tau) = s = 0.; q = exp(2*I*Pi*tau); q1 = 1.; \
          for (n = 1, 10500000, q1 *= q; s += v[n]/n*q1); s

(we do not need 10.5 million terms for the computation of $\phi(x_1)$, but we are not optimizing here), and we compute z (in 78 seconds) thanks to the commands

z1 = ph((-b + I*sqrt(39))/(2*N)); \
          z2 = ph((-b + I*sqrt(39))/(4*N)); z = 2*real(z1 + z2);

and we obtain

$$z = -5.6391112750083176600769616630731603632356240 6574706\ldots .$$

We write z += 27*om1 to make it as small as possible. Among the possible points to be studied in the algorithm, we see that $2z + 2\omega_1$ already does the trick. Since the nontrivial torsion points are $\omega_1/2$, $\omega_2/2$, and $(\omega_1 + \omega_2)/2$, it follows that $2z + 6\omega_1$, $2z + 2\omega_1 + 4\omega_2$, and $2z + 6\omega_1 + 4\omega_2$ would also work. Thus we compute $\wp((2z + 2\omega_1)/8)$ thanks to the command x1 = ellwp(e, (2*z + 2*om1)/8). To have x as a rational number we write rx = contfrac(real(x1)) (the imaginary part of x1 is zero to the accuracy of our computations). This gives a continued fraction for the approximation x, which has a large partial quotient toward the end, precisely after index 39. We thus write

    mx = contfracpnqn(vector(39, i, rx[i])); x = mx[1,1]/mx[2,1].

We check that the denominator of x is a perfect square, which is a good sign. To obtain the value of y, we can be lazy and use the built-in function y=ellordinate(e,x)[1], or else we write $x = n_x/f^2$, we compute $n_x^3 - 157^2 n_x f^4$, and we check that it is the square of some integer n_y, so that $y = n_y/f^3$. In any case, we finally find the rational point

$$\left(\frac{95732359354501581258364453}{2774877873292446 32169121},\ \frac{834062764128948944072857085701103222940}{1461725457917215265681 55259438196081}\right).$$

Remarks. (1) Using all the improvements to the basic method mentioned in Section 8.6.5, and in particular the fact that E is a CM curve, the computation time of the above example can be reduced from a couple of minutes to 7 seconds.

(2) See Exercises 32 and 33 for other examples of the use of Heegner points.

8.7 Computation of Integral Points

For this section I have closely followed the presentation of Chapter XIII of [Sma].

8.7.1 Introduction

Let E be an elliptic curve defined over $\mathbb{Q}$, but which we now assume to be given by a generalized Weierstrass equation

$$y^2 + a_1 xy + a_3 y = x^3 + a_2 x^2 + a_4 x + a_6 \,,$$

where the a_i are all integral. We want to compute the complete set $E(\mathbb{Z})$ of integral points on E, which by Siegel's theorem we know to be finite. Note that although $E(\mathbb{Q})$ does not depend on the chosen model (i.e., equation or system of equations) for E, on the other hand, $E(\mathbb{Z})$ does depend on the model.

We first describe the general strategy, and give details afterward. We assume that using one of the above methods we have completely computed the Mordell–Weil group $E(\mathbb{Q})$ (and not only the rank), in other words that we know the torsion subgroup $E_t(\mathbb{Q})$ of $E(\mathbb{Q})$ and an r-element basis $(P_i)_{1 \leqslant i \leqslant r}$ of the torsion-free part of $E(\mathbb{Q})$, where r is the rank of E. Thus any element P of $E(\mathbb{Q})$, and in particular any element of $E(\mathbb{Z})$, can be written in the form

$$P = T + \sum_{1 \leqslant j \leqslant r} p_j P_j$$

for some $T \in E_t(\mathbb{Q})$ and some $p_i \in \mathbb{Z}$.

A deep theorem generalizing the theorems of A. Baker on linear forms in logarithms to linear forms in *elliptic* logarithms implies that when $P \in E(\mathbb{Z})$, the $|p_j|$ are bounded by a (usually very large) constant. When we say very large, we mean something like e^{1000} for instance (see examples below). However, the sheer *existence* of this bound is sufficient to continue. Applying a now classical method used by several authors and systematized by Tzanakis and de Weger, using lattice reduction algorithms such as the LLL algorithm we can *drastically* reduce the upper bound on the $|p_j|$ (see Section 2.3.5). Thus typically after two passes of the LLL algorithm we often obtain bounds such as $|p_j| \leqslant 30$. It then becomes possible to do a systematic search on the possible p_j, and we thus obtain the set $E(\mathbb{Z})$ of integral points.

The search for integral points on an elliptic curve is an important Diophantine problem, and although it requires techniques of a different kind from those that we have studied up to now, its importance justifies a detailed study of the necessary tools. By nature the present section is more algorithmic than most of the rest of this book.

8.7.2 An Upper Bound for the Elliptic Logarithm on $E(\mathbb{Z})$

Let E be an elliptic curve defined over $\mathbb{Q}$ by a generalized (affine) Weierstrass equation $y^2 + a_1 xy + a_3 y = x^3 + a_2 x^2 + a_4 x + a_6$, where we assume $a_i \in \mathbb{Z}$. We assume (and this is of course the essential and difficult assumption) that we have explicitly computed the Mordell–Weil group $E(\mathbb{Q})$, in other words the finite group $E_t(\mathbb{Q})$ and independent points P_i for $1 \leqslant i \leqslant r$ such that

$$E(\mathbb{Q}) = E_t(\mathbb{Q}) \oplus \bigoplus_{1 \leqslant i \leqslant r} \mathbb{Z}P_i \ .$$

Consider the *height pairing matrix* $Q = (\langle P_i, P_j \rangle)_{1 \leqslant i,j \leqslant r}$, which we have already introduced, where as usual $\langle P_i, P_j \rangle$ is the bilinear form associated with the canonical height function $\widehat{h}$. Since by definition the P_i are independent this matrix is nonsingular, and it is the matrix of a positive definite quadratic form on $E(\mathbb{Q})/E_t(\mathbb{Q})$, so all its eigenvalues λ_i are strictly positive. In particular, $R(E) = \det(Q) = \prod_i \lambda_i > 0$ is the *regulator* of E, and is independent of the choice of the P_i.

Now let $P = (x(P), y(P)) \in E(\mathbb{Z})$ be an (unknown) *integral* point. We can write in a unique way $P = T + \sum_{1 \leqslant i \leqslant r} p_i P_i$ for some $T \in E_t(\mathbb{Q})$ and $p_i \in \mathbb{Z}$, and we set $H = \max_i(|p_i|)$. Our main goal is to find inequalities for $1/|x(P)|$ involving H and the elliptic logarithm $\psi(P)$ (see Sections 7.3.1 and 7.3.2). Since E^{gg} is compact, it is usually very easy to find all integral points on it, so we will always assume that $P \in E^0$ (although it would be a simple matter to generalize to the whole of $E(\mathbb{Q})$). We may also evidently assume that $x(P) \neq 0$.

Lemma 8.7.1. *Keep the above notation, let $c_2 = \min_{1 \leqslant i \leqslant r} \lambda_i$ be the smallest eigenvalue of Q, and set $c_1 = \exp(\mu(E) + 2.14)$, where $\mu(E)$ is defined in Theorem 8.1.18. If $P = (x(P), y(P)) \in E(\mathbb{Z})$ is any integral point with $x(P) \neq 0$ and $H = \max_i(|p_i|)$ as above then*

$$\frac{1}{|x(P)|} \leqslant c_1 e^{-c_2 H^2} \ .$$

Proof. Since $x(P) \in \mathbb{Z} \setminus \{0\}$ we have $h(P) = \max(\log(|x(P)|), 0) = \log(|x(P)|)$, so by Theorem 8.1.18, $\log(|x(P)|) \geqslant \widehat{h}(P) - \log(c_1)$, with c_1 defined as above. Let $X = (p_1, \ldots, p_r)^t$ be the column vector of the p_i. Since $\widehat{h}$ is a positive definite quadratic form on $E(\mathbb{Q})/E_t(\mathbb{Q})$ with matrix Q, we have $\widehat{h}(P) = X^t Q X$. A standard undergraduate exercise (see Exercise 34) shows that

$$\widehat{h}(P) \geqslant c_2 X^t X \geqslant c_2 \sum_{1 \leqslant i \leqslant r} p_i^2 \geqslant c_2 H^2 \ ,$$

proving the lemma. $\qquad\square$

We now want a link between $x(P)$ and the elliptic logarithm $\psi(P)$. After the standard transformations explained in the preceding chapter we can put our curve in the form $Y^2 = f(X)$ with $f(X) = 4X^3 - g_2X - g_3$, where $g_2 = c_4/12$ and $g_3 = c_6/216$, and where we recall that $X = x + b_2/12$. Denote as usual by e_1, e_2, and e_3 the complex roots of $f(X) = 0$.

The following lemma relates $\psi(P)$ to $x(P)$ for any real point $P \in E^0$, not necessarily integral.

Lemma 8.7.2. *Let* $P = (x(P), y(P)) \in E^0$ *be a real point, and assume that* $|x(P) + b_2/12| \geqslant 2 \max(|e_1|, |e_2|, |e_3|)$. *If we choose the (essentially) unique determination of* $\psi(P)$ *such that* $|\psi(P)| \leqslant \omega_1/2$ *we have*

$$\psi(P)^2 \leqslant \frac{c_3}{|x(P)|}, \quad where \quad c_3 = \frac{\omega_1^2 |b_2|}{48} + 8 .$$

Proof. Write $P = (X(P), Y(P))$. Since $P \in E^0$ we have $X(P) \geqslant e_3$. Since $X(P) = x(P) + b_2/12$ this implies that $X(P) > 0$, since otherwise $e_3 \leqslant X(P) \leqslant 0$, so $|X(P)| \leqslant |e_3|$, contrary to our assumption. Now as mentioned in Section 7.3.2, for $P \in E^0$ and the chosen determination we have the explicit formula

$$\psi(P) = \mathrm{sign}(Y(P)) \int_\infty^{X(P)} \frac{dt}{\sqrt{f(t)}} .$$

By assumption if $t \geqslant X(P)$ we have $t \geqslant 2|e_i|$ for $1 \leqslant i \leqslant 3$, so

$$|t - e_i| \geqslant t - |e_i| = \frac{t - 2|e_i|}{2} + \frac{t}{2} \geqslant \frac{t}{2} .$$

It follows that $|f(t)| = 4 \prod_{1 \leqslant i \leqslant 3} |t - e_i| \geqslant t^3/2$, so

$$|\psi(P)| = \left| \int_\infty^{X(P)} \frac{dt}{\sqrt{f(t)}} \right| \leqslant 2^{1/2} \int_{X(P)}^\infty \frac{dt}{t^{3/2}} \leqslant \frac{2^{3/2}}{\sqrt{X(P)}} ,$$

or equivalently, $X(P) \leqslant 8/\psi(P)^2$. We thus have

$$|x(P)| = \left| X(P) - \frac{b_2}{12} \right| \leqslant X(P) + \frac{|b_2|}{12} \leqslant \frac{8}{\psi(P)^2} + \frac{|b_2|}{12}$$

$$\leqslant \frac{8 + \psi(P)^2 |b_2|/12}{\psi(P)^2} \leqslant \frac{8 + \omega_1^2 |b_2|/48}{\psi(P)^2}$$

since $|\psi(P)| \leqslant \omega_1/2$, proving the lemma. $\qquad \square$

From now on, when $P \in E^0$ we will always assume that we choose the above *principal determination* of $\psi(P)$, i.e., such that $|\psi(P)| \leqslant \omega_1/2$.

Corollary 8.7.3. *Let c_i be the constants defined in the above two lemmas and set $c_5 = \sqrt{c_1 c_3}$. If $P = (x(P), y(P))$ is an integral point in E^0 with $x(P) \neq 0$ and if $|x(P) + b_2/12| \geqslant 2 \max(|e_1|, |e_2|, |e_3|)$ then*

$$|\psi(P)| \leqslant c_5 e^{-c_2 H^2/2} .$$

Proof. Clear by combining the two lemmas.   □

8.7.3 Lower Bounds for Linear Forms in Elliptic Logarithms

Now is the time to introduce high technology. This should be taken as an easy-to-use black box, but the reader should be aware that the mathematics and computations leading to lower bounds for linear forms in logarithms (elliptic or not), initiated by A. Baker, are one of the major advances in number theory in the second half of the twentieth century.

The following theorem is due to S. David, and we give only the special case of $\mathbb{Q}$. For the general case, as well as the corresponding statements for linear forms in complex or p-adic logarithms, I refer to [Sma].

First, we need some notation. Let E be an elliptic curve defined over $\mathbb{Q}$, and as above let $Y^2 = 4X^3 - g_2 X - g_3$ be the equation of E obtained after the standard changes of variable. Recall that if $P = (x_0 : \cdots : x_n) \in \mathbb{P}^n(\mathbb{Q})$ we have defined $h(P) = \max_{0 \leqslant i \leqslant n} \log(|y_i|)$, where $P = (y_0 : \cdots : y_n)$ is one of the two representations with $y_i \in \mathbb{Z}$ for all i and $\gcd(y_0, \ldots, y_n) = 1$, so in particular if $u = n/d \in \mathbb{Q}^*$ with $\gcd(n, d) = 1$, we have $h(u) = \max(\log(|n|), \log(|d|))$. We define the height $h(E)$ of the elliptic curve by the formula

$$h(E) = \max(1, h(1, g_2, g_3), h(j(E)))$$

and we set $c_7 = 3\pi/(\omega_1^2 \Im(\omega_1/\omega_2))$. If $P \in E(\mathbb{Q})$ we define a modified height function $h_m(P)$ by the formula

$$h_m(P) = \max(\widehat{h}(P), h(E), c_7 |\psi(P)|^2) .$$

Theorem 8.7.4 (David). *Let $P_1, \ldots, P_n$ be points in $E(\mathbb{Q})$, and set*

$$c_8 = \max(e h(E), \max_{1 \leqslant i \leqslant n} h_m(P_i)) \quad and \quad c_9 = \frac{e}{\sqrt{c_7}} \min_{1 \leqslant i \leqslant n} \frac{\sqrt{h_m(P_i)}}{|\psi(P_i)|} .$$

For $p_i \in \mathbb{Z}$ we set $H = \max(|p_i|)$, and $L = \sum_{1 \leqslant i \leqslant n} p_i \psi(P_i)$. Then if $H \geqslant \exp(c_8)$ and $L \neq 0$ we have

$$\log(L) > -c_{10}(\log(H) + \log(c_9))(\log(\log(H)) + h(E) + \log(c_9))^{n+1} ,$$

where

$$c_{10} = 2 \cdot 10^{8+7n} (2/e)^{2n^2} (n+1)^{4n^2 + 10n} \log(c_9)^{-2n-1} \prod_{1 \leqslant i \leqslant n} h_m(P_i) .$$

Remark. Recall that $\psi(P)$ is defined only modulo Λ. This theorem is valid for any determination of $\psi(P)$. In particular, we have $\psi(\mathcal{O}) \equiv 0 \pmod{\Lambda}$, so by choosing one of the P_i equal to $\mathcal{O}$ we can include any integral linear combination of ω_1 and ω_2 among the $\psi(P_i)$.

We now explain how to use this theorem combined with Corollary 8.7.3 to find integral points. As already mentioned, the fundamental assumption is that we have computed exactly the Mordell–Weil group as

$$E(\mathbb{Q}) = E_t(\mathbb{Q}) \oplus \bigoplus_{1 \leqslant i \leqslant r} \mathbb{Z}P_i \ ,$$

and we recall from Section 7.3.2 that we have a disjoint union $E(\mathbb{R}) = E^{gg} \cup E^0$, where E^0 is the connected component of the identity, and the possibly empty set E^{gg} is compact. I claim that we can assume that at most one of the P_i is in E^{gg}: indeed, if $P_i \in E^{gg}$ and $P_j \in E^{gg}$ with $i \neq j$, then in the Mordell–Weil basis we may replace $\{P_i, P_j\}$ by $\{P_i + P_j, P_j\}$, and since $P_i + P_j \in E^0$ by Section 7.3.2, we have one point fewer in E^{gg}, proving my claim. We may thus assume that $P_i \in E^0$ for $2 \leqslant i \leqslant r$.

We write our integral point P as $P = T + \sum_{1 \leqslant i \leqslant r} p_i P_i$ for some unknown $T \in E_t(\mathbb{Q})$ and $p_i \in \mathbb{Z}$. We may clearly assume that $P \in E^0$, that $P \notin E_t(\mathbb{Q})$ (so that the p_i are not all equal to 0), and that $x(P) \neq 0$, since all the points that we exclude in this way are easy to find. Since we want to restrict to E^0 we set $Q_i = P_i$ for $2 \leqslant i \leqslant r$, and $Q_1 = 2P_1$ if $P_1 \in E^{gg}$ and $Q_1 = P_1$ if $P_1 \in E^0$, so that $Q_i \in E^0$ for all i. We can thus write $P = T + U + \sum_{1 \leqslant i \leqslant r} q_i Q_i$, where $q_i = p_i$ if $i \geqslant 2$ or if $i = 1$ and $P_1 \in E^0$, and $q_1 = \lfloor p_1/2 \rfloor$ if $P_1 \in E^0$, and $U = P_1$ if $P_1 \in E^{gg}$ and p_1 is odd, $U = \mathcal{O}$ otherwise. Since P and the Q_i are in E^0 we also have $T + U \in E_0$, so that $T + U$ belongs to a finite set having at most $2|E_t(\mathbb{Q})|$ elements. We will write $Q_{r+1} = T + U$ (we can of course avoid this extra point if we know in advance that it will be equal to $\mathcal{O}$, for instance if $P_1 \in E^0$ and $E_t(\mathbb{Q}) = \{\mathcal{O}\}$).

By definition the elliptic logarithm is additive modulo Λ, and since on E^0 we have chosen the principal determination it is clear that if $P = \sum_{1 \leqslant i \leqslant n} R_i$ then $\psi(P) = m\omega_1 + \sum_{1 \leqslant i \leqslant n} \psi(R_i)$, where $|m| \leqslant \lfloor n/2 \rfloor$. In particular,

$$\psi(P) = m\omega_1 + \psi(Q_{r+1}) + \sum_{1 \leqslant i \leqslant r} q_i \psi(Q_i)$$

with $|m| \leqslant (1 + \sum_{1 \leqslant i \leqslant r} |q_i|)/2$, and even $|m| \leqslant \sum_{1 \leqslant i \leqslant r} |q_i|/2$ if $Q_{r+1} = \mathcal{O}$. This is a linear form in elliptic logarithms, so we can combine Corollary 8.7.3 with David's theorem. If we set $L = \psi(P)$ and

$$H_q = \max(1, \max(|q_i|)) \leqslant \max(|p_i|) = H \ ,$$

Corollary 8.7.3 tells us that when $|x(P) + b_2/12| \geqslant 2\max(|e_1|, |e_2|, |e_3|)$ we have

$$-\log(|L|) \geqslant c_2 H^2/2 - \log(c_5) \geqslant c_2 H_q^2/2 - \log(c_5) \,,$$

and David's theorem (applied to the Q_i and to $n = r + 2$ or to $n = r + 1$ if $Q_{r+1} = \mathcal{O}$) implies that

$$-\log(|L|) < c_{10}(\log(H_r) + \log(c_9))(\log(\log(H_r)) + h(E) + \log(c_9))^{n+1} \,,$$

where $H_r = rH_q + 1$, or $H_r = rH_q$ when $Q_{r+1} = \mathcal{O}$. Since the upper bound grows logarithmically in H_q and the lower bound grows like H_q^2, it is clear that these bounds are contradictory for H_q sufficiently large.

Since all the constants are explicit we can thus compute some bound B such that $H_q > B$ leads to a contradiction, so that we know that $H_q = \max(|q_i|) \leqslant B$. This bound will usually be extremely large, but now we apply the techniques explained in Section 2.3.5 (in particular Corollary 2.3.17 and Proposition 2.3.20) to the inequality $|L| \leqslant c_5 \exp(-c_2 H_q^2/2)$, possibly two or three times, to reduce the bound to something manageable, which we then enumerate by brute force.

8.7.4 A Complete Example

Since the above description contains a large amount of notation and may be hard to understand at first, it may be useful to give in detail a complete example. We will again consider the curve $y^2 + y = x^3 - 7x + 6$ studied in detail in Section 8.5.6. We have seen that it has no torsion, and that it has rank 3, where generators can be taken to be $P_1 = (2,0)$, $P_2 = (-1,3)$, and $P_3 = (4,6)$. The reduced Weierstrass equation of this curve is obtained by setting $Y = 2y+1$ and $X = x$, so that $Y^2 = 4X^3 - 28X + 25$. We thus compute that $b_2 = 0$, $g_2 = 28$, $g_3 = -25$, $\mathrm{disc}(E) = 5077$, $j(E) = 37933056/5077$, hence $h(E) = 17.45$, $\mu(E) = 3.90855$, $c_1 = 423.5$, $\lambda_1 = 0.3228$, $\lambda_2 = 0.4925$, $\lambda_3 = 2.623$, hence $c_2 = 0.3228$, $c_3 = 8$, $c_5 = 58.21$, $e_1 = -3.0124$, $e_2 = 1.0658$, $e_3 = 1.9466$. Thus if $P = p_1 P_1 + p_2 P_2 + p_3 P_3 \in E^0$ is an integral point such that $|x(P)| \geqslant 7$ we have the fundamental inequality $|\psi(P)| \leqslant 58.21 \exp(-0.1614H^2)$, where $H = \max(p_i)$. Now among the P_i only P_2 is in E^{gg}, so we set $Q_1 = P_1$, $Q_3 = P_3$, and $Q_2 = 2P_2 = (114/49, -720/343)$. Since there is no torsion we have $P = q_1 Q_1 + q_2 Q_2 + q_3 Q_3 + U$ with $q_1 = p_1$, $q_2 = \lfloor p_2/2 \rfloor$, $q_3 = p_3$, and $U = \mathcal{O}$ or P_2, but since we are only looking for $P \in E^0$, we have in fact $U = \mathcal{O}$, $q_2 = p_2/2$, and $Q_4 = \mathcal{O}$.

We are now ready to apply David's theorem to the form $L = m\omega_1 + q_1\psi(Q_1) + q_2\psi(Q_2) + q_3\psi(Q_3)$ (hence $n = 4$), where $m \leqslant (|q_1| + |q_2| + |q_3|)/2 \leqslant (3/2)\max(|q_i|)$. We obtain $c_7 = 1.5599$, $h_m(Q_i) = h(E)$, hence $c_8 = 47.4376$, $c_9 = 5.8503$, and finally $c_{10} = 2.97 \cdot 10^{107}$. Thus we have the inequalities

$$-\log(|\psi(P)|) \geqslant 0.1614H_q^2 - 4.064 \quad \text{and}$$

$$-\log(|\psi(P)|) \leqslant 2.97 \cdot 10^{107}(\log(1.5H_q) + 1.7665)(\log(\log(1.5H_q)) + 19.218)^5 \,,$$

the second one being valid only for $1.5H_q > \exp(c_8) = 4.01 \cdot 10^{20}$. We immediately find that these equations are incompatible for $H_q > 10^{60}$, so that we have a first basic upper bound $H_q \leqslant 10^{60}$. Note that it is completely unnecessary to take sharp bounds anywhere in this computation, since the next step, i.e., the use of the LLL algorithm, will drastically reduce the bound anyway. Since we are no longer going to use David's theorem, we can also forget the lower bound $(3/2)H_q > 4.01 \cdot 10^{20}$ necessary for the validity of his theorem.

After having used the above high technology, we can now use the magic of the LLL algorithm, more precisely Corollary 2.3.17 and Proposition 2.3.20 applied to the inequality

$$|m\omega_1 + q_1\psi(Q_1) + q_2\psi(Q_2) + q_3\psi(Q_3)| \leqslant 58.21\exp(-0.1614H_q^2) \ ,$$

where we now know that $H_q = \max(|q_i|) \leqslant 10^{60}$ and $m \leqslant 1.5 \cdot 10^{60}$. We first choose $C > (10^{60})^4$, say $C = 10^{250}$, and form the 4×4 matrix

$$\begin{pmatrix} 1 & 0 & 0 & 0 \\ 0 & 1 & 0 & 0 \\ 0 & 0 & 1 & 0 \\ \lfloor C\psi(Q_1) \rceil & \lfloor C\psi(Q_2) \rceil & \lfloor C\psi(Q_3) \rceil & \lfloor C\omega_1 \rceil \end{pmatrix} .$$

An application of the (integral) LLL algorithm shows that the first vector of an LLL-reduced basis of the lattice generated by the columns of B is an explicit vector whose entries have approximately 60 decimal digits. We easily compute that for $i = 1, 2, 3, 4$ we have $\|\mathbf{b}_1\|^2/\|\mathbf{b}_i^*\|^2 = 1, 0.718, 0.426, 0.338$ respectively, so that with the notation of Corollary 2.3.17 we have $c_1 = 1$, hence

$$d(L,0)^2 \geqslant \|\mathbf{b}_1\|^2/c_1 \geqslant 2.5 \cdot 10^{120} \ .$$

With the notation of Proposition 2.3.20 we have $Q = 3 \cdot 10^{120}$ and $T = (1 + 4.5 \cdot 10^{60})/2$, so $d(L,0)^2 \geqslant T^2 + Q$. Since the points Q_i are independent, we deduce from Proposition 2.3.20 that

$$|q_1\psi(Q_1) + q_2\psi(Q_2) + q_3\psi(Q_3) + m\omega_1| \geqslant 1.5 \cdot 10^{-184} \ .$$

Combining this with the inequality

$$|q_1\psi(Q_1) + q_2\psi(Q_2) + q_3\psi(Q_3) + m\omega_1| \leqslant 58.21e^{-0.1614H_q^2}$$

gives $H_q \leqslant 51$, which is much more manageable than our initial bound of 10^{60}.

Although 51 is now a reasonable number, it is worthwhile to iterate the whole LLL process using the new inequality $H_q \leqslant 51$. This time we must be a little careful with the choice of C so as to be able to obtain an improvement using Proposition 2.3.20. We choose $C = 10^9$, and after a similar computation to that performed above we find the new bound $H_q \leqslant 11$. By using still another LLL process with $C = 10^7$, we could still reduce this to $H_q \leqslant 10$, but there is not much point in doing so.

We now perform a direct systematic search: on E^{gg} we find the integral points $(-3,0)$, $(-3,-1)$, $(-2,3)$, $(-2,-4)$, $(-1,3)$, $(-1,-4)$, $(0,2)$, $(0,-3)$, $(1,0)$, $(1,-1)$, $(2,0)$, $(2,-1)$. The points on E^0 with $x(P) \leqslant 6$ are $(3,3)$, $(3,-4)$, $(4,6)$, and $(4.-7)$. All the others are on E^0 with $x(P) \geqslant 7$, hence of the form $q_1 Q_1 + q_2 Q_2 + q_3 Q_3$ with $|q_i| \leqslant 11$, and we may assume $q_1 \geqslant 0$ if we take care to compute also the opposites of the points that we find. Thus after searching through $12 \cdot 23^2 = 6348$ points we find (in seconds) the additional integral points (where we of course also include the opposites of the points found) $(8,21)$, $(8,-22)$, $(11,35)$, $(11,-36)$, $(14,51)$, $(14,-52)$, $(21,95)$, $(21,-96)$, $(37,224)$, $(37,-225)$, $(52,374)$, $(52,-375)$, $(93,896)$, $(93,-897)$, $(342,6324)$, $(342,-6325)$, $(406,8180)$, $(406,-8181)$, $(816,23309)$, and $(816,-23310)$, all corresponding to coefficients q_i with $H_q = \max(|q_i|) \leqslant 3$. We have thus found a total of 36 integral points, and we have *proved* (this was of course the main difficulty) that there are no others.

It should be remarked that 36 is a very large number of integral points for an elliptic curve, but it is a completely general and only partly understood phenomenon: if we choose an elliptic curve having the smallest or one of the smallest conductors for a given rank, it will have a large number of integral points. Indeed, it is known that our curve is the curve of rank 3 with the smallest conductor (see Exercise 35 for rank 2).

8.8 Exercises for Chapter 8

1. Prove Corollary 8.1.9.

2. Let R be a commutative ring, let $f \in R[X]$ be a monic polynomial, let $A = R[X]/(f(X)R[X])$, and let α be the class of X modulo $f(X)$, so that $f(\alpha) = 0$. Finally, set $g(Y) = (f(Y) - f(\alpha))/(Y - \alpha) = f(Y)/(Y - \alpha) \in A[Y]$.

(a) Prove that
$$\operatorname{disc}(g) = g(\alpha)^2 \operatorname{disc}(f) = f'(\alpha)^2 \operatorname{disc}(f) \ .$$

(b) Deduce the existence of polynomials U and V in $R[X]$ such that $U(X)f(X) + V(X)(f'(X))^2 = \operatorname{disc}(f)$, thus explaining the identity used in the proof of Theorem 8.1.10 (I thank H. W. Lenstra for this proof).

3. Generalizing Theorem 8.1.10, let E be an elliptic curve given by $y^2 + a_1 xy + a_3 y = x^3 + a_2 x^2 + a_4 x + a_6$ with $a_i \in \mathbb{Z}$, and let $T = (x, y)$ be a torsion point of order not dividing 2. Show the following:

(a) $(x, y) \in \mathbb{Z}^2$ and $(2y + a_1 x + a_3)^2 \mid 4 \operatorname{disc}(E)$.
(b) If $a_1 \in 2\mathbb{Z}$ then $(2y + a_1 x + a_3)^2 \mid \operatorname{disc}(E)$.
(c) If $a_1 \in 2\mathbb{Z}$ and $a_3 \in 2\mathbb{Z}$ then $(y + (a_1/2)x + (a_3/2))^2 \mid \operatorname{disc}(E)/16$.

4. Using Sections 7.3.6 and 7.3.7, generalize the Nagell–Lutz Theorem 8.1.10 to a general number field.

5. Prove Corollary 8.1.15 (2).

6. Let E be the elliptic curve $y^2 = x^3 - 2x$ and $P = (-1, 1) \in E(\mathbb{Q})$, considered in the text to illustrate the fact that the naïve height h behaves approximately quadratically. Prove that the sign of $x(kP)$ is equal to $(-1)^k$ and that

$\exp(h(kP)) \equiv 1 \pmod{168}$ if $k \not\equiv 2 \pmod 4$ and $\exp(h(kP)) \equiv 9 \pmod{168}$ if $k \equiv 2 \pmod 4$.

7. Using reductions to standard Weierstrass form, compute a Weierstrass equation for the hyperelliptic quartic curve $y^2 = 226x^4 - 1$ using the known rational point $(x, y) = (1, 15)$, and show that the rank of this elliptic curve is equal to 3.

8. (Bremner–Tzanakis.) This exercise is a sequel to Exercise 46 of Chapter 6, whose notation we keep.

(a) By setting $x = -P/Q^2$ show that for $n = 7$, the coprime pairs (P, Q) of the above-mentioned exercise are in one-to-one correspondence with rational points on the elliptic curve E whose equation is $y^2 = x^3 + 6x^2 + 5x + 1$.

(b) By using the methods of this chapter, show that this curve has no torsion, and that it has rank 1 generated by the point $(-1, 1)$.

(c) Give the first seven coprime pairs (P, Q) coming from the preceding question.

In their papers [Bre-Tza1], [Bre-Tza2], and [Bre-Tza3], the authors study the general case in detail.

9. Let E be the elliptic curve $Y^2 = X^3 - 34992$.

(a) Using mwrank or descent, show that E has rank 1 and no torsion, a generator being $P = (36, 108)$, which corresponds to the point $(x, y) = (1, 2)$ under the birational transformation of Proposition 7.2.3.

(b) Using the group law and that proposition, solve Fermat's challenge (which he knew how to solve) of finding *strictly positive* coprime integers x, y, and z such that $x^3 + y^3 = 9z^3$ other than $(1, 2, 1)$ and $(2, 1, 1)$ (the smallest answer has 12 decimal digits).

(c) Perform the same computation, but now using Exercise 10 (b) of Chapter 7.

10. Find all $x \in \mathbb{Q}$ such that $x^2 + 4$ is the square of a rational number and

$$2 + 2x - \frac{4x}{x^2 + 4} + 2\frac{x^2 + x + 2}{\sqrt{x^2 + 4}}$$

is also the square of a rational number (reduce to finding all rational points on an elliptic curve).

11. (J.-F. Mestre.) Let r_1, r_2, r_3, and r_4 be distinct rational numbers and let $t \in \mathbb{Q}$ be a parameter. Consider the 12th-degree polynomial

$$P(X) = \prod_{1 \leqslant i, j \leqslant 4, \; i \neq j} (X - (r_i + t r_j)) .$$

(a) By considering the Laurent series expansion of $Q^{1/3}$ show that for any monic polynomial Q of degree 12 there exists a unique polynomial $g \in \mathbb{Q}[X]$ such that $\deg(Q(X) - g^3(X)) \leqslant 7$, and show that in our special case we have in fact $\deg(P(X) - g^3(X)) \leqslant 6$.

(b) Show that there exists $q(X) \in \mathbb{Q}[X]$ and $r(X) \in \mathbb{Q}[X]$ such that $P(X) = g^3(X) + q(X)g(X) + r(X)$ with $\deg(q) \leqslant 2$ and $\deg(r) \leqslant 3$.

(c) Deduce from this that the equation $Y^3 + q(X)Y + r(X) = 0$ is the equation of a cubic with rational coefficients, and that the 12 points $(r_i + t r_j, g(r_i + t r_j))_{i \neq j}$ give 12 (not necessarily distinct) rational points on this cubic.

(d) Give explicit values of the r_i and t for which the cubic is nonsingular and the above 12 points are distinct, and in fact linearly independent for the group law on the cubic.

(e) Using the algorithm described in Section 7.2.4, find a Weierstrass equation corresponding to the cubic, and give explicitly an elliptic curve defined over $\mathbb{Q}$ whose rank is greater than or equal to 11 as well as 11 independent points on the elliptic curve (note that we have to "lose" a point in order to obtain an elliptic curve).

Remarks.

(i) To answer the last two questions of this exercise, the reader is strongly advised to use a package such as `mwrank`.

(ii) The largest known rank for an elliptic curve defined over $\mathbb{Q}$ is 28; see [Elk2].

12. (R. Schoof.) Define a *Cassels–Sansone number* (abbreviated to CS number) as an integer a of the form $x/y + y/z + z/x$ for some nonzero integers x, y, and z. Let C_a be the curve with projective equation $x^2 z + y^2 x + z^2 y = axyz$.

(a) Prove that C_a is an elliptic curve if and only if $a \neq 3$, and give a rational parametrization of the curve C_3. From now on assume that $a \neq 3$.

(b) Let E_a be the elliptic curve with affine equation $y^2 + axy + y = x^3$. Show that E_a and C_a are isogenous over $\mathbb{Q}$, and give explicitly the isogenies and their degrees.

(c) The point $T = (0,0)$ is trivially a point of order 3 on E_a. Prove that the torsion subgroup of E_a is strictly larger than $\langle T \rangle$ if and only if $a = -1$ and $a = 5$, and give the torsion subgroup in these cases, as well as the corresponding points on the curve C_a.

(d) Prove that a is a CS number if and only if $a = -1$, 3, 5, or if the rank of the elliptic curve E_a is greater than or equal to 1.

(e) Using Tate's algorithm, it can be shown that the sign of the functional equation of $L(E_a, s)$ is equal to $(-1)^{d-1}$ if $3 \nmid a$ or if $a \equiv 12 \pmod{27}$, and is equal to $(-1)^d$ otherwise, where d is the number of prime divisors of $a^3 - 27$ that are congruent to 1 modulo 3. Using BSD, deduce a sufficient condition for a to be a CS number.

(f) By computing numerically $L(E_a, 1)$ when the sign of the functional equation is equal to 1, make a small table of CS numbers.

(g) Using the Heegner point method, compute explicitly integers x, y, and z such that $x/y + y/z + z/x = -32$.

(h) Show that if a is a CS number the Diophantine equation $x^3 + y^3 + z^3 = axyz$ has a nontrivial solution, but that the converse is false. Does this remark simplify the Heegner point computation of the preceding question?

13. Using a software package such as `mwrank` or descent methods, show that the parabolic-type super-Fermat equations $x^2 + y^4 = 2z^4$ and $x^4 + 8y^4 = z^2$ have an infinity of integral solutions with x, y, and z pairwise coprime (see Section 6.5). Find all such solutions with $\min(|x|, |y|, |z|) \leqslant 10^{100}$.

14. Consider the hyperelliptic quartic equation $y^2 = (x - 1)x(x + 1)(x + 2)$.

(a) Using Proposition 7.2.1, find a generalized (not necessarily minimal) Weierstrass equation for this elliptic curve.

(b) Compute its torsion subgroup and its rank, using 2-descent.

(c) Deduce the following result due to Euler: the product of four integers in arithmetic progression can never be a nonzero square; in other words, the only solutions in $\mathbb{Z}$ to the Diophantine equation $n(n+d)(n+2d)(n+3d) = m^2$ have $m = 0$.

15. In this chapter, in Chapter 14, and at other places we needed to explicitly compute the Mordell–Weil group of a number of elliptic curves. Although we can use Cremona's mwrank package, it is instructive to do some of the calculations by hand. Perform explicitly the necessary computations as follows. For each given curve E, first transform it into the canonical 2-descent form $y^2 = x^3 + ax^2 + bx$ (when it has rational 2-torsion, which will usually be the case), and compute $\alpha(E(\mathbb{Q}))$ and $\widehat{\alpha}(\widehat{E}(\mathbb{Q}))$ by looking at the real, 2-adic, or 3-adic solvability of the necessary quartics.

(a) $y^2 = (x-1)(x^2 - 14x + 1)$: show that $|\alpha(E(\mathbb{Q}))| = 2$ and $|\widehat{\alpha}(\widehat{E}(\mathbb{Q}))| = 2$, hence $r(E) = 0$.

(b) $y^2 = (x+1)(x^2 + 14x + 1)$: show that $|\alpha(E(\mathbb{Q}))| = 4$ and $|\widehat{\alpha}(\widehat{E}(\mathbb{Q}))| = 2$, so that $r(E) = 1$, and give explicit generators.

(c) $y^2 = (x-1)(x^2 - 4)$: show that $|\alpha(E(\mathbb{Q}))| = 4$ and $|\widehat{\alpha}(\widehat{E}(\mathbb{Q}))| = 1$, so that $r(E) = 0$.

(d) $y^2 = x^3 - 8$: show that one can reduce to the equation studied in (a), hence that $r(E) = 0$.

(e) $y^2 = x(x^2 - 9)$: show that $|\alpha(E(\mathbb{Q}))| = 4$ and $|\widehat{\alpha}(\widehat{E}(\mathbb{Q}))| = 1$, so that $r(E) = 0$.

(f) $y^2 = 9x^4 + 18x^2 + 1$: show that this curve is isomorphic to $y^2 = x(x^2 - 9)$ hence that $r(E) = 0$.

(g) $y^2 = 12x^4 + 1$: setting $y = 1 + xY$ and using the algorithm described in Section 7.2.4 and Proposition 8.2.15 show that $r(E) = 0$.

16. Consider the elliptic curve defined by the affine equation $y^2 = f(x)$ with $f(x) = x^3 + 3x - 11$, let θ be a root of f, and set $K = \mathbb{Q}(\theta)$. With the notation of Corollary 8.3.7, show that $I(\theta) = 5$ and $\mathfrak{q} \mid f'(\theta)$ for one of the two prime ideals $\mathfrak{q}$ of K above 5, but that the Tamagawa number $c_5 = 3$ is odd so that, as mentioned in the text, the condition $c_\mathfrak{q}$ odd does not follow from the others.

17. Continue the study of 3-descent by first proving an analogue of Proposition 8.2.4 (3) and (4). You will need in particular to work with prime *ideals* of K_d. Prove also an analogue of Proposition 8.2.8.

18. Let E be an elliptic curve defined over a field K of characteristic 0. It is clear (and we have used this fact in 2-descent) that if E has a rational point of order 2 its equation can be taken of the form $y^2 = x^3 + ax^2 + bx$ (here and afterward the parameters a and b are implicitly assumed to be in K).

(a) Using Proposition 8.4.2, show that if E has a rational point of order 3 its equation can be taken of the form

$$y^2 + by = x^3 + a^2 x^2 + abx .$$

(b) By writing $2(0,0) = -2(0,0)$, show that if E has a rational point of order 4 its equation can be taken of the form

$$y^2 + 2aby = x^3 + (a + b^2)x^2 + 2ab^2 x .$$

(c) By writing $3(0,0) = -2(0,0)$, show that if E has a rational point of order 5 its equation can be taken of the form

$$y^2 + (2a - b)b^2 y = x^3 + (a^2 + 2ab - b^2)x^2 + (2a - b)ab^2 x .$$

(Hint: transform the polynomial relation between a_2, a_3, and a_4 obtained from $3(0,0) = -2(0,0)$ by setting $a_4 = ta_3$, dividing by a_3^6, setting $z = a_2 - t^2$, solving in t, and simplifying the resulting equation.)

(d) Using practically the same method as for order 5, show that if E has a rational point of order 6 its equation can be taken of the form
$$y^2 - 2a(a-b)(2a-b)y = x^3 - (2a^2 - 6ab + 3b^2)x^2 - 2ab(a-b)(2a-b)x \ .$$
Note that since these equations are not unique, your results may be a little different.

19. Using Legendre symbols or otherwise, prove that if E_1 is the elliptic curve with affine equation $y^2 = x^3 + 1$ we have $a_p(E_1) \equiv 0 \pmod 2$. Using Propositions 2.5.20 and 8.5.3, deduce Corollary 8.5.4.

20. The aim of this exercise is to study the decomposition of 22 as a sum of two rational cubes.

(a) Let x, y, z be pairwise coprime integers such that $x^3 + y^3 = 22z^3$ (since 22 is cubefree, it is clear that we can reduce to this case). Show that $66 \mid x + y$ and that x and y are odd.

(b) Using this and making a systematic search with $|y| \leqslant x$, show that 22 is the sum of two rational cubes by giving explicitly x, y, and z.

(c) Thanks to Proposition 7.2.3, we can also study the rational points on the elliptic curve E whose Weierstrass equation is $y^2 = x^3 - 432 \cdot 22^2$ or, more simply, $y^2 = x^3 - 27 \cdot 11^2$ by changing (x, y) to $(4x, 8y)$. Using Proposition 8.5.10 and the remark that follows, show that $\varepsilon(E) = -1$, hence that under the BSD conjecture the rank of E is odd, hence greater than or equal to 1, so that 22 is indeed a sum of two rational cubes.

(d) Show that $L'(E, 1) \neq 0$, hence that the analytic rank of E is equal to 1. By the *proved* results of Gross–Zagier et al., this shows that the rank of E is equal to 1.

(e) By performing a general 2-descent, find explicitly a rational point on E, and hence a decomposition of 22 as a sum of two rational cubes.

(f) Do the same, but now using the Heegner point method, since we know that E has rank 1.

21. Set
$$f(x) = xe^x E_1(x) = xe^x \int_x^\infty \frac{e^{-t}}{t} \, dt \ .$$

(a) Show that if we set $y_0(t) = f(1/t)$ then y_0 is a solution of the differential equation $t^2 y' + (1+t)y - 1 = 0$.

(b) Prove that y_0 is a C^∞ function around $t = 0$, and that it has the (nonconvergent) series expansion $y_0(t) = \sum_{n \geqslant 0} (-1)^n n! t^n$.

(c) Consider the Riccati differential equation $t^2 y' + (1 + at)y + bty^2 - c = 0$, where a, b, and c are parameters. Prove that there is a unique solution of this equation that is C^∞ around $t = 0$ and that its value at $t = 0$ is equal to c.

(d) Let y_n be the C^∞ function that is a solution of $t^2 y' + (1 + a_n t)y + b_n ty^2 - c_n = 0$. Prove that if we set $y_n = c_n/(1 + ty_{n+1})$ then y_{n+1} is the C^∞ function that is a solution of $t^2 y' + (1 + a_{n+1}t)y + b_{n+1}ty^2 - c_{n+1} = 0$ with $a_{n+1} = 1 - a_n$, $b_{n+1} = 1$, and $c_{n+1} = a_n + b_n c_n$.

(e) By proving the convergence of the continued fraction, deduce that
$$E_1(x) = \cfrac{e^{-x}}{x + \cfrac{1}{1 + \cfrac{1}{x + \cfrac{2}{1 + \cfrac{2}{x + \ddots}}}}} \ .$$

(f) By contracting this fraction, deduce finally that

$$E_1(x) = \Gamma_1(1,x) = \cfrac{e^{-x}}{x+1-\cfrac{1^2}{x+3-\cfrac{2^2}{x+5-\cfrac{3^2}{x+7-\cdots}}}}$$

22. (Continuation of the preceding exercise.)

(a) Denote by p_n/q_n the nth convergent of the continued fraction for $E_1(x)$ obtained at the end of the preceding exercises (so that $p_0 = 0$, $p_1 = e^{-x}$, $q_0 = 1$, $q_1 = x+1$). Show by induction that

$$q_n = \sum_{j=0}^{n} \binom{n}{j}^2 (n-j)!x^j .$$

(b) Show that the largest summand in this sum is obtained for j_0 equal to one of the two integers closest to

$$j_0 = -1 - x/2 + \sqrt{nx + x + x^2/4} ,$$

and using Stirling's formula show that as $n \to \infty$ this summand is asymptotic to

$$\frac{n!e^{2\sqrt{nx}}e^{-x/2}}{2\pi\sqrt{nx}} .$$

(c) By setting $j = j_0 + \lambda n^{1/4}$ and approximating the sum by an integral, show that as $n \to \infty$ we have

$$q_n \sim \frac{n!e^{2\sqrt{nx}}e^{-x/2}}{2\sqrt{\pi}(nx)^{1/4}} \quad \text{and} \quad p_n \sim \frac{n!e^{2\sqrt{nx}}e^{x/2}E_1(x)}{2\sqrt{\pi}(nx)^{1/4}} .$$

(d) Deduce that

$$E_1(x) - \frac{p_n}{q_n} \sim 2\pi e^{-4\sqrt{nx}} .$$

23. Prove completely similar results to those of Exercises 21 and 22 for the incomplete gamma function $\Gamma(s,x) = \int_x^{\infty} e^{-t}t^{s-1}\,dt$. In particular, show that

$$\Gamma(s,x) = \cfrac{x^s e^{-x}}{x+1-s-\cfrac{1(1-s)}{x+3-s-\cfrac{2(2-s)}{x+5-s-\cdots}}}$$

and that if p_n/q_n is the nth convergent of this continued fraction, then as $n \to \infty$ we have

$$\Gamma(s,x) - \frac{p_n}{q_n} \sim \frac{2\pi}{\Gamma(1-s)}e^{-4\sqrt{nx}}$$

(when $s \in \mathbb{Z}_{\geqslant 1}$ this means that the left-hand side is equal to 0, due to the fact that the continued fraction terminates).

24. Assume that E is an elliptic curve defined over $\mathbb{Q}$ of rank 1 and that one knows a point of infinite order in $E(\mathbb{Q})$. Explain how to find a generator for the torsion-free part of $E(\mathbb{Q})$.

25. Prove Proposition 8.6.2.

26. Prove Proposition 8.6.6.

27. (Atkin–Lehner.) Let $N \geqslant 1$ be an integer, and let Q be a (positive) divisor of N such that $\gcd(Q, N/Q) = 1$.

   (a) Prove that there exist x, y, z, and w in $\mathbb{Z}$ such that if we set

   $$W_Q = \begin{pmatrix} Qx & y \\ Nz & Qw \end{pmatrix},$$

   then $\det(W_Q) = Q$. Such a matrix W_Q will be called an Atkin–Lehner matrix for the divisor Q of N.

   (b) Prove that W_Q is unique up to left multiplication by an element of $\Gamma_0(N)$, in other words that if $\gamma \in \Gamma_0(N)$ then γW_Q is again an Atkin–Lehner matrix for Q, and conversely that if W_Q and W_Q' are two such matrices there exists $\gamma \in \Gamma_0(N)$ such that $W_Q' = \gamma W_Q$.

   (c) Prove that $W_Q^2 = Q\gamma$ for some $\gamma \in \Gamma_0(N)$, hence that the action of W_Q via linear fractional transformations is an *involution* on $\Gamma_0(N)$-invariant functions.

28. (Continuation of the previous exercise.) Let τ be a Heegner point of level N, and let $(A, B, C) = 0$ be the corresponding primitive positive definite quadratic form, hence by Proposition 8.6.3 such that $N \mid A$ and $\gcd(A/N, B, CN) = 1$ (or $\gcd(N, B, AC/N) = 1$). Let $W_Q = \begin{pmatrix} Qx & y \\ Nz & Qw \end{pmatrix}$ be an Atkin–Lehner matrix corresponding to some $Q \mid N$ such that $\gcd(Q, N/Q) = 1$. We set $\tau_1 = (Qx\tau + y)/(Nz\tau + Qw)$, which we write as $\tau_1 = W_Q(\tau)$, and let (A_1, B_1, C_1) be the corresponding primitive quadratic form.

   (a) Compute explicitly A_1, B_1, and C_1 in terms of A, B, and C (and of course of the matrix W_Q), and conversely compute explicitly A, B, and C in terms of A_1, B_1, and C_1. (Hint: this second computation is immediate from the first.)

   (b) It is immediately seen from the formulas that $N \mid A_1$, so the first condition for a Heegner point of level N is satisfied. Using $\gcd(A/N, B.CN) = 1$, prove that if p is a prime such that $p \mid N/Q$ then $p \nmid \gcd(A_1/N, B_1, C_1N)$.

   (c) Using now $\gcd(N, B, AC/N) = 1$, prove that if p is a prime such that $p \mid Q$ then again $p \nmid \gcd(A_1/N, B_1, C_1N)$, so that $\tau_1 = W_Q(\tau)$ is a Heegner point of level N.

29. Compute all the integral points on the elliptic curves $y^2 = x^3 + t$ for $t = -26$, -29, -38, -39, -53, -59, -61, -83, -89, and $t = 17$, which are needed for Theorem 6.7.13 and Exercise 40 of Chapter 6. For instance, first show that a basis of the Mordell–Weil group is given by $(P, Q) = ((5, 6), (153/4, 1891/8))$ for $t = -89$, $(P, Q) = ((4, 5), (10, 31))$ for $t = -39$, and $(P, Q) = ((-2, 3), (4, 9))$ for $t = 17$, and proceed similarly for the other values of t.

30.

   (a) Let P be the point found in the example of Section 8.6.6. Compute $P + T$ for the four 2-torsion points of E.

   (b) Note that the numerators of the x-coordinates of these four points are of the form $\pm 157^a u^2$ for some sign $\pm$ and $a = 0$ or 1, and in particular one of these four points has an x-coordinate that is a square in $\mathbb{Q}$. Explain and generalize this phenomenon.

31. Compute a rational point on the elliptic curve $y^2 = x^3 - 157^2 x$ considered in Section 8.6.6, but now using the 2-descent method. (Hint: cheat and start from the point found in the text to see on which quartics to look, and at what height.)

32. Using the Heegner point method, compute rational numbers u and v such that $u^3 + v^3 = 697$. For this, first show that the minimal Weierstrass model of the corresponding elliptic curve is $y^2 + y = x^3 - 3279211$. You will have to perform all the subsequent computations with reasonably small accuracy (66 to 75 decimal digits), but you will need several hundred million coefficients of the L-series, so these will have to be computed on the fly. The smallest denominators for u and v have 50 decimal digits.

33. (Bremner–Cassels [Bre-Cas].) Using the Heegner point method, find a nontorsion rational point on the elliptic curve $y^2 = x(x^2 + 877)$ (again, this is a large computation).

34. Let Q be the matrix of a positive definite quadratic form, and let λ be its smallest eigenvalue. By diagonalizing Q on an orthonormal basis of eigenvectors, show that $X^t Q X \geqslant \lambda X^t X$.

35. Prove that there are exactly 20 integral points on the elliptic curve with equation $y^2 + y = x^3 + x^2 - 2x$ (this is the curve of rank 2 having smallest conductor, equal to 389).

Bibliography

[Abou] M. Abouzaid, *Les nombres de Lucas et Lehmer sans diviseur primitif*, J. Théor. Nombres Bordeaux **18** (2006), 299–313.

[Abr-Ste] M. Abramowitz and I. Stegun, *Handbook of Mathematical Functions*, Dover publications (1972).

[AGP] R. Alford, A. Granville, and C. Pomerance, *There are infinitely many Carmichael numbers*, Ann. of Math. **139** (1994), 703–722.

[Ami] Y. Amice, *Les nombres p-adiques*, SUP/Le Mathématicien **14**, Presses Universitaires de France (1975).

[Ang] W. Anglin, *The square pyramid puzzle*, American Math. Monthly **97** (1990), 120–124.

[Ax] J. Ax, *Zeroes of polynomials over finite fields*, Amer. J. Math. **86** (1964), 255–261.

[Bac] G. Bachman, *Introduction to p-adic Numbers and Valuation theory*, Academic paperbacks, Acad. Press (1964).

[Bak1] A. Baker, *Linear forms in the logarithms of algebraic numbers*, Mathematika **13** (1966), 204–216.

[Bak2] A. Baker, *Transcendental Number Theory*, Cambridge University Press, 1975.

[Bak-Dav] A. Baker and H. Davenport, *The equations $3x^2 - 2 = y^2$ and $8x^2 - 7 = y^2$*, Quart. J. Math. Oxford Ser. (2) **20** (1969), 129–137.

[Bak-Wus] A. Baker and G. Wüstholz, *Logarithmic forms and group varieties*, J. reine angew. Math. **442** (1993), 19–62.

[BDD] R. Balasubramanian, J.-M. Deshouillers, and F. Dress, *Problème de Waring pour les bicarrés 1 : schéma de la solution, 2 : résultats auxiliaires pour le théorème asymptotique*, C. R. Acad. Sc. Paris **303** (1986), 85–88 and 161–163.

[Bal-Dar-Ono] A. Balog, H. Darmon, and K. Ono, *Congruences for Fourier coefficients of half-integral weight modular forms and special values of L-functions*, Proceedings of a Conference in honor of H. Halberstam **1** (1996), 105–128.

[Bar] D. Barsky, *Congruences de coefficients de séries de Taylor (Application aux nombres de Bernoulli–Hurwitz)*, Groupe d'Analyse Ultramétrique **3** (1975-1976), Exp. 17, 1–9, available on the NUMDAM archives.

[Bat-Oli] C. Batut and M. Olivier, *Sur l'accélération de la convergence de certaines fractions continues*, Séminaire Th. Nombres Bordeaux (1979–1980), exposé **23**.

[Bel-Gan] K. Belabas and H. Gangl, *Generators and relations for $K_2 \mathcal{O}_F$*, K-Theory **31** (2004), 195–231.

[BBGMS]   C. Bennett, J. Blass, A. Glass, D. Meronk, and R. Steiner, *Linear forms in the logarithms of three positive rational numbers*, J. Théor. Nombres Bordeaux **9** (1997), 97–136.

[Ben1]   M. Bennett, *Rational approximation to algebraic numbers of small height: The Diophantine equation* $| ax^n - by^n | = 1$, J. reine angew. Math. **535** (2001), 1–49.

[Ben2]   M. Bennett, *Recipes for ternary Diophantine equations of signature* (p, p, k), Proc. RIMS Kokyuroku (Kyoto) **1319** (2003), 51–55.

[Ben3]   M. Bennett, *On some exponential Diophantine equations of S. S. Pillai*, Canad. J. Math. **53** (2001), 897–922.

[Ben-deW]   M. Bennett and B. de Weger, *The Diophantine equation* $| ax^n - by^n | = 1$, Math. Comp. **67** (1998), 413–438.

[Ben-Ski]   M. Bennett and C. Skinner, *Ternary Diophantine equations via Galois representations and modular forms*, Canad. J. Math. **56** (2004), 23–54.

[Ben-Vat-Yaz] M. Bennett, V. Vatsal, and S. Yazdani, *Ternary Diophantine equations of signature* $(p, p, 3)$, Compositio Math. **140** (2004), 1399–1416.

[Ber-Eva-Wil] B. Berndt, R. Evans, and K. Williams, *Gauss and Jacobi Sums*, Canadian Math. Soc. series **21**, Wiley (1998).

[Bha1]   M. Bhargava, *Higher composition laws I, II, and III*, Ann. Math. **159** (2004), 217–250, 865–886, 1329–1360.

[Bha2]   M. Bhargava, *The density of discriminants of quartic rings and fields*, Ann. Math. **162** (2005), 1031–1063.

[Bha-Han]   M. Bhargava and J. Hanke, *Universal quadratic forms and the 290-theorem*, Invent. Math., to appear.

[Bilu]   Yu. Bilu, *Catalan's conjecture (after Mihailescu)*, Séminaire Bourbaki **909** (2002–2003), 1–25.

[Bil-Han]   Yu. Bilu and G. Hanrot, *Solving Thue equations of high degree*, J. Number Th. **60** (1996), 373–392.

[Bil-Han-Vou] Yu. Bilu, G. Hanrot, and P. Voutier, *Existence of primitive divisors of Lucas and Lehmer numbers*, with an appendix by M. Mignotte, J. reine angew. Math. **539** (2001), 75–122.

[Boe-Mis]   J. Boéchat and M. Mischler, *La conjecture de Catalan racontée à un ami qui a le temps*, preprint available on the web at the URL http://arxiv.org/pdf/math.NT/0502350.

[Bom]   E. Bombieri, *Effective Diophantine approximation on* $\mathbf{G}_m$, Ann. Scuola Norm. Sup. Pisa Cl. Sci. (4) **20** (1993), 61–89.

[Bor-Bai]   J. Borwein and D. Bailey, *Mathematics by Experiment*, A. K. Peters (2004).

[Bor-Bai-Gir] J. Borwein, D. Bailey, and R. Girgensohn, *Experimentation in Mathematics*, A. K. Peters (2004).

[Bor-Sha]   Z. I. Borevitch and I. R. Shafarevitch, *Number Theory*, Academic Press, New York (1966).

[Bre-Cas]   A. Bremner and I. Cassels, *On the equation* $Y^2 = X(X^2 + p)$, Math. Comp. **42** (1984), 257–264.

[Bre-Mor]   A. Bremner and P. Morton, *A new characterization of the integer 5906*, Manuscripta Math. **44** (1983), 187–229.

[Bre-Tza1]   A. Bremner and N. Tzanakis, *Lucas sequences whose 12th or 9th term is a square*, J. Number Th. **107** (2004), 215–227.

[Bre-Tza2]   A. Bremner and N. Tzanakis, *On squares in Lucas sequences*, J. Number Th., to appear.

[Bre-Tza3]   A. Bremner and N. Tzanakis, *Lucas sequences whose nth term is a square or an almost square*, Acta Arith., to appear.

[BCDT]   C, Breuil, B. Conrad, F. Diamond, and R. Taylor, *On the modularity of elliptic curves over $\mathbb{Q}$: wild 3-adic exercises*, J. Amer. Math. Soc. **14** (2001), 843–939.

[Bri-Eve-Gyo] B. Brindza, J. Evertse, and K. Győry, *Bounds for the solutions of some Diophantine equations in terms of discriminants*, J. Austral. Math. Soc. (Series A) **51** (1991), 8–26.

[Bru1]   N. Bruin, *Chabauty Methods and Covering Techniques Applied to Generalized Fermat Equations*, CWI Tract **133**, CWI, Amsterdam (2002).

[Bru2]   N. Bruin, *The Diophantine equations $x^2 \pm y^4 = \pm z^6$ and $x^2 + y^8 = z^3$*, Compositio Math. **118** (1999), 305–321.

[Bru3]   N. Bruin, *Chabauty methods using elliptic curves*, J. reine angew. Math. **562** (2003), 27–49.

[Bru4]   N. Bruin, *Primitive solutions to $x^3 + y^9 = z^2$*, J. Number theory **111** (2005), 179–189.

[Bru-Kra]   A. Brumer and K. Kramer, *The rank of elliptic curves*, Duke Math. J. **44** (1977), 715–742.

[Bug]   Y. Bugeaud, *Bounds for the solutions of superelliptic equations*, Compositio Math. **107** (1997), 187–219.

[Bug-Gyo]   Y. Bugeaud and K. Győry, *Bounds for the solutions of Thue–Mahler equations and norm form equations*, Acta Arith. **74** (1996), 273–292.

[Bug-Han]   Y. Bugeaud and G. Hanrot, *Un nouveau critère pour l'équation de Catalan*, Mathematika **47** (2000), 63–73.

[Bug-Mig]   Y. Bugeaud and M. Mignotte, *On integers with identical digits*, Mathematika **46** (1999), 411–417.

[BMS1]   Y. Bugeaud, M. Mignotte, and S. Siksek, *Classical and modular approaches to exponential Diophantine equations I. Fibonacci and Lucas perfect powers*, Annals of Math. **163** (2006), 969–1018.

[BMS2]   Y. Bugeaud, M. Mignotte, and S. Siksek, *Classical and modular approaches to exponential Diophantine equations II. The Lebesgue–Nagell equation*, Compositio Math. **142** (2006), 31–62.

[BMS3]   Y. Bugeaud, M. Mignotte, and S. Siksek, *A multi-Frey approach to some multi-parameter families of Diophantine equations*, Canadian J. Math., to appear.

[Buh-Gro]   J. Buhler and B. Gross, *Arithmetic on elliptic curves with complex multiplication II*, Invent. Math. **79** (1985), 11–29.

[BGZ]   J. Buhler, B. Gross, and D. Zagier, *On the conjecture of Birch and Swinnerton-Dyer for an elliptic curve of rank 3*, Math. Comp. **44** (1985), 473–481.

[Cal]   E. Cali, *Points de torsion des courbes elliptiques et quartiques de Fermat*, Thesis, Univ. Paris VI (2005).

[Can]   D. Cantor, *Computing on the Jacobian of a hyperelliptic curve*, Math. Comp., **48** (1987), 95–101.

[Cas1]   J. Cassels, *Local Fields*, London Math. Soc. Student Texts **3**, Cambridge University Press (1986).

[Cas2]   J. Cassels, *Lectures on Elliptic Curves*, London Math. Soc. Student Texts **24**, Cambridge University Press (1991).

[Cas3]   J. Cassels, *On the equation $a^x - b^y = 1$, II*, Proc. Cambridge Phil. Soc. **56** (1960), 97–103.

[Cas-Fly]   J. Cassels and V. Flynn, *Prolegomena to a Middlebrow Arithmetic of Curves of Genus 2*, LMS Lecture Note Series **230**, Cambridge University Press (1996).

[Cas-Frö]   J. Cassels and A. Fröhlich, *Algebraic Number Theory*, Academic Press, London, New York (1967).

[Cat]   E. Catalan, *Note extraite d'une lettre adressée à l'éditeur*, J. reine angew. Math. **27**, (1844), 192.

[Cha]   C. Chabauty, *Sur les points rationnels des variétés algébriques dont l'irrégularité est supérieure à la dimension*, C. R. A. S. Paris, **212** (1941), 1022–1024.

[Coa-Wil]   J. Coates and A. Wiles, *On the conjecture of Birch and Swinnerton-Dyer*, Invent. Math. **39** (1977), 223–251.

[Coh0]   H. Cohen, *A Course in Computational Algebraic Number Theory (4th corrected printing)*, Graduate Texts in Math. **138**, Springer-Verlag (2000).

[Coh1]   H. Cohen, *Advanced Topics in Computational Number Theory*, Graduate Texts in Math. **193**, Springer-Verlag (2000).

[Coh2]   H. Cohen, *Variations sur un thème de Siegel et Hecke*, Acta Arith. **30** (1976), 63–93.

[Coh3]   H. Cohen, *Sums involving L-functions of quadratic characters*, Math. Ann. **217** (1975), 271–285.

[Coh4]   H. Cohen, *Continued fractions for gamma products and $\zeta(k)$*, unfinished postscript preprint available on the author's home page at `http://www.math.u-bordeaux1.fr/~cohen/`.

[Coh-Fre]   H. Cohen and G. Frey, eds., *Handbook of elliptic and hyperelliptic curve cryptography*, Chapman & Hall/CRC press, 2005.

[Coh-Fri]   H. Cohen and E. Friedman, *Raabe's formula for p-adic gamma and zeta functions*, submitted.

[Coh-Len]   H. Cohen and H. W. Lenstra, *Heuristics on class groups of number fields*, Springer Lecture Notes in Math. **1068** (1984), 33–62.

[Coh-Mar]   H. Cohen and J. Martinet, *Class groups of number fields: numerical heuristics*, Math. Comp. **48** (1987), 123–137.

[Coh-Rhi]   H. Cohen and G. Rhin, *Accélération de la convergence de certaines récurrences linéaires*, Séminaire Th. Nombres Bordeaux (1980–1981), exposé **16**.

[Coh-Vil-Zag] H. Cohen, F. Rodriguez-Villegas, and D. Zagier, *Convergence acceleration of alternating series*, Exp. Math. **9** (2000), 3–12.

[Cohn1]   J. Cohn, *The Diophantine equation $x^2 + C = y^n$*, Acta Arith. **65** (1993), 367–381.

[Cohn2]   J. Cohn, *The Diophantine equation $x^2 + C = y^n$, II*, Acta Arith. **109** (2003), 205–206.

[Col]   R. Coleman, *Effective Chabauty*, Duke Math. J., **52** (1985), 765–780.

[Colm]   P. Colmez, *Arithmétique de la fonction zêta*, Journées mathématiques X-UPS (2002), Publications de l'Ecole Polytechnique, 37–164.

[Con-Sou]   J. B. Conrey and K. Soundararajan, *Real zeros of quadratic Dirichlet L-functions*, Invent. Math. **150** (2002), 1–44.

[Con]   J.-H. Conway, *The Sensual (Quadratic) Form*, Carus Math. Monographs **26**, MAA (1997).

[Con-Slo]   J.-H. Conway and N. Sloane, *Sphere Packings, Lattices and Groups (3rd ed.)*, Grundlehren der math. Wiss. **290**, Springer-Verlag, New York (1999).

[Cre1]   J. Cremona, *Computing the degree of the modular parametrization of a modular elliptic curve*, Math. Comp. **64** (1995), 1235–1250.

[Cre2]   J. Cremona, *Algorithms for Modular Elliptic Curves (2nd ed.)*, Cambridge Univ. Press (1996).

[Cre-Pri-Sik]   J. Cremona, M. Prickett, and S. Siksek, *Height difference bounds for elliptic curves over number fields*, J. Number theory **116** (2006), 42–68.

[Dar]   H. Darmon, *Rational Points on Modular Elliptic Curves*, CBMS Regional Conference Series in Mathematics **101** (2004), American Math. Soc.

[Dar-Gra]   H. Darmon and A. Granville, *On the equations $z^m = F(x,y)$ and $Ax^p + By^q = Cz^r$*, Bull. London Math. Soc. **27** (1995), 513–543.

[Dar-Mer]   H. Darmon and L. Merel, *Winding quotients and some variants of Fermat's Last Theorem*, J. reine angew. Math. **490** (1997), 81–100.

[Dem1]   V. Dem'yanenko, *О Суммах четырех кубов* (*On sums of four cubes*), Izv. Visch. Outch. Zaved. Mathematika **54** (1966), 64–69.

[Dem2]   V. Dem'yanenko, *Rational points on a class of algebraic curves*, Amer. Math. Soc. Transl. **66** (1968), 246–272.

[Dem3]   V. Dem'yanenko, *The indeterminate equations $x^6 + y^6 = az^2$, $x^6 + y^6 = az^3$, $x^4 + y^4 = az^4$*, Amer. Math. Soc. Transl. **119** (1983), 27–34.

[Den]   P. Dénes, *Über die Diophantische Gleichung $x^\ell + y^\ell = cz^\ell$*, Acta Math. **88** (1952), 241–251.

[DeW1]   B. de Weger, *Solving exponential Diophantine equations using lattice basis reduction algorithms*, J. Number Th. **26** (1987), 325–367.

[DeW2]   B. de Weger, *A hyperelliptic Diophantine equation related to imaginary quadratic number fields with class number 2*, J. reine angew. Math. **427** (1992), 137–156.

[Dia1]   J. Diamond, *The p-adic log gamma funnction and p-adic Euler constants*, Trans. Amer. Math. Soc. **233** (1977), 321–337.

[Dia2]   J. Diamond, *On the values of p-adic L-functions at positive integers*, Acta Arith. **35** (1979), 223–237.

[Dia-Kra]   F. Diamond and K. Kramer, *Modularity of a family of elliptic curves*, Math. Res. Lett. **2** (1995), No. 3, 299–304.

[Dok]   T. Dokchitser, *Computing special values of motivic L-functions*, Exp. Math. **13** (2004), 137–149.

[Duq1]   S. Duquesne, *Rational Points on Hyperelliptic Curves and an Explicit Weierstrass Preparation Theorem*, Manuscripta Math. **108:2** (2002), 191–204.

[Duq2]   S. Duquesne, *Calculs effectifs des points entiers et rationnels sur les courbes*, Thesis, Univ. Bordeaux I (2001).

[Edw]   J. Edwards, *Platonic solids and solutions to $x^2 + y^3 = dz^r$*, Thesis, Univ. Utrecht (2005).

[Elk1]   N. Elkies, *ABC implies Mordell*, Internat. Math. Res. Notices **7** (1991), 99–109.

[Elk2]   N. Elkies, *$\mathbb{Z}^{28}$ in $E(\mathbb{Q})$*, Internet announcement on the number theory listserver (May 3rd, 2006).

[Ell]   W. Ellison and M. Mendès France, *Les nombres premiers*, Hermann (1975).

[Erd-Wag]   P. Erdős and S. Wagstaff, *The fractional parts of the Bernoulli numbers*, Illinois J. Math. **24** (1980), 104–112.

[Eva]   R. Evans, *Congruences for Jacobi sums*, J. Number Theory **71** (1998), 109–120.

[Fal]   G. Faltings, *Endlichkeitssätze für abelsche Varietäten über Zahlkörpen*, Invent. Math. **73** (1983), 349–366.

[Fer-Gre]   B. Ferrero and R. Greenberg, *On the behaviour of p-adic L-functions at $s = 0$*, Invent. Math. **50** (1978), 91–102.

[Fly]   V. Flynn, *A flexible method for applying Chabauty's Theorem*, Compositio Math. **105** (1997), 79–94.

[Fly-Wet1]   V. Flynn and J. Wetherell, *Finding rational points on bielliptic genus 2 curves*, Manuscripta Math. **100** (1999), 519-533.

[Fly-Wet2]   V. Flynn and J. Wetherell, *Covering collections and a challenge problem of Serre*, Acta Arith. **98** (2001), 197–205.

[Fre]   E. Freitag, *Hilbert Modular Forms*, Springer-Verlag (1990).

[Frö-Tay]   A. Fröhlich and M. Taylor, *Algebraic Number Theory*, Cambridge Studies in Adv. Math. **27**, Cambridge Univ. Press (1991).

[Gel]   A. O. Gel'fond, *On the approximation of transcendental numbers by algebraic numbers*, Doklady Akad. Nauk SSSR **2** (1935), 177–182.

[Gou]   F. Gouvêa, *p-adic Numbers: An Introduction*, Universitext, Springer-Verlag (1993).

[Gra-Sou]   A. Granville and K. Soundararajan, *Large character sums: pretentious characters and the Polya–Vinogradov theorem*, Journal of the American Math. Soc., to appear.

[Gran]   D. Grant, *A curve for which Coleman's effective Chabauty bound is sharp*, Proc. Amer. Math. Soc. **122** (1994), 317–319.

[Gras]   G. Gras, *Class Field Theory: From Theory to Practice*, Springer monographs in mathematics (2003).

[Gre-Tao]   B. Green and T. Tao, *The primes contain arbitrarily long arithmetic progressions*, Ann. Math., to appear.

[Gri-Riz]   G. Grigorov and J. Rizov, *Heights on elliptic curves and the Diophantine equation $x^4 + y^4 = cz^4$*, Sophia Univ. preprint (1998).

[Gro]   B. Gross, *Heegner points on $X_0(N)$*, in Modular forms, edited by R. Rankin (1984), 87–105.

[Gro-Kob]   B. Gross and N. Koblitz, *Gauss sums and the p-adic Γ-function*, Ann. Math. **109** (1979), 569–581.

[Guy]   R. K. Guy, *Unsolved Problems in Number Theory (3rd edition)*, Problem books in math. **1**, Springer-Verlag (2004).

[Hal-Kra1]   E. Halberstadt and A. Kraus, *Sur les modules de torsion des courbes elliptiques*, Math. Ann. **310** (1998), 47–54.

[Hal-Kra2]   E. Halberstadt and A. Kraus, *Courbes de Fermat : résultats et problèmes*, J. reine angew. Math. **548** (2002), 167–234.

[Har-Wri]   G. H. Hardy and E. M. Wright, *An Introduction to the Theory of Numbers (5th ed.)*, Oxford University Press (1979).

[Hay]   Y. Hayashi, *The Rankin's L-function and Heegner points for general discriminants*, Proc. Japan. Acad. **71** (1995), 30–32.

[Her]   G. Herglotz, *Über die Kroneckersche Grenzformel für reelle quadratische Körper I, II*, Gesam. Schr. (ed. H. Schwerdtfeger), Vandenhoeck and Ruprecht (1979), 466–484.

[Hul]   W. Hulsbergen, *Conjectures in Arithmetic Algebraic Geometry*, Aspects of math., Vieweg (1992).

[Ire-Ros]   K. Ireland and M. Rosen, *A Classical Introduction to Modern Number Theory (2nd ed.)*, Graduate Texts in Math. **84**, Springer-Verlag (1982).

[Ivo1]   W. Ivorra, *Sur les équations $x^p + 2^\beta y^p = z^2$ et $x^p + 2^\beta y^p = 2z^2$*, Acta Arith. **108** (2003), 327–338.

[Ivo2]   W. Ivorra, *Equations diophantiennes ternaires de type $(p, p, 2)$ et courbes elliptiques*, Thesis, Univ. Paris VI (2004).

[Ivo-Kra]   W. Ivorra and A. Kraus, *Quelques résultats sur les équations $ax^p + by^p = cz^2$*, Can. J. Math., to appear.

[Iwa-Kow]   H. Iwaniec and E. Kowalski, *Analytic Number Theory*, Colloquium Publications **53**, American Math. Soc. (2004).

[Jan]   G. Janusz, *Algebraic Number Fields*, Pure and applied math. **55**, Academic Press (1973).

[Kap]   I. Kaplansky, *Ternary positive quadratic forms that represent all odd positive integers*, Acta Arith. **70** (1995), 209–214.

[Kat1]   N. Katz, *On a theorem of Ax*, Amer. J. Math. **93** (1971), 485–499.

[Kat2]   N. Katz, *The congruences of Clausen–von Staudt and Kummer for Bernoulli–Hurwitz numbers*, Math. Ann. **216** (1975), 1–4.

[Kea]   J. Keating, talk in Bordeaux, 2005.

[Kel-Ric]   W. Keller and J. Richstein, *Solutions of the congruence* $a^{p-1} \equiv 1$ (mod p^r), Math. Comp. **74** (2005), 927–936.

[Kna]   A. Knapp, *Elliptic Curves*, Math. Notes **40**, Princeton University press (1992)

[Ko]   Ko Chao, *On the Diophantine equation* $x^2 = y^n + 1$, $xy \neq 0$, Sci. Sinica **14** (1965), 457–460.

[Kob1]   N. Koblitz, *p-adic Numbers, p-adic Analysis, and Zeta-Functions (2nd edition)*, Graduate Texts in Math. **58**, Springer-Verlag (1984).

[Kob2]   N. Koblitz, *An Introduction to Elliptic Curves and Modular Forms (2nd edition)*, Graduate Texts in Math. **97**, Springer-Verlag (1993).

[Kra1]   A. Kraus, *Sur l'équation* $a^3 + b^3 = c^p$, Experimental Math. **7** (1998), 1–13.

[Kra2]   A. Kraus, *On the equation* $x^p + y^q = z^r$*: a survey*, Ramanujan Journal **3** (1999), 315–333.

[Kra3]   A. Kraus, *Majorations effectives pour l'équation de Fermat généralisée*, Can. J. Math. **49** (1997), 1139–1161.

[Kra-Oes]   A. Kraus and J. Oesterlé, *Sur une question de B. Mazur*, Math. Ann. **293** (1992), 259–275.

[Kul]   L. Kulesz, *Application de la méthode de Dem'janenko–Manin à certaines familles de courbes de genre 2 et 3*, J. Number Theory **76** (1999), 130–146.

[Lan0]   S. Lang, *Algebra*, Addison-Wesley, Reading, MA (1965).

[Lan1]   S. Lang, *Algebraic Number Theory (2nd ed.)*, Graduate Texts in Math. **110**, Springer-Verlag (1994).

[Lau]   M. Laurent, *Linear form in two logarithms and interpolation determinants*, Acta Arith. **66** (1994), 181–199.

[Lau-Mig-Nes] M. Laurent, M. Mignotte, and Yu. Nesterenko, *Formes linéaires en deux logarithmes et déterminants d'interpolation*, J. Number Theory **55** (1995), 255–265.

[Leb]   V. Lebesgue, *Sur l'impossibilité en nombres entiers de l'équation* $x^m = y^2 + 1$, Nouv. Ann. Math. **9** (1850), 178–181.

[Lem]   F. Lemmermeyer, *Kronecker–Weber via Stickelberger*, preprint.

[Ma]   D.-G. Ma, *An elementary proof of the solution to the Diophantine equation* $6y^2 = x(x + 1)(2x + 1)$, Sichuan Daxue Xuebao **4** (1985) 107–116.

[Man]   Yu. Manin, *The p-torsion of elliptic curves is uniformly bounded*, Izv. Akad. Nauk SSSR Ser. Mat. **33** (1969), 459–465; Amer. Math. Soc. Transl. 433–438.

[Mar]   J. Martinet, *Perfect Lattices in Euclidean Spaces*, Grundlehren der math. Wiss. **327**, Springer (2003).

[Marc]   D. A. Marcus, *Number Fields*, Springer-Verlag, New York (1977).

[Mart]   G. Martin, *Dimensions of the spaces of cusp forms and newforms on* $\Gamma_0(N)$ *and* $\Gamma_1(N)$, J. Number Theory **112** (2005), 298–331.

[Mat]   E. M. Matveev, *An explicit lower bound for a homogeneous rational linear form in logarithms of algebraic numbers. II*, Izv. Ross. Akad. Nauk Ser. Mat. **64** (2000), 125–180. English transl. in Izv. Math. **64** (2000), 1217–1269.

[Maz]   B. Mazur, *Rational isogenies of prime degree*, Invent. Math. **44** (1978), 129–162.

[McC]   W. McCallum, *On the method of Coleman and Chabauty*, Math. Ann. **299** (1994), 565–596.

[Mes-Oes]   J.-F. Mestre and J. Oesterlé, *Courbes de Weil semi-stables de discriminant une puissance m-ième*, J. reine angew. Math. **400** (1989), 173–184.

[Mig]   M. Mignotte, *A note on the equation $ax^n - by^n = c$*, Acta Arith. **75** (1996), 287–295.

[Mig-Weg]   M. Mignotte and B. de Weger, *On the Diophantine equations $x^2 + 74 = y^5$ and $x^2 + 86 = y^5$*, Glasgow Math. J. **38** (1996), 77–85.

[Mom]   F. Momose, *Rational points on the modular curves $X_{split}(p)$*, Compositio Math. **52** (1984), 115–137.

[Mon-Vau]   H. Montgomery and R. Vaughan, *Exponential sums with multiplicative coefficients*, Invent. Math. **43** (1977), 69–82.

[Mord]   L. Mordell, *Diophantine Equations*, Pure and applied Math. **30**, Academic Press (1969).

[Mori]   M. Mori, *Developments in the double exponential formula for numerical integration*, in Proceedings ICM 1990, Springer-Verlag (1991), 1585–1594.

[Morit1]   Y. Morita, *A p-adic analogue of the Γ-function*, J. Fac. Sci. Univ. Tokyo Sect. IA Math. **22** (1975), 255–266.

[Morit2]   Y. Morita, *On the Hurwitz–Lerch L-functions*, J. Fac. Sci. Univ. Tokyo Sect. IA Math. **24** (1977), 29–43.

[Nak-Tag]   Y. Nakkajima and Y. Taguchi, *A generalization of the Chowla-Selberg formula*, J. reine angew. Math. **419** (1991), 119–124.

[New]   D. Newman, *Analytic Number Theory (2nd corrected printing)*, Graduate Texts in Math. **177**, Springer-Verlag (2000).

[Pap]   I. Papadopoulos, *Sur la classification de Néron des courbes elliptiques en caractéristique résiduelle 2 et 3*, J. Number Theory **44** (1993), 119–152.

[Poo-Sch-Sto] B. Poonen, E. Schaefer, and M. Stoll, *Twists of $X(7)$ and primitive solutions to $x^2 + y^3 = z^7$*, Duke Math. J., to appear.

[Poo-Wil]   A. van der Poorten and K. Williams, *Values of the Dedekind eta function at quadratic irrationalities*, Canadian Jour. Math. **51** (1999), 176–224, corrigendum **53** (2001), 434–448.

[Rap-Sch-Sch] M. Rapoport, N. Schappacher, and P. Schneider, *Beilinson's Conjectures on Special Values of L-Functions*, Perspectives in Math. **4** (1988), Academic Press.

[Rib1]   K. Ribet, *On modular representations of $\mathrm{Gal}(\overline{\mathbb{Q}}/\mathbb{Q})$ arising from modular forms*, Invent. Math. **100** (1990), 431–476.

[Rib2]   K. Ribet, *On the equation $a^p + 2b^p + c^p = 0$*, Acta Arith. **LXXIX.1** (1997), 7–15.

[Rob1]   A. Robert, *A Course in p-adic Analysis*, Graduate Texts in Math. **198**, Springer-Verlag (2000).

[Rob2]   A. Robert, *The Gross–Koblitz formula revisited*, Rend. Sem. Math. Univ. Padova **105** (2001), 157–170.

[Rod-Zag]   F. Rodriguez-Villegas and D. Zagier, *Which primes are sums of two cubes?*, Canadian Math. Soc. Conference proceedings **15** (1995), 295–306.

[Ruc]   H.-G. Rück, *A note on elliptic curves over finite fields*, Math. Comp. **49** (1987), 301–304.

[Rud]   W. Rudin, *Real and Complex Analysis*, Mc Graw Hill (1970).

[Sam]   P. Samuel, *Théorie Algébrique des Nombres*, Hermann, Paris (1971).

[Sch]   E. Shaefer, *2-descent on the Jacobians of hyperelliptic curves*, J. Number Theory **51** (1995), 219–232.

[Sch-Sto]   E. Schaefer and M. Stoll, *How to do a p-descent on an elliptic curve*, Trans. Amer. Math. Soc. **356** (2004), 1209–1231.

[Scho]   R. Schoof, *Class groups of real cyclotomic fields of prime conductor*, Math. Comp. **72** (2003), 913–937 (see also the errata on Schoof's home page).

[Sel1]   E. S. Selmer, *The Diophantine equation $ax^3 + by^3 + cz^3 = 0$*, Acta Math. **85** (1951), 203–362.

[Sel2]   E. S. Selmer, *Completion of the tables*, Acta Math. **92** (1954), 191–197.

[Ser1]   J.-P. Serre, *Cours d'arithmétique*, P.U.F., Paris (1970). English translation: Graduate Texts in Math. **7**, Springer-Verlag (1973).

[Ser2]   J.-P. Serre, *Corps locaux (2nd ed.)*, Hermann, Paris (1968). English translation: Graduate Texts in Math. **67**, Springer-Verlag (1979).

[Ser3]   J.-P. Serre, *Sur les représentations modulaires de degré 2 de* $\mathrm{Gal}(\overline{\mathbb{Q}}/\mathbb{Q})$, Duke Math. J. **54** (1987) 179–230.

[Ser4]   J.-P. Serre, *Abelian ℓ-adic Representations and Elliptic Curves*, W. A. Benjamin, New York, 1968.

[Ses]   J. Sesanio, *Books IV to VII of Diophantus's Arithmetica in the Arabic Translation Attributed to Qusta ibn Luqa*, Sources in the History of Mathematics and Physical Sciences **3**, Springer-Verlag (1982).

[Shi]   G. Shimura, *Introduction to the Arithmetic Theory of Automorphic Functions*, Iwami Shoten (1971).

[Sho-Tij]   T. Shorey and R. Tijdeman, *Exponential Diophantine Equations*, Cambridge Tracts in Mathematics **87**, Cambridge University Press (1986).

[Sik]   S. Siksek, *On the Diophantine equation $x^2 = y^p + 2^k z^p$*, Journal de Théorie des Nombres de Bordeaux **15** (2003), 839–846.

[Sik-Cre]   S. Siksek and J. Cremona, *On the Diophantine equation $x^2 + 7 = y^m$*, Acta Arith. **109** (2003), 143–149.

[Sil1]   J. Silverman, *The Arithmetic of Elliptic Curves*, Graduate Texts in Math. **106**, Springer-Verlag (1986).

[Sil2]   J. Silverman, *Advanced Topics in the Arithmetic of Elliptic Curves*, Graduate Texts in Math. **151**, Springer-Verlag (1994).

[Sil3]   J. Silverman, *The difference between the Weil height and the canonical height on elliptic curves*, Math. Comp. **55** (1990), 723–743.

[Sil4]   J. Silverman, *Rational points on certain families of curves of genus at least 2*, Proc. London Math. Soc. **55** (1987), 465–481.

[Sil-Tat]   J. Silverman and J. Tate, *Rational Points on Elliptic Curves*, Undergraduate Texts in Math., Springer-Verlag (1992).

[Sim1]   D. Simon, *Solving quadratic equations using reduced unimodular quadratic forms*, Math. Comp. **74** (2005), 1531–1543.

[Sim2]   D. Simon, *Computing the rank of elliptic curves over a number field*, LMS J. Comput. Math. **5** (2002), 7–17.

[Sma]   N. Smart, *The Algorithmic Resolution of Diophantine Equations*, London Math. Soc. Student Texts **41** (1998).

[Sta1]   H. Stark, *Some effective cases of the Brauer–Siegel theorem*, Invent. Math. **23** (1974), 135–152.

[Sta2]   H. Stark, *Class numbers of complex quadratic fields*, in Modular Forms in One Variable I, Springer Lecture Notes in Math **320** (1973), 153–174.

[Sto]   M. Stoll, *Implementing 2-descent for Jacobians of hyperelliptic curves*, Acta Arith. **98** (2001), 245–277.

[Sug]   T. Sugatani, *Rings of convergent power series and Weierstrass preparation theorem*, Nagoya Math. J. **81** (1981), 73–78.

[Swd]   H.-P.-F. Swinnerton-Dyer, *A Brief Guide to Algebraic Number Theory*, London Math. Soc. Student Texts **50**, Cambridge University Press (2001).

[Tak-Mor]   H. Takashi and M. Mori, *Double exponential formulas for numerical integration*, Publications of RIMS, Kyoto University (1974), 9:721–741.

[Tay]   P. Taylor, *On the Riemann zeta function*, Quart. J. Math., Oxford Ser. **16** (1945), 1–21.

[Tay-Wil]   R. Taylor and A. Wiles, *Ring theoretic properties of certain Hecke algebras*, Annals of Math. **141** (1995), 553–572.

[Ten]   S. Tengely, *On the Diophantine equation $F(x) = G(y)$*, Acta Arith. **110** (2003), 185–200.

[Tij]   R. Tijdeman, *On the equation of Catalan*, Acta Arith. **29** (1976), 197–209.

[Tun]   J. Tunnell, *A classical Diophantine problem and modular forms of weight 3/2*, Invent. Math. **72** (1983), 323–334.

[Tza-Weg]   N. Tzanakis and B. de Weger, *On the practical solution of the Thue Equation*, J. Number Th. **31** (1989), 99–132.

[Vel]   J. Vélu, *Isogénies entre courbes elliptiques*, Comptes Rendus Acad. Sc. Paris Sér. A **273** (1971), 238–241.

[Wald1]   M. Waldschmidt, *Minorations de combinaisons linéaires de logarithmes de nombres algébriques*, Canadian J. Math. **45** (1993), 176–224.

[Wald2]   M. Waldschmidt, *Diophantine Approximation on Linear Algebraic Groups*, Grundlehren der math. Wiss. **326** (2000), Springer-Verlag.

[Wals]   P. G. Walsh, *A quantitative version of Runge's theorem on Diophantine equations*, Acta Arith. **62** (1992), 157–172.

[Was]   L. Washington, *Introduction to Cyclotomic Fields (2nd ed.)*, Graduate Texts in Math. **83**, Springer-Verlag (1997).

[Watk]   M. Watkins, *Real zeros of real odd Dirichlet L-functions*, Math. Comp. **73** (2004), 415–423.

[Wats]   G. Watson, *A Treatise on the Theory of Bessel Functions (2nd ed.)*, Cambridge Univ. Press (1966).

[Wet]   J. Wetherell, *Bounding the Number of Rational Points on Certain Curves of High Rank*, PhD thesis, Univ. California Berkeley (1997).

[Wil]   A. Wiles, *Modular elliptic curves and Fermat's last theorem*, Annals of Math. **141** (1995), 443–551.

[Yam]   Y. Yamamoto, *Real quadratic number fields with large fundamental units*, Osaka J. Math. **8** (1971), 261–270.

[Zag]   D. Zagier, *Modular parametrizations of elliptic curves*, Canad. Math. Bull. **28** (1985), 372–384.

Index of Notation

Page numbers in Roman type refer to the current volume, while italicized page numbers refer to the complementary volume.

Symbols

G

G	usually a group, also Catalan's constant, *127*
g	sometimes the genus of a curve, 90, *441*
g	sometimes the number of prime ideals above $\mathfrak{p}$, 134
G_0	group of points reducing to a nonsingular point, 507
G_1	group of points reducing to $\overline{\mathcal{O}}$, 508
$g_2(\Lambda)$	g_2-invariant of lattice Λ, 483
$g_3(\Lambda)$	g_3-invariant of lattice Λ, 483
γ	usually Euler's constant, *33*
$\Gamma_p(s)$	p-adic gamma function at s, *368*
$\Gamma_r(s, x)$	higher incomplete gamma function, 574
$\gamma(s)$	$\pi^{-s/2}\Gamma(s/2)$, *172*
$\Gamma(s, x)$	incomplete gamma function, 573
$\Gamma(x)$	gamma function at x, *78*
$\gamma_p(\chi),\ \gamma_p$	p-adic Euler constants, *308*
$\gcd(a, b)$, GCD	greatest common dnivisor, *viii*, viii
$\gcd(a, b^\infty)$	limit of $\gcd(a, b^n)$, *ix*, ix
G_N	group of points of level $\geqslant N$, 508
$G(\tau, s)$	nonholomorphic Eisenstein series, *211*

H

$H^\perp$	orthogonal of H in V, 286
$h(D)$	class number of quadratic order of discriminant D, 318
$h(E)$	height of the elliptic curve E, 603
H_k	$\sum_{1\leqslant j\leqslant k} 1/j$, harmonic sum, *110*
$h(K),\ h$	class number of K, 131
H_n	$\sum_{1\leqslant k\leqslant n} 1/k$, *85*
H_n	harmonic sum $\sum_{1\leqslant j\leqslant n} 1/j$, *128*
HNF	Hermite normal form, 16, 340
$h(P)$	naïve height of a point $P \in E(\mathbb{Q})$, 530
$\widehat{h}(P)$	canonical height of $P \in E(\mathbb{Q})$, 530
h_p	class number of $\mathbb{Q}(\zeta_p)$, 432
h_{p^k}	class number of $\mathbb{Q}(\zeta_{p^k})$, 148
$h_{p^k}^-$	minus class number of p^kth cyclotomic field, 149
$h_{p^k}^+$	class number of maximal totally real subfield, 148
$H(p, t),\ H(t)$	conditions for $y^2 = x^p + t$, 411

I

$\Im(s)$	imaginary part of s, *ix*, ix
$I(\mathfrak{P}/\mathfrak{p})$	inertia group of $\mathfrak{P}/\mathfrak{p}$, 134
$I_s(m),\ I_s$	Stickelberger ideal, 160

J

$J(\chi_1, \chi_2)$	Jacobi sum associated with two characters, 82
$j(E)$	j-invariant of elliptic curve E, 467
$J_k(\chi_1, \ldots, \chi_k)$	Jacobi sum, 79

K

K	usually a number field, x, x
$\mathcal{K}$	a general $\mathfrak{p}$-adic field, x, x, 235
$K_n(F)$	higher K-groups, 244
$K_\mathfrak{p}$	completion of K at the prime ideal $\mathfrak{p}$, x, x, 195
$K(T, 2)$	same as $S_T(K)$, 551

L

L	usually a number field, x, x
$\Lambda(\chi, s)$	completed L-function for χ, 172
$\Lambda(E, s)$	completed L-function of an elliptic curve, 521
$\Lambda(n)$	von Mangoldt's function, 159
$\lambda(N)$	Carmichael's function of N, 93
$L(a, s)$	Dirichlet series associated with a, 151
$\mathcal{L}$	a general $\mathfrak{p}$-adic field, x, x
$L(\chi, s)$	L-series of character χ, 162
$\mathrm{lcm}(a, b)$, LCM	least common multiple, $viii$, viii
$L(E_D, s)$	L-function of elliptic curve E twisted by D, 590
$\left(\frac{a}{p}\right), \left(\frac{m}{n}\right), \left(\frac{a}{b}\right)$	Legendre, Jacobi, or Kronecker symbol, 33
$\left(\frac{a}{b}\right)_m$	mth power reciprocity symbol, 166
$\mathcal{L}(f)$	Laplace transform of f, 108
$\mathcal{L}_F(E)$	space of F-linear maps from E to E, 118
$\mathrm{Log}\Gamma_p(x)$	Diamond's log gamma function for $x \in \mathcal{Z}_p$, 330
$\mathrm{Log}\Gamma_p(\chi, x)$	Morita's log gamma function for $x \in \mathbb{Z}_p$, 337
$\mathrm{Log}\Gamma(s)$	complex log gamma function at s, 81
L^H	fixed field of L by H, 104
Li_2	dilogarithm function, 278, 404
Li_k	polylogarithm function, 278
$\mathrm{Li}(x)$	logarithm integral, 257
$[L : K]$	the degree of L over K, or the index of K in L, 107
L_n	usually the Lucas sequence, 421
$\log_\mathfrak{p}(x)$	$\mathfrak{p}$-adic logarithm, 211
$L_p(\chi, s)$	p-adic L-function of character χ, 301

M

$\mathcal{M}(f)$	Mellin transform of f, 107
$\mu(n)$	Möbius function of n, 153
μ_n	group of roots of unity of order n, 18
$\boldsymbol{\mu}_n = \boldsymbol{\mu}_n(K)$	subgroup of nth roots of unity in K, 112
$\mu_\mathfrak{p}$	group of $(\mathcal{N}\mathfrak{p} - 1)$st roots of unity in $K_\mathfrak{p}$, 228

S

$S^1(X)$, $S^1(\mathbb{Z}_p)$	strictly differentiable functions, *277*
$S_2(E)$	2-Selmer group of E, 555
$S(a, b; p)$	Kloosterman sum, 100
$s\backslash p$	$(s - a_0(s))/p$, essentially $\lfloor s/p \rfloor$, *365*
$\sigma(n)$	sum of divisors of n, *157*
$\sigma_k(n)$	sum of kth powers of divisors of n, *157*, 317
$s(N)$	sometimes the squarefree part of N, 541
$s_p(n)$, $s(n)$	sum of the digits of n in base p, 155, 207
$S_T(K)$	T-Selmer group of number field K, 551
$\displaystyle\sum^{(p)}$	sum over integers prime to p, *302*

T

τ	often an element of the upper half-plane $\mathcal{H}$, 586
$\tau(\chi)$, $\tau(\chi, a)$	Gauss sum for multiplicative character χ, 31
$\tau(\chi, \psi)$	Gauss sum with additive character ψ, 75
$\tau(n)$	Ramanujan τ function, *159*
$\tau_q(r)$	Gauss sum associated with a Dwork character, *386*
$\theta(\chi, \tau)$	theta function of character χ, *170*
$t(n)$	product of factorials of digits of n in base p, 155

U

U_0	the group of $\mathfrak{p}$-adic units, 226
U_1	group of $\mathfrak{p}$-adic units congruent to 1 mod $\mathfrak{p}$, 228
U_i	group of $\mathfrak{p}$-adic units congruent to 1 mod $\mathfrak{p}^i$, 228
$U(K)$	unit group of K, 131
$U_T(K)$	T-unit group of K, 550
$\langle u \rangle$	distance from u to the nearest integer, 58

W

$W(\chi)$	root number of modulus 1, 49
W_Q	Atkin–Lehner operator, 596

Z

$\mathbb{Z}_{\geqslant 0}$	nonnegative integers, *ix*, ix
ζ	usually a primitive pth root of unity, 432
$\zeta_C(T)$	zeta function of a curve or variety C, 91
$\zeta_K(s)$	Dedekind zeta function of a number field, *216*
$\zeta_p(s, x)$	p-adic Hurwitz zeta function, *283*
$\zeta_Q(s)$	Epstein zeta function for the quadratic form Q, *215*
$\zeta(s)$	Riemann zeta function, *72*, *153*
$\zeta(s, z)$	Hurwitz zeta function, *71*, *168*, *190*
$\mathbb{Z}_{>0}$	strictly positive integers, *ix*, ix

Index of Names

Page numbers in Roman type refer to the current volume, while italicized page numbers refer to the complementary volume.

General Index

Page numbers in Roman type refer to the current volume, while italicized page numbers refer to the complementary volume.

Graduate Texts in Mathematics

(continued from page ii)